T0142195

Advances in Intelligent Systems and Computing

Volume 715

Series editor

Janusz Kacprzyk, Polish Academy of Sciences, Warsaw, Poland
e-mail: kacprzyk@ibspan.waw.pl

About this Series

The series “Advances in Intelligent Systems and Computing” contains publications on theory, applications, and design methods of Intelligent Systems and Intelligent Computing. Virtually all disciplines such as engineering, natural sciences, computer and information science, ICT, economics, business, e-commerce, environment, healthcare, life science are covered. The list of topics spans all the areas of modern intelligent systems and computing.

The publications within “Advances in Intelligent Systems and Computing” are primarily textbooks and proceedings of important conferences, symposia and congresses. They cover significant recent developments in the field, both of a foundational and applicable character. An important characteristic feature of the series is the short publication time and world-wide distribution. This permits a rapid and broad dissemination of research results.

More information about this series at http://www.springer.com/series/11156

Michael E. Auer · David Guralnick
Istvan Simonics
Editors

Teaching and Learning in a Digital World

Proceedings of the 20th International Conference on Interactive Collaborative Learning – Volume 1

Editors
Michael E. Auer
Carinthia University of Applied Sciences
Villach, Kärnten
Austria

Istvan Simonics
Obuda University
Budapest
Hungary

David Guralnick
Kaleidoscope Learning
International E-Learning Association
New York, NY
USA

ISSN 2194-5357 ISSN 2194-5365 (electronic)
Advances in Intelligent Systems and Computing
ISBN 978-3-319-73209-1 ISBN 978-3-319-73210-7 (eBook)
https://doi.org/10.1007/978-3-319-73210-7

Library of Congress Control Number: 2017962881

Printed on acid-free paper

This Springer imprint is published by Springer Nature
The registered company is Springer International Publishing AG
The registered company address is: Gewerbestrasse 11, 6330 Cham, Switzerland

Preface

ICL2017 was the 20th edition of the International Conference on Interactive Collaborative Learning. This interdisciplinary conference aims to focus on the exchange of relevant trends and research results as well as the presentation of practical experiences in Interactive Collaborative Learning and Engineering Pedagogy. This conference is at the same time the annual conference of the "International Society of Engineering Pedagogy (IGIP)."

ICL2017 has been organized in cooperation with the Óbuda University from September 27 to 29, 2017, in Budapest, Hungary.

This year's theme of the conference was "Teaching and Learning in a Digital World."

Again, outstanding scientists from around the world accepted the invitation for keynote speeches:

- **Dale A. Martin**, CEO of Siemens Hungary,
- **Michael K. J. Milligan**, ABET Executive Director and CEO, USA,
- **András Benedek**, Professor of Education at Budapest University of Technology and Economics, and DSc of the Hungarian Academy of Sciences, and
- **Greet Langie**, Vice-Dean of the Faculty of Engineering Technology at KU Leuven, Belgium.

Since its beginning this conference is devoted to new approaches in learning with a focus on collaborative learning and engineering education.

We are currently witnessing a significant transformation in the development of education. There are three essential and challenging elements of this transformation process that have to be tackled in education:

- the impact of globalization on all areas of human life,
- the exponential acceleration of the developments in technology as well as of the global markets and the necessity of flexibility and agility in education, and
- the necessity of a closer cooperation between the industry, academia, and governmental organizations, especially in Engineering Education.

Therefore, the following main themes have been discussed in detail:

- Collaborative Learning,
- Project-based Learning,
- New Pedagogies with a focus on Engineering Pedagogy,
- K-12 and Pre-university programs,
- Learning Culture, Diversity & Ethics,
- Lifelong Learning and Academic-Industry Partnerships,
- Mobile Learning Environments Applications,
- New Learning Models and Applications,
- Online Environments and Laboratories,
- Game-based Learning,
- Computer-aided Language Learning (CALL),
- Entrepreneurship in Engineering Education,
- Real-world Experiences and Pilot Projects, and
- Ubiquitous Learning Environments, Platforms, and Authoring Tools.

The following submission types were accepted:

- Full Paper, Short Paper,
- Work in Progress, Poster,
- Special Sessions, and
- Roundtable Discussions, Workshops, Tutorials.

All contributions were subject to a double-blind review. The review process was very competitive. We had to review 569 papers. A team of about 150 reviewers did this terrific job. My special thanks go to all of them.

Due to the time and conference schedule restrictions, we could finally accept only the best 190 submissions for presentation.

Our conference had again more than 270 participants from 47 countries from all continents.

ICL2018 will be held in Kos Island, Greece, and ICL2019 in Bangkok, Thailand.

Michael E. Auer
ICL General Chair

Organization

Committees

General Chair

Michael E. Auer

ICL2017 Conference Chair

Istvan Simonics, Hungary

International Chairs

Samir Abou El-Seoud, Africa
Neelakshi Chandrasena Premawardhena, Asia
John Sandler, Australia/Oceania
Arthur Edwards, Latin America
Alaa Ashmawy, Middle East
David Guralnick, North America

Program Co-chairs

Michael E. Auer, Villach, Austria
David Guralnick, New York, USA
Teresa Restivo, Porto, Portugal

Technical Program Chair

Sebastian Schreiter, France

IEEE Liaison

Russ Meier, Milwaukee, USA

Workshop and Tutorial Chair

Barbara Kerr, Canada

Special Session Chair

Peter Toth, Hungary

Demonstration and Poster Chair

Danilo Zutin, Austria

Awards Co-chairs

Andreas Pester, Austria
Agnes Toth, Hungary

Publication Chair

Sebastian Schreiter, France

Senior PC Members

Andreas Pester	CUAS Villach, Austria
Axel Zafoschnig	Ministry of Education, Austria
Doru Ursutiu	University of Brasov, Romania
Eleonore Lickl	College for Chemical Industry, Vienna, Austria
George Ioannidis	University of Patras, Greece
Samir Abou El-Seoud	The British University in Egypt
Tatiana Polyakova	Moscow State Technical University, Russia

Program Committee

Agnes Toth, Hungary
Alexander Soloviev, Russia
Christian Guetl, Austria
Cornel Samoila, Romania
Costas Tsolakis, Greece
Hanno Hortsch, Germany
Hants Kipper, Estonia
Herwig Rehatschek, Austria
Imre Rudas, Hungary
Istvan Simonics, Hungary
Ivana Simonova, Czech Republic
James Uhomoibhi, UK
Jürgen Mottok, Germany
Martin Bilek, Czech Republic
Nael Barakat, USA
Olga Shipulina, Canada
Pavel Andres, Czech Republic
Rauno Pirinen, Finland
Roman Hrmo, Slovakia
Teresa Restivo, Portugal
Tiia Rüütmann, Estonia
Viacheslav Prikhodko, Russia
Victor K. Schutz, USA

Contents

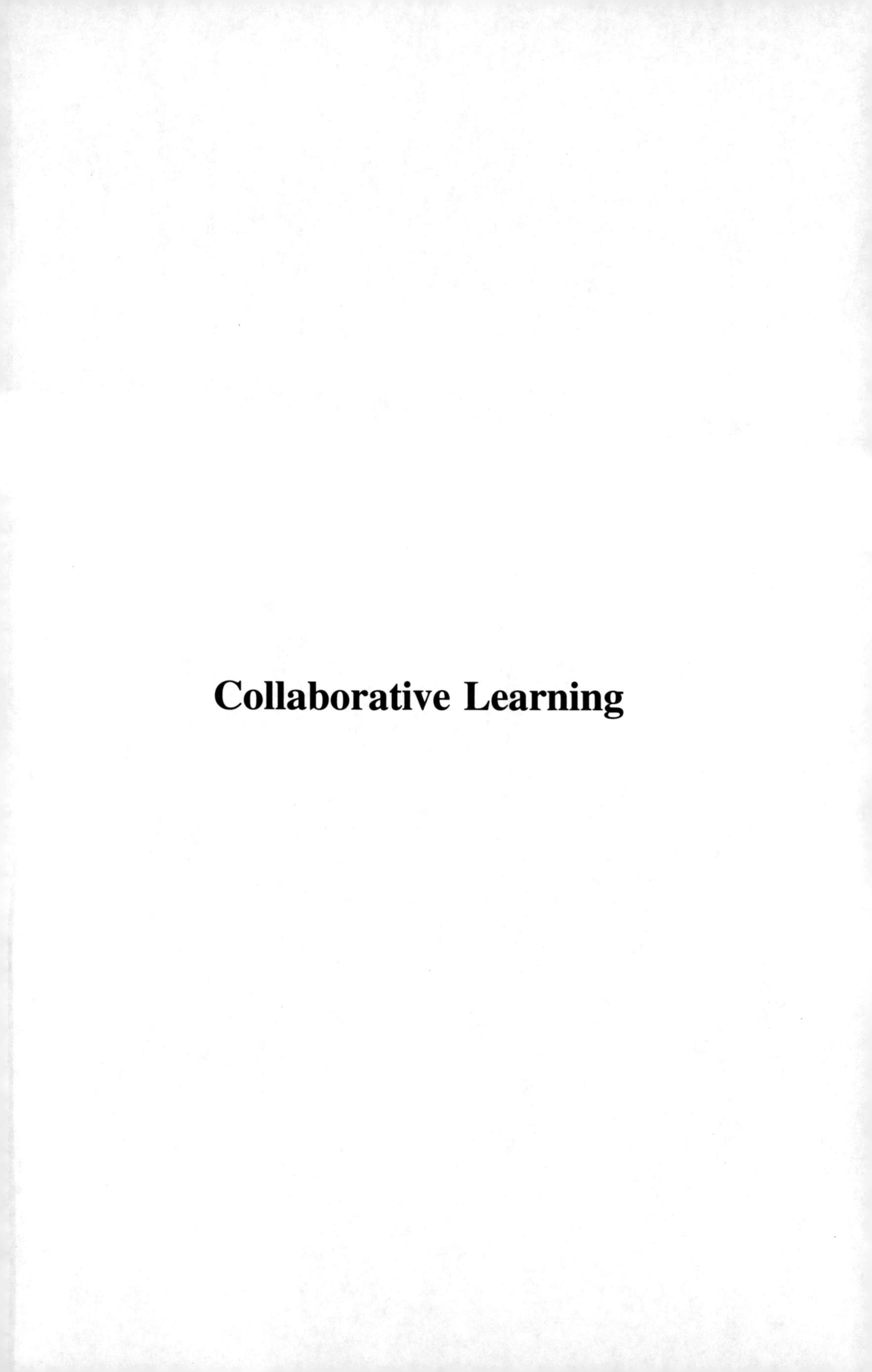

Collaborative Learning

Scratch as Educational Tool to Introduce Robotics

Pedro Plaza[1(✉)], Elio Sancristobal[2], German Carro[2], Manuel Castro[2], Manuel Blázquez[3], Javier Muñoz[1], and Mónica Álvarez[1]

[1] Plaza Robotica, Torrejón de Ardoz, Spain
{pplaza,javi,monica.alvarez}@plazarobotica.es
[2] Electrical and Computer Engineering Department, Spanish University for Distance Education (UNED), Madrid, Spain
{elio,mcastro}@ieec.uned.es, germancf@ieee.org
[3] Universidad Antonio de Nebrija, Madrid, Spain
manuel.blazquez.merino@gmail.com

Abstract. There are many necessities that need to be improved in STEM (Science, Technology, Engineering and Math) education. The robotics represents a promising educational tool. Nowadays, robotic education tools arise with the aim of promoting the innovation and the motivation of the students during the learning process. Robots are becoming more common in our daily life; thus, it is important to integrate robots at all levels of our society. The aim of this paper is to present the use of Scratch - a widely-used tool - in order to guide educational robotics as the first step in introducing students into robotics. The robotics requires several skills such as systems thinking, programming mindset, active learning, mathematics, science, judgement and decision making, good communication, technology design, complex problem solving and persistence. These skills can be easily developed using Scratch. The obtained outcomes from the educational robotic course demonstrate how children without previous experience in programming or robotics can start learning both through experiences in the classroom. The result of this work shows that it is better to make very easy challenges, to adapt the difficulty to each of the children. Furthermore, it is necessary to develop previous concepts. Moreover, it is necessary to work the design, instead of programming directly. Additionally, it is important to combine theory and practice with the aim of including fun tasks intertwined with the challenges that are posed to apply theory in problem solving.

Keywords: Programming · Robotics · Education · STEM

1 Introduction

STEM (Science, Technology, Engineering and Mathematics) education is a powerful tool which it is being more popular these days [1]. Additionally, the robotics also provides an attractive manner to transform boring concepts into an amusing learning process. Robotic kits facilitate the ease with which students can make connections among STEM disciplines [2].

M. E. Auer et al. (eds.), *Teaching and Learning in a Digital World*,
Advances in Intelligent Systems and Computing 715,
https://doi.org/10.1007/978-3-319-73210-7_1

Games and simple materials are a good combination with the aim of engaging students about computer science fundamentals. The article [3] describes a set of learning activities that were found suitable for non-formal learning environments. In [4], the authors present a programming education session through the production of the game program using Scratch as a programming environment and NanoboardAG as a sensor board.

The robotics is being used as the modernization and improvement for most of processes. This occurs as result of robots can be easily integrated within the current industrial processes [5]. Furthermore, the robotics has become an important tool for the students to be involved in STEM. Nevertheless, the introduction to the robotics is not a simple task. First, the robotics combines mechanics, electricity, electronics, and computer science with the purpose of designing and building robotics applications. Robots are programmable electromechanical machines that include sensors to make decisions and adapt to different situations. Hence, the first step in order to be introduced into the robotics should be the acquisition of programming skills.

The aim of this paper is to propose a tool which ease the learning process into the robotics. The chosen tool to accomplish this is Scratch. This paper is divided in five sections. Section 2 presents Scratch as a simple and powerful STEM tool. Section 3 details a learning course using Scratch to introduce the robotics to the children. A discussion related to the exposed content and the achieved results is provided in Sect. 4. Finally, the last section summarizes the inferred conclusions after the performed investigation. Moreover, Sect. 5 includes some proposed activities with the aim of promoting the robotic education within STEM education.

2 Scratch: Imagine, Program and Share

Scratch was developed by the MIT (Massachusetts Institute of Technology) Education division. The mission of the Massachusetts Institute of Technology is to advance knowledge and educate students in science, technology, and other areas of scholarship that will best serve the nation and the world in the 21st century. They are also driven to bring knowledge to bear on the world's great challenges [6]. With Scratch interactive stories, games and animations can be programed Additionally, it allows sharing the creations with others in the online community. Scratch helps young people learn to think creatively, to reason systematically, and to work collaboratively - essential life skills in the 21st century. The Scratch project has received financial support from the National Science Foundation, Scratch Foundation, Google, LEGO Foundation, Intel, Cartoon Network, Lemann Foundation, and the MacArthur Foundation [7]. Along this section, how Scratch can be obtained is detailed. Moreover, the Scratch programming environment is described too. Finally, the type of users and communities are included.

2.1 Getting Scratch and Reference Resources

Scratch is an easy-to-use programming environment. Its programming interface uses a block programming language which eases its use by children aged below 16 years. The educational robotics can be implemented easily with it due to the simple structure of the Scratch interface.

MIT Education Scratch website provides different ways of using Scratch:

- Online use in [7].
- Offline use for different operating systems:
  - Windows.
  - Debian/Ubuntu.
  - Mac OS X.

There are two online versions: the 1.4 version which can be downloaded in [8] and the 2.0 version which can be downloaded in [9].

The following minimum system requirements are needed for run Scratch properly:

- Windows, Mac, or Linux (32 bit) operating system,
- Adobe Air version 2.6+ (included in download),
- over 23 MB of free hard drive space.

MIT Education also provides two type of reference resources for Scratch programming projects. The first one is located in the Scratch website [10]. The second one can be found as part of the programming environment. Figure 1 shows the window to import projects with the aim of using as reference or in order to modify them.

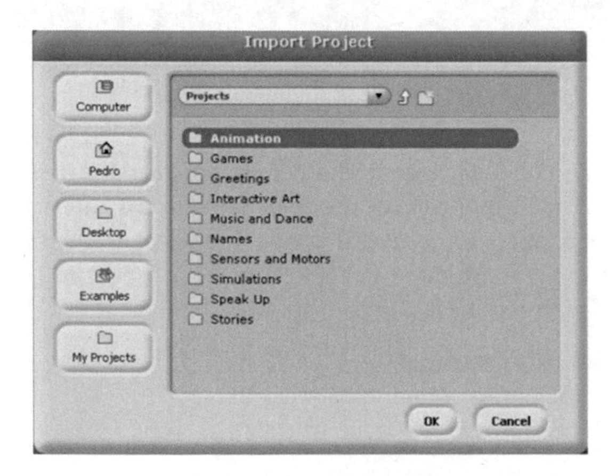

Fig. 1. Import projects in Scratch

2.2 The Programming Environment

The Scratch programming environment is divided into three main sections: the command panel, the programming panel and the visualization panel. The Fig. 2 depicts the Scratch environment. The Fig. 2 also identifies each panel: orange circle for the command panel, blue circle for the programming panel and grey circle for the visualization panel. Command panel is located on the left. It contains 8 categories of commands: motion, control, looks, sensing, sound, operators, pen and variables. Each category has a set of commands that are related to each other. The programming panel is located on the center. It contains controls on the Sprite, a space to build the script that will execute the Sprite, the costumes that can use the Sprite and the sounds that can cast the Sprite. The Sprites are the actors of the stories or the characters of the games that are created with Scratch. The display panel is located on the right. It is divided into two zones. The upper area allows the view of the Sprites while they run their scripts, and the backgrounds that have been programmed on the stage. The lower zone allows the selection of both the background and the Sprites that have been created in the open project.

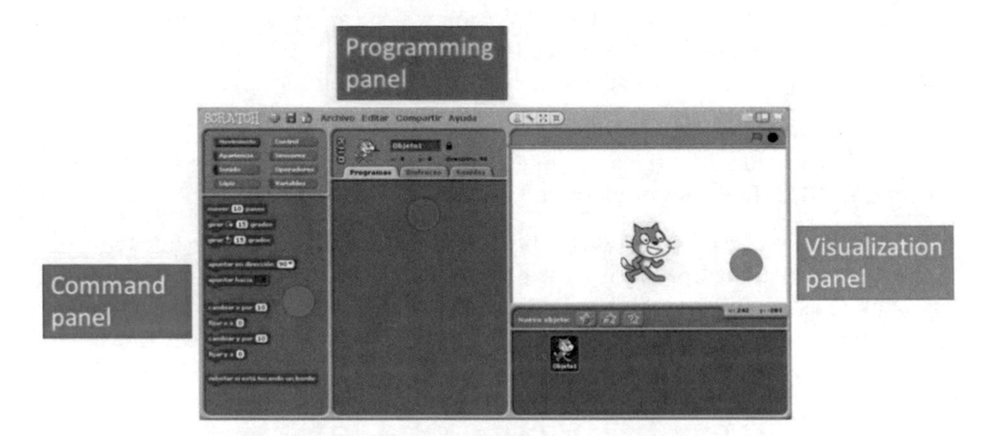

Fig. 2. Scratch environment and panels

At the top of the command panel, the command categories can be seen. At the bottom of this panel, the grouped commands in the selected category are showed.

On the other hand, at the top of the programming panel, the Sprite can be seen. To the left of the Sprite there are three buttons, from top to bottom have the following functions:

- Allows the character to rotate 360°.
- Allows the character to rotate on its vertical axis.
- Does not allow the character to rotate or rotate.

In this area, the Sprite's name is shown. Additionally, the location of the Sprite is defined as the coordinates x and y where it is. Furthermore, the direction the Sprite is pointing is included too. In this panel there are three tabs:

- Programs: place where the commands are added to build a script.
- Costumes: costumes for the Sprite, they can be imported or created.
- Sounds: sounds for the Sprite, they can be imported or created.

Finally, at the top of the visualization panel the background and the Sprites can be seen while they run their programs. At the bottom of the visualization panel the background and the different Sprites can be selected.

2.3 Scratch Users and the Communities

While Scratch was primarily designed for 8 to 16 years old children, it is also used by people of all ages, including younger children with their parents. The MIT Scratch Team works with the community to maintain a friendly and respectful environment for people of all ages, races, ethnicities, religions, sexual orientations, and gender identities [11]. The Scratch Team works each day to manage activity on the site and respond to reports, with the help of tools such as the CleanSpeak profanity filter [12]. Moreover, Scratch is widely-used in education [13]. The educators are using Scratch in a wide variety of:

- Setting: schools, museums, libraries and civic centers.
- Courses: Primary, Compulsory and Baccalaureate.
- Subjects: language, science, social, mathematics, computer science, foreign languages and arts.

There is a community for educators, ScratchEd [14] is an online community in which Scratch educators share stories, exchange resources, ask questions and meet people. ScratchEd is developed and supported by Harvard Graduate School of Education. Furthermore, there are meetings of Scratch educators who want to learn from each other, share their ideas and strategies to support computational creativity in all its forms [15].

3 Scratch Course

In the context of a course to introduce children into the robotics, the first step has been based on Scratch as a tool which provides students with skills such as systems thinking, programming mindset, active learning, mathematics, science, judgement and decision making, good communication, technology design, complex problem solving and persistence. Along this section, the location and used resources are detailed. Moreover, the course structure is described too. Finally, the course results are also included.

3.1 Course Location and Resources

The course was held in the center of La Estera [16], located in the town of Camarma de Esteruelas. Camarma de Esteruelas is located in the eastern area of the Community of Madrid. La Estera is an independent socio-cultural center, self-funded by neighbors and non-for-profit organization.

The classroom is formed by tables in isle. In one side of the isle the visual learning material is projected. In the other side, the teacher manages the computer, presents the session and interacts with the children.

The students have four personal computers (PC), each PC is used in pairs. These PCs run Ubuntu as the operating system. This operating system was chosen for its compatibility with Scratch and because both Ubuntu and Scratch do not require many resources such as microprocessor speed or RAM (Random-Access Memory) memory. Additionally, Ubuntu can be used for free. Ubuntu is a GNU/Linux based operating system that is distributed under a Free Software license and is based on a graphic system based on Unity. In which free software programs for education purposes can be installed.

The students group was formed by eight students aged from 7 to 15 years. The course duration was two months and the contents were provided during two hours each week.

3.2 Course Structure

The course was divided into two parts, the first one was focused on getting knowledge about the Scratch tool and the second part was aligned to application building by problem solving activities. The course comprises a total of 16 h.

The session contents were designed with the aim of combining theory and practice. The theory was deployed during the first quarter of the session and the rest of the time was devoted exclusively to problem solving in a practical way in order to apply the theoretical concepts explained previously.

The Table 1 lists the sessions which form the course. Additionally, the topics which are intended to be covered during students perform these sessions are included too in order to provide the distinctive features for each one of them.

Table 1. Session titles and objectives.

Session title	Session objectives
Knowing Scratch (I)	Know Scratch, notions of programming, draw geometric figures and create a story
Knowing Scratch (II)	Go into detail about Scratch, notions of programming, interact with the program and improve a story
Knowing Scratch (III)	Go into detail about Scratch, notions of programming, programming scenarios and improve a story
Knowing Scratch (IV)	Go into detail about Scratch, notions of programming, use of variables and improve a story
Programming a clock	Introduction to projects with Scratch, advanced programming, teamwork and work for objectives
Line follower	Introduction to sensors, advanced programming, teamwork and work for objectives
Obstacle avoider	Go into detail about sensors, advanced programming, teamwork and work for objectives
History of a robot	Go into detail about Scratch, advanced programming, teamwork and work for objectives

All activities shown in the Table 1 are composed by the following contents:

- Knowing Scratch (I): this session covers a brief of main characteristics of Scratch such as panels, command categories and Sprites. During this session, a simple task about programming without computer was done. Furthermore, geometric figures were painted using a script. At the end of the session the children were making simple programs for a Sprite, they used the pen command and they worked in a simple history programmed with Scratch.

- Knowing Scratch (II): firstly, the panels and the commands were reviewed. After this, new commands from looks, control and sensing categories were showed in order to make geometric figures using user interaction with the Sprite script. The simple story made in the previous session was improved too.
- Knowing Scratch (III): at the begin of the session, the commands used in the previous session were reviewed. After this, the use of backgrounds was explained. Moreover, the interaction of several Sprites and users was added. Hence, the students can make geometric figures using user interaction with the Sprite script as part of a group. Furthermore, they scaled the simple story made in the previous session was improved too.
- Knowing Scratch (IV): this session introduces variables and operators. Some activities such as operations with variables, variables with Sprites, drawing with variables and drawing geometric figures with variables were carried out.
- Programming a clock: projects, roles and objectives were presented in this session. After this, the students were requested to accomplish a simple project, make a clock using Scratch. Sprites edition and background edition was needed. Students programmed and tested their creations. After the project completion, they made simple modifications such as the clock speed.
- Line follower: an introduction to sensors was made. For this session, the objectives were: programming a line follower robot and the modification of the script in order to allow that two users can compete. Once students finished their projects, they were testing the projects of other students.
- Obstacle avoider: a going into detail about sensors knowing was presented. This session was similar than the previous one but students were challenged to get a robot which was able to avoid obstacles. They also modified their project with the aim of allowing that two users can compete. Once students finished their projects, they were testing the projects of other students.
- History of a robot: along this last session, the students created a storyline that took into account the characters and the environments which they had created. The project included conversations and movements of the Sprites and the backgrounds.

3.3 Course Results

As exposed in the sessions' contents, different kind of activities can be carried out with the aim of providing how STEM education can be covered using simple tasks and project-oriented activities.

Students had not any previous experience using Scratch. Additionally, they had not any previous experience in programming. Over the sessions, the instructor guided the students through the theoretical contents and the proposed challenges. Each day's session started with a short discussion of the previous session's content. The instructor also observed a change in the questions asked by some of the students, both in connecting the experiences of the activity with real-world examples as well as in trying to understand the underlying concepts.

During the first four sessions, students were dealing with Scratch environment. This environment is little complex, see Sect. 2.2. where a summary is included. Along these sessions, the instructor was conducting students in order to cover completely the content. Hence, all students completed the assigned tasks. Some of them without help the other needed some clarifications about the activity objectives and how to solve it. Table 2 includes information about activities carried out in each of the sessions Knowing Scratch (I), Knowing Scratch (II), Knowing Scratch (III) and Knowing Scratch (IV) respectively and how many students completed each one of these activities without help, with some help or required supervision by the instructor with the aim of complete the corresponding activity.

Table 2. Knowing Scratch (I) session activities and students which completed them.

Session activities	Completed without help	Completed some help	Not completed without supervision
S1. Programming without computer	0 (0%)	6 (75%)	2 (25%)
S1. Geometric figures	1 (12.5%)	5 (62.5%)	2 (25%)
S1. Simple programs for a Sprite	6 (75%)	0 (0%)	2 (25%)
S1. Simple history	0 (0%)	6 (75%)	2 (25%)
S2. Geometric figures using user interaction	0 (0%)	6 (75%)	2 (25%)
S2. Improving the simple history	1 (12.5%)	5 (62.5%)	2 (25%)
S3. Improving user and Sprites interaction	0 (0%)	6 (75%)	2 (25%)
S3. Improving the simple history	1 (12.5%)	5 (62.5%)	2 (25%)
S4. Operations with variables	0 (0%)	2 (25%)	6 (75%)
S4. Variables with Sprites	0 (0%)	2 (25%)	6 (75%)
S4. Drawing with variables	0 (0%)	6 (75%)	2 (25%)
S4. Drawing geometric figures with variables	0 (0%)	6 (75%)	2 (25%)

During the second four sessions, students were dealing with different Project-oriented activities using Scratch environment. Along these sessions, the instructor was conducting students in order to set project objectives and attaching surrounding theory about the corresponding session. Hence, all students completed the assigned tasks. Some of them without help the other needed some clarifications about the activity objectives and how to solve it. Table 3 includes information about activities carried out in each of the sessions Programming a clock, Line follower, Obstacle avoider and History of a robot respectively and how many students completed each one of these activities without help, with some help or required supervision by the instructor with the aim of complete the corresponding activity.

Table 3. Programming a clock session activities and students which completed them.

Session activities	Completed without help	Completed some help	Not completed without supervision
S5. Make a clock using Scratch	0 (0%)	2 (25%)	6 (75%)
S5. Simple modifications such as the clock speed	6 (75%)	2 (25%)	0 (0%)
S6. Programming a line follower robot	1 (12.5%)	5 (62.5%)	2 (25%)
S6. Allow that two users can compete	6 (75%)	0 (0%)	2 (25%)
S6. Testing the projects of other students	6 (75%)	2 (25%)	0 (0%)
S7. Get a robot which was able to avoid obstacles	6 (75%)	0 (0%)	2 (25%)
S7. Allow that two users can compete	6 (75%)	2 (25%)	0 (0%)
S7. Testing the projects of other students	8 (100%)	0 (0%)	0 (0%)
S8. Created a storyline with characters and environments	6 (75%)	2 (25%)	0 (0%)
S8. Include conversations and movements of the Sprites	8 (100%)	0 (0%)	0 (0%)
S8. Include conversations and movements of the backgrounds	8 (100%)	0 (0%)	0 (0%)

Students increased their motivation for programming and creating simple projects. At the beginning of the course, all students had knowledge about new technologies such as computers, tablets and smartphones but this knowledge was limited to internet queries or gaming. During the course, they were scaled their abilities about systems thinking, programming mindset, active learning, mathematics, science, judgement and decision making, good communication, technology design, complex problem solving and persistence.

4 Discussion

Along the previous sections, this educational robotic course demonstrates how children without previous experience in programming or robotics can start learning both through experiences in the classroom.

Educational robotics is a term widely used to describe the educational use of robotics as a learning tool. From 2006 as stated in [17] where LEGO Mindstorms NXT was used. This is an example of educational robotics. Currently, LEGO evolved the NXT to EV3. Nowadays, there are a wide variety of robotic tools such as Arduino [18].

Scratch is a good choice to introduce the robotics because of its cost, MIT Education branch provide the tool free of charge. Furthermore, the only one requirement is the use of a PC (Personal Computer), laptop or tablet. For Scratch to run properly, the following minimum system requirements are needed:

- Windows, Mac, or Linux (32 bit) operating system,
- Adobe Air version 2.6+ (included in download),
- over 23 MB of free hard drive space.

As exposed in the Scratch course sections, different kind of activities are exposed with the aim of providing how STEM education can be covered using simple and effective sessions. These sessions contain theory but practice is an important factor due to students' age requires that students are part during the learning process. Additionally, it is important that students find fun the activities in order to get their attention and motivation for learning. The results from each sessions have been summarized in Table 2 for the first part of the course and in Table 3 for the second part of the course. Along of the sessions, some students were able to complete the proposed activities without help from the instructor. Most of the students were able to complete the activities with some kind of help. In addition, some students needed support from the instructor in order to complete the activities assigned to them. The need for help was not related to the age of the students. That is, not because they were younger, they needed more help.

The students enjoyed their activities and learned a great deal from the experience. In addition, they enhance their abilities related to STEM.

5 Conclusions

The result of this work shows that it is better to make very easy challenges, to adapt the difficulty to each of the children. Furthermore, it is necessary to develop previous concepts. Moreover, it is necessary to work the design, instead of programming directly.

Additionally, it is important to combine theory and practice with the aim of including fun tasks intertwined with the challenges that are posed to apply theory in problem solving.

In conclusion, it has been shown that Scratch is a good choice when it comes to introducing the robotics in an inexpensive, simple and convenient way for teachers to develop scalable concepts. Moreover, students find programming with Scratch an enjoyable activity while there are acquiring STEM concepts.

Finally, the results will be integrated in an Open Hardware platform which promotes innovation and motivation for students during the learning process [19]. The platform which is being developed presents wirelessly connections such as Bluetooth and WiFi as enhancements [20]. This research continues the development described in [21]. The doctoral thesis is being carried out in the Engineering Industrial School of UNED (Spanish University for Distance Education) and the Electrical and Computer Engineering Department (DIEEC).

Acknowledgment. The authors acknowledge the support provided by the Engineering Industrial School of UNED, the Doctorate School of UNED, and the "Techno-Museum: Discovering the ICTs for Humanity" (IEEE Foundation Grant #2011-118LMF).

And the partial support of the eMadrid project (Investigación y Desarrollo de Tecnologías Educativas en la Comunidad de Madrid) - S2013/ICE-2715, IoT4SMEs project (Internet of Things for European Small and Medium Enterprises), Erasmus+Strategic Partnership nº 2016-1-IT01-KA202-005561), mEquity (Improving Higher Education Quality in Jordan using Mobile Technologies for Better Integration of Disadvantaged Groups to Socio-economic Diversity), Erasmus+Capacity Building in Higher Education 2015 nº 561727-EPP-1-2015-1-BG-EPPKA2-CBHE-JP, PILAR project (Platform Integration of Laboratories based on the Architecture of visiR), Erasmus+Strategic Partnership nº 2016-1-ES01-KA203-025327, the GID2016-17 Remote Electronics Practices in the UNED, Europe and Latin America with Visir - PR-VISIR G-TAEI Research Group and the Research Project 2017 IEQ 17 Industrial Communications Networks. School of Industrialists UNED.

The authors are also thankful to La Estera due to their collaboration.

References

1. Pickering, T.A., Yuen, T.T., Wang, T.: STEM conversations in social media: implications on STEM education. In: 2016 IEEE International Conference on Teaching, Assessment, and Learning for Engineering (TALE), Bangkok, pp. 296–302 (2016)
2. Fernandez, G.C., et al.: Mechatronics and robotics as motivational tools in remote laboratories. In: 2015 IEEE Global Engineering Education Conference (EDUCON), Tallinn, pp. 118–123 (2015)
3. Cápay, M.: Engaging games with the computer science underlying concepts. In: 2015 International Conference on Interactive Collaborative Learning (ICL), Florence, pp. 975–979 (2015)
4. Yoshihara, K., Watanabe, K.: Practice of programming education using scratch and NanoBoardAG for high school students. In: 2016 10th International Conference on Complex, Intelligent, and Software Intensive Systems (CISIS), Fukuoka, pp. 567–568 (2016)
5. Fernandez, G.C., Gutierrez, S.M., Ruiz, E.S., Perez, F.M., Gil, M.C.: Robotics, the new industrial revolution. IEEE Technol. Soc. Mag. **31**(2), 51–58 (2012)
6. MIT Education website. http://web.mit.edu/. Accessed 7 May 2017
7. MIT Education Scratch website. https://scratch.mit.edu/. Accessed 15 Apr 2017
8. Scratch version 1.4 website. https://scratch.mit.edu/scratch_1.4/. Accessed 7 May 2017
9. Scratch version 2.0 website. https://scratch.mit.edu/scratch2download/. Accessed 7 May 2017
10. Scratch resources website. https://scratch.mit.edu/explore/projects/all. Accessed 7 May 2017
11. Scratch for parents website. https://scratch.mit.edu/parents/. Accessed 7 May 2017
12. CleanSpeak profanity filter website. https://www.inversoft.com/products/profanity-filter. Accessed 7 May 2017
13. Scratch for educators website. https://scratch.mit.edu/educators/. Accessed 7 May 2017
14. ScratchEd community website. http://scratched.gse.harvard.edu/. Accessed 7 May 2017
15. ScratchEd Meetups website. https://www.meetup.com/pro/scratched/. Accessed 7 May 2017
16. La Estera website. http://laestera.blogspot.com.es/. Accessed 7 May 2017

17. Gale, R., Karp, T., Lowe, L., Medina, V.: Generation NXT. In: 2007 IEEE Meeting the Growing Demand for Engineers and Their Educators 2010–2020 International Summit, Munich, pp. 1–13 (2007)
18. Araújo, A., Portugal, D., Couceiro, M.S., Rocha, R.P.: Integrating Arduino-based educational mobile robots in ROS. In: 2013 13th International Conference on Autonomous Robot Systems, Lisbon, pp. 1–6 (2013)
19. Plaza, E.S., Fernandez, G., Castro, M., Pérez, C.: Collaborative robotic educational tool based on programmable logic and Arduino. In: 2016 Technologies Applied to Electronics Teaching (TAEE), Seville, pp. 1–8 (2016)
20. Merino, P.P., Ruiz, E.S., Fernandez, G.C., Gil, M.C.: A wireless robotic educational platform approach. In: 2016 13th International Conference on Remote Engineering and Virtual Instrumentation (REV), Madrid, Spain, pp. 145–152 (2016)
21. Merino, P.P., Ruiz, E.S., Fernandez, G.C., Gil, M.C.: Robotic educational tool to engage students on engineering. In: 2016 IEEE Frontiers in Education Conference (FIE), Eire, PA, pp. 1–4 (2016)

Design and Set-up of an Automated Lecture Recording System in Medical Education

Herwig Rehatschek(✉)

Department Organization of Teaching and Learning with Media, Medical University of Graz, Harrachgasse 21, 8010 Graz, Austria
Herwig.Rehatschek@medunigraz.at

Abstract. In classroom lectures with huge groups it is still time consuming for students to follow the teacher and accumulate all the information given on various media. Scripts provided in advance may support the learning and understanding process, however, this is – for various reasons - in many cases not done or done too short before the start of the lesson. So far we provided eLectures which contained the slides and the voice of the teacher. They were produced on an individual basis causing a lot of efforts. On our new Medical Campus we decided to provide a fully automated lecture recording system which is easy to operate by the teachers and provides a maximum of flexibility for the students. We introduce a concept and the implementation of this recording system which enables students to receive to full HD streams of the PC output and the whiteboard, which can be scaled to their individual needs.

Keywords: Automated lecture recording · Lesson streaming · Video streaming portal

1 Introduction

In many classroom lectures with huge groups it is still a tedious task for students to follow the teacher and accumulate all the information given on various media such as PC/projector, visualizer and white board in an appropriate way. Scripts and/or power point slides provided in advance may support the learning and understanding process, however, this is – for various reasons - in many cases not done or done too close before the start of the lesson. That's why we started at the Medical University of Graz to offer so called eLectures [1–3] which are slides with animations in connection with the voice of the teacher. This concept was well accepted by the students and we received good evaluation results. Since eLectures were produced on an individual basis together with the teachers they had the clear advantage to contain all the information and wishes of the teachers which content has to be shown to the students. This included also complex learning objects such as animations and virtual microscopes [4]. The disadvantage was the – depending on the complexity of the content - quite high effort it took to produce eLectures and that they could not be produced automatically or on live classroom lectures.

M. E. Auer et al. (eds.), *Teaching and Learning in a Digital World*,
Advances in Intelligent Systems and Computing 715,
https://doi.org/10.1007/978-3-319-73210-7_2

Since currently a new Medical Campus is built for our university which will be ready by October 2017 we decided to introduce a new concept for lecture recording in the teaching rooms which allows to record classroom lectures automatically and during their live presentation to the students. So students can have access on the material shortly after the lesson and can use it for preparation of the exam. Furthermore not only the PC/projector and the voice of the teacher shall be recorded but optionally also the whiteboard and/or the teacher himself talking and maybe showing essential objects or experiments. Last but not least the system shall be easy to use by the teachers by giving them as much flexibility as needed but as little technical details as possible. So teachers with no technical background should be able to use the recording system without the necessity of time intensive trainings and/or reading tons of manuals. Furthermore we want to provide a video portal solution to our students which enables them to comfortably watch the lessons according to their needs.

2 Related Work

Looking at the market someone can find many providers offering automated and fully integrated lecture recording systems including Extron, Panopto, Epiphan and StreamAMG. The Extron [5] solution was at the time we started the design (2016) not fully ready, meaning that there was no native software solution in order to store and manage the produced streams. Panopto [6] and StreamAMG [7] with its product Stream LC offer a software solution for recording and managing self recorded videos in connection with slides and a streaming portal solution. It does not natively offer recording hardware for lecture rooms but provides certified partners. Epiphan [8] provides hardware for live video production and streaming, but no software for management, organisation and broadcast of the recorded material.

3 Technical Approach

Having no experience so far with the various products on the market we decided to take the following approach for the design: we applied for a research project at the Austrian government together with two universities in our city (Karl Franzens University and University of Technology Graz), where the University of Technology in Graz already had an automated lecture recording system in place and hence had years of experience with the hardware, software and the technical and organisational workflow. That's why we decided to go for the Epiphan hardware solution in combination with Opencast Matterhorn [9], an open source lecture capture and streaming software.

The technical implementation mainly shall consist of two main steps: first step is the set up of the hardware recording solution, the second step is the set up of the content organization and streaming portal.

3.1 Set-up of Recording Hardware

The lecture classrooms will be equipped with a camera capable of filming the whiteboard and/or the teacher at the lectern. Furthermore the PC output/projector output can be recorded. Both streams shall be recorded in full HD resolution. The transmission will be done in a single stream containing both videos (PC output/projector, whiteboard) and the voice of the teacher. This is done due to easy and full synchronization of the PC output/projector and the video of the teacher/whiteboard. When transmitted in two streams experience showed that synchronization is always an issue and this is critical, because if e.g. the slides and the voice of the teacher are not synchronously students won't be able to understand the content anymore. Within the teaching rooms Crestron control units are provided with record and stop buttons in order to provide teachers an easy possibility to start/stop the recording. Furthermore we provide the teachers four recording pre-settings. The first setting will record PC output/projector and the white-board and the lectern, the second setting will record PC output/projector and the white-board, the third setting will record PC output/projector and the lectern and the fourth setting will record whiteboard and lectern but not the PC output/projector. These easy to understand recording scenarios together with the record and stop button will be the entire interface for the teachers in order to fully automatically record their lessons.

3.2 Set-up of the Content Organization and Streaming Portal

The second step is the set up of the video portal VITAL (Video Portal of Med Uni Graz), which we just started. As a software we decided to take OpenCast Matterhorn. We plan two main access possibilities for the students: first students shall be able to login via the video portal and access all recorded lessons directly. Second students can access the recorded lectures directly via our learning management system Moodle where also other virtual content can be found. Recorded lectures will be directly linked to the appropriate content. Technically we plan to utilize the LTI (Learning Tools Interoperability - a specification developed by IMS Global Learning Consortium) interface [10] of Moodle in order to directly access the lectures stored on OpenCast server.

4 Technical Workflow

All steps in connection with preparation, recording and playout will be integrated in an overall technical workflow consisting of 5 major steps – Pre-processing, data recording, data transmission, post processing and playout - which we will describe in more detail in the next sub chapters.

4.1 Pre-processing Steps

When planning the next semester all teachers will be contacted by E-mail in order to state if they want to record lessons and if yes they have to specify which ones. All these data will be collected by our technical Virtual Medical Campus team and integrated in one comprehensive recording calendar. Based on this calendar the team knows when recordings will take place and when they have to provide technical assistance and post processing.

4.2 Data Recording in Lecture Room

The data recording in the lecture room can be solely started by the teacher by selecting the preferred recording format from the four pre-settings (as described above) and by pressing record, i.e. no further technical assistance should be necessary. At the end of the lecture the teacher has to press the stop button.

4.3 Data Transmission

Recorded lessons are locally stored at the Epiphan storage and will be automatically transferred to a storage server for further processing. The Epiphan storage is a ring storage, when it is full the oldest data is deleted. Of course the storage is huge enough to store several recordings. Since data is usually immediately transmitted after the end of the recording to the storage server this storage is also not critical.

4.4 Post Processing

The post processing is triggered by the recording calendar. Hence each time a recording is expected the technical team checks whether data has arrived. If the stream is available the first step is a cropping. We transmit both videos in one stream, as visualized in Fig. 1.

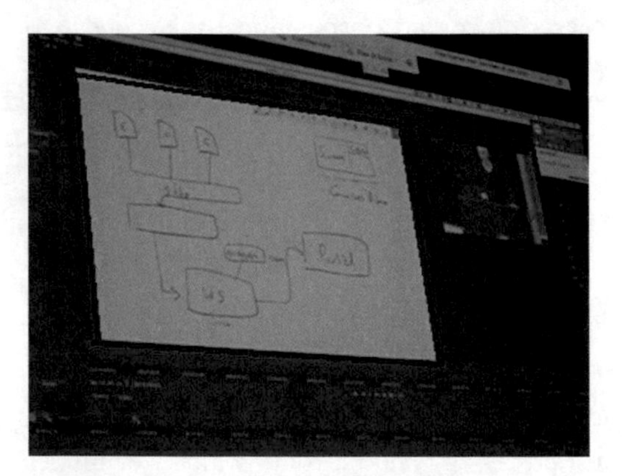

Fig. 1. Transmission of two videos in one stream

This is done in order to keep synchronization between the speaker/whiteboard video and the PC output stream. Hence in the first step we crop out the two streams. Then we enhance the audio by utilizing the OpenCast functionalities. Next we add an intro and check roughly the quality. Finally the video will be placed on the OpenCast server and linked with the appropriate lecture in Moodle.

4.5 Playout and Student View

At the student side we will provide a video portal solution. Students shall be able to easily search and view recorded lessons they need for learning. It is a clear objective that the video portal shall enable students full control over the arrangement and video size of the two recorded streams. Hence if there is at some stages shown something interesting on the whiteboard, students may view this stream in a large format on their screens. The planned student view is visualized in Fig. 2 which is an example taken from the existing video portal from the University of Technology, Graz.

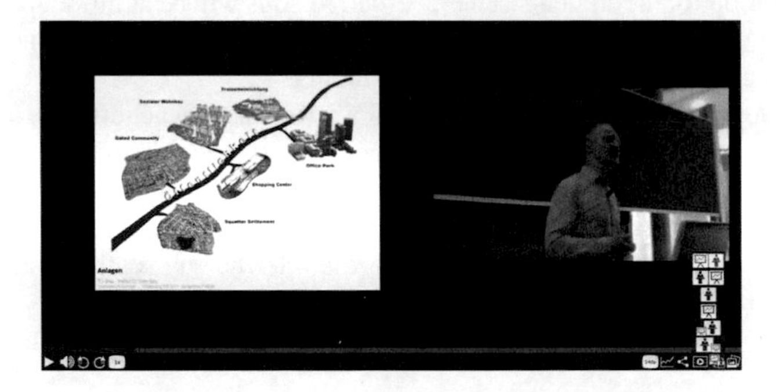

Fig. 2. Planned student view of recorded lectures

The same shall be true if the interesting content is shown on the PC. Furthermore there shall be an easy way to automatically structure the content (e.g. by slides) and to search the content (e.g. by automated OCR of the slides). Since most of the content is copyright protected and not freely public available the video portal will be only accessible by students and affiliates of the university.

Technically this is planned to be realized by the Paella player, a HTML5 multi-stream video player capable of playing multiple audio & video streams synchronously which is specially designed for lecture recordings with Opencast Lectures developed and provided as open source by the university politecnica de Valencia [12].

5 Summary and Conclusions

With the full automated recording solution in place we expect a significant improvement for students to prepare for the examinations. They also can now recap certain parts of their study later on when they happen to need it again. Additionally the recording also provides more flexibility to students in terms of time, in case a lesson cannot be attended due to e.g. children caring activities the content still is available for them and they can still get the most important information.

We introduced a concept and a technical workflow for automated lecture recording at a medical university. However, this concept is not restricted to medical education but can be applied at any higher education institution. The software solution is based on open source software in order to save costs, the hardware solution is based on many years of experience of a partner university.

We proposed to take advantage of the experience of a partner university who has already experience with lecture recording and initialized a funded research project in order to transfer the knowledge and exchange ideas, which will help also themselves to further improve their existing system.

The system will be applied in two stages. Stage one is the hardware based recording solution which is mounted into the lecture rooms (camera and recording hardware with storage). Stage two is the management software for the recorded lessons and the streaming solution for the students. The recorded lessons will be additionally seamlessly integrated in our open source LMS Moodle where they are linked to the appropriate modules and can be directly viewed by the students.

Furthermore we provide now an easy possibility even for teachers who are not affine with technical equipment. Every teacher shall now be able to record lessons without the need of time consuming technical trainings and reading manuals.

The student view will provide two HD streams of the PC output and the speaker/whiteboard. Students will have full control of the interface by scaling and moving the videos according to their needs.

References

1. Rehatschek, H., Aigelsreiter, A., Regitnig, P., Kirnbauer, B.: Introduction of eLectures at the Medical University of Graz - results and experiences from a pilot trial. iJet. Int. J. Emerg. Technol. Learn. **8**(1), 29–36 (2013). ICL 2012, ISSN 1868-8799
2. Rehatschek, H., Hruska, A.: Fully automated virtual lessons in medical education. In: Proceedings of the International Conference on Interactive Collaborative Learning (ICL), Kazan, Russian Federation, 25–27 September 2013, pp. 3–8. IEEE Catalog Number: CFP1323R-ART, ISBN 978-1-4799-0153-1
3. Rehatschek, H., Aigelsreiter, A., Regitnig, P., Kirnbauer, B.: Introduction of eLectures at the Medical University of Graz – results and experiences from a pilot trial. In: proceedings of the International Conference on Interactive Collaborative Learning (ICL), Villach, Austria, 26–28 September 2012. IEEE Catalog Number: CFP1223R-USB, ISBN:978-1-4673-2426-7
4. Rehatschek, H., Hye, F.: The introduction of a new virtual microscope into the eLearning platform of the Medical University of Graz. In: Proceedings of the 14th Conference on Interactive Collaborative Learning (ICL), Piešťany, Slovakia, 21–23 September 2011, pp. 10–15. ISBN 978-1-4577-1746-8
5. Extron Electronics, Interfacing switching and control, March 2017. http://www.extron.com
6. Panopto, March 2017. https://www.panopto.com
7. StreamAMG broadcast quality, March 2017. https://www.streamamg.com/
8. Epiphan, capture stream record, March 2017. https://www.epiphan.com/
9. Opencast Matterhorn, open source solution for automated video capture and distribution at scale, March 2017. http://www.opencast.org/matterhorn
10. Learning Tools Interoperability, IMS specification, Moodle Addon, May 2017. https://docs.moodle.org/32/en/LTI_and_Moodle
11. Tube, Video portal of the technical university of Graz, May 2017. https://tube.tugraz.at
12. Paella player, multistream player for lectures, May 2017. http://paellaplayer.upv.es/

Collaborative Learning Advancing International Students

A Multidisciplinary Approach

Ana Virtudes[1(✉)], Ilda Inacio Rodrigues[2], and Victor Cavaleiro[1]

[1] Department of Civil Engineering and Architecture, University of Beira Interior (UBI), Covilhã, Portugal
{virtudes,victorc}@ubi.pt

[2] Department of Mathematics, UBI, Covilhã, Portugal
ilda@ubi.pt

Abstract. Universities are making great efforts in order to improve their rates in terms of internationalization. With the rise of international mobility, among students, new challenges are emerging on the high education system. In this sense, this paper aims to show some key strategies in terms of teaching methodologies that Universities should follow forward to a better internationalization performance. It is based on the case of University of Beira Interior in Portugal, in a multidisciplinary approach, which involves scholars from the Department of Civil Engineering and Architecture, and from the Department of Mathematics, teaching at the Civil Engineering studies. The main research questions of this paper are the following: are the currently used teaching methodologies prepared to accommodate international students who have different standards of proficiency whether in terms of language or in basic knowledge of mathematics? Which are the challenges of high education system, in order to improve the performance of international students, engaging them as part of a collaborative learning approach? The conclusions show that there is the need of following new teaching methodologies in a more collaborative approach, in order to promote the integration of international students.

Keywords: Collaborative learning · International students
Multidisciplinary approach · Engineering studies

1 International Students at the Portuguese Universities

All over the world and in European countries in particular, after Bologna process, the universities are making great efforts in order to improve their rates in terms of internationalization. They are trying to be more attractive for students coming from abroad, in a multicultural context of diversity. In fact, one of the results of Bologna agreement, with the European Credit Transfer System

M. E. Auer et al. (eds.), *Teaching and Learning in a Digital World*,
Advances in Intelligent Systems and Computing 715,
https://doi.org/10.1007/978-3-319-73210-7_3

(ECTS), comprising 29 European countries, including Portugal, was the rise of international mobility, among students. Therefore, new challenges are emerging on the high education system, requiring a more collaborative learning in order to improve the performance of international students. In this sense, the main goal of this paper is to show some key strategies in terms of teaching methodologies that Universities should follow towards a better internationalization performance.

It is focused on the case study of the University of Beira Interior (UBI) in the Interior region of Portugal, in the city of Covilhã, which is not far away from the border with Spain. It regards to a multidisciplinary approach, involving scholars from different departments, the Department of Civil Engineering and Architecture (DECA), and the Department of Mathematics (DM), teaching at Civil Engineering studies. Consequently, the research questions for these scholars are the following: are the currently used teaching methodologies prepared to accommodate international students, with different standards of proficiency, whether in terms of language or in basic knowledge of mathematics? Which are the challenges of high education system, in order to improve the performance of international students, engaging them as part of a collaborative learning approach?

The first university was created in Portugal in 1290. Since then, the evolution of the higher education system comprises a set of two types of institutions: the polytechnics and the universities. There are both, public and private institutions, highly developed and well respected [1]. The Polytechnics are mainly focused on vocational and practical training. They offer several academic subjects such as accountancy, teaching or nursing. The Universities are mainly focused on research and theory skills. They offer academic subjects in sciences, law or architecture. Subjects such as civil engineering, are offered by polytechnics and universities. Despite the majority of the degrees are taught in Portuguese, in some cases the language is the English and the students are required to have a good level of proficiency to be admitted. If some years ago, the international students were mainly coming from Portuguese speaking countries, such as Angola or, Brazil currently they are coming from non-Portuguese speaking countries. Thus, they should have good language skills in English. An increasing number of international students is responsible for new approaches at the teaching context, focused on the way its teaching methodologies are related to the way how students are learning [2]. Such as in other countries [3], the literature reveals that this is a multi-faceted phenomenon.

2 Collaborative Learning as a Methodology

2.1 A Multidisciplinary Approach for the Integration of International Students

Several teaching methodologies are ongoing at the Master Degree in Civil Engineering (MIEC) at UBI, in order to improve the performance of international students. They are based on a multidisciplinary approach, engaging scholars of the DECA and the DM. Given that a proficient knowledge in maths is a crucial issue, in particular in the fields of engineering, the maths scholars are developing new strategies to improve the knowledge of international students, with a low level of skills in this domain. There is the use of ICT, *information and communication technologies*, such as e-learning platforms, designed to help the students to understand the maths calculations and exercises. The used software was designed as a mobile-accessible tool on tablets, compatible with a virtual course management environment (online), which is an e-learning platform.

The key outcomes of the new teaching methodologies at the MIEC, show that the main changes in teaching practice are the following: to have a reduced number of students working in each group, mixing the Portuguese speakers with the others, to get a better engagement of all; to have extra hours of classes in the domain of math, which is the base of several subjects in the research field of engineering; to have additional materials in terms of bibliographic references; to use ICT tools. The latter is related to e-learning platforms used to solve automatically maths exercises, showing with a detail all the steps of the exercises resolution (see two examples below) (Figs. 1 and 2):

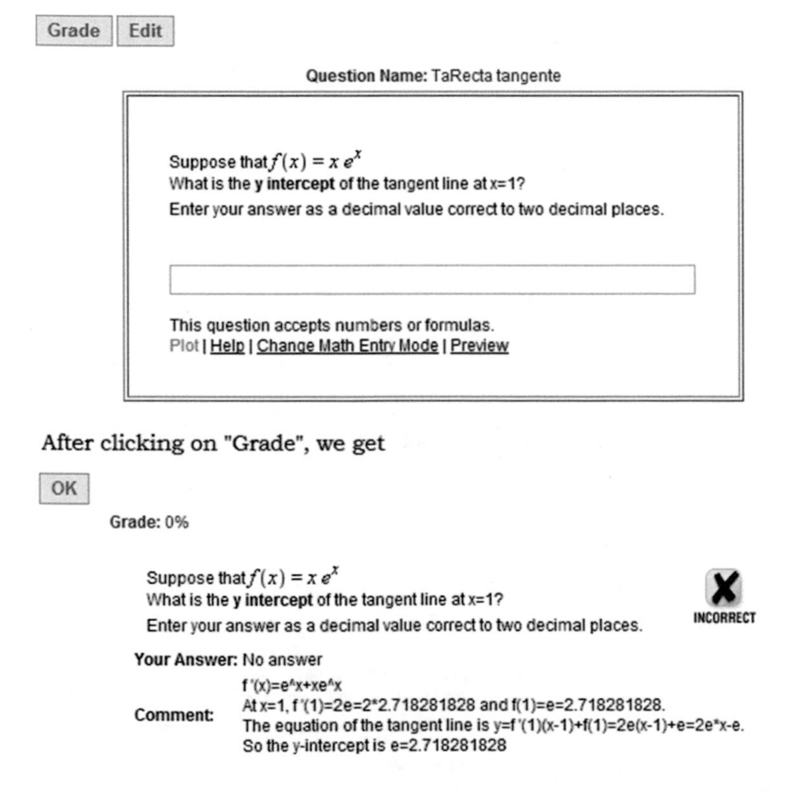

Fig. 1. An example of exercising the e-learning platform used in the DM (*MAPLE T.A.*)

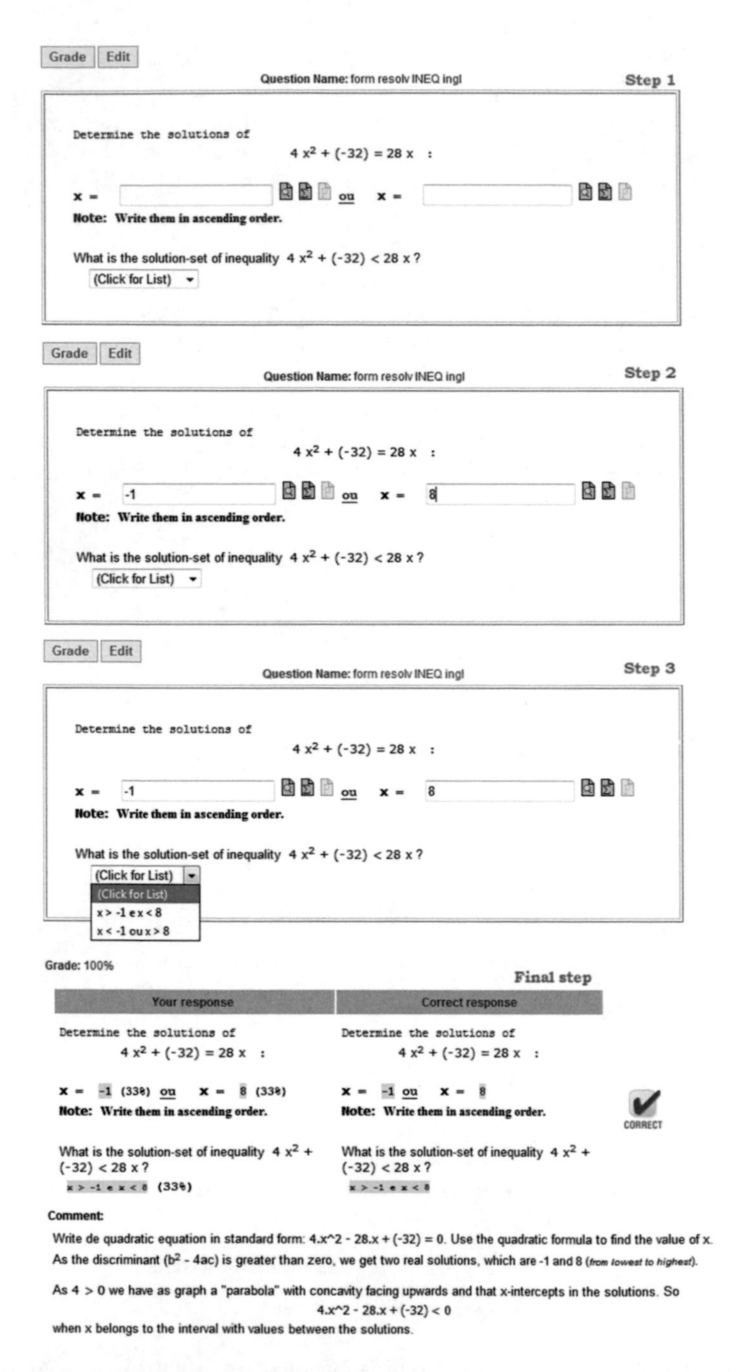

Fig. 2. Another example of exercising the e-learning platform used in the DM (*MAPLE T.A.*)

In Europe, there are several programmes and funds focused on the exchange of international students in between the countries, supported by a political and scientific interest in their mobility [4]. New demands in the teaching process are required to scholars in terms of good practices to integrate diverse groups of international students. Among these good practices there is the idea that students should be taught in a way that allows them to make decisions while surrounded by uncertainty [1], in order to lead teachers to adapt their teaching approaches to match with the students' background context [5]. These authors advocate that this goal requires turning from a traditional teacher-focused approach of teaching to a student-centred approach of teaching within a knowledge-centred learning framework [2]. The international students require adjustments from their home school 'culture of learning' or previous universities [1], by the higher education institutions, considering their different backgrounds, ethical standards and principles [6].

The Australian universities have a great experience in this type of teaching methodologies, based on their contact with international students coming from the Asia-Pacific countries. Examples like the Australian University of Victoria are promoting a collaborative process in order to adapt the teaching methodologies, by defining the teaching/learning framework regarding the international context or training the scholars to work in a multicultural framework and a praxis of international cooperation in research domain [7–10]. The experience at the MIEC with international students, reveals a diversity of experiences on the development of students' preparation for the globalized society, examining students' views and attitudes, is contributing to their intergroup attitudes and civic engagement. This result is consistent with the international literature [11–14], proposing that universities should be aware of the ways in which students can use their diversity as a strength. Thus, universities play a pivotal role in fostering high-quality in intercultural terms amongst their students, preparing them for a diverse and global society [11], alerting for an interdisciplinary approach [15].

2.2 The Pedagogic Triangle

Given that teachers are responsible for training human resources, they are present as active agents in all speeches about education. The result of this training is the basis for the economic development of the countries. Therefore, scholars are pivotal branches on the development of societies, being responsible for training the generations for the challenges of the 21st century, under a globalized and technological realm. The pedagogic triangle (see Fig. 3.), proposed by Jean Houssaye [16] refers that there are three vertices on the educational process, which are the following: the teachers, the students and the knowledge. According to this scheme the teachers and the students are responsible for the educative process; the teaching process is the result of the relationship in between teachers and knowledge; and the learning process results from the engagement in between students and knowledge.

According to the same reference, is possible to imagine three main pedagogic models: the connection in between teachers and knowledge which is focused

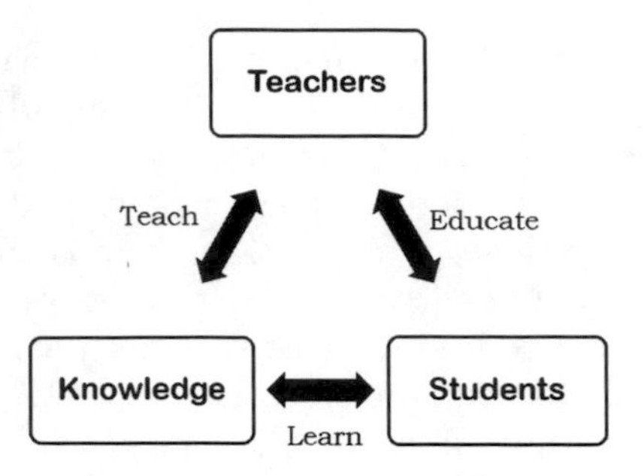

Fig. 3. The pedagogic triangle of the educational process

on the teaching process and on the knowledge transmission; the connection in between teachers and students focused on the valorization of the relational and formative processes; and finally, the articulation between the students and knowledge favoring a logic of (self) learning. At this point, there is the interest of highlighting the tendencies that point to a consolidation of the know-how of students, with teachers occupying the 'place of the dead'. The latter approach is not intending to criticize pedagogical situations that rely on self-training or self-management practices, but rather to alert the reappearance of movements that advocate a technology of teaching. Nowadays, the technological developments and the success of planetary expansion strategies for computing and telecommunications equipment, place the debate in a new perspective [17]. In this sense, what seems to be important is the way in which sometimes theoretical discourses are built that underlie a certain devaluation of the human relationship and the qualifications of teachers. The use of teaching technologies implies not only the acquisition of new skills, but also the reinforcement of traditional skills. Consequently, there is some difficulties in imagining an educational process that does not rely on the relational and cognitive mediation of teachers.

2.3 The Knowledge Triangles

The knowledge triangle of the educational process (see Fig. 4.a) is a translation of three great types of knowledge: the knowledge of experience, related to the teachers; the knowledge of pedagogy, related to experts in education sciences; and the knowledge of the disciplines, regarding several specialists from different fields of knowledge. Considering the European experience and its history, there are some authors [18] advocating that the free circulation of knowledge and ideas was stronger in the past times rather than currently. However, this situation depends on the country and its educational system. In any cases, there is a learning triangle (see Fig. 4.b) consisting of education, research and innovation, which is a key factor for the productivity growth [19].

In sum the universities have to substantially improve the ways how the knowledge is transferred throughout the educational process. The production of new knowledge through research, and the use and application of knowledge through

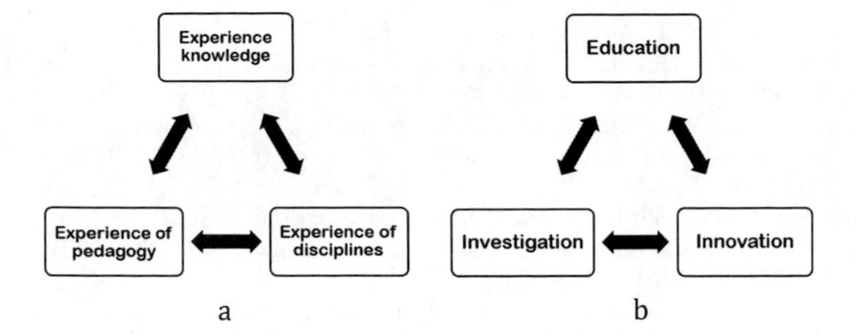

Fig. 4. The knowledge triangles of the educational process

innovation are key words on this process. In this sense, there is the need of encouraging the use of information and communication technologies/ICT which is the backbone of the knowledge economy, an area where Europe has not performed well.

3 Conclusions

As this article, has reveals, the key word for a better integration of international students is to involve them at the learning process, being part of the definition of the features of units'contents, in an open debate, considering case studies coming from their home backgrounds. This methodology requires a multidisciplinary approach of scholars of different scientific domains, gathering at the teaching of engineering domains.

A better way of facilitate this task, improving the performance of international students is the use of ICT, which has revealed as a crucial tool, in particular in the domain of math. In this sense, one of the challenges still ahead at the high education system is to have means to insure the functioning of these tools, requiring permanent uploads of the software and technical support.

Acknowledgment. This work is supported with Portuguese national funds by FCT - Foundation for Science and Technology, I.P., within the GEOBIOTEC - UID/GEO/04035/2013.

We would like to thank Professor Rogério Serôdio (UBI) for his help.

References

1. Mavor, S.: Socio-culturally appropriate methodologies for teaching and learning in a Portuguese university. Teach. High. Educ. **6**(2), 183–201 (2010)
2. Allendoerfer, C., Wilson, D., Kim, M.J., Burpee, E.: Mapping beliefs about teaching to patterns of instruction within science, technology, engineering, and mathematics. Teach. High. Educ. **19**(7), 758–771 (2014)

3. Skyrme, G., McGee, A.: Pulled in many directions: tensions and complexity for academic staff responding to international students. Teach. High. Educ. (2016). https://doi.org/10.1080/13562517.2016.1183614
4. Kratz, F., Netz, N.: Which mechanisms explain monetary returns to international student mobility? Stud. High. Educ. (2016). https://doi.org/10.1080/03075079.2016.1172307
5. Virtudes, A., Cavaleiro, V.: Teaching methodologies in spatial planning for integration of international students. Earth Environ. Sci. **44**, 1–6 (2016). https://doi.org/10.1088/1755-1315/44/3/032022
6. Arenas, E.: How teachers' attitudes affect their approaches to teaching international students. High. Educ. Res. Dev. **28**(6), 615–628 (2009)
7. Guruz, K.: Higher Education and International Student Mobility in the Global Knowledge Economy. New York State University Press, Albany (2008)
8. Kumar, M., Ang, S.: Transitional issues of induction into design education for international undergraduate students: a case study analysis of architecture. J. Educ. Built Environ. **3**(2), 10–32 (2008)
9. Jolley, A.: Exporting Education to Asia. Victoria University Press for the Center for Strategic Economic Studies, Victoria (1997)
10. Daniels, J.: Internationalisation, higher education and educators' perceptions of their practices. Teach. High. Educ. **18**(3), 236–248 (2013)
11. Denson, N., Bowman, N.: University diversity and preparation for a global society: the role of diversity in shaping intergroup attitudes and civic outcomes. Stud. High. Educ. **38**(4), 555–570 (2013)
12. Gurin, P., Nagda, B.A., Lopez, G.E.: The benefits of diversity in education for democratic citizenship. J. Soc. Issues **60**, 17–34 (2004)
13. Laird, T.F.N., Engberg, M.E., Hurtado, S.: Modelling accentuation effects: enrolling in a diversity course and the importance of social action engagement. J. High. Educ. **76**, 448–476 (2005)
14. Meaney, T., Rangnes, T.E.: Book review: how research fields change - the documentation of a process. In: Halai, A., Clarkson, P. (eds.) Teaching and Learning Mathematics in Multilingual Classrooms. Educational Studies in Mathematics, vol. 95, pp. 219–227 (2017). https://doi.org/10.1007/s10649-017-9753-8
15. Domíngues-Mujica, J. (ed.): Global Change and Human Mobility. Springer, Singapore (2016)
16. Nóvoa, A.: O lugar dos Professores: o terceiro excluído? (2017). http://www.apm.pt/apm/revista/educ50/educ50_3.htm
17. Misfeldt, M., Jankvist, U.T., Aguilar, M.S.: Teachers' beliefs about the discipline of mathematics and the use of technology in the classroom. Math. Educ. **11**(2), 395–419 (2016)
18. Barroso, J.M.D.: O triângulo do conhecimento: uma base sólida para o crescimento e o emprego (2005). https://www.publico.pt/espaco-publico/jornal/o-triangulo-do-conhecimento-uma-base-solida-para-o-crescimento-e-o-emprego-12319
19. Maassen, P., Stensaker, B.: The knowledge triangle, European higher education policy logics and policy implications. High. Educ. **61**(6), 757–769 (2011). https://doi.org/10.1007/s10734-010-9360-4

Collaborative Learning of DC Transients in Series Circuit with MATLAB as a Learning Aid

Gargi Basu[1] and Urmila Kar[2](✉)

[1] Jnan Chandra Ghosh Polytechnic, 7, Mayurbhanj Road, Kolkata 700023, India
himanibasu@gmail.com
[2] NITTTR, Kolkata, Block-FC, Sector-II, Salt Lake City 700106, India
urmilakar@rediffmail.com

Abstract. This paper proposes an active learning approach for study of DC transients in series circuits. Students, divided into small groups, are advised to simulate various circuits using MATLAB and study the waveforms and construct the equations characterizing them. This method helps in concept development through collaborative learning.

Keywords: Collaborative learning · Problem-solving ability
Energy storing elements · Simulation · Switching

1 Introduction

Transient phenomenon occurs when a circuit containing inductor or capacitor is suddenly connected or disconnected from the supply or there is a sudden change in the supply voltage or a component (resistor/inductor/capacitor) is varied. Transients may harm both electrical and electronic devices. Thus transient study is an essential part of curriculum of Electrical Engineering both at Diploma and Degree level.

Usually in a conventional teacher-centric classroom, transient is taught by deriving an expression for the circuit current or capacitor or inductor voltage through the classical method of solving differential equations or using Laplace transform, followed by plotting of these quantities. During these mathematical derivations, many of the students lose interest in this topic. It has been found that active learning is a better option [7] for the 21st century students. In this paper, a collaborative learning approach to study DC transients is suggested. This facilitates students' achievement and content literacy which is the goal of modern education system through cognitive thinking and comprehension [4]. It also helps them to work together, support each other in learning through sharing of ideas and discover [1] which is considered as a valuable skill by the employers.

Students are advised and guided by the instructor to build circuits and observe the waveforms first and arrive at the mathematical part on themselves

M. E. Auer et al. (eds.), *Teaching and Learning in a Digital World*,
Advances in Intelligent Systems and Computing 715,
https://doi.org/10.1007/978-3-319-73210-7_4

through experiential learning [5]. Doing these in real life is not only time consuming but also not economically feasible, due to limitations of available resources. Hence, simulation of the circuits seems to be more viable and is suggested in this paper.

2 Approach

A batch of sixty students are subdivided into three groups comprising of twenty students in each group.

Group-I Study of transients in R-L circuit.
Group-II Study of transients in R-C circuit.
Group-III Study of transients in R-L-C circuit.

Each group is subdivided into three sub-groups.

Group-A Simulate and comment on circuit current.
Group-B Simulate and comment on voltage drop across circuit resistance.
Group-C Simulate and comment on voltage drop across inductor/capacitor.

2.1 Assignment Modules of Group I

Students are advised to
Module 1 (Allotted time: 3 days): Use Simscape tool box of MATLAB to build models resembling the real life series R-L circuit and study the following cases.

Case-1 it is suddenly connected to a DC source.
Case-2 when supply voltage is suddenly increased.
Case-3 when supply voltage is suddenly reduced.
Case-4 it is suddenly disconnected from a DC source and shorted.

Module 2 (Allotted time: 2 days): Use the basic Simulink blocks to form mathematical model of the above circuit and study all the cases.
Module 3 (Allotted time: 2 days): Use Transfer function block and comment.
Module 4 (Allotted time: 2 days): Write Script file and study all the cases.

2.2 Getting Started [2]

Before they start, following suggestions were provided to them:

- To measure current and different voltage drops, as in real life systems, voltage and current sensors are required.
- To link the physical system (formed by Simscape) to Simulink, PS-S blocks are to be incorporated.
- While using the Simscape blocks, initial conditions of the circuit, if any, is to be assigned to the inductor or/and capacitor blocks.

- To keep the models flexible, the values of different parameters are to be provided from the MATLAB workspace.
- Inclusion of a switch is not required in these models. When the Start simulation button is clicked, the effect of sudden connection to or disconnection from supply is obtained.
- To form a mathematical model using the basic Simulink blocks, initial conditions of the circuit, if any, is to be assigned to the integrator blocks.
- Data forming the waveforms may be saved to workspace and plotted later using a small script file (Fig. 2).
  Steps to be followed:
  - Go to Scope window: *parameters>data history* menu (Fig. 1).
  - Check the *Save data to workspace* box.
  - Give a variable name and set format to *Structure with time.*

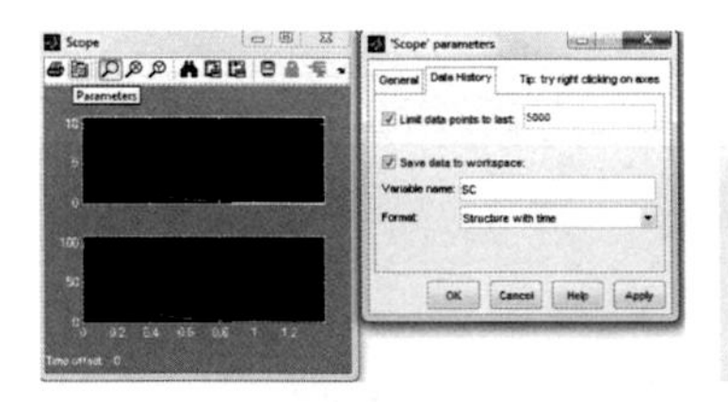

Fig. 1. Scope parameters window

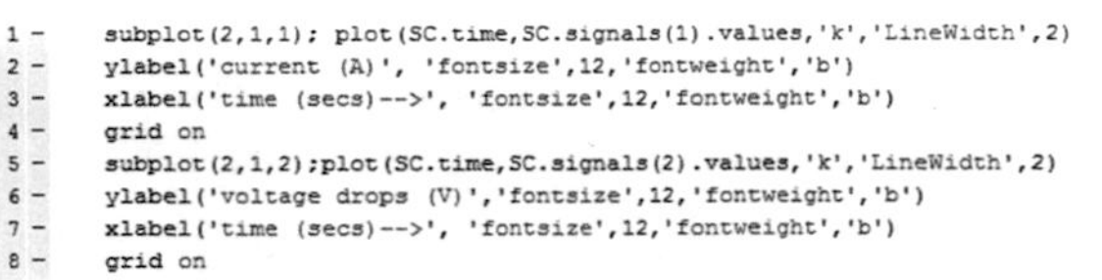

```
1 -  subplot(2,1,1); plot(SC.time,SC.signals(1).values,'k','LineWidth',2)
2 -  ylabel('current (A)', 'fontsize',12,'fontweight','b')
3 -  xlabel('time (secs)-->', 'fontsize',12,'fontweight','b')
4 -  grid on
5 -  subplot(2,1,2);plot(SC.time,SC.signals(2).values,'k','LineWidth',2)
6 -  ylabel('voltage drops (V)','fontsize',12,'fontweight','b')
7 -  xlabel('time (secs)-->', 'fontsize',12,'fontweight','b')
8 -  grid on
```

Fig. 2. Script file to plot waveforms

3 Activities of Group-I

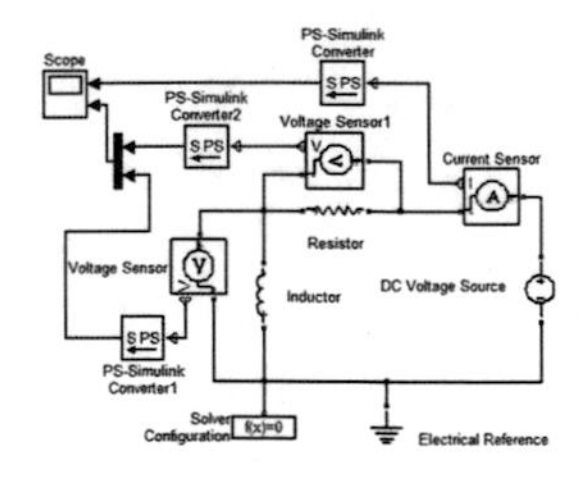

Fig. 3. Simscape model developed by Group-I

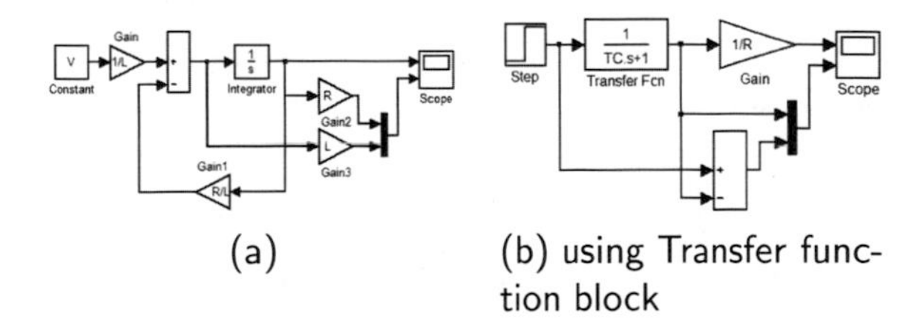

Fig. 4. Mathematical models developed by Group-I

The shared goal of Group-I is to simulate a series circuit comprising of resistance R and inductance L. Let V be the magnitude of DC supply voltage, i be the current flowing into the circuit and v_R and v_L be the potential drops across resistance and inductance. At the instant of switching, values of i, v_R and v_L are $I_0(=\frac{V_{R0}}{R})$, $V_{R0}(=I_0R_0)$ and $V_{L0}(=V-V_{R0})$ respectively.

3.1 Analysis of the Problem

Under the guidance of instructor, the members of Group-I, through group discussion and peer teaching, were able to analyze that Case 1, 3 and 4 are just variations of Case 2. For **case 1**, no need to provide initial value of circuit current, I_0 i.e. inductor current. For cases 2, 3 and 4, initial values should be provided. For **case 4**, supply voltage should be made zero. Hence, if a model for case 2 can be formed, it will help to study all the cases.

3.2 Models and Outputs

The model developed using Simscape is shown in Fig. 3. Initial value of current, I_0 (IL0 assigned to *Inductor* block) was supplied from the MATLAB workspace. Mathematical model (based on Eq. 1) using basic Simulink library is shown in Fig. 4a.

$$\dot{x} = \frac{V}{L} - \frac{R}{L}x \tag{1}$$

$$\text{where } i = x.$$

If the circuit is initially relaxed, i.e. $I_0 = 0$, for a series R-L circuit, admittance transfer function is $\frac{I(s)}{V(s)} = \frac{1}{R}.\frac{1}{1+\tau s}$ and voltage transfer function is $\frac{V_R(s)}{V(s)} = \frac{1}{1+\tau s}$. Developed model is shown in Fig. 4b.

While using transfer function blocks, even though the no. of blocks used in a model is reduced but it has been observed that only Case 1 can be studied as there is no provision to supply the initial conditions. Script file was written using Eq. 1.

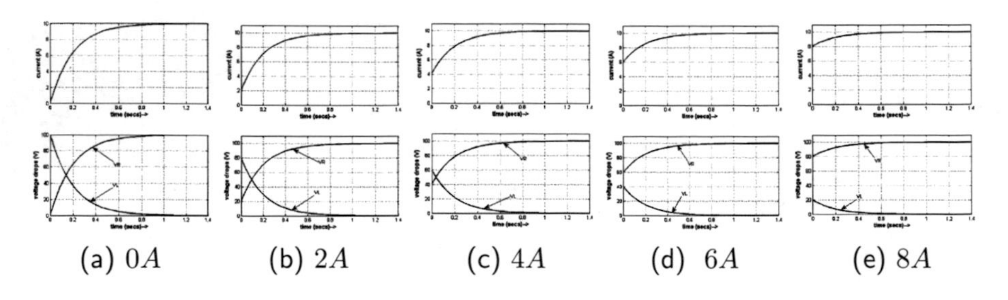

(a) $0A$ (b) $2A$ (c) $4A$ (d) $6A$ (e) $8A$

Fig. 5. Waveforms obtained Group-I: Case 1 & 2 for different I_0

These models were tested with $R = 10\,\Omega$ and $L = 2\,\mathrm{H}$. Waveforms from different models are studied and found that they are identical.

Case 1: With $V = 100\,\mathrm{V}$, the waveforms obtained are shown in Fig. 5a. Tangents were drawn to find the initial rates of change of different quantities.

Case 2: When circuit voltage suddenly increases, there is some initial value of circuit current. For different initial values of 2 A, 4 A, 6 A and 8 A, waveforms are shown in Figs. 5b, c, d and e respectively.
Case 3: Supply voltage is decreased from 200 V to 100 V. With initial value of inductor current 20 A, waveforms obtained are shown in Fig. 6.
Case 4: With $V = 0\,\text{V}$, initial values of circuit current 10 A, 15 A and 20 A the waveforms obtained are shown in Figs. 7a, b and c respectively. Tangents were drawn to find the initial rates of change of different quantities.

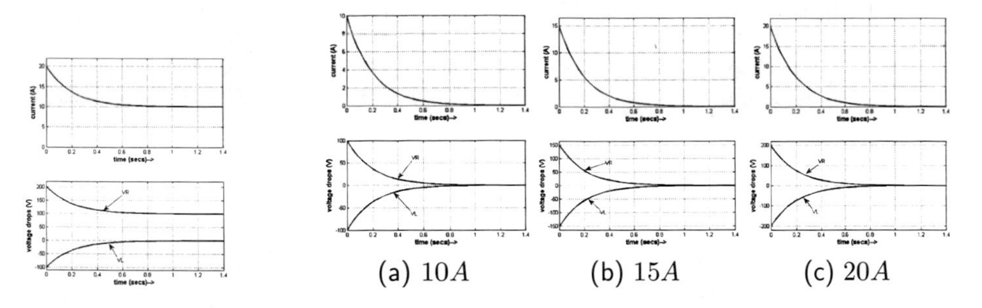

(a) $10A$ (b) $15A$ (c) $20A$

Fig. 6. Group-I: Case 3

Fig. 7. Waveforms obtained by Group-I: Case 4 for different I_0

Under the guidance of instructor, the members of subgroups listed their observations on different waveforms through group discussion (Table 1). Later three subgroups sat together, shared their views and prepared the Table 3 for case 1 and 2. Similarly, for Case 4, observations are listed in Table 2 and prepared the Table 4.

It was observed that students themselves from their observations could arrive to the expressions for circuit current and voltage drops without long mathematical derivations.

Remarks: If an R-L series circuit is suddenly connected to the supply, the transient phenomenon is observed due to presence of inductance. The steady state is achieved after a time period solely determined by the circuit parameters, L and R i.e. time constant of the circuit (L/R). Once, the transient is over,

Case 1, 2 & 3: inductor acts as a short-circuit, the entire supply voltage appears across the resistor.

Case 4: stored energy in the inductor entirely gets dissipated in the resistor.

4 Activities of Group-II

The shared goal of Group-II is to simulate a series circuit comprising of resistance, R and capacitance C. Let V be the magnitude of DC supply voltage, i be the current flowing into the circuit and v_R and v_C be the potential drops across

Table 1. Observations of Group-I: Case 1 & 2

Subgroup-A	*Subgroup-B*	*Subgroup-C*
Whatever be the initial value, 1. Circuit current i rises exponentially. 2. Initial value of current, $i = I_0$. 3. Steady state value obtained is $I_{ss} = V/R = 100/10 = 10A$. 4. At t=0.2sec, the circuit current reaches its steady state value (10A) if it was allowed to rise linearly with its initial rate of rise. This period of time is defined as time constant, τ (measured in seconds). Time constant $\tau = 0.2sec = 2/10 = L/R$. 5. Initial rate of rise of current depends on the difference of the final value and initial value ($I_1 = I_{ss} - I_0$) and circuit parameters($\tau = L/R$). 6. Current reaches its final value at t=1sec which is almost five times the circuit time constant . 7. Generalized expression for current, $i = I_{ss} - I_1 e^{-t/\tau}$ where $I_1 = I_{ss} - I_0$. 8. Generalized expression for rate of change of current, $\frac{di}{dt} = \frac{I_1}{\tau} e^{-t/\tau}$. 9. Higher the value of circuit time constant, longer it takes to reach the steady state.	1. Voltage across the resistance,v_R rises exponentially. 2. Initial value of voltage across the resistance, $v_R = V_{R0}$. 3. Steady state value obtained $V_R = V = 100V$ 4. At t=0.2sec, the v_R reaches its steady state value 100V if it was allowed to rise linearly with its initial rate of rise. 5. Initial rate of rise of v_R depends on the initial value of inductor voltage and circuit parameters (i.e. time constant). 6. v_R reaches its final value at t=1sec at about $t = 5\tau$. 7. Generalized expression for voltage across the resistance, $v_R = V - (V - V_{R0})e^{-t/\tau} = V - V_{L0}e^{-t/\tau}$	1. Voltage across the inductance, v_L falls to zero exponentially. 2. Initial value of voltage across the inductance, $v_L = V - V_{R0}$. 3. At steady state $v_L = 0$, i.e. inductance acts as a short circuit. The entire voltage appears across the resistance. Steady state current is solely determined by the circuit resistance. 4. At $t = \tau$, the v_L reaches its steady state value 0V if it was allowed to fall linearly with its initial rate of fall. 5. Initial rate of fall of v_L depends on the initial value of inductor voltage and circuit parameters (i.e. time constant). 6. v_L reaches its final value at t=1sec which is almost five times the circuit time constant. 7. Generalized expression for voltage across the inductance, $v_L = (V - V_{R0})e^{-t/\tau} = V_{L0}e^{-t/\tau}$.

resistance and capacitance. At the instant of switching, values of i, v_R and v_C are $I_0(= \frac{V_{R0}}{R})$, $V_{R0}(= V - V_{C0})$ and V_{C0} respectively.

Simscape model and mathematical model (based on Eq. 2) developed by Group-II are shown in Figs. 8 and 9a respectively.

$$\dot{x} = \frac{V}{RC} - \frac{1}{RC}x \quad (2)$$
$$\text{when}\, x = v_C.$$

Initial value of capacitor voltage, V_{C0} was assigned to *Integrator* block.

Table 2. Observations of Group-I: Case 4

Subgroup-A	*Subgroup-B*	*Subgroup-C*
1. Current falls exponentially. 2. Initial value of circuit current, $i = I_0$. 3. Steady state value obtained is $I_{ss} = 0A$. 4. Initial rate of fall of current, $\frac{di}{dt} = \frac{V_{R0}}{L}$ (Figure **7(a)**). 5. At $t = \tau = 0.2sec$, the circuit current reaches its steady state value (0A) if it was allowed to fall linearly with initial rate of fall of current. 6. At $t = 5 \times 0.2sec = 1sec$ which is five times the circuit time constant, current reaches to almost its final value. 7. Generalized expression for current, $i = I_0e^{-t/\tau}$. 8. Generalized expression for rate of change of current, $\frac{di}{dt} = -\frac{I_0}{\tau}e^{-t/\tau} = -\frac{V_{R0}}{L}e^{-t/\tau}$. The negative sign indicates that the current is falling.	1. v_R falls exponentially. 2. Initial value of voltage across the resistance, $v_R = V_{R0} = I_0R$. 3. Steady state value obtained $V_R = V = 0V$ 4. At t=0.2sec, the v_R reaches its steady state value 100V if it was allowed to fall linearly with its initial rate of fall. 5. Initial rate of fall of v_R depends on the initial value of resistor voltage and circuit parameters (i.e. time constant). 6. v_R reaches almost its final value at t=1sec which is $t = 5\tau$. 7. Generalized expression for voltage across the resistance, $v_R = RI_0e^{-t/\tau} = V_{R0}e^{-t/\tau}$	1. v_L rises to zero exponentially. 2. Initial value of voltage across the inductance, $v_L = -V_{R0}$. 3. At steady state $v_L = 0$. 4. At $t = \tau$, the v_L reaches its steady state value 0V if it was allowed to rise linearly with its initial rate of rise. 5. Initial rate of increase of v_L depends on the initial value of resistor voltage and circuit parameters (i.e. time constant). 6. v_L reaches almost its final value at t=1sec which is five times the circuit time constant. 7. Generalized expression for voltage across the inductance, $v_L = -v_R = -V_{R0}e^{-t/\tau}$

If the circuit is initially relaxed, i.e. $V_{C0} = 0$, voltage transfer function for a series R-C circuit, $\frac{V_C(s)}{V(s)} = \frac{1}{1+\tau s}$. A simple model of R-C series circuit is derived as shown in Fig. 9b. The script file was written using Eq. 2.

These models were tested with $R = 100\,\Omega$, capacitance, $C = 1\,\mu$F.

Case-1: Suddenly connected to a DC source: With V = 100 V, the waveforms obtained are shown in Fig. 10a.

Cases-2: Supply voltage is suddenly increased: Waveforms obtained are shown in Figs. 10b, c, d and e.

Students' observations are tabulated as in Table 5.

Cases-3: Supply voltage is suddenly decreased: Waveforms obtained are shown in Fig. 11.

Case-4: Disconnected from supply and shorted: Output waveforms for initial values of capacitor voltage 100 V, 150 V and 200 V are shown in Figs. 12a, b and c respectively. Students' observations are tabulated as in Table 6.

Table 3. Students' observations on waveforms of Group I: Case 1 & 2

At $t = 0$ (initial values)						At $t = \tau$						At $t = 5\tau$						At $t \to \infty$ (final values)					
i	v_R	v_L	$\frac{di}{dt}$	$\frac{dv_R}{dt}$	$\frac{dv_L}{dt}$	i	v_R	v_L	$\frac{di}{dt}$	$\frac{dv_R}{dt}$	$\frac{dv_L}{dt}$	i	v_R	v_L	$\frac{di}{dt}$	$\frac{dv_R}{dt}$	$\frac{dv_L}{dt}$	i	v_R	v_L	$\frac{di}{dt}$	$\frac{dv_R}{dt}$	$\frac{dv_L}{dt}$
A	V	V	A/s	V/s	V/s	A	V	V	A/s	V/s	V/s	A	V	V	A/s	V/s	V/s	A	V	V	A/s	V/s	V/s
0	0	100	50	500	-500	6.3	63.2	36.8	18.4	184	-184	9.93	99.3	0.7	0.34	3.4	-3.4	10	100	0	0	0	0
2	20	80	40	400	-400	7	70.6	29.4	14.7	147	-147	9.95	99.5	0.5	0.27	2.7	-2.7	10	100	0	0	0	0
4	40	60	30	300	-300	7.8	77.9	22.1	11.0	110	-110	9.96	99.6	0.4	0.2	2	-2	10	100	0	0	0	0
6	60	40	20	200	-200	8.53	85.3	14.7	7.35	73.5	-73.5	9.97	99.7	0.3	0.13	1.3	-1.3	10	100	0	0	0	0
8	80	20	10	100	-100	9.26	92.6	7.4	3.68	36.8	-36.8	9.99	99.9	0.1	0.07	0.7	-0.7	10	100	0	0	0	0
$I_{ss} - I_1 e^0$	$V - V_{L0} e^0$	$V_{L0} e^0$	$\frac{I_1}{\tau} e^0$	$\frac{V_{L0}}{\tau} e^0$	$-\frac{V_{L0}}{\tau} e^0$	$I_{ss} - I_1 e^{-1}$	$V - V_{L0} e^{-1}$	$V_{L0} e^{-1}$	$\frac{I_1}{\tau} e^{-1}$	$\frac{V_{L0}}{\tau} e^{-1}$	$-\frac{V_{L0}}{\tau} e^{-1}$	$I_{ss} - I_1 e^{-5}$	$V - V_{L0} e^{-5}$	$V_{L0} e^{-5}$	$\frac{I_1}{\tau} e^{-5}$	$\frac{V_{L0}}{\tau} e^{-5}$	$-\frac{V_{L0}}{\tau} e^{-5}$	$I_{ss} - I_1 e^{-\infty}$	$V - V_{L0} e^{-\infty}$	$V_{L0} e^{-\infty}$	$\frac{I_1}{\tau} e^{-\infty}$	$\frac{V_{L0}}{\tau} e^{-\infty}$	$-\frac{V_{L0}}{\tau} e^{-\infty}$

Table 4. Students' observations on waveforms of Group I: Case 4

At $t = 0$ (initial values)						At $t = \tau$						At $t = 5\tau$						At $t \to \infty$ (final values)					
i	v_R	v_L	$\frac{di}{dt}$	$\frac{dv_R}{dt}$	$\frac{dv_L}{dt}$	i	v_R	v_L	$\frac{di}{dt}$	$\frac{dv_R}{dt}$	$\frac{dv_L}{dt}$	i	v_R	v_L	$\frac{di}{dt}$	$\frac{dv_R}{dt}$	$\frac{dv_L}{dt}$	i	v_R	v_L	$\frac{di}{dt}$	$\frac{dv_R}{dt}$	$\frac{dv_L}{dt}$
A	V	V	A/s	V/s	V/s	A	V	V	A/s	V/s	V/s	A	V	V	A/s	V/s	V/s	A	V	V	A/s	V/s	V/s
10	100	-100	-50	-500	500	3.7	37	-37	18.4	-184	184	0.07	0.7	-0.7	0.34	-3.4	3.4	0	0	0	0	0	0
15	150	-150	-75	-750	750	5.5	55	-55	27.6	-276	276	0.1	1.0	-1.0	0.5	-5.0	5.0	0	0	0	0	0	0
20	200	-200	-100	-1000	1000	7.4	74	-74	36.8	-368	368	0.14	1.4	-1.4	0.67	-6.7	6.7	0	0	0	0	0	0
$I_0 e^0$	$V_{R0} e^0$	$-V_{R0} e^0$	$-\frac{I_0}{\tau} e^0$	$-\frac{V_{R0}}{\tau} e^0$	$\frac{V_{R0}}{\tau} e^0$	$I_0 e^{-1}$	$-V_{R0} e^{-1}$	$V_{R0} e^{-1}$	$-\frac{I_0}{\tau} e^{-1}$	$-\frac{V_{R0}}{\tau} e^{-1}$	$\frac{V_{R0}}{\tau} e^{-1}$	$I_0 e^{-5}$	$V_{R0} e^{-5}$	$-V_{R0} e^{-5}$	$-\frac{I_0}{\tau} e^{-5}$	$-\frac{V_{R0}}{\tau} e^{-5}$	$\frac{V_{R0}}{\tau} e^{-5}$	$I_0 e^{-\infty}$	$-V_{R0} e^{-\infty}$	$V_{R0} e^{-\infty}$	$-\frac{I_0}{\tau} e^{-\infty}$	$-\frac{V_{R0}}{\tau} e^{-\infty}$	$\frac{V_{R0}}{\tau} e^{-\infty}$

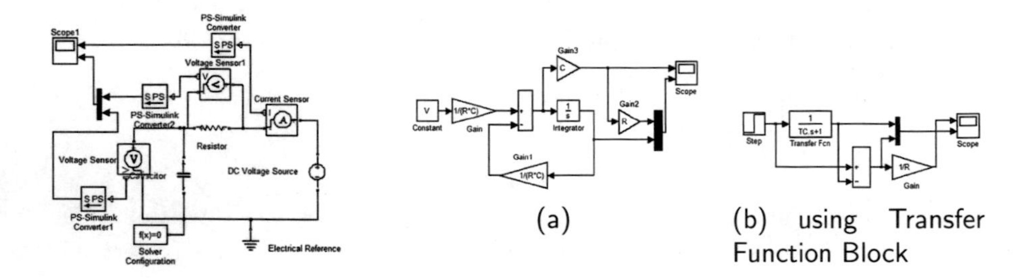

(a)

(b) using Transfer Function Block

Fig. 8. Complete Simscape model developed by Group-II

Fig. 9. Mathematical models developed by Group-II

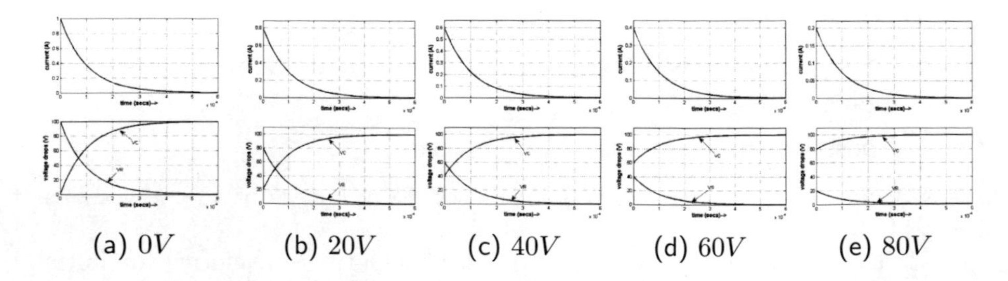

(a) $0V$ (b) $20V$ (c) $40V$ (d) $60V$ (e) $80V$

Fig. 10. Waveforms obtained by Group-II: Case-1 & 2 for different V_{C0}

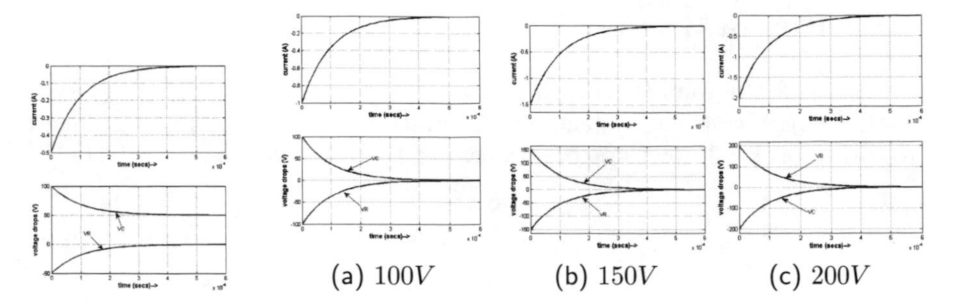

(a) $100V$ (b) $150V$ (c) $200V$

Fig. 11. Waveforms of Group-II: Case-3

Fig. 12. Waveforms of Group-II: Case-4 for different V_{C0}

Table 5. Students' observations on waveforms of Group II: Case 1 & 2

At $t = 0$ (initial values)						At $t = \tau$						At $t = 5\tau$						At $t \to \infty$ (final values)					
i	v_R	v_C	$\frac{di}{dt}$	$\frac{dv_R}{dt}$	$\frac{dv_C}{dt}$	i	v_R	v_C	$\frac{di}{dt}$	$\frac{dv_R}{dt}$	$\frac{dv_C}{dt}$	i	v_R	v_C	$\frac{di}{dt}$	$\frac{dv_R}{dt}$	$\frac{dv_C}{dt}$	i	v_R	v_C	$\frac{di}{dt}$	$\frac{dv_R}{dt}$	$\frac{dv_C}{dt}$
A	V	V	kA/s	MV/s	MV/s	mA	V	V	kA/s	MV/s	MV/s	mA	V	V	A/s	kV/s	V/s	A	V	V	A/s	V/s	V/s
1.0	100	0	-10	-1.0	1.0	368	36.8	63.2	-3.68	-36.8	36.8	6.7	0.7	99.3	-67	-6.7	6.7	0	0	100	0	0	0
0.8	80	20	-8	-0.8	0.8	294	29.4	70.6	-2.94	-29.4	29.4	5.4	0.5	99.5	-54	-5.4	5.4	0	0	100	0	0	0
0.6	60	40	-6	-0.6	0.6	221	22.1	77.9	-2.21	-22.1	22.1	4.0	0.4	99.6	-40	-4.0	4.0	0	0	100	0	0	0
0.4	40	60	-4	-0.4	0.4	147	14.7	85.3	-1.47	-14.7	14.7	2.7	0.3	99.7	-27	-2.7	2.7	0	0	100	0	0	0
0.2	20	80	-2	-0.2	0.2	73.6	7.4	92.6	-0.74	-7.4	7.4	1.3	0.1	99.9	-13.5	-1.35	1.35	0	0	100	0	0	0
$\frac{V_1}{R}e^0$	V_1e^0	$V - V_1e^0$	$-\frac{V_1}{R\tau}e^0$	$-\frac{V_1}{\tau}e^0$	$\frac{V_1}{\tau}e^0$	$\frac{V_1}{R}.e^{-1}$	V_1e^{-1}	$V - V_1e^{-1}$	$-\frac{V_1}{R\tau}e^{-1}$	$-\frac{V_1}{\tau}e^{-1}$	$\frac{V_1}{\tau}e^{-1}$	$\frac{V_1}{R}.e^{-5}$	V_1e^{-5}	$V - V_1e^{-5}$	$-\frac{V_1}{R\tau}e^{-5}$	$-\frac{V_1}{\tau}e^{-5}$	$\frac{V_1}{\tau}e^{-5}$	$\frac{V_1}{R}e^{-\infty}$	$V_1e^{-\infty}$	$V - V_1e^{-\infty}$	$-\frac{V_1}{R\tau}e^{-\infty}$	$-\frac{V_1}{\tau}e^{-\infty}$	$\frac{V_1}{\tau}e^{-\infty}$

Table 6. Students' observations on waveforms of Group II: Case 4

At $t = 0$ (initial values)						At $t = \tau$						At $t = 5\tau$						At $t \to \infty$ (final values)					
i	v_R	v_C	$\frac{di}{dt}$	$\frac{dv_R}{dt}$	$\frac{dv_C}{dt}$	i	v_R	v_C	$\frac{di}{dt}$	$\frac{dv_R}{dt}$	$\frac{dv_C}{dt}$	i	v_R	v_C	$\frac{di}{dt}$	$\frac{dv_R}{dt}$	$\frac{dv_C}{dt}$	i	v_R	v_C	$\frac{di}{dt}$	$\frac{dv_R}{dt}$	$\frac{dv_C}{dt}$
A	V	V	kA/s	MV/s	MV/s	A	V	V	kA/s	kV/s	kV/s	A	V	V	A/s	V/s	V/s	A	V	V	A/s	V/s	V/s
-1.0	-100	100	10	1.0	-1.0	0.37	-37	37	3.7	368	-368	-0.007	-0.7	0.7	67.40	6740	-6740	0	0	0	0	0	0
-1.5	-150	150	15	1.5	-1.5	0.55	-55	55	5.5	552	-552	-0.01	-1.0	1.0	101.10	10110	-10110	0	0	0	0	0	0
-2.0	-200	200	20	2.0	-2.0	0.74	-74	74	7.4	736	-736	-0.014	-1.4	1.4	134.75	13475	-13475	0	0	0	0	0	0
$-I_0e^0$	$-V_{C0}e^0$	$V_{C0}e^0$	$\frac{I_0}{\tau}e^0$	$\frac{V_{C0}}{\tau}e^0$	$-\frac{V_{C0}}{\tau}e^0$	$-I_0e^{-1}$	$-V_{C0}e^{-1}$	$V_{C0}e^{-1}$	$\frac{I_0}{\tau}e^{-1}$	$\frac{V_{C0}}{\tau}e^{-1}$	$-\frac{V_{C0}}{\tau}e^{-1}$	$-I_0e^{-5}$	$-V_{C0}e^{-5}$	$V_{C0}e^{-5}$	$\frac{I_0}{\tau}e^{-5}$	$\frac{V_{C0}}{\tau}e^{-5}$	$-\frac{V_{C0}}{\tau}e^{-5}$	$I_0e^{-\infty}$	$-V_{C0}e^{-\infty}$	$V_{C0}e^{-\infty}$	$\frac{I_0}{\tau}e^{-\infty}$	$\frac{V_{C0}}{\tau}e^{-\infty}$	$-\frac{V_{C0}}{\tau}e^{-\infty}$

Remarks: If an R-C series circuit is suddenly connected to the supply, the transient phenomenon is observed due to presence of capacitance. The steady state is achieved after a time period solely determined by the circuit parameters, C and R i.e. time constant of the circuit (RC). Once, the transient is over,
Case-1, 2 & 3: capacitor is charged to the supply voltage and no current flows into the circuit.
Case-4: energy stored in the electrostatic field of the capacitor entirely gets dissipated in the resistor.

5 Activities of Group-III

The shared goal of Group-III is to simulate a series circuit comprising of resistance, R, inductance L and capacitance C. Simscape model and mathematical model (based on Eq. 3) developed are shown in Figs. 13 and 14a respectively.

$$\ddot{x} = \frac{V}{LC} - \frac{R}{L}.\dot{x} - \frac{1}{LC}x \tag{3}$$
$$\text{where}\, v_C = x.$$

I_0 and V_{C0} were assigned to *Integrator* and *Integrator1* blocks respectively.

When $I_0 = 0$ and $V_{C0} = 0$, voltage transfer function of a series R-L-C circuit, $\frac{V_C(s)}{V(s)} = \frac{1}{LC} \times \frac{1}{s^2+\frac{R}{L}s+\frac{1}{LC}}$. Using a transfer function block, a simple model of R-L-C series circuit is implemented as shown in Fig. 14b. The script file was written using Eq. 3.

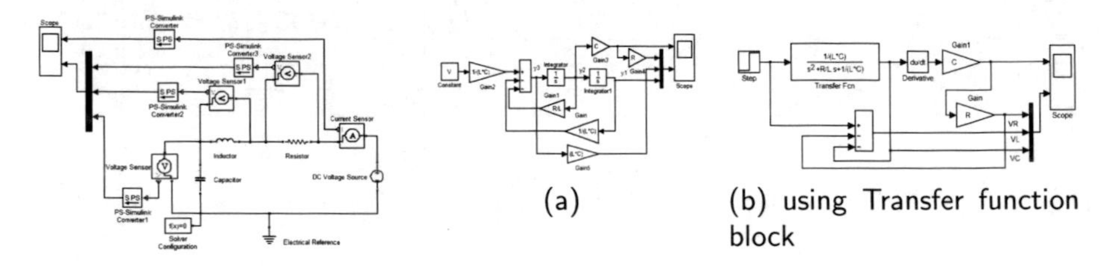

(a) (b) using Transfer function block

Fig. 13. Simscape model developed by Group-III

Fig. 14. Mathematical models developed by Group-III

These models were tested with L = 1 H and C = 1 F, and three different values of resistance: $R = 3\,\Omega$ ($\zeta = \frac{R}{2L} = 1.5$), $R = 2\,\Omega$ ($\zeta = 1$) and $R = 0.5\,\Omega$ ($\zeta = 0.25$).

Case 1: Suddenly connected to a DC source: With V = 100 V, waveforms obtained are shown in Figs. 15a, b and c.

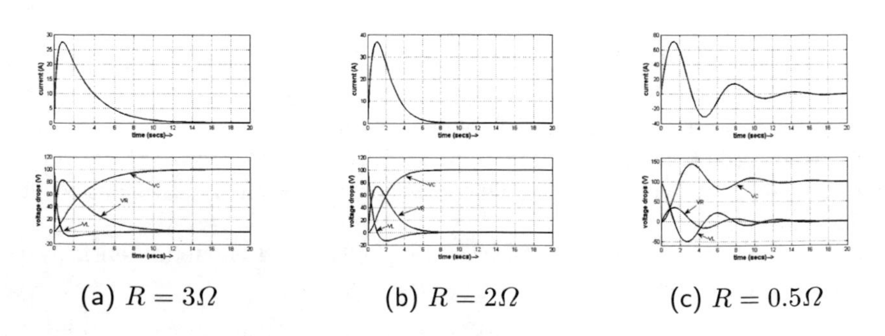

(a) $R = 3\Omega$ (b) $R = 2\Omega$ (c) $R = 0.5\Omega$

Fig. 15. Waveforms of Group-III: Case-1

Case-2 & 3: Supply voltage is suddenly changed: If DC supply voltage is changed to 5 V, when $V_{C0} = 2\,\mathrm{V}$ and $I_0 = 2\,\mathrm{A}$ [6], waveforms obtained for $R = 3\,\Omega$, $2\,\Omega$ and $0.5\,\Omega$ are shown in Figs. 16a, b and c respectively.

Case-4: Disconnected from supply and shorted: Considering a circuit comprising of L = 1 H and C = 1 F, if it is disconnected from DC supply when $V_{C0} = 100\,\mathrm{V}$ and $I_0 = 1\,\mathrm{A}$ [6], waveforms obtained for $R = 3\,\Omega$, $2\,\Omega$ and $0.5\,\Omega$ are shown in Figs. 17a, b and c respectively.

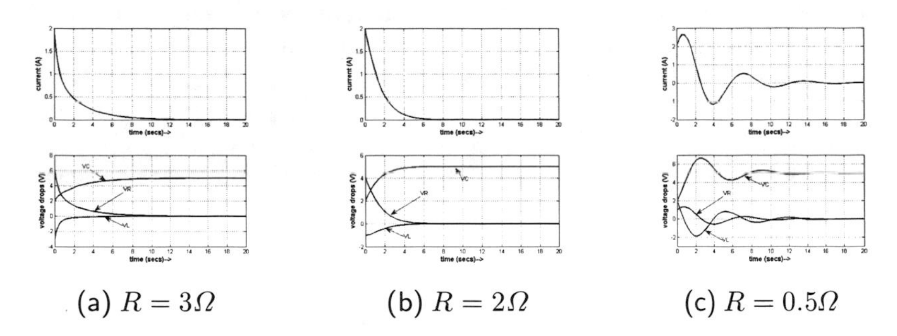

(a) $R = 3\Omega$ (b) $R = 2\Omega$ (c) $R = 0.5\Omega$

Fig. 16. Waveforms of Group-III: Case-2

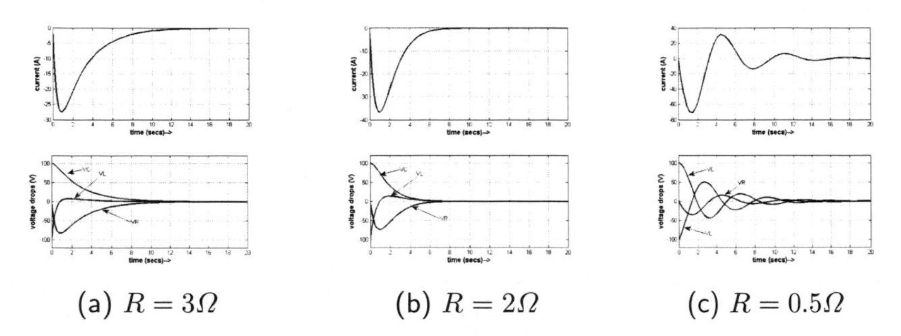

(a) $R = 3\Omega$ (b) $R = 2\Omega$ (c) $R = 0.5\Omega$

Fig. 17. Waveforms of Gruop-III: Case-4

Remarks: If an R-L-C series circuit is suddenly connected to or disconnected from the supply, the transient phenomenon is observed due to presence of inductance and capacitance. The behaviour depends on circuit parameters. Once, the transient is over,

Case 1, 2 & 3: It is observed that the capacitor voltage reaches the supply voltage and steady state is achieved and no current flows into the circuit. Capacitor acts as an open-circuit.

Case 4: Stored energy gets dissipated in the circuit resistance.

Effect of circuit resistance, R: If $R = 2\,\Omega$, the circuit takes less time to achieve the steady state i.e. system settles down quicker than the response with $R = 3\,\Omega$. When circuit resistance is further reduced to $R = 0.5\,\Omega$, an oscillation is observed before the system achieves the steady state. The first maximum overshoot occurs at 3.2 s. System settling time is about 16 s.

6 Conclusion

The proposed method of learning helped individual students to develop the concept in transient analysis at their suitable pace and time collaboratively [3]. It has been observed that by constructing the circuits and studying the waveforms, students are being able to make valuable observations and form the equations characterizing the transient behaviour of series circuits on their own through collaborative thinking and experiential learning.

They learnt how to develop mathematical models and compare the waveforms with those obtained from previous method. They could conclude less no of blocks are required if Transfer Function block is used when the circuits are initially relaxed. They also observed how MATLAB programming makes the study flexible for any sort of excitation and initial conditions. Using the simulated models students who are learning the topic in ICT mode, will have the feel how things work when an energy storing element is present in the circuit.

It has been observed that this creates an active, involved and exploratory learning environment which supports development of social skills of students. This method enhances students' problem formulation, analysis and solving abilities, improves student-teacher and student-student interactions and builds team spirit among students. It may be concluded that this method will certainly develop learners' lifelong learning skill and facilitates achievement and content literacy which is the goal of 21st century education.

References

1. Educause: Learning initiative. http://teambasedlearning.apsc.ubc.ca
2. User Manual. http://www.mathworks.com
3. Basu, G., Kar, U., Impact of instruction on binary multipliers using simulink to improve cognitive ability. In: Proceedings of IEEE EDUCON Global Engineering Education (2016)
4. Cooper, B.B.: The science of collaboration: how to optimize working together (2015). http://thenextweb.com/entrepreneur/2014/07/15/science-collaboration-optimize-working-together
5. Stice, J.E., Felder, R.M., Woods, D.R., Rugarciar, A.: The future of engineering education ii: teaching methods that work. Indian Science Cruiser (2000)
6. Soni, K.M.: Circuits and Systems. KATSON Books, New Delhi (2013)
7. Guthrie, R.W., Carlin, A.: Waking the dead: using interactive technology to engage passive listeners in the classroom. In: Proceedings of the Tenth Americas Conference on Information Systems, New York (2004)

The Practical Experiences with Educational Software for Modelling Interactive Collaborative Teaching

Stefan Svetsky(✉), Oliver Moravcik, Pavol Tanuska, and Iveta Markechova

Faculty of Materials Science and Technology,
Slovak University of Technology in Bratislava, Trnava, Slovakia
{stefan.svetsky,oliver.moravcik,pavol.tanuska,
iveta.markechova}@stuba.sk

Abstract. In comparison with the traditional learning, the CSCL (Computer Supported Collaborative Learning) represents a combination of didactics and informatics approaches. This requires researchers to solve an additional design of specific educational software and suitable ICT infrastructure as was presented at previous ICL Conferences by the authors. This contribution describes a continuous progress under the umbrella of the research on technology-enhanced learning; and how the ICT integration is modelled for the collaborative writing and activities when using an educational beta-software BIKE(E)/WPad. The practical experiences are demonstrated with some examples of how the WPad is applied to the CSCL issues, in off-line mode or shared virtual space.

Keywords: Technology enhanced learning
Computer supported collaborative teaching · CSCL · Educational software
Database applications

1 Introduction

Technology-enhanced learning (TEL) represents an interdisciplinary issue. As the key-term, TEL was used within the project calls of the FP7 EU funded research program. The research focus was more on the informatics aspects of computer support of teaching and learning than the didactics ones. There are many literature sources related to the TEL. One can find useful information especially in the "basic" books on TEL [1–3]. A specific computer support of academic teaching is described in [4], including the CSCL, which is presented from a historical perspective in [5].

Actually, the computer supported teaching and learning covers many specific areas of education. The global goal of authors' research is to computerize teaching processes on a personalized level with focus on the complex computer support without a need to use global software solutions (see e.g. [6–8] - interdisciplinary approach, or [9, 10] - CSCL/ICL Conferences). So, the educational software WPad was continually developed in the period of 2007–2016 (by one of authors). The specific goal is to test and modify the pilot version of the WPad also for collaborative teaching (WPad is a part of the beta in-house software BIKE(E)-Batch Information and Knowledge Editor and Environment). It supports various collaborative activities in engineering education, e.g. in the

M. E. Auer et al. (eds.), *Teaching and Learning in a Digital World*,
Advances in Intelligent Systems and Computing 715,
https://doi.org/10.1007/978-3-319-73210-7_5

framework of the courses of study: Background of environmental protection; Occupational health and safety; Basics of Chemistry; Programming languages.

Figure 1 illustrates the actual approach to the ICT integration into the teaching and learning.

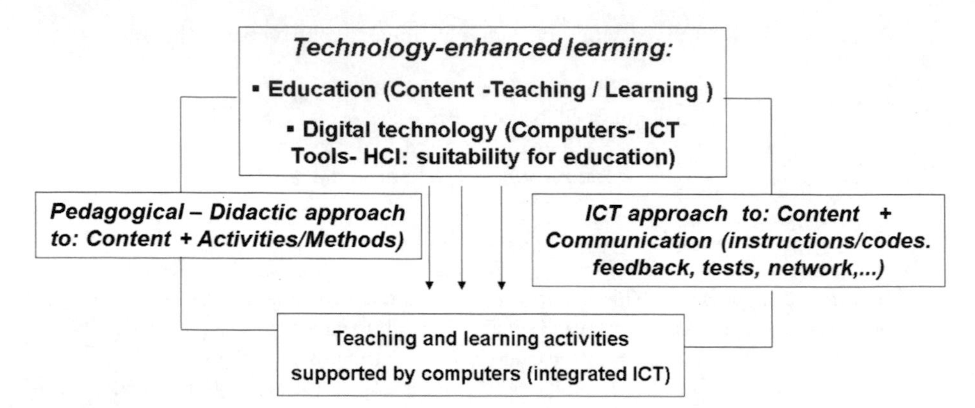

Fig. 1. Schema of the actual approach to the ICT integration into teaching and learning

The education issues cover not only the educational content but also the teaching processes. The digital technology issues cover the technology infrastructure and the ICT tools. In this context, the pedagogical-didactic approach relates to the content and activities, and the ICT approach to the content and communication issues. This general approach more clearly illustrates the CSCL infrastructure in Fig. 2.

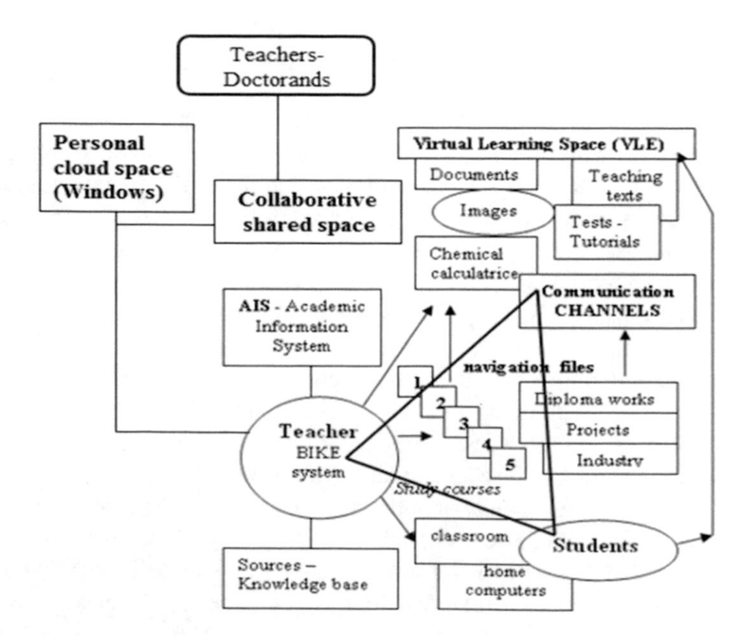

Fig. 2. CSCL infrastructure for the automation of teaching processes

In comparison with previous research presented in [7, 9, 10], the collaborative shared space and personal cloud space were implemented. The actual approach of authors' research on TEL is presented as the "automation of educational processes" [11]. It should be understood as the integration of three key areas: (1) didactics processes running in classrooms, (2) informatics processes running during teaching, and (3) the adaptation of individual's activities related to using the Windows operation system, Internet browsers, software, hardware, clouds, networks (this affects the Human-Computer-Interaction). Authors' long-term research showed that the computer support must be strictly tailor made to the teaching activities, so not vice versa (teachers usually applied only a general software, e.g. office packages).

2 Examples of Using WPad

Any integration of ICT into education requires one to use an interdisciplinary approach. The issue belongs to the applied informatics, or applications in the field of cognitive science. This should be clear from the introduction, especially from the schemas in the Figs. 1 and 2. In the context of this, the authors published their approach both from the didactics and informatics point of view (especially because one of them writes the educational software BIKE(E)/WPad).

For example, their actual IT approach is based on a theoretic background which was derived from Cybernetics approaches for automation of mental processes (teaching is considered as a knowledge based process). In our case, the WPad uses a so-called "virtual knowledge unit" which is both human and machine readable and isomorphically bridges mental processes with physical-computer processes [11]. So, the basic CSCL problems are solved by using the WPad in the framework of the combined off-line/online infrastructure (computers in classrooms/faculty's servers and open domain). This is illustrated by Fig. 3, i.e. the WPad, which is installed on computers in the classroom, virtual faculty's spaces and teacher's computers, represents a key IT tool for CSCL.

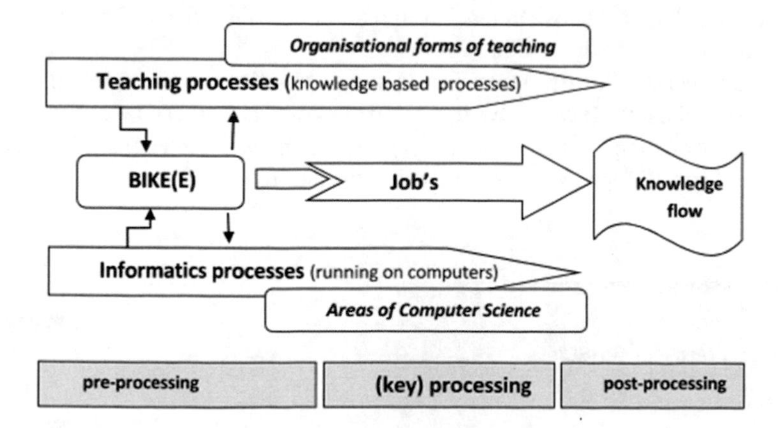

Fig. 3. The schema of the BIKE(E)/(WPad) function within the research on CSCL

As was mentioned above, the supporting ICT infrastructure was implemented for modelling CSCL. The outcomes represent both the implemented collaborative methods (e.g. collaborative writing semester works) and the supporting IT background infrastructure (database application WPad, virtual learning space). It should be emphasized that the WPad tables are used by authors, teachers and students as the platform for sharing the same domain knowledge and to manage and move it via informatics paths (i.e. between classroom or home computers and faculty's server or internet).

Figure 4 shows a screenshot of the WPad table. It represents an information exchange among teachers (see e.g. the note about Hopper's "Human are allergic to change", which was made from an e-mail).

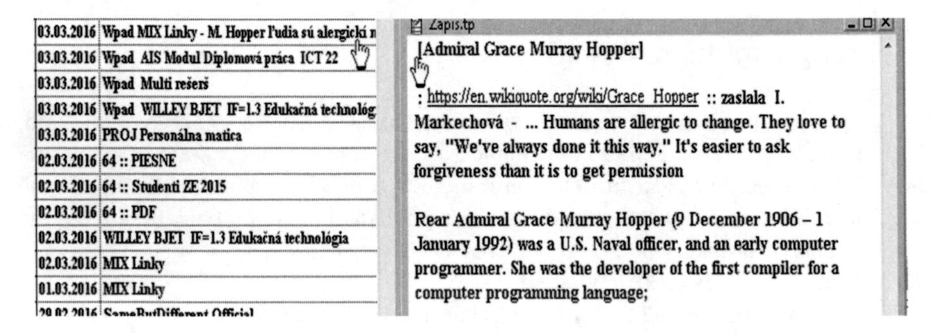

Fig. 4. Screenshot of the collaborative knowledge table for information exchange

Within the programming of WPad, an internet application was designed (the personalized TEL system) and placed on the virtual learning space. It functions as a communication channel for feedback and collaborative activities. So the collaborative teaching is performed in combination with the communication channels which connect the virtual learning space at the faculty's server with classrooms' computers, notebooks or mobiles. The communication channels work as a personal network where information, knowledge and instructions "travel" between teacher and students to be shared. They represents a sophisticated analogy to the Internet forums.

Figure 5 illustrates how the teacher or students write information (ASCII-data) into the channel (on the left). It can be written by using keyboard or teacher can use a simple HTML text. For example, there is a comparison of emission and absorption spectrum of hydrogen from Wikipedia in the screenshot on the right.

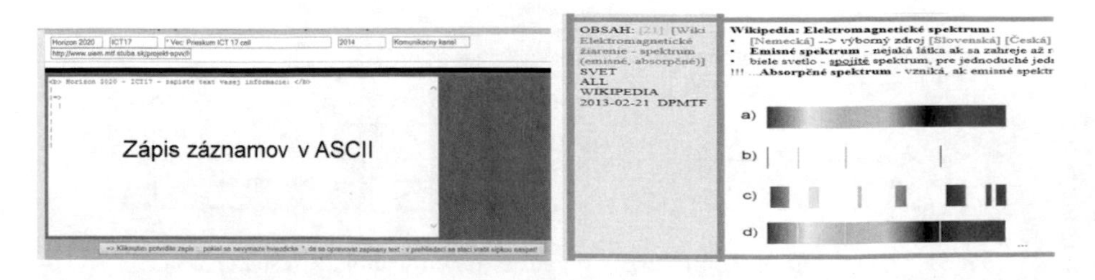

Fig. 5. Examples of screenshots of the communicational channel for Chemistry

While Fig. 5 represents one record (information) of the communication channel, as one can see, the screenshot in the Fig. 6 represents the more complex information exchange. In this case: (1) data related to the glucose oxidation, (2) links for downloading the study material prepared by students in previous year, (3) link for uploading files from the classroom computers to the virtual learning space, and (4) link to the Chemistry folder (a study material was produced by joining the knowledge tables, which were prepared by around fifty students using WPad, in order to eliminate the absence of basic knowledge from Chemistry).

Zdroj: Marenčík	**Oxidácia glukózy - príklad premeny energie na mechanickú prácu** [Viacej-More]
Materiály študentov a periodická tabuľka + výpočty /kliknite číslo záznamu/	|||||| Prekopírujte si do Zápisníka (WPad) na počítačoch v triede |||||| **1. Štúdium linkov študentov z roku LS 2014-2015** • a) http://www.uiam.mtf.stuba.sk/projekt-apvv/UBEI/upload/CHEM • b) http://www.uiam.mtf.stuba.sk/projekt-apvv/UBEI/upload/CHEM ??? Oxidácia / Redukcia [zmena oxidačného čísla,] 2. Stavová rovnica - výpočty pre ideálne plyny => p1 x V1 / T1 = p2 x V 3. Periodická tabuľka => Orbitály: http://www.ptable.com/#Orbital ale 4. Zapisovanie údajov do komunikačného fóra => http://www.uiam.mt [Viacej-More]
Upload	• **Kliknite => Vyberte súbor z počítača, ktorý ch** • Kliknite => Tu je adresár s nahranými súbormi D [Viacej-More]
Collaborative learning	**Výsledok spoločnej práce - linky a materiály z chémie, ktoré sprac** [Adresár CHÉMIA] [Viacej-More]

Fig. 6. Communicational channel containing the collaborative shared educational data

The research performed under the umbrella of technology-enhanced learning, the results of which are presented in this contribution, is considered by the authors to be an automation of traditional "face to face" teaching. The software, a database application BIKE(E)/WPad, was written by one of the authors (Svetsky) who is both teacher, designer and researcher. Case studies on BIKE(E) implementation within traditional teaching were presented in [6–11]. Basically, it works as an all-in-one tool used by him for producing a set of teaching aid tools, i.e. from simple off-line tutorials to the complex virtual learning environment functioning online on the faculty's server. It is not simple to explain in details because an atypical approach is used which was patented as an utility model [12]. The database application is also unique, however it could not be patented because in the EU it is not possible to patent any software (only technical solutions).

The BIKE(E)/WPad as a multipurpose personal tool is also used for modelling interactive collaborative teaching as was mentioned above. However, it is possible to perform only a limited number of collaborative activities during the teaching hours (due to lack of time, when one teaches ten to fifty students). So under the "automation of traditional teaching" one should understand that computer support - with BIKE(E)/WPad as teaching tool - enables a teacher to solve new methodological approaches that could not otherwise be solved within regular teaching time. One can imagine simply that the computer works as a teacher's assistant. For example, the content from a computer - as illustrated on Figs. 4, 5 and 6 - can be shared, evaluated, managed, i.e. used by the teacher for didactical purposes, assessment, writing collaborative works etc.

To the best knowledge of authors, no such solution has been described in the literature. For example, several years ago, the learning management system Moodle was evaluated as the most frequent technology-enhanced learning tool used at European universities [13]. However, in contrast to Moodle, BIKE(E) works both off-line and online, allows the teacher to produce eLearning materials, perform internet retrievals or launch external applications in Windows operation system. Being a personal knowledge management tool, as published previously by authors [13], BIKE(E) could even be used to build a Moodle-like personal learning management system.

3 Conclusions

The real computer support of teaching represents always a mix of various kinds of teaching and learning processes and areas of ICT. Some practical experiences with the in-house developed educational software BIKE(E)/WPad for modelling the interactive collaborative teaching were briefly described via examples from teaching bachelors students (e.g. how communication channels "teacher-students" function and how the computer support of collaborative writing is solved by using the WPad tables). From the collaborative point of view, both teacher and students can use the same knowledge, including transfer of the domain knowledge flow via digital paths. The teacher designs collaborative methods and the informatics background infrastructure is hierarchically subordinated to those methods. The communication channels were also tested when examining bachelors students (on their classroom or home computers they see the communication channels, located on the faculty's shared virtual space).

The interdisciplinary research on CSCL actually continues in the framework of an institutional project. For instance, teachers and PhD students test collaborative methods for sharing their personal data (in their WPad tables on their devices) with the shared central WPad table (localized on the faculty's server). A future work is aimed at the design of database applications for modelling the automation of mental processes (based on the utility model), including solving the CSCL visualisations. It should be also emphasized that modelling the computer supported collaborative teaching and learning requires the combined pedagogic (didactic) and informatics research (it is not possible to write programing codes if no teaching steps exist).

References

1. Goodman, P.S.: Technology Enhanced Learning: Opportunities for Change, 336 p. Lawrence Erlbaum Associates, Inc., USA (2001). Routledge 2002
2. Balacheff, N., Ludvigsen, S., Jong, T., Lazonder, A., Barnes, S. (eds.): Technology -Enhanced Learning. Principles and Products. Springer (2009)
3. Martens, A.: Software engineering and modeling in TEL. In: Huang, R., Kinshuk, C.N.S. (eds.) The New Development of Technology Enhanced Learning Concept, Research and Best Practices, pp. 27–40. Springer (2014)
4. Moursund, D.G.: A Faculty Member's Guide to Computers in Higher Education (2007). http://uoregon.edu/~moursund/Books/Faculty/Faculty.html
5. Stahl, G., Koschmann, T., Suthers, D.: Computer-supported collaborative learning: an historical perspective. In: Sawyer, R.K. (ed.) Cambridge Handbook of the Learning Sciences, pp. 409–426 (2006). http://gerrystahl.net/cscl/CSCL_English.htm
6. Svetsky, S.: The practical aspect of knowledge construction and automation of teaching processes within technology-enhanced learning and eLearning. Habilitation thesis. Slovak University of Technology (2012)
7. Svetsky, S., Moravcik, O.: The supportive system for the processing of human and non-human knowledge sources and combining mining techniques. In: AASRI Conference on Intelligent Systems and Control, Vancouver (2013)
8. Svetsky, S., Moravcik, O.: The automation of teaching processes based on knowledge processing. Trans. Mach. Learn. Artif. Intell. 2(5) (2014). http://scholarpublishing.org/index.php/TMLAI/article/view/568
9. Svetsky, S., Moravcik, O., Stefankova, J., Schreiber, P.: IT Support for knowledge management within R&D and Education. In: Proceedings of the 15th International Conference on Interactive Collaborative Learning, ICL 2012, Villach, Austria. IEEE, Piscataway (2012)
10. Svetsky, S., Moravcik, O.: The practice of CSCL in engineering education within the research on TEL. In: Proceedings of the WEEF-ICL, Florence, Italy (2015)
11. Svetsky, S., Moravcik, O.: The implementation of digital technology for automation of teaching processes. In: Presented at the Future Technologies Conference 2016, San Francisco (2016). http://ieeexplore.ieee.org/stamp/stamp.jsp?tp=&arnumber=7821632
12. UV 7340. The connection of an unstructured data processing system using a specific data structure. Industrial Property Office of the Slovak Republic. https://wbr.indprop.gov.sk/WebRegistre/UzitkovyVzor/Detail/45-2014
13. Matusu, R., Vojtesek, J., Dulik, T.: Technology-enhanced learning tools in European higher education. In: Proceedings of the 8th WSEAS International Conference on Distance Learning and Web Engineering, Santander, Cantabria, Spain (2008)

Exploring Student Interest of Online Peer Assisted Learning Using Mixed-Reality Technology

Sasha Nikolic(✉) and Benjamin Nicholls

University of Wollongong, Wollongong, Australia
sasha@uow.edu.au

Abstract. Supplementary Instruction, also known as Peer Assisted Study Sessions (PASS), is a popular program supporting the educational development of students in a collaborative setting. Flexibility of delivery has been explored for a number of reasons including: work and family commitments; distance from campus; and integrating regional and transnational satellite campuses. Previous studies have found attempts to undertake online delivery of PASS lacking in student interest and have been restrained by the technology. This study attempts to build upon this research by investigating student interest and the suitability of using a mixed reality technology called iSee, based on video avatars within a 3D virtual world. Consistent with previous studies student interest was low, converting a planned quasi-experimental study into a simulation. The simulation suggests that the technology was suitable for online collaboration, with effective communication of course content between participants and a good sense of presence. This suggests this trial may gain greater student interest if undertaken within institutions offering predominantly online, distance education.

Keywords: Collaborative learning · iSee · Online learning · PASS · Peer learning Mixed-reality · Supplementary instruction

1 Introduction

Government reforms, accreditation bodies and university strategic plans have continued to focus on the importance of universities to provide students a quality education. One approach used to enhance the learning experience is by providing Supplemental Instruction (SI), used by over 1000 higher education institutions across 29 countries [1]. The administration, structure and name of SI varies across institutions but the core focus of all programs is to improve student retention and learning outcomes through the facilitation of peer-learning; predominantly with senior students supporting junior students [2]. Within Australia, SI is known as Peer Assisted Study Sessions (PASS) with the benefits to student achievement well documented [3]. A review of research conducted by Dawson et al. [4] established three distinct benefits of PASS: students who participate in PASS receive higher grades; students regardless of ethnicity or prior academic achievement are more likely to succeed and at a higher rate; students persist at the institution at a higher rate. The key to a successful SI/PASS program are student leaders with strong technical, interpersonal and collaborative skills [5]; with the importance of

M. E. Auer et al. (eds.), *Teaching and Learning in a Digital World*,
Advances in Intelligent Systems and Computing 715,
https://doi.org/10.1007/978-3-319-73210-7_6

capable leaders also identified in other teaching roles [6]. While face to face learning experiences remain popular, advancements in information technology has led to the greater adoption and exploration of e-learning and blended learning possibilities [7].

Online forms of PASS are at a fairly experimental stage investigating a range of technologies and approaches and have been labelled with many names such as ePAL, OPAL, OPTEN, PALS Online and PTEN [8]. Online versions of PASS provide the opportunity to increase flexibility, especially for students that do not live close to campus or providing a link with regional and transnational campuses. Within the various studies, an interest has been to investigate the differences in student engagement and collaboration between the face to face and online versions. This study closely aligns with the work of Beaumont et al. [3] testing the implementation of an Online Peer Assisted Learning (OPAL) scheme at the University of Melbourne across the faculties of Business and Economics and Engineering. The study used Adobe Connect as the platform and found that due to technology issues video and voice was rarely used and communication centered on text based chat, using whiteboards, and uploaded documents. This approach allows for anonymity which gave students greater confidence to participate. However, this led to greater temptation towards distraction, less personal contact and reduced social aspects associated with PASS hampering collaboration. In turn, participation for OPAL was low and it was recommended that this approach might be more suitable for institutions offering predominantly online, distance education.

An alternative to classical 2D web based collaboration are new 3D video augmented technologies blending video avatars with virtual environments. A pilot study by Nikolic et al. [9] compared the differences in team meetings between Adobe Connect and iSee (a mixed-reality platform). The study found students thought iSee was easier to use due to its similarities with gaming and that team collaboration was better due to the focus of face to face communication. This was followed up by a number of other studies [10–13] that found that the video avatar substantially contributed to the acceptance and quality of engagement online; and provided a mechanism to facilitate transnational education. Therefore, this pilot study investigates if iSee is a suitable technology for running a virtual PASS (V-PASS) scheme at a traditional face to face university that could then be expanded to include students from other transnational satellite campuses.

2 Research Design

This pilot study was undertaken in a second year signal and systems engineering course at an Australian university that had been historically supported by PASS. The student cohort comprised of 86 local and 25 international students. The original intention was to undertake a quasi-experimental design as outlined in [3] to compare both student experience and learning between a PASS control group and a V-PASS experimental group.

The V-PASS sessions were conducted by the second author who had undertaken PASS classes as a student and worked together with the PASS leader to replicate the V-PASS experience. However, he was not an accredited PASS leader going through the required training programs. This was suitable for the small scale pilot and an accredited

PASS leader would be required if the study was expanded. A virtual world was created for use within the iSee platform. The virtual world was designed to conform to Jonassen's model [14] for constructivist learning environments (CLE) with three core components; context, representation and manipulation. Area one, marked out in Fig. 1, outlines the central hub consisting of multiple interactive boards for students use and one large board for the V-PASS leader to relay information. Areas two, three and four were designed as break out areas with interactive boards allowing students to work in small groups.

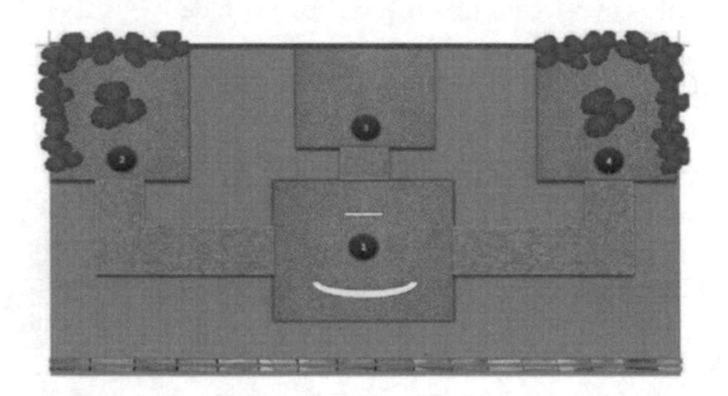

Fig. 1. Top down perspective of map concept design

The research study was initially promoted via a presentation during the lecture. From a pool of 111 students 34 registrations to participate in the study were received. The registered students were provided with all documentation required for ethics approval and instructions on how the V-PASS sessions would take place.

3 Results and Discussion

Despite the use of a technology built for enhanced collaboration, student participation was poor, similar to the findings in [3]. Only two students participated in the first week of scheduled V-PASS sessions. In contrast, PASS on average had 7 students attend each session. This was followed up with another presentation in the lecture, and advertising channeled through student social media and the PASS leader. Attempts were also made to schedule V-PASS at times most desirable for students and to run session throughout the week. This effort resulted in only another two students thus eliminating the possibility of a sound experimental design. It was revealed that the students participating were also attending PASS at the same time. A survey was conducted in a PASS session to try and understand the reason for such low participation. A total of 4 responses were received from 7 students. Three of the four students clearly labelled that their preference was for direct face to face contact. This confirmed informal feedback suggesting that regardless of the technology, students did not want to participate in an online PASS session when a direct face to face option was available. This reinforces the findings in [3] that such online approaches might be best suited to institutions offering predominately online, distance education compared to traditional face to face universities.

An alternative approach was required to test the suitability of using iSee. A simulated V-PASS session was advertised to all students across the four year engineering degree via social media. A simulation does not accurately represent V-PASS participation due to a change in dynamics of the participants. Self-selection and the impact this has on the findings is a further limitation. However, the simulation can provide an estimate of the possible effectiveness of the technology that could lead to further studies. A total of 15 students from third and fourth year accepted to participate in an hour long V-PASS simulation. An example of how the students appeared within the simulation can be seen in Figs. 2 and 3 representing how students see each other within the 3D world and how they can interact with learning materials. A survey instrument adapted from Bower et al. [15] was used to measure the simulated experience using a Likert scale from 1-Strongly Disagree to 5-Strongly Agree. The survey questions, mean responses and standard deviation are shown in Table 1.

Fig. 2. Student discussion within iSee

Fig. 3. Students working with interactive boards and reviewing notes

Table 1. Mean responses from simulated V-PASS participants

No.	Question	Mean	Standard Deviation
1	The chat tool was an effective way of communication with other V-PASS participants	4.07	0.27
2	Verbal discussion was an effective way of communication with other participants	4.40	0.63
3	Being able to see other participants via a video avatar gave me a strong sense of presence	3.93	1.34
4	Being able to manipulate (move) my avatat allowed me to interact with other participants more freely than a standard PASS class	3.33	1.29
5	The interactive whiteboards were useful in conveying course related information	3.87	1.13
6	The virtual learning space provided opportunities for discourse/collaboration with other V-PASS participants	4.07	0.70
7	The virtual learning space allowed me to find and source course related information effectively	4.07	1.03
8	I was able to effectively indicate my status to other (e.g. wanting attention, agreeing, being unsure, etc.)	3.60	0.91
9	I felt like I was present with my peers during V-PASS	3.93	0.96
10	The software provided clear and accurate representation of information and people	3.80	1.01
11	The software enable collaboration to occur	4.07	0.80
12	V-PASS classes are an effective method of teaching course related content	4.20	0.86
13	The idea of a V-PASS class is appealing	4.47	0.92

All 15 students participated in the survey and while no concrete findings can be determined by the simulation, the mean score across all questions was 3.98, indicating an 'agree' across the board suggesting that iSee may be a suitable technology for running V-PASS sessions. In particular Q2 indicates that the platform is effective for verbal communication, one of the key limitations in [3]. The responses show that the participants believed that V-PASS provided a good sense of presence, allowed collaboration to occur and was an effective way of teaching course related content, providing some agreement with the findings in [10–13]. This does provide some confidence in using iSee to trial a transnational V-PASS session if a way to encourage student participation could be found. The best investigation to understand if iSee can improve student interest in V-PASS would require repeating this study at a predominately online university.

4 Conclusion

This study implemented a pilot program to test the suitability of using a mixed reality technology, using video based avatars, to run a virtual PASS program. The study found that students had very little interest in considering a V-PASS opportunity, confirming the findings of similar studies using different technologies [8]. The evidence continues to suggest that students at traditional universities have a preference for face to face learning opportunities regardless of the benefits and flexibility associated with a V-PASS program [3]. Consequently, the study was transformed into a simulation activity to gain an estimate of the suitability of using the mixed reality technology, iSee. Students that did participate in the simulated activity expressed that the iSee technology was suitable for online collaboration, with effective communication of course content between participants and a good sense of presence. The next stage of the project will be to test the interest of running V-PASS between PASS leaders at the main campus with students from a satellite campus.

References

1. Power Ms, C.: Peer Assisted Study Sessions (PASS): through a complexity lens. J. Peer Learn. **3**(1), 1–11 (2010)
2. Birch, E., Li, I.: The impact of peer assisted study sessions on tertiary academic performance. In: Paper presented at the Quantitative Analysis of Teaching and Learning in Higher Education in Business, Economics and Commerce: Forum Proceedings (2009)
3. Beaumont, T.J., Mannion, A.P., Shen, B.O.: From the campus to the cloud: the online peer assisted learning scheme. J. Peer Learn. **5**(1), 1–15 (2012)
4. Dawson, P., van der Meer, J., Skalicky, J., Cowley, K.: On the effectiveness of supplemental instruction: a systematic review of supplemental instruction and peer-assisted study sessions literature between 2001 and 2010. Rev. Educ. Res. **84**(4), 609–639 (2014)
5. Stout, M.L., McDaniel, A.J.: Benefits to supplemental instruction leaders. New Dir. Teach. Learn. **2006**(106), 55–62 (2006)
6. Nikolic, S., Suesse, T., McCarthy, T., Goldfinch, T.: Maximising resource allocation in the teaching laboratory: understanding student evaluations of teaching assistants in a team based teaching format. Eur. J. Eng. Educ. (in Press)
7. Zhang, Y., Dang, Y., Amer, B.: A large-scale blended and flipped class: class design and investigation of factors influencing students intention to learn. IEEE Trans. Educ. **59**(4), 263–273 (2016). https://doi.org/10.1109/TE.2016.2535205
8. Watts, H., Malliris, M., Billingham, O.: Online peer assisted learning: reporting on practice. J. Peer Learn. **8**(1), 85–104 (2015)
9. Nikolic, S., Lee, M.J.W., Vial, P.J.: 2D versus 3D collaborative online spaces for student team meetings: comparing a web conferencing environment and a video-augmented virtual world. Paper presented at the 26th Annual Conference of the Australasian Association for Engineering Education, Geelong (2015)
10. Lee, M.J.W., Nikolic, S., Vial, P.J., Ritz, C., Li, W., Goldfinch, T.: Enhancing project-based learning through student and industry engagement in a video-augmented 3D virtual trade fair. IEEE Trans. Educ. **59**(4), 290–298 (2016). https://doi.org/10.1109/TE.2016.2546230

11. Nikolic, S., Lee, M.J.W., Goldfinch, T., Ritz, C.H.: Addressing misconceptions about engineering through student–industry interaction in a video-augmented 3D immersive virtual world. In: Paper Presented at the Frontiers in Education Conference (FIE) (2016)
12. Nikolic, S., Li, W.: Facilitating student and staff engagement across multiple offshore campuses for transnational education using an immersive video augmented learning platform. In: Paper Presented at the 2016 IEEE International Conference on Teaching, Assessment, and Learning for Engineering (TALE), 7–9 December 2016
13. Lee, M.J.W., Nikolic, S., Ritz, C.H.: Supporting the conceptualization of student innovation projects through peer and expert feedback on virtual pitches. In: Wilder, H.A., Ferris, S.P. (eds.) Unplugging the Classroom, pp. 119–131. Chandos Publishing, Cambridge (2017)
14. Jonassen, D.H.: Designing constructivist learning environments. Instr. Des. Theor. Models New Paradigm Instr. Theory **2**, 215–239 (1999)
15. Bower, M., Kennedy, G.E., Dalgarno, B., Lee, M.J.W., Kenney, J.: Blended Synchronous Learning: A Handbook for Educators (2014)

Monitoring the Knowledge Building Process in a CSCL Environment: A Case Study from Turkey

Gülgün Afacan Adanır(✉)

Ankara University Distance Education Center, Ankara, Turkey
gafacan@ankara.edu.tr

Abstract. The Computer Supported Collaborative Learning (CSCL) field is based on social learning theories which offer that knowledge is constructed through learners' interaction, knowledge sharing, and knowledge building as a community. Knowledge Building Theory presents a conceptual framework that allows researchers to investigate the social character of learning and its application with corresponding methods of CSCL area. This case study mainly aims to reveal learners' knowledge building process located in time-stamped logs of a VMT environment that covers chat, shared whiteboard, and wiki functions. The study was conducted in the setting of a graduate level course in a state university in Turkey. In order to examine learners' knowledge construction process in the VMT system, their verbal interaction related to course assignments were considered. Sentences of messages were applied as the unit of analysis and inspected them based on Progressive Knowledge Building Inquiry cycle (Hakkarainen 2003; White and Frederiksen 1998), which begins with a trigger activity and covers of four main stages; (a) idea generation, (b) idea connection, (c) idea improvement, and (d) rise above. In addition, we applied content analysis method to analyze learners' submissions to wiki environment.

Keywords: Computer supported collaborative learning · Knowledge building

1 Introduction

The Knowledge Building (KB) theory states that knowledge is constructed by mutual aims and negotiation of diverse ideas (Scardamalia and Bereiter 2003). The theory also makes a distinction between learning and knowledge building that sees the learning as an inside process which leads to changes in beliefs, attitudes, or skills. And, it views the knowledge building as the production or alteration of common knowledge that new cognitive artifacts are obtained as a result of mutual goals, group talks, and combination of ideas. According to the theory, knowledge building process enhances team members' existing understanding related to topics or tasks.

Knowledge building opportunities are provided to learners through different learning environments such as blogs, wikis, virtual worlds, CSCL environments, and discussion tools integrated to learning management systems. Computer-supported

M. E. Auer et al. (eds.), *Teaching and Learning in a Digital World*,
Advances in Intelligent Systems and Computing 715,
https://doi.org/10.1007/978-3-319-73210-7_7

Intentional Learning Environments (CSILE) project is respected as the first attempt that aims to equip schools with technology in order to achieve knowledge building communities (Scardamalia and Bereiter 1994). CSILE with its recent variation called as the Knowledge Forum was developed as an educational software that supports collaborative knowledge building activities, share of ideas in textual, audio, graphical, and video formats, and the organization of learning outputs. As the CSILE project, Learning through Collaborative Visualization (CoVis) Project attempted to transform traditional science learning by the help of networking technologies which provide learners with the opportunity of studying collaboratively with distant learners, teachers and scientists (Edelson and O'Neill 1996). The collaborative activities are satisfied through different tools, which include desktop video teleconferencing, shared software environments for remote/realtime collaboration, access to the World Wide Web resources, a multimedia scientist's notebook, and scientific visualization software.

Several methods have been proposed for the investigation of group knowledge in collaborative learning environments. Typically, such methods consider learning outputs (i.e. a report, a plan, a software application, a design artifact) for the assessment process. Yet, examining only delivery of group products is not sufficient for understanding the development of knowledge in individual and group levels. Therefore, online discussions are respected as the main instructional activity and corresponding logs are treated as the major products related to knowledge building. Although such a method is one important way to follow the group and individual's knowledge building trajectories, it results in some problems because of the unstructured form of the interaction. Alternatively, peer assessment methods are employed for the analysis of group learning, which consider learners' reflections about activities of their peers and eliminate the need for conducting a comprehensive log analysis (Strijbos and Sluijsmans 2010). Yet, this approach leads to additional burden on students and their assessment may not be sufficient for the investigation of knowledge building processes (Hong and Scardamalia 2014). As another assessment method, portfolios are utilized to acquire students' reflections on their learning during the collaborative work. Although portfolios provide researchers with detailed information regarding knowledge building in individual level, they are limited in understanding the progress of the group.

One major goal of knowledge building theory is to offer hands-on principles that could guide instruction in social environments. In this regard, for the investigation of activities in collaborative learning environments, it offers a set of principles rather than specific activity structures, procedures, or rules, (Scardamalia and Bereiter 2003). By considering this approach, we considered the Progressive Knowledge Building Inquiry cycle for the analysis of learners' knowledge building process in a CSCL environment called Virtual Math Teams (VMT). We provided the details of the Progressive Knowledge Building Inquiry cycle in the Methodology section. The remainder of the paper is organized as follows. In the Sect. 2, we presented our methodology. We dedicated the Sect. 3 for the results. In the final section, we presented the summary and implications of the results for researchers and practitioners.

2 Methodology

2.1 Research Design

With this study, we aim to examine how a group of learners achieve the knowledge construction process in an online CSCL environment called Virtual Math Teams. We utilized qualitative methods for the purpose of our study.

2.2 Context

We conducted the study in the context of a graduate level Research Methods course in one of the state universities of Turkey. In the context of this course, the instruction was provided in course settings and course assignments were collaboratively performed in Virtual Math Teams (VMT) environment. The VMT provides groups of learners to study on assignments while performing online discussion and collaboration (Stahl 2009). During their studies in assignments, teams firstly performed online chat meetings and then submitted their solutions as Wiki outputs.

2.3 Participants

Participants of the study were graduate students of Informatics Institute in one of the state universities in Turkey. In this paper, we considered experts from one chat session of one of the teams formed in the course. We provided demographic characteristics of students in the Table 1.

Table 1. Demographic characteristics of students

Subject handle	A_B	D_C	H_K
Gender	Male	Female	Male
Grade	PhD	Masters	PhD
Graduate major	Cognitive Science	Cognitive Science	Cognitive Science
Current GPA	3.50–4.00	3.00–3.50	2.50–3.00

2.4 Procedure

We explained expectations of the assignment in the following format: "A researcher is investigating the relation among the variables - IQ, VerbalIQ, PerformanceIQ, MRI_volume, height, and weight of participants. He collected corresponding data of participants and saved in Data.sav file. According to this dataset, on what scale are these variables measured (nominal, ordinal, interval, ratio) and why?".

2.5 Data Collection

When the learning group's online collaboration session terminated, we acquired data of chat log files which are automatically produced by the VMT tool. The chat log file is composed of date, start time, post time, duration, and event type about learning activities. Additionally, it consists of chat messages and other activities such as learners' joins to chat room, system messages, etc.

2.6 Data Analysis

Our overall research considered approximately 71% of the whole data. That is, we analyzed the ones generated by teams 1, 2, and 5, which involves 6978 chat messages in total. We didn't analyze the remaining data (i.e. data of teams 3 and 4) involving 2735 chat messages, mainly because of these teams' rare use of the VMT tool to work on assignments. In this paper, we considered experts from one chat session of one of the teams formed in the course.

In this study, our analysis followed the procedure of gathering the session's protocol data, detecting the core points of the data, and examining instructional benefits in the data (Inaba et al. 2002). In the context of collaborative learning, learners' level of benefits is parallel to their interaction in the group. In this regard, we applied qualitative interaction analysis methods in order to identify learners' processes of knowledge building. In order to examine learners' knowledge construction process in the VMT system, we considered their verbal interaction related to course assignments. We applied sentences of messages as the unit of analysis and inspected them based on Progressive Knowledge Building Inquiry cycle (Hakkarainen 2003; White and Frederiksen 1998), which begins with a trigger activity and covers of four main stages; (a) idea generation, (b) idea connection, (c) idea improvement, and (d) rise above. Trigger activity typically includes the question text that encourages learners to propose their thoughts and solutions. Idea generation refers learners' offers of ideas about the current theme or topic. Idea connection is related to comparison and compare of different ideas. Idea improvement demonstrates investigation of information and knowledge from sources. Rise above is the last phase that relates learners' reflections towards their learning. Then, we applied content analysis method to analyze learners' submissions to wiki environment. That is, we examined the solution and detected if it is wholly correct, partly correct or incorrect. Additionally, we reported the inadequate parts.

3 Results

We presented the results of learners' discussions according to the Progressive Knowledge Building Inquiry cycle. In line 80, the team summarized their former study and provided the solution that indicates types and scales of variables. The team stated that physiological parameters and IQ variables are in interval scale since there is equality of difference between values 9–10 and 11–12, and there is no absolute 0. However, one member (i.e. A_B) indicated his displeasure about this solution (line 171). Therefore, H_K proposed that they could work on the question again (line 174). A_B had confusion regarding the scale of the iq variable and offered that ordinal scale may be the appropriate scale (line 175–177), which illustrates the *idea generation* phase. Yet, H_K indicated the relevance of the interval scale because of the extinction of absolute 0 for the iq variable (179–182), which showed the *idea connection* phase since there is a comparison between ideas (Table 2).

Table 2. Learners' discussions between lines 171–185

Line	Date	Chat message/Whiteboard activity
80	11/16/2016	*The team shared the solution "In this research design, while brain volume, gender, height and weight are independent variables, all IQ variables are dependent variables. Intelligence changes according to the physiological factors. While physiological parameters and IQ variables are interval, gender variable is nominal. The difference between 9–10 and 11–12 is the same and we do not have an absolute 0. Therefore, physiological parameters and IQ variables are interval and since we only have male and female as gender, gender variable is nominal." in the whiteboard*
171	11/18/2016	A_B: I'm not very satisfied with the question 1 explanation
173	11/18/2016	H_K: ok
174	11/18/2016	H_K: lets do it again
175	11/18/2016	A_B: I'm not sure that IQ is interval
176	11/18/2016	A_B: what do you think?
177	11/18/2016	A_B: it feels like ordinal to me but then again it's like interval:)
178	11/18/2016	H_K: hmm.
179	11/18/2016	H_K: There is not true zero for IQ
180	11/18/2016	H_K: it feels ordinal
181	11/18/2016	H_K: No
182	11/18/2016	H_K: this is interval
183	11/18/2016	A_B: "However, these data tell us nothing about the differences between values"
184	11/18/2016	A_B: yes
185	11/18/2016	A_B: this explanation is for ordinal so its interval

Next, A_B considered an explanation from a source and provided the explanation of the ordinal scale as "these data tell us nothing about the differences between values" (line 183). Since he investigated the knowledge from sources, this demonstrates the idea improvement phase. After understanding the information, A_B offered the scale of iq as interval scale (line 184, 185), which showed the idea connection phase since there is a comparison between ideas.

The team initially classified physiological parameters in interval scale. Yet, H_K altered the solution by indicating height and weight in ratio scale, which was more proper classification and demonstrated the *idea generation* phase (line 187). A_B had confusion regarding the scale of the height variable. He indicated the real life and asked if there is an individual without height. Additionally, he stated that the dataset doesn't comprise "no height" situation. Hence, he proposed the height as in interval scale (line 189–191), which can be respected as *idea connection* phase since there is a comparison with real life example. For the explanation of the thought, H_K considered the weight variable and classified it in ratio scale by stating that 0 kg refers nothing (line 193–194), which demonstrated the *idea connection* phase. A_B stated her understanding (line 195). Next, A_B considered the scale of the MRI volume and offered that it should be in ratio scale because weight is ratio (line 197, 198), which demonstrated the *idea connection* phase. Yet, he experienced confusion regarding the interpretation for the scale of the MRI volume. Hence he asked whether 0 value refers no brain within the context of MRI volume (line 199). H_K provided her confirmation by indicating equality of 0 kg brain to no brain (line 203) (Table 3).

Table 3. Learners' discussions between lines 186–203

Line	Date	Chat message/Whiteboard activity
186	11/18/2016	H_K: gender is nominal
187	11/18/2016	H_K: height and weight are ratio
188	11/18/2016	A_B: yes
189	11/18/2016	A_B: well according to the description yes but in real life is this ok? no height in a person?
190	11/18/2016	A_B: and there is no "no height" situation in the dataset
191	11/18/2016	A_B: i think it's interval?
192	11/18/2016	H_K: 0 kg means
193	11/18/2016	H_K: nothing
194	11/18/2016	H_K: so it is ratio
195	11/18/2016	A_B: hımm, ok
196	11/18/2016	H_K: I have sent you an email
197	11/18/2016	A_B: ok weight and height are ratio so what about MRI_Volume?
198	11/18/2016	A_B: if weight is ratio then this should be ratio
199	11/18/2016	A_B: but does 0 mean no brain?
200	11/18/2016	H_K: right
201	11/18/2016	A_B::)
202	11/18/2016	A_B: well i'm confused so i trust you in this
203	11/18/2016	H_K: physiologically yes 0 kg brain = no brain:)

Wiki Reflection

Regarding scales of variables, the team shared the following output: "While physiological parameters and IQ variables are interval, gender variable is nominal. The difference between 9–10 and 11–12 is the same and we do not have an absolute 0. Therefore, physiological parameters (height, weight, volume) are ratio and IQ variables are interval and gender variable is nominal." This content demonstrated that the team correctly identified scales of iq variable. In addition, they provided correct reasoning for the interval scale by stating equal intervals and extinction of absolute 0. However, the scale of gender should be binary since it consists of two categories. For the scale of the physiological parameters, they both offered interval and ratio scales. Their second categorization was correct, but they should mention existence of ratios along the scales, and occurrence of a true and meaningful zero for the reasoning of ratio scale.

4 Discussion and Conclusion

In order to investigate how learners achieve knowledge building in their collaborative study, we examined their verbal interaction in the VMT system while they were discussing about the assignment. In our analysis we considered the phases of Progressive Knowledge Building Inquiry cycle. According to the results, learning groups' discussions usually started with the trigger activity which consists of the question statement. Next, learner activities continue with the phases of idea generation and idea connection. Yet, idea improvement and rise above phases were rare in learners' interactions. This may be because learners preferred to benefit from their peers' knowledge. Our finding is similar to the study of So et al. (2010), which employed content analysis to Knowledge Forum postings for examining student groups' enhancement in their ideas. Their results indicated that students had lack of ability to improve their ideas and mention sources in their solutions. Our future study could include more challenging questions, which will encourage learners to engage in idea improvement and rise above phases.

Our purpose was to examine how the learning group demonstrated progress in their chat activities while studying on "variables" concept. We provided the corresponding results belong to one team. More specifically, we demonstrated students' difficulties, explanations, ideas, and ultimate solutions related to the concept. Moreover, we investigated sufficiency of their final answer submitted as the wiki output. These results revealed the instructional benefits that students attained during their collaborative study.

The initial limitation of this study is that we analyzed data of one team among five teams registered to the course. The future study could investigate knowledge building process of whole teams and make comparisons among their improvements. The other challenge is related to the chat corpus that consists of non-English words and has a noisy structure. Therefore, we need to conduct data preprocessing before the major analysis. By the help of the interaction analysis, we investigated students' progress according to the variables concept. The forthcoming study could analyze knowledge building process based on whole concepts of the course.

References

Edelson, D.C., O'Neill, D.K.: The CoVis collaboratory notebook: computer support for scientific inquiry. In: Annual Meeting of the American Educational Research Association, New Orleans, LA, USA, April 1996

Hakkarainen, K.: Emergence of progressive-inquiry culture in computer-supported collaborative learning. Learn. Environ. Res. **6**, 199–220 (2003)

Hong, H.Y., Scardamalia, M.: Community knowledge assessment in a knowledge building environment. Comput. Educ. **71**, 279–288 (2014)

Inaba, A., Ohkubo, R., Ikeda, M., Mizoguchi, R.: An interaction analysis support system for CSCL: an ontological approach to support instructional design process. In: 2002 Proceedings International Conference on Computers in Education, pp. 358–362. IEEE, December 2002

Scardamalia, M., Bereiter, C.: Computer support for knowledge-building communities. J. Learn. Sci. **3**(3), 265–283 (1994)

Scardamalia, M., Bereiter, C.: Knowledge building. In: Encyclopedia of Education, 2nd edn., Macmillan Reference, New York, pp. 1370–1373 (2003)

Stahl, G.: Studying Virtual Math Teams. Springer, New York (2009)

So, H.J., Seah, L.H., Toh-Heng, H.L.: Designing collaborative knowledge building environments accessible to all learners: impacts and design challenges. Comput. Educ. **54**(2), 479–490 (2010)

Strijbos, J.W., Sluijsmans, D.: Unravelling peer assessment: Methodological, functional, and conceptual developments. Learn. Instr. **20**(4), 265–269 (2010)

White, B.Y., Frederiksen, J.R.: Inquiry, modeling, and metacognition: Making science accessible to all students. Cogn. Instr. **16**(1), 3–118 (1998)

Improving Online Interaction Among Blended Distance Learners at Makerere University

Harriet M. Nabushawo[1(✉)], Paul B. Muyinda[1], Ghislain M. N. Isabwe[2], Andreas Prinz[2], and Godfrey Mayende[1,2]

[1] Department of Open and Distance Learning, Makerere University, Kampala, Uganda
hnabushawo@gmail.com

[2] Department of Information and Communication Technology, University of Agder, Grimstad, Norway

Abstract. This article reports on a study done to improve interaction among distance learners offering the blended Bachelor of Education (B.Ed.) programme at Makerere University. The study attempts to answer the question: How can a Learning Management System be used to improve learner interaction on the blended B.Ed. programme at Makerere University? The study adopted the Affordance eLearning Design Framework. This study was done among 54 students studying a Policy Planning and Implementation course on the B.Ed. programme. The study employed qualitative approaches to data collection and analysis. These included semi-structured interviews and observation of the interaction logs within the groups and open forums. The results revealed that LMS affordances coupled with well-structured activities increased interaction among learners. Other factors that accelerated interaction and participation included grading of contributions and regular tutor presence. In conclusion, technology alone cannot bring about interaction among students; the way the activity is structured should be emphasized for interaction.

1 Introduction

Learner interaction is central in any teaching and learning processes. It assists in filling knowledge gaps through sharing of multiple perspectives on difficult concepts. In an Open and Distance Learning (ODL) mode of study, where the teacher and student are separated most of the time, interaction is even more encouraged. The distance created by the separation is bridged through interaction of the student with the tutors, peers and content (Anderson 2003; Moore 1993).

The B.Ed. programme at Makerere University is administered using the blended mode of study. Learners attend two-week face-to-face sessions and then do independent study. The programme has adopted group assignment as a way of bringing learners together for interaction. However, this is still limited because the learners are dispersed. This situation is likely to affect learning outcomes since learners miss out on the benefits of collaborative learning. Online connections can offer possibilities for learner interactions. However, online connection does not automatically bring about effective online interaction. This requires systematic planning of the online pedagogy and technologies.

M. E. Auer et al. (eds.), *Teaching and Learning in a Digital World*,
Advances in Intelligent Systems and Computing 715,
https://doi.org/10.1007/978-3-319-73210-7_8

Based on this background, the study attempts to find a way of improving interaction among B.Ed. learners at Makerere University through the use of the institutional Learning Management System (LMS) which is called MUELE. It was therefore anticipated that the utilization of the LMS can create possibilities of increasing interaction and hence learning outcomes. The study was therefore guided by the following research question: How can the LMS be used for improving interaction for learning among the B.Ed. students at Makerere University?

The rest of this paper is organized in four sections. Section 2 presents the literature. In Sect. 3, approaches and methods are presented. Section 4, presents the findings and discussions. Finally, the paper is concluded in Sect. 5.

2 Literature

The concept of interaction entails "reciprocal events that require at least two objects and two actions to influence each other" (Wagner 1994). Interaction may also be understood as the exchange of information, ideas and dialog between and among learners about the course. These interactions may be synchronous (real time) or asynchronous (over time) communication depending on the available channels. When learners interact with each other, difficult concepts are demystified as they share and negotiate meaning together. Similarly, Mayende et al. (2015) contends that learner interactions are effective in fostering "peer tutoring" and "peer collaborative learning" that involves learners working in pairs or small groups to discuss concepts, or find solutions to academic problems. Such educational experiences that are active, social, contextual, engaging, and learner-owned lead to deeper learning and development of higher-level thinking (Jonassen 1991).

Drawing from Anderson's theory/model of interactivity, interaction is valued for investigating and developing multiple perspectives among learners. Similarly, mainstream constructivists such as Piaget and social constructivists like Vygotsky contend that learners learn best in interaction with peers as opposed to interaction with teachers or other authorities (Anderson 2003). Bringing students together therefore helps them to create a learning community which is key in fostering constructivist learning approaches and enhances development of critical thinking skills. According to Vygotsky these social interactions increase competence through socio-cultural development as learners process, assess, generate and co-construct knowledge. These approaches lead to deeper learning and development of higher order thinking skills like evaluation, synthesis, analysis and application as interaction gives students opportunities to formulate their own understanding of the information read and shared.

Emerging technologies, such as LMS, hold a lot of promise for ODL as a cognitive tool to enhance interactive collaborative learning. The web affords interaction in many modalities like face to face, video, audio and computer conferencing (Anderson 2003). These technologies are ideal for ODL because of their capacity to support independent study in terms of time and place with their various affordances. Course resources and activities uploaded on the LMS can be accessed using mobile devices like phones, laptops, tablets, i-Pads etc. enabling learners to interact anytime anywhere. According to Knight (2009) creating a course on an LMS has advantages of extending learning beyond the classroom. Learners become active in managing their learning and gaining support from discussion forums and blog.

3 Approaches and Methods

The approaches are described in the following subsections: affordance analysis eLearning design framework, learning task/activity and affordances requirement of the task and tools.

3.1 Affordance Analysis eLearning Design Framework

This study used the affordance analysis eLearning design framework advanced by (Bower 2008). This was chosen because it could help in matching tasks with technologies to construct e-learning task designs. The following steps are followed while employing this framework:

1. Identifying the educational goal – In this study, our goal was to improve interaction among the distance learners offering a blended B.Ed. programme at Makerere University.
2. Postulate suitable learning tasks – These are tasks which are in line with the educational goal and could foster interaction among the learners. An example is a task that requires them to share individual ideas as well as comment on each other's submissions.
3. Determine affordance requirements of the task – In this case, to establish the affordances needed to provide the desired interactions, for instance media, temporal and navigational affordances.
4. Determine available affordances – Depending on the context, determine the technology affordances. In this case determine medias that can afford the interactions to support the learning task.
5. ELearning task design – Here we synergise the affordances of the learning tasks and the affordances of the tool. Different studies have shown that a mismatch between affordances of the task and those of the tool can lead to learner frustration.

3.2 Learning Task/Activity

Learners were randomly divided into groups of five. The lecturer posted the reading materials and the task on LMS. The task required each learner to share their understanding of the concept of social policy by posting their contributions to the group forum. The groups were expected to discuss the different opinions from members and thereafter make a one-page summary and post it in the open forum for other groups to access their ideas. It was a requirement that each student comments on at least two group submissions. The leader of each group finally made a one-page summary after getting feedback from other groups in the open forum and submitted it for assessment to the lecturer through the official submission system on the LMS. The activity structure was adopted from Mayende et al. (2016).

Given that this course is part of the courses required to be done in their programme, the facilitator, with emphasis, informed the learners that their submission will be marked. The marks are divided as follows. This group assignment contributed 20% of the final coursework mark. Students earned marks for individual submissions (5%) and groups also earned marks for their preliminary submissions (5%) and final submissions (10%). They later did a home assignment which contributed (20%). Therefore, the group assignment score of 20% plus the home assignment of 20% contributed 40% as a continuous assessment mark. The final course examination contributed 60%, which was then added to the 40% from the continuous assessment to give 100%. This therefore encouraged students to submit their individual contributions, which is a precursor to better interactions in the group processes.

3.3 Affordance Requirements of the Task and Tool

The study endeavoured to appreciate the learning context to carefully select and utilise the technologies for carrying out the given tasks. In this case, the task required the technology or tool that possessed media, temporal and navigational affordances to facilitate and foster student interaction. LMS could afford view-ability, read-ability and write-ability which enabled students to view and read the tasks and activities given by the lecturer and then be able to write back. This facilitation of knowledge exchange and interaction fostered learning among students. Specifically, the temporal affordances of LMS included accessibility to the LMS anytime anywhere, for students to interact and make their contributions to the group activities. This unlimited access to LMS enabled both synchronous and asynchronous interaction to take place, making the learning experience friendly and flexible.

The navigational affordances found with LMS included features like browse-ability, link-ability and searchability. These also enabled students to browse other sections of the resources by moving back and forth, linking to other sections within the resource or other resources. This easy access to information and resources enables meaningful interaction and richer discussions, which promotes deep learning. MUELE was therefore found to be an appropriate tool to improve interaction among the blended distance learners because it had the required affordances for the task.

The affordance requirement of the task was to improve interaction among the blended B.Ed. students at Makerere University. On the other hand, the affordance requirement of the tool was to enable B.Ed. students to interact online using MUELE. Given that all registered students had access to MUELE, it was found to be the most appropriate tool to use to improve interaction among distance learners. This enabled the students to access course materials and activities anywhere anytime.

3.4 Methods

This study was done among students offering a course titled policy planning and implementation on the B.Ed. programme of Makerere University. This course was offered in the second semester of 2016 to 54 students. The study used qualitative methods which included semi-structured interviews and observation of the interaction logs within the groups and open forums. Data was then transcribed and validated by a second person. This data was then analysed thematically to generate themes.

4 Findings and Discussion

This section describes the findings and discussions into 5 thematic areas that are learning activity organisation, facilitation, course design, assessment, and tutor presence and feedback.

4.1 Learning Activity Organization

The activity was organized in three levels. The first level was the mandatory submission by each learner. This discussion helped the individual learners to come in the group with their opinion to the activity. This helped in increasing interaction among the learners having attempted to answer the group assignment. They also assist them to engage in discussions with each other on a more personal level for effective scaffolding. One of the respondents alluded to this, that "… the initial activity helped me participate in the group discussion with confidence …". The second level was the discussion in the groups. This was enhanced interaction given that the learners had already made the initial submissions. Just before the third level the learners are required to give mandatory feedback to at least two submissions. This helped in improving interaction among the learners. The third level reconvened back to their groups for the final discussion and submission. At this level, they use feedback on their submission and other group's submission to improve and submit their final assignment for assessment. This kind of activity organisation helped learners to engage hence promoting deep learning. The way the activity was structured therefore promoted interaction among students.

4.2 Facilitation

The facilitation is a key role in online learning. The online tutors contribute a lot to enhancing and sustaining online interactions among learners. They should therefore have skills that promote interactions; for example, the nature of feedback given to students. The study revealed that students interacted more whenever the tutor prompted them with a question or substantiated on their submissions. This can only be done if the tutors are trained in the online pedagogy to provide questions that mediate learning rather than questions for assessment. This is because the tutors' questions or comments triggered their thinking, causing them to look at the activity from another angle. This is in line with Gallimore and Tharp (2002) assertion that assistive questions stretch students' thinking to a level they would not have attained. The tutor feedback helped in improving online interactions.

4.3 Course Design

For effective online interactions, the course study guide should have all the required information on activities and the relevant resources. The study guide for the course unit under study was provided to the students. It guided them on how to navigate to different pages of the LMS, specifically the discussion forum. This helped them to browse and locate the posted tasks and activities as well as the relevant resources for meaningful interactions. Activities were given to the students on the platform and they successfully engaged with them by reading the posted resource material, sharing their ideas on the forum as well as reacting to ideas given by the peers. This was made possible by operationalising chat rooms and discussion forums where students were engaged in both individual and group activities while the tutor monitored the progress.

4.4 Assessment

When assessment is integrated in the online activities, it is a precursor for motivation and sustained interaction. In this study, interactions were also enriched by embedding assessment in the activities. Every submission made by students whether at individual or group level contributed to the final mark. This made students participate actively in the activities. For clarity, the tutor provided a rubric on how marks would be awarded. The awarding of marks was based on the quality of participation. According to Mayende et al. (2014), motivation for online interaction is increased when the learners know the type of interaction which will give them higher marks. However, self- and peer-evaluation through the comments on other students' submissions is also useful for promoting critical reflection.

4.5 Tutor Presence and Feedback

This is constructive and specific information provided by the tutor to the learners while studying online. After the design of the course and corresponding activities, the tutor is charged with the task of monitoring, guiding and supporting the learning process. The study revealed that the presence of the tutor online encouraged many students to participate in the activities. The tutor provided appropriate feedback and posed questions that enriched the student's interactions. This was done through explicit prompts on how to move on, setting time frames, and controlling the coordination of activities among the group members. Some of these were either provided as instructions, or embedded in the LMS itself. As the tutor gives reminders, asks some questions, and guides on how to navigate the resources posted on the LMS, s/he will be indirectly scaffolding students' learning.

5 Conclusion

In conclusion, emphasis should be placed on the activity organisation to improve interaction among learners. This is important because it embeds interaction within the activity. In addition, facilitation was also identified as very important element in improving interaction. Learners interacted with the course materials, shared ideas about the activities, and gave and received peer feedback. The rich interactions were also facilitated by well-structured interactive activities posted by the instructor. The way the activities were set up and organised enhanced student engagement and enabled multiple feedback from both individuals and groups. The grading of the course motivated learners to participate in the online discussion forums. Additionally, the regular presence of teacher during the discussions was a greater motivator for learners since they received timely interventions and prompts for further discussions.

Acknowledgements. This work has been supported by the DELP project which is funded by NORAD. Special appreciation to the University of Agder and Makerere University, who are in research partnership, for their support.

References

Anderson, T.: Modes of interaction in distance education: recent developments and research questions. In: Moore, M., Anderson, G. (eds.) Handbook of Distance Education, pp. 129–144. Erlbaum, NJ (2003)

Bower, M.: Affordance analysis–matching learning tasks with learning technologies. Educ. Media Int. **45**(1), 3–15 (2008)

Gallimore, R., Tharp, R.: Teaching mind in society: teaching, schooling and literate discourse. In: Moll, L.C. (ed.) Vygotsky and Education: Instructional Implications and Applications of Socio Historical Psychology. Cambridge University Press (2002)

Jonassen, D.H.: Objectivism versus constructivism: do we need a new philosophical paradigm? Educ. Tech. Res. Dev. **39**(3), 5–14 (1991)

Knight, J.: Coaching. J. Staff Dev. **30**(1), 18–22 (2009)

Mayende, G., Isabwe, G.M.N., Muyinda, P.B., Prinz, A.: Peer assessment based assignment to enhance interactions in online learning groups. Paper Presented at the International Conference on Interactive Collaborative Learning (ICL), 20–24 September 2015, Florence, Italy (2015)

Mayende, G., Muyinda, P.B., Isabwe, G.M.N., Walimbwa, M., Siminyu, S.N.: Facebook mediated interaction and learning in distance learning at Makerere University. Paper Presented at the 8th International Conference on e-Learning, 15–18 July, Lisbon, Portugal (2014)

Mayende, G., Prinz, A., Isabwe, G.M.N., Muyinda, P.B.: Learning groups for MOOCs: lessons for online learning in higher education. Paper Presented at the 19th International Conference on Interactive Collaborative Learning (ICL2016), 21–23 September, Clayton Hotel, Belfast, UK (2016)

Moore, M.G.: Theory of transactional distance. Theor. Principles Distance Educ. **1**, 22–38 (1993)

Wagner, E.D.: In support of a functional definition of interaction. Am. J. Distance Educ. **8**(2), 6–29 (1994)

A Technological Proposal Using Virtual Worlds to Support Entrepreneurship Education for Primary School Children

Angela Pereira[1(✉)], Paulo Martins[2,4], Leonel Morgado[3,4], Benjamim Fonseca[2,4], and Micaela Esteves[5]

[1] CiTUR - Tourism Applied Research Centre, IPLeiria-Polytechnic Institute of Leiria, Leiria, Portugal
angela.pereira@ipleiria.pt

[2] UTAD - University of Trás-os-Montes e Alto Douro, Vila Real, Portugal
pmartins@utad.pt, benjaf@utad.pt

[3] Universidade Aberta, Coimbra, Portugal
leonel.morgado@uab.pt

[4] INESC TEC, Porto, Portugal

[5] CIIC - Computer Science and Communication Research, IPLeiria-Polytechnic Institute of Leiria, Leiria, Portugal
micaela.dinis@ipleiria.pt

Abstract. The importance of entrepreneurship education from elementary school through college is now recognized as an important aspect of children's education. At the level of basic education, the development of entrepreneurial activities using Information and Communication Technologies, specifically three-dimensional virtual worlds, is seen as an area with potential for exploration.

The research presented herein is a model that allows the development of entrepreneurial activities in virtual worlds with children attending primary education. This model allows the preparation, monitoring and development of entrepreneurship education activities in virtual worlds, including safe interaction in virtual worlds between the children and the community. For this, we identified a set of requirements that would allow the teaching and learning of entrepreneurship in virtual worlds, from which a technological model was implemented through an application, EMVKids (after the Portuguese expression "*Empreendedorismo em Mundos Virtuais com Crianças*", entrepreneurship with children in virtual worlds).

Keywords: Entrepreneurship · Primary education
Three-dimensional virtual worlds · Collaborative learning
Educational virtual environment · User-Centered Design

M. E. Auer et al. (eds.), *Teaching and Learning in a Digital World*,
Advances in Intelligent Systems and Computing 715,
https://doi.org/10.1007/978-3-319-73210-7_9

1 Introduction

Entrepreneurship is widely recognized as one of the basic skills to be acquired through life-long learning. Educational systems can make a significant contribution towards this, encouraging the development of entrepreneurial attitudes and skills, starting with youths and school-aged children.

Changes in educational practices have come to recognize the potential of Information Communication Technology (ICT) in teaching and learning. One of these technologies are the three-dimensional online virtual worlds that can promote playful, immersive and interactive learning experiences, allowing users to test, explore and interact through a visual learning environment, as well as real-time communication and collaboration with other users [1–3]. These three-dimensional virtual worlds can be projected to support the design and implementation of educational activities and not just for the reproduction of traditional educational content.

To explore the potential of entrepreneurship education in virtual worlds with children attending primary schools we conducted an early empirical study [1]. To accomplish this activity, we needed a virtual world that allowed the following: the development of technological solutions; access to educators and children under 10; create and edit objects by users and support for collaborative learning. From the range of available Massively Multiplayer Online Social Games (MMOSG), the Active Worlds Educational Universe (AWEDU) was selected to develop this activity because it has the possibility of obtaining an isolated world accessible only to authenticated/professional users, in this case educators and children.

From the previous study developed by Pereira et al. [1] several technological problems for the use of virtual worlds in entrepreneurship education were identified and had to be solved manually. We classify these problems in three phases according to the evolution of activities: First phase - Virtual World preparation for a new educational activity; Second phase - children development activities related to entrepreneurship inside the Virtual World; and Third phase - Connection with adult virtual world community.

This paper reports the architectural model to support the development of entrepreneurial education activities in AWEDU virtual world for elementary school children and its prototype implementation, in order to validate it. We began by situating the study in the current literature, discussing what is known about entrepreneurship education practices, defining what virtual worlds consist of and how they can be used for educational purposes, and discussing studies that have been conducted in this field. Next, the architecture model and its components are described. Finally, we present the conclusions and the future work.

2 Related Work

Entrepreneurship must be seen as a dynamic process that involves the perception, the conception and the concretization of a business opportunity, which presupposes the involvement of people and processes that, together, lead to the transformation of ideas into opportunities. Its importance has been recognized for more than 50 years in the development of economies and studied by many authors in various parts of the world [4, 5]. Education can make an important contribution to entrepreneurship by stimulating the development of entrepreneurial attitudes and skills, starting with young people and with schooling [6–8].

Regarding to initiatives in the field of entrepreneurship education, a documentary research of good practices implemented in some countries has been carried out on specific programs related to entrepreneurship education in the 1st cycle of basic education.

In all basic schools of Luxemburg the program of 6th grade French class entirely integrates a unit devoted to the creation of companies. This unit is based on the comic book of the same name Boule et Bill créent une entreprise [9]. That comic book explains how well-known personalities became successful in the business world. This program was also adopted by schools in the French region of Nord-Pas-de-Calais.

"Una empresa en mi escuela" (EME) – is a program of the Astúrias region (Spain) directed to basic education (5 to 12-year-olds). This program is managed by Valnalón, a public company of the Ministry of Labour and Industry, which works in cooperation with the Ministry of the Education for the creation and implementation of programs in the area of entrepreneurship for different grades. The learners carry out various tasks related to the creation and operation of a company, the aim being the creation of business skills and to establish relations between schools and companies [9, 10].

In Scotland, an infrastructure was created to provide education for the development of entrepreneurship in basic education [11]. The programs are aimed at school-aged children from the age of five and involve them in the creation of mini-enterprises and in the development and commercialization of for-profit products or services. There are many other examples of good practice in stimulating entrepreneurship since an early age like in some countries of the European Union. After Estonia, Luxembourg has the 2nd highest Total Early entrepreneurship Activity (TEA) among European countries. Luxembourg's neighbors have a TEA rate of about 6%. In 2015, Belgium reported a TEA of 6.2%, Netherlands 7.2% and Germany 4.7% [16].

A growing number of researchers have used virtual worlds as educational platforms, typically referring to them as educational Multi-User Virtual Environments (MUVEs). Some authors [12, 13] point out that three-dimensional virtual worlds have the potential to promote various educational and collaborative activities, allow-ing their users to communicate and collaborate with others in a virtual shared space that is created by the users themselves. This generates opportunities to develop creativity, the creation of new ideas and experiences that are not always possible to achieve in the real world. Educators have been including three-dimensional virtual worlds as an alternative to improve students' learning experiences, from primary education to higher education.

A study carried out by Bers and Chau [14] involved children in the construction and social organization of a virtual city in a three-dimensional multi-user virtual environment, with the aim of learning about moral and civic values. The results of this study show that virtual environments allow access to a wide range of information and resources, as well as communication tools and support for collaboration, and are platforms with potential for the development of civic education programs.

Quest Atlantis (QA) is a teaching and learning project developed by Barab et al. [15] that uses the virtual environment ActiveWorlds for the development of educational activities with children between the ages of 9 and 12. The results revealed an increase in students' participation in the curriculum, progress in explaining processes and conclusions, and a significant increase in knowledge about content.

In this context we decided to use three-dimensional virtual worlds in entrepreneurship education for children. Three-dimensional virtual worlds can offer several contributions specifically to entrepreneurship education of children, by allowing the creation of learning activities where children may reproduce the operation of a business or the organization of a social activity [1].

An exploratory study was conducted to verify the possibilities of using virtual worlds for education purposes for children, which revealed some constraints on their use.

Thus, the need arose to develop a technological solution that allows the use of these worlds in an effective way. The next section presents the proposed solution.

3 Architecture Model

From the previous study [1] we identified several technological problems in the use of virtual worlds in entrepreneurship education that had to be solved manually. We classified these problems in three phases: First phase – Preparation of the educational activities related with the organization of the virtual world for a new activity; Second phase – Development of the activities related with the implementation of the entrepreneurship activities by children inside the virtual world; and the Third phase – Connection with the community. A number of specific requirements to promote the learning activities was identified. A requirements summary are presented in Table 1.

Table 1. Requirements

Phase	Found requirements
Preparation and monitoring of the educational activities	Management of information about schools, classes, learners, groups and avatars; Division of the virtual learning space; Monitoring the learner's activities
Development of the activities	Presentation of business information; Inserting images (logos and products); Supporting the creation of objects; Providing products/element information
Connection with the community	Secure interaction between entrepreneurs and targets; Monitoring entrepreneurial activities

These requirements drive the recognition of three key areas, which are:

1. Management system: prepare, monitoring and show to the community the educational entrepreneurship activities.

2. Automate processes: assist students in the development of entrepreneurship activities in a virtual world.
3. Online ordering: allows purchases, orders and statistical data about the projects (businesses).

The technological model key areas that interact with the virtual world are presented in Fig. 1.

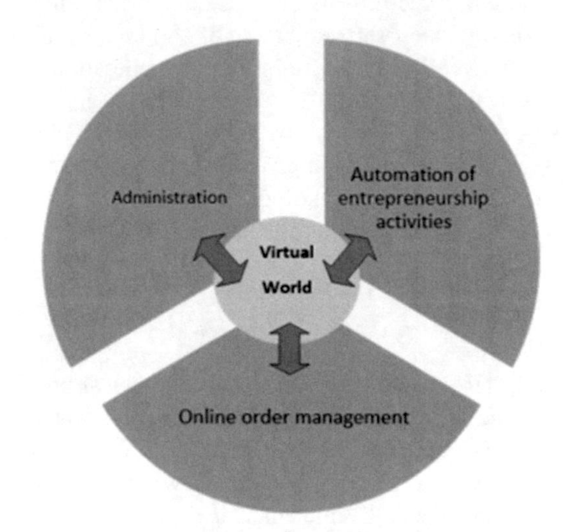

Fig. 1. Model key areas

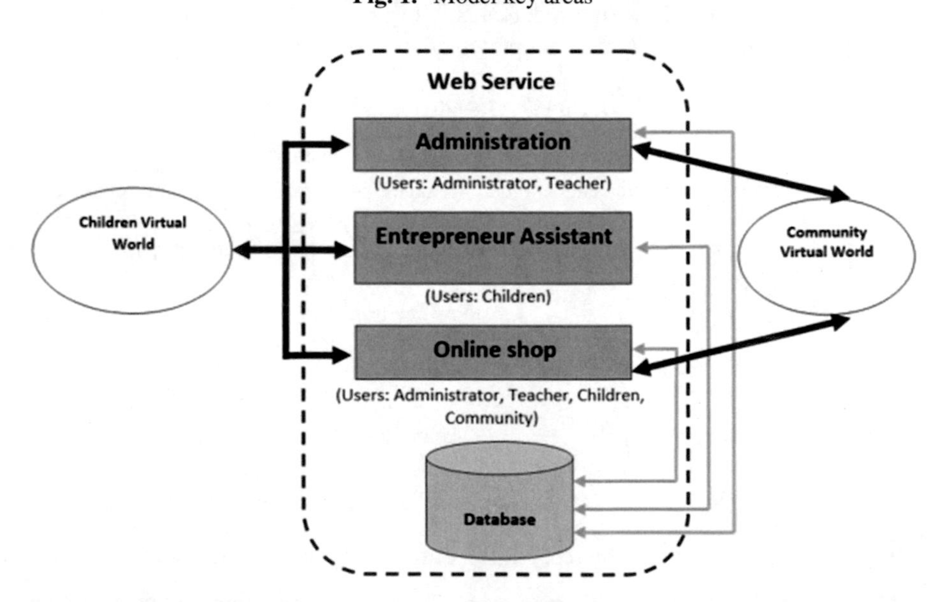

Fig. 2. Architecture to support the activities of teaching entrepreneurship to children in virtual worlds

The next step was the creation of the architectural basis (Fig. 2) that allowed the development of technological solutions for the entrepreneurship education activities with primary school children inside the virtual world AWEDU. The architecture supports different users' profiles; each of them has access to a specific part to manage their activities.

The interaction takes place with two virtual worlds that have different purposes. A virtual world reserved for students (children) for the development of their educational activities. Another virtual world with free access to the community (adults) in which the activities developed by the children will be presented. This virtual world's separation was to ensure the safety of children against malicious users. Since it is important that business developed by the children is available to all community, it was necessary to create a safety way to make the business public. The solution found consisted in the creation of a replication mechanism. This mechanism copies all the businesses from the children's virtual world to the community virtual world and allows children to manage all their business inside the safety of their virtual world.

3.1 Administration

The Administration component consists of two modules: **Resources management** – enables the development of entrepreneurial educational activities in the virtual world, such as creating the users' profiles, register the virtual world server, configuration of the world such as the division of virtual space in areas for each group of students work, define avatars, schools, classes and groups; **Virtual world management** – allows the administrator or teacher to initiate entrepreneurship activities, monitor the progress of these activities and replicate to the virtual world community.

3.2 Entrepreneurship Assistant

The Entrepreneur Assistant component is used only by students through their avatar. This component helps the students develop their business inside the world. The assistant consists in four modules which are: Project Information, Building Construction, Product Management and Advertising. **Project Information** – helps the students structure all their business, such as: name, logo, products information, among others. **Building Construction** – allows the students to build and personalize their shops. **Product Management** – helps the students to manage all their business specially defining the prices, quantities and manage the sales. **Advertising** – enables the creation of advertising for the community's virtual world.

3.3 Online Shop

The Online shop was designed to manage the orders from the clients and consists of three modules: Promotions, Businesses and Administration. **Promotions** – shows to the community all the promotions that exist in the children's businesses. **Business** – displays the products from the children's businesses to the all community. **Administration** –

manages the orders from the clients for each business. The next step consists in the creation of the design prototype in order to validate the architecture model.

4 Prototype Design

In order to validate the architecture proposed we developed the EMVKids prototype. In the prototype development a User-Centered Design (UCD) process was used, which is an effective approach to satisfy users' needs and to improve the interface. More specifically, the UCD includes three major activities: design, evaluation, and re-design. These three activities were repeated until users were fully satisfied with the functionalities and inter-face provided by the system. This prototype was tested with primary school teachers and students. The prototype made possible a better organization of the virtual world and activities, such as: help improve students' performance in performing their tasks in the virtual world and allow teachers to monitor the activities developed by the students. This technological solution allowed the use of a virtual world for education purposes with young children.

5 Conclusions

The developed model allowed to realize several results: the development of real economic activities in small scale through the production and sale of real products; the development of entrepreneurial skills, personal qualities and transversal skills of students; the execution of students' ideas and the development of creativity; team-work in a business project; the secure interaction of entrepreneurs with the community through the asynchronous data replication system.

Some recommendations to be implemented in the future are that of an alert system to register clients who visit the virtual world (virtual world of community access) and the possibility to send a warning (via email or SMS) to the teacher. Another one is the integration of activities in social networks such as Twitter and Facebook so that the teacher can spread the business and new products.

As future work the researchers intend to validate the prototype developed in a real context.

References

1. Pereira, A.M., Martins, P., Morgado, L., Fonseca, B.: A virtual environment study in entrepreneurship education of young children. J. Virtual Worlds Res. **2**(1) (2009)
2. Esteves, M., Fonseca, B., Morgado, L., Martins, P.: Improving teaching and learning of computer programming through the use of the second life virtual world. Br. J. Educ. Technol. **42**(4), 624–637 (2011)
3. Hew, K.F., Cheung, W.S.: Use of three-dimensional (3-D) immersive virtual worlds in K-12 and higher education settings: a review of the research. Br. J. Educ. Technol. **41**(1), 33–55 (2010)
4. Schumpeter, J.A.: The theory of economic development. H.U. Press, Cambridge (1949)

5. Reagle, C., Maggioni, V., Boicu, M., Albanese, M., Joshi, M., Sklarew, D., Peixoto, N.: From idea to prototype: introducing students to entrepreneurship. In: 2017 IEEE Integrated STEM Educational Conference (ISEC), pp. 71–75. IEEE (2017)
6. Child, J., Mcgrath, R.G.: Organizations unfettered: organizational form in an information-intensive economy. Acad. Manage. J. **44**(6), 1135–1148 (2001)
7. Kirby, D.: Entrepreneurship. McGraw-Hill Education, London (2002)
8. Kirby, D.A., Ibrahim, N.: Entrepreneurship education and the creation of an enterprise culture: provisional results from an experiment in Egypt. Int. Entrepreneurship Manage. J. **7**(2), 181–193 (2011)
9. Pereira, Â., Morgado, L., Martins, P., Fonseca, B.: The use of three-dimensional collaborative virtual environments in entrepreneurship education for children. In: Proceedings of the IADIS International Conference WWW/Internet, pp. 319–322 (2007)
10. Marina, J.A.: A competição para realizar competência empreendedorismo. J. Educ. **351**, 49–71 (2010)
11. Wiggins, A., Cowie, M., Tymms, P.: Evaluation of the Whole School Approach to Enterprise in Education. Department of Lifelong Learning. Scottish Executive (2007)
12. Eschenbrenner, B., Nah, F.F.H., Siau, K.: 3-D virtual worlds in education: applications, benefits, issues, and opportunities. J. DB Manage. **19**(4), 91–110 (2008)
13. Dalgarno, B., Lee, M.J.: What are the learning affordances of 3-D virtual environments? Br. J. Educ. Technol. **41**(1), 10–32 (2010)
14. Bers, M.U., Chau, C.: Fostering civic engagement by building a virtual city. J. Comput. Mediated Commun. **11**(3), 748–770 (2006)
15. Barab, S.A., Thomas, M., Dodge, T., Carteaux, R., Tuzun, H.: Making learning fun: quest atlantis, a game without guns. Educ. Technol. Res. Dev. **53**(1), 86–107 (2005)
16. Peroni, C., Riillo, C., Sarracino, F.: Entrepreneurship and immigration: evidence from GEM Luxembourg. Small Bus. Econ. **46**(4), 639–656 (2016)

Social Life in Virtual Universities

Jozef Hvorecký[1(✉)] and Monika Dávideková[2]

[1] Vysoká škola manažmentu (School of Management), Bratislava, Slovakia
jhvorecky@vsm.sk
[2] Faculty of Management, Comenius University, Bratislava, Slovakia
monika.davidekova@fm.uniba.sk

Abstract. Teaching and learning at virtual universities have their own obstacles. The isolation of students separated from the rest of their "classmates" results into specific problems non-observable in traditional classrooms. The students miss their social contacts which naturally evolve in traditional educational environments. They contribute to cohesion in the group and support team building. The feeling of belongingness is an important motivational factor and has a positive impact on the capability of an individual to overcome problems (in our discussed case, the ones connected to his/her study). This paper aims to study the ways in which the "university spirit" can be enhanced.

Keywords: Quality of tertiary education · Isolation of online students Belongingness to Alma Mater · Facilitating the motivation to study University spirit

1 Introduction

The emergence and development of information and communication technology (ICT) enabled real-time communication and collaboration across any distance [1] as well as sharing of massive knowledge with wide public on instant. Academic institutions significantly changed as a consequence of developments associated with globalization and informatization; and adapted by demonstrating the capacity to continuously re-invent themselves, whilst continuing to pursue their fundamental mission and sustaining their core values [2]. The increased deployment of resource-based learning enabled by ICT is accompanied by shortcomings and barriers that can be mitigated by applying appropriate models of learning, teaching and course design [3]. Among obstacles that are not observable in traditional class rooms, but are significantly influencing the learning process in virtual classrooms belong the feeling of isolation of students who are separated from the rest of their "classmates". These students lack their social contacts which naturally evolve in traditional educational environments of locally assembled attendees during activities that are performed in teams through collective effort.

The feeling of belongingness is an important motivational factor [4] with a significant positive impact on the capability of an individual to overcome obstructions influencing the cohesion of the whole group attending given course(s). It results into team spirit – the feeling of pride and loyalty among the members that makes them want their team to

M. E. Auer et al. (eds.), *Teaching and Learning in a Digital World*,
Advances in Intelligent Systems and Computing 715,
https://doi.org/10.1007/978-3-319-73210-7_10

do well, to succeed or to be the best [5]. It is built and facilitated among the academic community by solving problems collectively, by supporting collaborators to find solutions for them and by sharing experience gained from tackling those [6]. Individuals are learning from experiences of their peers who usually struggle with similar impediments and describe their victorious and unfortunate attempts in a language that is close for them. All these encounters build bonds among them. Relationships between former classmates influence the environment in which they collaborate and develop their business. Their existence at virtual universities therefore is highly desired.

In this paper the ways of building team spirit at virtual universities are considered. Enhancing the "university spirit" can be done by various ways, in particular through activities belonging to following two categories:

(a) Activities directly connected to teaching and learning processes;
(b) Out-of-school events and happenings.

One can easily imagine the forms of application of above mentioned activity categories at traditional universities: The first group includes direct contacts of students with their educators, their mutual collaboration with their classmates on group projects, their discussions during workshops, etc. Simply said: students build bonds by creating mutually shared collective experiences. Socializing of students is fostered by group activities in official academic working environment. These connections often survive their graduation and may later lead to long lasting relationships and/or cooperation in business activities.

The same frequently repeats with informal contacts formed during out-of-school activities like sport events, parties, summer courses and others – the other out of the above categories. In these environments, the contacts cover even wider groups expanding the particular class and/or year of study. Such encounters often lead to creating new (out-of-the-study-field) interactions and to synergic effects when students with diverse specializations and types of knowledge join together. Their mutual participation on informal group events supports building of their mutual understanding and cohesion between/among partners and may help overcoming obstacles and improve overall relationships.

The contacts built during their study years represent an important capital for the graduates. They denote a base network of connections spread in diverse companies at various positions specializing in different narrow fields of know-how. Also in cases when the students are working in close fields like their acquaintances, such contact networks help to reach the right persons and/or arrange the opportunities to be "in the right time on the right place". Those discussed activities are implicitly present for "face-to-face" daily students. Obviously, many of the above ways of teambuilding are difficult to transfer into virtual universities. In this contribution, the aim is to analyze both (internal and external) encounters in order to enhance the students' feeling of affiliation to their virtual university and towards fostering cohesion building within course groups.

This paper addresses tools reinforcing the feeling of belongingness by applying a qualitative approach in form of structured and semi-structured interviews with students of selected virtual classrooms to find additional ways of enhancing their feeling of "being a member of the community" which may lead to forming long-term partnerships among graduates of the same virtual academic institution. The study conducted and described in this paper intends to analyze to what degree the existing ICT supports the formation of social connections and establishment of relationships among attendees that may turn into future valuable partnerships (if at all); whether students are utilizing available ICT means to build and establish social bonds and linkages and what could be done to facilitate their formation. Our aim is to expand the portfolio of the offerings which may facilitate the current state of social life in virtual university courses and expand the offer "contact-friendly" virtual environments for online students. Properly built environments could diminish many shortcomings present in today's on-line settings of virtual universities to the benefits of both – the universities and their attendees sustainable contributing to building relationships.

This paper is organized as follows: Sect. 2 describes the current status of students' social life *inside* virtual classrooms. It is then followed in Sect. 3 by their social life *outside* classrooms. In the end, Conclusions summarize our findings addressing the actual situation of social life within an online course of a virtual university and propose a framework for its enhancement.

2 Social Life Inside Virtual Classrooms

There are various ways of enhancing social cohesion inside online courses.

First, every course should start with teacher's and students' introduction by providing some knowledge about them. They should say a bit about their character, their previous study and professional experience(s). References to their additional personal data like family, origin and hobbies are welcome, too. In the ideal case, the students share also their photos, videos and addresses to their profiles in social networks, personal websites and/or other Internet connections. This all contributes to the opportunity to be perceived by others as an individual in a wider digital space – to the building a perception of being appreciated not only as a classmate but also as a human. It also fosters initiation of communication between individuals who can start to talk about any topic of interest that is known to both communication parties. This principle is known for quite long [7, 8]. For its eternal validity, it still has to be recommended for first encounters with students in each course.

Everyone's introductory information must be acknowledged by his/her teacher to demonstrate to the student that he/she is a valued partner. In particular, if the instructor and/or students met the individual in their previous courses, the welcome message has its specific value. It is an indirect proof of the person's unique value and respect of the sender of the welcome message towards the individual. This welcome greeting and similar informal messages enhance his/her self-confidence and strengthen the feeling of membership and belongingness to the team. The students should also be asked to respond to (at least some of) their classmates' introductions in order to get known each other and start building their first partnerships.

The virtual classroom should contain a room for general chats among classmates. As the teacher can access all components of their virtual classroom, the students should be advised to keep a universal character of their discussed topics: working experience, family news, local public events, interesting stories etc. The small talks, gossips and slanders should be avoided. In principle, all interpersonal communication should be general enough not to offend or discriminate anyone and to reflect high level of tolerance and respect towards human dignity.

Naturally, more private communication is also a part of standard contacts and supports the feeling of partnership. However, due to its possibly sensitive character, it should run outside classroom and reach just preselected addressees. The reason is obvious. Due to their content, some messages and news may have negative consequences. The instructors should eliminate them from their "official territory" or at least try to reduce their presence. Also, the distribution of post-truth, "alternative facts", not proven and questionable speculations and hoaxes should be excluded to the maximum degree (unless they themselves are the subject of course). In such occasions, references and guidance of officially accepted and proven sources should be provided.

Education itself offers various social activities. As the online education is primarily based on discussions, the argumentation style during the idea exchange becomes a trait helping to identify every individual – to what degree he/she simply repeats the learned materials, whether he/she searches additional sources over the internet to widen his/her knowledge, which type of information and which sources of information he/she prefers and tends to believe, how he/she elaborates it, the influence of previous experiences, the basic threats for taking a standpoint, tolerance against other viewpoints etc. The individual sense of humor and his/her capability to incorporate the jokes into his/her texts also play an important role in recognizing individuals among the masses. In addition to that, they relax the atmosphere and imprint a human touch into the course content in the impersonal virtual environment.

Group assignments have a similar supportive function with a more intensive effect. As they are usually solved by small teams [9], the success of a team member depends on the effective communication and collaboration with the rest of the entire team. The team members chose the communication channel that may be implicitly the most suitable for their effective collaboration: exploiting videoconferencing (either built-in in provided e-learning management system (LMS) or a free tool like Skype) or initiate separate chat (inside or outside the LMS via mobile applications like WhatsApp or social networks like Twitter/Facebook by mutually exchanging contact information), etc. However, the teacher should monitor the state of collaboration within a group to eliminate passive groups with "one man" working and to foster tackling communication in groups which chose an impropriate communication channel.

The geographical distribution of the students plays its role as well. When the distances are small enough, the instructors and/or students can occasionally meet with each other and build the social bonds alike physically attended universities. Personal encounters are significantly impacting the building of a relationship. However, the current ICT provides a reality close alternative through transmitting audio and visual characteristics through a videoconference as well as its personality expressed in spoken and written messages and through behavioral interactions.

3 Social Life Outside Virtual Classrooms

Outside the class room, students implicitly exploit communication channels in social networks, mobile messenger applications and/or emails. Talks and messages outside the class room are no longer overseen by any official academic entity that might try to mitigate or denote an undesired audience to sharing of sources like books, elaborated works and tests with correct answers from previous years etc. This information not only represents a valuable source of knowledge for students – it is also a sign of their desire and readiness to form a community. The sharing of such resources enhances the importance and value of the connection among individuals and contributes to their perception as a unique and special group based on their common knowledge and access to shared sources.

There can be two types of external channels fostering community formation:

- *A community including the lecturer* for example the university web site and its incorporated communication services: Here, the rules identical to those in the virtual classroom should apply. The importance of community lies in the fact that it helps the students finds supplementary contacts according to their additional expectations: their region of origin, individual interests, hobbies etc. Access to these shared sources can e.g. contain information about their past, current and future courses, to help them to recognize how well (or shallow) they are prepared and how to acquire wider and deeper additional knowledge. Public events like used books markets are also welcome.
- *A community excluding the lecturer:* At the same time, students love exchanging their experiences and know-how not only on the content of one course. Social networks provide a place for posting threads on topics of common interest including previously mentioned small talks, gossips and slanders. The existence of such places can substantially facilitate the social life of the geographically dispersed groups as it is often the sole way of their informal contact. ICT provides the place that is accessible to everyone at any time from everywhere and provides the access flexibility for community members to participate in activities without the necessity of being physically assembled. The possibility to attach a file that is suitable for whole course/group announcements and sharing of books, works, tests and pictures etc. is also available.

In order to learn the current situation, students of our online courses were asked about the existence of groups excluding the instructor. Interviewed students confirmed the existence of such private groups in online social network sites (Facebook in particular) to share information and resources within their group. Due to the sensitivity of issue, they were not asked to disclose their content.

Our results also show that a typical student is a member of multiple groups. As a rule, their study fellows from their year and study program form one of those groups. In others, all study fellows as well as graduates of given university are accepted (independently on year of study and study program). This more open group was also utilized by graduates for offering internships and part time jobs of current students and integrating them into the business world on that way. It also provided the place for market places of used books etc. While the former one usually excludes instructors, the latter on is far more open.

Most of interviewed students were also members of official group created through the official profile of the university [10] expressing their connectedness towards and belonging to their Alma Mater.

Mobile communicating applications and instant messaging clients were more used for instant messaging through cell phones for immediate response as not every person followed all posts in a social network. By creating a group in a messenger, the chat room for whole group is being created and all group members can be informed at once.

The mobile communicator is also a more utilized communication channel for occasional accidental arrangements agreed only a short time before the actual happening, e.g. going out for a cup of coffee when being near to somebody's place.

Email communication was exploited by smaller project groups that elaborated assignments. Students working in teams exchanged their versions of common team project and other additional resources. Team members utilized emails mostly for sharing of resources with smaller group and communicated mostly via instant messaging clients.

All the above mentioned "outside class room" communication channels including the history of activities conducted through these channels also during the study remain to study fellows also after finishing the course or/and the study at the virtual academic university. These connections remain open for future deepening relationships and business collaborations.

4 Building Social Environment

The prolegomenous findings from interviews indicate room for fostering building social relationships in virtual environment and the necessity of informal communication for the perception of human touch and building of relationships.

To evolve a connection to a common topic and discussion are basic attributes initiating communication and fostering the transfer and sharing of tacit knowledge. The socialization of individuals through discussions, brainstorming, listening to other opinions and opposing common opinions contribute to mutual collection of experiences and fosters relationship building. At the same time, it builds the base and first phase for transfer of tacit knowledge [11]. Therefore, a group assignment by the lecturer gives the topic and may start the discussion among students. At the same time, the bonding assignment does not need to be directly connected to the course content. It should rather support the interpersonal ties and give to everyone an opportunity to express his/her uniqueness. It shall denote a topic that anybody can talk about to create relaxed atmosphere.

Assigning the bonding assignment to a team forces its members to combine their skills and know-how in order to succeed. It often leads to combining approaches and lateral thinking that may result in new innovative solutions. A good assignment that may foster the building of relationships should disclose some knowledge about members and also open room for their creativity to let them express their handling. A good example may be a virtual "group backyard garden" where each group member pastes his/her flower, tree, herb or any other plant, sitting or picnic area, family members, etc. The selection will exemplify the value system of the person. For example the choice of a particular plant should be accompanied by an explanation why he/she decided for the particular one. On that way, participants show knowledge about plants, tell something about their country and reveal their own characteristics through the way of explaining and expressing. Each individual could contribute with an endogenous tree or a flower that is typical for his/her own surroundings. Then the students create their visions of the group garden by combining those plants. At the end, they compare their solutions and can vote for the most favorite one.

The virtual garden could be also realized in physical reality, e.g. planted by the lecturer in the university garden, where the group members could trace and sight their community garden by online cameras. The group will have something that makes them special and perceived as a team that belongs together. It will be a neutral common topic that will strengthen the connection bonds.

5 Conclusions

Our survey was very informal and done with a small number of students. Thus, its results can only have an orientation character. They still gave us interesting outcomes that must be surveyed and analyzed more systematically.

- The out-of-classroom communication among classmates of an online course is minimal and rarely continues after the course completion.
- Today, there is a barrier between on-site and on-line students as well as among on-line students themselves.
- The online students would welcome certain support from the side of university that would enhance their out-of-the-classroom social life.

Our preliminary findings show undisputable importance of building a social environments specific for needs of online students. Their results also create room for our further research which will lead to forming solid hypotheses and to their verification using a larger numbers of students and formats of their online study.

Acknowledgements. This work was supported by the Faculty of Management, Comenius University and School of Management (Vysoká škola manažmentu), both in Bratislava, Slovakia.

References

1. Dávideková, M., Hvorecký, J.: Collaboration tools for virtual teams in terms of the SECI model. In: Auer, M., Guralnick, D., Uhomoibhi, J. (eds.) Interactive Collaborative Learning, ICL 2016. Advances in Intelligent Systems and Computing, vol. 544, pp. 97–111. Springer, Cham (2017)
2. Robins, K., Webster, F.: The Virtual University? Knowledge, Markets, and Management. Oxford University Press, Oxford (2002)

3. Ryan, S., Scott, B., Freeman, H., Patel, D.: The Virtual University: The Internet and Resource-Based Learning. Kogan Page Ltd., London (2013)
4. Maslow, A.H.: A theory of human motivation. Psychol. Rev. **50**(4), 370–396 (1943)
5. Cobuild: Team spirit. Advanced English Dictionary. https://www.collinsdictionary.com/dictionary/english/team-spirit. Accessed 22 May 2017
6. Erdem, F., Ozen, J.: Cognitive and affective dimensions of trust in developing team performance. Team Perform. Manage. Int. J. **9**(5/6), 131–135 (2003)
7. Salmon, G.: E-Moderating – The Key to Teaching and Learning Online. Kogan Page, London (2000)
8. Paloff, R.M., Pratt, K.: Lessons from the Cyberspace Classroom – The Realities of Online Teaching. Jossey-Bass, San Francisco (2001)
9. Hvorecký, J.: Team projects over the Internet. In: Auer, M., Guralnick, D., Uhomoibhi, J. (eds.) Interactive Collaborative Learning (ICL 2006). ICL, Villach (2006)
10. Dávideková, M., Greguš, M.: A case study analyzing social network integration and its opportunities to selected academic institution. In: The 5th International Conference on Serviceology (ICServ2017) (2017, in press)
11. Hvorecký, J., Šimúth, J., Lichardus, B.: Managing rational and not-fully-rational knowledge. Acta Polytechnica Hungarica **10**(2), 121–132 (2013)

Evaluating Collaborative Learning Using Community of Inquiry Framework for a Blended Learning Formal Methods Course

Saad Zafar, Naurin Farooq Khan(✉), and Seema Hussain

Riphah International University, Islamabad, Pakistan
{saad.zafar,naurin.zamir}@riphah.edu.pk, seema@gmail.com

Abstract. A Formal Methods course was taught using a blended-learning pedagogy at graduate level. The blended-learning environment was designed with the objective of improving the students' learning experience and to address some of the inherent challenges of teaching FM. This study presents results of evaluating the contents of online discussion forums that were used in the course. The Community of Inquiry (CoI) framework was used to assess the level of collaborative learning by measuring social presence, teaching presence and cognitive presence of the participants. The results of the study show healthy levels of participation on all the three CoI dimensions. More importantly, there was no or little difference between mandatory and non-mandatory discussions. However, the discussion in mandatory forum was more open and organized on social front indicating an evolved sense of community and trust for time-critical and mission critical tasks within the course.

Keywords: Formal Methods · Blended learning · Community of Inquiry Discussion forum

1 Introduction

In computing field, teaching Formal Methods (FM) is not an easy task. Apart from the inherent difficult nature of the subject [1], the methodologies adopted for teaching are traditional that do little to encourage students to learn [2]. Moreover, there is a lack of pedagogical practices in the field that can help students better understand the subject [3]. We adopted blended learning pedagogy [4] in the delivery of graduate level FM course. Students were required to take face-to-face lecture as well as do project activities online. For collaboration, students used Moodle [5] discussion forums. The discussion forums were differentiated as mandatory and non-mandatory in terms of marks awarded to students on participation in the former type. For evaluating collaborative learning that happened in the course using blended learning pedagogy, content analysis of the student discourse (retrieved from the discussion forums) was performed. The cognitive presence, social presence and teaching presence were measured by coding the student discourse using Community of Inquiry (CoI) framework [6] by the help of indicators' list that CoI offers. Statistical analysis was performed to find out the effect of mandated

M. E. Auer et al. (eds.), *Teaching and Learning in a Digital World*,
Advances in Intelligent Systems and Computing 715,
https://doi.org/10.1007/978-3-319-73210-7_11

and non-mandated discussion on the cognition, social and teaching presence in the course. It was found out that in terms of cognition and teaching presences, there was no significant difference observed for mandated and non-mandated discussion forums. However in the social presence, there was a significant difference in the open communication and organization categories between mandated and non-mandated discussion forums. The mandatory discussion forums were responsible for increased open communication and organizational posts by the student.

We present the background information in Sect. 2. Section 3 describes the research questions followed by Sect. 4 where the detailed methodology is discussed. The analysis and results are presented in Sect. 5 while Sect. 6 concludes the paper.

2 Background

2.1 Formal Methods

Computing curriculum has a due place for Formal Methods and its criticality cannot be denied. Formal Methods teaching is not an easy task. The mathematical nature of Formal Methods pose students with the problems of understanding, reading, writing and mastering the FM notations [1, 7, 8]. Both teachers and students take a considerable amount of time at the start of the semester in understanding the basic mathematical terminology. Moreover, the students' and teachers' negative attitude towards the subject exacerbates the problem. Students refrain from studying theory of Formal Methods [9] since they find it dry and boring [10–12] and hence show low participation and interest during the lectures [13]. The teaching methods adopted for Formal Methods fall short in motivating the students to learn [14]. Teachers mostly rely on traditional methods of using paper and pencil in the class room [2, 15] that do little to encourage students to participate. There are few studies that try to address the challenges of teaching FM using active learning methods [16] and web based environment. However, what is missing are the pedagogical practices [3] in the area that can play an important role in better understanding of the subject. To the best of our knowledge, Formal Methods teaching reported in the literature do not exploit the advantages of blending traditional face-to-face teaching with online e-learning technologies.

2.2 Blended Pedagogy

The collaborative learning by blending traditional face-to-face teaching with online contents has recently boomed [17]. It is most likely that it may become the dominant model in education due to the fact that it incorporates the best elements of both teaching methods [4]. The live synchronous environment of face-to-face method is considered to have high fidelity experience while the asynchronous material in an online setting is self-paced and offers flexibility [18]. Blended learning environment is composed of richness of face-to-face interaction as well as the flexibility of online environment. This effective integration of two methods [19, 20] plays a lead role in enhancing the learning experience of the students, the student engagement and learning [4].

Online discussion forums play an important part in blended learning pedagogy and have proliferated recently [21]. Students can actively engage in sharing information and perspective with other students [22]. The interactions in discussion forums are archived which allows for reflection and debate. As a result of which creation of knowledge takes place [23] and learning gets transformed from one way instructional strategy towards highly interactive learning [24]. Moreover, it allows for students to take into consideration different perspective to a particular problem which in turn foster new meaning construction [25]. This exposure to diverse views is missing in face-to-face interaction and is responsible for creation of a more complex perspective on a particular topic [26]. The student's discussions can later be analyzed and evaluated to find out the quality of learning that took place in the blended learning environment. One method is to perform content analysis of the online discussion of the students which qualitatively evaluates the contents. Many content analysis models exist in literature that can be used to qualitatively assess the electronic discourse. In this study we use Community of Inquiry model for content analysis of the discussion forums used in the blended learning course. The model is popular in the research community with a lot of empirical validations and replication of the studies [27, 28]. Another reason to choose the model is its robustness and its validity due to high level of inter-rater reliability of its coding scheme.

2.3 Community of Inquiry Framework

The Community of Inquiry (CoI) framework represents the mostly researched model that is focused on three major constructs of online learning depicted as presences. The three presences – cognitive, social and teaching are central to the learning processes that occur in an online community [6, 29] and work together to support deep and meaningful learning [30]. Each presences' are further subcategorized. The first element cognition is further decomposed into (1) a triggering event (2) exploration, (3) integration and (4) resolution. These subcategories actually present the cognition level with 1st being the lowest while 4th the highest. Similarly the social presence is further divided into (1) interpersonal communication, (2) open communication, (3) cohesive communication and (4) organization that shows the supportive context for expression of emotion, open communication and group cohesion respectively. The third element is the teaching presence that is categorized as (1) direct instruction, (2) design and organization and (3) facilitating discourse.

We use CoI model to qualitatively assess the electronic discourse of the students. The CoI list of indicators are used to give correct category to the posts. Furthermore, statistical analysis is performed in order to achieve the objective of the study i.e. the cognition, social and teaching presences' in mandatory and non-mandatory discourse. In accordance to our study objectives, the following section presents the research questions.

3 Research Questions

RQ1. What is the level of cognition observed in electronic discourse of mandatory and non-mandatory discussion forums?

RQ2. What level of social interaction is observed in mandatory and non-mandatory discourse?
RQ3. Is there a difference of teaching presence in mandatory as compared to non-mandatory discussion forums?

4 Methodology

4.1 Study Context

Advanced Software Requirements Engineering (ASRE) course was offered in Spring (2015) to graduate students in the Software Engineering discipline at a medium sized university located in Islamabad. The objective of the course was to develop understanding of requirements engineering and to engineer requirements using Formal Methods (FM). The activities involved class lectures, videos, projects, project workshops and extensive use of discussion forums. Two instructors carried out the design and delivery of the course. The course was designed in such a way that the class met face-to-face with the instructors after every fortnight, with projects workshops assigned to them during the time in between. The students were expected to carry out the tasks assigned to them by collaborating among themselves and solving the related problem via use of discussion forums. We differentiated between the discussion forums as mandated and non-mandated in terms of 5% of the total assessment reserved for the later. It should be noted here that some of the discussion forums were made compulsory due to unfamiliarity of students with blended learning pedagogy and doing so allowed the students to acknowledge the participation in the forums and consider them a part of teaching methodology in this course.

ASRE course was divided into two major projects which were in turn broken down into milestones. Project 1 was assigned to students so that they can develop knowledge and skills of FM tools and techniques in order to engineer requirements. The project was divided into different activities, each related to the different phases of requirements engineering. Activities contained the background information, task, time and deadlines as online workshops. In project 2, students were required to carry out literature review involving modelling of systems using Formal Methods. They were instructed to find the application of formal methods in different industrial domains. The databases that students had to search included IEEE Xplore, ScienceDirect, ACM digital library and SpringerLink.

4.2 Subjects

There were 22 students who enrolled in the ASRE course. Eight of the students were female. The students came from varied backgrounds, some currently working full time in the industry. Moodle Learning Management System (LMS) was used for setting up the course. Instructors uploaded documents to set up the participation rules in discussion forums. To familiarize students with the discussion forum environment in Moodle and to overcome the initial inertia, icebreaker discussion forums were set up where participants were asked to introduce themselves. The activities related to the two projects were carefully designed and were given as workshops in Moodle. Each workshop was properly structured to contain objective, task description, the outcome and the due date and time. One discussion forum for each activity was also initiated where students were instructed to participate to solve problems related with the tasks. Students replied to the already initiated threads as well as initiated new thread for different problems. For mandatory discussion forums, a time period of 1 week was allotted after which students' posts were not considered for marks awards.

4.3 Data Collection

There were 16 discussion forums that were set up over the course of 16 weeks. The students and instructors initiated threads as well as replied to the problems posed by the students. The record of these discussions was downloaded as pdf document and the third author performed content analysis. There were total (N = 191) threads initiated in which 89 were from mandatory and 102 threads were from non-mandatory discussion forums. The three presences – cognitive, social and teaching were measured using the instrument developed by [6, 31]. For coding of messages, each message was considered in terms of its meaning. The thematic unit was preferred over whole message as a single message can contain sub-messages which can belong to multiple categories of single presence as well as multiple presences. The coding of messages was performed in chronological order and the data was coded in Nvivo software. Table 1 presents the total coded messages in each presences. The coded data was imported into SPSS v.21 and statistical analysis was performed. For coding in teaching presence, the course main page and discussion forum main pages from Moodle were also selected to be included in the sample since they contain the teaching presence in terms of structured workshops and discussion forums set up.

Table 1. Total number of coded messages in the three presences

	Triggering Event	Exploration of Ideas	Integration of Ideas	Resolution of Ideas	Total	Interpersonal Communication	Open Communication	Cohesive Communication	Organization	Total	Design and Organization	Facilitating Discourse	Direct Instruction	Total
Mandated	36	95	201	28	360	125	82	6	46	259	8	12	24	44
	6%	15%	31%	4%	56%	19%	13%	1%	7%	40%	1%	2%	4%	7%
Non Mandated	49	173	228	15	465	82	46	8	17	153	9	5	20	34
	8%	27%	35%	2%	72%	13%	7%	1%	3%	24%	1%	1%	3%	5%

4.4 Data Analysis

In the imported data, each sub category for each presence becomes a quantitative variable. The sample consisted of threads that were initiated in each discussion forums with their frequency of occurrence recorded for each sub category of CoI model as shown in Table 1. The frequency analysis of cognition presence reveals that the total number of posts in non-mandatory discussion forums were greater as compared to mandatory, with 8% posts depicting triggering events, 27% exploration messages and 35% integration of ideas messages. The resolution of ideas posts were greater in mandatory discussion forums with 4%. On the other hand, the social presence was observed to be greater in term of no of posts for mandatory discussion forums with 19% interpersonal communication, 13% open communication and 7% organization. Cohesive communication was lesser in number in mandatory as compared to non-mandatory discussion forums. The teaching presence for mandatory and non-mandatory discussion forums was almost the same for design and organization category with 1% while for facilitating discourse mandatory discussion forums had 2%, a little higher than that of non-mandatory discourse. On the similar terms direct instruction was a little bit higher in mandatory discourse with 4% as compared to 3% for non-mandatory discourse.

5 Results and Discussion

For performing quantitative analysis, independent sample t-test was administered on the data obtained from qualitative coding using CoI model. In order to answer RQ1, the dependent variable is the cognition level and the independent variable is the type of discussion forum (2 groups: mandatory and non-mandatory). The t-test was run four times for each sub category of cognition level. The results of levene's test (Table 2) $F = 0.804$, $p = 0.371$ for triggering event category and $F = 0.166$, $p = 0.684$ and for Integration sub category indicate that the variance between the two population are assumed to be approximately equal therefore the standard t-test results are considered. On the other hand, $F = 4.507$, $p = 0.03$ for exploration category and $F = 9.132$ with $p = 0.003$ for resolution indicate that the variance between the two population are not assumed to be equal hence t-test for equal variances not assumed is considered. The results of the independent t-test (Table 3) for the four sub categories of cognitive presence were not significant, with values $t(189) = -0.709$ with $p = 0.479$, $t(130.713) = -1.739$ with $p = 0.084$, $t(189) = 0.045$ with $p = 0.964$ and $t(117.1)$ with $p = 0.140$ for triggering event, exploration, integration and resolution sub categories respectively. The p values for each sub category is greater than 0.05 which shows that there is no significant difference between level of cognition for mandatory and non-mandatory type of discussion.

Table 2. Group statisitcs and Levene's test for equality of variance

	Discussion forums	N	Mean	Std D	F	Sig
C-triggering event	Mandatory	89	.40	.669	.804	.371
	Non Mandatory	102	.48	.793		
C-exploration	Mandatory	89	1.07	1.241	4.507	.035
	Non Mandatory	102	1.70	3.400		
C-integration	Mandatory	89	2.26	2.810	.166	.684
	Non Mandatory	102	2.24	4.103		
C-resolution	Mandatory	89	.31	.984	9.132	.003
	Non Mandatory	102	.15	.431		
S-interpersonal communication	Mandatory	89	1.40	3.063	5.239	.023
	Non Mandatory	102	.80	1.372		
S-open communication	Mandatory	89	.92	1.517	9.813	.002
	Non Mandatory	102	.45	1.114		
S-cohesive communication	Mandatory	89	.07	.252	.327	.568
	Non Mandatory	102	.08	.305		
S-organization	Mandatory	89	.52	1.188	19.371	.000
	Non Mandatory	102	.17	.646		

Table 3. Independent sample T-Test Results

	t	df	Sig. (2-tailed)	Mean dif	Difference	
					Lower	Upper
C-triggering event	−.709	189	.479	−.076	.107	−.287
C-exploration	−1.739	130.713	.084	−.629	.361	−1.344
C-integration	.045	189	.964	.023	.516	−.996
C-resolution	1.487	117.123	.140	.168	.113	−.056
S-interpersonal communication	1.707	118.348	.091	.601	.352	−.096
S-open communication	2.413	159.506	.017	.470	.195	.085
S-cohesive communication	−.270	189	.788	−.011	.041	−.092
S-organization	2.479	131.640	.014	.350	.141	.071

For answering RQ2, the dependent variable is social presence's sub categories. The levene's test for interpersonal communication (Table 2) has the F values of 5.239 with $p = 0.02$, for open communication, $F = 9.813$ with $p = 0.002$ and for organization sub category, $F = 19.371$ with $p < 0.001$ indicate that the variance between the two population are not assumed to be equal hence t-test for equal variance not assumed is considered. On the other hand, for cohesive communication sub category standard t-test results are considered with levene's test F values of 0.327 and $p = 0.568$. The results of the independent t-test (Table 3) for the social presence were mixed. For interpersonal communication and cohesive communication the results were not significant with values $t(118.3) = 1.707$ with $p = 0.091$ and $t(189) = -0.270$ with $p = 0.788$. The results for open communication and organization sub categories of social presence were statistically different for mandatory and non-mandatory type of discussion. The values $t(159.5) = 2.413$ with $p = 0.01$ and $t(131.64) = 2.479$ with $p = 0.01$ for the two sub categories. In each case the p value (0.01) is less than 0.05. The open communication and organization of social presence for mandatory discussion forums are greater with mean values of (M= 0.92, Std dev = 1.517) and (M = 0.52, Std dev = 1.188) respectively.

In order to answer RQ3, frequency analysis of the teachers' posts was performed as shown in Table 1. Due to smaller data set that is available, statistical analysis was not viable. The percentages of teaching presence sub categories; design and organization, facilitating discourse and direct instruction for both mandatory and non-mandatory discourse was almost the same. Therefore, there was no difference in the teaching presence for the two types of discussion forums.

The statistically significant results for open communication depict that the stringent time line in mandatory discussion forums to post within a week provided a learning climate in which students asked questions to clarify the subject matter. The students also complimented on their peers posts and expressed agreements to the ideas they deemed worthy, thereby creating an environment in which they trusted each other to reveal about their thoughts. Moreover, the greater organization presence with statistically significant results in the mandatory discussion forums reveal that students were concerned about the deadlines and posted postings related to management and time out of the tasks. These results depict that an evolved sense of community was realized for time critical tasks in the course.

6 Conclusion

The challenge of teaching Formal Methods was addressed by incorporating blended learning pedagogy. The course was designed in such a way that face-to-face lectures were augmented by using online discussion forums. In mandatory discussion forums students were required to take part and as a result their participation contributed towards 5% of their final grade while in non-mandatory discussion forums the participation was left to the students' choice. Content analysis of the students' posts were carried out to see the level of cognition that took place in the two types of discussion forums. Moreover, social and teaching element in the electronic discourse were also measured. The frequency analysis revealed that non-mandatory discussion forums allowed for greater number of cognition while the mandatory discussion forums allowed for increased social presence. The teaching presence almost remained the same. However, the statistical analysis showed that there was no significant difference between the cognition and teaching levels of mandatory and non-mandatory discussion forums. On the other hand, open communication and organization sub categories of social presence was greater in mandated as compared to non-mandated discussion forums. We conclude that the time critical nature of tasks and the prospect of marks made students to communicate effectively with their peers in terms of asking questions to clarify the task at hand, agreeing to their peers' point of view, giving compliments as well as managing class activities and time.

Acknowledgement. This research is part of a larger research project at Riphah International University WISH campus. The contributions of Dr. Aslam Asadi, Dr. Saba Riaz and Ms. Sadia Nadir are also acknowledged.

References

1. Schreiner, W.: The RISC ProofNavigator: a proving assistant for program verification in the classroom. Form. Asp. Comput. **21**(3), 277–291 (2009)
2. Dony, I., Le Charlier, B.: A tool for helping teach a programming method. ACM SIGCSE Bull. **38**(3), 212–216 (2006)
3. Gopalakrishnan, G., et al.: Some resources for teaching concurrency. In: Proceedings of the 7th Workshop on Parallel and Distributed Systems: Testing, Analysis, and Debugging, p. 2 (2009)
4. Watson, J.: Promising Practices in Online Learning: Blended Learning: the Convergence of Online and Face-to-Face Education. North American Council for Online Learning (2008)
5. Cole, J., Foster, H.: Using Moodle: Teaching with the Popular Open Source Course Management System. O'Reilly Media Inc., Sebastopol (2007)
6. Garrison, D.R., Anderson, T., Archer, W.: Critical inquiry in a text-based environment: computer conferencing in higher education. Internet High. Educ. **2**(2), 87–105 (1999)
7. Habrias, H.: Teaching specifications, hands on. In: Formal Methods in Computer Science Education (FORMED), pp. 5–15 (2008)
8. Barbu, A., Mourlin, F.: Enhancing student understanding of formal method through prototyping. Formal Methods in the Teaching Lab, p. 85 (2006)
9. Paige, R.F., Ostroff, J.S.: Specification-driven design with Eiffel and agents for teaching lightweight formal methods. In: International Conference on Technical Formal Methods, pp. 107–123 (2004)
10. Henz, M., Hobor, A.: Teaching experience: logic and formal methods with Coq. In: International Conference on Certified Programs and Proofs, pp. 199–215 (2011)
11. Walther, C., Schweitzer, S.: Verification in the Classroom. J. Autom. Reason. **32**(1), 35–73 (2004)
12. Ahrendt, W., Bubel, R., Hähnle, R.: Integrated and tool-supported teaching of testing, debugging, and verification. In: International Conference on Technical Formal Methods, pp. 125–143 (2009)
13. Miller, A., Cutts, Q.: The use of an electronic voting system in a formal methods course. In: Workshop on Formal Methods in the Teaching Lab (FM-Ed 2006), pp. 3–8 (2006)
14. Lau, K.-K., Bush, V.J., Jinks, P.J.: Towards an introductory formal programming course. ACM SIGCSE Bull. **26**, 121–125 (1994)
15. Rodger, S.H., Wiebe, E., Lee, K.M., Morgan, C., Omar, K., Su, J.: Increasing engagement in automata theory with JFLAP. ACM SIGCSE Bull. **41**, 403–407 (2009)
16. Lau, K.-K.: Active learning sheets for a beginner's course on reasoning about imperative programs. ACM SIGCSE Bull. **39**, 198–202 (2007)
17. Graham, C.R.: Blended learning systems. In: The Handbook of Blended Learning, pp. 3–21 (2006)
18. Garner, R., Rouse, E.: Social presence–connecting pre-service teachers as learners using a blended learning model. Stud. Success **7**(1), 25–36 (2016)

19. Pelliccione, L., Broadley, T.: RU there yet? using virtual classrooms to transform teaching practice. In: Curriculum, Technology and Transformation for an Unknown Future. Proceedings Ascilite Sydney, pp. 749–760 (2010)
20. Stacey, E.: Effective Blended Learning Practices: Evidence-Based Perspectives in ICT-Facilitated Education: Evidence-Based Perspectives in ICT-Facilitated Education. IGI Global, New York (2009)
21. Chen, W., Looi, C.-K.: Incorporating online discussion in face to face classroom learning: a new blended learning approach. Australas. J. Educ. Technol. **23**(3), 307–326 (2007)
22. On-line education: a new domain*. http://www.bdp.it/rete/im/harasim1.htm. Accessed 05 May 2017
23. Gay, G., Sturgill, A., Martin, W., Huttenlocher, D.: Document-centered peer collaborations: an exploration of the educational uses of networked communication technologies. J. Comput. Commun. **4**(3) (1999)
24. Vrasidas, C., McIsaac, M.S.: Factors influencing interaction in an online course. Am. J. Distance Educ. **13**(3), 22–36 (1999)
25. Ruberg, L.F., Moore, D.M., Taylor, C.D.: Student participation, interaction, and regulation in a computer-mediated communication environment: a qualitative study. J. Educ. Comput. Res. **14**(3), 243–268 (1996)
26. Prain, V., Lyons, L.: Using information and communication technologies in English: an Australian perspective. In: English in the Digital Age. Cassell Education, London (2000)
27. Garrison, D.R., Anderson, T., Archer, W.: The first decade of the community of inquiry framework: A retrospective. Internet High. Educ. **13**(1), 5–9 (2010)
28. Szeto, E.: Community of Inquiry as an instructional approach: what effects of teaching, social and cognitive presences are there in blended synchronous learning and teaching? Comput. Educ. **81**, 191–201 (2015)
29. Garrison, D.R., Anderson, T., Archer, W.: Critical thinking, cognitive presence, and computer conferencing in distance education. Am. J. Distance Educ. **15**(1), 7–23 (2001)
30. Morueta, R.T., López, P.M., Gómez, Á.H., Harris, V.W.: Exploring social and cognitive presences in communities of inquiry to perform higher cognitive tasks. Internet High. Educ. **31**, 122–131 (2016)
31. Rourke, L., Anderson, T., Garrison, D.R., Archer, W.: Assessing social presence in asynchronous text-based computer conferencing. Int. J. E-Learn. Distance Educ. **14**(2), 50–71 (2007)

The "Architectural Jewels of Lublin" Game as a Tool for Collaborative Interactive Learning of History

Marek Milosz(✉) and Jerzy Montusiewicz

Lublin University of Technology, Lublin, Poland
{m.milosz,j.montusiewicz}@pollub.pl

Abstract. The city of Lublin in Poland is a place of many overlapping cultures and religions: Polish, Russian, Ukrainian, Armenian, Jewish, German, etc. Remains of these cultures are present in the city space in the form of architectural monuments. Unfortunately, the awareness of this fact disappears in society, especially among the school youth.

The paper presents the results of a survey of the level of knowledge about the architectural monuments of the city of Lublin, Poland. The study was conducted among K14 youngsters with the use of questionnaires and 3D models. The results of the research show a low level of knowledge about monuments and their history.

The next part of the article presents the game "Architectural Jewels of Lublin", which was developed as a board game, using modern information technologies such as RFID systems, 3D modelling and printing, gamification and real time systems programming. The technical layer of the game is an electronic board with the city plan of Lublin and models of architectural monuments, a sensory network using RFID technology, and a palmtop with game control software. The game allows the implementation of collaborative interactive learning of history. The article also presents the logic of the game and the elements of gamification used in it. The directions of its development and dissemination are also presented.

Keywords: Interactive learning · Board game · Knowledge level
Architectural monuments

1 Intoduction

The city of Lublin is the largest Polish city on the eastern side of the Vistula River. Slavic settlement in Lublin is dates back to the end of the 8th century. The favourable geographical location on the numerous hills above the Bystrzyca and Czechówka valleys, at the crossroads of trade routes, allowed for a rapid development of the castle in the following centuries and the arrival of representatives of many different nationalities. In 1317, in order to reorganise the existing state of affairs and strengthen further development, the city was re-located on the Magdeburg law, well-known and commonly used in Western Europe.

At present the inhabitants of Lublin celebrate the 700th anniversary of this very important event. Over the centuries, Lublin and the whole Polish state were an area in which Poles, Ruthenians, Russians, Jews, Armenians, Germans, Ukrainians and other

M. E. Auer et al. (eds.), *Teaching and Learning in a Digital World*,
Advances in Intelligent Systems and Computing 715,
https://doi.org/10.1007/978-3-319-73210-7_12

ethnicities lived side by side in symbiosis and the spirit of tolerance. Alongside each other, there were opportunities for different cultures and religions to develop: Catholic, Orthodox, Uniate (Greek-Catholic), Protestant and Jewish. This situation persisted until Poland's loss of independence in the 18th century, when the Uniates were persecuted by the Russian authorities. During the Second World War, the German occupation authorities exterminated the Jewish population, destroying many of its historical objects.

Multinationality, multiculturalism and multiconfessionality are the standard image of eastern Poland, as exemplified by Lublin. Remnants from those centuries are numerous preserved architectural monuments, which testify to both the specific character of the merchant town and the wealth of its inhabitants, and the foundations made by the Polish kings and surrounding nobles.

Unfortunately, as experience shows, young people (especially in the younger age bracket, K14), have little historical awareness of Lublin's rich history. This is reflected in the poor recognition of monuments (including architectural ones) that remain after the people once living in the city.

In view of the fact that in 2017 the city of Lublin celebrates 700 years of its location, in the Institute of Computer Science of the Lublin University of Technology arose the idea of building a board game entitled "Architectural Jewels of Lublin", which in a modern way would allow to raise the level of pupils' knowledge about the architectural monuments of Lublin [1]. The developed game realises the idea of gamification, which assumes that in addition to the competition phase, winning points and prizes by participants. It also allows the transmission of strictly defined content. In preparation of the game, advanced 3D modelling and printing technology was used, as well as automatic recognition of the 3D model's location on the map of the old city. The whole process of competition is managed by real-time systems programming.

Games are a very popular educational tool used in different areas and levels of education [2–12]. The effectiveness of their use is confirmed by numerous scientific studies [13, 14] and the opinions of participants, which indicate [15, 16] their interest and motivation, the intensification of learning, the approximation to reality, the interdisciplinarity, the development of practical skills without the usual risks associated with such activity, interaction and error learning.

Boards have been used to organise games for thousands of years, examples being as chess, sudoku, or war games. Also, attempts to automate board games have been made since ancient times. The development of board game automation technology led from mechanical, through electromechanical to electronic one, using a variety of component bases and the use of various physical phenomena to build sensory networks [17–19]. Contemporary computer technologies provide a wide range of opportunities for automated board games [20–22], which was used in the "Architectural Jewels of Lublin" game [23].

The article consists essentially of two main parts. Section 2 presents the research on the level of knowledge about architectural monuments, conducted among young people of K14 from the city of Lublin and the region. The research was carried out in a classical format: defining goals, formalising them, conducting surveys, processing results, and discussing them. Section 3 presents the game "Architectural Jewels of Lublin" in terms of: game objectives, technical layer and game execution logic. The article ends with conclusions from the achievements so far and indicates further work.

2 Study of the Level of Knowledge About the Architectural Monuments of the City of Lublin Among Young People

2.1 Purpose and Scope of Research

The aim of the study was to assess the level of knowledge of K14 pupils of the architectural monuments of the city of Lublin. The study covered two groups of pupils: those living in Lublin (1st group) and those outside (2nd group).

The level of knowledge was assessed on the basis of the results of the test, consisting of questions about the 7 most distinctive architectural monuments of the city of Lublin, selected by experts. For each monument the questions concerned:

P1. Name of the monument.
P2. The time at which the monument originated.
P3. The architectural style in which the monument was built.

For the level of knowledge about architectural monuments of the city of Lublin the following absolute indicator (K) was adopted:

$$\mathrm{K} = \sum\nolimits_{i=1}^{7} \sum\nolimits_{j=1}^{3} w_j P_{ij} \tag{1}$$

where:

i – number of the monument,
j – number of the question for the given monument,
w_j – the weight of the j-th question for the monument,
P_{ij} – the answer to the j-th question for the i-th monument (0 - incorrect, or 1 - correct).

The following weights for individual questions (P1… P3) were used for each monument: P1 – 1, P2–2 and P3–3. These scales (1… 3) in the opinion of experts mirror the scale of the difficulty of the questions.

The relative level of knowledge (k) is the ratio of a student's absolute knowledge index to its maximum value – K_{max}:

$$k = \frac{K}{K_{max}}, \%. \tag{2}$$

For 7 monuments and 3 questions of different difficulty levels, $K_{max} = 42$.

2.2 Thesis and Research Hypotheses

The analysis of the research objective allowed us to formulate the following research thesis:

Pupils from the K14 group living in the Lublin province have little knowledge about the architectural monuments of the city of Lublin.

In order to confirm this thesis, two hypotheses were formulated:

H1. The average level of knowledge of K14 pupils living in the Lublin province about the architectural monuments of Lublin is very low.
H2. The level of knowledge of K14 pupils living in the Lublin province about the architectural monuments of Lublin practically does not depend on whether the student lives in or outside Lublin.

2.3 Research Methodology

The research group consisted of pupils who participated in a visit to the Lublin University of Technology. All participants were aged K14.

In order to carry out the research, a survey was prepared, consisting of a questionnaire (containing answers classifying the respondent) and questions on the architectural monuments of Lublin. For each of the seven monuments, 3 questions were asked in Sect. 2.1.

The study was conducted on two groups of pupils living in the Lublin province. The pupils from the first group of respondents were permanent residents of Lublin from birth, while those from the other group were not. The size of the two groups was the same: 12 pupils. Survey participants were randomly selected – each from a different school in Lublin and the region.

The research was carried out using paper surveys and printed 3D models of architectural monuments on a scale of 1:200. After showing a model selected from the questionnaires of all models, the pupils answered 3 questions (P1... P3) about the model shown. After the last question the polls were collected, and the researcher gave correct answers about each of the monuments, along with its extended historical background.

Fig. 1. Implementation of the research on pupils' knowledge about the architectural monuments of Lublin

This presentation was a reward for the pupils and was an element of building historical knowledge (Fig. 1).

2.4 Results

The results of the questionnaires were transformed according to Formulas (1) and (2) – for each student a relative level of knowledge was determined. In Table 1 the basic statistics of the survey results are presented.

Table 1. Results of questionnaires on the statistical parameters of the relative level of respondents' knowledge about architectural monuments of Lublin

No.	Name	Value k for all pupils	Value k for pupils from outside Lublin
1	Average	23.1%	20.2%
2	Median	26.2%	20.2%
3	Min.	0.0%	0.0%
4	Max.	50.0%	40.5%
5	Standard deviation	13.2%	13.7%

A more accurate statistical analysis of the results of the study was made by defining the series of distributions and constructing histograms of the distribution of the values of the relative level of knowledge of pupils about the architectural monuments of Lublin. Its results are presented in Fig. 2.

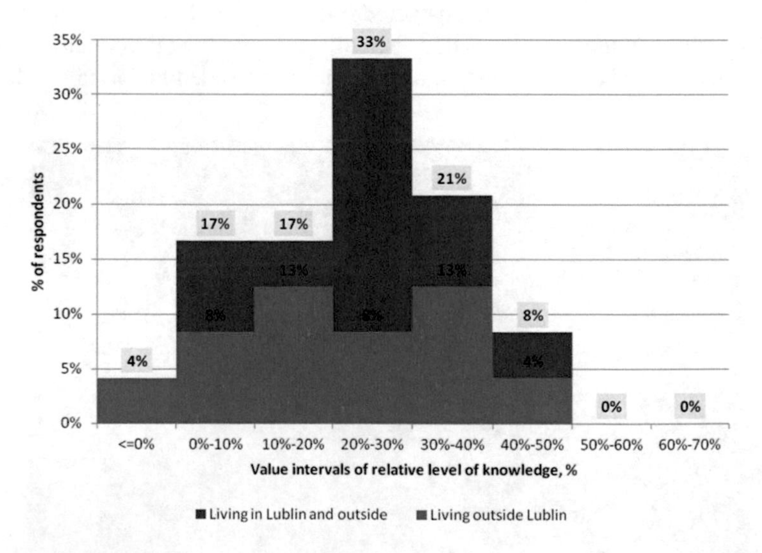

Fig. 2. Histogram of distribution of the relative level of knowledge of pupils living in Lublin and those from outside it

2.5 Discussion

The very low mean (and median) values of the relative knowledge level of pupils, attaining a little over 20% (Table 1), clearly indicate a low level of pupils' knowledge of Lublin's architectural monuments. This is also confirmed by the fact that no student exceeded the 50% threshold, usually required to pass the course.

Hypothesis H1 has been positively verified
The mean of the pupils' self-esteem of their level of knowledge about the monuments of Lublin amounted to 1.96 in the scale of 1… 4 (where 1 – very bad, 4 – very good), whereas for the pupils from outside Lublin this value was 1.41, and for those living in Lublin – 2.50. Nevertheless, the comparison of the average level of knowledge of all pupils with the average level of knowledge of non-Lublin pupils (Table 1 and Fig. 2) indicates that these levels are not much different from one another.

Therefore hypothesis H2 can be considered as positively verified
The results of the K14 pupils' performance showed a low level of knowledge about Lublin's architectural monuments and their history.

3 The Game "Architectural Jewels of Lublin" – Presentation

3.1 Idea of the Game

The game "Architectural Jewels of Lublin" is to provide participants with the most distinctive and famous monuments of Lublin, their history and location in urban space. In its basic form, the game is designed for two players at the same time. Players aim to score as many points as possible and thus defeat a direct opponent. Points can be earned for proper positioning of the object on the board and correct answers to questions. Competitors have the following interactive features:

- 3D models of Lublin monuments,
- the "I answer" button is used to take over question 2,
- buttons to answers (1/2/3).

In addition, there is a palmtop on the board that displays questions, possible answers, historical information about the object being held, current game results and the final score, as well as the loudspeakers giving the relevant messages.

The large board and 3D models, combined with a palmtop with control software, facilitate the collaborative and interactive capabilities of the participants in the game.

3.2 Technical Layer

The game board is a prepared and printed 125 × 142 cm plan of the old city, placed on a specially designed support structure. The visible part of the board contains a simplified layout of streets and squares of the historical centre of the city with places where 3D models of architectural monuments in Lublin are marked. At these locations, a LED

RGB diode will light up in green when the model is placed correctly, or in red in the case of a mistake, Fig. 3. The board also has a palmtop responsible for managing the entire game, retrieving information about the current state of layout of the models on the board. It displays current attendee messages, questions and historical information, and a set of loudspeakers to generate sound messages.

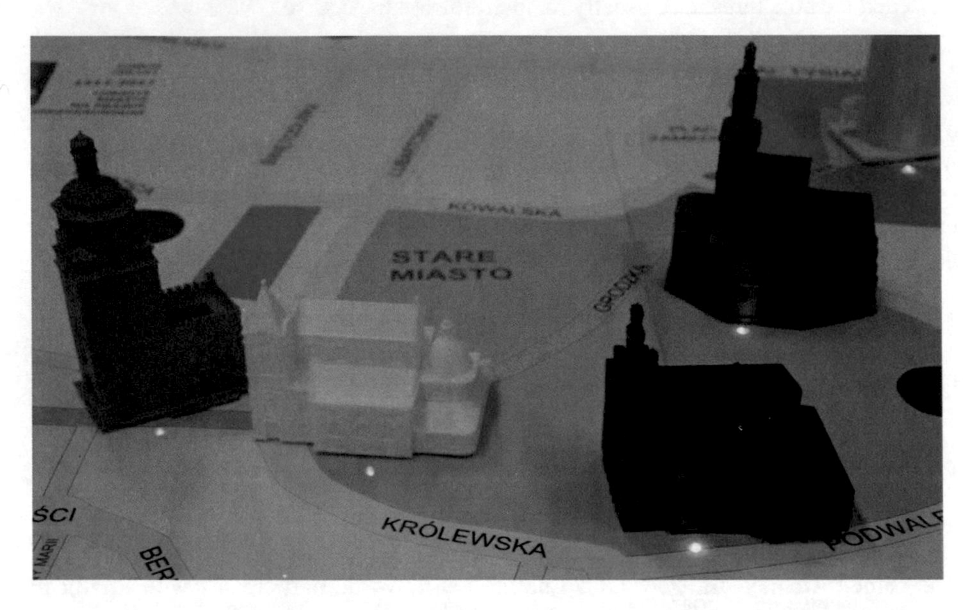

Fig. 3. The game board with examples of 3D models of Lublin's architectural monuments placed as a demonstration

Invisible parts of the board are RFID readers, whose task is to identify the correctness of 3D model placement, Arduino boards responsible for controlling readers and RGB LEDs, and for transferring their current state to the software controlling the whole game, wiring (over 130 m), power strip and power supplies.

3.3 The Logic of the Game Allowing for Collaborative-Interactive Learning of History

The game is turn-based and consists of repeated cycles sequentially played by players. Each cycle includes:

- retrieving a 3D model from the bank,
- placing it on the board,
- answering question 1 (concerning the object),
- answering question 2 (about the city's history). This question can be taken over by the opponent.

The gameplay and its course are dynamic, controlled by the real-time software developed. Depending on the situation, the messages displayed on the screen change and the course of the current turn is appropriately modified. In the course of the game different situations are possible (Table 2).

Table 2. The scheme of the course of a single turn of the game

No.	Occurrence	Action	Consequences
1	Correct placement of the model	The turn is continued	Display of question 1
2	Incorrect placement of the model or exhausting the time limit	Placement of a model into the object bank	No question 1. The opponent starts a new cycle
3	Wrong answer to question 1	Starting a new cycle	The opponent starts
4	No answer to question 1 – time out	The answer is given by the opponent	The opponent starts a new cycle
5	Wrong answer to question 2	Starting a new cycle	The opponent starts
6	No answer to question 2 – time out	The answer is given by the opponent	The opponent starts a new cycle

The number of points scored by a player depends not only on models correctly placed on the board and correct answers, but largely on the number of mistakes made by players. When a model is placed incorrectly and must return to the object bank, it becomes "more valuable" – it gets more points than it had. It is assumed that it is more difficult to recognise. The opponent starting the next cycle of the game can take a chance and choose it, or download another model with an initial score. The number of potential places for placing models on the board is much higher than the number of available models. This is to prevent the last model from being placed correctly merely by default.

4 Conclusions and Further Work

Survey research on a selected research sample has shown that K14 youths have a very low knowledge of the history of the city of Lublin, and in particular of the historical monuments built in the past. To improve this situation, the computerised board game "Architectural Jewels of Lublin" was developed. The large size of its board and the need to manually handle large 3D architectural monuments by many participants makes this game interactive and collaborative. The rules of gamification were used to create algorithms for individual turns and the whole game.

Further work related to the "Architectural Jewels of Lublin" game will focus on:

- Development of software for game control, mainly for elements that increase the dynamics of the turn and increase the educational value of the game.
- Study of the effectiveness of learning by using games – this efficiency, understood as the increase of knowledge in the participants of the game, can be analysed by conducting an active experiment with the participants (experiment: survey-game-play-survey).
- Increasing game difficulty by introducing models of non-existent monuments.

- Improving the realism of 3D models of monuments by applying realistic colours to their surfaces.
- Examination of the ergonomics of the game control interface software.
- Development and implementation of adaptive algorithms, modifying the way points are calculated depending on the course of the game.
- Increasing the number of players from the current 2 to 4, which will allow to develop completely different rules of the game, increasing its collaborative value on the basis of playing in pairs (similar e.g. to bridge).

The technical and programming solutions developed in the creation of the game "Architectural Jewels of Lublin" can be easily used to build analogous games related to the geolocation of monuments, cities, rivers, national parks, etc. or historical events (e.g. battles). The content layer of such games is of course unique, but the solutions developed on this occasion can easily be used in new circumstances.

References

1. Montusiewicz, J., Barszcz, M., Dziedzic, K., Kęsik, J., Milosz, M., Tokovarov, M.: The concept of a 3D game board to recognise architectural monuments. In: 11th International Conference of Technology, Education and Development (INTED 2017), Valencia, Spain, pp. 8665–8674 (2017)
2. Dicheva, D., Dichev, C., Agre, G., Angelova, G.: Gamification in education: a systematic mapping study. J. Educ. Technol. Soc. **18**(3), 75–88 (2015)
3. de la Guía, E., Lozano, M.D., Penichet, V.R.: Educational games based on distributed and tangible user interfaces to stimulate cognitive abilities in children with ADHD. Br. J. Educ. Technol. **46**(3), 664–678 (2015)
4. Forman, H.: Implementing a board game simulation in a marketing course: an assessment based on "Real World" measures. J. Acad. Bus. Educ. **13**, 41–54 (2012)
5. Huizenga, J., Admiraal, W., Akkerman, S., Dam, G.: Mobile game-based learning in secondary education: engagement, motivation and learning in a mobile city game. J. Comput. Assist. Learn. **25**, 332–344 (2009)
6. Kordaki, M., Gousiou, A.: Digital card games in education: a ten year systematic review. Comput. Educ. **109**, 122–161 (2017)
7. Kujawski, J., Januszewski, A.: Business simulation game as an educational method for postgraduate studies in management and finance. In: 11th International Technology, Education and Development Conference (INTED 2017), Valencia, Spain, pp. 8944–8951 (2017)
8. Milosz, M., Milosz, E.: Small computer enterprise on competitive market decision simulation game for business training of Computer Science specialist. In: 7th International Conference of Education, Research and Innovation (ICERI 2014), Seville, Spain, pp. 1831–1838 (2014)
9. Plechawska-Wojcik, M., Schmidtke, M., Skublewska-Paszkowska, M.: Applying of gamification in rising social awareness of urban residents. In: 9th International Conference of Education, Research and Innovation (ICERI 2016), Seville, Spain, pp. 8390–8396 (2016)
10. Popescu, M., Romero, M., Usart, M.: Serious games for serious learning using SG for business, management and defense education. Int. J. Comput. Sci. Res. Appl. **3**(1), 5–15 (2013)

11. Salvador-Ullauri, L., Luján-Mora, S., Acosta-Vargas, P.: Development of serious games using automata theory as support in teaching people with cognitive disabilities. In: 9th International Conference of Education, Research and Innovation Conference (ICERI 2016), Seville, Spain, pp. 4508–4516 (2016)
12. Simões, J., Redondo, R., Vilas, A.: A social gamification framework for a k-6 learning platform. Comput. Hum. Behav. **2**(29), 345–353 (2013)
13. Borys, M., Mitaszka, M., Pudło, P.: Why game elements make learning so attractive? A case study using eye-trackin technology. In: 10th International Technology, Education and Development Conference (INTED 2016), Valencia, Spain, pp. 2774–2780 (2016)
14. Guillen-Nieto, V., Aleson-Carbonell, M.: Serious games and learning effectiveness: the case of "It's a Deal!". Comput. Educ. **58**(1), 435–448 (2012)
15. Laskowski, M., Borys, M.: The student, the professor and the player: usage for gamification and serious games in academic education – a survey. In: 8th International Conference on Education and New Learning Technologies (EDULEARN 2016), Barcelona, Spain, pp. 2933–2941 (2016)
16. Milosz, M., Milosz, E.: Business simulation games during the summer schools – the case study. In: 6th International Conference of Technology, Education and Development (INTED 2012), Valencia, Spain, pp. 1577–1583 (2012)
17. Ferguson, D.: Board game move recording system. US patent number 3843132, 1974/10/22 (1974)
18. Ryan, P., Tse, E.K.Y., Lo, C.K.L.: Sensory games. US patent number 5082286, 1992/01/21 (1992)
19. Gilboa, P.: Computerized game board. US patent number 5853327, 1998/12/29 (1998)
20. Barredo, A., Garaizar, P.: Flow paths: a standalone tangible board system to create educational games. In: 2015 IEEE Global Engineering Education Conference (EDUCON 2015), pp. 301–304 (2015)
21. Melero, J., Hernández-Leo, D.: A model for the design of puzzle-based games including virtual and physical objects. J. Educ. Technol. Soc. **17**(3), 192–207 (2014)
22. Taspinar, B., Schmidt, W., Schuhbauer, H.: Gamification in education: a board game approach to knowledge acquisition. Procedia Comput. Sci. **99**, 101–116 (2016)
23. Montusiewicz, J., Milosz, M., Kesik, J., Barszcz, M., Dziedzic, K., Tokovarov, M., Kopniak, P.: Developing an educational board game using information technology. In: 9th International Conference on Education and New Learning Technologies (EDULEARN 2017), Barcelona, Spain (2017)

Acquiring the History of the City with Collaborative Game Based Learning

Dariusz Czerwinski(✉), Marek Milosz, Patryk Karczmarczyk, Mateusz Kutera, and Marcin Najda

Lublin University of Technology, 38A Nadbystrzycka Str, 20-618 Lublin, Poland
{d.czerwinski,m.milosz}@pollub.pl, mateusz.kutera@pollub.edu.pl
karczmarczyk.patryk94@gmail.com, marcinio.najda@gmail.com

Abstract. The paper presents the 3D maze collaborative game for learning the history of the city. The idea of the game was described as also the main functionalities. Collective gameplay consists in helping the lost participant to get between two points of the 3D maze. The research on the influence of joint gameplay on historical facts acquiring was made. The results of interview and survey were described. The study focused on game design and gamebased learning results.

Keywords: Game based learning · Collaborative gaming · History

1 Introduction

The most effective learning, deepened and permanent, takes place when is initiated and continued by the learner [1]. To make this possible, we need to create a learning environment, an indispensable space of freedom and an autonomous, creative action where digital natives have a chance to show their strengths and develop their potential. Teaching history, especially the home town history is a difficult task [2].

On the Internet, one can find a lot of games that are created solely for entertaining purposes, taking valuable time to the players who could devote themselves to developing their knowledge or skills [3]. One of the best ways to learn is learning through fun, games are motivating, accelerate user response time and simulate real situations. In decision-making games, the primary task is not simply to provide entertainment, but to solve a problem or goal [4].

There are many historical games where the players can acquire the knowledge about the history. For games of this type can be included among others: What in the World is That?, Gladiator, The Battle of Waterloo, Iron Age Life, Vikings, Mummy Maker, Battlefield Academy. What is unique about these games is that the gameplay is one-player and the knowledge is acquired by solving the given tasks (Fig. 1a).

Cooperative games focus on promoting collaboration, learning work in a team and making common decisions [5]. These are autonomous games in which entertainment is only a subordinate goal. They offer an effective approach to learning

M. E. Auer et al. (eds.), *Teaching and Learning in a Digital World*,
Advances in Intelligent Systems and Computing 715,
https://doi.org/10.1007/978-3-319-73210-7_13

and developing skills [6] These games combine elements of teaching, training, communication and information with entertainment elements [7]. Cooperative games are designed to achieve tangible, lasting changes in the behaviour of their users [8]. The sample of such game is "Keep talking and nobody explodes", where one player is trapped in a virtual room with the bomb. The other players, which cannot see the bomb, have to help him to defuse the bomb using defusal manual and information given by trapped user (Fig. 1b).

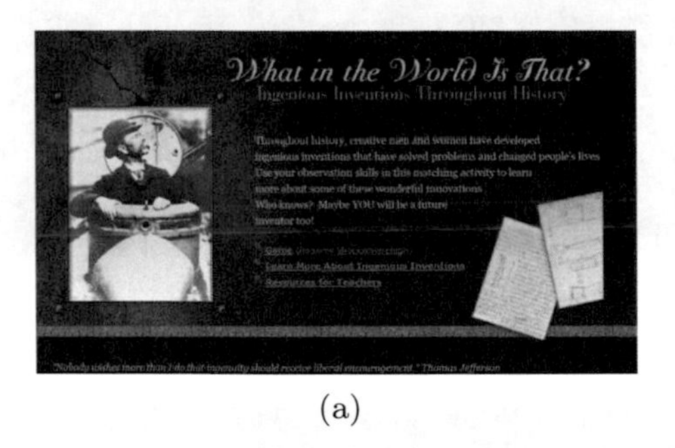

(a)

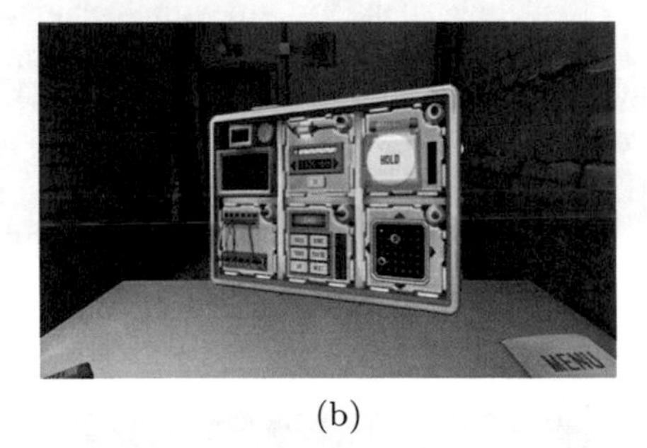

(b)

Fig. 1. Sample games: (a) historical "What in the World is That?" [9], (b) cooperative "Keep talking and nobody explode" [10]

There are many games in the network that use similar mechanics and the way of players connection, to that presented in the article. However, one can not find an identical solution that uses the world as a 3D maze game with quests and enables players to cooperate. This makes the described in article solution innovative.

In this paper the authors studied the history teaching path with the use of collaborative 3D maze game. The results show the influence on gathering the historical facts about the home town with the use of designed serious game. The remainder of the paper is organised as follows: Sect. 2 describes the game idea and the way of building. Section 3 describes the research way in the form of interviews and survey. Results were also presented in this chapter. Conclusions are presented in Sect. 4.

2 The Game

When designing the games for history it is important to consider the pedagogical goals [11]. The teacher can focus more on historical facts and data or (and) resource management and decision making [12]. The balance between the retaining historical accuracy and gameplay has to be considered. Well designed team games allow students to effectively assimilate historical facts and understand the atmosphere of history [13].

According to the definition the labyrinth is a complicated route with many passages, which is virtually impossible to navigate without a guide [14].

In ancient times, the maze served as protection against unauthorized persons entering the tomb or vault. Today is used in pop culture or as a tourist attraction. The form of labyrinth (3D maze) that has been used in the game draws direct inspiration from the book "Harry Potter and the Goblet of Fire" by Rowling [15]. In this book, the exit from the labyrinth does not directly relate to finding a path that leads to the way out, but to the particular point where the artefact is located. This artefact is symbolizing the successful completion of the quest.

Learning situation, in which the described game can be used is that, the teacher wants to learn young people some historical facts about home city or like in presented paper to gather more historical knowledge because of Lublin citys 700th anniversary. The anniversary is an good opportunity for city residents to play cooperative web game to learn more facts from the history of the city. The content that is useful for learners are unknown home city historical facts gathered through quests and riddles which are randomly picked up from database. In described case the teaching equipment and learning environment are PC with web browser and 3D maze collaborative game.

The game is a turn-based, collaborative team game between its participants. Players, in order to achieve the goal and win, should help each other. The obvious goal of the game is to get through the maze as quickly as possible from the starting point to the destination and to find a trophy cup. The hidden objective of the game is to learn the history of the city of Lublin. For this purpose, in the maze there are riddles/quests that the player must solve. Riddles are related to the history of the city. The player should correctly solve all the quests - only then the trophy cup will appear somewhere in the maze and you will be able to finish the game successfully. The game will end itself (without the trophy-cup, that is with the loss) after a certain time. Such limitation makes the game more dynamic.

The cooperation of many (at least two) players was organized as follows. Exactly one player (called Lost) is supposed to move around the maze and solve riddles/quests. This player is randomly selected from the group of players who have signed up for the game. The other players (called Helpmates) are supposed to suggest the Lost player the solution. They can communicate with each other, as well as receive a quest from Lost player (in the form of scene picture which Lost see), and send suggestions to him (or consult each other). Communication between the Lost and Helpmates players is limited by the time (type: not more than) and by the number (max. 50 during the game). The victory and time of its achievement in the game is assigned to all its participants and is the basis for the statistics of players.

An essential element of the game, apart from creating a situation involving cooperation between players, is its educational element. It was made using the riddles/quests depicted on the walls of the maze. These quests concern the history of the city of Lublin. They are randomly drawn from a large quests database. Four types of riddles/quests and their graphical interfaces are used in the game:

- Indicate appropriate colors (e.g. city coat colors).

- Enter the date (e.g. the date of signature of the Polish-Lithuanian union in Lublin).
- Indication of the correct answer (e.g. choosing one of the 3 answers to the question).
- Pressing the buttons in the specified order (this type is used for scheduling, e.g. serialized during an event).

After giving the correct answer, Lost is rewarded, and in the opposite case, riddle/quest is reappearing. Image of the quest interface the Lost player can send to the Helpmates for consultation.

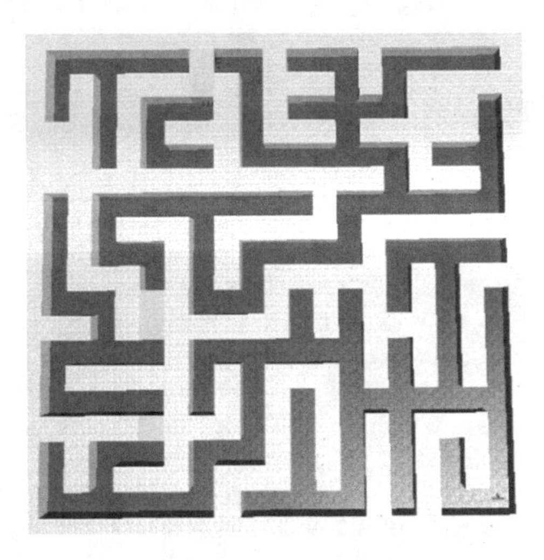

Fig. 2. Sample look of generated 3D maze

Fig. 3. Sample maze scenes with quests and message

The 3D maze is generated randomly in the created game room. At the stage of game development that was possible to look at the maze from above. The sample top look of 3D maze is shown in Fig. 2. During the game the communication between Lost player and helping team is carried out with the use of short text messages appearing on th screen (Fig. 3).

2.1 Functional and Non-functional Requirements of the Game

The main functional requirements for the web technology based 3D maze game are:

- Anyone can sign up for the service.
- The logged in users can edit their account, create or join existing game rooms.
- When creating a new game room, user gives the room name, the number of players, the game time and difficulty, and the optional password. Once created, the creator is the administrator of the room.
- An administrator of the room can discard or block a player, delegate administration to another player, and edit room options.
- Roles of players are awarded randomly.
- Players are communicating using text messages.

Non-functional requirements of the game can be described as follows:

- You can take part in one game at a time and join one room at the same time.
- The user can not block other users' accounts and suspend the game.
- An unregistered person can not participate in the game.
- Play time is limited to the following three options: 5, 10 and 15 min.
- There must be a minimum of two players per room.
- The person using the site must have a browser supporting JavaScript and minimal requirements are: Chrome version 30, Firefox version 30, Internet Explorer 11, Opera version 12.10 or later,
- The server on which the site will be hosted should have a PHP 5.5 or later interpreter and a PostgreSQL 9.3 or later server.

2.2 Data Model and Game Logic Design

The game consist of several quests. First of all the lost player needs to find the all places in the labyrinth to solve the riddles. Whole data structure of the game is presented in the Fig. 4.

It can be noticed that all riddles/quests are stored in the tables: *puzzle*, *puzzle_group* and *puzzle_element*. The riddles (quests) are divided into the four main groups and several subgroups presented in Table 1. Each question in the database has assigned difficulty level which is used during room creation.

The main idea of the functionality of the game application is presented in Fig. 5 in the form of a use case diagram. This diagram shows the roles of the application users and the basic use cases. The application user roles (any person becomes user as a result of the application registration process) were defined as follows (actors shown in Fig. 5):

- Player - any person who registered in the web application.
- Game Administrator - a player who initiated another (named) turn of the game, resulting in a creating maze, starting and ending points, a collection of riddles-quests, difficulty of the game and other gameplay parameters.

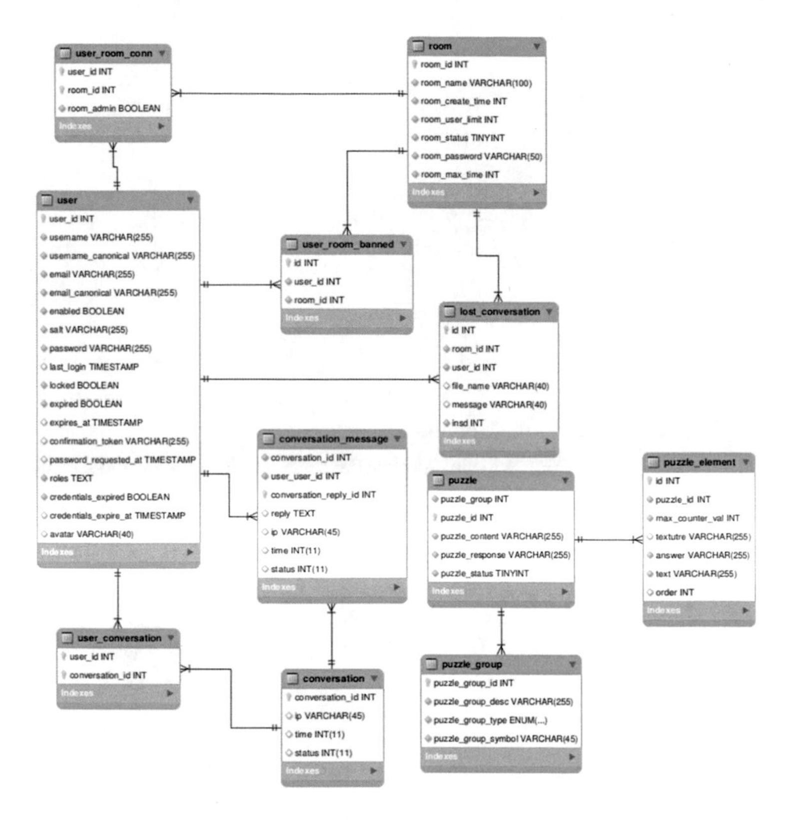

Fig. 4. Game data structure in the form of ERD diagram

- Lost - a player who moves through a virtual maze in a given game.
- Helpmate - a player or their group who works with Lost and prompts him to solve riddles/quests.
- Time - an actor who maps the passage of time (e.g., to end the game after the elapsed time specified by the Game Administrator).

Figure 5 (use cases) describe the main functionalities of the application. These are:

- Edit account data - manage your player profile data.
- Create game room - allows the Player to create a new game, in which the Player is the Game Administrator. The application generates a new maze, sets the start and end points, and generates (gets from the Database) and arranges riddles-quests in the maze.
- Join the game - gives you the ability to join a specific (created and named) game as a participant. The application randomly assigns the role of participant (Lost or Helpmate).
- Manage the game - allows the Game Administrator to set its parameters (e.g. duration), manage players (e.g., remove a player from the game) and start the game.

Table 1. Types of quests

Riddle/quest type	Subtype	Number of quests
Set order	Universities	3
	Districts	4
	Historical facts	3
	Annexation	2
Set date	City	2
	World War II	2
Set colours	Lublin City	2
	Voivodeship	2
Answer select	Monuments	2
	Local authorities	2

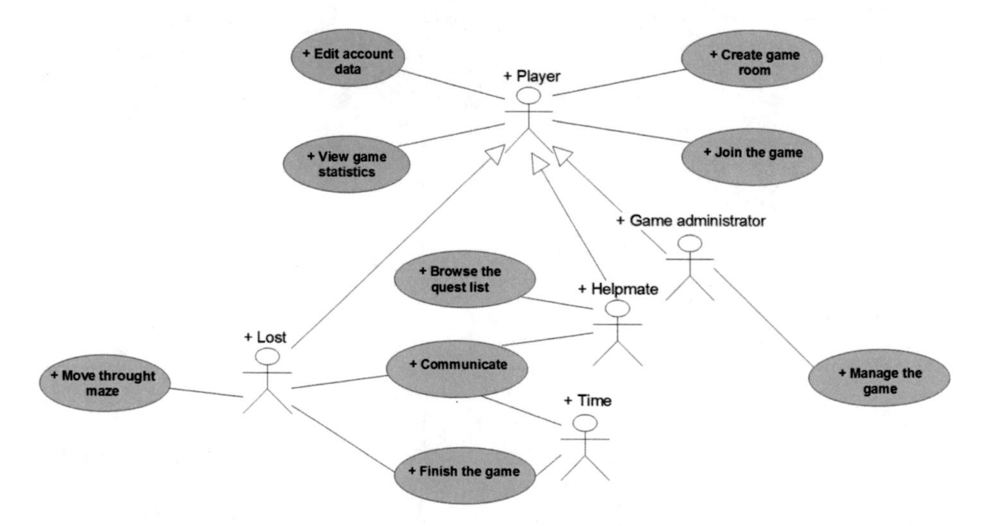

Fig. 5. Use case diagram of the application

- Move trough maze - change the position in the maze by Lost and answer the riddles-quest. While moving, Lost has the ability to communicate with Helpmates.
- Communicate - allows communication over the network (typical chat) between Lost and Helpmates and between Helpmates. Communication is limited by the time (Time actor) and quantitative.
- Browse the quest list - browse through the list of riddles/quests to identify an extended description of the riddle/quest. Helpmate gets from Lost the look of the riddle/quest and needs to refine it based on a list of riddles-quests, to work out in group answers sent to Lost player.

- Finish the game - finish the game and memorize its outcome. Completion of the game can be successful (Lost finds the cup) or without (passage of time).
- View game statistics - view statistics of past games, including the top players list.

3 Research Results

In this study a series of interviews with current students were performed. They were students of computer science who had completed the education of history at the secondary school level. During the interviews the students were examined with their reference to games, their motivations for playing games, their attitude to game based learning and knowledge about historical facts of home town. After playing the cooperative game the students filled out a survey questionnaire. The 30 postgraduate students were examined, with the intention of how they acquired the new historical facts about the home city, which they did not know before, with the use of 3D maze collaborative game.

The series interviews were conducted with students, who were known to the researchers. There were 21 male and 9 female participants, with the ages of 21–23 years. Seventeen persons were people who considered themselves as a playing games in their leisure time and the rest of them said, that they prefer the other type of amusement.

A short survey of ten questions was designed to examine the gaming preferences of the students and their attitudes in the collaborative gaming. Interviewees were asked whether they played computer collaborative games often, regularly or never. The next question was about the experience of playing collaborative game for learning and feeling positively motivated, not motivated or negatively motivated. At this stage of the research this three-point scale was sufficient to differentiate between positive and negative perceptions.

In the survey there was also a question how the students prefer to learn and acquire new knowledge, independently, in 2–3 persons group or group over 3 persons. The rest of the questions in the survey concerned the knowledge gained during the game.

The interviews showed that the participants who considered themselves to be game players had different motivations. Among these students the primary motivations for playing games were mental challenge and social experience. In the case of people who did not consider themselves gamers they could play games in the case of boredom or social events. However most students (23) said that acquiring historical knowledge with the use of 3D maze game could be interesting.

The results of the survey showed, that the 40% of the students played the collaborative games often, over 53% who said they were playing regularly and almost 7% who said that never played. It can be seen that in this group over 93% of interviewees played in the collaborative games. Although only 34% of the group have ever come across earlier with game based learning. The results of being motivated were as follows: over 73% were positively motivated, almost 17% were not motivated and 10% was negatively motivated.

The question about the acquiring new knowledge in group gave an interesting results, 54% of students prefer to learn independently, 33% want to learn in 2–3 persons groups and 13% like to learn in larger group. This shown an interesting trend that students are more likely to cooperate in the group and acquire new knowledge when playing the game.

An interesting results were obtained with the questions about historical facts absorbed during the 3D maze game. In the case of "Set colours" quest type 70% of students memorised the proper order, however the "Set date" quests were memorized only by almost 27% of students. In the "Set order" riddle type most problems raised questions about historical facts. Only almost 17% of students remembered the proper order of historical facts after playing the game. However in the case of other subtypes of "Set order" type the results were much better. Average percent in these subtypes was over 51%. "Answer select" riddle type gave an 40% result in acquiring the new knowledge.

4 Conclusions

Historical space since the birth of the games has been the background for their scenarios, and the presence of characters, events or historical symbols has become inherent attribute of the games. Historical games are characterized by, on the one hand, attempts at a realistic, mimetic representation of the past, and on the other, allow for a fictional story. Realism involves the need for consistency with the knowledge of the structure of the past presented and the experience of the players.

It should be noted that in historical games one can talk about the disappearance of the distinction between fiction and reality, the real world and the virtual world. The image of the past created in the game is not an imitation, representation or representation of past reality - it is a cultural construct, attractive primarily because its creator is a player.

At the same time it is important to emphasize that historical games are not only creative tools, but they also require knowledge or familiarity of historical realities. Some of them try to remain faithful to factographies and are educational or educational aided games - or, alternatively, to counter the current knowledge of historical decision, remain faithful to the historical image of the epoch scenery (e.g. Europa Universalis) [16].

The results of the research clearly showed that the students can acquire the home town historical facts much faster when they are playing collaborative game. However there were persons which were not motivated by the game based learning. An immersive game environment in the form of 3D maze connected with the collaborative game enabled players to learn some knowledge during gameplay.

References

1. Robinson, K.: Out of Our Mind: Learning to Be Creative. Capstone Publishing, West Sussex (2011)
2. Schrier, K.: Using digital games to teach history and historical thinking. Learning, Education and Games. ETC Press Pittsburgh, pp. 73–91 (2014)
3. Oliver, M.B., Bowman, N.D., Woolley, J.K., Rogers, R., Sherrick, B., Chung, M.Y.: Video games as meaningful entertainment experiences. Psychology of Popular Media Culture, Educational Publishing Foundation, vol. 5, no. 4, p. 390 (2016)
4. Laamarti, F., Eid, M., Saddik, A.E.: An overview of serious games. Int. J. Comput. Games Technol. **2014**, 11 (2014). Hindawi Publishing Corp
5. Mortara, M., Catalano, C.E., Bellotti, F., Fiucci, G., Houry-Panchetti, M., Petridis, P.: Learning cultural heritage by serious games. J. Cult. Heritage **15**(3), 318–325 (2014). Elsevier
6. Backlund, P., Hendrix, M.: Educational games-are they worth the effort? a literature survey of the effectiveness of serious games. In: 2013 5th International Conference on Games and Virtual Worlds for Serious Applications (VS-GAMES), pp. 1–8. IEEE (2013)
7. Milosz, E., Milosz, M.: Business simulation games during the summer schools–the case study. In: INTED2012 Proceedings, IATED, pp. 1577–1583 (2012)
8. Milosz, M., Pastuszak, Z., Milosz, E.: Decision-making simulation games as a tool for verifying enterprise logistics strategies. In: INTED2015 Proceedings, IATED, pp. 3558–3567 (2015)
9. Library of Congress, What in the World is That? June 2016. http://www.loc.gov/teachers/classroommaterials/presentationsandactivities/activities/science/learn_more.html
10. Keep Talking And Nobody Explodes, June 2016. http://store.steampowered.com/app/341800/Keep_Talking_and_Nobody_Explodes/
11. Romiszowski, A.J.: Designing Instructional Systems: Decision Making in Course Planning and Curriculum Design. Routledge, Boca Raton (2016)
12. Perrotta, C., Featherstone, G., Aston, H., Houghton, E.: Game-based learning: latest evidence and future directions. NFER Research Programme: Innovation in Education. NFER, Slough (2013)
13. Li, M.C., Tsai, C.C.: Game-based learning in science education: a review of relevant research. J. Sci. Educ. Technol. **22**, 877–898 (2013). Springer
14. Scaligero, M., Santarcangeli, P.: Il libro dei labirinti. Storia di un mito e di un simbolo, JSTOR (1969)
15. Rowling, J.K.: Harry Potter and the Goblet of Fire (2000)
16. Admiraal, W., Huizenga, J., Akkerman, S., Ten Dam, G.: The concept of flow in collaborative game-based learning. Comput. Hum. Behav. **27**(3), 1185–1194 (2011)

Engineering Pedagogy

Work-in-Progress: Lean Education/Lean Innovation

Andrei Neagu[1(✉)], Doru Ursutiu[2], and Cornel Samoila[3]

[1] University “Transilvania” Brasov, Brasov, Romania
andrei.neagu@unitbv.ro
[2] University “Transilvania” Brasov, AOSR Academy, Brasov, Romania
udoru@unitbv.ro
[3] University “Transilvania” Brasov, ASTR Academy, Brasov, Romania
csam@unitbv.ro

Abstract. This article analyzes strategies that aim at creating innovative solutions or business concepts on a user-centered approach: lean startup. Those approaches involve customers, potential users, or other stakeholders into their development process. Although there are significant differences between the traditional approach and the lean approach, in our future work we plan to investigate the strategies of applying tailored structures to guiding students in in Higher Education Institutions, in creating market-demanded innovation.

Keywords: Lean startup · Innovation · Lean manufacturing
Higher education institutions

1 Introduction

Lean principles were developed in the early seventies by Toyota in Japan, called lean manufacturing, to optimize production processes. As Singo Shigeo is mentioning in his book: A Study of the Toyota Production System -“Production is a network of processes and operations, transforming material into product”. [1] The idea of lean principles is to make the production process more efficient by reducing any sort of waste in the process—this could mean either the reduction of resources (human or material) or the elimination of needless or redundant activities or expenses. This strategy revolutionized production processes in the automotive industry. By now, lean principles have become also important for general management, and other disciplines like IT development, which make use of lean concepts but transfer them also to non-manufacturing contexts.

2 Lean in Higher Education Institutions Context

Higher Education institutions (HEIs) are the perfect case of non-manufacturing organizations that can benefit by implementing Lean tools and principles, at a strategic level and also as operational and administrative point of view. [2] Observing the “World University Rankings 2016-2017” published by timeshighereducation.com, the Educational institutions now days are facing groundbreaking competition. Even more the

M. E. Auer et al. (eds.), *Teaching and Learning in a Digital World*,
Advances in Intelligent Systems and Computing 715,
https://doi.org/10.1007/978-3-319-73210-7_14

competition for students, research funds, prestige, quality ratings, incubated companies, fundraising, academicians, etc. forced those institutions to consider new modes of operation. [3]

According to the case study published by University of Windsor to demonstrate how the application of lean principles can assist in improving the quality of an engineering design course, they identified 5 principles that can make a parallel between lean philosophy and the process of education [4] (Fig. 1).

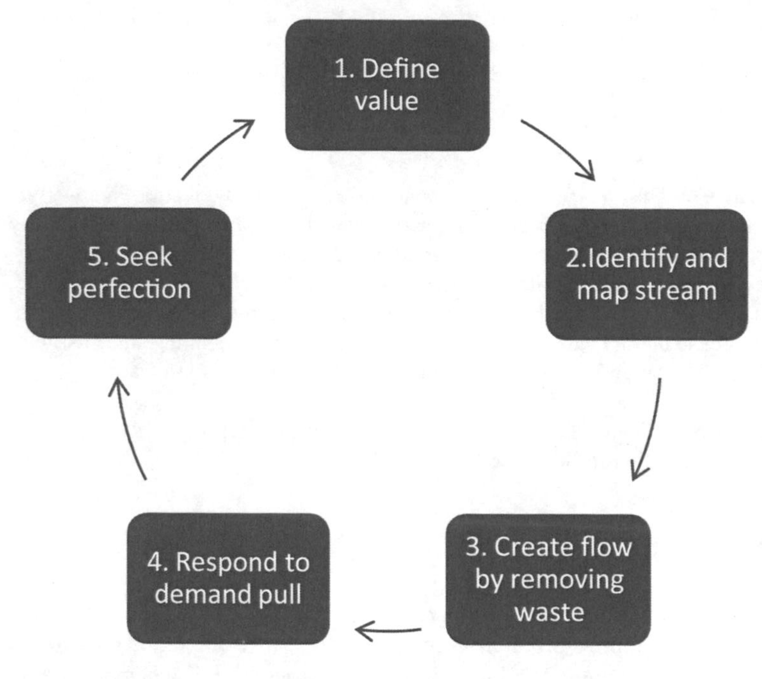

Fig. 1. Lean principles

3 Lean in Higher Education Institutions Entrepreneurial Context

Having in mind the fourth principle mentioned above, "respond to demand pull", one topic of interest, that HEIs focus more and more, is bringing entrepreneurial spirit among students.

HEInnovate, representing the collaboration between European Commission and the OECD, is one of the initiatives that support collaboration on matters of entrepreneurial and innovation between HEIs. Till now the programs which were implanted, support in build new relationships and synergies across the Institution; engaging and recruiting individuals with entrepreneurial attitudes, behaviour and experience; invests in staff development to support its entrepreneurial agenda. According to the case study:

"Entrepreneurial Behaviours and Organisation Culture" one conclusion was clearly stated: "it is a truism that 'culture beats strategy'. This implies that if strategies are to be effective they must be supported by appropriate behaviours, attitudes and cultural pre-disposition." [5]

On the other hand, if we analyze the entrepreneurs and their needs, starting new business ether is a local business or global one, whether it's a tech start-up or a services business, has always been a hit-or-miss proposition. Respecting the well-known, old formula: writing a business plan, assemble a team, create an MVP, pitch it to investors, introduce a product, and start selling as hard as you can. Even so, in this sequence of events, most probably just 25% might be successful according to the research conducted by Shikhar Ghosh - Harvard Business School. [6]

In the last years the methodology called "lean startup" started to bring the process of developing a business closer to the potential customer, focusing on "validation" and less on "intuition".

4 Lean Startup in University Education

"Lean startup" is considered an innovation method by startup companies that claims "the most efficient innovation is the one for which there is an actual demand by the users". [7] In other words: the biggest waste is creating a product or service that nobody needs. This concept is highly relevant for any strategy or method that aims at creating innovations inside or outside HEIs.

The "lean startup" nowadays is more than a method is a trend that created an entrepreneurial movement. If we take a look at the European Hi-educational environment, more and more teachers apply those methods, guiding students in their entrepreneurial careers. One of those environments is the Entrepreneurship department at Gent University Belgium. Through the "Dare to start" program, students are guided in validating their business ideas by applying lean startup principles and in the same time learning to think lean. More than that, in the last years, several international students went through this process, and started successful businesses in their countries, and in Belgium.

During 12 weeks, as the program lasts, we were able to participate and to observe the process. The program is based on the "Launchpadcentral" an online tool developed by Steve Blank. [8] According to the inputs that we collected, we were able to create a scenario of how the students manage to understand and to apply the lean principles in validating their ideas [9].

As a brief conclusion, from our observation, the process of understanding and implementing lean thinking, in the process of validating their ideas, is different for each student and depends on his entrepreneurial experience gained in the past. [10] Even more, one aspect that actually got our attention is the involvement and dedication of the team for the process, after their idea is pivoted according to the potential market. More information regarding our observation we will provide in our future publications (Fig. 2).

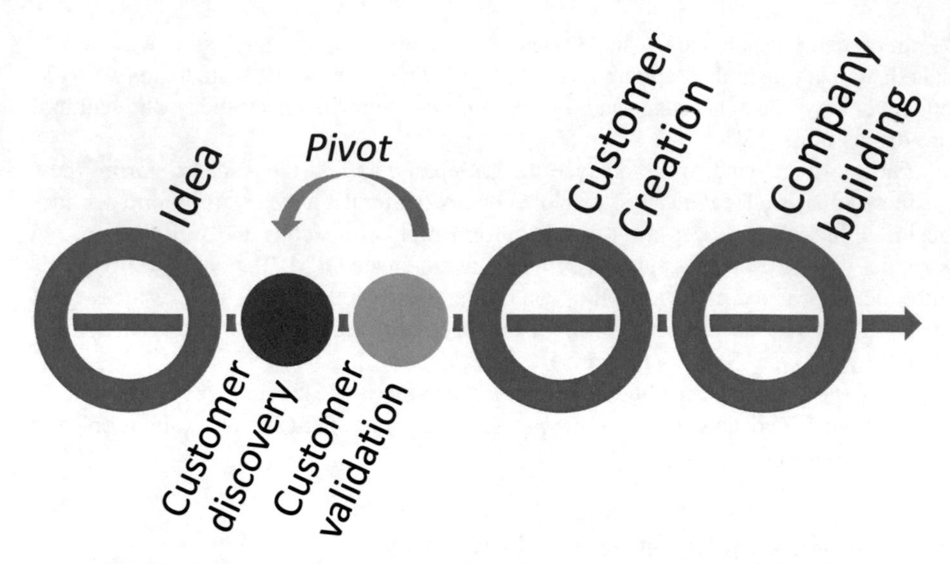

Fig. 2. Steps in creating a business according to lean startup principles [11]

5 Further Work

We intend, based on this study case that we conducted in the University of Gent, Belgium and their support, to develop and start a similar program at "Transilvania" University, tailored to the cultural and economic aspects of the region.

Acknowledgment. We thank the "Durf Ondernemen" team – Gent University for the support in developing the study.

References

1. Shingo, S.: A study of the toyota production system from an industrial engineering viewpoint (1989)
2. Vukadinovic, S., Djapan, M., Macuzic, I.: Education for lean & lean for education: a literature review. Int. J. Qual. Res. **11**(1), 35–50 (2016). ISSN 1800-6450
3. World University Rankings. www.timeshighereducation.com
4. Pusca, D., Northwood, D.O.: Can lean principles be applied to course design in engineering education? Glob. J. Eng. Educ. **18**(3), 173–179 (2016)
5. Richard, T.: In collaboration with staff members of Dundalk Institute of Technology, "Entrepreneurial Behaviours and Organisation Culture – A Case Study", Dundalk Institute of Technology
6. Shikhar, G., Farre-Mensa, J., Payton, C.: "Entrepreneurial Finance Lab: A Founding Story (A)." Harvard Business School Case 815-119, March 2015
7. Ries, E.: The Lean Startup (2011)
8. Blank, S.: The four steps to the epiphany: successful strategies for startups that win (2nd ed.), San Francisco (2006)

9. Baden-Fuller, C., Morgan, M.S.: Business models as models. Long Range Planning (2010)
10. Doganova, L., Eyquem-Renault, M.: What do business models do? innovation devices in technology entrepreneurship (2009)
11. Osterwalder, A., Pigneur, Y.: Business Model Generation: A Handbook for Visionaries, Game Changers, and Challengers (2010)

Using PBL and Rapid Prototyping Resources to Improve Learning Process

Jovani Castelan(✉) and Rosemere Damasio Bard

SATC College, Criciuma, Brazil
{jovani.castelan,rosemere.bard}@satc.edu.br

Abstract. This paper aims to demonstrate the modeling and implementation of innovative teaching-learning practices in Higher Education, based on Active Learning Methodologies, by highlighting especially, PBL – Problem Based Learning using Rapid Prototyping devices. In order to apply PBL's methodology to the graduation courses, it was developed an implementation model, in which, there are four levels of implementation. Each level has four class attributes, which are classroom space-time, student autonomy, teaching role and problem's scope. The obtained results show that the students demonstrate higher levels of interest, participation, and involvement with classmates, motivation and content's perennial assimilation. With the application of these methodologies, skills required by job market, such as teamwork, relationship, collaboration, proactivity and entrepreneurship are also developed.

Keywords: PBL · Learning model · Active learning · Rapid Prototyping

1 Introduction

Introducing active learning in Higher Education is an important challenge because it brings disruptive changes in the way that the teaching-learning process traditionally occurs. Around the world, many universities identified the need for a cultural shift, in order to rescue the societal relevance, nature and protagonism of undergraduate engineering courses that are based on traditional curriculum [1]. The shift involves the transition from an educational system based on teaching, to a system based on learning, making the student the center of the educational process [2].

Until the advent of the Internet and its massification in the 1990's, the traditional method remained successful for centuries. In the 2000's onwards, we saw the closure of Information Era and the start of the Knowledge Era. Currently, we are experiencing the Fourth Industrial Revolution; that is, the merging of digital, physical and biological technologies in a cybernetic world. The 4.0 Industry with the Internet of Things (IoT), cloud computing and manufacturing information are already part of our daily lives. However, the teaching model that persists up to this day is similar to the year of 1088, the year of the foundation of Bologna's University.

Problem-Based Learning (PBL) is an educational approach that is student-focused. The focus changes: rather than having a teacher-driven approach that leads the students,

M. E. Auer et al. (eds.), *Teaching and Learning in a Digital World*,
Advances in Intelligent Systems and Computing 715,
https://doi.org/10.1007/978-3-319-73210-7_15

PBL aims to empower students to develop self-directed and perennial learning, developing also their cognitive and metacognitive skills [3]. At its higher level, the student is "mentored and encouraged to conduct research, integrate what is learned, and apply it to develop a viable solution to an ill-defined problem" [4].

This methodology engages students in active learning, and in addition, it promotes and increases students' cognitive and practical abilities, as well as developing other important skills to professional life, such as collaboration, teamwork, creativity and proactivity to solve problems and face challenges. At this point, the materialization of solutions, made via Rapid Prototyping (RP) resources, becomes the class' synthesis, promoting a perennial and meaningful learning.

Therefore, searching for innovation and reform of higher education in engineering courses via Active Learning implementation starts with the need to develop important abilities and skills, widely discussed in national and international scope [5, 6]. Furthermore, once problems have become more and more complex, requiring professionals of several fields to solve them, it is indispensable that engineers are able to work in multidisciplinary teams. Therefore, it is important that teachers experience and develop among their students' creativity, teamwork, decision-making, communication and problem solving.

Thus, the present work shows the development and application of a PBL model in engineering courses from SATC College, in Santa Catarina State, Brazil. The aim is to integrate theory and practice, by integrating university and enterprises. However, the methodological changes needed to attend the demands of the job market, depend on a design that, considers the need for a cultural transformation. Therefore, taking into the consideration how complex Higher Educational settings are, and how difficult it is to implement new learning models, we propose to implement the PBL-based learning model gradually.

2 Problem-Based Learning Model

Based on research of Problem-Based Learning (PBL) applications in engineering courses [4, 7–9], as well as institutional visits to American universities (MIT and Olin College on November, 2016) and also through experiences we have had in our own institution, it was developed a PBL implementation model for engineering courses' curriculum (Fig. 1). For the development of this model, we have considered teacher and students' behavior, infrastructure and the integration between academy and industry.

These are the four levels developed: Level 1 as the most basic, and the first, which is applied; Levels 2 and 3 are intermediate ones and Level 4 is the most advanced, thus the last one to implement. Each level is composed of four attributes and each attribute as a four-degree scale. Figure 1 below shows PBL's levels to be achieved.

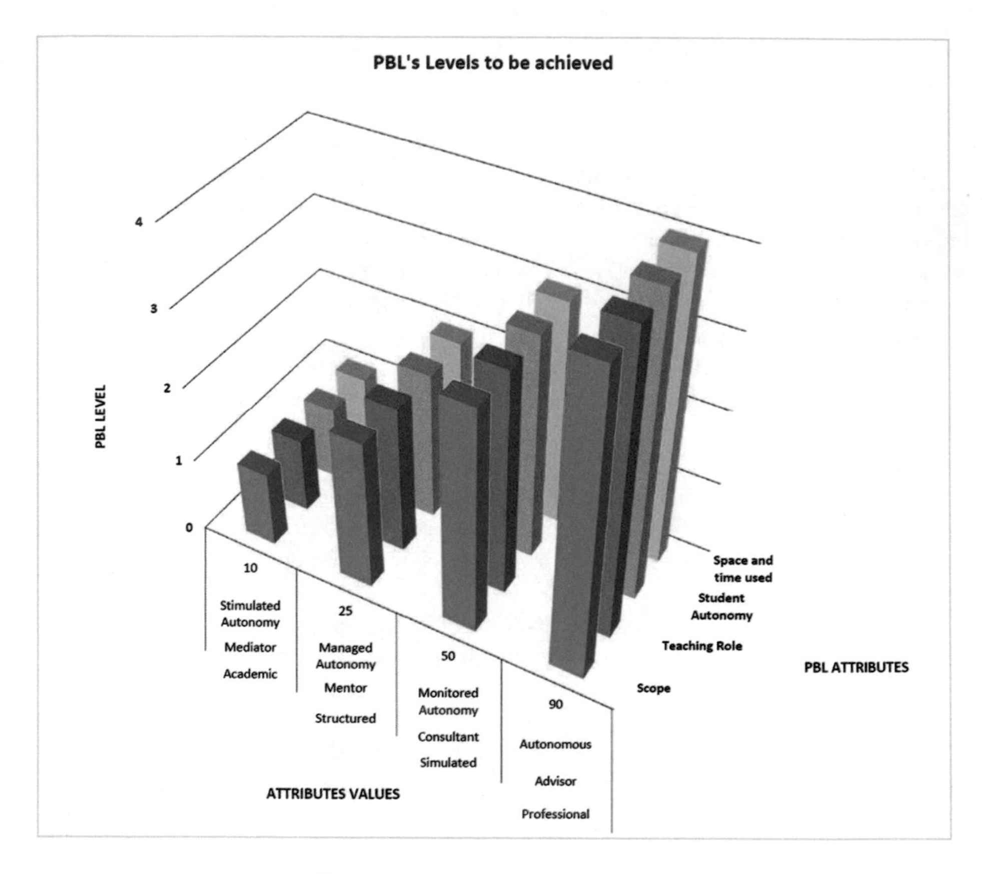

Fig. 1. PBL's levels of development

2.1 Level 1: Solving an Academic Problem Inside the Classroom

At Level 1, the space-time attribute has a value of 10. That means that in a discipline with a semester of 60-h workload (20 week-meetings with 3 h for each class), it would be used 6 h, 10% of time which is equivalent to two meetings, for the development of PBL's activity. Due to the short period available for this level, which is expected to be 2 meetings during the semester, the workspace is likely to be the classroom; however, other academic learning spaces, such as the library, computer labs and workshop could also be used.

The Student Autonomy (Stimulated) takes into consideration that students, on their first contact with PBL, have not experienced PBL yet, but the traditional one. Therefore, autonomy is encouraged and problem solving is constantly stimulated. Teacher Role attribute is that of a Mediator who ensures that learning takes place by interacting with students, asking meaningful questions and constantly challenging them, as well as, recommending research sources and leading students in the process of finding solutions. This role may also requires that professor-mediator to give short lectures and intervene during the PBL activity. Thus, continually nurturing the learning process by monitoring and leading the teams of students.

The problem falls into the academic scope, which is likely to focus on a specific discipline's topic. The students work in groups and the solutions are likely to be similar. It is less likely, therefore, to produce a work that is unique due to a few variables and the low complexity of the problem itself.

2.2 Level 2: Solving a Structured Problem

The space-time attribute shows a value of 25, that is, we may need 25% of the discipline's total workload for the PBL activities. Considering again a discipline that has 60 h in a semester (20 meetings of 3 h each one), that would be 15 h – equivalent to 5 meetings. With a longer time, there are more possibilities to extrapolate the classroom's space and using other academic spaces (library, computing labs and workshops). There is also more flexibility. The professor can plan one activity, based on level 2 criteria, with five meetings or two activities of two and three meetings, respectively. The first one, perhaps, at level 1 and the second one, more elaborated, according to the criteria of level 2. The intention is to provide students and teachers the opportunity to become more familiar with the methodology, allowing a judicious evaluation of the progress as well as of the failures that occur during the implementation.

Regarding student autonomy (Managed Autonomy), considering that stimulation has occurred in the previous experiences, students should at this level show a discreet skill to self-learning and proactivity. Thus, rather than constant, the stimulation becomes frequent.

A mentor, according to the dictionary, is an individual considered wise and inspiring, that drives, leads and encourages someone. The propositions presented in this level might be less structured and in an intermediate complexity. In the mentoring role, the professor will answer questions that students might have by pointing out possibilities ("and if…") and showing previous cases, nudging students to search and make new discoveries. Teaching through short lectures to small groups will still occur on demand, and lecturing the whole group only when needed, but likely to be less frequent than in level 1. Even though this level aims to foster self-directed learning, constant group monitoring will occur, like in the first level.

The scope of the problem becomes "Structured". In this one, the resolution attends the medium complexity of the problem and requires content integration of two or more disciplines that are concomitants or pre-requirements. PBL's activities, in some cases, can go beyond the academic and classroom spaces. In loco visits, where the problem is happening, is a real possibility, but not expected at this level. The final product, presented commonly in class, could be presented to the external community (liberal professionals, representatives and enterprise's CEOs). In this level, solutions to the problems must be validated through a scientific approach, which must include references, justification, methods, results, discussion and conclusion.

2.3 Level 3: Simulating a Problem's Solution

At this level, 50% of total workload may be done through PBL's activities. Thereby, of the 60 reference-hours, 30 will be for implementing active learning. Students become protagonists of the learning process by taking more responsibilities and, the outcomes become more elaborated, as learning situations become more complex as well. The situations proposed in this level come from the observation of professionals in several places: shops, offices, agencies, factories, farms, inside a coalmine, means of public transportation, inside a car, hospitals, at the bank, other schools, in their own house, etc.…).

Based on the assumption that, the designed problems in this level is embedded in situations that are part of professional or personal students' lives, the engagement is expected to be spontaneous, without the need to tap into students' intrinsic motivation or emphasize how meaningful the activity is. The professor, therefore, is not obliged to motivate constantly the student, since, it is expected that students have already developed some skills by level 3. Thus, student's autonomy is monitored, as occasional stimulation to avoid deviations from the task might be needed. In this sense, the professor acts like a consultant, acting on demand.

The simulated scope means that the obtained solution to a real problem exposed at the start of PBL activity is validated and presented, but it is not in fact applied in a real context. An engineer designing a crane bridge can simulate and validate it using real data (constructive materials, dimensions, friction, lubrication, safety factor, energy consumption, ultimate tensions, etc.…) without, in fact, the need to build one. Likewise, a discussion about drivers' aggressive behavior in sociology can be synthesized in an advertising campaign or a toll planning about defensive driving, without having to produce them.

Anyway, the resolution via problem's simulation will require concepts' integration and, consequently, will develop skills and abilities that are important for the job market.

2.4 Level 4: Solving a Professional Problem

In the last level, while practically all the classroom time (90%) is dedicated to PBL activities, the classroom space itself is minimally occupied, and restricted to meetings with professor-advisor. The learning itself develops on the space-time where the phenomena that is investigated occurs. At this level, students are the main actors and become responsible for driving and accomplishing the pre-established goals and achieving their full autonomy. In this context, the crane bridge and the toll examples, given above, would be developed and the results evaluated, according to a professional criteria (costs, technical viability, ethics and safety), to attest or not, the students' ability to solve a problem. There is no lecturing and the conversations that occur, between advisor and the student, happens on an individual basis. This could be exemplified by the Course Work expected to be accomplished at the end of the engineering course (called TCC in Portuguese), which is similar to senior capstone projects where a student engages in a project, as part of their senior year, and is completed in close consultation of a faculty mentor. However, it is tacit and explicit that a great number of graduating students, educated in a traditional context, when enrolled in the last semester, does not show the skills, the abilities and the attitudes needed to deal with the highest level of PBL. Consequently, the professor, who should act like an advisor, returns to the role discussed in level 1, which is mediating the process, stimulating the autonomy and assuming responsibilities about deadlines, fulfillment of goals and outcome's analysis. This is the reason by which, TCC causes stressful situations and is uncomfortable for the students. We educate them during all the academic cycle in the passive, traditional, unilateral, and non-autonomous form and focuses mainly on theory. Suddenly, in their last year of graduation, we insert them in an active process, which is contemporaneous, multilateral, fully autonomous and free to obtain knowledge based on real experiences. Consequently, the work presented to a faculty board council at the end of the course (TCC) is disorganized, lacks originality, texts bypassing the theme and oral presentations are discouraging, not to say disastrous. It is the right formula to create embarrassment to everyone involved, especially for the student.

2.5 The Use of Rapid Prototyping (RP) Resources in the PBL Levels

According to Orey [10], who based his levels of thinking in the 'Learning Pyramid' (Fig. 2) of Bloom's Taxonomy, in order to ensure cognitive development, it is important to work with all levels of thinking, from lower to higher order. Therefore, lecturing followed by memorization exercises are examples of a low-level learning; the listen-read-write-look activities can generate some medium-level learning but still not effective as an excellent student learns– and does not forget – around 40% to 50% of what is taught. Furthermore, desirable skills are not contemplated (teamwork, collaboration, creativity and proactivity). On the other hand, high levels of thinking are achieved, when a professor provides discussion-evaluation moments and activities that foster creativity and hands-on work. Because of that, it is not enough for students to listen to a lecture or understand a text. By creating, drawing and manufacturing a new product, students are working with a range of cognitive levels of thinking, especially higher-order thinking. Hence, using RP in the

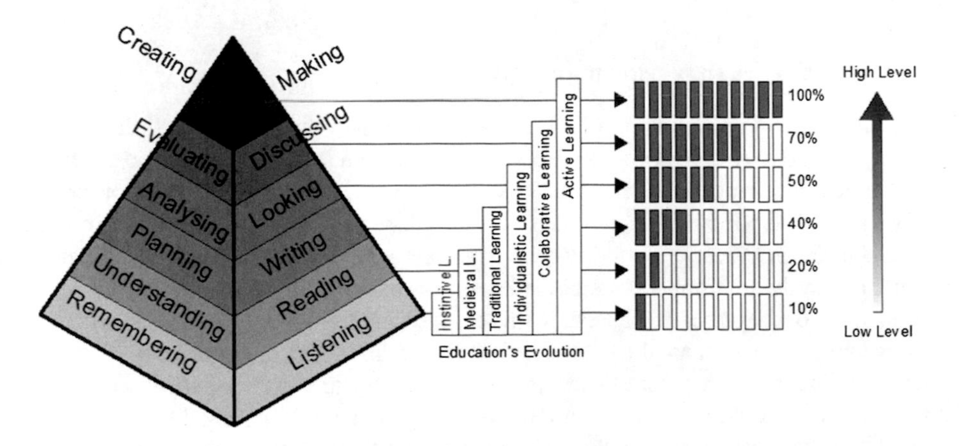

Fig. 2. Relation between the Bloom's Taxonomy, academic actions, education's historic evolution, perennial learning percent and thinking levels. Adapted and modified from [10].

classroom provides the opportunities for students to analyze, to evaluate and to apply knowledge.

PRONTO 3D – Laboratory of Prototyping and Digital Manufacture Oriented to 3D is part of a Brazilian network of labs linked to the FAB Foundation, associated to MIT – Massachusetts Institute of Technology. The Fab Lab Network is an open, creative community of fabricators, artists, scientists, engineers, educators, students, amateurs, professionals, who have the mission to share access to the tools for technical invention. This community is simultaneously a manufacturing network, a distributed technical education campus, and a distributed research laboratory working to digitize fabrication, inventing the next generation of manufacturing and personal fabrication. In each unit of PRONTO 3D, there are 3D printers, router milling, laser cutting machines, computers and software (Fig. 3). There, it is developed CAD modeling, print 3D physical models, manufacture of prototypes, final products, complex structures, assemblies and installations. PRONTO 3D attends several areas, such as, architecture, civil, mechanical, mechatronics and electrical engineering, industrial and graphic design, among others. PRONTO 3D unit, from SATC, is composed by a coordinator (Professor-researcher) and students who receive scholarship, and provides services to internal customers (SATC undergraduate, high school and technical education courses) and external customers (companies and others PRONTO 3D units from Santa Catarina State).

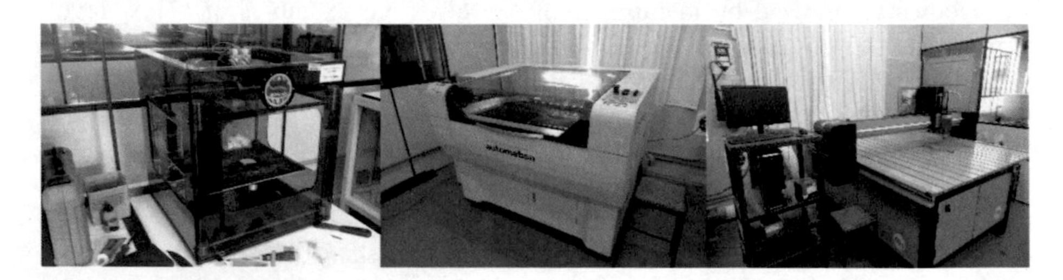

Fig. 3. RP equipment to support PBL classes (3D Printer, Laser cutting machine and Milling Router).

2.6 PBL's Classes Development in Level 1

The following is an example of a PBL's class, which was applied Level 1, during the 2nd semester 2016 (Fig. 4). Groups of four students solved a problem related to a subject in Technical Drawing. The PBL's task was to develop a Mini Baja's prototype by applying the discipline basic contents (dot, line, planes and graphics process to obtain true distances and areas). Normally when implementing PBL, classroom's layout is different from the traditional rows of desks. In this experience, tables for four or more students, to work together, were used. In this case, several tools (scissors, pliers, stilettos, screws and other mechanical tools) were made available and raw materials (fine and round wood bars, filament for 3D printer, wood and acrylic thin sheets for laser cutting machine) were provide to manufacture the prototypes. After designing the set, individuals plan of the parts of the cowl were drawn. From them, it was generated 2D CAD drawings that were sent to Laser Cutting Machine Software, which, then, performed the cuts on wood sheet. From 3D CAD models, car seats, wheels and steering wheel were manufactured on 3D printer. The professor (indicated in detail in Fig. 4) does not stay seated in his chair - he remains very close to the teams, helping them to solve the proposed problem. There were short lectures, but most of the class time was spent mediating the groups. Not all groups delivered a final piece, incomplete assemble occurred, but all groups performed the drawings and used them to build the prototype. As it shows, to plan and execute PBL class requires important changes, such as, in the classroom layout, equipment, planning, and adapting infrastructure. However, the main and most difficult change, which is paramount, is in the professor's mindset, who needs to leave their comfort zone and adapt to provide the education of the XXI century.

Fig. 4. Left side: Active Learning classes (Level 1); 2nd semester 2016, SATC College, Mechanical Engineering Course. Subject: Technical Drawing. Right side: manufacture of prototypes using the basic concepts of the subject and rapid prototyping resources (laser cut machine and 3D printer).

3 Conclusions

It is important to consider that, the applicability of this PBL-based learning model to different disciplines needs to take into consideration the nature of each discipline, due to limitations imposed by course's current structure. Thus, it is plausible to expect professors to achieve level 2 when implementing in one discipline and level 3 in another, without moving up to next level. On the other hand, the incapacity to apply levels 1 and 2 indicated that, there were structural problems in the disciplines, in which prevented defining the objective, importance, nature, protagonism and utility of the discipline itself. Furthermore, each attribute cannot occur in the same intensity, even if it is in the same level of implementation. A problem of simulated scope (Level 3), can be solved through a mediator professor (Level 1). Actually, it is a possible situation but unlikely, in according to case studies observed. The integral application of PBL, contemplating every learning unit and every level, requires a revision of the curriculum for all the courses. However, the current proposition does not see this as a possibility. PBL's curricular implementation requires change of teachers' consciousness towards teaching-learning process, the steeped application of each level, the radical change from content-based curriculum to skills-based curriculum, the immersion of universities on professional world and vice versa. This immersion can be achieved via partnership between enterprise and the university, providing and fomenting research projects; scholarship and extra-curricular internship; university learning units inserted into the enterprises and enterprise laboratory units inserted into university.

References

1. Hasna, A.M.: Problem Based learning in engineering design. In: SEFI 36th Annual Conference, European Society For Engineering Education (2008)
2. Rodríguez, J., Laverón-Simavilla, A., Del Cura, J.M., Ezquerro, J.M., Lapuerta, V., Cordero-Gracia, M.: Project based learning experiences in the space engineering education at technical University of Madrid. Adv. Space Res. **56**(7), 1319–1330 (2015)
3. Jamaludin, M.Z., Yusof, K.M., Harun, N.F., Hassan, S.A.H.S.: Crafting engineering problems for problem-based learning curriculum. Procedia Soc. Behav. Sci. **56**(Ictlhe), 377–387 (2012)
4. Hitt, J.: Problem-Based Learning in Engineering, p. 8. States Military Academy, West Point (2010)
5. Belhot, R.V.: Reflexões e propostas sobre o 'ensinar Engenharia' para o século XXI, Escola de Engenharia de São Carlos, Universidade de São Paulo (1997)
6. Drake, B.E., et al.: Teaching and Learning in Active Learning Classrooms (2014)
7. Ríos, I.D.L., Cazorla, A., Díaz-Puente, J.M., Yagüe, J.L.: Project-based learning in engineering higher education: two decades of teaching competences in real environments. Procedia Soc. Behav. Sci. **2**(2), 1368–1378 (2010)
8. Esteban, S., Arahal, M.R.: Project based learning methodologies applied to large groups of students: airplane design in a concurrent engineering context. IFAC-PapersOnLine **48**(29), 194–199 (2015)
9. Hassan, S.A.H.S., Yusof, K.M., Mohammad, M., Abu, M.S., Tasir, Z.: Methods to study enhancement of problem solving skills in engineering students through cooperative problem-based learning. Procedia Soc. Behav. Sci. **56**(Ictlhe), 737–746 (2012)
10. Orey, M.: Emerging Perspectives on Learning, Teaching and Technology. Jacobs Foundation, Zurich (2010)

Influence of Study Skills on the Dropout Rate of Universities: Results from a Literature Study

Nilüfer Deniz Bas[1(✉)], Robert Heininger[1], Matthias C. Utesch[1,2], and Helmut Krcmar[1]

[1] Chair for Information Systems, Technical University of Munich (TUM), Munich, Germany
nilufer.bas@tum.de, {robert.heininger,utesch,krcmar}@in.tum.de
[2] Staatliche Fachober- und Berufsoberschule Technik München, Munich, Germany

Abstract. A high dropout rate from universities has been a topic of interest in educational research for more than a decade. The withdrawal from the university on the one hand is frustrating for the students and their families, and on the other hand mostly means a waste of time and money. Tailoring every higher education program to each students' learning styles and needs is difficult. However, mastering the requisite study skills at the pre-university phase (K-12) is considered one of the solutions to diminish university dropouts. By conducting a literature study, we identified relevant study skills and their influencers as well as the appropriate period to improve these skills. We created a model illustrating the relationship between five study skill categories, their influencers and their effects on dropout from university. This study aims to increase awareness and active stakeholders such as families, teachers, and universities for cooperating to strengthen the study skills of school students.

Keywords: Study skills · Study habits · Learning skills · Dropout
Academic success · Intervention studies · K-12

1 Introduction

Education plays a crucial role equipping generations who will shape their future directly and the society that they live in indirectly [1]. Therefore, the formal education is important because skills and strategies are acquired during this period. Learners will use these skills and strategies during a life-long learning period outside the classroom starts. One of the aims of the Bologna Process is to make higher education accessible for people from different backgrounds and qualification [2]. This is one reason that the number of the students enrolling to higher education has been increasing gradually in recent years, and thus, student groups are becoming more heterogeneous [3]. In this sense, heterogeneous group means consisting of students from different academic and family backgrounds along with different learning styles and paces. As it might be difficult to tailor higher education for each of the students' learning styles and needs, dropouts from the study programs might occur. Thus, the higher enrolment rate to university does not always result in a proportional number of graduates from the study programs. Especially in the natural sciences, the failure rate is noticeably higher compared to other fields of

M. E. Auer et al. (eds.), *Teaching and Learning in a Digital World*,
Advances in Intelligent Systems and Computing 715,
https://doi.org/10.1007/978-3-319-73210-7_16

science. A report published by Brugger et al. [4] related to German universities and the period between 2002 and 2010 shows a dropout rate of 34.6% in natural sciences and 27.3% in engineering sciences. For this reason, educational researchers have focused on the topic of dropouts from higher education during the last decades. Higher education targets successful completion of study programs to reinforce the well-being of their individuals in society. The Europe 2020 strategy plans the action of reducing dropout and increasing completion rates of 30-34-year-olds by at least 40%. This goal is important to create high-level skills for fostering innovation, productivity, and social justice in societies [5].

The high dropout rates have a series of repercussions on different stakeholders and contributors. "*Retention not only has an impact on the individual and his/her family but also produces a ripple effect on the postsecondary institutions, the work force and the economy*" [6]. By preventing university dropouts, more students will achieve secondary education as a successful entry to their working life, and families, teachers, and universities reap the rewards of their invested time, effort, and money on students' achievement. Literature shows us several factors that might help to improve the study success [7]. One of these factors is having good study skills which would be seen as a part of students' personal, academic, and professional development [8]. "*As a general term, study skills include all those activities necessary to manage learning and to handle tests*" [9]. Mastering these skills is very important, crucial for students. Improving them in an early phase of life is important to have a smooth transition to university and to graduate successfully from the bachelor's program. Early identification and intervention of deficiencies in skill set during students' pre-university phase – also known as K-12[1] [10] – reinforces student success in educational career, and students continue their educational track without the fear of failure [11]. The K-12 students should be supported to become self-regulated learners so that they can take active control of their own learning process, direct their own efforts to acquire knowledge and skills rather than depending on teachers or parents [12].

The motivation of this contribution is to explore the literature and to unearth during which phases of lives the school students' study skills should be improved. Furthermore, to find out who supports students to gain these skills and whether there is a strong correlation between the study skills of students and their dropout rates from higher education are the motivation of this study. Currently, the influence of study skills on the dropout rate of universities is only partly researched and there is still a lack of knowledge which offers a holistic view of the correlation of improving study skills and reducing dropout rate. Furthermore, it is also interesting to investigate the complexity and the details of the question 'who is able to strengthen study skills': e.g. the K-12 students themselves, their families, the schools, or even the universities (so-called influencers). Therefore, this paper aims to answer to following research questions:

1. *What is the relationship between study skills and the dropout rate at universities?*
2. *Which influencers can best promote these study skills?*

[1] "The K to 12 Program covers Kindergarten and 12 years of basic education [...]" GOVPH. (n.d., 04.04.2017). *WHAT IS K TO 12 PROGRAM?* Available: http://www.gov.ph/k-12.

Most educators would agree that students' dropout problem from schools or higher education institutions is a problem. In addition, families might read news about the latest school dropout numbers. Therefore, the implications of this study will increase consciousness and activate stakeholders such as families, teachers, and universities for cooperating to strengthen the study skills of school students. Moreover, they will help K-12 students and their families, as well as schools and universities to find both the requisite study skills and the appropriate starting point to improve study skills of K-12 students. In the long term, our research aims at reducing dropout rate at universities. Figure 1 shows the research model of our research showing the described relationships.

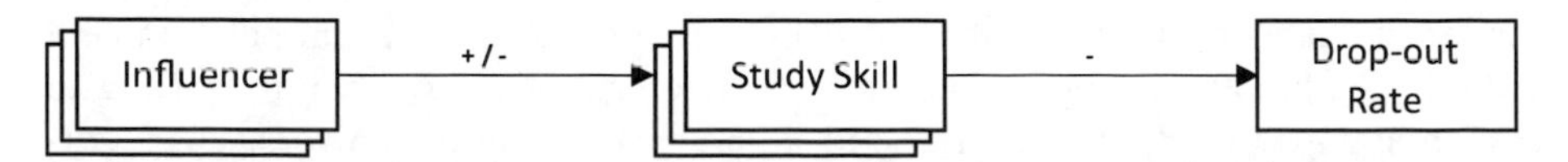

Fig. 1. Research model (own illustration)

This review article started by explaining the current situation of higher education regarding the higher dropout rates and the possible solution of implementing study skills into the K-12 education. In the second part, we provide a brief overview of the current state of the art concerning the enhancement of study skills for further academic achievement. Next, we explain the literature research methods adopted for this study. In the last part, we present the results of the literature research and the model that we developed from these results. Then we conclude this article with presenting the implications of this study.

2 Study Skills and Academic Achievement-State of the Art

One of the preventive alternatives is improving the study skills of the students before starting an academic program [7]. Study skills indicate students' knowledge of applicable study strategies and methods, their competence of time management, and other sources to complete academic tasks [13]. A research conducted by Asikainen et al. [14], which aimed to improve study skills of bioscience program students, showed that when organizing and time management skills strengthened, study success and grades become higher. Basically, study skills include all those activities necessary to manage learning and to handle tests at the university level [15]. Robbins et al. defined study skills as all activities that are directly related to productive class performance at school. And this productive class performance determine the ability to pursue and successfully graduate a university program [15]. Therefore, it can be said that academic success requires good usage of study skills. On the other hand, students with low academic achievement generally exhibit ineffective study skills and they accept a passive role which shows dependency on teachers or parents [16]. Furthermore, Gettinger and Seibert [16] also explained features of studying in their literature review study. Two of these features are that studying requires skills and practical training to help learners how to get, organize, process and use information. Secondly, studying is volitional activity which needs special time and effort by learners.

Much research shows the importance of study skills in academic success. For instance; in their meta-analysis, Crede and Kuncel [13] found that study habits and skills foresee academic performance more than any other affective individual difference variable investigated until now and could be regarded as the third pillar of academic success. Robbins et al. [15] presented an influential meta-analysis in Psychological Bulletin examining the incremental contribution of what they called psychosocial and study skills factors (PSFs) to the prediction of academic success. One of these factors are academic-related skill factors, which include study skills and habits. The results of this meta-analysis showed moderate relationship between academic success and academic-related skills (r = .366). Literature foresees that academic achievement comes when academic learning skills are acquired. Although some of the literature supports the idea of teaching study skills before starting to academic program, other supports the idea that first year of the university is a critical time for the student engagement and study skills can be boosted at the beginning of the university [17].

Besides prime importance of teaching study skills to the students regardless of before or during academic program, teaching approaches also deserves attention. Integrating study skills in separate courses or seminars has been changed to integrating it into subject courses because the terms knowledge and subject content have close interrelation. The instructors can teach students the 'skills of learning' or 'learning to learn' approach by embedding these skills' instruction methods within their subject matter courses [18]. In his meta-analysis, Hattie [19] proposed that courses in study skills alone can be effective on a surface level of understanding but combining study skills course with a content to have far better influence on learners' a deeper level of understanding. In their review article from 51 intervention studies on study skills, Hattie et al. [20] also supported the idea of training study skills within a context and using task-related activities to develop learner activity and their metacognitive awareness. In this sense, training study skills in metacognitive level requires metacognitive intervention courses. Differently from cognitive interventions which focus on task-related skills such as note-taking or summarizing, metacognitive interventions focus on self-management of learning when, where, why and how to use specific strategies in appropriate contexts [20].

We would like to find out that when these study skills, either in cognitive or metacognitive level, implemented into instruction courses before entering to university, the students' learning becomes more direct and effective by using these skills. In the next section, we explain the methodology of this literature review study and subsequently the results that we had.

3 Method

For this study, we adopted literature searching approaches following the guidelines of [21]. Back and forward searching methods as suggested by [22] were also utilized. We used the database EBSCOhost, especially the psychology and educational databases PsycARTICLES, Education Resource, ERIC, and SocINDEX with the option 'scholarly peer-reviewed academic journals'. Additionally, the IEEE Xplore database and Google Scholar were included. In a first search, we used the following keywords: *study skills* or *study habits* or *learning habits* and *dropout* or *drop out* or *drop-out*. Doing so, we received 407 hits. Next, we limited the results to publications related to preschool age (2–5 years), school age (6 to 12 years), or adolescence (13–17 years) and the number of the relevant articles for our study dropped to 165. We screened the titles and abstracts of these publications looking for their study designs and their relevancy to our search criteria of university dropouts. In this way, we reduced the amount of relevant publications to 8. Most of the eliminated publications are related to open and online education, high school dropouts, and behavioral problems of students. However, the article found during back and forward searches was also included in this study, following Webster and Watson's advice about writing a literature review [22]. Thus, we finally analyzed 9 publications.

Our database research showed that although the relevant literature selection criteria were set within the ages of K-12 students, the results include university students and university preparatory year students. Besides, our results were restricted to accessible full-text articles provided by our institutional library

4 Results

Because this study only investigated contributions to the relationship between study skills and dropout from university, only a small number of relevant publications could be found. E.g., only three studies [9, 23, 24] focused on successful student transition from high school to university. Showing that this specific subject has so far not been researched extensively. Additionally, none of the publications provide a holistic view of the subject matter. Thus, our model showed in Fig. 2, provides tentative first results.

Next, we present brief summaries of the relevant publications, followed by a first version of our model showing the relationships between study skills, dropout rate, and some influencers. Finally, this chapter describes the individual components of the model.

4.1 Summary of the Relevant Literature

A study by Lowe and Cook [23] explored high school students' transition to the university. In this study, a single group of students was surveyed both before entering to university as well as two months after starting their studies. The results revealed that at the end of the first two months, 22% of the students had experienced lacked study skills they expected to have in the pre-test. Furthermore, this study evaluated how much 'extended support' students need for the development of academic skills in nine study skills areas. The results revealed that students require more support from academic staff in formal lecture, presentation, academic writing, IT, and time-management skills. Hence, the authors concluded that students experiencing academic difficulties are more prone to dropping-out of their studies.

Creating a learning atmosphere in which students learn and practice their study skills can inspire them for better study and successful completion of their programs in the future. Therefore, integrating study skills courses with subject content, as we mentioned in the second section of this study, can be a good approach. To give an example, in recent research, a study skills course was integrated into an interactive educational game. The SAP ERP simulation game, ERPsim is an enterprise game which helps to strengthen 11th graders study skills. The participating students visited a university in Germany; the focused skills were time management and teamwork. They collaborated with university students during the game while being informed about studying at a university by these senior students. The main aim of this study was to support students' smooth transition to university. Although it was only a one-day training, it significantly increased in pupils' awareness about university as well as their time management and teamwork skills [9].

Another example of embedding technology with study skills is the program 'Study-MATE' [25]. This project was implemented by a university's study support services aiming to inform 'at risk students' who are adjusting to the university and to boost their awareness of the student learning services. The participants received informative study skills hints and tips via their student e-mail account and via SMS. The results revealed that this program increased the students' awareness about study skills and supported their habitual study throughout semester. However, only 30% of the participants agreed that this program increased their planning and preparation for exams and assignments [25].

More than 1.000 freshmen attended 10 h of metacognitive-skills courses which aimed to improve their study skills, study habits, and self-knowledge necessary to succeed at university. The course sessions were guided by teachers, educational counselors, and psychologists. The aim was to increase these competencies to reduce failure rates (dropouts). The results demonstrated that the course improved the students' perceived ability to use the study strategies significantly [26].

A study by Loyens et al. [27] measured students' conception of constructivism at the beginning and end of the academic year. The students' learning activities, self-study times, and cooperative learning activities were observed and rated by tutors. One of the results of this study was that when students have higher conceptions of learning, they cooperated with fellow classmates. Another finding was that when students spare more time to studying, they were more likely to complete their degree.

Another alternative to support students' teamwork skills would be learning communities. A pilot study by Cuevas et al. [28] split first year university students into two groups: learning community and non-learning community. The aim was to gather students in the learning community to discuss the subject contents, to cooperate during studying, and to meet outside the classroom for preparation of exams. The instructor guided and engaged students for discussions and for learning activities. Therefore, these meetings let them study together regularly and they broke last-minute studying habits for exams. At the end of the pilot study, the differences between groups were significant. The learning community received higher grades and improved both their time management and teamwork study skills.

A study by Wernersbach et al. [29] focused on teaching academically underprepared students the following study skills: note taking, time management, learning strategies, and test preparation skills. These study skill courses were introduced by instructors during seven weekly sessions. At the end of the course, a post-test survey was sent to the participants. Academically underprepared students improved their study skills significantly more than the control group. This highlights the importance of identifying these students in the early stages of their formal education.

A case study [30] focused on teaching academic literacy skills such as essay writing, complex problem solving, and active reading to preparatory school students. The successful students were awarded admission to the bachelors' program. However, students coming from poorly-resourced schools couldn't get admission to the program.

Our backward research resulted in one study, which aimed to prepare school students for the university life as well as to strengthen their study skills for smooth transitions to university. The Pupil's Academy of Serious Gaming project focused on improving time management, communication and presentation skills, and problem solving strategies of 11th graders. The participant students were invited to play a business game in teams, helping them to develop their cooperation skills. The results displayed positive perspectives of students after attending Pupil's Academy. Most of them stated that they now have a clear idea what studying really means. Beyond these results, this study compared participant students' and the control group's final examination (the German 'Abitur') results, and found that participants students achieved lower failure rate than control group [24].

4.2 Development of a Model

By analyzing the aforementioned publications, we were able to develop a preliminary model showing the relationship between the study skills, their influencers, and the dropout rate, in line with our research model shown in Fig. 1. Furthermore, in her study, Wingate [8] integrated each study skill into the QCA key skills framework [31]. This framework inspired us to categorize our identified study skills into these five categories. Overall, we identified *support services*, *university students*, *instructors*, *counselors*, *psychologists*, and *teachers* as influencers. We also identified twelve study skills affecting the dropout rate at university and grouped them into the following categories: *communication*, *working with others*, *problem solving*, *self-learning and evaluation*, and *information technology numeracy*. Figure 2 illustrates our preliminary model showing the relationship between the study skills, their influencers, and the dropout rate from university.

However, due to few relevant publications we assume that this model is incomplete. It must be borne in mind that on the one hand, additional study skills could influence dropout rate at university, and on the other hand, additional influencers or additional relations between the known influencers and the study skills must be considered. In the next section, we define our findings of study skills under the five skill categories followed by a section defining the six identified influencers and their relations to the study skills.

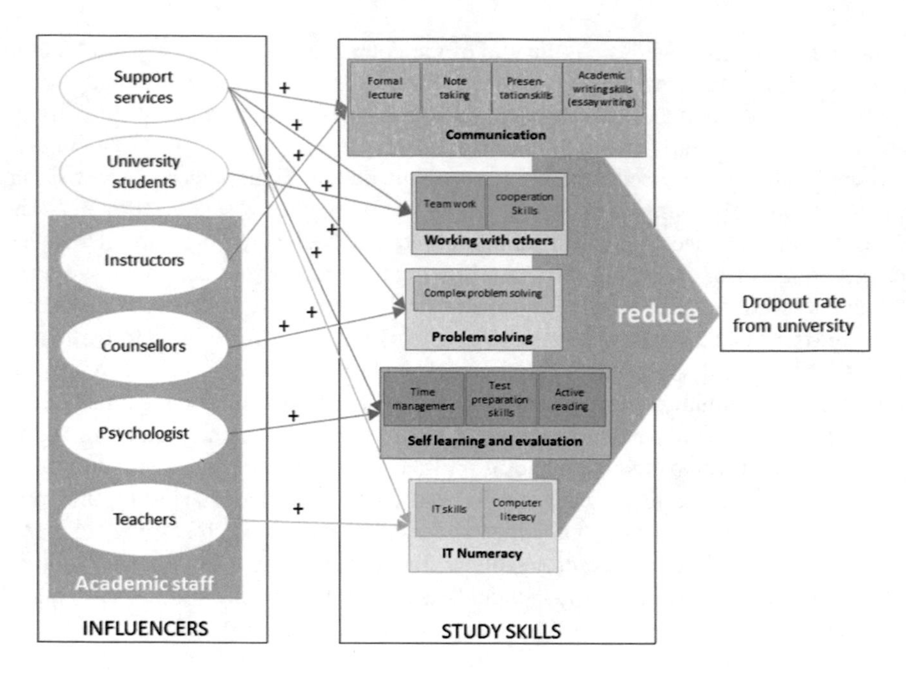

Fig. 2. Model showing the relationship between the study skills, their influencers, and the dropout rate from university (own illustration)

4.2.1 Study Skill Categories and Study Skills

Communication [23, 24, 29] is one of the study skills categories and includes *formal lecture* [23], *note taking* [29], *presentation skills* [23, 24], and *academic writing skills* [23]. Communication skills are vital both in speaking and in writing to convey own thoughts, to present and analyze complex information, as well as, in listening, comprehending and responding to others [32]. In this sense, formal lecture and note taking are important skills for students' success, because taking good lecture notes means achieving in recording information from professors' lectures [33]. In oral communication, presentation skills keep a prominent importance as it addresses all audience. Participant students improved their presentation skills as a team in a business game in which they presented the business activities of their companies [24]. However, students might feel necessity of extra support in certain communication skills branches. For instance, in the study of Lowe and Cook [23] the participating students expressed that they require more support in giving a verbal presentation to other students and writing a practical report.

The working with others skill [9, 24, 28] encompasses *teamwork* and *cooperation.* Cooperative working requires students to use interpersonal skills appropriately, valuing others' contribution, but at the same time taking responsibilities for their own mission and aims [32]. Students with good teamwork skills should be able to enjoy working with other group members on a same mission, ready to take responsibilities, and feel comfortable with presentations in front of a group [9]. Therefore, the difference between teamwork and cooperation is that teamwork requires group members to work on a common task altogether, however in cooperation, each group member has their own tasks but works collaboratively both to improve their skill set and the other group members'.

The problem solving skill [24, 30] has three stages and it resolves issues or problems in personal, social, and work contexts. The first stage is critical thinking in which students use reasoning skills to make the best decision, then comes planning resources and managing tasks to completion, and the last stage is reflecting on outcomes of the tasks [32].

The self-learning and evaluation [9, 29, 30] skill includes the skills of *time-management* [9, 24, 29], *test preparation* [29], and *active reading* [30]. These skills focus on individual tasks. For instance, students with good *time management skills* can organize their time efficiently, complete their tasks on schedule without delay, prioritize tasks, and make responsible decisions [9]. By creating learning communities, Cuevas et al. [28] focused on teaching group of students the *test preparation skills* for their biology exam. The results showed learning community scored higher in every test than the non-learning community. Active reading and reading comprehension are crucial study skills as science students needs extensive academic vocabulary and complex language skills to understand the script [30].

The IT numeracy skill covers *information management* and *computer literacy*. These skills enable students to search for the most appropriate sources of information in the library or on the internet [23]. Students can also perform more complex tasks i.e. using Microsoft Excel to deepen their practical knowledge in statistics [9].

4.2.2 Influencers' Roles in Improving Study Skills

Through our analysis, we identified six influencers who have direct impact on students gaining, improving, and strengthening skills. All the analyzed publications deal solely with positive effects of the influencers on the study skills.

Support services [25] play an effective key role for new students' integration and offer a wide range of induction courses such as study skills courses to increase students' awareness of learning services. In the study of Jayde et al. [25] support services took the active role of informing the first year students with the available skills courses. Furthermore, they built up the Study-MATE messaging system to give study tips and clues to students. One of the messages aiming to give tips about success examples is: *"Message Week 6 - So glad da break is here! Was starting 2 stress:0 Latest tips for success are here* http://tiny.cc/StudyMATE *or BBoard Study-MATE"*. The result of this study shows that the support services' action in creating such interactive services increased students' awareness of the study skills courses.

University students [9] might be efficient influencers and role models for the school students in many ways. In the study of Utesch et al. [9], school students visited a university where they played a simulation game and explored the university atmosphere. The university students took an active role of mentoring these students and answered their inquisitive questions, thus giving them insight into university life. The outcomes of this study were positive, as university students' companionship helped school students to break barriers and to develop an idea of being a university student. Furthermore, with the simulation game, they strengthened their time management and teamwork skills [9].

Academic staff [26, 28, 29] includes *instructors*, *counsellors*, *psychologists,* and *teachers*. In their study, instructors took an active role of teaching study skills to the underprepared students during seven sections of the course [29]. First forming a learning group, then engaging them with discussions and test preparation strategies, instructors performed well by proving the significant results between learning and non-learning community students [28]. School counselors take active and direct roles of designing structured lessons and providing students the knowledge and skills, which are fitting for their developmental level [34]. In a school environment, counselors, psychologists and teachers work collaboratively to improve academic achievement. School psychologists give direct support and interventions to students and consult with teachers, families and school counselors to improve the academic achievement of students as well as their engagement and learning [35].

5 Conclusion and Outlook

As we mentioned in Sect. 1 of this paper, a higher university enrolment rate does not always result in a proportional number of graduates from these universities. One of the ways preventing university dropouts is to prepare current K-12 students for their future academic life by practicing study skills. Therefore, this study aimed to look for academic publications which address this issue. Our database research resulted in nine publications which showed some relation between K-12 students' study skills and their influences on the future university dropout probabilities. Despite the number of these studies, their findings shaped our preliminary model of the relationships between influencers, study skills, and dropouts. Identifying underprepared students early and providing them with relevant study skills before entering university are the key points to reduce the probability of university dropouts.

However, the presented model was developed based on nine publications, and we must conclude that this model is incomplete. It must be borne in mind that, on the one hand, additional study skills could influence dropout rates at university and, on the other hand, additional influencers or additional relations between the known influencers and the study skills must be considered. There is therefore a need for further research to complete and to validate the model.

With this study, we aim to increase awareness in families, teachers, and universities. Thus, more students will hopefully achieve education as a successful transition to their working lives. Additionally, families, teachers, and universities reap the rewards of the time, effort, and money they have invested in students' success.

References

1. Heininger, R., Seifert, V., Prifti, L., Utesch, M., Krcmar, H.: The playful learning approach for learning - how to program: a structured lesson plan. In: 30th Bled eConference: Digital Transformation – From Connecting Things to Transforming Our Lives, Bled, Slovenia, pp. 215–230 (2017)
2. Bologna Declaration: Joint declaration of the European Ministers of Education (1999)

3. Hanft, A.: Heterogene Studierende – homogene Studienstrukturen. Bundesministeriums für Bildung und Forschung, 13–28 (2015). http://www.hof.uni-halle.de/web/dateien/pdf/Hanft_Zawacki-Richter_Gierke_Open_Access.pdf#page=15
4. Brugger, P., Threin, M., Wolters, M.: Hochschulen auf einen Blick - Ausgabe 2012, Statistisches Bundesamt, Wiesbaden (2012)
5. European Commission: Dropout and Completion in Higher Education in Europe, European Commission Education and Culture (2015)
6. Hagedorn, L.S.: How to define retention: a new look at an old problem. In: Transfer and Retention of Urban Community College Students Project (TRUCCS), ed. Transfer and Retention of Urban Community College Students Project (TRUCCS). Lumina Foundation (2006)
7. Prifti, L., Heininger, R., Utesch, M., Krcmar, H.: Analysis and evaluation of tools, programs, and methods at German University to support the study skills of school students (accepted). Presented at the IEEE Global Engineering Education Conference (EDUCON 2017), Athens, Greece (2017)
8. Wingate, U.: Doing away with 'study skills'. Teach. High. Educ. **11**, 457–469 (2006)
9. Utesch, M., Heininger, R., Krcmar, H.: Strengthening study skills by using ERPsim as a new tool within the Pupils' academy of serious gaming. In: 2016 IEEE Global Engineering Education Conference (EDUCON), pp. 592–601 (2016)
10. GOVPH. (n.d., 04.04.2017). WHAT IS K TO 12 PROGRAM? http://www.gov.ph/k-12
11. Seidman, A.: College Student Retention: Formula for Student Success. ACE/Praeger Publishers, Westport (2005)
12. Zimmerman, B.J.: A social cognitive view of self-regulated academic learning. J. Educ. Psychol. **81**, 329–339 (1989)
13. Crede, M., Kuncel, N.R.: Study habits, skills and attitudes- the third pillar supporting collegiate academic performance. Assoc. Psychol. Sci. **3**, 425–453 (2008)
14. Asikainen, H., Parpala, A., Lindblom-Ylänne, S., Vanthournout, G., Coertjens, L.: The development of approaches to learning and perceptions of the teaching-learning environment during bachelor level studies and their relation to study success. High. Educ. Stud. **4**, 24–36 (2014)
15. Robbins, S.B., Lauver, K., Le, H., Davis, D., Langley, R., Carlstrom, A.: Do psychosocial and study skill factors predict college outcomes? A meta-analysis. Psychol. Bull. **130**, 261–288 (2004)
16. Gettinger, M., Seibert, J.K.: Contributions of study skills to academic competence. School Psychol. Rev. **31**, 350–365 (2002)
17. Tinto, V.: Research and practice of student retention: what next? Res. Theor. Pract. **8**, 1–19 (2006)
18. Pressley, M., Woloshyn, V.: Cognitive Strategy Instruction that Really Improves Children's Academic Performance. Brookline Books, Cambridge (1995)
19. Hattie, J.: The contributions from teaching approaches - part 1. In: Visible Learning A Synthesis of Over 800 Meta-Analyses Relating to Achievement ed. SAGE Publications (2009)
20. Hattie, J., Biggs, J., Purdie, N.: Effects of learning skills interventions on student learning: a meta-analysis. Rev. Educ. Res. **66**, 99–136 (1996)
21. vom Brocke, J., Simons, A., Niehaves, B., Riemer, K., Plattfaut, R., Cleven, A.: Reconstructing the giant: on the importance of rigour in documenting the literature search process. Presented at the European Conference for Information Systems (ECIS), Verona, Italy (2009)

22. Webster, J., Watson, R.T.: Analyzing the past to prepare for the future: writing a literature review. MIS Q. **26**, 13–23 (2002)
23. Lowe, H., Cook, A.: Mind the gap: are students prepared for higher education? J. Further High. Educ. **27**, 53–76 (2003)
24. Utesch, M.C.: The pupils' academy of serious gaming: strengthening study skills (iJEP). Int. J. Eng. Pedagogy **5**, 25 (2015)
25. Jayde, C., Elaine, H., Boris, H., Justin, D., Mark, N.: Study-MATE: using text messaging to support student transition to university study. Youth Stud. Aust. **31**, 34 (2012)
26. Costabile, A., Cornoldi, C., Beni, R.D., Manfredi, P., Figliuzzi, S.: Metacognitive components of student's difficulties in the first year of university. Int. J. High. Educ. **2**, 165 (2013)
27. Loyens, S.M.M., Rikers, R.M.J.P., Schmidt, H.G.: The impact of students' conceptions of constructivist assumptions on academic achievement and drop-out. Stud. High. Educ. **32**, 581–602 (2007)
28. Cuevas, M., Campbell, K., Lowery-Hart, R.D., Mallard, J., Andersen, A.: Using faculty learning communities to Link FYE and high-risk core courses: a pilot study. Learn. Commununities Res. Pract. **1**, 1–12 (2013)
29. Wernersbach, B.M., Crowley, S.L., Bates, S.C., Rosenthal, C.: Study skills course impact on academic self-efficacy. J. Dev. Educ. **37**, 14–16 (2014)
30. McKay, T.M.: Academic success, language and the four year degree: a case study of a 2007 cohort. South Afr. J. High. Educ. **30**, 190 (2016)
31. Qualifications and Curriculum Authority: Religious Education (2004)
32. Scottish Qualifications Authority: SQA, Core Skills Framework: An Introduction Communication. Scottish Qualifications Authority (2013)
33. Van Blerkom, D.L.: College Study Skills: Becoming a Strategic Learner. Cengage Learning, Boston (2011)
34. Grabowski, C., Bjerring, L., Peykarimah, S., Sorensen, L.B., Bracht, R., O'Hara, J., et al.: Integration architecture of multi-technology management systems. In: Proceedings of the Sixth IFIP/IEEE International Symposium on Integrated Network Management. Distributed Management for the Networked Millennium, Boston, MA, USA, pp. 955–956 (1999)
35. National Association of School Psychologists. Who Are School Psychologists (2014). https://www.nasponline.org/about-school-psychology/who-are-school-psychologists

Electrical Engineering Students' Achievement in Measurement Accuracy of Digitized Signals – Work in Progress

Nissim Sabag(✉) and Elena Trotskovsky

ORT Braude College of Engineering, Karmiel, Israel
{nsabag, elenatro}@braude.ac.il

Abstract. The current paper presents of a continuous longitudinal research dealing with different aspects of accuracy and the way the students understand it. Previous studies have shown that understanding the accuracy concept is not obvious. A previous study exposed students' difficulties in calculating measurement accuracy of digitized signals. After writing a special study unit on measurement accuracy and measurement error and adding it to the curriculum of Digital Electronics course, two questions questionnaire entered to the final test on digital electronics course. The students' grades showed an improvement in understanding the accuracy concept, but not sufficient. The study unit included explanation, tutorial and homework. A replication of the previous experiment in Fall semester 2015 showed a substantial improvement in the students' achievements regarding the accuracy concept. This time the students got explanation and example but no homework concerning the accuracy concept. Nevertheless, they had the chance to see and solve the previous questionnaire so the questions regarding accuracy did not surprise them. Solving old test questionnaires is a common mode of learning among our students therefore this may be the explanation of the improvement in the students' grades.

Keywords: Misconception · Accuracy · Error · Digitizing

1 Introduction

The ability to conduct measurements is of the basic engineer's skill [1]; this, of course, is accompanied with assessing the results accuracy. Therefore, it is extremely important for the engineering students to acquire a conceptual knowledge and understanding of the accuracy concept [2]. One definition of the measurement accuracy is "the closeness of agreement between a test result and the accepted reference value" [3]. The researchers of the current study carried out a series of studies concerning students' perception of the accuracy concept. In [4], students of electrical and electronic engineering expressed misconceptions about the concept of accuracy while using engineering models. Following this a wide population of different engineering disciplined answer a questionnaire about accuracy concept. The results demonstrated insufficient understanding and a confusion between measurement accuracy and error [5]. Paper [6] exposes serious misunderstanding concerning the accuracy of digitized signals. Students misperceived the essence of quantization of signal and noise, tended to confuse

M. E. Auer et al. (eds.), *Teaching and Learning in a Digital World*,
Advances in Intelligent Systems and Computing 715,
https://doi.org/10.1007/978-3-319-73210-7_17

quantization and noise errors, and confused the relationships between noise and resolution. The study described in [7] deal with integration a micro-study unit (about 1 h) about principles of accuracy into the course of Digital Electronics. It contained explanation, tutorial and homework. At the end of the semester, two questions about accuracy inserted into the final exam of Digital Electronics. The results show that 25 out of 61 students did not answer the questions at all. Only 27.8% of the students answer a reasonable or perfect answer that is 16.4% of the complete group.

The current study is an obvious interest of the researchers to keep on investigating the issue of accuracy concept, until reducing students' misunderstanding about measurement accuracy in the area of Digital Electronics dramatically.

2 The Research Setting

The main research tool was a questionnaire of three open problems concerning the concept of accuracy; the problems were developed and incorporated into the final exam in the course of Digital Electronics. The written solutions and explanations were then analyzed.

The Participants are forty-three students of electrical and electronics engineering at an academic college of engineering, who took the Digital Electronics course in Fall semester 2015 participated in the study.

Definitions of Error and Accuracy:

$$Absolute\ Error = |A_{MEASURED} - A_{TRUE}| \quad Relative\ Error = \frac{|A_{MEASURED} - A_{TRUE}|}{A_{TRUE}} \times 100\% \tag{1}$$

Where A_{TRUE} is a true value of a measured parameter and $A_{MEASURED}$ is a measured value.

As shown, there are different definitions of accuracy parameters. In the course, the relative measurement error and accuracy were defined in percentage as follows:

$$Accuracy = 100\% - Relative\ Error \tag{2}$$

The Questionnaire: Questions about measurement error and accuracy inserted into the final exam of digital electronics course. In their answers, the students had to take to consideration the sensor's accuracy (or error), Analog to Digital Conversion (ADC) and the effect of display method on the accuracy. All the questions were open problems that required calculations and explanations. The detailed questionnaire appears below.

1. An analog voltage of 200 mV was inserted to a 14 bit – Analog to Digital Converter (ADC), with a reference voltage of 0.5 V. Given that the environment radiates a 60 μV noise on the measured signal, calculate the measurement error as percentage of the measured voltage.

2. The reference voltage of a Digital to Analog Converter (DAC) is not stable; its range is $V_{REF} = 5 \pm 0.1$ V. Calculate the accuracy measurement for the digital input that present the decimal number N = 33.
3. Given a Digital temperature meter with the following specifications:

   - The equipment will consist of a temperature sensor, signal-conditioning circuit, 7-bit-ADC with $V_{REF} = 5$ V, binary to 7-segment display converter and a decimal display. See a block diagram below.
   - The measured temperature range is 0–99.9°C
   - The temperature sensor outputs 39 mV for 1°C with ±2% error.

   Calculate the error of a measured temperature of 20°C.

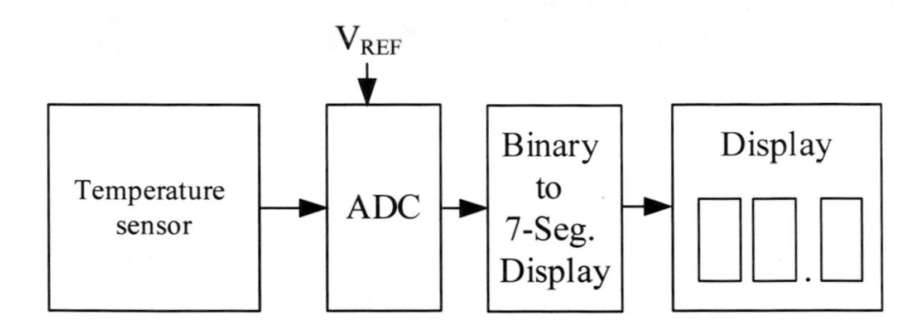

3 Results

Forty-three students answer the questionnaire. The questions about accuracy was graded in three categories: 1. Wrong answer or non-answer, 2. Partial correct answer and 3. Correct answer. The distribution of the students' answer is presented in Table 1.

Table 1. Distribution of students' answers to the questionnaire

	Q1		Q2		Q3		Average	
N = 43	N_1	%	N_2	%	N_3	%	Nav.	%
Correct answer	**23**	**53.5**	**16**	**37.2**	**9**	**20.9**	**16**	**37.2**
Partial answer	**7**	**16.3**	**6**	**14**	**14**	**32.6**	**9**	**20.9**
Wrong answer	**13**	**28.3**	**21**	**45.7**	**20**	**46.5**	**18**	**41.9**

4 Conclusion

The students got one lecture hour including explanation and an example of measurement using digitized signal in non-noisy environment as the 2014 students did. At the final test of the Digital Electronics course, the questionnaire included three questions about measurement accuracy of digitized signals in noisy environment, different from the 2014 questionnaire but on the same difficulty level. Among the 2015 students 37.2% answered completely correct answers and 20.9% partial correct answers, whereas only 16.4% of the 2014 students answered correct or partial correct answers. It is well known that our students use to solve older questionnaires as their main learning method towards final examinations. Therefore, the substantial improvement in the 2015 students' achievements can be attributed to the fact that they solved older questionnaires, while their predecessors did not have older questionnaires to train themselves.

The encouraging results of the 2015 students led us recommend on integrating the accuracy concept issue into the curriculum of more engineering studies.

Appendix – Questionnaire Solution

Answer 1. The noise amplitude can be either 0.06 mV or −0.06 mV. Therefore, signal + noise range is between N + S(max) = 200.06 mV to N + S(min) = 199.94 mV. The digital output for 200.06 mV is: $N\text{max} = \left[\frac{200.06}{500} \times 2^{14}\right] = 6555$ and for 199.94 mv is: $N\text{min} = \left[\frac{199.94}{500} \times 2^{14}\right] = 6551$.

The digital output for 200 mV is: $N = \left[\frac{200}{500} \times 2^{14}\right] = 6553$. The Error for both extremes is: $Err = \frac{6555-6553}{6553} \times 100\% = \frac{6553-6551}{6553} \times 100\% = 0.03\%$.

References

1. Sydenham, P.H., Thon, R. (eds.): Handbook of Measurement Science. Elements of Change, vol. 3, p. 1428. John Wiley & Sons, New York (1992)
2. Streveler, R.A., Litzinger, M.A., Miller, R.L., Stief, P.S.: Learning conceptual knowledge in the engineering sciences: overview and future research directions. J. Eng. Educ. **97**(3), 279–294 (2008)
3. Accuracy (trueness and precision) of measurement methods and results – Part 1: General principles and definitions. ISO 5725 – 1, First edition, 15 December 1994, p. 2 (1994)
4. Trotskovsky, E., Sabag, N., Waks, S., Hazzan, O.: Students' Achievements in Solving Problems Using Models in Electronics. IEEE Trans. Educ. **58**(2), 104–109 (2014). https://doi.org/10.1109/TE.2014.2331918
5. Trotskovsky, E., Sabag, N.: Students' misconception of accuracy. In: Joint International Conference on Engineering Education & International Conference on Information Technology, ICEE/ ICIT 2014, Riga, pp. 148–155 (2014)
6. Trotskovsky, E., Sabag, N.: How do engineering students misunderstand the concept of accuracy? – Work in progress. World Trans. Eng. Technol. Educ. **12**(4), 1–5 (2014)
7. Trotskovsky, E., Sabag, N.: How electrical engineering students understand the accuracy concept concerning digitized signals. In: International Conference on Engineering Education ICEE, Zagreb, Croatia, July 2015

How Does Indirect Feedback Affect the Attitude in Higher Software Engineering Education?

Martina Kuhn(✉)

University of Applied Sciences and Arts Coburg, Friedrich-Streib-Strasse 2, D-96450 Coburg, Germany
Martina.kuhn@hs-coburg.de

Abstract. Specialized programming knowledge and further generic competencies are needed to be able to develop software adequately. Therefore, it is essential to foster communication skills in higher software engineering education. One aspect of communication – feedback – is used rather frequently in agile software development as well as in educational settings, but: How does this indirect feedback affect the attitude of students in higher software engineering education? This paper describes a research proposal in order to detect how and to what extent indirect feedback has an impact on the individual's attitude.

Keywords: Feedback · Attitude · Higher education · Software engineering

1 Introduction

Today nearly every technical product uses some kind of software. Large software-systems are complex and so it its creation. Therefore specialized programming knowledge, but also different generic competencies are needed. In higher software engineering education it is essential to foster mainly communication skills, as they are one of the most important non-technical competencies for working life [39, p. 21ff., 15, p. 73ff.]. A special form of communication is giving and receiving feedback, but only little research has been done so far on the impact of indirect feedback in software development on the person itself.

Due to the present gap in research, this paper depicts some preliminary considerations to lay the theoretical foundation of the effect or influence of indirect feedback on a person's – here undergraduate's – attitude concerning university education in software engineering.

Therefore, the paper initiates definitions of several central terms in Sect. 2; e.g., (direct/indirect) feedback, software engineering, and attitude. Subsequent, some related work is in the center of attention (see Sect. 3) before the research proposal is presented in Sect. 4, that covers questions, design, and a didactical setting for the pretest. The paper concludes with a status of the research and an outlook on future activities in Sect. 5.

M. E. Auer et al. (eds.), *Teaching and Learning in a Digital World*,
Advances in Intelligent Systems and Computing 715,
https://doi.org/10.1007/978-3-319-73210-7_18

2 Definitions

This Section defines the terms (direct & indirect) feedback (see Sects. 2.1 and 2.2), software engineering (and its feedback methods; cf. Sects. 2.3 and 2.4), and attitude (see Sect. 2.5).

2.1 Definition: Feedback

In German language feedback has two different meanings[1] [6]. Therefore, feedback can be defined with a technical point of view, so to find in the definition from Ramaprasad [34], who defines feedback as follows: "Feedback is information about the gap between the actual level and the reference level of a system parameter which is used to alter the gap in some way" [34, p. 4ff.]. A definition of feedback in the context of organizations, referring in a non-technical sense, can be found by [33], who defines feedback as an intended verbal communication to a person about behavior or effects of behavior have been perceived or experienced [33, p. 6].

- As seen from the definition above, feedback can be seen from different *use cases*:
  - From a *technical point of view* feedback is, if in a control looped system, the modification of an output value that leads to an adjustment of an input value [31, 34].
  - In a *non-technical sense* the meaning of feedback is a special form of communication, in which the recipient of some information gives feedback about the received message to the original sender [8, 33].
- Feedback can also be divided by *roles*:
  - One or more persons – the feedback provider(s) – are *giving feedback* concerning a certain attitude, product or process to one or more persons.
  - One or more persons – the feedback receiver(s) – are *receiving feedback* about a given event (e.g., attitude, product or process).
- Feedback can also be distinguished referring to hierarchical standing (see Fig. 1).
  - In case the feedback receiver is lower in status than the provider of the feedback, the feedback goes *vertically downwards*; e.g., lecturer to student, supervisor to employee.
  - In contrast, *vertically upwards* means that the feedback provider has a lower status than the feedback receiver. This constellation will be found rarely in assessment of employees or in evaluation settings at universities.
  - If the feedback provider and feedback receiver are on the same level, they are peers and most of them are not in a dependent relationship; i.e. the feedback can be called *horizontally*.

[1] In German language the words "Rückmeldung" and "Rückkopplung" are used synonymous [6].

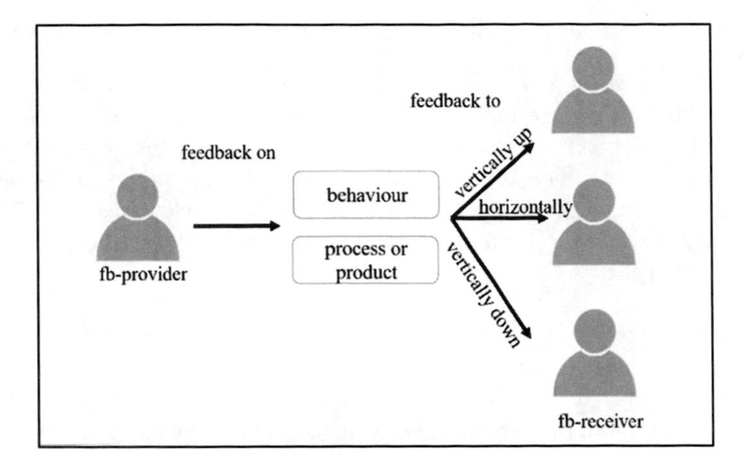

Fig. 1. Kinds of feedback

In all kinds of feedback an existing mixture on behavior, process or product feedback[2] is given, which cannot be eliminated to potentially avoid these effects. Therefore it is vital that clear feedback rules are formulated and met.

2.2 Definition: Direct and Indirect Feedback

Direct Feedback. In this paper direct feedback means feedback focusses mainly on behavior. These kinds of feedback are often used in learning situations. Here the goal is a change of learner's behavior.

Indirect Feedback. Here, indirect feedback means that the feedback focusses on a process or a product. The aim is an improvement of the process or the product.

2.3 Definition: Software Engineering

Software engineering (SE) is the goal-oriented deployment and systematic utilization of principles, methods and tools for a collaborative, engineering development and usage of large software systems [2, S. 36].

The definition in the IEEE-standard 610.12 (1990) [18] for software engineering: "(1) The application of a systematic, disciplined, quantifiable approach to the development, operation, and maintenance of software; that is, the application of engineering to software. (2) The study of approaches as in (1)" [1, p. 67].

Software engineering covers the following areas: requirements, design, implementation, testing, operation and maintenance.

[2] *For example: During a Pair Programming situation, the navigator says to the driver: "Ey, there must be a closing bracket! This fact, I have already told you three times".*

2.4 Exemplary Methods of (Indirect) Feedback in Software Engineering

During the software development process different opportunities for feedback situations are given. In software engineering many different artefacts (e.g., specifications, source code, and documentations) are created. Exemplary, to improve the quality of program code different methods – such as code reviews or Pair Programming – are applied and both of them use feedback on a product; i.e. in an indirect way (see Sect. 2.2). The participants who give and receive feedback are mainly from the same level (cf. horizontal feedback, Sect. 2.1).

Pair Programming. One method in agile software engineering – namely Extreme Programming (XP) [5] – is Pair Programming. Two persons go through a development process, while each person has a specific role; one is the driver, one the navigator, who cooperates. The goal is to increase the quality of software, to speed up the development process and distribute necessary knowledge in the team (or group) [5].

Reviews. Another method in software engineering to check results manually is the review. Each artefact will be checked by different persons, depending on priorities, based on checklists or perspectives [12, p. 6]).

2.5 Definition: Attitude

There are also different existing definitions about attitude; most of them are common that the attitude is a summary assessment of social situations [41]. A complete definition about attitude is given by Eagly and Chaiken [11, p. 1]: Attitude is a psychological tendency, which is expressed by the fact that a defined entity is valued with a certain degree of consent or rejection. This is determined by psychological factors like ideas, values, beliefs, perception, etc. All these have a complex role in determining a person's attitude [28].

3 Related Work

In the scientific community research about **feedback mechanisms in general** has a nearly 70 year old history. In 1946 Lewin [27] was the first to carry out some studies how feedback can lead to extended learning situations during group-processes. Most of the studies are in the context of situations in schools or in the field of teacher training or educational sciences (e.g., [13, 16, 22, 24, 25]). Mostly the feedback situations in these studies are vertical (as shown in 2.1). The results of the studies are inconsistent [3, 22], some of them are in contradiction [30, 32]. Narciss supposed that additional multiple situational or personal factors have to be included to reduce these inconsistences [10, p. 42]. Different theoretical approaches[3] have been raised.

[3] *Response-Certitude-Model [26], Mindful-Processing-Model [3], Feedback for self-regulated learning [7], Feedback Intervention Theory [22].*

Comparatively, little literature is available about mechanisms of feedback in a horizontal situation, often named peer feedback. Some studies are located in the context of school [13, 16, 20, 37] and higher education [24, 37]; fewer papers have been found in the context of working or organizational processes [23].

Concerning **feedback in conjunction with SE**, a systematic literature research like described from Kitchenham et al. [21] was undertaken. The selection of databases covers central important journals and proceedings in the field of SE[4]. The results are that investigations about feedback mechanisms in software engineering are in the topics of user feedback [e.g., 17, 29, 39, 40], on cost-benefit-ratio during Software development [e.g., 9, 23], or using feedback methods focused on software quality [36]. Only one paper was be found about the impact of feedback in software engineering on motivation [35], focusing on motivation and job satisfaction in a non-educational environment.

4 Research Questions and Research Design

Interest in the role of indirect feedback, the following questions arise (see Sect. 4.1), which are planned to be answered during the proposed research design (cf. Sect. 4.2). As a pretest, a didactical setting is conceived where Pair Programming, as one possible software engineering feedback method concerning the artefact "code", can be used to observe and analyze the effect of indirect feedback on the attitude of students' in higher software engineering education (see Sect. 2.5).

4.1 Research Questions

The central research question is: How does indirect feedback affect the personal attitude? This issue can be broken down to the following queries:

- Does indirect feedback change the attitude towards the given problem?
- What happens regarding the person's meta-cognition?
- Does indirect feedback have a motivational aspect?
- How important is the sender's expertise for the acceptance of the feedback?

4.2 Research Design

Since there is little known about the correlations between feedback in the software development process and its impact on the attitude of the people involved in the feedback, an exploratory, iterative research design seems to be preferable [14, p. 126]. Starting with the research question (see Sect. 4.1) and their several sub-questions (*pre-assumptions*) different feedback situations in the field of SE will be *investigated* and used for *data collection*. The result of such various *scenarios* will be *evaluated* and the results *compared* with each other. Relevant theoretical *pre-assumption* and different viewpoints will be reformulated or formulated more exactly. Through evaluation the research will be promoted. The core of the iterative research is the opportunity to develop

[4] *ACM [1], Base [4], IEEE [19], Science Direct [38], WebofScience [42].*

a *theory* and at the same time a base for decisions to collect additional data. The fundamental model of the research design is shown in Fig. 2.

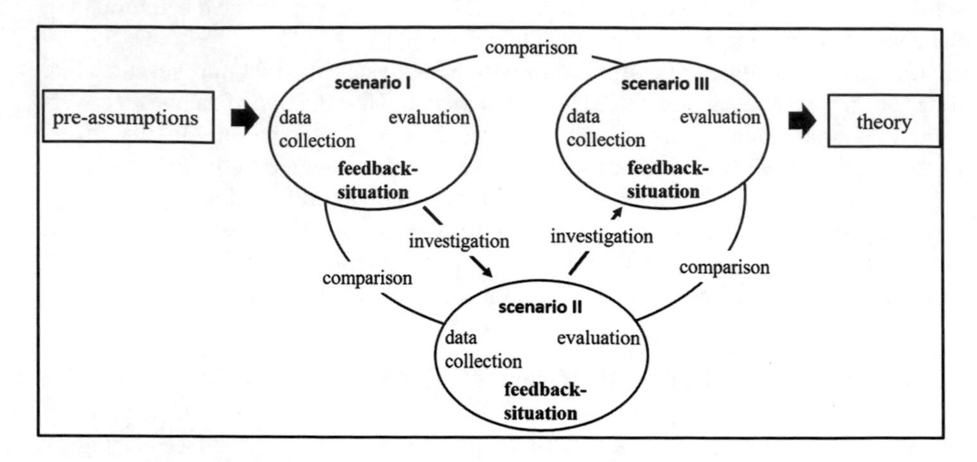

Fig. 2. Circular model of the research design (adapted from [14, p. 128])

4.3 Didactical Setting

At the moment exercises with Pair Programming in a programming course in a Bachelor's degree program – i.e., informatics/computer science non-majors – are taking place. During PP (see Sect. 2.4), two persons are working together on a solution for an assigned coding task. They have different roles, one is the navigator and one is the driver. The functions of the driver are to produce program code and to explain what he/she is doing. The function of the navigator is to give permanent indirect feedback about the produced artefact (e.g., program code). Sequentially they are changing their roles in pre-defined temporal intervals. So each of them provides (navigator) and receives feedback in a horizontal way (see Sect. 2.1).

5 Status and Outlook

The descripted feedback scenario (see Sect. 4.3) will be observed and at the end of the lesson a formative survey shall be filled out to catch the learners' experiences. Since the course is not yet completed, the formative surveys are still to be completed. At the end of the semester the impact of the feedback experience will be inquired with a summative survey. After the analysis of data, a final assessment will be given. On that basis, the detected results are suitable to gain further more detailed and focused systematical surveys or observations in other related feedback scenarios.

Acknowledgement. The present work as part of the EVELIN project was funded by the German Federal Ministry of Education and Research (Bundesministerium für Bildung und Forschung) under grant number 01PL17022A. The author is responsible for the content of this publication.

References

1. ACM (2017). http://dl.acm.org. Accessed 12 May 2017
2. Balzert, H.: Lehrbuch der Software-Technik [German]. Bd.1. Software-Entwicklung. Spektrum Akademischer Verlag, Heidelberg (2001)
3. Bangert-Drowns, R., Kulik, C., Kulik, J., Morgan, M.: The instructional effect of feedback in test-like events. Rev. Educ. Res. **61**, 213–238 (1991)
4. BASE: Bielefelder Academic Search Engine (2017). https://www.base-search.net. Accessed 12 May 2017
5. Beck, K.: Extreme Programming Explained. Embrace Change, 1st edn. Addison Wesley, Boston (2000)
6. Bibliographisches Institut GmbH – Duden Verlag (ed): Duden. Feedback [German] (2017). http://www.duden.de/node/714633/revisions/1382751/view. Accessed 19 May 2017
7. Butler, D.L., Winne, P.H.: Feedback and self-regulated learning: a theoretical synthesis. Rev. Educ. Res. **65**, 245–281 (1995)
8. Clark, H.: Using Language. Cambridge University Press, Cambridge (1996)
9. Curtis, B., Krasner, H., Iscoe, N.: A field study of the software design process for large systems. Commun. ACM **31**(11), 1268–1287 (1988)
10. Ditton, H., Müller, A.: Feedback und Rückmeldungen. Theoretische Grundlagen, empirische Befunde, praktische Anwendungsfelder [German]. Waxmann, Münster, New York (2014)
11. Eagly, A.E., Chaiken, S.: The Psychology of Attitudes. Harcourt Brace Jovanovich, Fort Worth (1993)
12. Fagan, M.E.: Design and code inspections to reduce errors in program development. IBM Sys. J. **15**(3), 182–211 (1976)
13. Fengler, J.: Feedback geben: Strategien und Übungen [German]. Beltz, Weinheim (2004)
14. Flick, U.: Qualitative Sozialforschung. Eine Einführung [German]. Rowohlts enzyklopädie, Reinbek bei Hamburg (2012)
15. Gold-Veerkamp, C.: Erhebung von Soll-Kompetenzen im Software Engineering: Anforderungen an Hochschulabsolventen aus industrieller Perspektive [German]. Springer-Verlag, Wiesbaden (2015)
16. Hattie, J., Timperley, H.: The power of feedback. Rev. Educ. Res. **77**, 81–112 (2007)
17. Hilbert, D.M., Redmiles, D.F. (n. a.): Collecting User Feedback and Usage Data on a Large Scale to Inform Software Development. http://citeseerx.ist.psu.edu/viewdoc/summary?doi=10.1.1.46.8257. Accessed 20 May 2017
18. IEEE (ed.): 1028-2008 - IEEE Standard for Software Reviews and Audits (2008). http://standards.ieee.org/findstds/standard/1028-2008.html. Accessed 11 May 2017
19. IEEE-Xplore digital library (2017). http://ieeexplore.ieee.org/Xplore/home.jsp. Accessed 12 May 2017
20. Jakobs, G., Curtis, A., Braine, G., Huang, S.-Y.: Feedback on student writing: taking the middle path. J. Second Lang. Writ. **7**(3), 307–317 (1998)
21. Kitchenham, B., Brereton, O., et al.: Systematic literature reviews in software engineering – A systematic literature review (2008). http://www.cin.ufpe.br/~in1037/leitura/meta-systematic-reviews-kitchenham-jan09ist.pdf. Accessed 11 May 2017
22. Kluger, A.N., DeNisi, A.: The effects of feedback interventions on performance: a historical review, a meta-analysis, and a preliminary feedback intervention theory. Psychol. Bull. **119**(2), 254–284 (1996)
23. Kohli, A.K., Jaworksy, B.J.: The influence of coworker feedback on salespeople. J. Mark. **58**(4), 82–94 (1994)
24. Krause, U.-M.: Feedback und kooperatives Lernen [German]. Waxmann, Münster (2007)

25. Krause, U.-M., Stark, R., Mandl, H.: Förderung des computerbasierten Wissenserwerbs durch kooperatives Lernen und eine Feedbackmaßnahme [German]. Zeitschrift für Pädagogische Psychologie **18**(2), 125–136 (2004)
26. Kulhavy, R.W., Stock, W.A.: Feedback in written instruction: the place of response certitude. Educ. Psychol. Rev. **1**, 279–308 (1989)
27. Lewin, K.: Feldtheorie in den Sozialwissenschaften [German]. Huber Verlag, München (2012). Ausgewählte Schriften
28. Maio, G., Olson, J.: Attitudes in social behavior. Part Three. Social psychology. In: Willey, J. (ed.) Handbook of psychology. Klumer, Academic/Plenum Publishers, New York (1998)
29. Morales-Ramirez, I., Perini, A., Guizzardi, R.: Providing foundation for user feedback concepts by extending a communication ontology. In: International Conference on Conceptual Modeling, pp. 305–312. Springer International Publishing. http://link.springer.com/content/pdf/10.1007%2F978-3-319-12206-9_25.pdf. 20 May 2017
30. Mory, E.H.: Feedback research revisited. In: Jonassen, D.H. (ed.) Handbook of research for educational comunications and technology, pp. 745–783. Lawrence Erlbaum Associates, Mahwah, NJ (2004)
31. Narciss, S.: Informatives tutorielles Feedback [German]. Waxmann, Münster (2006)
32. Narciss, S.: Feedback in instructional contexts. In: Seel, N. (ed.), Encyclopedia of the Learning Sciences, Volume F, pp. 1285–1289. Springer Science & Business Media, LLC, New York (2012)
33. Oberhoff, B.: Akzeptanz von interpersonellem Feedback. Eine empirische Untersuchung zu verschiedenen Feedback-Formen [German]. Dissertation (1978)
34. Ramaprasad, A.: On the definition of feedback. Behav. Sci. **28**(1), 4–13 (1983)
35. Sach, R., Petre, M.: Feedback: how does it impact software engineers? In: Proceedings of the 5th International Workshop on Co-operative and Human Aspects of Software Engineering, pp. 123–134. IEEE Press (2012)
36. Salinger, S.: Ein Rahmenwerk für die qualitative Analyse der Paarprogrammierung [German]. Dissertation (2013). http://www.diss.fu-berlin.de/diss/receive/FUDISS_thesis_000000094430. Accessed 8 May 2017
37. Schulz, F.: Peer Feedback in der Hochschullehre hilfreich gestalten – Onlinegestütztes Peer Feedback in der Lehrerbildung mit der Plattform Peer Gynt. [German]. Dissertation (2013). https://kluedo.ub.uni-kl.de/frontdoor/index/index/year/2013/docId/3629. Accessed 11 May 2017
38. ScienceDirect (2017). http://www.sciencedirect.com. Accessed 12 May 2017
39. Sedelmaier, Y., Landes, D.: SWEBOS – The Software Engineering Body of skills (2015). https://www.researchgate.net/publication/275646070_SWEBOS_-_Software_Engineering_Body_of_Skills. Accessed 20 May 2017
40. Sherief, N., Abdelmoez, W., Phalp, K.T., Ali, R.: Modelling Users Feedback in Crowed-Based Requirements Engineering. An Empirical Study (2015). http://eprints.bournemouth.ac.uk/22518/. Accessed 20 May 2017
41. Six, B.: Einstellungen. Lexikon der Wissenschaft [German] (2000). http://www.spektrum.de/lexikon/psychologie/einstellungen/3914. Accessed 20 May 2017
42. WEBOFSCIENCE (2017). https://www.webofknowledge.com. Accessed 20 May 2017

"We Don't Want to Know Their Names!"

A Long Way to Go from Engineering Versus Pedagogy to Engineering Pedagogy

Evgenia Sikorski[1,2(✉)], Cristina Urbina[1,2], Michael Canz[1,2], and Evelyn Großhans[1,2]

[1] Offenburg University of Applied Sciences, Badstr. 24, 77652 Offenburg, Germany
{evgenia.sikorski,michael.canz, evelyn.grosshans}@hs-offenburg.de

[2] URV (Universitat Rovira i Virgili), Carrer de l'Escorxador, s/n, 43003 Tarragona, Spain
cristina.urbina@urv.cat

Abstract. In the course of the last few years, our students are becoming increasingly unhappy. Sometimes they stop attending lectures and even seem not to know how to behave correctly. It feels like they are getting on strike. Consequently, drop-out rates are sky-rocketing. The lecturers/professors are not happy either, adopting an "I-don't-care" attitude.

An interdisciplinary, international team set in to find out: (1) What are the students unhappy about? Why is it becoming so difficult for them to cope? (2) What does the "I-don't-care" attitude of professors actually mean? What do they care or not care about? (3) How far do the views of the parties correlate? Could some kind of mutual understanding be achieved?

The findings indicate that, at least at our universities, there is rather a long way to go from "Engineering versus Pedagogy" to "Engineering Pedagogy".

Keywords: Demotivated students · Overburdened professors
Non-pedagogical engineering education

1 Context

In Germany, like in many other European countries, there are two types of institutions for higher education – research universities and teaching universities. The latter ones are called Universities of Applied Sciences (UAS). Their former name Fachhochschule, highly appreciated just a decade ago, is now considered somewhat inferior and is not supposed to be in use anymore. The most distinctive features of Fachhochschulen were

Some paragraphs in this paper may give the impression to be anecdotes but authors reassure the readers that all the described attitudes and statements are not only genuine but also typical ones. Furthermore two out of four authors were part of the target group. They were supposed to gain some understanding of their own attitude partly by conferring with two other authors. In order to comprehensively describe this process of grasping due to conversational conferring a conversational style was retained in some parts of this paper.

M. E. Auer et al. (eds.), *Teaching and Learning in a Digital World*,
Advances in Intelligent Systems and Computing 715,
https://doi.org/10.1007/978-3-319-73210-7_19

"professional" (as opposed to "purely academic") education, extended practical training (two semesters, now shortened to one semester) and smaller academic groups (up to 40 students). The main task of professors/lecturers at Fachhochschulen was teaching, indicated by the so-called "primacy of teaching"[1].

The Bologna reform brought about new degree titles (Bachelor's and Master's degrees replacing much-loved Diplomas). Professors at UAS were assigned an additional duty of conducting applied research. The raising of third-party funds became the main distinctive feature; teaching is not spoken of anymore, let alone the "primacy of teaching".

It is not like nobody noticed the tiny contradiction: If professors are to fulfill a new additional task on a regular basis, the task or tasks they have fulfilled before will suffer. To put it straight: Teaching will suffer. German Council of Science and Humanities stated in 2010: "An increase in effort towards third-party research funding brings along a danger that professors at Fachhochschulen will possibly become overburdened which could in turn result in suffering of the quality of teaching" (Behrenbeck 2013).

It would be all right for the UAS professors becoming more potent, able and capable with every educational reform assigning them a new additional duty. The trouble is that, obviously, teaching has suffered indeed. Our students are getting increasingly kind of unhappy. A lot of students don't bother to attend lectures (there is no compulsory attendance in Germany). The code of conduct (showing up late, surfing on the Internet during lectures, being noisy and even almost childishly nasty) becomes inappropriate. Sometimes, it feels like the students are going on strike! There are also hard facts, not just a feeling: Drop-out rates are sky-rocketing (up to 40% or even 60%). The students clearly demand a change. In the eyes of students, that's for sure, lecturers are the ones who have to deal with the problem.

The lecturers are also far from being happy. They are getting unnerved, many of them adopting an "I-don't-care" attitude. Most of them are engineers teaching quite a number of different subjects and working long hours preparing and refreshing all the material. They do not even consider the task of making students want to learn as their task – never this! They take the view that university students are adults, have to behave as such and to bring along the motivation to learn as adults do. Whenever students don't do all this (increasingly often), it is their (the students') problem – and that is that. In the eyes of professors, that's for sure as well, students are the ones who have to deal with the problem.

Nothing of this helps: Whoever has to deal with the problem – both sides are not happy with it and, come to think of it, this "unhappiness" comes as no surprise. It feels like we have known for a long time that it is coming this way and like there is absolutely nothing we can do about it.

[1] The regular teaching load at UAS is 18 SWS (9 lectures of 90 min. per week) as opposed to 9 or 10 SWS at universities.

2 Purpose and Approach

The authors are members of an interdisciplinary team: two professors of Mechanical Engineering, a pedagogue and a study counselor[2] with expertise in the field of procrastination and exam-nerves.

The goal was to find a rationale for such "functional unhappiness" and verify it with students and lecturers:

1. Students: What are the students unhappy about? Why it is becoming so difficult for them to cope?
2. Lecturers/professors: What does the "I-don't-care" attitude actually mean? What do they care or not care about? How do they address/treat "problematic students" or rather a "problematic" students' group?
3. How far do the views of the parties correlate? Could some kind of mutual understanding be achieved just by letting one party know the position of the other?

A questionnaire for interviewing students and professors was set up and adopted several times as needed. This mutual questionnaire was confined to merely 8 questions. The language of the questionnaire was kept informal. Questions were worded the way people talk in an informal environment only. The authors also refrained from using "he or she", "his or her" etc. for the same reason. As well over 90% of our students (Mechanical and Energy Systems Engineering programs only) and most of our professors are male, "he", "his" etc. were used instead[3].

Quite a number of interviews were conducted – with individual students and student groups (both "still-happy" and "troubled" ones/6 groups in total, all but one "still-happy") as well as with lecturers/professors (both in Germany and abroad, 15 and 5 respectively).

All interviews were conducted face-to-face. This enabled the authors to make sure a particular question was understood the way it was meant and to clarify how the answers were meant. Most importantly, this allowed for an informal way of questioning.

Very soon it became obvious that students are an almost easy-to-grasp target group, completely astonishingly so.

Professors, quite on the contrary, proved a highly intricate, sophisticated and demanding target group, which in reality was not a group at all. Practically each interview delivered a different, even if always (!) comprehensive, point of view. Surprising was that all professors asked for the interview showed interest and spoke openly during these one-to-one talks. (Later, when presenting some of the findings at a faculty workshop, the authors saw the other side of it.) Especially illuminating were talks with foreign colleagues.

[2] Offenburg UAS appointed a pedagogue and a study counselor some 8 yr ago. It took a couple of years for professors getting used to it. The general attitude is still We-are-not-in-school-here but having a pedagogue and a study counselor "at hand" proved quite a good thing after all.

[3] Please note that no disrespect for female actors is meant by this – given that three of four authors are female, two of them professors of Mechanical Engineering! The described peculiar behavior has been shown exclusively my male students so far. The behavior of our extremely few female students follows complete different patterns.

3 Students Who Shouldn't Be There

A noticeable part of German engineering freshmen have no serious ambitions to graduate. They get enrolled just for the lack of alternatives! Some of them would rather have done an apprenticeship but their school grades were too poor for any company to accept them. Or they would rather have studied something else (with no math and physics involved) but were not accepted for the same reasons. Or they would prefer doing nothing at all but their parents kept pushing them. OK, then, let's get enrolled[4] into an engineering program, they accept everyone who fulfills formal requirements! (Yes, they do, at least at the UAS in question.)

No one in their right mind would want such engineering freshmen, unless there is a reason – and yes, there is one! Universities are happy to enroll such freshmen, in fact any freshmen – very much so! – for most of the funding is allocated[5] according to the number of freshmen (Bachelor programs only).

As far as the university is concerned, all freshmen – even those with no prospect to graduate – are welcome to get enrolled. As far as professors are concerned, the latter ones are also welcome to never show up at the lectures.

But they do show up at the lectures – for different reasons:

- Some of them don't see why they couldn't graduate just by caring on being enrolled. After all, "this is how it worked at high school"!
- Some other students have extremely poor school grades (4.0 in both math and physics, which is the last passing grade in Germany). You would think that they (and at least their parents) should know beforehand that academic education is for (academic) elite and extremely poor school grades are no basis for elite! But they REALLY don't know this, they would never accept this view, they argue that their 4.0s are "good 4.0s"[6] !
- And graduate they all certainly can (they think) – why else would the university accept them in the first place? In the end, it was the university who wanted them (not vice versa), so it is just right to expect it to see them through!

They (will) get expelled eventually – after two, three or four semesters. But before they get expelled, some of them show up in the lectures after all, trying to follow for a chance and failing, getting frustrated and noisy, drawing the attention of the whole class to themselves and their unhappiness. These particular students are not about to go quietly!

[4] There are no tuition fees in Germany. By getting enrolled, some financial benefits are to be gained (e.g. a low-priced health insurance, tax deductions for parents, even a small study allowance). On the "Parking Enrollment" see (Sikorski and Canz 2016).

[5] In Germany, universities are funded by the federal state (except for a few private ones).

[6] There is no point in explaining that there is no such a thing as "a good 4.0" (the very last passing grade)! The opposite of it would be "a poor" 1.0" (the best grade achievable), which would be clearly a nonsense!.

One of the authors tried to explain this to a student and failed at it. Yet it could still do some good to a struggling "kid" in a not-so-distant future. For the "kid" is going to fail very soon. And might recall the conversation, at least to some extent, and might think about it ...

4 Who Is Responsible for Creating a Positive Learning Atmosphere?

With a lot of the shouldn't-be-there students around, it is difficult to create "a positive learning atmosphere" for "regular" students, i.e. students with a serious ambition to graduate. This is especially the case if all the teaching is done by extremely overworked engineers with no connection to pedagogy[7].

The first part of the questionnaire dealt with this issue.

4.1 Students

Question #1: How often do troubles occur during your lectures?

- Group #1 (referring to typical "still-happy" groups, first or second study semester): less than 10%
- Group #2 (two students of a group widely known as troubled and deeply unhappy probably from the very beginning; at the time of the interview the regular study period was already over[8]; the students interviewed were learning for exams they skipped or failed in previous semesters): 25%

Question #2: Who do you think is responsible for creating a proper learning atmosphere during the lecture?

- Group #1: 90% lecturer + 10% students[9]
- Group #2: 60% lecturer + 40% students

Question #3: What do you think shall the lecturer do, when students are getting too noisy?

- Group #1: Send them (the trouble-makers) packing! (In this context "to send them packing" means "to make them leave the classroom".)
- Group #2: Send them (the trouble-makers) packing, yes. But not till the lecturer feels EXTREMELY disturbed/annoyed.
- An additional comment on the proposal for the lecturer to speak up/talk in a lower voice: "It might be a nice idea to speak in lower voice but it wouldn't help any, since those (students) who don't want to learn wouldn't mind not to be able to listen."

[7] Author #1 (engineer and professor) admits having heard the term "a positive learning atmosphere" for the very first time after having some troubles with students and consulting the pedagogue (Author #3).

[8] Very few Engineering students (Bachelor) in Germany graduate within the regular study time of 7 semesters. The reasons for this do not only lie in overloaded study programs but also in the hiring procedure of employers as well as in the Master courses starting mainly in October.

[9] The students' groups still intact were nevertheless aware of how easily they can lose a nice learning atmosphere (probably from previous learning experience). They argued that the responsibility rests mainly with the lecturer and were not prepared to assume more responsibility for themselves. "How could I get them (noisy co-students) quiet? I can ask them once or twice but if they don't do so, there is no way for me to quiet them down!".

Question #4: Is the lecturer ALLOWED to send a trouble-making student packing?

- The answers varied from "Yes" to merely "I suppose he is".
- On some other occasion a student showed that he actually thinks the lecture is not allowed to do so (see the conversation "You can't make me leave … you are not allowed!"). Students know this rule from school where the teacher is indeed not allowed to "send them packing" and they obviously "forget" that they are not at school anymore.

Question #5: Is the following statement correct? "The lecturer shall simply ignore the trouble-makers. They will get bored and stop it."

- All students and all groups: "Never!"

Especially surprising are the answers of the Group #2:

- The students admitted having had a tremendous amount of trouble during the whole study (even 25% were clearly an understatement). But at the same time the students think that an enormous part of responsibility rests with themselves (40%). They obviously are aware of themselves at least having co-caused a large amount of troubles and of having failed their part in dealing with trouble-makers.
- During quite a few attempts to sort out the matter with this particular group (some 4 semester prior to the interview) the students had argued quite differently: "Why does it bother/unnerve/trouble you (when students are late/noisy etc.)? Why don't you just speak up?" But in the interview they say what generally all other students say "Send the trouble-makers packing!"
- Indirectly, they admitted of making quite a couple of lecturers feeling "extremely disturbed/annoyed". (During the interview it felt like, looking back, they were regretting having wasted their time on annoying lecturers instead of using it to actually LEARN.)

All in all, the students seem to be aware that, to create a good learning atmosphere, the asymmetric communication between the professor and the student group is of fundamental importance.

4.2 Professors

Question #1: Students are causing troubles during a lecture whenever they show up late, use gadgets like smartphones, talk to each other etc. Substantial troubles are those which negatively affect the process of efficient learning. How often do such troubles occur in your lectures?[10]

[10] This is the last version of the question. The first version ''How often do troubles occur during your lectures?'' was fine with students but not accepted by professors. The comparison shows that eventually the language became very formal (and quite boring at that).

Professor WG + BJ: 0% (!)/Additional comments:

- "Well, it can happen that I send someone (a student) packing once in a while."[11]
- "Just recently (when they were too noisy), I left the classroom; 5 min later they asked me to return (so I returned)."

All other professors interviewed gave 15% to 35% of troubled lectures for an answer.

Striking is that on average, professors reported distinctively more troubled lectures than the students did.

Question #2: Who do you think is responsible for creating a proper learning atmosphere during the lecture?

Professor WG + BJ: 100% professor (Note: These two professors, and only these two, said that they have zero trouble in their lectures!)

The other answers varied from 90% professor + 10% students to 50% professor + 50% students.

Striking is that the professors are willing to accept clearly less responsibility for dealing with this issue than the students expect them to do. Vice versa this means that the professors expect the students accepting more responsibility that the latter are willing to accept and probably more that they are capable of bearing.

Question #3: What do you think shall a lecturer do, when the students are getting too noisy?

Listed answer #3.1: He shall speak up

- "I often speak up unconsciously. But then I realize it and turn the volume down, for it never helps to speak up."
- "It never helps."

Listed answer #3.2: He shall speak in a lower voice

- "It helps much better than speaking up."
- "I speak in a lower voice for a little while first. If it doesn't help, I send them (the trouble-makers) packing!"

Listed answer #3.3: He shall send them (the trouble-makers) packing!

- "Yes, he shall indeed!" (the most prominent answer)
- I don't know, I wouldn't have the heart to do so!" (one single answer)

[11] Comments of the authors (very typical ones):
Author #1 (engineer & professor): Yes, exactly! – This is why he has no trouble in his lectures! I tried this one (move) too – it worked out brilliantly! You do this just once and you have it for a couple of years coming.Author #2 (engineer & professor): I don't know, it might be, but I haven't had such a case so far.Author #3 (pedagogue): You shouldn't do this – it is not right! You should make clear to him that he is welcome in your lecture!Author #4 (study counselor): It would be better to talk to him, dear ...Author #1 going again: Oh yeah, 'welcome in my lecture'! Why him getting so sensitive at once? This is him who wants to graduate, not me, so he will be back soon enough! And if he's going to go funny again, well, then we will certainly need a talk ... (Authors #1 and #4 had three such talks with troubled students within seven years. All of them settled the matter.)

Other answers:

- "There is no line of action which would be always appropriate. The noise level increases, for instance, if students are not able to follow (the explanation of the lecturer). In this case it helps to repeat some of the material."
- "(It helps to) just stop talking and wait until the students quiet down. If it happens for the second time (immediately afterwards), stop talking again and tell them 'Just once more – and I am going to leave (the classroom)'. It has never gone so far that I had actually had to leave."
- "It helps to stop talking or rather to pause."
- "If students have difficulties to concentrate, it helps to open the window and let some fresh air in." (In older campus buildings the windows can be opened.)
- "It is never too noisy in my class!"
- "I would neither speak up nor lower my voice! I just ask them to be quiet, if needed repeatedly so. It has always worked so far."
- "To deal with trouble-makers in a more or less appropriate way, you need to know who they are. Knowing their names would probably help. When sending them packing, I can't ask them to spell their names prior to leaving the classroom – it would sound ridiculous!"

Question #4: Is the lecturer allowed to send a trouble-making student packing?

- "Yes, he is definitely allowed/has the authority to do so!" (the answer reflecting the general attitude)
- "I don't know if he is!" (one single answer)

Question #5: Is the following statement correct? "The lecturer shall simply ignore the trouble-makers. They will get bored and stop it."

- Note: The question was included in the questionnaire due to one very persistent advice of a colleague the other day: "It would be much better to just ignore them, they will eventually get bored and stop it on their own. This is how it always works with my children!"
- Almost all of the answers were "Never!" One answer was "Rather not."
- Opinion of the pedagogue (Author #3): "To ignore a trouble-maker means to ignore a person who is unhappy and needing help. Making trouble sometimes is the only way a needy student is able to act."[12]

[12] Comment of the Author #1: They are adults and therefore have to act like adults do, i.e. keep their troubles for themselves!.

5 These Gadgets!

And last but not least: Them gadgets! – Surfing on the Internet during the lectures, looking who has just posted what on Facebook, Twitter, and/or other social platforms![13]

Question #6 of the Questionnaire dealt with the gadgets.

Question #6: Is the following statement correct? To make students learn efficiently, it is better when the lecturer forbids using gadgets (smartphones and the likes) during the lecture.

Answers/Students:

- The vast majority of the students with one voice: "Definitely yes!"
- Additional remark: "Yes, but only on classes like Mechanics, not in all classes! Using smartphone keeps the students quiet. If I can't use it, I will start talking to the neighbor."
- Additional remark: "Yes, definitely, but you (Author #1) were the only one who did so. It would be better – in terms of this and other issues too – if all the lecturers acted in a similar way. A way predictable for the students. Now we have a different set of rules with every lecturer!

Answer/Professors:

- "I haven't had such problems yet so I haven't spent a though on it so far."
- "Yes, it would be better but I do not forbid using them. I want the students to decide for themselves!"
- "Difficult to decide. Look – in our faculty meetings a lot of professors are busy with their gadgets, so it would be sort of dishonest to set for students a rule which professors wouldn't follow themselves.[14] I would rather say that if students are surfing on the Internet during a lecture, then something must be wrong with the teaching."
- "Yes, unless there are just one or two (students doing so) for just a few seconds."
- "Not really, probably just marginally."
- Author #2:"Yes, unless, like in my Mechanics lectures (in Spain), students need their smartphones or other Internet devices to download problems, to complete tasks, to

[13] A typical conversation between a lecturer and a student who (repeatedly) is busy with gadgets during the lecture: "It looks to me like you are on the Internet again!" – "I just wanted to look up ..." – "As I said before what feels like for a dozen of times: You can't do this during the lecture! One more time and you will have to leave the classroom!" – "You can't make me leave!" – "Try me!" – "You are not allowed to!" – "Oh yes, I certainly am!" (After this conversation, Author #1 realized that she actually did not know for sure whether or not a professor is allowed to make a student leave. So she went to see a prorector for teaching and was told that yes, a professor is allowed to do so, as long as he/she does not do it on a daily basis.).

[14] The colleague is talking about his Department (Business Administration and Industrial Engineering). This is not the case in the Department of Mechanical and Process Engineering where most of the interviews were conducted.

run mechanical simulation programs, to do short tests etc." (This is not done in Offenburg yet.)

6 What Do Professors Care or not Care About?

The most controversial question (#7) read in the first version "If a lecturer says that he doesn't care whether the students show up at his lecture or not, whether they show up on time for the lecture or are late, whether they surf on the Internet, talk to each other, eat their sandwiches etc. during the lectures: Do you believe this/him?" The possible answers: Yes, No idea, No.

Upon an objection of a professor ("You can't ask students, WHETHER they BELIEVE the lecturer!") the wording "If a lecturer says that he doesn't care whether …" was replaced with "Do you believe that a lecturer doesn't care whether …"; no possible answers were given in this version. Please note what we are actually at: *No one objected to the wording "…the lecturer doesn't care whether the students show up at his lecture or not, whether they show up on time for the lecture or are late, whether they surf on the Internet, talk to each other, eat their sandwiches etc. during the lectures …"!*

After a few more rounds of objections from professors, the wording of the Question #7 was softened even further and the Question #8 was added.
Question #7: Some lecturer tend to say to the students that they do not care whether the students attend the lectures or not, whether they show up on time or are late, whether they surf, talk to each other, eat or the like. (Yes, some lecturers DO say all this to the students!) Do you think they really do not care?

Answers/Students:

- "We haven't had such a case so far but I can easily imagine it. No, I do not think they do not care!"
- "Directly like this, no. We were told this just once – by professor XY …" Please note: This was a group of freshmen at the beginning of their very first semester!
- "We are told this all way along! – We don't care either!" (This was the a.m. troubled group.)

Answers/Professors:

- "No, I do not believe they do not care!" (several answers)
- "I would care either!"
- "I don't care either!"
- "You can add my name to the list! This is how I feel."
- "I don't care if they don't show up for the lecture. But if they do (show up) they have to do so on time!"
- "Oh yes! I don't care either! And I tell them so! I tell them no one needs to come here. If they want to talk (to each other) they'd better go to the cafeteria where they wouldn't be disturbed by an old babbler talking Mechanics … But the exam – they would better pass the exam at the first attempt, otherwise I will need to grade their horrible exam papers several times!"
- "It is not possible for the lecturer not to care!"

- "Some students prefer to learn home using a script. I was such learning type when I was a student. In this case I don't mind if they don't show up at the lecture."
- Author #2: "If they say such things, they are not telling the truth. In reality, they care a lot! But yes, some of my colleagues (in Spain) act like that and tell this the students. The trouble is that in this case the students would think that the lecturer doesn't like his job."
- Author #3 (pedagogue): "If you talk like this, students will think that you don't care about them, them like persons. They would feel like disregarded, ignored and rejected, unwelcome in your lecture. You can't create a good learning atmosphere this way."

Question #8: If you think these lecturers DO CARE, then what is probably behind this way of talking?
Listed answer #8.1: The lecturer is just not prepared (does not want) to deal with the trouble-makers.
Students: "I think so.", "No idea!"
Professors:

- "I am (always) dealing with it!"
- "Yes, probably in 50% of cases."
- "I don't want to deal with this! They are grown-ups and that's that!"
- "I would be prepared!"
- "Probably so."
- "If a professor has troubles in his lecture, he MUST do something about it!"
- "Yes, he is not prepared (does not want) to! Or he is doing it reluctantly for this is NOT HIS TASK!"

Listed answer #8.2: The lecturer would actually prefer to create a good learning atmosphere, but he/she would not know how to do this.

- "And I don't feel unsure at it."
- "Yes, probably so (to 50%)".
- "This is not a task of a professor!"
- "Yes, definitely so"!
- "Probably so. But you know, before I started teaching here, I did a didactic course of some 100 h. It probably helps me now."
- "I would know how but, again, this is NOT MY TASK!"

7 "We Don't Want to Know Their Names!"

The findings were summarized and presented in a faculty workshop (Mechanical and Process Engineering).

The question was put forward whether we would want to have a mutual approach in addressing/treating the student group, beginning with a names-and-photos sheet to allocate students' faces to their names as it is common practice in at least some other European countries (contribution of Author #2, extremely illuminating for other authors).

After a somewhat embarrassing non-discussion, the participants agreed that a UAS does definitely not have an educational mandate (with our students being adults after all). Given this, it is preferable to keep them anonymous for the lecturers and therefore we wouldn't even want to know their names! And there is no need in any further discussion, no need at all …

Nothing doing with pedagogy there, let alone Engineering Pedagogy … And it definitely doesn't look like it is going to change a bit any time soon.

Nevertheless, for whatever reasons, the atmosphere changed for the better afterwards (during a one-year period of time). It might be that all the talks have had a positive effect after all.

8 Conclusions

It is certainly not a single case, at least in Germany: Professors at a UAS (a teaching university!) with a supposedly excellent student-to-teacher ratio and the "primacy of teaching" wouldn't want to know even the names of their students, let alone anything about them! If pedagogy is education of a person, then keeping teaching impersonal is quite an opposite of pedagogical and there is a long way to go from "Engineering versus Pedagogy" to "Engineering Pedagogy" – but it obviously helps to get going.

References

Sikorski, E., Canz, M.: A cheer-and-challenge approach in teaching mechanics to demotivated freshmen. In: 5th IEEE International Conference on Teaching, Assessment, and Learning for Engineering, Bangkok, Thailand, 12/17, p. 6 (2016)

Behrenbeck, S.: Profilierung durch Exzellente Lehre: Institutionelle Strategien von Fachhochschulen. Gleichartig – aber anderswertig? Zur künftiger Rolle der (Fach-) Hochschulen im Deutschen Hochschulsystem, p. 59. Bertelsmann, Bielefeld (2013)

Internet of Vehicles Demo System for Autonomous Driving Applications Used in Engineering Education

Gerald Kalteis(✉), Gabriele Schachinger(✉), Bernhard Miksovsky(✉), Martin Potocnik(✉), and Rezek Michael(✉)

Higher Technical School, TGM Vienna, Wexstrasse 19-23, 1200 Vienna, Austria
{gkalteis,gschachinger}@tgm.ac.at

Abstract. This paper shows two best practices examples to bring Internet of Things, especially Internet of Vehicles in a classroom (i.e. at engineering educational institutions). It includes a proposal on how to incorporate the most complex topic "Internet of Things" in particular "Internet of Vehicles" with small exercises on a model system that students themselves have designed and built. The model system was created as a diploma thesis at a higher technical school TGM in Vienna at the Department of Mechanical Engineering with the focus on automotive engineering. Using this practical model system, the networking of things can be more easily understood, or even problems can be easily shown.

Keywords: Autonomous driving · Internet of Things · Internet of Vehicles

1 Modular Demo System for Learning Experiences with Autonomous Vehicles

The aim of one diploma thesis project in the department of mechanical engineering at TGM in Vienna was to design a modular demo system (using a 1:24 scale model car), which allows experiments and learning experiences with autonomous vehicles in real traffic like conditions. The modular design allows different scenarios. It is easy to expand and transport. Active elements like traffic lights support realistic representation. Each module of the whole demo system has microcontrollers which are connected via an intelligent bus systems and each module can be exchanged with others to get different traffic situations. The modules and cars are equipped with different embedded sensors and actuators. This demo system can be used in the classroom for simple exercises and thus approach step by step the complex topic internet of vehicles [1].

1.1 Practical Implementation

The following section gives a brief overview of the practical implementation written by students:

After receiving the requirements, we began to think on how to make the parcour for the vehicles. Also, the characteristics like height, surface area and lightning of the room where the parcour is going to be were considered. The first idea for the parcour was to

M. E. Auer et al. (eds.), *Teaching and Learning in a Digital World*,
Advances in Intelligent Systems and Computing 715,
https://doi.org/10.1007/978-3-319-73210-7_20

make a fixed route spread over 12 different modules. Later our supervisors suggested us this parcour should be more modularized by providing every module with its own logic and connecting with the other modules via a BUS-System. The individual segments are now designed in such a way that they consist only of road sections such as straight lines, curves, intersections, etc., to allow for variations and arbitrary combinations. The following figure shows examples of the top side of basic modules, which will be linked (hand sketch) (Fig. 1).

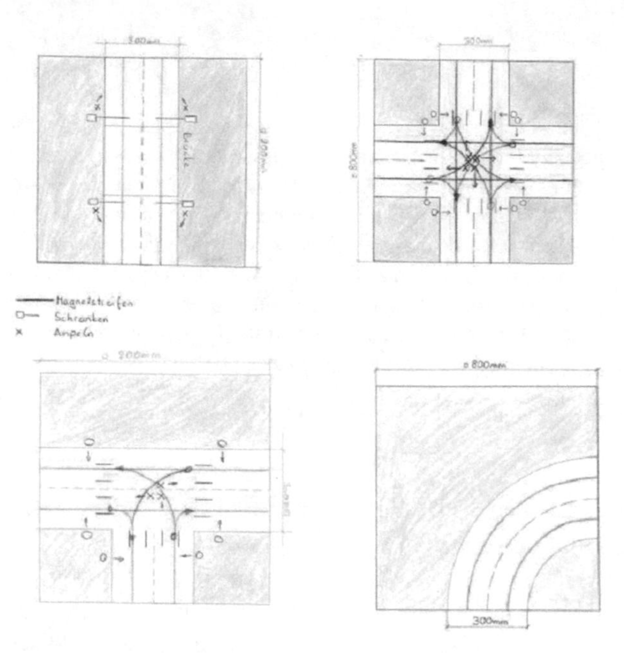

Fig. 1. Four basic modules (800 mm x 800 mm) of the parcour (hand sketch).

As every module had to be independent and to comprise its own logic we had to think about how to connect every module in an efficient, stable and easy way. Also the connection had to be designed in such a way making it possible to connect every module to another module in every direction possible. As for the plugs, we decided to use banana plugs as they are very robust and cheap (Fig. 2).

For the modules, we decided to produce them as wooden elements because it is easy to work with, cheap and more than stable. We decided to use a "Eurolight"-board with a honeycomb core made of recycled paper between thin layers. The advantage of such wood is that it is produced as a sandwich element with a paper structure between two wooden panels making it extremely light. By cutting out a cross in the middle we have created a storage compartment for the electronics within the wooden panel. The following figure shows the bottom view of one module (Fig. 3).

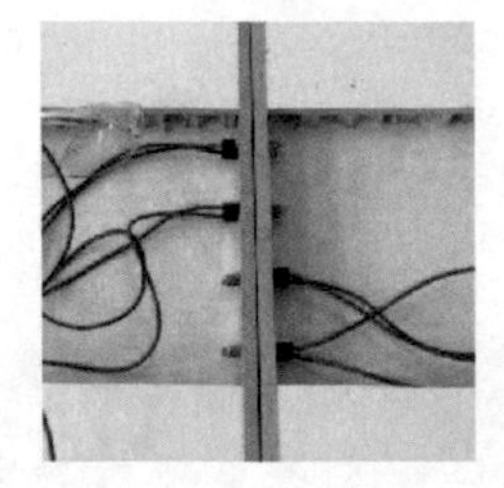

Fig. 2. Connection pattern allowing a universal connection (example of two connected modules on the right side)

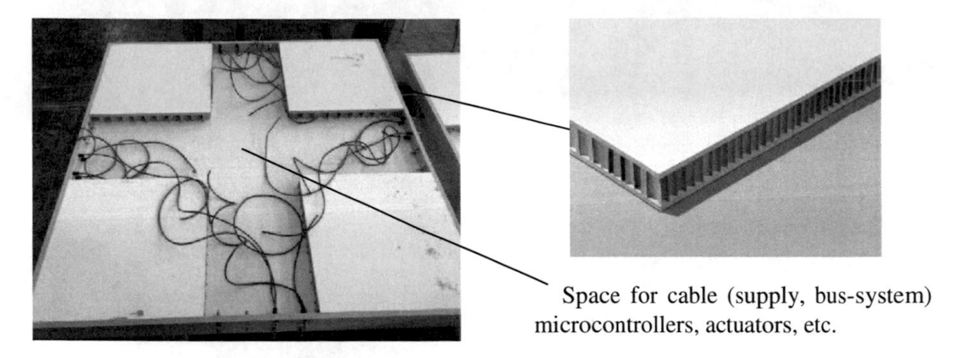

Fig. 3. Module with cross cut-out and connections (button view)

Each module should represent for a special exercise about autonomous driving for example or different scenario examples:

- various traffic crossings
- movement within a roundabout
- traffic sign recognition
- automatic parking
- etc.

This flexible model unit is easily expandable and new sensors, actuators and communication systems can be indirectly tested.

1.2 Model Cars

Before designing the actual model car, we prepared a prototype to see what parts are needed to be changed to provide an autonomous vehicle. As it turned out everything had to be changed because the original car was much smaller than expected and so we had to make room for all the components which had to be part of the car like the following components (Figs. 4 and 5):

- Arduino controller or Raspberry pi
- Batteries

- Different sensors (ultrasonic sensor for distance measuring, potentiometer for steering, etc.)
- DC motor
- Servo motor

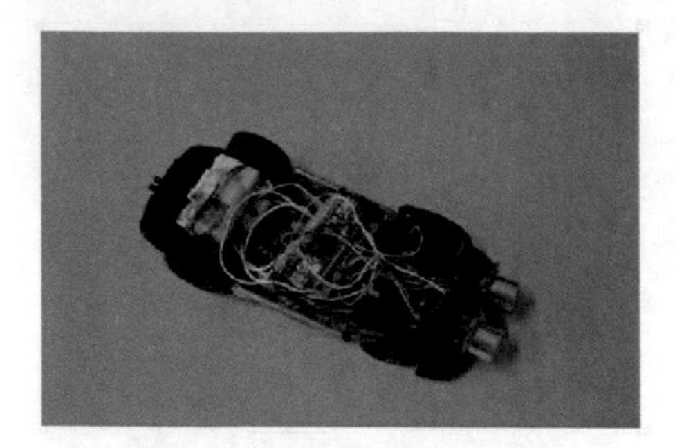

Fig. 4. Prototype of the model car

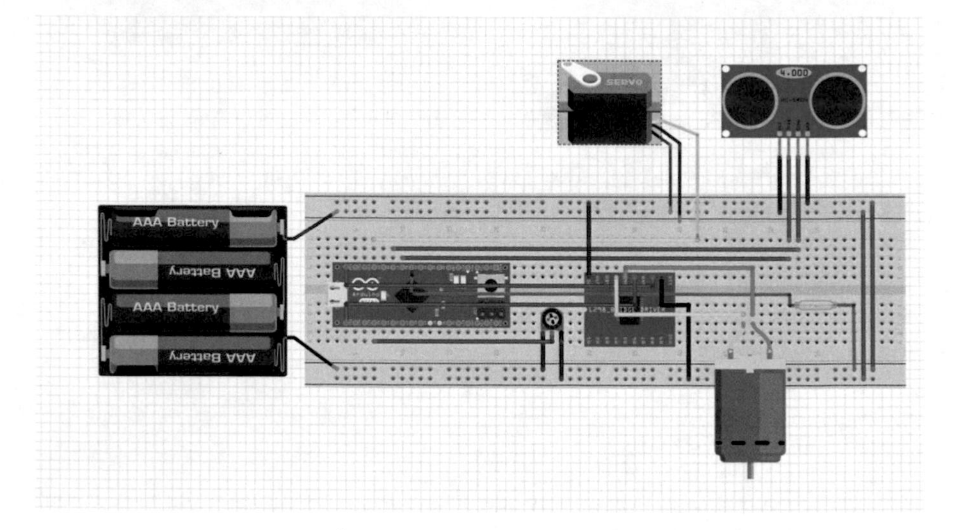

Fig. 5. Connection of all components of the model car

1.3 Exercise Topics

The purpose or goal was to design a modular demo system, which allows experiments and learning experiences with autonomous vehicles in real traffic like conditions. The modular design allows different scenarios on each module. A list of exercises is proposed.

- obstacle detection
- traffic lights detection
- vehicle to vehicle communication

- vehicle to mobile devices communication
- black or white line detection principle
- line following
- navigation in a known environment
- collision detection
- traffic sign recognition
- test of driver-assistance systems

Fig. 6. Demo system for autonomous driving

Based on the knowledge of the model unit, students are then able to implement the following project for example.

The second example was also a diploma thesis with the aim to reconstruct an ordinary kart with a combustion engine to be able to drive a lap on a course on its own. As control unit two Arduinos and one Raspberry were used. These can communicate via CAN bus (internal communication). To identify the centre of the boundaries of the course a webcam is used. These data are transmitted via CAN bus to the other microcontrollers. One of the Arduinos operates an H-bridge to determine the direction of rotation and the velocity of an electric motor for the steering. The second operates a servomotor to adjust the throttle valve and a direct current gear motor for the break. It is also possible to control the kart via radio communication (external communication). Therefore, a third

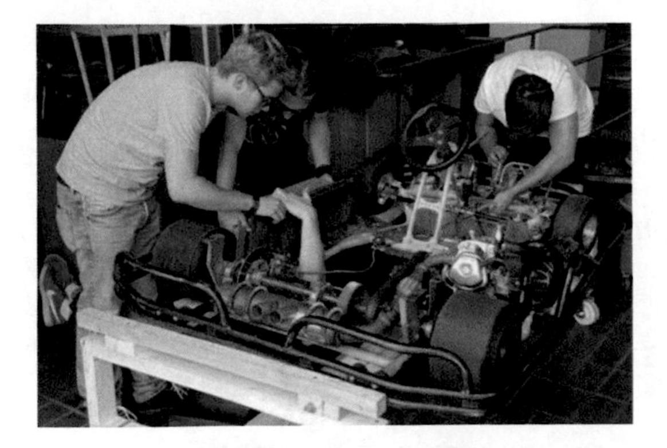

Fig. 7. Disassembling of unnecessary parts

Arduino, a 2.4 GHz radio transceiver and a joystick shield for the Arduino are linked together. The data from the joystick gets transmitted to the Raspberry on the kart.

To be able to fix any sensors or actuators, some modifications had to be made at the kart. As can be seen in Fig. 7 all components that were redundant for mounting sensors have been removed initially.

To reduce weight, the original seat mount and the seat adjustment were replaced by a simple steel tube construction. Furthermore, a bracket for the steering motor was welded. To be able to transmit the engine torque to the steering rod, the construction of two gears was necessary. In addition, small brackets for the second brake cylinder, the camera and warning light, the accumulators and the brake motor had to be manufactured. A 12 V DC windshield wiper was used as breaking motor to ensure safe braking. This motor is controlled by an Arduino MEGA with an H-bridge. Via the CAN - BUS connection between the microcontrollers, the Arduino is getting commands from the Raspberry Pi to brake or to release the brake. To brake safely during a fault or power failure, a second brake cylinder was installed. As shown in the following Fig. 8 the cylinder is biased by a spring and actuated by a magnet. Whenever the voltage drops, the magnet gets turned off and the kart will brake safely. In case of an error, the magnet can be turned off by any microcontroller.

The implementation of the automated steering was more complex. First, it had to be determined which force influences the wheels and any elements of the steering rod, when the kart moves through a curve. During the calculation, it turned out that an electric motor, applying a torque of at least 27.5 Nm, is required.

Additionally, we acquired a suitable worm gear and a pulse generator. The worm gear is served to minimize the speed, to increase the torque and to enable a self-locking function. A further Arduino MEGA and a further H-bridge are used to control the motor. In order to end the steering process once the maximum steering angle is reached, on the left and the right side a push-button was attached. Additionally this feature allows a calibration of the steering at the beginning.

To drive the kart fully autonomously, it needs be able to recognize the road. A Raspberry Pi 3 and a USB webcam as a hardware are used to detect the road. Our software has the following tasks:

- Recognition of the roadside
- Calculation of the laterally distance up to the kart
- Calculation of the steering angle
- Calculating the acceptable speed
- Transmission of the data to the relevant microcontroller

A library called "OpenCV" is used to process the recorded images. With this library, it is possible to process the images of the video in such a way that the lateral road markings are recognised and their position is determined. In Fig. 9 you will see the recognised roadsides (marked in green). The reticle (marked in red) marks the horizontal and vertical screen centre. The vertical marking lines (black) tagged both horizontal positions of the intersection between the edges of the road and the reticle as well as the horizontal positions of those points with a selected distance are also positioned on the edges of the road. From the increase of inter-related points, the steering angle is calculated. By means of a steering angle and an acceleration sensor the speed is identified.

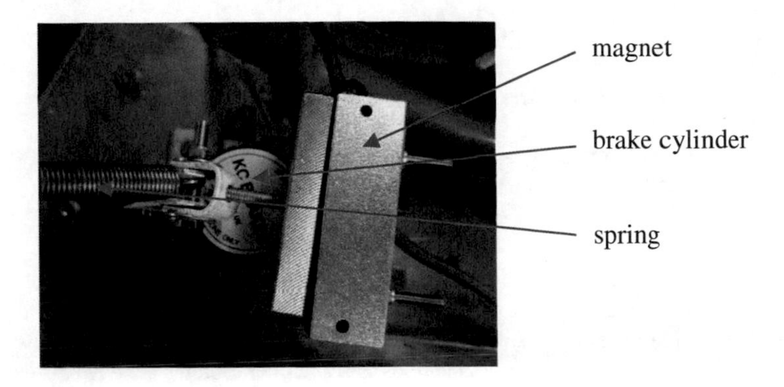

Fig. 8. Emergency brake cylinder

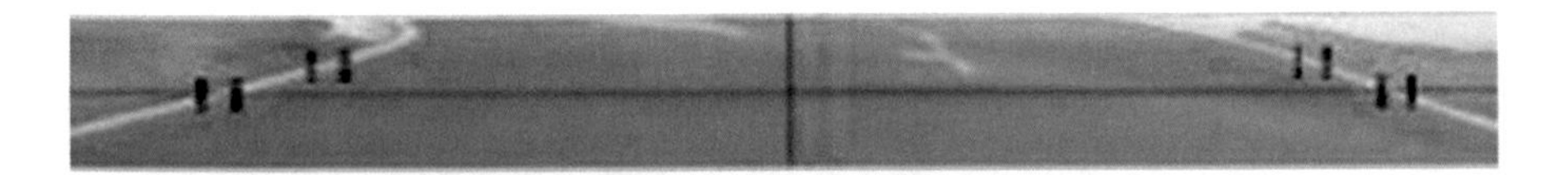

Fig. 9. Image after contour detection with reticle and intersection determination

2 Conclusions

Both best practice examples, the experimental autonomous model car unit and the autonomous kart support teachers to convey the content in the form of exemplary situations. This demo system can be used in the classroom for simple exercises and thus approach step by step the complex topic of internet of vehicles. First experiences at a small training facility with model cars can then be used to be able to move real vehicles autonomously as shown in the paper.

Reference

1. Gerla, M., Lee, E.-K., Pau, G., Lee, U.: Internet of Vehicles: from intelligent grid to autonomous cars and vehicular clouds. In: 2014 IEEE World Forum of Internet of Things (WF-IoT) (2014)

New Concept of Learning Mathematics

Anil Baburao Satpute(✉)

GSM Services, MTNL, Mumbai, India
anil@7pute.com
http://anil7pute.blogspot.in
http://www.7pute.com/

Abstract. To study mathematics, many students face problem to understand the concept of question and the method to solve it. Here the purpose is to develop simple methods to solve them. Looking at the method of the students solving math problems, simplified methods to solve the problems are developed. These methods can save the time and take less effort to solve the problems.

Read, Think, and Learn (RTL) is the main course of studies for the students. Once one acquires it, he/she will enjoy mathematics.

Keywords: RTL · Simple method · Basic concept of mathematics

1 Practical Mathematics

The paper's interest lies in establishing and developing 3-D Mathematical Lab to improve the learning technique.
One simple example is given below.

Consider the sum of all angles of a triangle. Here, their sum is always going to be 180°. Practically this can be shown by the following method. Let's start by marking angles X, Y and Z of a triangle with YZ as base (Fig. 1). Then fold top angle X to the base YZ (Fig. 2). Then fold angle Z and Angle Y to meet to angle X (Figs. 3 and 4). Here all three angles of the triangle form a linear pair. So, the sum of the measures of these angles is 180° which is also the sum of all angles of a triangle.

This will greatly improve the thinking level of students and they will understand the concept better and they will start applying this technique wherever required.

2 Read, Think, and Learn Concept

Read, Think, and Learn **(RTL)** is the main course of studies for the students wherein, at least once in a week, they should choose any one topic of their interest which is one level higher for their grade. Then they must read **(R)** and think **(T)** on the same topic so that they will learn **(L)** it in due course. In the same manner, they need to understand the basic concept of the problem and then try to apply the simple method to solve it. We also insist the students that while reading or solving the problem; they need to understand the technical meaning of the words. Then read the first line of the topic with proper understanding and subsequently the paragraph of that topic which follows. In this manner, they will understand the complete topic.

M. E. Auer et al. (eds.), *Teaching and Learning in a Digital World*,
Advances in Intelligent Systems and Computing 715,
https://doi.org/10.1007/978-3-319-73210-7_21

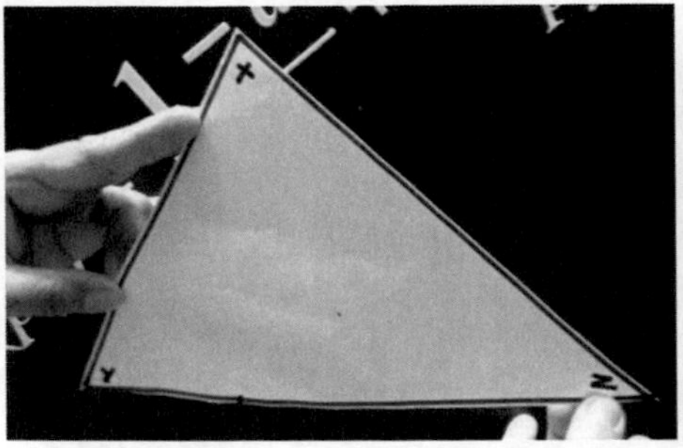

Fig. 1. Triangle XYZ [1].

Fig. 2. Angle X joins the base YZ at a point (as shown in the diagram) [1].

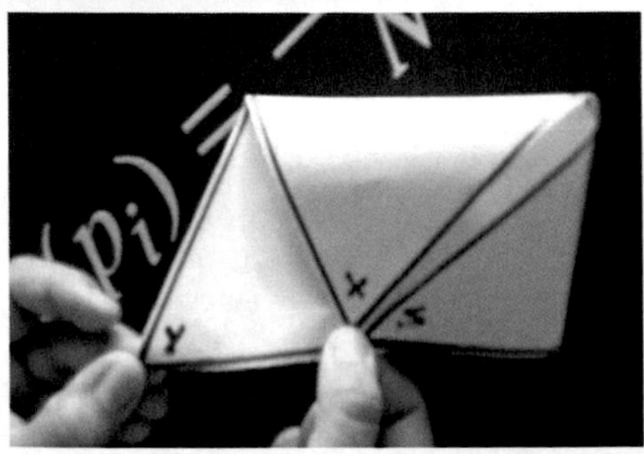

Fig. 3. Angle Z joins the same point on base YZ as Angle X [1].

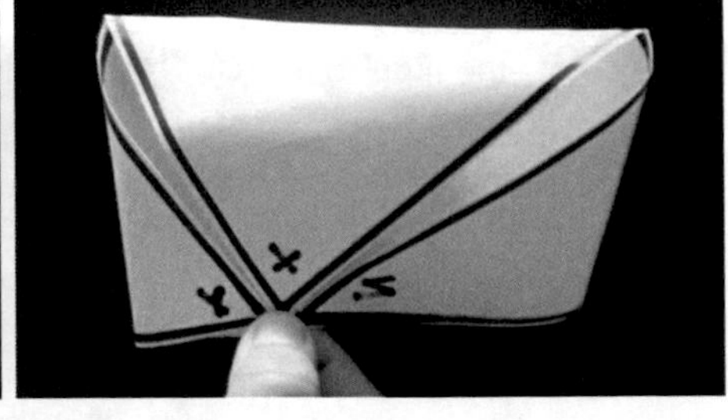

Fig. 4. Angle Y also joins the same point on base YZ as Angle X and Angle Z [1].

3 Simple Methods

Some examples are given below with simple method along with remarks. This will help the students to solve the problems using these simple techniques.

3.1 Example-1

Which number is to be subtracted from 49, 143, 61, and 179 so that the numbers so obtained are in proportion? [3]

General Method Opted by Many Students

(1) Let x be the number which is to be subtracted from 49, 143, 61 and 179.
(2) So (49 – x), (143 – x), (61 – x) and (179 – x) are in proportion.

$$
\begin{aligned}
\therefore\quad \left[\frac{(49-X)}{(143-X)}\right] &= \left[\frac{(61-X)}{(179-X)}\right]\\
\therefore\quad [(49-x)(179-x)] &= [(61-x)(143-x)]\\
\therefore\quad [8771-49x-179x+x^2] &= [8723-61x-143x+x^2]\\
\therefore\quad [8771-228x+x^2] &= [8723-204x+x^2]\\
\therefore\quad [8771-228x] &= [8723-204x]\\
\therefore\quad [8771-8723] &= [228x-204x]\\
\therefore\quad [48] &= [24x]\\
\therefore\quad x &= 48/24\\
\therefore\quad x &= 2
\end{aligned}
$$

(3) Hence the number to be subtracted from 49, 143, 61 and 179 is 2.

Here while doing cross multiplication; we need to expand the brackets. Multiplication of two larger numbers is very difficult. It can be done in the following simple method.

Simple Method

(1) Let x be the number which is to be subtracted from 49, 143, 61 and 179.
(2) So (49 – x), (143 – x), (61 – x) and (179 – x) are in proportion.

$$
\begin{aligned}
\therefore\quad \left[\frac{(49-X)}{(143-X)}\right] &= \left[\frac{(61-X)}{(179-X)}\right] && \rightarrow \text{By Alternendo}\\
\therefore\quad \left[\frac{(49-X)}{(61-X)}\right] &= \left[\frac{(143-X)}{(179-X)}\right] && \rightarrow \text{By Invertendo}\\
\therefore\quad \left[\frac{(61-X)}{(49-X)}\right] &= \left[\frac{(179-X)}{(143-X)}\right] && \rightarrow \text{By Dividendo}\\
\therefore\quad \left[\frac{(61-X-49+X)}{(49-X)}\right] &= \left[\frac{(179-X-143+X)}{(143-X)}\right]\\
\therefore\quad \left[\frac{(61-49)}{(49-X)}\right] &= \left[\frac{(179-143)}{(143-X)}\right]\\
\therefore\quad \left[\frac{(12)}{(49-X)}\right] &= \left[\frac{(36)}{(143-X)}\right] && \rightarrow \text{Dividing by 12}\\
\therefore\quad \left[\frac{(1)}{(49-X)}\right] &= \left[\frac{(3)}{(143-X)}\right]\\
\therefore\quad [(143-x)] &= [3\times(49-x)]\\
\therefore\quad [(143-x)] &= [147-3x]\\
\therefore\quad [(3x-x)] &= [147-143]\\
\therefore\quad 2x &= 4\\
\therefore\quad x &= 2
\end{aligned}
$$

(3) Hence the number to be subtracted from 49, 143, 61 and 179 is 2.

3.2 Example-2

The radius and height of a Cylinder are in the ratio of 4:9 and its volume is 19386.36 cm^3. Find its radius (π = 3.14) [2].

(Note: Use of calculator is prohibited)

Here the Simple method of calculation is used. Please note the concepts/steps given in box Brackets [ ].

Given

(1) The ratio between radius and height is r:h = 4:9. (So, h = 9 r/4)
(2) The volume of the cylinder V = 19386.36 cm^3.
To Find: The radius of the cylinder.

Solution

(1) We know that the volume of the cylinder is $V = \pi r^2 h$

$\therefore V = \pi r^2 (9 r/4)$ as h = 9 r/4.
$\therefore V = (9 \pi r^3)/4$
$\therefore r^3 = 4 V/9 \pi$, put V = 19386.36 and $\pi = 3.14$
$\therefore r^3 = (4 \times 19386.36)/(9 \times 3.14)$
[Multiply numerator & denominator by 100]

(2) So $r^3 = (4 \times 1938636)/(9 \times 314)$
[Here 1938636 is divisible by 9, so divide numerator & denominator by 9 and we get]
(3) So $r^3 = (4 \times 215404)/314$
[As we need cube term in RHS, we take 2 out from 215404. So, 4 and 2 will give us 8, which is perfect cube of 2]
(4) So $r^3 = (4 \times 2 \times 107702)/314$
[Here 107702 is divisible by 2, so divide numerator & denominator by 2 and we get]
(5) So $r^3 = (4 \times 2 \times 53851)/157$
[Here 53851 is not divisible by 2, 3, 4, 5, 6, we will see for 7, so take out 7 from 53851]
(6) So $r^3 = (4 \times 2 \times 7 \times 7693)/157$
[For getting perfect cube, we need to take 2 more 7's from 7693 one after another]
So $r^3 = (4 \times 2 \times 7 \times 7 \times 1099)/157$
(7) So $r^3 = (4 \times 2 \times 7 \times 7 \times 7 \times 157)/157$
[Now an answer is in your hand. Simply divide numerator & denominator by 157 we get]
(8) So $r^3 = 4 \times 2 \times 7 \times 7 \times 7$

$\therefore r^3 = (2 \times 2 \times 2) \times (7 \times 7 \times 7)$
$\therefore r = (2 \times 7)$
$\therefore r = (14)$
Therefore the radius of the cylinder is 14 cm.

3.3 Example-3

In the figure given bellow, PR = 6 units & PQ = 8 units. Semicircles are drawn taking Sides PR, RQ & PQ as diameters. Find the area of the shaded portion RxPsQtPyR ($\pi = 3.14$) (Fig. 5) [2].

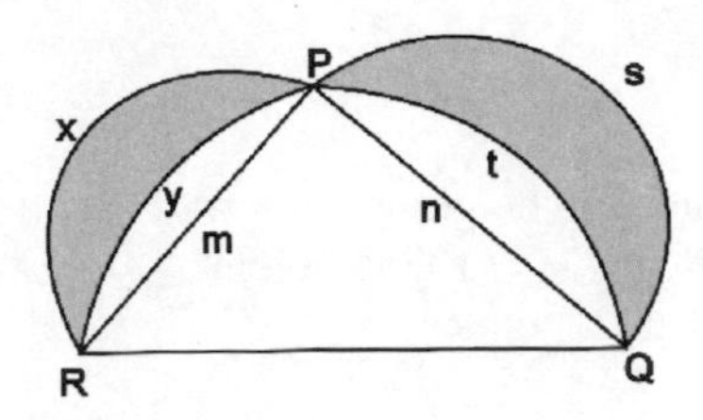

Fig. 5. Shaded portion RxPsQtPyR.

Solution

Given: PR = 6 units, PQ 8 units
To Find: Area of the shaded Portion

(1) In Δ RPQ, < RPQ = 90° [Angle inscribed in Semicircle RyPtQ is right angle]
(2) In Δ RPQ,

$$RQ^2 = PR^2 + PQ^2$$
$$RQ^2 = 6^2 + 8^2$$
$$RQ^2 = 36 + 64$$
$$RQ^2 = 100$$
$$RQ = 10 \text{ Units.}$$

(3) Let radius of semicircle RxPm be r_1, PsQn be r_2 & RyPtQ be r_3
(4) Let Base & Height of Δ RPQ be PR = m = 6 & PQ = n = 8.
(5) Area of the shaded portion RxPsQtPyR = Area of semicircle RxPm + Area of Semicircle PsQn + Area of Δ RPQ – Area of semicircle RyPtQ $= \left[1/2\,\pi\, r_1^2 + 1/2\,\pi\, r_2^2 + 1/2\,m\,n\right] - \left[1/2\,\pi\, r_3^2\right]$
Note: Here many students calculate the values separately and then find their answers. Basic concept is to be applied on these problems to reduce the burden taken by troublesome calculations.
Rearranging the terms, we get,

$$= \left[1/2\,\pi\, r_1^2 + 1/2\,\pi\, r_2^2 - 1/2\,\pi\, r_3^2\right] + [1/2\,m\,n]$$
$$= 1/2\,\pi\left[r_1^2 + r_2^2 - r_3^2\right] + [1/2\,m\,n]$$
$$= 1/2\,\pi\left[3^2 + 4^2 - 5^2\right] + [1/2 \times 6 \times 8]$$
$$= 1/2\,\pi\,[9 + 16 - 25] + [3 \times 8]$$
$$= 1/2\,\pi[0] + [24]$$
$$= [0] + [24]$$
$$= [24]$$

(6) So, area of the shaded portion RxPsQtPyR is 24 square units.

There are many other methods for solving these problems, but above mentioned methods will lessen the burden taken by troublesome calculations. Like this, there are so many new techniques to solve the problems. Here, our main interest is to help the students all over the world to improve their thinking capabilities and use this knowledge for the application of the same.

An interest can be created to learn mathematics in students by teaching them the fun of mathematics. Mathematical topics are more easily and effectively learnt by developing basic ideas of the topic among themselves.

4 Fun with Mathematics

Every student must work on some type of new inventions to develop an interest in their fields. Here few diagrams of magic cube are shown.

In Fig. 6, these types of magic cube can be prepared with the help of magic square and magic sketch [4].

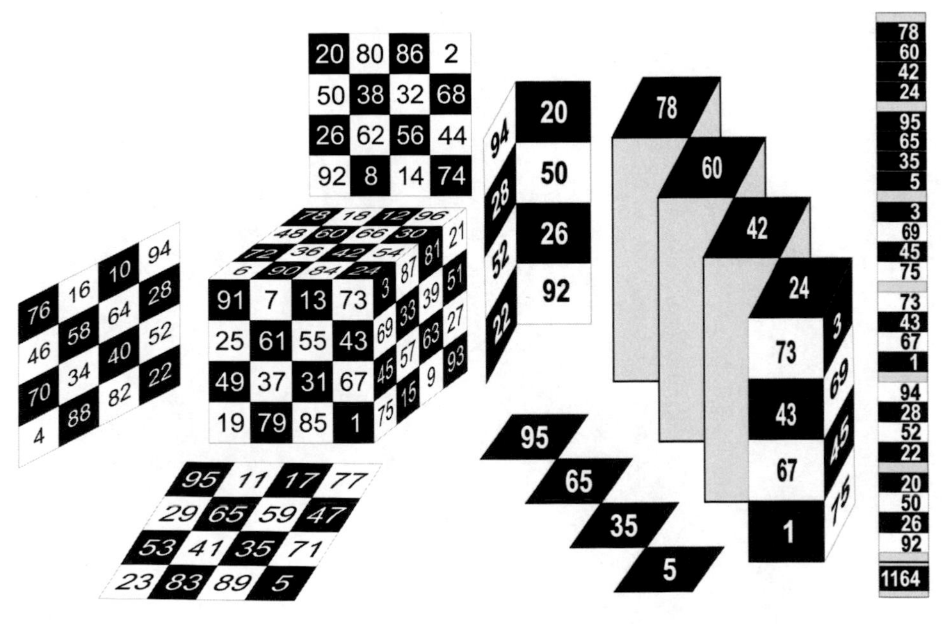

Fig. 6. Magic cube

Fig. 7. 4 × 4-Top-Bottom-Diagonal-Front-Right

In Fig. 7, "4 × 4-Top-Bottom-Diagonal-Front-Right" addition pattern is given.

Here, the motive is to develop different types of interesting concepts across all the students worldwide.

References

1. "Paper folding" concept. http://www.paperfolding.com/math/, https://mathigon.org/origami
2. Geometry text books of 8th, 9th and 10th grade of Maharashtra state board of secondary & Higher Secondary Education, Pune 411004
3. Algebra text books of 8th, 9th and 10th grade of Maharashtra state board of secondary & Higher Secondary Education, Pune 411004
4. http://anil7pute.blogspot.in/2013/01/new-concept-about-magic-cube.html

How to Bring a Graduate Program Closer to Employers' Needs?

A Case Study in Qatar University in the Field of IT

Laurent Veillard[1(✉)], Stéphanie Tralongo[1], Catherine Galli[1], Abdelaziz Bouras[2], and Michel Le Nir[1]

[1] Université Lumière Lyon 2, Lyon, France
{laurent.veillard,stephanie.Tralongo, michel.lenir}@univ-lyon2.fr, catherinegalli@laposte.net
[2] Qatar University, Doha, Qatar
abdelaziz.bouras@qu.edu.qa

Abstract. Since the mid of the 1990s, Qatar is engaged in an economic diversification strategy to move away from an economy strongly based on gas and oil. The national authorities consider that high-quality education, especially at the tertiary level, is absolutely essential to succeed in this strategy. Despite of repeated reforms since the middle of the nineties, there are still some important difficulties in the Qatari higher education system, and especially in Qatar University (QU), the main national university. One of them is related to academic programs not really preparing students for professional integration. The Pro-Skima project aims at finding possible pedagogical modifications of the curriculum of the Master of Science in Computing of QU, to better meet the needs of highly skilled workers in the IT professional sector in Qatar. Part of this project is a sociological study of the main characteristics of the existing curriculum of this master program. Results highlight a curriculum based on an Anglo-Saxon model. In this model, several internal and external accreditation institutions have a strong influence on any evolution of the curriculum. They put different constraints that give little space to discuss more flexibly with local employers in order to change the curriculum. Further developments of the project consist in finding what could be some concrete solutions to overcome these difficulties and bring the curriculum closer to employers' needs. On possibility is to experiment a curriculum that includes more workplace learning periods.

Keywords: Curriculum development · Sciences of computing
Academic-employers partnerships · Workplace learning

1 Introduction

Since the mid of the 1990s, Qatar is involved in an economic diversification strategy. Mainly based on hydrocarbons, the country's economy is now moving to technological innovations activities and high-end services in areas such as cultural tourism, sporting and diplomatic events, etc. [1]. The national authorities consider that high-quality

M. E. Auer et al. (eds.), *Teaching and Learning in a Digital World*,
Advances in Intelligent Systems and Computing 715,
https://doi.org/10.1007/978-3-319-73210-7_22

education, especially at the tertiary level, is absolutely essential to succeed in this strategy, strongly based on a so-called *knowledge society* [2]. The aim is to build a high-quality academic system that can both produce innovations in terms of products and services and locally train the highly skilled workers needed for a rapid economic development. There is also a stake of national sovereignty: better and faster train nationals in order to escape the country's heavy dependence on expatriate experts and foreign companies.

In order to reach this objective, Qatar Authorities have chosen to set up a higher education system based on the Anglo-Saxon academic model, considered to be the best according to international rankings. On the one hand, a new campus, (*Education City*) has been built at the end of nineties and several very renowned foreign universities (mostly US) have been encouraged (including strong funding incitation) to establish local branches. On the other hand, the main national university, Qatar University (QU) has been reformed in the recent years, and is now organised very similarly as Northern American Universities. Furthermore, smaller elite University (Hamad Ben Khalifa University - HBKU) has been launched recently, which also replicates this Anglo-Saxon model. This important transformation of the tertiary education provision was followed by an extremely strong and rapid increase in the number of students: 83% increase in 5 years according to the figures provided by the Ministry of development planning[1] (15352 students in 2010/2011; 28106 in 2014/2015).

Nevertheless, several studies show there are still some important difficulties in this new Qatari higher education system [3–6]:

- an important imbalance between male and female students, the former being almost 3 times more numerous and the latter being much less motivated to pursue their studies in the higher educative system;
- a low high-school level of some Qatari students, in mathematics, sciences and English language in particular, that makes harder their access to academic institutions (especially in elite universities like foreign university branches and HBKU) and/or complicate their academic success; QU is particularly concerned with this difficulty giving the fact that national authorities ask this national institution to accept all native students, even if they lack basic knowledge;
- Additionally, intercultural problems leading to misunderstandings between teachers and students (and sometimes their parents) and even rejection of a pedagogical model based on American *liberal arts*; this problem is especially important when academics come from western universities and teach humanities or social sciences;
- a lack of links between universities and employers to design training courses; a consequence is that a lot of academic programs not really equip students with the necessary knowledge, skills and attitudes required by employers.

The Pro-Skima project focuses more particularly on this last and important problem. Its general aim is to reflect on possible pedagogical modifications within the Master of Science in Computing of Qatar University, to better meet the growing

[1] http://www.mdps.gov.qa/.

needs of highly skilled workers in the IT professional sector in Qatar. This is a 3-year interdisciplinary research project, funded by the Qatar National Research Fund (QNRF), and carried out since April 2015 in partnership between QU and a French University (Lumière Lyon 2). It is based on a collaboration between academics from different disciplines (computer scientists, sociologists of education, researchers in vocational didactic) and from different countries (Qatar, France, Switzerland, Canada).

The project involves several stages: (1) Studying different models of curriculum development at tertiary level (France, Switzerland, Canada); (2) Analyzing the existing curriculum of the master of science in Computing at QU (design process and pedagogical characteristics); (3) Examining and proposing possible adjustments, in collaboration with both academics and some representatives of IT companies, to better prepare students to professional situations. We presented the work carried out during the first stage of the project in a previous publication [7]. This paper is focused on the second step, i.e. the analysis of the existing curriculum and its development process, considered in a sociological way. Our research questions are: how was this curriculum built and how (and when) is it updated? What are the different actors (including both persons and institutions) concerned by the curriculum evolution process? What are the main constraints (rules, criteria, etc.) to respect during this process and who sets them? Considering these constraints, what is possible to change in a reasonable and concrete way to better prepare students for the workplace?

2 Theoretical and Methodological Approach

We carried out our analysis by using concepts from sociology of education. Specifically, we use two main notions. The first is *curriculum*, which refers to everything that is supposed to be taught, learned and assessed in a given order and progression, within a certain cycle of studies [8]. In our case, we are interested in the curriculum of the Master of Science (Msc) in Computing of QU. This curriculum can be considered either in its formal or real versions. The formal curriculum is the planned programme of objectives, content, resources, teaching situations and assessment defined by the educative institution. This version can generally be inferred from the written texts published internally or externally (ex: internet site) by this institution. The real version refers to what is actually taught and practiced in the classrooms or other learning sites (workplace in case of internship periods), which can be sometimes very different from the official version of the curriculum. In our research, we focused our attention on the official curriculum.

We also use the notion of *recontextualization* proposed by Bernstein [9, 10], which characterises the whole process of curriculum production. This process involves different actors, belonging to several institutions (ex: teachers from a specific academic department; staff from the office of academic affairs; experts from external accreditation agencies; etc.). During curriculum creation or change, they interact with each other in different places and at different times, sometimes in a conflict way. Conflicts can arise because actors not defend the same interests and have not the same goals, constraints and problems. For instance, teachers of specific theoretical subjects can be very dissatisfied with the proposition from the head of the department to reduce the number of hours of their disciplines, in order to give a longer duration to an internship period. They can consider these types of theoretical knowledge absolutely essential for students to succeed both at University and in the workplace. Consequently, some can fight against this possibility, by using more or less official ways. Additionally, *recontextualisation* generally leads to use and produce different types of written documents (minutes of meeting, reports, syllabus, study plan, etc.). Draft versions of these documents are especially interesting in a sociological perspective, because they give traces of issues, agreements, or conflicts at different steps of the *recontextualisation* process.

Methodologically, the Pro-Skima project is globally based on a collaborative approach, involving researchers in social sciences (sociology, vocational didactic), researchers and teachers in IT (especially those who teach in the Master degree, or manage this training course), and representatives from companies that have growing recruitment needs in this field of competence (IT) and want to develop a partnership with QU. This interdisciplinary and collaborative approach is necessary considering the aim of the project, which is to evolve an academic curriculum in the field of computer sciences, towards a stronger professionalization perspective and more collaboration between academics and professionals. Such a goal requires not only academics in the technical field of computer sciences, and others in the research field of education, but also workplace practitioners and representatives from companies.

For the sociological analysis, we have constituted a corpus composed of different types of data, documenting our case study on the Msc in Computing: internal documents of QU (ex: study plan of the master course; university specifications for developing or modifying a curriculum, etc.); external ones (ex: ABET accreditation criteria); interviews with several actors involved in the construction or evaluation of the master curriculum; local observations during 3 workshop sessions with the participation of academics (from the Msc in computing), representatives from ministries (IT, education) and IT experts from different national and foreign companies based in Qatar. We had also the objective of conducting interviews with students to study their perception and experience of the curriculum. However the project conditions (several stays in Doha since the beginning of the project, each stay had of short duration) did not allowed us to make all of them for now.

The methodology of the study was based on content analysis of documents, interview transcriptions and field notes. We combined these different sources as much as possible in triangulation process, and through a collective and progressive strategy of hypotheses refinement and consolidation [11, 12]. Our analysis aimed at:

1. Reconstructing the different steps of the curriculum change process.
2. Identifying actors and institutions who played a role during this process and understanding their goal, motivation and professional culture. In order to ensure this, it was sometimes necessary to go beyond our initial corpus, firstly focused on what was going on inside Qatar University. This was the case when we identified the very important role played by a US accreditation agency (ABET) in the recontextualisation process. Then, we started to collect additional information about this external institution: history, strategy, methods, members' profile, etc.

3. Analysing what were constraints, conditions or criteria asked or imposed by these actors or institutions during the *recontextualisation* process, and identifying where some flexibility was possible within the curriculum to propose some changes.

3 Main Results

Our sociological analysis highlights a long and complex *recontextualisation* process, started both for pedagogical and social issues. Our work also shows that the curriculum revision process was strongly constrained, not only by several internal actors and institutions of Qatar University, but also an external one: ABET (Accreditation Board for Engineering Education).

3.1 Why This Curriculum Has Been Updated?

The previous version of the curriculum has been designed in 2009, when the Msc of computing was created. Its initial objective (still valid today) was to train students in applied computer science, either for a thesis (thesis option) or to prepare them for a professional position (project option). According both to the head of the current master and several other teachers, the experience of running this curriculum over several years has led to identify several difficulties.

First, compulsory courses (core courses) were refering to three quite different specializations in computer science: *database systems*, *operating systems*, *architecture and design of computer systems*. Teachers experienced great difficulties with these courses because they were requiring prior knowledge that many students not yet acquired. In the new curriculum, the decision was taken to make these courses optional and to propose 3 new compulsory courses: *algorithm design and modeling*, *applied research methodology*, *seminar in computing*. These courses are now more general and considered by teachers as a common and necessary knowledge base for all subsequent specializations.

Second, the 2009 version was offering a choice of elective courses in two areas: *information science* and *network systems*. For each area, students had to choose 3 amongst 6 offered courses. There were few students who chose courses in the network systems area. By questioning students, teachers realized that they were misrepresenting the content of this area, seeing it as less prestigious and interesting. Nevertheless, this field corresponds to important research and company needs. Depending on the year, some courses in this area could not be opened, or taught with only a very small number of students. Thus, it became more and more difficult to keep them with regard to the university. To solve this problem, academics from the Computer Science and Engineering (CSE) department proposed broader and more comprehensible labels, respectively *computer science* and *computer engineering*. According to the words of head of the master program, the first one corresponds to a more 'software' specialization (*database systems*, *software engineering*, *machine learning*, etc.) and the second has a more 'hardware' orientation (i.e. IT developments for specific electronic devices like networks, robotics, processors, etc.). We can see here a desire to propose more intelligible terms for students, closer to their categories of representations.

These broader titles have also allowed introducing new courses, closer to students' interests, like advanced robotics. This new course was added in the *computer engineering* area. A lot of students are very interested with this topic. Since they have to choose 2 additional elective courses, this is a good way to fill other less popular courses (for example in *network systems*). These new courses also come from requests of teachers of the department who complained that they did not participate in the master's program. These teachers only taught at the Bachelor level, which was less prestigious and did not allow recruiting future PhD students.

At this stage of our analysis, it appears that the main reasons for curriculum revision are not related to skills or knowledge expressed by IT employers, but mainly related to pedagogical and social issues within the department.

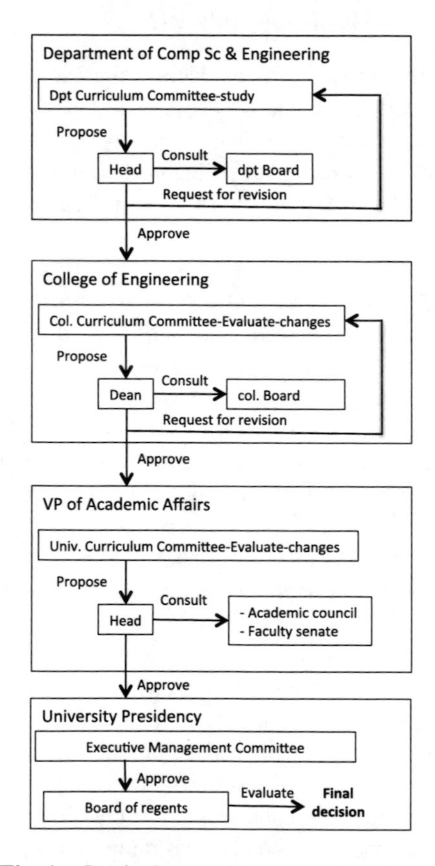

Fig. 1. Curriculum revision procedure in QU

3.2 Numerous Internal Constraints

Who are actors and institutions concerned by the *recontextualisation* process? In Qatar University, any curriculum creation or modification of a curriculum must follow a precise and formalized circuit. Thus, the process of revising the curriculum of the Msc of Computing involved bodies located at 3 levels of the university organisation: CSE department; college of engineering (including IT department); Vice-Presidency for academic affairs (which tackles the various colleges). This process was initiated by a *department curriculum committee-study*, composed of several academics. Before making a written proposal, they consulted their colleagues and students through various means (meetings, surveys, etc.). Afterwards, the process followed the detailed circuit below (see Fig. 1). This figure highlights numerous instances and actors who have given their opinion or validation. The head of the master program, who managed this review, explained us that it was a long and complex process (3 years). There were several back and forth between these bodies before a final agreement could be found. Why is such a process so long and involves so many different actors?

A first explanation lies in the interdependencies between the different training programs of QU. First, curricula must respect a common general structure, taken over from the Anglo-Saxon universities. Every academic year must be divided into two semesters (Fall and Spring), with a total of 16 teaching weeks. Courses must be accounted for in Credit Hours (corresponding roughly to a number of hours per week during a semester), with a minimum number of CHs to validate a course. For masters, it is a minimum of 30 CHs. And for the Msc in computing in particular, it is 31 CHs. As already mentioned, there must be both compulsory (required) courses and optional (elective) courses. In the case of the Msc of computing, some of these optional courses are not specific to this training: students who choose them can be grouped with students from another master of the college of Engineering. Thus *applied research methodology* is a same course for students from two master programs of the college of engineering (Msc in *computing* and in *electrical engineering*), because teachers considered it as very transversal. At the undergraduate level, some courses are even offered to students of all the bachelors of the university. When such pooling is possible, it is set up to reduce training costs.

Here we can see the complexity of this curriculum revision process, for administrative (availability of teachers, students' interest and demand for a given course), pedagogical (avoiding students with too heterogeneous levels within a course) and financial (limiting the number of courses open among all possible ones) reasons. Any modification of a curriculum can have consequences on several others. It is therefore not surprising that the university seeks to secure the approval of several actors and bodies at different levels of the organization before agreeing to a modification of a curriculum.

But we can also offer a second explanation, which should nevertheless be further substantiated by a more thorough investigation: historical legacy of a relatively bureaucratic and centralized organizational culture in this University [13]. According to several teachers we interviewed, this culture seems to be still in place today. As a consequence, colleges and departments have little leeway to make faster and more specific educational choices.

3.3 A Strong External Constraint: ABET Accreditation

There were not only internal, but also external constraints on the *recontextualisation* process. CSE department decided to ask for an international accreditation for the Msc of computing. This choice is part of the general strategy of QU, which is to ask as much as possible this type of accreditation for all training courses (especially bachelor, but for some years also master programs). QU selected agencies that are most recognized in each field of study, mostly Northern American (US, Canada). QU legitimizes this choice of external accreditation by the aim of reaching international quality standards. Another important explanation factor is that other universities within the Gulf Area (Saudi Arabia, United Arab Emirates, etc.) have made a similar choice. It is important for QU to gain similar accreditations to compete in this regional academic market. As a sign of the institutional importance of this type of accreditation, we can find (in May 2017) 'proud to be accredited'on the first page of the website of the college of engineering.

ABET (Accreditation Board for Engineering and Technology) is the main international accreditation agency for degrees in the engineering field. Bachelor degrees of QU engineering college are accredited by this agency for several years. The CSE department decided to ask for an accreditation for the Msc in Computing in 2015 (and got it one year later) to gain international credibility at this level. Gaining this accreditation required both writing a very big and formatted report with multiple topics (ex: background information, program description, students selection process, curriculum description, etc.), and then be visited by ABET international experts (in this case, two academics from 2 different Northern American universities). In particular, it was necessary to demonstrate how the curriculum could respect several criteria fixed by ABET. One of them is the necessity for the master program to generate 11 learning outcomes. For instance, each graduate student has to develop: "an ability to design, implement, and evaluate a computer-based system, process, component, or program to meet desired needs."; or" an ability to analyse the local and global impact of computing on individuals, organizations, and society."

In addition to these outcomes, ABET also asks for necessary inputs, i.e. specific teaching contents which have to be implemented within the curriculum. For instance, "Students must have the following amounts of course work or equivalent educational experience: a. Computer science: One and one-third years that must include: (1) Coverage of the fundamentals of algorithms, data structures, software design, concepts of programming languages and computer organization and architecture; (2) An exposure to a variety of programming languages and systems, [...]. A detailed analysis of all the norms fixed by ABET would except the scope of this paper. But we can conclude that gaining the accreditation requires a strong normalisation process, imposed by an external institution to QU. In particular, the eleven learning outcomes define a generic profile of what must be a graduate student in computer science and computer engineering, whatever higher education institution and its cultural, political and socio-economic environment.

4 Conclusion: What Is Possible to Change?

Our sociological analysis shows that different internal and external institutions have a very important influence on any evolution of the curriculum. They put numerous and strong constraints that give little space to discuss more flexibly with local employers and to make some rapid curricular changes. In particular, respecting ABET criteria requires academics to align the learning outcomes of the curriculum onto a generic computer science expert profile. Of course, the ABET report requires to make some connections with employers (surveys, joint committees, etc.). But this is not to build the curriculum in collaboration with them: only to ask if they are satisfied with the existing curriculum or if they are interested with some proposed changes. These diffcrent constraints or limitations constitute a strong brake to collaborate more closely with experts and representatives from professional fields, especially when they are in very rapid evolution, like the IT sector. For instance, cyber security issues are stronger and stronger in Qatar. It would be important to be able to develop quickly a new option of the Msc of computing, specialised in this vocational field. Learning to be an expert in cyber-security requires to be closely and strongly connected to rapid evolutions in the real world.

As said previously, the Pro-Skima project consists in finding concrete solutions to make the learning outcomes of the curriculum of the Master of Sciences in computing closer to the local employers' needs. An important difficulty is the big difference between this Anglo-Saxon type of curriculum design process, strongly based on local procedures and international accreditations, and some alternative models presented in the project. For instance, in some European countries such as France, designing or updating a curriculum at Bachelor or Master level is a much more local process, which can associate closely representatives from local companies, and is not constrained by external criteria like learning outcomes or required courses in specific subjects. In Switzerland, curriculum design in higher vocational colleges or universities of applied sciences is more centralised, but very based on a DACUM approach [14], which gives a great place to national vocational experts in the curriculum design process.

When we presented these alternative models to the faculty of the Master of Science in computing (workshop sessions), they were interested but very sceptical concerning the possibility to introduce short term noticeable changes in their curriculum, given the strong constraints coming from both internal and external institutions. Discussions are in progress with both academics of QU and representatives from several IT companies to find and experiment some possible changes. A possibility is to create an experimental group, with some few (10–15) volunteer students and interested companies, to test a new organisation of the curriculum, without making some changes at the level of the teaching courses. Students who will participate to this experimental group will have to manage a project for companies. In the current version of the master program (project option), students already have to manage a project, but this latter is rather academic, with specifications coming from teachers. In the experimental group, students' projects will start from real needs of companies. Students will work on their project during the day, and follow existing teaching courses of the master program during the evening (17 h–19 h). This organisation is possible without modifying the teaching part of the master program (very important to avoid new validations from the different university committees and from ABET), because courses are already done during the evening, to allow participations of salaries from companies. Students will stay 3 days a week in the workplace, managed by a mentor (IT expert from the company), and will spend 2 days at University, helped by teacher who will play the role of a supervisor for the project. This double supervision should contribute to bring CSE department closer to local companies.

Acknowledgement. This publication was made possible by NPRP grand # NPRP 7-1883-5-289 from the Qatar National Research Fund (a member of Qatar Foundation). The statements made herein are solely the responsibility of the authors.

References

1. General Secretariat for Development Planning. Qatar National Vision 2030, Doha, Qatar (2008)
2. Berrebi, C., Martorell, F., Tanner, J.C.: Qatar's labor markets at a crucial crossroad. Middle East J. **63**(3), 421–422 (2009)
3. Rostron, M.: Liberal arts education in Qatar: intercultural perspectives. Intercult. Educ. **20**(3), 219–229 (2009)
4. Said, Z.: Science education reform in Qatar: progress and challenges. Eurasia J. Math. Sci. Technol. Educ. **12**(10), 2253–2265 (2016)
5. Ellili-Cherif, M., Alkhateeb, M.: College students' attitude toward the medium of instruction: Arabic versus English dilemma. Univ. J. Educ. Res. **3**(3), 207–213 (2015)
6. Lemke-Westcott, T., Johnson, B.: When culture and learning styles matter: a Canadian university with Middle-Eastern students. J. Res. Int. Educ. **12**(1), 66–84 (2013)
7. Veillard, L., Tralongo, S., Bouras, A., Le Nir, M., Galli, C.: Designing a competency framework for graduate levels in computing sciences: the Middle-East context. In: Auer, M., Guralnick, D., Uhomoibhi, J. (eds.) Interactive Collaborative Learning, ICL 2016. Advances in Intelligent Systems and Computing, vol. 544. Springer, Cham (2017)
8. Forquin, J.-C.: Sociologie du Curriculum. PUR, Rennes (2008)
9. Bernstein, B.: Pedagogy, Symbolic Control and Identity. Theory, Research, Critique. Taylor & Francis, London (1996)
10. Stavrou, S.: Negotiating curriculum change in the French university. The case of regionalising social scientific knowledge. Int. Stud. Sociol. Educ. **19**(1), 19–36 (2009)
11. Patton, M.: Enhancing the quality and credibility of qualitative analysis. HSR Health Serv. Res. **34**(5), 1189–1208 (1999)
12. Olivier de Sardan, J.P.: La politique du terrain. Enquête **1**, 71–109 (1995)
13. Moini, J.S., Bikson, T.K., Richard Neu, C., DeSisto, L., Al Hamadi, M., Jabor Al Thani, S.: The Reform of Qatar University, Doha (2005)
14. Norton, R.E.: DACUM Handbook, 2nd edn. The Ohio State University, Columbus (1997)

A Constructivist Approach to the use of Case Studies in teaching Engineering Ethics

Diana Adela Martin(✉), Edward Conlon, and Brian Bowe

School of Multidisciplinary Technologies,
College of Engineering and Built Environment,
Dublin Institute of Technology, Dublin, Ireland
{dianaadela.martin,edward.conlon,brian.bowe}@dit.ie

Abstract. Our paper aims to explore the effectiveness of a constructivist approach to the teaching of engineering ethics through case studies, by putting forward a contextualization of the much discussed case study "Cutting Road Side Trees" [12] in light of the constructivist frame suggested by Jonassen [8]. First, we briefly analyse how the use of case studies for the teaching of engineering ethics eludes the complexity of the engineering professional environment before arguing that constructivism is a learning theory that can help to address this complexity. The final section proposes a constructivist reworking of the case method in a manner that aims to correct the deficiencies identified, followed by a discussion of the results of applying the contextualized exercise to First Year group of engineering students. The key findings reveal that the contextualized scenario enhances, in some respects, students' understanding of the social dimension of the engineering profession.

Keywords: Engineering education · Engineering ethics
Social dimension of engineering · Case studies · Role-playing · Constructivism

1 Case Studies in the Teaching of Engineering Ethics

Case studies presenting moral dilemmas faced by engineers have been a dominant teaching method in engineering ethics, and as Colby and Sullivan [3] highlight, discussion of cases is still the most prevalent means of teaching ethics in engineering colleges in US. This approach has recently attracted criticism pointing to its weakness in capturing the dynamics and realities of the work place [2, 4, 11]. The method appears to elude the metaphysical characteristics of the engineering profession, related to the nature of the artefacts produced [7, 14], engineering practice [1, 13] and the professional environment [5, 6]. Colby and Sullivan ([3], p. 330) further note that "few schools had instituted systematic programs to educate for this broad sense of professional responsibility […] and engineering ethics is not usually taught with this kind of scope," with engineering programmes lacking the integration of "technique with the social meaning and broader ethical context of engineering practice." It is thus crucial to enquire what learning theory could support such a broad pedagogical approach. In light of this line of criticism, the teaching of engineering ethics needs to make students aware that:

M. E. Auer et al. (eds.), *Teaching and Learning in a Digital World*,
Advances in Intelligent Systems and Computing 715,
https://doi.org/10.1007/978-3-319-73210-7_23

(i) the artefacts created incorporate social and political values
(ii) the decision and design process of creating an artefact is also a social process
(iii) even if identifying the moral thing to do is a necessary first step for being a socially responsible engineer, acting upon it depends on wider structural factors.

Constructivism appears to be a learning theory that responds to the need of broadening the teaching of engineering ethics beyond the individualistic outlook, in a way that reflects the context in which engineering is practiced and the metaphysical complexity of the profession. In what follows we explore how social constructivism addresses the first two metaphysical characteristics of engineering practice, seen from an ethical lens.

As the Challenger case shows, the meaning of the values an engineer operates with is fluid, subject to modifications brought by different factors related to one's past experience, organizational culture and structure, or the personal characteristics of individuals. Constructivism is focused on meaning making. Knowledge is "constructed, negotiated, propelled by a project and perpetuated for as long as it enables its creators to organize their reality in a viable fashion" ([9], p. 8). Knowledge and learning are thus active social processes resulting from the interaction of different subjectivities.

A constructivist teaching frame encourages students to see how their views about the meaning of engineering values change as an outcome of their interaction. Considering the *Cutting Roadside Trees* exercise, the solution proposed and the method to reach it reflect a particular understanding of what counts as an acceptable compromise or what students understand by sustainability. Working in groups, as students enrolled in the *Professional Practice* course do, "requires the learner to produce an output by acting on the world in some way […] It demands more than discussion, argument, question and answer: it demands also group consensus on producing an output" ([10], p. 57). This is a first step in familiarizing students with the way in which a collective represents a structure that can affect the outcome, and also encourages them to devise a solution given the constraints of their own micro-professional structure.

Jonassen ([8], p. 220) notes that the physical, organizational, and sociocultural context in which the problem is set should always be included. According to Jonassen [8], rich contexts for group projects or role playing exercises can raise students' awareness of the many factors that contribute and influence the output of their task. This helps in drawing attention to the way in which the different subjectivities of the agents encountered in the workplace will affect the decision making process and the creation of engineering artefacts as future professionals.

Thus, by situating learning within a social context and considering engineering decision making and artefact creation as a collective endeavour, constructivism manages to reflect *some* of the major metaphysical characteristics of the professional environment of the engineering profession.

2 Cutting Roadside Trees: A Constructivist Contextualization

At Dublin Institute of Technology, we have put in practice for the course *Professional Practice* a contextualization of the much discussed case study "Cutting Road Side Trees" [12] in light of the constructivist frame suggested by Jonassen [8]. Thus, to the scenario designed by Pritchard [12], we have added a contextual description for the three main characters of the case study, which highlighted their professional experience and status within the organization and community, their values and feared outcome (see Box 1), followed by a set of questions related to the scenario (see Box 2).

Box 1. The contextualized "Cutting Road Side Trees" case study

"Kevin Clearing is the engineer for the Verdant County Road Commission (VCRC). VCRC has primary responsibility for maintaining the safety of county roads. Verdant County's population has increased by 30% in the past 10 years. This has resulted in increased traffic flow on many secondary roads in the area. Forest Drive, still a two lane road, has more than doubled its traffic flow during this period. It is now one of the main arteries leading into Verdant City, an industrial and commercial center of more than 60,000 people.

For each of the past 7 years at least 10 persons have suffered a fatal automobile accident by crashing into trees closely aligned along a 3 mile stretch of Forest Drive. Many other accidents have also occurred, causing serious injuries, wrecked cars, and damaged trees. Some of the trees are quite close to the pavement. Last year two law suits have been filed against the road commission for not maintaining sufficient road safety along this 3 three mile stretch. Both were dismissed because the drivers were going well in excess of the 45 mph speed limit.

Members of VCRC have been pressing Kevin Clearing to come up with a solution to the traffic problem on Forest Drive. They are concerned about safety, as well as law suits that may someday go against VCRC. Clearing now has a plan – widen the road. Unfortunately, this will require cutting down about 30 healthy, longstanding trees along the road.

Clearing's plan is accepted by VCRC and announced to the public. Immediately a citizen environmental group forms and registers a protest. Tom Richards, spokesperson for the group, complains, "These accidents are the fault of careless drivers. Cutting down trees to protect drivers from their own carelessness symbolizes the destruction of our natural environment for the sake of human 'progress.' It's time to turn things around. Sue the drivers if they don't drive sensibly. Let's preserve the natural beauty and ecological integrity around us while we can."

Many letters on both sides of the issue appear in the Verdant Press, the issue is heatedly discussed on local TV, and Tom Richards presents VCRC with a petition to save the trees signed by 150 local citizens." [12]

Correspondingly, the three character description we added are:

-The young engineer

You are Kevin Clearing, an engineer who graduated 6 years ago and is now working as an engineer at the Verdant County Road Commission. You enjoy your work and hope to get in the next 6 months a promotion as engineering manager, knowing that your professional trajectory within VCRC depends on the board and how satisfied they are by your decisions. In your work you value practical solutions and you take pride in considering sustainability in your decisions.

-The top manager

You have worked for VCRC for more than 30 years. Back in the days Verdant was a small community with light traffic, which saw one lethal car accident every few years. The current traffic brings new challenges for your line of work and you fear that losing a lawsuit would have disastrous consequences for the public image of VCRC and for an already strained budget. You value decisions that protect the image of VCRC and are cost effective.

-The influential environmentalist

You are Tom Richards, and have been living in the Verdant County since you were born, 60 years ago. You appreciate its natural scenery, and consider that the forest lining up besides each side of the road, home to so many wild species, is an invaluable part of the city Verdant. The recent urban developments have already led to some of the county's green areas get torn down to make room for industrial buildings, and now you fear that the new deforestation plans of VCRC will continue such a trend. The group you represent aims to protect the natural habitat that makes Verdant County unique. You value nature and want your grand-children to enjoy the same landscape and quality of air that you have benefitted from.

Box 2. Assignment questions

CONTEXTUALISED CASE STUDY (3 STUDENT DIVISIONS)

(Q1) What is the main problem you (Kevin Clearing/ VCRC/TR) have to solve?
(Q2) What do you (Kevin Clearing/VCRC/TR) consider to be the best solution?
(Q3) What do you (Kevin Clearing/VCRC/TR) think are the main barriers for achieving this solution?

After answering these questions by yourself, discuss the scenario with your group by adopting the stance of the character you represent and agree on a solution or line of action (20 min). Then answer the following questions (15 min):

(Q4) What solution was reached following the discussion?
(Q5) What criteria or values have been considered to reach this solution?

(Q6) Was the solution agreed by all or did it result from one person imposing their views on others? Why do you think this happened?
(Q7) Do you personally agree with the solution reached? Why/Why not?

ORIGINAL CASE STUDY (CONTROL GROUP: 1 STUDENT DIVISION)

1. Discuss how Kevin Clearing should proceed at this point. Think about the following questions for 5 min by yourself:
   What is the main problem that Kevin has to solve?
   What is the best solution?
2. Discuss the two questions in your groups and arrive at agreement on an answer (15 min)
3. Pick one member of your group to report back.

The sample group to which this exercise was applied consisted of 112 first year students enrolled in four divisions of the *General Engineering* programme, in the course *Professional Practice*, during the spring semester of the academic year 2016–17. The control group consisted of one division of 23 students, split into groups of 4–5, who received the original case study without the contextualisation. The other three divisions totalling 89 students were split into groups of 3–5 and given different roles, of a young engineer, a top manager or an influential environmentalist, as described in Box 1. The exercise required students to discuss and propose a solution to the dilemma informed by the contextual information provided. They had to answer in writing a set of questions, as seen in Box 2, which invited participants to reflect on what they consider to be the problem presented by the scenario given their character description, a solution and possible barriers to implement it, then following a joint discussion to name the solution agreed, the values behind it and how it was reached. The answers were collected and represent the data on which this paper draws. The key findings reveal that the contextualized scenario enhances students' understanding of the *social dimension of the engineering profession*.

As such, in the pre-discussion stage, when the control group with no role assigned and the three divisions of students were asked about their preferred solution given the scenario, there was no marked difference in the solution proposed between the control group and the top manager and young engineer typologies, while there was a difference between them and those who took on the role of the influential environmentalist.

As can be seen from Table 1, the role adopted by students informed their preferred approach. The students assigned an environmental role were more focused on solutions for changing drivers' behaviour through constructing speed barriers or implementing sanctions, that would thus avoid cutting trees. The assignment of differential roles thus generated a greater variety of solutions based on the students' perceptions of what values might inform the approach of those occupying the roles they were assigned. This created then the need to engage with each other, as bearers of different values, to arrive at a solution on which all could agree.

Table 1. Student solution to the "Cutting roadside trees" scenario

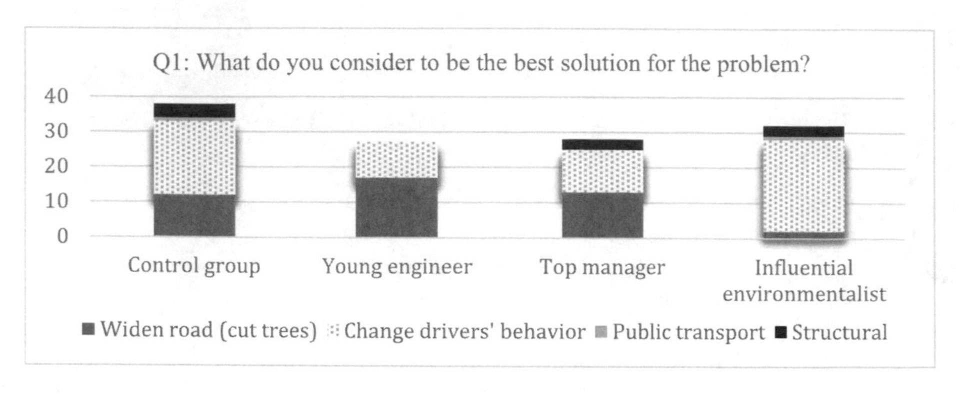

The final solution was reached after discussions in small groups comprised of the three different roles. As such, as seen in Table 2, given the number of students who had indicated the need to cut the trees in the pre-discussion stage, there was a diminished number of proposals to cut the trees. This suggest that the process of discussion and persuasion led to changes in some students perspectives. It is notable that the groups with assigned roles developed a greater variety of solutions than the control groups. Thus the assignment of roles has the potential to increase awareness that engineering design and decision making is also a social process, and that the characteristics of the different actors involved in the stages of design and decision making are ultimately embedded in engineering artefacts.

Table 2. Students' solution for the "Cutting roadside trees" scenario

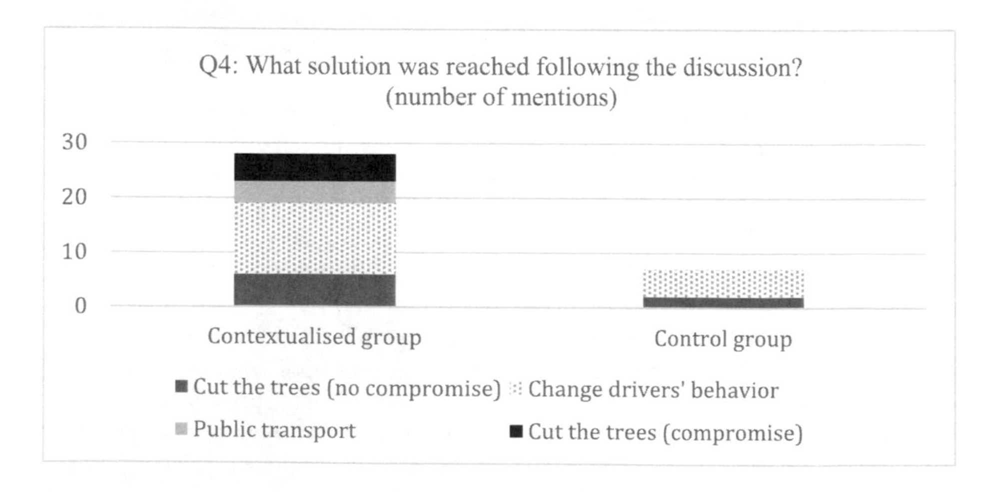

Further evidence for this is derived from a question about what the students considered to be the barrier to achieving their desired solution in the pre-discussion stage. Across all three roles the most mentioned barrier is one of the other roles. So, of those playing the role of top manager, 65% said environmentalists were the main barrier. Of those playing the role the young engineer 67% identified the environmentalist as the main barrier. The environmentalists were less likely to name the other two parties and more prone to identify several different actors (such as the public, drivers or big companies). While students assigned top manager roles were more likely to identify resource barriers. We can also note that there is a lower concern with structural factors, such as "population increase" or a "decision making style oriented towards quick solutions," and more on how different actors can affect the solution to the problem.

One way in which different actors are seen to contribute and affect an engineering solution is suggested by the answer to the question about the values and criteria which contributed to the final decision that followed each discussion. Table 3 shows a wide spread of values, with numerous mentions each, revealing that there was no single value or perspective imprinted on the solution. The engineering solution and the proposed artefact that resulted from the discussion is the outcome of the different values brought in by different actors. The results show that the role playing scenario conveys to students awareness of the *social values embedded in engineering artefacts and decisions*.

Table 3. Values and criteria influencing the solution to the scenario

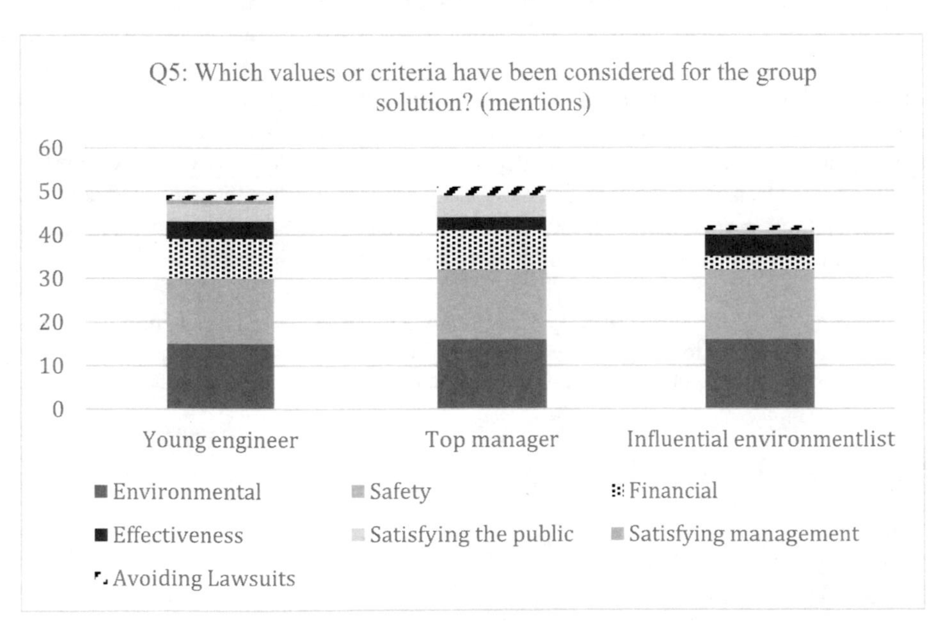

According to the data collected, the answers of the students who were presented a contextualized scenario were diverse and complex, reflecting an awareness of how the interplay between agents with different goals and values influences the engineering process. By receiving a description of different actors involved in the design and decision-making process of the engineering profession, students can see that the practice of engineering contains a strong social component, namely that *engineering artefacts contain social values* and that *the process of design and decision-making are social processes*. This leads us to suggest that to better capture the metaphysical characteristic pertaining to the social dimension of the engineering profession, case studies can benefit from more contextual information that details the actors' characteristics.

A shortcoming of the exercise is that while students gained insight into engineering practice as a social process (aims 1–2 in Sect. 1) there was less of a focus on the social structural dimension of engineering practice (aim 3). This may be to expect too much from a "small case" [11] used with first year students who may lack wider knowledge about the social and organizational context in which they may work in the future. "Larger cases" may be needed to explore the wider constraints on engineering practice and the possibilities for addressing them [11].

3 Conclusion

The pedagogical exercise of contextualising the case study "Cutting Roadside Trees" is informed by a macroethical outlook driven by the ideal of enabling engineers to change the economic and social context in which they work as to promote the development of sustainable and safe solutions. A prerequisite for achieving this is to increase students' awareness about the constraining or enabling factors present in the workplace, the inherent imbalance of power and institutional dynamic, and how different subjectivities interact and shape the decision making process in the workplace. We believe that the pedagogical exercise proposed manages to convey *some* of the metaphysical characteristics of the engineering profession, those related to its social dimension, and to increase student awareness of the complexity of engineering practice.

Acknowledgment. The authors want to thank Kevin Kelly and the DIT - School of Multidisciplinary Technologies, for the support in the implementation of the contextualized "Cutting Roadside Trees" exercise and facilitating the presentation of key findings at ICL 2017.

References

1. Beder, S.: Beyond technicalities: expanding engineering thinking. J. Prof. Issues Eng. Educ. Pract. **125**(1), 12–18 (1999)
2. Bucciarelli, L.: Ethics and Engineering Education (2007). http://dspace.mit.edu/bitstream/handle/1721.1/40284/ethics_20_talk.pdf?sequence. Accessed 13 May 2017
3. Colby, A., Sullivan, W.M.: Ethics teaching in undergraduate engineering education. J. Eng. Educ. **97**(3), 327–338 (2008)

4. Conlon, E., Zandvoort, H.: Broadening ethics teaching in engineering: beyond the individualistic approach. Sci. Eng. Ethics **19**(4), 1589–1594 (2010)
5. Conlon, E.: Marco, micro, structure, agency: analysing approaches to engineering ethics. In: SEFI Annual Conference, Lisbon, Portugal, 27–30 September 2011
6. Davis, M.: Thinking like an engineer: the place of a code of ethics in the practice of a profession. Philos. Public Aff. **20**(2), 150–167 (1991)
7. Feenberg, A.: Questioning Technology. Routledge, London (1999)
8. Jonassen, D.H.: Designing constructivist learning environments. In: Reigeluth, C.M. (ed.) Instructional-Design Theories and Models, vol. II, pp. 215–239. Lawrence Erlbaum Associates, New Jersey (1999)
9. Larochelle, M., Bednarz, N., Garrison, J. (eds.): Constructivism and Education. Cambridge University Press, Cambridge (1998)
10. Laurillard, D.: Teaching as a Design Science. Routledge, New York (2012)
11. Lynch, W., Kline, R.: Engineering practice and engineering ethics. Sci. Technol. Hum. Values **25**(2), 195–225 (2000)
12. Pritchard, M.: Cutting roadside trees. In: Teaching Engineering Ethics: A Case Study Approach. Center for the Study of Ethics in Society National Science Foundation, University of Michigan, Michigan (1992)
13. Vaughan, D.: The Challenger Launch Decision. University of Chicago Press, Chicago (1996)
14. Winner, L.: Do artifacts have politics? In: Winner, L. (ed.) The Whale and the Reactor: A Search for Limits in an Age of High Technology, pp. 19–39. University of Chicago Press, Chicago (1986)

Measuring the Increase in Students' Comprehension in a Flipped Introductory Calculus Course

Rahmad Dawood[1,2(✉)], Mohd. Syaryadhi[1], Muhammad Irhamsyah[1], and Roslidar[1]

[1] Teknik Elektro dan Komputer, Universitas Syiah Kuala, Banda Aceh, Aceh 23111, Indonesia
{rahmad.dawood,syaryadhi,irham.ee,roslidar}@unsyiah.ac.id

[2] Telematics Research Center, Universitas Syiah Kuala, Banda Aceh, Aceh 23111, Indonesia

Abstract. Introductory Calculus is one of the foundational courses in any engineering curriculum. Many innovations have been introduced in the teaching of this course to enhanced students' understanding of its material. One new innovation that is increasingly being adopted is the Flipped Classroom method. This paper reports on an experiment that measures the increase in comprehension of students who took an Introductory Calculus course taught using the Flipped Classroom method that was specially designed for a developing country setting. Results from the experiment shows that students who were taught using this Flipped Classroom method had better comprehension than students who were taught using the traditional classroom method. Specifically, students who were taught using the Flipped Classroom method had, on average, 17.45 points higher exam scores.

Keywords: STEM education · Calculus · Flipped Classroom · Developing country

1 Introduction

A good command of calculus is required for almost all advanced engineering subjects. Due to this requirement, Introductory Calculus is one of the foundational courses in any engineering curriculum and numerous innovations have been introduced in the teaching of this course, such as: using software tools to help students understand complex topics, employing real world examples to ground students on how a particular topic applies in their fields, and adopting new pedagogy methods to help students learn better.

One new pedagogy method that is increasingly being adopted in the teaching of Introductory Calculus is the Flipped Classroom method [1, 2, 5]. In the traditional classroom method, a student will learn a new topic in the classroom under the instruction of a teacher and then given a problem set to be solved at home to increase their understanding of the taught topic. Whereas in the Flipped Classroom method, the traditional method is flipped, a student will learn a new topic in his or her home through a set of provided materials and then she or he will be given a problem set to solve in the classroom together with the teacher. Provided materials tend to be lecture videos, online examples, and reading assignments.

M. E. Auer et al. (eds.), *Teaching and Learning in a Digital World*,
Advances in Intelligent Systems and Computing 715,
https://doi.org/10.1007/978-3-319-73210-7_24

Several Introductory Calculus courses have adopted the Flipped Classroom method and reported positive results in increasing students' comprehension of the taught materials [6–8, 10, 11]. These courses were all carried out in developed countries, however, and direct adoption of the Flipped Classroom method from these courses in a developing country setting, in our opinion, will not be feasible. This will be primarily because all of these courses rely heavily on online videos as its main material for students to review in their home, which will require access to high speed Internet service. In developing countries, access to high speed Internet service is still not widely available, cost of service is relatively expensive, and the quality of service is still unreliable, thus making direct adoption of the Flipped Classroom method from these courses not feasible.

This paper reports on a study to adopt the Flipped Classroom method for a developing country setting and measures whether it will produce the same positive results as in previous Flipped Classroom courses in developed countries. In addition, this study further advances findings from these case studies by quantitatively measuring the amount of improvement in students' comprehension who are taught using this Flipped Classroom method.

2 Method

A 1×2 natural experiment was conducted on two classes of an Introductory Calculus course. First year electrical engineering undergraduate students in *Universitas Syiah Kuala* took these classes. The experiment had two treatments randomly assigned to one of these two classes. In the first treatment, the class was taught using the Flipped Classroom method (we termed 'Flipped Class'). In the second treatment, the class was taught using the typical classroom method (we termed 'Control Class'). This study did not randomly assign students to one of these two classes but instead randomly assigned the treatment to the classes.

Both classes were taught using the same syllabus, slide deck, reading materials, examples, problem sets, exam questions, and on the same time in the same day as well as with the same length of time for the face-to-face session (around 150 min). Each topic in the syllabus was taught during the same week for both the Control Class and the Flipped Class. Even though each class had a different lecturer, each lecturer was experienced in teaching the course, with at least five years of Introductory Calculus teaching experience. Moreover, these lecturers would hold a weekly meeting to synchronize their teaching for the week.

A typical session in the Flipped Class is described in the "A Flipped Calculus Course for Developing Country" section of this paper. Whereas in the Control Class, each session starts by students handing in their completed problem set that was assigned from the previous session. The lecturer would then start teaching by explaining that session topic using a slide deck together with working out several examples in the whiteboard. During each session, students can freely ask questions to the lecturer regarding the presented topic and examples. The session ends with the lecturer handing out a problem set relevant to the just taught topic and a reading assignment to prepare students for next week's session. The problem set consisted of ten questions and will be collected in the next session. The slide deck, examples, problem sets, and solutions were developed together by both lecturers of the Control Class and the Flipped Class.

To properly understand Calculus, students are required to have a basic understanding of Algebra and Geometry. Due to this requirement, we suspect that one source of variation in our study will be students' previous knowledge of Algebra and Geometry. Because in Indonesia basic algebra and geometry are taught as part of the national high school curriculum, as a proxy for this previous knowledge we collected each student's score from their Mathematics National Exam. To graduate from high school, Indonesia requires each student to past a nationally administered exam where one of the tested subjects is Mathematic.

As a measure of students' comprehension, this study uses students' exam score, which was administered in the 9th week of the semester, and was a closed-book exam. Students were told the exam would make up 25% of their final grade. Students were given a problem set consisting of ten questions where grading was both on correctness of the answer and on steps required to derive the answer. Each lecturer conducted grading based on the same solution set. Any discrepancies or confusion in grading were discussed and resolved together between the two lecturers.

All analysis was conducted using the R software package [9].

3 A Flipped Calculus Course for Developing Country

In the Flipped Class, all materials for a given session are handed out to students at the end of the previous session, which is a week before the current session. Students are expected to have reviewed these materials during the week before the start of the session. These materials consisted of: reading assignments, a slide deck, and several worked out examples. All of these materials are identical to the one given to students in the Control Class.

Each session in the Flipped Class starts with a quick review of the provided slide deck and worked out examples, which typically takes between 20 and 30 min. Students are encouraged to ask any questions they might have about the session's topic and materials during this review. This review is intended not only to remind students but also to prime students on the topic to be taught in the current session.

After the quick review, a problem set consisting of ten questions is distributed to students to be answered in class. For each question, both the correct answer and how the answer was derived must be written down on the student's own sheet of paper. This is intended so that students will have a sense of ownership of their answer sheet, will take better care of it, and more willing review it later. All problem sets are identical to the one given to students in the Control class.

While working on the problem set, the lecturer and one teaching assistant will move around the class to see if students needs help in understanding the problem set or any other aspect of the session's topic. Students are also encouraged to ask for help from their classmates. The lecturer and the teaching assistant never provide the answer to any of the questions but instead direct students to discover the answer themselves. A teaching assistant was recruited for the Flipped Class because we realized, after the first two sessions, that one lecturer would not be able to serve all of the students during the allocated class time.

Once a student has completed one question in the problem set, it is brought to the lecturer to be graded. All answer sheets are newer collected and are returned to students so they will be available in the future if needed. If the solution is incorrect, the student is asked to try again, and most likely either the lecturer, the teaching assistant, or one of their classmates are asked to point out the issue with the solution. This is repeated until the student can correctly answer the question. With this approach, students ultimately will receive a perfect score of 100 for each question. We believe this approach will build student's self-esteem and confidence in answering future problem sets.

Before the end of the session, materials for next week's session are assigned.

4 Results

The profile for each class can be seen in Table 1. In the Flipped Class, there were 22 students with only 4 female students. While in the Control Class, there were 15 students but with 9 female students. The average Mathematics National Exam score in the Flipped Class is 55.44 while for the Control Class is 57.93. A two sample t-test shows there is no difference between the average score for the Flipped Class and for the Control Class, specifically with $t(35) = 0.36$ and p-value $= 0.72$.

Table 1. Class profile.

	Control class	Flipped class
N	15	22
Gender		
Male	6 (40.0%)	18 (81.8%)
Female	9 (60.0%)	4 (18.2%)
Mathematics National Exam Score		
Mean	57.93	55.44
SD	24.06	17.88

A linear model was developed to measure whether the developed Flipped Classroom method had a positive impact on students' comprehension and to measure the increase in this comprehension. As stated in the Method section, students' comprehension is measured through a closed-book exam that was administered in the 9th week of the semester. Thus, the linear model tries to see whether students' exam score could be explained by what class they were in and their Mathematics National Exam score. Specifically, a linear model that consists of the exam score as the response and two predictors: student's Mathematics National Exam (MATH) score and the categorical variable whether the student was assigned to the Flipped Class or not (FLIPPED). Results from this linear model are shown in Table 2.

Table 2. Results.

	Coefficient	Standard coefficient	SE	*t*-score	*p*-value
Intercept	16.68	0.00	14.00	1.19	0.24
MATH	0.43	0.31	0.21	2.00	0.05
FLIPPED	17.45	0.31	8.67	2.01	0.05

N = 37, *F*-score = 3.80 with *p*(F) = 0.03, R^2 = 0.18 and Adj. R^2 = 0.13, Residual SE = 25.85.

The developed linear model is significant with an *F*-score = 3.8, with the corresponding *p*-value = 0.03, and an R^2 = 0.18. To validate the model, a diagnostic was carried out as suggested by [4]. Results from the diagnostic showed that: the error assumptions for the model are upheld (errors were independent, have equal variance, and normally distributed); the model had no outliers or observations with high leverage but had one influential observation, which we decided to include in the model; and the model did not have any structural problems.

Based on the developed linear model, students in the Flipped Class, holding all other variables constant, on average will have 17.45 points higher course exam score than students in the Control Class. This result is significant with a *p*-value = 0.05. A student's Mathematics National Exam score is also predictive of his or her exam score. Specifically, holding all other variable constant, on average an increase of 1 point in the Mathematics National Exam Score will increase the student's course exam score by 0.43 points. This result is significant with a *p*-value = 0.05.

Due to the wide difference in the gender composition between the Control Class and the Flipped Class, we also tested whether there would be a gender effect in our result by creating a new model that included the student's gender. An ANOVA test comparing our initial model and the new model, which included gender, was not significant with an *F*-score = 0.00 and a *p*-value = 0.96. This leads us to conclude that there is no gender effect in our result.

5 Discussion and Conclusion

In this study, we developed a Flipped Classroom method for teaching Introductory Calculus that do not rely on videos but instead rely on textual materials, which we perceive to be more appropriate for a developing country setting. To determine whether this method is effective, we trial the method in teaching Introductory Calculus for first year Electrical Engineering students at *Universitas Syiah Kuala*. To measure the effectiveness of the method in increasing students' comprehension of the course materials, a linear model was developed that tries to see whether students' exam score in the course can be explained by the application of the developed Flipped Classroom method.

The significance of both the linear model and the categorical variable FLIPPED in the model provided evidences that the developed Flipped Classroom method had a positive effect on student's comprehension in the Introductory Calculus course. Specifically, students in the Flipped Class had a higher exam score by 17.45 points on average and over students in the Control Class when holding all other variables constant.

Moreover, because the model has a Cohen's $f^2 = 0.22$ [3], this points to the application of the developed Flipped Classroom method as having a medium effect size on improving students' comprehension of the course material.

A student's prior knowledge of Mathematics also plays a big role in increasing comprehension of Introductory Calculus. This role is pointed out by the significance of the MATH variable where the higher a student's Mathematics National Exam score is the higher their course exam score will be. In fact, looking at the standardized coefficients for the linear model, the influence of a student's prior knowledge of Mathematics is at par with the influence of our developed Flipped Classroom method, where both variables had a standardized coefficient of 0.31.

Based on these positive results, we are further tweaking the course to make it more effective. Specifically, we are looking into ways to further optimize the reading assignment. In the exit survey, many students commented that reading assignments were not effective and perceived the worked out examples to be more useful. While we disagree with the student, reading is still needed to introduce students to the underlying theory, but we are searching for alternative methods to make the reading assignment more interactive. We also highly suspect that some students might be skimming through, or even not doing, the reading assignments at all thus we are also looking for easily implemented mechanism to guarantee that students are doing the readings.

We are also investigating whether our method can be applied on courses that are less mathematical, such as Introduction to Programming and Human–Computer Interaction.

References

1. Bergmann, J., Sams, A.: Flip Your Classroom: Reach Every Student in Every Class Every Day. International Society for Technology in Education, Washington (2012)
2. Bishop, J.L., Verleger, M.A.: The flipped classroom: a survey of the research. In: Proceeding of the 120th ASEE Annual Conference & Exposition, Atlanta, GA (2013)
3. Cohen, J.: A power primer. Psychol. Bull. **112**(1), 155–159 (1992)
4. Faraway, J.J.: Linear Models with R. CRC Press, Boca Raton (2014)
5. Herreid, C.F., Schiller, N.A.: Case studies and the flipped classroom. J. Coll. Sci. Teach. **42**(5), 62–66 (2013)
6. Jungić, V., et al.: On flipping the classroom in large first year calculus courses. Int. J. Math. Educ. Sci. Technol. **46**(4), 508–520 (2015)
7. McGivney-Burelle, J., Xue, F.: Flipping calculus. Primus Probl. Resour. Issues Math. Undergrad. Stud. Phila. **23**(5), 477–486 (2013)
8. Palmer, K.: Flipping a calculus class: one instructor's experience. PRIMUS **25**(9–10), 886–891 (2015)
9. R Development Core Team: R: A Language and Environment for Statistical Computing. R Foundation for Statistical Computing, Vienna, Austria (2009)
10. Sahin, A., et al.: Flipping a college calculus course: a case study. J. Educ. Technol. Soc. **18**(3), 142–152 (2015)
11. Ziegelmeier, L.B., Topaz, C.M.: Flipped calculus: a study of student performance and perceptions. PRIMUS **25**(9–10), 847–860 (2015)

Training Using Professional Simulators in Engineering Education: A Solution and a Case Study

Dorin Isoc(✉)

Technical University, Cluj-Napoca, Romania
dorin.isoc@utcluj.ro

Abstract. Continuing previous researches on engineering education, this paper presents how to build and use a professional training simulator. The training simulator is specified and then, it is implemented as a set consisting of a technical simulator, basic concepts and techniques and a basis for professional regulation documents. An example of usage shows that, through the construction of the training simulator, the individual performance of student work increases significantly, professional skills are acquired in an integrated way and cooperation between students becomes a permanent practice. In conclusion, recommendations are being made to extend professional simulators to a larger coverage of professional skills. This extension will be at the expense of accumulation of theoretical knowledge with a potential and probable future application.

Keywords: Professional skills · Simulator · Training
Professional regulation · Engineering education

1 Introduction

Engineering school is not an extension of classical university. Engineering can not be reduced to a few theoretical courses or, in general, to theory. Engineering means a way to think specifically, a way to act specifically. This way of thinking is formed in time and the school must offer it every moment, along with knowledge. Whatever, the skills to be learned, the basic technique remains of systematic repetition. For engineering, systematic repetition is effectively possible with simulators. These are simulators of activities associated with the engineering profession.

Simulators existing on the market are regarded as simple tools, equipment, or, more generally, technical means that avoid effort or investment during training [7,9,10]. It is noteworthy that the medical simulators [3,5] present outstanding results. We hold this finding to highlight the very applicative nature of medical simulators. This is to say that they are used either after theoretical training or with a very little theoretical training.

M. E. Auer et al. (eds.), *Teaching and Learning in a Digital World*,
Advances in Intelligent Systems and Computing 715,
https://doi.org/10.1007/978-3-319-73210-7_25

This paper aims to present a solution for the specification, developing and using of a professional simulator for training in the engineering school.

Section 1 introduces the features of professional training simulators in school. Section 2 is dedicated to specifying and developing the professional simulator and, further describing its mode of using.

The conclusions follow how the professional training simulator can be integrated into the engineering school and the consequences of this integration.

2 About Professional Training Simulators

The novelty of the approach and the subject obliges us to define the concept more strictly.

A professional simulator appears as a multivalent training tool. Its purpose is to provide the opportunity to develop activities and sequences of activities that are found in the practice of the engineering profession and which need to be learned. A professional simulator is all the more valuable as the number of activities he permits is more significant.

The use or exploitation of the professional simulator is a didactic activity that the instructor has to coordinate. Through the simulator, the instructor is able at all times dose the constituent elements of professional skills in such a way that they are relevant and useful to the formation of the future professional. A simulator is a concrete way to integrate knowledge and expand professional skills. In this way, the instructor gets to define the content of the skill, to evaluate it, and then to make right corrections in relation to the concrete requirements of reality.

Professional procedural activities are those activities that develop on the basis of standards, norms, instructions, regulations. Despite the fact that these professional documents impose restrictions, each of them does so specifically. In relation to all professional activities, the student acquires the elements of the future responsibility of the professional, the future elements of the possibility to act in a given framework and, especially, in relation to various requirements and tasks.

The design of the professional simulator assumes in this way: (a) the choice of the covered field of competencies, (b) the identification and classification of the activities, (c) the implementation of the simulator, (d) the selecting or building of the set of professional regulation documents, (e) the chaining of activities and the construction of the evaluation-conditioning system.

3 Building the Professional Training Simulator

In the context of this paper, the Professional Training Simulator (PTS) consists of a technical simulator, a knowledge base at the user's disposal and a set of professional regulatory documents. Its role is to train in a school, in relation to a set of professional skills and in relation to a particular field of application.

3.1 Specifying of the Training Simulator

In order to build a training simulator, it is taken into account that it is meant to form professional skills.

Forming a professional skill requires full knowledge of the skill. For this, the skill is decomposed into steps that allow understanding, learning, repeating and evaluating the sequence. The skill appears as a chain of activities or actions. Each activity is conditional upon activation, each activity is conditioned in the context of the set of activities, and each activity presents a certain deliverable. From a teaching point of view, each activity is evaluable and allows analysis both by the student and by the instructor.

The detailed example involves developing the following skills:

i. the ability to build and run a project and then interpret independently the results of an engineering project;
ii. the skill to learn and know a technical means that will be used professionally;
iii. the skill to manipulate and use at each moment a given volume of professional regulatory documents;
iv. the ability to report fully and responsibly the results of a project;
v. the skill to work in a hoc team set up for peer review both as executor and verifier.

3.2 Technical Simulator

The technical simulator is specialized equipment, oriented to the training of the personnel in order to know or study a certain reality.

In the present study, the reality as in Fig. 1 to which the simulator is oriented, is the design of automated control systems from the position of a practicing engineer. In the block diagram shown in Fig. 1, as in [6], are found essential elements of a model of a general automatic control system. Such a system is developed around three constructive parts: the differentiation or comparison element, D, in order to determine at any moment the value $e(t) = r(t) - y(t)$; the controller, C which determines at each moment the command $u(t) = C(e(t)$ and the controlled process, P.

Component parts are connected to a feedback loop and provide for the processing of generated or measured signals that are: $r(t)$ - the reference signal of the system; $e(t)$ - the error signal; $u(t)$ - the command signal; $p_f(t)$ - additive phenomenon perturbation; $u_p(t)$, $y_p(t)$ - the input, i.e. the output signal of the controlled process so that $y_p(t) = P(u_p(t))$; $y(t)$ - the output signal of the control system; $z_m(t)$ - the additive measurement noise signal.

In Fig. 1 we used the notation that at one given moment, $y(t) = F(x(t))$ with the meaning that F is the mathematical model of the information processing block that supplies $y(t)$ when $x(t)$ is known. In this way, the controller C can be any of the types $\{C_c, C_d, C_m\}$ i.e. C_c - continuous controller; C_d - discrete controller; C_m - multi-positional controller (with bi-and -three-positional

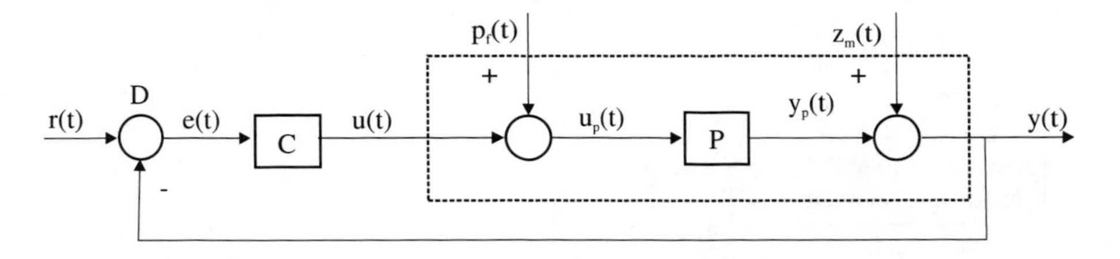

Fig. 1. The block diagram of simulated control system.

variants, with or without hysteresis, with three positions, with or without hysteresis, with or without zone of insensitivity) and P - controlled process with continuous compliant model (1), with continuous simplified model (2) or with discrete model (3).

In order to exploit the reality in the destination of the technical simulator, the structure of Fig. 2 is developed.

The representative situations provided by the simulator will be specified. These situations are related to simulator modules that are configured by human operator action.

Module $S0$**.** Simulating of a control system of the continuous model of the controlled process P under the action of the model of a continuous controller, C_c.

Module $S1$**.** Simulation of a set of models of autonomous continuous systems, S_1, respectively S_2.

Module $S2$**.** Simulation of a set consisting of two models of autonomous continuous systems S_1, respectively S_2 and a set of experimental measured signals, $u_{exp}(t)$, respectively $y_{exp}(t)$.

Module $S3$**.** Simulation of a control system of the continuous model of the controlled process P under the action of the model of the discrete controller, C_d.

Module $S4$**.** Simulation of a control system of the continuous model of the controlled process P under the action of the model of a multi-position controller, C_m.

Modules put at the user's reach are tailored to specific requirements that include situations from dynamic systems analysis, dynamic system identification and analysis, design and evaluation of control systems.

It is noteworthy that simulator operation does not involve any qualified intervention in the simulator, as a realization although the simulator is an open software platform. The only user interventions refer to configuration and setting of parameter values. In this way, the simulator can not identify to other products for the analysis of dynamic systems, such as $MATLAB$ or $SIMULINK$ [1].

In order to respond to specific training objectives, it was chosen that the entire simulator should operate under the action of certain selection switches, such as Ks, and configuration sequences.

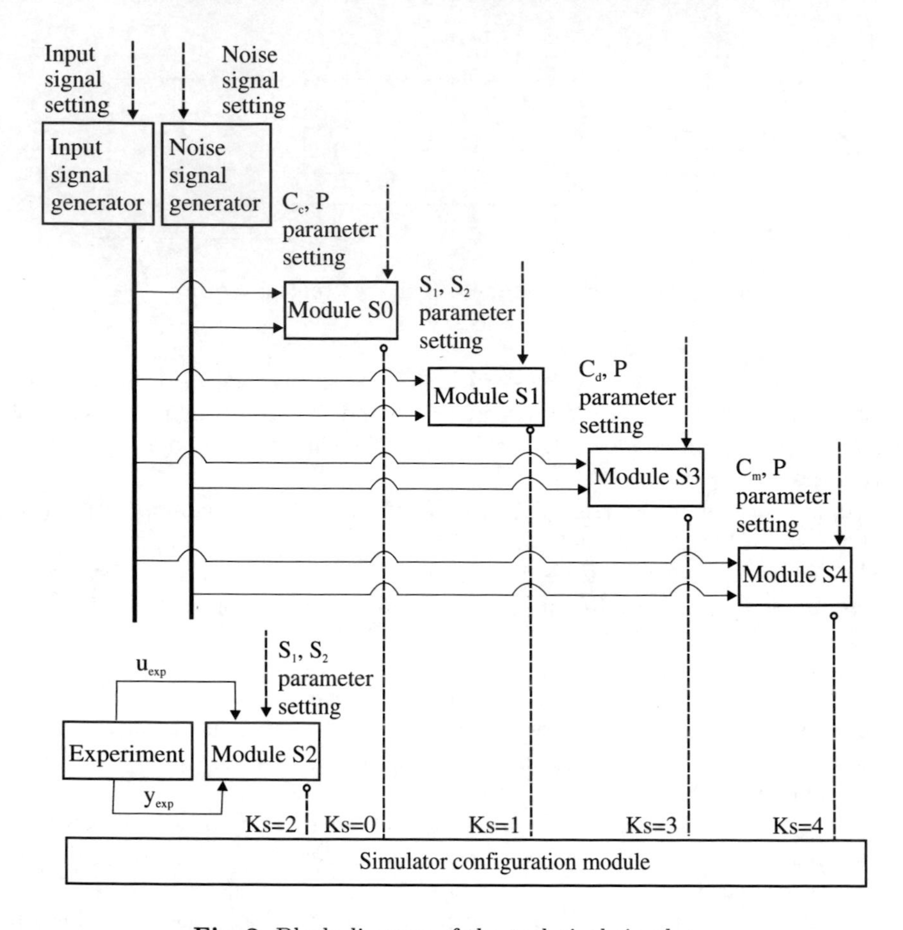

Fig. 2. Block diagram of the technical simulator.

$$P(s) = \frac{B(s)}{A(s)} = \frac{b_{nb}s^{nb} + b_{nb-1}s^{nb-1} + \ldots + b_1 s + b_0}{a_{na}s^{na} + a_{na-1}s^{na-1} + \ldots + a_1 s + a_0} \tag{1}$$

$$P(s) = \frac{B(s)}{A(s)} = \frac{(b_{nb}s^{nb} + b_{nb-1}s^{nb-1} + \ldots + b_1 s + b_0)e^{-T_m s}}{a_{na}s^{na} + a_{na-1}s^{na-1} + \ldots + a_1 s + a_0} \tag{2}$$

$$P(z) = \frac{B(z)}{A(z)} = \frac{(b_{nb}z^{-nb} + b_{nb-1}z^{-(nb-1)} + \ldots + b_1 z^{-1} + b_0)z^{-m}}{a_{na}z^{-na} + a_{na-1}z^{-(na-1)} + \ldots + a_1 z^{-1} + a_0} \tag{3}$$

3.3 The Knowledge Base Available to the User

The simulator described above is used in design and training. Its use is to customize the component parts as signals and parameters after the configuration of the work system structure.

As the simulator's destination is the design, it is imperative that the user always have a minimum of theoretical knowledge but sufficient for the application he/she is working on.

For this, the professional simulator has at its disposal a knowledge base consisting of application notes, theoretical details or presentations of working techniques.

It is noteworthy that this knowledge is not the basis of a course but only details that are necessary for the immediate use of the simulator.

If the user wishes to deepen their theoretical knowledge, it is his responsibility to appeal to other bibliographic sources, some of which are recommended by the author of the simulator base itself.

3.4 Set of Professional Regulatory Documents

The specific element of the training simulator is that it addresses not only the technical aspects that are found in the literature for most of the simulators [2,8–10] but also the action in a professional field, in the present case, in the professional field of the engineer.

The professional field is the entire context of a professional's activity. For an engineer, the professional field is provided with a necessary volume of regulatory documents. These documents are the foundation of professional skills.

Professional regulatory documents include standards, user manuals, working rules, and internal regulations, working instructions, standardized or dedicated forms.

In the specific case, the design of the professional field took into account:

Instructions for management and coding the electronic information. The regulatory document takes into account that design, experimentation and evaluation work involves a significant amount of electronic information. Hence the need for systematic coding is obvious.

Norms of technical reporting. The regulatory document adds the knowledge of professional technical communication for a consistent reporting of the results of the solved projects.

Rules of management and evaluation of work. The regulatory document defines all the details regarding the organization of work. It defines the activities, the deliverables, and the way of interaction to ensure self-control, mutual control, quality, and reporting in the professional hierarchy of the working group.

Rules for the management of assisted design works. The regulatory document deals with the specific details of the assisted design work. It insists on the implementation of certain defined procedures but also the iterations and the supposed checks within them.

Operating manual of dynamic systems simulator. The regulatory document contains all the technical details that relate to the use of the technical simulator. All modes of operation are described. Each sequence that is important for complete and rational exploitation is exemplified.

Along with the regulatory documents, a series of working sheets summarizing representative information on the working steps and the activities carried out are introduced. Here are the self-checking and peer-check sheets, the experiments configuration sheets for each simulator module.

4 Using of the Professional Simulator. Case Study

4.1 Roles Setting

The defined, designed and implemented professional simulator has been widely applied in training. Obviously, it is also necessary to establish the role of the instructor in using the simulator.

In order for the student to be able to correctly acquire all the planned professional skills, the instructor has to define his role and the way of intervention. All these details are to be announced at the beginning of the activity.

In the specific case, the following rules for the role of the instructor were defined:

i. In the working phase, the instructor plays the role of the team leader and plays also the role of the client. This role is one with a high dose of professional passivity but with a maximum dose of responsibility. The functions of the team manager were included in the regulatory documents. Compliance with regulatory documents is ensured by their transparency and general character for all members of the student group.
ii. Each student receives a project that differs by definition within a common theme.
iii. Within the working team, the control teams, which are made up of verification pairs [4], are established by default.
iv. The assessment of each student's work is carried out in relation to the way in which he/she performs his/her tasks on taught deliverables, compliance with the delivery terms and compliance requirements, defined in a uniform manner. The role of the instructor is to censor definitively and drastically, in the end, the work of each student by requesting to go through some segments of work stages in which the simulator is involved, under the conditions set.

4.2 Authorization for Operation

The scenario to be achieved through the use of the technical simulator will begin by operating authorization.

In spite of all the habits enacted in school, access to the technical simulator will only occur after the student's individual authorization.

Authorization involves knowing, through assisted execution, all stages of work with a particular simulator module. Assisted execution means once again the action of performer-verifier pairs, running a configuration sequence to a palpable result.

A mandatory component part of the authorization activity is to complete a written report specific to any simulator operation.

4.3 Exploitation of the Professional Simulator

In the manner in which the simulator modules have been defined, it is possible to make a consistent set of situations where the simulator needs to be used.

In principle, the exploitation of the professional simulator involves a succession comprising:

i. dimensioning at least one simulation experiment;
ii. crossing at least one design sequence;
iii. crossing at least one simulation situation;
iv. interpreting a certain number of experiments related to each situation leading to optimization;
v. full written reporting of the project according the specific communication standards announced.

5 Interpretations, Results

The professionally designed simulator is built on the MATLAB environment. The implementing mode envisages that the student never comes to achievement aspects other than those specific to the activities he has to accomplish.

The character of a professional tool is ensured as a software tool, through sobriety, that is, by the very small number of items left to the user's discretion. All user options are governed by manuals and usage procedures. Interaction between the student and the simulator occurs only after an authorization process through which the user proves that he is able to use all available facilities.

Any interaction of the student with the simulator is done by anticipating a use project in which all conditions, parameters and information are specified. There is an endorsement phase, that is to say, admission to use in a collegiate relationship within the group. When all working conditions are assured, the student switches to using the simulator according to the technical tasks received.

The access to the simulator, the work and the presentation of the results are done in accordance with a working plan with time scales.

The introduction of the concept of a professional simulator in the engineering school has as a first effect a new redistribution of the didactic tasks. Working on the simulator involves a significant amount of knowledge that must be permanently integrated into real technical activities. Integrating knowledge into well-identified professional activities is not optional, and this becomes a formative pressure. By this detail, the principle of linking the acquiring of knowledge specific to the conventional school is diluted.

School activity becomes piloted by professional activities and of the engineer profession, organized for practice training, under conditions similar to those in reality.

Both activity-oriented work and the common background of professional regulation documents are an intrinsic motivation to stimulate teamwork. It is the team of professionals that exists because of working relationships and the team of students, as a group formed by criteria of individual acceptance of the received tasks.

The knowledge, typically included in classical courses, is now focused, first of all, on professional regulatory documents. Regulatory documents are designed to be action guides. This new reality obliges the teacher to insist on the framework created by the simulator, to prioritize activities so their chaining leads to the intended goals.

Using professional simulators does not eliminate classical courses but makes them more applicative. This tendency adds to the possibility that sources of information and knowledge acquisition are external to the school but accessible, such as on the Internet. The novelty lies in the fact that professional regulation documents become a filter both directly and through the new optics that the student assimilates day by day.

Building simulators is not limited to the curriculum's current disciplines. Building simulators is a way with multiple consequences. The first consequence is that of enhancing the integration of knowledge and skills. Another consequence is a practical one. One can imagine that such an engineering school has a number of professional simulators. The provision of simulators should cover a competency and hence the number of simulators is always finite and limited. This number of simulators is dictated by the number of activities a profession, whether engineer or engineer, requires them to be covered with skills.

From a quantitative point of view, the use of professional simulators, as they have been defined, has led to an increase in the actual individual workload of students from an average of about 30–40% under conventional conditions through direct work coordinated by the instructor at about 81–94%.

By using the professional simulator, that is, setting up the professional environment, in each quota, about 12 ... 15% no longer coped with the pace of work and called for administrative forms of recovery.

One particular aspect is the drastic censorship exerted by the instructor. The role of this censorship is to reduce the proportion of violations of professional ethics, plagiarism and false data, but also to increase individual responsibility. As a result, through this censorship the students of the band were filtered out with the elimination of incorrectly performed works, in the proportion of about 25 ... 30% of the group's staff, only in the first two promotions of training with professional simulators. Later on, probably because of a better knowledge of the way they work, the students responded correctly to the work requirements.

The relationship between the number of working-class students and the use of professional simulators is favorable because the instructor's effort is greatly diminished and his interventions during work are reduced to scheduled sessions of about 10 min per working hour.

6 Concluding Remarks

Using professional simulators in engineering school is a new way of training. Unlike other methods such as project-based training or problem-based training, this type of training is based on the set of professional activities, but not only. The paper reveals that the engineering school forms some people with a certain way of thinking. This way of thinking results from the formation of complex skills. In this way, the simple definition of skills in other fields, especially psychology and sociology, becomes obsolete.

The analysis phase in the development of the professional simulator provides a schematic for integrating knowledge and other skills. The efficiency of the simulator is all the more significant as the set of identified activities is better profiled with reference to the engineer profession to which it tends.

The mechanism remains for other phases of training-including after graduation. Thus training based on professional simulators proves to be a technique of changing the vision of the engineering school. The essence of this technique is given by the fact that adjustments that are required or accepted can be done without directly affecting the students, but only the means of their training.

A remarkable conclusion shows that the whole training technique speaks of a professional simulator, but this training simulator is actually made up of a technical simulator and a set of professional regulatory means: standards, norms, regulations, instructions in which are integrated knowledge and working techniques.

References

1. Board, M.: Control algorithm modeling guidelines using MATLAB, Simulink, and Stateflow. V2.0. Technical report, The Mathworks, Inc. (2007)
2. Dozortsev, V.: Development of computer-based training simulator for industrial operators: main participants, their roles and communications. Autom. Remote Control **71**(7), 1476–1480 (2010)
3. Gordon, J., Oriol, N., Cooper, J.: Bringing good teaching cases "to life": a simulator-based medical education service. Acad. Med. **79**(1), 23–27 (2004)
4. Isoc, D., Isoc, T.: Practice of peer-review and the innovative engineering school. In: 2015 9th International Symposium on Advanced Topics in Electrical Engineering (ATEE), Bucharest, Romania (2015)
5. Joe, R., Otto, A., Borycki, E.: Designing an electronic medical case simulator for health professional education. Knowl. Manag. E-Learn. Int. J. **3**(1), 63–71 (2011)
6. Lazăr, C., Vrabie, D., Carari, S.: control systems using PID controllers. Matrixrom, Bucharest (2004). (In Romanian)
7. McGrattan, K., Hostikka, S., Floyd, J., Baum, H., Rehm, R., Mell, W., McDermott, R.: Fire Dynamics Simulator (Version 5) Technical Reference Guide. NIST Special Publication, Gaithersburg (2004)
8. Reinhardt, E., Crookston, N.: The fire and fuels extension to the forest vegetation simulator. General Technical report RMRS-GTR-116, Ogden (2003). (Technical Editors)
9. Rogalski, T., Tomczyk, A., Kopecki, G.: Flight simulator as a tool for flight control system synthesis and handling qualities research. Solid State Phenom. **147**, 231–236 (2009)
10. Scalese, R., Issenberg, S.: Effective use of simulations for the teaching and acquisition of veterinary professional and clinical skills. J. Vet. Med. Educ. **32**(4), 461–467 (1982)

Engineering Students' Solutions to Accuracy Problems in Analog Electronics Course

Elena Trotskovsky(✉) and Nissim Sabag

ORT Braude College of Engineering, Karmiel, Israel
{elenatro,nsabag}@braude.ac.il

Abstract. In the course of the last three years, a longitudinal study researching engineering students' understandings of the concept of accuracy and error has been carried out. Previous studies of the researchers investigated general misunderstandings of the concept of accuracy and error among engineering students from different programs, and specific misunderstandings of those concepts among electrical and electronic engineering students in Digital Electronics course. The studies showed that students' understanding of these important engineering concepts is insufficient. The current study researches students' achievements in solving problems that relate to the concepts of accuracy and to basic analog electronics concept of non-linearity and saturation. The research was carried out in April - May 2017 during Analog Electronics course with 38 participants. A mixed methodology was applied; in the quantitative stage the students solved problems relating to the concept of accuracy, and in the qualitative stage 15 open interviews were carried out. It was found that engineering students with inadequate understanding of basic accuracy concepts struggle with interpreting the concepts of analog electronics while solving problems related to the issue. The research intends to be continued with the aim to show that consistent treatment of the issue will benefit students' understanding of basic electronics and accuracy concepts.

Keywords: Misconception · Accuracy · Error

1 Introduction

Accuracy is considered to be a basic science and engineering concept. There are various definitions of accuracy in science literature. One definition [1, p. 21] is: "Accuracy of measurement is the closeness of agreement between a quantity value obtained by measurement and the true value of the measurand" (measured parameter).

The concept of accuracy is well established in everyday engineering practice. The authors of [2] emphasize the importance of a conceptual knowledge in academic education. Therefore, it is crucial for engineering students to understand the concept of accuracy and to apply it in different engineering courses.

The paper presents the results of the advanced stage of a longitudinal study that referred to engineering students' understanding of accuracy. Various aspects of the issue were investigated in previous studies. In [3], students of electrical and electronic engineering expressed misconceptions about the accuracy concept while solving simple

M. E. Auer et al. (eds.), *Teaching and Learning in a Digital World*,
Advances in Intelligent Systems and Computing 715,
https://doi.org/10.1007/978-3-319-73210-7_26

problems using engineering models. [4] pointed out that engineering students from different engineering programs—mechanical engineering, electrical and electronics engineering, and industrial engineering and management—demonstrated insufficient understanding of the concept of accuracy and the relationship between accuracy and measurement error. Studies [5, 6] describe misunderstandings concerning the accuracy of digitized signals exposed by the electrical and electronics students in Digital Electronics course. Thus, in [5] it was shown that a large part of students misinterpreted the essence of quantization of signal and noise, confuse quantization and noise errors, and the relationships between noise and resolution. Paper [6] deals with the integration of a micro-study unit into the Digital Electronics course program for the purpose of explaining accuracy terms.

The aim of the current study is to broaden the research frame and to investigate students' understandings of accuracy in Analog Electronics course.

2 The Research Setting

2.1 Methodology

The research was carried out in April - May 2017 during the course of Analog Electronics. The students take the course in the fourth semester of their study program. A mixed methodology was applied in the research.

The main research tool in the quantitative stage was a questionnaire of three problems concerning operational amplifier. The two last problems relate to the concept of accuracy. The students solved the problems during a lecture after learning the appropriative subject matter. The written solutions and explanations were analyzed. In the qualitative stage of the research open interviews with the students who demonstrated misunderstandings of accuracy concept were carried out.

2.2 Questionnaire Set up

The first problem (P1) in the questionnaire refers to the subject of saturation and non-linearity of transfer characteristic of non-inverting amplifier based on real operation amplifier. In order to provide correct answers on two following questions the students had to solve P1 correctly. While solving the second problem (P2), the student had to calculate absolute or relative error of the output voltage of the real amplifier in relation to ideal amplifier, and in the third problem (P3) he or she rated the answers of P2, according to the accuracy of the given results.

2.3 Questionnaire

The amplifier circuit shown in the Fig. 1 uses a real operational amplifier. The supply voltage is $\pm V_{CC} = \pm 12V$.

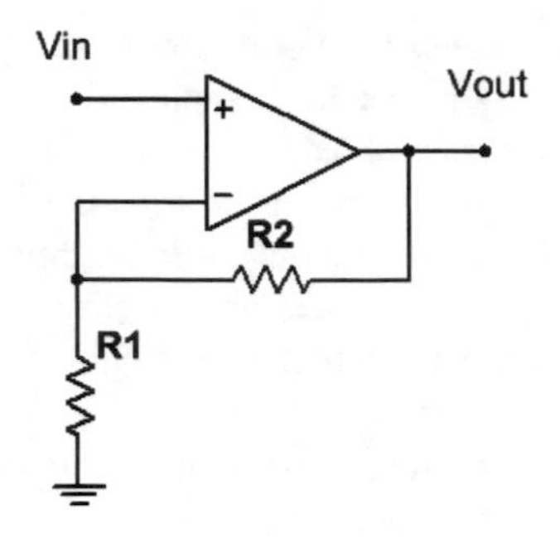

Fig. 1. Amplifier.

Transfer characteristic of the circuit is presented in Fig. 2.

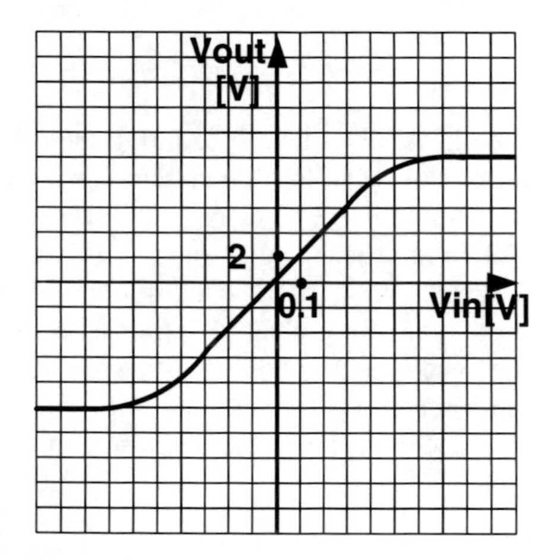

Fig. 2. Transfer characteristic of the amplifier.

1. Find the output voltage of the amplifier in each case.
a. $V_{IN} = 0.2V$ $V_{OUT} = ?$
b. $V_{IN} = 0.5V$ $V_{OUT} = ?$
c. $V_{IN} = 0.6V$ $V_{OUT} = ?$
2. Calculate the error of the output voltage of the real amplifier relative to an ideal amplifier.
a. $V_{IN} = 0.2V$
b. $V_{IN} = 0.5V$
c. $V_{IN} = 0.6V$
3. Complete the following table based on the results of question 2.

Accuracy	V_{IN}
Low	
Middle	
High	

2.4 Research Population

38 students of electrical and electronics engineering at an academic college of engineering, who took the Analog Electronics course in the 2017 spring semester participated in the study. 15 of them were interviewed during the qualitative stage of the research.

3 Results

The distribution of the students' answers to the whole questionnaire is presented in Table 1.

Table 1. Distribution of students' answers

	P1		P2		P3	
N = 38	N_1	%	N_2	%	N_3	%
Correct answer	21	55.3	14	36.8	9	23.7
Wrong answer	17	44.7	8	21.1	17	44.7
No Answer	0	0	16	42.1	12	31.6

21 out of 38 students correctly solved P1. All of the students who provided wrong answers to the problem did not refer to the non-linearity of transfer characteristic, and calculated the output voltage of ideal operational amplifier. By observing the interviews with the students who incorrectly solved the problem it shows that they found the amplifier gain in the first point on the linear part of characteristics with the parameters $Vin = 0.1V, Vout = 2V, A = \frac{Vin}{Vout} = 20$, and did not notice that two additional points with $Vin = 0.5V$ and $Vin = 0.6V$ belong to the non-linear part of the characteristics. Thus, they found the output voltage by multiplying input voltage by the gain in the linear part of the transfer curve.

The distribution of the students' answers to the second problem is shown in Table 2.

Table 2. Distribution of students' answers to the second problem

P2				
N = 38	Correct answer – absolute error	Correct answer – relative error	Partial correct answer	No answer
	7	7	8	16
%	18.4	18.4	21.1	42.1

It can be seen that only 14 out of 38 participants gave a correct answer, whereas half of them (seven out of 14) found an absolute error, the other half – a relative error, and eight students provided a partial correct answer – to section a. only. The students who gave a partial correct answer are included in the category "wrong answer" in Table 1.

The distribution of the students' answer to the third problem is presented in Table 3.

Table 3. Distribution of students' answers to the third problem

P3				
N = 38	Correct answer – inverse relationship between error and accuracy	Wrong answer – direct relationship between error and accuracy	Other wrong answer	No answer
	9	9	8	12
%	23.7	23.7	21.1	31.6

The same number of students – nine out of the 38 – rated both correctly and wrongly the relationship between error and accuracy. Additional eight students provided other incorrect answers, and 12 students did not give any answer.

4 Discussion

It is apparent that most of the students who cannot solve P1 are unable to provide correct answers to the two following questions as well. Most of the interviewed participants emphasized that they did not understand what these questions were about. However, after the lecturer's explanation about their error in P1, they solved P2 and P3 successfully.

It can be noticed in Table 2 that the same number of students (seven), who solved P2 correctly, found an absolute and relative error. The similar result was obtained while interviewing students who answered P2 after the lecturer's explanation; about half of them settled for the absolute error and the other half tried to calculate the relative error. But most of them did not remember the accurate formula of the relative error and proposed different erroneous formulas such as a relationship between a real and an ideal result, or a relationship between an absolute error and a sum of real and ideal results. The usage of incorrect formula for the calculating relative error is known from a previous study [5].

It can be seen in Table 2 that eight students gave partially correct answers to section a. −0 or 0% and no answers to sections b. and c. During the interviews they explained that they intuitively felt that in case a. the error of the output voltage is zero, but they could not provide any proof to this.

When answered P3 (Table 3), nine students provided wrong rating, which shows a direct relationship between error and accuracy. These results are in line with the outcomes of a previous study [4] that shows that approximately quarter of the students misconceive the relationship between error and accuracy.

In the interviews the lecturer asked the students in which engineering courses they have met the accuracy concept. 14 out of 15 participants approved that they encountered the concept in a physics lab course only one year ago. One student solely, who took the Analog Electronics course for second time in his eight semester and answered all of the questionnaire correctly, noted that he met it in Digital Electronics course, which he learned in his fifth semester.

5 Conclusions

The results of the current study reaffirm the results of the previous studies, which show that the accuracy concept is not sufficiently treated in the engineering program. As it was expected, engineering students with inadequate understanding of basic accuracy concepts struggle with interpreting the concepts of analog electronics while solving problems related to the issue. The authors of the current paper intend to continue the research during the Analog Electronics course and to offer the students to answer a similar questionnaire about another subject matter – diode - following a lecturer's explanation about accuracy and errors. The authors expect improving the results and deepening students' understanding of basic electronics and accuracy concepts.

References

1. BIPM International vocabulary of metrology - Basic and general concepts and associated terms in (VIM). 3rd Edn., JCGM (2012). http://www.bipm.org/en/publications/guides/vim.html
2. Streveler, R.A., Litzinger, M.A., Miller, R.L., Stief, P.S.: Learning conceptual knowledge in the engineering sciences: overview and future research directions. J. Eng. Educ. **97**(3), 279–294 (2008)
3. Trotskovsky, E., Sabag, N., Waks, S., Hazzan, O.: Students' achievements in solving problems using models in electronics. IEEE Trans. Educ. (2014). https://doi.org/10.1109/TE.2014.2331918
4. Trotskovsky, E., Sabag, N.: Students' misconception of accuracy. In: ICEE/ ICIT 2014, Joint International Conference on Engineering Education & International Conference on Information Technology, Riga, pp. 148–155 (2014)
5. Trotskovsky, E., Sabag, N.: How do engineering students misunderstand the concept of accuracy? – work in progress. World Trans. Eng. Technol. Educ. **12**(4), 1–5 (2014)
6. Trotskovsky, E., Sabag, N.: How electrical engineering students understand the accuracy concept concerning digitized signals. In: International Conference on Engineering Education ICEE, Zagreb, Croatia, July 2015

The Necessity of Competency Development in Engineering Informatics Education in the Light of Students' Characteristics

Zita Tordai(✉) and Ildikó Holik

Ágoston Trefort Centre for Engineering Education, Óbuda University, Budapest, Hungary
tordai.zita@tmpk.uni-obuda.hu

Abstract. Current technological innovations and continuous change in the labor market have generated new challenges for higher education, and universities now underline the importance of competency development. The purpose of this paper is to examine the competencies of undergraduates attending Engineering Informatics education, and to identify their personal needs for development in the light of workplace demands. The characteristics of the motivation and personality profiles of Engineering Informatics students are also addressed in this research in order to reveal possible ways of motivating them by implementing new methods or approaches in Engineering Informatics education.

Keywords: Engineering education · Competency development
Students' characteristics

1 Introduction

There is a growing interest in the attributes and skills required by engineers in our rapidly changing information society [1–4], with particular regard to examining the characteristics of engineering undergraduates from different perspectives [5–7].

Globalization has intensified the demands for flexible, socially adept and communicative engineers [4]. The engineers' weaknesses in soft skills such as effective communication, cooperation, teamwork, project management, lifelong learning have been noted by industry and various professional organizations [3, 6]. Other studies highlight the wider social context of engineers' work, arguing that there is a need for new engineers who are not only equipped with employability skills but who are also socially and environmentally responsible [2]. It is also reported in the literature that personality traits are important predictors of job performance and satisfaction, and are also key characteristics for the engineering profession. Engineers differ from members of other occupations as they are less likely to be assertive, extravert, emotionally stable and optimistic, yet they show more intrinsic motivation and tough-mindedness [7].

Current technological innovations and continuous change in the labor market have generated new challenges for higher education. Solid evidence indicates significant differences between the competencies developed by institutions of higher education,

M. E. Auer et al. (eds.), *Teaching and Learning in a Digital World*,
Advances in Intelligent Systems and Computing 715,
https://doi.org/10.1007/978-3-319-73210-7_27

perceived by graduates and expected by employers. Competence has become a key concept in higher education, regarding the combination of knowledge, skills, abilities and personal attributes that contribute to enhanced academic performance and success in the workplace [8]. Higher education in engineering primarily focuses on the development of professional competencies and technical skills, and students are not appropriately prepared for the demands of the workplace and lack social, communicative or personal competencies [9].

Several studies emphasize that along with the technical changes, conventional methods of engineering education should be re-evaluated and non-technical, generic competencies, including the interpersonal skills of the future engineers also need to be developed in order to reflect the changing demands of working life and industry [6].

2 Method

The study was designed to measure a wide range of attributes and competencies (e.g. learning style, logical thinking, emotional intelligence, personality traits, achievement motivation and self-efficacy) at three time points among engineering informatics students during their academic careers. This paper focuses on the perceived workplace competencies of students, and on the motivational and personality aspects of competency development based on the first measurement.

The following research questions are addressed in the study:

1. How do undergraduates rate their proficiency in a range of competencies at the beginning of their academic career?
2. How do they rate the importance of these competencies for future employment?
3. How high is their level of achievement motivation? and
4. What is the profile of Engineering Informatics students' personalities?

2.1 Sample

A sample of 188 first-year undergraduate Engineering Informatics students attending a Hungarian university participated in the study, of whom 166 males (88.3%) and 22 were females (11.7%). The ages of participants ranged from 18 to 26 years (Mean = 20.07, SD = 1.459). Sixty percent of the participants took their final secondary school examinations in 2016, 30% of them in vocational schools, 28% in high schools. Most of the participants (76%) have work experience, including summer work or student work, and 7% of them are working students.

2.2 Instruments

A set of questionnaires was used to assess different attributes and preferences of Engineering Informatics students before starting their academic career. The data was collected in September 2016.

The competencies of Engineering Informatics students were measured with a self-rating list of competencies consisting of 24 items which was constructed by using job vacancy advertisements and the results of previous competency assessments in higher education in Hungary. We examined, on the one hand, the extent to which these competencies are required for future employment according to the opinion of Engineering Informatics students, and, on the other hand, to the extent to which they possess these competencies at the beginning of their studies. Participants were asked to rate the importance of each competency in future employment using a 5-point Likert scale rating from (1) of minimum importance to (5) of maximum importance, and to self-evaluate their level of proficiency in the same competencies at the present moment rating from (1) at a minimum level to (5) at a maximum level.

Regarding the motivation of the students, the reasons for entering the selected institution of higher education and the motives for obtaining a degree were investigated by a series of questions. Achievement motivation was measured by 12 items derived from Helmreich's and Spence's [10] three-factor model [11]. Each item should be rated by a 5-point Likert scale rating from (1) 'strongly disagree' to (5) 'strongly agree' and a high score reflects a high degree of work, mastery, and competitiveness. The work orientation scale represents the effort a person makes to work hard and complete a task well. The mastery of needs scale reflects a preference for difficult and challenging tasks and excellent performance, and the competitiveness factor describes the desire to win and be better than others [10].

The personality traits of Engineering Informatics students were assessed with the Hungarian version of Big Five Questionnaire [12, 13]. The BFQ measures five personality factors, namely energy, friendliness, conscientiousness, emotional stability and openness. The questionnaire consists of 132 items, with five dimensions and ten sub-scales, and a social desirability scale. Each BFQ factor is measured by 24 items, 12 of which are positively phrased and 12 of which are negatively phrased. Table 1 shows the factors and sub-scales with examples of items. Participants were asked to rate their responses on a 5-point Likert scale, (1) meaning 'does not apply at all' and (5) meaning 'does apply entirely'. For the BFQ, norm data of a representative sample of Hungarian population was available [13].

Table 1. Factors, sub-scales and sample items of the BFQ, based on [12]

Factors	Sub-scales	Sample items
Energy	Dynamism Dominance	I am an active and vigorous person.
Friendliness	Cooperativeness Politeness	I hold that there's something good in everyone.
Conscientiousness	Scrupulousness Perseverance	I always pursue the decisions I've made through to the end.
Emotional Stability	Emotion control Impulse control	Usually I don't lose my calm.
Openness	Openness to culture Openness to experience	I'm fascinated by novelties.

3 Results

3.1 Perceived Competencies

Table 2 presents means and standard deviations of all 24 competencies, and the discrepancies between the self-reported proficiency and the importance level of competencies perceived by the Engineering Informatics students. As the distributional assumptions of parametric statistics were not met for the competency list, the Wilcoxon test was used to determine whether there is an association between the two set of variables. The two sets of competencies show acceptable internal consistency (Cronbach's alpha: 0.863 and 0.871).

Table 2. Means, standard deviations (SD) and differences between the perceived importance and self-reported proficiency level of competencies

	Importance Mean (SD)	Own level Mean (SD)	Difference	Wilcoxon (Z)
1. Oral communication	3,87 (.752)	3.24 (.989)	.622	−6.523**
2. Problem solving	4.87 (.407)	3.79 (.674)	1.080	−11.213**
3. Ability to work precisely	4.78 (.496)	3.73 (.798)	1.053	−10.495**
4. Cooperation	4.17 (.733)	3.87 (.883)	.303	−3.866**
5. Teamwork ability	4.23 (.792)	3.84 (.973)	.388	−4.780**
6. Working independently	4.56 (.568)	3.87 (.811)	.691	−8.489**
7. Analytical thinking	4.68 (.589)	3.73 (.810)	.947	−9.542**
8. Learning ability	4.66 (.567)	3.58 (.871)	1.080	−10.163**
9. Innovation	4.66 (.575)	3.70 (.905)	.963	−10.125**
10. Conflict resolution	3.24 (1.046)	3.49 (.989)	−.250	−2,169*
11. Organization	3.48 (.967)	3.18 (.951)	.309	−3.863**
12. Persistence	4.20 (.885)	3.64 (.831)	.559	−6.171**
13. Written communication	3.01 (1.065)	3.29 (1.015)	−.277	−3.237**
14. Openness	3.94 (1.048)	3.82 (.986)	.117	−1.768
15. Goal orientation	4.62 (.614)	3.97 (.824)	.644	−7.863**
16. Self-knowledge	3.27 (1.212)	3.61 (.973)	−.340	−3.376**
17. Stress tolerance	4.27 (.892)	3.55 (1.046)	.713	−7,123**
18. Responsibility	4.36 (.779)	3.92 (.820)	.441	−5.889**
19. Adaptation to change	4.39 (.734)	3.98 (.821)	.415	−5.561**
20. Concentration	4.80 (.464)	3.64 (.824)	1.154	−10.710**
21. Understanding causal relationships	4.84 (.386)	3.87 (.690)	.963	−10.586**
22. Appling knowledge	4.81 (.417)	3.97 (.701)	.846	−10.241**
23. Flexibility	4.31 (.739)	3.85 (.807)	.457	−5.853**
24. Evaluation and self-evaluation	3.59 (1.033)	3.57 (.865)	.016	−.129

* Significant at the 0.05 level.
** Significant at the 0.01 level.

The findings regarding the importance of competencies indicate that in the opinion of Engineering Informatics students the most necessary skills for their future profession are problem-solving skill (M = 4.87, SD = .407), the ability to work precisely (M = 4.78, SD = .496), understanding causal relationships (M = 4.84, SD = .386), and the ability to apply knowledge (M = 4.81, SD = .417). The students attributed less importance to self-expression and writing ability (M = 3.01, SD = 1.065), conflict management (M = 3.24, SD = 1.046) and self-knowledge (M = 3.27, SD = 1.212). Furthermore, four other competencies did not receive a 4.00 rating, namely self-evaluation, openness, the ability to organize, and the ability to connect and communicate effectively with others. These results show that first-year students consider cognitive skills and profession-related competencies much more important for employment than soft skills.

Regarding the proficiency level of these competencies, higher ratings were revealed for adaptation to change (M = 3.98, SD = .821), goal orientation (M = 3.97, SD = .824), applying knowledge (M = 3.97, SD = .701) and taking responsibility (M = 3.92, SD = .820). Lower ratings were found for organization (M = 3.18, SD = .951), oral communication (M = 3.24, SD = .989) and written communication (M = 3.29, SD = 1.015). The mean ratings of the self-reported levels of competencies ranged from 3.18 to 3.98, indicating that development of these competencies would be necessary during academic courses.

In accordance with previous studies (e.g. [6]), students rated the importance of the competencies higher than their actual level of proficiency in most cases. The Wilcoxon signed rank test was used to analyze the median difference. Significant differences were found between ratings of all competencies, except for "openness" and "self-evaluation", so there is a gap in 22 of the 24 analyzed competencies. The most evident differences were found for concentration, learning ability, problem-solving skill, working precisely and innovation, and the following eight other competencies obtained mean differences greater than 0.50: analytical thinking, understanding causal relationships, working independently, ability to apply knowledge, stress tolerance, goal orientation, oral communication, and persistence.

There are three competencies for which the Engineering Informatics students estimated their proficiency higher than the perceived importance for future work, namely self-knowledge, written communication and conflict resolution. However these skills also received lowest ratings regarding the importance of competencies.

3.2 Achievement Motivation

Regarding achievement motivation a preference for work orientation was detected among the students, in that 32% of the sample is motivated by obtaining a sense of satisfaction from work and pursuing self-realization and growth. Although the proportion of the three types of motivation do not show sizeable differences, mastery and competition seem to be less important factors (24% and 22% of the sample), indicating that challenges and excellent performance or the desire to win and be better than others are less motivating factors for these students. The remaining 22% of the students show preferences for two or all the three motivational factors. The mean ratings for the subscales are the following: work orientation (W): M = 14.11 (SD = 1.888), mastery of needs (M): M = 13.73 (SD = 2.321) and competitiveness (C): M = 12.69 (SD = 3.285). The reliability test for the 12 items was considered acceptable (Cronbach's alpha: 0.648). Gender differences were also revealed, as the women obtained higher scores on the mastery scale than the men and the men scored higher on the competitiveness scale, although the difference was not significant.

3.3 Personality Traits

Table 3 contains the means and standard deviations of the participants on the five personality dimensions and the social desirability scale. For comparison, average scores of the Hungarian norm group [13] are also provided. Note that the scores of the BFQ scales arc represented in raw scores although standardized T scores were used for further analysis.

Table 3. Means and standard deviations (SD) of Engineering Informatics students in five personality factors and social desirability scale compared with the Hungarian norm group (raw scores)

	Engineering informatics students Mean (SD)	Hungarian norm group (N = 774) Mean (SD)
Energy	74.74 (12.20)	77.51 (11.85)
Friendliness	79.22 (10.43)	82.25 (10.09)
Conscientiousness	82.65 (10.43)	81.34 (11.11)
Emotional stability	74.12 (12.39)	68.60 (15.83)
Openness	79.83 (10.78)	85.52 (6.88)
Social desirability	34.49 (5.00)	29.54 (6.88)

Engineering Informatics students obtained lower scores on energy, friendliness and openness dimensions, but higher scores on conscientiousness and emotional stability than Hungarian norm group. Higher scores were found on the lie (validation) scale indicating that the participants tend to distort their profile in a positive way, and to present themselves in a better light than in reality. Regarding factor scales, gender differences were revealed using an independent samples t-test (the normal distribution of the data was confirmed by the Kolmogorov-Smirnov test). Its result shows that men obtained significantly higher scores on emotional stability compared with women ($M_{men} = 75.23$, SD = 11.92, $M_{women} = 65.73$, SD = 12.86, t = 3.479, p = 0.001), but lower scores on openness ($M_{men} = 79.15$, SD = 10.53, $M_{women} = 84.95$, SD = 11.52, t = −2.401, p = 0.017).

The internal consistency coefficients of the dimensions ranged from .78 (openness) to .84 (energy), and .621 for the lie scale (social desirability) which is also feasible.

Based on the standardized T scores (provided in the BFQ Manual, [12]) personality profiles may be portrayed. T scores for each factors and sub-scales may be categorized as low (below 45), average (between 45 and 55) and high (above 55). Engineering Informatics undergraduates in general achieve average scores for energy (M = 48.26), friendliness (M = 48.88), conscientiousness (M = 52.6) and emotional stability (M = 52.81), but low scores for openness (M = 44.01), although they achieve high scores for social desirability (M = 54.8).

A correlation analysis was performed to determine the relation between personality traits and the achievement motivational characteristics of the students. Table 4 shows the Pearson correlation coefficients between variables.

Table 4. Correlation between BFQ factors and achievement motivation scales

	Work orientation	Mastery	Competition
Energy (E)	**,359****	**,410****	**,402****
Friendliness (F)	**,340****	,027	−,179*
Conscientiousness (C)	**,510****	**,398****	**,263****
Emotional stability (S)	,068	,107	−,186*
Openness (O)	**,384****	**,468****	,056

* Correlation is significant at the 0.05 level.
** Correlation is significant at the 0.01 level.

It was revealed that persons with a higher level of work orientation tended to rate themselves higher in all personality traits, except emotional stability. Undergraduates with a higher level of mastery showed higher ratings of personality traits related to energy, conscientiousness and openness. Furthermore, students who have a strong degree of competitive motivation can be described as active, extravert, dedicated and persistent.

4 Discussion

Our findings based on perceived competencies and personality traits suggest that Engineering Informatics students have several attributes, skills and competencies before starting their academic career which are relevant to the selected specialty in the university and crucial for their future profession as Engineering Informatics, for example they are capable of working both independently and dependably, meeting with high standards, controlling their emotions and showing consistent and coherent behavior without oscillating because of emotional states. However, they show a lack of social competencies such as effective communication, self-expression, organization, learning ability and cooperation which are important aspects for successful employment.

Based on the results related to the Big Five personality traits, the Engineering Informatics students display a low level of social skills, activity and dominance, implying that most of them are introvert and inactive, prefer working independently and tend to avoid interactions with others. They are characterized by moderate friendliness, helpfulness, tolerance and cooperation, but in contrast, they obtained higher than average scores for conscientiousness, meaning that they show accuracy and precision in various activities, have great respect for order and discipline, hence responsibility and reliability may be considered as their strengths. The participants also show emotional stability, indicating that they are rather calm and relaxed, and capable of controlling anxiety and impulses and coping with associated emotions.

These findings are in line with other studies that confirm existing evidence on emotional stability and conscientiousness among engineers together with lower levels of agreeableness compared with a norm population [5], and found that engineering students are more orderly, tough-minded and conventional than students in other disciplines [14].

Surprisingly, Engineering Informatics students obtained very low scores on the openness scale, indicating that they are neither interested in cultural and intellectual activities and events, nor open to change and to innovation, and that rather they have a traditional and conventional way of thinking.

These findings are in accordance with Holland's vocational personality types [15], which are based on the assumption that correspondence between key personality characteristics and work environments leads to important vocational outcomes (such as satisfaction, performance, etc.). Although it was not possible to identify clear types, 32% of the sample of Engineering Informatics undergraduates could be labeled as realistic or conventional types. Realistic individuals prefer to work with things rather than ideas or people, they enjoy physical activities, and they are skilled in mechanical and physical activities. Typical realistic careers include electrician and engineer (in their BFQ profile they score low in E and F, high in C, low in S and O factors). Conventional individuals are more likely to be conformist, they are organized and conscientious, they prefer organized, systematic activities and well-defined instructions, and are persistent and reliable in carrying out tasks. Typical conventional careers include secretary, accountant and banker (in their BFQ profile they score low in E, average in F, high in C, average in S and low in O factors). It seems that only a third of the first-year Engineering Informatics students selected the university specialty in accordance with their personality traits.

The findings of the study suggest that many of the Engineering Informatics students are not aware of the expectations of the selected profession, and of a mismatch of the attributes required for the job and their personal needs and skills. Practical implication of our findings for engineering education would suggest that the competency development of engineering informatics students should include the development of self-knowledge and several personal and interpersonal skills which mainly contribute to the students' academic career and future success in the workplace. The improvement of learning abilities, concentration and problem-solving skills are also crucial for preventing dropout among the students. Reforms in Engineering Informatics education are necessary, emphasizing the key role of the students' activity in the learning process. There is solid evidence that self-regulated learning [16] has a strong positive impact on students' achievement [17]. It has also been demonstrated by other research that achievement motivation and self-regulated learning strategies have an indirect influence on undergraduates' academic accomplishment [18].

Considering the motivational characteristics, personality traits and skills of the Engineering Informatics students, diverse learning organization methods and appropriate pedagogical techniques within the framework of student-centered learning approaches (e.g. problem-based learning, case studies, students' presentations and projects) [19], need to be implemented in engineering education in order to prepare the students for the employment challenges of today.

References

1. Markes, I.: A review of literature on employability skill needs in engineering. Eur. J. Eng. Educ. **31**, 637–650 (2006). https://doi.org/10.1080/03043790600911704
2. Conlon, E.: The new engineer: between employability and social responsibility. Eur. J. Eng. Educ. **33**(2), 151–159 (2008). https://doi.org/10.1080/03043790801996371
3. Lappalainen, P.: Communication as part of the engineering skills set. Eur. J. Eng. Educ. **34**, 123–129 (2009). https://doi.org/10.1080/03043790902752038
4. Kolmos, A.: Future engineering skills, knowledge and identity. In: Christensen, J., et al. (eds.) Engineering Science, Skills, and Bildung, pp 165–186. Aalborg University, Aalborg (2006)
5. Van Der Molen, H.T., Schmidt, H.G., Kruisman, G.: Personality characteristics of engineers. Eur. J. Eng. Educ. **33**, 495–501 (2007). https://doi.org/10.1080/03043790701433111
6. Direito, I., Pereira, A., Olivera Duarte, A.M.: Engineering undergraduates' perceptions of soft skills: relations with self-efficacy and learning styles. Procedia Soc. Behav. Sci. **55**, 843–851 (2012). https://doi.org/10.1016/j.sbspro.2012.09.571
7. Williamson, J.M., Lounsbury, J.W., Hanc, L.D.: Key personality traits of engineers for innovation and technology development. J. Eng. Tech. Manage. **30**(2), 157–168 (2013). https://doi.org/10.1016/j.jengtecman.2013.01.003
8. Wheeler, P., Haertel, G.D.: Resource Handbook on Performance Assessment and Measurement: A Tool for Students, Practitioners, and Policymakers. The Owl Press, Berkeley (1993)
9. Schomburg, H.: The professional success of higher education graduates. Eur. J. Educ. **42**, 35–57 (2007). https://doi.org/10.1111/j.1465-3435.2007.00286.x
10. Spence, J.T., Helmreich, R.L.: Achievement-related motives and behavior. In: Spence, T. J. (ed.) Achievement and Achievement Motives: Psychological and Sociological Approaches, pp. 10–74. Freeman, San Francisco (1983)
11. Nguyen Luu, L.A., Kovács, M., Frieze, H.I.: Values and ambivalence towards men and women: A study in Hungary and the United States. Alkalmazott Pszichológia (Applied Psychology, English edition) 2003–2004/3–4:7–19 (2004)
12. Caprara, G.V., Barbaranelli, C., Borgogni, L., Perugini, M.: The "big five questionnaire:" a new questionnaire to assess the five factor model. Pers. Individ. Differ. **15**, 281–288 (1993). https://doi.org/10.1016/0191-8869(93)90218-R
13. Rózsa, S., Kő, N., Oláh, A.: Rekonstruálható-e a Big Five a hazai mintán? (Is it possible to reconstruct Big Five in a Hungarian sample?). Pszichológia (Psychology) **26**, 57–76 (2006)
14. Kline, P., Lapham, S.L.: Personality and faculty in British universities. Pers. Individ. Differ. **13**, 855–857 (1992)
15. Holland, J.L.: Making Vocational Choices: A Theory of Vocational Personalities and Work Environments, 2nd edn. Prentice-Hall, Englewood Cliffs (1985)
16. Boekaerts, M.: Self-regulated learning: where we are today. Int. J. Educ. Res. **31**, 445–457 (1999)
17. Zimmerman, B.J.: Self-regulated learning and academic achievement: an overview. Educ. Psychol. **25**, 3–17 (1990). https://doi.org/10.1207/s15326985ep2501_2
18. Yusuf, M.: The impact of self-efficacy, achievement motivation, and self-regulated learning strategies on students' academic achievement. Procedia Soc. Behav. Sci. **15**, 2623–2626 (2011). https://doi.org/10.1016/j.sbspro.2011.04.158
19. Mykrä, T.: Learner-centered Teaching Methods. Indiana University Bloomington, Bloomington (1995)

Assessing the Needs of Technical Intelligentsia for Professional Development

Ekaterina Makarenko, Larisa Petrova(✉), Alexander Solovyev, and Vjatcheslav Prikhodko

Moscow Automobile and Road Construction State Technical University (MADI), Moscow, Russia
makarenko_madi@mail.ru, petrova_madi@mail.ru, soloviev@pre-admission.madi.ru, prikhodko@madi.ru

Abstract. In the paper we studied the results of sociological surveys conducted in 2 target groups including representatives of the so called "technical intelligentsia" social stratum - engineers and top-managers employed in industry and scientific research. Both groups are involved in creation and R&D of intellectual products for the development of science and technology. The analysis of respondents approach on the following problems is given herein: attitude to the national scientific and technical policy, skills necessary for professional career, evaluation of personal competences and needs to forge qualification, look into their own knowledge of current regulating documents for educational and technological activities, perception of interconnections between engineering education and the labor market. The needs within target groups are identified by areas of activity and within the framework of educational programs of professional development.

Keywords: Technical intelligentsia · Additional education · Sociological survey Scientific and technology policy · Professional development

1 Introduction

The new national scientific and technical policy of the Russian Government focuses on forming of high-qualified engineering staff capable to readily meet the challenges in economics and labor market. According to the Government conceptual strategy, it is necessary to motivate technical intellectuals to acquire new competences. Consequently, educational programs for professional development should reflect the corresponding needs of technical intelligentsia.

The purpose of the investigation is to identify different needs of technical elite in the contents of study programs on professional qualification improvement. We have studied what technical intellectuals think about the declared scientific and technical policy, the attitude to their own competences and their own needs in professional career, the viewpoint on their knowledge of current regulating documents for educational and technological activities, their perception of interconnections between engineering education and the labor market, etc.

M. E. Auer et al. (eds.), *Teaching and Learning in a Digital World*,
Advances in Intelligent Systems and Computing 715,
https://doi.org/10.1007/978-3-319-73210-7_28

2 Background

2.1 "Technical Intelligentsia": General Concept and Approaches

The origin of the word "Intelligent" is Latin and it means knowledgeable, and clever. We can meet similar words in many languages but having not the same meanings. The Russian writer P.D. Boborykin firstly used the term «intelligentsia» in the 60-ies of the XIX century.

In the paper we shall use the word "intelligentsia" to determine the social stratum of people professionally engaged in mental (especially difficult) work and, as a rule, graduated higher education [1]. Technical intelligentsia includes specialists with higher technical education engaged in the production, as well as in the development of the intellectual products in science and technology.

We use several methodological approaches for description of the place and the role of intelligentsia within the society and, in particularly, in science and technology [2]. The philosophical approach does not divide intelligentsia to professional groups and suborders. Sociological approach investigates qualitative and quantitative characteristics of the intelligentsia, and of its specific groups. Socio-Economics approach pays more attention to the practical exploring of intelligentsia (or intellectuals). Independently of the approach used, there are common features of the social group that we call "technical intelligentsia":

- Higher technical education;
- Mastering technical specialty, professionalism;
- Real-time employment in the field of intellectual labor.

So technical intelligentsia is recognized as independent social class [1].

2.2 Specifics of the Russian National System of Qualification Improvement and Professional Retraining

In the Russian Federation there exists the "system of additional professional education" (APE) for postgraduate persons [3]. APE bases on specific study programs: programs for qualification improvement designed for formation of new competences within the same specialty, and programs for professional retraining forming competences for the new profession or for the acquisition of new skills.

Additional professional programs are suggested for persons with an average vocational education or with higher education, i.e. for persons, who are attributed to a certain level of technical intelligentsia.

Experts in the sphere of the transport complex are in demand in the labor market [4]. In MADI University, Moscow, there exists a special qualification improvement institute - the Institution for Professional Development (IPD). Additional professional education in the IPD gives the opportunity to implement training programs and professional retraining according to the profile of basic educational programs of the University. Scientific-pedagogical personnel of the University have the opportunity to improve their skills not only in the theory of engineering education and pedagogical technologies, but also in the professional profile of activities. Teachers are involved in the development of additional professional programs and methodologies, as well as in the organization of feedback with students and employers [5].

3 Sociological Surveys: Results and Discussion

3.1 Method of Study

To assess the possible needs of different groups of technical intelligentsia in special training two sociological surveys were carried out. The study approach consists in the comparative analyses of surveys outcomes. (1) The large-scale survey "Modernization of economics and technical intelligentsia" was carried out at enterprises of Moscow, the Moscow Region, and in Tolyatti, Samara Region, with city-forming AVTOVAZ car factory, and in Pavlovo city, Nizhny Novgorod Region, having large bus plant. The developed questionnaire was distributed to 910 technical specialists - engineers of manufacturing enterprises selected by representative quota sampling. (2) An expert survey was carried out among the trainees of IPD MADI and to delegates of the Russian Association for Engineering Education Congress. Among 74 respondents selected by random sampling there were staff members of transport enterprises; 44% of them are top-managers, and so they can be considered as employers for respondents of the mass survey.

Processing of surveys' results was made by ranking method. A certain rank was appropriate to a question depending on the percentage of respondents marked corresponding answers. The results of the analysis of this feedback allow responding to the real needs of industries in the development of educational programs both basic and additional [6].

3.2 Ranking of the National Scientific and Technical Policy Priorities

Results of respondents' opinions concerning the national scientific and technical policy priorities are represented in the Table 1 in hierarchical development order.

The analysis shows that among the announced priorities, both surveys' participants gave the first rank to the priority "Innovations stimulating, introduction of inventions and new technologies". We can state that the modern technical intelligentsia expects innovations, and is ready to participate in moving them forward.

According to the respondents' opinion, during formation of scientific and technological policy it is necessary to take into account the experience of national economical projects of the past. For example, organization of industrial construction in the Tsarist Russia, "new economic policy" of the Soviet Russia in 20-ies of the former century, "industrialization" of 30-ies, "scientific and technological revolution" of 50-ies and 60-ies, etc. Among the examples of the positive modernization, experience of the past respondents gave the first rank to the development of the technical education system in the USSR. We can conclude that engineering personnel training must respect the traditions of the systemic education.

Table 1. Ranking of the national scientific and technical policy priorities

Priority	Rank in the mass survey	Rank in the expert survey
Innovations stimulating, introduction of inventions and new technologies	1	1
Increase of existing manufacturing efficiency	2	3
Increase of management efficiency	3	5
Development of engineering education system	4	2
Increase of energy efficiency	5	4
Support of military enterprises	6	4
Development of IT-technologies	7	4
Development of space technologies and communications	8	5
Development of nuclear energy	9	6

3.3 Ranking of Necessary Engineer's Skills and Needs in Professional Development

The surveys make it possible to determine prevailing forms of professional development training for engineers and experts. There are three popular forms indicated by respondents among suggested 14 variants: qualification improvement courses, additional certification courses, and self-education on acquaintance with actual professional news and publications. Majority of the mass survey participants (1-st rank) and a lot of experts (2-nd rank) passed courses of qualification improvement for the last 5 years.

The question "How do you evaluate your own competence level in your professional sphere today?" received the following answers: 20.7% of engineers – participants of the mass survey, and only 8% of experts, are satisfied with their professional competences. So it is reasonable, that more respondents of the expert survey feel the lack of professional knowledge (26.5%) comparing to the respondents of the mass survey (16.6%). We suppose that this factor motivates top-managers and chiefs of different levels to the professional development training.

The question is what target group members expect from additional study programs on qualification improvement? "What skills are necessary for engineering intellectuals to come up to the modern industry?" In the Table 2 the ranking of respondents' answers are summarized.

The analysis shows that members of both target groups indicated priorities of the first five positions in the list among the suggested 12 variants. The common rule follows from this hierarchy: for better adaptation at the labor market an engineer has to be motivated for professional development. Respondents are unanimous in their confidence: an engineer should have fundamental technical knowledge. Fundamental training of specialists is traditional for the Russian higher technical school. Engineers with deep knowledge in physics, mathematics, mechanics, materials, etc. are flexible in selection and changing of their activity direction. This gives him/her more adaptability both in professional and in social spheres.

Table 2. Ranking of the necessary skills and properties of an engineer of modern industry

Skills and properties	Rank in the mass survey	Rank in the expert survey
Mastering the fundamental knowledge	1	2
Motivation to permanent professional development	2	3
Ability to organize one's work	3	5
Knowledge in technics fundamentals	4	4
IT-skills	5	1
Knowledge of foreign languages	6	8
Adaptability to new economic conditions	7	6
Knowledge of the financial basis of an enterprise	8	9
Knowledge of corporate culture	9	9
Understanding of society development regularities	10	10
Adaptability for the labor market	11	7
Readiness to professional changes	12	10

There are some discrepancies in opinions of experts and mass surveys' respondents. Comparative analysis identified that experts are better oriented in science and technical policy, and in regulating documentations; they have strategic vision of future economic development. Experts gave the first rank to the IT-skills necessary for engineers, whereas the mass survey respondents consider work organization skills more important. We see that such specifics should be taken into account during additional professional programs development for different target groups. The surveys' results are used as the recommendations for development of study programs for additional professional education [7].

4 Conclusions

The study has shown that in the stratum of technical intelligentsia a significant demand exists on mastering qualification improvement programs and programs of professional retraining. For integration of educational services with the labor market further modernization of the additional educational system is urgent.

The surveys determined the following foreground requirements to the necessary competences and capacities of technical intelligentsia: having fundamental technical knowledge, mastering IT-technologies, acquaintance with actual regularity documentation, ambitions on permanent qualification improvement. Development of qualification improvement institutions in Universities consists in realization of new study programs specially formed according to the requirements of different target groups.

References

1. Sillaste, G., Pavlov, N., Georgiev, N.: Creating new economic intelligentsia of Russia and Bulgaria in a market economy (sociological analysis) (2008)
2. Makarenko, E.: Different approaches to the concept of "Technical Intelligentsia" in Russian and western sociology. In: 2013 International Conference on Interactive Collaborative Learning, Kazan, Russia, pp. 369–370 (2013)
3. The Order of MECRF on 1.07.2013 No. 499 (edition of 15.11.2013). On validation of the Order of organization and realization of educational activity on additional professional programs
4. Ushakov, V., Petrova, L., Silyanov, V., Shkitskiy, Yu., Makarenko, E.: Training of high-qualified staff for road branch at the international level. Sci. Eng. Highw. **3**, 1–3 (2012)
5. Ivanov, V.: Training of teachers and the system of APE as a basis for development of research university. High. Educ. Russia **5**, 63–68 (2015)
6. Solovyev, A., Petrova, L., Prikhodko, V., Makarenko, E.: Quality of study programmes or quality of education. In: Interactive Collaborative Learning, ICL 2016. Advances in Intelligent Systems and Computing, vol. 544, pp. 362–366 (2017)
7. Prikhodko, V., Petrova, L., Solovyev, A., Makarenko, E.: Recommendations and instructions on application of the methodic of quality evaluation of educational professional programs of skills improvement and of professional retraining of educational institutions teachers for high-technological economics branches (2010)

Plagiarism Deterring by Using a Sociocybernetic System

Dorin Isoc(✉)

Technical University, Cluj-Napoca, Romania
dorin.isoc@utcluj.ro

Abstract. After more than four years of research and assessment of a sociocybernetic system, the present paper makes the first coherent and concrete reporting. The purpose of the sociocybernetic system is to deter plagiarism in a society clogged with this scourge. The technical features of identification of the act are given, and further the sociocybernetic tool functions included in the sociocybernetic system are specified in the form of the Plagiarized Works Index. The functionality of the Index is placed on the detailed structure, organized on the presentation of records and databases. From the operation of the Index, information is collected so that one builds up statistics that are detailed interpreted. In conclusion, the way of action on historical plagiarisms is made so that they represent an exemplification in the training of the young students, but also a way in which the authors of the published plagiarized works are subject to public rebellion.

Keywords: Plagiarism · Civil society · Intellectual pollution Sociocybernetic system · Academic integrity · Plagiarism deterrence

1 Introduction

Plagiarism is a social phenomenon that is more and more present in higher education and research in many countries. It has become, over time, a serious premise for the intellectual pollution of the environment. All fields are targeted, but the emphasis has shifted significantly towards applied sciences and specialized schools, such as engineering school.

Literature concerning plagiarism excels in ascertaining and, much less, by action or initiative.

First category of works are those concerning methods and forms of diminishing of the plagiarism incident [1–3, 13]. The reported solutions range from organizing teaching to discouragement by severe punishment. There is a tendency to isolate the plagiarism phenomenon. This trend is not intended and does not include professional ethics in the becoming of the future graduate and professional.

M. E. Auer et al. (eds.), *Teaching and Learning in a Digital World*,
Advances in Intelligent Systems and Computing 715,
https://doi.org/10.1007/978-3-319-73210-7_29

A next uniform category of subjects is the one regarding the use of electronic means of detection of clues of plagiarism suspicion [8,10,14]. These detection methods are oriented primarily for didactic activities and relate to school work. The abstract analysis of incidence detection methods through digitized text analysis avoids integrated treatment of plagiarism as a social phenomenon.

All the works reported, as above, are based on the assumption that plagiarism is a violation of academic ethics. This violation involves students, and each university has its own experience and events.

Recently they began to appear syntheses concerning plagiarism in a broader context, sometimes internationalized [4,12], including the predictable connection between plagiarism, dishonesty and corruption [11].

As a whole, the reported literature has a unitary view of plagiarism as to its actual, unique and isolated character. In our view, however, this vision is easily distorted and does not take into account the realities of the current world, despite the emergence of electronic means of plagiarism detection.

In contrast to what we mentioned, our approach was fundamentally different. The starting point is the idea of active deterring approach of plagiarism using a way where the whole academic community - community is regarded as a cybernetic system subjected to an internal disturbance and voluntary action of civil society.

The theoretical basis is that associated with sociocybernetic systems and the development of these systems [5,9]. By nature of the application, it will be taken into account that the sociocybernetic system to be accomplished corresponds to a cybernetics of the third order, i.e. to a cybernetics in which, simultaneously, the observed agent coexists with the observing agent.

The present paper aims to approach the historical plagiarism, i.e. plagiarism identified in previously published works. This plagiarism can not be neglected and produces effects on current and future written production. Essential to this research is that it was requested by a non-governmental organization and that the work of this organization is entirely funded by private sources.

To begin with, the concept of historical plagiarism has been introduced and the need for it to be defined and regarded as an act to be proven is justified. One details the features that make historical plagiarism identifiable.

It is further defined the sociocybernetic system to be implemented through structure, variables and mode of operation. As a particular case, the sociocybernetic machine of the Index of Plagiarized Works is described as a means of dealing with historical plagiarism.

The conclusions of the paper insist that the treatment of historical plagiarism is a way to deter the plagiarism phenomenon and should be gradual, tenacious, permanent and public.

2 Historical Plagiarism: Act and Reporting

2.1 About Historical Plagiarism

Plagiarism is an academic ethics infringement of great complexity and with special consequences. First of all, it is necessary to specify that plagiarism can be treated at an early stage, where there is no publication, such as school themes, or where there are review procedures, applied before publication.

We will further use the name of *historical plagiarism* for any plagiarism that is identified after a public disclosure process.

A brief analysis highlights the consequences of historical plagiarism as a social act:

i. Historical plagiarism affects works that, once published, become accessible to readers.
ii. Historical plagiarism distorts the natural growth of the written cultural heritage of the community, be it in general, even society.
iii. Historical plagiarism has outcomes in justifying social hierarchies in which their authors become favored at the expense of other honest members of the community.
iv. Historical plagiarism is the means of propagating and encouraging other acts of dishonesty such as making untrue data and information, and attempts to fraudulently spend of public money.

For historical plagiarism, identification of the act can occur long after publishing and inserting into the patrimony. The existence of the paper, in paper format, is only one of the technical reasons for which the plagiarism can be identified late.

2.2 Definitions and Reporting Principles of the Plagiarism

The analysis of the reported literature shows unequivocally that the identification and treatment of plagiarism as a deed are rarely approached systematically.

We keep from Isoc [6] the hypotheses about the existence of plagiarism and the character of the act, and we introduce here two issues.

The first issue is the deepening of the content of plagiarism based on concepts brought about by the practice of intellectual creation. In this way, the subject of plagiarism is emphasized as the work that is:

> *...the written work, done and published by a person or of a group of persons who declare themselves in this way. Plagiarism is a work that includes, in whole or in part, a work of written intellectual creation, previously performed and published by another person or by others. ...*

Each pair, a plagiarized work - authentic work, forms a plagiarism or a plagiarism act. It follows that one plagiarized work can identify or identify one or more plagiarisms.

A second systematized issue is of the features of the plagiarism as an act:

i. Plagiarism is a complex act and with many outcomes that manifest itself during the existence of a work. Plagiarism can be detected with evidence at any time, by anyone and to any extent.
ii. Plagiarism can not be prescribed, obviated or disappear.
iii. Plagiarism does not depend on the length of the intellectual creation piece, not attributed, from the work of the authentic author.
iv. Once it exists, plagiarism is incompatible with any kind of reconciliation or re-entry following its public disclosure.

A creative work is a plagiarized or plagiarized work in relation to another work considered genuine if:

i. The two works deal with the same subject or related topics.
ii. The authentic work was disclosed by publication before the suspicious work.
iii. The two works contain common identifiable creative pieces that each, have a well-defined subject and form of presentation.
iv. For common creative pieces, that is, in the authentic work and in the suspicious work, there is no explicit mention of origin. The mention of origin is made through a citation that allows identification of the creation piece taken from the authentic work.
v. Creative pieces taken from the authentic work are used in constructions made by juxtaposition without being treated by the author of the suspected work by his explicit position.
vi. In the suspicious work one identifies a logical thread or several logical threads of argumentation and treatment linking the same premises with the same conclusions as in the authentic work.

The essential issue that appears in this definition is the concept of "*creative piece*" introduced in [6] after which:

> *...the creative piece is an issue of communication presented in written form, as text, image or combined, possessing a logical subject, organization or logic and argumentation that presupposes some premises, a reasoning and a conclusion. The creative piece necessarily presupposes a form of expression specific to a person. The creative piece can be associated with the entire authentic work or with a part of it ...*

2.3 Plagiarism as a Matter of Prevention and Deterrence

Plagiarism is not an act of an individual. It is noted that plagiarism can be an act involving more individuals as coauthors.

By extension, one can conclude that plagiarism exists only if society allows it. In this way, two generic participants are introduced to any plagiarism act that affects the cultural heritage of a community.

The enhancing agent is any legal person or assimilated person, together with the persons who represent it or any of the persons empowered by the author of a plagiarized work, in connection with the occurrence or spread thereof, in any way and anyhow in the society. As an example, for a doctoral dissertation, plagiarism facilitators, and so enhancing agents, are the university or institution awarding the scientific title, together with its employees, empowered in connection with the plagiarized thesis, including the head of the doctoral dissertation and the thesis analysis committee.

It should be noted that the position of an enhancing agent of plagiarism has, not necessarily, an explicit guilt, but a moral, social one.

The integrity guarantor is the person who enjoys notoriety and who, in a direct or indirect form, is brought by the author of a plagiarized work or his enhancing agent to be associated with the purpose of increasing society's trust in them and in their actions. For a university, examples of integrity guarantors are the honoris causa doctors. Despite the fact that they have nothing to do with a particular plagiarism, these people can contribute, through their position, to the incentives for sanctioning a plagiarized thesis case by its retracting and, further, by withdrawing the scientific title.

3 The Sociocybernetic System to Plagiarism Deterring

In the specific literature [5], sociocybernetic systems are analyzed in a sociological context. This time we will consider that the sociocybernetic system has the structure as in Fig. 1. The built system [7] is a functional set, including a community where the individuals appear, at the same time, as observer and observing agents and a technical construction as a set of some information processing resources together with the related connections.

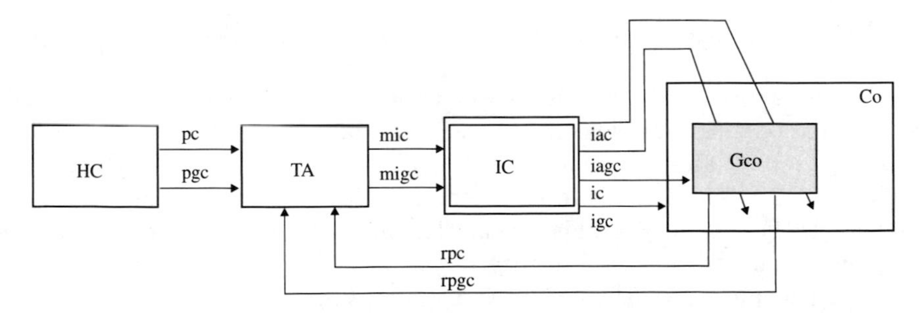

Fig. 1. The structure and associated information with the developed sociocybernetic system.

If we look at the structure in Fig. 1, we find that in relation to first order sociocybernetic systems, the system being managed is a community, Co, and a GCo group inside this community. Unlike technical systems, GCo group and the

Co community can not be separated and they consist of human individuals. In this case, the Co community is associated with a country and the GCo group is associated with its academic community. Plagiarism has the role of an internal disturbance that is about to destabilize the community of the point of view of its integrity.

The sociocybernetic system will build on the basic principles of cybernetics with respect to the negative feedback. A human symbolic decision-maker, HC is who sets the performance of *pc*, respectively *pgc*, required by the governed system that is the Co community and the GCo group, respectively. It is obvious that the set $\{pc, pgc\}$ represents a possible strategy.

Referring to this strategy, also defined by the values of the real performance assessment $\{rpc, rpgc\}$, a TA, *tactical analyzer*, will define the intervention actions of $\{mic, migc\}$ on the Co community and the GCo community group, respectively.

Intervention actions are converted by an IC, *intervention controller*, into several control and adapting actions on the driven system. A first category of interventions $\{ic, igc\}$ acts directly on the Co community and the GCO group respectively. Another category of interventions $\{iac, iagc\}$ acts for the dynamic adaptation of the Co community and the GCo group to change.

There are a number of differences from the technical systems, as systems of a cybernetics of first order. Without insisting, it is noted that the mathematical models of the community and the community group are not technically available. It is further to be noted that both the community and the group are considered as assemblies of individuals, despite the fact that individuals are the observed subject or the observing subject that lead to a situation of third class sociocybernetic system.

Finally, the CI intervention controller is built by rules, some of which are implemented in the structure of an IT application.

4 Embodiment of the Intervention Controller

The intervention controller, IC, is an application that resulted from a nearly 10-year research to which this work refers and it has been materialized on the www.plagiate.ro website and is known as the "Plagiarized Works Index in Romania".

This is the outcome of the research and it is focused on the plagiarism acts that have authors from Romania, no matter where they published, or authors from other countries who publish together with authors from Romania or who publish in Romanian publishing houses or publications.

4.1 Principles

The Plagiarized Works Index is based on a well-defined set of principles.

i. The Index is public utility computer application.
ii. The Index is built of plagiarism acts, but it is always possible to identify a plagiarized work that includes each of the identified plagiarism acts.

iii. Plagiarism acts for which there are well-established clues are included in the Index without being ever removed.
iv. The Index includes in its structure also the social relations that contribute to the promotion and propagation of indexed plagiarism.

Thus, the Plagiarized Works Index is not a simple database but a dedicated database available to a user who wishes to be informed and who wishes to know everything about a complex social phenomenon that affects him.

In the current organization, the Index of Plagiarized Works allows more activities to be described later with reference to its component parts.

Direct information on plagiarism acts. The activity can be done by directly inspecting the plagiarism acts included in the Index. Each record in the table shown as in Fig. 2 represents a plagiarism act.

Informing about the phenomenon of plagiarism in society. The plagiarism phenomenon is initiated by a man or a group of people but is supported by institutions or entities called social agents. Through its policy, the Plagiarized Works Index assumes that social agents are the universities or institutions employing the authors of the plagiarized works, the publications that allow the spreading of plagiarized works and publishing houses that put the plagiarized works into circulation.

4.2 Functional Block Diagram

The solution in Fig. 2 was built and implemented in order to meet the specification resulting from the principles definition of Plagiarized Works Index.

The whole solution is a web page accessed by selection chains and made for users of the Romanian language as a language of origin and English as a language of circulation and impact.

The user operates the technical system as application via a man-machine interface (1).

Once entered into the operating mode, the user has direct access to the full record of plagiarism acts (2), ordered in chronological order as well as to the record of plagiarism acts organized by domains of Universal Decimal Classification (UDC) [15], to the evidence of institutional agents (A) and to the evidence of information related to plagiarism (B). For reasons of operation principle, the two categories of records, A and B, are "frozen" in relation to the time when the plagiarism records were included in the Index.

The evidence of plagiarism acts assumes implicit access to information related to plagiarism acts, that is, to lists of social agents, (9) and to evidentiary documents (10).

Each plagiarism act has a list of enhancing and integrity guarantor agents.

Separately, the user may have access to the record of plagiarism acts distributed on employing universities (4), academies (5) or research institutions (6) at the time of the acts, to the record of plagiarism acts, distributed on the publications that public disclosed the work (7), to the recording of plagiarism acts on publishing houses (8).

Evidential documents of a plagiarism act are represented by the suspicion sheet, plus a number of pages that come to link copies of pages of the plagiarized work with pages of the authentic work. To these can be added documents justifying the situation in which a plagiarized work has been brought to a state of retraction.

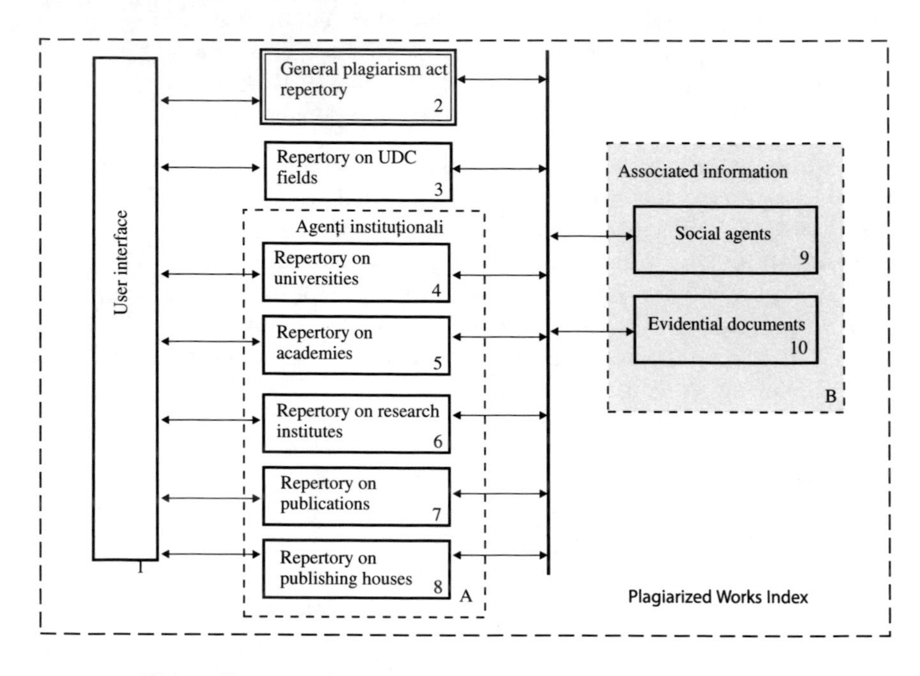

Fig. 2. Functional block diagram of Plagiarized Works Index.

4.3 Direct Informing on Identified and Indexed Plagiarism Acts

In evidence of plagiarized acts, the information is organized into records, as in Fig. 3, which are always arranged in chronological order so that the first is the newest record.

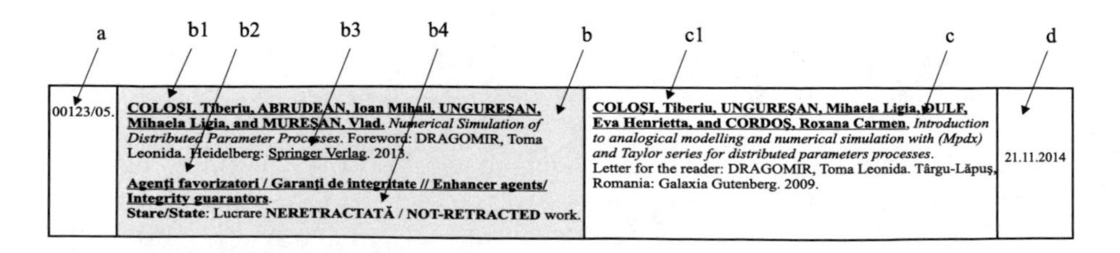

00123/05.	**COLOŞI, Tiberiu, ABRUDEAN, Ioan Mihail, UNGUREŞAN, Mihaela Ligia, and MUREŞAN, Vlad.** *Numerical Simulation of Distributed Parameter Processes*. Foreword: DRAGOMIR, Toma Leonida. Heidelberg: Springer Verlag. 2013. **Agenţi favorizatori / Garanţi de integritate // Enhancer agents/ Integrity guarantors**. **Stare/State**: Lucrare **NERETRACTATĂ / NOT-RETRACTED** work.	**COLOŞI, Tiberiu, UNGUREŞAN, Mihaela Ligia, DULF, Eva Henrietta, and CORDOŞ, Roxana Carmen**, *Introduction to analogical modelling and numerical simulation with (Mpdx) and Taylor series for distributed parameters processes*. Letter for the reader: DRAGOMIR, Toma Leonida. Târgu-Lăpuş, Romania: Galaxia Gutenberg. 2009.	21.11.2014

Fig. 3. Record of Plagiarized Works Index with all significant details.

As principle, the information includes the registration number in the Index, a, the information on the plagiarized work, b and the description of the authentic work, c.

The indexing Number (00123/05) a, is made up of the serial number (00123) and the universal decimal codification indication associated with the plagiarized work. The description information for the plagiarized work, b contains the bibliographic description of the plagiarism act area, $b1$ is a link to the suspicion sheet, then a link, $b2$, to the group of persons who are the enhancing agents and integrity guarantors of plagiarism, a link, $b3$ to the web page of the publishing house that placed the plagiarized work on the market and an indication of the plagiarism status in relation to the retraction of the work, $b4$. In the field of the authentic work c, there is description information and the field of the author group, $c1$, is a link to the suspicion sheet. Field d includes the date of publication of the suspicion of plagiarism. This date is particularly significant as it is the time to enter to publicity stage. This time has nothing to do with the moment of publication, but it is significant in relation to a conventional but certain moment.

5 Results and Interpretations

The Plagiarized Works Index, as intervention controller of a sociocybernetic system is working on $February$ 6, 2012 with successive upgrades and an evolution that has transformed it into a moral authority to treat plagiarism in the academic world.

In the following, only some information resulting from the operation of the intervention controller will be presented. With details of the global functioning of the sociocybernetic system, it will return to separate works.

5.1 About the Authors of Plagiarized Works

The first point of divergence of our analysis with those reported in literature is that it is not possible and fair that plagiarism be attributed exclusively to students.

The results presented in Table 1 show that a sample of 247 plagiarized works identifies 68.53% of authors as professors.

Table 1. Detailing by relevant professional categories of the authors the information volume of the Plagiarized Works Index (350 plagiarism facts).

	Professional categories of the authors of plagiarized works		
	Professors	Doctoral students	Other
Physical	379	27	147
Reported [%]	68.53	4.88	26.59

Even if it does not enter the substance of this information for a detailed analysis in terms of the number of authors/work published, it is obvious that the information may be shocking. On the other hand, it can be explained why, for many universities, it is preferable that plagiarism is not questioned and, in any case, not treated as a worrying social phenomenon.

5.2 About the Form of Publishing of Plagiarized Works

In literature, it is often insisted on accrediting the idea that plagiarism is associated, mainly, with scientific articles. Our study, even partial and incomplete but representative because it is based on a significant amount of actual information synthesized in Table 2, shows that there is significant plagiarism as a volume, and when it comes to monographs or books.

Table 2. Detailing by the way of publishing the plagiarized written works contained in the information volume of the Plagiarized Works Index (350 plagiarism acts).

	How to publish the written work			
	Articles	Books	Thesis	Other
Physical	151	105	87	1
Reported [%]	43.94	30.92	24.85	0.29

5.3 About the Specialized Field of Plagiarized Works

If we analyze the specialty field of plagiarized works, by the definitions of our colleagues, philologists or philosophers, it would result that plagiarism would be an art that excels through the subtlety of unspent paraphrases.

Table 3. Detailing according to the specialized field of plagiarized works included in the information volume of the Plagiarized Works Index (247 plagiarized works/350 plagiarism facts).

	Specialty fields						
	Engineering	Fundamentals	Economics	Medicine	History	Religion	Other
Physical	99	28	20	24	11	4	61
Reported [%]	40.08	11.34	8.10	9.72	4.45	1.62	24.70

An analysis on real data, as in Table 3, indicates, in the first instance, superficially, that plagiarism has a particular incidence in engineering. If the reference is made to the Universal Decimal Classification (UDC) then it is found that plagiarized works in engineering and medicine (6th UDC class) are almost equal to the share that proves the plagiarism economists' appetite (Table 4).

Table 4. Detailing by specialized field of plagiarized works described in the *UDC* classes and included in the information volume of the Plagiarized Works Index (247 plagiarized works/350 plagiarism facts).

	Class by *UDC* (Universal Decimal Codification)									
	0	1	2	3	4	5	6	7	8	9
Physical	14	0	7	125	-	54	128	1	7	14
Reported [%]	4.00	0.00	2.00	35.71	-	15.43	36.57	0.29	2.00	4.00

5.4 About Attitudes and Reactions to Evidences

The factual reporting, certified by concrete results, must be added, even partially, to some attitudes and reactions of those involved, directly or indirectly, in the plagiarism.

All references will be made in relation to the current situation in Romania, but some ideas may also be encountered in other countries or environments:

i. University institutions and academies, as well as publishers and publications, have yet not understood the dangers of plagiarism and the need for all the levers of society to stop it. It proves to me, unquestionably, the fact that, as a result of the published information, there are about 10 retractions to 350 facts of plagiarism reported.
ii. A number of authors, either directly or through "*representatives*", have begun to claim more and more that each profession has the right to define its plagiarism and the specific forms in which it exercises its right to take without indicating the source and without to delimit the takeover.

The only thing they omit each time the authors of plagiarized works is that their ethics violation has consequences and that their written works pollute, by publishing, a cultural patrimony they have reached.

The Internet makes pollution more evident, but also more deterrent the intervention.

6 Concluding Remarks

Plagiarism is a profoundly detrimental social act which, like other similar acts, will not disparage.

The implications of plagiarism are particularly serious and, under these conditions, it is reasonable to act gradually and with extreme toughness. The danger of plagiarism is all the more so as it particularly affects "*white collar*", and so the path of corruption seems to be particularly close.

As always, the possible ways are not unique and are never the most effective. Overall, however, plagiarism can be deterred by a continuous process of education.

The active way to deter the plagiarism is a technical solution materialized in a sociocybernetic system. The key element of the built system is the structure and intervention controller. It is also important to note that the sociocybernetic system acts directly on the community and the target group that produces the plagues, but also ensures their adaptation over time, by building the interventions.

Along with education, a process of maintaining written cultural heritage appears to be necessary in order to identify by all means the plagiarism and dishonesty in their elaboration, and the systematic obedience to public offense of the facts and the perpetrators.

One particular aspect is the incidence between the violation of academic ethics and the criminal offense. The implications are multiple and go from similarities with the phenomenon of counterfeiting to acts of corruption. Any action to discourage plagiarism is a way of anticipating a much greater degree of serious crime. Like other situations, plagiarism is not the effect of an individual's action but the consequences of a community's lack of reaction.

As the danger is a major one, society can not wait for the unique and effective solution of the institutions, whatever they may be. It is obvious, through the results of this research, that a well-thought-out and developed action of civil society representatives can be a pillar of stability for society.

References

1. Berlinck, R.: The academic plagiarism and its punishments - a review. Revista Brasileira de Farmacognosia **21**(3), 365–372 (2011)
2. Born, A.: How to reduce plagiarism. J. Inf. Syst. Educ. **14**(3), 223 (2003)
3. Burke, M.: Deterring plagiarism: a new role for librarians. Libr. Philos. Pract. **6**(2), 10 (2017). (e-journal)
4. Glendinning, I.: Responses to student plagiarism in higher education across Europe. Int. J. Educ. Integrity **10**(1) (2014)
5. Hornung, B.: Principles of sociocybernetics. In: Symposium on Sociocybernetics, Social Complexities from the Individual to Cyberspace, ISA–RC51, Paris (2005)
6. Isoc, D.: User Action Guide Against Plagiarism: Good Conduct, Preventing, Combating (in Romanian). Ecou Transilvan, Cluj-Napoca (2012)
7. Isoc, D.: Notes on the design of socio-cybernetic systems (2017, to be published)
8. Keuskamp, D., Sliuzas, R.: Plagiarism prevention or detection? The contribution of text-matching software to education about academic integrity. J. Acad. Lang. Learn. **1**(1), A91–A99 (2007)
9. Krippendorff, K.: The cybernetics of design and the design of cybernetics. Kybernetes **36**(9/10), 1381–1392 (2007)
10. Martin, D.: Plagiarism and technology: a tool for coping with plagiarism. J. Educ. Bus. **80**(3), 149–152 (2005)
11. Osipian, A.: Corruption in higher education: does it differ across the nations and why? Res. Comp. Int. Educ. **3**(4), 345–365 (2008)
12. Pupovac, V., Bilic-Zulle, L., Petrovecki, M.: On academic plagiarism in Europe. An analytical approach based on four studies. Digithum **10**, 13–19 (2008)

13. Stabingis, L., Šarlauskienė, L., Čepaitienė, N.N.: Measures for plagiarism prevention in students' written works: case study of ASU experience. Procedia-Soc. Behav. Sci. **110**, 689–699 (2014)
14. Sutherland-Smith, W., Carr, R.: Turnitin. com: teachers' perspectives of anti-plagiarism software in raising issues of educational integrity. J. Univ. Teach. Learn. Pract. **2**(3), 94–101 (2005)
15. UDC: Universal decimal classification (2017). http://www.udcc.org

Systemic Approach to In-house Training in English for Technical University Staff

Inga Slesarenko, Irina Zabrodina, and Maria Netesova(✉)

National Research Tomsk Polytechnic University, Tomsk, Russia
{slessare, zabrodina, netesova}@tpu.ru

Abstract. The necessity to design and implement training programmes in English is conditioned by their relevance to current goals of the university development. The training programmes for TPU staff since 2013 incorporate three pathways: courses for research and teaching staff; courses for professional services staff; courses for university senior managers. The programmes are complemented by joint programmes run with international partner. Verification of courses accomplishment is being monitored by independent testing procedures run by another TPU structural unit in order to confirm the correspondence of the level of language proficiency of the courses graduates to the results announced at the beginning of the studies.

Keywords: University internationalisation · Training in English
University staff · English medium

1 Context

Universities are international by nature; they strive for leadership in academic and research performance globally. International character of all activities of universities allows for true competition and advancement. According to the Times Higher Education World University Rankings 2016–2017, 980 universities worldwide are acknowledged as leading international institutions due to their core missions – teaching, research, knowledge transfer and international outlook [1].

Russia has launched Academic Excellence Programme (Project 5-100) to enhance leading positions of Russian research intensive universities in the world [2]. In this context National Research Tomsk Polytechnic University (TPU) as one the leading participants of Russian Academic Excellence Project 5-100 was the first of Russian universities that developed and implemented the in-house training in foreign languages for its students and staff.

2 Purpose and Approach

The design and implementation of training programmes in English is conditioned by their relevance to current goals of the university development first stated in TPU Competitiveness Enhancement Programme (TPU Roadmap) back in 2013 [3]. According to the university roadmap among others TPU effectiveness indicators are

M. E. Auer et al. (eds.), *Teaching and Learning in a Digital World*,
Advances in Intelligent Systems and Computing 715,
https://doi.org/10.1007/978-3-319-73210-7_30

focused on the following priorities that require fluency in English as a foreign language:

1. Increase in the number of articles in Web of Science and Scopus databases
2. Increase in percentage of international professors, teachers and researchers among research and teaching staff, including Russian citizens - foreign universities PhD holders
3. Increase in percentage of international students enrolled in main educational programmes
4. Increase in the number of educational programmes delivered in English
5. Increase in income from international programmes and grants in total amount of income-generating activities
6. Increase in the number of research and teaching staff participated in programmes of international and Russian academic mobility
7. Increase in the number of specialists with experience gained at leading foreign and Russian universities and research organisations [3].

Proficiency in foreign languages, such as English and German, becomes one of the key factors of successful academic and professional performance influencing the university internationalization. The effectiveness indicators mentioned above allow decomposing the goals of English learning and using it at workplace for different staff categories.

According to the priorities in the university development at the current stage, TPU research and teaching staff is engaged in activities that require proficiency in English for the following purposes:

1. English as medium of instruction (EMI) and content courses design performed in English and in accordance with international standards
2. Developing scientific writing skills and writing for publication
3. Developing socializing skills, conferencing, networking with international colleagues, presenting research, one`s own professional portfolio, building professional e-presence, etc.
4. Communicating to professional community and translating research results to public.

To be able to achieve TPU effectiveness indicators stated in the university roadmap other staff categories are also engaged in learning English or enhancing their foreign language proficiency. Thus, university senior managers are to be able to hold meetings, host international delegations, network, and execute written communication, read and prepare documents in English, produce HR documentation and online news.

According to the functional responsibilities of TPU staff work place naming that this TPU staff category becomes the target audience for offering training in a foreign language, the system of TPU in-house training in English was revamped and redesigned in accordance with the TPU development goals. Thus, the training programmes and courses for TPU staff since 2013 incorporate three pathways:

1. Research and teaching staff
2. Professional services staff
3. University senior managers, leadership and administration.

The thorough investigation has been performed [4, 5] to analyse competences groups for each staff category that allowed to formulate the content of workplace duties that is related and to be performed in English in spoken and written communication as presented in Table 1.

Table 1. Learning objectives in mastering English for university staff

University staff category	Learning objective for mastering English
Research and teaching staff	1. Enhancement of English and German language proficiency for the purposes of professional communication and activities 2. Training for exams to certify the level of foreign language proficiency 3. Course delivery and course design in English 4. Research in international collaborations and communicating research results in research papers in English
Senior managers and other university leaders and administration	1. Enhancement of English and German language proficiency for the purposes of professional communication and activities 2. Training for exams to certify the level of foreign language proficiency 3. Network and collaboration with international partners 4. Presenting the university experience and learning from international best practices 5. Business visits 6. Emailing correspondence and documentation in English
Professional services staff	1. Enhancement of English and German language proficiency for the purposes of professional communication and activities 2. Paperwork for international students and staff; front-desk service 3. Providing document flow in English

The approach discussed further in the paper also enabled the authors to validate through practical training provided that the view on teaching and learning English when proficiency in a foreign language is considered a means and instrument for performing professional activities not the final purpose of studies significantly influenced the methodology of needs analysis, teaching materials design, courses design and delivery, planned and measured outcomes and follow-up on professional performance.

The architecture of the training programmes allows for pathways for different staff categories; it also allows for individual pathways for each staff member. The pathways for different staff categories reflect their professional duties. The levels of foreign language proficiency are introduced in compliance with the Common European Framework of Reference for Languages: Learning, Teaching, Assessment (CEFR) [6].

Depending on the entry test results each staff member can choose the relevant programme for studies and is not obliged to take the programmes from the first level on; there is an open access to the programmes to join the training depending on the level of foreign language proficiency and further progress sequentially.

TPU programmes are complemented by joint courses designed and delivered jointly with international partner – University of Southampton (UoS), UK, for those staff categories whose level of language proficiency is required to be advanced for the specifics of professional activities.

Target audience	Research and teaching staff. Learning goal: - enhancement of English and German language proficiency for the purposes of professional communication and activities; - training for exams to certify the level of foreign language proficiency.				Research and teaching staff. Learning goal: professional activities and performance in English	
	Programmes					internships within joint training courses
	Level 0-A1	*Level A1-A2*	*Level A2-B1*	*Level B1-B2*	*Level B2-C1*	*Level B2-C1*
Research and teaching staff	English. Module 1	English. Module 2	English. Module 3	English. Module 4		
			German. Module 2	German. Module 3		
	Research and teaching staff. Learning goal: - course delivery and course design in English; - research in international collaborations and communicating research results in research papers in English					
			Educational technologies for teaching foreign languages	Didactic Competence for Professional Performance in English	Teaching through the Medium of English	Delivering Through the Medium of English (joint course TPU-UoS)
						E-Learning Production and Delivery (joint course TPU-UoS)
				English for Science and Engineering	Writing for Publication	
				Grants and Conferencing in German speaking countries		

Fig. 1. Path ways for training English for university staff target groups. Part 1.

Currently TPU in-house training incorporates 21 courses for mastering the levels of foreign language proficiency and is designed for three main target groups of university staff. The system allows for three pathways into administrative language, language for teaching and research, professional services communication in written and spoken interaction with the option of choosing the corresponding level of foreign language proficiency. Thus each category of university staff members is offered its own clusters of programmes presented in Figs. 1 and 2.

Professional Services Staff	Professional Services Staff. Learning goal: enhancement of English and German language proficiency for the purposes of professional communication and activities					
	Programmes					
	English for Professional Services Staff (beginners)	English for Professional Services Staff (intermediate)		English for PR staff		
			English for HR staff			
Senior managers and other university leaders and administration	Senior Managers. Learning Goal: enhancement of English and German language proficiency for the purposes of professional communication and activities					
	Programmes					internships within joint training courses
		English for Senior Managers. Module 1	English for Senior Managers. Module 2	English for Senior Managers. Module 3 *in progress of design*		Professional Development Through the Medium of English for University Managers (joint course TPU-UoS)

Fig. 2. Path ways for training English for university staff target groups. Part 2.

Module based programmes presented in Fig. 2 are the core of the system and designed for 18 weeks delivery and have more than 200 or 400 contact hours. By the end of 2016/17 academic year the implementation of the discussed here system of TPU in-house training in English and German for university staff is envisioned for the planned anticipated results:

1. From 350 to 400 course participants this academic year who will successfully accomplish their training
2. More than 80% of research and teaching staff among course participants
3. Majority of programmes' graduates will verify their level of achieved foreign language proficiency basedon independent certification exams
4. About 250 TPU staff members will demonstrate B1–C1 level of English language proficiency in 2017
5. About 30 complex teaching aids (UMKD) in English to be developed this academic year within studies in EMI training courses at TPU
6. About 30 e-courses to support content disciplines delivered through the medium of English are to be designed within joint course realization run by TPU and UoS in spring term 2017
7. 14 TPU graduates from Writing for Publication course with 14 research papers submitted for publication in international scientific journals indexed in Web of Science or Scopus data bases in 2017
8. 16 senior managers to graduate from English training courses in 2017
9. 30 professional services staff to graduate from English training courses of various levels of English language proficiency in 2017.

The reengineering of the existing system of training takes place at the end of academic year depending on the university needs analysis and changes that can be introduced to the development goals stated in TPU roadmap [3].

3 Learning Outcomes Verification

Annually the system of English and German training undergoes changes and new programmes are offered instead of those that are not in demand. The review of the programmes takes place twice a year and is secured by needs analysis, questionnaires, follow-up interviews and inquires on the results implementation into workplace activities. There are two kind of assessment inquiries performed to investigate if the programme offered is successfully delivered.

One type of inquires is designed by the staff members of the department responsible for the programmes implementation. The second type of inquires is performed by another structural division within the university responsible for quality assurance in learning process. The purpose of this inquiry is to investigate if the programme is chosen by the appropriate target audience; if the results are meaningful for professional performance of the presumed staff category.

All the programmes are finalized by credit tests and exams. The programmes of advanced level are also finalized by final paper which incorporates the results of studies embodied in development of teaching aids in English; accumulating materials for research paper or analytics, including the following issues for consideration:

1. Investigation into principles of teaching aids design for content disciplines to be delivered through the medium of English; delivering Russian as foreign languages; teaching in multinational groups, organizing independent studies, project work; etc.
2. Presentation of research results for prospective publication – preparation of a structural part of a research paper, e.g. methodology or discussion of research results, etc.
3. Comparative qualitative or quantitative analysis of theories and practical approaches in the fields of university management, modern trends in higher education, educational technologies, advances in science, etc.

The described fields for final paper presented above are not limited to the areas stated but rather indicate the directions in the information research and investigation performed in foreign language by university staff while taking the advanced level training courses in English or German.

Language portfolio as another means to verify the learning results has been introduced in 2017. The portfolio is envisioned as additional verification of learning results; it allows to accumulate the course participants achievements in written and spoken communication, including final paper findings, and serves as evidence of a learner's progress in training [7, 8]. Language portfolio is also a tool for learning process self-management and self-evaluation. The tentative results of portfolio implementation will be available for discussion at the end of 2017.

It is worth mentioning that 94% of TPU staff who took TPU training courses in foreign languages verified their foreign language proficiency by the results of taken certificate exam in 2016. TPU in-house system for certification of the university staff in foreign languages proficiency is also designed in compliance with the Common European Framework of Reference for Languages: Learning, Teaching, Assessment (CEFR) [6].

Among professional enhancement issues for staff responsible for in-house training design and delivery it is worth mentioning that design and implementation of certain pathway for each staff category within the university allows for development of new methodology of teaching and learning. This way within the research and teaching staff pathway the technology of pedagogical design of content courses was developed and successfully implemented with TPU in-house programmes and joint courses run jointly with TPU international partner - UoS. The technology is in compliance with the logic of EMI teaching as introduced in [8, 9]. The training results are actual content courses developed by TPU staff members for bachelor, master degree students enrolled in English speaking educational programmes.

At average from 350 to 450 course participants successfully accomplish training per one academic year. Their results allow university staff members to advance to courses of higher foreign language proficiency level and perform more complex tasks at their work place through the medium of English and/or German.

It is also worth underlying that in 2016 TPU in-house training in English as Medium of Instruction was acknowledged among European best practices of teaching university staff to deliver their content subjects in English [10].

4 Conclusion

The presented system of TPU in-house language training courses design discussed in this paper has been implemented successfully since 2013. Regular reengineering of TPU language training system for university staff has resulted in higher motivation of programmes participants and increased number of staff using English and German as foreign languages for professional purposes followed by TPU staff engagement in university internationalization in different fields of university activities. TPU experience is worth to be disseminated over other universities in countries where educational programmes in universities are designed and delivered in English however English is not official language of a country. The transfer of the technology of language training courses design to other national research universities in Russia has been presented and shared with other Project 5-100 and federal universities since 2014 in form of all-Russian research and methodology seminars and conference devoted to EMI implementation as well as in the form of professional training courses for universities' staff responsible for languages training, working with international students and university internationalization. All events are held on TPU premises and/or delivered online.

Acknowledgment. We would like to express our great appreciation to Tomsk Polytechnic University project partners from European Universities and organizations within the framework of funds and programmes, TEMPUS, TACIS, DAAD, INTAS, Marie Curie Fellowship, FP6 (INCO) etc. The research is funded from Tomsk Polytechnic University Competitiveness Enhancement Program grant.

References

1. World University Rankings 2016–2017. https://www.timeshighereducation.com/world-university-rankings/2017/world-ranking#!/page/0/length/25/sort_by/rank/sort_order/asc/cols/stats. Accessed 19 May 2017
2. Russian Academic Excellence Programme Project 5-100. http://5top100.com. Accessed 17 May 2017
3. TPU Action Plan on the Implementation of National Research Tomsk Polytechnic university Programme for Promoting the Competitiveness ("Roadmap") for 2013–2020. TPU Publishing House, Tomsk (2013)
4. Zabrodina, I., Slesarenko, I., Sivitskaya, L.: The system of training in English for university staff. In: Proceedings of III All-Russian Research and Methodology Seminar-Conference Technical University Students Training in a Foreign Language: Methodological Awareness of Teaching Staff, Tomsk, Russia, 21–23 April 2016, pp. 147–148 (2016)
5. Benson, G., Slesarenko, I., Schamritskaya P.: Technology of language training courses design for university research and teaching staff. In: Proceedings of INTED 2017 Conference, Valencia, Spain, 6th–8th March 2017, pp. 3729–3733 (2017)
6. Common European Framework of Reference for Languages: Learning, Teaching, Assessment (CEFR). http://www.coe.int/t/dg4/linguistic/cadre1_en.asp. Accessed 10 May 2017

7. Reynolds, C., Patton, J., Rhodes, T.: Leveraging the e Portfolio for Integrative Learning: A Faculty Guide to Classroom Practices for Transforming Student Learning. Stylus Publishing, Sterling (2015)
8. Baker, W., Huettner, J.: English and more: a multisite study of roles and conceptualisations of language in English medium multilingual universities from Europe to Asia. J. Multiling. Multicult. Dev. **38**(6), 501–516 (2016)
9. Leung, C., Jenkins, J., Lwekowicz, J.: English for academic purposes: a need for remodeling. Engl. Pract. **3**(3), 55–73 (2016)
10. Lingua Franca Learning: The Growth of EMI. Pie Rev. **12**(4), 60–65 (2016)
11. Slesarenko, I.V., Zabrodina, I.K.: Internet technologies as pedagogical condition for forming foreign language skills of content teachers (at Technical University). In: Proceedings of the 2nd ASSHM, China, vol. 12, pp. 96–99, December 2014

Inbound International Faculty Mobility Programs in Russia: Best Practices

Artem Bezrukov[1], Julia Ziyatdinova[1(✉)], Phillip Sanger[2], Vasily G. Ivanov[1], and Natalia Zoltareva[3]

[1] Kazan National Research Technological University, Kazan, Russian Federation
uliziatd@gmail.com
[2] Purdue University, West Lafayette, USA
[3] Russian State Social University, Moscow, Russian Federation

Abstract. The paper aims at analyzing the best practices of government funded inbound international faculty mobility programs in Russia through the experience of a leading national research engineering university. The paper gives an overview of the global inbound international faculty mobility programs, and then narrows down to the programs funded by the Russian government institutions. Analyzing a specific case of a visiting foreign professor, the study reveals several short-term and long-term outcomes of the inbound mobility practices: from enhancing motivation of students to be internationalized and further developing other academic mobility initiatives such as student mobility, joint research activities, and dual degree programs to a positive impact on a regional economy.

Keywords: Internationalization · Academic mobility · Exchange faculty

1 International Academic Mobility in the Context of Engineering Education

The phenomenon of internationalization has attracted a sustainable interest of the global engineering education community. This process is motivated by internationalization of all the engineering activities in recent decades: from standardization of technologies to migration of engineers through continents as a part of their professional career [1, 2].

Academic mobility is an integral part of education internationalization [3]. Millions of students study abroad today and a foreign student is a common symbol of academic mobility for the world higher education community [4]. Despite students in humanities, who pursue study abroad to become more competitive at the national or global labor market, overseas academic mobility for engineering students is often the only way to obtain competencies their home country education system cannot offer. Such students will then be hired by international companies which need graduates with world level skills or return to their home country to participate in international industrial projects.

Higher education internationalization results in student and faculty exchange programs funded by independent organizations or governments. Major attention is again given to student mobility and exchange of youth. There is little data on faculty mobility,

M. E. Auer et al. (eds.), *Teaching and Learning in a Digital World*,
Advances in Intelligent Systems and Computing 715,
https://doi.org/10.1007/978-3-319-73210-7_31

and every university has its unique experience in this field. It is quite interesting, however, to focus on the benefits of faculty mobility for engineering education and discuss its outcomes in the short or long run.

2 From Student to Faculty Academic Mobility

Much less attention is paid to the faculty mobility as the necessary internationalization activity for enhancing other academic mobility initiatives. Despite the mobility of students, the mobility of faculty is usually expected to develop face-to face faculty cooperation and nurture global student mindset [5].

Is it necessary to develop faculty mobility for engineering education? For centuries, foreign professors were invited to universities to give lectures within their area of expertise and to do research for months or for years. Student and faculty mobility were considered as separate activities what they actually were.

Nowadays, the world research and education community share a common opinion that the universities are in the era of internationalization. Faculty mobility can be considered to be a necessary prerequisite for further development of other academic mobility initiatives such as student mobility, joint research activities, and dual degree programs [6].

A foreign professor travelling to universities abroad now represents the global education community and expertise. A more important role of faculty for internationalization is that they bring mobility opportunities with them. A travelling professor (not a postdoc) is usually a tenured scientist representing academic infrastructure through his/her own research group, connections with other university departments, granting agencies and business. A key distinction of engineering professors from their colleagues in humanities from the viewpoint of internationalization, is that they bring connections with foreign industry that usually cannot be reached by their colleagues or students abroad.

Face-to-face contacts between foreign professors will, therefore, bring two overseas infrastructures together. While mobility of a single student ends with a diploma, mobility of a professor may initiate other forms of internationalization. An engineering educator is the key contact point for a foreign engineering university with the focus on internationalization.

3 "Glocalization" of Faculty Mobility Programs in Russia

3.1 What is "Glocalization"?

There are various ways of funding academic mobility programs. In many countries there are foundations and granting agencies offering grants to faculty and students to study abroad. Such grants are usually subdivided by category and age of grantees: programs for Master's students, PhD students, postdocs and faculty. Grants are also subdivided by country or a world region. European Union offers mobility grants as a part of its Erasmus programs.

Many countries offer their own granting programs, such as Fulbright in the USA and DAAD in Germany. Originally developed to satisfy mobility needs of its citizens, such programs have matured well beyond a domestic level. They now offer many grants to foreign students and faculty in nearly all major world countries.

For such programs, a term "Glocalization" is suitable. This word originally appeared to describe customization of a global product or service on a regional level so it can be sold or commercialized successfully considering peculiar features of different countries or even regions within countries. Having become a global phenomenon, academic mobility faces different local demands in different countries.

3.2 An Example of a Regional Program for International Academic Mobility

For engineering education internationalization, it is interesting to analyze the programs that intensify mobility of engineering students and educators. In this paper, we describe the engineering faculty mobility case funded by the regional program in Russia in one its industrial regions – the Republic of Tatarstan. Being number 3 among its most industrially active regions after Moscow and St. Petersburg, Tatarstan academic infrastructure is focused on engineering education [7].

The regional government developed its own grant program titled "Algarysh" intended to train best specialists for the regional economy [8]. This program is a part of the regional strategy of economic development [9]. As engineering branches of industry, such as chemical engineering, mechanical engineering and petrochemistry are the local economic drivers, the local academic mobility program is "glocalized" for engineering education. This program offers grants for the following categories of applicants:

- Master's students;
- PhD students;
- Secondary School Teachers;
- University Faculty;
- Project Groups.

In the last two years, it offered a new grant to invite top-class foreign scientists to regional universities.

It is interesting that the board of experts of this program is often formed by the CEOs of local engineering companies and the key attention at the grant interview is thus given to applications of engineering students and engineering educators as outbound and inbound visiting faculty.

4 A Foreign Faculty Benefits for Engineering Students: Facts and Figures from a Specific Example

4.1 Grant Environment

In this paper we discussed the experience of the Engineering Education department at one of universities of Tatarstan: a national research university [10]. As in many world countries, national research universities in Russia were created as centers of research and education activities especially in the engineering area [11]. The department won the inbound academic mobility grant to invite and engineering professor from a US engineering university.

The total duration of grant was 4 months in the fall 2017. The grant was given to invite this professor to give lectures and organize workshops in engineering education. About 100 engineering students and 30 faculty members attended the course of lectures. The courses were also organized within the framework of ING PAED IGIP – IGIP International Engineering Educator Program and the professor shared the experience with several engineering faculty members and increased their expertise as future engineering educators.

4.2 Visiting Professor Audience Survey

During this training a survey (involving around 100 engineering students) was carried out to track changing attitude of students to internationalization. The survey consisted of three major groups of questions: intercultural tolerance [12], attitude to internationalization and general desire to learn English as a foreign language [13] to further use it as a part of internationalization activities.

The survey was carried out prior the start of the training course and short time after the course completion. The survey revealed that before the course, the students had moderate motivation to internationalize. Around 25% of students wanted to add international components to their curriculum. An intercultural tolerance component was strong (over 70%), while less than 40% of students considered learning English necessary for their future career.

The course completion resulted in significant changes in these statistics. The amount of students willing to internationalize and learn English for their future professional communication almost doubled (50% and 76% respectively). Intercultural tolerance values remained high (78%). The initially high level of intercultural tolerance, however, can be explained by regional traditions, as the Republic of Tatarstan is among top Russian regions with national and religious diversity and national and cultural tolerance are historically high here.

4.3 Contribution to Engineering Education Internationalization and Globalization of a Region

From the point of longer term outcomes, this grant contributed to the overall development of academic mobility between two universities. Internationalization activities initiated by this grant involved joint research and publications in international peer-reviewed journals, joint participation in international IGIP and ASEE conferences and grant applications for Erasmus+ programs. This project suited the general university internationalization strategy assuming progress from separate globalization initiatives to planned projects resulting in creation the network of partners from different world countries with sustainable and long-lasting joint activities [14, 15].

Final long-term outcomes of such programs for the regional economy will be seen in 10–15 years. Certain statistics, however, can be already gathered. The alumni association of “Algarysh” grantees (created in 2007) includes over 3,500 graduates with around 70% of student graduates. The program graduates are employed by nearly all regional large-scale engineering companies with foreign investment or global outreach including Mitsubishi, Honeywell, Yokogawa, Gazprom, and etc. The project groups which win this program grants for study abroad activities also involve engineers from local companies who then implement their international experience to improving their business.

5 Conclusions

Thus, this study revealed several short-term and long term outcomes for engineering education internationalization from the discussed inbound mobility practices.

The short-term outcome of a locally funded program of a foreign professor visit to an engineering university is found to be the motivation of students to globalize their education by applying for international grants, publishing papers abroad and using English for professional communication.

A mid-term benefit of inbound mobility is boosting student exchange and joint research in engineering education through direct professor-to professor contacts that will, in their turn, intensify other academic mobility programs such as student exchange, attraction of international funding as grants and industrial contracts.

A final long-term outcome is that the university will train engineers with international competitiveness which make a positive impact on a regional economy by forming a pool of qualified engineering specialists for industrial projects with foreign investment and adding a global outreach to a region.

Acknowledgements. The research was funded by Russian Foundation for Basic Research, project #15-26-09001 “Development and Implementation of a Network Interaction Model for Regional Universities of Vietnam and Russia for Internationalization of Engineering Education”.

References

1. de Wit, H.: Who Owns Internationalisation? University World News, **350** (2015)
2. Ziyatdinova, J.N., Osipov, P.N., Bezrukov, A.N.: Global challenges and problems of Russian engineering education modernization. In: Proceedings of 2015 International Conference on Interactive Collaborative Learning, ICL 2015 (2015). https://doi.org/10.1109/ICL.2015.7318061
3. Shageeva, F.T., Erova, D.R., Gorodetskaya, I.M., Prikhodko, L.V.: Socio-psychological readiness for academic mobility of engineering students. In: Proceedings of 2016 International Conference on Interactive Collaborative Learning, ICL 2016. Paper ID 1743 (2016)
4. Knight, J.: Student mobility and internationalization: trends and tribulations. Res. Comp. Int. Educ. **7**(1), 20–33 (2012)

5. Altbach, P.G., Yudkevich M.: The Role of International Faculty in the Mobility Era, **444** (2017)
6. Marginson, S., Van der Wende, M.: Europeanisation, international rankings and faculty mobility. Three cases in higher education globalisation. In: Higher Education to 2030, vol. 2, pp. 109–144 (2009)
7. Ziyatdinova, J., Bezrukov, A., Sanger, P.A., Osipov, P.: Best Practices of Engineering Education Internationalization in a Russian Top-20 University. Paper presented at 2016 ASEE International Forum, New Orleans, Louisiana, June 2016. https://peer.asee.org/27236
8. Kvon, G.M., et al.: Features of socio-economic development of the republic of Tatarstan and conditions for the implementation of the investment and cluster policies of the region. Int. J. Environ. Sci. Educ. **11**(17), 10553–10567 (2016)
9. Strategy of Social and Economic Development of the Republic of Tatarstan till 2030. http://invest.tatarstan.ru/advantages/strategy-2030/. Accessed: 21 May 2017
10. Altbach, P.G.: Advancing the national and global knowledge economy: the role of research universities in developing countries. Stud. High. Educ. **38**, 316–330 (2013)
11. Oleynikova, O.N., Zolotareva, N.M.: Conceptual Aspects of the National Qualifications Framework. Vestnik Tverskogo Gosudarstvennogo Universiteta. Bulletin of Tver State University, No. 4, pp. 113–125 (2014)
12. Gorodetskaya, I.M., Romani, P.M., Sanger, P.A.: Cross cultural learning motivations for engineering students. Paper presented at 2016 ASEE Annual Conference & Exposition, New Orleans, Louisiana. Paper ID14960 (2016)
13. Valeeva, E.E., Bezrukov, A.N.: New Teaching Methods for Professional English Language Courses to Follow Internationalization of Engineering Education. Sovremennye Problemy Nauki I Obrazovaniya. Modern Problems of Higher Education. No 1–1, pp. 430–434 (2016) (In Russ., abstract in Eng.)
14. Ziyatdinova, J., Bezrukov, A., Sukhristina, A., Sanger, P.A.: Development of a Networking Model for Internationalization of Engineering Universities and its Implementation for the Russia-Vietnam Partnership Paper presented at 2016 ASEE Annual Conference & Exposition, New Orleans, Louisiana. 10.18260/p. 26808, June 2016
15. Sukhristina, A.S., Ziyatdinova J.N., Kochnev A.M.: Networking as a Form of Internationalization in Education: Case Study of KNRTU. Vyssheeobrazovanie v Rossii. Higher Education in Russia. No.11, pp. 103–110 (2016) (In Russ., abstract in Eng.)

Effect of an Emotional Video on Skin Conductance Response of Respondents in Dependence on Personality Type

Martin Malcik(✉) and Miroslava Miklosikova

Department of Social Sciences, VSB-Technical University of Ostrava, Ostrava, Czech Republic
{martin.malcik,miroslava.miklosikova}@vsb.cz

Abstract. This paper reports on the results of research whose aim was to find out whether and how the respondents' skin conductance change due to emotions experienced while watching emotional videos and whether there is a relation between a type of personality and a type of "emotional" curves we measured (the respondents were the teachers of vocational subjects). Emotions arise as an immediate response to a current situation with regard to the individual's experience, interests and goals, comprising both subjective experiences and physiological changes, changes in attention and readiness, and changes in motoric expressions. The paper aims to show how secondary school teachers experience emotions and whether there is a relation between the measured values of skin conductance and types of personality. We found out that the skin conductance method reliably measures the intensity of experiencing emotions. It became apparent that the intensity of experiencing emotions is connected with a personality type in certain cases.

Keywords: Emotions, skin conductance response
Personality test Myers–Briggs type indicator · Secondary school teacher

1 Introduction

Emotions permeate the whole psyche of an individual [1]; they influence memory, learning [2], motivation [3] and the system of values [1]. Goleman [4] argues that an individual's success and satisfaction with him/herself depends on the way he/she can handle his/her emotions and use them for his/her own benefit. He assumes that young individuals experience what he called "the paradox of young generation". It can be characterized by the rise of rational, and the decrease of emotional intelligence so they experience to a larger extent the feelings of loneliness, nervousness, aggression, anxiety, impulsivity and dependency, alienation, depression etc. The connection between emotions and a physical response has always been in the centre of attention because, given the changes they cause, they quickly became the subject of various physiological research [5]. In 1890, James [6] stated that physical changes, which a person feels, come after the perception of an exciting reality, and he called the subsequent feeling an emotion. When searching for specific physiological profiles of various emotions, a small variability of responses of target organs appeared as a problem, i.e. the same

M. E. Auer et al. (eds.), *Teaching and Learning in a Digital World*,
Advances in Intelligent Systems and Computing 715,
https://doi.org/10.1007/978-3-319-73210-7_32

physiological response (rapid heartbeat when experiencing happiness, anger or fear) was caused by a greater number of stimuli [7]. The research of emotions continued and it proved that a remarkable and separate variable which shows stability, even though only of rough differentiation of emotional states (sadness, fear, anger, disgust), is skin conductance. Among current opinions, complex models of parallel processing of emotional stimuli prevail. An important source of new information is neuropsychological knowledge, which is enhanced by modern imaging techniques such as emission tomography, magnetoencephalography, functional magnetic resonance imaging etc. For our purpose, we used the method of Galvanic skin response (GSR) and measured the skin conductance of respondents in a situation when they watched an emotional video. We are aware of the fact that respondents could knowingly suppress emotional reactions to a certain degree because they knew that they were participating in research. Thompson [8] claims that emotional regulation is related to processes by which people influence what emotions they have, when they have them, how they experience them and express them, which can influence physiology and skin conductance. On the other hand, we believe that skin conductance can be influenced less than for instance facial expression.

Naturally, we expected that respondents' responses to an emotionally-tinged stimulus would differ and we decided to find out whether there is a connection between respondents' measured values of skin conductance and their personality type. Personality as a set of characteristics, a typical compact individuality of a given person, is manifested by a quality of experience as well as by the character of behaviour.

1.1 The Influence of Emotions on Physiological Changes in an Organism

Emotions are complex phenomena which are characterised by enormous sensitivity and changeability. The sensitivity of emotions related to changes in personal and situational circumstances is manifested by the fact that they can change without apparent changes in objective circumstances and only based on subjective evaluation. This is manifested in a way where in a given situation, an emotion is triggered, but in a different situation, equally typical, it is not. Such an amount of sensitivity is not common in other mental processes [5].

Emotions are closely related to physical responses, which is evidenced in many psychological approaches which solely focus on research of peripheral physiological changes or brain processes when experiencing emotions. The quality of physiological changes in an organism is significantly influenced by individual's interpretation of what s/he is feeling since apart from the present situation, s/he also includes prior experience and expectation resulting from it. Individuals who realise this fact can understand the processes that are going on within themselves better, and also regulate their emotions better. One way how to regulate emotions is a change in the evaluation of a situation. This modification can lead to an increase or decrease in intensity of a given emotional response. However, LeDoux [9] states that re-evaluation is effective only from the social point of view and not from the physiological one since cognitive evaluation does not influence physiological response very much. The above-mentioned finding led some theorists to study a mutual relationship between a mental and physiological state and to try to explain the opposite better, i.e. how changes in a physiological state influence a mental state. According to [5], one of the possibilities are the techniques balancing on the edge of behavioural and physiological intervention, for example, biofeedback or focusing. By showing that there is a way how to get emotional experience under control, we wanted to point out the possibility that our respondents could slightly knowingly modify their physiological response.

1.2 Emotional Experience in the Context of Personality Types

Personality consists of an individual unity of biological, psychological and social factors; it is created and manifested in relationships between people [10]. It is an individuality, individual's distinctiveness from other individuals, mainly individuals of the same age and culture [11]. Types are not people, but only precise configurations of characteristics, a hypothetical construct. It is necessary to respect the fact that a majority of descriptive characteristics of an individual have continuity, for instance in the descriptive dimension of introversion - extroversion, there is only a small number of clear introverts or extroverts. Empirical research shows that an ideal type appears very rarely, while mixed types appear regularly [12].

As for temperament, it means a consistent emotional tendency, habitual dispositions, which typically characterise behaviour of an individual. It has innate dispositions and it is very important in the dynamics of individual's adaptation to circumstances. Its relation to emotions is apparent. Buss, Plomin (in [13]) mention dimensions of temperament, and one of them is sociability. It is manifested, for example, by a preference to be in a company of other people and sensitivity to social situations. In our opinion, the amount of sensitivity in the perception of social situations could influence our respondents' responses to emotional videos selected by us. To find out whether our assumption is true, we first determined the type of respondents' temperament using Myers-Briggs Type Indicator and then we compared it with measured values of their skin conductance.

1.3 MBTI

Myers-Briggs Type Indicator works with these personality traits:

- introversion versus extroversion,
- intuition versus senses,
- thinking versus feeling,
- perception versus judging.

By combining these possibilities, we can get 16 groups, which can be divided into four types of temperament:

- Dionysian SP (39% population; characteristics: based on the combination of sensory perception and awareness of the moment - impulsiveness, action, copes well under pressure, stress, risk-taking, need for audience, optimism, entertainer, aversion to monotony).

- Epimethean SJ (38% population; characteristics: based on the combination of sensory perception and logical judging - willingness to comply with rules, obedience, a need to belong somewhere, caring for others).
- Promethean NT (12% population; characteristics: based on the combination of performance and skill - competence, self-control and self-criticism, a tendency toward perfectionism, sometimes arrogance, inability to perceive the complexity of human relationships).
- Apollonian NF (11% population; characteristics: based on the combination of intuition and feeling - yearning for the meaning of life, self-actualisation, raised sensitivity, selflessness, genuineness, empathy, creativity, emphasis on human relationships) [14].

2 Material and Methods

The aim of our research was to measure the intensity of experiencing emotions in secondary school teachers using the galvanic skin conductance (GSC) method and to find out whether there is a connection between the measured values of skin conductance and their personality types.

2.1 Respondents

The research sample consisted of tertiary-educated people, prospective teachers, who broaden their qualifications in pedagogy within further education. There were 29 people, of which 20 were women and 9 men aged 25 to 45.

2.2 Questionnaire Determining the MBTI

To determine the MBTI, we used David Keirsey's questionnaire, which consists of 70 dichotomous closed questions [14]. Acquired data were assessed according to given instructions. To measure respondents' personality type, we used Myers and Briggs' type indicator (MBIT), which serves as a personality test working with sixteen behaviour patterns. To acquire necessary data, we used Keirsey's test of temperament.

2.3 Electrodermal Activity (EDA)

Emotional excitation affects the skin of a tested person. Electrodermal activity refers to electrical changes, measured at the surface of the skin, that arise when the skin receives innervating signals from the brain [15]. During this excitation, in accordance with the sympathetic response, sweat glands in the skin filled with sweat, a weak electrolyte and good conductor. This results in many low-resistance parallel pathways, thereby increasing the conductivity of the skin [16]. EDA is divided into tonic and phasic phenomena, which can roughly be thought of as "the smooth underlying slowly-changing baseline" vs. "the rapidly changing peaks." EDA can be measured in many different ways electrically including skin potential, resistance, conductance, admittance, and impedance [17]. The Edlab provides a way to capture electrical conductance across the skin (GSC).

2.4 Skin Conductance Response

Unless an investigator is speciffficaly interested in comparing his work with the literature os skin potencial, skin conductance measurement are to be preferred [17]. Skin Conductance Responses (SCRs) are abrupt increases in the conductance of the skin. The big SCRs that are circled in Figs. 2, 3, 4 and 5 illustrate their most common form: they usually have a faster rise time than decay time. These SCR's are examples of "phasic activations".

2.5 Eduction Laboratory Board (EdLab)

The edlab measuring device is an interface for connection up to 2 digital and 6 analogue sensors to a computer using USB interface. Edlab uses 12-bit A/D converter with 2×30 kHz sampling rate. For recording and evaluating the measurement, there is the Edlab software, which enables to export measured data in a csv format. As a sensor, we used the sensor of skin conductance, which uses two electrodes placed on fingertips (see Fig. 1) and works as if they were terminals of a resistor (skin resistance). This resistance is one line of the voltage divider. Voltage recorded in this way is filtered and increased and then evaluated by EdLaB measuring system through equalising algorithm.

Fig. 1. Placement of sensors on respondent's hand. Source: EdLab manual.

2.6 Video as a Stimulus

To find out how respondents respond to an emotionally escalated situation, we prepared a stimulus in form of a video. This video consisted of two parts, the first part was 1.5 min long and contained 5 scenes showing dangerous movement and working at heights. The second part was 1.6 min long and contained completely different sections, where the first section lasting 1 min was supposed to create a relaxing atmosphere and pleasant experience. The second section beginning at 124 s into the whole video contained drastic experience of a violent death and lasted 62 s.

2.7 Questionnaire

For research purposes, we created a non-anonymous questionnaire, which contained 4 items related to the present mental state of a respondent, the quality of his/her concentration on a given video and a question whether s/he is afraid of heights. In the last open item, respondents were supposed to note which moment in the film was emotionally strongest for them. Acquired data were compared with MBTI and GSC measured values.

2.8 Research Process

At the beginning of the research, we determined respondents' types of temperament using MBTI and divided them into groups based on this. Then, we played the video to each respondent during which we measured galvanic skin conductance (GSC). At the end, the respondents filled in a short questionnaire, in which they wrote about the items related to their own subjective experience when watching the video. Then, we processed acquired data.

3 Results and Discussion

3.1 Evaluating MBTI

First, we evaluated respondents' MBTI and we found out that there are 9 people with ESFJ combination, 8 people with ESTJ, 3 people with ISTJ, 3 people with XSTJ, 2 people with ESXJ, and 1 person with ISXJ, INTJ, ESTP and ENFP respectively. These combinations show that the sample consists of 26 respondents (90%) of the SJ Epimethean type and one respondent in SP, NT and NF. This proportion does not correspond with the issued distribution of types in the population which is caused by the size of the sample; nonetheless, this fact does not influence other findings.

3.2 Evaluating Data from GSC Measuring

In GSC charts, we can observe several regularities. In all bellow-mentioned cases (Figs. 2, 3, 4 and 5), it is possible to identify the moment when the dramatic event happened, without regard to respondent's personality type - it was approximately 130 s into the video (marked red in Figs. 2, 3, 4 and 5). The amount of intensity of the experience is apparent from individual charts. Respondents from F (feeling) group showed a higher intensity of experience than respondents from T (thinking) group.

After dividing respondents into two groups according to their personality type, FJ (feeling, judging) and TJ (thinking – judging), we found out following information:

- In the first group of respondents, there was a big difference between the lowest and highest GSC value - more than 500 units (see Figs. 2 and 3).

- In the second group of respondents, there was a small difference between the lowest and highest GSC value - about 200 units (see Figs. 4 and 5). We did not take into consideration the amount of the GSC value itself since it is dependent on the physiological parameters of the skin.

According to the different monotony of charts (increasing, decreasing, constant chart), we observed that some respondents respond to the stimulation video by higher impulsiveness, which was manifested by a more frequent change in the monotony of the chart. In our research sample, we were not able to confirm the dependence between a personality type and a frequency of changes in monotony, but we suppose that this dependence exists to a certain degree.

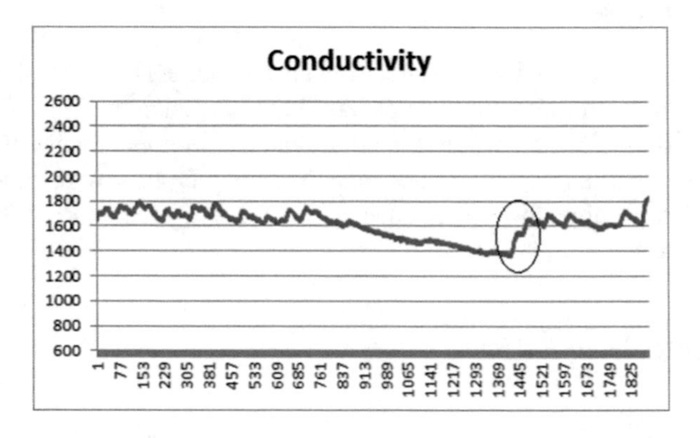

Fig. 2. GSC chart for the respondent A of the ESFJ type. Source: authors' measurement. Legend: horizontal axis - tenths of a second, vertical axis - calculated quantity based on the skin conductance ranging from 0 to 5000.

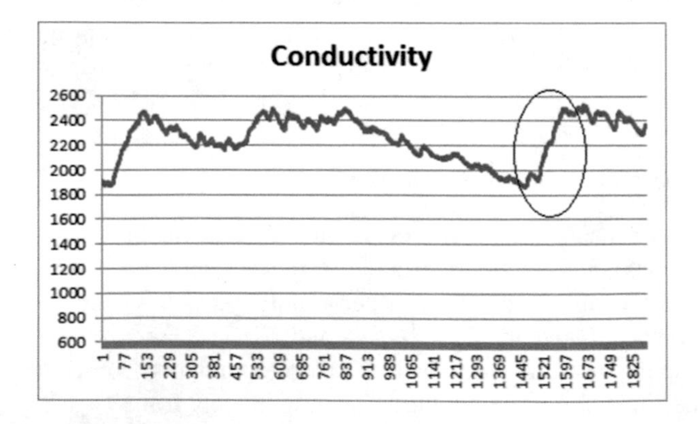

Fig. 3. GSC chart for the respondent B of the ESFJ type. Source: authors' measurement. Legend: horizontal axis - tenths of a second, vertical axis - calculated quantity based on the skin conductance ranging from 0 to 5000.

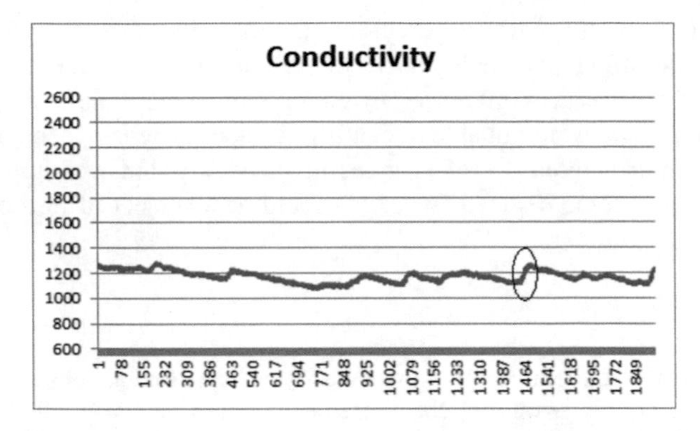

Fig. 4. GSC chart for the respondent C of the ESTJ type. Source: authors' measurement. Legend: horizontal axis - tenths of a second, vertical axis - calculated quantity based on the skin conductance ranging from 0 to 5000.

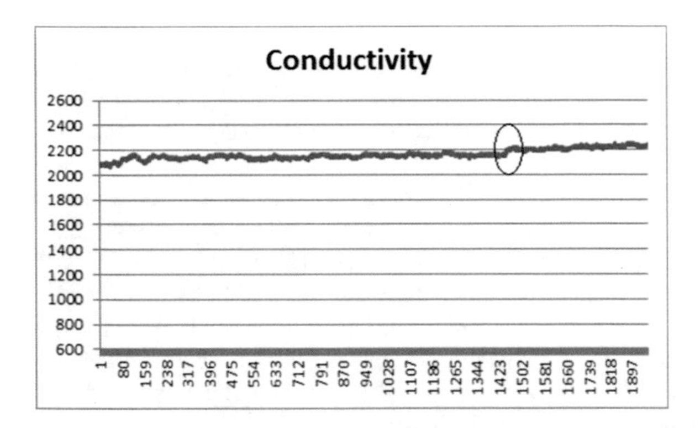

Fig. 5. GSC chart for the respondent D of the ISTJ type. Source: authors' measurement. Legend: horizontal axis - tenths of a second, vertical axis - calculated quantity based on the skin conductance ranging from 0 to 5000.

3.3 Evaluation of the Questionnaire

As for their present mental state, 22 respondents noted very good or good, 4 respondents average and 3 respondents rather bad or bad. 15 respondents were totally immersed in the film, others focused only 'partly'. 15 respondents are afraid of heights, which influenced their experience when watching the first part of the video.

4 Conclusion

The aim of our research was to measure the intensity of experiencing emotion in secondary school teachers using the GSC method and to examine whether there is a connection between teachers' measured values of skin conductance and their personality types. We found out that the GSC method reliably measures the intensity of experiencing emotions. It became apparent that the intensity of experiencing emotions is connected with MBTI personality type in certain cases. We are aware that the intensity of emotions was also influenced by the present mental state of respondents. In our research sample, we were not able to confirm whether there is dependence between a personality type and frequency of changes in monotony, but we suppose that this dependence exists to some degree. This fact should be a subject of further research.

References

1. Schmidt, L., Atzert, L.: Lehrbuch der Emotionspsychologie, Berlin, Köln (1966). 312 s
2. Reykowski, J.: Experymentalna psychologia emocji, Warszawa: PWN (1968). 579 s
3. Leeper, R.W.: A motivational theory of emotion to replace "emotion as disorganized response.". Psychol. Rev. **55**, 5–21 (1948)
4. Goleman, D.: Práce s emoční inteligencí, Praha, Columbus (2000). 366 s
5. Stuchlíková, I.: Základy psychologie emocí, p. 227. Portál, Praha (2007)
6. James, W.: Principles of Psychology. Henry Holt, New York (1890)
7. Qigley, K.S., Berntson, G.G.: Autonomie irigins of cardiac responses to nonsignal stimuli in the rat. In: Behavioral Neuroscience, s. 104, pp. 751–762 (1990)
8. Thompson, R.A.: Emotion and self-regulation. In: Thompson, R.A. (ed.) Nebraska Symposium on Motivation Socioemotional Development. University Of Nebraska Press, Lincoln (1990)
9. LeDoux, J.E.: Cognitive-emotional interactions in the brain. Cogn. Emot. **3**, 267–289 (1989)
10. Nakonečný, M.: Psychologie osobnosti, ACADEMIA, Praha (1995). 336 s
11. Říčan, P.: Psychologie osobnosti. Geada, Praha (2010). 208 s
12. Šnýdrová, I.: Psychodiagnostika. Grada, Praha (2008). 144 s
13. Blatný, M., et al.: Psychologie osobnosti. Grada, Praha (2010). 304 s
14. Keirsey, D., Bates, M.: Jaký jste typ osobnosti?. Grada, Praha (2006)
15. Picard, R.W., Fedor, S., Ayzenberg, Y.: Multiple arousal theory and daily-life electrodermal activity asymmetry. Emot. Rev. **8**, 1–14 (2015)
16. Malmivuo, J., Plonsey, R.: Bioelectromagnetism. The Electrodermal Response. Oxford University Press, New York (1995)
17. Boucsein, W.: Electrodermal aktivity. Springer, New York (2012)

Motivation of Adolescents to the Study of Technical Branches as a Priority of the Czech Education System

Pavel Andres(✉), Alena Vališová, Petr Svoboda, and Lenka Mynaříková

Department of Pedagogical and Psychological Studies, Masaryk Institute of Advanced Studies, Czech Technical University in Prague, Prague, Czech Republic
{pavel.andres, alena.valisova, petr.svoboda, lenka.mynarikova}@cvut.cz

Abstract. The project is at the theoretical level oriented to the possibilities of enhancement of the adolescents' motivation to the study at technically oriented secondary schools. Innovative methods and forms of teaching are important means that put emphasis on the use of information and communication technologies and elements of practice learning. The project proceeds from the combination of extensive (theoretical analyses, observation) and intensive way of research (moderated discussions, focus groups, pedagogical experiment). The teaching methods and techniques using ICT will be proposed on the basis of the respective school documents and professional literature analysis. The attention will be concentrated on outlining of a motivation programme for the secondary school students that will be realized and verified in teaching at the secondary schools. At the same time the interest of adolescents in science and technology will be systematically developed. The project anticipates cooperation with the distinguished science and technology specialists and with experts from technical workplaces. The project is based on the following fundamental premises: the need of study at technical schools as a priority of the Ministry of education in 2016 and the important demands of the labour market, i.e. the fulfilment of graduates at the technical professional posts.

Keywords: Motivation · Adolescent · Technical branches
Technical education · Innovative teaching methods
Education system · Information and communication technologies
M-learning · Information ethic

1 Introduction

The objective of the project is based on the previously achieved results of the preceding projects, primarily on the results of the project "Preparation of technicians for education and human resource management under the conditions of the integrative Europe" solved in the period 2006 – 2008, registration No. 406/06/1414. The goal of this project was a thorough theoretical and empirical analysis of the existing system of preparation

M. E. Auer et al. (eds.), *Teaching and Learning in a Digital World*,
Advances in Intelligent Systems and Computing 715,
https://doi.org/10.1007/978-3-319-73210-7_33

of technicians for the pedagogical a managerial activity, its optimization in accordance with the up-to-date knowledge and the advancing trends in the integrative Europe and in the Czech Republic. Further to the new legislation, the project followed the questions of the readiness of technicians, as to the professional, pedagogical and managerial aspects, for implementation of the fundamental programme education documents, In: Semrád J., Škrabal M. "Preparation of the technicians for education and human resource management under the conditions of the integrative Europe". The final report of the grant solution GAČR. ISBN 978-80-01-04287-8.

Further we refer to the project of the Education Politics Fund of the Ministry of Education (MŠMT) titled "Management of motivation readiness of the technically educated students to the practice of teaching profession, realized in 2014 and 2015, the investigation object of which was to enhance the students' motivation by a focused educational activity, and professionally taking part in the pedagogical activity at the technically oriented regional schools in the Czech Republic. The first phase of our research was devoted to the bibliographic search of sources, from our country and from abroad, related to the motivation theory with respect to the choice of employment (with emphasis on the technical professions), especially of the Vroom's theory of "job involvement", proving importance of the detailed knowledge of concrete conditions of the work, with respect to the probability of its acquirement and maintaining. At the subsequent phase two following research investigations were carried out:

1. The intensive form of investigation, by way of an in-depth analysis, carried out at a selected group of first-year students, took place in the form of moderated discussions with 10 students of both study branches. At this phase of research was ascertained the motivation readiness of students to the practice of teaching profession, and for this purpose the postural scales of a Likert type were used for verification of the modality and intensity of stances.
2. Extensive form of investigation via questionnaires was carried out at the basic group of first-year students of both study branches "Teaching of technical subjects and "Teaching of practical education and vocational training, the objective of which was to describe the attitudinal orientation of students connected with the first study semester (cognitive, emotional and conative aspects in the choice of a study branch, expectations, problems, etc.) and the potential choice of the teacher profession oriented to technical subjects/practical education and vocational training.

On the basis of the above mentioned preliminary research probes, the research team initiated at the end of 2014 an empirical investigation of stands of the first-year and third-year students by means of a semantic differential.

Important for the selected theme seems to be also the grant project "Authority and the qualitative changes of its conception in the educational environment" (reg. No. P407/10/0796-R2010-2012), the main output of which is the publication of a team of authors titled "Authority in the education and social work", University in Pardubice 2012, p. 411, ISBN 978-80-73-95-507-6.

The research team further follows the outputs (of the conference organized on the university ground ČVUT MÚVS) which have been published and have a relation to the issue of motivation and its management in the framework of the educational system, for example ANDRES, Pavel and Alena VALIŠOVÁ "Introduction of electronic devices into education, the phenomenon of the present time". In: Pavel Andres, Alena Votišová and the team (ed.) Interdisciplinary relations between the technical, humanistic a social sciences. Collection of papers on CDROM from the international scientific conference themed "Technical, Human and Social Sciences: Is It Possible to Dialogue in the Pedagogical Process?" Czech University of Technology in Prague, Masaryk's Institute of Higher Studies, 2013, ISBN 978-80-01-05287-7.

The presented project intention extends the present knowledge in the sense of looking for basic answers to the influence of motivation factors in relation to the study of technical branches. We are looking for basic relations between the cause (technological innovations, trends in the sphere of information and communication technologies) and the effect in the form of varying requirements for preparation of teachers oriented on the technical subjects. On the basis of the content analysis, school documents, critical searches and other theoretic-empirical approaches we are looking for an answer, how to incorporate conceptually the possibilities of motivation to the study of technical branches.

In looking for this causes and effects it is necessary to accept the present situation but also take a look further ahead on the development of new technologies that are suitable and meaningful for application into the education reality. From the report "Innovating Pedagogy 2013 and 2014" (Sharples, Adams, Ferguson and the team, 2014) it is possible to estimate that expansion concerns primarily the m-learning personal education environment, MOOC, new objects in the distance education, wiki, blogs, RSS, use of the licence Creativecommons, sharing of electronic study supports in the cloud, u-learning, t-learning, educasting, seamlesslearning, social webs, omnipresent clever telephones and tablets, extended mobile reality, and in general a shift towards mobile education technologies. To the fore come new skills, frequently called as skills for the 21st century.

It is also necessary to be aware of the fact that the present-day time brings new technologies called as e-technologies. Communication in the traditional school was and is concentrated on the direct verbal and nonverbal contact of the communicating persons. Currently enter into the education space the electronic communications. To the best known belong for example: E-mail, Chat, ICQ, Skype, WhatsApp, Viber, Linkedln, Facebook, Cloud, LMS systems, Webinars, Educasting, Podcasting. They are an effective and prospective support of education and also a positive supportive mean of upbringing.

In the context of the issue related to the teachers of technical subjects and their readiness to meet the accelerating requirements of the information society, the presenter of this project solves in cooperation with the partners from the Czech Republic and from abroad (Estonia, Slovakia, Poland), also the sphere of the growing influence of technological innovations, not only in upbringing and education but also in broader pedagogical and psychological contexts. The solution of both the issue of the contemporary preparation of teachers and their skills to motivate students to the study of technical branches and also the role of information and communication technologies is beyond their traditional sight and comes to the new contexts, relations and relevance.

The entire project takes into account the role of information ethic that tries to find an answer to the question what is the ethically correct way to work with information (for example their origin, dissemination, searching and using). Nowadays is the information ethic a very relevant theme, it is fundamental for taking into account the ethical principles and their observance. The theme of information ethic in this project will contribute to the positive influence of information on the moral development of adolescents and to the prevention of incorrect use of information. Information ethic is necessary besides other things also in the context of adolescents' motivation to the study of technical branches.

2 Problem Field and Questions Resulting from the Broader Social Contexts

Information and communication technologies underwent over the last few years a rapid development with consequences also for the sphere of education and upbringing. It is necessary to emphasize that this happened not only at the level of technological thinking.

The reflection of the modern didactical trends takes place on the boundaries of the pedagogical, psychological and also sociological disciplines. The development is considerably accelerating in consequence of the ongoing technological changes and innovations. What is the real impact on the key and professional competences of the secondary school graduates? What does these competences follow? Do the education programmes reflect the demand of competences required by the industry and the practice generally? Wherein is the present education reality determined in the sense of a broader conception of education technologies? How is the problem of adolescents' motivation to the study of professional technical subjects interpreted from the theory point of view – i.e. from the analysis of professional publications and studies in our country and abroad? The development of motivation to the study of professional technical branches is thus a priority of the contemporary education system in the Czech Republic.

The presenter of this project, as well as other members of the research team have rich experiences in this sphere from the previous partial projects.

The presented project, in cooperation with the school-teachers of primary and secondary schools, sets up to enhance the interest in the study of technical branches, mathematic, physics and chemistry. The impact of the lack of interest and insufficient education of the secondary school students in the technical sense has also an impact on their choice of the study branch at the university. If the student chooses a technically oriented university, he often disposes with a low quality and insufficient knowledge of the mentioned study subjects. A considerable percentage of students accepted to the study therefore leave the university already in the course of the first semester. The project endeavours to prevent such situations and eliminate concern about the study of these subjects at the university and motivate to a greater interest in the study at the technically oriented universities.

In connection with the interpretation of the term of motivation to the technically oriented study it is necessary to take into account the need of a scientific interpretation. If we take a look into the various conceptions of motivation and motivation factors we realize that in this sense there does not exist an unequivocal consensus as to the content. The individual authors, according to their specialization, orient themselves to a wide scale of the content determination and definitions (for example Maslow, McClelland, Herzberg, Vroom, Skinner and others). Consequently it can come to certain terminological (semantic) inaccuracies.

Properly used motivation techniques can to a considerable extent influence the fact how the students are learning and how they approach the school subjects and also the education as a whole. Sufficient motivation enables to lead the behaviour of students towards the specified education goals, increases efforts and also persistence in its achievement and can positively influence also the cognitive processes, necessary for learning. A higher degree of motivation to the study of a certain branch thus can lead to better study results. An important role in the motivation of students represents the shift from the traditional way of teaching centred round the person of the teacher to a greater focus on the student himself and practical use of his knowledge. This approach enables a greater involvement of students into teaching with the assistance of exercises, experiments, discussion groups and use of modern technologies.

The attention will be focused on outlining of a motivation programme for the secondary school students that will be realized and verified in teaching at the secondary schools, and the interest of adolescents in science and technique will be systematically developed. The project also anticipates realization of discussions with distinguished professionals in the sphere of science and technique and with experts from the technical workplaces.

The future of the education system is at the present social paradigm based primarily on openness, becoming evident in the content of education, technological solutions and consequently also in the openness of education for everybody. The contemporary school must project its school education programmes with respect to the future social fulfilment of students (the profile of the student). The dynamics of social changes and technological innovations brings of course also a change in the form of a rapidly changing structure of the labour market. Some professions cease to exist and on the contrary increases the demand for knowledge-based workers. In sociology we speak about the flexibility that is necessary to project into the systematic preparation aimed for the teachers of technical subjects, who can subsequently use their professional competence to the motivation of students' interest in the technical education.

Currently we are able to react flexibly to the requirements of the labour market and to the requirements of the school practice thanks to the feedback from the secondary school teachers and pedagogues. This feedback is usually acquired in the course of realization of a number of education programmes of the lifelong education and further education of school-teachers in the sphere of modern technologies, and at the same time provides new research questions for solving of the problem at the basic research level.

3 The Goals of the Project and the Time Schedule of Solution

Our objective at the theoretical level is to emphasize the importance of the motivation to the study of technical branches. The research problem will be analyzed in the context of the social sciences where first of all will be described the situation and importance of the present paradigm of the information society with respect to the level of historical development.

The goal of the presented project, especially at the theoretical level and also at the application level, is to increase the interest of students in the study of technically oriented secondary schools and universities. It concerns the fundamental priority of the current education system. The reason for presentation of the project and a long-term aim to which the project should contribute is to increase the number of students at the technical secondary schools and universities. As a result it will be possible to satisfy the requirements of the labour market that to a considerable extent feels the lack of technically educated graduates.

The proposed research investigation (extensive and intensive research design) will be realized at a group of students, pedagogues and experts in the Czech Republic, Slovakia, Poland and Estonia.

As partial goals we specify:

- To analyze the school documents and the education programme with respect to the motivation to the study at the professional technical schools.
- To elaborate theoretical grounds of the conception of motivation and motivating students and continue in the theoretical studies from the outputs of the preceding grant projects.
- To analyze the content specification of motivation and its importance for orientation of the technically aimed study from the viewpoint of the present days and future prospects in our country and abroad.
- To compare the results of the research investigation with the interested colleagues from abroad and with their experiences from the professional institutions, solving similar projects in the framework of the Czech Republic and EU.
- To analyze the influence of new methods and forms of education oriented to the use of modern technologies in motivation of adolescents.
- To propose didactical aids and curriculum for teaching and independent preparation focused on the support of adolescents' interest in the technical branches.
- To increase the interest of the public, state and non-state professional enterprises and technically oriented institutions, as potential employers, in motivation of adolescents to the study of technical branches.
- To support mutual long-term cooperation between the secondary schools and universities.

3.1 Time Schedule

Key activities for the first year (2017)

- To analyze the present state of knowledge in our country and abroad (theoretical studies, collection of disposable data and findings in our country and abroad).
- To analyze the pedagogical documentation (RVP, ŠVP), to compare the education systems in the Czech Republic and abroad.
- To create networks of technical secondary schools and universities with the possible involvement of the present partner institutions CTU MIAS.

- To analyze the problem by means of qualitative methods of research (group of respondents teachers, students and experts), and identify the motivation factors for the technical a creative development of the adolescents.

Key activities for the second year (2018)

- To design the methodology of empiric research aimed at the use of ICT elements in teaching of technical subjects.
- To carry out the observation of stimulating methods and organizational forms of technical subject teaching.
- To carry out a pedagogical experiment (to compare the application of stimulating methods and forms of teaching at the experimental and control classes) at the technical schools.
- To design an innovative education programme using the ICT elements in support of adolescents' interest in the study of technical branches.

Key activities for the third year (2019)

- To propose didactical aids and curriculum for teaching based on the analyses in the first and second year of the project.
- Pilot employment of modern methods and forms of teaching in selected schools, including the evaluation and reflection.
- To evaluate results of the empirical investigation (extensive and intensive forms of research).
- To interpret the research data of the quantitative and qualitative part of the research.
- To compare the results of the empiric investigation in our country and abroad.
- To publish the results of theoretical analyses and research investigations in the professional magazines, publications and monographs.
- To present the results of the grant project at the international level (conferences, seminars, workshops).
- To draw up the final report of the three year lasting grant project.

4 Conceptual and Methodological Approach Proposed for the Solution of the Project

The conception of project solution is based on the knowledge and experiences that the authors gained on the basis of earlier solved assignments (see the bibliography).

The focal point of the presented project consists in research investigation, its evaluation and interpretation of results – further also on the comparison of research results realized in the Czech Republic and abroad (Slovakia, Poland, and Estonia). The research investigation will be carried out mainly by the method of questioning which will be supplemented with selected research methods of an intensive form of empiric investigation at a smaller research (selected) group.

Methodological approaches will comprise a comparison research (using the following complex of research methods:

- Attitudinal scales – stands of students and teachers to the importance of modern information and communication technologies and its didactical importance in the system of technical education.
- Comparison of opinions of the secondary school students and teachers on the possibilities of increasing interest in the study of technical subjects.
- Moderated discussions with teachers, students and experts; focused on the use of modern technologies in teaching and respecting the information ethic.
- Analysis of the school and education documents – RVP, ŠVP, and other things.
- The technique of questioning for which will be used a standardized questionnaire in order to find out the value orientation and motivation factors influencing the study performance (questionnaire HO-PO-MO).

The relevant statistical methods will be used for the determination and verification of relations and contexts between the individual variables, considered as important.

4.1 Anticipated Outputs

- A specific theoretical study on the issue of adolescents' motivation to the study of information and communication technologies.
- A study on the comparison of empirical findings from the research in our country and abroad.
- A study on the research investigation results carried out in the Czech Republic, Estonia, Poland and Slovakia.
- Conception of a special programme for development of adolescents' motivation to the study of technical branches.
- Curriculum for students of the secondary schools.
- Professional contributions at conferences in our country and abroad, active participation and presentation of research results at these conferences.
- Partial research reports from the individual phases of the research investigation.
- Final report and organization of an international conference.

5 The Social Significance of the Solved Grant

As far as we sum up the most important findings we can say that as decisive characteristic of the so called society of knowledge can be considered the following items:

- A great proportion of the population achieves higher education;
- The absolute majority of the population has access to the information technologies and internet;
- A great proportion of the population form the so called "knowledge workers", who need university education and extensive experiences for successful performance of their work (knowledge worker, In-: Drucker, Peter F. Management Challenges for the 21st Century);
- Individuals and also states massively investing into education, research and development;

- Private firms and organizations of the public sector are forced to permanent innovations.

The conception of the knowledge society is necessary to dynamize. The role of knowledge in the present-day society cannot be understood statically but as a continual process. Actually, we should rather speak about a society of intensive knowledge processes instead of a knowledge society. Indeed, everything that can be observed is a huge and unprecedented dynamic of processes connected with the knowledge. Permanent innovation becomes a necessary condition for the survival of a firm on the market, the lifelong learning is an unavoidable prerequisite of employment, the continual recombination of a huge quantity of findings produces new and new knowledge, the unexpressed knowledge is codified so that it can be distributed by information technologies, the produced knowledge is subject to a permanent reflection and recombination with another knowledge and thereby just comes to further production of knowledge and so on. Innovation depends on the transmission of various type of knowledge that frequently is not codified but forms a part of the social networks, in consequence of which increases also the importance of relations and relationships between the participants who produce the knowledge.

Modern technologies are becoming significant assistants in education. It is possible to state that the schools so far do not offer much possibilities of a meaningful use of modern technologies in connection with the motivation to the study of technical branches. It is important to call attention to the fact how these means can be effectively used and open a space to the teachers for a creative approach to this issue at the individual schools. The way, how the adolescents can be motivated to the study of technical branches and work with information in an ethical way, is a very desirable theme under the current conditions of the Czech education system. The requirements of the labour market and the requirements of the school practice correspond with the development of the technique, information and communication technologies in the Czech Republic and in the world.

One of the key goals of the project is to increase the interest of the secondary school students in technical branches and increasing their motivation to the study of these branches at the universities. This goal thus responds to the persistent trends on the labour market and to the need of increasing the number of technically educated professionals.

An important element of increasing the student's motivation is the opportunity to show them the possibilities of practical use of theoretical information, acquired by the study, and lead them to independent application of this knowledge into the practice. Methods and forms of teaching used in the project enable the students to be acquainted with new approaches to the acquirement of knowledge in technical subjects as are lectures held by the academicians using modern information technologies, emphasis on the independent study and also a team work in solution of common projects and a critical analysis of information searched on the internet.

The activities of the project consist primarily in elaboration of new methodical materials and a proposal of new innovative aids necessary for the new concept of teaching which will be supported by technical and material equipment enabling introduction of interactive forms of teaching and also professional preparation of school teachers. For the fulfilment of the mentioned goals are important not only the activities incorporated directly into the teaching but also professionally led excursions and lectures for the school teachers and the students themselves, aimed at the assignment competences and at the interesting themes from the sphere of technical branches and information technologies. A narrow cooperation of the secondary schools and universities will enable both the introduction and application of the described methods and also to acquaint the students with the course of university studies and thus to support the fulfilment of the main goal of the project i.e. increasing of the students' motivation.

The grant assignment is conceptually based on the strategic and conception documents of the Ministry of Education (MŠMT) where to the priority goals belongs primarily opening to the new methods and ways of teaching by means of digital technologies, improving of competences of students in the sphere of work with information and digital technologies and last but not least also development of students' informatics thinking. We consider the theme of secondary school students' motivation to the choice of technical branches as a priority direction of the contemporary education system in the Czech Republic.

Acknowledgement. This paper was supported by the Fund of educational policy of the Ministry of Education, Youth and Sports of the Czech Republic: Readiness of technically educated students for the teacher profession, management and motivation.

References

1. Adair, J.: Efektivní motivace. Alfa, Praha (2004)
2. Andres, P., Dobrovská, D.: Dilemmas of student technical and social science thinking. In: 44th IGIP International Conference on Engineering Pedagogy, World Engineering Education Forum WEEF, Florence, Italy, pp. 99–101 (2015)
3. Andres, A., Měřička, J., Vališová, A.: The characteristic dilemmas of engineering education. Radom, Rocznik Pedagogiczny Komitetu Nauk Pedagogicznych Polske Akademie Nauk, nr 36, pp. 105–111 (2013). ISSN:0137-9585
4. Andres, P., Svoboda, P., Vališova, A.: Snižování studijní neúspěšnosti a celoživotní vzdělávání na ČVUT v Praze. In: Sborník příspěvků z mezinárodní vědecké konference EDUCOM. Pedagogica actualis VIII. 1. vyd. Trnava: Univerzita sv. Cyrila a Metoda v Trnave (2015)
5. Andres, P., Svoboda, P.: Vybrané aspekty celoživotního vzdělávání učitelů - techniků. In: Danielova, L., Schmied, J. (eds.) Sborník příspěvků ze 7. mezinárodní vědecké konference celoživotního vzdělávání Icolle 2015. 1. vyd. Mendelova univerzita v Brně, Křtiny, s. 17–34 (2015). ISBN:978-80-7509-287-8
6. Andres, P., Vališová, A.: Elektronizace ve vzdělávání - fenomen současné doby. In: Interdisciplinární vztahy mezi technickými, humanitními a společenskými vědami. MÚVS ČVUT Praha, Praha (2013). ISBN:978-80-01-05287-7
7. Andres, P., Vališová, A.: Mezinárodní konference Technické, humanitní a společenské vědy: Je možné vest v pedagogickém procesu dialog? Praha: AULA, č. 1, roč. XXI, s. 126–129 (2013)
8. Andres, P., Vališová, A.: Institucionální vzdělávání učitelů-techniků. Lifelong Learning - celoživotní vzdělávání. roč. 3, č. 3, s. 8–22 (2013). http://www.vychova-vzdelavani.cz/ll1303.pdf. ISSN:1805-8868

9. Holeček, V.: Psychologie v učitelské praxi. Grada Publishing, Praha (2014)
10. Hrabal, V., Pavelková, I., Man, F.: Psychologické otázky motivace ve škole. Státní pedagogické nakladatelství, Praha (1989)
11. Kvítek, L. (ed.): Možnosti motivace mládeže ke studiu přírodních věd: sborník recenzovaných příspěvků. Univerzita Palackého v Olomouci, Olomouc (2008)
12. Novotná, J.: Motivace nadaných žáků a studentů v matematice a přírodních vědách. Masarykova univerzita, Brno (2012)
13. Pavlas, I.: Výkonová motivace a interpersonální potřeby. Pedagogická fakulta Ostravské univerzity v Ostravě, Ostrava (2011)
14. Semrád, J., Vališová, A., Andres, P., Škrabal, M., et al.: Výchova, vzdělávání a výzvy nové doby. Brno, Paido (2015). ISBN:978-80-7315-258-1
15. Svoboda, P.: Digitální kompetence. In: Trojan, V., Svoboda, P. (eds.) Sborník příspěvků z II. mezinárodní vědecké konference školského managementu. 1. vyd. Univerzita Karlova, Centrum školského managementu PedF UK v Praze, Praha (2013). ISBN:978-80-7290-696-8
16. Svoboda, P.: M-learning – využití moderních technologií ve výuce. In: Sborník příspěvků z výroční XX. mezinárodní vědecké konference České asociace pedagogického výzkumu: Kvalita ve vzdělávání. 1. vyd. Univerzita Karlova, Praha, s. 693–701 (2012). http://capv2012.pedf.cuni.cz/wp-content/uploads/2013/09/Sbornik_FINAL.pdf. ISBN:978-80-7290-620-8
17. Svoboda, P.: M-learning - využití mobilních technologií ve výuce pro další vzdělávání pedagogických pracovníků (audiovizuální tvorba, elektronický dokument, blended learning), Výzkumný ústav pedagogický (2011). http://elearning.rvp.cz/katalog-kurzu/informace-o-kurzu?k=37. ISSN:1802-4785. Accessed 30 Nov 2011
18. Svoboda, P.: Využití m-technologií v modelovém scénáři aktivity pedagoga. In: Media4u Magazine, 7. ročník, 3/2010, s. 125–127 (2010). http://www.media4u.cz/aktualvyd.pdf. ISSN:1214-9187. Accessed 28 Sept 2010
19. Svoboda, P.K.: využívání m-learningových technologií v současné škole s příkladem z výuky fyziky. In: Sborník příspěvků z mezinárodní konference – Modernizace vysokoškolské výuky technických předmětů. 1. vyd. Hradec Králové: Gaudeamus, s. 158–160 (2010). ISBN:978-80-7435-014-6
20. Svoboda, P.: Modelové scénáře aktivit s využitím m-learningu. In: Media4u Magazine, 6. ročník, 4/2009, s. 33 – 35 (2009). http://www.media4u.cz/aktualvyd.pdf. ISSN:1214-9187. Accessed 29 Dec 2009
21. Svoboda, P.: M-learning a příklady využití mobilních technologií se vztahem k výuce technických předmětů (2009). http://www.media4u.cz/mvvtp2009.pdf. ISSN:1214-9187. Accessed 10 May 2009
22. Svoboda, P.: M-learning ve výuce technických předmětů. In: Sborník příspěvků z mezinárodní konference – Modernizace vysokoškolské výuky technických předmětů. 1. vyd. Hradec Králové: Gaudeamus, s. 172–175 (2008). ISBN:978-80-7041-154-4
23. Šmahaj, J., Cakirpaloglu, P.: Význam motivace v pojetí osobnosti: teoretický, výzkumný a aplikační rozměr. Univerzita Palackého v Olomouci, Olomouc (2015)
24. Řehulková, O., Osecká, L.: Výkonová motivace ve škole. Brno: Akademie věd České republiky, Psychologický ústav (1996)
25. Vališová, A., Bratská, M., Sliwerski, B.: Rozvoj české společnosti v Evropské unii. Jedinec a společnost v procesu transformace a globalizace. In: příspěvky z konference konané ve dnech, 21–23 October 2004, 218 s. Matfyzpress, Praha (2004). ISBN:80-867-3235-5
26. Vališová, A., Dobrovská, D., Tureckiová, M., Andres, P.: Společnost a vzdělávání – vzájemné zrcadlení. Zpráva o mezinárodní konferenci Knowledge and its Communities. AULA, č. 4, s. 52–54 (2006). ISBN:1210-6658

27. Vališová, A., Andres, P., Šubrt, J.: E-learning from the technical university students' point of view. In: Computers and Advanced Technology in Education. Globalization of Education through Advanced Technology, Beijing (Anaheim, Calgary, Zurich), pp. 185–190. ACTA Press (2007). ISBN:978-0-88986-699-7
28. Vališová, A.: Application of electronics in the formative educational process. In: Studia z teorii wychowania. Tom III: 2012, nr 2(5). Warszawa: Wydawnictwo Naukowe CHAT, pp. 114–122 (2011). ISSN:2083-0998 (nr 02/2012)
29. Vališová, A., Andres, P.: Myth of an ideal teacher? Prepossessions and reality. In: 44th IGIP International Conference on Engineering Pedagogy, 2015 World Engineering Education Forum WEEF, Florence, Italy, pp. 181–183 (2015)
30. Vališová, A., Andres, P., Šubrt, J.: Teacher specialists: authority in relation to social competence. In: 44th IGIP International Conference on Engineering Pedagogy, 2015 World Engineering Education Forum WEEF, Florence, Italy, pp. 188–190 (2015)

Psychology and Pedagogical Maintenance of Formation of Career Competence of Future Engineers

Khatsrinova Olga(✉) and Vasily G. Ivanov

Kazan National Research Technological University, Kazan, Russia
khatsrinovao@mail.ru, mrcpkrt@mail.ru

Abstract. For ensuring quality of engineering training it is necessary to use in educational process of a technique, providing this quality. Each subject matter has to play a role in formation of professional competences of future experts. The humanitarian discipline "Management of training processes" in the contents considers questions of creation of professional career.

During classroom occupations by masters methodical recommendations "Planning of career as an advance element on an office ladder" have been developed. During an educational semester they actively participated in research, educational and other actions, and then investigated personal changes.

Keywords: Career · Master · Vocational training · Professionalism

1 Introduction

In quickly changing social and economic conditions the main objective of the state strategy of development of education in Russia till 2020 has defined achievement of new quality of education. The changes happening in the country involve changes of practice of the educational organizations. Finding by the person of professional competence of process of training represents a complex problem which is characterized by such questions as obtaining professional qualification, employment, creation of professional career. Professionalism is defined as the characteristic of the person which provides quality of activities for the effective solution of professional tasks in new conditions. This research shows the experience got with masters in the Technosphere safety direction when studying humanitarian discipline. Undoubtedly, realization of this approach in training of future experts improves student's motivation and increases their participation during the work in audience. The initial objectives have been achieved successfully.

1.1 The Purpose

The purpose consists in that in the course of studying of discipline "Management of training processes" to create an orientation of masters on formation of career competence. It is expressed in formation of career potential, a subject position of masters in

M. E. Auer et al. (eds.), *Teaching and Learning in a Digital World*,
Advances in Intelligent Systems and Computing 715,
https://doi.org/10.1007/978-3-319-73210-7_34

karyerooriyentirovanny study, professional personal development on the basis of realization of the subject focused training models.

The research of success of entry of masters into professional communities has shown that from activity, ability to find application for the intellectual and professional qualities, manifestations of flexibility and a non-standard in the solution of professional tasks how there will be future professional career in many respects depends. Carrying out a research will allow to claim that the problem of formation of career competence of future experts is very urgent both in practical, and in the theoretical plan. Process of formation of career competence of students demands existence of the competitive environment of university.

2 Approach

The problem of formation of professionalism is closely connected with a question of resources of mental development. B. G. Ananyev wrote that for social forecasting special scientific knowledge of reserves, resources, potential of development of the personality which is insufficiently used by society [1] is required.

These provisions haven't lost the relevance and today. Professionalism is defined as such characteristic of the person which provides new quality of activities for the effective solution of professional tasks in new conditions. In formation of professionalism the professional self-determination assuming active search of opportunities of the development, formation of system of valuable orientations is of great importance. According to N. F. Talyzina the expert meets social expectations if development of his personal, general and professional culture is carried out by the rates advancing in relation to other members of society [2].

3 Pedagogical Context of "Education for Career"

Now many researchers specify that the world of professions in modern society is characterized by exclusive dynamism, intellectualization, the increasing combination of different types of activity within one profession. Therefore inclusion in programs of preparation of questions of "education for career" promote formation of a set of the adaptation abilities and skills allowing the individual in the course of vocational training and further professional development to effectively use the personal potential [3]. The psychology and pedagogical perspective of consideration of a phenomenon of career allows to see her not only at an angle future advance on an office ladder, and is much wider - in the context of personal and professional formation and functioning of the individual in society.

The analysis of contents which is carried out by the author "educations for career, has shown that it can be structured in the form of several educational blocks, filling of each of which can be in any traditional subject matter. By means of functional descriptions it is possible to allocate the following blocks which substantially provide full support of development of career of students.

- Providing the person with educational system of motivation to career activity, the analysis of own professional trajectory regarding degree of a realizovannost of the inclinations, professional requirements.
- Reflexive planning of career. Training of the person in design of career steps which need to be undertaken for formation of the long-term program of own actions, formation at the studying abilities of a reflection of the result achieved at a certain stage and a way of his achievement.
- Mastering effective behavior models and thinking. In the course of implementation of the choice the individual is faced by a problem of preservation of integrity of the personality.
- Work with information resources. Information on the professional sphere has to contain the facts of all sphere of professional activity, possible for the individual, an opportunity to get acquainted with a big range of opportunities of the choice, without restriction with tightly professional aspects.
- Structure of knowledge. Assimilation of knowledge focused on understanding of dynamics and regularities of development of career during the different age periods, knowledge of the main vital stages of development of career, understanding of gender aspects of development of career, importance of integration in the course of realization of career, knowledge of the theory of development of career.
- World outlook component. Formation of valuable representations and outlook.
- Adaptation component. Formation of concrete adaptation skills: communicative skills, skills on increase in efficiency of work, skills on decision-making, the arch of the choice of educational opportunities, skills on management of changes and adaptation to them.

We consider it expedient to present in "education for career" three fundamental elements. It is a substantial component (support of the choice of field of the chosen professional activity, studying of the generalized characteristics of professional activity). On the second place we have put a methodical component (studying of future professional environment and mental "occurrence" this Wednesday with the subsequent reflection).

The third place was allocated to an organizational component. It is expressed in identification of an opportunity to acquire additional professional knowledge, qualifications, to master adjacent professions. Rendering the consulting help in formation of career, support of development of career.

3.1 Methods of Practical Realization

We will allocate steps of passing of career by the master in the Technosphere safety direction. The first step – presence of requirement at the student to self-development. The second step – interest in concrete specialty and its development, understanding of values of future profession. The third step – responsibility of future expert for results of work on the basis of the level of the professional skill. The fourth step – creation and distribution of own experience, the new ideas.

On the basis of generalization on classroom occupations occupation of joint researches in the field of career and features of professional activity of specialists in the

Technosphere safety direction methodical recommendations for masters "Planning of career as an advance element on an office ladder" have been developed. Each person in modern conditions can and has to be the manager of own career and own personality as ability to influence other people begins with ability to operate himself, effectively reconstructing their interaction and relationship. At the same time, to strengthen creative approach to the solution of the problems facing the modern expert, to increase his independence, to increase his readiness for expansion of system of life and professional priorities, it is necessary to use new opportunities for the career development. It has been defined that throughout all educational semester masters will build soy cognitive activity, being guided by elaborated provisions of methodical recommendations.

In the course of training we approved set of the basic principles of psychology and pedagogical maintenance of process of formation of career competence of masters. The principle of social activity and subjectivity assumes development of an initiative of the student as independent source of knowledge and transformation of reality through their inclusion in process of psychology and pedagogical design, joint search and opening of the truth, values, meanings; understanding by each master of the fact that "I" can't develop without other people who are objectively creating a collision situation but also subjectively ready to help search of a consent and correlation.

The principle of dialogue and productive interaction allows each student to have equal opportunities in the solution of various professional and educational tasks and problems on the basis of dialogue and focused on receiving, both a collective product, and individual. The principle of a continuity and openness provides to all subjects of educational process in engineering higher education institutions freedom in the choice of the direction of continuous creative development. This principle means variety of zones of development and situations of their choice; freedom of self-determination; freedom of discussion of problems. The principle of systemacity and integrity means interrelation and interaction of all substantial and active components in the course of training. The main criterion of level development of career competence is the solution of "career tasks". They include: self-determination in a career problem; definition of the purpose and type of career; self-organization in space of career development; reflection of the opportunities; definition of inquiries of masters taking into account requirements of higher education institution and branch.

4 Actual Results

Creation of the career by the master as subjectivity parameter, has been analysed on the basis of special interviews. Justification of the subject attracting interest of students is significant: "how to find work", "How to pass an interview with the employer", "How to keep a workplace", "Work and free time". During interviewing by the examinee it was offered to tell about perception of such concepts as "professional and personal self-determination", "self-updating", "self-realization in a profession"; to define what place these phenomena take in their life; what they are guided at achievement of the professional and personal plans by that can disturb or promote their realization both in life, and in a profession. We will present versions of the analysis of separate diagnostic inspections. The master Alexander K. is capable to estimate really the merits and demerits and to accept himself it what is, out of character of an assessment of the advantages.

Describing in an interview the understanding above the presented concepts, Alexander has noted their special value in development of as subject of different types of activity, despite sensibleness or not sensibleness of presence of these processes not only in his life, but also in life of other people. In this regard he notes importance not of simple statement and achievement of the objectives "for himself", and abilities to make non-standard decisions, to create something new and useful to people around: "… I consider that if self-realization goes only for itself and doesn't bring any benefit to people around, then this wrong opinion that the person realizes himself and the opportunities".

This understanding of the respondent is conformable to the statement that "self-realization goes by means of personalisation process". At the same time he considers himself not completely self-actualized personality though some fragments of professional and personal self-determination as it seems to it, have high rates. Generalization of results of interviewing has shown that at masters considerable changes on an indicator "career participation" was observed, the characterized importance for the master of a personal contribution to achievement of the objectives of the professional organization, his readiness to work in the complicated circumstances. In a methodical grant the following subjects have been considered: main stages of development of career, technology of self-diagnostics of problems of development of individual career.

Literature and questions for discussion is without fail offered to students. A summarizing task is the essay "Determination of Career Prospect and Identification Significant Actions Which to This Student Are Expedient for Carrying Out". Students describe the professional activity in ten years and reveal personal qualities which have helped to realize professional plans.

From total number of respondents, over 30% were lifted to higher level. The "career intuition" indicator - only at 22% of masters has least of all changed. During time of studying of discipline (for a semester) 15 masters have taken part in various projects, conferences, seminars.

5 Conclusion

The psychologist - pedagogical maintenance is considered as process of familiarizing of masters, teachers with productive interaction and is based on a subject - the subject relations. It is implemented at two interdependent levels: at the level of the process including unity of activity and the relations of teachers and masters for detection of regularities of their development as subjects; at the level of the specific master who is actively developing in himself the self-actualizing ability to the analysis of creative limits and opportunities in own professional and personal growth.

References

1. Ananyev, V.G.: People as knowledge subject. SPb: piter (2001). 288 pages
2. Talyzina, N.F.: A technique of drawing up the training programs. Mosk Publishing House. un-that, (1980). 176 pages
3. Khatsrinova, O.: Career-building training as a component of talent management. In: Proceedings of 2015 International Conference on Interactive Collaborative Learning (ICL), Florence, Italy, 20–24 September 2015. IEEE (2015). ISBN:978-1-4799-8706-1/15/$31.00

Collaborative Project-Based Learning in Training of Engineering Students

Gulnara F. Khasanova[1(✉)] and Phillip A. Sanger[2]

[1] Kazan National Research Technological University, Kazan, Russia
gkhasanova@mail.ru
[2] Purdue University, West Lafayette, USA
psanger@purdue.edu

Abstract. This paper describes an implementation of project based learning to a class of Russian students studying psychodiagnostics. A survey of the students after the experience was compared to a separate class of BS students in professional education. Based on these surveys and interviews with students, the results suggest that explicit training in project management tools and exercises in skills for team dynamics might enhance the positive experience.

Keywords: Project based learning · Psychodiagnostics · Russian education

1 Introduction

Training engineers in universities needs to focus on 21st century competencies such as collaboration, communication and teamwork, learning to learn, information and ICT literacy, adaptability, group responsibility, activities of Project-Based Learning (PBL), planning co-working strategies. However, dominating pedagogical strategies do not provide sufficient contribution to meeting these goals.

This paper describes an experience on the implementation of collaborative projects by students of engineering at Kazan National Research Technological University. An approach to engineering education implemented in the research combined project-based learning, social constructivism and connectivism theories. Student teams developed diagnostics tools and used them to conduct a psychodiagnostics survey. Activities were aimed at developing skills of enhancing learning processes in connected activities and using Web 2.0-based social technologies in teamwork.

2 Types of Teaching Approaches

2.1 Project-Based Learning

Project-based learning (PBL) is a pedagogical approach developed in the 1970s and initially applied to early childhood education. Lately project based learning has resurfaced and is being re-invented internationally as a path to relate professional training to real world experience. The value of project-based learning is in training the individual with life experiences, and, in the process, mastering new ways of solving

M. E. Auer et al. (eds.), *Teaching and Learning in a Digital World*,
Advances in Intelligent Systems and Computing 715,
https://doi.org/10.1007/978-3-319-73210-7_35

problems and generating new knowledge. Universities in many parts of the world are adopting PBL to develop professional graduates capable of being the practical, application oriented professionals needed in today's global society. This pedagogical approach is well established and has been reviewed extensively [1–3].

PBL is being implemented in a variety of different ways depending on the curriculum and the surrounding economic climate. Essential characteristic of projects within PBL are that the projects are central to the content being taught and not peripheral to the course, projects are focused on a driving question, the projects require transforming acquired knowledge, the projects are largely student controlled, and finally the projects are real world problems.

With the introduction of projects into the learning process, students investigate problems and propose solutions over an extended period of time to acquire a deeper understand of the techniques and approaches being taught. The learner is actively engaged in the project, feels responsibility for the results and recognizes the trust placed in him. The PBL approach is often described as "learning by doing". An additional benefit to PBL is that many of these projects are team based requiring the acquisition of and practice of interpersonal skills and increases an awareness of the complexity of interdisciplinary work.

2.2 Social Constructivism Theory

According to social connectivism theory developed by L. Vygotsky, learning and development are socially shaped and need social context. Cognitive functions first appear in person's social interaction with someone with higher levels of knowledge and skills and then are internalized on the intrapsychological level. Communication with "more knowledgeable other" (MKO) is a key condition for learning, and the role of MKO can be played by not only instructor but by peers as well.

Knowledge is actively constructed by a learner, and it is not simply constructed but is co-constructed. Cooperation and collaboration with peers in solving problems and decision-making enhance learning. Learning should take place in a context being meaningful for a learner. Application of knowledge and experience to meaningful tasks, real-life problems intensifies the learner's internal drive to understand and promote the learning process. So teaching assignments should locate in the zone between two levels – the level of actual development and that of potential development – that Vygotsky named the "zone of proximal development" [4, 5].

2.3 Connectivism Theory

Connectivism considers knowledge as distributed, connected, emergent, adaptive, composed of networked entities. So learning process should not restrict to an internal, individualistic activity but have such properties as diversity of opinions, connections between ideas, fields, and information sources, social interaction, constructing networks, exchanging knowledge and experience in practice, being student-initiated, intentional and situated. Learning takes shape of not only knowledge consumption but knowledge creation process [6–8]. It is important for the perspective of team projects that connectivism considers decision-making a key component of learning process.

In the context of changing reality, "[c]apacity to know more is more critical than what is currently known" [6].

Being a response to growing public demand for teams and cooperativeness, connectivism-oriented approach helps students obtain experience in cooperation, mutual aid, supporting teamwork, and co-creation. It also enables teams obtain competencies that do not appear from simple adding those of separate individuals but have synergetic effect.

2.4 Combination of the Three Approaches

Combination of the three approaches resulted in the list of requirements to projects executed in our experiment: (1) close connection to the content of the course; (2) challenging degree of complexity, fitting into the zone between "too simple" and "too complex"; (3) establishing communication networks not only within the team but with external environment as well; (4) interdisciplinary character; (5) being student controlled; (6) focusing on real-life problems.

Learning outcomes resulted from students' engagement in experimentation and activity in communication with others, distributed group work in social networks and Web 2.0-based social technologies. Project activities were constituted by series of interactions in which participants changed roles of a student and more knowledgeable other.

Learning outcomes included both subject-related skills and those focused on teamwork: (1) an ability to plan and implement connected activities; (2) an ability to self-monitor and fulfill an assigned task in due time; (3) an ability to use programming skills for constructing digital psychodiagnostics tests; (4) an ability to use statistics to process survey data; (5) an ability to analyze data sets and use them for psychological assessment and evaluation of personality; (6) an ability to communicate using social networks, Web 2.0-based social technologies.

3 Project Implementation

3.1 Psychometry Project in the Human-Machine Interaction Course

PBL activities were focused on teaching students how to develop automatized diagnostics tools and use them in a psychometry survey. Sophisticated and easy-to-use electronic versions of various psychological tests are currently broadly available for conducting surveys. Total delegation of routine procedures to intellectual systems, however, makes logic of a survey and data processing less comprehensible for users. It was assumed that activities on construction of tools for automatic processing of data may help students "internalize" scenarios of data evaluation and interpretation.

Three-week projects were implemented in groups. Teams were formed on a voluntary basis, without any restrictions except one regarding team's size that should not exceed 6 members. Teams determined for themselves how they distribute responsibilities of team members. Students broke down the assignment according to their own preferences, planned approaches and checked their efficiency, revised and searched for better solutions. Major part of tasks required collective activities, and even the types of work students implemented separately were crucial for teammates to complete successfully the entire project. No teaming and interpersonal skill training was given to these students. For most of the students this was the first experience that they have working on a team project.

A set of deliverables included electronic versions of a questionnaire and a data processing tool, a PowerPoint presentation, and a paper copy of a report covering course and results of a project. Teams were supposed to use Web 2.0-based instruments to conduct a survey without direct contact.

In the course of presenting the results, students pointed at the complexity of the assignment. Certain part of learning occurred not only within teams but as well in the "inter-team" space. A kind of "meta-cooperation" between teams was established, something that was not supposed and prescribed in the project design. First, teams exchanged experience in resolving similar problems in constructing digital tools. Second, they elaborated a report template that was finally used by the most of teams.

Though no representativeness requirement was applied to samples of test-takers, learners were interested in including respondents of different ages, education degrees and other characteristics. One student, who had previously volunteered in Sochi Olympic Games, commented that she contacted to volunteers throughout Russia.

Contents of paper reports showed that several of teams did not restrict themselves with mandatory tasks in project implementation but went further, having conducted research on the foundations, theories, history and contexts related to the tests they employed. In the report summaries, teams expressed their "satisfaction with the atmosphere of free discussion, experimenting, sharing experiences that reigned during the whole work".

3.2 Survey of Attitudes

The research also included a survey in which the course participants, now graduates, expressed their opinion on their collaborative project experiences. The survey tool included following issues:

1. If I did the project again, I would spend more time doing research about the project background.
2. I was able to apply my computer science knowledge to this project.
3. I was very pleased to work with my teammates.
4. All of my teammates did their fair share of the work.
5. Working to a deadline was difficult.
6. Our team created a schedule for the tasks to do.
7. Making a decision was difficult.
8. Our team had to learn about topics that we had not had in class.
9. Helping and supporting my teammates was fun.
10. Afterwards I used skills obtained during the work on this project in other activities.

Nine of 28 course graduates contacted through the internet have provided feedback. 77% of respondents expressed their appreciation of the chance to help and support teammates during the work and recognized that they faced difficulties in developing strategies and making decisions to solve the problems. 66% reported that they enjoyed the collaborative format of the project and said that afterwards they used skills obtained during the work. 55% recognized they developed no schedule to implement the project tasks while 44% accepted that they faced difficulties meeting a deadline.

Distribution of the responses is shown in Fig. 1.

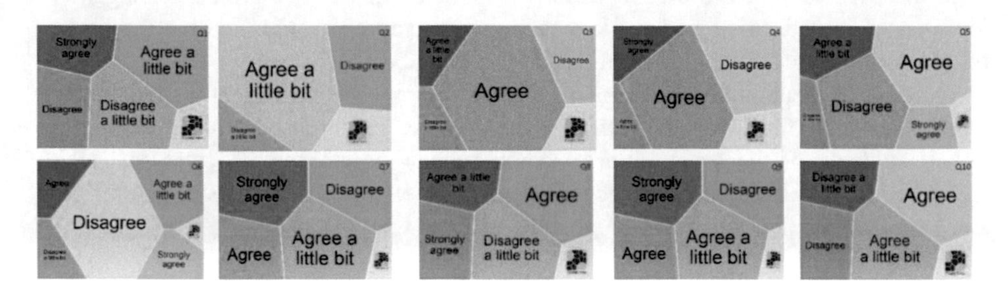

Fig. 1. The results of a survey on perceptions by participants of PBL activities.

3.3 Comparison of Results

The survey results from this class is compared a class similar in age and in stage of academic level but which were was following a course dedicated primarily to project management and also using the PBL approach. This class of professional development students is full described in another paper in this conference [8]. This class received extensive training in project management techniques and several exercises to prepare them for the dynamics of a team project. The results of the survey for each class and compared to each other are shown in Fig. 2.

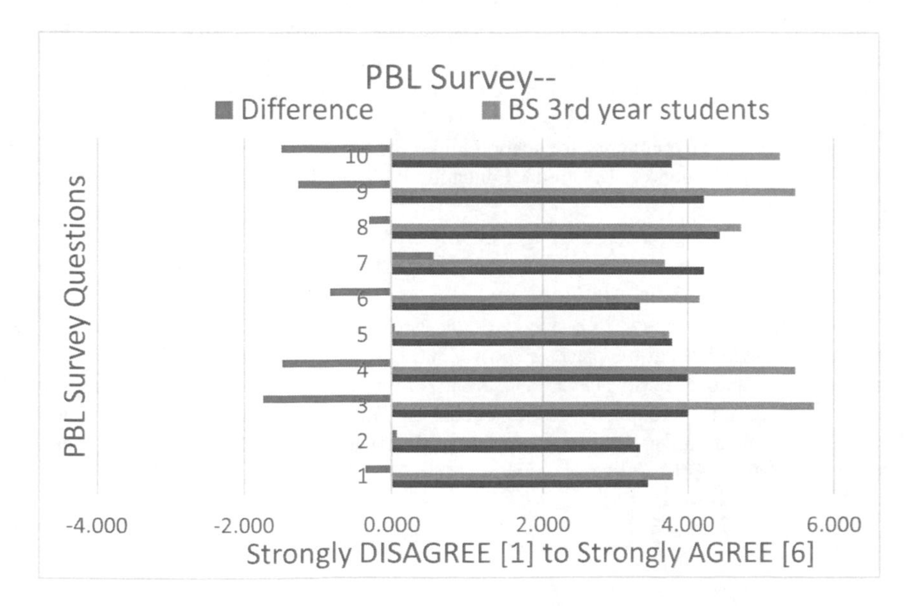

Fig. 2. Comparison of two classes using PBL – with and without specific teaming and project management training.

Both teams felt that they did adequate research and adequate planning for the project. In addition reaching decision appeared to ok as well. The biggest difference between the teams is in the area of team dynamics – questions 3, 4, and 9. Working in a team was significantly more satisfying for the BS students, may be due to the lack of specific training in teaming and project management in how to operate effectively in a team that was provided to the BS team. In addition, using good project management tools just as Gantt charts and work breakdown structures should increase the probability of more uniform contribution from team members.

Notable to mention that participants in the psychometry experiment were contacted three years after finishing the course and even after graduation from the university, unlike students of the second class, so the results from these two classes can be compared to a limited degree. Particularly, question 10 could be understood differently since graduates might mean usage of skills in their working places.

4 Conclusions

This paper describes the example of collaborative project on computerized psychometry instruments development that affected positively engineering students' engagement. Our findings indicate as well that special time management instruction for students that missed in our experiment could be recommended.

References

1. Thomas, J.W.: A review of research on project-based learning (2000). http://w.newtechnetwork.org/sites/default/files/news/pbl_research2.pdf
2. Helle, L., Päivi, T., Erkki, O.: Project-based learning in post-secondary education–theory, practice and rubber sling shots. High. Educ. **51**(2), 287–314 (2006)
3. Bell, S.: Project-based learning for the 21st century: skills for the future. Clearing House **83** (2), 39–43 (2010)
4. Vygotsky, L.S.: The dynamics of the schoolchild's mental development in relation to teaching and learning. In: Vygotsky, L.S. (ed.) Educational Psychology, pp. 391–410. Pedagogika, Moscow (1991)
5. Vygotsky, L.S.: Thought and Language. Labirint, Moscow (1999)
6. Siemens, G.: Connectivism: a learning theory for the digital age. Int. J. Instr. Technol. Distance Learn. **2**(1), 3–10 (2005)
7. Downes, S.: An introduction to connective knowledge, Stephen's Web, 22 December 2005. http://www.downes.ca/cgi-bin/page.cgi?post=33034. Accessed 20 May 2017
8. Sanger, P.A., Pavlova, I., Shageeva, F.T., Khatsrinova, O., Ivanov, V.G.: Introducing project based learning into traditional russian education. In: 20th International Conference on Interactive Collaborative Learning and 46st International Conference on Engineering Pedagogy (ICL & IGIP), Budapest, Hungary (2017)

Work in Progress: Real-Time Annotations of Video-Lectures

Marco Ronchetti(✉)

DISI, Università di Trento, Via Sommarive 14, 38123 Povo di Trento, Italy
Marco.ronchetti@unitn.it

Abstract. We present our work about the development of a system, which allows taking notes during a lecture, and having the notes integrated in the video recording of the lecture itself. The notes become semantic markers into the video, and could possibly be shared with peers and teacher.

1 Introduction

The work we present is an attempt to put together several areas: capturing of lectures in the form of videos, annotation sharing and multimedia annotations.

Recording traditional lectures in the form of videos is a practice that dates back to the end of the nineties, and has been widely deployed in recent years (see [1, 2] for a review).

The idea of sharing annotations is not new: already in 1999 Davis et al. [3] discussed the possibility to allow group members easy access to each others experiences through their personal notes. The system allowed group members to share the notes they were taking through a shared repository. Robertson et al. [4] observed users behavior in three different scenarios: No Notes, Private Notes, and Shared Notes. A more learning-centered approach to note taking was developed by Miura et al. [5]. Their AirTransNote was an interactive learning system augmented by digital pens and PDAs for each student. Notes taken by the students were transmitted to teacher's PC to generate feedback. Miyake and Masukawa too [6] investigated a note-sharing scenario in a university setting.

Also the idea of annotating multimedia for learning purposes has been explored: see for instance Chu et al. [7].

Our work starts from a video-recording scenario, and enriches it with the possibility to take on-line notes (either during the actual lecture, or while watching the recorded lecture). These notes are attached to the video, and become a personal tool for marking relevant passages in the video itself, for later reference. Also, these notes can be shared with peers, and become part of lecture metadata, which can be deployed for learning analytics purposes.

In the following we shortly describe our scenario and how we achieve our goals before coming to discussion and conclusions.

M. E. Auer et al. (eds.), *Teaching and Learning in a Digital World*,
Advances in Intelligent Systems and Computing 715,
https://doi.org/10.1007/978-3-319-73210-7_36

2 Scenario and System Description

We deal with a setting, in which lectures are (video) recorded and made available over the lecture shortly after their end. We assume here that lectures are mostly frontal, methodology that, in spite of being highly deprecated, is still the quite commonly encountered in academy.

The practice of recording lectures and making them available via web, either in open form or with protected access in a Learning Management System, is obviously useful for students who cannot attend classes. However, it has been demonstrated that even students who were in class use the videos to selectively review portions of lectures, for instance for checking their notes, clearing some doubts, resolving interpretation conflicts with their peers [8, 9]. In general, to identify the relevant portion in the recording, students can seek by using a time bar, but some video-recording systems also provide certain semantic markers such as indicator of slide transition, slide titles or text and/or slide thumbnails. We thought that some other types of markings could be generated, and devised a system that allows students to annotate the (video) lecture.

In our scenario, the teacher uploads on a web site some lecture notes before the beginning of the lecture. An example of such handout could be a PDF containing the slides that the teacher will use during the lecture. In absence of such resource, even a white PDF (representing a blank sketchpad) can be provided (it will be clear later why). Students are given access to the handouts in a browser, and a Single Page web Application (SPA) allows them to annotate their own (virtual) copy of the lecturer's notes in class, while the lecture is given. The SPA allows adding notes in the form of typed post-it-like sheets, by highlighting text and by free hand sketching. Figure 1 shows the current aspect of the prototype.

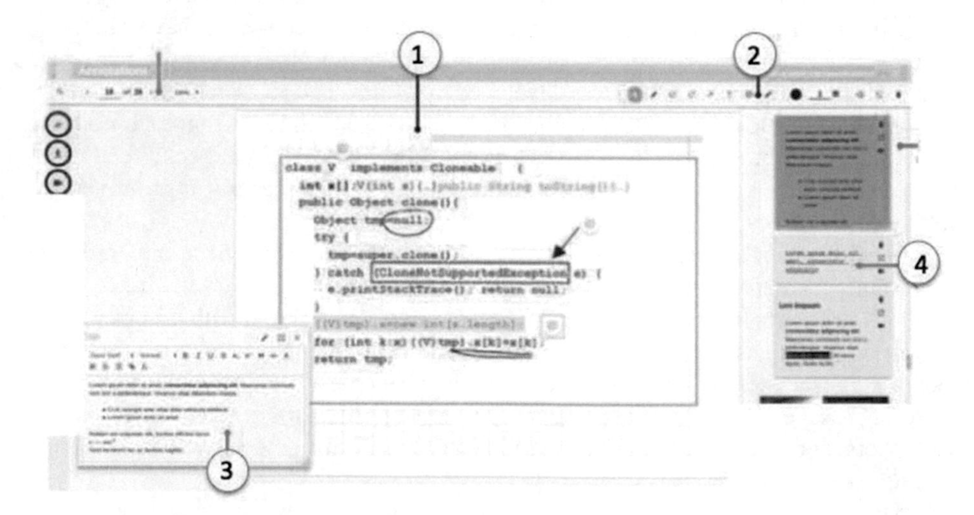

Fig. 1. Look of the prototype SPA for annotating the lecture in class.

In the figure, 1 indicates the current page of the PDF (in general it should coincide with what the teacher is projecting on the classroom); 2 is the menu from which students can select the tool for annotating the PDF, 3 shows the area where a typed annotation takes place and 4 shows the already typed notes for the current page.

Free-hand writing is not very natural when using a mouse o a touchpad on a laptop, but is very effective when using a tablet. Hence, students using laptops will mostly type, while those using tablet will prevalently hand-write.

Student notes are saved in a database. Each student has a personalized view of her/his annotated handout, so that s/he only sees her/his own notes: access to the SPA is individualized via a login gate. Apart of the content and the reference to the user, the database also saves a timestamp, which records when the note was jotted.

At home, students use a second SPA, which allows viewing the recorded lecture together with the notes taken in class. Figure 2 shows the second SPA.

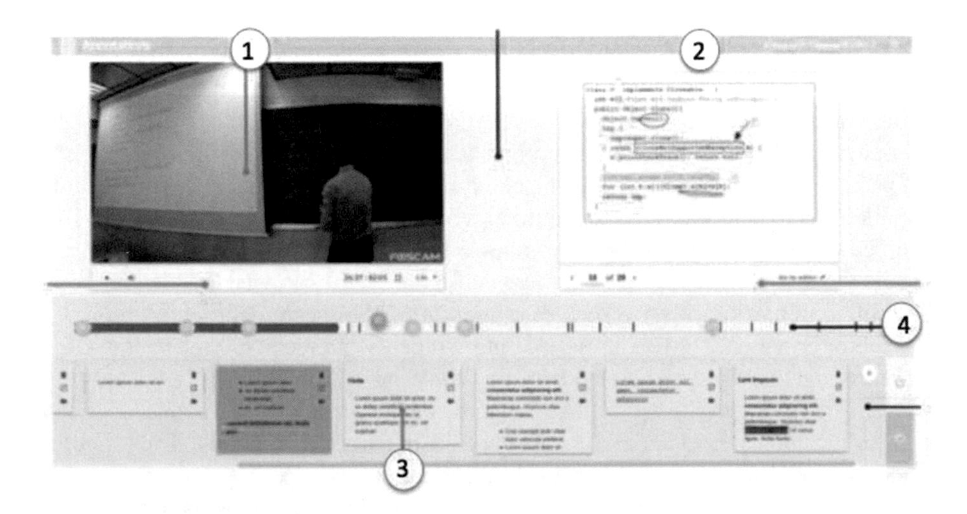

Fig. 2. Look of the prototype SPA for reviewing the lecture after class.

In the figure, 1 and 2 respectively show the video and the annotated slide. The system looks similar to many video-lecture playing tools, but here the slides contain the markings superimposed by the student. In fact, the time-bar (4) contains markers showing the time-location of the annotations, and the lower region (3) shows a (scrollable) series of notes so that students can quickly jump to a note (which lets the video jump at the time, when the note was taken.

Of course the synchronization of video and notes is allowed by the timestamps we saved in the database Hence the notes become a personal index for the video-lecture.

The system also supports a different type of scenario: a variant of the first SPA allows taking notes when watching a video-lecture. This covers the case, when a student not present in class at lecture time takes the lecture in its recorded form: still, s/he has the option of taking notes while watching the recorded lecture. In that case the system is able to detect that notes are taken in a deferred scenario, and not during the real lecture.

The synchronization mechanism, aware of this fact, is able to mark the time stamp of the note as if it was taken in class, hence making notes taken in class equivalent to those taken when watching the video. After that, the student is still able to review her/his notes together with the synchronized video using the second SPA.

All students' actions are monitored, so that learning analytics can be employed and students' behavior can be tracked for later analysis of their learning processes. Apart the note-taking activity, also recalling notes, watching videos and all the operation done in this case (such as pausing, stopping, jumping ahead or backward, or playing video at higher speed) are recorded.

An overview of the software architecture which is at the basis of our system has been presented elsewhere [10].

3 Discussion and Conclusions

Our preliminary work makes it possible to allow students to take on-line notes while in class or when watching a recorded video-lecture, and automatically aligns these notes with the video-lecture itself.

An advantage offered by this system is that if, during the lecture, the student finds that a passage is unclear or difficult, s/he can quickly take a note and later use the note as a marker to position the video to the time, where the portion to be re-watched is.

Sometimes it also happens that one does not have the time to mark down all the things he would be interested in (e.g. there can be references to URLs which are long or difficult to transcribe): also in such case annotating the specific portion of the video where the information is given is useful. Later students can recall the relevant portion and retrieve the information which otherwise would have been lost.

Also, it may happen that one takes a note, but when reviewing it one does not remember the context, and hence the note itself loses part of its meaning. The synchronization of notes and video allows reconstructing the lost context.

Given such role played by the annotations, even taking them on a blank sketchpad (in absence of teacher's handout) has a value. Hence, in absence of a PDF pre-loaded by the teacher, students can still take notes on such virtual sketchpad, with the advantage that these notes are synchronized with the video.

Yet another dimension is the social one: students could share notes with all their peers, or only with a group they belong to, or with (e.g.) their Facebook friends.

At the time of the writing of the present paper, a prototype of the system has been created and deployed. Experimentation has started, for the moment only with four pilot users: each of them uses a different client, so as to test the system is various environment. The test is being performed with two different laptops (an Apple Macintosh and a MS Windows 10 machine) and two different tablets (an Apple iPad and an Android tablet), to check for possible problems when using different machines (although the SPAs run into a browser, and hence should not be affected by operating system nuisances) and to verify the user-friendliness with two different device classes (laptop vs. tablet). The social aspects, although made possible by our software architecture, have not yet been investigated. Our work is in progress, so we are planning to have soon a more extensive and comprehensive validation.

The small-scale experimentation with real students showed that (in spite of some glitches due to the immature user interface) the idea is valuable and interesting to them. A possible but hitherto unexplored potential is the possibility of sharing the notes among the students, so we do not know yet how much this will be considered useful by them: this is the subject of a study we just started. As we mentioned, the system fully traces the students' actions, both when taking notes and when watching videos. This generates a wealth of data, which will be used for learning analytics. The density of notes, for instance, will be an indicator that teachers can use to monitor either the interestingness or the difficulty of certain passages in the lecture, and hence will become a tool for improving their teaching. The design of the logs still has to be refined, so as to allow extracting the maximum possible information out of them. Also this part of the study is presently in progress.

Videos could be used in a flipped classroom scenario, as suggested in [11]. In such case, students would use the version of the first SPA, which allows taking notes while watching the videos. In such scenario, notes taken by the students will contribute to the preparation of the following in-class phase. For instance, by formally asking students to mark the lecture points where they have troubles in understanding, or to add questions in-place in the lecture, the teacher will have rich material for the preparation of such phase. Also, the logs, which indicate for instance that the students re-watched certain portions of the videos, will be indicators of difficulties to which the teacher can later respond in class.

In summary, although the work we report here is only at a preliminary phase, we envision a set of interesting research areas that we intend to pursue in the near future.

Acknowledgments. This work has been performed in the context of the MIUR project CTN01_00034_393801 "Città Educante".

References

1. Ronchetti, M.: Video-lectures over Internet: the impact on education. E-Infrastructures and Technologies for Lifelong Learning: Next Generation Environments, pp. 253–270. IGI Global, New York (2011)
2. Ronchetti, M.: Perspectives of the application of video streaming to education. Streaming Media Architectures, Techniques, and Applications: Recent advances, pp. 411–428. IGI Global, Hershey PA, (2011). Information Science Reference
3. Davis, R.C., Landay, J.A., Chen, V., Huang, J., Lee, R.B., Li, F.C., Schilit, B.N.: NotePals: lightweight note sharing by the group, for the group. In: Proceedings of the SIGCHI Conference on Human Factors in Computing Systems, pp. 338–345. ACM, May 1999
4. Robertson, S.P., Vatrapu, R., Abraham, G.: Note taking and note sharing while browsing campaign information. In: 42nd Hawaii International Conference on System Sciences, HICSS 2009, pp. 1–10. IEEE, January 2009
5. Miura, M., Kunifuji, S., Sakamoto, Y.: Airtransnote: an instant note sharing and reproducing system to support students learning. In: Seventh IEEE International Conference on Advanced Learning Technologies, ICALT 2007, pp. 175–179. IEEE, July 2007

6. Miyake, N., Masukawa, H.: Relation-making to sense-making: Supporting college students' constructive understanding with an enriched collaborative note-sharing system. In: Proceedings of the 4th International Conference of the Learning Science, pp. 41–47, April 2013
7. Chiu, P.S., Chen, H.C., Huang, Y.M., Liu, C.J., Liu, M.C., Shen, M.H.: A video annotation learning approach to improve the effects of video learning. In: Innovations in Education and Teaching International, pp. 1–11 (2016)
8. Ronchetti, M.: Using the Web for diffusing multimedia lectures: a case study. In: Proceedings of EdMedia: World Conference on Educational Media and Technology. Association for the Advancement of Computing in Education (AACE), pp. 337–340 (2003)
9. Amendola, D., Perali, A., Vitali, D.: Using video-lectures in e-learning platform to improve physics teaching at university level. In: Proceedings of EDULEARN 2017 Conference, Barcelona, Spain, 3rd–5th July 2017, pp. 10018–10024 (2017). ISBN 978-84-697-3777-4
10. Ronchetti, M., Lattisi, T., Zorzi, A.: Architecture for a video–lecture annotation system. In: Proceedings of EDULEARN 2017 Conference, Barcelona, Spain, 3rd–5th July 2017, pp. 2496–2504 (2017). ISBN 978-84-697-3777-4
11. Ronchetti, M.: Using video lectures to make teaching more interactive. Int. J. Educ. Technol. **5**(2), 45–48 (2010)

The Perception of the One-Semester International Academic Mobility Programme by Students of Computer Science

Marek Milosz(✉) and Elzbieta Milosz

Institute of Computer Science, Lublin University of Technology, Lublin, Poland
{m.milosz,e.milosz}@pollub.pl

Abstract. Increase in student mobility is one of the objectives of the European Union (EU) and many other countries of the world. This mobility is meant to increase the ability and flexibility to acquire knowledge and skills by students, get them used to the freedom of movement of people and services, and educate them in the area of international and multicultural cooperation. The EU, through a number of programmes (especially those grouped in Erasmus+), supports such mobility organisationally and financially. The article presents the results of research among students of Computer Science (CS), aimed at detecting their perception of the opportunity to participate in a one-semester international academic mobility, i.e. to study abroad for one semester. The survey covered almost all undergraduate CS students of the Lublin University of Technology at their fifth semester, when they take decisions to apply to go abroad for one semester. We analysed their overall assessment of this type of travel, the level of the perceived usefulness of the experience, the concerns and risks associated with it and the students' interest. Some elements of the research were differentiated for Polish and foreign students. The results of the tests are not conclusive, because despite the generally positive perception of the programme, a relatively small percentage of students declare their willingness to take part in it.

Keywords: International student mobility · Perception of usefulness
Risks and concerns

1 Introduction

Student mobility programmes are playing an ever greater role in the area of higher education, leading to its internationalisation. The Erasmus programme has been operating for 30 years and includes in its range not only European countries but also partner countries in Africa, Asia and America. More than half of the budget (about 1.2 trillion euros) of the Erasmus+ programme in 2015 was devoted to Key Action 2, i.e. student, school and professional mobility. This allowed the financing of more than 559.000 student mobilities [1]. The main destinations of mobility were Spain (12%), Germany (11%) and the United Kingdom (10%). Student flows depend on many factors, such as cost of living, distance, educational background, university quality [2, 3].

M. E. Auer et al. (eds.), *Teaching and Learning in a Digital World*,
Advances in Intelligent Systems and Computing 715,
https://doi.org/10.1007/978-3-319-73210-7_37

Studies show that participants in exchange programmes “are satisfied about the programme (96%), improved their skills (94%) and feel better prepared for finding a job (80%)” [1]. Other studies point out an increase in students general perceived self-efficacy through international academic mobility [4].

Research [5] suggests that many EU countries are not willing to finance the education of students from abroad, which is why such mobility is financed by international programmes, such as Erasmus+.

EU employers do not always take into account the fact of education abroad during recruitment [6]. Nevertheless, the findings indicate that international education is valued particularly when employers need graduates with good foreign language and decision-making skills [6, 7].

An important success factor in various student exchange programmes is a proper selection of students for exchange [8]. Student exchange does not always have positive effects [9], but is generally considered as positive [1].

Poland joined the international educational programmes of the EU in 1990 (the first edition of the Tempus programme). In 2004 Poland entered the EU and became a country actively participating in academic exchange programmes. These programmes have a positive impact on the careers of graduates – former Erasmus students [10].

The Lublin University of Technology (Lublin, Poland) is a participant of international educational programmes since 1990 [11]. At the moment it is an active participant in Erasmus+. To handle outgoing and incoming students within the Erasmus+ exchange at the Lublin University of Technology (LUT) the Office of International Studies was established. Despite the organisational efforts of the Office, in the academic year 2014/15 almost 280 students came to the LUT for a semester academic mobility, and only 54 left [12]. The coming students were numerically dominated by those from Turkey, Spain and Portugal. In turn, the countries most frequently chosen by LUT students included Spain, Denmark and Italy [12].

The LUT, particularly in the field of Computer Science (CS), has for many years pursued a policy of curriculum adaptation to the needs of local and EU industry [13–18]. One-semester international academic mobility programme is a part of this adaptation.

In the academic year 2016/17 none of the students of CS has left for a semester international academic exchange, despite the fact that in previous years the number of students ranged from 2 to 7 (which is from 4% to 14% of the total mobility of the LUT). At the same time, individual and group discussions with students indicate a declining interest in semester trips abroad among students of CS, despite their good preparation for studies in English [19]. This raises the question about the reasons for this state of affairs.

2 Definition of the Research Problem and Hypotheses

The goal of this study is to analyse the perception of the semester-long international academic programme of mobility by students of CS as well as their intentions to participate in it.

In the formulation of the research process elements have been used of the familiar Technology Acceptance Model – TAM [20]. In this model, the acceptance of new elements (i.e. technology) stems from its perceived usefulness and perceived ease of use, which affect the attitude toward using and the behavioural intention to use it. This model describes the reasons for the decision to use new technologies.

The use of international mobility is new for most students, maybe not a technological one, but this decision is mentally comparable. Therefore, the study used elements of the TAM model.

The subjects of the research are 5th semester CS undergraduate students of the Lublin University of Technology (LUT) in Lublin, Poland. The students were informed by the LUT Office of International Education about the possibility of mobility within Erasmus+ and the relevant procedures. They are thus a natural target audience for this type of project.

The following hypotheses were formulated in the area of the research problem:

H1. CS students have a positive attitude towards the mobility programme.
H2. CS students want to take part in academic exchanges.
H3. CS students perceive the international exchange programme as an opportunity to gain knowledge and skills, and enhance their future position in the labour market.
H4. Family reasons and a sense of terrorist threat are the primary causes of the students' lack of desire to travel.

3 The Research Methodology

In order to verify the hypotheses a questionnaire was prepared and research conducted among full-time undergraduate students of CS, semester 5. The surveys were conducted using a standard paper technique under conditions ensuring complete anonymity of respondents.

The survey contained mostly closed-ended questions with the possibility to indicate other answers. In the preamble to the questionnaire the situation was explained as follows:

> *"The Lublin University of Technology, within the Erasmus+ programme, offers its students an exchange which consists in sending a University student for one semester of training at a specific institution of the European Union. The student sent gets a special grant to cover the costs of study abroad.*
> *Universities agree on a programme of study in such a way that the student has not lost a semester of teaching at his/her alma mater. The details of the agreement are quite flexible.*
> *The procedures of sending students on the exchange is dealt with by the Office of International Studies of the Lublin University of Technology and departmental representatives for the Erasmus + programme. Most of the subjects realised abroad are performed in English."*

The survey form included questions about the student's gender, nationality (Polish vs foreign), age, years of work, self-evaluation of the command of English, and the fact of being abroad. Most of these elements may indeed strongly influence the perception of such trips by students. The above introduction was followed by a proper questionnaire aimed at obtaining data for verification of the research hypotheses. The questionnaire results were subjected to a simple quantitative statistical analysis.

4 Research Results

The survey was conducted in February 2017. The research included 105 5th-semester full-time first-degree students in CS. This corresponds to approximately 85% of all students of this field and semester.

Among the respondents were 12.4% women, which is the norm for CS at the LUT. Exactly 20% of respondents were foreign students, mainly from Ukraine (the few 1–2 students per year from Belarus were not singled out).

Study abroad is carried out in English. Research has shown that only less than 5% of students rated their level of English as weak (Fig. 1). Meanwhile, good and very good knowledge of English was declared by nearly 68% of students (Fig. 1).

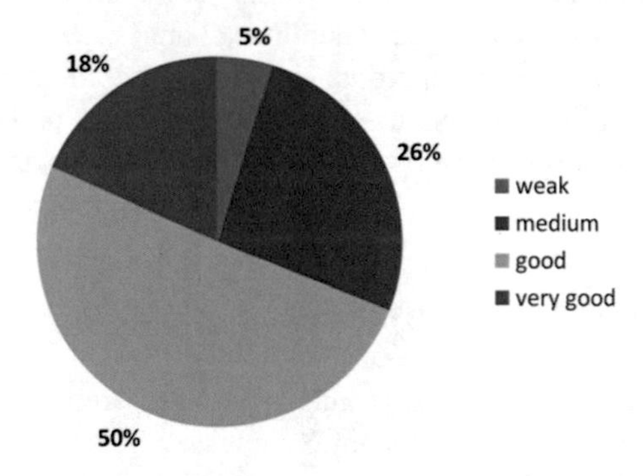

Fig. 1. Knowledge of English according to the students' self-evaluation

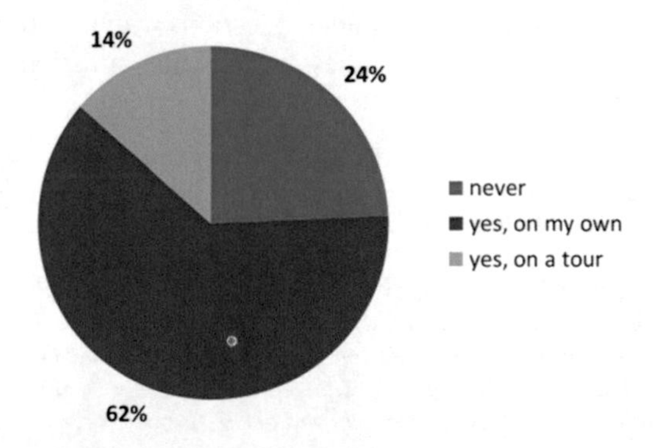

Fig. 2. The results of answers to the question: "Have you been abroad and how?"

The perception of international mobility may affect the existing experience of travelling abroad. It may be of some interest that over 24% of respondents had never been abroad (Fig. 2). If we take into account that foreign students are already abroad, the percentage of Poles who have never been outside of Poland amounts to nearly 30%. In most cases, students organised their stay abroad independently – Fig. 2.

The programme of the semester-long international academic exchange was rated very negatively and negatively by only about 10% of the respondents (Fig. 3). The others assessed it positively (69.5%) and very positively (19.0%).

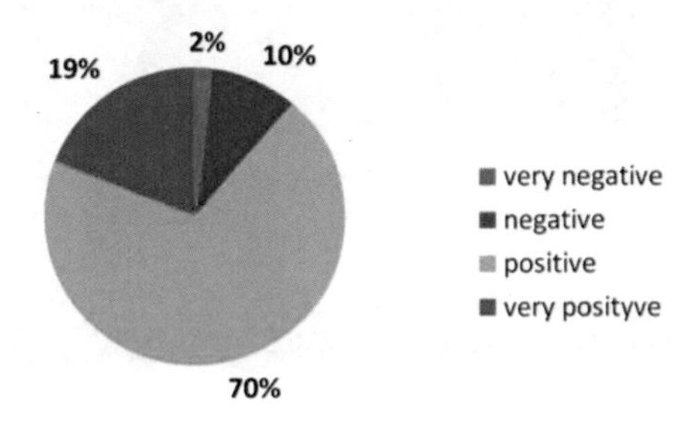

Fig. 3. The evaluation of the semester-long international academic exchange by students in Computer Science

Despite the generally positive assessment of the exchange programme, reluctance to participate in it was shown by more than 32% of the respondents (Fig. 4), and a relatively unequivocal desire – just over 24%. Many of the students merely "considered" the opportunity to participate – nearly 43% (Fig. 4).

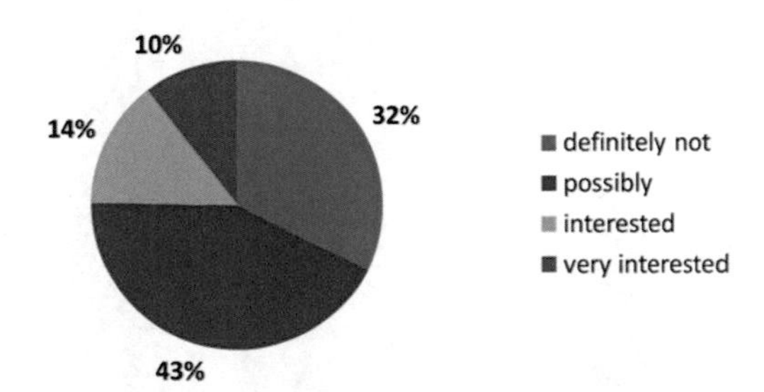

Fig. 4. Interest among Computer Science students in participation in the international academic exchange

The situation related to the interest in the academic exchange programme is different in the two groups of students – Polish and foreign – Fig. 5. In this division a clear willingness to participate (interested and really interested) is expressed by only 20% of Poles and up to 43% of foreign students (i.e. more than twice as many).

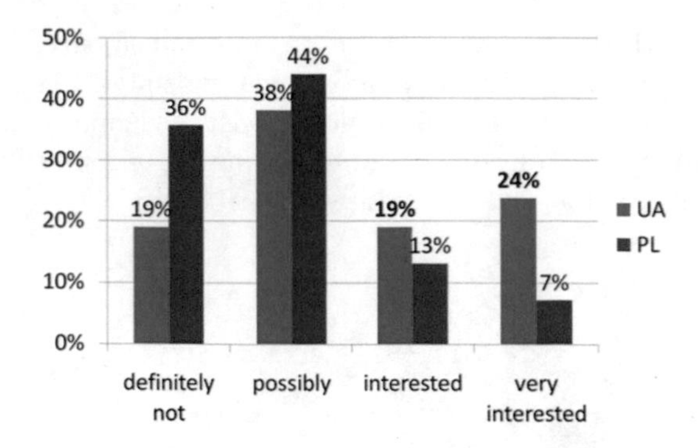

Fig. 5. The interest of Computer Science students in participation in the international academic exchange by nationality (students: UA – Ukrainian, PL – Polish)

Students asked about the perceived usefulness of departure for one semester of study abroad mainly pointed to improving their language skills (74%), tourism implemented in the course of study (68%) and obtaining international work experience (55%) – Fig. 6. They evaluated the experience very low in the aspects of strengthening their position in the labour market (21%) and acquiring new knowledge and skills (36%).

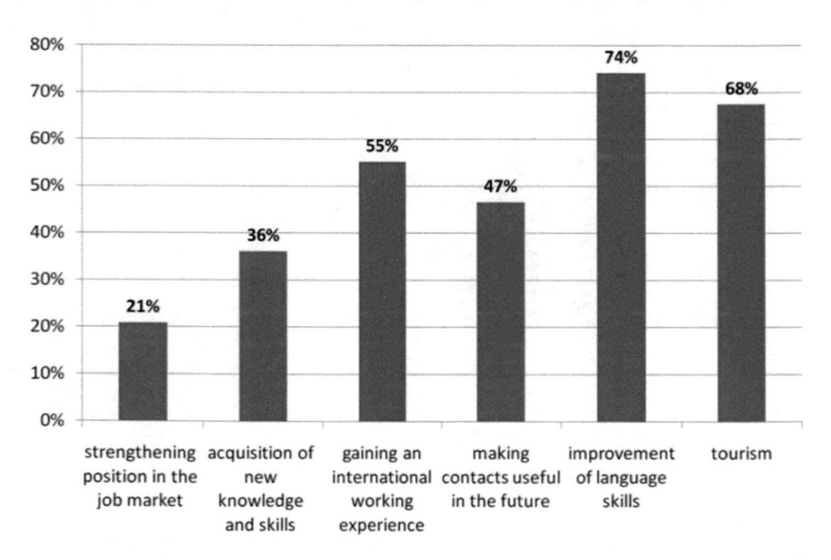

Fig. 6. Perception of the usefulness of the experience of one-semester study abroad

Among the major causes of concern about leaving for a semester abroad, students indicated financial problems (45%) and leaving family and friends in the country (44%) – Fig. 7. In contrast, the biggest perceived threat associated with the departure is the anxiety about the completion of the mobility programme abroad (53%) and fear of terrorism (44%) – Fig. 8.

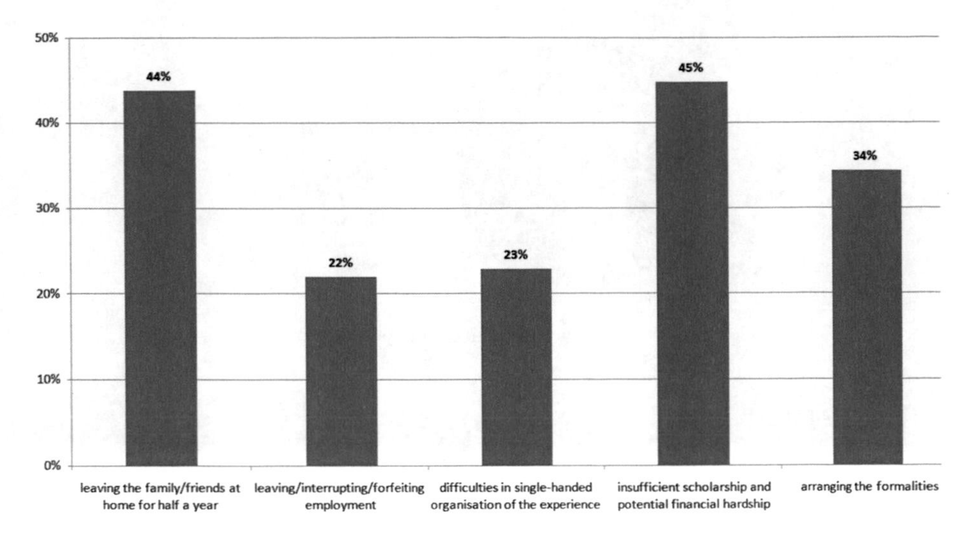

Fig. 7. Reasons for concern before going abroad for one semester

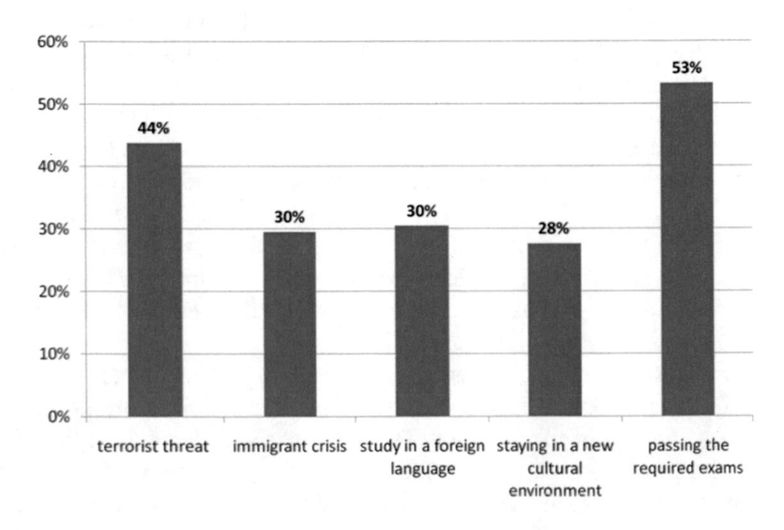

Fig. 8. Risks associated with the experience

Most students (50%) are considering a trip to EU countries. One quarter of them (25%) indicate countries in South Asia (Korea, China), and only 1% – the Eastern Partnership countries (Ukraine, Russian Federation, Belarus and Kazakhstan). Among other countries, the most popular are the United States and Australia.

Students (Fig. 9) mostly have a general knowledge about the programme of academic mobility (81%). Nearly 25% of them also have information from people who participated in this type of trips. Only 10% of respondents indicate knowledge of the detailed procedures of the programme.

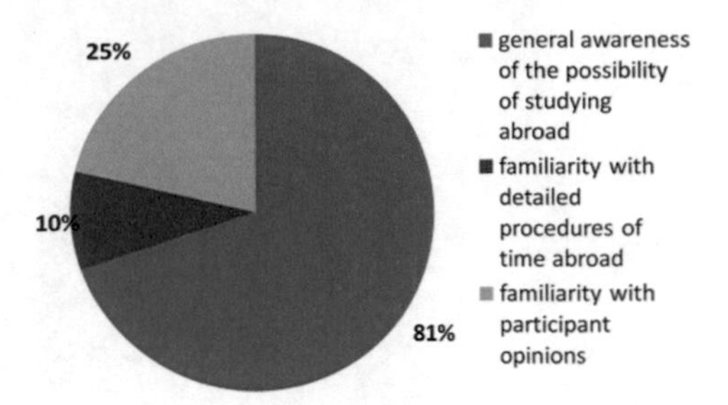

Fig. 9. Knowledge of the international academic exchange programme

5 Discussion

The research results presented in Sect. 4 allow to verify the hypotheses and draw conclusions. Some of the hypotheses have been verified positively (or partly positively), and some not. In particular:

H1. The hypothesis "CS students have a positive attitude towards mobility programme" has been completely verified. The programme of a semester-long international academic mobility by students of CS has been assessed negatively and very negatively by an insignificantly small number of students – only 11.4%. The rest, i.e. more than 88% assessed the programme positively and very positively.
H2. The hypothesis "CS students want to take part in academic exchanges" was confirmed only partially, because only 24.8% of respondents were willing and very willing to take part in a student exchange. Among the foreign students (mostly from Ukraine), this ratio is over 43%, while among the Poles only about 20%. Nevertheless, the 20% is not so low in comparison to the proposion of students actually taking part in academic exchange programme (app. 10%).
H3. The hypothesis "CS students perceive the international exchange programme as an opportunity to gain knowledge and skills, and enhance their future position in the labour market" has not been confirmed. The utility of these elements of the exchange was perceived in the last positions. The first position was occupied by raising the knowledge of a language, the next by tourism and the following – by getting international working experience.
H4. The hypothesis that "family reasons and a sense of terrorist threat are the primary cause of a lack of desire to travel by students" was partially confirmed. Both family reasons and the terrorist threat were indicated by 44% of respondents. The misgivings also included low scholarships and the related potential financial problems (45%) and the anxiety to do with completion of the study programme abroad (53%).

Only 10% of the respondents were familiar with the detailed procedures of arranging an international academic exchange.

6 Conclusions and Future Works

Most of the research hypotheses were partially confirmed (one completely), and one of them rejected.

Despite the very positive attitude towards the mobility programme (H1), a substantial portion of students do not plan to, or simply do not want to use the programme (H2). The percentage of negative attitudes is higher among Polish students than in foreigners. Given a not too high knowledge about Erasmus+, this indicates a need to intensify information activities by the Office of International Studies of the Lublin University of Technology.

The perception of the benefits of the exchange programme, despite an only partial verification of hypothesis H3, should be considered positive. The study indicates that the knowledge and experience of working internationally are important advantages in undertaking employment. Tourism, understood as learning about other countries and cultures, is one of the objectives of exchange programmes. One can therefore conclude that the students want to pursue this objective.

A significant barrier before participating in the exchange programme Erasmus+ is the fear of financial problems during a stay abroad (indeed, the financial situation may be to the main reason for the lack of experiences abroad in nearly 30% of the Polish respondents and in all students from Ukraine) and the fear of completing the requirements of the programme of studying abroad. Assistance in this area can be obtained from the Office of International Studies of the LUT, and possibly by applying for an increase in the level of mobility funding.

The research carried out allows to diagnose the perception of the programme of one-semester international academic mobility by CS students of the Lublin University of Technology and develop the areas and methods of interaction with students in order to increase the popularity of Erasmus+ and the number of student mobility.

The presented research will be repeated next year. This will allow to assess the situation in the dynamic. It is also planned to expand research into students of other disciplines and different universities using the presented methodology.

References

1. Erasmus+ Programme Annual Report 2015, European Commission (2017). https://ec.europa.eu/programmes/erasmus-plus/sites/erasmusplus/files/erasmus-plus-annual-report-2015.pdf
2. González, C.R., Mesanza, R.B., Mariel, P.: The determinants of international student mobility flows: an empirical study on the Erasmus programme. High. Educ. **62**(4), 413–430 (2011)
3. Beine, M., Noël, R., Ragot, L.: Determinants of the international mobility of students. Econ. Educ. Rev. **41**, 40–54 (2014)
4. Petersdotter, L., Niehoff, E., Freund, P.: International experience makes a difference: effects of studying abroad on students' self-efficacy. Personal. Individ. Differ. **107**, 174–178 (2017)
5. Haussen, T., Uebelmesser, S.: Student and Graduate migration and its effect on the financing of higher education. Educ. Econ. **24**(6), 573–591 (2016)
6. Van Mol, C.: Do employers value international study and internships? A comparative analysis of 31 countries. Geoforum **78**, 52–60 (2017)

7. Koziel, G., Milosz, M.: IT studies and IT industry – a case study. In: Proceedings of the 6th International Conference on Education and New Learning Technologies (EDULEARN 2014), pp. 5796–5802. IATED, Barcelona (2014)
8. Crescenzi, R., Gagliardi, L., Orru', E.: Learning mobility grants and skill (mis)matching in the labour market: the case of the 'Master and Back' programme. Reg. Sci. **95**(4), 693 (2016)
9. Aresi, G., Moore, S., Marta, E.: Italian credit mobility students significantly increase their alcohol intake, risky drinking and related consequences during the study abroad experience. Alcohol Alcohol. **51**(6), 723–726 (2016). Oxford
10. Bryla, P.: The impact of international student mobility on subsequent employment and professional career: a large-scale survey among Polish former Erasmus students. In: Procedia – social and behavioral sciences. International Educational Technology Conference, IETC 2014, vol. 176, Chicago, IL, USA, 3–5 September 2014, pp. 633-641 (2015)
11. Milosz, M., Milosz, E., Grzegórski, S.: Tempus programme and its influence on development of EU Universities. In: Proceedings of the 9th International Conference of Education, Research and Innovation, (ICERI 2016), pp. 8615–8623. IATED (2016)
12. The Report on the Activities of the Office of International Exchange for the Academic Year 2014/15, p. 5 (2016). http://www.bwm.pollub.pl/pic/4242.pdf (in Polish)
13. Lukasik, E., Skublewska-Paszkowska, M., Milosz, M.: Meeting the ICT industry needs by universities. In: Proceedings of the 9th International Technology, Education and Development Conference (INTED 2015), pp. 3135–3140. IATED (2015)
14. Milosz, M.: Social competencies of graduates in Computer Science from the employer perspective – study results. In: Proceedings of the 7th International Conference of Education, Research and Innovation (ICERI 2014), pp. 1666–1672. IATED (2014)
15. Borys, M., Milosz, M., Plechawska-Wójcik, M.: Using Deming cycle for strengthening cooperation between industry and university in IT engineering education programme. In: Proceedings of the 15th International Conference on Interactive Collaborative Learning (ICL), pp. 1–4. IEEE (2012)
16. Milosz, M., Plechawska-Wójcik, M., Borys, M., Luján-Mora S.: International seminars as a part of modern master Computer Science education. In: Proceedings of the 6th International Conference on Technology, Education and Development (INTED 2012), pp. 1494–1500. IATED (2012)
17. Plechawska-Wójcik, M., Milosz, M., Borys, M.: Contribution of international seminars on Computer Science to education adjustment on the European IT industry market. In: Proceedings of the 15th International Conference on Interactive Collaborative Learning (ICL), pp. 1–7. IEEE (2012)
18. Milosz, M., Lukasik, E.: Additional trainings of students as a way to closing the competency gap. In: Proceedings of the 10th International Technology, Education and Development Conference (INTED 2016), pp. 8590–8596. IATED (2016)
19. Milosz, M., Milosz, E.: Computer Science studies in English from the perspective of students and business. In: Wrycza, S. (ed.) Information Systems: Development, Research, Applications, Education. SIGSAND/PLAIS 2016. LNBIP, vol. 264, pp. 167–178. Springer, Cham (2016)
20. Davis, F.D.: Perceived usefulness, perceived ease of use, and user acceptance of information technology. MIS Q. **13**(3), 319–340 (1989)

Mobile Learning Environments and Applications

Influence of the Mobile Digital Resources (MDR) Conceptual Model in Motivation of Disadvantaged Students

María José Albert Gómez[1], Clara Pérez Molina[2], María García Pérez[1], Isabel Ortega Sánchez[1], and Manuel Castro[2(✉)]

[1] School of Education, Spanish University for Distance Education (UNED), Madrid, Spain
{mjalbert,mgarcia,iortega}@edu.uned.es
[2] Industrial Engineering Technical School (ETSII), Spanish University for Distance Education (UNED), Madrid, Spain
{clarapm,mcastro}@ieec.uned.es

Abstract. The aim of the European research project mRIDGE (Using mobile technology to improve policy Reform for Inclusion of Disadvantaged Groups in Education, PROJECT Number 562113-EPP-1-2015-1-BG-EPPKA3-PI-FOR WARD) is the design of digital learning resources for mobile devices, for improving the educational integration of disadvantaged students, i.e., groups at risk whose ethnical and cultural features, special needs or socioeconomic status significantly constrain their possibilities to receive a suitable education.

The mRIDGE project aims at evaluating the suitability and development of mobile applications with augmented reality (MDR Model) for improving motivation of students in the educational process, according to the Quality Assurance Plan designed for this project.

Keywords: Educational digital resources · Motivation · Mobile devices

1 Introduction

Throughout these last years, different technological means have gradually appeared; they are present in all areas of society nowadays, as well as in the educational field. The integration of information and communication technologies (ICT) into educational practices can improve the teaching-learning process and increase the educational success of students.

Motivation is one of the key elements for improving learning and for educational success. Several researches [6] have found a direct relation between motivation and the use of the technology in the classroom as well as other factors that have a direct influence in the improvement of learning, such as self-efficiency. Some authors [1] state that technology applied to education favors students' learning, increases their interest and creativity, improves their ability for solving problems, fosters team work and reinforces their self-esteem, thus contributing to increase motivation of students.

M. E. Auer et al. (eds.), *Teaching and Learning in a Digital World*,
Advances in Intelligent Systems and Computing 715,
https://doi.org/10.1007/978-3-319-73210-7_38

However it is important to point out that academic motivation is a complex process where many interacting factors converge, so it would be necessary to open a debate in the educational community on its scope [2] and use a single approach when dealing with this key factor.

Under this assumption, the European project mRIDGE (Using mobile technology to improve policy Reform for Inclusion of Disadvantaged Groups in Education, PROJECT Number 562113-EPP-1-2015-1-BG-EPPKA3-PI-FORWARD) aims at creating educational digital learning resources for mobile devices in order to improve a series of elements, among which is motivation of disadvantaged students; i.e., groups at risk whose ethnical and cultural features, special needs or socioeconomic status significantly constrain their possibilities for receiving a suitable education.

Therefore, the specific objectives of this project are the following:

- To analyze users' needs in different contexts, as well as the current curricula in Bulgaria and Romania. This study aims to evaluate to what extent specific groups of students use mobile technology, thus creating new chances for its inclusion in the learning process.
- To design a Mobile Digital Resources (MDR) Model to support the educational features of mobile technology in order to adapt them to the learning conditions of disadvantaged groups of people.
- To develop and adapt mobile applications and educational digital resources. This objective addresses the need of applying m-Learning in different subjects to meet the educational needs of the disadvantaged groups which the project is destined for.

The mRIDGE project aims at fulfilling the educational needs of persons at risk of exclusion in Bulgaria and Romania, and follows the guidelines of the partners involved in it, which are the following educational institutions:

- Plovdiv University "Paisii Hilendarski", Bulgaria
- UNED, Spanish University for Distance Education (Universidad Nacional de Educación a Distancia), Spain
- Ravensbourne Higher Education Institution RAVE, United Kingdom
- The University of Craiova, Romania
- Primary school "Geo Milev", Sadovo, Bulgaria
- Secondary Vocational School for Children with Hearing Disabilities "Prof. Stoyan Belinov", Plovdiv, Bulgaria
- Special gimnazial school "Sf. Mina", Craiova (Mina), Romania
- Regional Industrial Association – Smolyan, Bulgaria
- Plovdiv Municipality – The "Education Department" of Plovdiv, Bulgaria

2 The MDR Model

This research focuses on the MDR Model - which is based on the development and use of digital educational resources, mobile devices and augmented reality - and how it affects the teaching process and the variable studied: motivation in education.

The use of mobile devices and digital resources in education offers innovative solutions to meet the needs of the groups at risk taking part in this research:

- They provide teachers with resources for making complex presentations to improve understanding and solutions to problems according to the educational needs of students.
- They give freedom and autonomy to students by overcoming limitations related to time, place and the amount of school materials. Students can use this technology both in the classroom and outside at the most convenient time, and they can have access to a wealth of information resources.
- The use of this technology does not require great knowledge. Augmented reality (AR) is based on intuitive perceptions and personal preferences, and offers many opportunities to raise interest and motivation in the long term.
- This innovative learning makes students proactively and enthusiastically accept, by using everything related to technology, the subject to be studied [3]. It leads to greater motivation for actively taking part in the learning process, as well as to a substantial improvement in memorizing school materials, and - due to students using more senses - facilitates learning of disadvantaged students, making education more effective.

The MDR model is based on active learning where students discover, process and apply the information received. As Green and Casale-Giannola [4] point out, for meaningful learning to occur there must be an active learning where students operate, judge things on their own, test their skills, reflect on their own knowledge. Therefore, active learning allows students to participate in their own learning process and increase their motivation [5].

However, it is also important to add that the MDR model is used in an electronic learning platform, DIPSEIL (http://env.dipseil.net/v3), which allows designing, developing and offering resources for the educational process, being one of its features to ensure support to students when and as much as needed, for them to deal with real assignments in a problem-based educational model. Thus, according to the needs and abilities of students, educational modules/courses have been developed and adapted in the electronic learning platform DIPSEIL.

3 Methodology

As aforementioned, one of the hypotheses of the project is based on discerning and evaluating how the MDR model improves motivation in disadvantaged students. Students' motivation is a variable that has to be measured from the point of view of the student interacting with the educational resources of the MDR model, as well as from the teachers who guides the teaching and learning process. Therefore, in the research a quasi-experimental method has been used and this variable is measured with several online questionnaires for all those involved (teachers and students) to have immediate access to them, being data visual and easily classified.

This variable is measured in two different phases in order to give answer to the aforementioned objectives:

- The "Pre" Phase, before students use the educational digital resources in the teaching-learning process.
- The "Post" Phase, after students use the educational digital resources.

The questionnaires that have been used are different for teachers and students taking part in the pilot experience, although the same variables are measured. They consist of open and closed questions. The closed questions are measured by a four-point scale where each participant gives their opinion (1 = strongly disagree, 2 = disagree, 3 = agree, 4 = strongly agree) and open questions have been created for each participant to describe what they would improve.

An intentional non-probabilistic sampling has been used for selecting the sample. The sample is made up of students from disadvantaged groups such as people with sensory impairment, musculoskeletal disorders and unemployed, as well as their teachers. The sample of the research has involved a total of 49 people, 40 students and 9 teachers.

4 Results

The analysis of the results shown hereunder has been carried out with the information collected with the questionnaires used in this research. In general, data shows that participants are satisfied with the MDR model, obtaining in all cases an average of more than 3 on a scale of 4 in the closed questions.

As aforementioned, the mRIDGE project aims to create digital educational resources for mobile devices to improve motivation of disadvantaged students, among other elements. For this reason, the questions used in the questionnaires not only gave meaning to motivation as a single element but also gave it to academic performance and improvement in teaching with the MDR model, factors with a direct influence in motivation in education.

In the results collected, teachers consider that the MDR model has been very satisfactorily designed according to the activities to be carried out and taking into account the features of the groups at risk, a factor which has influence in improving their teaching abilities and increasing motivation within their students. Teachers have also pointed out that this model allows specific approaches based on pedagogical theories and their practical application. As per data obtained, teachers think that the MDR model has promoted socialization and personal satisfaction of their students.

Students as well have scored high on how this model provides new knowledge that otherwise would be difficult to understand. They highlight that this system promotes interaction in education as well as an active participation of all those people involved. According to data collected, students consider that the MDR model has stimulated their interest in studying and class attendance.

Both teachers and students score high when directly asked if they consider that this MDR model has been of influence for improving motivation of disadvantaged students and if it enhances the educational potential of mobile devices. Another question yielding good results in students and teachers has been the one related to how this system promotes interactive communication between them.

Data collected in the different questionnaires points out that the MDR model fosters self-esteem and motivation among disadvantaged groups, and in turn, it encourages the development of skills and abilities in different areas of specialization. In addition, results show that the respondents considered that the training activities and objectives which follow the MDR model have been properly scheduled and designed.

5 Conclusions

To conclude we can say that the participants in the mRIDGE project consider the MDR model to be appealing and useful, as well as a motivating element in the teaching and learning process of this group of teachers and students.

The results of this research reinforce the idea that the MDR model fosters motivation among disadvantaged groups as well as it promotes the development of competences and skills in different areas of educational specialization.

Teachers who took part in the project highly appreciated that the design and application of the programs through the MDR model were adapted to the different areas of study, which has facilitated their work and improved their own motivation.

Moreover, a sound pedagogical design is essential for students to acquire knowledge thus favoring high quality learning processes. It is not only a matter of teaching multidisciplinary contents, which is of a clear importance, but also the pedagogical design must ease the development of efficient learning strategies.

The MDR model was designed and planned as a balanced tool to ensure that the content of the subjects was properly taught as well as for creating learning scenarios which encourage different strategies and skills. According to the results obtained – reviewed in detail in previous sections of this paper – the objective has been fulfilled in this phase of the project, although this same objective is to be achieved in the following phases of the research.

As per data collected this model and especially the technology used in the educational field, has favored learning in students, fostered its interest and improved their problem-solving skills. Thus their self-esteem has been reinforced, something that has contributed to increase motivation of students.

This study expects to contribute for removing barriers and prejudices in the integration of technology for improving education and educational motivation of people at risk of exclusion which are the basis of this project. It is sought that all students, regardless of their personal circumstances and abilities, can benefit from a better education and that teachers see technologies as a set of resources and possibilities.

Acknowledgments. We would like to acknowledge the work and contribution of all the people taking part in the European Research Project mRIDGE (Using mobile technology to improve policy Reform for Inclusion of Disadvantaged Groups in Education, PROJECT Number 562113-EPP-1-2015-1-BG-EPPKA3-PI-FORWARD). We would like to acknowledge as well UNED's *ETS de Ingenieros Industriales* (School of Industrial Engineers) for the funding received for the research project in e-learning through the Call for Aid to support teaching and research activities of the Departments of the School in the 2017 Call.

References

1. Amar, V.: Planteamientos críticos de las nuevas tecnologías aplicadas a la educación en la sociedad de la información y de la comunicación. PíxelBit. Revista de medios y Educación **27**, 1–6 (2006)
2. Broc, M.A.: Motivación y rendimiento académico en alumnos de Educación Secundaria Obligatoria y Bachillerato LOGSE. Revista de Educación **340**, 379–414 (2006)
3. Díez, E.J.: Modelos socioconstructivistas y colaborativos en el uso de las TIC en la formación inicial del profesorado. Revista de Educación **358**, 175–196 (2012)
4. Green, L.S., Casale-Giannola, D.: 40 Active Learning Strategies for the Inclusive Classroom, Grades K 5. Corwin Press, Thousand Oaks (2011)
5. Montes, A.H., Vallejo, A.P.: Efectos de un programa educativo basado en el uso de las TIC sobre el rendimiento académico y la motivación del alumnado en la asignatura de tecnología de Educación Secundaria. Educación XX1 **19**(2), 229 (2016). https://doi.org/10.5944/educXX1.14224
6. Zenteno, A., Mortera, F.J.: Integración y apropiación de las TIC en los profesores y los alumnos de educación media superior. Apertura, Revista de Innovación Educativa **3**(1), 142–155 (2011)

The Math Trail as a Learning Activity Model for M-Learning Enhanced Realistic Mathematics Education: A Case Study in Primary Education

Georgios Fessakis(✉), Paschalina Karta, and Konstantinos Kozas

University of the Aegean, Rhodes, Greece
{gfesakis,psemdt140162,psed150073}@aegean.gr

Abstract. Seeking a systematic combination of the pedagogical model of m-learning with the Realistic Mathematics Education (RME) approach, this study concerns the use of math trail as a learning activity model that can take the advantages of mobile computing devices for the design of effective learning experiences in an authentic context. The paper presents the design and the study of the first pilot implementation of a math trail, using mobile devices for primary school students. In this math trail, the students are guided, through a digital map, to a sequence of preselected sites of a park where they solve specially designed math problems using data from the environmental context. The students measure real objects' dimensions either with conventional instruments or by measurement applications of their tablet. According to the findings of the study, students solved the puzzles by applying mathematical knowledge, discussion and collaboration. The students applied and reinforced their knowledge through an effective and engaging learning activity. Moreover, the students were puzzled about the differences of the measurements by conventional and digital instruments and this confusion triggered social negotiation. Further research is needed for a grounded theory development about m-learning design for RME.

Keywords: Learning design · M-learning · Realistic mathematics education
Math trails

1 Introduction

The widespread use of various kinds of mobile devices (e.g. Tablet PC, Smartphones, iPad) and their integration into the daily life of children have created new facts on the potential of the pedagogical model of mobile learning (m-learning). Due to the extensive use of technology inside and outside the school setting, today's children are classified as "i-generation students" [1]. The broad use of mobile devices creates new challenges for the learning design. Learning technology researchers and teachers are trying to integrate mobile devices into schools in a meaningful way. In this paper, we focus on the learning design of meaningful m-learning activities for elementary school mathematics. The authors are interested in the idea that m-learning is consistent with the principles of the Realistic Mathematics Education (RME) approach, and it can

M. E. Auer et al. (eds.), *Teaching and Learning in a Digital World*,
Advances in Intelligent Systems and Computing 715,
https://doi.org/10.1007/978-3-319-73210-7_39

improve its applications. This paper proposes the use of the math trail as a learning activity model for systematic mobile learning design for mathematical education [2] in elementary school students. The following sections of the paper present a brief introduction of m-learning and math trails and then the design and the analysis of the case study of the math trail that was designed for this research.

2 Theoretical Framework

2.1 Mobile Learning

Widely spread portable computing devices and wireless internet are now radically transforming the notions of discourse and knowledge [3] and make possible new learning models such as m-learning [4]. M-learning, which is defined as the use of wireless handheld devices in order to engage participants in some form of meaningful learning [5], as a component of formal or informal education [6]. During m-learning experiences, the learners have access anytime and anyplace to information in order to perform authentic activities [7]. In other words, m-learning constitutes a relatively new pedagogical model in which students learn as they move, interacting with each other, their environment as well, through the mediation of applications that run on varying types of mobile digital devices [8]. As it is supported by UNESCO m-learning gives literal meaning to the principle that "the world is a classroom" [9]. Until recently, the context "sensitive location depended learning" has been applied mainly to the informal learning because of the cost of the devices and the requirement to move outside the school environment [10]. With the broad use of the mobile devices and the wireless internet access, mobile learning functions bridges the gap between formal and informal learning [8].

2.2 Mobile Learning and Realistic Mathematics

Freudenthal [11], the main representative of RME movement, has argued that mathematics had to be taught in such a way to become useful for solving everyday-life problems. He was a strong supporter of "Mathematics for All", trying to make mathematics accessible to everybody and he also advocated problem-solving as a teaching method. According to RME, learners should find the starting point of their learning in rich, complex structures of the real world and afterward they must continue to the abstract structures of the world of symbols. Other RME researchers consider important to view concepts as problem-solving tools and discover them through their application in authentic contexts [12]. With the use of digital technology, learners can develop deeper understanding of mathematics, because technology as a mindtool can support learners' inquiry, decision making, reflection, reasoning, and problem-solving capacity [13]. The principles of RME constitute a context in which digital technology could provide significant assistance in teaching and learning [14], because digital technology itself influences the kind of mathematics that are taught and enhances students' learning [13]. Furthermore, networked mobile devices contribute to mathematics education by fostering the collaboration between participants [15].

2.3 Research Review for Mobile Learning and Realistic Mathematics

Research review reports that only 1.9% of research published between 2009 and 2014 had as a subject the investigation m-learning for Mathematics education [16]. Available research on RME and m-learning is even more limited [17–20]. The results of a systematic research review regarding m-learning applications in mathematics education have shown positive results concerning: performance in basic mathematical concepts and arithmetic in Kindergarten [14, 21] and the first grade [22], in engagement and confidence of secondary education students [19] and finally in team collaboration and understanding of graphs [23]. The studies above do not concern key distinctive features of the pedagogical model of m-learning, such as students' movement and their interaction with the natural environment, but the use of mobile devices as a more flexible type of electronic computers. Moreover, the search of educational applications, for elementary school mathematics, in mobile devices (e.g. in app store and iTunes services) showed that most the available apps are educational applications concern arithmetic drill and practice instructional games. In other words, they do not require moving in space and do not utilize the sensors of portable devices. Hence, a research gap is ascertained regarding the combination of mobile learning with RME.

2.4 Mobile Learning, Realistic Mathematics Education and Math Trails

In an attempt to explore models for the use of space and context sensitive applications of mobile devices in learning designing in general and for mathematics learning more specifically, the authors consider various models of activities such as treasure hunt games [24] and math trails. As stated in Ref. [25], math trail is a walk (a tour) towards the discovery of Mathematics. More typically, the math trail includes a pre-planned route, which is defined by a sequence of stops in which students examine the environment in a mathematical way [2]. A remarkable collection of mathematical trails is available from the MathCityMap project of the University of Frankfurt (https://mathcitymap.eu). The math trail constitutes a learning activity model that combines: *movement, communication, collaboration, problem-solving, links of school knowledge with the real world and other school subjects, practical application of knowledge and skills in the conceptual and natural environment* [26]. The compatibility of the mathematical trails features with the principles of RME is quite evident. Mobile devices support the advanced interaction between the user and the environment. The above statements highlight the need to search the possibility of the m-learning application in order to improve mathematics trails and realistic mathematics.

3 The Mathematical Trail "The Fairy of the Waterfalls"

For research purposes, a math trail was designed, which was named "The fairy of the waterfalls". The site chosen is the waterfalls' park of Edessa city in Greece. The estimated duration of the trail was approximately 1–2 h. To make it more attractive, a story was integrated into the math trail. Students are requested to solve a series of math problems in order to find the fairy of waterfalls. As soon as the students solve a problem, they earn some pieces of the fairy's puzzle. In the end, after they provide correct answers to all problems, they can assemble the puzzle and see the image of the famous fairy. There are six problems in the trail located in corresponding stops of the park area. In order to facilitate the children to find these six places, a Google Map was made for the specific mathematical trail. The math trail map that was given to the students is available on: http://goo.gl/iMeu53. Each stop of the math trail is marked by a corresponding pinpoint on the map. Arriving at each stop of the trail, students can see on the digital map, a photo of the stop and a text with instructions about the problem. At each stop, the students are requested to solve the problems in Table 1.

Table 1. The math trail's stops and problems

Stop's name	Photo	Problems
No. 1 ***"The circle"***		Calculate the perimeter of the circular square: 1. With the tablet apps and 2. With the measuring tape (Work in 2 groups)
No. 2 ***"The church"***		1. Calculate the height of the church. 2. Calculate the height of the door of the church 3. Calculate the surface of the door (Work in one group)
No. 3 ***"The watermill"***		1. Calculate the radius of the circular pulley, 2. Calculation of the diameter of the pulley, 3. Calculation of the angle in degrees of each section of the pulley (Work in one group)
No.4 ***"The stair"***		1. Calculate the height of the stairs. 2. Calculate the height of each step, 3. Calculate the steps going up 2-2 stairs (Work in one group)
No. 5 ***"The shelter"***		1. Calculate the area of the covered rectangular square with tape, 2. Calculate the same area as the Object Height embodiment (Work in 2 groups)
No. 6 ***"The waterfall"***		Open-ended problem. Recommended ways to calculate the height of the Karan's waterfall.

For the required measurements, students were provided with a conventional tape measure and the application "Object Height" (Fig. 1). The measurement applications usually require the user's height and then they permit the measurement of the user's distance from a specific target point on the camera of the device. They also measure the height of objects on the camera, by applying the Pythagorean Theorem for a given angle of inclination which they measure with the help of the tilt sensor of the device.

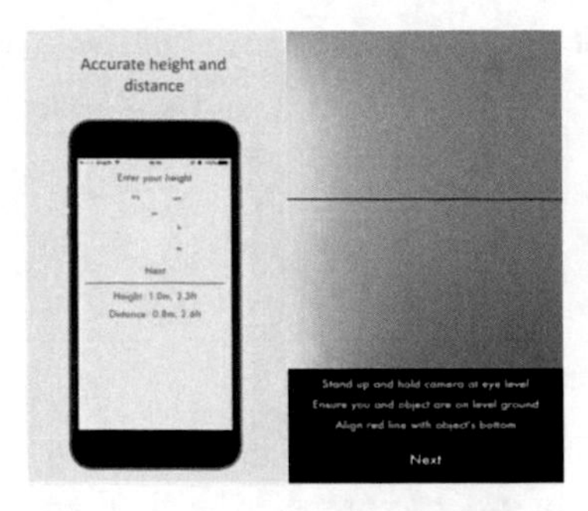

Fig. 1. Object Height app's user interface screenshots

The "Object Height" is a specific mobile application for measuring both height and length. In order to use the app, the user has to follow three steps: Step 1. Insert his/hers height, Step 2. Take a photo so that the red line that appears on the camera (Fig. 1) is at the base of the object to be measured and Step 3. Take a photo so that the red line of the camera is on the top of the object. This will show the height of the object and the user's distance from it. The application works by using the same principles as a simple theodolite, or as a more modern geodetic measurement station. This measuring apps accuracy is very sensitive to the user's alignment and skills. As a result, the digital measurements will vary among student groups as well as those made with the tape measure.

4 Methodology, Research Questions, and Research Conditions

The following research questions were posed before the study:

RQ1: Can math trails, enhanced by mobile technologies, help primary school students in the development of mathematical concepts, such as the length, the circumference, the area and their measurement?
RQ2: Can math trails, enhanced by mobile technologies, be effective, attractive and feasible for primary school students?
RQ3: Can math trails, enhanced by mobile technologies, foster collaboration among the learners?

The methodology chosen is a combination of the design experiment [27] and exploratory case study [28]. For the collection of research data questionnaires, the students' worksheets, the interviews and the researcher's observation log have been used. The second author participated in the implementation of the trail as the teacher. The participants of the research were four children of the sixth grade of a public Elementary school in Edessa. The details of the participants are shown in Table 2. The rules of research ethics were followed the children's and their guardians' consent was given after they were informed about the purpose and the research process. All children knew how to use a tablet while children who had their own tablet mentioned that they use it 1–3 h per day for gaming and web browsing.

Table 2. The participants of the research

A/A	Alias	Gender	Age	Owns tablet
1	P.A.	M	12	Yes
2	E.K.	F	12	Yes
3	X.A.	M	12	Yes
4	L.K.	F	12	No

Children implemented the math trail on Sunday, 18/12/2016 and the total duration of the process was 75 min. In order to prepare the children, a three-hour meeting took place, one day before the trail implementation. In this meeting, the students became familiar with the use of the map on tablet and the measuring tape. In addition, students' previous knowledge of the concepts of the trail was examined by written exercises using paper and pencil. The examination showed that the children knew the formulas for rectangles dimensions but they faced problems with the formulas for the cycle. During the preparation of the students, previous knowledge was recalled, questions were answered and misconceptions were worked out.

5 Findings

The students completed the trail and solved the problems that corresponded to all six stops. They also kept their interest and enthusiasm during the entire activity. The students collaborated, discussed and made common decisions and initiatives. All the students used the tablet and the applications, Google Maps and Object Height. The difficulties that appeared in the solution of the problems were expected for this level of education. More specifically, children mixed up the concept of the radius with the diameter during the calculations related to the circle (stops 1 and 3). However, they came up with a solution, by discussing the problem and referring to the preparation material. Students were familiar almost with the whole park area and they found some of the stops without using the map. Nonetheless, in some cases (e.g. stops No. 3 & No. 4) they used the map and their GPS position trace to find the point that they were looking for. The automatic navigation of Google Maps was not used by the students even they were allowed to. Students noticed discrepancies in the representation of the GPS position on the map, when the internet connection was not strong enough and they critically interpreted these discrepancies based on the position of the known landmarks. The dialogues that took place in stop 4 (the staircase) are characteristic:

L.K.: *"Miss, the map shows that we are moving, it doesn't show where we are".*

L.K.: *"Miss, we are here [at stop No. 4] but the sign on the map for stop No. 4 is farther away".*

As it was expected, students came up with different results in measurements with the measuring tape and the Object Height App. Table 3 shows characteristic comments of the students on the differences in measurements in various stops. Generally, the students trusted more the measurements with the measuring tape instead of the tablet, since for them this kind of a measuring procedure is like a black box.

Table 3. Vignettes from the math trail implementation

<table>
<tr><th>Stop</th><th>Students Comments</th><th>Answers</th></tr>
<tr><td>No.1
«The circle»</td><td>P.A. (Group B): "Miss, why did we measure 5,10m while the girls (Group A) found 4,77m? Did not we measure correctly?"
X.A. (Group B): "Miss, maybe we should measure it again to be certain"</td><td><table><tr><th>Group A</th><th>Group B</th></tr><tr><td>r = 4.77m.</td><td>r = 5.1 m.</td></tr><tr><td>c = 29.96m.</td><td>c = 31.4 m.</td></tr></table></td></tr>
<tr><td>No.3
«The watermill»</td><td>P.A.: "Miss, this cannot be 3m, it seems to be much less"</td><td>a) circle, b) diameter = 1,10m., c) radius = 0.55m.
d) 6 parts, e) 360°/6 = 60°</td></tr>
<tr><td>No.5
«The shelter»</td><td>E.K.: "Miss, a minus sign appears in front of the result of the measurement. Is this because it is dark in here and the tablet did not measure it well? We should do it again".
E.K.: "Aaah!, now the result is similar to the result of the other group so it must be ok".</td><td><table><tr><th>Group A</th><th>Group B</th></tr><tr><td>a = 11.7m</td><td>a = 11m</td></tr><tr><td>b = 8.7m</td><td>b = 8.9m</td></tr><tr><td>A = 101.7 m²</td><td>A = 97.9m²</td></tr></table></td></tr>
<tr><td>No.6
«The waterfall»</td><td>P.A.: "Maybe we can only measure the point that the waterfall starts until the point we are standing now"
E.K: "We cannot do it with the measuring tape either"</td><td>They concluded that:
1. It was not possible to measure the height with the measuring tape.
2. Only one part of the height could be measured with the tablet application.
3. Maybe the measurement of the height can be achieved with the use of other instruments that they are not aware of.</td></tr>
</table>

In the case of discrepancies, they tried to re-measure so that tablet measurements approximate the values of the measuring tape. The cases of stops 3 and 6 (Table 3) are more interesting where the children could only use the tablet. In the case of stop 3, the students were asked to measure from a distance the diameter of a circular metal structure. The first measurement was rejected as false by the children because they empirically considered it to be very large. Also in the case of stop 6, students concluded that the height of the waterfall cannot be measured with the measuring tape so they just estimated a part of it by using the tablet. Children were amazed when the teacher told them that this was possible using analogy in photos, in the discussion after the trail walk.

6 Results

This section, based on the findings, gives answers to the research questions.

RQ1: Can math trails, enhanced by mobile technologies, help primary education students in the development of mathematical concepts, such as the length, the circumference, the area and their measurement?

The students practiced in measuring length, the distinction of the radius from the diameter and the circumference from the area. They also applied all the knowledge acquired in school to realistic situations of everyday life in the natural environment in the area calculation of a rectangle and a circle transferring. In addition, the students became familiar with the concept of accuracy and measurement error. With the use of the digital map, the students practiced in reading the map and its use to navigate in space. Math trail and the use of the tablet can contribute to the strengthening of the knowledge of mathematical concepts. Furthermore, the students realized the limits of their ability to make accurate measurements and calculations. This experience improved significantly their mathematical and computational thinking.

RQ2: Can math trails enhanced by mobile technologies be effective, attractive, and feasible for primary school students?

As far as efficiency is concerned, the experiment results are positive since the learning objectives of the math trail, such as the application of mathematical concepts in authentic environment, the use of a tablet as a learning tool, the radius distinction from the diameter, the comparison of the measuring tape with the application Object Height, the use of a digital map, as well as the collaboration were achieved satisfactorily by all students.

The application of the mathematics trail seems to be feasible and worth of the extra effort and time required. This is not only due to its effectiveness but also due to its impact on the quality of the students' interaction. Moreover, the math trail that was applied seemed to be particularly attractive to the students. On the question *"how they characterize the mathematical trail"* their answers are: *"good, fun, fantastic, and different"*. Finally, the students stated that they would wish to participate again in such an activity in the future. The following statement of P.A is characteristic.: *"Miss, if you repeat this experiment, please choose us again to do it together!"*.

RQ3. Can math trails, enhanced by mobile technologies, foster collaboration among the learners?

The math trail with the specific script of action has fostered collaboration among the students for map reading and navigation as well as for the solving of math problems. The students had intensive on task discussions about navigation, measurements and problem-solving. Students overcame the difficulties that appeared in the solution of the problems through their collaboration. Table 3 contains characteristic episode of the quality of interaction among students as well as students and the teacher.

7 Summary-Discussion

M-learning is a modern pedagogical model which can make learning more enjoyable and effective as it utilizes mobile devices which are very popular. M-learning can support the construction of knowledge in the context of its application. These characteristics make mobile learning compatible with RME. However, the applications of mobile technologies in RME are fairly a few and they do not take full advantage of the pedagogical model. In order to facilitate the design of m-learning application for RME, the enhancement of a math trail with mobile technologies was explored. The math trail had a positive impact on the students learning and contributed to the transfer of school knowledge to authentic situations. The results of this research are in alignment to those of other researchers [18, 19, 24]. M-learning fosters collaboration, facilitates the math trail implementation, and makes it more efficient and attractive. The students applied and improved their map skills with the support of GPS and Google Maps application. Moreover, by the comparison of the measurements with a tape measure to the ones of the Object Height application, the students understood better the concepts of measurement accuracy, error and approximation. The students reached the limits of the potential practical applications of measurement with measuring tape and a tablet. Thus, they laid the foundations for acquiring new knowledge such as the applications of the Pythagorean Theorem and measurements through ratios in photos. Generally, the researchers believe, according to the findings of this research is that mobile learning can contribute to the improvement and the enhancement of realistic mathematics education for primary school students. The research can be continued in the future with more trails, a larger sample of children and more detailed records of the interactions during implementation. It could also be the basis for developing an online community of practice that will share math trails with mobile technologies.

References

1. Bouck, E.C., Flanagan, S., Miller, B., Bassette, L.: Technology in action. J. Spec. Educ. Technol. **27**(4), 47–57 (2012)
2. Cross, R.: Developing math trails. Math. Teach. **158**, 38–39 (1997)
3. Traxler, J.: Defining mobile learning. In: Isaías, P., Borg, C., Kommens, P., Bonanno, P. (eds.), Proceedings of the IADIS International Conference on Mobile Learning, Qwara, Malta, pp. 261–266 (2005)
4. Zhang, Y.: Design of mobile teaching and learning in higher education: introduction. Handbook of Mobile Teaching and Learning, pp. 3–10 (2015)
5. Traxler, J.: Current state of mobile learning. Mob. Learn Transform. Deliv. Educ. Train. **1**, 9–24 (2009)
6. Stevens, D., Kitchenham, A.: An analysis of mobile learning in education, business, and medicine. In: Kitchenham, A. (ed.) Models for Interdisciplinary Mobile Learning: Delivering Information to Students, pp. 1–25. IGI Global, Hersey (2011)
7. Martin, F., Ertzberger, J.: Here and now mobile learning: an experimental study on the use of mobile technology. Comput. Educ. **68**, 76–85 (2013)
8. Kukulska-Hulme, A., Traxler, J.: Mobile teaching and learning. In: Kukuluska-Hulme, A., Traxler, J., (eds.) Mobile Learning: A Handbook for Educators and Trainers, pp. 25–44 (2005)
9. Kraut, R.: UNESCO Policy Guidelines for Mobile Learning. UNESCO, France (2013)
10. Markouzis, D., Fessakis, G.: Interactive storytelling and mobile augmented reality applications for learning and entertainment—a rapid prototyping perspective. In: International Conference on Interactive Mobile Communication Technologies and Learning (IMCL) 2015, pp. 4–8 (2015)
11. Freudenthal, H.: Why to teach mathematics so as to be useful. Educ. Stud. Math. **1**(1), 3–8 (1968)
12. Van den Heuvel-Panhuizen, M., Drijvers, P.: Realistic mathematics education. In: Encyclopedia of Mathematics Education, pp. 521–525. Springer, Dordrecht (2014)

13. National Council of Teachers of Mathematics. In: Principles and Standards for School Mathematics, vol. 1 (2000)
14. Clements, D.H.: From exercises and tasks to problems and projects: unique contributions of computers to innovative mathematics education. J. Math. Behav. **19**(1), 9–47 (2000)
15. Koole, M.L.: A model for framing mobile learning. Mob. Learn. Transform. Deliv. Educ. Train. **1**(2), 25–47 (2009)
16. Soykan, E., Uzunboylu, H.: New trends on mobile learning area: the review of published articles on mobile learning in science direct database. World J. Educ. Technol. **7**(1), 31–41 (2015)
17. Daher, W.: Students' perceptions of learning mathematics with cellular phones and applets. Int. J. Emerg. Technol. Learn. **4**(1), 23–28 (2009)
18. Baya'a, N., Daher, W.: Students' perceptions of mathematics learning using mobile phones. In: Proceedings of the International Conference on Mobile and Computer Aided Learning, vol. 4, pp. 1–9 (2009)
19. Bray, A., Oldham, E., Tangney, B.: Technology-mediated realistic mathematics education and the bridge21 model: a teaching experiment. In: Proceedings of the Ninth Congress of the European Society for Research in Mathematics Education, Prague, pp. 2487–2493 (2015)
20. Cahyono, A.N., Ludwig, M.: MathCityMap: exploring mathematics around the city. Presented at the 13th International Congress on Mathematics Education (ICME-13), Hamburg (2016)
21. Zaranis, N.: Does the use of information and communication technology through the use of realistic mathematics education help kindergarten students to enhance their effectiveness in addition and subtraction? Presch. Prim. Educ. **5**(1), 46–62 (2017). https://doi.org/10.12681/ppej.9058
22. Zaranis, N., Baralis, G., Skordialos, E.: The use of ICT in teaching substraction to the first grade students. In: Proceedings of Fourteenth the IIER International Conference, Paris, France, pp. 99–104 (2015)
23. Widjaja, Y.B., Heck, A.: How a realistic mathematics education approach and microcomputer-based laboratory worked in lessons on graphing at an Indonesian Junior High School. J. Sci. Math. Educ. Southeast Asia **26**(2), 1–51 (2003)
24. Fessakis, G., Bekri, A.-F., Konstantopoulou, A.: Designing a mobile game for spatial and map abilities of kindergarten children. In: 10th European Conference on Games Based Learning, ECGBL 2016, Scotland, pp. 183–192 (2016)
25. Shoaf, M.M., Pollak, H., Schneider, J.: Math Trails. COMAP, Lexington (2004)
26. Richardson, K.M.: Designing math trails for the elementary school. Teach. Child. Math. **11** (1), 8–14 (2004)
27. Cobb, P., Confrey, J., diSessa, A., Lehrer, R., Schauble, L.: Design experiments in educational research. Educ. Res. **32**(1), 9–13 (2003)
28. Yin, R.: Case study Research. Design and Methods. Sage Publications, New Delhi (2014)

Teachers' Feedback on the Development of Virtual Speaking Buddy (VirSbud) Application

Radzuwan Ab Rashid[1](✉), Saiful Bahri Mohamed[2], Mohd Fazry A. Rahman[3], and Syadiah Nor Wan Shamsuddin[4]

[1] Faculty of Languages and Communication, Universiti Sultan Zainal Abidin, Kuala Nerus, Malaysia
radzuwanrashid@unisza.edu.my
[2] Faculty of Innovative Design and Technology, Universiti Sultan Zainal Abidin, Kuala Nerus, Terengganu, Malaysia
saifulbahri@unisza.edu.my
[3] Kolej Komuniti Shah Alam, Shah Alam, Malaysia
fazryrahman@kksa.edu.my
[4] Faculty of Informatics and Computing, Universiti Sultan Zainal Abidin, Kuala Nerus, Terengganu, Malaysia
syadiah@unisza.edu.my

Abstract. This study aims at identifying the ideal characteristics of a virtual speaking buddy (VirSbud), as perceived by the teachers, so that the current VirSbud application can be further developed to meet the users' expectation. Employing a qualitative approach, an in-depth group interviews was conducted with five English language teachers teaching in Standard Three (nine-year-old pupils). Prior to the interview, the teachers used the application as a teaching aid for eight weeks. The interview reveals the ideal characteristics of an application for virtual speaking buddy. The teachers expected a more sophisticated application which enables online participation, two different interfaces accessible by learners and teachers, looping in the prompts given to the learners, and the use of voice recognition which enables automatic response to the learners. The findings point towards the need of semi-formal learning, instead of formal and informal learning, in developing learners' speaking skills with the teachers being virtually present at a distance.

Keywords: English as a second language · Speaking skill
Virtual speaking buddy · Technology enhanced learning in Malaysia

1 Introduction

English is given a special emphasis in many countries across the globe as it is among the most widely spoken languages in the world. It is ranked third with 335,000,000 speakers [5]. In Malaysia, learners are required to learn English for 11 years before pursuing their tertiary education. It is compulsory to learn the language in the six years of primary education and the five years of secondary education [6]. Beginning from year 2017, it is made as a must pass subject for enrolment in the tertiary education [1].

M. E. Auer et al. (eds.), *Teaching and Learning in a Digital World*,
Advances in Intelligent Systems and Computing 715,
https://doi.org/10.1007/978-3-319-73210-7_40

Despite being exposed to English for 11 years in the school system, majority of the learners have poor language skills especially in terms of speaking [13]. It is common for teachers in the country to come across students who have sufficiently mastered the writing and reading skills but are unable to express themselves verbally. This is mainly because speaking skill receives little attention in the classroom as it is not tested in examination. Besides that, the teaching materials to develop the skill are scarce. Addressing this problem, we developed an interactive application named Virtual Speaking Buddy (VirSbud), which can be used as a collaborative source to develop learners' speaking skill. This paper discusses the ideal characteristics of a virtual speaking buddy, as perceived by the teachers, in the attempt to further develop the current VirSbud application to meet the users' expectation.

The research questions addressed in this study are: (1) What are the ideal characteristics of a virtual speaking buddy as perceived by the teachers? (2) How can the current VirSbud application be further developed so that its effectiveness is optimized? It is hypothesized that learners engagement with a virtual speaking buddy will develop their speaking skill as the buddy serves as an interactive partner for learners to practice the language.

2 Literature Review

2.1 A Sociocultural Perspective on Learning to Speak

Socio-cultural theory proposed Lev Semenovich Vygotsky (1896–1934) 'systematically synthesizes' the notions of culture, development, and learning [3]. The main tenet of this theory is that human action, on both the social and individual planes, is mediated by tools and signs [12]. As explained by [11] through the concept of semiotic mediation:

> language; various systems of counting; mnemonic techniques; algebraic symbol systems; works of art; writing; schemes, diagrams, maps and mechanical drawings; all sorts of conventional signs and so on' are all important in mediating social and individual functioning, and connecting the social and the individual (p. 137).

Vygotsky's semiotic mediation thus suggests that knowledge is developed through the use of socially-created 'psychological tools' [3]. Knowledge is not something that is directly internalized [7]. The notion of semiotic mediation contributes to our understanding that students' confidence to speak in English explored in this study is mediated by the discourse that the participants engage in. VirSbud provides a platform for the students to engage in the discourse by talking to their virtual speaking buddy. The engagement in the discourse will help the students to gradually develop their confidence level to orally use the language. Once they have greater control of their confidence, the students will no longer feel anxious to use the language in face-to-face setting.

2.2 Related Recent Studies

A search in Scopus database using the keyword "Speaking AND Technology" results in 1,143 documents from the last five years (2013–2017). The screening of the abstract of these documents reveals that only a handful of studies has explored the use of technology to enhance speaking skill. One of the most recent studies was conducted in the context of Turkey by [2]. The study involved two intact classes of Grade 3 Turkish learners of English and lasted for four months. The experimental group was engaged in recorded communicative exercises for their asynchronous speaking practice homework. The students were able to develop their speaking ability where they speak more confidently with minimal pauses and hesitations compared to the students in the control group. The study concludes that computer-mediated activities for homework speaking practice is particularly useful for students who come from non-English speaking family background.

Another recent study integrating technology for the development of speaking skill was conducted by [10]. They designed a virtual auditorium with audience who exhibit physical and vocal cues to be used as a training platform to reduce public speaking anxiety. The study recommends that this simulator is used as a tool for identifying cues to which speakers are more sensitive to.

[4] are developing an interesting innovation named SamiTalk. It is a sami-speaking robot which is linked to samiwikipedia. SamiTalk allows users to have spoken dialogues with a humanoid robot which speaks and recognizes North Sami. The robot has the ability to access information from the Sami Wikipedia, talk about topics requested by the users using the Wikipedia texts, and make smooth topic shifts to related topics using the Wikipedia links. The prototype has not yet been tested but this is one of the latest developments in the use of technology to enhance learners' speaking skill.

3 The Current Version of VirSbud Application

VirSbud is designed based on two interrelated concepts: Fun Learning and Autonomous Learning. Based on the fun learning concept, VirSbud is designed to be appealing to young learners by having an animation as its main character. The animation serves as the virtual speaking buddy to the learners. This speaking buddy invites the learners to speak to him by asking several questions to the learners. The questions are arranged according to three level of difficulties: beginner, intermediate and advanced. This application is made interactive by having the virtual speaking buddy asking follow up questions to the learners. The follow up questions are smartly designed so that they are responsive to the answers of the earlier questions given by the learners. This is done by designing the follow up questions as broad as possible but within the scope of the earlier questions.

In its current form, there are several ways that the teachers can integrate VirSbud application in the teaching process. Firstly, they can use the application as a teaching material in the classroom as the prompts in VirSbud are designed based on text book syllabus. Secondly, the learners can be assigned to use VirSbud as a homework as learners can easily access the application through their computer, which has been installed with the software, regardless of their geographical distribution. In addition, the learners can use the application as an informal learning tool without the teachers' presence. This is because the recorded version can be accessed by any 'more-knowledgeable others', such as parents, brothers, or friends who can give them feedback on their recorded conversation (see [8, 9] for previous research on VirSbud).

4 Methodology

This study employed a qualitative approach to gain in-depth feedback of the ideal characteristics of a virtual speaking buddy. Prior to the study, a VirSbud application was developed using Adobe Flash as the main software. The supporting software of this application includes (i) Adobe Photoshop and Illustrator to incorporate the objects and images in the application; and (ii) PHP and MySQL to record the input, such as the users' voice, profiles and marks. The application was introduced to five English language teachers teaching in Standard Three (nine-year-old pupils). The teachers came from three primary schools located in an urban area in the east coast of Malaysia. They were asked to use VirSbud application as a teaching aid for eight weeks. The teachers were purposively sampled as they were technology savvy. Towards the end of the week eight, the teachers were invited for a group interview to gauge their perceptions towards the application. During the interview, the teachers were asked about the weaknesses and strengths of the application and how it can be improved so that the application serves as an ideal virtual speaking buddy for the development of speaking skill.

5 Findings and Discussion

The interviews with the teachers reveal that they hope for a more sophisticated application to be used as a tool to improve speaking skill. This can be manifested in four themes: (i) Online database storing users' recorded speeches; (ii) Different interfaces accessible by learners and teachers; (iii) Looping in the prompts given to the learners; and (iv) The use of voice recognition technology. These themes are discussed in the paragraphs that follow.

The teachers expressed that any recorded speeches produced by their students should be accessible online so that they can evaluate the speeches at any time and any where. In the current VirSbud application, the students' recordings were installed in the computer/laptop through the application's database. Thereafter, the students need to show the laptop to the teachers. As revealed during the interview, this process slows down the teachers' feedback and the smoothness of the learning process in the sense that the students need to see the teachers face-to-face to get their speeches marked.

The two different interfaces for teachers and learners requested by the teachers are closely related to the first theme described above - the online participation. In teachers' words:

> You know it is good to have something like when the students log on, they see different things on the screen and when the instructor log on they will see different things. You know something like what we get when we use the Turnitin software. The instructor is more powerful than the students that they have the opportunity to control what the students can do or they can do something that the students cannot do. For example, when we log on to VirSbud, we should be presented with marking rubrics, and we can edit the topics that will appear on the students' screen.

The teachers' response suggests that they want an autonomy for them in designing the content of VirSbud application so that they can personalize the learning topics to cater for the need of their students. This is important as different students have different needs. This finding is useful for the software developer in the sense that it shows that it is not enough only to design the content of the software based on the standardized text-book topic. They should design the software in a way that affords the teachers to regulate the content.

Following the request for different interfaces, the teachers also requested for the looping in the prompts that appear on the students' screen. This is to enable learners to have a variety of experience every time they log on to VirSbud in the sense that they are exposed to different topics within the same theme randomly. The teachers perceived that the looping will make the students' learning more enjoyable as they cannot anticipate in advance what topics they will talk about with their animated speaking buddy.

Last but not least, the teachers requested for the integration of voice recognition technology. The teachers expressed that they hope for the animated speaking buddy to be able to respond to the students if they mispronounce any word. In the teachers' words:

> It is good if the speaking buddy can detect the words which are not pronounced clearly. You know something like *Siri* in the Iphone. And then the buddy responds to the students to suggest a better way of pronouncing the words.

6 Conclusion

The findings point towards the need of semi-formal learning, instead of formal and informal learning, in developing learners' speaking skills with the teachers being virtually present at a distance. The need for the semi-formal learning is reflected in the teachers' four requests discussed in the previous section. Instead of engaging students in informal learning on their own without any teachers' regulation, the teachers hope that they are able to monitor the students' learning. However, the teachers hope that this monitoring can be done virtually to enable more flexible learning experience, instead of them being physically present as in a formal class setting. In addition, the findings also suggest that there is a need for a more sophisticated application to facilitate the development of speaking skill. At present, there is no comprehensive application that manages to fulfil the ideal characteristics of a virtual speaking buddy portrayed by the teachers in this study.

Acknowledgment. This research has been funded by Universiti Sultan Zainal Abidin (UniSZA) under the DPU Research Grant UniSZA/2016/DPU/03.

References

1. Albury, N.J., Aye, K.K.: Malaysia's national language policy in international theoretical context. J. Nusant. Stud. **1**(1), 71–84 (2016)
2. Buckingham, L., Alpaslan, R.S.: Promoting speaking proficiency and willingness to communicate in Turkish young learners of English through asynchronous computer-mediated practice. System **65**, 25–37 (2017)

3. John-Steiner, V., Mahn, H.: Sociocultural approaches to learning and development: a Vygotskian framework. Educ. Psychol. **31**(3), 191–206 (1996)
4. Jokinen, K., Hiovain, K., Laxström, N., Rauhala, I., Wilcock, G.: Digisami and digital natives: interaction technology for the North Sami language. In: Lecture Notes in Electrical Engineering, vol. 999, pp. 3–19 (2017)
5. Lane, J.: The 10 Most Spoken Languages in the World (2017). https://www.babbel.com/en/magazine/the-10-most-spoken-languages-in-the-world
6. Rashid, R.A., Rahman, M.F.A., Rahman, S.B.A.: Teachers' engagement in social support process on a networking site. J. Nusant. Stud. **1**(1), 34–45 (2016)
7. Rashid, R.A.: Responding to nurturing global collaboration and networked learning in higher education. Res. Learn. Technol. **24**, 31485 (2016)
8. Rashid, R.A., Mohamed, S.B., Rahman, M.F.A, Wan Shamsuddin, S.N.: Developing speaking skills using virtual speaking buddy. Int. J. Emerg. Technol. Learn. **12**(5), 195–201 (2017)
9. Rashid, R.A., Mohamed, S.B., Rahman, M.F.A., Wan Shamsuddin, S.N.: VirSbud: Key characteristics, applications, and its future. Int. J. Interact. Mobile Technol. **11**(6), 158–163 (2017). SCOPUS
10. Söyler, E., Gunaratne, C., Akbaş, Mİ.: Towards a comprehensive simulator for public speaking anxiety treatment. Adv. Intell. Syst. Comput. **481**, 195–205 (2017)
11. Vygotsky, L.S.: The instrumental method in psychology. In: Wertsch, J. (ed.) The Concept of Activity in Soviet Psychology, pp. 134–143. Sharpe, New York (1981)
12. Wertsch, J.: Voices of the Mind: A Sociocultural Approach to Mediated Action. Havard University Press, Cambridge (1991)
13. Zaid, S.B., Zakaria, M.H., Rashid, R.A., Ismail, N.S.: An examination of negotiation process among ESL learners in higher institution. Int. J. Appl. Linguist. Engl. Lit. **5**(6), 228–234 (2016)

Mobile Learning Environments Application M-Tais Timor: A Study of East Timor

Hariyanto Santoso[1,2], Ofelia Cizela da Costa Tavares[1,2], and Suyoto[1(✉)]

[1] Universitas Atma Jaya Yogyakarta, Yogyakarta, Indonesia
haiyantosantoso89@gmail.com, ofelia_tavares@yahoo.com, suyoto@staff.uajy.ac.id

[2] Dili Institute of Technology, Dili, Timor-Leste

Abstract. With the passage of time that is modern, technology has been very advanced in various places so that humans are very easy to meet human needs quickly and efficiently. In this social network or often called an online store with all the needs of people ranging from small to very easy to get. The M-Tais Timor is an initiative to create a mobile application learning environment. In this mobile learning application is more helpful with the process of learning Tais Timor and Tais Timor's webbing process. So by looking at human needs that more and more, then the process of learning Tais Timor webbing is very helpful to be able to keep traditional art mats and Tais Timor can be processed into any outfit according to needs. The tutorial on Tais Timor fabric and the online system created in East Timor Tais.

Keywords: Impact of globalization · Cost affectivities · Real word experience

1 Introduction

In today's modern world smartphones and mobile devices are needed by the public, and with the widespread use smartphones against multimedia, among others Facebook, web and YouTube because is part of the integral internal [1]. In the learning process currently facing is the process of learning how to process or woven fabric Tais Timor especially for art lovers who want to know the process of woven Tais Timor. Tais itself has existed since antiquity before the Portuguese colonization in East Timor in the 14th century Tais already exists. The tais is a sacred traditional garb, which in the 14th century was used only for important people like Liurai (King) and used in traditional Timorese ceremonies. Tais Timor is basically a fabric that is processed by traditional looms and not machines, the tool is in abouth called as Soru or Songket. With the weaving process finally became a tais with various motifs that show the culture or identity of each region in East Timor. By seeing the Tais learning process indispensable for art and culture lovers, the mobile devices are now more resilient and portable, a lot of equipment to help people who experience everyday life, and with the advancement of mobile technology, mobile issues have been thought Widely researched in E-learning research. Many studies are considered important for. Development and learning [2]. In terms of in-flight learning, how to find or know the art of East Timorese, the mobile multimedia

M. E. Auer et al. (eds.), *Teaching and Learning in a Digital World*,
Advances in Intelligent Systems and Computing 715,
https://doi.org/10.1007/978-3-319-73210-7_41

application that will be used is to build a collaborative learning model for East Timorese society, which is revealed from positive learning. Results and can help who just access to the app [3]. In the development of this application using android Studio 2.3.2 and tools used in the development of Java application development kit (JDK). The database uses SQLite, SQLite is one of the most popular software, in combination SQL interface and very little memory usage at very fast speed.

In today's modern world smartphones and mobile devices are needed by the public, and with the widespread use smartphones against multimedia, among others Facebook, web and YouTube because is part of the integral internal.

2 Literature Review

In this writing, the writing will take several scientific papers, which check the same as the writing that will be made, including:

Namely discussing mobile technology applied in the decision-making process and an alternative device that is unsatisfactory so it cannot be fixed from time to time to be able to cope with dynamic decision situations, and this research explores the prototype that can combine a series of alternatives that can be juxtaposed Occurring during the decision-making process [4].

Which discusses the use of basic information about mobile applications thinking incoming applications is the most important in planning and design and large systems. Design on information about the rational need to be captured from the sharing of resources and context [5]. And hypermedia that facilitates the retrieval of various types of rational knowledge using multiple media, and system maintenance in changing the requirements of display displays, and meeting the requirements of multimedia information such as mobile online application is done for the Tais Timor learning process.

Which writes that collaboration interactions for video on mobile devices are based on multiple attitudes. It requires a pretty smart technique to drive, navigate and videos on mobile devices that are collaboratively interesting in the mobile environment, so a study proves that users do show that the flexibility and collaborative access of video exporting is increasing in public circles [6].

This research will have a concept to be able to understand the conceptual concept for mobile learning applications that offer support for mobile lifelong learning design [7]. In another study also said that mobile learning can be applied using lifelong learning, the proposed framework provides advanced engineering support for the success of future mobile lifelong learning system design [8].

This research will have an inner conceptual Mobile learning (m-learning) as a kind of learning model that anytime can be used mobile technology and internet. The deployment of mobile efficient and ideal so that mobile learning will be successful and the implementation is efficient, research discussing the theory-based research is needed to understand the basic motivation that leads the system academically to adopt the elements and learning mobile [9].

In the same way, this research will apply application learning to mobile learning learning process to mobile has been appealing that can generate many independent and android-based mobile applications, and the goal is to give personal service that helps other similar learning [10].

Said that the mobile gets the spotlight as an alternative to payments with payments in mobile applications, the research says that knowing the behavior and acceptance factors of mobile wallet technology users [11]. With the same research in this paper, it proposes the design and deployment of mobile applications with an interactive multimedia approach - called "m-leadership". This application is used for services not directly from the guidance and counseling that runs on the mobile device that is mobile phones, and uses methods to develop an approach that is with a combination of interactive multimedia and educational psychology [12].

In this paper will discuss the new development of m-psychology for the three types of applications used as guidance counseling in three applications used as direct guidance counseling that runs on mobile, and the method used in this application is a combination of alternative media and educational psychology [13, 14]. In the research results [15]. In this paper proposes the design and application of mobile leadership applications with an interactive multimedia approach. This application is used for service and development with interactive multimedia on learning about Tais,. in that research helps the community learning process about Tais Timor webbing.

Discusses how mobile applications learn something related to mobile learning system for learning process to what will be discussed, the goal is to build mobile outdoor learning activity by using the latest wireless technology, so that mobile learning Applications benefit from mobility, portability and individualization of mobile learning devices [16].

Along with the rapid development of mobile technology, then there is research discussed, that the need to learn mobile learning is increasingly important, and gaining important benefits in the expansion of new learning needs [17, 18].

3 Propose Method

A. System

This study has several concepts to understand the many learning processes that exist in the Timorese Society on the Tais Timor webbing process, therefore the existence of a system built with online learning concepts based on android, so as to develop the process of learning about Tais Timor webbing well [19], The use of computers for online computer-based systems that help the community of East Timor can install on hand phone respectively for free so that boost the learning knowledge about Tais both domestically and abroad. The lesson understands that people's understanding through the Android-based online multimedia application trains to know the Tais Timor weaving process from beginning to end, and the application is not only used for the people of East Timor and also as training for the easterners or for tourists who want to know traditional wear East Timor [20].

In Fig. 1 shows the process stages can be described in the application of Tais Timor mobile learning in the process of learning Tais Traditional East Timor, the purpose of this plot to make it easier to know the way of application made in helping the learning process of Tais Timor webbing. At this stage can be explained as below:

1. Users: users who will be logged in or who use the application must login at login first. Download the app and install on the smartphone and open and make sure online.
2. Login: when the User already has the application then the user must log in and register a new account, to be able to see the application process in the system. Register a personal account consisting of Full Name, email and Payment, after getting user account.
3. Process: The process will take the user to enter the registration after that user has operating rights for the mobile application. The learning process in this application is a stage of a good collaborative learning process that can use mobile anywhere with their respective internet packages.
4. Output: In this stage is the output stage which is also the stage that becomes the final stage of the mobile application, so that stage is the final result and the output makes a change in someone in knowing the way or the process of making traditional Tais, without to the production of the Tais.

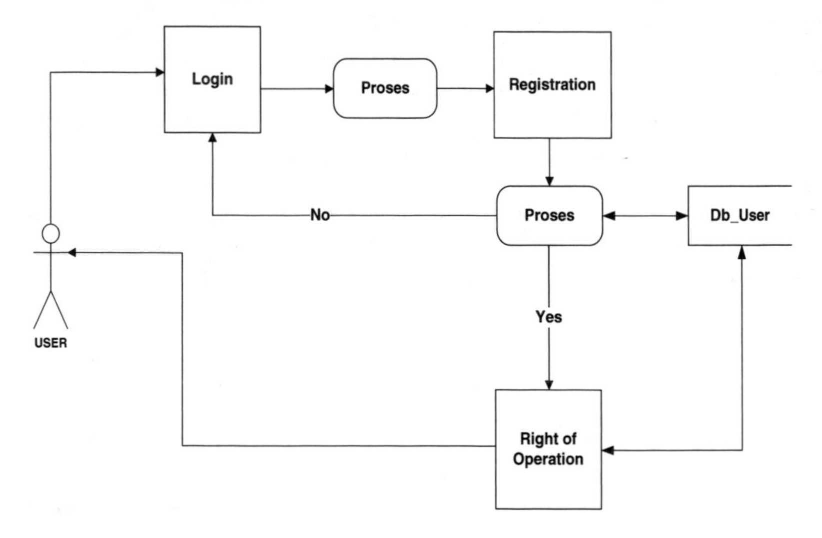

Fig. 1. The process of mobile learning application flow Tais Timor

Proses: Checking the user's original
Registration: Register to be a user consisting of:

- Full name
- Email
- Payment: Visa, ATM and others used as a transaction tool
- Password

Proses: If all the above provisions have been filled with true and clear then go

- Yes: Then the user registered tan user data stored in the database.

- No: Then the user must re-registration from the beginning.

Operation rights: Users are offered an existing offer in an existing tutorial application
Option 1: Users can learn the Tais-Timor learning process.
Option 2: Users can access bids on Tais-Timor training election.

B. Application Architecture

In the application architecture tells about the work process of the application there are some menus such as Home, Registration, about and Help as in the following picture:

In Fig. 2 above the home menu displays the Tais results display and order list display, which displays the learning offered. On the Registration menu displays the Login and Registration view and on the registration view there is full name, payment model and password to get user account. On the menu tells a short story about application development. The Help menu shows help about the application.

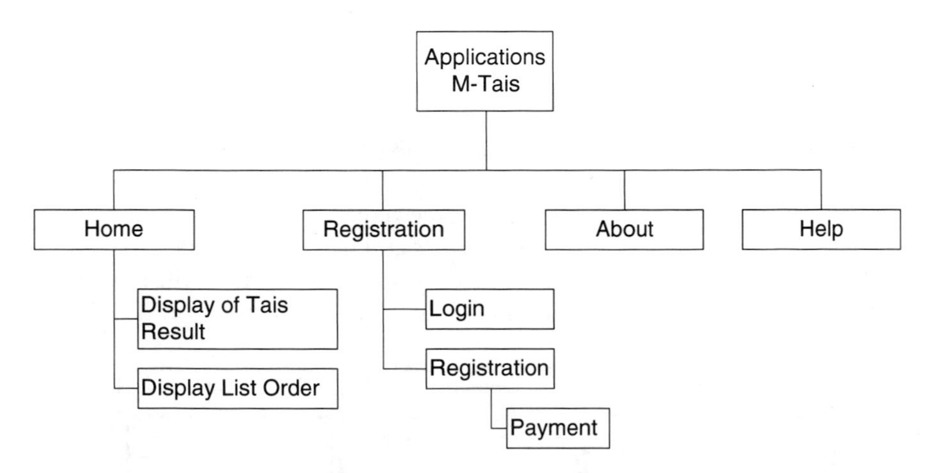

Fig. 2. Application architecture

4 Result and Discussion

As previously discussed, the process of this system can be used based on the results of other research, which is compatible with this method, and this application is derived from the research of traditional Tais making which is very closely related to multimedia approaches such as, sound, video and color. This online mobile app is built, with some interesting menus available that can attract users with color and sound, the application can be seen in the picture of the menu content.

In the Fig. 3(A) Users who will be logged in or who use the application must Register (B) or Login (C) first. When users have to sign in and register a new account, to be able to see the application process in the system. Register a personal account consisting of Full Name, payment and Email, after obtaining user account.

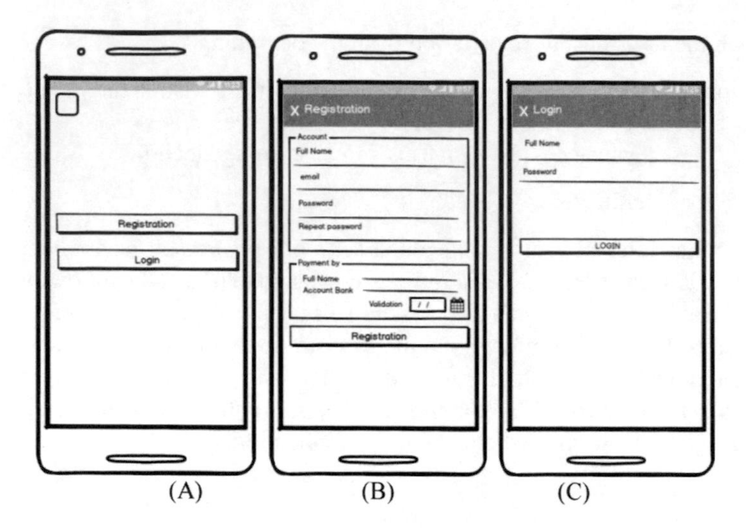

Fig. 3. The login and registration display

In the Fig. 4(A) shows the order of the finished Tais plaited products, from each region according to the model and characteristics of the 13 districts of East Timor. Figure 4(B) lists the order of Tais Motive from the culture of each district in East Timor and briefly tells the cultural history of each District. Figure 4(C) displays the selected motif and displays an explanation of the origin and history of the motif, if the manufacturing process button automatically clicks directly to the Payment screen.

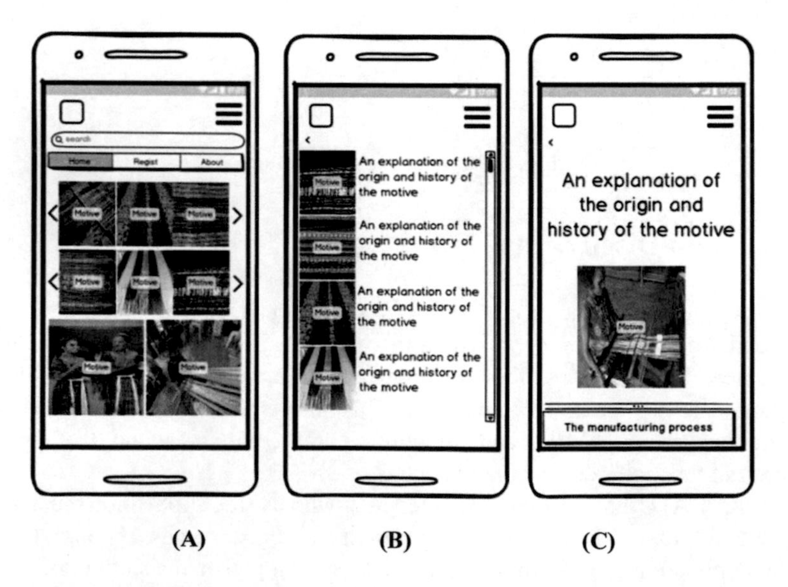

Fig. 4. Display of Tais result from each district

Figure 5(A) displays the payment model for each selected lesson, if the payment is successful then the process button is activated immediately and if clicked it automatically enters in the learning view. Figure 5(B) displays a learning process screen where each user can learn how to make woven Tais from the selection of yarn according to the color for each different motifs, and what tools are used in the manufacturing process to complete.

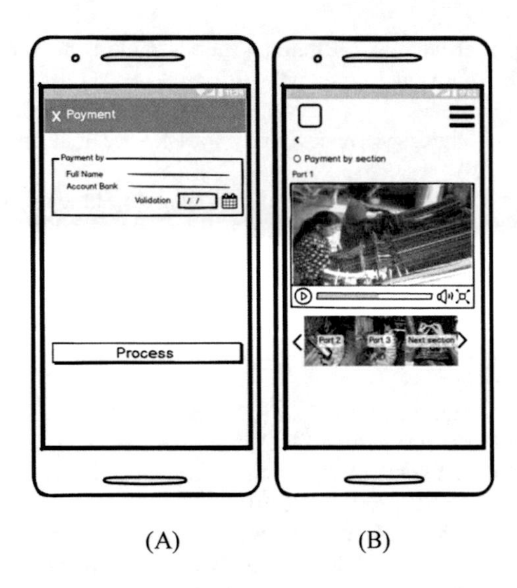

Fig. 5. Display payment model and process of making Tais

The superiority of this application can help and train for anyone who wants to learn to create Tais Timor, with simple steps and can be learned for all ages and does not need much time and cost by simply installing and doing the step-step registration to have their own account, The lack of this learning application is the lack of video quality and the size of the screen or the resolution is too small.

5 Conclusion and Recommendation

In this paper, the reviewers present in several studies on learning about traditional tais woven process with good cost affectivity, real word experience, to aid the learning process of East Timor Traditional Tais. Mobile Learning Tais East Timor is more focused on the Tais woven process in more detail from the results of the application for the users of this application, to better know the process of woven Tais East Timor. The results of the application have shown the image that produced the woven video process, and has performed an analysis that yields good results on the application, so that the users better understand the results of the Traditional Tais process well.

Acknowledgments. At the end of this writing the author would like to say many thanks to the University of Atma Jaya Yogyakarta, Indonesia which is very supportive of the guidance process in making this paper well, and can be presented with good results in the participants of the conference.

References

1. Kovachev, D., Cao, Y., Klamma, R.: Building mobile multimedia services: a hybrid cloud computing approach. Multimed. Tools Appl. **70**(2), 977–1005 (2014)
2. Jeng, Y.-L., Wu, T.-T., Huang, Y.-M., Tan, Q., Yang, S.J.H.: The add-on impact of mobile applications in learning strategies: a review study. Part A Spec. Issue Innov. Des. Mob. Learn. Appl. **13**(3), 3–11 (2010)
3. Huang, Y.M., Jeng, Y.L., Huang, T.C.: An educational mobile blogging system for supporting collaborative learning. Educ. Technol. Soc. **12**(2), 163–175 (2009)
4. Perez, I.J., Cabrerizo, F.J., Herrera-Viedma, E.: A mobile decision support system for dynamic group decision-making problems. IEEE Trans. Syst. Man Cybern. Part A Syst. Hum. **40**(6), 1244–1256 (2010)
5. Ramesh, B., Sengupta, K.: Multimedia in a design rationale support system. Decis. Support Syst. **15**, 181–196 (1995)
6. Zhang, J.K., Ma, C.X., Liu, Y.J., Fu, Q.F., Fu, X.L.: Collaborative interaction for videos on mobile devices based on sketch gestures. J. Comput. Sci. Technol. **28**(5), 810–817 (2013)
7. Nordin, N., Embi, M.A., Yunus, M.M.: Mobile learning framework for lifelong learning. Procedia Soc. Behav. Sci. **7**(C), 130–138 (2010)
8. Wang, Y.S., Wu, M.C., Wang, H.Y.: Investigating the determinants and age and gender differences in the acceptance of mobile learning. Br. J. Educ. Technol. **40**(1), 92–118 (2009)
9. Ozdamli, F., Cavus, N.: Basic elements and characteristics of mobile learning. Soc. Behav. Sci. **28**, 937–942 (2011)
10. Gavalas, D., Kenteris, M.: A web-based pervasive recommendation system for mobile tourist guides. Pers. Ubiquit. Comput. **15**(7), 759–770 (2011)
11. Megadewandanu, S., Suyoto, Pranowo: Exploring mobile wallet adoption in Indonesia using UTAUT2: an approach from consumer perspective. In: Proceedings of 2016 2nd International Conference on Science and Technology (ICST 2016), pp. 11–16 (2017)
12. Paper, C., et al.: Multimedia, Computer Graphics and Broadcasting, vol. 262 (2011)
13. Suyoto, Prasetyaningrum, T., Gregorius, R.M.: Design and implementation of mobile leadership with interactive multimedia approach. In: Kim, T., et al. (eds.) Multimedia, Computer Graphics and Broadcasting. Communications in Computer and Information Science, vol. 262. Springer, Heidelberg (2011). https://doi.org/10.1007/978-3-642-27204-2_27
14. Suyoto, Suselo, T., Dwiandiyanta, Y., Prasetyaningrum, T.: New development of m-psychology for junior high school with interactive multimedia approach. In: Kim, T., et al. (eds.) Multimedia, Computer Graphics and Broadcasting. Communications in Computer and Information Science, vol. 262. Springer, Heidelberg (2011). https://doi.org/10.1007/978-3-642-27204-2_28
15. Abas, Z.W., Lim, T., Kaur, H., Singh, D., Shyang, W.W., Lumpur, K.: Design and Implementation of Mobile Learning, vol. 20, no. 1, pp. 13–15 (2009)
16. Chen, Y.S., Kao, T.C., Sheu, J.P.: A mobile learning system for scaffolding bird watching learning, pp. 347–359 (2003)

17. International Educational and Technology Conference, Need For Mobile Learning: Technologies and Opportunities **103**, 685–694 (2013)
18. Wu, W.H., Jim Wu, Y.C., Chen, C.Y., Kao, H.Y., Lin, C.H., Huang, S.H.: Review of trends from mobile learning studies: a meta-analysis. Comput. Educ. **59**(2), 817–827 (2012)
19. Alqahtani, M., Mohammad, H.: Mobile applications' impact on student performance and satisfaction. Tojdel Online J. Distance Educ. e-Learn. **14**(4), 102–112 (2015)
20. Mlitwa, N.W.B., Wanyonyi, D.W.: Towards interactive open source based m-Learning solution: the m-Chisimba framework. J. Eng. Des. Technol. **13**(3), 463–485 (2015)

Mobile Learning Zoo: A Case Study of Indonesia

Jourgi Epardi[1], Oktavianus Teguh Prayitno[1,2], and Suyoto[1(✉)]

[1] Universitas Atma Jaya Yogyakarta, Yogyakarta, Indonesia
juliusjourgi@gmail.com, otegoohp@gmail.com,
suyoto@staff.uajy.ac.id

[2] Akademi Maritim Nusantara Cilacap, Cilacap, Indonesia

Abstract. Technology development for the purpose of learning virtually now is much needed. This research integrating knowledge virtual based on education and learning at the zoo especially in Indonesia based on mobile applications, in this case, will integrate application mobile use technology of QR Code and Image Recognition. Android system and web used for implementation in this research. Android runs a web application and to display the information in the form of text, sound, pictures, and video. QR Code technology used with capturing the barcode 2D (2 Dimensions). While image recognition used to catch photo the animals and do image processing, the results second this method of information for of text, sound, pictures, and video in an extract from a web. With this application, visitors can learn and also it can increase in learning about animals and the zoo in Indonesia through the application of mobile.

Keywords: *Mobile* learning zoo · Educational virtual environments
Mobile learning environments applications

1 Introduction

In most zoos in Indonesia in particular information about the animals in the zoo still very less and impress visitors only see the animals there without reading the board information about animals that exist in every animal cages. Technological progress, today also affect pattern the lives of the community that tends to want all the things in practical, for example in seeking for information about various for which has been available on the internet. The use of a mobile device for the media in informal learning has been fished attention of researchers in the context of technology in the field of education [1, 2]. With the development of communications technology and information in this case of the internet and technology mobile, Wu et al. did the research on learning based mobile get attention gradually by an increase of positive [3]. Therefore, the developments of application mobile now are very a lot of innovation and the varied. But, in the development of research on a mobile device who assisted with the process of learning informal, besides having to research and evaluate a system that is on a mobile device itself and do effectively for the use of in media learning.

A method of learning in the zoo constituting a topic of research is important in the development of a method of learning informally. Now, research on technology-assisted

M. E. Auer et al. (eds.), *Teaching and Learning in a Digital World*,
Advances in Intelligent Systems and Computing 715,
https://doi.org/10.1007/978-3-319-73210-7_42

in the zoo focus on the development of a method of learner-centered and apply this technology to help the visitor to conduct exploration and learn at the zoo. But, the zoo, especially in Indonesia now is still lacking in provides information complete and easily accessed by visitors the zoo and lack of reading interest visitors on the information board that is. Therefore, to research is made an application that helps visitors to obtain information about fauna that is seen as is in the zoo and assisting increase the public learning about animals.

This study attempts to develop a method based on the learning environment virtual, in other words, use integration between technologies with education. In a case, that we choose the integrated technology of QR Code and Image Recognition. The application of this technology uses a base Android and web. The application using platform Android to run it and use the web as a data base of text information, sound, pictures, and video. The final result on the application this is when visitors do scanning QR Code or scanning the animals, so application to show information taken from the web a picture, text, sound, and video.

2 Related Literature

Of development of information technologies which is very fast has been a supporter of main in obtaining information easy and fast. Technology other supporting technology called mobile, with a combination of most delivery of information can be undertaken faster. Application of technology based mobile has been done a number of studies capable of to develop the ability of leadership approach in multimedia [4] and development psychology mobile based [5]. Other research, discuss in the primacy of learning based mobile [6] and influence short message mobile to the process learning collaborative in South Korea [7]. In addition, there is research discuss the influence of two learning model, of using tablet PC and tabletop [8] and research on increased capacity asked by an approach based learn in mobile problem [9]. Researchers who choose a topic same has made several analysis and research. Previous research revealed that learning that applied to areas of tourism is needed because there are a lot of information that is, while information provided very limited.

The use of QR Code technology has been a need of now because by using QR Code all data will readily in governance and minimize error. The technology it has been used in the health sector [10], in the field of libraries to help the promotion of the book collection [11] and in the tourism museum [1, 2, 12, 13]. Integration between technology, education, and tourism, will certainly facilitate the visitors of tourist attractions to find out information about objects that exist, adding insights into your visitors and new learning methods. Development of a multiplatform mobile application offers a service to its visitors. By using the integrated technology of QR Code and web services, the mobile application to visitor's tourism place can provide complete information to the number. Then from it, a literature review that in proposing in this research to design applications that can help visitors in obtaining information, particularly visitors the zoo in Indonesia. This program is also innovation in the learning that combination with technology mobile, so information about those who are at the zoo capable of was obtained through easy and comfort.

3 Method and Design

The analysis is an important step before application made. Frequent application desirable made less well in the analysis. The analysis aims to make application established in accordance with expected. So, a mistake while implementation can be minimized by interviewing to stakeholders and field studies to seek information from people about application designed.

Usually, visitors, the zoo visit the zoo just to see a collection of an animal that there has been no want to know information about the animal. In this study, application learning in the zoo especially in Indonesia generally provides learning about a collection of fauna that is seen as for being at the zoo or can also provide information about the zoo in Indonesia. Visitors can obtain information on for only with through the telephone they through the application of M-Learning Zoo.

Methods used to collect data are interviewing and carry out literature. The interview how was conducted to obtain a collection of the zoo. In addition, we use also analysis of previous research, for example, journal, paper, and books.

In a case that we chose we apply 2 technologies that are integrated technology of QR Code and image recognition. The application of this technology uses a base Android to deploy and use the web services as a data base of text information, sound, pictures, and video. With the existence of mobile application is expected to attract visitors to go to the zoo. The application is also used to facilitate visitors aware of information about all animals that there is, as well as expected capable of resources to add the interest of all people learning and technology learning being fun.

In the implementation of the later on this program will support technology of QR Code and image recognition. Technology of QR Code is used with capturing the of barcode 2D that there had been in luminance the information board in every kennel and the result of information for of text, sound, pictures and video in extract from a web, while image recognition used to catch photo the animals and do image processing and the result of information for of text, pictures and video in extract from a web.

Application by means of technology barcode 2D recently is found growing very rapidly in conjunction with the camera telephone. Al-Khalifa (2008) use barcode 2D technology to develop marker products, in the health, press and a lot of other areas [14].

One sort of barcode 2D is QR Code. QR Code capable of encodes all kinds of data including symbol, multimedia binary and so on. The capacity of data maximum of the numeric, alphanumeric and binary is of 7089 characters, 4296 character and 2953 bytes for each data [15]. QR Code formed of two parts, namely the area and the pattern of encoding function that in it including tracer (finder), splitters (separators), a pattern of time, the alignment of the pattern, quiet zone, the format of information, version of information and data and an area of error. Four sides bounded by quiet zone. Figure 1 shows a pattern of QR Code. Function pattern would read all directions with detecting the position of QR Code and will fix distortion code [16].

Also need attention, similar to the use of QR Code that encodes apply a URL, is the stability of a URL itself, size content in web, not which limit user and something unique about this application [11].

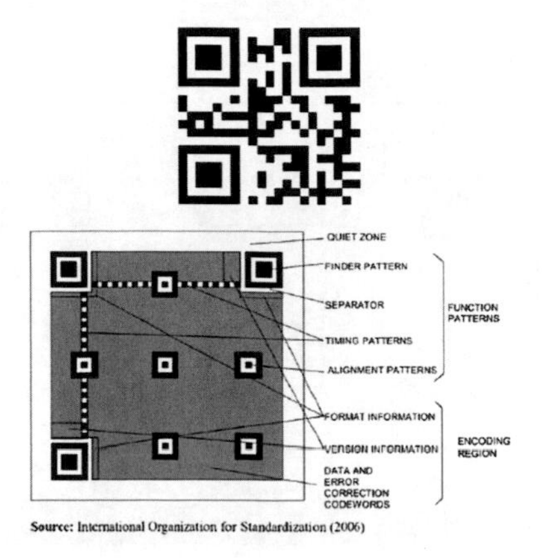

Fig. 1. Pattern of QR Codes

4 Result and Discussion

This application gives two services to get the information in the application and visitors would not have to do authentication to get into application M-Learning Zoo.

First services are to use the technology of QR Code to get information about animals. Users application use this technology simply by making a tap in QR Code available on the information board in every kennel or on brochures, QR Code on board information and brochure kept a URL or address the web, the web services here used for storing information on for of text, sound, pictures, and video. After a tap on QR Code, the application will display information from animals to find know the information.

For the second services applications, them-learning zoo provides technologies using image recognition. This technology in their (image recognition) implementation is nearly the same with the technology of QR Code. Users take the photographs of the animal's, from a photograph obtained the system in comparison to the data on a database uses the image processing, the photo has a common to the data on the base hence application to show the same information to what is on technology QR Code.

Figure 2 show Flow Chart of M-Learning Zoo. Showing use application process is flows m-learning zoo. Visitors the zoo can choose a way to an understanding of animals with three options, do scanning QR Code, scanning the face of live animals or read and see the provided the zoo. Technology and information used and displayed on the application M-Learning Zoo:

1. QR Code
   a. Animal information of text
   b. Animal information of picture
   c. Animal information of sound
   d. Animal information of video

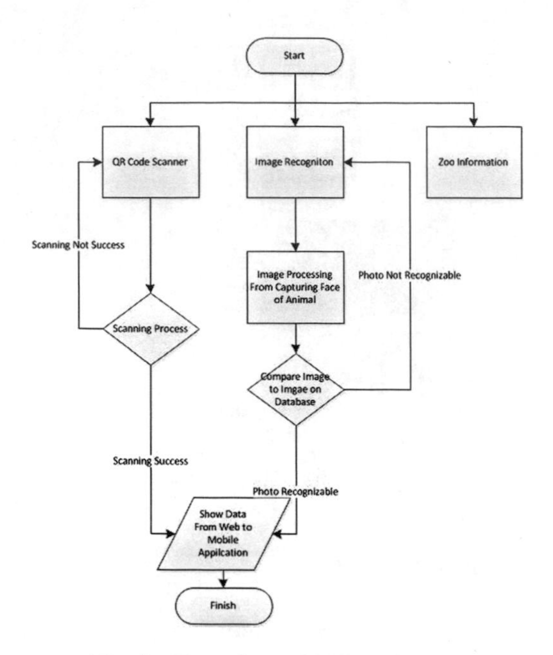

Fig. 2. Flow chart of M-Learning zoo

2. Image Processing
   a. Animal information of text
   b. Animal information of picture
   c. Animal information of sound
   d. Animal information of video
3. Information about the zoo in Indonesia

In this service application showing all information about the zoo in Indonesia, namely schedule, the ticket cost an event that is at the zoo and collection of animals.

In the development of the application there are three pieces of a menu that provided, among others, can be seen in Fig. 3 show a homepage of application. In this session, visitors live pressed the button "Mulai" if you are going to do the next process. Figure 4 show page main menu. In this session users application provided the three menus the menu scan QR Code, Scan Photo the animals and information about the zoo. On Fig. 5 shows page scan QR Code. This page appears when the user chooses buttons "Scan QR Code", this page contains a display QR Code scanner, if scan QR Code successful in this area and will continue to next session. Figure 6 show page scan photo the animals. On page emerged as the user choose buttons "Scan From a Photo", this page contains display the photos to the animals, after were the beast taken so will continue to next session. Figure 7 shows page animal information. On page is the result of the process of scan QR Code and the photo, the information displayed in the text, pictures, of sound and video. Figure 8 shows a page of information about the zoo in Indonesia. This session is showing all the information about the zoo.

Fig. 3. Main page

Fig. 4. Page menu

Fig. 5. Page scan QR Code

Fig. 6. Page scan from photo

Fig. 7. Page information about animal

Fig. 8. Page information about zoo

Based on screenshots application mobile above, this mobile application Facilitate visitors to know information about the fauna there was in the zoo with technology, and display information that interesting, so that visitors do not feel bored in visiting the zoo. This mobile application also gives information concerning the various fauna that is seen as is in the zoo. Information in a show on the application it will be text, a picture, of sound and video. In addition, this application can also be innovation in the education sector, where this application provides virtual learning about for the visitors wrapped in an application learning based mobile. So that this application can increase power public learning, especially study of animals.

5 Conclusion

In this study, we offer an application for learning virtually the zoo based application mobile. Visitors, the zoo will get information about fauna that is seen as is in the zoo with do scanning QR Code or by doing scanning photo the face an animal. In addition, an application is expected to increase the public learning and was one of innovation in the methods of learning.

References

1. Hou, A.H., et al.: A blended mobile learning environment for museum learning. Int. Forum Educ. Technol. Soc. **17**(2), 207–218 (2015)
2. Sung, Y.T., Hou, H.T., Liu, C.K., Chang, K.E.: Mobile guide system using problem-solving strategy for museum learning: a sequential learning behavioural pattern analysis. J. Comput. Assist. Learn. **26**(2), 106–115 (2010)
3. Wu, W.H., Jim Wu, Y.C., Chen, C.Y., Kao, H.Y., Lin, C.H., Huang, S.H.: Review of trends from mobile learning studies: a meta-analysis. Comput. Educ. **59**(2), 817–827 (2012)
4. Suyoto, Prasetyaningrum, T., Gregory, R.: Design and implementation of mobile leadership with interactive multimedia approach. In: Kim, T., et al. (eds.) Multimedia, Computer Graphics and Broadcasting. Communication in Computer and Information Science, vol. 262, pp. 217–226. Springer, Heidelberg (2011). https://doi.org/10.1007/978-3-642-27204-2_27
5. Suyoto, Suselo, T., Dwiandiyanta, Y., Prasetyaningrum, T.: New development of M-psychology for junior high school with interactive multimedia approach. In: Kim, T., et al. (eds.) Multimedia, Computer Graphics and Broadcasting. Communication in Computer and Information Science, vol. 262, pp. 227–236. Springer, Heidelberg (2011). https://doi.org/10.1007/978-3-642-27204-2_28
6. Flintoff, K.: Advancing mobile learning in formal and informal settings via mobile app technology: where to from here, and how? J. Educ. Technol. Soc. **19**(3), 16–26 (2016)
7. Kim, H., Lee, M., Kim, M.: Effects of mobile instant messaging on collaborative learning processes and outcomes: the case of South Korea. J. Educ. Technol. Soc. **17**(2), 31–42 (2014)
8. Hwang, W.Y., Shadiev, R., Tseng, C.W., Huang, Y.M.: Exploring effects of multi-touch tabletop on collaborative fraction learning and the relationship of learning behavior and interaction with learning achievement. Educ. Technol. Soc. **18**(4), 459–473 (2015)
9. Pi-Hsia, H., et al.: A problem-based ubiquitous learning approach to improving the questioning abilities of elementary school students. J. Educ. Technol. Soc. **17**(4), 316–334 (2014)
10. Cai, Y., Li, X., Wang, R., Yang, Q., Li, P., Hu, H.: Quality traceability system of traditional Chinese medicine based on two dimensional barcode using mobile intelligent technology. PLoS ONE **11**(10), 1–13 (2016)
11. Semenza, J.L., Gray, C.J., Fons, T., Penka, J., Wallis, R.: The Zombie Library: Books Reanimated via QR Codes OCLC's Linked Data Initiative: Using Schema. org to Make Library Data, vol. 32, pp. 32–35 (2013)
12. Haworth, A., Williams, P.: Using QR codes to aid accessibility in a museum. J. Assist. Technol. **6**(4), 285–291 (2012)
13. Chianese, A., Piccialli, F.: Designing a smart museum: when cultural heritage joins IoT. In: Proceedings of the 2014 8th International Conference on Next Generation Mobile Applications, Services and Technologies, NGMAST 2014, pp. 300–306 (2014)
14. Al-Khalifa, H.S.: Utilizing QR code and mobile phones for blinds and visually impaired people. Lect. Notes Comput. Sci. (including Subser. Lect. Notes Artif. Intell. Lect. Notes Bioinformatics), LNCS, vol. 5105, pp. 1065–1069 (2008)
15. Kato, H., Tan, K.T., Chai, D.: Novel colour selection scheme for 2D barcode. In: Proceedings of the 2009 International Symposium on Intelligent Signal Processing and Communication System, ISPACS 2009, pp. 529–532 (2009)
16. Xu, F.: QR codes and library bibliographic records. Vine **44**(3), 345–356 (2014)

Learning to Read and Identifies the Level of Hearing Disability Early Age Using a Mobile Learning Application

Paulus Haba Lena, Wayan Rupika Jimbara, and Suyoto(✉)

Universitas Atma Jaya Yogyakarta, Yogyakarta, Indonesia
paul.habalena@gmail.com, Jim.rupika@gmail.com,
suyoto@staff.uajy.ac.id

Abstract. In recent Decades, a lot of applications are designed to assist communities in carrying out their activities from, whether applications for education, services or games. Notice from the significant development of technology in this world, especially in the smart phone application. Some of Reviews These developments are also being felt by the disabilities, some applications for education are designed to provide education from an early age. The aim of this study is to assist or to Facilitate Reviews those who have problems with Reviews their hearing. There are many factors that can cause problems with Reviews their hearing. They need an education about the problems that occur at their hearings. With the help of computer, information about what happen in their hearing can be quickly intervening and Easily Obtained. Moreover, the presence of a diagnosis system at hearing loss, they will get certainty about what already happen at their hearings. In diagnosing of Hearing Loss, the writer uses Mobile application Learning. And in writing, the writer Also uses forward chaining method and certainty factor method. But there are Also people who have problems with Reviews their hearing from birth. People with disabilities on their hearing are named the deaf. Also this system contains of the education about sign language for the deaf and the mute. The writer multimedia designs by using a smartphone application so it will be easy to access anywhere. The designer of this system was equipped by visualization and audio so it will be more easy to be understood by the user.

1 Introduction

As the development of information technology today, many applications are created in order to facilitate the process of communication, one application in particular for persons with disabilities *(Disable)* [1], it is because of the difficulty of establishing communication with people with disabilities in society, so that the disability is less active in the social life, then the disabilities are difficult to follow the educational process as normal people [2, 3]. Therefore, of course, it would impede the movement for the disabled. In particular many who do not understand about people with disabilities? [4]. Here, the things that will be discussed are the lessons for the disability part of the Deaf – the mute (Cannot hear and cannot speak). This creation offers interactive multimedia applications

M. E. Auer et al. (eds.), *Teaching and Learning in a Digital World*,
Advances in Intelligent Systems and Computing 715,
https://doi.org/10.1007/978-3-319-73210-7_43

solutions for activities that allows people to learn interact with others using sign language and diagnosing hearing loss in early childhood [5–8].

The skill of Communication by using with sign language have not only addressed to the disability but it also can be learned by people who want to communicate with the disability. It would have a big impact on persons with disabilities they will feel very appreciated. It will show not only body movements hand gestures but also show facial expression [9]. As suggested by several studies, people who communicate through sign language has a lot of difficulties in access to e-learning platform for SL is the language of visual-spatial (gesture recognition and hand/arm movements are not sufficient for the correct translation [10, 11]. It required the interpretation of facial expression) there is the fact that it is not the language of "universal", but some are independent, with different symbols, grammar, and syntax [12, 3].

Currently, the problem of hearing loss is a hearing process that happens when the eardrum to vibrate causes sound waves enter the ear canal. And continuing into the middle ear. Then proceed to the cochlea, from "*kolkea*" sends a signal via the auditory nerve to the brain. Regarding the disruption of the hearing usually develop gradually. Not a lot of people who do not know the problems that occurred in the hearing. There are so many factors that cause interference with hearing bias. Then it will be discussed Diagnosing hearing loss computerized according to the degree of hearing loss experienced by patients. Diagnostic methods used is by *forward chaining and certainty factor (CF)* by looking at the existing symptoms will be issued degrees of hearing loss and the percentage of hearing loss are experienced.

2 Literature Review

Several studies using forward chaining method approach and certainty factor by reference solve a known problem of some common symptoms of the disease, Several studies using the approach forward chaining and certainty factor is in resolving a problem diagnosing young children, In these cases forward chaining perform the processing of collecting the symptoms - symptoms that occur resulting in diagnosis and methods of certainty factor (CF) in diagnosing the disease in infants, it will get the value of certainty in determining the contents of disease in infants. By using a reasoning system, shows the percentage of the truth of what has happened. System testing is done on the diagnosis based - symptoms - symptoms that are fed ate out the results of the symptoms - symptoms and directly defined on diseases that occur and ways of treatment [13]. The principle of the method certainty factor (CF) is by reducing the size of the hypothesis distrust childhood illnesses with symptoms - symptoms that cause disease in infants. So it is equal to the diagnosis conducted on the symptoms - symptoms that experienced by the hearing impaired and what degree of hearing loss experienced by sufferers of - the symptoms - symptoms. In addition, some references are used in learning for disabled people (the deaf) that the problem is also common to hearing problems do increase understanding of language acquisition and use of mobile applications [14, 15].

The method Forward Chaining is advanced tracking methods that match the fact that starting from the IF. In other hand, reasoning starts from the facts first to test the truth of the hypothesis resulting in a conclusion of the symptoms - symptoms of the problems occurred. And Certainty Factor Method is a method to look at certain factors to describe the level of confidence of experts on the problems being faced. After diagnosing the disease beyond will continue into the learning system, it is shown by learning multimedia audio and video [16–18]. The use of the app allows them to learn to interact between hearing impaired people with the language of gestures. Sign language skills between individuals to express themselves, to help cognitive skills and thought processes then make an android-based learning system to learn sign language [19]. The strength of this application is shown in two aspects: first normal person's ability to communicate with people - people in the target without having knowledge of sign language.

3 Research Methodology

3.1 Classifications of Hearing Loss

At normal human hearing levels are able to hear sounds between 10 to 25 dB. Hearing loss is classified according to a person's ear level or ability to hear sounds or sounds commonly referred to as a reduction in the ear's ability to hear which the drop is usually categorized by its level of mild, moderate hearing loss and severe hearing loss. Table 1 shows that each stage of a person's hearing rate is measured by the number of disables and the frequency of sound waves [22, 21].

Table 1. Classification of hearing loss [22]

Classification of hearing loss	ISO	ATA
Normal hearing	10–25 dB	10–15 dB
Mild hearing loss	26–40 dB	16–29 dB
Moderate hearing loss	41–55 dB	30–34 dB
Hearing loss a little heavy	56–70 dB	45–59 dB
Severe hearing loss	71–90 dB	60–79 dB
Deaf	>90 dB	>80 dB

3.2 Diagnosis of Hearing Loss

Physical examination: The doctor checks hearing loss to seek such hearing loss, disorders of earwax, infection or damage to the eardrum.

Tuning fork test: In addition to detecting hearing loss, a tuning fork test can also determine where the damaged inner ear.

Pure tone audiometry test: A machine to produce sounds with different volume and frequency will be heard by the patient through the mobile phone.

Table 2 shows degree of hearing loss experienced patients and traits – characteristics.

Table 2. Diagnosis classification of hearing loss

Classification of hearing loss	ISO	ATA
Normal hearing	10–25 dB	10–15 dB
Mild hearing loss	26–40 dB	16–29 dB
Moderate hearing loss	41–55 dB	30–34 dB
Hearing loss a little heavy	56–70 dB	45–59 dB
Severe hearing loss	71–90 dB	60–79 dB
Deaf	>90 dB	>80 dB

Rule base *(Forward Chaining)* and the percentage level of confidence *(certainty factor* method) a diagnosis is the percentage weight of each characteristic of the level of hearing loss. If the symptoms are all characteristic of the level of hearing loss the confidence level to a maximum percentage of 100%.

From the foregoing analysis of decision-making through a series of analysis steps degree of hearing loss, the analysis methods of tracking and tracing, can be obtained weighting rule base.

3.3 Flowchart System

Figure 1 shows the design of the process flow system that will be built.

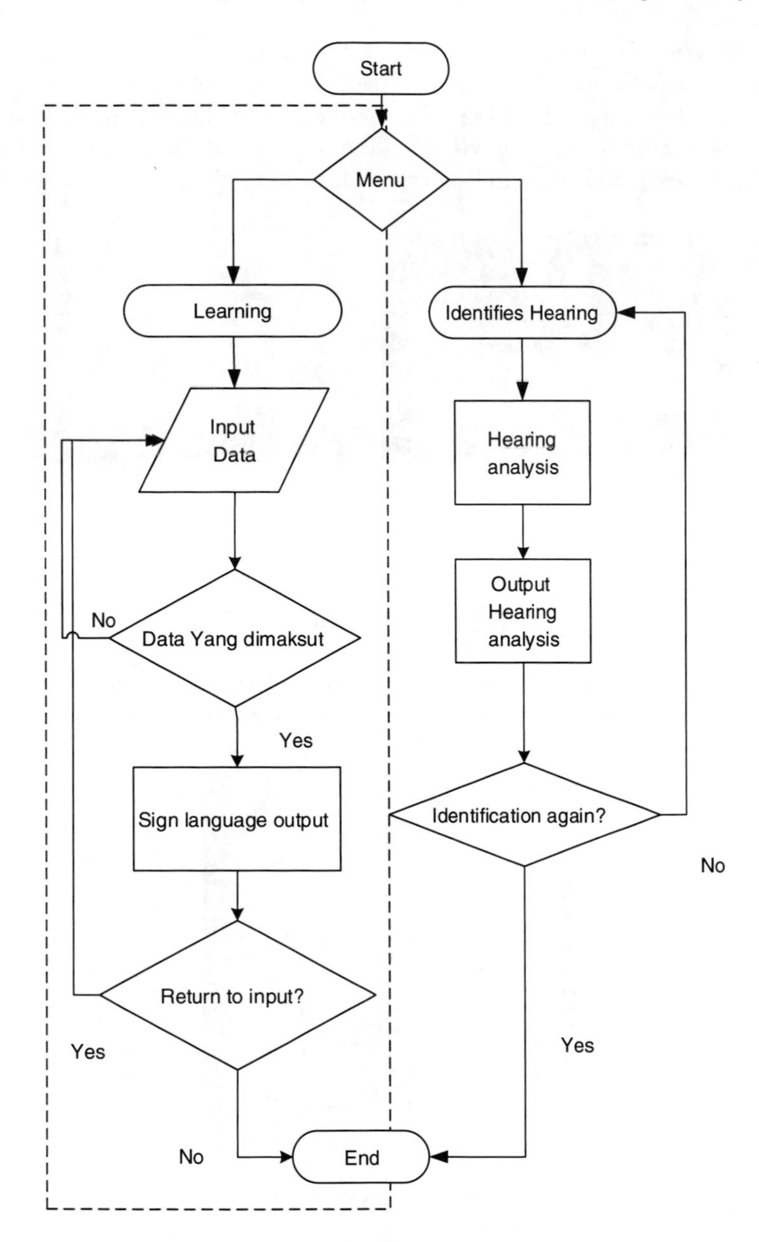

Fig. 1. Flowchat incorporation method.

4 Virtual Education for Early Childhood

In previous studies that have been conducted to conclude that the learning process easier on the thigh, I when the learning material is also implementing the use of images, audio and video [20–22].

Here are some features of the application are designed.

Figure 2 Shows the main menu page of the system, on the page there is the option to go to a page "Learning and hearing tests". On the page the user can make your choice learning such types of learning activities in the home, animal names, types of sports so much easier for users to learn sign language materials based on the classification.

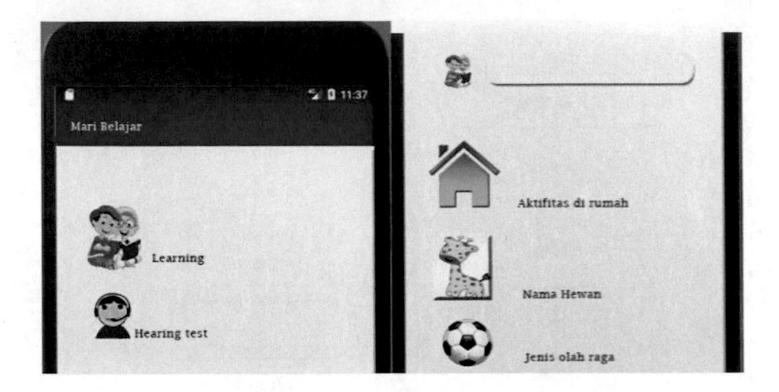

Fig. 2. The main menu

Figure 3 shows how to learn with the application. "Learning" menu aims to present audio and video of the users to make it easier for users to learn sign language.

Fig. 3. Learning

Figure 4 show "Hearing test" menu. The menu serves as a tool for users to train the level of hearing, in this application provided audio and video based on a predetermined frequency. In video that is provided is set based on the frequency in which the volume of the audio, it is designed to stimulate nerve make it easier to identify the level of hearing level of the hearing impaired. However, its implementation must use tools (headset).

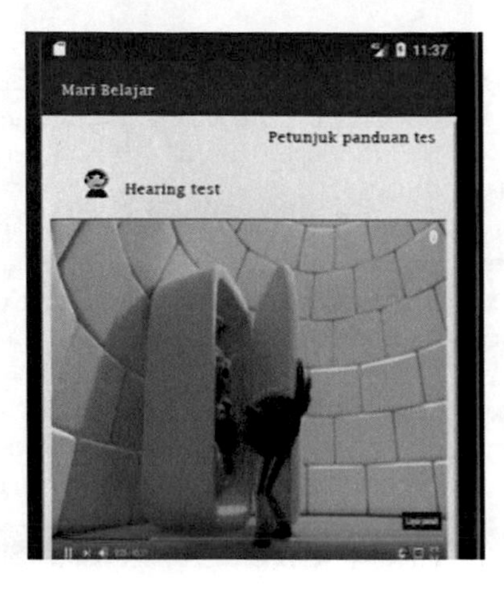

Fig. 4. *Hearing test*

The design of the system as can be seen from Fig. 4 above is a brief summary of the application is intended as a medical-based learning android. System is reserved for persons with disabilities in order to make learning sign language from an early age by using audio and video, it is in because the child more quickly capture or study a state of image, audio and video [21].

5 Conclusion

That mobile learning application gives some contributions for the disabilities especially for the deaf and the mute to learn how to communicate from the early age. This application is designed using audio and visual equipment and it will be cover with attractive appearance so it will be easy to be understood by the disabilities in early age. For the deaf, this application can help them to learn about sign language, besides that, the surplus that differentiate it from the other application, this application can classify the level of the problems that happen to the deaf, meanwhile, the mute can use this application to learn sign language from an early age.

References

1. Alghabban, W.G., Salama, R.M., Altalhi, A.H.: Mobile cloud computing: an effective multimodal interface tool for students with Dyslexia. Comput. Hum. Behav. **75**, 160–166 (2017)
2. Bjeki, D., Obradovi, S., Vu, M., Bojovi, M.: E-teacher in inclusive e-education for students with specific learning disabilities. Procedia Soc. Behav. Sci. **128**, 128–133 (2014)
3. Mary Drozd, R.N.: The experiences of orthopedic and trauma nurses. Int. J. Orthop. Trauma Nurs. **22**, 13–23 (2015)
4. Laabidi, M., Jemni, M., Ayed, L.J.B., Brahim, H.B., Jemaa, A.B.: Learning technologies for people with disabilities. J. King Saud Univ. Comput. Inf. Sci. **26**(1), 29–45 (2014)

5. Suyoto, Suselo, T., Dwiandiyanta, Y., Prasetyaningrum, T.: New development of M-Psychology for junior high school with interactive multimedia approach. In: Kim, T. et al. (eds) Multimedia, Computer Graphics and Broadcasting. Communications in Computer and Information Science, vol 262. Springer, Berlin, Heidelberg (2011). https://doi.org/10.1007/978-3-642-27204-2_28
6. Suyoto, Prasetyaningrum, T., Gregorius, R.M.: Design and implementation of mobile leadership with interactive multimedia approach. In: Kim, T., et al. (eds) Multimedia, Computer Graphics and Broadcasting. Communications in Computer and Information Science, vol 262. Springer, Berlin, Heidelberg (2011). https://doi.org/10.1007/978-3-642-27204-2_27
7. Mich, O., Pianta, E., Mana, N.: Interactive stories and exercises with dynamic feedback for improving reading comprehension skills in deaf children. Comput. Educ. **65**, 34–44 (2013)
8. Cano, M.D., Sanchez-Iborra, R.: On the use of a multimedia platform for music education with handicapped children: a case study. Comput. Educ. **87**, 254–276 (2015)
9. Amrutha, C.U., Davis, N., Samrutha, K.S., Shilpa, N.S., Chunkath, J.: Improving language acquisition in sensory deficit individuals with mobile application. Procedia Technol. **24**, 1068–1073 (2016)
10. Suyoto, Megadewandanu, S., Pranowo: Exploring the mobile wallet adoption in Indonesia using UTAUT2. In: 2016 2nd International Conference on Science and Technology-Computer, ICST, pp. 11–16 (2016)
11. Nugraha, N.B., Suyoto, Pranowo: Mobile application development for smart tourist guide. Adv. Sci. Lett. **23**: 2475–2477 (2017)
12. Li, J., Yin, B., Wang, L., Kong, D.: Chinese Sign language animation generation considering the context. Multimed. Tools Appl. **71**(2), 469–483 (2014)
13. Rasal, I.: Rule based expert system for diagnosing toddler disease using certainty factor and forward chaining, pp. 215–220 (2011)
14. Abdallah, E.E., Fayyoumi, E.: Assistive technology for deaf people based on the android platform. Procedia Comput. Sci. **94**(FNC), 295–301 (2016)
15. Sue, B.: Snapshots of interactive multimedia at work across the curriculum in deaf education: implications for public address training. J. Educ. Multimed. Hypermed. **15**(2), 159 (2006)
16. De Araújo, T.M.U., et al.: An approach to generate and embed video sign language tracks into multimedia contents. Inf. Sci. **281**, 762–780 (2014)
17. Munson, E.V., Pimentel, M.D.G.C.: Issues and contributions in interactive multimedia: photos, mobile multimedia, and interactive TV. Multimed. Tools Appl. **61**(3), 519–522 (2012)
18. Zydney, J.M., Warner, Z.: Computers & education mobile apps for science learning: a review of research. Comput. Educ. **94**, 1–17 (2016)
19. Malo, F.X.K., Joko, U.S., Pranowo: Mobile base least significant bit method for steganography. Adv. Sci. Lett. **23**: 2223–2227 (2017)
20. Atma, U., Yogyakarta, J.: Multimedia learning (2016). (in Bhs.)
21. Su, C., Cheng, C.: A mobile game insect-based learning system for improving the learning. Procedia Soc. Behav. Sci. **103**, 42–50 (2013)
22. Zirzow, N.K.: Signing avatars: using virtual reality to support students with hearing loss. Rural Spec. Educ. Q. **34**(3), 33–36 (2015)

Mobile Pocket KalDik: Dynamic and Interactive Academic Calendar

Amaya Andri Damaini[1], Ginanjar Setyo Nugroho[1,2], and Suyoto[1(✉)]

[1] Universitas Atma Jaya Yogyakarta, Yogyakarta 55281, Indonesia
amayaandridamaini@gmail.com, ginanjarsetyonugroho@gmail.com, suyoto@staff.uajy.ac.id

[2] Sang Timur Senior High School Yogyakarta, Yogyakarta 55161, Indonesia

Abstract. Academic calendar serves as a reference in planning and management of academic activities at a University. Many universities that include the academic calendar on web pages or in the form of a specific file format to be accessible. The trend of smartphone use is growing rapidly, making the academic calendar potentially developed into a mobile application that can help the management of academic activities become more effective and efficient. Pocket KalDik that incorporates the principles of the calendar and social media, allows the user to create learning schedule virtually and dynamically distribute the information of activities at the University in a real-time. This can indirectly help increase the productivity of education in the academic community. Pocket KalDik is a mobile application implemented with the Representational State Transfer (REST) architecture model. Pocket KalDik is expected to improve productivity and better time management for students and other academic community.

Keywords: Mobile application · REST
Mobile learning environments applications · Educational virtual environments

1 Introduction

Academic calendar is educational tools that provide information such as schedules in learning activities per school year. Educational institutions such as schools and universities implement learning activities based on the academic calendar. In learning activities, the academic calendar has a function to assist the effectiveness and efficiency of the teaching process, to harmonize the provisions on effective and holidays, the guideline in preparing the program of learning activities and the annual program, semester, and create the syllabus and teaching units of events. For people who are not directly involved in the learning activities, the academic calendar can be used as a reference to draw up a specific agenda. For example, parents of students who want to know the date of payment of tuition can find out in the academic calendar. Another example is student community that will hold a particular event on campus can use the academic calendar as a reference to decide the date of the event.

M. E. Auer et al. (eds.), *Teaching and Learning in a Digital World*,
Advances in Intelligent Systems and Computing 715,
https://doi.org/10.1007/978-3-319-73210-7_44

Today many universities that have worked on the academic calendar for easy access by providing a download link to the academic calendar file or display it directly on the University website. With the rapid development of mobile cloud applications nowadays, the common academic calendar has the potential to be developed into mobile applications and explored for its use.

In this paper, the authors proposed Pocket Kaldik to facilitate students and academic community to access and manage the agenda of academic activities anytime and anywhere. Thus, Pocket Kaldik can help distribute information about the schedule of the universities more effectively and efficiently. Another goal is to help organize personal schedules related to academic activities in a real-time.

2 Related Work

Trends in the use of smartphones today resulted in many developments from various fields towards mobile applications [1]. Many web-based systems such as news portals websites or e-commerce evolved into mobile applications to serve the needs of users in terms of ease of access [2]. Currently, the development of mobile technology has a large impact [3] and one of the important things in education because mobile technology can help improve access to educational resources to make it affordable for everyone [4–7]. Wannous et al. [8] utilizing cloud technology for e-learning system during fight situation in Syria to keep the learning activities going.

Many application systems and technology architectures have been developed for teaching and learning environment [9, 10]. One of them is mobile optimized application architecture using Ionic framework to provide a new tool for M-Learning application [11]. This architecture covers the problem of hardware accessibility in web applications and cross-platform running issues in native applications and combines the advantages of mobile applications and web applications. Several mobile application for learning and counseling guidance subject for Junior High school student have been developed by Suyoto et al. [12, 13]. One of which is an application called "m-NingBK: Social GC" [14] that developed for interactive social service counseling for Junior High school student.

3 Proposed System

Pocket KalDik is an Android based mobile application that integrates with web services [15, 16] to manage the resources. The web services implement architectural model Representational State Transfer (REST) [17]. Pocket KalDik application is divided into two sides, namely the client and server side. Application of the client side is an interface that relates directly to the user, while the web server provides the web service for storing and processing data. The client application is built using the Java programming language. REST architectural models, usually run over HTTP, involves a process of reading a specific web page that contains an XML file or JSON. In this case, the web server communicates with the client using JSON (JavaScript Object Notation) as a means of data exchange. Some of the systems developed using the REST architecture are online geo-information service [18] and Nevada Energy-Water-Environment Nexus Project [19].

REST web services are implemented with web standards (HTTP, XML, and URI) and REST principles that are addressability, uniformity, connectivity and stateless [20, 21]. The architectural design of the Pocket Kaldik Application can be seen in Fig. 1.

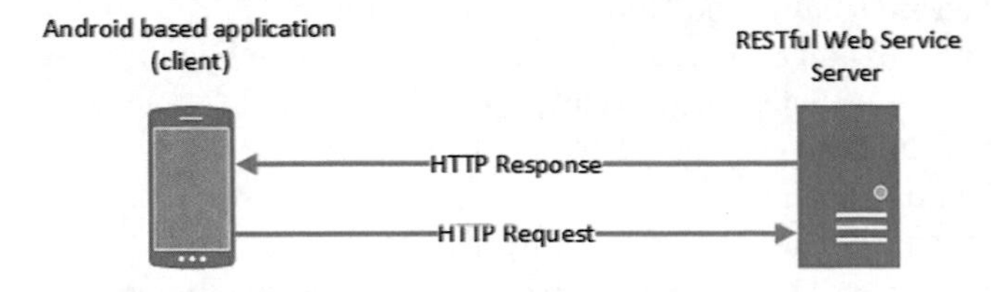

Fig. 1. Architecture design of Pocket KalDik system

Pocket KalDik built with the principle of combining the calendar and social networking [22]. By using the 'following' system among users, the users can find out the public schedule information that created by other users. For example, when a University administrator posts public events schedule, then the schedule will be automatically integrated with other users who follow the calendar of the University. Thus it can help students and lecturers to know that there are schedule changes in real-time. Figure 2 shows the general features of Pocket KalDik.

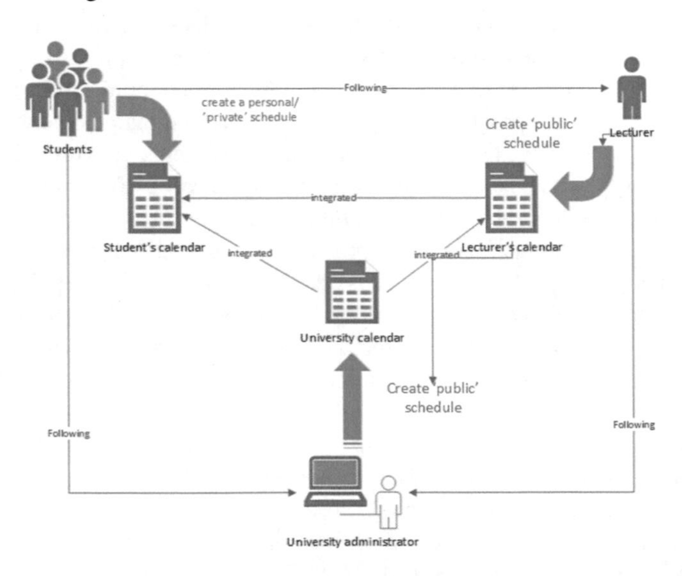

Fig. 2. General features of pocket KalDik

Schedule of events in Pocket KalDik consists of three types of schedules, namely "events", "personal schedule", and "course". If a lecturer posted a course schedule in the calendar, then the students get notifications question whether he follows the class. Students who join the class will measure the level of mastery of the course by giving a rate. If user gave "weak" rate, then the system will recommend private study schedule. Some other features found in Pocket KalDik are a reminder to schedule lectures and campus activities, and integration with the national calendar.

Features of Pocket KalDik are:

- Users can follow the schedule of the University and lecturers; there are three types of users: student, lecturer and University administrator.
- Schedule changes and events can be received in real-time by other users.
- Three types of schedules, the "event", "personal schedule", and "course"; for the "course" schedule can only be created by the teacher.
- Integrated with the national calendar.
- Integrated with the University's schedule and course schedule.
- Recommendation of personal study plan for students who did not master the course that they followed. It is useful for students who are often lazy to make a personal study plan, so can help them have a structured study schedule.

Through this features, Pocket KalDik is expected can help the management of academic activities for students, lecturers and other academic community within the University.

4 Result and Discussion

Currently, Pocket KalDik application has been successfully designed and implemented. Pocket Kaldik is one of the mobile learning environment applications. Implementation is done on the client and server side. On the client side, the application is developed using Android based Java programming language. Resource management is done by a web service that implements the REST architecture. Web services with REST architecture are known to process data transfer on mobile applications faster and lighter [23]. Users access the Pocket Kaldik app via mobile phone.

All academic community has mobile phones and most of their mobile phone using Android operating system. To be able to use the features of the Pocket Kaldik application, users must register using email first. When a user accesses Pocket Kaldik after being registered as a user, the main interface that appears is a calendar view. Pocket Kaldik integrates with national calendar automatically. Therefore, any national event or national holiday will appear on the Pocket KalDik calendar.

When the user touches a certain date, then at the bottom of the calendar displays the schedule that existed on that date. Users can set what schedule they want to display through the combo box on the top right. There is an option "All", "Event", "Course", and "Personal". For example, if the user only wants to display an event schedule only on the calendar, then the user can select "Event". When one user creates a new public schedule, then the schedule will automatically be integrated with the Pocket KalDik followers. Figure 3 shows Home interface of Pocket KalDik.

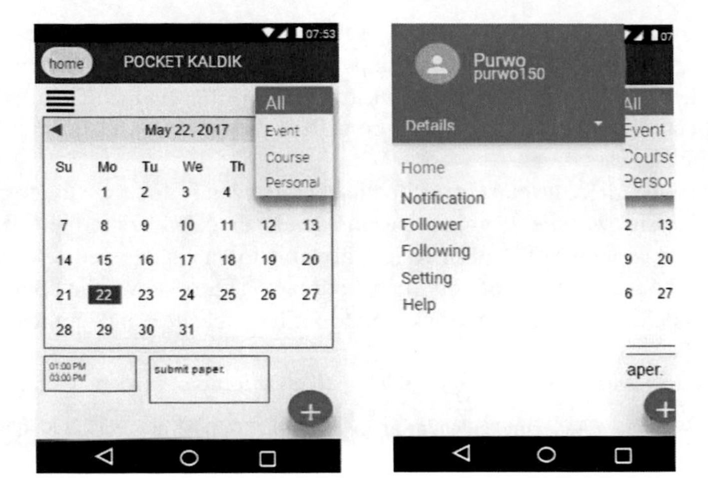

Fig. 3. Home interface of pocket KalDik

Pocket KalDik has six lists of menus on Sliding Menu, that is "Home", "Notification", "Following", "Follower", "Setting", and "Help".

- "Home": Directing users to the main form.
- "Notification": Display notifications; For example, there are new followers, there are new schedules that appear on a certain date, or there is a certain schedule changed.
- "Following": Displays a list of followers.
- "Follower": Displays the list of people they follow.
- "Setting": Feature to make settings in the application.
- "Help": Feature to provide assistance to users.

Generally, academic calendar is separate from lecture schedule, but Pocket Kaldik can be used as lecture schedule. When lecturers create a new course schedule, students will get notices (Fig. 4). If the student chooses to join the class, then the student is

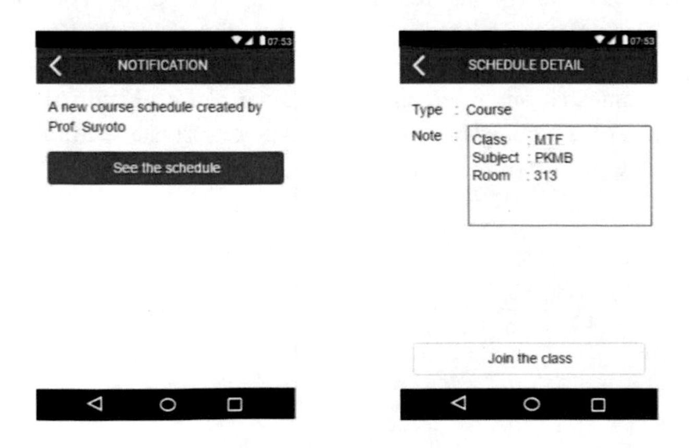

Fig. 4. Notification detail

required to give the rate of mastery of the subject. When a student chooses a "weak" rate then Pocket KalDik will recommend a personal study schedule on a date that does not have a schedule. With the Pocket KalDik application, users can schedule and manage educational activities virtually. This feature distinguishes Pocket Kaldik with regular calendar applications.

As explained earlier, the current academic calendar can be found in the form of files or on the university website. Compared to the academic calendars in the form of files and websites, the features of PocKet Kaldik are able to maximize the use of academic calendars for the management of activity schedules. Table 1 shows the comparison of academic calendars in the form of Pocket Kaldik files, websites, and applications.

Table 1. Comparison of Pocket KalDik with other forms of academic calendar

Form	Academic calendar in file form	Academic calendar in website form	Pocket KalDik
How to access	Open the file; must be downloaded first	Open the university website address	Open the application
Reminder	No reminder	No reminder	Has reminder alarm
Schedule update & information distribution	Schedule is not up to date, files must be edited and re-uploaded if there is a schedule change	Schedule is updated by the website administrator	Schedule is up to date; can be changed and distributed in a real-time

5 Conclusion

Pocket KalDik is an Android-based interactive app that uses RESTful Web Service to manage resources. RESTful Web Service in Pocket KalDik application manages data dynamically so it can provide information in real-time. The schedule created by the user is managed by the Pocket Kaldik system so it integrates on the user's phone right away. This causes information about academic activities to spread in real-time.

Through the Pocket KalDik features, it can help the management of academic activities for students, lecturers and other academic community within the University. Pocket Kaldik will be effective if students are active in making their personal study plans. Through the recommendation of personal study given Pocket Kaldik, it is expected to increase awareness of the importance of students to do private learning to master the subject that they are weak in it.

References

1. Onomza Victor, W., Abah, J., Sunday Adewale, O., et al.: Institute of advanced engineering and science a survey on mobile cloud computing with embedded security considerations. Int. J. Cloud Comput. Serv. Sci. **3**, 53–66 (2014)
2. Barroca Filho, I.M., Aquino Júnior, G.S.: Development of mobile applications from existing web-based enterprise systems. Int. J. Web. Inf. Syst. **11**, 162–182 (2015). https://doi.org/10.1108/IJWIS-11-2014-0041
3. Khaddage, F., Müller, W., Flintoff, K.: Advancing mobile learning in formal and informal settings via mobile app technology: where to from here, and how? Educ. Technol. Soc. **19**, 16–26 (2016)
4. Ally, M., Prieto-Blázquez, J.: What is the future of mobile learning in education? RUSC Univ. Knowl. Soc. J. **11**, 142–151 (2014). https://doi.org/10.7238/rusc.v11i1.2033
5. Ulfa, S.: Mobile technology integration into teaching and learning. IEESE Int. J. Sci. Technol. **2**, 1–7 (2013)
6. Fasae, J.K., Adegbilero-Iwari, I.: Mobile devices for academic practices by students of college of sciences in selected Nigerian private universities. Electron. Libr. **33**, 749–759 (2015). https://doi.org/10.1108/EL-03-2014-0045
7. Sung, Y.-T., Chang, K.-E., Liu, T.-C.: The effects of integrating mobile devices with teaching and learning on students' learning performance: a meta-analysis and research synthesis. Comput. Educ. **94**, 252–275 (2016). https://doi.org/10.1016/j.compedu.2015.11.008
8. Wannous, M., Amry, M.S., Nakano, H., Nagai, T.: Work-in-progress: utilization of cloud technologies in an E-learning system during campus-wide failure situation. In: 2014 International Conference on Interactive Collaborative Learning, pp. 13–16. IEEE (2014)
9. Chung, S.H., Khor, E.T.: Development of interactive mobile-learning application in distance education via learning objects approach. In: Studies in Computational Intelligence, pp. 373–380. Springer, Cham (2015)
10. Ramachandran, N., Sivaprakasam, P., Thangamani, G., Anand, G.: Selecting a suitable cloud computing technology deployment model for an academic institute: a case study. Campus Wide Inf. Syst. **31**, 319–345 (2014). https://doi.org/10.1108/CWIS-09-2014-0018
11. Wang, N., Chen, X., Song, G., et al.: Design of a new mobile-optimized remote laboratory application architecture for M-learning. IEEE Trans. Ind. Electron. **64**, 2382–2391 (2017). https://doi.org/10.1109/TIE.2016.2620102
12. Suyoto, Suselo, T., Dwiandiyanta, Y., Prasetyaningrum, T.: New development of M-psychology for junior high school with interactive multimedia approach. In: Communications in Computer and Information Science, pp. 227–236. Springer, Heidelberg (2011)
13. Suyoto, Prasetyaningrum, T., Gregorius, R.M.: Design and implementation of mobile leadership with interactive multimedia approach. In: Communications in Computer and Information Science, pp. 217–226. Springer, Heidelberg (2011)
14. Suyoto, Prasetyaningrum, T.: Development of mobile application social guidance and counseling for junior high school. Int. J. Comput. Electr. Autom. Control Inf. Eng. **7**, 1398–1404 (2013)
15. Tuli, A., Hasteer, N., Sharma, M., Bansal, A.: Exploring Challenges in Mobile Cloud Computing: An Overview (2013)
16. Thu, E.E., Aung, T.N.: Developing mobile application framework by using RESTFuL web service with JSON parser. Adv. Intell. Syst. Comput. **388**, 177–184 (2016). https://doi.org/10.1007/978-3-319-23207-2_18
17. Verborgh, R., van Hooland, S., Cope, A.S., et al.: The fallacy of the multi-API culture. J. Doc. **71**, 233–252 (2015). https://doi.org/10.1108/JD-07-2013-0098

18. Shi, S.: Design and development of an online geoinformation service delivery of geospatial models in the United Kingdom. Environ. Earth Sci. **74**, 7069–7080 (2015). https://doi.org/10.1007/s12665-015-4243-8
19. Lee, S., Jo, J., Kim, Y., Stephen, H.: A framework for environmental monitoring with arduino-based sensors using restful web service. In: 2014 IEEE International Conference on Service Computing, pp. 275–282. IEEE (2014)
20. Jiang, H., Yu, Q., Huang, K.: Design and implementation of an improved cloud storage system. In: 2016 12th International Conference on Natural Computation, Fuzzy Systems and Knowledge Discovery, pp. 1816–1823. IEEE (2016)
21. Sundvall, E., Nyström, M., Karlsson, D., et al.: Applying representational state transfer (REST) architecture to archetype-based electronic health record systems. BMC Med. Inform. Decis. Mak. **13**, 1 (2013). https://doi.org/10.1186/1472-6947-13-57
22. Al-Aufi, A., Fulton, C.: Impact of social networking tools on scholarly communication: a cross-institutional study. Electron. Libr. **33**, 224–241 (2015). https://doi.org/10.1108/EL-05-2013-0093
23. Arroqui, M., Mateos, C., Machado, C., Zunino, A.: RESTful web services improve the efficiency of data transfer of a whole-farm simulator accessed by Android smartphones. Comput. Electron. Agric. **87**, 14–18 (2012). https://doi.org/10.1016/j.compag.2012.05.016

Learning Style and Consciousness Factors in E-Learning System on Information Security

Kiyoshi Nagata[1(✉)], Yutaka Kigawa[2], and Tomoko Aoki[3]

[1] Faculty of Business Administration, Daito Bunka University, Tokyo, Japan
nagata@ic.daito.ac.jp
[2] Faculty of International Communication, Musashino Gakuin University, Saitama, Japan
yutaka.kigawa@u.musa.ac.jp
[3] Faculty of Law, Heisei International University, Saitama, Japan
aoki@hiu.ac.jp

Abstract. An E-learning system, which is considered as personalized, interactive, self-studying system, and free from time and place constraints, is an important factor for the flipped teaching. In Japanese higher educational institutions as in universities, many studies and practices on e-learning have been reported, but the state of trial and errors are still continuing. Thus, we propose the construction of an e-learning system incorporating not only with learner's consciousness but also with their learning style. The system is consistent of three individual parts and one integration phase. The first part is composed of tools, methods, and facilities used to provide proper e-learning environment. The second part is composed of theoretical or practical result on learning styles. The third part is composed of learners' consciousness factors on information security which we have studied by conducting questionnaire surveys to some Asian countries' university students. In the integration phase, the components of e-learning discussed in the first part are evaluated from learning style related psychological aspects. Anticipated result in the paper is that we propose a prototype which includes several types of content with proper links to information security factors and to learning styles. By clarifying the relationship among some constituent elements related to the e-learning which helps flipped teaching, we propose a total integrated system of e-learning.

1 Introduction

In the early 2000s, just after the e-Japan was proposed, research and argument on e-learning have been actively conducted in Japan. Learning systems using e-learning has been established to some extent in several areas such as education of employees in companies, video learning materials in language studies, so-called preparatory schools, etc. In higher educational institutions as in university's education, many studies and practices on e-learning have been reported, but the state of trial and error are still carried out. The difference between the situation

M. E. Auer et al. (eds.), *Teaching and Learning in a Digital World*,
Advances in Intelligent Systems and Computing 715,
https://doi.org/10.1007/978-3-319-73210-7_45

of companies, preparatory school and language teaching material learning and that in university is thought to be the learning motivation of learners. In Japan, universities or junior colleges students select classes from various motives, and not everybody is interested in the lecture contents nor enthusiastic to acquire knowledge, and it is difficult to deal with learners' motives and learning styles with conventional e-learning systems.

As one of methods of the active learning which have been actively advocated recently in Japan, flipped or inverted teaching is adopted and practiced in some educational institutions. E-learning systems, which are considered as personalized, interactive, self-studying system, and free from time and place constraints, are important factor for the flipped teaching. Learners can select subject or item to study at any time and from any place, and an important advantage of e-learning is that it can help to respond the diversity of each learner's motivation and his/her learning preference.

We have been investigated an e-learning system on information security which takes in account learners' consciousness and educational experience. In our previously proposed system, questionnaire on information security issues are initially posed, then calculate some fixed factors from which priority of items to study is determined. However, the subject and each item on information security issues are not so familiar to many students especially in social science studies, and it is not easy to cheer them up for self-learning. Thus, we try to propose constructing an e-learning system incorporating not only with leaner's consciousness but also with their learning style.

Our proposing e-learning system is consistent of three individual parts and one integration phase. The first part is composed of tools, methods, and facilities used to provide proper e-learning environment for learners. Here we discuss the managing cost, practical efficiency by reviewing several research papers or reports. The second part is composed of theoretical or practical result on learning styles. In this part, we discuss the learning style in psychological aspects, and focus especially on MBTI (Myers-Briggs Type Indicator), Kolb's four knowledge types, 4MAT. The third part is composed of learners' consciousness factors on information security which we have studied by conducting questionnaire survey to some Asian countries' university students during several years. In the integration phase, referring the table illustrating the relationship between components of inverted classroom and students' learning styles by Lage et al. [8], components of e-learning discussed in the first part are evaluated from learning style related psychological aspects. The consciousness factors are also linked to preferred learning styles. We consider the learning style preference can be combined with learning tools, methods, facilities, etc., and consciousness factors is mainly combined with content which learners should study. Thus, the system leads learner to his/her recommended contents to study in the learner's preferable situation suggested by learning style aspects. For this tasks, we need some functions describing the relationship among e-learning components (tools, methods, and facilities etc.) and learning style types and consciousness factors.

The ultimate goal of our research project is to develop the useful and effective total e-learning system helping learners to improve their knowledge and ethical sensitivity on information security related issues. At the time being, in order to implement our proposing system as a computer program, we need to prepare many types of educational content discussed in the first part of our paper on information security issues. Some text based contents with scores related to the information security consciousness factors are already made, and a small application program was completed. Anticipated result in the paper is that we create a prototype which includes several types of content with proper links to the information security factors and to the learning styles. At the same time, for the improvement of the system, data on media preference in relation to psychological aspects of learning style should be accumulated. We have some plans to ask for the test usage by university students in some Asian countries.

2 E-Learning Environment -Facilities, Tools, Methods-

In recent years, e-learning has enabled communication between teachers and learners, among learners, even among teachers through the advancement and spread of remarkable information technology. Here we list up various things related to e-learning, and classify them from several aspects such as the environment, tools, methods, and the other by taking into account of their merit and demerit from learners' point of view.

2.1 Merit and Demerit

When we consider from various point of view, there are many merits and demerits of e-learning. In this paper our aim is to propose an e-learning system accompanied with learner's learning style and his/her consciousness factors especially on information security items. Thus, we focus on the merit and demerit only in learner's point of view.

Major Merit and Demerit. E-learning has a merit and a demerit because it may cause a change in the relationship between a conventional teacher and a learner. When considering them on the learner's side, the main items are listed below.

- Merit for Learner
  - Free from time and place constraints
  - Learn according to individual's skill level and at one's own pace
  - Available to get uniformed or standardized lectures
- Demerit for Learner
  - Difficult to sustain learning motivation
  - Difficult to solve problems on the spot such as questions
  - Difficult to interact with teachers and other learners
  - Difficult to concentrating on thinking or learning

It is said that the "blending" education of face-to-face and e-learning should be performed in a way to compensate for demerits and take advantage of merits.

2.2 E-Learning Environment

Several learning environment for e-learning, such as Computer-Assisted or -Aided Instruction (CAI), Computer-Based Training (CBT) and Web-Based Training (WBT), have been delivered. Learners can access them not only to get learning materials but also to see their performance level or what they should study. They also provide many systems or functions such as the attendance management, the performance evaluation and grading support, report distribution and collection function, distribution function of various information on the lecture. Some systems also provide e-portfolio or Rubric for evaluating learner's performance.

These days, SNS (Social Networking Service) system including simple blogs becomes very popular especially among younger people, and real time remote communication technologies such as Skype also becomes available with lower cost. These systems release learners from time or place constrains.

As hardware tools for e-learning, PC (on-line or off-line), tablet, and mobile device, equipped with web-cam, microphone, or speaker will be considered. Although an e-learning system accessible from tablet or mobile devices provide the freedom from time and place constrains, it might have disadvantage in view of concentrating on learning. As software tools or methods for distributing learning contents, there are many types of application related files such as audio, video, text, web page, picture, animation, presentation (with or without audio, video, etc.), work sheets. Quizzes, reports, collaborative learning, on-line office hours, real or non-real time question receiving and responding system are also considered as methods for educations.

3 Learning Styles -Theoretical or Practical Works-

By using technologies of e-learning, it is becoming possible to provide learners with individualized environments. In this section, we try to look at individual differences in learning styles aiming to provide students with optimum learning environments or to have learners aware of their own learning styles so that they can choose their own optimum learning environment. There are many ways in classification of learning styles based on different theories such as theories of intelligence, experiential theories by Kolb [7], sensory modalities (the VARK model [4]), cognitive styles, or psychological types by Briggs Myers, see [2]. Here we refer Kolb, 4MAT in the middle layer, "Information Processing Style", of the Lynn Curry's onion model of learning style [3], and MBTI (Myers-Briggs Type Indicator) in the inner layer, "Cognitive Personality Style".

3.1 Kolb's Learning Style

Kolb's learning style (1984) is one of the best-known and widely used learning style theories. He believed that our individual learning styles emerge due to our genetics, life experiences, and the demands of our current environment. In addition to describing four different learning styles, Kolb also developed a theory of experiential learning and a learning style inventory. Thus, for example, a person with a dominant learning style of "doing" rather than "watching" the task, and "feeling" rather than "thinking" about the experience, will have a learning style which combines and represents those processes, namely an "Accommodating" learning style, in Kolb's terminology.

Table 1. Kolb's Learning Styles

	Doing (AE) Active Experimentation	Watching (RO) Reflective Observation
Feeling (CE) Concrete Experience	Accommodation	Diverging
Thinking (AC) Abstract Conceptualization	Converging	Assimilating

3.2 4MAT

Based on Kolb's learning style theory, Bernice McCarthy developed the framework for designing learning activities named 4MAT System [9]. There are four quadrant clock-wisely turning around in the order of "Diverging", "Assimilating", "Converging", and "Accommodating" in the Table 1, and each quadrant has right- and left- brain modes of processing techniques. These quadrants also correspond to learners stages as follows.

- Imaginative Learner:
  Demand to know 'why'. Likes to listen, speak, interact, and brainstorm.
- Analytic Learner:
  Want to know 'what' to learn. Likes to observing, analyzing, classifying, and theorizing.
- Common-sense Learner:
  Want to know 'how' to apply the new learning. Likes experimentation, manipulation, improvement, and tinkering.
- Dynamic Learner:
  Want to ask 'what if'. Enjoy modification, adapting, taking risks, and creating.

Huitt describes the 4MAT system for a Web-based Instruction in [5] like as the Table 2. Where the learning process proceeds from 1 to 8. The first type of learner, the imaginative of Concrete-Random learner, demands to know "Why" he/she should be involved in the activity. The second type of learner, the analytic or Abstract Sequential learner, wants to know "What" to learn. The third type of learner, the Common-sense or Concrete Sequential learner, wants to know "How" to apply the learning. The last type of learner, the Dynamic or Abstract-Random learner asks "If" this is correct how can one actively modify it to make it work for one-self. He also indicates each type of learner to the MBTI, and we describing them in the last part of each low on the first column of the Table 2.

Table 2. Instructional Events in the 4MAT System (by William G. Huitt)

	Left-Brain Activity	Right-Brain Activity
WHY? (Motivate and Develop Meaning) (Sensing/Perceiving)	2. Analyze/reflect about the experience (EXAMINE)	1. Create an experience (CONNECT)
WHAT? (Reflection and Concept Development) (iNtuitive/Thinking)	4. Develop concepts/skills (DEFINE)	3. Integrate reflective analysis into concepts(IMAGE)
HOW? (Usefulness and Skill Development (Sensing/Judging)	5. Practice defined "givens" (BY)	6. Practice and add something of oneself (EXTEND)
WHAT IF? (Adaptations) (iNtuitive/Feeling)	7. Analysis application for relevance (REFINE)	8. Do it and apply to more complex experience (INTEGRATE)

Source: http://www.edpsycinteractive.org/papers/4matonweb.html

3.3 MBTI

As a prominent psychological type theory based on Jung's psychological types, we focus on the MBTI [10] developed by Katherine Cook Briggs and Isabel Briggs Myers. Here, let's take a look how a person's preferences in terms of Jung's and Briggs Myers' approach to personality type may influence learning styles. MBTI is the preference of general attitude with four dimensions such as Extroverted (E) vs. Introverted (I), Sensing (S) vs. iNtuition (N), Thinking (T) vs. Feeling (F), and Judging (J) vs. Perceiving (P).

The E/I dimension which is very similar to Kolb's AE/RO dimension reflects the direction of an individual's general interest and, indicates where one's interests and motivation lie. Person high on extroversion tends to be doer, while one high on introversion tends to be watchers. An introvert's motivation and interests primarily stem from and are driven by their inner world, whereas an extrovert is primarily motivated by the world outside of oneself, and most of interests are externally focused. This pair of styles is concerned with the direction of one's energy. If one prefers to direct the energy to deal with people, things, situations, or the outer world, then his/her preference is for extroversion.

The N/S dimension reflects the preference perceive for the world and thinking. Person high on the intuition has the preference perceive for the world and thinking in broader categories, whereas one with sensing have that in a more concrete and direct way. This pair concerns the type of information or things that one process. If one prefers to deal with facts, what you know, to have clarity, or to describe what you see, then his/her preference is for sensing. If one prefers to deal with ideas, look into the unknown, to generate new possibilities or to anticipate what isn't obvious, then the preference is for intuition.

The F/T dimension is very similar to Kolb's CE/AC dimension. Person high in the feeling and concrete experience areas tends to be more focused on the here-and-now, while one high in the areas of thinking and abstract conceptualization prefers to focus on theoretical concepts. Person with high on feeling preference tends to judge and respond to events based on his/her feeling, whereas one with high thinking preference tends to do it based on reason and logic. This pair reflects one's style of decision-making. If one prefers to decide on the basis of objective logic, using an analytic and detached approach, then his/her preference is for thinking. If one prefers to decide using values - i.e. on the basis of what or who one believes important - then the preference is for feeling.

The J/P dimension reflects the preference for learning style. Person high on the judging preference comprehends information in a more structured way and is likely to prefer a more systematic and structured learning process, whereas one high on the perceiving preference might favor a less rigid, more heuristic approach to learning and might prefer a trial and error method of comprehending information. This final pair describes the type of lifestyle one adopts. If one prefers his/her life to be planned, stable and organized then the preference is for judging (not to be confused with "Judgmental", which is quite different). If one prefers to go with the flow, to maintain flexibility and respond to things as they arise, then his/her preference is for perceiving.

The MBTI was specifically designed as a tool to categorizes individual's personality type in general, and their approaches to relationships with others. For this reason, the MBTI differs in tone from other influential personality trait theories, by being more positive or neutral in its descriptors. This aspect may account for its influence in the learning styles' field, where theorists who have drawn upon it have tended to emphasize descriptors of normal behavior and actions, rather than the identification of pathological traits or tendencies.

Combining the letters associated with each preference, we get the Myers Briggs personality type. For example, having preferences for E, S, T and J gives a personality type of ESTJ. Thorne and Gough [11] picked out 10 most common types by mentioning with some positive and negative traits as follows.

INFP Artistic, reflective, sensitive ↔ Careless, lazy
INFJ Sincere, sympathetic, unassuming ↔ Submissive, weak
INTP Candid, ingenious, shrewd ↔ Complicated, rebellious
INTJ Discrete, industrious, logical ↔ Deliberate, methodical
ISTJ Calm, stable, steady ↔ Cautious, conventional
ENFP Enthusiastic, outgoing, spontaneous ↔ Changeable, impulsive
ENFJ Active, pleasant, sociable ↔ Demanding, impatient
ENTP Enterprising, friendly, resourceful ↔ Headstrong, self-centered
ENTJ Ambitious, forceful, optimistic ↔ Aggressive, egotistical
ESTJ Contented, energetic, practical ↔ Prejudiced, self-satisfied

4 Learners' Consciousness Factors on Information Security

We have proposed an e-learning system based on research on information ethics of foreign students since 2003. In many Japanese universities and colleges, there are lectures on information ethics, and many of them are carried out on the premise of students who grew up under Japanese culture, education and social environment. However there is no information ethics education in consideration of the social environment and the current situation of education in the countries where international students are from. Based on this idea, we thought that the information security education for each international student with different cultures and social background should be conducted.

Therefore, in 2008, we started "Developing Information Ethics Education Teaching Materials Considering Education and Legal System in the Countries of International Students". By conducting a questionnaire survey on information ethics for students from China, Taiwan, South Korea, Philippines and Singapore from 2008 to 2010 in order to know the characteristics of international students [6].

During our research activities, we could notice that we should take the personal characteristics and the educational experiences into account for the education of the contents of a certain field such as information ethic and security. Since the human resource for education are not so rich to correspond to each student, an e-learning system with the potential for personalization is needed as a complementary educational tools which might cope with even one's learning style [1].

4.1 User Consciousness Oriented E-Learning System (Prototype Version)

Our having proposed e-learning system in [6] consists of two major parts, one is the data part such as database of stories categorized according to information security related issues, the other is user's interface representing stories, explanations, and so on. This system presenting some stories on each topic such as copy right, personal information etc. according to the priority values calculated from learners consciousness factor scores.

The flow of the system is described as follows,

1. Login the system using the learner's ID and password.
   (a) If it's first time to login, then ID is issued and a password is set
   (b) Choose the country or language from the selection box.
2. Show the initial page of the e-learning system with selected language.
   (a) Three buttons are disposed in the "Menu Area" on the left
   (b) The white box on the right, called "Presentation Area", is used for presenting user's characteristics and learning materials.

3. Choose one of three buttons.
   (a) If the "Attribute Q" button is selected, the attribute questionnaire sheet is displayed as an individual window, and the attribute file of the user is created or recreated
   (b) If the "Consciousness Q" button is selected, the consciousness questionnaire sheet is displayed as an individual window, and the consciousness file of the learner, in which the learner's individual scores for each of three consciousness factors are recorded is created or recreated
   (c) If the "Recommended Study Stories" button is selected, learner's consciousness scores along with rating comments by comparing each score value to the average value of students in the same country, and some buttons for recommended stories appear in the "Presentation Area".
4. Learning is repeated by showing each story and its explanation.
   (a) When learner first selects the "Best Story" button, the first story that seems to be the best for the current user appears in the box just below the button, and the "Best Story" button disappears
   (b) If learner selects the "Next Story" button, the next best story will be displayed in that area
   (c) If learner selects the "Show description" button, a description about the story currently displayed in the story area will be displayed in the white box.

One of unique points of the system is reflecting learner's individual and educationally experienced characters when choosing learning subjects. In this system, the fuzzy out-ranking method was applied to put each story in order. It can happen that several stories have the same order, and we can give them a dynamic order using time-dependent random number system. The data part consists of several data files, such as ID file, learner's attribute file, user's consciousness factor files, story file, and explanation files. The procedure of inner process is as follows.

1. ID file stores user ID with corresponding hash value.
2. Attribute file is created when the attribute questionnaire is completed, and used to determine the preferable degrees of story categories such as "Personal Information", "Copyright", and so on.
3. Consciousness factors' score file is created when the consciousness questionnaire is completed. Story file involves several stories in each category, and a vector with three value components corresponding to consciousness factors is assigned to each story. These values should have been given carefully considering the meaning of story.
4. There are number of stories explanation files in which each story is explained and commented in several perspectives such as legal, ethical, Netiquette, safety, financial, etc. Especially in legal point of view, differences between laws in Japan and user's original country are taken up and explained.

5 Integration of Three Parts

In our proposing system, there are two independent concepts, one is learning style and the other is learner's consciousness on information security. Figure 1 describes the image of the framework of the e-learning system.

The upper part is on content choosing process according to the learner's consciousness factors, and the lower part is constructed by integrating e-learning environments and learning style theories we referred in Sect. 3. Lage et al. gave an table describing the learning methods with some of learning style theoretical relations [8]. Following them, we give evaluation scores to the tools or methods in the Sect. 2.2. On the other hand, after earning learner's MBTI type, the learner is mapped on the 4MAT circle according to the value of the indicator. Then, the system determine the preference degree of each tool or method. About the content on information security, we adopted the system described in the Sect. 4. Thus, some recommended contents will be proposed to learn in the preferred tools or method in order with uncertainty.

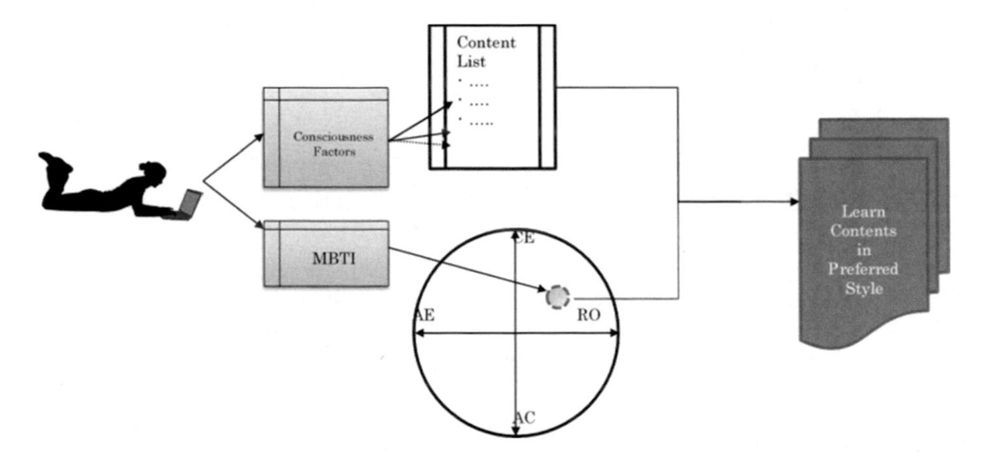

Fig. 1. Framework Image of the Proposed System

For implementing our system, we need to prepare the followings,

- Learner's consciousness extracting system
- MBTI generating system
- 4MAT circle where all tools and methods is put on
- Contents list on information security issues to be learn
- Different types of tools and methods for each content item.

Huitt pointed out that "Instructors must remember that Web-based instruction is more dependent on materials and activities than is classroom-based instruction." Thus the critical part of our e-learning system is the corresponding function between Kolb's four learning styles or MBTI and learning material or method.

Table 3 is a trial version of the correspondence where we quote most of MBTI checking parts from the table in [8]. We give the indication of learner type from the Table 2 by considering MBTI and brain types.

Table 3. Table Between Learning Tools and MBTI, Learner Type (Trail Version)

	MBTI				Brain Type	Leaner Type in Table 2
	E/I	N/S	F/T	J/P	Left/Right	
Tools						
PC	I				Left	
Tablet	E			P		1,2
Mobile device	E		F	P	Right	1,8
Methods						
Video	I	N		P	Right	1,3,8
Text	I		T	J	Left	4,5
Picture		N	F			3,4,7,8
Animation		N	F	P	Right	1,3,8
Audio			F			7,8
Reports		S	T	J	Left	2,4,5
PowerPoint		S		P		1,2,5.6
WorkSheet(group work)	E	S	F	J	Left	2,5,7
Quizzes			T	P	Left	2,4
Office hours(on-line)	I	S		J	Right	1,6
Chat room	E			J	Right	6
SNS	E	N	F	P	Right	1,3,8

6 Conclusion

We proposed the total system of e-learning which chooses recommending contents according to learner's information security related consciousness factors in learner's preferred learning style. In order to make a program, we have many things to clear described in the Sect. 5. We need to complete the Table 3 by considering much more details of tools and methods, and make the correspondence more proper.

Since the inner process programming is not difficult, so on the assumption that the MBTI patent issues are resolved, we can construct a prototype system with relatively small database of learning contents in several contents type such as text, video, etc.

References

1. Aoki, T., Kigawa, Y., Nemenzo, F., Nagata, K.: E-learning system based on user's consciousness and characteristic. In: Proceedings of 2015 International Conference on Computer Application Technologies, pp. 108–113 (2015)
2. Coffield, F., Moseley, D., Hall, E., Ecclestone, K.: Learning styles and pedagogy in post-16 learning: a systematic and critical review. Learning and Skills Research Centre (2004)
3. Curry, L.: An organization of learning styles theory and constructions. ERIC Doc. **235**, 185 (1983)
4. Fleming, N.D., Mills, C.: VARK a guide to learning styles (1992), http://www.vark-learn.com/English/index.asp. Accessed 17 May 2017
5. Huitt, W.: Using the 4MAT system to design web-based instruction. In: Delivered at the 8th Annual Conference Applied Psychology in Education, Mental Health, and Business, http://www.edpsycinteractive.org/papers/4matonweb.html. Accessed 17 May 2017
6. Kigawa, Y., Nagata, K., Aoki, T.: Multilingual e-learning system for information security education with users' consciousness. In: Advances in Web-Based Learning - ICWL 2014. LNCS, vol. 8613, pp. 201-206. Springer (2014)
7. Kolb, D.A.: Experiential Learning: Experience as the Source of Learning and Development. Prentice Hall, New Jersey (1984)
8. Lage, M.J., Platt, G.J., Treglia, M.: Inverting the classroom: a gateway to creating an inclusive learning environment. J. Econ. Educ. **31**(1), 30–43 (2000)
9. McCarthy, B.: Using the 4MAT System to bring learning styles to schools. Educ. Leadersh. **48**(2), 31–37 (1990)
10. Myers, I.B., McCaulley, M.H., Quenk, N.L., Hammer, A.L.: MBTI manual: a guide to the development and use of the Myers-Briggs Type Indicator, 3rd edn. Consulting Psychologists Press, Inc., Palo Alto (1998)
11. Thorne, A., Gough, H.: Portraits of type: an MBTI research compendium, 2nd edn. Center for Applications of Psychological Type, Inc., Gainesville (1999)

Mobile Device or Personal Computer for Online Learning – Students' Satisfaction in Yemeni Universities

Daniela Tuparova[1(✉)], Abdul Rahman Al-Sabri[1], and Georgi Tuparov[2]

[1] South-West University "Neofit Rilski", Blagoevgrad, Bulgaria
daniela.tuparova@gmail.com
[2] New Bulgarian University, Sofia, Bulgaria

Abstract. The widespread use of mobile technologies allows today's students and teachers to work and study independent of place and time. Moreover, in some developing countries mobile access to internet is more predominant than other Internet access methods due to poor or destroyed communication infrastructure caused by political conflicts; therefore, mobile devices, like smartphones, are frequently used for Internet access. The aim of our research is to outline student satisfaction and preference regarding the usage of different devices – personal computers, tablets, and smartphones to access various types of learning content in higher education in the Republic of Yemen. In this study, 72 students were involved, of which, 51 of participated in the survey for evaluation of satisfaction and usability.

Keywords: Mobile learning · Usability · Cloud delivering · LMS Middle East

1 Introduction

Nowadays smart devices take a significant place in the educational area. The widespread use of mobile technologies allows today's students and teachers to work and study independent of place and time. Moreover, in some developing countries mobile access to Internet overruns other Internet access methods due to poor or destroyed communication infrastructure caused by political conflicts; therefore mobile devices, like smartphones, are frequently used for Internet access.

Another reason for the increase in the number of mobile users is described in [1] "Mobile devices are cheaper than a personal computer and are used by many because the devices are more affordable and in the case of mobile phones, is almost a necessity to have."

However, wide acceptance of mobile devices is not enough for the successful e-learning process. As [2] states, "when educational institutions, such as universities, design and develop their m-learning systems, they need to consider students' expectations of m-learning. In other words, they should develop their m-learning services based on students' suggestions, to better meet their performance expectations."

M. E. Auer et al. (eds.), *Teaching and Learning in a Digital World*,
Advances in Intelligent Systems and Computing 715,
https://doi.org/10.1007/978-3-319-73210-7_46

According to [3], students at Yemeni universities are familiar with the use of smartphone facilities in daily activities, but they do not use actively smartphones for learning. Moreover, Yemeni students have a positive attitude towards the use of e-learning and m-learning, but suffer from a lack of e-learning and m-learning resources.

The aim of our research is to outline student satisfaction and preferences regarding the use of various devices – personal computers, tablets, and smartphones to access different types of learning content in higher education in the Republic of Yemen.

2 Methodology

We are forced to use free cloud hosting to provide e-learning content due to the lack of working e-learning infrastructure in Yemen due to political conflict in the country. Moreover, the Yemeni universities are not able to conduct classes regularly and students must access internet sources using personal devices – mobile phones, tablets, and personal computers. Therefore the learning management system (LMS) used for content delivering, must support responsive interface.

Many contemporary LMS support a responsive interface, and also have an Arabic localization. We have chosen Moodle, due to its rich functionality and worldwide experience accumulated during its long-term usage. Also, there are two good options for free hosting of e-learning content – moodlecloud (https://moodlecloud.com) and Gnomio (https://www.gnomio.com). Due to the limitations of the free option in moodlecloud – up to 50 users and 200 MB storage, we have chosen Gnomio, which has no such limitations.

The course "Introduction to PHP programming" was designed to be accessed through smartphone, tablet, and personal computers. Learning content was divided into 6 modules. In each module, learning content was presented as MS PowerPoint presentation, pdf file, MS Word file, video file generated from MS PowerPoint presentation, and web page. One of modules contained an interactive simulation training exercise, developed as a SCORM package. At the end of the course, a questionnaire was given to collect students' evaluations learning content.

The course "Introduction to PHP programming" was delivered to undergraduate students from the Faculty of Computers and Information Technology at Sanaa University, the Faculty of Computers and Information System at Taiz University, the Faculty of Computers and Information Technology in Hodeidah University, and one of the private universities in Sanaa, during March and April 2017. The experimental training was attended by 72 third and fourth year undergraduate students. The students were asked to complete the course for six weeks and, after that, complete a questionnaire. We received feedback from 51 of 72 students – 14 females and 37 males.

We developed the questionnaire for the evaluation of usability and students' adoption and satisfaction of proposed learning content and activities, accessed via different devices. The terms pedagogical and technical usability of multimedia learning materials and e-learning environments are discussed in the following papers [4–7] and include basic constructs such as understandability and motivation. The questionnaire consisted of 8 parts: 1. description of respondents with two indicators - gender and university; 2. use of smartphone, tablet, computer, and printed materials (four indicators); 3, 4, 5 - three parts for usage of smartphone, tablet, and computer, each of them consisting of two subparts – one for usability and accessibility of content, quizzes, and simulation training (four indicators) and another for level of usage of different delivering formats; 6. usability of design (two indicators); 7. pedagogical usability with construct of understandability of whole content – four indicators.

All items, except the first two (gender and university) were evaluated using a 5-point Likert Scale (1 for "strongly disagree"/"very low level", 5 for "strongly agree"/ "very high level"). The statement – "I do not use" is noted with a missed value.

3 Findings and Results

The data were processed with SPSS v.19. The Cronbach's Alpha for Reliability of the survey with listwise excluded missed values was 0.905.

In our study, we set several research questions regarding preferences to use of different devices to learn and different formats of learning materials.

Due to small numbers of female's respondents and type of data (Nominal and Ordinal), we applied nonparametric tests with level of confidence $\alpha = 0.05$.

RQ1: Is there difference in preferred type of device to learn? Is there difference in preferred type of devices regarding gender?
About 9.8% of respondents do not use smartphones, 25.5% do not use tablets, 11.8% do not use computers, and 13.7% do not use printed materials. There is a significant difference in way learning materials from the course are used. The Nonparametric Friedman Test was applied. The Friedman Test's Statistics are $\chi 2 = 10.831$, df = 3, $p = 0.013 < 0.05$. The descriptive statistics of variables are presented in Table 1. To find the preferred device, the Wilcoxon Signed Rank Test was performed. The use of tablets vs smartphones and personal computers was statistically significant. There was no significant difference among preferences to the use of smartphone, personal computers, or printed materials (Table 2). For the factor of gender, there was a significant difference observed for the variable "I am using smartphone to read course materials" (Table 3). Kruskal-Wallis Test have been performed. About 72% of the males read learning materials with smart phone and 33% of the females respond that they use smart phone to read learning materials.

Table 1. Descriptive statistics

Variables	N	Mean	Std. Dev.
I am using smartphone to read course materials	51	3.5098	1.43349
I am using Tablet to read course materials	51	2.7843	1.87951
I am using computer to read course materials	51	3.7255	1.58844
I am using printed files pdf, ppt and doc to read course materials	51	3.2549	1.57281

Table 2. Wilcoxon signed rank test statistics

Null hypothesizes: The median of differences between next variables equals 0	Z	p
I am using Tablet to read course materials. - I am using smartphone to read course materials	**2.217**	**0.027**
I am using computer to read course materials. - I am using smartphone to read course materials	−1.061	0.289
I am using printed files pdf, ppt and doc to read course materials. - I am using smartphone to read course materials	−1.020	0.307
I am using printed files pdf, ppt and doc to read course materials. - I am using Tablet to read course materials	−1.450	0.147
I am using computer to read course materials. - I am using Tablet to read course materials	**−3.041**	**0.002**

Table 3. Kruskal - Wallis test statistics, grouping variable - gender

Null hypothesis: The distribution of next variables is the same across categories of Gender	Chi-Square	df	p
I am using smartphone to read course materials	**4.010**	**1**	**0.045**
I am using Tablet to read course materials	0.067	1	0.795
I am using computer to read course materials	0.029	1	0.632
I am using printed files, pdf, ppt and doc to read course materials	2.740	1	0.098

RQ2. Is there difference in usability of video demonstration, training simulation exercise, and quizzes regarding to use of device?
Regarding the usability of video demonstration, training simulation exercise, and quizzes, there is no significant difference in any of the users' groups (smartphones, tablets, or personal computers). There were 46 students that mentioned use of smartphones for course study. The Friedman Test's Statistics are $\chi2 = 3.796$, df = 3, p = 0. 284 > 0.05.

For the users of tablets (33 total) the observed Friedman Test's Statistics are as follows $\chi2 = 3.614$, df = 3, p = 0.302 > 0.05.

The results for the respondents that used personal computers (45 users) for learning of the course are similar, $\chi2 = 3.201$, df = 3, p = 0.362 > 0.05.

The statistically significant differences observed in comparisons of answers from students that use smartphone, tablet, and personal computer (total 32) regarding the usability of whole learning material and simulation training xampp exercise (Table 4).

Table 4. Friedman test statistics for usability of learning resources regarding the used devices.

Null hypothesis	χ2	df	p
The distributions of "When I am using smartphone easy I get the course materials", "When I am using tablet I get the course materials" and "When I am using computer easy I get the course materials" are the same	**6.164**	**2**	**0.046**
The distributions of "When I am using smartphone the visualization of video demonstration is sufficient", "When I am using tablet the visualization of video demonstration is sufficient" and "When I am using tablet the visualization of video demonstration is sufficient" are the same	2.425	2	0.297
The distributions of "When I am using smartphone I use training exercise of xampp very easy", "When I am using tablet I use training exercise of xampp very easy", "When I am using computer I use training exercise of xampp very easy" are the same	**6.576**	**2**	**0.037**
The distributions of "When I am using smartphone the filling of quizzes is appropriate", "When I am using tablet the filling of quizzes is appropriate", "When I am using computer the filling of quizzes is appropriate" are the same	3.937	2	0.140

The students reported that it was easier to learn course material by use of computer (84% agree or strongly agree) than tablet (68% agree or strongly agree). (Wilcoxon Signed Ranks Test, $Z = -2.456$, $p = 0.014$).

The significant difference occurred in use of training exercises of xampp between smartphone and computer (Wilcoxon Signed Ranks Test, $Z = -2.289$, $p = 0.022$), and between tablet and computer (Wilcoxon Signed Ranks Test, $Z = -2.858$, $p = 0.004$). About 76% of respondents agree or strongly agree that xampp training exercise is easy for use with personal computer versus about 56% for use with smartphone and tablet.

RQ3. What kind of learning content format delivery is preferred when students use smartphone, tablet and computer?

Significant differences have been observed regarding the used formats of learning content for users of smartphones. The null hypothesis: "The distribution of "I am learning from video clips", "I am learning from pdf files", "I am learning from ppt presentations", "I am learning from doc files", "I am learning from web pages integrated in e-learning environment" are the same" could confidently be rejected. The observed Friedman Test's Statistics are as follows: $\chi 2 = 17.548$, $df = 3$, $p = 0.002 < 0.05$.

When students used smartphones to learn course materials, 2% of them did not prefer (strongly disagree or disagree) to use video clips, 6.52% did not prefer to learn from pdf files, 13% did not prefer to learn from MS PowerPoint presentations, 28% did not prefer to learn from MS Word files, and 20% did not prefer to use web pages integrated into the e-learning course.

Students who used tablets for learning course content also indicated a significant difference in preferred usage of delivery formats. The observed Friedman Test's

Statistics were: $\chi2 = 13.556$, $df = 3$, $p = 0.009 < 0.05$. When students used a tablet to learn course materials 6% of them did not prefer to use video clips, 9% did not prefer to learn from pdf files, 11% did not prefer to learn from MS PowerPoint presentations, 14% did not prefer to learn MS Word files, and 29% did not prefer to use web pages integrated into the e-learning course.

There was no significant difference regarding format delivery preferences for students who use a computer for course learning. The observed Friedman Test's Statistics were: $\chi2 = 7.098$, $df = 3$, $p = 0.131 > 0,05$.

RQ4. Is the graphical design of learning materials acceptable for the students? The students enjoyed colors for video clips and presentations. For colors in presentations, 6% of the students strongly disagreed or disagreed, 31% were neutral, and 63 agreed or strongly agreed. Acceptance of colors in video clips was similar. There was no significant difference regarding acceptance of graphical design of the presentations and video clips.

RQ5. Is the content understandable for students (pedagogical usability)? There was no significant difference observed regarding the understandability of learning content through presentation, quizzes, pdf files, and video. The Friedman Test's Statistics are $\chi2 = 5.122$, $df = 3$, $p = 0.163 > 0.05$. Only 2% of the respondents disagree that quizzes helped them easily understand learning content, and nobody disagreed that pdf and video files helped them to easily understand learning content.

RQ6. What are the preferences regarding work in offline mode and choice opportunity given by delivering of different formats of learning content? 96% of students work in offline mode that is provided by Moodle. 4% reported that they are not satisfied with different formats of learning content delivery. Furthermore, 6% strongly agree, and 34% agree that different delivering formats may overload the course.

4 Conclusions

In this research study we presented results from the experimental application of responsive learning content conducted with undergraduate computer science students in Yemen. Students prefer to learn with smartphones, computers, and printed materials with learning content. It was found that students do not use tablets frequently. For students, it is easier to learn course material by use of computer than tablet. They met some difficulties with a simulation training exercise with a tablet and smartphone. When students use smartphones and tablets, they prefer to use video clips and pdf files. For the students who use computers, there is no difference regarding the delivery format of learning content. The proposed activities and resources as quizzes, pdf files, and videos helped students to understand learning content more easily. Finally, the students liked being given various choices for learning resources, but the course designers have to reduce non-preferable formats for content delivery as web pages integrated in e-learning environment and MS Word documents.

Acknowledgments. This paper is partially supported by University Research Fund at South-West University "Neofit Rilski", Blagoevgrad, Bulgaria.

References

1. Alawi, G.A.A.A., Shwal, M., Nasreen, N.: Interaction Triangle of Mobile Learning & ELearning and Computer Tools (CUAELML) in the Basic Class: Attitudes & Opinions of Pre-Service Teachers, Multimedia Technology (MT), vol. 4 (2015). https://doi.org/10.14355/mt.2015.04.001, www.seipub.org/mt
2. Alksasbeh, M.Z.S.: Integrating mobile technology quality service, trust and cultural factors into technology acceptance of mobile learning: a case of the Jordan Higher Education Institution. Ph.d. thesis, Universiti Utara Malaysia (2015). http://etd.uum.edu.my/3370/
3. Alsabri, A.A.A, Tuparov, G., Tuparova, D.: Students' readiness for mobile learning in Republic of Yemen - a pilot study. In: International Conference on Interactive Mobile Communication Technologies and Learning (IMCL), Thesaloniki, Greece (2015). https://doi.org/10.1109/IMCTL.2015.7359584
4. Brodahl, C., Smestad, B.: A taxonomy as a vehicle for learning. Interdisc. J. E-Learn. Learn. Objects **5**, 111–127 (2009)
5. Hadjerrouit, S.: Developing web-based learning resources in school education: a user-centered approach. Interdisc. J. E-Learn. Learn. Objects **6**, 115–135 (2010)
6. Nokelainen, P.: An empirical assessment of pedagogical usability criteria for digital learning material with elementary school students. Educ. Technol. Soc. **9**(2), 178–197 (2006)
7. Teoh, B.S., Neo, T.: Interactive multimedia learning: students attitudes and learning impact in animation course. Turk. Online J. Educ. Technol. TOJET 2007, **6**(4).http://www.tojet.net/articles/643.pdf. Accessed 28 Nov 2011. Article 3, ISSN: 1303-6521

Cryptography Applied to the Internet of Things

Javier Sanchez Guerrero[1,2(✉)], Robert Vaca Alban[1,2(✉)], Marco Guachimboza[1,2(✉)], Cristina Páez Quinde[1,2(✉)], Margarita Narváez Ríos[1,2(✉)], and Luis Alfredo Jimenez Ruiz[1,2(✉)]

[1] Dirección de Tecnología de la Información y Comunicación, Facultad de Ciencias Humanas y de la Educación, Universidad Técnica de Ambato, Ambato-Ecuador, Ecuador

[2] Facultad de Contabilidad y Auditoría, Universidad Técnica de Ambato, Ambato-Ecuador, Ecuador

{jsanchez,rvaca,marcovguachimboza,mc.paez, mm.narvaez,la.jimenez}@uta.edu.ec

Abstract. Concepts and technologies that have made it possible for devices to interconnect to real-world objects, have existed for some time. The Internet of things can be defined as a pervasive and ubiquitous network that allows the monitoring and control of the physical environment by collecting, processing and analyzing data generated by sensors or smart objects. As the number of devices connect to the Internet increases, it is necessary to provide a safe and reliable operation of such devices. This paper emphasizes the use of cryptographic algorithms that can be implemented in devices that can connect to the internet of things, in order to provide security for the transmission of information on the vast network of sensors, a network that is increasing in size and number. This paper describes the problem of security on networked devices and describes the main algorithms and processes that provide security to this technology. It is concluded that since the devices that form the Internet of things have little memory and processing, it is necessary to apply algorithms that use the least amount of bits, with encryption algorithms that are based on elliptic curves. The advancement of technologies and algorithms that provide security should be considered, because every day we are more dependent on the internet of things.

Keywords: Internet · Security · Encryption

1 Introduction

The Internet has so far been a communication network among humans. Computers learn, and with the right software they can become better than humans, or at least help humans in their day-to-day activities. Remote monitoring is now an ordinary activity, it can be performed from anywhere in the world. The internet of things links the chips of things with advancements in information technology. Computers that can reason and learn, this means that you can make something intelligent. A red light on the dashboard of a vehicle indicates that something is

M. E. Auer et al. (eds.), *Teaching and Learning in a Digital World*,
Advances in Intelligent Systems and Computing 715,
https://doi.org/10.1007/978-3-319-73210-7_47

happening. A good owner will take it to the shop and have it repaired. It means that connected things will be able to anticipate problems [15]. You can find the problem before it occurs, or at least you can identify these problems and act on them. What will happen is that eventually any industrial product will have the desired connectivity. If you have appliances to monitor the temperature at home, such as a thermostat, why not use them to protect the house?. The internet of things is a global phenomenon that includes human machine communication, radio frequency identification, location - based services, Lab-on-a-Chip sensors, augmented reality, robotics and vehicle telematics. Its common feature is its ability to combine embedded objects with sensory capabilities and communication intelligence, exchanging data through a combination of wired and wireless networks. Companies can analyze patterns of behavior, so this can create new business value by simply monitoring things [17].

The amount of computer attacks is increasing day by day. This can lead to severe breaches of sensitive information. The security of communication is an aspect that is causing concern, since the number of devices connected to the internet grows by the hour. Although Internet of Things (IoT) devices do not appear to be critical devices, they can become critical devices if they are not used properly [29].

An open security issue is the distribution of passcodes between devices. Risks vary depending on the criticality of the device either by their function or dependence that people may have on them. Threats to which devices are exposed will have a negative effect on accessibility, integrity, identity, availability and confidentiality. This last aspect is very important since it must exist in the device, and even more if this information is transmitted over the Internet. GPS positioning can become a threat, because the user's location can be recorded on a website [20]. In many cases you can access these devices via internet, but if a third party has obtained a passcode, then privacy issues appear. Remote control of the devices by third parties, whose non legitimate use can affect the computer and physical security of the users, is something that must be avoided. Since devices send information via the Internet, in most cases an extensive use of communication networks is needed [1]. All these communications that propagate via public networks are sensitive to being compromised.

In this paper, we describe the means by which the attacks are performed, and cryptographic solutions that prevent their development. Elliptical curves are considered, and we argue that the use of such elliptical curves is one of the best methods of information encryption. The aim is to publicize the importance of IoT and the security that can be provided to the links created between the devices from the beginning to the end of the transmission of information. This is very important, since by the year 2020 many devices will be connected to the internet and it is necessary to disclose how you can provide adequate security on these devices. The following section of related work problems and recent developments in relation to cryptography for IoT is described, thus explaining the computer algorithms and procedures needed to provide the highest security of communications that take place on the Internet of things. This description allows for conclusions and future work that will contribute to improving safety on the IoT.

1.1 Theoretical Framework

Security must remain throughout the device's lifecycle, from initial design to its deployment. This implies that there must be credentials that allow secure access to networks in the IoT. It is necessary that the devices have unique protocols, which must be verified, this being the only form of safe and proper operation. Many security systems need to deploy patches that reduce bandwidth consumption or reduce the possibility of attacks when someone could steal the credentials to access the device's operation. This implies that a connected device that maintains limited bandwidth and intermittent network must authenticate itself before receiving data from another device. Then, security must exist from end to end, both at the device and network levels; the intelligence that allows the devices to carry out their tasks, should also allow them to recognize and repel threats [28]. The Man in the Middle attack means that the receiver's data is intercepted by the attacker, who pretends to be the originator of the message that is being sent to the receiver, acting as an intermediate point between communications, remaining invisible during the process [26]. Reduced versions of operating systems are used in certain devices, which means lowering software costs. This implies a security risk, since this weakens the security of the entrance doors, compromising the information transmitted by the devices. Web interfaces that are used in IoT, allow their administration from another enabled device, which implies that these interfaces allow the administration of a device from the network, so that when they are attacked, the attack is not only made at the interface level but in the hijacked devices as well. Some devices like Smart tv's allow the installation of third-party applications just like Smartphones, in those cases, the applications can be used as an entry platform to the device [4].

In some cases, developers or users do not follow an appropriate security policy, this means that most devices enable many of their functionalities in their configurations, which means having a security gap. The attacks against hardware are carried out through the Internet, as well as through an analysis of its structure and functioning. An example of these situations are attacks on electrical components or components of network traffic. Depending on the equipment, attacks such as monitoring interfaces, reverse engineering or manipulation of internal sources can take place. Social engineering consists of exploiting user behavior as a gateway for fraud, that's what is known as phishing. An attack much more elaborate, is to study the internet victim when the victim publishes personal information on the Internet, making him more vulnerable. The most elaborate attacks are aimed at more profitable targets [22].

The principles of reliable computing are accessible for embedded systems, but IoT devices with limited resources and low energy make their implementation a major challenge. A comparison of the LPC interface, I2C is most commonly included in embedded microcontroller and requires fewer E/S pins. For handheld devices or other applications where space is limited, the Atmel ATSHA204A provides client and host security capabilities in a small 3-pin SOT23 package with a single cable interface to the host. Other possible uses include images of encryption code to prevent tampering, session key exchange, secure data storage and verification of user passwords. Faced with the reality of millions of devices connected to each other and the Internet, our real challenge is to provide security solutions that meet the needs of heterogeneity and scalability that imposes the Internet of Things [13].

Cryptographic techniques such as preservation and homomorphic encryption use procedures that increase the processing time by 25% and also work with proxies which are expensive to implement in electronic circuits that work with IoT. By contrast, Talos is a system that allows the storage of secure information and allows encrypted queries. It focuses on the partially homomorphic encryption that provides a high level of security when working with a reasonable data overload [25]. When working in a client - server model, the device is connected via LAN, WLAN or WPAN. The device communicates certain localized information or service requests to a network concentrator of a cloud-based service. An IoT device connects a physical device to the cloud for processing additional data or services. Architectural considerations for CPU performance for IoT devices depend on the scope of what the CPU has to do as well as hardware security provisions contained in the CPU hardware. A key requirement for IoT applications is security. Devices connected in networks open a variety of threats as they are more connected to the network and finally to the cloud. Among the elements to be considered for safety we should consider: (a) secure boot, (b) updating secure code, (c) password protection, (d) controlling access to secure resources, and (e) DMA (Direct Memory Access) with data encryption for critical functions such as session authentication [11]. An interesting trend that contributes to the growth of IoT is the transition from IPv4 to IPv6 which allows for interactions at a machine-machine level. IPv6 is one of the most important enablers of IoT since it is not possible to add thousands of millions of devices over IPv4. Security Considerations and Implications of IPv6 are critical to secure IoT [30].

Existing technologies and security solutions can take advantage of a network architecture, especially through the central layers and data centers in the cloud, there are unique challenges in the field of IoT. Methods must be implemented to ensure that the authenticity of the data, the path from the sensor to the collector, the authentication parameters between the initial installation/configuration of the device and its possible presence in the IoT infrastructure are not compromised. Man-in-the-middle is the means by which the attacker can successfully create a connection between two points and spy on their conversation while capturing the data. Hence, IPv6, a function of the IoT, is subject to the same threats of attack as IPv4, such as smurfing, recognition, spoofing, fragmentation attacks, sniffing, neighbor discovery, rogue devices, man-in-the-middle, and others [2]. When IoT/M2M devices are connected, they need access to the infrastructure, a secure connection is started based on the identity of the device. In this domain, many devices may not have enough memory to store certificates or may not even have the CPU power needed to run cryptographic operations to validate X.509 certificates. Data generated by IoT devices is only valuable if appropriate analysis algorithms or other processes are defined to identify the threat. While the security implications for construction of IO/M2M are vast, the deconstruction of a security framework/M2M viable IO can be the basis for implementing security in production environments [6].

Identity Based Encryption (IBC) is one of the most important applications of cryptography based matching, which is used in low memory devices. The use of another class of elliptical curves must be further extended in order to increase efficiency. The safety multiplier refers to the size of the base field on which signals in the elliptical curve are manipulated [19]. Hence, if a large security multiplier is needed, it means that it must apply elliptical or hyper-elliptic curves over a smaller base field, with the advantages of efficiency that come with it [3].

Elliptic curves over a finite Fp field, are a set of (x, y) that satisfy the equation y^2 mod p = x^3 + ax + b mod p, with 43 + 27b 2 <> 0, such that a, b, x, y belong to a special point Fp plus a special point known as not infinite point O. Another important feature of the elliptic curve cryptography is that a point G can be selected, which will have an order n and a cofactor h [18]. An elliptic curve can be found in major modern technologies such as the Web and technologies that rely upon TLS, PGP and SSH. RSA, DSA, and DH algorithms are combined with ECC in order to increase security. The RSA algorithm implements asymmetric cryptography, however this algorithm uses very large keys. In IoT it is necessary to use simpler computational processes especially in computational calculations, as is the case with the use of elliptic curves, especially in devices with lower computing power, shorter memory and limited energy resources. Symmetric cryptography algorithms may be simple to decode or decrypt [7]. Finite fields are a set consisting of a finite number of elements. Given a finite field Fp, is a set of integers with modulo p, where p is a prime number. A set of module p integers consists of all the integers from 0 through p − 1 and mathematical operations are performed with modular arithmetic [8]: a = b mod n, being a the residual of the division by n times b. For example 2 = 38 mod 12, as 38/12 = 3 with a residue of 2. An elliptical E curve on the body K, is a cubic curve, irreducible and non-degenerated, defined on the projective plane P2 (k) with a point O which belongs to the curve. The curve will have points in the same plane, but it also has points in infinity [23]. An example of elliptic curve is

$$y^2 = (x)^3 + 2x + a, over K = Xs \tag{1}$$

To calculate the points at infinity curve we have y = y/z x = x/z, replacing in the equation we have:

$$\frac{y^2}{z^2} = \frac{x^3}{z^3} + 2\frac{x}{z} + 1 \tag{2}$$

Given that we are looking for points in the infinite we convert z = 0, therefore x^3 = 0 and x = 0.

Therefore the infinite points of the curve are: (0:Y:0) = (0.1:0)

This means that the normal Weierstrass formula can be located:

$$y^2 = (x)^3 + ax + b \tag{3}$$

The number of points are calculated by defining the elliptic curve:
E = EllipticCurve (GF (5), [2,1]), * Previous curve, over the body z(5)
If E.points () is located, we will notice a set of points calculated for 7 elements [(0:1:0),(0:1:1),(0:4.1),...]

In this case, the number of points is 7. However, in other cases the curves have many points which are calculated with E.Cardinality () function.Curve points are operable. If there are two points P, Q belonging to the curve, one can define the P + Q operation, resulting in another point of the curve (this sum differs from the coordinate plus coordinate sum). This sum has the commutative and associative properties and has an element O which together with P, yields as a result the same point P. For every point of the curve there is a point −P, such that P + (−P) = O [16].

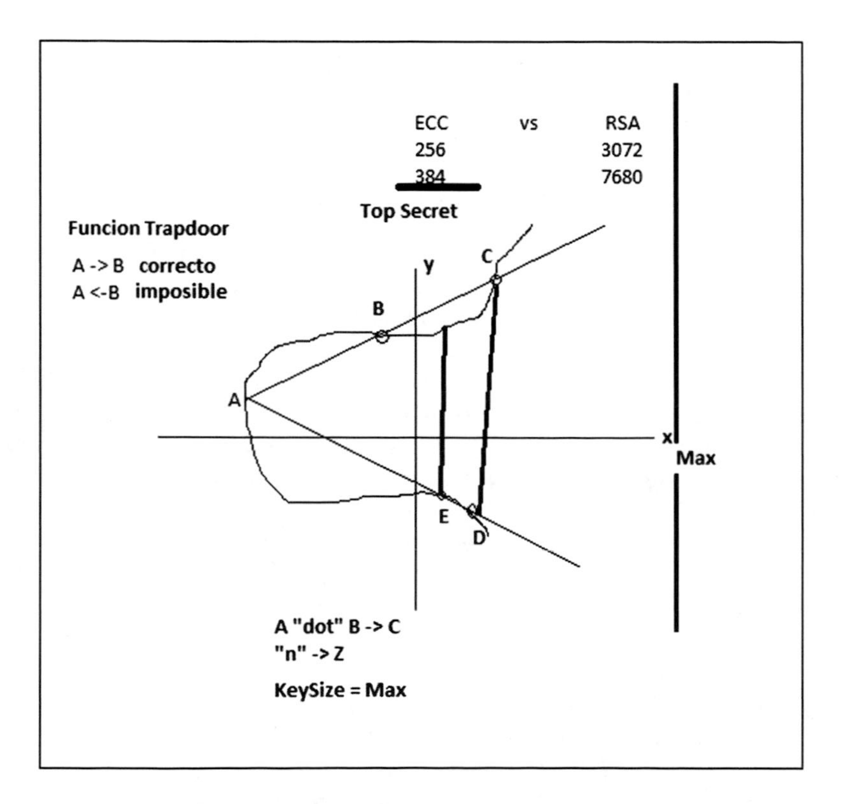

Fig. 1. Elliptic curves in cryptography: the number of bits used to encrypt is much smaller than the RSA algorithm

Elliptic curves have a direct application in cryptography, and in the use of public and private key. The Trapdoor function makes it possible to send information from A to B, which reaches its destination but is very difficult to decipher from B to A [10]. The ECC outweigh the RSA, because fewer bits are used in the encryption process (top secret level information). The characteristics of the elliptic curve are symmetrical with the axis x. In Fig. 1, the three elliptic points (A, B, C) in cryptography, the idea is to apply the function A "dot", for example A "dot" B "dot" C, and as there is symmetry there is a point on the other side of the curve called D, and therefore an E point, and so on until all elliptical points are completed. Through this process we can obtain the peak that can be connected from A. These points (values) are taken into account for the public key, KeySize = Max for the private key Priv = "n" dot, in order to find the symmetrical points that allow it to return to the original message (E, D) [27].

Systematic analysis of information is a challenging problem. A hacker can compromise the privacy and integrity of the transaction process. Being able to encrypt messages is vitally important especially if encryption processes can reside in IoT devices which in 2020 will be connected and producing large amounts of data. This section described the processes hitherto known to encrypt information, and has provided details regarding elliptic curves as one of the best elements to achieve an increase in information security. The following section describes the conclusions achieved in this research work.

1.2 Conclusions, Limitations and Future Work

This study is a journey through the state of the art in the field of encryption and cryptography and algorithms. Overall, it is noted that emphasis on using public and private keys, with the main issue being its encryption (Abboud 2015). Soon, smart automobiles (which are controlled by touchscreens, recognize driver's gestures and detect parking spaces) will move by smart cities where everything will be controlled by sensors interconnected in order to optimize traffic, reduce pollution and manage video surveillance systems. The Internet of things is already a reality in trends as wearables, accessories and clothing that offer all kinds of information about the wearer. The good news is that there are several procedures and solutions that can be taken to prevent malicious attacks; Cryptography Data; Anti-Malware; Identity and Access control and device management. Investing in security solutions for the growing array of increasingly sophisticated threats is critical to maintaining data security. In addition to investing in the "digital education" of people, it is also necessary to create a strategy for Information Security with constant investments, seeking to eliminate the vector of attacks on your network and applications. That's because multiple devices are simultaneously connected; computers, notebooks, tablets, smartphones, TVs, among many others. Considering the speed with which the Internet of Things advances, there is still much to do to ensure data security. All integrated services must follow best security practices, from solution design, through all stages of implementation of IT solutions. The devices that we see today are electronic devices with different names and purposes. These devices process data and therefore can be classified as computers. The growth of connectivity facilitates daily activities, but also generates a great concern: Information Security. This work describes the main algorithms to provide security for IOT transactions, especially elliptic curves, which, when using less bits than RSA, allow better performance with IoT devices. It is very difficult to decrypt a message based on IBC and elliptic curves [9]. However, in order to apply the asymmetric algorithmic processes and their relationship in the elliptic finite space, an appropriate knowledge of mathematics is required, hence it is important to describe with more examples the computational processes of elliptical encryption, which Is a task that will be covered in future research topics. An example is used for this paper, in which we describe the handling of elliptic curves, where; the management of core operations differs from traditional handling, highlighting the fact that it is necessary to opt for other mechanisms of decomposition. This implies that there may be difficulty in applying encryption methods in IoT, so it is important to conduct tests on the functioning of algorithms in real world scenarios. Furthermore, progress is made on encryption and security processes on a daily basis. Changes in algorithms and their different ways of applying them are on the agenda, and therefore requires updating and patching process in IoT devices. Quantum computing has been proposed by IBM, such technology would allow decrypting of security algorithms but at the same time would strengthen encryption methods (Noor-ul-Ain et al. 2015). Hence, it is very important to make a comparison of the cost-benefit ratio of the application of the algorithms that provide security for the IoT devices especially since they have little memory and processing capacity.

References

1. Allen, N.: Weaknesses cybersecurity to make smart cities threaten more costly and dangerous than analog their prdecessors. USAPP
2. Botta, A., Donato, W., Persico, V., Pescapé, A.: Integration of cloud computing and internet of things: a survey. Future Gener. Comput. Syst. **56**, 684–700 (2016)
3. Breezely George, M., Igni Sabasti Prabu, S.: Secured key sharing in cloud storage using elliptic curve cryptography. In: Proceedings of the International Conference on Soft Computing Systems (2015)
4. Chaitanya, B., Sekhar, C., Ramesh, N.: IR IOT based smart device using CC3200. Sci. Technol
5. Chen, S., Maode, M., Zhenxing, L.: With an authentication scheme for identity-based cryptography M2M security in cyber-physical systems. Secur. Commun. Netw. **9**(10), 1146–1157 (2016)
6. CISCO.: Securing the Internet of Things: A Proposed Framework, http://www.cisco.com/c/en/us/about/security-center/secure-iot-proposed-framework.html
7. Delfs, C., Galbraith, S.: Computing isogenies supersingular between elliptic curves over Fp. Des. Codes Cryptogr. **78**(2), 425–440 (2016)
8. Galbraith, S., Gaudry, P.: Recent progress on the elliptic curve discrete logarithm problem. Des. Codes Cryptogr. **78**(1), 51–72 (2016)
9. Geetha, D., Sakthivel, D.: Orient stream cipher based service key management scheme for secure data access control using elliptic curve cryptography in wireless broadcast networks. Am.-Eurasian J. Sci. Res. **11**(1), 63–71 (2016)
10. Hoffman, L.: Q & A: finding new directions in cryptography. Commun. ACM **59**(6), 112–ff (2016)

11. Imagination Technologies: Internet of Things-device Opportunities for differentiation. Desing & Reuse
12. Kamboj, D., Gupta, D., Kumar, A.: Efficient scalar multiplication elliptic curve over. Int. J. Comput. Netw. Inf. Secur. **8**(4), 56–61 (2016)
13. Kanase, P., Gaikwad, S.: Smart hospitals using Internet of Things. Int. Res. J. Eng. Technol. **3**(3), 1735–1737 (2016)
14. Khan, M., Deen, S., Jabbar, S., Gohar, M., Ghayvat, H., Mukhopadhyay, S.: Context-aware intelligent low power SmarHome based on the Internet of Things. Comput. Electr. Eng. **52**, 208–222 (2016)
15. Kopetz, H.: Internet of Things. In: Real-Time Systems, pp. 307–323. Springer (2011)
16. Kumar Das, S., Sharma, G., Kumar Kevät, P.: Integrity and authentication using elliptic curve cryptography. J. Interdisc. Res. Imperial (2016)
17. Lechelt, Z., Rogers, Y., Marquardt, N., Shum, V.: Connectus: a new toolkit for teaching about the Internet of Things. In: Proceedings of the CHI 2016 Conference on Human Factors Extended Abstracts in Computing Systems, p. 4. ACM, New York (2016)
18. Maitin-Shepard, J., Tibouchi, M., Aranha, D.: Elliptic curve multiset hash. Cryptogr. Secur. (2016)
19. Malina, L., Hajny, J., Hosek, J.: On perspective of security and privacy-preserving solutions in the Internet of Things. Comput. Netw. **102**, 83–95 (2016)
20. Nazim, M., Ali Shah, M., Kamran Abbasi, M.: Analysis of embedded web resources in web of things. In: Proceedings of the International Conference on Communication IOARP and Networks. IOARP, London (2015)
21. Noor-ul-Ain, W., Atta-ur-Rahman, M., Nadeem, M., Abbasi Ghafoor, A.: Quantum cryptography trends: a milestone in information security. In: Advances in Intelligent Systems and Computing, pp. 25–39. Springer, Cham (2016)
22. Powell, A.: Hacking in the Public Interest: Authority, Legtimacy, Means, and Ends. New Media & Society, New Haven (2016)
23. Rao, M., Rao, B.: Paper reduction and total inter-carrier interference detection in canceled multiuser CDMA-OFDM CFO corrected elliptic curve cryptography with. Int. J. Appl. Eng. Res. **11**(8), 5880–5888 (2016)
24. Scott, M.: Pairings faster using an efficient elliptic curve with an endomorphism. LNCS, pp. 258–269 (2005)
25. Shafagh, H., Duquennoy, S., Hu, W.: Talos: encryupted query processing for the Internet of Things. In: Proceedings of the 13th ACM Conference on Embedded Networked Sensor Systems, New York (2015)
26. Sicari, S., Rizzardi, A., Miorandi, D., Cappiello, C., Coen-Porisini, A.: A secure and quality-aware prototypical architecture for the Internet of Things. Inf. Syst. **58**, 43–55 (2016)
27. Smart, N.: Elliptic curves. In: Cryptography Made Simple. University of Bristol, London (2016)
28. WIND: Security in the Internet of Things: Lessons from the Past for the Future Connected. WIND, USA (2013)
29. Zambonelli, F.: Towards a general software engineering methodology for the Internet of Things. Softw. Eng. (2016)
30. Palattella, M., Dohler, M., Grieco, A., Rizzo, G.: Internet of Things in the 5G era: enablers, architecture, and business models. IEEE J. Sel. Areas Commun., 4 (2016). IEEE

Be Active and Explore

An Interactive Online Walk as an Example of Engaging Place-Based Learning

Magdalena Brzezinska(✉)

WSB University, Poznan, Poland
magdalena.a.brzezinska@gmail.com

Abstract. This paper focuses on cross-curricular instruction that encompasses place-based learning (as defined by Sobel, Elder and Lane-Zucker), location-based learning (one mediated by mobile platforms; as understood by Yehiel and Ziv, followers of Walter Benjamin and his distinction between the concepts of a tourist and a wandering man), problem-and-quest-based learning (particularly in its task-based aspect embracing tacit knowledge and supporting student choice) and project-based learning. It also shows how this type of "eclectic instruction" can be used to boost teenage student engagement and make young adults genuinely interested in local culture and history, examined through the tinted spectacles of a foreign language.

Keywords: Place-based learning · Location-based learning
Project-based learning

1 Introduction to the Study

1.1 The Goal of the Study

There were several equally important goals of the study. Firstly, the author undertook to examine whether (and/or prove that) it is in fact possible to change the attitude and approach of teenagers to (frequently underappreciated or mocked) local culture and history by engaging them in a group project: designing an interactive online walk for their peers around the city that they live in.

Secondly, as the project constituted a part of a course on English as a Foreign Language (EFL), and the walk was to be drafted in a language that was not native to the students, the students' active and passive knowledge of English was to broaden and improve.

Thirdly, the author was interested in finding out whether such a project, also developing teenagers' ICT skills and competences, could be successfully used in cross-curricular foreign language instruction in a variety of contexts and whether it could serve to develop the desired 21st century skills.

1.2 The Scientific Approach Adopted

The six-month cross-curricular project combined place-based, location-based, problem-based, quest-based and project-based learning.

M. E. Auer et al. (eds.), *Teaching and Learning in a Digital World*,
Advances in Intelligent Systems and Computing 715,
https://doi.org/10.1007/978-3-319-73210-7_48

Place-based learning and location-based learning, as defined by Sobel [10], Elder [3] and Lane-Zucker and Sahn [8], in which proficiency and understanding of local history, traditions and culture are valued, and in which links between an individual and his place on earth are formed, have been among the author's scholarly interests for several years. She is also interested in how this kind of learning can be mediated by mobile platforms, such as, e.g. ExperienCity.

As for problem-based learning, the author tried to observe all the criteria listed by Gallagher et al. [4] or Hmelo and Evensen [6]: the learning process was student-centred, even though the topic was selected by the author; learning occurred in a small group (there were only 5 students in the class); the author was not a lecturer, but rather a guide and/or a facilitator; problem-solving was an essential part of the learning process; and students' learning was mostly self-directed. Some aspects of quest-based learning, as seen by Haskell [5], were adopted too, namely orientation on the task (one embracing tacit knowledge), ending the tasks with a deliverable goal, and providing students solely with formative feedback.

The project aimed at involving students in a variety of ways. It encompassed vast research conducted by the students (which included examining both offline and online resources); investigation of local environment through all the senses; inventing attractive and interactive activities to be performed by the students' peers "in situ"; and finally, structuring all the formerly mentioned elements as an action-packed BYOD hike around the city, created on the ExperienCity platform, at http://beta.experien.city/index.php?return=myExperiencieCity.php#.

2 The Process of Creation of the Interactive Walk

2.1 Inventing a Persona

Inspired by the information given in the Creative Box course offered by Prof. Sadik-Rozsnyai of the Ecole Supérieure des Sciences Commerciales d'Angers Ecole de Menagement, whose Creative Path the author successfully completed, she decided that creating a walk aimed at teenagers in general would result in cliché and uninteresting choices. Instead, she asked her 13–17-year-old students to come up with a detailed description of the persona of the user who was going to tour their city. She wanted the students to be able to identify with the invented character, and thus create a personalized and exciting experience.

The students invented a persona named Filippo: a 17-year-old Italian teen who liked to visit new places and interact with the locals, was mildly interested in monuments and art, and whose favourite activities were: sports, dancing and tasting good local food.

Once the persona was constructed, the teenagers brainstormed a catchy name for their walk to make "Filippo" interested in their final product. It led to the origination of "Be Active and Explore", appealing to their persona's energy and desire for investigation.

2.2 Selecting Optimal Localizations

Following the introductory stage presented above, there took place vast research and fervent discussions as to which spots in their city could be of interest to "Filippo".

The students wanted their persona to be a wanderer rather than a tourist, as differentiated by Benjamin (see [1]) (and promoted by Talila Yehiel, museum expert and educator, and Shani Ziv, the creator of the ExperienCity online platform) or by Jenks, who claims that

> "A 'psychogeography' depends upon the walker 'seeing' and being drawn into events, situations and images by an abandonment to wholly unanticipated attraction. (…) In the *derive* the explorer of the city follows whatever cue, or indeed clue, that the streets offer as enticement to fascination." (see [7])

The teenagers examined city maps (both online and offline) and critically read guidebooks, brochures and promotional leaflets, meeting the objectives specified by Bloom and his followers (see [2]), while recalling facts and explaining concepts through designing and producing original work – in English.

Apart from addressing their persona's needs, an important criterion adopted during the selection of locations was the places' – relative – significance in the city, and also their vicinity (by the standards of a city whose area is 261.85 km^2).

The students' choices were partly influenced by the material read, but all the activities (e.g. dancing or taking selfies) to be performed at the designated venues were adapted to the curiosity and energy of a teen, while some locations (e.g. the largest shopping mall in the region) were chosen specifically for the persona.

Finally, seven sites were selected:

- Malta Thermal Baths
- ICHOT/Gate of Poznan Interactive Museum
- The Unrecognized, an outdoor collection of sculptures created by the late Polish artist Magdalena Abakanowicz in the year 2000
- The so-called "Bamberka" Fountain from the year 1915
- Old Brewery – a center of art, business and trade
- Kontenery Outdoor Cultural Center
- Posnania – the largest shopping mall in the Greater Poland Region.

2.3 Creating a Map of the Senses

The students were given coloured cardstock paper to make the activity more appealing to their senses. Next, they were asked to use that paper to re-create an approximation of their city centre map featuring just the places they had selected (for that, photographs and drawings were used). Finally, they were to brainstorm what could be seen, heard, tasted, smelled or touched in each location. The activity resulted in a lengthy and heated exchange of opinions – in English. All the students' ideas were then listed beside the appropriate locations, on the "Map of the Senses."

Once that stage was completed, by means of Socratic questioning, and specifically by asking the "why" questions, the author prompted her students to select one specific activity that they viewed as the most unique to each place (but not obviously so), based on the Map of the Senses they created. The approach forced the students to revise their stereotypes about and change their perspective on the city they live in and the places they visit often.

Unsurprisingly, most of the activities were focused on the "see". The teenagers decided the suitable action for the thermal baths would be looking for the figure at the front of the pirate ship and then taking a selfie of oneself in the pose of a pirate. For the ICHOT Interactive Museum, full of glazed surfaces, the students decided it would be interesting to look for a "place with glass" and take a photo to show how the past connects with the present "through the glass." Standing in front of the Unrecognized, depicting a group of armless and headless figures, the "wanderer" was to choose one of them and recreate it, adding arms and a head. At the Bamberka fountain, the visitor was to "modernize" the figure, drawing the sculpture dressed in contemporary clothes. In the Old Brewery, he/she should find a huge chessboard and stand positioned as the Black King at the beginning of the game.

At the Kontenery Cultural Center, where there is always some music playing, the teen was to ask someone their age to dance with them and have the dance recorded. It was only at this point that the "hear" and "touch" rather than the "see" was addressed. Finally, at the Posnania shopping mall, after an eventful day, the tired but satisfied visitor was to indulge in a meal, ordering a traditional Polish treat, pierogi. Here, eventually, the "taste" came to focus.

2.4 Reconstructing the Map of the Senses Online

Upon finishing the offline activities, the stopping places needed to be reconstructed online. To optimize and simplify the process, only the author's ExperienCity profile was used for this purpose. Following the cues and prompts from the platform, online stations were created and edited.

The four main tabs were: Go, Do, Discuss and Info. The required tasks were: to direct the visitor to an individual place, to ask him/her to perform an activity there, to make him/her comment or upload some media that would reflect the "do", and to give him/her additional information that would be relevant and interesting.

When all the stations were complete, the "Create an Experience" option was chosen. Then, following the prompts from the platform, the appropriate area was fenced in on the map, and the relevant stations were selected. The stations were named in what the students presumed to be an appealing way:

- Have a Swim with the Pirates!
- History through the Glass
- Make a Man!
- A Woman with Yokes
- Searching for the Black King
- Dance with a Posnanian!
- Bon Appetit!

The experience was given the same name the students had brainstormed initially: Be Active and Explore. It was described as "Poznan for teens": http://beta.experien.city/viewExperienceDetailsUser.php?toEdit&idexp=3696&saved.

Recreating the walk online may have been the most arduous and time-consuming stage of the project, but it was also very effective towards improving foreign language written communication skills, including grammar and vocabulary, but also style. Additionally, the students' ICT skills and competences were developed.

2.5 Testing the Draft

As soon as the interactive online walk was drafted, it was advertised on the students' English CLUB blog, again in English. The teenagers asked visitors for feedback and suggestions to further improve their walk. Several pieces of advice appeared, which led to the revision and edition of the stroll. One of them was adding a prompt in case the visitor did not know where the position of the Black King was (he/she should ask a passer-by, thereby interacting with the locals). Another one was adding less impersonal information about the thermal baths, which resulted in the teens' specifying which water slides they liked the most.

Each remark was carefully read and responded to by the students, which resulted in further perfection of their foreign language skills.

3 The Outcomes of the Project

The cross-curricular instruction described in the article and the project itself turned out to be a daring, yet very attractive, engrossing and effective, shift in the author's EFL classroom. The students successfully completed all the project tasks, and the arising problems and difficulties, such as temporary exhaustion or lack of motivation, were solved. The greatest motivator in this respect was the fact that the students' blog posts describing the process of crafting the walk were read and commented on, and the teenagers' work had not gone unnoticed.

The level of student creativity and engagement throughout the project was high. The teenagers also changed their attitude towards local culture and history, conducting vast and profound research on the several places of interest they selected. They started perceiving both, the culture and the history, as relevant elements of their heritage. They also believed it could be made attractive to their peers from abroad.

The expected additional benefit was the fact that the students improved their English skills (both passive and active ones) and their ICT competencies.

4 The Conclusions

The author's findings, based on the extensive and comprehensive project presented in the article, are that cross-curricular instruction combining place-based, location-based, quest-based and project-based learning is very attractive and effective in teaching teenagers who are a demanding group of learners. This kind of teaching can be successfully employed in EFL to boost teenage student creativity and engagement; to enhance their project- and group-work skills; and to make their learning process a "process of enculturation, emphasizing the socio-cultural setting and the activities of people within the setting" (Yehiel 2017).[1] In other words, "it is bringing a more social aspect to the classroom material to bring it alive" (Badwal, as quoted by Raths [9], 2015).

References

1. Benjamin, W.: The Arcades Project. The Belknap Press of Harvard University Press, Cambridge (2002). https://archive.org/stream/BenjaminWalterTheArcadesProject/Benjamin_Walter_The_Arcades_Project_djvu.txt
2. Bloom, B.S., Engelhart, M.D., Furst, E.J., Hill, W.H., Krathwol, D.R.: Taxonomy of Educational Objectives: The Classification of Educational Goals. Handbook I: Cognitive Domain. David McKay Company, New York (1956)
3. Elder, J.: Teaching at the edge. In: Stories in the Land: A Place-Based Environmental Education Anthology. Nature Literary Series No. 2. The Orion Society, Great Barrington (1998)
4. Gallagher, S., Sher, B., Stepien, W., Workman, D.: Implementing problem-based learning in science classrooms. Sch. Sci. Math. **95**(3), 136–146 (1995)
5. Haskell, Ch.: Understanding Quest-Based Learning (2013). https://issuu.com/loraevanouski/docs/qbl-whitepaper_haskell-final
6. Hmelo, C.E., Evensen, D.H.: Problem-Based Learning: A Research Perspective on Learning Interactions. Routledge, London and New York (2000)
7. Jenks, Ch. (ed.): Visual Culture, p. 154. Routledge, London and New York (1995)
8. Lane-Zucker, L., Sahn, J.: Stories in the Land: A Place-Based Environmental Education Anthology. Nature Literary Series No. 2. The Orion Society, Great Barrington (1998)
9. Raths, D.: Bringing Location-Based Learning to Life (2015). https://campustechnology.com/Articles/2015/07/08/Bringing-Location-Based-Learning-to-Life.aspx?Page=2
10. Sobel, D.: Place-Based Education: Connecting Classrooms and Communities. Nature Literary Series No. 4. The Orion Society, Great Barrington (2004)

[1] Unpublished slides of the course *Trees, buildings, signs, statues: Public space as a mobile learning environment.*

New Learning Models and Applications

PDM Field Study and Evaluation in Collaborative Engineering Education

Andreas Probst[1(✉)], Detlef Gerhard[2], and Martin Ebner[3]

[1] HTL Ried, Ried im Innkreis, Austria
Andreas.Probst@eduhi.at
[2] Technische Universitaet Wien, Vienna, Austria
[3] Graz University of Technology, Graz, Austria

Abstract. Collaboration in general but especially between students of Austrian Federal Secondary Colleges of Engineering (HTL) is becoming more and more important. Therefore, the joint diploma thesis has been introduced into the curriculum. Furthermore, joint student projects have become a crucial topic within the subject of mechanical engineering design. Because of the worldwide activities of most companies, being able to collaborate within a huge team is seen as an essential for future jobs. To support the collaboration process in engineering education, a product data management (PDM) system was introduced to several Austrian HTLs within projects carried out in the Sparkling Science program of the Austrian Federal Ministry of Science and Research. In the academic year 2016/17 a field study was started to figure out how to enhance collaboration between students by using this kind of software and methods. The outcome of this study will be to find out about the level of collaboration within students' design projects.

Keywords: Engineering education · Engineering collaboration PDM · CAD

1 Introduction

There are several major techniques and methods in engineering design education which are being tested in pilot projects and will be incorporated into student design lessons at Austrian HTLs in the near future:

- Mechanical design education in interdisciplinary teams [1]
- PDM and PLM techniques [2]
- 3D printing, scanning and reverse engineering
- Internet of Things (IoT) [3]
- Systems engineering [4]

This paper gives an overview of the introduction of PDM into mechanical engineering design lessons at Austrian HTLs as well as the conducting of a field study to gather information on the level of collaboration within students' projects.

M. E. Auer et al. (eds.), *Teaching and Learning in a Digital World*,
Advances in Intelligent Systems and Computing 715,
https://doi.org/10.1007/978-3-319-73210-7_49

2 Research Design

The purpose of this research paper is to describe a field study which is being conducted to measure the quantity and quality of collaboration within students' engineering design projects. The first part of the study, called "HTL internal" will be done by approximately 100 students and their corresponding teachers of several HTLs throughout Austria. The main task is a short collaborative engineering design task by designing shafts and gear-wheels into a given gearbox by groups of up to four students. During these tasks all students have to make notes concerning frequency of arising difficulties such as interface issues, as well as the necessary time to solve these problems. Additionally, their teachers are documenting the design task and the occurring problems such as non-working computers. After the study is completed, there will be a measurement of the degree of collaboration by evaluating the PDM database. The second part is a survey amongst HTL students, teachers and industry staff with collaboration and PDM background to find out about the importance of PDM for collaboration in engineering. Qualitative interviews are carried out to get a clear picture of industry's demands of collaborative education. The research results will be presented in this research paper and additionally will be discussed with members of the nationwide HTL working group for mechanical engineering design in Austria (http://www.3d-cad.at). The results may influence the next generation of the HTL curriculum.

3 Activities to Introduce PDM to Mechanical Engineering Design Education

3.1 Research Activities

Since 2006 two research projects concerning PDM have been carried out with TU Wien. Figure 1 shows the two projects BLUME (Basis PDM Lern- und Projekt Umgebung für ganzheitliche Mechatronische Produktentwicklung in German; in English, the learning and education environment for mechatronic product engineering) [5] and PDM-UP (UP – Umweltgerechte Produktentwicklung, in English the "Design for Environment – DfE", also referred to as Ecodesign or green design) [6] together with a survey about the past research projects. 68 people participated in the survey, answering 18 questions overall. The main research question asked "Do the projects conducted by the Sparkling Science program influence teaching practices in engineering education at participating Austrian HTLs and the project members' perception of the program itself?". The participants were classed under different groups, consisting of current and former students, HTL teachers and TU researchers; both those that had and hadn't taken part in either one or more of the previous projects [4]. For further research projects it is planned to evaluate the participants' knowledge of project topics at the beginning and at the end of upcoming projects to get an idea about the change of their knowledge.

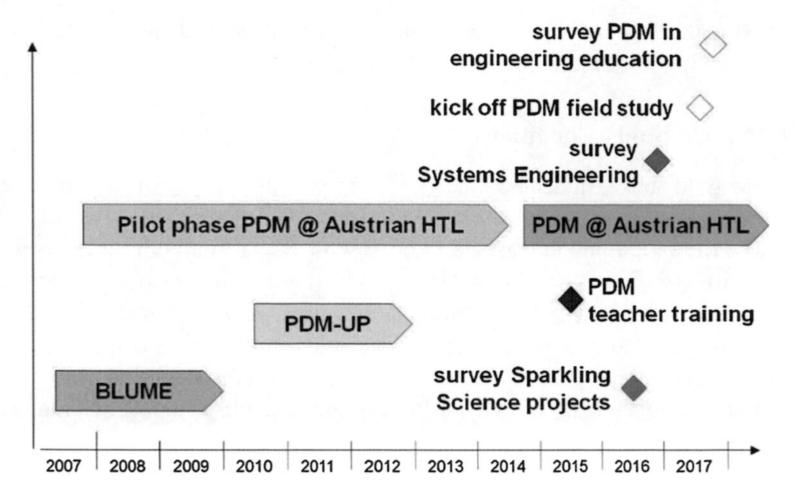

Fig. 1. Research activities and introduction PDM at Austrian HTL

3.2 Introducing PDM to Austrian HTL

After conducting two research projects in 2015, teacher training for using PDM in classes was organized with the aim of encouraging the use of PDM within mechanical engineering design (see Fig. 1). Some HTLs use PDM very often in their design lessons, mainly those who have attended the research projects BLUME and PDM-UP. Because persuading people of the benefits of using PDM tools is quite difficult, it was decided to conduct a field study to find out more about the usage of PDM systems, as well as to figure out the benefits of using PDM for mechanical engineering design at Austrian HTLs. A project group consisting of HTL teachers had chosen an example of a single stage transmission and adopted it for a field study. Additionally, the conditions of testing were elaborated which are described in this paper.

4 PDM Field Study Setting and Boundaries

The experimental design is strongly following a field-study-design; in with other words, we let research happen as realistically as possible. Therefore, a typical environment has been chosen where PDM is used amongst students. The main characteristic of a field study [7] is that the experiment itself is placed in a real-life scenario and is not influencing it in any way [8]. On the other side we have to bear in mind that unexpected situations can happen and change the outcome instantly. In this research work schools have been chosen where PDM is or will be used in near future and teachers were instructed to create a real-life situation as well as possible.

4.1 PDM Field Study at Austrian HTL

The field study is being conducted within ten HTLs spread all over Austria, and for each HTL there are different groups of students: there are students and teachers who have worked with PDM for several years, and there are other students and teachers who have been working with PDM for less than one year.

4.2 PDM Field Study Conditions

For the PDM field study, students in the class are split into groups of four and together must solve the assignment using their CAD program in combination with the PDM software. The time for doing the tasks is defined with a maximum of four hours without any information on how to collaborate. The given time seems to be short but the students are used to designing and calculating a multistage gear transmission, which is much more complicated compared to the PDM field study task of a single stage gear transmission. This ensures that students can concentrate on the field study task and how to collaborate together on it. Within the task students have to build 3D models and production drawings within their CAD program and save everything into the PDM database.

4.3 Design Task and Deliverables

The overall design task for students is to design a single stage gear transmission (see Fig. 2 [9]). The gearbox housing is already given as a 3D model and stored within the PDM project database.

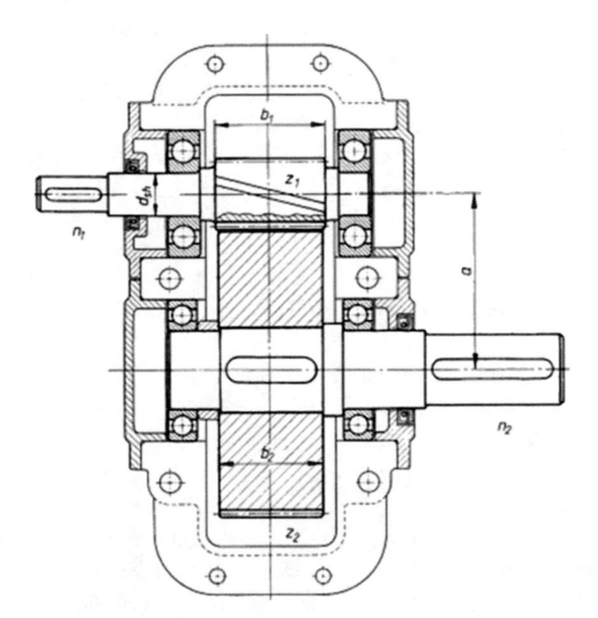

Fig. 2. Design task single stage transmission [9]

The deliverables for the PDM field study are displayed in Table 1.

Table 1. Deliverables of PDM test project

	3D model	2D drawing
Pinion shaft	Yes	Yes
Gear wheel	Yes	Yes
Output shaft	Yes	Yes
Bearing cover	Yes	Yes
Gear assembly	Yes	Yes

To ease the design process, the number of gear teeth as well as the gear module is given. Students do not have to calculate the parts concerning stress. Interface information like gearwheel bore and the corresponding output shaft diameter has to be figured out between the members of a test group, on the basis of given geometric information of the gearbox. During their group work, students have to split up the tasks between themselves as well as decide on the level of collaboration. The students are also informed that their project will be marked like a standard mechanical engineering design project. All field study projects get graded; the basis for the assembly grades are that the assembly function is given. For the grading of the parts the basis is:

- Parts are producible
- Dimensioning of the parts are complete
- Expedient surface symbols and tolerances are on the drawings
- No geometrical dimensions and tolerances are needed in the drawings

4.4 PDM Setting for Field Study

For the field study specific PDM project structure and settings are created within the PDM software Windchill from PTC, which is used for all projects. All PDM study projects are named “2017-06_PDM-Testung-HTL-nnn”, where “nnn” is the organization, therefore “2017-06_PDM-Testung-HTL-Zeltweg” means the test project of HTL Zeltweg in Styria. Concerning file structure there are data folders defined within each PDM project numbered from “Schüler_01” to “Schüler_10”, where the folder “Schüler_01” contains all the data of the first test group and “Schüler_04” contains all the data of the forth test group. The folder structure is shown in Fig. 3.

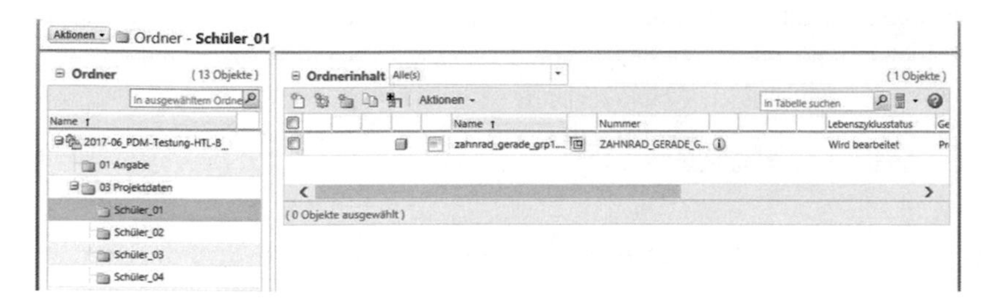

Fig. 3. Project structure of PDM test project

To ensure that every member of each test group has only access to their own data the security function of Windchill is used. The students of each test group are connected to a role, therefore the members of the role "Schüler01" only have access to the folder "Schüler_01". This setting is displayed in Fig. 4.

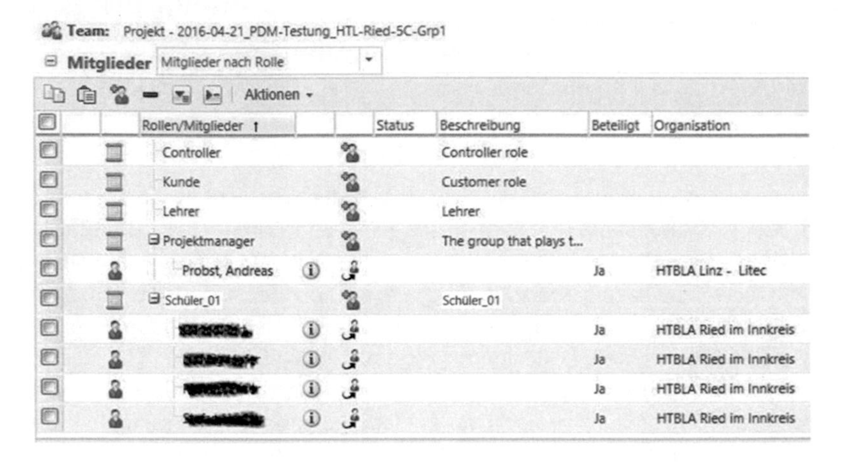

Fig. 4. Team definition of PDM test project

4.5 Students' and Teachers' Notes

During the PDM field study the students have to note down several effects which are evaluated afterwards, as shown in Table 2. Despite the raw database data, this gives the opportunity to get some additional information from the field study.

Table 2. Overview pattern of students' notes for PDM field study

No.	Measurable effect	Collected data	Remarks
1.	Improvements within data conflicts	Number of occurrence	
		Time for correction	
2.	3D interface problems	Number of occurrence	3D parts and assembly do not match
		Time for correction	
3.	Coordination concerning 2d working draft	Number of discussions	Students have to carry out and adapt a working draft
		Time for each discussion	
4.	Working on the same parts or assemblies	Number of occurrence	Due to a PDM concept only one person can work on one part at one time
		Time for correction	
5.	Individual students notes	Effects and remarks	

Teachers also have to note down several effects to be evaluated during the PDM field study which are shown in Table 3. This gives additional information to the evaluated data, for example if a test group generates no data, there could be different causes for this: perhaps the network did not work, or some students were ill and could not attend the field study.

Table 3. Overview pattern of teachers' notes for PDM field study

No.	Measurable effect	Collected data
1.	Set up PDM field study project	Durance
2.	Invite students to test project	Durance
3.	Start PDM field study project	Time
4.	CAD program started	Time
5.	Individual students workspace is generated	Time
6.	Start of coordination concerning 2d working draft	Time
7.	End of coordination concerning 2d working draft	Time
8.	End of PDM field study	Time
9.	Individual teachers notes	Effects and remarks

5 PDM Field Study Evaluation

5.1 PDM Database Evaluation of Collaboration

From the beginning of the summer term 2017 until the end of the fall term 2017/18, ten HTLs with approximately 120 students are participating in the field study. Every HTL class is spilt into groups of four students, with each group working on one PDM project. The generated CAD and PDM data of each student group is collected and all the data together is evaluated anonymously. Table 4 shows an evaluated example of a PDM project of a test group of four students. For instance, the third row shows the students have created the deliverable of the gear assembly and finished the task

Table 4. Evaluation of collaboration with file versions from PDM database

HTL-nnn – Group-x		Task % finished	File version								
Deliverable	File type		1	2	3	4	5	6	7	8	9
Gear assembly	Assembly	100	B	B	B	B	B	B	B	B	
	Drawing	100	B	B	B	B					
Output shaft	3D part	100	D	D							
	Drawing	100	D								
Gear wheel	3D part	100	B	B	B	B	B	B			
	Drawing	0									
Pinion shaft	3D part	100	A	A	A	A	A	A	A	B	B
	Drawing	100	A	A	A	A	A	A			
Bearing cover 1	3D part	100	C	C							
	Drawing	90	C	C							
Bearing cover 2	3D part	100	C	C	C						
	Drawing	100	C								

completely. In the table “A” means student A, “B” student B etc., so obviously only student B has worked with this file.

5.2 Research Questions

There are three main research questions which are investigated:

1. Can collaboration be identified in the student groups?
2. What difficulties occurred during the field study?
3. How many tasks were completed in the given time?

For the first research question, the file versions of the PDM database are collected. Looking at Table 4 collaboration occurred only within creating the 3D part of the pinion shaft, student A generated seven versions of the file while student B generated the last two file versions. In contrast the 3D part and drawing of the bearing cover are created only by student C. This student group did not collaborate that much. Evaluating all data of all student groups, this will give a clear picture about collaboration and usage of the PDM system. Adding the information on how long the students are working with PDM for will complete the picture.

The second research question concerning difficulties, answers if collaboration did or did not happen due to several causes. For example, if one student did not attend the field study for example because of illness, looking only at the PDM database it seems that only three out of four students were working with the field study. Therefore, the teachers’ observation and notes will give additional information for the evaluation results.

The third research question about the completed tasks within the given time will determine, together with the evaluated PDM data, whether there is a correlation between collaboration of students and percentage of task completion.

6 Conclusion and Outlook

In this research study the settings and evaluative methodology for a PDM field study have been presented. Because the field study is running during the writing of this paper, the first results will be presented afterwards.

We would like to thank to Sparkling Science program for making it possible for, students to use PDM for collaboration tasks in their engineering design lessons. PDM was introduced at several Austrian HTLs and due to the increasing complexity of engineering design tasks and the industry’s demand, more HTLs will follow suit. PDM will become more and more important as a centralized database for CAD data as well as for all Internet of Things data. Therefore, the effort of adapting and maintaining a PDM system as well as conducting a field study is significant. Based on the results and experiences made so far, the authors of this paper are convinced that the quality of design lessons and education in mechanical engineering will improve and will be closer to demands of the industry’s needs.

References

1. Bitzer, M., Burr, H., Eigner, M., Vielhaber, M.: Integrated concepts of design education. In: DS 46: Proceedings of E&PDE 2008, the 10th International Conference on Engineering and Product Design Education, Barcelona, Spain, 4–5 September 2008
2. Gerhard, D., Grafinger, M.: Integrative engineering design using product data management systems in education. In: Global Engineering Alliance for Research and Education (GEARE) - A Comprehensive Study & Intern Abroad Program for Engineering Students (2009)
3. Köhler, M., Wörner, D., Wortmann, F.: Platforms for the internet of things–an analysis of existing solutions. http://cocoa.ethz.ch/downloads/2014/02/1682_20140212%20-%20Bocse.pdf. Accessed 24 May 2017
4. Probst, A., Gerhard, D., Bougain, S., Nigischer, C.: Continuous research and development partnership in engineering education. In: Proceedings of the 19th ICL Conference. Springer (2016)
5. Probst, A., Gerhard, D.: PDM supported engineering design education. In: Proceedings of 2014 International Conference on Interactive Collaborative Learning (ICL), pp. 117–120 (2014)
6. Ostad-Ahmad-Ghorabi, H., Rahmani, T., Gerhard, D.: Implementing PDM systems in design education to enhance design collaboration. In: Buck, L. (ed.) Design Education for Future Wellbeing: Proceedings of the 14th International Conference on Engineering and Product Design Education. DS/Design Society, vol. 74. Artesis University College, Antwerp, Belgium, pp. 53–57. Design Society, Glasgow, 6–7 September 2012
7. Fößl, T., Ebner, M., Schön, S., Holzinger, A.: A field study of a video supported seamless-learning-setting with elementary learners. Educ. Technol. Soc. **19**, 321–336 (2016)
8. Ebner, M., Schön, M., Neuhold, B.: Learning analytics in basic math education – first results from the field. eLearning Pap. **36**, 24–27 (2014)
9. Matek, W., Muhs, D., Wittel, H.: Roloff/Matek Maschinenelemente Aufgabensammlung, 6th edn. Friedr. Vieweg & Sohn, Braunschweig/Wiesbaden (1984)

Soundcool Project: Collaborative Music Creation

Elena Robles Mateo[✉], Jaime Serquera, Nuria Lloret Romero, and Jorge Sastre Martínez

IDF-iTEAM, Universitat Politècnica de València, Valencia, Spain
elrobma@bbbaa.upv.es

Abstract. This paper addresses four criteria that the Soundcool project meets: to "be sustainable", "be future-oriented", "be transformative" and "be innovative". Soundcool is a pedagogical and technological project. A brief description of the technology behind Soundcool will be useful for the reader before addressing the four criteria. Soundcool is like a "Lego" for sound; Soundcool is composed of a series of software modules that run on a central computer, or host computer. Each module is sort of a musical instrument; it could be a synthesizer, a sampler, a sound effect processor, etc. these modules can be interconnected in different ways allowing the users, i.e. the students, to create their own arrangements, as we call the module creations and interconnections. Then, each module can be controlled either with the mouse or, what is more interesting, with a mobile device through WiFi. This way, every student can control one or several modules of the whole arrangement from their mobile device contributing to a collaborative and participative experience.

Keywords: Collaborative creation · Sound creation · Modules · Innovative Transformative · Sustainable · Future-oriented education · Music education Pedagogic · Kinect · MAX-MSP · Mobile devices · Tablets

1 Soundcool Project

1.1 Soundcool is Sustainable

Soundcool is sustainable from two perspectives, the technical and the pedagogical. Regarding software, sustainability can refer to different concepts, one of which is being a long-lasting program. This is the case of Soundcool for several reasons. The project began in 2013 and today, in 2017 we are still doing amazing activities with students outside the classrooms. For example a recent Soundcool performance was carried out at the TEDxUPValència event held in February 2017 in Valencia, Spain. The Soundcool software will be maintained in the future because, in spite of being a free software, the technical university where it is developed, the Universitat Politècnica de València (UPV, Spain), can provide a great number of motivated students to carry out their final projects or internships within the Soundcool team. We always have young people interested in doing technological developments for this project. We have managed to maintain the software for Windows and Mac, adapting it to new versions every year. Apart from maintenance, we are also constantly including new features.

M. E. Auer et al. (eds.), *Teaching and Learning in a Digital World*,
Advances in Intelligent Systems and Computing 715,
https://doi.org/10.1007/978-3-319-73210-7_50

Since music is a universal language, Soundcool is also global, and its characteristics allows it to be used by any community in the World. In the future, when more users will adopt Soundcool, we are planning to adopt a similar sustainable model to that of the famous audio software Audacity (http://www.audacityteam.org/). And we are confident we will be able to implement it successfully because the co-creator, Prof. Roger Dannenberg, is an active collaborator of Soundcool from the very beginning of the project. Soundcool was born in 2013 with a UPV research project and a Spanish ministry grant for Dr. Jorge Sastre to visit CMU as a researcher at Dannenberg's group. In this visit Dr. Sastre discussed with Dannenberg about the conception of Soundcool and gave the first talks about the system at CMU (Pittsburgh, USA).

Soundcool is sustainable also because it does not require expensive equipment. Any computer can run Soundcool on its own with a mouse and speakers. The use of smartphones and tablets is optional, and they can be inexpensive Android phones or tablets which the students already own, and iOS devices are also allowed.

1.2 Soundcool is Future-Oriented

The school of the 21st century opens the doors to the technological progress more and more every day. This advancement has to go hand-in-hand with pedagogical approaches and didactic strategies that allow our kids and teens to integrate the learning process in an innovative, functional, social and competent way. In this sense, Soundcool is a tool capable of developing creative and artistic thinking, and thus, the students can manage skills at the level of divergent thinking, which is essential when our aspirations are, for example, to develop active citizens as well as to train the future entrepreneurs of our society.

The incorporation of new technologies for music education at primary and secondary levels opens up great possibilities to improve the teaching and learning processes. New interfaces for human-computer interaction such as multi-tactile surfaces, smartphones, tables and also systems for collaborative creation represent promising tools to improve the motivation and the interest of the students, to develop their cognitive skills and to support the educational process. However, still nowadays in many countries, the most extended kind of music education continues to be entrenched in the traditional musical language and the conventional use of instruments commonly found in the classroom, such as the recorder, within a system where classes are mainly traditional lectures with little discussion or participation of the students. In this sense, Soundcool is future-oriented since it is based on the use of mobile devices by the students. Tactile devices allow direct manipulation of objects and controls by means of multi-tactile interactions with excellent results in electronic musical instrument design and also in educational projects in general. Additionally, these devices are in fact already part of our daily life. The worldwide mobile phone market has followed a remarkable trend and it will continue growing in the following years according to reports of the International Data Corporation (IDC). The number of mobile devices has already overcome the number of all humans.

Soundcool is also future-oriented because it includes pedagogical work and technical developments for people with special educational needs. It is believed that the level of development of a society can be estimated by how it deals with disabled people. The societies of the future will be more and more inclusive every day. In this sense, Soundcool team's aim is the training of music teachers, therapists, relatives and everyone who works and shares their lives with persons with functional diversity using music as an educational and motivational tool. This is why a research group has emerged inside Soundcool's team. It is called Emosons (from emotion and sound) and it focuses on pedagogical and music-therapy developments and also aims at addressing the usability and implementation of the Soundcool application in groups with functional diversity, to introduce the application as a therapeutic tool in the set of playful and creative activities of the "La Torre Occupational Centre" (Valencia, Spain), where the tests are currently being conducted.

1.3 Soundcool is Transformative

Soundcool has an enormous potential to transform education worldwide, in a universal way, by offering an education for all. Its easiness of transference allows for its replication in different parts of the world. Soundcool can be used by any person regardless of their cultural and socio-economical level, by using low-cost smartphones and PCs or recycled ones. Ongoing investigations show that the use of Soundcool is equally satisfactory by both participant groups of people with music knowledge and people with no musical training. Therefore, the system allows any user to express creatively without a previous specialized training.

Soundcool has the capacity of breaking physical barriers and transforms the spatial sense of places for sound-art creation; beyond the traditional classroom, Soundcool encourages experimentation transcending boundaries, global on-line work, and the use of "lab" environments that generate spaces for learning based on experimentation. This technology will generate significant changes in the music education area. On the Soundcool website we are designing a collaborative environment (http://network.soundcool.org) to facilitate the exchange of ideas and creations of all the users.

We are seeing by ourselves how teachers are transforming their view towards music education in the courses and workshops that we are delivering across all of Spain and the countries that participate in the Erasmus + European project (explained below) such as Portugal, Italy and Romania. We know that apart from the technical aspects of Soundcool, a change in the mentality is necessary regarding the didactic methodologies and teaching and learning approaches. In relation to teacher training, our interest is focused on the development of creative strategies that boost a new approach from a new paradigm that sees the students as creators and not simply as information receptors. In this sense, Soundcool has been presented by the authority responsible for education in the Valencia Community (Spain) as a promising system that can help develop the competencies needed for the education of our students in the foreseeable future.

Soundcool is also transforming the practical activities of the students at Primary and Secondary levels. With Soundcool they take part in concerts outside the classroom in auditoriums and other public venues like the Principal Theatre or the Palau de les Arts in Valencia (Spain), which is a great complementary activity for their education. In the TEDxUPValencia event on Feb. 17, 2017 Soundcool was presented and the students performed the piece Metropolis, based on the sounds of the city. In addition, a large-scale Opera entitled "La Mare dels Peixos" ("the Mother of the Fish") has been premiered at the Opera House of Valencia, "Palau de les Arts", on 16th Dec., 2016.

The above mentioned Erasmus + European project has also contributed to transform the education using Soundcool. The project is a European strategic partnership for school education KA2: "Technology for learning and creativity: weaving European networks through collaborative music creation" involves seven organizations in the field of education and culture from Spain, Portugal, Italy and Romania (primary, secondary and music schools). This project emerges from a vision to use arts education and especially music education to transform the classroom by incorporating new methodologies based on Soundcool system in order to enhance the development of skills needed for the twenty first century society such as creativity, digital literacy, cultural and linguistic competence and team work.

1.4 Soundcool is Innovative

The Soundcool innovative contribution to education has been recognized with several prizes. We have to say that, although what we propose is a total paradigm shift and involves a profound transformation in the ways of teaching music in classrooms, it is not our intention to break with tradition, but rather to expand the possibilities and the sound palette and thus the music education.

Pedagogically, Soundcool is based on placing students at the center of learning, giving them the opportunity to be the creators of their own creations. These ways of "learning from doing" and learning between equal individuals are in perfect agreement with the latest advances in theories of learning. The teacher's role as a totem shifts towards a much more horizontal view of learning. Now the teacher's role has to be that of a facilitator and builder of creative spaces.

Expanding teaching music based on music history and playing traditional instruments, Soundcool allows the creation of an entire orchestra of the 21st century through mobile devices. The ability to imagine new sound universes is now infinite.

Acknowledgements. This work is supported by Daniel & Nina Carasso Foundation, project 16-AC-2016 and Cátedra Telefónica-UPV 2016-17.

References

Sastre, J., Cerdà, J., García, W., Hernàndez, C.A., Lloret, N., Murillo, A., Picó, D., Serrano, J.E., Scarani, S., Dannenberg, R.B.: New technologies for music education. In: Proceedings of the 2nd International Conference on e-Learning and e-Technologies in Education (ICEEE) Ed., pp. 149–154. IEEE (2013)

Wright, M.: The Open Sound Control 1.0 Spec., V. 1.0, 26 March 2002. http://opensoundcontrol.org/

Regelski, A.T.: Music and music education: theory and praxis for 'making a difference. In: Lines, D.K. (ed.) Music Education for the New Millennium: Theory and Practice Futures for Music Teaching and Learning. Blackwell Publishing, Malden (2005)

Nicolls, S.: Seeking out the spaces between: using improvisation in collaborative composition with interactive technology. Leonardo Music J. **20**, 47–55 (2010)

Miranda, A., Santos, G., Stipicich, S.: Algunas características de investigaciones que estudian la integración de las TIC en la clase de ciencia (Some characteristics of research exploring the integration of ICT in science class). Revista Electrónica de Investigación Educativa, **12**(2) (2010). http://redie.uabc.mx/index.php/redie/article/View/259

Almirón, M.E., Porro, S.: Las Tic en la enseñanza: un análisis de casos (ICT in education: an analysis of cases). Revista Electrónica de Investigación Educativa **16**(2), 152–160 (2014)

Zabala, V., Arnau, L.: Métodos para la enseñanza de las competencias (Methods for teaching of skills), Barcelona, Graó, pp. 59–81 (2014)

The Application of Action Research to Review Modern Techniques in Manufacturing Systems

Sungyoul Lee(✉)

Department of Engineering Management, College of Engineering, Prince Sultan University, Riyadh, Kingdom of Saudi Arabia
lsungyoul@psu.edu.sa

Abstract. This is a case study of action research project applied to review modern techniques in manufacturing systems. It is often considered as one of learning outcomes in engineering courses for the engineering student to have insights regarding recent technologies applied in manufacturing fields. However, it is a challenge for the instructor to achieve the goal in a given semester hours while delivering scheduled subjects. Therefore this study proposes a teaching approach using an action research concept as a tool which motivates and engages more to the students as well as achieves the learning outcome to review modern technologies in limited semester hours. In other words, the same subject can be exposed to the students two times; once when they review it through searching recent journals by themselves and again when it was covered by the instructor according to syllabus schedule. Consequently, the findings from the action research have been summarized. The proposed approach shows how action research application in engineering disciplines is an effective tool to motivate learner's abilities to review recent technologies.

Keywords: Action research · Modern technology · Formative learning Saudi Arabia

1 Introduction

Due to rapid technological development, it is worth to review recent technological trends in manufacturing systems. In the Engineering Management program of study, students are frequently asked to review recent technologies in the field of their specific course subject. Ever since fall 2014, I have taught a course entitled Production Planning and Control (PPC) which is a junior level standing course at P University in Saudi Arabia. This course addresses basically fundamentals of production planning and control in manufacturing systems. One of the course learning outcomes (CLO) in this specific course is to review modern techniques in the field of manufacturing systems. The class size is usually 5 to 20 students. The teaching method that I have applied to this specific course is a classical way which a teacher deals with the subjects chapter by chapter. Looking back on teaching this course for last four semesters, I had literally two difficulties to achieve this specific CLO; limited resources to cover from the textbook and limited times to address various technologies. Although it was one of course outcomes

M. E. Auer et al. (eds.), *Teaching and Learning in a Digital World*,
Advances in Intelligent Systems and Computing 715,
https://doi.org/10.1007/978-3-319-73210-7_51

to be covered during the semester, I always don't feel comfortable at the end of the semester because I couldn't achieve this outcome as expected. It was difficult to achieve the goal with the current setting of delivering subject chapter by chapter.

While I was thinking to overcome these obstacles, I realized that the best way to achieve this outcome is to review most recent journals of the corresponding subject areas as students' project. Once I decided to take action research with this topic, I explained the students about the action research we are going to go through. The students had interests and agreed to take part. The PPC course has enrolled 9 students in this semester. There are five broad areas of production planning and control based on the current syllabus. Four groups which have three groups of two students and one group of three students have been formed. Thus, the four groups chose four areas out of five according to their preferences as possible. In order to review each of four fields in relatively limited time, it is desirable to share the areas by student groups so that each group is in charge of surveying one area. Another important reason why I choose and plan this way is to expose the students the same subjects two times through the semester. What that means the four subject areas that each group investigates will be addressed again based on the syllabus schedule by the instructor.

I believe that the formative learning approach learning by themselves first will engage and motivate more to the students when they learn the subject later. The students present their findings after surveying recent journals in the fields of the selected subjects. Through the student's presentation and following discussion, the CLO can be achieved completely in terms of covering various selected areas in relatively short time.

2 Action Research Practice

2.1 An Overview of the Project

The first problem we faced was that the student is lacking how to search their material efficiently. Their primary source to search was the internet. However, they could not find any journal paper surfing from the internet. To cope with this issue, I showed them how to log in the central library of P University and potential relevant sites such as ProQuest, Sage Journals, and ScienceDirect. The small size class like this was suitable to do this walk-through method. After that, every student could bring the required material next class. The second problem faced that I expected was that the students have no basic knowledge about their topic. For this issue, I tutored briefly about their topic to every four group in my office using office hours. This gives us to reduce a gap between the teacher and the student. While in tutoring, we could talk about some personal issues such as the reason for being late and questioning other subjects.

The focus of this action research (AR) is on two things; achievement in one of the course learning outcome and improvement in the students' self-study ability, motivation, and engagement. The CLO stated as 'Research the modern techniques in managing manufacturing systems' was not properly achieved since the last four semesters that I have taught the PPC course. It was because the traditional way of lecturing chapter by chapter could not handle the various technologies applied in the fields of the selected subjects within limited times. Hence, the goal of this action research is to achieve the proposed outcome efficiently in a given time through investigating associated recent journals by the students' project. A collaborative approach is the best way to proceed when practicing action research [2]. Thus, this idea was also confirmed by one of my colleagues who had AR experience last year as a peer reviewer.

Instead of simply getting knowledge from the instructor, the student gets a good chance to study by himself for a new subject. It means that the student gets two times of exposure to the same subject firstly by himself in advance before it is addressed by the teacher and secondly covered by the instructor through the semester. This way gives the student more motivation to get engaged on the subject and ultimately a better chance for the student to know the subject more thoroughly.

2.2 Instructor's Intervention

In this course, I used to address more than five major subjects such as forecasting, aggregate planning, inventory control, supply chain management, and operations scheduling. After I decided to do action research on reviewing modern techniques in manufacturing systems, I designed the four inter-related stages; plan, act, observe and reflect according to Coats [1].

I prepared four topics for the students to choose to review the recent technology in regard to the topic since there were four groups of students formed by themselves. The three groups of two students and one group of three students have been formed out of 9 students enrolled in this course this semester. Each group has to choose one subject by their own interest. The task given to the group is to do recent literature review regarding new technology prevailing in their subject area. Once surveyed, every group has to present their findings and discuss any issues together with the rest of students.

The project period to review and present their findings was around two months. During the period of the project, we have around 10 min Q and A session at the beginning of each class hour to proceed and guide the project in right direction. The session gave students a good opportunity to share and discuss the similar problems they faced. In this way, students have gradually achieved the required CLO successfully.

2.3 Outline What Happened

Before starting AR, I explained to the students why we are doing this and what we are going to do and received their consent. As previously mentioned, the issue is to achieve successfully reviewing modern techniques as various as possible in manufacturing systems through the course work. The action research has been performed step by step based on the following four step inter-related activities:

1. Plan: to solve the issue, I planned two months period project where four groups of students investigate their assigned subject area respectively. I assigned a subject area incorporated with their interests for each group. The plan includes pre-decided progress report due dates and how to review and summarize the journals.
2. Act: instead of teacher's lecture type of reviewing some limited areas, four groups of students review four areas mainly from the digital journal sources of the central

library website in P University and share their findings through presentation and discussion at the end of the project.

3. Observe: initially, the students showed a lack of searching ability where to find the appropriate journal materials of their concerns. With a slight guidance along with a demonstration by the teacher, they could find the relevant source from proper journals. Based on their presentation and summary reports, the results were encouraging in terms of various topics they found except for difficulty understanding the concepts of some new technologies.
4. Reflect: when they were asked to summarize their findings in journal format, no students prepared as required for the first time. This issue needs to be resolved as a further action research subject. Other than that, I could complete this action research successfully in terms of achieving the course learning outcome saying that review modern techniques in manufacturing systems within limited semester hours. I am willing to use the same approach in other courses as well.

2.4 Results

I believe that educators should take the role of facilitator rather than material deliver when working with their students. In such a role, we would provide information and guidance aimed at supporting students to participate actively. While conducting my own action research project, I encountered that the students were lacking at searching relevant material relating to their specific topics. They were asked to find three journal papers related to their topics from the recent 5 years publication. I realized that all of 9 enrolled students tried to use the internet as their only source of information and failed to find the right ones. To overcome this issue, I decided to apply the experiential learning method, by which students learn by doing, in other words, they learn through reflection when researching relevant material [3]. I showed them the procedure of searching for some example journals step by step in central library website of the P University. I also encouraged them to work together since I observed they were facing similar issues often. It helped them to share their knowledge and resolve the issues relatively in short time. After that, they could have hands-on experience to find their own concerned journals successfully by themselves. By doing so, students would develop self-confidence and would be equipped to find the information needed from the relevant periodicals throughout their careers.

Upon completion of the project, a questionnaire was distributed to the students asking them how they felt about their experience and further improvement. The questionnaire was developed with help of my colleague as a peer reviewer. As shown on the Table 1, most of them agreed positively that they got some insights of modern techniques applied in manufacturing systems through this project and were willing to use the same approach when they need to review any other topics in future. In overall comments, most students agreed that the project was useful to connect between subjects we cover in class and real life as well as get insights about modern technologies.

Table 1. Students' response after implementing the Action Research project

No.	Questions	Number of students who answered yes	Percentage who answered yes
1.	Did you have some insights about modern techniques applied in manufacturing systems through this project?	9	100
2.	Is this project helpful for you to engage in the same subject when it is covered again by the teacher?	9	100
3.	Are you going to use the same methodology in other courses?	8	89
4.	Are you going to share this approach with your friends?	9	100
5.	Did you meet any difficulties during the project? If yes, what are they?	8	89
6.	Overall any comments regarding the project?	N/A	N/A

3 Conclusions

I decided to apply the experiential learning method [4], by which students learn by doing, in other words, they learn through reflection when searching relevant material. I showed the students how to log into the central library website and search for the relevant material relating to specific topics, in the same way, those students would if they would need to search other topics.

Based on their summary reports and presentations, the results were encouraging in terms of reviewing what modern technologies applied in specific areas except for difficulty to understand some details of the new technologies which I had to supplement in the discussion. Also, in free talking after the presentation, most students agreed how they motivated and engaged they felt after this project compared to typical lecture-type of learning.

Before doing action research I had little knowledge about educational theories. Through the experience of this action research, I found new ways of enhancing my practice. In the future, I intend to apply the concept of action research to any issue that I face in teaching since I believe in continuous self-improvement as an educator.

Acknowledgement. The author would like to thank Prince Sultan University for supporting this research.

References

1. Coats, M.: Action Research: A Guide for Associate Lecturers. The Open University (2005)
2. Carr, W., Kemmis, S.: Becoming Critical: Education, Knowledge and Action Research. Falmer Press, Lewes (1986)
3. Felicia, P.: Handbook of Research on Improving Learning and Motivation, p. 1003. IGI Global, Hershey (2011). ISBN: 1609604962
4. Kolb, D.A.: Experiential Learning: Experience as the Source of Learning and Development. Prentice Hall, London (1984)

Analysis of Behaviors of Participants in Meetings

Eiji Watanabe[1(✉)], Takashi Ozeki[2], and Takeshi Kohama[3]

[1] Konan University, Kobe 658-8501, Japan
e_wata@konan-u.ac.jp
[2] Fukuyama University, Fukuyama, Hiroshima 729-0292, Japan
[3] Kindai University, Kinokawa, Wakayama 649-6493, Japan
http://we-www.is.konan-u.ac.jp

Abstract. In this paper, we analyze the behaviors of participants in two types of meetings (brainstorming and decision-making). First, we introduce the use of participant behavior based on facial movement. Next, we propose a method for modeling the behaviors of participants based on multi-layered neural networks. Lastly, based on our experimental results, we discuss the relationships between the meeting phase, participant behaviors, and the model parameters in these two types of meetings. Our results show the parameters in the above models to be strongly related to the behaviors and ideas of the participants in these two types of meetings.

Keywords: Meeting · Participant · Behavior · Image processing Modeling · Neural network

1 Introduction

Meetings can be roughly categorized into the two style types-one whose focus is information sharing (communication and report) and the other problem-solving (decision-making and brainstorming). When brainstorming, participants must adhere to the following rules: (i) make no comment or judgement, (ii) emphasize quantity more than quality. As such, participants must evaluate ideas by monitoring the behaviors of fellow participants. Shinnishi et al. proposed a system for evaluating meeting activity based on the movement and speech of participants [1]. McCowan et al. discussed the recognition of group actions using the hidden Markov model (HMM) [2]. On the other hand, in decision-making meetings, some participants give opposing and others approving opinions with respect to those of the speaker. Otsuka et al. discussed the estimation of the conversation structure based on the non-verbal behaviors of participants during conversation [3]. However, at the same time, some participants indicate their options in their expressions and facial movements without voicing their opinions.

Therefore, it is important for meeting participants to analyze the behaviors of their fellow participants while considering their decisions regarding their own

M. E. Auer et al. (eds.), *Teaching and Learning in a Digital World*,
Advances in Intelligent Systems and Computing 715,
https://doi.org/10.1007/978-3-319-73210-7_52

opinions and to assess the progress of the meeting. If we can develop models to represent the behaviors of meeting participants, we can evaluate the relationships between behaviors and the influences of meeting participants. Moreover, we can easily apply models of participant behaviors to the analysis of the behaviors of students in collaborative learning.

In this paper, we analyze the behaviors of participants in two types of meetings (brainstorming and decision-making). First, we introduce the use of participant behavior based on their facial movements. Next, we propose a method for modeling participant behaviors based on multi-layered neural networks. Lastly, based on our experimental results, we discuss the relationships between the meeting phase, participant behaviors, and the model parameters with respect to these two types of meetings.

2 Detection of Participant Behaviors

2.1 Decision-Making Meetings [4]

In decision-making meetings, we can observe participant behaviors with respect to eye contact and facial movements and we can observe the facial regions of participants via video images recorded by a camcorder in the center of the room [5]. By counting the number of pixels in each facial region (upper-left, upper-right, lower-left, lower-right), as shown in Fig. 1, we can obtain the ratios of each region [6] as follows.

$$R_{UL} = \frac{1}{S} \sum_{x,y \in UL} f(x,y), \quad R_{UR} = \frac{1}{S} \sum_{x,y \in UR} f(x,y), \tag{1}$$

where S denotes the size of the facial region. When the object pixel $f(x,y)$ is skin-colored, $f(x,y)$ is set to 1.

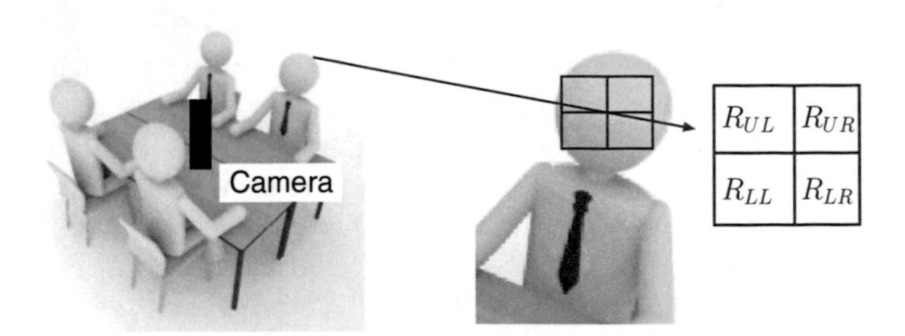

Fig. 1. Observation of facial movements based on the number of pixels in facial regions.

We define the feature R^{DM}_{face} concerning the facial movements of participants by $R^{DM}_{face} = \sum |R_{UL} - R_{UR}|$. Since participants often touch their chins, we focused on the upper facial region of each participant. Here, R^{DM}_{face} denotes facial movement in the horizontal direction and does not apply to eye movement.

2.2 Brainstorming Meetings [7]

We consider brainstorming to be when participants write their ideas on a whiteboard (WB) whereby we can observe the participants writing their ideas on a WB and their corresponding reactions. Therefore, we evaluated participant behaviors in a brainstorming meeting based on their WB writing behavior and their subsequent facial movements (reactions).

Figure 2 shows a schematic of the process of detecting various writing behaviors in video-image time sections, where f_i denotes the ith image. Therefore, we can detect writing behavior in section T by a participant by the difference between images. Moreover, we can evaluation the ratio WB^{BS} of the writing behavior by $WB^{BS} = S_{Body}/S_{WB}$, where S_{Body} and S_{WB} denote the participant's body size in front of the WB and the size of the WB, respectively.

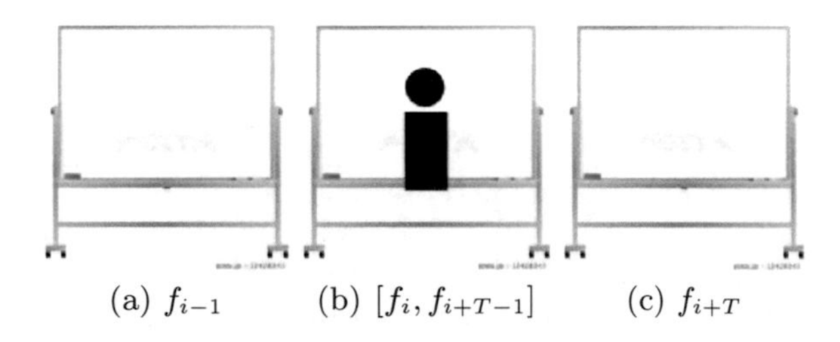

(a) f_{i-1} (b) $[f_i, f_{i+T-1}]$ (c) f_{i+T}

Fig. 2. Detection of video-image time section for writing on WB by the difference between images.

Furthermore, as described in Sect. 2.1, we can detect the facial region based on the flesh-colored pixels and we can define the feature R^{BS}_{face} for participant facial movement behavior by $R^{BS}_{face} = \sum |R_{UL} - R_{UR}|$. Therefore, we can evaluate participant behaviors in brainstorming meetings by the sum of the ratio WB^{BS}, as indicated by the WB writing behavior, and denote the participant facial movements (reactions) R^{BS}_{face}.

3 Modeling Participant Behaviors

In this section, we describe our method of modeling the behaviors of meeting participants. Here, we consider facial movements to comprise participant behavior.

First, we introduce our non-linear time-series model for the participant feature $R^L(t)$, in which we define $x_p(t) = R^{BS}_{face}$ of the p-th participant for the brainstorming meeting and $x_p(t) = R^{DM}_{face}$ for the decision-making meeting:

$$x_p(t) = \sum_{q}^{Q} \alpha_{p,q} f(\sum_{\ell=1}^{L} w_{p,q,\ell} x_q(t-\ell)) + e(t), \quad (2)$$

where $\alpha_{p,q}$ denotes the influence of the non-verbal behavior $x_q(t)$ of Participant-q on Participant-p, $w_{p,q,\ell}$ denotes the correlation with the non-verbal behavior $x_q(t)$ of other participants, $f(\cdot)$ denotes the sigmoid function, and $e(t)$ denotes Gaussian noise. Here, we can represent the non-linear time-series model defined in Eq. (2) by a neural network [8], as shown in Fig. 3. As we see in the figure, we can use the same neural network model for both meeting types (decision-making and brainstorming).

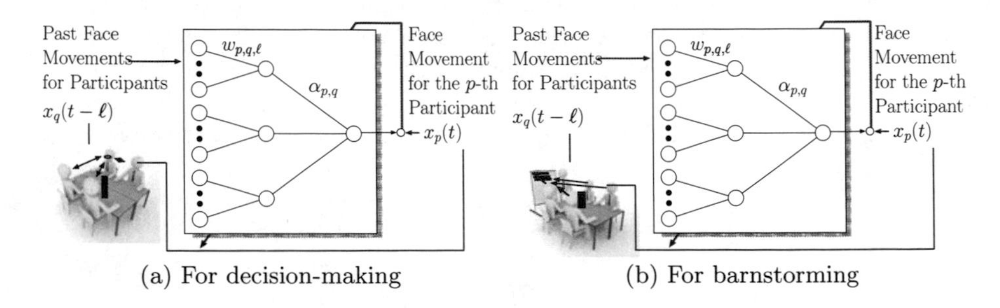

Fig. 3. Neural network model for (2).

The learning objective for this neural network model is to minimize the error function $E = \sum_{t=1} E_t = \sum_{t=1}(x_p(t) - \hat{x}_p(t))^2$. Here, $\hat{x}_p(t)$ denotes the prediction value for $x_p(t)$. We can express the learning law for weights $\alpha_{p,q}$ by the following:

$$\alpha_{p,q} = \alpha_{p.q} - \eta \frac{\partial E_t}{\partial \alpha_{p,q}}, \tag{3}$$

where η denotes the learning coefficient. We can calculate the differential coefficient $\partial E_t / \partial \alpha_{p,q}$ as follows:

$$\frac{\partial E_t}{\partial \alpha_{p,q}} = \frac{\partial E_t}{\partial \hat{x}_q(t)} \frac{\partial \hat{x}_q(t)}{\partial \alpha_{p,q}} = (x_p(t) - \hat{x}_p(t)) f(\sum_\ell w_{p,q,\ell} x_q(t-\ell)). \tag{4}$$

Moreover, we can express the learning law for weights $w_{p,q,\ell}$ by the following:

$$w_{p,q,\ell} = w_{p,q,\ell} - \eta \frac{\partial E_t}{\partial w_{p,q,\ell}}. \tag{5}$$

The differential coefficient $\partial E_t / \partial w_{p,q,\ell}$ can be calculated by

$$\begin{aligned} \frac{\partial E_t}{\partial w_{p,q,\ell}} &= \frac{\partial E_t}{\partial \hat{x}_q(t)} \frac{\partial \hat{x}_q(t)}{\partial w_{p,q,\ell}} = \frac{\partial E_t}{\partial \hat{x}_q(t)} \frac{\partial \hat{x}_q(t)}{\partial o_q} \frac{\partial o_q}{\partial w_{p,q,\ell}} \\ &= (x_p(t) - \hat{x}_p(t)) f'(\sum_\ell w_{p,q,\ell} x_q(t-\ell)) \alpha_{p,q} x_q(t-\ell), \end{aligned} \tag{6}$$

where $o_q = \sum_\ell w_{p,q,\ell} x_q(t-\ell)$. We can remove redundant weights by using a forgetting learning algorithm [9].

4 Experimental Results for Decision-Making Meetings

We held a decision-making meeting under the following conditions; (i) Theme: Determination of food menu (Fig. 4(a)) and the amount of food to be provided for a party, (ii) Participants: four persons (undergraduate students), (iii) Camera: MR360 (King Jim Co. Ltd.).

Figure 4(b) shows the timing of the speaking by participants. We can summarize the characteristics of the comments of each participant as follows; (i) Participant-A: comments were different than those that characterize the flow of the meeting, (ii) Participant-B: followed the lead of Participant-D, (iii) Participant-C: made only a few comments, (iv) Participant-D: commented on the management of the meeting flow.

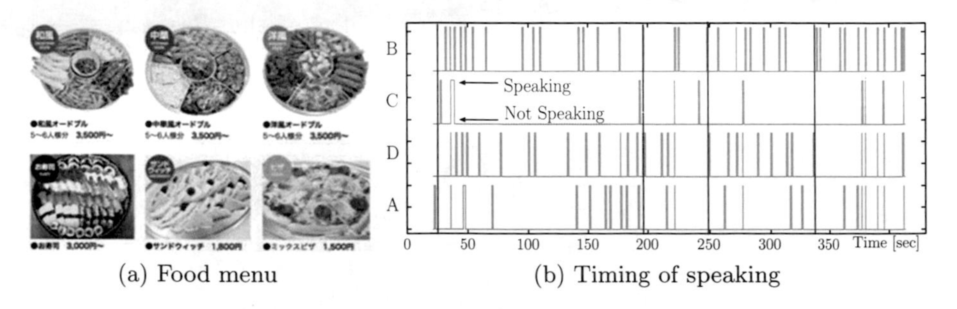

(a) Food menu (b) Timing of speaking

Fig. 4. Food menu and timing of speaking (red lines indicate a change in the order (food menu and amount)).

Figure 5(a) shows the feature $R_U = R_{UL} - R_{UR}$ and we can see that the facial movements of Participant-A and C are large. Figure 5(b) shows the sum $R_{face}^{DM} = \sum |R_U|$ of the features R_U for all participants and we can see that the facial movements R_{face}^{DM} become comparatively larger with respect to the timing of the change in the order (food menu and amount).

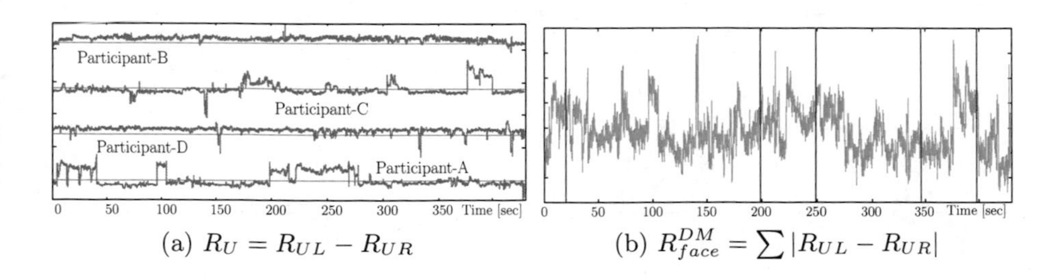

(a) $R_U = R_{UL} - R_{UR}$ (b) $R_{face}^{DM} = \sum |R_{UL} - R_{UR}|$

Fig. 5. Features R_U and R_{face}^{DM} regarding participant facial movements (red lines indicate a change in the order (food menu and amount)).

4.1 Relationship Between the Phase of Meeting and Weights $\alpha_{p,q}$

We define the phases of the meeting by the video-image sections between the times of change in the order (food menu and amount), as shown in Fig. 4(b).

Next, Fig. 6 shows the weights $\alpha_{p,q}$ in Eq. (2). Here, the weight $\alpha_{p,q}$ denotes the influence on Participant-p by Participant-q. We can describe the relationship between the meeting phase and the change of weights $\alpha_{p,q}$ as follows:

- From 200 [s], weights $\alpha_{A,C}$, $\alpha_{B,C}$ and $\alpha_{D,C}$ show great change. Therefore, the change in these weights shows that Participants-A, -B, and -D were influenced by Participant-C. At 193 [s], Participant-C made the statement "We do not need two dishes of Japanese cuisine." Participant-C had offered few comments throughout the meeting so the other participants were strongly influenced by Participant-C's comment.
- From 270 [s], weights $\alpha_{p,A}$ change greatly. We confirmed that Participants-A, -B, and -C are nodding in agreement. Thus, we can see that Participants-A, -B, and -C are influenced by Participant-A.

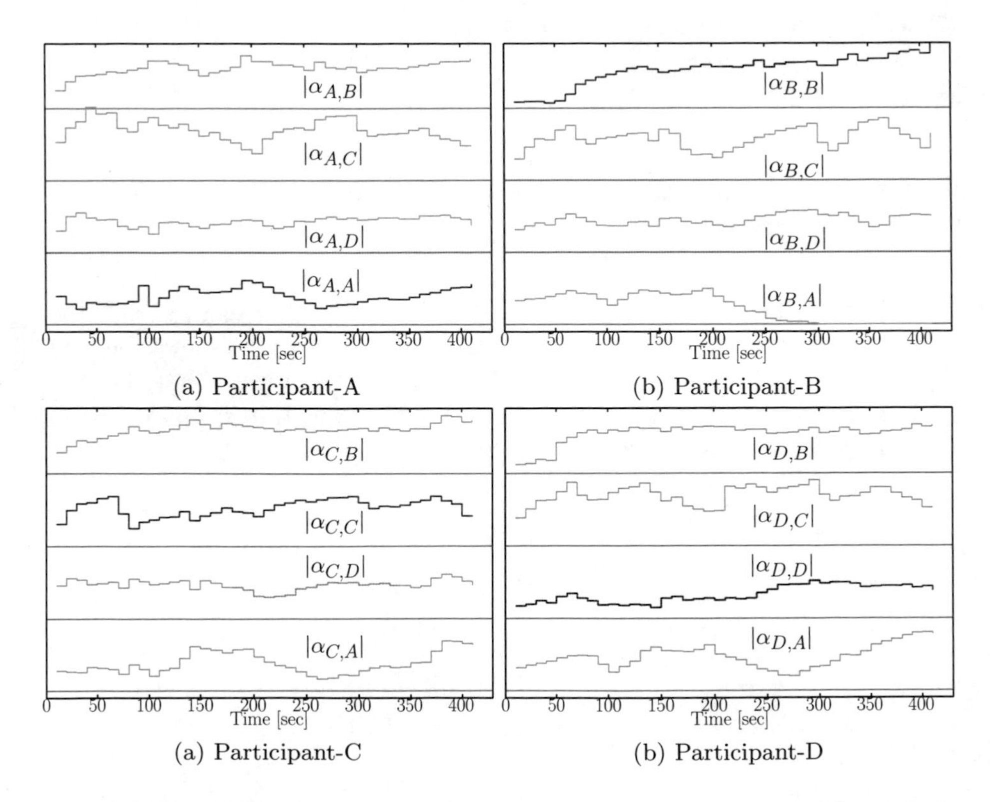

(a) Participant-A (b) Participant-B

(a) Participant-C (b) Participant-D

Fig. 6. Weights $\alpha_{p,q}$: influence of Participant-q (q = A, B, C, D).

4.2 Evaluation of the Meeting by Other Individuals

We showed our evaluation results from the meeting video to three other individuals (undergraduate students) who had not attended for which the evaluation item was the detection of characteristic behaviors and comments. Figure 7 shows their evaluations of the non-verbal behaviors and comments of participants.

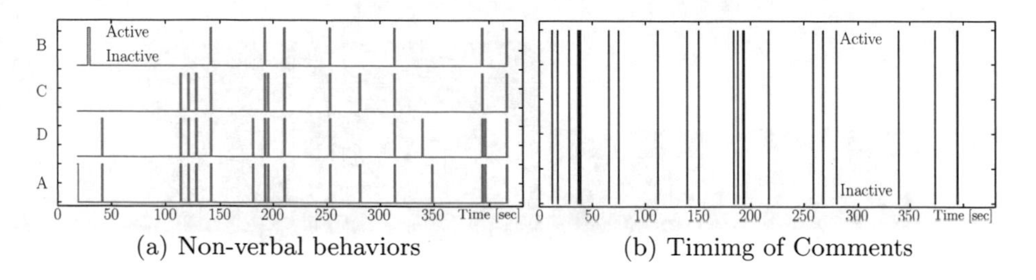

(a) Non-verbal behaviors (b) Timimg of Comments

Fig. 7. Evaluation of non-verbal behaviors and comments by other individuals.

Evaluation of behaviors: Figure 7(a) shows the evaluation results for the meeting video without speech, in which the green bar denotes the time at which the other individuals detected characteristic behaviors.

- In section [109,136] [s], they determined that some participants were nodding and looking at the menu.
- In section [190,200] [s], they determined that Participant-D exhibited the following behavior: showing his three fingers while commenting "We need three types of food." (Participant-D exhibited behavior corresponding to this comment.)
- In section [350,390] [s], the meeting was stagnating. However, the meeting ended based on the comment by Participant-C.
- In section of [390,395] [s], the other individuals determined that one participant exhibited the following behavior; Participant-C pointed to the menu and made the comment, "We need one more food."

Evaluation of comments: Figure 7(b) shows the evaluation results for the meeting audio recording (speech only) without any video images.

- In section [28,40] [s], the other individuals considered the comment concerning "pizza," but did not consider any of the characteristic behaviors exhibited in section [28,40] [s] shown in Fig. 7(a).
- In section [180,195] [s], they considered the comment "Japanese, Western-style, and Chinese foods." Similarly, they considered the characteristic behaviors of section [180,195] [s] shown in Fig. 7(a).
- In section [260,280] [s], they considered the comments "Sandwich" and "For breakfast if it were left over," but did not consider any characteristic behaviors shown in Fig. 7(a).
- In section [390,395] [s], they considered the comment "We need one more Western-style food" as well as the characteristic behaviors shown in Fig. 7(a).

5 Experimental Results for Brainstorming

We held a brainstorming meeting under the following conditions; (i) Theme: What can you obtain by a part-time job? (ii) Participants: four persons (undergraduate students). Figure 8 shows the meeting scene recorded by the two types of cameras.

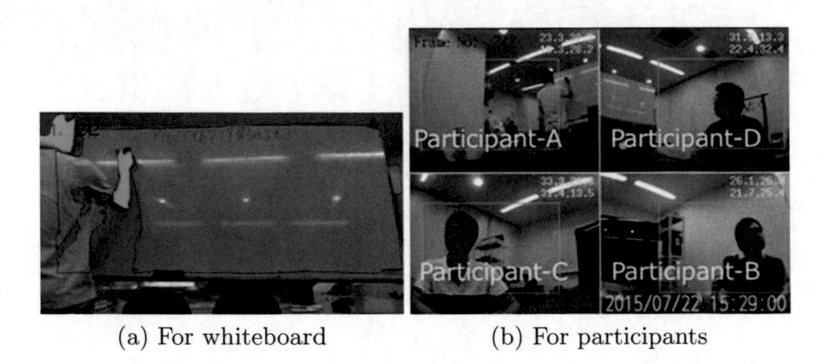

(a) For whiteboard (b) For participants

Fig. 8. Recorded images of whiteboard and participants (Camera for participants: MR360 (King Jim Co. Ltd.) and for whiteboard: MacBook Air (Apple Inc.))

In this brainstorming meeting, the four participants wrote 38 ideas on the WB over 720 [s], with the average time required for each idea being about 19 [s]. Table 1 shows four categories that contain similar ideas and the ratio *Agree* of the participant's agreement with this idea. From the table, we can see that many participants agreed with Category-2 (development of skills in interpersonal relations).

Table 1. Idea categories and the ratio *Agree* (Category-1: Business and time skills, Category-2: Interpersonal relations, Category-3: Income, Category-4: Self-esteem)

Category	Idea	*Agree* [%]
1	Efficient use of time, Scheduling, Planning	0.0
2	Communication, Interaction with peers and those from different generations, Cooperativeness, Learn how to contact customers	50.0
3	Income, How to manage money	37.5
4	Self-esteem, Accomplishment	12.5

5.1 Writing Behaviors WB^{BS} and Reactions R^{BS}_{face} by Participants

- **Writing behaviors** WB^{BS}: Figure 9(a) illustrates the participant writing behaviors WB^{BS} for which when the value of WB^{BS} becomes large, it indicates that the participants are writing their ideas on a whiteboard. We can see that the value of WB^{BS} becomes small from 450 [s] and the frequency of ideas decreases.
- **Reactions** R^{BS}_{face} **by participants for written ides**: Figure 9(b) shows the two facial movement features $R_U = R_{UL} - R_{UR}$ and $R^{BS}_{face} = \sum |R_U|$. Here, when a participant was writing ideas, R_U becomes 0. This figure shows that the R_U values of Participants-C and -D are larger than those of the other participants. As a result, the feature R^{BS}_{face} of facial movement by all the participants becomes large. We confirmed that participants-C and -D moved their facial expressions to a great degree and the videos show them to be exhibiting happy emotions.

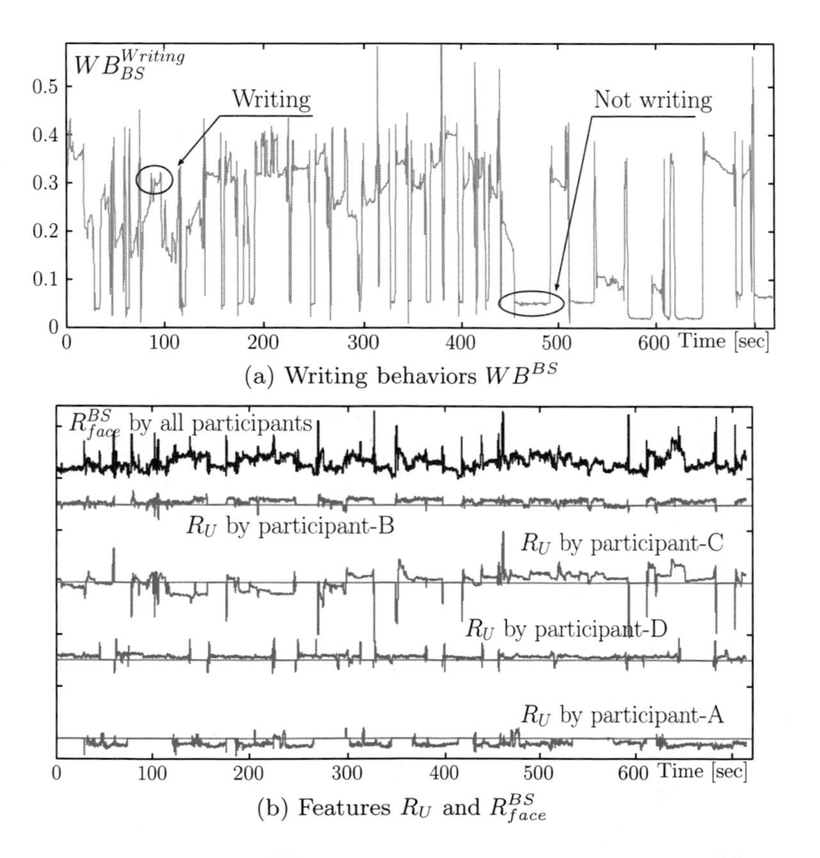

Fig. 9. Writing behaviors WB^{BS} and Features $R_U = R_{UL} - R_{UR}$ and $R^{BS}_{face} = \sum |R_U|$ for facial movements.

5.2 Modeling Results of Participant Behaviors and Relationships Between Behaviors and Ideas

- **Modeling results of participant behaviors**: Figure 10(a) shows the sum $\sum_q |\alpha_{p,q}|$ of weights $\alpha_{p,q}$. Here, $\alpha_{p,q}$ denotes the influence of the behavior of Participant-q on Participant-p, as calculated by Eq. (2) and $\sum_q |\alpha_{p,q}|$ denotes the influence by all participants on Participant-p. When $\sum_q |\alpha_{p,q}|$ becomes large, the behavior by Participant-p is influenced by the other participants.
- **Relationships between behaviors and ideas**: Figure 10(b) shows the timing of writing ideas on the WB. Here, "Agree-n" denotes the category shown in Table 1. We can see that many ideas written by Participant-B and C are categorized as agreed ideas.

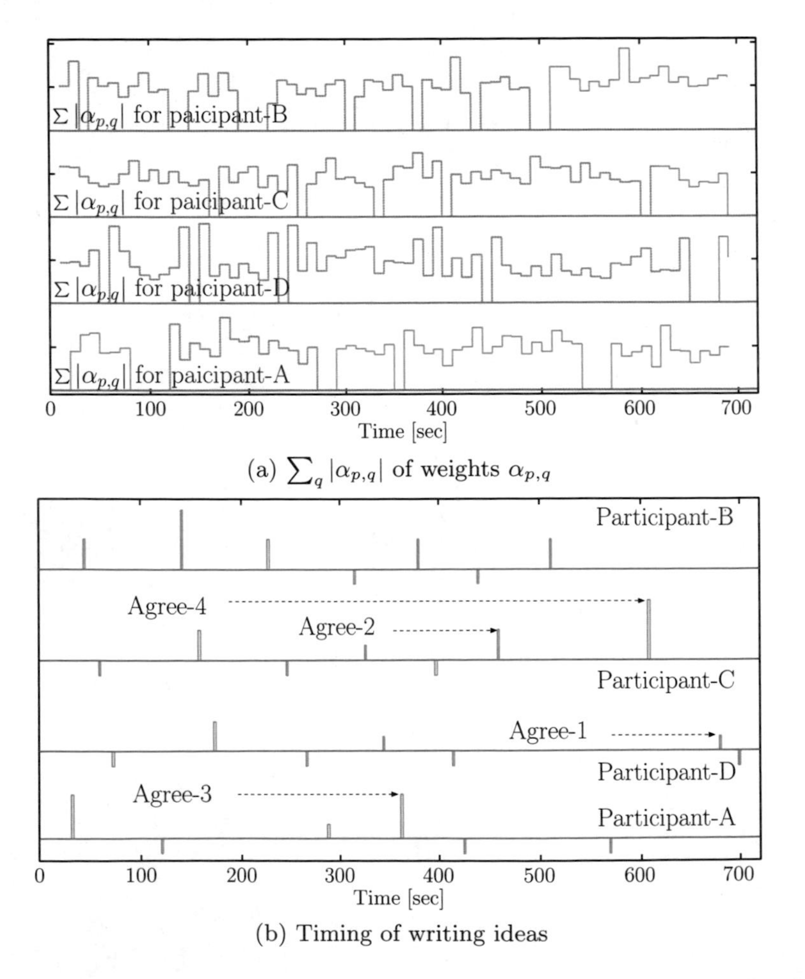

(a) $\sum_q |\alpha_{p,q}|$ of weights $\alpha_{p,q}$

(b) Timing of writing ideas

Fig. 10. $\sum_q |\alpha_{p,q}|$ of weights $\alpha_{p,q}$ and timing of writing ideas on whiteboard.

We can summarize the relationships between participant behaviors (Figs. 10(a) and (b)) and ideas (Table 1) as follows;

- The idea (Category-3) by Participant-A at 30 [s]: $\sum |\alpha_{p,q}|$ for Participant-B and C become large.
- The idea (Category-3) by Participant-A at 350 [s]: $\sum |\alpha_{p,q}|$ for Participant-B, C and D become large.
- The idea (Category-2) by Participant-B at 45 [s]: $\sum |\alpha_{p,q}|$ for Participant-C becomes large.
- The idea (Category-2) by Participant-B at 220 [s]: $\sum |\alpha_{p,q}|$ for Participant-D becomes large.
- The idea (Category-2) by Participant-B at 380 [s]: $\sum |\alpha_{p,q}|$ for Participant-A, C, and D become large.
- The idea (Category-2) by Participant-B at 510 [s]: $\sum |\alpha_{p,q}|$ for Participant-A and C become large.

With respect to the relationships between behaviors and ideas, our results show a clear relationship between having the agreement for an idea and the value of $\sum |\alpha_{p,q}|$.

6 Conclusions

In this paper, we analyzed the behaviors of participants in two types of meetings (brainstorming and decision-making). We introduced the use of participant behavior based on facial movement. Next, we proposed a method for modeling the behaviors of participants based on multi-layered neural networks. Lastly, based on our experimental results, we discussed the relationships between the meeting phase, participant behaviors, and model parameters for the two meeting types. Our results show that $\alpha_{p,q}$ (the influence of the behavior by Participant-q on Participant-p) is strongly related to the behaviors and ideas of participants in brainstorming and decision-making meetings.

In future work, we plan to discuss nodding agreement behavior and the relationship between the modeling results and what the participants hear. Our proposed modeling method can be easily applied to collaborative learning analyses and we plan next to analyze the behaviors of students engaged in collaborative learning.

Acknowledgements. This work was supported by JSPS KAKENHI Grant Number JP16K00499.

References

1. Shinnishi, M., Kasuya, Y., Inamoto, H.: Wi-Wi-Meter: a prototype system of evaluating meeting by measuring of activity, IEICE Technical report, HCS2014-63, pp. 19–24 (2014)
2. McCowan, I., Gatica-Perez, D., Bengio, S., et al.: Automatic analysis of multimodal group actions in meetings. IEEE Trans. PAMI **27**(3), 305–317 (2005)
3. Otsuka, K., Araki, S., Ishizuka, K., et al.: A realtime multimodal system for analyzing group meetings by combining face pose tracking and speaker diarization. In: Proceedings of the 10th International Conference on Multimodal Interfaces, pp. 257–264 (2008)
4. Watanabe, E., Ozeki, T., Kohama, T.: Analysis of behaviors by participants in meetings for decision making (second report), IEICE Technical report, HCS2016-1, pp. 1–6 (2016)
5. Watanabe, E., Ozeki, T., Kohama, T.: Extraction of relations between lecturer and students by using multi-layered neural networks. In: Proceedings of IMAGAPP 2011, 6 p. (2011)
6. Watanabe, E., Ozeki, T., Kohama, T.: Analysis of non-verbal behaviors by students in cooperative learning. In: Proceedings of International Conference on Collaboration Technologies, 9 p. (2016)
7. Watanabe, E., Ozeki, T., Kohama, T.: Analysis of behaviors by participants in brainstorming (second report), IEICE Technical report, ET2016-6, pp. 81–86 (2016)
8. Rumelhart, D.E., McClelland, J.L., The PDP Research Group: Parallel Distributed Processing. MIT Press, Cambridge (1986)
9. Ishikawa, M.: Structural learning with forgetting. Neural Netw. **9**(3), 509–521 (1996)

Educational Effects for University Students Through Multiple-Years Participation in Out-of-Curriculum Project Activities

Makoto Hasegawa(✉)

Chitose Institute of Science and Technology, 758-65 Bibi, Chitose, Hokkaido 066-8655, Japan
hasegawa@photon.chitose.ac.jp

Abstract. The out-of-curriculum project team "Rika-Kobo" organized by university student members has actively performed various activities for over 10 years in local community that mainly aim to stimulate interests of children and other generations into various fields of sciences and technologies. The activities of the project team are in out-of-curriculum basis, which means the student members are voluntarily involved in. Thus, they are likely to have the same level of motivations and enthusiasms, leading to generally high qualities in the activities. Through several questionnaires for the student members, participation experiences into the activities of this student project team has been found to be very effective for the student members to achieve and/or improve various skills and abilities such as communication skill, collaboration, leadership, scheduling ability as well as problem finding & solving skill. In addition, their participations can also serve as desirable opportunities for allowing them to realize the fact that certain skills/abilities are required to be achieved and improved. Thus, although their out-of-curriculum basis, the activities of this student project team can become an effective educational and training scheme for university students as future researchers and engineers. In this paper, for the purpose of further investigating any possible educational effects achievable from the activities, based on the results of questionnaires for the student members conducted over several years, some build-up effects for achievements and improvements of their skills and abilities will be discussed, which can be realized through participation into the activities over multiple years.

Keywords: Project-Based-Learning (PBL) · Student program
Science education · Engineering education · Career development

1 Introduction

Several attempts have been tried over recent years to realize innovation in university and other higher-level education. Significant emphasis has been placed on how to encourage students to be actively involved in educational experiences. For that purpose, various new schemes have been tried, in place of traditional classroom-style teachings, and Project-Based-Learning (PBL) style activities are one of very popular schemes, as reported in various papers [1–3]. In typical PBL-style activities, students are required

M. E. Auer et al. (eds.), *Teaching and Learning in a Digital World*,
Advances in Intelligent Systems and Computing 715,
https://doi.org/10.1007/978-3-319-73210-7_53

to actively participate in activities so as to find out and/or set a certain goal to be achieved, and to work, often as a team.

However, in order to realize effective PBL activities, some problems have to be overcome. For example, the team members in the PBL activities are required to have the same level of motivation so that successful and desirable team-working atmospheres can be achieved. When the PBL activities are provided as a mandatory class in an official education curriculum, this might be difficult to realize. Even when the PBL activities are provided as a non-mandatory class, similar undesirable situation might happen, for example, when some students take the PBL class because they think it may be easier for them to earn their credits than normal classes. As another problem, when their outcomes at the end of the PBL activities are not so desirable, the students may not be able to achieve successful mental conditions. In order to avoid such situations, any re-challenging opportunities should be desirably given. However, this may not be possible when the PBL activities are incorporated into the official educational curriculum.

As one of possible effective ways to avoid such undesirable situations in the PBL activities, the author has been involved in the out-of-curriculum project activities to be performed by a project team "Rika-Kobo" that is organized by university student members [4–8]. This project team has actively performed various activities in local community that mainly aim to stimulate interests of children and other generations into sciences and technologies for over 10 years. The activities of the project team are in out-of-curriculum basis, which means the student members are voluntarily involved in. Thus, they are likely to have the same level of motivations and enthusiasms, leading to generally high qualities in the activities. As a result, their activities have been warmly welcome in the local community, and the total number of their activities for each Japanese fiscal year reached at around 60 to 70 since 2010 (see Fig. 1 as shown later).

The author already reported that participation experiences into the activities of this student project team are very effective for the student members to achieve and/or improve various skills and abilities such as communication skill, collaboration, leadership, scheduling ability as well as problem finding & solving skill [5–8]. Their participations can also provide them with desirable opportunities for allowing them to realize that certain skills/abilities are required to be achieved and improved by themselves. Thus, the activities of this out-of-curriculum project team can become an effective educational and training scheme for university students as future researchers and engineers.

In this paper, the author has tried to further analyze the educational effects of the activities. Specifically, based on the results of questionnaires for the student members, some buildup effects for achievements and improvements of their skills and abilities will be discussed, which can be realized through participation into the activities over multiple years.

2 Outlines of the Project Team Activities

The project team was first organized over ten years ago. At that time, the originally intended task was to perform science experiment classes at local schools (elementary and junior-high schools) for the purpose of stimulating interests towards various fields of sciences and technologies. Since then, the team has actively performed various activities each year, and further the student members of the project team have gradually extended their activities by themselves. Recently, it is in general unnecessary for the author as the project supervisor to give them specific instructions in their daily activities, and the student members can make appropriate decisions on what and how they should do towards coming events. In that sense, the activities have been self-disciplined and self-controlled by the student members.

The current activities performed by the project team can be categorized as follows:

(1) Science experiment classes at elementary and junior-high schools

Currently, the project team offers regular science experiment classes for two elementary schools and one junior-high school in the local community. For each of those classes, the student members do necessary preparation works such as preparing original experimental tools and presentation slides. During the class, one member acts as a teacher or an instructor and the others as assistants.

(2) Science experiment classes at other educational institutes

Some science museums and other educational institutes offer the project team to perform science classes, typically 30-90 min long each. Main target audience is usually children in elementary-school age and their parents, but in some cases, people in elder generations also attend such classes. Similar to the classes at schools, the student members do necessary preparation tasks for realizing attractive classes, and perform instructor's and assistants' roles.

(3) Participation to various events

The project team is often invited to participate in various events and provide science demonstration. Those events in general have a wider variety of audiences in their ages as well as in the degrees of their interests in science, and thus, atmospheres of these activities are quite different from those of classes at schools and other educational institutes. In order to make demonstrations more attractive, the student members need to exhibit more talented communication skills.

(4) Activities in response to local organizations' requests

Local organizations also offer requests of participations in their various activities. In order to meet their expectations, the student members are required to organize and perform the best-matched contents of activities.

(5) Others

In addition to the activities in the local community, the project team is sometimes asked by the university administration to work for supporting student education in the first-grade students. In addition, the team is also asked to assist recruitment activities towards high school students.

Figure 1 shows the total number of the activities performed by the student project team for each Japanese fiscal year (from April 1 through March 31 of the next year). Significant increase in the number is clearly recognizable.

Each of these activities is mainly based upon request from various institutes and organizations in the local community, including elementary school, junior high school, science museum and other social education institutes, as well as Parent and Teacher Association of local schools.

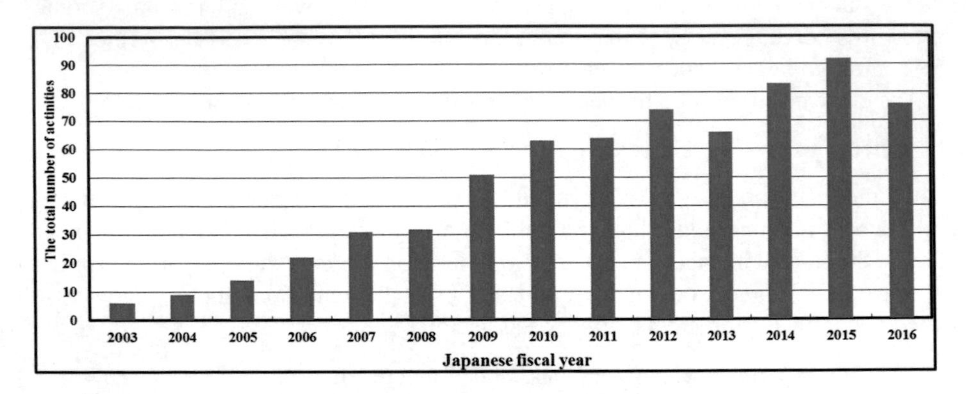

Fig. 1. The total number of the activities performed by the student project team for each Japanese fiscal year.

3 Results of Questionnaires

3.1 Questionnaires for the Student Members

Quantitative evaluations of educational effects for the student members of the project team to be achieved through their participation in the project activities are not so easily conducted. The author prepared some questionnaires for the student members of the project team to see if any advantageous educational effects are realized or not through their participation experiences in the activities of the project team [5–8]. More specifically, in the questionnaires, several skills and abilities were listed up, and the student members were asked to select any skills and abilities in the list that, through participation in the activities of the project team: (i) they thought achieved and/or improved; and (ii) they realized lacking or having but only with insufficient level so that it would be necessary for them to achieve and/or improve. In the responses, they were allowed to pick up skill and abilities without numerical limitation.

Although the similar questionnaires were also conducted in the past, the results of such previous ones were simply summed up without considering any factors such as differences in their grades (in other words, differences in the time period of their participations). In contrast, in this paper, the author tried to analyze their results in view of differences in the time periods of the student members' participations in the activities of the project team to see any buildup effects over multiple years.

The questionnaires were performed on December of the year 2015. All of the 32 student members involved in the project activities at that time responded.

3.2 The Overall Results

Figure 2 shows the responses from all of the student members. More specifically, Fig. 2(a) shows the results on which skills and abilities they thought achieved and/or improved through participation in the activities, and Fig. 2(b) shows the results on those skills and abilities they realized lacking or having but only with insufficient levels. In each of the results, the numbers of percentages of the student members who selected the respective skills/abilities are shown.

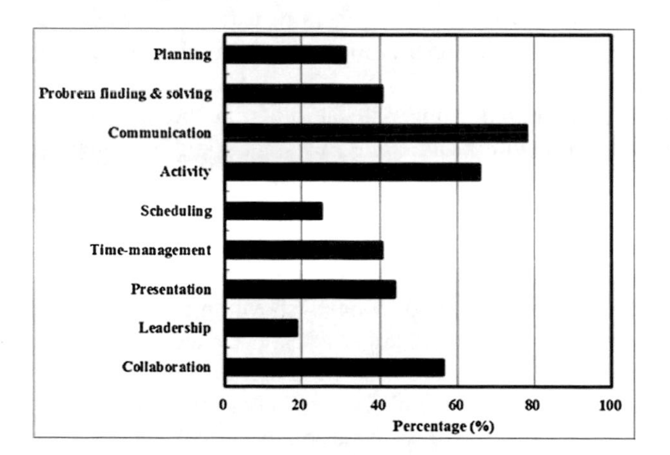

(a) skills and abilities that the student members thought achieved and/or improved through their participation in the activities of the project team.

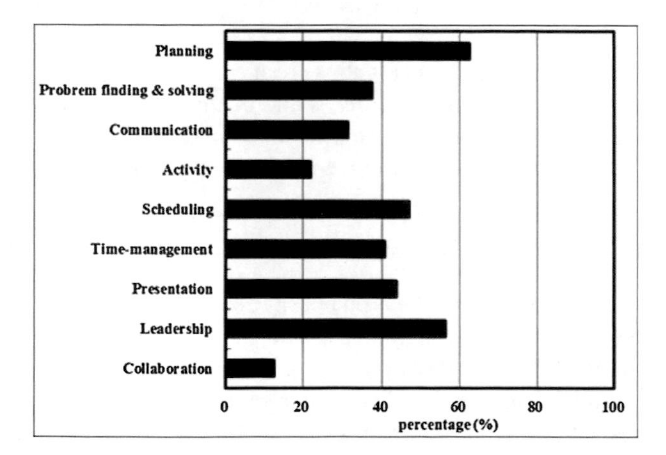

(b) skills and abilities that the student members realized lacking or having only with insufficient levels and need to be improved through their participation in the activities of the project team.

Fig. 2. The responses to the questionnaires from the student members.

The results in Fig. 2(a) show that communication skill, activity or active attitude (this was intended to indicate whether they have been becoming actively, not passively, involved in the various activities), as well as collaboration attitude were the top three categories selected. Such tendencies of selection by the student members imply that their active and voluntary engagement to the PBL-style activities of the project team seems effective for achievement and/or improvement of those skills/abilities. With respect to achievement/improvement of their collaboration attitude, the fact that they are always required to work in a team with close collaboration among each other is believed to exhibit a desirable effect. Such team-working with collaboration is necessary in order to meet tough schedules while providing their clients (such as teachers or staffs of local schools or other educational institute, members of parent-and-teacher associations or other organizations, who offered opportunities of the activities to the project team) with sufficient satisfactions.

In the results in Fig. 2(b) indicating skills/abilities that the student members selected as lacking or having only with insufficient levels, the top three categories selected as achieved/improved in Fig. 2(a) are placed as the least top three. In contrast, planning, scheduling, and time-management skills (which are normally required to be improved toward coming events) are likely to be selected. These tendencies can be explained in view of the fact that the total number of the activities of this project team per each year reaches around 80 in recent years, as previously shown in Fig. 1. The schedule of each of these activities normally cannot be changed after once fixed. This means that planning, scheduling, and time-management are very important between events to make coming events successful, and the student members are likely to strongly feel that they need to improve the related skills through experiences of such tough schedule demands.

Another category likely to be selected is leadership skill. Since the student members work in a team, someone in a team is required to serve as a leader of that team. Such an experience often makes the student member feel that improved leadership skill is desirable.

Moreover, a further important aspect in those results is that the student members can realize, based on their own experiences, that some of their skills/abilities have to be improved/achieved. Such achievement or improvement may be required for their career after graduation, and thus important as their career development.

3.3 The Results Based on the Time Periods of Participation

For the purpose of making further analysis on possible educational effects for the student members that can be obtainable through their participation experiences in the activities of the project team, the results as shown in Fig. 2 are split based on the time period of participation for each student member, and shown in Figs. 3 and 4. More specifically, Fig. 3 show the results of the questionnaires obtained from the student members who have two or more years of participation experiences in the activities (in other words, the student members in the second or higher grade), and Fig. 4 shows the results obtained from the first-grade student members who have experiences of only less than one year.

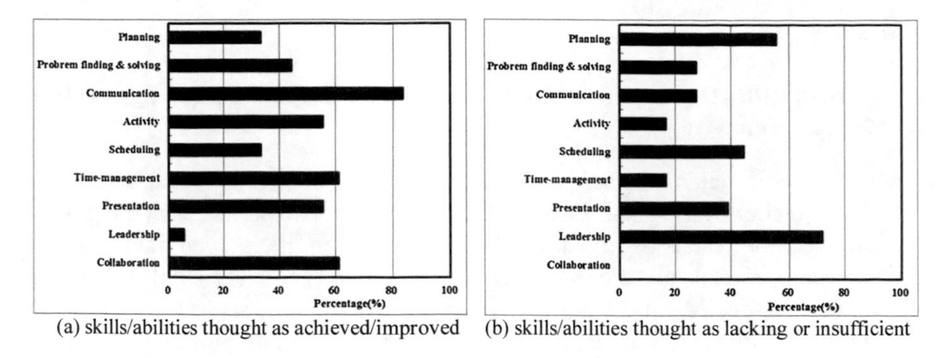

(a) skills/abilities thought as achieved/improved (b) skills/abilities thought as lacking or insufficient

Fig. 3. The responses to the questionnaires from the student members who have two or more years of participation in the activities (in the second or higher grade).

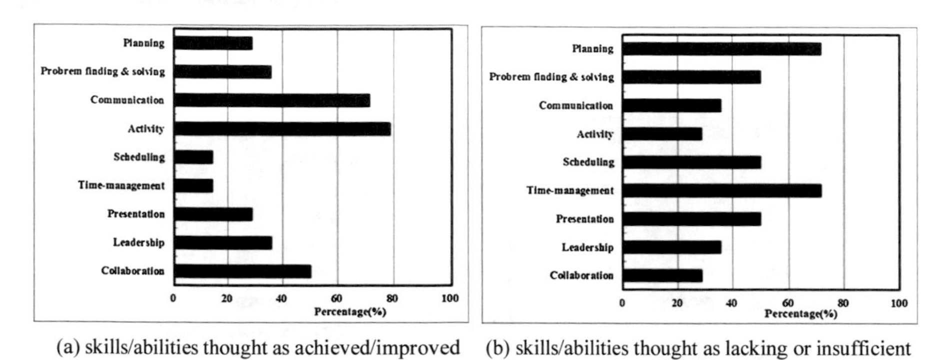

(a) skills/abilities thought as achieved/improved (b) skills/abilities thought as lacking or insufficient

Fig. 4. The responses to the questionnaires from the student members who have less than one year participation experience in the activities (in the first-grade).

The results in Fig. 3 show in general the similar tendencies as the results from all of the student members shown in Fig. 2. In Fig. 3(a), communication skill as well as collaboration attitude are likely to be selected. In Fig. 3(b), leadership is the most likely skill that the student members of two years or longer participation experiences realize lacking or having but only with insufficient levels. In contrast, everyone believed they have already achieved collaboration attitude of sufficient level, and thus, no one selected collaboration as lacking.

An interesting aspect in general tendencies obtainable from comparisons between these figures is that the student members of two years or longer participation experiences, shown in Fig. 3, are likely to select "more" skills/abilities as achieved and improved than those as lacking or having only with insufficient levels. In contrast, the first-grade student members shown in Fig. 4 are likely to select "more" skills/abilities as lacking or having only with insufficient levels than those achieved or improved. These tendencies indicate that educational effects for the student members can be built up while they spend longer time periods in the activities of the project team.

4 Discussions

4.1 Skills/Abilities Development Effects Through Participation Experiences Over Multiple Years

For the purpose of making further analysis on possible buildup educational effects for the student members over longer years of their participation experiences in the project team activities, the results of responses obtained only from the second-grade student members were picked up and are summarized as shown in Fig. 5. The similar questionnaire was also conducted in February of the year 2015 when those second-grade student members were in their first-grade. Among the results in this previous 2015 February questionnaire, the responses of the first-grade student members at that time (roughly corresponding to the second-grade student members in the 2015 December questionnaires) were picked up and summarized as shown in Fig. 6. Although the student members who responded to the respective questionnaires are not strictly the same, some tendencies are obtainable from comparisons of those two results.

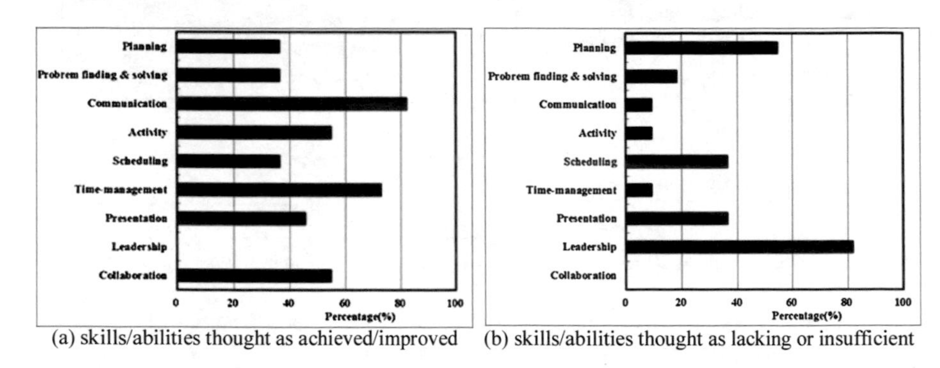

(a) skills/abilities thought as achieved/improved (b) skills/abilities thought as lacking or insufficient

Fig. 5. The responses to the questionnaires from the student members in their second-grade in the 2015 December questionnaires.

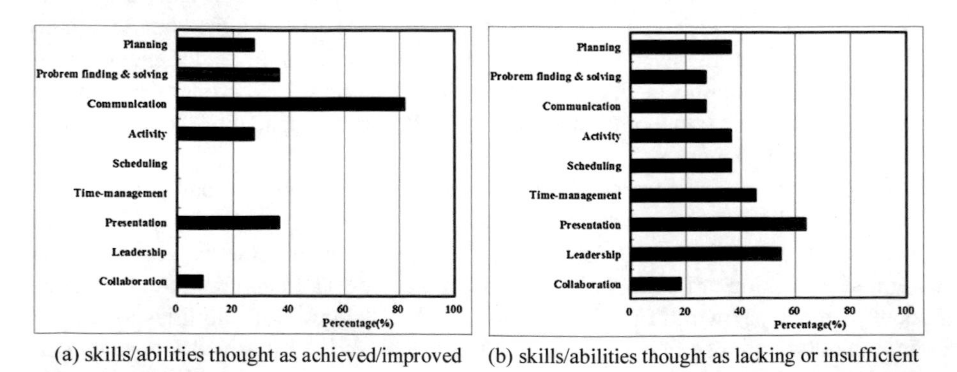

(a) skills/abilities thought as achieved/improved (b) skills/abilities thought as lacking or insufficient

Fig. 6. The responses to the questionnaires from the student members in their first-grade in the 2015 February questionnaires.

In Fig. 6 showing the responses for the 2015 February questionnaires from the student members in their first-grade (in other words, with participation experiences with less than one year), they selected only a few skills/achieved as achieved or improved, and selected "more" skills/abilities as lacking or having only with insufficient levels. Those tendencies are quite similar to the responses from the first-grade student members in the 2015 December questionnaires as previously mentioned with reference to Fig. 4. Those tendencies indicate that participation experiences in their first year strongly make them realize insufficient levels of their various skills/abilities.

In contrast, after they spent more time period for participation in the project team activities, they responded to the 2015 December questionnaire in much more positive ways in which they selected more skills/abilities as achieved or improved, as shown in the results in Fig. 5. Thus, although based on self-evaluation by the student members and not statistically or quantitatively confirmed, the student members can realize skills/abilities developments through participation experiences in the second year.

Further interesting aspect in the results as shown in Fig. 5 is that no student members in their second-grade responded that they already have had sufficient level of leadership skill, as can be seen in Fig. 5(a). When they had new student members after becoming the second-grade students, they are likely to face several scenes in which they have to act as a leader of a working group. Such experiences can serve as desirable opportunities for allowing them to realize lack of their leadership skills.

The similar questionnaires were also surveyed on December 2016. Figure 7 shows, among the results, the responses from the student members in their third-grade who are corresponding to the response group in Figs. 5 and 6 (not strictly the same members). In contrast to the results in Figs. 5(a) and 6(a), they responded that they could achieved and/or improved their leadership skills. Additional one more year experiences in the activities seem to work preferable for them.

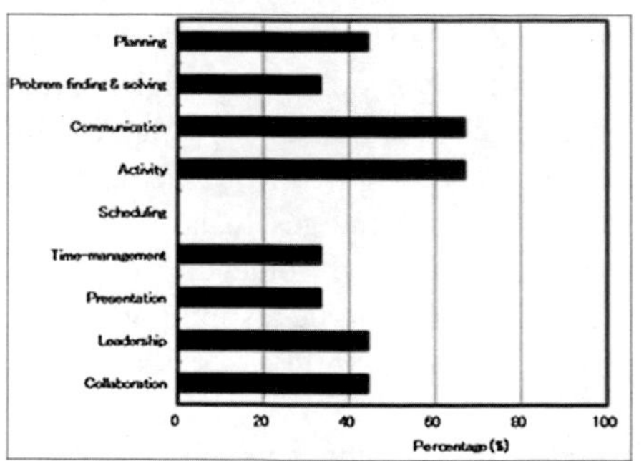

(a) skills/abilities thought as achieved/improved

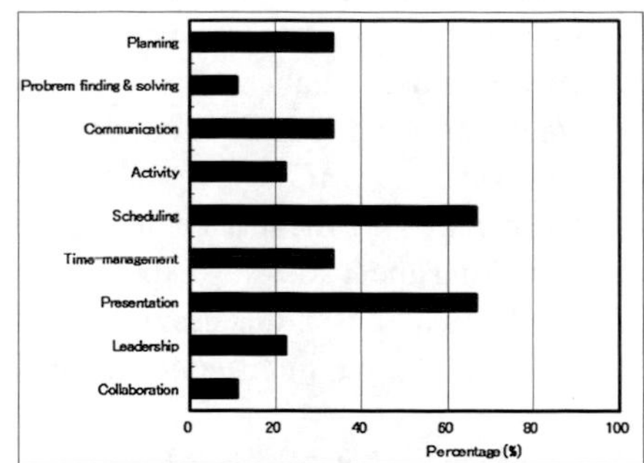

(b) skills/abilities thought as lacking or insufficient

Fig. 7. The responses on skills/abilities thought achieved/improved to the questionnaires from the student members in their third-grade in the 2016 December questionnaires.

Another interesting aspect in Fig. 7 is that no members selected scheduling skill as achieved/improved in Fig. 7(a). In contrast, they selected it as lacking or insufficient skill, as shown in Fig. 7(b). They also realized that their presentation skills should be further improved. Although any exact reasons for such results are not clear, the fact that those student members had to often act as leading persons among the lower-grades members might cause them to again realize insufficient levels of their certain skills.

4.2 Desirable Aspects of the Out-of-Curriculum Style Activities

For allowing the student members to realize certain skills/abilities development through their participation in the project team activities, participation experiences over multiple years are important and effective, as mentioned in the above. In their first-year, the student members are forced to realize their situations in which several skills/abilities are lacking or only at insufficient levels. Then in their second and later year, development (achievement and/or improvement) of those skills/abilities become possible. The fact that the activities of the project team are on out-of-curriculum basis is desirable for realizing such multiple years' participation [6–8].

Specifically, the student members have usually voluntarily joined in this out-of-curriculum project team. When asked the reasons why they decided to join in the activities, they typically answered that they would like to have advantageous experiences and/or improve some skills and abilities through participation in the activities. This means that the student members are likely to be internally motivated, instead of external motivation (such as considering earning some credits or improving their GPA scores as their rewards). Therefore, they can be very challenging in general. Different from in-curriculum activities, the time period of participation in the out-of-curriculum activities is not limited, and the student members can be actively involved in over multiple years. Moreover, the total number of the activities of the project team is quite large as mentioned previously, and the student members can attend similar activities many times. Thus, even when their performances are not satisfactory at one occasion, they can get another chance of re-challenging.

The out-of-curriculum style of the project team activities can also realize inter-grade collaboration among the student members because students in different grades can work in a team with close collaboration. This aspect is especially beneficial for higher-grade student members. They can have opportunities to improve their leadership-related skills through interactions with lower-grade student members, as implied from the results of the questionnaires as mentioned in the above. Similar advantages to be obtainable through teaching and leadership experiences can also be found in the other case of project activities [3]. In addition, discussions among each other seem to be also important and effective for realizing improvements of various skills and abilities of the student members.

As an interesting aspect to be found in attitudes of the student members, most of them are likely to think that the activities should remain in the out-of-curriculum basis. They are afraid that if the activities were incorporated into official education curriculum, less-motivated students might come in, which would down-grade the qualities of the activities. Since the student members were led to the activities by their internal motivations, realization of satisfactory achievements should be their rewards in participation into the activities. For that purpose, relatively highly motivated atmospheres among the student members in the activities have to be maintained, and therefore, the student members are likely to think that the activities should be in the out-of-curriculum basis.

5 Conclusions

Through participation in the activities of the out-of-curriculum student project team "Rika-Kobo" mainly aiming to stimulate interests of children and other generations towards sciences and technologies, certain educational effects for the student members can be achieved in which the student members can be allowed to achieve and/or improve their various skills/abilities through participations in the activities over multiple years. In general, the student members can realize, during their first year, that several skills/abilities are lacking or having but only as insufficient levels, and in the second or later year, achievement and/or improvement of those skills/abilities become possible. Thus, participation experiences in the activities of the project team over multiple years are important for realizing such buildup effects for developments (in other words, achievements and improvements) of skills/abilities.

As mentioned previously, the originally intended task of this students' project was to stimulate interests towards sciences and technologies in local community through science experiment demonstrations. Although that is still an important task, providing the student members with some chances for allowing them to grow up has become another important goal of the project activities.

References

1. Li, K.F., Gebali, F., McGuire, M.: Teaching engineering design in a four-course sequence. In: Proceedings of the 2015 IEEE International Conference on Teaching, Assessment, and Learning for Engineering (TALE 2015), Session 3B, pp. 288–293, December 2015
2. Katayama, H., Takezawa, T., Tange, Y.: Education of practical engineering skills aiming for solving real problems related to local area. In: Proceedings of the 2015 IEEE International Conference on Teaching, Assessment, and Learning for Engineering (TALE 2015), Session 3A, pp. 138–143, December 2015
3. Reith, D., Haedecke, T., Schulz, E., Langel, L., Gemein, L., Groβ, I.: The BRSU race academy: a tutored peer-teaching learning approach. In: Proceedings of the 19th International Conference on Interactive Collaborative Learning (ICL 2016), Session 6C, pp. 584–591, September 2016
4. Hasegawa, M.: Physics and science education through project activities of university students and regional collaboration. In: 12th Asia Pacific Physics Conference (APPC 2012), no. F-PTu-1. JPS Conference Proceedings, vol. 1, July 2013, pp. 017016-1–017016-4 (2014)
5. Hasegawa, M.: Roles and effects of activities of a student project team in engineering education for university students in lower grades. In: Proceedings of the 2013 IEEE International Conference on Teaching, Assessment, and Learning for Engineering (TALE 2013), no. 140, pp. 87–90, August 2013
6. Hasegawa, M.: New education scheme for college students through out-of-curriculum project activities. Int. J. Mod. Educ. Forum (IJMEF) **4**(3), 120–123 (2014)
7. Hasegawa, M.: Case study on educational effects for university students of their out-of-curriculum project activities. In: Proceedings of the 2015 2nd International Conference on Educational Reform and Modern Management (ERMM 2015), no. ERMM2015-E040, pp. 205–208, April 2015
8. Hasegawa, M.: Engineering educational effects for undergraduate students through out-of-curriculum. In: Proceedings of the International Conference on Electrical Engineering (ICEE 2016), no. 90064, July 2016

Presentation Skills of Mentor Teachers

Istvan Simonics(✉)

Trefort Agoston Centre for Engineering Education, Obuda University, Budapest, Hungary
simonics.istvan@tmpk.uni-obuda.hu

Abstract. Since 2011, we have been educating mentor teachers in four semester further training. Mentor teachers can support the preparation process of our engineering teacher students in secondary vocational schools by coaching their teaching practice. They have high standard expectations to learn the newest knowledge based on the most modern educational technology. It was important and interesting to get acquainted with presentation skills of mentor teachers, their knowledge about preparedness and success on that field. In 2016 we had organized a survey to measure presentation skills of mentor teachers and how they can use it in teaching learning process.

For the survey we had elaborated a questionnaire. The purpose of questionnaire was to survey in four parts the presentation skills of mentor teachers. The goal of survey was to define how frequently they use presentations in their work, what kind of lack they have in editing process and how can we support their application of presentations in mentor and teaching work.

In this study we described only the basic elements of survey. Deeper evaluation of data is in process. The survey raised our attention that further development of presentation skill needs methodology upgrading of our teaching process.

Keywords: Presentation skills · Mentor teaching · Theoretical training
Learning pedagogy

1 Introduction

Since 2011, we have been educating mentor teachers. Mentor teachers can support the preparation process of our engineering teacher students in secondary vocational schools by coaching their teaching practice and they help their trainee colleagues in the first two years in beginning period of career. This postgraduate course was new for our colleagues as well. It was a challenge to prepare ourselves as best as possible, to give lectures to our Mentor teacher students, who are practical teachers in schools and have good experience in training of their subjects. They have high standard expectations to learn the newest knowledge based on the most modern educational technology. The best Mentor teacher can accelerate changes in teaching-learning process which needs creativity.

Past 5 years the content and methodology of mentor teacher training courses have changed [1]. In the preparation process of course design we had to take into account the changing learning styles as well [2].

In 2013 we introduced at our university to use Moodle eLearning Management System to support our students in learning process.

M. E. Auer et al. (eds.), *Teaching and Learning in a Digital World*,
Advances in Intelligent Systems and Computing 715,
https://doi.org/10.1007/978-3-319-73210-7_54

Our mentor teacher students have several practical lessons. They have to work together in groups, find and collect information process and share with their colleagues [3–5]. In this way they have to use Information and Communications Technology – ICT, edit presentations and give them to their colleagues. Presentation skills for teachers were surveyed nationally and internationally as well. Alshare and Hindi emphasized “Presentation skills in the classroom are a complex tasks” [6]. “Competence in using ICT in teaching is required for Qualified Teacher Status and considerable attention” [7]. There are several books and online courses about presentation techniques. But researches about effectiveness of presentations are only few.

2 Background, Goal and Methodology of Research

Mentor teachers have high standard expectations to learn the newest knowledge based on the most modern Educational Technology. It was important and interesting to get acquainted with presentation skills of mentor teachers, their knowledge about, preparedness and success on that field. In 2016 we had organized a survey to measure presentation skills of mentor teachers and how they can use it in teaching learning process.

For the survey we had elaborated a questionnaire. The purpose of questionnaire was to survey in four parts the presentation skills of mentor teachers. In the first part we measured their personal data: age, sex, area of teaching, type of school of work. In second part we involved questions about preparation of presentation skills, did they learn Educational Technology at university in teacher training period, how did they learn making of presentations, did they know the basic rules of editing presentations. In the third part we were interested in management of resources for editing presentations. In the fourth part we measured the application of presentations in their education process.

The goal of survey was to define how frequently they use presentations in their work, what kind of lack they have in editing process and how can we support their application of presentations in mentor and teaching work.

We elaborated a hypothesis: *We have to emphasize the development of presentation skills of mentor teachers in our training and support them to learn the effective information process.*

The survey happened in November 2016. We elaborated a questionnaire for mentor teachers. These target groups involved mentor teacher students from three different semesters. The questionnaire consisted of four different parts and contained 32 questions.

We could ask 16 from first semester, 18 from second semester and 75 from third semester, altogether 109 mentor teacher students. In this study we focus only the results of part of questionnaire. Results were analyzed on level of significance, correlation, factors tests and used SPSS software.

In mentor work teacher needs several competencies and skills which help them in preparation and development of their students or colleagues. These competences are various e.g. knowledge of pedagogy, professional subjects, methodology, teacher training, psychology etc. They had to accept the requirement of self-development. Application of presentations in pedagogical work is essential for mentors. Cooperation and communication are also very important factors in mentors' work, so we support activities which can accelerate them.

3 Personal Data and Professional Knowledge

3.1 Gender Ratio of Mentor Teachers

In the first part we asked them about personal data. Answers about sex mirrored the ratio in general of educators' society. 84% were women and 16% were men. In Hungary the ratio of men are low in school teachers. After secondary school, majority of girls prefers teacher training because the prestige of teachers and salary are not so high both in primary and secondary schools Fig. 1.

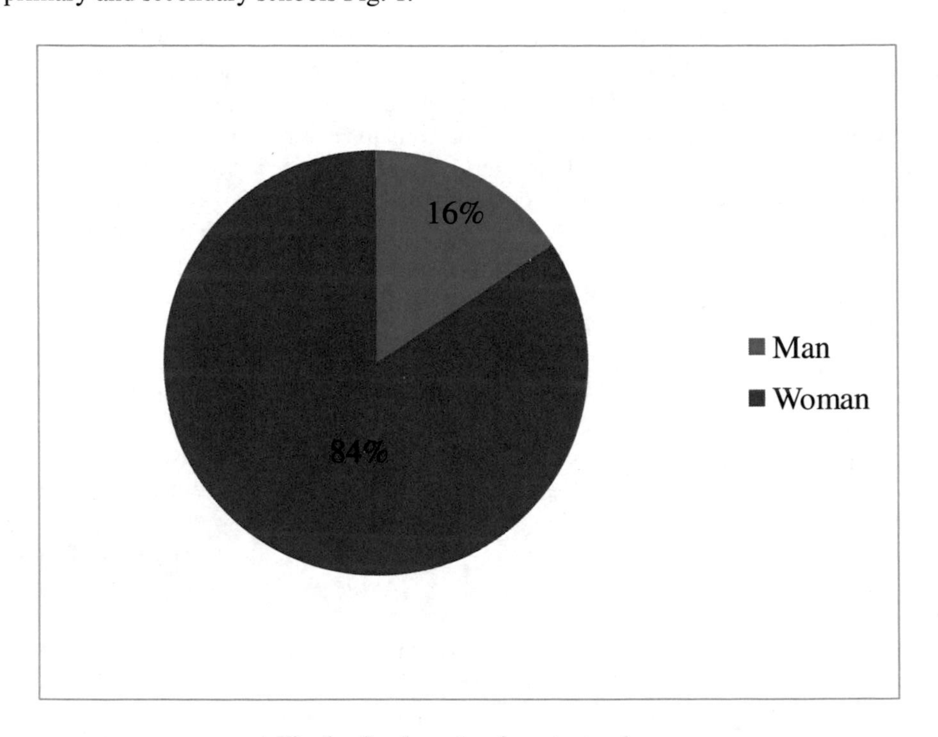

Fig. 1. Gender ratio of mentor teachers

3.2 Age of Mentor Teachers

69% of mentor teachers were between 36 and 50 ages. It can be explained with two reasons. To be a mentor teacher needs minimum three years experiences in teaching, so below 29 nobody can be mentor teacher. Mentor works need extra energy besides teaching so above 50 it is not so interesting and popular to select this kind of professional further training. Our mentor teacher students were working several kinds of schools from primary to secondary vocational schools Fig. 2.

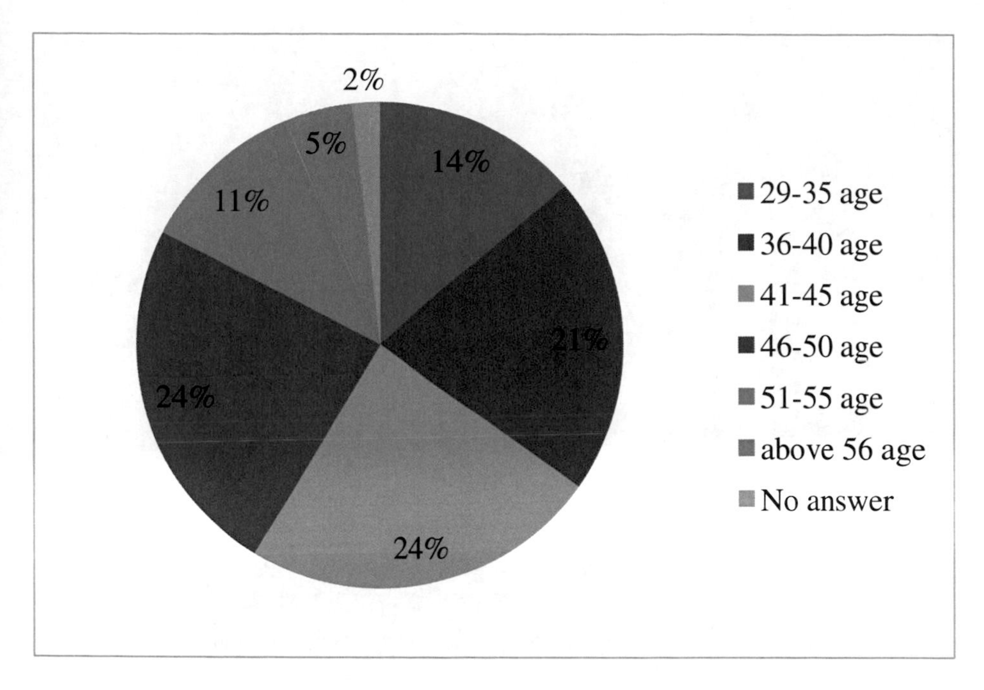

Fig. 2. Age of mentor teachers

4 Previous Knowledge for Presentation Skills

4.1 Learning of Educational Technology

In the second part of questionnaire we measured the previous knowledge for presentation skills. First we asked our students about learning of Educational Technology. Knowing bases of Educational Technology is essential for any kind of demonstration, application of rules and equipment is integrated in that subject generally. Mentor teacher students in age 46–55 indicated definitely they had subject such as Educational Technology in their teacher training study. Mentors age 29–35 and above 56 stated they did not learn Educational Technology Fig. 3.

4.2 Learning of Presentations' Preparation

In second question of this section we measured how they could learn the preparation of presentations. More than ¾ of mentor teacher students learnt alone as a self-preparation Fig. 4.

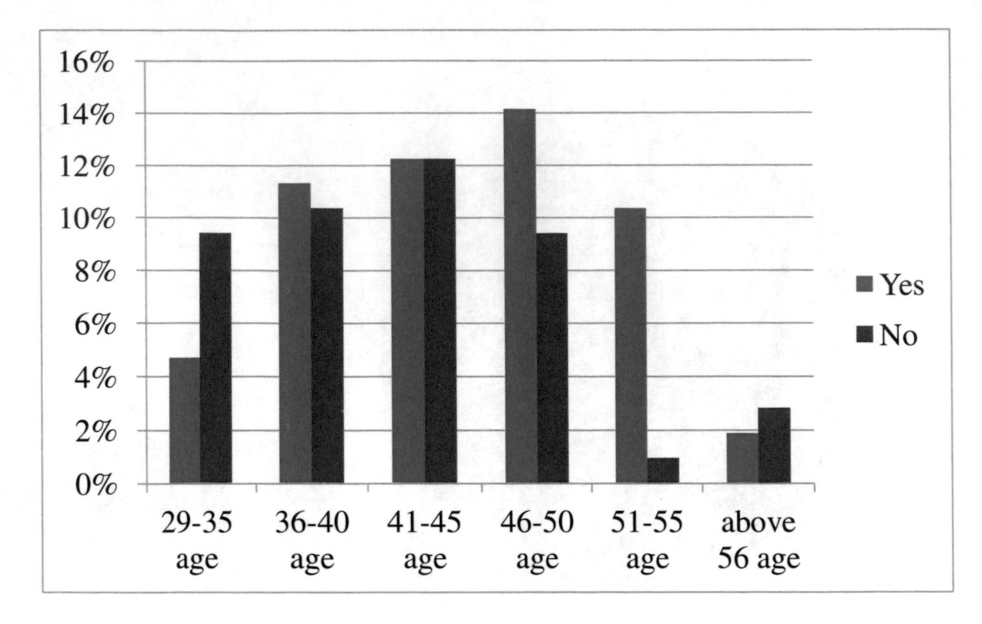

Fig. 3. Learning of Educational Technology subject in teacher training study

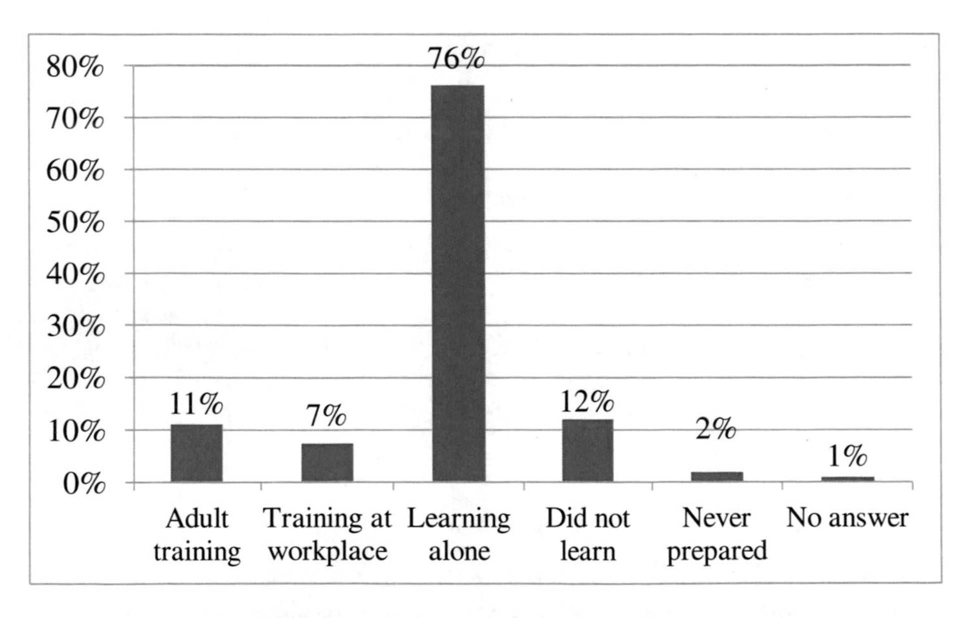

Fig. 4. How they could learn the preparation of presentations

12% of answerers did not learn the preparation of presentations. 11% of Mentor teacher students learnt in Adult training, 7% was acquainted with preparation in course and one did not answer.

4.3 Basic Rules of Presentations' Preparation

In this question we measured the knowledge of basic rules about layout, design, colours, letters, embedding pictures, sounds and video.

They could select from the following list:

- Layout planning
- Design
- Using colours
- Size of letters
- Proportion of layout
- Page numbers
- Sizing of pictures
- Insert of pictures
- Insert of voices
- Insert of video

It is important to mention, they could select, more than one possible equipment Fig. 5.

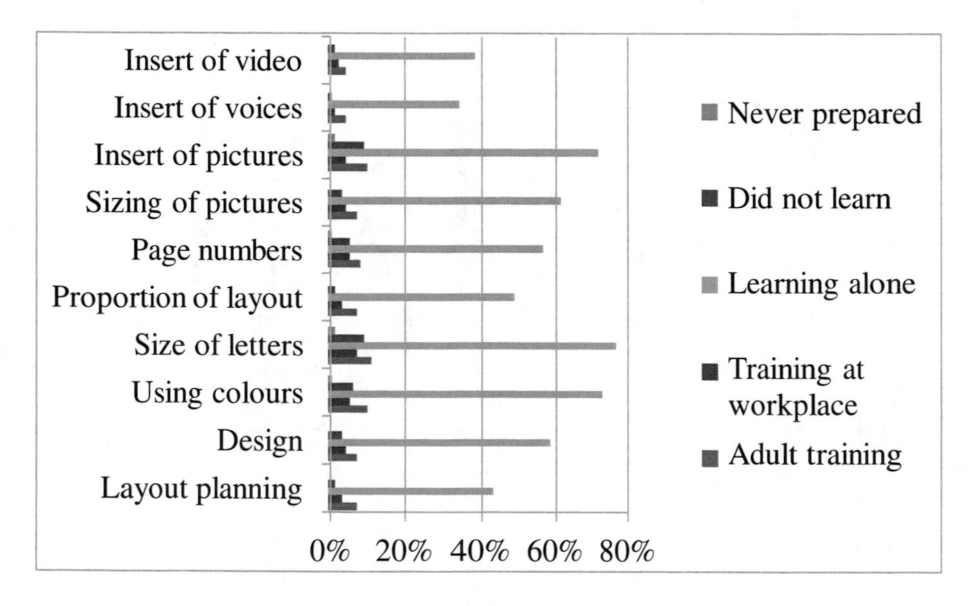

Fig. 5. Basic rules of presentations' preparation

We analyzed the results connecting the previous question. The results were similar than previous question. According to the figures majority of Mentor teachers learnt most of the rules alone: 34–76%. Probably 10% was the proportion of adult training and more than 10% who did not learn any rules.

That effects can explain, why our mentor teachers have problems with editing, using strange design, small letters, awful colours etc.

5 Planning and Preparation of Presentations

In the third part of questionnaire we have dealt with planning and preparation of presentations. We tried to measure how the mentor teachers use resources for planning and preparation of presentations.

The possible answers were as follow: printed material, data from internet, Wikipedia, articles from internet, professional pages from internet, information broadcasted by radio or TV, information from colleagues, relatives, information from pupils, other. They could select more than one items Fig. 6.

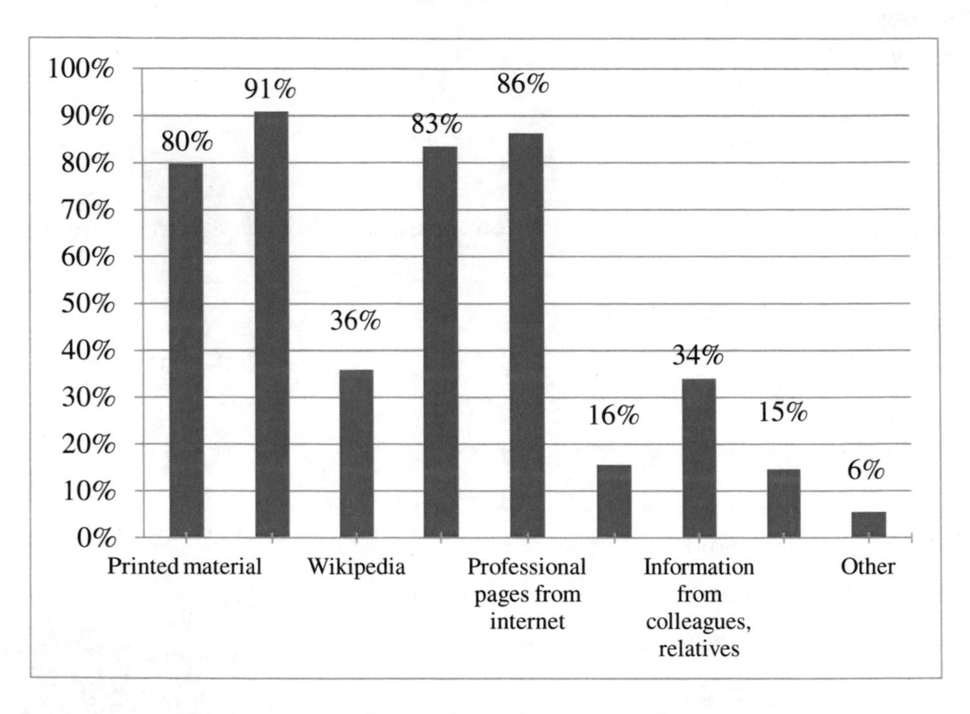

Fig. 6. Resources for planning and preparation of presentations

More than 80% they use printed materials, data, articles and professional pages from internet. Only 36% used Wikipedia and 34% information from colleagues, relatives. All the other resources were below 20%. These results mean, they generally believe the information from printed material and internet, reliability of information from Wikipedia and colleagues, relatives is more or less acceptable and all the other not or not useful. Printed materials in most cases checked by publisher's readers. All the other information is not so reliable. When our students giving presentations, we had to recognize they believe in data, never check the reliability of them.

We have to teach them to manage information carefully and wise!

6 Application of Presentations in Their Work

In the fourth part we measured the application of presentations in their work. In this study we evaluated only two questions. First question dealt with frequency of application of presentation Fig. 7.

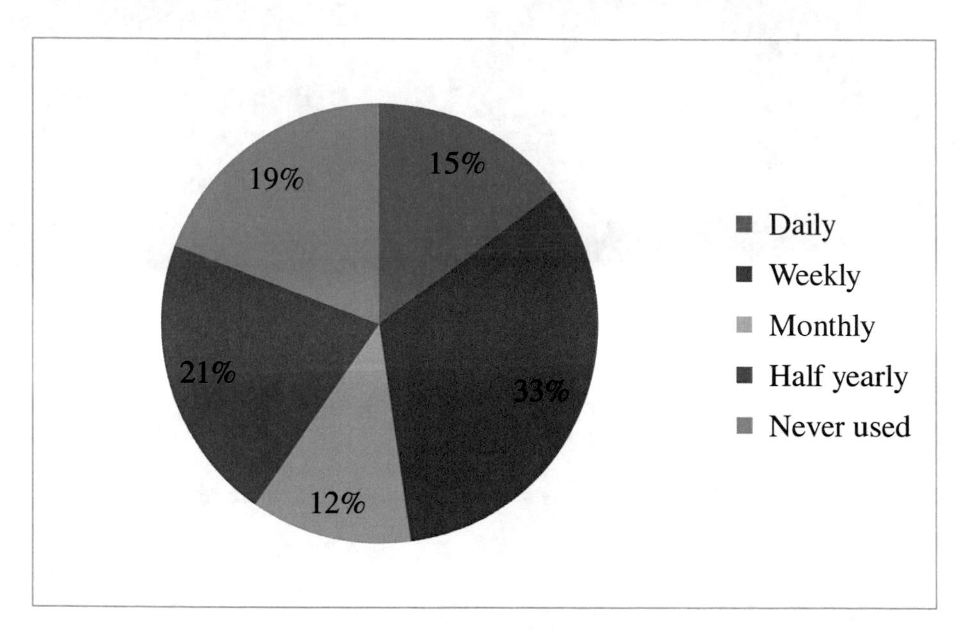

Fig. 7. Frequency of application of presentation

1/3 of answerers used presentations weekly, 1/5 used only half yearly, 15% daily, 12% monthly and 19% never used presentations in their work.

Last question is in this study, how they used presentations in education Fig. 8. 39% of mentor teacher students used presentations in theoretical lessons, 10% used in practical lessons, 19% used both theoretical and practical lessons. 22% never used and 10% did not answer.

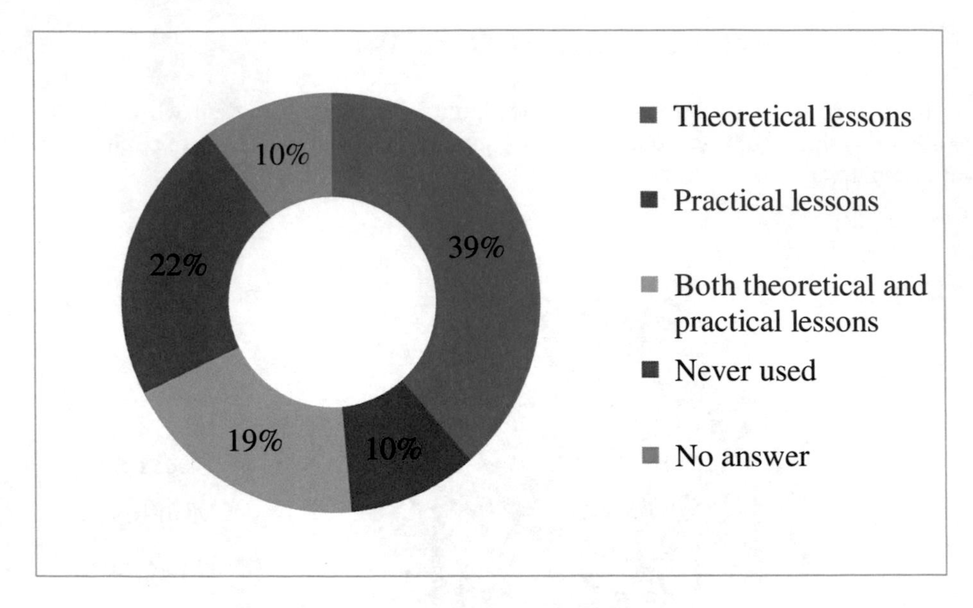

Fig. 8. Application of presentations in education

Fortunately, two third of mentor teachers used presentations in their educational work.

7 Results and Conclusions

The survey was organized for Mentor teacher students at our university. The purpose of questionnaire was to study in four parts the presentation skills of mentor teachers.

The goal of survey was to define how frequently they use presentations in their work, what kind of lack they have in editing process and how can we support their application of presentations in mentor and teaching work.

In this study we described only the basic elements of survey. Deeper evaluation of data is in process. However these analyses have proved our hypothesis, we have to emphasize the development of presentation skills of mentor teachers in our training and support them to learn the effective information process.

The survey raised our attention that further development of presentation skills needs methodology upgrading of our teaching process.

References

1. Toth, P., Rudas, I.J.: New methods for course development and management in engineering education. In: Ali, A., Abab, M., Santosa, B. (eds.) 4th International Conference on Industrial Engineering and Operations Management, Indonesia, Bali, pp. 1865–1874. Curran Associates, Inc. (2014). ISBN: 978-0-9855497-1-8

2. Toth, P.: New possibilities for adaptive online learning in engineering education. In: Ali, A., Abab, M., Plaha, R., Rahim, A. (eds.) International Conference on Industrial Engineering and Operations Management, United Arab Emirates, Dubai, pp. 2242–2246. Curran Associates, Inc. (2015). ISBN: 978-0-9855497-2-5
3. Simonics, I.: Preparation for information processing of mentor teachers In: International Conference on Interactive Collaborative Learning, Florence, Italy, Piscataway, pp. 536–539. International Society for Engineering Pedagogy (IGIP) (2015). ISBN: 978-1-4799-8706-1
4. Holik, I.: Experience and possibilities of information processing in training of mentor teachers. In: Szakál, A. (ed.) IEEE 13th International Symposium on Applied Machine Intelligence and Informatics, SAMI 2015, Herlany, Slovakia, pp. 219–222. IEEE Hungary Section (2015). ISBN: 978-1-4799-8220-2; 978-1-4799-8221-9
5. Molnár, G.: The impact of modern ICT-based teaching and learning methods in social media and networked environment. In: Turčáni, M., Balogh, Z., Munk, M., Benko, L. (eds.) 11th International Scientific Conference on Distance Learning in Applied Informatics, Štúrovo, Slovakia, pp. 341–351. Wolters Kluwer, Nitra (2016). ISBN: 978-80-7552-249-8
6. Alshare, K., Hindi, N.M.: The importance of presentation skills in the classroom: students and instructors perspectives. J. Comput. Sci. Coll. **19**(4), 6–14 (2004)
7. Darling-Hammond, L.: Evaluating Teacher Effectiveness – How Teacher Performance Assessments Can Measure and Improve Teaching, p. 36. Center for American Progress, Washington (2010)

A Two-Sided Approach of Applying Software Engineering Perspectives in Higher Education

Rebecca Reuter[1(✉)], Martina Kuhn[2], and Jürgen Mottok[1]

[1] OTH Regensburg, 93053 Regensburg, Germany
{Rebecca.reuter, Juergen.mottok}@oth-regensburg.de
[2] Coburg University of Applied Sciences and Arts, 96450 Coburg, Germany
Martina.kuhn@hs-coburg.de

Abstract. Using the Perspective Based Reading technique opens doors to new and enriching opportunities to design new teaching/learning arrangements. We set up a new approach that implements a perspective based task and (additionally) perspective based feedback. Therefore we provide the theoretical basis and elaborate more on Perspective Based Reading and Peer Feedback as well as on the needed perspectives, which we extract from the software development life cycle phases. We also define a new type of task that we called perspective based task with a perspective based feedback. For a better understanding we present an example scenario for a design pattern unit in a software engineering course.

Keywords: Perspective Based Reading · Peer Feedback
Software engineering education · Perspective based task
Perspective based feedback

1 Introduction

The landscape of higher education didactics has already changed towards a student centered, activating and thus, an inductive one. Although we are well on track, in our opinion learning obstacles are still existing. Thereby we assume that learning obstacles are of technical[1] and non-technical nature[2].

By offering different perspectives on the intended learning object respectively the exercise, we hope to overcome these existing learning obstacles. Our target audience are students of software engineering (SE) courses.

This paper contributes with the first steps to implement a new type of learning arrangement. It mainly bases on two techniques that are Perspective Based Reading (PBR) and Peer Feedback. PBR is a technique to inspect a software artefact from different perspectives. In our view, Peer Feedback, as we apply it, is a didactical method that fits to PBR in the case, that PBR as well as Peer Feedback inspect and evaluate an artefact.

[1] A learning obstacle concerning technical nature is for example a misconception regarding to a specific learning content.

[2] A learning obstacle concerning non-technical nature is for example a cooperation-related: If a learner is not able to learn with his/her assigned collegues.

M. E. Auer et al. (eds.), *Teaching and Learning in a Digital World*,
Advances in Intelligent Systems and Computing 715,
https://doi.org/10.1007/978-3-319-73210-7_55

We present the theoretical preparatory work needed for the implementation and the elaboration of the approach, as well an example scenario.

The remaining paper is structured as follows: First of all we explain the theoretical foundation that is necessary for our approach. Thereby we present the techniques of PBR and Peer Feedback, we base on. Further, we present the software lifecycle process and its roles per life cycle stage. After that we describe our approach in detail and give a first example for an exercise dealing with design patterns. Finally, we conclude and give an outlook on our further work.

2 Theoretical Foundation

To apply PBR (see: Sect. 2.1) in its origin purpose, it is necessary to choose roles, respectively perspectives, the inspectors have to take in order to detect errors in an artefact. Thus, we extract roles for each stage of the software life cycle (see: Sect. 2.2). The extracted perspectives shall then function as basis for the creation of tasks for our course, since all of the software development steps are handled in a SE course.

Furthermore, we explain Peer Feedback and its role in our approach, since this is the didactical method that is applied in the second step, when students evaluate their work taking different roles, respectively perspectives (see: Sect. 2.3).

2.1 Perspective Based Reading

Our approach for a new learning setting leans mainly on the idea of PBR. PBR is a technique that inspects a software artefact from the perspectives of different stakeholders, which are implemented in roles [1–3]. The main goal of PBR is to improve the defect detection coverage of a software artefact [4].

The perspectives depend on the existing roles within a software development life cycle. In PBR, they are derived from the stakeholders from a software artefact [5]. These stakeholders are "the most relevant people that [...] use the [...] artefact during its life cycle [5, p. 21]". In Table 1 we list the roles that are applied for PBR. The join of the three named references might be "the typical stakeholders used in perspective based reading [6, p. 501]" [5].

Table 1. Roles used in PBR

Reference	User	Designer	Implementer	Tester	Requirements engineer	Maintainer
[6]		x		x	x	x
[4]	x	x	x			
[7]	x	x		x		

Regarding the process, PBR provides a scenario for each inspector on the reading technique and on how to examine the document [4, 5]. The scenario is divided into an introducing section, instructions and questions. The introducing section "describes the quality requirements, which are most relevant to this perspective [4, p. 3]". The instructions section "describe what kind of documents to use, how to read, how to extract the necessary information [4, p. 3]". The section about the questions provide a set of questions, the "inspector has to answer during the inspection" [4, p. 3], [1, 6].

2.2 The Software Lifecycle Process and Its Roles Per Life Cycle Phase

There are many different characteristics of software development lifecycles. These show up through different process models, mainly differentiable by agile and classical approaches. All of them offer different roles, but our aim was to identify the general or "most important" roles (i.e. stakeholders) in order to derive necessary perspectives for one development step.

The extracted perspectives respectively roles shall function as basis for the creation of tasks for our course, since all of the software development steps are handled in a SE course.

The lifecycle of a software engineering process in general consists of five steps that have different emphasis depending on the chosen process model: requirements analysis, design, development, test, and operation/maintenance.

The roles for each phase are derived from the phase itself: this is a requirements analyst, a designer, a developer, a tester and a maintainer. Depending on the naming of the phases there might be individual changes such as "analyst" instead of "requirements analyst". Literature that provides a role description independent from process model lists similar roles but also some more, that are not of interest to us. Bittner and Spence present nine roles as common project roles: Development Team, Quality Assurance, Architect, Requirements Manager, Iteration Lead, User, Customer, Sponsor and Project Manager [8]. Murray defines also nine roles: Program Manager, Project Manager, Tech Lead, Sys Admin, Architect, Business Analyst, Design, UX and Programmers [9]. We see a potential usage in two more roles for our design: user, customer, since they function as interface to textual, unstructured formulations in natural language.

Concluding, we identify the following roles: Requirements Analyst, Designer, Developer, Tester, Maintainer, User and Customer.

A requirements artefact should be inspected using the following perspectives: user, designer, tester [7, 10]. Architecture artefacts should be reviewed by: user, designer, implementer [4] and/or requirements engineer, designer, tester, operator/maintainer [6].

The code should be reviewed by an implementer (one, who has not implemented the code). Artefacts of the test phase should be verified using the perspective of a requirements engineer, the implementer, and the tester itself. Currently, we do exclude the operation/maintenance phase, since we do focus on this in our courses. The operation and maintenance phase ought to be reviewed by the customer and the user and in case of maintenance also a developer. Using augmented reality this could be a possible approach to solve the operation/maintenance perspectives in future. We assume that this extracted roles are the relevant ones for our perspective-based design.

2.3 Feedback

Giving and taking feedback in an adequate way is a technique that students have to learn for their future work [11]. Especially in SE, that is a quite subjective discipline it is important to have these competencies, mainly for the review process. Besides, software development happens in teamwork, thereby it is a necessary skill that especially software engineers are able to give and take feedback from colleagues. During the software development process different opportunities for feedback situations are given, since many different artefacts (e.g., specifications, source code, and documentations) are created. Exemplary, to improve the quality of program code different methods – such as Pair Programming or PBR – are applied and both of them use feedback on a product; i.e. in an indirect way (see: Sect. 4). The participants who give and receive feedback are mainly from the same level (cf. horizontal feedback, see: Sect. 4), so called Peer Feedback. Thus, it is necessary to train this skill in our courses too. This section provides an overview about feedback mechanisms, especially peer feedback. We focus on Peer Feedback, since, in our view, this is the needed form of feedback for our approach.

Types of Feedback

Definitions about feedback can be found in several literature and also from different views: at a very high level, there are existing definitions from feedback in technical [12, 13] and non- technical senses [14, 15]. Further, definitions consider feedback regarding the concerned roles that are necessary: feedback provider[3] and feedback receiver (see Fig. 1): The feedback provider(s) – are giving feedback concerning a certain attitude, product or process to one or more persons. The feedback receiver(s) – are receiving feedback about a given event (e.g., attitude, product or process).

The last definition concerns a hierarchical view (see Fig. 1).

- In case the feedback receiver is lower in status than the giver of the feedback, the feedback goes vertically downwards; e.g., lecturer to student, supervisor to employee.
- In contrast, vertically upwards means that the feedback giver has a lower status than the feedback receiver. This constellation will be found rarely in assessment of employees or in evaluation settings at universities.
- If the feedback giver and feedback receiver are on the same level, they are peers and most of them are not in a dependent relationship; i.e. the feedback can be called horizontally.

Peer Feedback

After a definitions about feedback from different views we want to explain and define peer feedback as we use it.

Peer Feedback is defined by the relation between feedback provider and receiver. Thereby, the preposition “peer” indicates that provider and receiver of feedback have a similar status regarding to the learning process [16, 17] (e.g. provider and receiver are participants of the same course). Relationships between the tutor and student (feedback provider: tutor, receiver: student) are explicitly excluded. This is also the case if the student as the feedback provider has a more advanced level in the learning process. Therewith he/she is then more assigned to role of a teacher/tutor because of the fix role

[3] Terms in italics can be found in Fig. 1.

definition and not on the same level anymore [16, 18]. Peer Feedback is feedback that is given in a horizontally (and symmetric) way. Provider of the feedback and receiver on the feedback are on the same hierarchy [18] (see: Fig. 1).

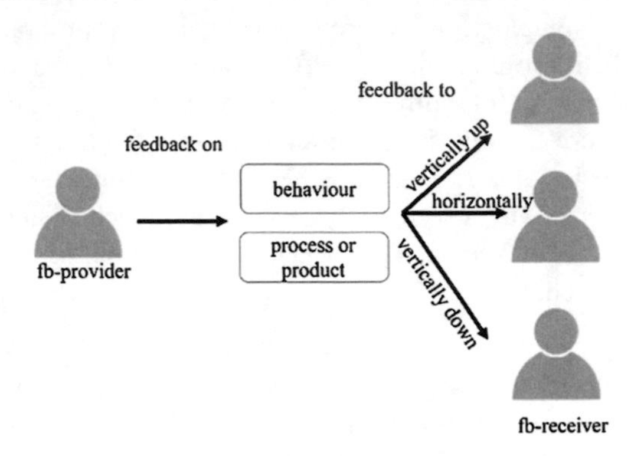

Fig. 1. Types of feedback

3 Approach

In this section we describe our suggestions for new teaching and learning arrangement using software engineering perspectives. In our point of view students entail learning obstacles such as a misconception when they are asked to solve a specific task. Our thesis is that our approach helps to overcome the existing technical obstacles and overcome the non-technical obstacle of giving and taking feedback. Of course, this feedback can be a cause for further learning obstacles (e.g. a new misconception) that might entail the process of further exercises.

In contrast to the PBR technique where the identification of mistakes is to the fore, we want our students to solve an apparently complex task but thereby taking a different perspective and further getting additional information from a different perspective. The students can choose their preferred perspective before they start to solve the task. During task solving they can order additional information from a different perspective using augmented reality, exemplarily. This arrangement of perspective based task and perspective based feedback are also implementable separately.

3.1 Perspective Based Task

The perspective based task can be structured of three ways:

1. As individual task, then additional perspective based information for solving the task is provided.
   In this case, we offer students additional perspective based information dependent on the identified perspectives (see: Sect. 2.2), in order to support the learning process. Since the student can choose the perspective on his/her own this can be

assigned to the adaptable learning approach. During the adaptable learning process, in contrast do adaptive learning where a system makes the choice, a student can choose for example the kind of additional information he/she needs to complete a task.

2. As task solved in group work:

(a) Group members solve a complex task dependent on work of team members (real life dependencies during software engineering), (see: Sect. 3.3).
The students solve a more complex task but are dependent from each other's work. The problem solving is then proceeded as in real live software engineering. The student is dependent on the partial result of his/her peer to solve his/her task.

(b) Group members solve a task independent from team members.
In this case, the students all get the same task but all take a different perspective. Here, the comparison of created results/artefacts is important.

One main goal of this practice is to proceed the whole software engineering process and to show that software engineering is subjective in many cases and different solutions are possible.

3.2 Perspective Based Feedback

After solving a task students have to provide feedback a task from a certain perspective. The perspective depends on the perspectives that were either taken in the perspective based task (see: Sect. 3.1) or if this method is applied independently, from the available perspectives according to the software life cycle phase (see: Sect. 2.2). The task to give feedback is very similar to the PBR technique. They get a predefined scenario that includes introducing section, instructions and questions (see: Sect. 2.1).

3.3 Example

Students in a group of three get the task to model a concrete implementation of the Abstract Factory design pattern. They take the role of the user (Student A), the designer (Student B) and the implementer (Student C). As a user, student A has the task to formulate the requirements for the design pattern Abstract Factory as described by the Gang of Four. We therefore provide a description of the design pattern in natural language. As a designer, student B has to model a UML class diagram that shows the Abstract Factory pattern as described by student A. If this fails or he/she has problems or questions, he/she can look up the predefined requirements by the provided additional information. As an implementer, student C has the task to implement the Abstract Factory pattern as modelled by student B. If this fails or he/she has problems or questions, he/she can look up the predefined class diagram requirements by the provided additional information again perspective based. Having solved the task, the second phase starts: Student C provides feedback on the model he/she got for the implementation to student B. Student B provides feedback on the requirements he/she got for the model to student A. Student A provides feedback has to check whether all requirements are fully implemented in the code and functions as required.

After the feedback cycle, the students can improve their work in a second iteration. In an evaluation, we want to measure whether and if yes what additional information was useful, and how students handle the feedback.

4 Outlook and Conclusion

We presented a new teaching/learning setting using an adapted Perspective Based Reading approach. The new setting is also implementable with single use of the perspective based task and the single use of perspective based feedback. Therefore, we presented the necessary theoretical foundation of PBR, Peer Feedback and the extraction of the needed perspectives using the common software life cycle process.

In our view three possible task types are conceivable for us at the moment. As the individual task provides additional perspective based information it should help to address individual needs and therewith help to overcome possible existing learning obstacles regarding the learning content. The second two options should address two obstacles that software engineering entails by nature: it is abstract and subjective and often there is no explicit solution in many cases. Therefore, we want our students to get an idea of how SE happens in real life and to proceed as many steps as possible in the SE life cycle by one task.

By implementing the perspective based Peer Feedback we want to train students using a communication skill that quite necessary for their further industry work.

In future work we have to test the new approach and we have to define and validate the detailed task description for the concrete course implementation. Thereby we have to consider structure and concrete formulations of the task [19].

Acknowledgement. The present work as part of the EVELIN project was funded by the German Federal Ministry of Education and Research (Bundesministerium für Bildung und Forschung) under grant numbers 01PL17022A and 01PL17022F. The authors are responsible for the content of this publication.

References

1. Laitenberger, O., Atkinson, C., Schlich, M., El Emam, K.: An experimental comparison of reading techniques for defect detection in UML design documents. J. Syst. Softw. **53**(2), 183–204 (2000)
2. Basili, V.R., Green, S., Laitenberger, O., Lanubile, F., Shull, F., Sørumgård, S., Zelkowitz, M.V.: The empirical investigation of perspective-based reading. Empirical Softw. Eng. **1**(2), 133–164 (1996)
3. Biffl, S., Halling, M.: Investigating the influence of inspector capability factors with four inspection techniques on inspection performance. In: Proceedings. Eighth IEEE Symposium on Software Metrics, pp. 107–117. IEEE (2002)
4. Sabaliauskaite, G., Matsukawa, F., Kusumoto, S., Inoue, K.: An experimental comparison of checklist-based reading and perspective-based reading for UML design document inspection. In: Proceedings of 2002 International Symposium Empirical Software Engineering, pp. 148–157. IEEE (2002)

5. Lahtinen, J.: Application of the perspective-based reading technique in the nuclear I&C context. CORSICA work report (2011)
6. Laitenberger, O., Atkinson, C.: Generalizing perspective-based inspection to handle object-oriented development artifacts. In: Proceedings of the 1999 International Conference on Software Engineering, pp. 494–503. IEEE (1999)
7. Regnell, B., Runeson, P., Thelin, T.: Are the perspectives really different?–further experimentation on scenario-based reading of requirements. Empirical Softw. Eng. **5**(4), 331–356 (2000)
8. Bittner, K., Spence, I.: Managing Iterative Software Development Projects. Pearson Education, Boston (2006)
9. Murray, A.: The Complete Software Project Manager: Mastering Technology from Planning to Launch and Beyond. Wiley, Hoboken (2016)
10. Shull, F., Rus, I., Basili, V.: How perspective-based reading can improve requirements inspections. Computer **33**(7), 73–79 (2000)
11. Sedelmaier, Y., Landes, D.: Swebos – the software engineering body of skills. Int. J. Eng. Pedagogy (iJEP) **5**(1), 20 (2015)
12. Ramaprasad, A.: On the definition of feedback. Behav. Sci. **28**(1), 4–13 (1983)
13. Narciss, S.: Informatives tutorielles Feedback: Entwicklungs- und Evaluationsprinzipien auf der Basis instruktionspsychologischer Erkenntnisse. Volume 56 of Pädagogische Psychologie und Entwicklungspsychologie. Waxmann, Münster and New York and München and Berlin (2006)
14. Oberhoff, B.: Akzeptanz von interpersonellem Feedback: Eine empirische Untersuchung zu verschiedenen Feedback-Formen (1978)
15. Clark, H.H.: Using Language. Cambridge University Press, Cambridge (1996)
16. Schulz, F.: Peer Feedback in der Hochschullehre hilfreich gestalten–Onlinegestütztes Peer Feedback in der Lehrerbildung mit der Plattform PeerGynt. Ph.D. thesis, Kaiserslautern, Technische Universität Kaiserslautern, Dissertation 2012 (2013)
17. Topping, K.: Peer assessment between students in colleges and universities. Rev. Educ. Res. **68**(3), 249–276 (1998)
18. Topping, K.J.: Trends in peer learning. Educ. Psychol. **25**(6), 631–645 (2005)
19. Figas, P., Hagel, G.: Task is not a task - empirical results about the quality of instructional tasks in higher education. In: 8th Proceedings IEEE Global Engineering Education Conference (EDUCON 2017), Athens, Greece. IEEE (2017)

Peer Review as a Tool for Person-Centered Learning: Computer Science Education at Secondary School Level

Dominik Dolezal[1,2(✉)], Renate Motschnig[3], and Robert Pucher[1]

[1] University of Applied Sciences Technikum Wien, Vienna, Austria
{dominik.dolezal,robert.pucher}@technikum-wien.at
[2] TGM – Vienna Institute of Technology, Vienna, Austria
[3] University of Vienna, Vienna, Austria
renate.motschnig@univie.ac.at

Abstract. Using peer assessment in the classroom in order to increase student engagement by actively involving them in the assessment process has been practiced and researched for decades. The literature suggests using peer review for project-based exercises. This paper analyzes the applicability of peer assessment to smaller exercises at secondary school level and makes recommendations for its use in computer science courses. For this purpose, two secondary school classes consisting of a total of 57 students were introduced to the peer assessment method within the scope of the same software engineering course. Two of 13 exercises were assessed using peer reviews via the Moodle workshop activity. The students were asked to evaluate these two exercises using an anonymous online questionnaire. At the end of the course, they were asked to rate all of the 13 exercises regarding their motivation to learn.

Overall, the anonymous feedback on the peer review exercises was very positive. It has shown that the students not only obtained more feedback, but also received it in a timelier manner compared to regular teacher assessment. The results of the overall rating of all 13 exercises regarding the motivation to learn revealed that the two peer reviewed exercises have been rated distinctly better than the average of the other eleven exercises only assessed by the teacher. Evidence therefore suggests that peer reviews are a viable option for small- and medium-sized exercises in the context of computer science education at secondary school level under certain conditions.

Keywords: Peer review · Person-centered learning · Computer science education Moodle workshops

1 Introduction

Peer assessments have been used in the classroom as a tool to increase student engagement by actively involving them in the assessment process for decades. Carl Rogers, who is seen as the inventor of the person-centered approach as a result of his research in psychotherapy, named self-assessment and peer-assessment elements of significant learning in 1983 [1]. Motschnig et al. iteratively introduced person-centered principles to a computer science course in higher education, making it the best-rated bachelor-level

M. E. Auer et al. (eds.), *Teaching and Learning in a Digital World*,
Advances in Intelligent Systems and Computing 715,
https://doi.org/10.1007/978-3-319-73210-7_56

course of the university rated by more than 5 students [2], which indicates that introducing person-centered approaches such as peer review to computer science courses leads to a positive impact on the students' perception of the course as well as the learning effect.

Dochy et al. performed an extensive literature review [3] and depicted positive effects of peer assessment on students' learning as they become more involved in both the learning and assessment process. Gibbs analyzed reports and studies regarding students' experience of feedback in his article "How assessment frames student learning" in 2006 [4] and found indicators that suggest an increase in student performance if their work is peer reviewed by other students. Gibbs stated that it is not the quality of the feedback which increases student engagement, but rather the instantaneousness of feedback and the fact that it is peer reviewed. He derived eleven "conditions under which assessment supports student learning", six of which address feedback, depicting the importance of the quality, quantity and timing of feedback.

Bauer et al. conducted an empirical study which analyzed students' opinion on online peer reviews in the context of higher education, implementing a peer review system for a scientific writing course [5]. They concluded that students appreciate online peer reviews and highlight the importance of the review criteria. A computer science course addressing Unix shell programming in higher education was evaluated by Sitthiworachart and Joy [6]. They concluded that students appreciate peer reviews as they realized their own mistakes by looking at the work of others and start thinking about their own work more deeply. In addition, most of the students were satisfied with the mark from the peer assessment. Gehringer used peer reviews in three computer science classes, an undergraduate one and two graduate ones, and evaluated the students' perception of peer reviews [7]. He found that students perceive peer reviews as being helpful to the learning process and value the feedback on their work.

However, peer reviews do not need to be done online: Figl et al. compared online peer reviews with face-to-face peer reviews in 2006 [8], focusing on collaboration aspects. They found out that face-to-face reviews improve communication as they promote discussions, but may be more effort for the students to conduct, which is why they recommend combining both methods. Standl developed a framework of educational patterns [9] to be applied to computer science courses at secondary school level, including the peer check as one of the assessment methods in person-centered learning. He suggests that students assessing each other learn more than students who only get assessed by the teacher. However, he recommends using this pattern primarily for projects as it is a time-consuming task.

This paper builds on the existing literature and analyzes the applicability of peer assessments to smaller exercises in computer science classes at secondary school level. Courses bound to a strict curriculum as an external requirement may require students to learn certain topics through exercises, which is not typical for person-centered classrooms. However, introducing person-centered methods such as peer review to a traditional classroom setting may still provide the benefits of peer assessments, which is analyzed in this paper. Standl suggests using peer reviews after a project phase, which raises the question of whether peer reviews are also useful for small- and medium-sized exercises. Based on the reviewed literature, it seems natural that the advantages of peer review can also be observed when they are used for regular exercises in traditional classrooms. Specifically, this paper deals with the following research questions and makes recommendations for using peer reviews for regular exercises in computer sciences courses at secondary school level:

1. To what extent do peer reviews promote student-centered learning?
2. Is the feedback quality of students comparable to the feedback of a teacher?
3. Do students receive feedback in a timelier manner using peer reviews?
4. Does grading become more or less transparent?
5. Are peer reviews a reasonable alternative to teacher assessments?
6. What is a reasonable number of exercises to be assessed by peers?
7. Are students overall satisfied with peer reviews?

2 Methodology

Test Setup. Two secondary school classes of the same year of study, hereafter referred to as "A" and "B", consisting of 29 and 28 students respectively were introduced to the peer-assessment method within the scope of the same software engineering course. Both classes had 13 exercises to be assessed throughout the software engineering course. Two of these 13 exercises were assessed by peers, while the other eleven exercises were assessed by one of the two teachers. The students were asked to evaluate those two exercises using an anonymous online questionnaire. At the end of the course, they were asked to rate all of the 13 exercises regarding their motivation to learn. The software engineering course used a blended learning concept, i.e. the lessons were supported by an online Moodle course. The peer reviews have been conducted using the Moodle workshop activity, which consists of five phases [10]: setup, submission, assessment, grading evaluation, and closing. One of the most important tasks of the instructor is defining clear and concise criteria for the assessment phase, which should be mutually agreed upon with the students.

Assignments. The students were asked to work on each of the 13 exercises for one to three weeks. The peer assessments have been conducted using an iterative approach: students' feedback on the first peer assessment and the lessons learned have been discussed in the classroom and incorporated into the second one. The first peer reviewed exercise involved implementing a simple Java Enterprise web application for a person database, while the second assignment asked students to develop a graphical interface for a route planner using a RESTful API.

After the students read the instructions and review criteria, they were given two weeks to solve the Java Enterprise exercise and one week to solve the RESTful client. After the submission, the students were randomly assigned five to six reviewers and five to six reviewees for the peer review. The whole process was designed to be anonymous: the reviewees did not know their reviewers and vice versa. This is untypical for person-centered approaches; however, it helps to reduce prejudice.

Evaluation. In order to quantitatively assess the students' attitude towards peer reviews, both classes were asked to voluntarily fill out an anonymous questionnaire to give feedback on both peer assessments and estimate the impact on several factors of learning. They were asked to compare the peer assessment with regular teacher assessment and rate their level of agreement on a semantic differential scale from "disagree"[1] (1) to "agree" (5) on several statements assessing different aspects of feedback, learning, grading, and satisfaction. Furthermore, the students were asked to define their preferred number of exercises to be reviewed by peers: 100%, 75%, 50%, 25%, or 0% of all exercises. Finally, they were able to give positive feedback and ideas for improvement through two open questions.

At the end of the school year, the students were asked for another rating of all 13 exercises regarding their motivation to learn. They were asked to rate each exercise from "not motivating" (1) to "motivating" (5). The students were shown the results directly after all of them filled out the form and gave feedback in a final discussion.

3 Results

Figure 1 presents the reported feedback quantity and timing of the first and second peer reviewed exercise. Students agreed in the second iteration with an average of 3.9 that they received more feedback when the exercise was peer reviewed, which is explainable by the number of reviewers: five assessments provide more feedback than a single assessment by the teacher. The measured increase in feedback quantity from the first iteration to the second is explainable by the revised feedback modalities. Written feedback was optional in the first iteration, which led to students giving only ratings, but no suggestions for improvement. This issue was discussed with the students and they agreed on giving written feedback on each submission in the second iteration. The students therefore report that they receive more feedback on peer reviewed exercises than on exercises assessed by the teacher.

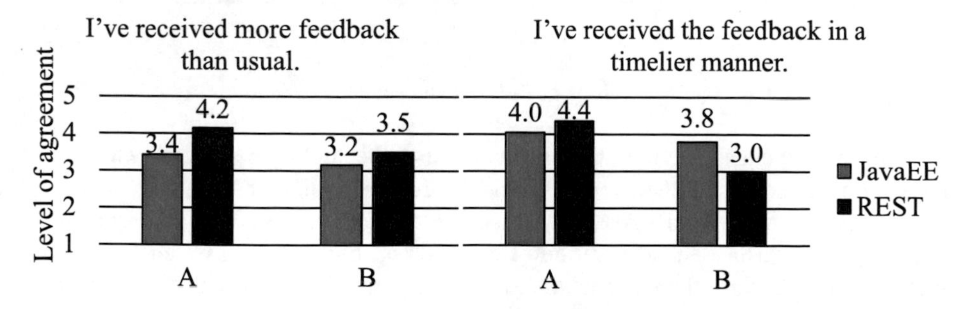

Fig. 1. Feedback quantity and timing

[1] The rating scales, feedback questions, and quotes have been translated from German.

It has also shown that students perceived a measurable improvement in the timing of feedback with an average level of agreement of 3.9 for both iterations. Receiving feedback on an exercise within one week seems to be faster than the average assessment time required by the teacher. Due to administrative work and late submissions, the second iteration needed two weeks for the grading phase in class B, which explains the outlier. Verbal answers regarding what the students liked about the quantity and timing of the feedback were:

"Mostly more feedback than by a teacher"
"Fast feedback"
"Exercises were graded within one week. You don't have to wait for a month for a feedback"
"Guarantee that the exercise is graded within one week"
"Feedback within one week"

Figure 2 depicts the feedback quality as well as the self-reported student engagement. It shows that students were not too pleased with the quality of the feedback in the first iteration as they rather disagreed with this statement. After the mentioned change from an optional written feedback to a mandatory feedback, there was a distinct increase in feedback quality to a total response average of 3.2, indicating that the quality of the feedback by students is indeed comparable to the feedback of a teacher. This suggests that students give reasonable feedback like a teacher would.

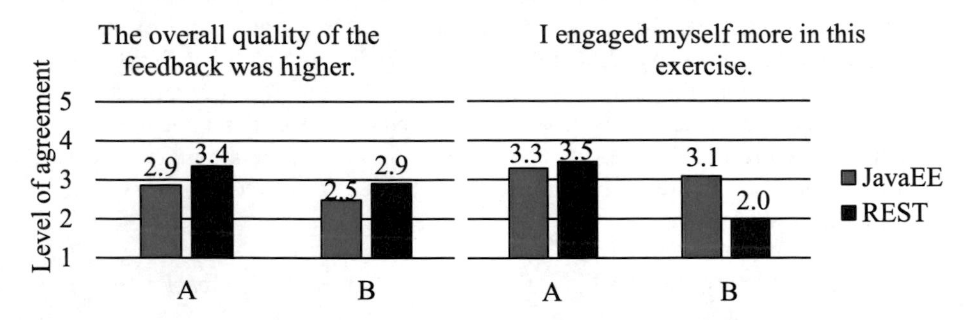

Fig. 2. Feedback quality and student engagement

The reported engagement of the students seems to be slightly better than when only assessed by the teacher: the average level of agreement of both iterations was 3.1, whereas the response of class A reached 3.5 in the second iteration. The noticeable drop of engagement in the second iteration of class B can be explained by analyzing the qualitative feedback of the students: three students who disagreed with this statement wrote that they experienced an unfair deduction of points, resulting in frustration and a drop of engagement. Therefore, it is important to leave the final grading to the teacher in order to minimize unfair penalties. Furthermore, the REST exercise was less open and less creative, which seems to have a limiting effect to student engagement. Some answers to what they liked about feedback quality and engagement were:

"Feedback of others is indeed helpful"
"Receiving hints which are not given by teachers in some cases"
"Altogether, better feedback than usual"
"Finally useful feedback!"
"Detailed feedback"
"Suggestions for improvement"
"More in-class communication about problems and solutions"
"Students engage themselves more"

The results of the evaluation of the own learning effect gained from the reviewing process as well as the estimation of the learning effect on others can be seen in Fig. 3. Most of the students reported to have learned something in the first iteration, while they rather disagreed with this statement in the second iteration. The reason again lies within the assignment: While the JavaEE exercise allowed many different solutions (free choice of database, user interface, validation etc.) and was more creative, the REST exercise had more predefined elements and contained a suggested user interface as a screenshot. Therefore, the solutions of the REST exercise were similar to each other, which is why students could not really learn new and different techniques. This suggests that peer reviews are especially suitable for more creative and open exercises. Verbal feedback on what they liked about the learning effect includes:

"You can see what others did better/worse"
"Making sure that everything works on different computers"
"Many hints"
"A reasonable, informative comment. I am happy!"
"I could give other students reasonable feedback, maybe even more than a teacher due to his limited time"
"Grading good and bad solutions (Dos and Don'ts)"
"You learn different coding styles"
"Seeing different approaches"
"Seeing how others solved the exercise"
"Through the assessments, you see how others solved the exercise. This improves the learning effect."

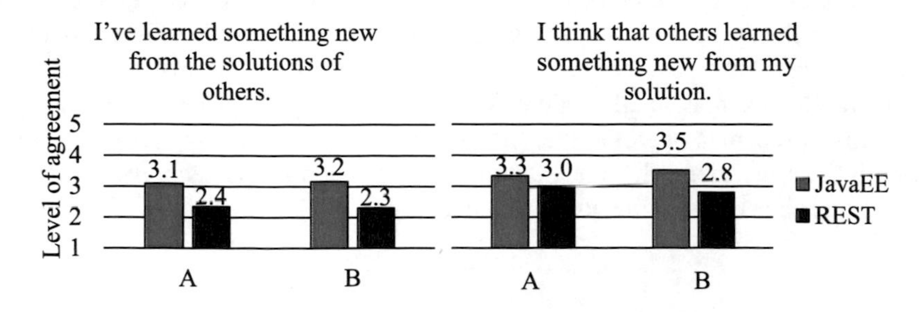

Fig. 3. Learning effect

Figure 4 shows the reported transparency of grading as well as the preferred number of exercises being peer reviewed. Students report that peer assessment follows a more transparent grading scheme than teacher assessment. The average level of agreement on both iterations was 3.6; only about 12% disagreed with this statement. Five reviews seem to give a better estimate of the grade than a single assessment by the teacher as seen by the students.

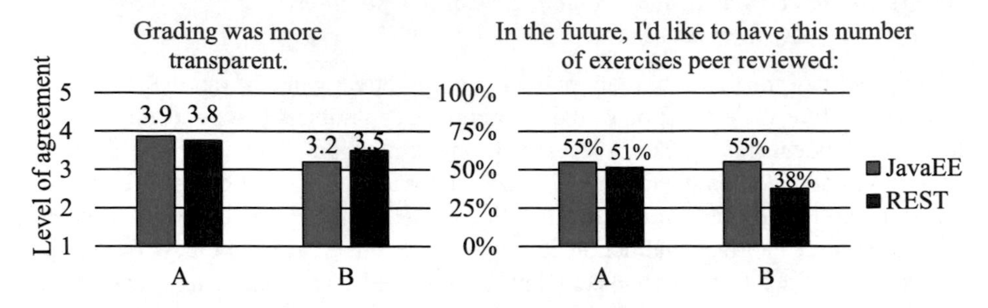

Fig. 4. Transparency of grading and amount of peer review exercises

On average, the students would prefer to have about half of the exercises (52%) peer reviewed. The noticeable drop in class B can again be explained by the frustration of some students due to reported unfair assessment. Qualitative feedback on what they liked regarding grading included:

"Criteria clear and understandable"
"You know what to focus on and what is important for grading"
"A wide selection for a precise grading"
"Mainly fair and understandable feedback"
"The criteria were more precise this time"
"More transparent"

The self-reported satisfaction with teacher assessment and peer assessment is depicted in Fig. 5. It can be seen that students still like to be assessed by the teacher. Although they report to like the peer assessment method with a total average of 3.4, students still prefer the opinion of the teacher as an expert. One of the students stated: "Teachers are always the best at grading. It's simple and reasonable." This can also be seen as a compliment to the teacher as they are very satisfied with his or her teaching and grading. However, although students value high-quality feedback of a teacher, peer assessments still seem to trigger higher student engagement, as also suggested by Gibbs' research [3]. Furthermore, the students themselves stated in the second iteration that student feedback is of the same quality as teacher feedback.

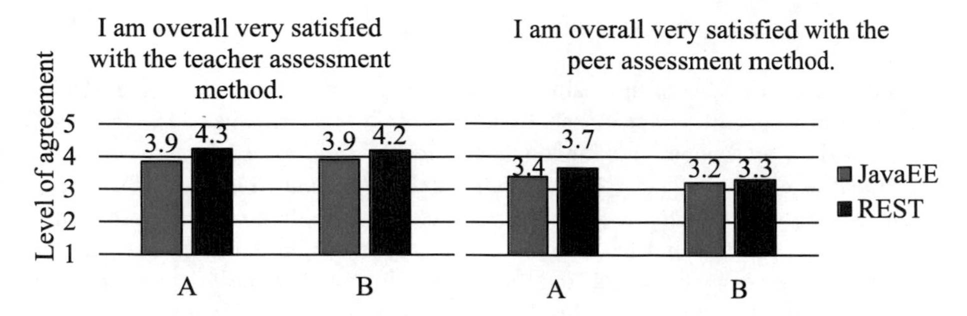

Fig. 5. Satisfaction with teacher assessment and peer assessment

Figure 6 shows the results of the final rating of all 13 exercises regarding their motivation to learn the respective topic. In three of four cases, the peer reviewed exercise received a rating notably better than the average of all other exercises. Interestingly, class B rated the second peer reviewed exercise distinctly better than the first one. This seems to contradict their initial feedback right after the peer reviewed exercises were completed as they reported more positive effects in the first iteration. In the discussion after the final rating of all 13 exercises, the students stated that the JavaEE exercise did not meet their field of interest, resulting in a low overall rating for the JavaEE exercise.

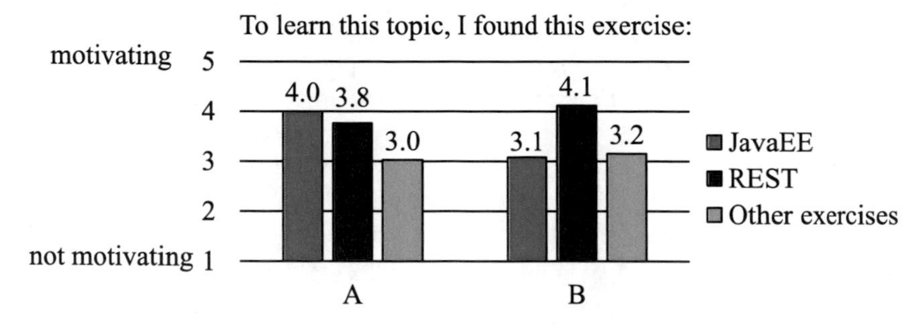

Fig. 6. Motivation to learn

4 Recommendations

Based on the introduced empirical results, the following recommendations for promoting student-centered learning using peer reviews in computer science courses at secondary school level in traditional classrooms can be formulated:

- **1. Qualitative feedback.** An agreement to provide written feedback should be made with the students. By default, giving qualitative feedback is optional in the Moodle workshop activity.
- **2. Anonymous feedback.** Although this is unusual for person-centered approaches, reviewer and reviewee should be anonymous to reduce prejudices. This maximizes transparency of grading and prevents "upvoting" and "downvoting". This may sound

easy at first glance, but students may be used to putting their name on their submissions.

- **3. Fast feedback.** One of the main advantages of peer assessment is fast feedback. In order to be useful, feedback should be given in a timely manner [4]. One week seems to be a reasonable time for computer science exercises.
- **4. Black-box testing.** The rating criteria should be formulated in a way that every student – regardless whether he or she was able to solve the exercise – can assess them. Some students and teachers raised the objection that students may not be qualified to assess the assignments of others. However, if the criteria focus on features, students can rate the submissions from a user's point of view.
- **5. Transparent criteria and conditions.** The rating criteria as well as the general conditions for giving feedback need to be communicated and agreed upon. This ensures that students use the same rating scale and grading becomes reproducible.
- **6. Final grading by the teacher.** Although most of the peer review grades did not need to be changed, the teacher always needs to do the final grading of the exercise. Mistakes happen and should be corrected by the teacher.
- **7. Shared level of basic knowledge.** Students need a certain level of expertise in order to give good feedback on the solutions of others. If they still struggle with computer science basics, it is questionable whether they are able to thoroughly test a program, even if the criteria are formulated from a user's perspective.
- **8. Exercises allowing creativity.** Peer reviews seem to be especially useful if there are multiple possible solutions for the exercise as the students seem to learn more from each other and are more engaged. This corresponds with the findings of Standl [9], who recommends using peer reviews for project-based assignments. The more behavioristic an exercise is, the less powerful peer reviews become.
- **9. Do not overuse it.** Students report that they would be fine with a peer review on every other exercise. However, peer reviews were a refreshing alternative to teacher assessments in this case. They are still time-consuming, both for the teacher and the students, and could quickly lose their charm if they are overused.

5 Conclusion

This paper analyzed the applicability of peer reviews to small- and medium-scaled exercises in computer science courses at secondary school level to introduce person-centered approaches to traditional classrooms. Based on empirical evidence collected over one year, the following answers to the research questions have been found:

1. To what extent do peer reviews promote student-centered learning?
   Peer reviews seem to promote student-centered learning if they are used correctly. The results indicate a clear improvement in feedback quantity and timing as well as student engagement and motivation to learn. In addition, the students liked the peer reviews and regarded them as a refreshing alternative to predominating teacher assessments.
2. Is the feedback quality of students comparable to the feedback of a teacher?

Yes. Students report that feedback quality is indeed comparable to the feedback of a teacher if written feedback is given.
3. Do students receive feedback in a timelier manner using peer reviews?
   Yes. The students perceived faster feedback on their assignments as they received it within one week or two weeks respectively.
4. Does grading become more or less transparent?
   The students stated that grading became more transparent, which is explainable by the higher number of persons who assess the submission. Furthermore, teachers have to pay special attention to the definition of the rating criteria for peer reviews, so that the grading scheme of exercises is sufficiently transparent.
5. Are peer reviews a reasonable alternative to teacher assessments?
   Yes. Peer reviews seem to be a reasonable alternative to teacher assessments in computer science courses. Nevertheless, some constraints need to apply in order to make them a useful tool for teaching and assessing.
6. What is a reasonable number of exercises to be assessed by peers?
   Students report that about every other exercise could be reasonably peer reviewed. However, peer reviews should not be overused. The exact number of exercises depends on the type of exercise.
7. Are students overall satisfied with peer reviews?
   Yes. Students are satisfied with peer reviews as an assessment tool. However, they report that they still value the high-quality feedback of an expert.

To conclude, the overall feedback on the peer review exercises was very positive. However, it was found that some additional constraints such as open assignments as well as obligatory and fast feedback need to apply to make peer reviews practicable and reasonable for small- and medium-scaled exercises in traditional classrooms.

References

1. Rogers, C.: Freedom to Learn for the 80's, 2nd edn. Merrill, Columbus (1983)
2. Motschnig, R., Sedlmair, M., Schröder, S., Möller, T.: A team-approach to putting learner-centered principles to practice in a large course on human-computer interaction. In: 2016 IEEE Frontiers in Education Conference, pp. 1–9. IEEE, Eire (2016)
3. Dochy, F., Segers, M., Sluijsmans, D.: The use of self-, peer and co-assessment in higher education: a review. Stud. High. Educ. **24**(3), 331–350 (1999)
4. Gibbs, G.: How assessment frames student learning. In: Bryan, C., Clegg, K. (eds.) Innovative Assessment in Higher Education, pp. 23–36. Routledge, Oxfordshire (2006)
5. Bauer, C., Figl, K., Derntl, M., Beran, P.P., Kabicher, S.: The student view on online peer reviews. In: SIGCSE Bulletin, vol. 41, no. 3, pp. 26–30. ACM, New York (2009)
6. Sitthiworachart, J., Joy, M.: Web-based peer assessment in learning computer programming. In: 3rd IEEE International Conference on Advanced Technologies, pp. 180–184. IEEE, Athens (2003)
7. Gehringer, E.F.: Strategies and mechanisms for electronic peer review. In: 30th ASEE/IEEE Frontiers in Education Conference, vol. 1, F1B/2-F1B/7. IEEE, Kansas City (2000)
8. Figl, K., Bauer, C., Mangler, J.: Online versus face-to-face peer team reviews. In: 36th ASEE/IEEE Frontiers in Education Conference, pp. 7–12. IEEE, San Diego (2006)

9. Standl, B.: Conceptual modeling and innovative implementation of person-centered computer science education at secondary school level. Doctoral thesis, University of Vienna (2013)
10. MoodleDocs: Using Workshop, 25 February 2017. https://docs.moodle.org/32/en/Using_Workshop. Accessed 25 Apr 2017

Can Pair Programming Address Multidimensional Issues in Higher Education?

Marco Klopp[1(✉)], Carolin Gold-Veerkamp[1(✉)], Martina Kuhn[2(✉)], and Joerg Abke[1(✉)]

[1] University of Applied Sciences Aschaffenburg, Wuerzburger Strasse 45, 63743 Aschaffenburg, Germany
{marco.klopp,carolin.gold-veerkamp,joerg.abke}@h-ab.de

[2] University of Applied Sciences and Arts Coburg, Friedrich-Streib-Strasse 2, 96450 Coburg, Germany
martina.kuhn@hs-coburg.de

Abstract. To handle heterogeneity within students, to foster needed generic competencies, to motivate them, and to increase their employability, a didactical method to teach and learn programming in non-major degree programs shall be found. Therefore, this paper covers strategies and the theoretical underpinning concerning these four challenges and gives solutions how to cope with them. Finally, Pair Programming is presented as a method that addresses the issues introduced here.

Keywords: Programming · Heterogeneity · Generic competencies · Motivation Employability

1 Introduction

In today's digital society software is omnipresent. For this reason, it is obligatory to compromise programming in scientific and technical subjects in higher education, although students often are not aware of it. Based on initial survey questionnaires, summative evaluations, and experiences some issues in teaching programming have been identified, which should be addressed in addition to specialized programming knowledge and skills. These challenges are: Students' heterogeneity, generic competencies, motivation of learners, and their future employability.

This might be due to the fact that the degree program considered is a non-major informatics or computer science course of study, which means that informatics and programming are rather unpopular.

The following chapters therefore cover a detailed look at the issue and define goals how to tackle them (see Sect. 2). The method found to address these issues will be displayed in Sect. 3 and an answer to the question: How can it solve the challenges?

To conclude the paper, Sect. 4 compromises a summary and an outlook, which covers first ideas for the setting and a strategy for evaluating the method with regard to the defined issues.

M. E. Auer et al. (eds.), *Teaching and Learning in a Digital World*,
Advances in Intelligent Systems and Computing 715,
https://doi.org/10.1007/978-3-319-73210-7_57

2 Challenges in Teaching Programming

The purpose of this paper is to find a method, which improves teaching and learning programming and addresses the presented issues simultaneously. The following subsections take a closer look at each issue described.

2.1 Students' Heterogeneity

Students have different entry qualifications, which vary from expert to novice level (cf. initial survey), but: Can this heterogeneity be used in a positive way? How?

Heterogeneity includes different approaches, which can be used in learning situations (see Fig. 1). The following classification is based on Tillmann [19] and is also used by Krüger-Baser et al. [11, p. 166f.].[1]

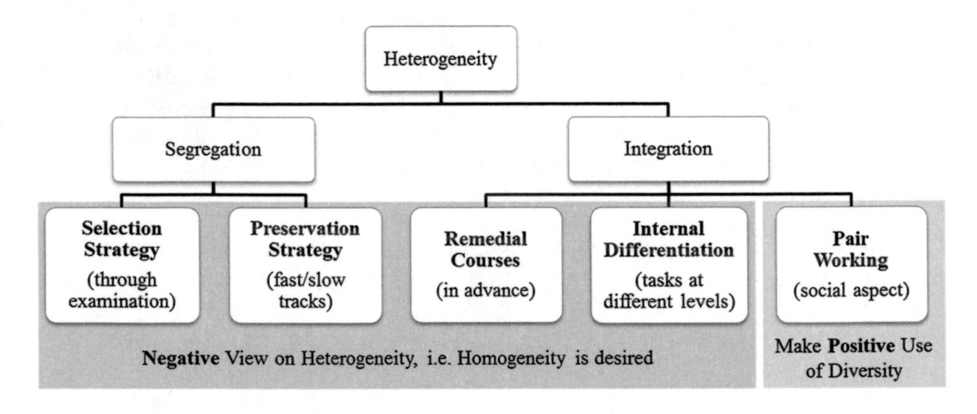

Fig. 1. Strategies to handle heterogeneity [cf. 11, p. 167f.]

Only Pair Working (see Fig. 1, most right item) uses the diversity of individuals in a positive way to create a learning arrangement. The other strategies regard heterogeneity as a deficit or handicap and try to achieve homogeneity [11, p. 166f.].

It appears sensible to look at it in a positive light, i.e. to use diversity in education and therefore apply the integrational approach with Pair Working.[2]

2.2 Mediation of Generic Competencies

In addition to specialized knowledge, which has been the core of university education for centuries, various generic competencies (GCs) are necessary. Concerning programming, these cover personal (PCs), methodological (MCs) as well as social skills (SCs). Combined with professional competency these generate job-related competency.

[1] Another structure of strategies to deal with heterogeneity is given by Weinert [20].

[2] One should be clear about the fact that the students also have to write the same exam at the end of the semester, which guarantees a minimum level of expertise. This – of course – contradicts the idea of heterogeneity in some way, but is an institution and accreditation issue.

Which GCs are at the centre of attention? This question should be viewed in the following part. Therefore three different perspectives (informatics[3] [1, 2, 6, 8], labour market[4] [3, 12, 23], and four lecturers[5]) are taken into consideration to define the generic competencies that should be trained in informatics[6] during studies.

The grey-marked competencies represent those ones that are of high importance as they are part of all three perspectives. Regarding a didactical method, which is looked for, it occurs meaningful to use a collaborative method in order to foster social interaction and the GCs (see SCs in Table 1).

Table 1. Competencies for programming from different perspectives (It has to be noted that no definitions are given; i.e., it is not sure whether the same competencies are implied. That is why clusters have been built and generic terms are used.)

Competencies		Discipline Informatics	Labour Market	Lecturers
MCs	Analytical Skills	[6]	[3, 12, 23]	X
	Learning Skills	[1, 2, 8]	[3, 12]	X
	Various Ways of Thinking	[6]		X
	Media Competence	[2, 8]		
	Working Techniques	[2]	[3]	
	Scientific Working	[2, 8]	[3, 12, 23]	
SCs	Communication	[1, 2, 6, 8]	[3, 12, 23]	X
	Intercultural Skills		[3]	
	Team Skills	[2]	[3]	X
PCs	Stamina	[8]	[23]	X
	Empathy	[8]		
	Concentration			X
	Creativity		[3]	X
	Competence to Reflect	[6]	[3]	
	Self-Management Competency	[8]		
	Self-Confidence		[23]	
	Professional & ethical responsibility	[1]		

2.3 Motivation of Students

Because of the non-major subjects, the motivation for learning programming is relatively low, i.e.: In which case can students' motivation be improved?

In university education the types of motivation are interesting, which apply to a sustainable learning so the students identify themselves with their subject [10, p. 32].

Based on reports of lecturers it is known that students do not have a lot of motivation for subjects, which are not central to the field of study – in their view [17, p. 88].

[3] Perspective "informatics" from ABET [1], ASIIN [2], a dissertation [6], and GI [8].
[4] Perspective "labour market" from BDA [3], KMK [12], and Wissenschaftsrat [23].
[5] The "lecturers" asked are professors in informatics in higher education.
[6] Due to a lack in literature on generic competencies for programming; focus on informatics.

The well-known Self-Determination Theory of Deci and Ryan [5] differentiates between three fundamental psychologic requirements: Autonomy, experience of competence and social integration. Regarding motivation, the differentiation between intrinsic and extrinsic motivation has become common. Deci and Ryan [5] prove in their theory that extrinsic motivation can function as an intrinsic motivation through fostering the three aspects mentioned above; i.e. these are essential for the development and maintenance of extrinsic motivation with an intrinsic character that is established from extraneous factors.

In the following contemplation these three central goals [5] will be focussed. Also the use of a method from/applied in industry (see Sect. 2.3) plus strengthening relevant soft skills (see Sect. 2.2) could help to motivate students.

2.4 Employability

Through the Bologna Process the focus is on competencies in order to emphasize employability and job readiness [9]; paradigm shift towards outcome-orientation.

For this reason, a didactical approach of universities, which use a method from industry or one that is at least used in the working life, seems to be reasonable.

2.5 Solution Approach

In the following, the purpose is to find a method, which is suitable to achieve the defined goals and four linked challenges. The theoretical view on the issues (in Sects. 2.1–2.4) and the handling of these aspects show a connection concerning the social dimension of learning: The strategy of "Pair Working" as a solution of **heterogeneity**, the need of autonomy, "Social Integration", and the experience of competencies to increase **motivation**, the **generic social competencies** working and communicating in a team are of central importance for programming and its training, as well as the promotion of **employability**, because of the outcome orientation and focus on SCs.

Regarding research and a literature review for adequate methods in informatics covering the desired requirements, only "Pair Programming" [4] seems to be suitable. Therefore, this method is displayed in more detail as follows. Additionally, a closer look at the challenges described above in connection to Pair Programming is given.

3 Pair Programming

Pair Programming (PP) [4] is a method for agile software engineering – namely Extreme Programming (XP). Two collaborating persons go through a development process, while each person has a specific role; one is the driver, one the navigator. The goals are to increase the quality of software, to speed up the development process and distribute necessary knowledge in the team (or group) [4].

Certainly, there are more positive effects possible, which apply to heterogeneity, generic competencies, motivation, and employability, which are shown hereafter.

3.1 Related Work

Some surveys already deal with PP and its benefits in higher education. Most surveys research whether the course achievements of students who use PP are superior to those of students who program on their own.

One result of Nagappan et al. [13] is that "student pair programmers were more self-sufficient, generally perform better on projects and exams, and were more likely to complete the class with a grade of C or better than their solo counterparts" [13, p. 359]. Fronza et al. [7] give a collection of literature that covers different positive facets of PP, such as: Quality of code, "facilitates the introduction of new team-members while not lowering the overall productivity" [7, p. 225] etc. Salinger [16, p. 27 ff.] also shows some related work concerning an increase in quality, concentration, trust in results, and the use of PP in education. What all these publications and their references lack is the combinations of aspects as presented in this paper.

3.2 Heterogeneity and Pair Programming

While using PP both partners are completely equal, irrespective of their age or their own programming skills [15, p. 54]. Usually the paired developers are differently socialized, have different experiences and different knowledge; they use different problem solving strategies and tools [15, p. 63]. Considering the approach of "Pair Working" (Sect. 2.1) and comparing it to "Pair Programming", it can be seen that PP is a domain-specific special form of integrating heterogeneous individuals.

3.3 Motivation in Pair Programming

Nosek [14, p. 57] and others describe that the users of PP have more fun at work [cf. 16, p. 3; 6, p. 227; 19; 20][7] and PP leads to greater confidence in one's own work [cf. 6, p. 227; 13]. The constant reciprocal feedback leads to an increase in motivation [6, p. 227; 13]. According to Deci and Ryan [5] (see Sect. 2.3), the factors "Autonomy", "Experience of Competence" and "Social Integration" are essential.

- The feedback, which happens while working together during PP, allows learners to gain more confidence in their own work (**self-confidence**) and recognize that they are able to master certain tasks effectively (**self-efficacy**).
- During PP the learners will not be restricted from the outside in the solution process; the couple works **self-determined/autonomous**.
- Through mutual real time feedback during the development process it is possible for the learners to **experience their own professional and general competencies**. For example, by explaining certain procedures to their partner and defending them.
- In the pair they are **socially involved** and also have to use their social competencies (see point above).

[7] "Pair programmers are happier" [18, p. 3]; "more enjoyable at statistically significant levels" [7, p. 227; cf. 21, 22]; "enjoy the problem-solving process more" [7, p. 227; cf. 14].

3.4 Generic Competencies in Pair Programming

On the one hand, certain key competencies are required to use the method effectively. But on the other hand, the method can provide a test/learning environment in order to acquire these competencies. The addressed key competencies, which can be extracted from the elaboration of Reinhardt [15], are the following:

All competencies which are important for PP (see Table 2) are also part of the summary of the perspective of informatics, labour market and lecturers (see Table 1, grey marked); only "Concentration" (Table 2) and "Stamina" (Table 1) are listed once. A difference can be recognized when looking at the MC "Problem Solving Skills" in Table 2, in Table 1 this competency has been subsumed in "Analytical Skills".

Table 2. Generic competencies in pair programming [cf. 15, p. 56 ff.]

MCs	SCs[a]	PCs[b]
Problem Solving Skills[c]	Team Skills[d]	Concentration
Learning Skills	Communication[e]	

[a] Also: "Conflict resolution competency" [18, p. 4]; "empathy and antipathy" [18, p. 4];
[b] Also: "Self-confident appearance" [18, p. 4]; "Developing critical facility" [18, p. 4];
[c] Also: "problem-solving" [7, p. 227; cf. 14]; "problem solving skills" [7, p. 227; cf. 21; 22];
[d] Also: "Pair Programming [...] improves teamwork" [18, p. 3]; "communication skills" [18, p. 4]; "Learning and understanding different patterns of team working" [18, p. 4]; "improves team communications" [7, p. 227]; "teamwork" [7, p. 227; cf. 21; 22];
[e] Also: "communication" [7, p. 227; cf. 21; 22]; "discuss [...] solutions" [18, p. 4]

As a conclusion, PP seems to be reasonable to help fostering these GCs.

3.5 Employability Concerning Pair Programming

In order to distribute knowledge in a team and thus ensure that many participants are knowledgeable, PP is used in the industrial practice [4] besides other factors (see Sect. 3 and Subsections); e.g., fewer code defects, more fun, etc. By embedding PP in higher education, the employability of future graduates is enhanced by understanding the importance of PP in industry and also the motivation is increased as students use a method that is relevant for their future profession. In addition, a well-known method gives the graduate self-confidence for entering the labour market.

3.6 Conclusion

Because of the method PP covering the competency requirements of informatics, the labour market, and the lecturers, the conclusion is to use PP as a teaching method in programming education. In addition, it also uses the heterogeneity inside pairs and has the ability to motivate students in terms of Self-Determination Theory. It has the advantage that students can try a method from agile software development, which is also applied in the real working world.

4 Summary and Outlook: Didactical Setting and Evaluation

Starting with the four challenges (heterogeneity, generic competencies, motivation, and employability), theoretical preliminary considerations – underlaid with literature – to handle them individually are given. Depending on these findings, a literature research was undertaken to find a suitable method to address these issues simultaneously. The only suitable method encountered was Pair Programming.

In prospect, a didactical setting is being designed, which is able to focus on the four issues (see Sect. 2) using PP; i.e. an operationalization of Sects. 2 and 3 in a beginners programming course. The implementation in this course is planned in the second half of this summer term in a Mechatronics Bachelor's degree program. Therefore, the conceptual design is started, but not yet finalised; this could be a central part of further publications – besides evaluation analyses.

To be able to evaluate this teaching and learning arrangement regarding the specified goals, various forms of formative and summative surveys will be carried out accompanied by additional observations to inspect the method concerning: Heterogeneity, generic competencies, motivation, and employability.

Acknowledgement. The present work as part of the EVELIN project was funded by the German Federal Ministry of Education and Research (Bundesministerium für Bildung und Forschung) under grant numbers 01PL17022B and 01PL17022A. The authors are responsible for the content of this publication.

References

1. ABET: Accreditation – Criteria for accrediting engineering programs 2016–2017 (2015). http://www.abet.org/wp-content/uploads/2015/10/E001-16-17-EAC-Criteria-10-20-15.pdf. Accessed 03 Apr 2017
2. ASIIN: Fachspezifische ergänzende Hinweise – Zur Akkreditierung von Bachelor- und Masterstudiengängen der Informatik (2011). [German]. http://www.asiin-ev.de/media/feh/ASIIN_FEH_04_Informatik_2011-12-09.pdf. Accessed 03 Apr 2017
3. BDA (Bundesvereinigung der Deutschen Arbeitgeberverbände (BDA): Memorandum zur gestuften Studienstruktur (Bachelor/Master) (2003). [German]. http://ids.hof.uni-halle.de/documents/t699.pdf. Accessed 03 Apr 2017
4. Beck, K.: Extreme Programming Explained, 1st edn. Addison Wesley, Boston (2000)
5. Deci, E.L., Ryan, R.M.: Handbook of Self-Determination Research. University of Rochester Press, Rochester (2002)
6. Dörge, C.: Informatische Schlüsselkompetenzen – Konzept der Informationstechnologie im Sinne einer informatischen Allgemeinbildung, Dissertation (2012). [German]. http://oops.uni-oldenburg.de/1426/1/doeinf12.pdf. Accessed 25 Apr 2017
7. Fronza, I., Sillitti, A., Succi, G.: An interpretation of the results of the analysis of pair programming during novices integration in a team. In: International Symposium on Empirical Software Engineering and Measurement (ESEM), Lake Buena Vista, FL, USA, pp. 225–235. IEEE (2009)

8. Gesellschaft für Informatik e.V. (GI): Empfehlungen für Bachelor- und Masterprogramme im Studienfach Informatik an Hochschulen (2016). [German]. https://www.gi.de/fileadmin/redaktion/empfehlungen/GI-Empfehlungen_Bachelor-Master-Informatik2016.pdf. Accessed 03 Apr 2017
9. Greinert, W.-D.: Beschäftigungsfähigkeit und Beruflichkeit – Zwei konkurrierende Modelle der Erwerbsqualifizierung? (2008). [German]. https://ww.bibb.de/veroeffentlichungen/de/publication/download/1365. Accessed 24 Mar 2017
10. Hawelka, B., Hammerl, M., Gruber, H.: Förderung von Kompetenzen in der Hochschullehre. Theoretische Konzepte und ihre Implementation in der Praxis. Kröning. Asanger Verlag (2007). [German]
11. Krüger-Basener, M., Ezcurra Fernandez, L., Gößling, I.: Heterogenität als Herausforderung für Lehrende der angewandten Technikwissenschaft im Teilprojekt Nord. In: Bülow-Schramm, M. (ed.) Erfolgreich studieren unter Bologna-Bedingungen – Ein empirisches Interventionsprojekt zu hochschuldidaktischer Gestaltung. Bertelsmann Verlag, Bielefeld, pp. 162–190 (2013). [German]
12. Kultusministerkonferenz (KMK): Qualifikationsrahmen für Deutsche Hochschulabschlüsse (2005). [German]. http://www.kmk.org/fileadmin/Dateien/veroeffentlichungen_beschluesse/2005/2005_04_21-Qualifikationsrahmen-HS-Abschluesse.pdf. Accessed 03 Apr 2017
13. Nagappan, N., Williams, L., Ferzli, M., Wiebe, E., Yang, K., Miller, C., Balik, S.: Improving the CS1 experience with pair programming. In: SIGCSE Technical Symposium on Computer Science Education, pp. 359–362. ACM Press, New York (2003)
14. Nosek, J.T.: The case for collaborative programming. Commun. ACM **41**(3), 105–108 (1998)
15. Reinhardt, W.: Einfluss agiler Softwareentwicklung auf die Kompetenzentwicklung in der universitären Informatikausbildung – Analyse und Bewertung empirischer Studien zum Pair Programming (Diplomarbeit – Universität Paderborn) (2006). [German]
16. Salinger, S.: Ein Rahmenwerk für die qualitative Analyse der Paarprogrammierung; Dissertation (2013). http://www.diss.fu-belin.de/diss/servlets/MCRFileNodeServlet/FUDISS_derivate_000000013531/ssr_thesis_vertical.pdf. Accessed 04 May 2017
17. Schmolitzky, A.: Zahlen, Beobachtungen und Fragen zur Programmierlehre. In: Bruegge, B., Krusche, S. (eds.) SEUH 2017, Software Engineering im Unterricht der Hochschulen. Hannover, Deutschland (2017). [German]. http://ceur-ws.org/Vol-1790/paper10.pdf. Accessed 05 Apr 2017
18. Schumm, M., Joseph, S., Schroll-Decker, I., Niemetz, M., Mottok, J.: Required competences in software engineering: pair programming as an instrument for facilitating life-long learning. In: International Conference on Interactive Collaborative Learning (ICL), Villach, Austria, pp. 1–5. IEEE (2012)
19. Tillmann, K.-J.: Separierung und Integration. Oder: was will Integrative Pädagogik? In: Pädagogik (Weinheim), vol. 47, no. 10, pp. 6–9 (1995). [German]
20. Weinert, F.E.: Notwendige Methodenvielfalt: Unterschiedliche Lernfähigkeiten erfordern variable Unterrichtsmethoden. In: Wege zur Selbstständigkeit. Seelze, pp. 50–52 (1997). [German]
21. Williams, L.: The collaborative software process. Ph.D. dissertation. The University of Utah (2000)
22. Williams, L., Kessler, R.R., Cunningham, W., Jeffries, R.: Strengthening the case for pair programming. IEEE Softw. **17**(4), 19–25 (2000)
23. Wissenschaftsrat: Empfehlungen zum Verhältnis von Hochschulbildung und Arbeitsmarkt – Zweiter Teil der Empfehlungen zur Qualifizierung von Fachkräften vor dem Hintergrund des demographischen Wandels, Bielefeld (2015). [German]. https://www.wissenschaftsrat.de/download/archiv/4925-15.pdf. Accessed 03 Apr 2017

PTD: Player Type Design to Foster Engaging and Playful Learning Experiences

Johanna Pirker[1], Christian Gütl[1,2](✉), and Johannes Löffler[1]

[1] Graz University of Technology, Graz, Austria
jpirker@iicm.edu, c.guetl@tugraz.at,
j.loeffler@student.tugraz.at
[2] Curtin University, Perth, Australia

Abstract. In this paper we present a design model, *PTD (Player Type Design)*, to create engaging gaming and non-gaming experiences for attracting different types of players to learning settings. Based on Bartle's four player types, elements grounded on game design theory are introduced to design collaborative, competitive, explorative, and rewarding learning experiences. We illustrate the use of the framework on two different experiences. The main contribution of this paper is the design model "PTD", which can be used to create and also analyse engaging experiences in different contexts (gaming and non-gaming) based on different player types as known from game design theory. The model is evaluated with two different experiences: (1) a blended learning experience, (2) a mobile game with purpose.

Keywords: Game-based learning · Design guidelines · Engagement
Player types · Education · Computer games

1 Introduction

Designing engaging experiences, in particular in a non-gaming context is a challenging task. Strategies based on game design theory introduce ways to make this task easier. In recent years, the use of video games, game design theory, or single game elements has attracted interest as a powerful tool to make different non-gaming tasks and experiences more engaging and "fun" [9, 11, 15]. One form of incorporating game elements in a non-gaming context is *gamification*. Gamification strategies describe the use of game design elements, which can be used to engage users in non-gaming contexts [6]. These game design elements can be used to make different non-gaming tasks more attractive and engaging. The gamification of domains such as learning, training, fitness, business applications, or health in particular has become increasingly popular in recent years. Gamification strategies, however, are also often criticized as being used to design experiences which are not meaningful (e.g. giving points for meaningless actions, using external rewards to control behavior) [13]. One of the reasons for this issue is that many designers do not consider that not all players are engaged for the same reasons and by the same engagement elements. All players do not have the same playing behavior, the same reason for playing, nor are they attracted by the same game design elements [1, 7], such as various forms of points, badges, and achievements. Bartle described in [2] four

M. E. Auer et al. (eds.), *Teaching and Learning in a Digital World*,
Advances in Intelligent Systems and Computing 715,
https://doi.org/10.1007/978-3-319-73210-7_58

main player types (in multi-user-dungeons), each of which is engaged by different interactions with the environment or other avatars. While different forms of achievement, such as points, badges, and awards, engage some players, others are more engaged by interacting with other users, or exploring the game environments. Also in the non-gaming context, simply adding points to reward specific actions is not engaging for every user. Some users would rather enjoy taking their time to explore the experience (e.g. website), or enjoy the experience shared with others and are engaged and rewarded by interactions of these kinds with the environment or other users. Gamification elements are used in the educational applications to increase the learners' engagement and interest in the learning content by adding game-based elements such as points, rewards, or badges. When looking closer at pedagogical theory, however, it is apparent that all learners do not learn in the same way: learners have different methods and styles of learning [8, 12]. This also applies in the issue of how to integrate game elements to engage learners: for example it is not every learner in game-based or gamified scenarios who can be engaged by winning points and badges for completing assignments, or seeing leaderboards and ranking information. Competitive elements in particular can even be stressful and frustrating for some learners, while by contrast, cooperative strategies very often achieve better learning outcomes [10, 17].

In this article we intend to introduce a model for designing and evaluating non-gaming experiences and add game-based elements to these strategies in order to attract and engage learners. The remainder of this paper is organized as follows: we will first take a closer look at various game-based learning design strategies and then discuss player types as they are known from game design theory. This is followed by the introduction of the Player Type Design (PTD) model and followed by investigated this model by two case studies in a learning, but also non-learning context.

2 Background

2.1 Game-Based and Gamification Strategies for Designing Experiences with a Purpose

While educational games or games with a purpose are usually designed in a process similar to that in the design of traditional games, gamification is the process of integrating game elements in non-gaming environments [6]. Different frameworks and design guidelines have been provided to design educational games or educational experiences based on gamification strategies.

Zichermann and Cunningham [21] describe different game mechanics to support gamification processes. These include elements for scoring (e.g. points), illustrating progress (e.g. levels, progress bars), indicating competition and rankings (e.g. leaderboards, high scores), or badges (to allow collecting and surprise elements). Additionally, they describe the importance of designing minor activities with clear goals, such as challenges, missions, or quests and also activities supporting social engagement as well as different onboarding strategies (helping user learning of how to play the game/interact with the system). Linehan et al. [15] introduce guidelines for designing educational games. They propose 'Applied Behavioral Analysis' as an educational framework, which can be aligned with the principles of the game design and the pedagogical strategies and goals: first, the target behavior students ought to improve is defined; second, the performance is measured; third, the performance is analyzed; fourth, feedback is presented. Following on from this the learner is located in a loop where performance is measured again or the learner is rewarded. Learning takes place in iteration cycles and learners are awarded based on these cycles. Kotini and Tzelepi [13] introduce a framework based on Kumar's player-centered design [14]. This supports the design of educational experiences based on gamification strategies and focuses on three categories of elements: behavior (elements focusing on human behaviors such as open-type problems, freedom of choice, imaginary, creating emotions, team cooperation), feedback (elements giving feedback, if possible immediate if the goals have been accomplished), and progression (progression elements give a sense of structure and advancement). Annetta [1] describes a framework for serious educational game design. The author presents six (nested) main elements for educational game design: identity (identification with the environment), immersion (feeling of presence and engagement with the content, success in achieving goals, feeling of flow), interactivity (social interactions and communication), increasing complexity (level, increasing difficulty), informed teaching (feedback and assessment), and being instructional (learning as goal).

While these frameworks use different elements, general design principles can be observed in all of these frameworks: clear goals, fast feedback, and a sense of control. These characteristics and design goals are also used by Csikszentmihalyi [4, 5] in describing the experience of flow. This is a state where people are fully immersed in and concentrated on a task. This state is very typical for immersive video games. The optimal goal of different game design strategies is to achieve this state also in the non-gaming tasks (e.g. learning) to fully immerse and engage users in activities. Csikszentmihalyi describes three main elements of flow: (1) clear goals and sense of progress, (2) clear and immediate feedback, and (3) balance between skill-level and perceived challenge of the task.

Based on these observations we define three principles for successful game design and gamification in learning experiences: (I) clear goals, (II) clear feedback and reward description, and (III) interaction possibilities and freedom of choice.

2.2 Player Types

Based on observations of different aspects of player engagements in MUDs (Multi-User Dungeons), the game designer Richard Bartle [2, 3] identified four main player types. In his '*Taxonomy of Player Types*' the following types are introduced based on their interactions with the environment or other players: (1) *achievers*, who are engaged by achieving goals in the game (e.g. rising levels, getting points), (2) *explorers*, who like to discover the game and try out different things in the environment (e.g. discover treasures, explore the maps), (3) *socializers*, who are interested in interacting with others players and building relationships (e.g. joking, chatting), and (4) *killers*, who are engaged by beating others or showing their 'higher in-game status' to others (e.g. rankings, helping others as reputation booster). While these player types represent Bartle's observation of players in MUDs, these or similar types can be observed in all sort of environments and situations, where several people interact, such as in learning situations [10, 17]. Different authors have explored and discussed Bartle's player types. Yee [20] explored the four player types and found three main principles summarizing the activities and preferences of the types: (1) achievement: advancement, mechanics, competition, (2) social: socializing, relationship, teamwork, (3) immersion: discovery, role-playing, customization [7]. While different models cover a more general version of engagement, Bartle's model is one of the earliest and simplest models and well known in the game design theory [16, 19].

Since Bartle's player types model is one of the best known and most widely recognized models, we have also adopted it as a basis for the player type design strategy in the context of learning experiences.

3 Player Types Design (PTD)

People are engaged by different elements. Bartle's taxonomy of player types [2, 3] helps us to identify game design elements suitable for different types of players. However, this taxonomy was originally designed especially for MUDs (multi-user-dungeons) and hence needs to be used with care. Using the different player types as design strategy gives designers the possibility to include different forms of engagements in an experience, in the context of this paper in a learning experience. In the following, we propose *Player Type Design (PTD)*, a design strategy based on the four player types. PTD incorporated the four player types and engagement elements, which the different player types might be likely to enjoy. Additionally, various typical game elements inspired by gamification literature are identified, which can help attracting and engaging different player types.

3.1 Engagement Activities and Elements

We identify four broader categories of engaging activities and design elements based on the four player types and their interactions with the environment or other users (see Fig. 1). When designing activities in non-gaming context, such as in learning settings, designers should think of specific tasks and engagement elements in the form of verbs. More specifically, designers can think of tasks in line with the following action verbs:

A: Achieving, Gaining, and Producing. To please the player type *achiever*, it is essential to design elements, which suggest the user/learner that something has been achieved. Typical game elements here include elements suggesting performance (points, progress bars, levels, etc.) or special visible rewards (badges, achievements). Achievers need clear goals and objectives to be completed, and also feedback on their current progress towards this goal.

E: Exploring, Researching, and Testing. The main goal for *explorers* is a depth exploratory experience featuring lots of freedom through discovery, experimentation, finding secrets, and surprise elements. Furthermore it is important to reward this behavior in a visible way. The real reward here is the possibility provided for interacting in an explorative way with the environment.

S: Socializing, Collaborating, and Joining. Interactions with other users, collaborations, discussions, and building relationships and friendships are the most important reward factors for *socializers*. Sharing information, completing tasks together, or working together towards a goal are activities to attract and engage them.

K: Competing, Challenging, and Bragging. The gamer type *killer* seeks ways to compete with others. Typical elements supporting this group of users are special rewards, leadership information, or rankings. However, the activities are not only limited to obvious competitions. Killers can also be engaged by activities, which might be helpful, such as sharing information or gift, just to make others aware of their higher status or simply bragging (demonstration of superiority over fellows). The personal reputation and the recognition of skills and levels are important to this gamer type.

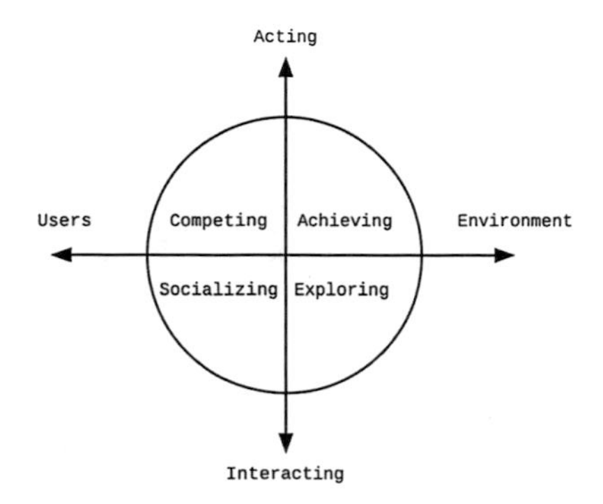

Fig. 1. Engagement activities and elements based on Bartle's player types.

3.2 Design Goals

As outlined in Sect. 2.1, we can define three main design principles to create an engaging playful experience, which can even create a flow experience: (I) clear goals, (II) informative and immediate feedbacks such as reward descriptions, and (III) possibilities to interact with the environments and other users and giving a freedom of choice.

Based on the type of feedback and interaction possibilities different player types can be engaged. This framework should help to design and analyze learning activities and engagement elements in learning platforms to understand what types of players are already motivated by the platform.

The core of the framework is built by engagement elements. An engagement element is an interaction with the system (e.g. finishing an assignment to get points) or an element provided by the system to engage (e.g. leaderboard).

Every engagement element should have a clear goal, an optional reward, and some extend of freedom. For the design of every engagement element the goal and the rewards should be clearly described. Additionally, different design strategies/elements should give players a sense of control and of their interaction possibilities. Since not every player type is attracted by every engagement element, some elements should be identified and designed as optional element (e.g. only showing ranking information on request instead of making it a part of the main site).

3.3 How to Use It

Table 1 illustrates a design framework for PTD. Game-based activities are listed and mapped to the engaged player type. A clear goal and feedback description should be added for each engagement element. Freedom refers to other choices as part of this engagement element. Additionally, the designer can indicate if an activity is optional.

An important point to mention is that Bartle's player types were originally described only for MUDs (Multi-User Dungeons). This framework adopts the player types in non-gaming contexts. It thus merely provides design inspirations on how to attract and engage different kind of users (in the specific context learners), but it is definitely not a complete and definitive guideline. PTD provides designers with a new method for designing game-based and gamified experiences that will engage different users. It can be also used to evaluate existing systems.

Table 1. PTD design framework: engagement elements are mapped to player types; goal, rewards and the possibility of freedom and interactions are described for each activity; additionally activities, which are optional, are marked

Engagement elements	A	E	S	K	Goal description	Feedback/ Reward	Freedom/ Interaction	O
1…								
2…								
…								
…								

The following sections describe and discuss two case studies as a means of evaluating applicability in learning settings.

4 Case Studies and Discussion

4.1 Case 1: Designing a Playful Blended Learning Environment

Motivational Active Learning (MAL) is a pedagogical model designed as a hybrid of the interactive learning model TEAL (Technology-Enabled Active Learning) and gamification strategies [17, 18]. TEAL uses mainly interactive engagement strategies including constant interactions with the students, collaborative assignments, and hands-on experiences (e.g. hands-on physics experiments). To make it more engaging for students, we combined this approach with game elements. The following main features of MAL were introduced:

- Small learning units (typically lectures are split in several activities in, before and after class, and the current learning progress of the students is steadily assessed), alternative task can be chosen
- Collaborative learning (many assignments, such as calculation problems, research activities, or discussions are designed as collaborative activities)
- Constant interactions (between the theoretical learning units given by the teacher, students' are asked to complete assignments, discuss the content with peers, or have some other form of interaction with the learning content as well as with other students or the instructor)
- Immediate feedback (for many interactions students receive immediate feedback on their performance through the lecturer, or the e-learning systems)
- Motivational feedback (the feedback is also enhanced by different forms of engaging feedback types such as points, ranking information, or badges; these feedback types are also designed to engage different player types)
- Flexible and adaptive class design (through the constant assessment in form of interactions between the small learning units, the current learning progress of the students can be assessed through the e-learning system)
- Errors are allowed (students can repeat assignments, quizzes, or other interaction types to improve, gain more points, step up in the ranking (Table 2))

Table 2. Examples of PTD framework for MAL

Engaging elements	A	E	S	K	Goal description	Feedback/Reward	Freedom	O
1. Small learning tasks in e-learning system	X				Complete learning unit	Feedback in form of progress-bar	Different/alternative task can be chosen	
2. Finishing research assignments in groups		X	X		Find answers to specific questions in a team	Get to know solutions from other groups and discuss different aspects	The extend of collaboration can differ	
3. Answering concept questions about learning progress with visible feedback and overall in-class statistics	X			X	Answer a question	Get feedback and see statistics what the rest of the class answered		
4. Work on clearly defined assignments	X			X	Finish an assignment;	Points, Leaderboard for points		
5. Working on clearly structured and defined assignment series	X			X	Finish an assignment series	Badge	Due to bonus assignments this activity is voluntary and the series can be chosen	X
6. Points are used for leaderboard information	X			X	Points influence the in-class ranking	Good ranking	The leaderboard is hidden on a subpage and must not be looked at; students can constantly improve assignments to get more points to enhance the ranking	X

4.2 Case 2: Designing an Engaging Mobile Application

In a second project we developed a playful and educational mobile app with the goal of engaging and motivating the user to walk and run more and learn about concepts of the city environment. The main idea was to develop an android application or game, which rewards the users for every active "own" movement. As current implementations of location aware games (e.g. Ingress, Resources, etc.) very often do not take into account the mode of transportation was used and also travelling e.g. by car is a legitimate action when playing the game and we needed to find a way to prevent this. Another common problem with current games is the fact that it sometimes suffices to stay still on one point to achieve certain game goals. These applications are focused on being games, without the addition of extrinsic motivation for getting people to be more fit and more on the move. We wanted to make movement the core element of our application. We tried to achieve this with carefully chosen game elements and making use of the smartphone sensors data (e.g. activity recognition with the help of the acceleration sensor). When designing the game we did not initially think of a story or the whole game it would be when complete. Instead of this our approach was to start designing the game with our focus on the player types. On the one hand we wanted to reach as many users as possible with this approach. On the other hand we did not want to be constrained in the possibilities by the rigidity of an initial fixed concept about what the completed game would need to be. Instead we approached the problem bottom up by adding game elements targeted to the player types, the limitations and possibilities of smartphones and the broadgoal of achieving fitter users. Not until when this task was completed did we plan further on what to implement to make this a single unified application that would add up to a game, instead of a collection of random game elements that do not fit together.

In the resulting game the players are separated into two opposing teams and the world is the playground. We separated the globe into trapezoids serving as areas, which can either be conquered for the own team or taken from the opposing team. Furthermore these areas can be leveled and thereby strengthened against being taken by going to the area more often. Furthermore those trapezoids are hidden for each individual player from the beginning. The players need to go to these areas to reveal what is happening there. We took this element from strategy video games where this "fog of war" is a very common. As the players use this core element of the areas, which is solely done by moving, points are earned and energy acquired. As in many games these points are an instant indicator of progress and lead to a level-up of the players. Energy is a consumable resource and as such leads to more possibilities in the game. Currently three options are available how to use this energy: *Plant a bacteria on an enemy area, Cure a bacteria on an friendly area, Reveal an area (which potentially may not be reachable e.g. restricted property).* Those areas affected by bacteria will spread to neighborhood areas every 4 h and as a result downgrade the area by one level or make it neutral ground again. This serves two purposes: on the one hand players are given another challenge; on the other hand this behavior should balance the problem of non-equal team sizes. Furthermore, we implemented elements, which are expected in nearly every multiplayer game. The game contains badges, a leaderboard to compare with other individual players, and a team rating. It also includes a world log which shows some of the actions of other players and in which area the actions happened. Independent from the game, the application also contains most of the functionality of classic sports tracking applications such as current speed, the distance run during the current session, duration of the current session, or average speed. It also implements a variety of statistics of past sessions to help the users keep track of their fitness development. Those statistics and the feedback of the current performance could also be interpreted as gamification elements targeted at achievers (Table 3).

Figure 2 shows the elements we introduced in the game correlated with the player types we tried to address with the specific elements

Table 3. Examples of PTD framework for Sportinate.

Engaging elements	A	E	S	K	Goal description	Feedback/Reward	Freedom	O
1. Discovering areas	X				Player should discover new areas	Area gets marked as discovered	Area can be chosen	
2. Uncovering fog of war		X			Area is uncovered and the fog disappears	Area is visible and usable now also from distance	Area can be chosen	
3. Infecting areas with bacteria				X	Area levels can be changed	Area loses enemy-levels	Area can be chose, activity is on choice	X
4. Seeing other player activities in World Log			X	X	See interactions with others early to help or intervene	Interactions with others	Player can decide to interact with the others	X
5. Getting a badge when completing specific challenges	X	X		X	Finish specific tasks	Badge		X
6. Seeing leaderboard information	X			X	Points influence the personal ranking	Good ranking		X
7. Seeing team scoring information	X		X	X	Points influence the group ranking	Good ranking, "better than others"		

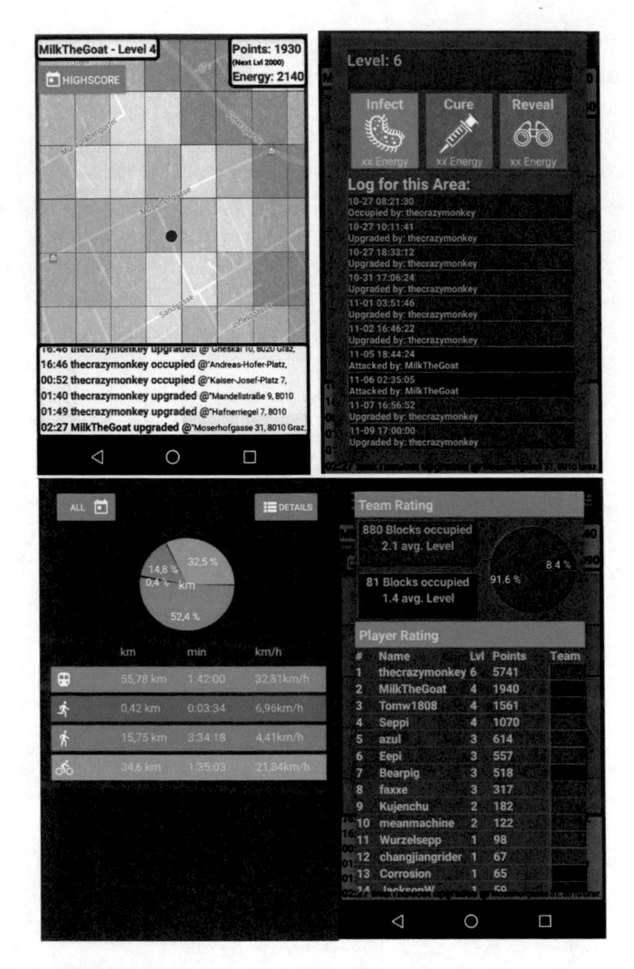

Fig. 2. Screenshots of the fitness app "Sportinate" (a) Fog of war, (b) Item: reveal areas, (c) Item: destroy enemy elements and statistics, (d) team ranking and player ranking.

5 Conclusions and Future Work

In this article we have proposed PTD (player type design), a design strategy to design experiences and activities in gaming, but also a in non-gaming context, such as educational environments to engage different player styles. As a means of making design activities aimed at different player types easier we presented each of the four following activity descriptions for each player type: (1) Achievers: Achieving, Gaining, and Producing, (2) Explorers: Exploring, Researching, and Testing, (3) Socializers: Socializing, Collaborating, and Joining, and (4) Killers: Competing, Challenging, and Bragging.

The crucial issues in the creation of an engaging experience are to design clear goals, think of direct or indirect rewards, and leave players (users or learners) interaction possibilities and freedom to interact with the system. This strategy cannot only be used to design new experiences, but also to evaluate existing experiences. The first usage of the design strategy reveals that the model is helpful in analyzing and designing applications and pedagogical models with a specific focus on different engagement types. As a follow-up project we are planning a user study to evaluate the effectiveness of the model as design tool with stakeholders in the context of learning applications.

References

1. Annetta, L.A.: The "I's" have it: a framework for serious educational game design. Rev. Gen. Psychol. **14**(2), 105–112 (2010)
2. Bartle, R.: Hearts, clubs, diamonds, spades: players who suit MUDs. J. MUD Res. **1**(1), 19 (1996)
3. Bartle, R.: Designing Virtual Worlds. New Riders, Indianapolis (2004)
4. Csikszentmihalyi, M.: Beyond Boredom and Anxiety: Experiencing Flow in Work and Play. Jossey-Bass, San Fransisco (1975)
5. Csikiszentmihalyi, M.: Flow: The Psychology of Optimal Experience. Harper & Row, New York (1990)
6. Deterding, S., Dixon, D., Khaled, R., Nacke, L.: From game design elements to gamefulness: defining gamification. In: Proceedings of the 15th International Academic MindTrek Conference: Envisioning Future Media Environments, pp. 9–15. ACM (2011)
7. Dixon, D.: Player types and gamification. In: Proceedings of the CHI 2011 Workshop on Gamification (2011)
8. Felder, R.M., Silverman, L.K.: Learning and teaching styles in engineering education. Eng. Educ. **78**(7), 674–681 (1988)
9. Gee, J.P.: What Video Games Have to Teach Us About Learning and Literacy. Palgrave Macmillan, New York (2007)
10. Johnson, R.T., Johnson, D.W., Stanne, M.B.: Comparison of computer-assisted cooperative, competitive, and individualistic learning. Am. Educ. Res. J. **23**(3), 382–392 (1986)
11. Kapp, K.M.: The Gamification of Learning and Instruction: Game-Based Methods and Strategies for Training and Education. Wiley, New York (2012)
12. Kolb, D.A.: Learning Styles Inventory. McBer and Company, Boston (1976)
13. Kotini, I., Tzelepi, S.: A gamification-based framework for developing learning activities of computational thinking. In: Gamification in Education and Business, pp. 219–252. Springer, Cham (2015)
14. Kumar, J.: Gamification at work: designing engaging business software. In: International Conference of Design, User Experience, and Usability, pp. 528–537. Springer, Heidelberg (2013)
15. Linehan, C., Kirman, B., Lawson, S., Chan, G.: Practical, appropriate, empirically-validated guidelines for designing educational games. In: Proceedings of the SIGCHI Conference on Human Factors in Computing Systems, pp. 1979–1988 (2011)
16. Pirker, J., Gütl, C.: Educational gamified science simulations. In: Gamification in Education and Business, pp. 253–275. Springer, Cham (2015)
17. Pirker, J., Riffnaller-Schiefer, M., Gütl, C.: Motivational active learning: engaging university students in computer science education. In: Proceedings of the 2014 Conference on Innovation & Technology in Computer Science Education, pp. 297–302. ACM (2014)

18. Pirker, J., Riffnaller-Schiefer, M., Tomes, L.M., Gütl, C.: Motivational active learning in blended and virtual learning scenarios: engaging students in digital learning. In: Handbook of Research on Engaging Digital Natives in Higher Education Settings, vol. 416 (2016)
19. Stewart, B.: Personality and play styles: a unified model. Gamasutra, 1 September 2011. http://www.gamasutra.com/view/feature/6474/personality_and_play_styles_a_.php
20. Yee, N.: Motivations of play in MMORPGs. In: Proceedings of DiGRA (2005)
21. Zichermann, G., Cunningham, C.: Gamification by Design: Implementing Game Mechanics in Web and Mobile Apps. O'Reilly Media, Inc., Sebastopol (2011)

The Use of New Learning Technologies in Higher Education Classroom: A Case Study

Micaela Esteves[1(✉)], Angela Pereira[2], Nuno Veiga[1], Rui Vasco[1], and Anabela Veiga[3,4]

[1] CIIC - Computer Science and Communication Research - ESTG, Polytechnic Institute of Leiria, Leiria, Portugal
{micaela.dinis,nuno.veiga,rvasco}@ipleiria.pt

[2] CiTUR - Tourism Applied Research Centre - ESTM, Polytechnic Institute of Leiria, Leiria, Portugal
angela.pereira@ipleiria.pt

[3] Geosciences Center of the University of Coimbra, Coimbra, Portugal

[4] Polytechnic Institute of Leiria, Leiria, Portugal
anabela.veiga@ipleiria.pt

Abstract. We have conducted a study with higher level education students, in lecture classes of three Undergraduate Courses and one Professional Higher Technical Course that involved six different subjects with a total of 324 students. In this research the use of Game-Based Learning platform was analysed in order to encourage the students' participation, increasing motivation and keeping them motivated and committed during lessons, therefore, increasing their learning skills.

Based on these results, we recommend that Kahoot is used in lectures in order to help students develop their performances and abilities and at the same time be more successful and prepared to have an active participation in society.

Keywords: Learning · Collaborative learning · Game-Based Learning Higher education

1 Introduction

During the recent years, there has been an increase in the participation rates of students in higher level education, mainly due to the Bologna process, and inevitably therefore an overall lowering of academic standards as universities and student populations have become more diversified [1]. There are still dedicated hardworking students in universities, however many of them have a low profile for higher education [2]. These changes have brought many challenges for teachers who have had to adapt their teaching methods according to students different characteristics and backgrounds.

In Portugal, as in other countries, the engineering areas deserve special attention from higher education institutions due to the low success rates and also because of the high dropout rate that is compared to other course rates. According to Paura and Arhipova [3] the main reasons for students' abandoning their studies is due to poor teaching and advising as well as the high difficulty level of the engineering curricula.

M. E. Auer et al. (eds.), *Teaching and Learning in a Digital World*,
Advances in Intelligent Systems and Computing 715,
https://doi.org/10.1007/978-3-319-73210-7_59

The difficulties of the engineering curricula are associated with the studied topics such as mathematics, physics and computer programming. Different studies show that the students' dropout rate depends on the studied subject at the university as well as their pre-college academic qualifications [4, 5]. On the other hand, today's students bring a rich and different set of literacy practices and background that is often unacknowledged or underused by educators, namely their use of Information and Communication Technology (ICT). All these issues lead to a high dropout rate and failing, especially, in the engineering areas.

In these times of rapid technological changes, the challenge for today's teachers is to build a bridge between the technological world that students live in and the classrooms in which teachers expect them to learn in, especially in higher education. These students are being referred to as the Millennials [6] once they are the first generation to be immersed in ICT for their entire lives. Considine et al. [7] argue that "*to develop a curriculum that is relevant to this generation, educators need to acknowledge and respect the skills, attitudes, and knowledge that students bring with them to school and build on those to ensure success in the academic disciplines. Thus, students will become engaged and connected to the traditional curriculum while developing crucial technological skills.*"

Due to the computerization society, education has been accompanying technological innovations. In this context, several digital applications for Game-Based Learning (GBL) purposes have emerged in higher education classrooms. These games should be aligned with the learning styles and needs of this current generation, the Millennials.

We propose the use of a GBL platform through mobile devices in the classroom to allow students to collaborate with each other, in order to encourage and increase their participation and motivation, during the lessons, therefore increasing their learning effectiveness. This paper includes a reflection on how using the game-based learning platform during lessons could improve learning experiences for students in higher education, mainly in engineering courses. In this context, this article presents the usage of Kahoot with higher education students from School of Technology and Management, Polytechnic Institute of Leiria, during the academic year 2016/2017 from September to May.

The rest of the article is organised as follows. Section 2 our motivation and related work is outlined. Section 3 describing the methodology of the project. In Sect. 4 we follow with the findings and the discussion. Finally, some conclusions of this study are presented in Sect. 5.

2 Related Work

The usage of Game-Based Learning (GBL) has been explored in recent years for educational purposes due to the ability to commit students to the task that they are working on and pushing them to work to the edge of their capabilities [8–11].

According to Whitton and Moseley [9] GBL has the ability to provide a scaffold for learners to gradually increase their knowledge and ensure that they are motivated to pursue additional content. Furthermore, games provide immediate feedback to the learners and can reinforce content and help students with the retention of concepts [12, 13].

The study carried out by Cheng and Su [14] makes a comparative study between traditional teaching methods and the Game-based Learning (GBL) approaches. The results show that when using games, the students' motivation has a significant impact on learning, allowing us to assert that GBL can achieve the learning goal effectively.

On the other hand, GBL stimulates the collaboration between students. Collaborative environments can offer important support to students in their activities for learning. According to Guzdial et al. [15], collaborating in problem-solving provides not only an appropriate activity but also promotes reflection, a mechanism that enhances the learning process. Students that work in groups need to communicate, argue and give opinions to other group members, encouraging a reflection that leads to learning.

One educational tool that can be used to promote the collaboration in class is Kahoot. Kahoot is a game-based learning platform that has been used in university field studies in different areas such as Mathematics, Physics, Languages and others. Nevertheless, some authors [16] argue that using mobile technology does not guarantee effective learning.

Kahoot is a free tool that has gained popularity amongst teachers for it being easy to use and its ability to establish dynamics of active work in the classroom. This application allows teachers to create quizzes, jumbles, surveys, discussions and also to obtain feedback from students in real time.

Kahoot has been used in higher education context. Studies [17, 19] evaluating the usage of Kahoot, show that this is a good GBL tool for classroom activities and also helps to improve student participation by promoting a positive relationship between groups. On the other hand, Cerro Gómez [18] emphasizes that the usage of Kahoot has led to an increase in the number of students attending classes. Buchanan et al. [20] argue that the existence of an alignment between the learning goals and the game design is important, which is possible to achieve whilst using the Kahoot GBL.

3 Methodology

This case study was carried out in the current school year, 2016/2017, in lecture classes of three Undergraduate Courses and one Professional Higher Technical Course and involved six different subjects with a total of 324 students. All subjects clearly fit into technological courses, with the exception of the undergraduate course, Health Information Sciences, which has also an interconnection with healthcare. All lecture classes lasted for 100 min, with the exception of one lasting only 50 min (Networks Laboratory I) and also had a different student attendance average.

Several Kahoot quizzes were applied in lecture classes and all of them were carried out with student teams. All Kahoot quizzes were always held at the end of the lecture, with the exception of Computational Systems (CmpS) subject. This allowed to validate the degree of attention and the consolidation of the contents from the students. In CmpS subjects with classes lasting 100 min, Kahoot was randomly interleaved as the lecture evolved. In addition to the previous objectives, this also allowed time to pause with the objective of regaining the students power of concentration and also to keep them motivated.

Three assessment methodologies (AM1, AM2, AM3) were chosen for evaluating the Kahoot effectiveness in the classroom (Table 1). The Comparison between the traditional approach versus Kahoot (AM1) was only performed on CmpS subject and was done by comparing the present subject student grades with those achieved in the preceding school year (without the use of Kahoot). This was possible because the students profile was identical in both years.

The analysis of the Kahoot approach acceptability and applicability was undertaken in two ways: team feedback (AM2) and an individual Kahoot survey (AM3). The individual survey included closed questions and open questions with regard to suggestions (Table 3).

Table 1. Overview of the case study

Subject acronym	Subject	Course	Average number of students in class	Assessment methodologies
CmpAT	Computer Architecture and Technology	Undergraduate in Health Information Sciences	20	AM2 and AM3
CmpNT	Computer and Network Technology	Undergraduate in Games and Multimedia	41	AM2 and AM3
CmpS	Computational Systems	Undergraduate in Computer Engineering	177	AM1 and AM2
NLab	Networks Laboratory I	Computer Networks and Systems Professional Higher Technical Course	25	AM2 and AM3
SoilMF	Soil Mechanics and Foundations	Undergraduate in Civil Engineering	14	AM2
VirtT	Virtualization Technologies	Undergraduate in Computer Engineering	47	AM2 and AM3

4 Results and Discussion

4.1 Comparison Between the Traditional Approach Versus Kahoot

In the subject of CmpS, the average theoretical grade has improved by 6.4%, from 8.86 out of 20 in the 2015/2016 school year (without Kahoot) to 9.57 in 2016/2107 (with Kahoot). There were no other significant changes in the operation of the subject, with contents and the teaching team remaining the same.

It should also be noted that there is a higher impact on lower grades with the failing students' scores improving 12%, from 6.17 out of 20 to 6.91, closer to success. The positive average rose 2.8% (Fig. 1).

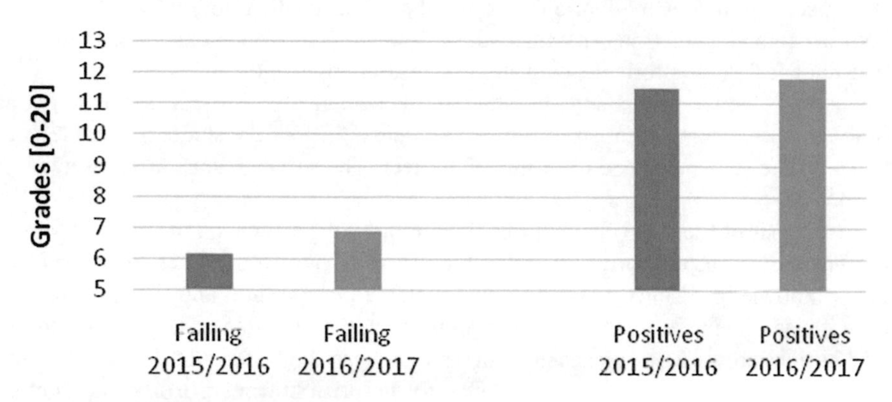

Fig. 1. Student average results [0–20] without and with Kahoot

These results are in line with the study carried out by Cheng and Su [14] showing the effectiveness of the tool in the learning process. Teachers could observe the positive impact on students' motivation. As for a possible cognitive influence assessment, more studies will be necessary in the future.

4.2 Analysis of the Kahoot Approach Acceptability and Applicability

With regard to the analysis of Kahoot applicability (Table 2) the results show that there is a high percentage average of correct answers of the three best teams, despite the fact that they have been evaluated right after the presentation of the contents, therefore without prior study. Thus, we believe that Kahoot can help students focus in class.

Some teams revealed difficulties in accessing Kahoot, either due to Wi-Fi problems or lack of hardware capacity, causing some teams to be unable to answer a considerable amount of the questions, being evaluated as "wrong" answers. This may partly explain the significant difference of the overall average (61%) in item 1 (Table 2) "Correct answers" to the 1st place average (86%) in item 1.1.

Table 2. Results of Kahoot acceptability and applicability

Item	Average	Standard deviation	Median
1. Correct answers	61%	13%	61%
1.1. Correct answers of 1st place	86%	19%	89%
1.2. Correct answers of 2nd place	79%	19%	75%
1.3. Correct answers of 3rd place	73%	23%	73%
2. How fun was it? (1 to 5)	4.3	0.5	4.4
3. Did you learn something?	92%	13%	100%
4. Do you recommend it?	94%	10%	100%
5. How do you feel?			
5.1. I feel positive	75%	20%	78%
5.2. I feel neutral	11%	11%	11%
5.3. I feel negative	15%	16%	13%

The results also emphasise the high average values of positive feedback, revealing a clear impact on student motivation. We highlight the medians of 100%. About the Kahoot acceptability the majority of students approved and appreciated it. However, 15% do not feel so comfortable with the use of this application.

In order to better understand the students' acceptance of Kahoot, we underwent satisfaction survey that was answered by 84 students, 23% of the students were female and 24% of the students attend the CmpAT subject. The survey results are presented in Table 3.

The analysis of the ease of technology handling by the students, reflects their digital skills. The survey results corroborate that the use of Kahoot contributes to consolidate students' knowledge. Concerning the time available per question and the quantity per Kahoot it was considered adequate. The results about the importance of podium and the reward attribution were not consensual once they present high standard deviations. This reveals that students do not give importance to the prize but rather to the pleasure of playing.

Table 3. Results of individual survey

Questions	Average (1 to 5)	Standard deviation	Median
1. Do you feel comfortable using electronic gadgets (smartphones, tablets, laptops, …)?	4.60	0.56	5
2. Does Kahoot contribute to the consolidation of the subject contents?	4.12	0.66	4
3. Will Kahoot contribute to better individual grades?	4.05	0.58	4
4. How fun was it?	4.37	0.77	4,5
5 Was the Kahoot question response time adequate?	3.02	0.38	3
6. Was the Kahoot number of questions adequate?	2.94	0.32	3
7. How important is the scoreboard?	3.68	0.97	4
8. Do you agree with the reward?	3.43	1.00	3
9. Do you recommend Kahoot?	4.30	0.70	4

Having only a few students answer the comments section. Those who did, were in general enthusiastic about using Kahoot and encouraged the continuation of its usage, as can be seen in the following answers examples: "*I loved Kahoot!*"; "*At the beginning Kahoot seemed like a joke, but then I realized its great benefits. Thank you*!"; "*The subject contents are hard and very theoretical. Kahoot helped a lot lightening the burden.*"; "*A good interaction moment and a fun way of reflexion on the contents.*"; "*Strengthening of team work and relationship between fellow students*".

5 Conclusions and Suggestions

Currently, educators are facing difficult times with a whole generation that was born surrounded by technology and that are subject to completely different stimuli from their teachers and parents. Researchers have recognized that these students learn differently thus teaching should adapt to avoid abandonment and improve the success of these digital natives.

The result of this study shows that students are more committed to learning in the classroom. Moreover, the number of students per class increased, softening a problem in our institution. Consequently, the learning results were better therefore it is our intention to spread the use of Kahoot in classrooms of other courses.

In this study, Kahoot was used as a way to assess and consolidate contents, as well as a way to regain power of concentration. However, we suggest other forms of using it: in the beginning of the next class to access and consolidate the contents of the previous class or simply as a diagnostic test, allowing the teacher to know the students' previous knowledge about the subject and motivating them to the learning process.

Based on our experience in teaching in higher education, we believe that using this type of applications could reduce the gap between student's way of life and the classroom environment, once mobile technology is intricately interwoven in their lives.

References

1. Altbach, P.G., Reisberg, L., Rumbley, L.E.: Trends in global higher education: tracking an academic revolution. In: Report for the UNESCO World Conference on Higher Education, 5–8 July 2009 (2009)
2. Biggs, J., Tang, C.: Teaching for Quality Learning at University, 4th edn. McGraw-Hill Education, New York City (2011)
3. Paura, L., Arhipova, I.: Student dropout rate in engineering education study program. In: Proceedings of 15th International Scientific Conference Engineering for Rural Development, Jelgava, Latvia (2016)
4. Arulampalam, W., Naylor, R.A., Smith, J.P.: Effects of in-class variation and student rank on the probability of withdrawal: cross-section and time series analysis of UK universities students. Econ. Educ. Rev. **24**(3), 251–262 (2005)
5. Smith, J., Naylor, R.: Schooling effects on subsequent university performance: evidence for the UK university population. Econ. Educ. Rev. **24**, 549–562 (2005)
6. Howe, N., Strauss, W.: Millennials and the Pop Culture. Life Course Associates, Great Falls (2006)
7. Considine, D., Horton, J., Moorman, G.: Teaching and reaching the millennial generation through media literacy. J. Adolesc. Adult Lit. **52**(6), 471–481 (2009)
8. Mayer, I., Bekebrede, G., Harteveld, C., Warmelink, H., Zhou, Q., Ruijven, T., Wenzler, I.: The research and evaluation of serious games: toward a comprehensive methodology. Br. J. Educ. Technol. **45**(3), 502–527 (2014)
9. Whitton, N., Moseley, A.: Using Games to Enhance Learning and Teaching: A Beginner's Guide. Routledge, Abingdon (2012)

10. Esteves, M., Fonseca, B., Morgado, L., Martins, P.: Improving teaching and learning of computer programming through the use of the Second Life virtual world. Br. J. Educ. Technol. **42**(4), 624–637 (2011)
11. Bodnar, C.A., Clark, R.M.: Can game-based learning enhance engineering communication skills? IEEE Trans. Prof. Commun. **60**(1), 24–41 (2017)
12. Kapp, K.M.: The Gamification of Learning and Instruction: Game-Based Methods and Strategies for Training and Education. Wiley, Hoboken (2012)
13. Hsu, W.C., Lin, H.C.K.: Impact of applying WebGL technology to develop a web digital game-based learning system for computer programming course in flipped classroom. In: International Conference on Educational Innovation through Technology (EITT), pp. 64–69. IEEE (2016)
14. Cheng, C.H., Su, C.H.: A game-based learning system for improving student's learning effectiveness in system analysis course. Procedia-Soc. Behav. Sci. **31**, 669–675 (2012)
15. Guzdial, M., Kolodner, J., Hmelo, C., Narayanan, H., Carlson, D., Rappin, N., Newstetter, W.: Computer support for learning through complex problem solving. Commun. ACM **39** (4), 43–46 (1996)
16. Nguyen, L., Barton, S.M., Nguyen, L.T.: Ipads in higher education—hype and hope. Br. J. Educ. Technol. **46**(1), 190–203 (2015)
17. Fuentes, M., del Mar, M., Carrasco Andrino, M.D.M., Jiménez Pascual, A., Ramón Martín, A., Soler García, C., Vaello López, M.T.: El aprendizaje basado en juegos: experiencias docentes en la aplicación de la plataforma virtual" Kahoot" (2016)
18. Cerro Gómez, G.M.D.: Aprender jugando, resolviendo: diseñando experiencias positivas de aprendizaje (2015)
19. Zarzycka-Piskorz, E.: Kahoot it or not? can games be motivating in learning grammar? Teach. Engl. Technol. **16**(3), 17–36 (2016)
20. Buchanan, L., Wolanczyk, F., Zinghini, F.: Blending Bloom's taxonomy and serious game design. In: Proceedings of the 2011 International Conference on Security and Management (2011)

Innovation of CAD/CAE System Teaching at Upper Secondary Education

Peter Kuna[1], Alena Hašková[1], Miloš Palaj[1], Miloslav Skačan[1], and Ján Záhorec[2(✉)]

[1] Department of Technology and Information Technologies, Faculty of Education, Constantine the Philosopher University, Nitra, Slovak Republic
pietro.kuna@gmail.com, ahaskova@ukf.sk, milos.palaj@gmail.com, miloslav.skacan@gmail.com

[2] Department of Education and Social Pedagogy, Institute of Educational Sciences and Studies, Faculty of Education, Comenius University, Bratislava, Slovak Republic
zahorec@fedu.uniba.sk

Abstract. In the traditional methodology of CAD system teaching, one starts with acquirement of technical drawing principles and design of technical schemes and drawings based on manual drawing. Only consequently students learn to draw the schemes and technical drawings in some of the CAD systems. In their paper the authors present a new, by them created methodology of teaching modelling and simulation in CAD/CAE systems at which students (ISCED 3 level) start immediately with a machine element proposal (draft) and modelling. A key question related to the use of this innovative teaching methodology is whether the reduction of the technical drawing teaching will not have a negative impact on the students' achievements. The answer to this question should result from a pedagogical experiment which the authors are going to carry out and a methodology of which, together with the research hypothesis and expected results, they present in the paper.

Keywords: CAD/CAE systems · Innovative methods of teaching · ISCED 3

1 Context of the Solved Issues

Computer aided systems are systems dedicated foremost to support activities in the area of graphical outputs projection and creation [1]. Maybe most often utilized they are in the engineering industry, where they are used in every production phase – from the machine part design, through production planning, from production up to the assembling, storing and exporting [2]. Nowadays they are also used in various others industry sectors as well as at different management levels. They enable to perform different engineering activities like drawing, constructing, dimensioning, projecting, but also many administrative activities like archiving, searching, reproduction etc. faster and more easily.

Right from the start it is important to define and explain, which computer aided systems are defined as CAD, i.e. Computer Aided Design (Drawing) systems and which are defined as CAE, i.e. Computer Aided Engineering systems. As CAD are defined systems, which mostly support the part of drawing and the creation of the technical

M. E. Auer et al. (eds.), *Teaching and Learning in a Digital World*,
Advances in Intelligent Systems and Computing 715,
https://doi.org/10.1007/978-3-319-73210-7_60

documentation and as CAE are defined systems, which support 3D modelling, simulations in computer environments of virtual reality. CAD/CAE is referred to mark a complex of CAD and CAE systems together, i.e. it refers to the meaning "everything together" [3].

The role of the CAD system is to aid the designer by providing him/her a possibility:

- to create and modify a graphical representation of the product,
- to view the actual product on screen,
- to make any modifications to the graphical representation of the product,
- to present his/her ideas on screen without any prototype, especially during the early stages of the design process,
- to perform complex design analysis in a short time,
- to store the whole design and processing history of a certain product, for future reuse and upgrade [4].

Nowadays we can't imagine an effective design of new products without the use of CAD/CAE systems. These systems are solving its own pre-construction part of the production, but also add the option of modification of existing products (AutodeskInventor). They are becoming something like a world of virtual reality, in which engineers and designers have the option to apply their creativity and inventiveness. CA technologies enable to create geometry of the models, to design further technological parameter and enable even to design the production process. The defined models can be in a very easy way adjusted, modified and also their mechanical properties can be very easy derived. The advantage of a computer created design is his close continuity on the next technological activity. The geometries of objects created in this way can be used for example as material for programming of machine tools. A separate chapter is the linking of the created objects to larger assemblies and computer simulations. (AutodeskInventor) CA technologies enable to test and prove functionality of a designer's ideas, before his product becomes an object in the real world. This substantially expands the possibilities of the design and the effective creation of a bigger amount of ideas and modifications, which otherwise would be financially very demanding [5].

The economic and technical aspects of the use of CA systems in design and manufacturing are generally known. In the last decade, we have witnessed the retreat of CA systems only as support tools for creating technical documentation. Current trends in their use emphasize the importance of human creativity. For example, CA systems today have very elegant and efficient tools for the automated creation of technical documentation. So the designer is not burdened by the lengthy and routine creation of technical drawings. Today trend in technical communication is to prefer virtual 3D models rather than drawing documentation. In any case, the CA systems are trying to relieve the constructors of any clichéd and time-consuming workflows. High emphasis is placed just on the inventiveness and creativity of human thinking, which we cannot replace by computer algorithmization.

In recent years, the Slovak Republic has become a European big power in car production. Availability and quality of skilled labor is one of the biggest problems of suppliers and manufacturers in the automotive industry. In Slovakia this problem was stated by 71% of respondents, which is comparable to the results of the global survey.

This is also a serious challenge for the reform of the education system in the Slovak Republic. It is understandable that there is a high demand for skilled workers from technical professions [6].

The conducted survey only confirms how very important it is for the education system to reflect the needs and interests of industrial practice. We consider teaching in the field of CA technology as one of the key elements of the successful application of graduates of technical schools in practice. The content of education in this field must also reflect the current trends and requirements of the industry.

2 Traditional Methodology of CAD System Teaching

Computer Aided Design (Drawing) has been a part of study program curricula from secondary schools to university level already for a long time [7–9]. In regard to the constantly increasing importance of CAD education more and more attention is being paid to the question how to teach CAD in practice [10–13]. Consequently the answer to this question, i.e. the general concepts, curricula for different stages of CAD knowledge, new teaching methods and didactical principles, new roles of teachers, has to be reflected also in relevant teacher training programs [14–16]. Teaching CAD is a very complex topic. A good teacher should understand the difficulties of learning CAD and should be able to counteract with these problems in various ways. Teaching CAD systems should provide students a wide diversity of problems and tasks and in this way by problem and process oriented sequences, the teacher should promote students' autonomous learning [10]. When a teacher starts to acquaint the students with the work with CAD systems, it is important to ensure that the students do not develop a dislike for these systems, or do not develop a feeling that their study is too difficult and complicated. A student must be offered education to that s/he enjoys the journey of the acquiring the new knowledge and skills, while picking up the principles of the issue naturally [8].

A traditional methodology of CAD system teaching rises from the historical technical development in the field of CAD systems. The first phase is acquirement of technical drawing principles. Students design schemes or technical drawings based on manual drawing. Only during the next phase the drawing board is substituted by computers and the students learn to draw the schemes and technical drawings in some of the CAD systems. In this case process of education is focused on acquirement of the practice (technique) at work with the software assigned for technical documentation creation, included technical drawings. The third stage is focused on computer assisted design of technical layouts and constructions. Students design machine elements by the means of CAD/CAE systems and compose them into integrated technical constructions.

Today's CAD-CAE systems are able to generate automatically a complex technical documentation following design and simulation test phase of the components or of the functional units. Through the survey of the labor market requirements, we have found that the emphasis on graduate requirements is precisely on the field of modelling and simulation. At present, in the technical practice, the creativity and inventor's invention is more valuable than the precise knowledge of rules and standards for the production of technical documentation. Of course, we do not want to claim that accurate and standard drawing of technical drawings is not a necessary part of the work of technical workers. However, current developments and trends in the use of CAD/CAE systems unambiguously points to the trend of retreat of these skills, in favor of the creativity and invention of the designer. In the final phase, the designer can generate accurate technical documentation through a computer within seconds (Fig. 1).

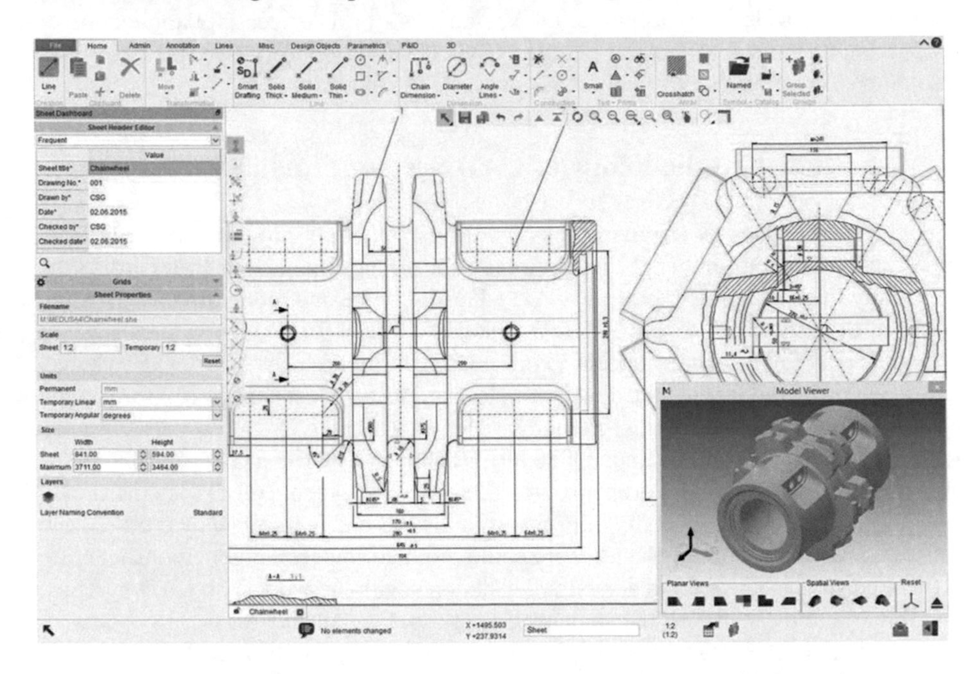

Fig. 1. Example of a 3D virtual model and to it related automatically generated technical drawing (Source: http://www.cad-schroer.com/products/medusa4.html)

In the current education system, the creation of technical documentation is the main content of the teaching. The standard teaching process devotes two thirds of the time allocation to technical drawing - theory, drawing on paper, or drawing on a PC in a CAD system. Only one third of the total time allocation is dedicated to the CAE system and deals with the topic of computer supported design, i.e. to the processes and tools that support the invention and the creativity of the constructor or designer (results of the analysis of the State Educational Program for Secondary Vocational/Technical Schools [17] done by the authors).

Up to date CAD/CAE systems can automatically generate complete technical documentation resulting from a draft/plan and simulated tests of the machine elements or machine function units [18]. Just this creation of the technical documentation is the subject of teaching, which precedes modelling and simulation methods. In the traditional ways of teaching a great part of time (lesson) allocation is devoted to technical drawing – theory, manual drawing on paper or computer assisted drawing in CAD systems. As a consequence then only one third of the total time allocation in the study programs (subject curricula) is devoted to CAD/CAM systems (as results from the analysis of the State educational program for upper secondary technical schools). But on the other hand a survey of the labour market requirements showed out that the labour market accentuates just the requirements on the graduates' skills in the modelling and simulation field [19].

One possibility how these requirements of the technical practice could be reflected at schools (in education), is to increase time allocation for CAD/CAE system teaching. This would result in a necessary modification of the relevant school documentation and a time demanding decision making process on the area of the technical education from which the relevant number of lessons would be relocated to gross up number of lessons needed for CAD/CAE system teaching. The authors of the paper have proposed a new, unconventional method to teach CAD/CAE systems, which on the one hand reflects current needs of both technical practice and labour market and on the other hand keeps the current range of this issue time allocation. In this unconventional method, students (ISCED 3 level) start immediately with a machine element proposal (draft) and modelling. In the next stage the students create necessary technical documentation to the proposed and modelled element. A key question related to this method (teaching methodology) is whether the reduction of the technical drawing teaching will not have a negative impact on the students' achievements.

3 Innovative Methodology of CAD/CAE System Teaching

The most effective way to reflect the requirements of technical practice in education is to increase the time allocation for CAE systems. However, such a procedure would entail a necessary adaptation of school documents and a lengthy process of deciding on which field of technical education we would take the time from to increase the time needed. That is why there was prepared by us a new, non-traditional procedure for CAD/CAE systems teaching, which reflects the requirements of the practice - increase of the time allocation for modelling and simulation, but with unchanged summed time support for CAD/CAE systems. In the proposed methodology, the students begin with the design and modelling of the components. After that a simulations phase follows and finally, creation of technical documentation for the proposed 3D model of the component is included. We believe that this by us proposed innovative approach to CAD/CAE system teaching follows better the practices established in technical practice. Today, the idea or intention of the constructor is at first "materialized" in the form of a virtual 3D model, behavior of which can be simulated by the constructor and subsequently modified in the virtual world of CAD/CAE systems. And only in the final the technical documentation for the production is produced (generated). The given situation is best visible in the catalogues of construction companies in which one has the opportunity to choose one of the offered homes. A customer is given a possibility to go through a virtual 3D home model and to specify to the designer parameters for any modifications. The designer uses the CAD/CAE system to design changes according to customer's requirements and to generate a new virtual 3D model which is once again at disposal to the customer. This may be repeated several times. Only at the end of the entire design process, the technical documentation is being developed. As it can be seen from this example, production of the technical documentation is also in practice only the final activity.

However the main idea of the proposed methodology is a significant reduction (by half) of the time to be allocated to teaching creation of technical documentation. It is natural that such an intervention will negatively affect the achievements of the students. In order to assess the extent to which this intervention will influence the learning results of the students, it is necessary to define the performance standard from the subject area and then establish a methodology for verification of the achieved results. Required knowledge and skills related to the area of technical documentation creation were divided as follows.

Theoretical knowledge

- Knowledge of standards in the field of technical documentation
- Knowledge of the functions and tools of the selected CAD system to create 2D technical documentation
- Theoretical knowledge of the drawing procedures (manual/PC) of the drawing documentation

Practical skills

- Reading of technical drawings with comprehension
- Creating technical documentation in accordance with the standard - manually
- Creating technical documentation in accordance with the standard - using a CAD system

When preparing new curricula, we want to achieve time savings made at the expense of mastering practical skills, especially when creating technical documentation. As mentioned, the current CAD/CAE systems can automatically generate a technical drawing from a virtual 3D model. Therefore, in this part of teaching we rather focus on reading comprehension of technical documentation and theoretical knowledge of procedures for the creation of technical documentation. We further define this knowledge and skills as the key ones. We assume that the students will be able to eliminate their handicap of the absence of practical skills in drawing technical drawings through their application practice. At the same time today's developments and trends in the use of CAD/CAE systems just relieve constructors of this activity.

By increasing the time allocation for CAE systems (twice), we expect, of course, that student achievements in the area of design and simulation will be improved. Currently, curriculum design for the entire issue of CAD/CAE system teaching at secondary vocational schools is under processing (Fig. 2).

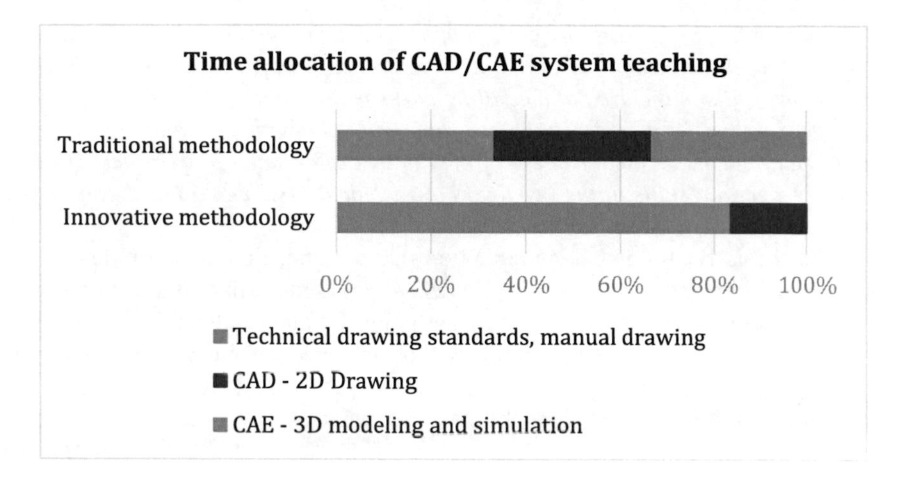

Fig. 2. Time allocation breakdown graph

4 Validation of the New Methodology of CAD/CAE System Teaching

A key question related to the new methodology of CAD/CAE system teaching is whether the reduction of the technical drawing teaching will not have a negative impact on the students' achievements. To find answer to this question and to validate the proposed new way of CAD/CAE system teaching, a project of a pedagogical experiment is under a construction at the Department of Technology and Information Technologies of the Faculty of Education at Constantine the Philosopher University in Nitra (Slovakia).

At the first phase the attention will be paid to the definition of the performance standard for the graduates of secondary vocational schools (ISCED 3) related to the area of CAD/CAE systems. The intention is to define this standard according to industry practice requirements. The relevant requirements should result from a large-scale questionnaire survey.

The prepared pedagogical experiment will be carried out with a research sample of students from selected relevant secondary technical schools (ISCED 3). There will be created two representative equivalent samples, one (a reference, control group) in which the CAD/CAE system teaching will be carried out in a traditional way and one (the experimental group) in which the experimental method will be applied, i.e. CAD/CAE system will be taught following the new methodology. Both groups will have the same total time allocation for the technical drawing and modelling lessons, but the ratio of technical drawing lessons to modelling lessons in case of the reference (control) group will be 2:1 while in case of the experimental group it will be 1:2. Intended duration of the experiment is 2–3 academic years. Duration of the pedagogical experiment is estimated from 2 upto 3 years. Currently the project of the experiment is in its preparatory phase. The research hypotheses which are going to be tested within the research are the following ones:

H1: *The members of the experimental group will achieve significantly better learning achievements as the members of the control reference group in acquirement of the knowledge and skills in the area of modelling and simulation in CAD/CAE systems.*

H2: *The members of the experimental group will achieve approximately the same learning achievements as the members of the control reference group in acquirement of the knowledge and skills in the key areas of technical documentation creation and reading.*

The hypothesis H1 follows the assumption that a higher time location devoted to teaching modelling and simulation in CAD/CAE systems will influence students' learning achievements in a positive way, what is logical and predictable. But the key point of this hypothesis is that fact that this will be achieved despite the decreased time allocation for training devoted to manual creation of technical documentation, i.e. we do not consider these skills (except the basic knowledge) as necessary to develop students' skills to work with CAD/CAE systems. Contrary to the hypothesis H1, the hypothesis H2 assumes that the decreased number of lessons devoted to teaching technical drawing and creation of technical documentation will not have any significant negative impact on students' knowledge and skills.

Very important factor to verify or falsificate the stated hypotheses will be development of an appropriate right evaluation methodology, as it is not easy to measure exactly degrees of the skill to express one's own thoughts nor the skill to perceive thoughts of others expressed through the technical graphical communication. Currently the relevant available resources dealing with this issue have been studied and analysed to create a platform for the relevant methodology development.

5 Anticipated Outcomes

In the event that our experiment confirms the established hypotheses, we can declare that our proposed non-traditional teaching process prepares the students better in the area of CAD/CAE systems for the needs of technical practice. Such an untraditional process would have a clear advantage in not imposing any extended material or time requirements. It would be enough to modify the teaching process with the change of the time allocation for the partial parts of the education in the given area.

If the carried out experiment confirms the given hypothesis, it will thus be proved that the above-presented unconventional methodology of teaching CAD/CAE systems at upper-secondary schools, which was designed by us, prepares their students better to meet the requirements of technical practice regarding their knowledge and skills related to modelling and simulation in CAD/CAE systems. Moreover an undisputed advantage in this unconventional method is the fact that this way of teaching the concerned issues is not connected with any increased or additional material or time needs. The only thing which it requires is to modify the way of teaching and to change the numbers of lessons devoted to the particular parts of the taught issues in the subject area.

References

1. Ye, X., Peng, W., Chen, Z., Cai, Y.: Today's students, tomorrow's engineers: an industrial perspective on CAD education. Comput.-Aided Des. **36**, 1451–1460 (2004)
2. Fořt, P., Kletečka, J.: Autodesk Inventor. Computer Press, Brno (2004)
3. Vláčilová, H., Vilímková, M., Hencl, L.: SolidWorks. Computer Press, Brno (2006)
4. Bilalis, N.: Computer Aided Design CAD. INNOREGIO: dissemination of innovation and knowledge management techniques. Technical University of Crete (2000)
5. Chang, K.H.: Product Design Modeling Using CAD/CAE. Academic Press, Cambridge (2014)
6. PWC: Prieskum dodávateľov automobilového priemyslu/Survey of automotive industry providers (2014). https://www.pwc.com/sk/sk/publikacie/assets/2014/prieskum-dodavatelov-automobiloveho-priemyslu-2014.pdf
7. Belmans, R., Geysen, W.: CAD-CAE in electrical machines and drives teaching. Eur. J. Eng. Educ. **13**(2), 205–212 (1988)
8. Kaminaga, K., Fukuda, Y., Sato, T.: Successful examples in CAD/CAE teaching. In: Sun, Q., Tang, Z., Zhang, Y. (eds.) Computer Applications in Production Engineering. IFIP Advances in Information and Communication Technology, pp. 214–221. Springer, Boston (1995). https://doi.org/10.1007/978-0-387-34879-7_23
9. Arthur, P. (ed.): CADCAM in Education and Training. Proceedings of the CAD ED82 Conference. Kogan Page Ltd., London (1984). https://doi.org/10.1007/978-1-4684-8506-6
10. Asperl, A.: How to teach CAD. Comput.-Aided Des. Appl. **2**(1), 459–468 (2005). https://doi.org/10.1080/16864360.2005.10738395
11. Ru-Xiong, L., Jiao, S.-H. : Teaching technique innovation on CAD/CAM/CAE of mold course. IERI Procedia **2**, 137–141 (2012). https://doi.org/10.1016/j.ieri.2012.06.064k
12. Xin-fang, Y., Ji-feng, L.: Probe on the teaching innovation of CAD/CAM for application-oriented course. J. Jiangsu Teachers Univ. Technol. **10**(2), 94–96 (2004)
13. Wang, H.: Research on the teaching system based on the characteristics of the three-dimensional CAD/CAM. New Technol. New Process **27**(2), 16–18 (2012)
14. Wittmann, E.C.: The mathematical training of teachers from the point of view of education. J. für Math.-Didaktik **10**, 291–308 (1989)
15. Humenberger, H., Reichel, H.C.: Teaching student teachers: various components of a complex task. Teach. Math. Comput. Sci. Debrecen **1**(1), 55–72 (2003)
16. Lantada, A.D., Morgado, P.L., Munoz-Guljosa, J.M., Otero, J.E., Muňoz Sanz, J.L.: Comparative study of CAD-CAE programs taking account of the opinions of students and teachers. Comput. Appl. Eng. Educ. **21**(4), 641–656 (2010). https://doi.org/10.1002/cae.20509
17. Jakubová, G.: Vzdelávacie programy/Educational programs. ŠIOV, Bratislava (2013). http://www9.siov.sk/statne-vzdelavacie-programy/9411s
18. http://www.uiam.mtf.stuba.sk/predmety/gms/studijne_texty/GMS10-09.pdf
19. Palaj, M., Skačan, M.: Vyučovanie technických predmetov na stredných školách z pohľadu začínajúcich učiteľov/Teaching technical subjects at secondary schools from the point of teacher novice's view. Technika a vzdelávanie/Technol. Educ. **5**(2), 43–45 (2016)

The Playful Approach to Teaching How to Program: Evidence by a Case Study

Matthias C. Utesch[1,2], Victor Seifert[1], Loina Prifti[1], Robert Heininger[1(✉)], and Helmut Krcmar[1]

[1] Chair for Information Systems, Technical University of Munich (TUM), Munich, Germany
{utesch, prifti, robert.heininger, krcmar}@in.tum.de, victor.seifert@tum.de

[2] Staatliche Fachober- und Berufsoberschule Technik München, Munich, Germany
utesch@igip.org

Abstract. Programming has become an important and popular skill in our economy. However, the digital age affects not only our economy but our daily lives, our free time, our habits, and our education systems as well. The way we perceive and especially access information has changed drastically in the last 20 years. IT-based learning has become a widespread approach to educate anyone who wants to learn, regardless of gender, age, or cultural background. Many new approaches have been made possible by the technological advancement in latest years. One of these approaches is the playful approach. As a part of self-regulated and personalized learning strategies, it focuses on the student's interaction with the subject. In this contribution, we connect the engagement students experience in video games with the educational content of standard curricula. By the means of a case study, we provide a scientific basis for future playful learning approaches.

Keywords: Teaching programming · Playful approach
Evaluated lesson structure · Programming course · Upper vocational school

1 Introduction

The digital age has changed the way we perceive and access information. Boundaries such as financial background, time, location, and availability of information are non-existent anymore – a smartphone with internet access is enough to access most of the documented knowledge of human history. In today's society, it has become normal for students to have constant access to the internet and the immediate on-the-spot knowledge acquisition opportunities provided through it. Students use digital means to learn, communicate, access knowledge, and even for professional application [1]. The importance of digital mediums as a form of education is increasing [2] and many academic researchers (e.g. [3–8]) put increasing effort into the exploration and validation of teaching techniques using the playful approach as a core concept.

M. E. Auer et al. (eds.), *Teaching and Learning in a Digital World*,
Advances in Intelligent Systems and Computing 715,
https://doi.org/10.1007/978-3-319-73210-7_61

The term 'Educational Technology' is defined as "*the study and ethical practice of facilitating learning and improving performance by creating, using, and managing appropriate technological processes and resources*" [9]. IT-based learning approaches gain momentum in our digital economy as countless numbers of software projects, for computers as well as for smartphones, are realized with the goal to educate students on different topics. E-learning, web-based learning, online learning, and IT-based learning are just a few examples for technological processes and resources based on Information Technology (IT).

In our digital economy, there is a huge demand for skilled IT workers, not only in the IT departments but across entire organizations [10, 11]. Fullan [12] asserted that the moral purpose of education is to equip students with the skills that enable them to be productive citizens after completing their education. Thus, in the age of the digital economy improving the digital skills must be one of the topics students learn in school. Trilling and Fadel [13] called education in information, media, and technology skills vital for the 21st century job market, differentiating between the traditional education subjects and upcoming subjects such as information transformation and coding skills which are much needed in today's economy. IT is a constantly evolving sector influencing many areas of all our lives and offering opportunities for change; education is one of these aspects changing under the influence of the digital age.

In some countries, computer science classes have already become a part of the standard education. On September 16, 2015, the mayor of New York City announced a 10-year deadline for all schools in New York City to offer computer science classes in their curriculum [14]. Another example is the UK whose government has published a new curriculum in 2013 that includes learning how to program [15]. In Germany, the specific curricula are under supervision of the individual federal states, some states have adapted their curricula to the digital age, while others have not included information technology in their education yet. An example, of those who have included it already, are the technical vocational schools in Bavaria. In Germany, vocational schools offer an alternative way to achieve the qualifications needed to obtain a university education or other types of postsecondary education next to secondary schools in Germany. Bavarian technical vocational schools' curricula include programming, as a basic skill to digitalization, which is implemented in the curriculum as part of an obligatory course in multiple modules, such as *basics of modern programming languages*, *programming techniques and data structures*, and *object-oriented programming* next to a variety of voluntary modules [16].

We conducted a case study at a Bavarian vocational school, implementing the playful approach and the lesson plan described in Heininger et al. [17]. As such, this paper aims at evaluating the results of the case study and to validate the success factors for playful teaching, which were identified in Heininger et al. [18]. The lesson structure was developed specifically for the playful approach to learning how to program.

2 Playful Approach

Playful learning, often also called edutainment, is a recent trend in academic education which focuses on the hands-on practice of learning instead of on the sit-and-listen approach, spanning between free play (in which the students play independently), and guided play (where an overseer directs their play) [19]. Many of these playful approaches use self-regulated or personalized learning as a way to transfer knowledge to students, as the usage of the self-regulatory processes has shown strong correlations with high academic achievement [20]. Pivec and Moretti [6] pointed out that games offer didactical-educational opportunities: games offer an 'ice-breaking' introduction to a new topic and inspire interest in the learner. Games also help to establish a dialogue, break social boundaries, and even encourage personal development and improvement [6].

Experiencing fun and joy while handling a given task can positively influence the learner's motivation to be involved and engaged in activities even if they have little or no previous experience with the subject matter [21]. Most games demand both engagement and active focus on the gameplay at every second, perhaps with the exception of some turn-based games. The motivation behind this thesis is therefore the combination of the engagement during playing video games paired with the educational aspects of standard curricula of schools in Germany.

Case studies implementing the playful learning approach in all kinds of fields of education are widespread, although most them supply no scientific basis concerning the structure and on how to implement such a playful teaching model. We aim to fill this gap by validating a lesson structure that applies a playful learning approach, which we developed in previous work [17]. The structure as well as the implementation of our case study is based on the success factors for teaching how to program with a playful approach. These were identified by a state of the art literature review and include: the motivation, integration & involvement, the audience-centered focus, giving feedback and enhancing interaction, and the fluent integration of the educational content into the gameplay [18]. Hence, in this paper we aim to answer the following research question to validate the developed approach:

> ***What are the advantages and disadvantages of the playful learning approach for teaching how to program?***

The goal of this paper is to highlight the advantages and disadvantages of the playful learning approach as well as to recognize benefits, errors, and problems that occurred during the lessons. A basis for this are the identified success factors [18] and their fulfillment by the developed structured [17] in the pilot class. Further on we will present beneficial changes to the developed structure, based on the results of the pilot class.

3 Case Study

The lesson structure, developed earlier by Heininger et al. [17], was implemented in a practical course in a classroom at a Bavarian upper vocational school in Munich, over the course of two months. Two classes took part in the case study, each of which was split into two groups, following the official lesson plan of the school. The students of each of the four groups attended six regular weekly lessons. The duration of each lesson was 45 min as is standard at Bavarian schools. The course was an introductory course to programming; only one student out of 44 had extensive previous experience with coding, while 21% of students mentioned to have some previous experience.

A limitation of the case study was the male/female ratio of the students, with 39 male students, but only 5 female students. This did not allow for a valid comparison of differences in the effect of the playful approach on male, respectively female students. The bigger part of the students (58%) were between the age of 16 and 18, with 37% between the age of 19 and 21 and 5% being 22 to 23 years old.

3.1 Organization

The classes for the individual groups were held on Tuesdays, Thursdays, and Fridays. A sufficient number of laptops for the students were provided by the school, but students were still encouraged to bring their own laptops in order to be able to take notes and to code in their own environment as they would on their own at home. Every student played and coded independently, following the self-regulatory setup of the approach. To control and assess the results of the practical experiment we carried out two surveys with the students in the class. The surveys took place at the beginning of the first and at the end of the last lesson. The purpose of the surveys was to acquire an overview of the skills and opinions about the subject of the students as well as to provide a scientific basis for the analysis and evaluation of the approach. Furthermore, the teacher's experiences as well as specific events and incidents during the pilot class were logged and analyzed in order to gain further insights regarding the experiment.

Note the fact that the teacher does not approach students but lets them work independently to allow them the freedom to solve problems on their own. This is in consent with the playful approach which shifts the role of the teacher from lecturer to a supportive, facilitating role [18]. The goal of this endeavor is fostering students' motivation, self-value and increasing their confidence in their own skills. A solution to a problem should not merely be offered by the teacher but worked towards together. Therefore, feedback to the students on their code is mainly given by the educational game. The teacher intervenes if difficulties come up and students cannot solve a problem based on the feedback of the game or coding error messages alone. By observing the students play, the teacher is also able to identify complications in the gameplay or during the execution of coding tasks. The feedback by the game is an important tool to give students a feeling of coding on their own and to teach them about identifying errors in their code by the standard error messages. To keep track of the students' progress, homework is assigned at the end of each unit. The assignments are given to students on paper to take home. The submission is in the form of python scripts on USB drives, which the students bring to class in the following unit.

3.2 Curriculum

The curriculum of the pilot classes was the curriculum of technical vocational schools in Bavaria [16]. It includes three major parts of learning how to program: basics of modern programming languages, programming techniques and data structures and object-oriented programming. The emphasis lies on students learning to independently break a problem down into smaller parts and create advanced pieces of software by structured thinking. In a previous work [17] we suggested the programming language Python as a first introductory language to learn for students because of its structured design, clear error message handling, shallow learning curve and high industry-relevance.

In the first five minutes of each unit, a short repetition of the knowledge learned in the last unit will be given by the teacher, combined with an outlook to what students learn in the current unit and how it is connected to previous knowledge. As students will have progressed at different speeds and hence be at different levels of knowledge the repetition is based on students' remarks and input. Following the self-regulatory environment of the playful approach means not setting a learning goal for the students for each unit but clearing up eventual difficulties with previous contents. Interim learning goals are set by the students themselves. The teacher provided handouts for each stage of the game to give students something tangible to study and base their learning effort on before the obligatory exam after the end of the course.

The last five minutes of the units will be used for a short repetition of the current content in order to gather more information on the difficulties the students experienced and how they felt about the learning process by playing a game. The time should be used for self-reflection by the students. After the units, the teacher himself will self-reflect and record how his insight differed from the students' remarks about their experience.

3.3 Survey Development

The questions in the surveys were of various types, dependent on the nature of the question. As seen below, most of the questions were based on a numerical, symmetric Likert scale with clearly defined linguistic qualifiers and a range of one (strongly disagree) to five (strongly agree). As this was a pilot class, the students answered a variety of given questions, for instance about their background in IT, about their readiness to learn how to program, their initial motivation and many more in order to cover as many aspects as possible and to be able to adequately compare the questionnaire results of both surveys. The first survey was conducted before the beginning of the first lesson; the second survey was conducted after the end of the last lesson and before an exam about the subject had taken place (Table 1).

Table 1. Survey question types

Question type	Survey 1	Survey 2	Example
Dichotomous	1, 3, 7	1, 3, 7	Gender
Likert scale (numerical 1–5)	4 and 8–26	4 and 8–35	Agreement to statement
Ordinal-multivalued	2, 5, 6	2, 5, 6	Age
Free text	None	36, 37, 38	Open Questions

3.4 The Educational Game

During the development, main design principles were derived from the iterative processes of conducting case studies by other authors about evolving and improving an educational game: interaction and iteration of those interactions, adaptive and personalized feedback, clear winning criterion and no or few opportunities to cheat the game [22]. Those criteria have been confirmed by another often cited study on educational games proposing practical design steps [23]. A good game is easy to learn but hard to master [24] meaning the handling of the basic game is easy, but to be good at it becomes more and more difficult the further you advance.

The style of the educational game used in our pilot class is live-action, demanding constant input and effort by the player. Turn based gameplay was considered but it would interrupt the flow of the game by initiating pauses, breaking the student's focus, active integration and participation. The authors agree with the idea that great graphics on their own are never enough to make for a good game [24]. Games are mostly about the gameplay, about the challenges, obstacles and following "success moments" [3]. The game was developed as a 2D jump-and-run, side-scrolling adventure game, a similar style like the immensely popular Mario games (Fig. 1).

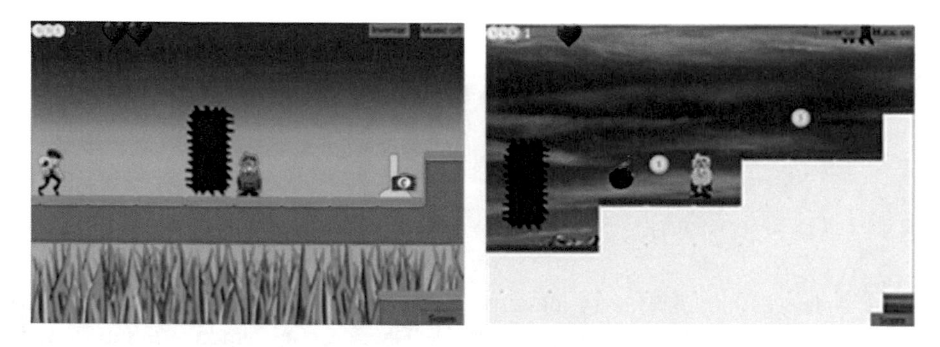

Fig. 1. Exemplary screenshots of the gameplay (own illustration)

The playful approach aims at motivating the students to become more interested in their own education and willingly sacrifice time for it, rather than presenting the knowledge that they are required to learn on a silver platter. The game achieves this by making students work for their knowledge by beating the game by means of defeating numerous enemies and mastering obstacles. Students playing next to each other will be able to compare their own progress to the progress of their classmates. Statistics and level progress are implemented to create a friendly basis of competition between the students, motivating them to put more effort into the game.

The educational aspect of the gameplay was represented by the implementation of a coding environment in which the students had to solve tasks in order to advance in the game while acquiring additional skills. It was connected to the gameplay by the means of tasks which results were reflected in the gameplay. An example is the introduction of variables and value assignment by students having to change the value for the jump-height of the in-game character from 1.5 to 3.5 – resulting in the game character being able to jump over obstacles. Another example is the implementation of arrays, which allowed the character to collect apples in the game which in turn enabled him to acquire additional 'life-points'. The coding interface contained an instructional text, a coding area as well as an output window in the style of system consoles (Fig. 2).

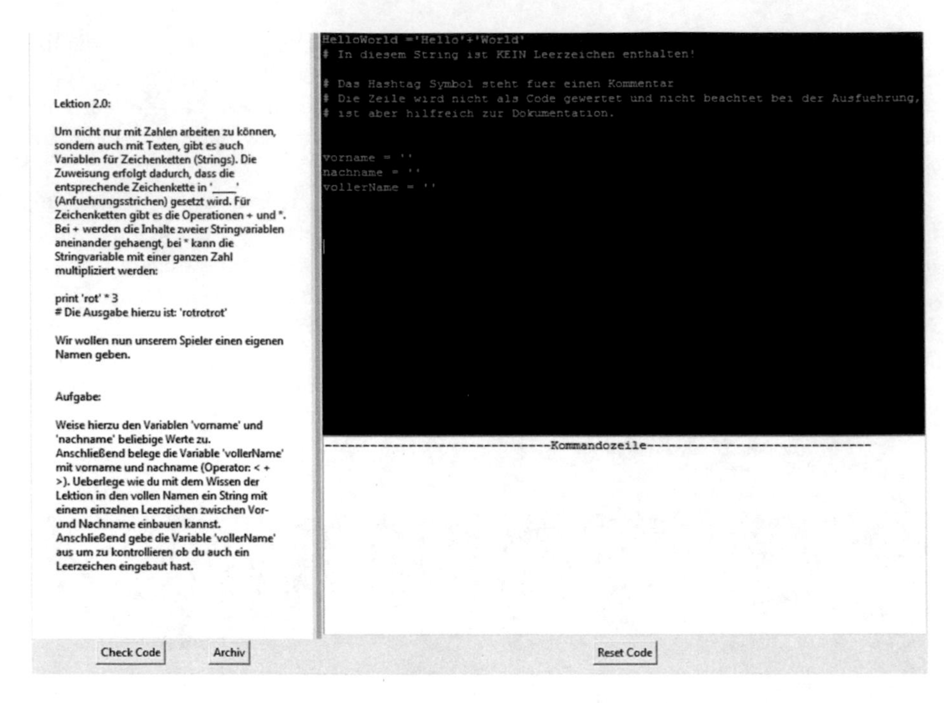

Fig. 2. Exemplary screenshot of the coding interface (own illustration)

The style was minimalistic to allow the students to focus on coding rather than having difficulty understanding the interface. A 'Check Code' button was introduced allowing students to check output and results of their code. Furthermore, an "Archive" button was added to enable students to return to older lessons during a coding exercise to refresh their memory or to look up previous content of the curriculum. Additionally, the students were able to reset their code or to continue to the next exercise, respectively return to the gameplay.

4 Survey Evaluation

Two surveys were carried out, including all students of all four groups. The first survey took place at the beginning of the first lesson, the second survey after the last lesson. The following chapter shows highlights as well as problems uncovered by the analysis and evaluation of students' answers to survey questions.

4.1 First Survey

43 Students filled out the first questionnaire, 41 students filled out the second. The discrepancy was caused by ill or otherwise absent students. The majority of the students (58%) was between the age of 16 and 18, with 37% between the age of 19 and 21 and 5% being between 22 and 23 years old. 61% of the students owned their first

smartphone between the age 14 and 15, with a few students having one earlier or later as seen in the figure below. The 'last' student to get a smartphone was at the time between the age of 18 and 19. None of the attending students did not have a smartphone. 16% of the students even had a smartphone before the age of ten. The students who stated that at least one member of their family worked in the IT sector were on average more interested in programming than those whose parents did not work in the IT sector. The authors' hypothesis is that students with a family background in IT have a better image of IT in general. The ratings of their image of IT showed that the difference of students who have a family member with an IT background and of those who do not was only 0.13 Likert points in average and as such too small to allow for a valid comparison. The first group had an agreement rating of 4.00 Likert points, the second achieved a rating of 3.87. The average of all 43 students was at 3.91 Likert points.

It was not possible to reject or prove the authors' hypothesis that students who think being able to code is useful in life are also interested in coding themselves. The test's independent variable was based on the previous question "*I think being able to program is useful in life*". The share of students who disagreed or strongly disagreed with this statement was too small to draw conclusions, hence not allowing for a valid comparison of the groups. Only 4.7% of students disagreed or strongly disagreed that coding is useful in life; the average rating was 4.2 in a Likert scale ranging from 1 (strongly disagree) to 5 (strongly agree). 95.3% of all students strongly agreed that they learn easier if the subject is interesting to them, which enhances the focus on the engagement aspect of the playful approach in order to foster initial curiosity about the subject to become intrinsic motivation.

4.2 Second Survey

85.4% of the students agreed or strongly agreed with the statement "*during the course in the last weeks I learned a lot of new things about programming*". 7.3% of students disagreed or strongly disagreed with this statement. The results hence show that the teacher was able to transfer knowledge by implementing the playful teaching approach – even for some of the students who already had prior knowledge. The authors' hypothesis that students were able to acquire new knowledge by attending the class which was previously unknown to them is therefore not rejected.

90.2% of the students agreed or strongly agreed with the statement that the teacher helped them to correct their mistakes. Furthermore, 95.1% of the students agreed or strongly agreed that the teacher addressed their issues. In another question, 95.1% of the students also agreed or strongly agreed that the teacher helped them if they encountered problems. This shows that the role of the teacher as a facilitator and supporter, rather than a lecturer, encompassed the overwhelming majority of the class and hence fulfills the success factor of feedback and interaction [18] in the approach of our pilot class.

41.5% of students agreed with the statement to have enjoyed themselves playing the game and another 41.5% of students even strongly agreed with this statement. Only one student strongly disagreed, accompanied by two students opting for the 'neutral' answer option. This confirmation of the approach in combination with the students' self-assessment about their success to learn something new in class, gives first indication of success concerning the implementation of the playful approach. It further validates the success factors of motivation, integration and involvement in the approach.

An assumption of the authors was that students liked the playful approach solely for the gameplay elements. However, when students were asked whether they felt like the educational game should have had a greater share of 'play'-content 34.2% of the students opted for the neutral answer option with the rest evenly divided between agreement and disagreement. This allows for the assumption that students did indeed like the game not only for its gameplay aspect, but for its educational value too. In another two questions the students' answers indicated that the balance between educational and playing content was perceived well, some students even wanted more educational content in the game. However, a fault by the authors and developers of the game was uncovered in a later question – students felt that connection of the educational content and the gameplay could have been closer. The fluent integration of gameplay and educational content was hence not completely achieved.

The majority of the students (90.2%) agreed or strongly agreed to have experienced the lessons as fun. The remaining 9.8% were neutral towards the statement. Furthermore, 90.2% of the students agreed or strongly agreed with the statement that the class was interesting to them, which fulfilled the criteria of the audience centered success factor of the implementation. After the course, 63.4% of the students agreed or strongly agreed that they were further interested in programming and wanted to learn more about it. In contrast, only 17.1% disagreed or strongly disagreed. 90.2% of the students stated to have enjoyed the playful approach, only one student disagreed with the statement and the rest choose the neutral answer option.

5 Observations During the Pilot Class

An extra-curricular voluntary lesson to deepen students' knowledge and skills, which was offered by the teacher, was not attended by the students. However, it was discovered later that the offered lesson overlapped with an obligatory school trip, which made it impossible for the students to attend.

Based on the teacher's observation, the students enjoyed the class. Nevertheless, a problem was uncovered by the teacher's assessment that students would need more time with the educational game. Students seemed to need 5 to 10 min at the beginning of each of the regular lessons to find back into the game. Some students even wanted to stay and keep playing during the school recess. A possible solution would be to reserve a timeslot of two weekly back-to-back lessons, each of 45 min, while reducing the weeks planned for this part of the curriculum from six to three, in order to enhance the continuous interaction with the educational game. This could be part of a future case study.

The most noteworthy observation during the case study, confirming the success of the playful approach and the achievement of motivating and engaging students further, was that students who never programmed before started to bring small pieces of code into the class without prior incentive or extrinsic motivation by the teacher.

An example of this was a student, who had coded a version of Tic Tac Toe based on console input and output in his free time, asking the teacher for suggestions and putting his code up for discussion in class with his peers. This goes to show that the approach was successful in motivating students to learn further by themselves and to be intrinsically motivated and interested in programming. Other students followed his example with other pieces of code without the teacher suggesting it.

6 Conclusion and Recommended Changes to the Approach

The authors were able to transfer the knowledge and inspire students to look outside the box. Students began to code independently, enjoyed learning, and wanted to learn more about programming. Students showed enthusiasm to play and learn by wanting to play during the breaks and additionally learning at home as well. The students voluntarily brought small pieces of code and software into class without prior incentive by the teacher in order to get feedback and to put it up for discussion with their peers. This provides evidence that the approach was able to capture students' interest and motivate them further. Students expressed that they experienced great support and feedback during the course in the questionnaires and showed their appreciation in class. Feedback was very important to the students during this case study – not in the form of grades but by the game and the teacher. A framework for structure and observation was still needed in order to guide the students. This was achieved by the structure developed in Heininger et al. [17] and the supervising teacher figure as a guide and facilitator of the students' learning progress.

The authors suggest two changes to the case study. The first change is class-related and concerning the success factor of feedback and interaction: the assessment by the teacher showed that especially continuous interaction with the educational game could benefit the students. Hence, we suggest a longer continual period as weekly lessons for the future in exchange for a shorter overall timeframe. The second change is related to the educational game: the integration of the educational content to the gameplay (one of our success factors) in a (to the students) satisfactory way was not achieved. This calls for more research on the design and development of educational video games in the future.

Letting students play a game in class and trying to motivate them to discover knowledge on their own and taking responsibility of their own progress did not result in students trying to distract themselves or not wanting to learn, but them being enthusiastic about the topic at hand. Especially upon seeing that their learning environment was created with the same tools, which they would possess at the end of the course. This was an incentive for the students to tackle problems that are more difficult and to come into closer contact with the subject. Therefore, the pilot class was able to confirm the success factors of Heininger et al. [18] and even suggest improvements to the approach of Heininger et al. [17] which was well perceived by the students as well as from a teacher's viewpoint.

Acknowledgements. The authors would like to express special thanks to the upper vocational schools at Bavaria represented by Günter Liebl, Werner Maul, Konrad Maurer, and Thomas Ondak.

References

1. Baggia, A., Žnidaršič, A., Borštnar, M.K., Pucihar, A., Šorgo, A., Bartol, T., et al.: Factors influencing the information literacy of students: preliminary analysis. In: 29th Bled eConference – Digital Economy, Bled, Slovenia, pp. 617–631 (2016)
2. Blamire, R.: Digital Games for Learning: Conclusions and Recommendations from the IMAGINE Project. European Schoolnet, Brussels (2010)
3. McGonigal, J.: Reality Is Broken: Why Games Make Us Better and How They Can Change the World. The Penguin Press, New York (2011)
4. Gee, J.P.: What video games have to teach us about learning and literacy. Comput. Entertain. (CIE) **1**, 20 (2003)
5. Fabricatore, C.: Learning and videogames: an unexploited synergy. In: Annual Convention of the Association for Educational Communications and Technology (AECT), Long Beach, CA, USA (2000)
6. Pivec, M., Moretti, M.: Game Based Learning: Discover the Pleasure of Learning. Pabst: Science Publishers, Lengerich (2008)
7. Prensky, M.: Engage me or enrage me: what today's learners demand. Educause Rev. **40**, 60 (2005)
8. Utesch, M., Hauer, A., Heininger, R., Krcmar, H.: An IT-based learning approach about finite state machines using the example of stock trading. In: Interactive Collaborative Learning (ICL) 2016, Belfast, UK (2016)
9. Robinson, R., Molenda, M., Rezabek, L.: Facilitating learning. In: Januszewski, A., Moleda, M. (eds.) Educational Technology: A Definition with Commentary, Lawrence Erlbaum Associates, New York (2008)
10. Capgemini Consulting: The Digital Talent Gap, Capgemini Consulting, Digital Transformation Research Institute - Capgemini (2013)
11. Prifti, L., Knigge, M., Kienegger, H., Krcmar, H.: A competency model for "industrie 4.0" employees. In: 13th Internationalen Tagung Wirtschaftsinformatik (WI 2017), St. Gallen, pp. 46–60 (2017)
12. Fullan, M.: The New Meaning of Educational Change, 4th edn. Teachers College Press, New York (2007)
13. Trilling, B., Fadel, C.: 21st Century Skills: Learning for Life in Our Times. Wiley, San Francisco (2009)
14. Taylor, K., Miller, C.C.: De Blasio to announce 10-year deadline to offer computer science to all Students. In: The New York Times (ed.) www.nytimes.com. The New York Times (2015)
15. Stuart, D.: Coding at school: a parent's guide to England's new computing curriculum. In: The Guardian (ed.) www.theguardian.com. The Guardian (2014)
16. Bayerisches Staatsministerium für Unterricht und Kultus. In: Lehrpläne für die Fachoberschule und Berufsoberschule; Unterrichtsfach: Technologie/Informatik, 20 March 2016 (2016). https://www.isb.bayern.de/download/9223/lp_fos_bos_technologie.pdf
17. Heininger, R., Seifert, V., Prifti, L., Utesch, M., Krcmar, H.: The playful learning approach for learning how to program: a best practice for a lesson structure. In: 30th Bled eConference: Digital Transformation – From Connecting Things to Transforming Our Lives, Bled, Slovenia, pp. 215–230 (2017)
18. Heininger, R., Prifti, L., Seifert, V., Utesch, M., Krcmar, H.: Teaching how to program with a playful approach: a review of success factors. In: IEEE Global Engineering Education Conference (EDUCON2017), Athens, Greece, pp. 189–198 (2017)
19. Lillard, A.S.: Playful learning and Montessori education. Am. J. Play **5**, 157 (2013)

20. Everson, H.: Barry Zimmerman, 8 September 2016 (n.d). http://learningandtheadolescentmind.org/people_04.html
21. Bisson, C., Luckner, J.: Fun in learning: the pedagogical role of fun in adventure education. perspectives. J. Exp. Educ. **19**, 108 (1996)
22. Tillmann, N., De Halleux, J., Xie, T., Gulwani, S., Bishop, J.: Teaching and learning programming and software engineering via interactive gaming. In: 2013 35th International Conference on Software Engineering (ICSE), pp. 1117–1126 (2013)
23. Linehan, C., Kirman, B., Lawson, S., Chan, G.: Practical, appropriate, empirically-validated guidelines for designing educational games. In: Proceedings of the SIGCHI Conference on Human Factors in Computing Systems, pp. 1979–1988 (2011)
24. Prensky, M.: The motivation of gameplay or, the REAL 21st century learning revolution. On Horiz. **10**, 5–11 (2002)

New Approach to Architectural Design Education

Yunus Turan Pekmezci(✉) and Taybuğa Aybars Mamalı(✉)

Nisantasi University, Istanbul, Turkey
{yunusturan.pekmezci,aybars.mamali}@nisantasi.edu.tr

Abstract. Architectural design is a creative activity whose aim is to determine the formal qualities of objects produced by art. These formal qualities are not only the external features but principal those structural and functional relationships which convert a system to a coherent unity from the point of view of user. Architectural design extends to embrace all the aspects of human environment which are conditioned by architectural production, besides designer occupies a fairly wide band on the spectrum of human experience. The arranging of a creative collaboration between different professional groups has never been easy and constitutes a classic problem of reconciliation. Design is perhaps a special case as architects have been on divergent educational and professional paths.

In the direction of all these current debates and statements, this investigation suggests a survey conducted to provide a better understanding through a case study of Nisantasi University's Architecture Program. The design considerations and general background must be related to the actual means of production available for the specific application. In this paper it is explained that a complete harmony is required in this relationship for real success.

Keywords: Architecture · Architecture design education
University and industry

1 Introduction

Architecture education differs based on the country it is presented. It is the cultural, sociological, economical, political and historical differences that cause the differentiation. The problematic area of architectural education is multi layered and bears the need for a historical etude. It is also parallel to many "histories" such as architecture, culture, education and globalization.

University as an institute started to take root after the Enlightenment due to the advancement on sciences. The Art Academies of Italy in 16th century were the pioneers of formal architectural educations. The Industrial Revolution caused to standardization on a global scale.

Movement of people, money, technology on a global scale is the a priori problem. Globalization rises the need for standardization that is problematic in terms of the differentiations that were mentioned above. This problem opens doors the debate of architectural education on a comperative study.

M. E. Auer et al. (eds.), *Teaching and Learning in a Digital World*,
Advances in Intelligent Systems and Computing 715,
https://doi.org/10.1007/978-3-319-73210-7_62

2 A Historical Approach to the Situation of Architectural Education in the Contemporary Age

Académie Royal d'Architecture of France was established in 1671, and later in 1819, it was rearranged as Ecole des Beaux-Arts, where the formal architectural education was established and institutionalised. The architectural education methods were erected in this school. Students were to study in ateliers, participate in at least 2 competitions per year. It eventually became the symbol of classic academic architectural mind that was openly criticised and even harshly protested by the Modernists of 19th and 20th century. It is possible to say that this opposition is still up to date since various methods such as the architectural design studio organization as the core of architectural education system, the jury system as the evaluation of student works owes its origins to Ecole des Beaux-Arts.

In the problematic era of Industrialization, Modern Architecture was born – but not so easily and without problems. The first reactions of the academic wing that was mentioned above to industrialization was to imitate the Western Architectural styles from its origins-Ancient Greek Architecture, through the Gothic of Middle Ages, Renaissance, to Baroque. These revivalist movements – Neoclassic, Neogothic, Neorenaissance, Neobaroque found strong protests from the progressive avant-gardes that will form the Modernist movement in architecture. The Modernists were also deeply interested in the architectural education.

Bauhaus was an institutional school representing this wing in architecture. Bauhaus school finds its origins in the Arts&Crafts movement. It is based on the integration of artistic disciplines and craftsmanship, and also different design disciplines therefore brings the multi-disciplinary education in many levels [1]. It also resonates with Art Nouveau in terms of total design. The system is merely a creative innovation in an age that is troubled by the problems of over-population and the stress on economy; and a philosophical question that seeks answer to produce a cultural language to the new age. Inter-disciplinary design studio permits to create this new language in an age determined by a cultural loss, standardization in production but also variation of different cultures in the over-growing city. This search resulted in a new approach: Basic Design. The aim of Basic Design education was to teach the basic principles of all disciplines of design in general. It was a multi-disciplinary studio that was being executed by worldwide known designers and artists such as Gropius, Moholy-Nagy, Kandinsky, Klee, Breuer et caetera… Today, basic design is one of the basic courses that is presented to the student in the first year of architecture education. It is accepted by all different wings in contemporary architecture education.

Another case in architectural education was exceptional, like its founder: Frank Lloyd Wright's Taliesin West Architecture School. Wright did not have the diploma of architecture and was approaching to the formal architectural education in universities cynically. He established an architecture school in one of his projects. Taliesin West was isolated in desert. There, he created a space for his students to live collectively and his main practice was to build the designs of his students' works collectively. His belief was that a student must experience and learn something by doing it rather than just learning from in-class academic education.

All these experiments in modern architectural education echoes today; the main formal basis of studio-jury system that was shaped by Ecole des Beaux Arts, Basic Design of Bauhaus, learning by doing in studios or special workshops of Taliesin West [2]. But constant searches for new methods and experimentations show that this basis is simply not enough. Also it seems that the method of pre-institutionalised architectural education cannot be abandoned: the archaic practice of master-apprentice is never outdated.

3 Developments from Past to Today

When the economic, cultural and political structures of societies are taken into consideration, it is observed that the universities have basic missions for the future of their communities. When we look at the historical development of universities, the transfer of information such as research and development as well as the production of information is emphasized by progressive universities of the 19th century, thus forming the second generation universities [3].

As a result of developments in science and technology, classical learning methods have begun to lose their influence on students. With the shift of expectations from universities, technology production has turned into a value that cannot be denied in terms of the prestige of universities both materially and socially. Along with that second generation universities have begun to suffer from the persistency problem.

The recent debates on the architectural education in the contemporary age has some main focus areas such as interdisciplinary practice, open studio for a multi-disciplinary learning, teamwork rather than a solo production, the raising presence of social sciences in course plans. This contemporary approach finds its basis in the globalization discourse. The contemporary architectural education is engaged to the international and multi-disciplinary business model. Especially in the global cities and architectural schools in the European Union are connected in a network for collaborations and exchange students, yearly workshops and so on. The open studio consists architects from different study areas, design professionals and engineers. This is basically the academic micro cosmos of the global world. The increasing effect of social sciences such as sociology, economy and philosophy may be interpreted as the result of globalization as well since the main critics against globalization is produced by these disciplines.

Today, knowledge has crossed the boundaries of the university campus and it has become important to transform knowledge into a product. This has brought the close cooperation of universities with the business community and non-governmental organizations. For this reason, changes and transformations in the education and research approaches of universities have become inevitable. This has emerged as the most important element in the formation of third generation universities.

In the present case it is assumed that the universities have two basic functions that are education and research, and that the purpose of existence is to generate social benefit. The concept of university is evolving and changing, depending on science, technology and ever-changing needs of today. With globalization, the 21st century has begun a new era in which information is liberalized in all dimensions today, and individuals and institutions can participate in the cycle of knowledge without a university. This new conjuncture brings new challenges for the universities and forces them to change and demonstrate a new education system for their students.

"Third Generation University Right" book written by Dutch academician Professor Dr. Hans Wissema notes that in the 21st century universities are undergoing a fundamental change, progressing from a science-based, single-disciplinary institution to becoming a global information center. Today, universities need to provide students with the opportunity to integrate into the global world.

Third generation universities are changing their curriculum from national to global, from theoretical to practical, from mono disciplinary to interdisciplinary. This change affects both students and academicians.

Entrepreneurship and innovation are the main entities third generation universities undertake by integrating education and research and development into the society, and instead of leaving it to third parties to transform the information they produce into social, economic, technological and social values. This way, third generation university as a knowledge hub, stand out as an institution that integrates research and education [4]. Everything from institutional organization to academical roles is redefined from the local to the universal, according to the principles of autonomy and freedom. Apart from traditional research-training and publishing activities, in the third generation universities, time is spent to solve the basic problems of the society and global issues, competition, practice, consulting and the market which echoes in the architectural projects strongly engaged in the mentioned issues. The search for form, structure and function has left its importance to the social aspects of architecture.

Universities which have strong cooperation with industry are no longer merely closed systems that produce information, instead they are open systems that are the centers of economic efficiency [5].

In the direction of all these current debates and statements, this investigation suggests a survey conducted to provide a better understanding through a case study of Nisantasi University's Architecture Program.

4 A Specific Assesment over Nisantasi University

Applied to develop the collaboration of industry and university in many countries of Europe and the USA, third generation university model has started to create an important agenda in Turkey as well. The first foundation university that was established as a 3rd generation university in order to provide more qualified professionals to the business world by a practical course program is Nisantasi University. The integration of education and the business world is the main aspect for the students in Nisantasi University.

The main difference of Nisantasi University form the other universities is that in the foundation of the university, the course programs were organized by academicians and professionals collaboratively in order to meet the demands of the professional world and that the education is formed by the collaboration of industry.

The architecture education in the university is formed in 8 semesters. Undergraduate program of Architecture demands an intense knowledge of basic mathematics, geometry, visual arts and philosophy. The education in the program is presented in four basic fields as follows: architectural design and building information, history of architecture, physical environment control and building technology, project and construction management. The production of projects from different scales and functions by the student in the studios and laboratories every semester defines the student activity based, discipline practice engaged part of the undergraduate education. The formation of building design, spatial organization, production of objects needed by the users, the

ARCHITECTURE - CURRICULUM

I. SEMESTER

#	Course Code	Course Name	T	U	K	AKTS
1	ARC101	Principles of Ataturk and History of Turkish Revolution I	4	0	4	4
2	ARC103	English I	2	0	2	3
3	ARC105	Architectural Technical Drawing	1	2	2	3
4	ARC107	Building Information I	2	0	2	2
5	ARC109	Introduction to Architectural Design	0	8	5	10
6	ARC111	Research and Presentation Techniques	2	0	2	3
7	ARC113	Mathematics	2	0	2	2
8	ARC115	Basic Design I	1	2	2	3
		TOTAL	16	10	21	30

II. SEMESTER

#	Course Code	Ders Adı	T	U	K	AKTS
1	ARC102	Building Information II	2	0	2	2
2	ARC104	Building Elements	2	0	2	3
3	ARC106	Basic Design II	1	2	3	3
4	ARC108	Architectural Design I	0	8	4	10
5	ARC110	Building Materials	2	0	2	4
6	ATA102E	Principles of Ataturk and History of Turkish Revolution II	2	0	2	4
7	ARC112	English II	2	0	2	4
		TOTAL			17	30

III. SEMESTER

#	Course Code	Course Name	T	U	K	AKTS
1	ARC201	Architectural Design II	0	8	4	10
2	ARC203	European Architecture History	2	0	2	2
3	ARC205	Computer-Aided Design I	1	2	3	3
4	ARC207	Building Elements II	2	0	2	2
5	ARC209	English III	2	0	2	3
6	ARC211	Structural Analysis	2	0	2	2
7		Elective Course (ITB)	3	0	3	4
8		Elective Course (MT)	3	0	3	4
		TOTAL	15	10	21	30

IV. SEMESTER

#	Course Code	Course Name	T	U	K	AKTS
1	ARC202	Architectural Design III	0	8	4	10
2	ARC204	Modern Architecture History	2	0	2	2
3	ARC206	Computer-Aided Design II	1	2	2	2
4	ARC208	English IV	2	0	2	2
5		Elective Course (ITB)	3	0	3	4
6		Elective Course (MT)	3	0	3	4
		TOTAL	11	10	16	30

Elective Courses

#	Course Code	Course Name	T	U	K	AKTS
1	ARC213	Theories of Architectural Dsgn	3	0	3	4
2	ARC215	Housing in Developng Countries	3	0	3	4

Elective Courses

#	Course Code	Course Name	T	U	K	AKTS
1	MIM210	Urban Interiors	3	0	3	4
2	ARC212	Real Estate Economics	3	0	3	4
3	ARC208	Consv of A Wrld Hrtg Site-İst	3	0	3	4
4	ARC214	Desg.Draw.:Pract.&Concep.Appr.	3	0	3	4

V. SEMESTER

#	Course Code	Course Name	T	U	K	AKTS
1	ARC303	Architectural Design IV	0	8	4	10
2	ARC305	Enviromental Control Systems I	2	0	2	2
3	ARC307	Structural System Design	2	0	2	2
4	ARC309	Turkish Architecture History	2	0	2	2
5	ARC311	Urbanism and Planning Law	2	0	2	2
6		Elective Course (ITB)	3	0	3	4
7		Elective Course (MT)	3	0	3	4
8						
		TOTAL	14	6	18	26

VI. SEMESTER

#	Course Code	Course Name	T	U	K	AKTS
1	ARC302	Architectural Design V	0	8	4	10
2	ARC304	Enviromental Control Systems II	2	0	2	2
3	ARC306	Urban Planing and Zoning	2	0	2	2
4		Elective Course (ITB)	3	0	3	4
5		Elective Course (ITB)	3	0	3	4
6		Elective Course (MT)	3	0	3	4
7						
		TOTAL	13	8	17	26

8

Elective Courses

#	Course Code	Course Name	T	U	K	AKTS
1	ARC311	Logic and Theory of Design	3	0	3	4
2	ARC315	Structural Systems in Arch.	3	0	3	4
3	ARC317	Archit.&Interdisciplinary Std.	3	0	3	4

Elective Courses

#	Course Code	Course Name	T	U	K	AKTS
1	ARC308	Drawing in Architecture	3	0	3	4
2	ARC110	Garden Des. in History&Arch.	3	0	3	4
3	ARC312	Acoustical Problms in Archtctr	3	0	3	4
4	ARC314	Visual Culture and Design	3	0	3	4

VII. SEMESTER

#	Course Code	Course Name	T	U	K	AKTS
1	ARC403	Architectural Design VI	0	8	4	10
2	ARC405	Conservation and Restoration	2	0	2	2
3	ARC407	Construction Management And Economy	2	0	2	4
4	ARC409	Installation Knowledge	2	0	2	2
5	ARC411	Relievo	2	0	2	2
6	ARC413	Construction Project	2	0	2	2
		Elective Course (ITB)	3	0	3	4
		Elective Course (MT)	3	0	3	4
		TOTAL	16	8	20	30

VIII. SEMESTER

#	Course Code	Course Name	T	U	K	AKTS
1	ARC404	Professional Practices at Work	0	4	2	30
		TOTAL	0	4	2	30

	KRED	AKTS
TOTAL	132	240

Elective Courses

#	Course Code	Course Name	T	U	K	AKTS
1	ARC415	Cities and Architecture	3	0	3	4
2	ARC419	Restoration of Cultural Prop	3	0	3	4

Fig. 1. Nisantası University Bachelor of Architecture Programme Curriculum

technical know-how of planning and coordination of construction process; and lifelong personal development based on arts and culture thus the knowledge and skill for solving the practical problems that they will face is presented to the students [6].

When the course plan of Nisantasi University Bachelor of Architecture degree program is examined, the main difference from the formal university education concerning the 3rd generation university is that the students have to pass the Field Practice Course successfully in addition to the Graduation Project. In this innovative program, students practice their disciplines and develop their Graduation Projects simultaneously. The aim in Field Practice Course is to provide students the first step to the sector and encourage them to acknowledge their professional field with the conglomeration of the theoretical knowledge of 7 semesters. (See Fig. 1).

5 Conclusions and Suggestions

Architectural education institutions are reforming the educational organizations and teaching-learning methods for the health, welfare and security of the society and the right to live in a high quality built environment concerning the European Union criterias since they are the institutions to train the professionals and service providers that shape our habitats form the smallest scale to the biggest. Also the fields that are influenced by architecture which consists indistinctive work professions such as arts, technology, social sciences, history, design are expanding. In this context, it is inevitable to adopt an approach that is an interdisciplinary teaching model, especially because that basic architectural field is in theory and practice integrity and is defined by applied sciences, technology and arts.

In this perspective, it is essential that the day-long master-apprentice based architectural education must be reformed in a way that open to contact with other disciplines, participative, transparent, open to every kind of scientific development and innovation, aiming to channel towards personal development and be encouraging to student.

References

1. Gürdallı, H.: Mimarın Formasyonunda Formel Mimarlık Eğitiminin Yeri, Doktora tezi, İstanbul Teknik Üniversitesi, İstanbul (2004). (in Turkish)
2. Nalçakan, H.: Küreselleşen Dünyada Mimarlık Eğitimi ve Türkiye, Yüksek lisans tezi, Yıldız Teknik Üniversitesi, İstanbul (2006). (in Turkish)
3. Nalçakan, H., Polatoğlu, Ç.: Türkiye'deki ve Dünya'daki Mimarlık Eğitiminin Karşılaştırmalı Analizi ile Küreselleşmenin Mimarlık Eğitimine Etkisinin İrdelenmesi. YTÜ Arch. Fac. E-J. **3**, 79–103 (2008). (in Turkish)
4. TMMOB Mimarlar Odası: Mimarlık ve Eğitim Kurultayı V – Kasım 2009, Ankara, Mart 2010. (in Turkish)
5. Mimarlık ve Eğitim Kurultayı–3: Mimarlık ve Eğitimi Yeniden Yapılanırken, TMMOB Mimarlar Odası, İstanbul (2006). (in Turkish)
6. https://www.nisantasi.edu.tr/bolumler-95/mimarlik-20

An Iterative Approach for Institutional Adoption and Implementation of Flipped Learning: A Case Study of Middle East College

Dhivya Bino(✉), Kiran Gopakumar Rajalekshmi, and Chandrasekhar Ramaiah

Middle East College, Muscat, Oman
{dhivyabino,kirangr,chandrashekar}@mec.edu.om

Abstract. This paper presents an iterative framework for institutional level adoption of flipped learning based on the experience of implementing it at Middle East College, Oman. The framework is designed in three phases with each phase spread out over 2 years. A combination of models: the 'Lewin's Change Management Model' and the quality assurance 'ADRI Model' has primarily inspired the design of the framework. The phased approach of implementation ensured that at any point in time all the modules undertaken by a student is taught in the flipped mode whereas all modules of a program is progressively flipped semester wise. Elements of work and evaluation of each phase were delineated into three categories as strategy, structure, and support. It was found that the phased cyclical approach helps to systematically and progressively implement flipped learning at a monolithic scale with minimal disturbance to the core functioning of the institution managing the associated risks effectively. The project also proved that contextual pedagogical definitions and E Learning content availability supported by curriculum with assessments suitable for flipped learning are required for ensuring quality of implementation.

Keywords: Flipped learning · Institutional framework for implementation ADRI model · Lewin's Change Management Model

1 Context

This paper presents a framework for institutional level adoption of flipped learning based on the experiences gained after implementation at Middle East College (MEC), Oman. The framework is built using a combination of two models: the classic 'Lewin's Change Management Model' and the quality assurance ADRI (Approach, Deployment, Results, and Improvement) model. MEC is a leading private higher education institution in the Sultanate of Oman consisting of more than 5000 Omani and international students and approx. 200 full-time faculty members. The institution follows modular credit based curriculum of 8 semesters delivered over 4 years in case of a Bachelor program. Generally, regular students after higher secondary education join the Bachelor program at semester 1 and the students with a diploma enter the Bachelor program at semester 5. The lecture based teaching methodology that was practiced in the institution was not suitable for attaining the revised vision of the institution. The revised Teaching and

M. E. Auer et al. (eds.), *Teaching and Learning in a Digital World*,
Advances in Intelligent Systems and Computing 715,
https://doi.org/10.1007/978-3-319-73210-7_63

Learning strategy formulated in 2015 to reflect the changed vision and mission identified Flipped Learning as the institutional approach to teaching and learning. The College Board of the institution hence directed all academic departments to implement Flipped Learning as a mainstream teaching and learning methodology across all programs at MEC. This directive was also based on the results of a pilot study undertaken in 2015 and encouraged by the success stories worldwide. After the first phase of implementing it across the institution, it is seen that the framework adopted is helpful in bringing change of practice in a systematic manner. The theoretical framework of the project is detailed in Sect. 2, the methodology adopted is discussed in Sect. 3, results, findings and conclusions are documented through Sects. 4, 5 and 6 respectively.

2 Theoretical Framework

2.1 Need for Flipped Learning

The K-12 and higher education scenario of the country is generally didactic and hence there is a strong need for evolving teaching and learning practices that will develop independent learning skills [1]. There is consensus among the different definitions in the literature that flipped learning requires the student to learn foundational subject content by themselves, prior to the class, so that the class contact hours can be utilized for wider and deeper learning guided by the teacher. This teaching and learning methodology puts students in the lead and teachers on the side [2]. It also has the potential to encompass all other socio constructivist methodologies like Activity Based Learning, Problem Based Learning, and Case Based Learning [3]. MEC's educational philosophy is based on the principles derived mainly from socio-cultural and cognitivist perspectives on teaching and learning. The cultural aspects of Oman by which people excel when tasks are attempted in a collaborative atmosphere have inspired the formation of its educational philosophy. Accordingly, MEC strategy emphasizes that learning should be active and engaging process where learning experiences are enhanced by students taking the lead in the process.

2.2 Institutional Implementation Models

There is a call to implement flipped learning at the institutional level due to the amount of benefits it offers in developing 21st century graduates [4]. A quick search of available sources however does not result in any relevant guidelines that can help an institution in implementing flipped learning at a wider scale. ADRI model is employed at MEC to ensure quality in all initiatives as advised by the regulatory bodies [5]. This model promotes systems that are sustainable and helps to facilitate structured and systematic realization of goals through deliberate interventions [6].

It is also seen that incremental models fare better in comparison to episodic models in terms of disturbance and risk management when implementing change in core business aspects [7]. Porter et al. [8] proposes a three stage model by which blended learning can be implemented institutionally. They identified Strategy, Structure and Support as the key areas to be explored through the three stages of (1) awareness (2) early implementation (3) mature implementation. Interestingly, these three stages are also mentioned in the work of Ping and Tianchong [9] as (1) Applying/Emerging (2) Infusing (3) Transforming. All these are basically the three stages of the Lewin's Change Management Model [10] and the key areas are inspired from the 7 factors detailed in the McKinsey 7 S Model [11].

3 Methodology - Approach and Deployment

The framework for implementation established on the ADRI Model was implemented using Lewin's Change Management Model. Approach focused on (a) determining the strategic purposes and defining the implementation plan suitable for the purposes (b) establishing governance structures and (c) instituting the support mechanisms to effectively deploy and evaluate the plan.

3.1 Strategic Purposes

(a) To ensure that all the modules undertaken by a student are taught in the flipped mode to provide a similar learning experience across all the modules - Student focus
(b) To systematically and progressively flip teach all the modules at the institution to ensure that implementers' load is manageable - Faculty focus
(c) To create a framework which provides for planning, implementing and reviewing systematically - Institutional/System focus.

3.2 Implementation Plan

To achieve the above mentioned aims, an incremental and iterative implementation strategy was developed to execute in 4 phases over the period of 2015–2021. The first phase, pilot project consisting of few sample modules from all the UG and PG programs was also planned to be undertaken to check the feasibility and to understand the implications of changing practice across all the departments of the institution. Each of the remaining three phases spread out over 2 years corresponded to the various phases of the Lewin's Change Management Model. Therefore, the second phase during the period of 2015–2017 was planned for 'unfreezing'-bringing in awareness and climate for change, third phase during the period of 2017–2019 was meant for the actual change and the fourth phase during 2019–2021 was aimed at 'refreezing'-reinforcing the changed practices. Each phase further included 4 stages focusing on the 4 main semesters of the 2 year period. The four stages were meant to progressively flip teach modules belonging to the semester in focus while also ensuring that every student will undertake all the modules of the curriculum in the flipped mode. The plan is detailed in Fig. 1.

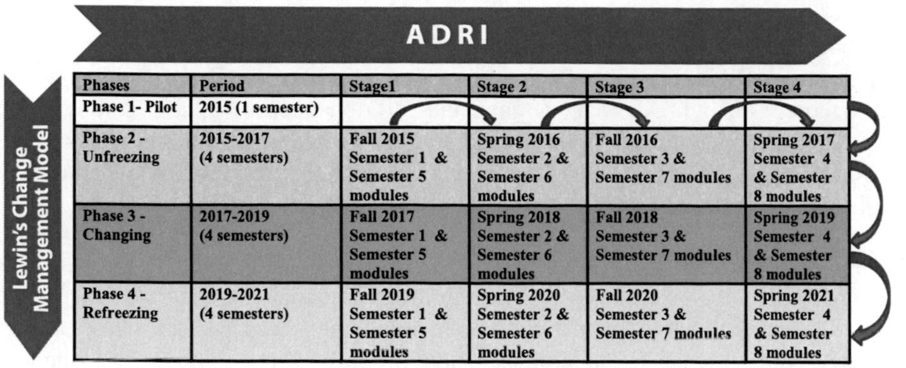

ADRI

Lewin's Change Management Model

Phases	Period	Stage1	Stage 2	Stage 3	Stage 4
Phase 1- Pilot	2015 (1 semester)				
Phase 2 - Unfreezing	2015-2017 (4 semesters)	Fall 2015 Semester 1 & Semester 5 modules	Spring 2016 Semester 2 & Semester 6 modules	Fall 2016 Semester 3 & Semester 7 modules	Spring 2017 Semester 4 & Semester 8 modules
Phase 3 - Changing	2017-2019 (4 semesters)	Fall 2017 Semester 1 & Semester 5 modules	Spring 2018 Semester 2 & Semester 6 modules	Fall 2018 Semester 3 & Semester 7 modules	Spring 2019 Semester 4 & Semester 8 modules
Phase 4 - Refreezing	2019-2021 (4 semesters)	Fall 2019 Semester 1 & Semester 5 modules	Spring 2020 Semester 2 & Semester 6 modules	Fall 2020 Semester 3 & Semester 7 modules	Spring 2021 Semester 4 & Semester 8 modules

Fig. 1. Flipped learning institutional implementation plan

3.3 Structures and Support Mechanisms

The Centre for Academic Practices (CAP) under the leadership of Associate Dean, Academic Affairs was given the responsibility to support, monitor and review the effectiveness of the implementation process. Support was planned in the form of (a) staff development workshops on pedagogy and technology (b) one to one consultations (c) infrastructure expansion (d) incentives and awards. Evaluation of the project was planned to be undertaken at interim exit points including the pilot, every stage of each phase and end of every phase. Multiple mechanisms like semester wise reports from program managers, semi structured interviews of faculty and questionnaires of faculty and students, review meetings and class room observations were established for effective evaluation. The review reports presented to Academic Affairs Committee, the Committee that oversees the academic matters of the institution enabled to decide on further course of action upon approval from College Board.

3.4 Deployment

Phase 1 - Pilot Project

The pilot project consisted of 11 level 2[1] modules belonging to seven academic departments. Evaluation of pilot was undertaken by CAP using semi structured interviews, and questionnaires of faculty and students and lesson observations, attendance and module grades. Evidence from the pilot project suggested that students' confidence in problem solving by themselves improved through the flipped learning approach. Module teaching teams and students attributed this improvement to better interaction between peers and greater engagement with learning materials. A considerable majority of students were positive about flipped learning citing benefits like easy access of learning materials at any time. The development of employability skills such as team working and presentation skills were also rated high by both module leaders and students [12].

[1] UK Level 5.

However, one of the main feedbacks for improvement from the students and faculty was to introduce the new practice from the first semester onwards so that students get familiar with the approach right from the beginning of their academic life. They pointed out that having most of the modules being taught in the traditional approach led to demotivation among the students in accepting the flipped practice in the few modules in which it was implemented. The need for a separate server for streaming videos was also identified at the end of the pilot study.

Phase 2 - Unfreezing
The objective of the phase 2 was to bring in awareness of flipped learning across the institution and build the foundation for changing the practice. This phase further involved 4 stages. Stage 1-The modules belonging to semester 1 and 5 were taught in the flipped mode to enable a student who joins the College after the school or after diploma to learn in this new approach. Stage 2-Along with Semester 1 and Semester 5 modules, modules belonging to Semester 2 and Semester 6 were flipped so that the student who joined in stage 1 progressed to study in the flipped mode. During stage 3, the student continued to study in the flipped approach as semester 3 and 7 modules were then flipped. The student completed his/her diploma or bachelor during stage 4 by continuing to learn all the modules in semester 4 or 8 through the flipped approach.

4 Results

As part of ongoing evaluations, semester wise review reports were submitted by all academic departments to the Academic Affairs Committee and review meetings were held by CAP with the academic Heads. At the end of the phase 1 of institutional implementation of Flipped Learning, surveys were also administered to the entire faculty members and a sample set of students using Survey Monkey. Purposive sampling was followed for the student survey and consisted of 900 students to ensure representation from all programs as well as diploma and bachelor levels. 94 (50%) faculty members and 118 (13%) students responded to the survey.

4.1 Student and Faculty Feedback

The student and faculty responses to the lickert scale survey questions are noted in Tables 1 and 2 respectively. Some of the student responses to the question on "How do you think we can make flipped learning better?" are also listed below.

> "no need to make it better, actually is good".
> "Teachers should be experienced in how to interact with students and making the classroom a healthy environment for the students to learn and participate in. Currently, the staff is unaware of how to keep the students active and interactive in the classroom".
> "The flipped learning should be improved by making some changes. First of all, we are asked to use flipped learning but the materials (videos) are too long and the total time may reach 45 min for one module. For part-time people 45 meaning a long time. Also, when flipped learning applies the attendance in the class should not be a matter. Another important thing, the videos should

be clear and the accent of the teacher who is speaking should be understood for all. We are suffering when we watching the videos of non-understandable and unlikeable voice".
"Created E-learning library available online all time from anywhere".
"Apply it in all modules".

Table 1. Student survey questions and responses

Survey questions and responses (in %)	SA[a]	A[b]	N[c]	D[d]	SD[e]
All the modules that I undertake at MEC are being delivered using flipped learning	15	25	42	14	4
I regularly engage with the preparatory materials before coming to class	10	31	35	19	5
I am able to learn the topic by myself using the materials provided	14	26	36	20	4
In the flipped classroom, my class spends more time doing experiments/exercises and activities than in traditional classrooms	13	31	29	24	4
The preparatory materials are sufficient for getting ready for classroom activities like experiments and advanced exercises	11	29	32	20	8
Activities undertaken during the class help me to practice and deepen my knowledge on the topic of study	17	35	30	12	7
Flipped learning gives my teacher more time to work one on one with me during class	15	31	31	15	8
I have more opportunity to study collaboratively with my friends because of flipped learning	22	27	26	15	9
I like videos made by my teacher than videos made by other teachers	19	35	24	14	8
Videos help me to catch up on lessons that I miss	28	36	23	7	7
Flipped learning gives me opportunity to learn with the help of technology	21	39	28	7	5
My learning experience is better in a flipped classroom than in a traditional classroom	21	22	30	18	9
I would rather study in a traditional teacher-led lesson than in a flipped one	25	25	26	18	6

[a]Strongly Agree
[b]Agree
[c]Neutral
[d]Disagree
[e]Strongly Disagree

Table 2. Faculty survey questions and responses

Faculty survey questions and responses in %	SA	A	N	D	SD
I am confident in making the necessary pre sessional content and activities for a flipped learning session	29	42	10	6	1
I know how to best utilize the class contact time during a flipped classroom	16	50	17	2	4
I create my own videos for the flipped classrooms when required	11	39	23	13	3
I find the videos online and use them for my flipped classrooms when required	24	53	8	2	2
Because of flipped learning initiative I have started doing a lot differently in my teaching practice	25	40	16	4	3
Flipped learning promotes activity based learning	31	39	14	3	2
Flipped learning methodology can be used for developing skills for lifelong learning in students	25	38	18	5	3
Flipped learning enables me to integrate technology into teaching and learning	32	44	8	2	3
Flipped learning can be used for creating collaborative learning environments	28	44	11	1	4
I am able to give individual attention to students because of flipped learning	17	34	24	8	6
Achievement of ILOs by students is more when taught using flipped methodology	18	24	33	7	8

5 Findings

The survey results indicate that all the faculty is aware of the new methodology envisioned and majority have flipped at least 40–60% of the content. They indicated that the incremental methodology has been helpful in changing practices efficiently. As expected, most of the faculty was very vocal in expressing that lack of student motivation and unpreparedness with the pre sessional materials have been the major constraints in successful implementation of flipped teaching. Technology related issues have also been pointed out by some as a hindrance to effective implementation.

Students who undertook the survey were positive of the approach even though the number of respondents was very less. The student feedback for improvement has only been in terms of improving the quality of videos uploaded, the duration of the videos and better training to the faculty in delivering a flipped lesson.

It was hence found that the strategic approach adopted was helpful to implement flipped learning at a monolithic scale with minimal disturbance to the core functioning of the college. There were systematic revisions in the implementation strategy, structure and support mechanisms. The initial resistance to a complete change of teaching practice gradually declined with every stage of first phase. Interim evaluations after every stage of the phase also showed increased popularity of the approach among the students and faculty.

6 Conclusions and Future Plans

Lack of adequate guiding frameworks and the uniqueness of flipped learning as an umbrella approach with the scope for employing a variety of instructional methodologies make it complicated for implementation at a massive scale. An incremental approach to implementing flipped learning at the institutional level seems effective in this regard. However institutional teaching and learning strategy should be clear about the approach for eLearning also. This will ensure that relevant technological infrastructure and pedagogical support mechanisms are created. Student motivational factors in case of flipped learning implementation need to be explored by promoting senior/alumni student involvement in content preparation and in educating current students. Curriculum redesign with specific focus on assessments for flipped learning should also be undertaken for a complete change of pedagogic practice.

References

1. Alsaadi, H.: Giving voice to the voiceless: learner autonomy as a tool to enhance quality in teaching and learning in higher education. In: Oman National Quality Conference, Research Gate, Muscat (2012)
2. Lage, M.J., Platt, G.J., Tregalia, M.: Inverting the classroom: a gateway to creating an inclusive learning environment. J. Econ. Educ. **31**, 30–43 (2000)
3. Baepler, P., Walker, J.D., Driessen, M.: It's not about seat time: blending, flipping, and efficiency in active learning classrooms. Comput. Educ. **78**, 227–236 (2014)
4. Forsythe, E.: Integrating recent CALL innovations into flipped instruction. In: Flipped Instruction: Breakthroughs in Research and Practice, pp 160–167. IGI Global, Japan (2017)
5. Razvi, S., Trevor-Roper, S., Goodliffe, T., Al-Habsi, F., Al-Rawahi, A.: Evolution of OAAA strategic planning: using ADRI as an analytical tool to review its activities and strategic planning. In: Proceedings of Seventh Annual International Conference on Strategic Planning for Quality Assurance and Accreditation of Universities and Educational Arab Institutions, Cairo (2012)
6. Sakhtivel, A.M.: Promotion of research culture in sur university: a case approach. In: Handbook of Research on Higher Education in the MENA Region: Policy and Practice. IGI Global (2014)
7. Pennigton, R.: Organizational Change and Development: An African Perspective. Knowres Publishing, Johannesburg (2003)
8. Porter, W., Graham, R., Spring, A.: Blended learning in higher education: institutional adoption and implementation. Comput. Educ. **75**, 185–195 (2014)
9. Ping, L.C., Tianchong, W.: A framework and self assessment tool for building the capacity of higher education institutions for blended learning. In: Blended Learning for Quality Higher Education: Selected Case Studies on Implementation, UNESCO (2016). http://unesdoc.unesco.org/images/0024/002468/246851E.pdf. Accessed 29 June 2017

10. Lewin, K.: Frontiers of group dynamics: concept, method and reality in social science, social equilibria, and social change. Hum. Relat. **1**, 5–41 (1947)
11. Waterman Jr., R.H., Peters, T.J., Philips, J.R.: Structure is not organization. Bus. Horiz. **23**, 14–26 (1980)
12. Rajalekshmi, K.G., Chikwa, G., Bino, D.: The end of the "sage on the stage": assessing the impact of flipping classrooms on teaching and learning practice across multiple disciplines at a higher education institution in Oman. In: ICERI 2015 Proceedings, Seville, pp. 3916–3922 (2016)

Evaluation of Early Introduction to Concurrent Computing Concepts in Primary School

Eleni Fatourou[1(✉)], Nikolaos C. Zygouris[1], Thanasis Loukopoulos[2], and Georgios I. Stamoulis[1,3]

[1] Computer Science Department, University of Thessaly, 2-4 Papasiopoulou st., 35100 Lamia, Greece
{efatourou, nzygouris, georges}@uth.gr
[2] Computer Science and Biomedical Informatics Department, University of Thessaly, 2-4 Papasiopoulou st., 35100 Lamia, Greece
luke@dib.uth.gr
[3] Electrical and Computer Engineering Department, University of Thessaly, 37 Glavani st., 38221 Volos, Greece

Abstract. Learning computer programming is a basic literacy in the digital age, which helps children develop creative problem solving, logical thinking and mental flexibility. Many countries have introduced computer science in their curriculum. For example, in the educational system of United Kingdom, pupils are introduced to computer science topics from the age of six, while in Greece the teaching of computer programming commences at the age of eleven. Given differences in culture, available infrastructures, as well as the age pupils are introduced to computer science, the challenge of forming a computer science curriculum that not only offers basic background but expands the cognitive horizon and cultivates the imagination of students, still remains a challenge. Towards this end, this study focuses on exploring the potential merits of introducing concurrent programming concepts early in the learning process. Results indicate that uninitiated to programming pupils at the age of eleven were able to comprehend basic concurrency topics, while pupils at the age of twelve with some programming familiarity were able to understand more advanced concepts and use them successfully for problem solving.

Keywords: Concurrent programming · Constructivism · Scratch

1 Introduction

Learning computer programming is a basic literacy in the digital age, which helps children develop creative problem solving, logical thinking and mental flexibility. As indicated by European Schoolnet in [5], only ten European countries have fully integrated computer programming in their primary school curriculum, as of 2015. Given differences in culture, available infrastructures, as well as the age pupils are introduced to computer science, the challenge of forming a computer science curriculum that not

M. E. Auer et al. (eds.), *Teaching and Learning in a Digital World*,
Advances in Intelligent Systems and Computing 715,
https://doi.org/10.1007/978-3-319-73210-7_64

only offers basic background but expands the cognitive horizon and cultivates the imagination of students, still remains a challenge.

Early computer science courses typically focus on structured programming concepts such as: control/selection constructs and iterations. The primary educational effort (particularly in the case of Greece) is tailored towards applying these building blocks into single execution thread scenarios, overlooking concurrent multiple thread execution issues. However, concurrency rises more than often in educational programming platforms such as Scratch. For instance, when two independent entities, e.g., sprites, move and act in a labyrinth, it is not uncommon that racing conditions appear, whenever the entities require simultaneous access to a common resource, e.g., some treasure object. As a result, pupils might experience "unexpected" program behavior and occasional program crashes. The teacher has then two practical options: (i) either overlook the problem, diminishing its importance, and continue focusing on single thread correctness criteria, or, (ii) attempt to give a thorough explanation of the reasons of such "unexpected behavior", thus, introducing concurrency issues to pupils, albeit in an ad-hoc, unstructured and unplanned manner which might discourage them.

Motivated by: (i) the apparent "knowledge gap" concerning concurrency that exists on many typical early computer programming syllabuses, (ii) the fact that pupils are accustomed to multitasking in their everyday life and (iii) the importance of multi-threading and multitasking in modern software and hardware, this study investigates the enrichment of a typical syllabus with multithreading concurrency issues. The goal is to introduce the pupils to the basic challenges of concurrent programming in a systematic manner, without sacrificing the level of detail contained on a typical syllabus as far as simple single thread structural programming is concerned. The developed syllabus was tailored and evaluated for the education system of Greece, whereby pupils are introduced to programming concepts at the age of eleven using Scratch.

The rest of the paper is organized as follows. Section 2 summarizes the related work from the particular standpoint of the pedagogical approach followed (constructivism). Section 3 illustrates the methodology together with the proposed syllabuses, while Sect. 4 describes the evaluation setup and discusses the results. Finally, Sect. 5 provides the concluding remarks.

2 Related Work

The core approach of the study adheres to the constructionist theory, according to which the learning process is not only transmitted from teacher to pupil, but rather constructed in the mind of the pupil in the form of active learning [13, 15] Constructivism theorists such as Piaget and Papert view children as the builders of their own cognitive tools, as well as their external realities [1]. Moreover, Papert believes that programming has a tremendous potential to improve classroom teaching [11]. Thus, the dominant theory of learning, supports that knowledge is actively constructed by the pupil and not passively absorbed from text books and lectures [1].

While constructivism has been intensively studied by researchers of science and mathematics education [1], in the field of computer science education fewer efforts were devoted [2]. During the 80s, many schools introduced in their curriculum computer programming projects using Logo. Nevertheless, the empirical observations on student learning did not always match the powerful theoretical claims. The ideal vision of students' performing better in science courses due to hands on Logo learning, collided with the documented reality of students' difficulties to learn even fundamentals of Logo [9]. Nowadays, research has entered into a new phase of multidisciplinary theory based protocols [4, 9]. The initial vision of teaching and learning computer programming has been altered. The focus of current researches is on understanding the conditions under which the skills that are learned in programming can translate to cognitive development of learners [3]. For instance in [8] it was concluded that programming in pairs (a common situation due to laboratory restrictions in schools) has limitations, while in [6] it was pointed out that despite its original limitations, the newer versions of Logo with enhanced graphics and interface might find applications in pre-school ages.

Visual programming languages such as Scratch have been widely adopted recently as the means for early introduction to programming concepts. Scratch uses blocks, which the pupils drag and drop to form their scripts. The first block of a script, called hat block, states the event that triggers it. An example of a sprite and a script written for this sprite is demonstrated in Fig. 1. An avid research interest exists on how to fine tune the learning process with Scratch in order to achieve the best pedagogical results in primary schools [7, 10, 14], but also in elementary ones [12]. Towards, this end the research presented in this paper aims at filling a tutoring gap that often appears when following a classic introductory syllabus to Scratch programming, namely the teaching of concurrency concepts.

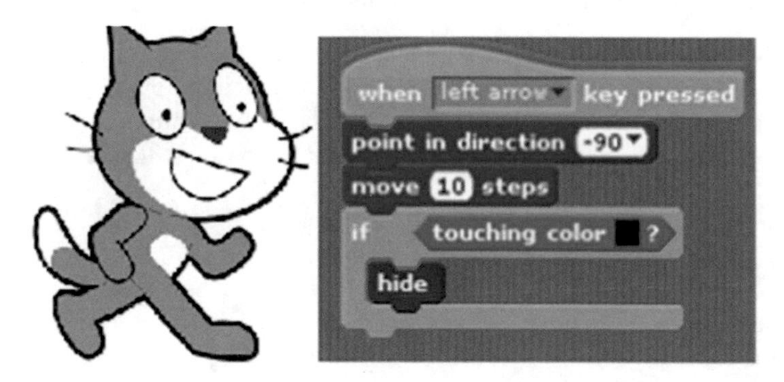

Fig. 1. Scratch sprite and script. When left arrow key is pressed, the sprite is programmed to turn left and move 10 steps. If it touches black color, it is hidden.

3 Methodology

3.1 Educational Context and Targets

The application of the methodology was performed in the Greek primary school whereby computer programming is taught at the last two years of the school (ages eleven and twelve). The official curriculum involves a total of 12 weeks of hourly laboratory lessons per year at each class (fifth and sixth grade). Applying the constructionist theory in the present study required the design of programming challenges that are incremental in nature and led after a certain point to concurrency problems that were self evident. It is straightforward that the proposed syllabuses should adhere to official curriculum constraints (for evaluation reasons). The rather limited timeframe for the computer programming courses offered a serious challenge in defining the educational goals and design a subsequent plan to achieve them.

The research conducted involved both classes that were already familiar with basic structured programming concepts and classes with no prior programming experience. As a result, it was decided that two different projects should be implemented. Instructional scaffolding was used for the learning process. Each project was split into equally hourly tasks. Pupils worked on the same file, extending game functionality. The teaching approach followed, was to introduce the notion of multiple running threads early on and incrementally build knowledge on concurrency issues according to the assigned tasks. A Scratch player often executes "simultaneous" scripts, so when different scripts are triggered by the same event, pupils must check if race conditions apply and if so, learn a way to tackle the problem. Building around this feature, tasks were escalated in order for pupils to:

1. Implement concurrent moves of a single sprite;
2. Implement concurrent moves of two or more sprites;
3. Synchronize sprites using time primitives;
4. Synchronize sprites using messages;
5. Distinguish local, sprite-level variables from global variables;
6. Synchronize sprites using condition variables.

The first four educational targets concerned both the beginners and the more advanced classes, while the last two were only attempted by pupils with prior programming knowledge. It should be noted that the aforementioned six educational targets were on top of the classic targets related to basic structural programming constructs.

3.2 Beginners' Syllabus

The beginners' project concerned a maze game whereby a hero sprite tries to capture a trophy, while chased by enemy sprites. Maze games are very common Scratch projects and are suitable for beginners. As a testament a google search for "maze", performed on 16/5/2017 on the site https://scratch.mit.edu returned roughly 140.000 results. The syllabus presented in Table 1 is designed to incrementally build fundamental programming knowledge, while introducing concurrency concepts and solutions, in a gradual self evident manner. It also contains a midterm and a final project presentation.

3.3 Advanced Syllabus

The more advanced classes were already introduced (previous year) to basic programming concepts with Scratch. However, the syllabus used for the introduction differed from the one in Table 1, focusing only on structured programming constructs

Table 1. Beginners' project outline

	Plan	Objectives	Concurrent issues addressed
1	Draw a maze stage Make a new sprite (hero)	Draw a maze Distinguish sprite from stage	N/A
2	Use commands to move the sprite from the beginning of the maze to the end	Execute a sequence of blocks	N/A
3	Make the sprite move using arrow keys	Use a trigger event to start a script	Concurrent scripts on a single sprite Pupils explain what happens if accidentally the same key is used for up and down movement
4	Draw two or more sprites as enemies. Use command forever to make the enemies move around constantly when green flag is clicked	Use iteration (forever)	Two sprites move at the same time. But no concurrent issues apply yet
5	Create trophy sprite for the hero and use command if to make trophy disappear when touched by the hero	Use iteration and condition Use hide command	A scripts loops until a condition applies but no concurrent issues apply yet
6	Add a script to the hero, to make it disappear when touched by an enemy	Use iteration and condition Use hide command	Concurrent scripts of two or more sprites. Place an enemy on food and make hero touch them
7	Midterm presentation		
8	Make nice guy, enemies and food appear in certain places when game starts	Initialization Use show command	Concurrent scripts of two or more sprites. If show command proceeds move, then race conditions apply
9	Make stage present an introductory message	Use wait command	Synchronize sprites using time primitives. Upon the game starts all sprites wait for a few seconds for the salutation to disappear
10	Make stage present a winning or losing message	Use message passing	Synchronize sprites using messages
11	Add any functionality to the project	Self assessment	Self assessment
12	Final project presentation		

using a single thread view of the executed scripts. Therefore, it was deemed that knowledge on concurrency concepts should be built from scratch, albeit at a faster pace compared to the beginners. Table 2 illustrates the syllabus for the advanced classes. As it can be observed, apart from the heaviest workload on concurrency topics (compared to Table 1), it contains two hours (instead of one) of free project additions and self-assessment. This served two purposes. Firstly, it encouraged a self-motivation attitude and secondly it served as a means of equalizing the effects of "missed hours" between classes that completed the 12 h schedule and those that only completed 11 h (due to national holidays).

Table 2. Advanced project outline

	Plan	Objectives	Concurrent issues addressed
1	Create two sprites representing players Create a second costume for each one holding a die Make them change costume on space pressed	Distinguish sprites and costumes	Concurrent scripts of two or more sprites
2	Change previous game, make them change costume every 2 s	Use timer	Synchronize sprites using time primitives
3	Delete costumes with the die. Create a die sprite. On click, the die goes to the other player. Use a variable to hold the dice owner	Use variables	N/A
4	When a player receives the die, says "I got it"	Use messages	Synchronize sprites using messages
5	Draw a die and make it turn randomly when clicked	Use random function	N/A
6	Make the die turn for 2 s until it shows the result when space is pressed	Use wait	Concurrent scripts of a single sprite. The script will be interrupted and start from the beginning again
7	Midterm presentation		
8	Create a local variable pocket for each player and a global for the die. Initialize them. Each time the die turns, it adds one to the players' wallet	Declare and use local and global variables	Distinguish local, sprite-level variables from global variables
9	Give or take money from the players depending on the value of the die and the player's turn		Synchronize sprites using condition variables
10	Add any functionality to the project	Self assessment	Self assessment
11	Add any functionality to the project	Self assessment	Self assessment
12	Final project presentation		

4 Evaluation

4.1 Subjects

A total of 74 pupils (age range 11–12 years old M = 11.26, SD = 0.440) participated in this study. All students presented typical academic performance according to their teachers' ratings. It is worth to notice that nineteen of students had already been familiar with the basic structured programming concepts and followed an advanced programming project. Fifty five of the children had never experienced programming on computer, so they followed the beginners' project.

4.2 Results

Statistical analysis is presented as follows: one – way analysis of variance (ANOVA) was conducted comparing the students' performance in the same skills of the beginners' and advance level project according to their age, namely the first four educational objectives as presented in Sect. 3.1 (abbreviated here as T1–T4). The results of the analysis is shown in Table 3.

Table 3. Differences in educational objectives according to age group (*$p < 0.05$, **$p < 0.01$).

	F
T1	6.312**
T2	4.617*
T3	0.00
T4	29.213**

As it is presented in Table 3, children performance had significant differences according to their age. In more details children twelve years old presented statistically significant better scores in the three out of four objectives of the project compared to children aged eleven years old. Figure 2 shows the mean results for the two age categories (1 = success, 2 = failure). Again age and previous acquaintance with programming concepts is shown to play role. Nevertheless, it is particularly encouraging for the methodology of this paper, that more than half of the beginners were able to achieve the first three educational objectives.

Next, a correlation analysis was calculated in order to evaluate childrens' performance in all computer programming targeted objectives. Table 4 presents the correlation analysis of children aged eleven years old.

As it is presented by the correlation analysis the concurrent moves of a single sprite (T1), the concurrent moves of multiple sprites (T2) and the sprite synchronization using time primitives (T3) had a strong correlation ($p < 0.01$). However, using messages for synchronization (T4) did not present any correlation compared to the other tasks.

Lastly a correlation statistical analysis was used in order to estimate the connection between tasks completed by children that attended the sixth grade of the primary school (12 years old). Table 5 presents the correlation analysis.

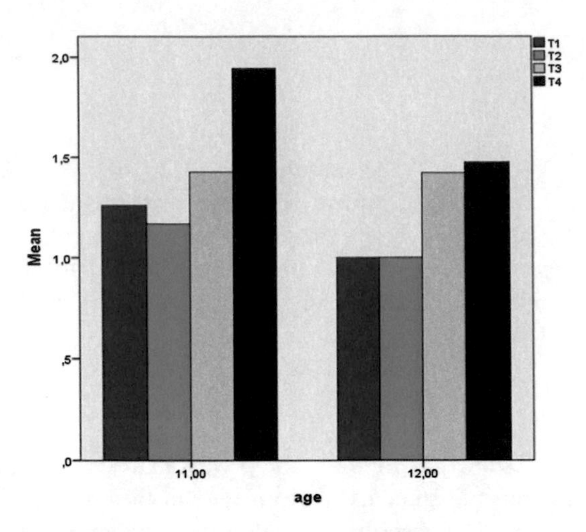

Fig. 2. Performance of children in the first four objectives (1 = success, 2 = failure).

Table 4. Correlation analysis for eleven years' old children (**p < 0.01).

	T1	T2	T3	T4
T1	1	0.757**	0.605**	0.143
T2	0.757**	1	0.522**	0.108
T3	0.605**	0.522**	1	0.209
T4	0.143	0.108	0.209	1

Table 5. Correlation analysis for twelve years' old children (*p < 0.05, **p < 0.01).

	T1	T2	T3	T4	T5	T6
T1	1	0.513**	0.81	0.053	0.119	0.27
T2	0.513**	1	0.198	0.061	0.092	0.03
T3	0.081	0.198	1	0.11	0.70	0.248
T4	0.053	0.061	0.11	1	0.42**	0.474**
T5	0.119	0.092	0.070	0.42**	1	0.379*
T6	0.027	0.030	0.248	0.474**	0.379*	1

The evidence suggests that concurrent move of a sprite (T1) correlated with concurrent move of multiple sprites (T2). Concurrency using messages (T4) correlated with distinguish local, sprite-level variables, from global variables (T5) and synchronizing sprites using condition variables tasks (T6). Distinguishing local, sprite-level variables from global variables correlated with control using messages (T4) and synchronize sprites using condition variables (T6). Lastly, synchronizing sprites using condition variables correlated with control using messages (T4) and comprehending the difference between, local and global variables (T5).

5 Discussion and Conclusions

The main purpose of this study was to investigate the introduction of multiple threading and related concurrency concepts into a typical early computer programming syllabus. Learning tasks were built in a structured approach so that pupils incrementally build knowledge on concurrency issues. By the end of the 12-week course most of the additional educational targets were achieved by the majority of the pupils. More importantly, the pupils demonstrated for the largest part an ability to "think concurrently". This was also manifested by the fact that no "unexplained" program behavior was reported as such at the end demonstration, but was rather attributed correctly to racing conditions. It is worth to notice that children aged 12 years old that had already followed an introduction to Scratch, presented significantly better results in commonly assigned tasks compared to children 11 years old. Nevertheless, the majority of beginners achieved three out of the four educational objectives originally posed, a result that is particularly encouraging for the research direction of the paper. Also, correlation analysis highlights the educational targets that should be addressed and focused by the tutor in order for pupils to complete all learning objectives.

Summarizing, we can state that the results of the study illustrate the usefulness of introducing concurrent programming concepts in primary school education. On the other hand, not all educational targets were successfully accomplished by all pupils, due to timetable and hardware failures. Thus, it is recommended that the proposed syllabuses are fine tuned depending on whether timetable and infrastructure conditions are favorable.

References

1. Ackermann, E.: Piaget's constructivism, Papert's constructionism: what's the difference. Future Learn. Group Publ. **5**(3), 438 (2001)
2. Allen, J.P., Pianta, R.C., Gregory, A., Mikami, A.Y., Lun, J.: An interaction-based approach to enhancing secondary school instruction and student achievement. Science **333**(6045), 1034–1037 (2011). AAAS
3. Brennan, K., Resnick, M.: New frameworks for studying and assessing the development of computational thinking. In: Proceedings of the 2012 Annual Meeting of the American Educational Research Association, Vancouver, Canada, pp. 1–25 (2012)
4. Clements, D.H., Sarama, J.: Research on logo: a decade of progress. Comput. Sch. **14**(1–2), 9–46 (1997)
5. European Schoolnet: Computing our future computer programming and coding - priorities, school curricula and initiatives across Europe. http://www.eun.org/c/document_library/get_file?uuid=3596b121-941c-4296-a760-0f4e4795d6fa&groupId=43887. Accessed 10 May 2017
6. Fessakis, G., Gouli, E., Mavroudi, E.: Problem solving by 5–6 years old kindergarten children in a computer programming environment: a case study. Comput. Educ. **63**, 87–97 (2013)
7. Franklin, D., Hill, C., Dwyer, H.A., Hansen, A.K., Iveland, A., Harlow, D.B.: Initialization in scratch: seeking knowledge transfer. In: SIGCSE 2016, pp. 217–222. ACM (2016)

8. Lewis, C.M.: Is pair programming more effective than other forms of collaboration for young students? Comput. Sci. Educ. **21**(2), 105–134 (2011)
9. Mayer, R.E.: Teaching and Learning Computer Programming: Multiple Research Perspectives. Routledge, Abingdon (1988)
10. Meerbaum-Salant, O., Armoni, M., Ben-Ari, M.: Learning computer science concepts with scratch. Comput. Sci. Educ. **23**(3), 239–264 (2013)
11. Papert, S.: Mindstorms: Children, Computers, and Powerful Ideas, 2nd edn. Basic Books, New York (1993)
12. Sáez-López, J.-M., Román-González, M., Vázquez-Cano, E.: Visual programming languages integrated across the curriculum in elementary school: a two year case study using "Scratch" in five schools. Comput. Educ. **97**, 129–141 (2016)
13. Tsihouridis, C., Vavougios, D., Ioannidis, G.: The effect of switching the order of experimental teaching in the study of simple gravity pendulum-a study with junior high-school learners. In: International Conference on Interactive Collaborative Learning, pp. 501–514. Springer, Cham (2016)
14. Wilson, A., Moffatt, D.C.: Evaluating scratch to introduce younger schoolchildren to programming. In: 22nd Annual Workshop of the Psychology of Programming Interest Group, pp. 64–74 (2010)
15. Zygouris, N.C., Vlachos, F., Dadaliaris, A.N., Oikonomou, P., Stamoulis, G.I., Vavougios, D., et al.: The implementation of a web application for screening children with dyslexia. In: International Conference on Interactive Collaborative Learning, pp. 415–423. Springer, Cham (2016)

Flipped Classroom Method Combined with Project Based Group Work

Ilona Béres(✉) and Márta Kis

Methodological Department, Budapest Metropolitan University, Budapest, Hungary
{iberes,mkis}@metropolitan.hu

Abstract. The purpose of this paper is to present in practice how the flipped learning approach works in the instruction of high number of higher education students. This article presents the developed flipped classroom method combined with project based group work in seminars. The participants were students majoring in field of Communication, Business and Tourism. In order to examine the efficiency of the implemented method and to confirm the positive or negative learning experiences, students were asked to complete a feedback questionnaire. In this paper we present the analysis of the opinions of 544 higher education students.

Keywords: Active learning · Project work · Flipped classroom

1 Introduction

Our past experience in higher education is that students are difficult to achieve in class student activity during practical courses. The expectation of higher education institutions and employers, on the other hand, is to improve students' collaborative, problem solving and communication skills. Higher education institutions widely use blended learning method to enhance traditional lecture-homework learning effects, where face-to-face instruction is combined with technological supported learning activities outside the classroom.

The latest research indicates that an active learning method can be an effective approach for present age higher education students [6, 10, 13]. The literature suggests that students' engagement is essential in effective teaching/learning process. Students who are more involved will be more active and challenged learners [1, 12]. Innovative approaches in higher education are shifting away from teacher centered instruction to student centered learning. One of the recently used technology enhanced learning models is known as a flipped classroom or an inverted classroom method. This model provides an opportunity for teachers to use a range of innovative teaching/learning approaches. These approaches are described in the next section.

M. E. Auer et al. (eds.), *Teaching and Learning in a Digital World*,
Advances in Intelligent Systems and Computing 715,
https://doi.org/10.1007/978-3-319-73210-7_65

2 Theoretical Framework

The active learning techniques integrate the student centered learning methods such as cooperative learning, problem-based learning, project based learning and peer assisted learning. These learning approaches mean that students work in groups in order to develop and reach their learning goals. The active learning process increases effectiveness of students [6, 10, 13]. In problem or project based learning environment, acquiring knowledge and the learning process are based on problem solving. Students themselves decide the type of knowledge that is necessary to solve the given real life problem and task [3]. This method develops (1) flexible knowledge, (2) problem solving skills, (3) self-directed learning skills, (4) collaboration skills [4]. In cooperative and collaborative learning environment, students communicate more about the issued presented by educators than they do when teachers present learning content [2].

In a flipped classroom, traditional lectures are moved outside the classroom. Knowledge construction is disseminated through online learning content. This content is processed by students individually before in class activities. Classroom time is then used to apply acquired knowledge in complex problems. In the last years numerous educational researches focused on this flipped classroom method. In this respect, there are several definitions describing flipped learning. "In the Flipped Learning model, teachers shift direct learning out of the large group learning space and move it into the individual learning space, with the help of one of several technologies" [7]. The pre-recorded lectures can be made available to students as homework, leaving class time open for interactive learning activities — activities that cannot be automated [11]. Figure 1 shows the difference between the traditional non-flipped classroom approach and flipped classroom method.

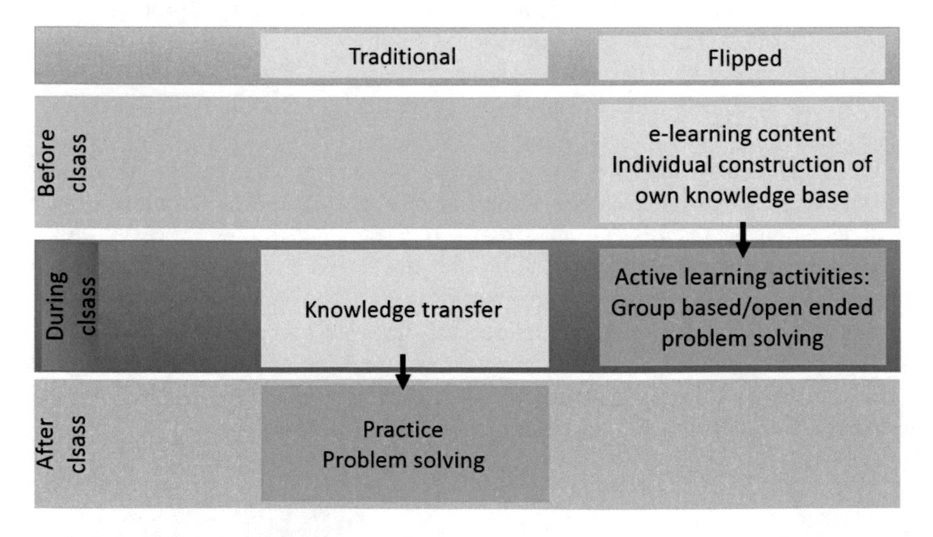

Fig. 1. Traditional classroom vs. Flipped classroom (based on [8])

We define the flipped classroom as an active teaching/learning method, which is comprised of two parts: (a) computer-based individual, self-directed learning outside the classroom; (b) open ended problem based collaborative work and activities inside the classroom.

Based on literature there are many studies focused around the efficiency of flipped classroom method. Studies have been developed in different courses such as engineering, information systems, statistics [4, 5, 9, 14].

Our study presents a flipped classroom model that attempts to achieve learning goals through the involvement and positive learning experience of students.

3 Research Goals

Our aim was to increase students' activity and problem solving skills during the course with an output of an active, flexible learning model that can be well adapted within higher education. Our main goals were:

1. to increase students responsibility and motivation in their own learning process
2. to increase students activity during the course
3. to support effective collaboration between students
4. to develop project tasks which increase individuality, responsibility
5. to analyze students attitudes and behavior in a project work situation and opinions about the applied method

In our experimental course, we had implemented flipped classroom method combined with project based group work in seminars. This research focuses on student behavior, involvement and activity during the course.

4 Course Design

This method was developed for Information Technology I (IT I) course in the 2016/17 fall semester. The participants were 700 first year students majoring in business administration and management, commerce and marketing, communication and media science, finance and accounting, human resources, international business economics, tourism and catering.

To achieve our goals a course model was developed that combines the flipped method with project-based learning and a collaborative evaluation. Figure 2 presents the weekly tasks in our applied model.

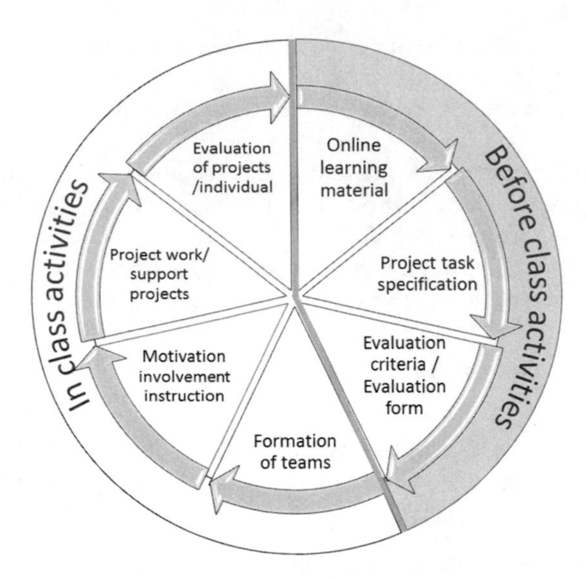

Fig. 2. Weekly tasks in developed course model

Before class objectives: Development of effective e-learning content. The online learning material, examples were uploaded on university Learning Management System (LMS) weekly. Students were required to process the learning material independently. A lot of work was invested in generating open ended project tasks all geared toward motivating students. The assessment method is important part of our model. The form of evaluation was developed for every project task.

In class objectives: All students had to take part weekly in interactive seminars. During these seminars, students in groups of three were asked to develop a complex task based on gathered knowledge. During these in-class activities, they worked as a project team and implemented project work. Their task was to understand the problem, divide it into sub-tasks, and apply the mastered tools, theoretical and practical knowledge to arrive to a solution.

As part of project work teams collected data, selected an appropriate technology and examination method independently. Different teams arrived at different project solutions; finally, they presented their own work and results.

The student teams' project work was evaluated weekly based on evaluation form. Projects were then awarded a total number of points, which were then divided amongst themselves by students. Students were asked to take into account the individual value added by each team member in the collective team work when allocating points.

Performance of students was measured by the weekly project works (50%) and an individual final examination (50%).

The quality and effectiveness of the used method was ensured by the students' feedback. Students submitted an online questionnaire about their opinion, they evaluated their own activities products and how the aims of the course were fulfilled.

5 Analyses and Results

In order to examine the efficiency of the implemented method and to gather students' experiences, we carried out a questionnaire analysis in the second semester of the 2016/2017 school year, during the first practice course of IT II classes. As this subject is only available to those students who have already successfully completed IT I., therefore this survey shows the opinions of students who have already been efficient in the IT I course. 544 students have filled in the questionnaire. Out of them, 80 have already completed the prerequisite course (IT I), and 464 students have learned via the new method. These 80 students can be considered as a control group when analysing our questionnaire.

This IT subject is learnt by students of the CBT area (Communication, Business and Tourism). The diagram below (Fig. 3) shows the different majors of the students who filled in the questionnaire.

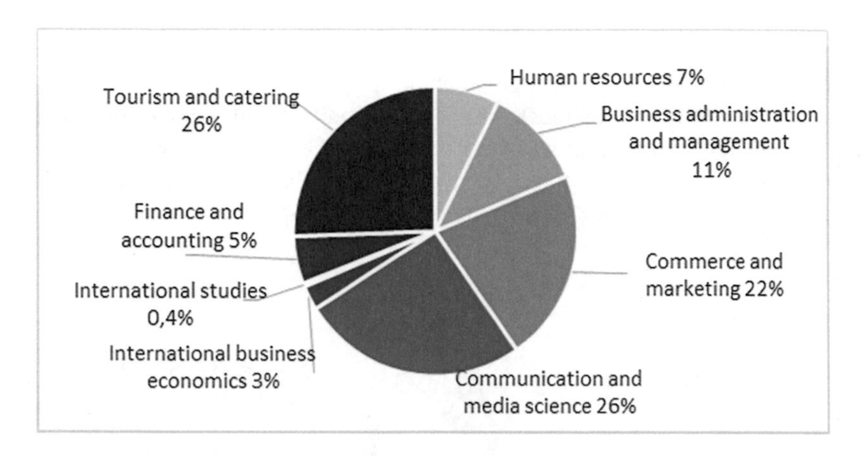

Fig. 3. Distribution of students by major (N = 544)

A very important element of the new method is to provide a possibility for those with limited basic knowledge to prepare by using online materials. Therefore we were curious to find out the extent to which this opportunity was used. The result is quite mixed: it is clear that half of the students using the new method have prepared for at least 50% of the tasks (Fig. 4). In comparison with the results of previous years, there is a significant increase among those who have prepared separately for each class (for 70–100% of the classes). The number of students in this category has tripled (has increased from 6% to 20%).

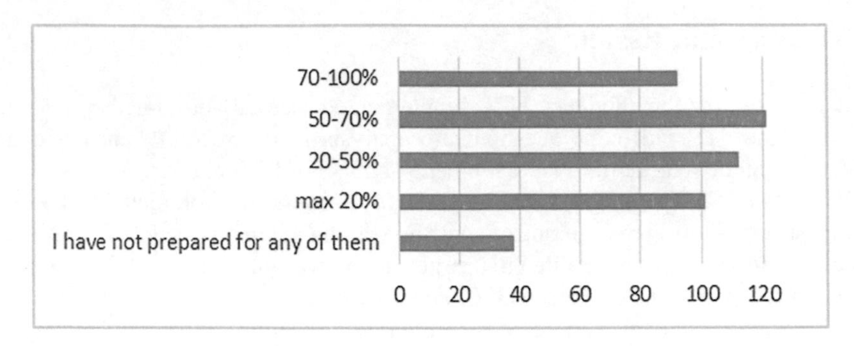

Fig. 4. Percentage of classes students prepared for (N = 464)

Out of the 464 interviewed students, 426 claimed to have prepared for a certain percentage of the classes. We have examined the periods of average preparation of those who studied the supporting materials previously. This time spent for preparation was mostly between half an hour and one hour. We can see from the diagram (Fig. 5), those students, who spent little time with preparation, have only prepared for a maximum of 20% of the classes (i.e. 1–2 classes), while there is quite a high number of those, who have spent less than half an hour with preparation. On the other hand, when considering he most diligent students, who have prepared for more than 70% of the classes (i.e. at least 9 classes out of 12), there is a significant increase of those who spent more time with preparation at home.

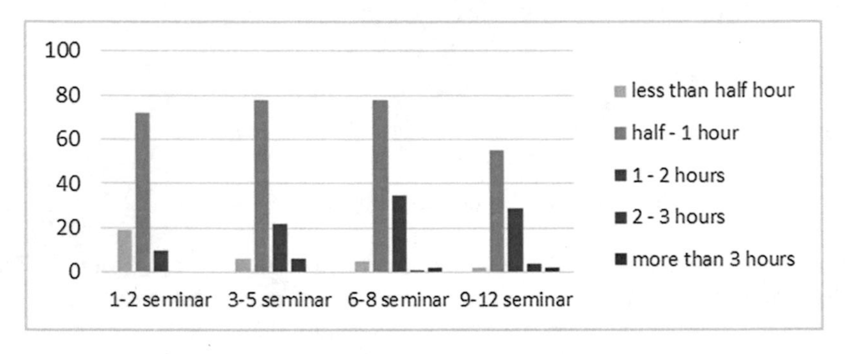

Fig. 5. How many classes have you prepared for and how much time have you spent with preparation? (N = 426)

When we examine the relation between the number of classes students prepared for, the amount of time spent with preparation and the effectiveness of the student, the result is the following.

There is a medium positive relationship (R = 0.43) between the number of occasions the student has prepared for classes and the amount of time spent on preparation. There is a weak positive relationship (R = 0.15) between the number of occasions spent with preparation and the achieved result. Surprisingly enough, though, the amount of time spent on preparation and the achieved result do not show (R = 0.01) a stochastic connection in the case of interviewed students. The main reason for this can be that in the case of this subject, the final result is greatly influenced by previous knowledge. Therefore someone with a solid basic knowledge did not need home preparation for good results, but those with weaker previous knowledge spent more time on preparation, which may have resulted in a medium grade result only.

The new method was launched in 38 groups of 20 students, with the participation of nine teachers, so continuous and coordinated work was very important. We paid attention to the feedback and effectiveness of the students during the semester, as well as the extent to which we could put our previous plans into practice. We shared our experiences and refined the method to make it more effective.

We have received more feedbacks from students and it was hard for us to decide whether those are general and refer to everyone or not. Someone would have needed a short presentation of the day's curriculum at the beginning of the class, others have asked for a more detailed task description, or more active teacher's participation in solving task during the classes. However, more have given feedback that everything is fine as it is and they were happy to work.

One of the goals of having a questionnaire filled out at the end of the year is to examine the individual feedbacks from students and see to what extent these are general or not, and what was the most difficult for them when studying using the new method.

They had to grade several statements on a scale from 1 to 5, showing how true those statements are in relation to themselves. The following chart (Table 1) shows these statements and the average results:

Table 1. Some statements of questionnaire and an average score of answers

To what extent are the following statements true? (1-not at all, 5-very much so)	
We had to solve tasks on our own: it was more difficult, but I understood the material better	3,1
In case I did not deal with the material at home, it was very difficult to do the tasks during class	2,8
It was good to work in teams, because we could help each other and could work quicker	4,2
I enjoyed having more creative tasks where we could find out what to examine	3,6
It was difficult to understand what the task was	2,3
It would have been good to see a sample of what to do in a given class	3,5
I think the evaluation system is fair	3,9
I received a grade that I deserved	4,0
This form of education requires a lot of energy from the students	2,5
This method is more effective to learn with than the usual frontal education	3,6
I find the things I studied during IT I. useful	3,8
I am satisfied with the available supporting materials	3,8

We can see from the values, students basically like the new way of learning. What they enjoyed most was teamwork (Fig. 6), which was not characteristic at all of this subject before.

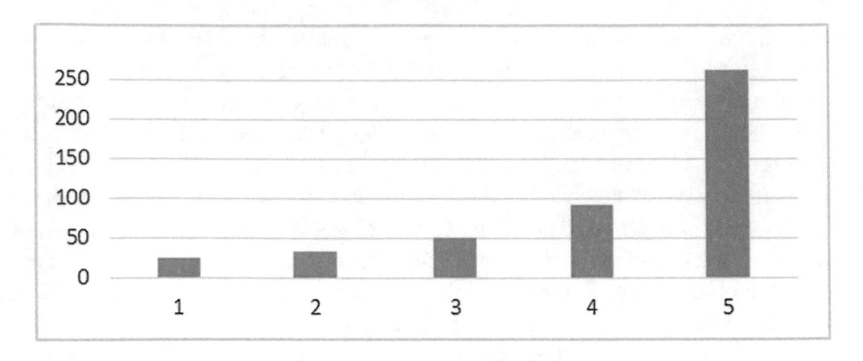

Fig. 6. Opinion about teamwork (N = 464)

A major element of the flipped classroom method is home preparation, so it is important to measure the extent to which they can use the available supporting materials for individual learning. We can see from the feedbacks (Fig. 7) that basically students are satisfied with them, but this element of the method can surely be developed. The analysis and perfection of this is a task for the future.

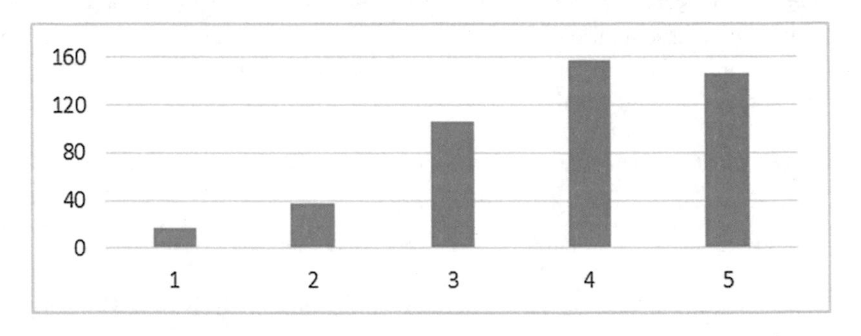

Fig. 7. Satisfaction with available supporting materials (N = 464)

It is useful to examine how effective the new education method was considered in relation to the achieved grade (Fig. 8). So whether in the future they would rather learn by the traditional method used earlier, or whether they would choose the flipped classroom method combined with project based group work, which demands a higher level of independence?

As you can see from the diagram (Fig. 8), every group (regardless of the achieved grade) was happier to learn with the new method. Obviously, the greatest difference was among students with the highest grade (5), in favour of the new method.

The questionnaire analysis has reinforced what we have already experienced in the classrooms: it was worth to change and we are on the right track, in spite of the high level of investment. Naturally, refinement is still necessary, but the education model we set up is effective, and gives more joy for both the teachers and the students.

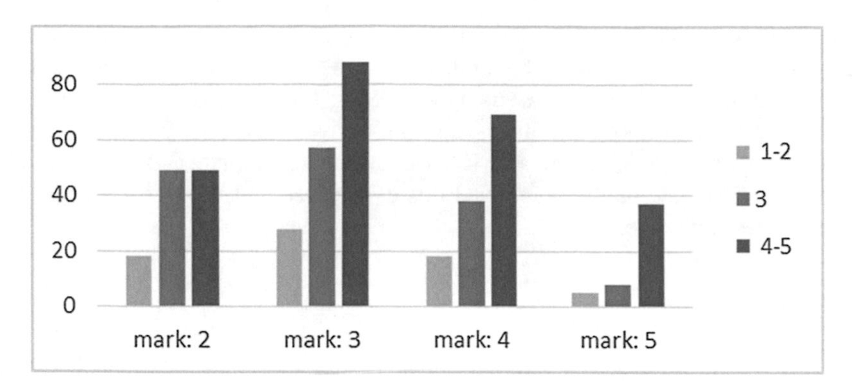

Fig. 8. Is it more effective to learn with this method than with the traditional education? (N = 464)

6 Conclusions

Developing a flipped classroom required a lot of effort on the part of teachers and new learning behavior on the part of students. Instructors have to develop weekly (a) web based learning content which support individual learning, (b) quality project tasks which are motivating and involve students, (c) evaluation of the students' projects required more effort on the part of the instructors.

One important advantage of this flipped classroom model is the enhancement of students' responsibility in their own learning process. They were required to learn the material before classroom activities, thus students had to take an active part in their own learning. This approach is new and was sometimes uncommon for our students. The prior knowledge of our students is quite diverse. As a result of this approach, students who were lacking in prior knowledge or are slower learners had an opportunity to spend more time with material; they could acquire the learning content at their own pace. Flopped classroom technique increases the 'a priori' involvement of students. Students were actively involved in their own knowledge building process; they could make sure that they had understood things correctly on a weekly basis by solving complex project tasks based on gathered knowledge. During the in-class activities it was a real workshop atmosphere. When instructors were asked of their opinion, they noted that the activity of students during the course had increased significantly. Students felt their projects tasks and their results as their own property. An interesting case that demonstrates students' engagement: a student who is a sport competitor (who did not attend all of the classes), whenever he could attend the lecture, he actively took part in the group work. During one lesson he was abroad in a training camp, when his teammates had contacted him on Facebook in order to help the team in problem solving, when they couldn't. Everyone enjoyed the situation, and teacher supported their ingenuity.

Implemented evaluation method provides an opportunity to evaluate students own and peers value added in project work. This is an effective tool for developing critical thinking skills. The marks were shared by the students taking into account their individual value added. We can conclude that most students were satisfied with their own grades. No student was protesting against the grade given to him by his team.

The final examination results shows that performance of students in flipped classroom approach is not significantly better than previous years traditional lectures-homework learners' results. However based on the students opinion and instructors observation inverting the classroom and integrating project work brought positive change in students' in-class behavior, their activity was undoubtedly better. Based on student feedbacks we can state that they enjoyed this experience and think it supported their learning process effectively.

References

1. Barkely, E.: Student Engagement Techniques: A Handbook for College Faculty. Josses-Bass, Hoboken (2009)
2. Boud, D., Cohen, R., Sampson, J.: Peer Learning in Higher Education: Learning from and with Each Rutledge (2014). https://books.google.hu/
3. Béres, I., Kis, M., Licskó, L., Magyar, T.: Technological support of web based project work in higher education. In: 2011 14th International Conference on Interactive Collaborative Learning (ICL), pp. 209–213. IEEE (2011)
4. Bishop, J.L., Verleger, M.A.: The flipped classroom: a survey of the research. In: ASEE National Conference Proceedings, Atlanta, GA, vol. 30, no. 9 (2013)
5. Davies, R.S., Dean, D.L., Ball, N.: Flipping the classroom and instructional technology integration in a college-level information systems spreadsheet course. Educ. Technol. Res. Dev. **61**(4), 563–580 (2013)
6. Freeman, S., Eddy, S.L., McDonough, M., Smith, M.K., Okoroafor, N., Jordt, H., Wenderoth, M.P.: Active learning increases student performance in science, engineering, and mathematics. Proc. Natl. Acad. Sci. **111**(23), 8410–8415 (2014)
7. Hamdan, N., McKnight, P., McKnight, K., Arfstrom, K.M.: A review of flipped learning (2013). http://flippedlearning.org/wpcontent/uploads/2016/07/LitReview_FlippedLearning.pdf. Accessed 15 Jan 2017
8. Jensen, J.L., Kummer, T.A., Godoy, P.D.D.M.: Improvements from a flipped classroom may simply be the fruits of active learning. CBE—Life Sci. **14**, 1–12 (2015). http://www.lifescied.org/content/14/1/ar5.full.pdf+html/. Accessed 15 Jan 2017. Spring
9. Mason, G.S., Shuman, T.R., Cook, K.E.: Comparing the effectiveness of an inverted classroom to a traditional classroom in an upper-division engineering course. IEEE Trans. Educ. **56**(4), 430–435 (2013)
10. Michael, J.: Where's the evidence that active learning works? Adv. Physiol. Educ. **30**(4), 159–167 (2006)
11. Mok, H.N.: Teaching tip: the flipped classroom. J. Inf. Syst. Educ. **25**(1), 7–11 (2014). http://ink.library.smu.edu.sg/cgi/viewcontent.cgi?article=3363&context=sis_research. Accessed 29 Sept 2016
12. O'Flaherty, J., Phillips, C.: The use of flipped classrooms in higher education: a scoping review. Internet Higher Educ. **25**, 85–95 (2015)
13. Prince, M.: Does active learning work? A review of the research. J. Eng. Educ. **93**(3), 223–231 (2004)
14. Strayer, J.F.: How learning in an inverted classroom influences cooperation, innovation and task orientation. Learn. Environ. Res. **15**(2), 171–193 (2012)

"Let's Go… Kahooting" – Teachers' Views on C.R.S. for Teaching Purposes

Marianthi Batsila[1(✉)] and Charilaos Tsihouridis[2(✉)]

[1] Directorate of Secondary Education, Ministry of Education, Larissa, Greece
marbatsila@gmail.gr
[2] University of Thessaly, Volos, Greece
hatsihour@uth.gr

Abstract. The present study constitutes the first part of a study about the use of the online game-based of Kahoot as a tool for teaching practices and ways to do this. This first part, described in this paper, focuses on investigating teachers' views on the use of Kahoot. A number of 149 secondary education teachers participated for this reason in workshops where they were introduced to Kahoot and were asked to design their own tasks. Upon completion of the workshops a questionnaire was delivered to them to evaluate the tool and focus group discussions were conducted to detect their in-depth thoughts. The teachers' opinions were positive as they considered Kahoot a motivating tool for teaching and assessment purposes which can make learning fun and a creative process. The teachers revealed their intention to use the tool in their future teaching practices with their classes to a great extent.

Keywords: ICT · Classroom Response Systems (C.R.S.) · Kahoot
Game-based learning

1 Introduction

Today we are living in a world where everything around us is changing rapidly due to continuous technological developments and new inventions in many fields of life. Education cannot do otherwise but follow along with this digital advancement which affects educational methods, tools and materials facilitating teaching and learning [1, 2]. Among the many modes employed for teaching purposes is the use of "Classroom Response Systems" (C.R.S.), else known as "Audience Response Systems" (A.R.S.) or "Students Response Systems" (S.R.S.) [3]. Classroom Response Systems are generally portable devices which are used by learners in the classroom allowing them to respond very quickly and anonymously to questions posed by their teacher/s. In other words, they are designed to improve questioning and answering in the classroom and to provide instant feedback to teachers and learners [4].

According to literature, C.R.S. are believed to enhance students' interaction and communication in the classroom, [5] and foster cooperation between teacher and learners, promoting their active participation in the teaching process. It is argued that

M. E. Auer et al. (eds.), *Teaching and Learning in a Digital World*,
Advances in Intelligent Systems and Computing 715,
https://doi.org/10.1007/978-3-319-73210-7_66

C.R.S. provide opportunities for student engagement [6], thus increasing their interest in attending their school lessons [7]. Additionally, C.R.S. are believed to promote the development of relationships between students and teachers and turn the educational procedure into an enjoyable activity resulting in an effective student participation in the classroom activities [8, 9]. What is more, C.R.S. have been found to motivate learners, thus facilitating their learning, and consequently offering positive educational results [10]. It has also been shown that C.R.S. enhance classroom communication, increase attention and interest in the lesson [11] whereas at the same time help instructors to generate discussion in the classroom, reduce stress, enhance understanding of the lesson and use them for formative assessment purposes [12]. Similarly, according to literature C.R.S. have been used for attendance, gauging comprehension and testing purposes but have also been used to overcome limitations of traditional lectures or improve students' attitudes [13]. What is more, it is suggested that when C.R.S. are included in curriculum design, they can provide a new dimension for interactivity in the classroom and allow student and teacher interaction to a great extent [14].

Kahoot is one of those classroom response systems that is considered an effective way to introduce new concepts or assess the extent to which those have been already mastered by learners [15]. This well-known game-based on-line platform of Kahoot presents the above C.R.S. features. Kahoot is used to enhance classroom participation and assess learners' cognitive level in the form of a game. It is a free online tool that allows the implementation of quick quizzes to assess students' knowledge in real time. The quizzes are displayed on a computer for the whole classroom or teams or pairs, and students respond to the application downloaded on their computers, smart phones, or tablets. The teacher can even add videos or pictures to accompany the questions and facilitate learning.

The quizzes are answered by learners in a form of a game and students answer them in real time through an easy-to-use interface, enabling the teacher to assess their progress and the extent to which they have reached the desirable cognitive level. Kahoot is a game-based learning platform that allows students to approach the lesson more as a fun game rather than a boring or obligatory process within school duties [16]. Additionally, Kahoot is considered both an enjoyable and educational platform which impresses learners through the use of quizzes and questions relevant to their learning. Similarly, Kahoot is a challenging tool for learners which may be exploited by the teacher to introduce new content or assess previously taught knowledge. Kahoot requires simple and basic computer skills to use and navigate through its platform.

2 Rationale as to the Method and the Topic Chosen

The classrooms of today are becoming increasingly demanding. The majority of today's students are efficient users of technology and their capability to surf the digital world and use the many applications to inform themselves of all the latest technological developments too often surpasses that of their own teachers'. These amazing digital capabilities and skills of students have some impact in teaching and learning. On one hand, they enrich learners' background knowledge on a specific subject they address in class. On the other hand, this plethora of digital skills raises their expectations for more exciting and innovative teaching methods and tools as opposed to traditional and indifferent or unattractive teaching methods and materials. To this end, teachers strive to find ways to attract their students' attention in order to make the lesson more interesting, perhaps turning it into a game but at the same time enabling them to increase their learners' cognitive level and efficiency. Thus, they seek ways to motivate them but also address the students' learning gaps, facilitate their understanding and guide their way into "learning how to learn", increase class interaction and gain new knowledge.

ICT and especially C.R.S. are believed to be among these digital suggestions that can help students become more active especially because learners view the lesson as a game, without however deviating from the learning goals and desired outcomes. Nevertheless, to achieve such goals it is necessary that teachers are trained to use such applications so as to lead their classes into joyful moments and better cognitive results. Based on the aforementioned points and on the need to help teachers supplement their knowledge with new applications and tools for more efficient and successful teaching practices we decided to conduct workshops for teachers of secondary education regarding the on-line C.R.S. platform of Kahoot for three basic reasons: introduce the platform and its educational use, explore teachers' views on its usage and features as well as their intention to apply Kahoot in their future teaching practices.

3 Methodology

3.1 Research Questions

Our research interest was to explore teachers' views on the use of Kahoot after its introduction to them in a two phase workshop. To this end, our effort emphasized the following: 1. Introduce the on-line C.R.S. Kahoot platform to teachers, 2. Investigate their views on its use and characteristics for teaching 3. Explore their intention to use Kahoot in the future. Therefore, our main research questions are as follows: 1. What is teachers' opinion on the features of Kahoot platform? 2. What is teachers' opinion on the educational use of Kahoot platform? 3. Do teachers intend to use Kahoot in their future classes and why?

3.2 The Sample

A total of 149 teachers of secondary education participated in the research. A number of 81 teachers taught subjects of theoretical studies (i.e. language, history, literature) and 68 of them taught science courses (physics, biology) or math. They were all formally certified by the Ministry of Education in basic ICT skills but this was not a prerequisite for their participation as some of them might had ICT skills but not necessarily certified and could still participate if they wished.

3.3 The Research Tools

The research employed both quantitative and qualitative research methods. Particularly, semi-structured questionnaires were delivered to teachers after their training and this decision was based on the fact that questionnaires are preferable for larger samples, as they are considered less time consuming and are easier to be quantified. However, and in order to have more analytical data and in-depth answers on teachers' views, a semi-structured focus group discussion was conducted with 25 randomly selected teachers. For validity purposes both questionnaires and focus group discussion questions were piloted with 18 (for the questionnaires) and 10 (for the focus group discussion questions) teachers respectively. The data taken were analyzed in order to modify, supplement or redesign the final versions for the appropriateness of the tools and clarification purposes. However, the researchers also observed and took notes throughout the whole process of teachers' training about Kahoot as they wished to have an opinion on their attitudes and responses regarding the new tool.

Structure of the Questionnaires
The purpose of the questionnaire was to collect data about Kahoot and consisted of eleven questions. These referred to issues such as easiness of use, clarity of instructions, environment interface, motivational features, ability to support teachers' work in class, ability to assess learners' knowledge, ability to enhance interactivity, accessibility, pedagogical implications, ability to enhance learning, its design features, and the extent to which they would use it in their future teaching. The answers were given on a five point Likert scale that ranged from "very much", "a lot", "quite", "a little" to "not at all".

Structure of the Focus Group Discussion
The focus group discussion was conducted in order to have an in-depth analysis of teachers' answers and clarify any vague points. The semi-structured questions were formed basically based on the questionnaire questions but explanations and clarification questions were also asked in order to provide us with more details for the purposes of the research questions.

Research Phases
Due to the fact that the number of the participants exceeded the capacity of the school laboratory (25 computers maximum) the research comprised six successive groups of 25 teachers each. The research was conducted in four phases as follows:

1st phase: The first phase involved teachers' invitation to participate in the workshop for the purposes of the research and their training on the use of Kahoot. After they had been fully explained the purpose of the workshop and their voluntary participation in the research was received we proceeded with the second phase.

2nd phase (3 h): This entailed a workshop for each one of the six groups during which, the first hour, the teachers were introduced to the tool as regards its use and characteristics with examples given by the trainers. During the last two hours, teachers were asked to implement quizzes and/or survey questions on topics of their interest or on topics they would like to introduce to their students. The aim was to allow teachers to work on their own quizzes to familiarize themselves with the steps needed and understand the ways they could use Kahoot and integrate it in their lesson activities.

3rd phase (3 h): During the third phase, the teachers presented their questionnaires/quizzes/surveys to the rest of the co-trainees for everyone to see how it worked, receive feedback and perhaps improve the way they would like to introduce it to their own classes. At the same time they had the opportunity to exchange ideas and cooperate with one another to expand on their knowledge both on the features of Kahoot and methodological suggestions/ways to integrate it in their classes.

4th phase (1 h – questionnaire completion and focus group discussion): Upon completion of the workshop, the teachers were given a questionnaire to answer regarding their opinion on the use of Kahoot and its application in their future teaching. Additionally, a focus group discussion of 25 teachers, randomly selected, was formed to provide the researchers with more in-depth answers regarding their views for Kahoot in education.

4 Data Analysis

4.1 Analysis of the Focus Group Discussion

For the analysis of the focus group discussion the "content analysis" method was employed. After the transcription, the repeated listening and reading of the discussion content, the most significant parts that linked to the research questions were isolated and recorded and the data analysis units/key words were determined. The key words were "easy", "clear", "environment", "motivating", "supportive", "helpful", "difficult", "access", "pedagogical", "design", that were included in the participants' answers, in relation to the aims of the research and research questions.

The thematic analysis of the focus group discussion results revealed teachers' positive opinion regarding Kahoot as a tool: "… quite different, new and fun to use!". According to the teachers, Kahoot can provide them with an enjoyable way to diagnose or assess learners' knowledge: "I want to see how it works in class too… I think they will like it… I'll use it in the exam after the tasks…". As they explained, what amazed them was the fact that it is a game-based tool, easily accessible and they considered this very important for their learners' motivation which they constantly try to enhance: "If I go now [means to school]…with it [means to use Kahoot]… we'll stay in the classroom…there won't be a break, I'm sure… they'll love it". For this reason they appeared to be quite confident that this would be an effective tool in class: "I feel it works all right… it can certainly give me a clear idea of their answers very quickly… which is necessary …".

Additionally, according to the teachers' views, Kahoot was considered an inspiring tool that can increase anticipation and interaction in class and boost learners' self-esteem because it is a digital tool that learners can understand and learn very easily and thus, make their learning a fun game: "We always need to get away from serious things… kids just enjoy games… this can be a fun game…". Moreover, as they explained, they felt it would allow all learners, even the weaker or shy ones to participate: "…those I have [means students] that don't work much, they'll want to be in [means participate], they will like it when their names are not shown… this is good for them… those that are shy… you know… with other students [means they won't feel embarrassed in front of others]".

Teachers also considered the possibility to insert videos very useful because learners are accustomed to them through YouTube mainly and they like their use in the lesson "I would particularly like to add videos… my students like this anyway. I'll use them in quizzes". However, the teachers also expressed their concern regarding the availability of a lab or a smart board as some of them (though a small percentage) had only one smart board in their schools which all teachers needed to use "I worry about it… there's only one smart board in school…and we all want to use when possible…". Regardless of this however, the majority seemed to be enthusiastic and suggested to those worried that all they needed was internet access, students' mobile phones and that it would not be difficult to overcome the lack of a smart board or a lab adding that they could just use their computer and a simple projector: "No worries! Just bring your laptop at school and use the wall… That will do it!… the rest leave it to them! [means the students]".

4.2 Analysis of the Questionnaires

The analysis of the teachers' questionnaires was conducted with the SPSS statistical package and some interesting results have been revealed (Table 1). According to the analysis the majority of the teachers (87.9%) answered that they found Kahoot a very easy tool to use with a percentage of 69.1% stating that the instructions that it offers are clear. A percentage of 74.5% of the teachers believed that Kahoot is a motivating tool for learning whereas an even bigger percentage of 81.9% admitted that it can support teachers' instructive work in class. Furthermore, 76.5% of the teachers answered that Kahoot has the ability to enhance interactivity and 51% believed that it also has pedagogical features. As regards the ability of Kahoot to enhance learning, 47% were positive with 65.8% of them admitting that it is an interesting tool for teaching. However, in the question about its future use the answers seemed to balance between 36.2% replying "a lot", 31.5% saying "quite" and 18.1% expressing "very much" as part of their intention to use it in the future. In detail, the answers are presented on Table 1 below.

Table 1. Results of teachers' questionnaires about Kahoot

Questions	Mean		%				
	Statistic	Std. error	Not at all	A little	Quite	A lot	Very much
Kahoot is an easy to use game-based platform	4.81	.049	0.7	1.3	2.0	8.1	**87.9**
The instructions of Kahoot are clear	3.81	.063	2.7	3.4	14.8	**69.1**	10.1
The environment of Kahoot is well presented and designed	3.37	.073	2.0	8.7	**52.3**	24.2	12.8
Kahoot is a motivating tool for learning	4.63	.062	1.3	1.3	4.7	18.1	**74.5**
Kahoot can support teachers' instructive work in class	4.01	.039	0.0	1.3	6.7	**81.9**	10.1
Kahoot can assess students' knowledge	4.64	.062	0.7	2.0	6.7	14.1	**76.5**
Kahoot has the ability to enhance interactivity	3.59	.080	4.0	8.7	25.5	**47.7**	14.1
Kahoot has pedagogical features	3.68	.073	2.7	6.0	26.2	**51.0**	14.1
Kahoot has the ability to enhance learning	3.61	.083	5.4	7.4	24.2	**47.0**	16.1
Kahoot is an interesting tool for teaching	4.45	.077	2.0	4.7	5.4	22.1	**65.8**
I intend to use Kahoot in my future teaching	3.50	.091	8.1	6.0	31.5	**36.2**	18.1

5 Discussion and Conclusions

ICT have had a very long journey since they last appeared suggesting an enormous number of ideas and applications to facilitate our daily lives. Especially for education, their integration in the teaching process has supported teaching and learning and has offered better educational results. Within the framework of ICT, Classroom Response Systems have been introduced for some time now as another way of approaching learning. Motivated by the necessity to assist teachers with their everyday practices, provide them with new ideas and offer them new tools to use in the classroom the researchers have presented in this paper their attempt to introduce the online game-based platform of Kahoot to secondary education teachers aiming to inform them about its features and suggest their exploitation as an instructive tool in the classroom.

According to the teachers' answers, Kahoot was found to be an interesting tool which can motivate learners, thus, turning a teaching session and especially assessment into a fun game rather than a boring process. As the teachers explained they enjoyed the ability to upload their questionnaires or quizzes on a platform which is well designed, easy and simple in its use with very clear instructions and objectives. Based on their answers, Kahoot has the ability to support their teaching and enhance interactivity among learners, a fact which, as they admitted, is very crucial because they want their students to be active participants rather than passive and indifferent learners. As the majority admitted they would like to use it in their future classes, though a percentage of 6% answered "a little" and 8.1% "not at all", when asked if they intended to use it in the future. However, everything new has the label of the "unknown" and sometimes this implies some sort of fear or it takes quite a long time for someone to get well acquainted with it, take the first step and "try it out".

Based on the focus group discussion results, the teachers would mostly like to use Kahoot for assessment purposes to turn this process into a more enjoyable one. As they admitted, assessment is quite a "painful" and boring process for learners which they feel very uncomfortable with and almost never participate positively. They are usually negative and many of them even refuse to study because the traditional types of written tests are not attractive to them, some even fear them and therefore they lack any interest in them because they see them as punishment and not as feedback. Therefore, as teachers exclaimed, they are hoping to change this situation and turn those indifferent learners into more active and interested students for a number of reasons. One would be to motivate them "to revise for their tests more willingly"; another would be to change their thinking of tests and turn them from being a terrible process to a genuine interest of students to discover all about their progress in a form of a fun game which would take their stress away and "release them from their fear for tests".

Additionally, and as the teachers revealed, their training on Kahoot as another tool to use in the classroom was considered important. As they explained, updating their knowledge on new methods and tools is useful for two reasons: new ideas are necessary because they can help them meet the needs and the challenges of today's demanding classrooms; their learners are a young community whose digital capabilities exceed those of their teachers', a fact which is quite stressful and/or sometimes "intimidating" as described by one of the participants. Therefore, to be able to have "arrows in their quiver", as a participant eloquently stated, and use them when and where accordingly is a good solution to their lesson plans design; and this availability can offer them opportunities of varying their methods and tools to support their teaching and students' learning based on each teaching purpose and each student/s' specific needs.

Being a teacher is perhaps one of the most challenging professions today. If we really wish to make a difference in education, we should decide with no fear to take the initiative and meet the challenges without being afraid to experiment and try new ideas and methods. Though a magical process, teaching is also an everlasting and stressful fight both inside and outside the classroom aiming however at new and wonderful paths to knowledge which never seize to amaze us. As the years go by, technology enhances the effort for innovation in education and what all of us educators need to do is simply be present in these efforts, open-minded and willing to try; and then, maybe we could let our passion lead us to different but exciting and more efficient perspectives of approaching our learners and our methods and support their wish to be successful in gaining knowledge. All it takes after all is that we, educators, should be the ones to first believe in ourselves and our efforts and keep on trying.

The second part of the research, which is beyond the scope of this paper, due to the bulk of the data collected, is about the views of those teachers who finally decided to use Kahoot in their classes and the ways they finally chose to do this. Additionally, the students' opinions are drawn with very innovative suggestions regarding the educational exploitation of the tool of Kahoot, a fact which is quite interesting, given that their views are very essential as students are the actual target of all our efforts as educators.

References

1. Mumtaz, S.: Children's enjoyment and perception of computer use in the home and the school. Comput. Educ. **36**, 347–362 (2001)
2. Leone, S.: The use of new technologies in advanced Italian classes. In: Proceedings of Emerging Technologies Conference, pp. 18–21. University of Wollongong (2008)
3. Muncy, J.A., Eastman, J.K.: Using classroom response technology to create an active learning environment in marketing classes. Am. J. Bus. Educ. **2**(2), 213–218 (2012)
4. Siau, K., Sheng, H., Nah, F.F.-H.: Use of a classroom response system to enhance classroom interactivity. Manag. Dep. Fac. Publ. **49**(3), 398–403 (2006)
5. Milner-Bolotin, M., Fisher, H., MacDonald, A.: Modeling active engagement pedagogy through classroom response systems in a teacher education course. LUMAT **1**(5), 523–542 (2013)
6. Heiss, B.M.: The effectiveness of implementing classroom response systems in the corporate environment. A Thesis submitted to the Graded College of Bowling Green (2009)
7. Abramson, D., Pietroszek, K., Chinaei, L., Lank, E., Terry, M.: Classroom response systems in higher education: meeting user needs with NetClick. In: IEEE Global Engineering Education Conference (EDUCON), March 2013, pp. 840–846 (2013)
8. Beatty, I.D., Gerace, W.J., Leonard, W.J., Dufresne, R.J.: Designing effective questions for classroom response system teaching. Am. J. Phys. **74**(1), 31–39 (2006)
9. Duncan, D.: Clickers in the Classroom: How to Enhance Science Teaching Using Classroom Response Systems. Pearson/Addison Wesley, San Francisco (2005)
10. Lucke, T., Keyssner, U., Dunn, P.: The use of a classroom response system to more effectively flip the classroom. In: IEEE Frontiers in Education Conference, October 2013, pp. 491–495 (2013)
11. Eastman, J.K.: Enhancing classroom communication with interactive technology: how faculty can get started. Coll. Teach. Methods Styles J. – First Quart. **3**(1), 31–38 (2007)
12. Owusu, A., Weatherby, N., Otto, S., Kang, M.: Validation of a classroom response system for use with a health risk assessment survey. In: Poster Session at 2007 AAHPERD National Convention and Exposition, Baltimore, Maryland (2007)
13. Fies, C., Marshall, J.: Classroom response systems: a review of the literature. J. Sci. Educ. Technol. **15**(1), 101–109 (2006)
14. Siau, K., Sheng, H., Nah, F.: Use of classroom response system to enhance classroom interactivity. IEEE Trans. Educ. **49**(3), 398–403 (2006)
15. Diaz, C., Trejo, C.: Kahoot: the student-teacher interactive classroom tool (2015)
16. Meijen, C.: Kahoot: using a game based classroom response system in teaching. School of Sport and Exercise Science, University of Kent (2015)

Introducing "Kodu" to Implement Cross Curricular Based Scenarios in English for K-12 Learners

Marianthi Batsila[1(✉)], Charilaos Tsihouridis[2(✉)], and Anastasios Tsichouridis[3(✉)]

[1] Directorate of Secondary Education, Ministry of Education, Larissa, Greece
marbatsila@gmail.gr
[2] University of Thessaly, Volos, Greece
hatsihour@uth.gr
[3] Democritus University, Komotini, Greece
tsabtsih@gmail.com

Abstract. In the present study the programming language of Kodu was used as a tool to implement a cross-curricular series of lessons combining the subjects of English and Computer Science. The purpose was to see the extent to which Kodu can be used as a creative and effective tool to enhance learners′ English language skills. A number of 74 Junior High School learners participated in the research, with a control group of 35, and an experimental group of 39 learners. The latter were introduced to the tool and implemented programming tasks with Kodu. A pre and post-test was delivered to the learners to detect their level before and after the intervention and informal discussions were conducted with them. The results revealed that Kodu made the lessons for the experimental group more vivid, creating a lively atmosphere which kept them active in class enhancing their use of English which they employed to follow the instructions, work with one another and implement the tasks assigned to them.

Keywords: ICT · Kodu · Scenarios · Cross-curricular

1 Introduction

We live in a world that everything around us is changing rapidly. This is due to the augmenting growth of technology, new inventions and devices. The consequences of all this is a change in the way people think and act; they have become more autonomous and efficient now in their learning than ever before in almost all walks of life. Within this framework of change ICT have come a long way and have played a very important role in man's effort to improve the personal, social, academic and professional life [1]. Their influence has been found to be of high importance especially in education; computers, the World Wide Web, software and an extensive variety of applications provide educators and learners with many advantages and assist learning to a great extent. Their benefits range between the facilitation of learning and transmission of knowledge to differentiated and personalized learning (learning difficulties, special needs, analphabetism, distance learning and so on). In addition to the above,

M. E. Auer et al. (eds.), *Teaching and Learning in a Digital World*,
Advances in Intelligent Systems and Computing 715,
https://doi.org/10.1007/978-3-319-73210-7_67

communication and group work are also issues which are promoted through the use of ICT for the success of which, lessons are no more teacher centered but learner centered, as instructors cooperate with learners and are mainly facilitators of learning rather than mere messengers of knowledge [2].

Thus, and for the teaching goals to be implemented, teachers need to organize their teaching in such a way in order for learners to benefit from the use of ICT, experiment, participate, focus, play and learn [3]. For learners to be motivated however, it is essential for teachers to focus not just on the use of ICT but on the appropriate use and exploitation of computers and their applications, in order for learners to arrive in pedagogical outcomes [4]. To this end, learners need to be given the opportunities to exploit learning, take initiatives and through a process of experimentation achieve the life skill of "learning how to learn", explore, question and take action [5]. Autonomous learning can be gained through creativity and design. Learners can acquire such skills through the world of programming which will allow them to create their own programs, through fun and exploration [6]. Thus, instead of simple users of a game for instance they can become its designers, define the behavior of the heroes of their games, what and when to do something, or what to do to win. Programming refers to specific instructions which one needs to follow on a computer in order to achieve the pre-set steps and targets. The instructions are understood by a computer through a programming language, in other words, the way we talk to the computer and form a series of guidelines and orders which learners need to know [7].

The above actions can be realized through programming environments [8]. Such an environment is the MS Kodu which Microsoft offers freely to the users. It is an environment which was designed in order to inspire young people and adults to create exciting games in a quick and easy manner. MS Kodu allows the users to create their own games, robots, stories, drawings etc. It was initially designed as a learning tool for young people, who used Xbox 360 and was officially in the market in January, 2009. It is easy to use and visually appealing to learners [9, 10]. The coding which is used is simple and intuitive but allows a high degree of computational thinking and basic principles of programming. It is easily integrated into any curriculum, like in math, science or geography. Furthermore, there is a broad support through resources and online communities to assist its use.

According to literature, Kodu has been found to have helped students demonstrate understanding of lawfulness by predicting the behavior of Kodu characters, applying idioms, and reasoning about rules. Additionally, it has been found that the power and simplicity of Kodu facilitate teaching students to reason formally about program behavior [11]. Kodu has been used to show that users express and explore fundamental computer science concepts by pairing language constructs with fundamental concepts in computer science. It has been shown that users spend more time programming and configuring their programs than they do playing them, which indicates that the Kodu environment has reached its goal of making programming accessible to all users [12]. Other researchers customized the mechanics available to users of Kodu Game Lab so it would appeal more to middle school girls but have not yet received feedback [13]. Similarly, researchers have shown that pupils interacted very well using Kodu [14] whereas others have analyzed the environment of Kodu to show that the diagrammatic analysis using Kodu provides opportunities for easy and quick comparison of essential dimensions of the digital storytelling environments [15]. Moreover, the environment of Kodu has been used as a game making tool for learning programming and have provided a supportive and productive learning environment that engaged a significant percentage of the students in general mainstream schools [16].

Within the framework of ICT skills and their integration in the teaching practice, the term "interdisciplinarity" (cross-thematic integration) is a form of teaching which means a multifaceted study of concepts in which knowledge is seen as a whole rather than fragmentary and linked to real life. In such an approach of knowledge mainly learner-centered methods are used (projects, group work) as they provide possibilities for cooperation in the teaching process. Students are free to explore the issues investigated, driven by their interests, whereas the role of the teacher is mainly auxiliary, yet decisive ensuring the quality of learners' work [17]. During such an approach learners interact with one another with team spirit developing their full potential (physical, social, cognitive and emotional) exploiting their interests and background knowledge and experiences. Thus, learners are in the center of teaching, while the teacher himself guides them discreetly, encourages and helps them when they have to confront difficulties. This interdisciplinary approach was trialed in this research integrating English into ICT skills and vice versa through the use of Kodu programming language environment with Junior High School students.

2 Rationale of the Present Study

Learning is quite a challenging procedure and is highly influenced by many factors, such as motivation, participation, communication and so on. Schools have undertaken a very difficult role and a lot need to be done if they wish to be successful in their goals. Especially today, with the constant upcoming and development of new technologies and their latest applications, learners are always informed about all kinds of newly emerging ICT applications and tools and quite often they are even a step ahead of their own teachers. To motivate them seems all the more a hard task and teachers need to strive if they wish to find ways to attract their attention. Furthermore, as learners have become so familiar with technology they demand a better quality of teaching and learning and traditional methods which keep them passive and inactive, such as the book centered one, seem to be ineffective. On the contrary, students feel very enthusiastic and interested in using ICT, surfing the net or playing electronic games and their integration in the lesson has become a tool in the hands of the "desperate" teacher. Driven by the need to motivate learners, enhance their active participation and improve their English language skills competence, the authors of this paper decided to use the "Kodu" programming language as a tool to implement a cross-curricular series of lessons combining the subjects of English and Computer Science.

3 The Research

3.1 The Purpose of the Research and Research Questions

The purpose of the research was to investigate the extent to which the use of the programming language of Kodu can be used as a teaching tool to improve learners' language skills in the English language. Additionally, the purpose was to detect learners' attitude towards the use of this tool in the lesson and the extent to which it can motivate them into using more English, activate them and enhance their class participation.

To this end, our research questions are as follows: 1. To what extent can "Kodu" improve learners' skills in English? 2. To what extent can the use of "Kodu" enhance learners' active participation in the classroom? 3. What do learners think of the use of "Kodu" in the lesson?

3.2 The Sample

In this research, 74 Lower Secondary School learners, aged 14 participated in a comparative study. They formed two groups: an experimental one of 39 (G1) and a control of 35 learners (G2). They all came from a city school and four classes, randomly selected. They were all taught the same book and syllabus and followed the same curriculum guidelines in both subjects, English and Computer Science. Their level of English was officially expected by the state to be B1 according to the Common European Framework of References for Languages, though they are usually mixed ability classes.

3.3 The Research Tools

The research was conducted with both quantitative and qualitative research methods. Learners' skills competence in English was measured with the use of the National Foreign Language Exam System (KPG test), which comprises four modules (reading comprehension and language awareness, writing and written mediation, listening comprehension and speaking and oral mediation) (Table 1). KPG is a stabilized test, used to assess learners' skills competence in foreign languages. The maximum possible score at B1 level candidates can gain is 100 whereas the minimum score required to gain in order to pass the test is 60 (30% of the maximum possible marks in Modules 1–3); there is no minimum mark required in Module 4, although the marks candidates receive are included in their total score (Table 1). Additionally, for the purposes of the second and third question the authors were based on their observations, note taking in the classroom and informal conversations with the learners in the end of the teaching interventions.

Table 1. KPG exam specifications

LEVEL B (B1 + B2)
KPG graded test in English

Module	Types of questions/tasks	Number of questions/tasks		Coefficient	Rating					Time (in min.)	Number of words in texts	
					max		min				Provided	Produced
1	Choice	25 + 25 = 50	60	0.8	40	50		B1	8	85	1.500–2.000	Not defined
	Completion	5 + 5 = 10		1.0	10			B2	15			
2	Semi-free production	4		–	60			B1	9	85	150–200	350–400
								B2	18			
3	Choice	7 + 8 = 15	25	2.0	30		50	B1	8	20–30	Not defined	Not defined
	Completion	5 + 5 = 10		2.0	20			B2	15			
4	Semi-free production	3		–	40					20–25	Not defined	Not defined

3.4 Research Stages

The research lasted for two months and 19 h. It comprised the following phases:

1st phase: During the first phase, the school and the classes were randomly selected in the research location. After having received the state permission, parents' and principals' consent the authors discussed with the teachers regarding the implementation of the procedure. All teachers had already been officially trained on the use of ICT and were familiar with various applications of Web 2.0 tools and software in the English classroom. However, before the intervention they had also been introduced to "Kodu" and had participated in a three week in-service training program for teachers of foreign languages, implemented by a Computer Science School Advisor regarding the use of "Kodu" and other tools in the classroom.

2nd phase (2 h): In this phase and before the intervention, all learners took a pre-KPG B1 level test to determine their level of English.

3rd phase (1 h): As a next step, in the beginning of the intervention, the students were announced that they would participate in a project about ICT, programming tools, and programming languages and especially the programming language of Kodu (this material was also part of their Computer Science syllabus). The project for both groups entailed, presentation of the material, watching videos, making presentations, filling in forms and tables, writing essays and answering relevant worksheets and all lessons for both groups (G1 and G2) were in English. However, the experimental group was also going to actually use the programming language of Kodu in order to program certain tasks themselves while the control group would be introduced to it as a tool but would not actually use it for programming.

4th phase (12 h): In this phase the control group worked on the steps of the project as set by the teachers, while the experimental group, apart from that, was also introduced to the language of Kodu in the lab and was asked to explore its features and get acquainted with its environment. While working on the project, the G1 learners were asked to implement specific tasks and actions with Kodu, acting as programmers themselves. In the end of every series of tasks they had with "Kodu" they had to complete a worksheet regarding terms and actions in English and the process of the games and discuss it with the teacher. All instructions were given in English and learners were obliged to use only this language in order to communicate with one another and respond to the tasks. They were also suggested to watch demonstration videos in YouTube for a better understanding on the use of "Kodu" and clarification purposes and use this information in order to proceed with their tasks in the lab. For whatever questions or explanations they discussed with the teachers of English and Computer Science who were present in the classroom. Learners were encouraged to work in pairs (of both groups), while their teachers contributed with their help when needed. The last 10 min of every lesson were dedicated to feedback in the English language, regarding the process and steps of the project and their design tasks with Kodu. Throughout the whole process the authors were observing all learners and their tasks.

5th phase (2 h): By the end of the interventions, students had to take another KPG B1 level post-test. The test had the same tasks and level as the previous one but learners had not been provided with the answers of the pre-test. The test was taken the same day by all students, who answered it with exactly the same time allocated by the test specifications. The G1 students also took part in informal conversations regarding their opinion on "Kodu".

6th phase (2 h): Three weeks after the second test, all learners took a similar follow up test, again of the same level to determine any differentiation in their answers.

4 Results

4.1 Statistical Analysis-Data Analysis

The use of the appropriate checking criterion (parametric or not) between research hypotheses depends mainly on the plan of the research, the commitment of the level of data, and the type of the indices of the measurement of the variables. To analyze the data obtained presently, the IBM-SPSS statistical package was used, and a t-test for independent and dependent samples were performed. For the purpose of the present study, the level of significance was set at 5%. The research hypotheses are:

- H_o: Null hypothesis: The participant groups of learners display the same performance after the teaching intervention.
- H_1: Alternative hypothesis: The participant groups of learners have displayed different performance between them, after the teaching intervention.

It should be noted that, in H1, there is no intrinsic attempt to predict which group displays the best or worst performance. Therefore, a two-sided checking of hypotheses is formulated. The results are presented below.

Checking groups G1 and G2 before and after the teaching intervention (Pre/Post – testing) for all individual teaching objectives.

The following table (Table 2) represents the t – test of independent samples.

Table 2. Results of t-test for independent samples at PRE/POST testing level for all individual teaching objectives t-test for equality of means

	t-test for equality of means			
	t	Sig. (2-tailed)	Mean difference	Std. error difference
Efficiency_Reading_PreTest	1.19	0.236	1.61	1.35
Efficiency_Listening_PreTest	0.43	0.666	0.69	1.59
Efficiency_Writing_PreTest	−0.32	0.747	−0.47	1.45
Efficiency_Speaking_PreTest	−0.28	0.777	−0.39	1.37
Efficiency_Reading_PostTest	6.11	**0.000**	8.75	1.43
Efficiency_Listening_PostTest	0.50	0.622	1.11	2.24
Efficiency_Writing_PostTest	0.30	0.768	0.50	1.68
Efficiency_Speaking_PostTest	2.89	**0.004**	5.50	1.90

This is based on the t-test, obtained for all individual teaching objectives which correspond to a (pre-determined) non-significant statistical ($p > 0.05$) result. This leads to the acceptance of the null hypothesis, meaning that the performance of learners in group G1 does not differ from that of the learners in group G2, before the teaching intervention ($\mu_{01TOTAL} = \mu_{02TOTAL}$), for each individual teaching objective. One can therefore proceed with the rest of the comparisons. Further statistical analysis using the t-test of the results, showed a statistically significant difference at the instructive goal of reading skills, i.e. [($t(65) = 6.11, p = 0.001$] between the groups and in the instructive goal of speaking skills, i.e. [$t(65) = 2.89$, $p < 0.004$]. This allows one to assume that the use of Kodu has improved learning of the specific skills, and for this specific age group, when compared to the group who did not work with Kodu (Fig 1).

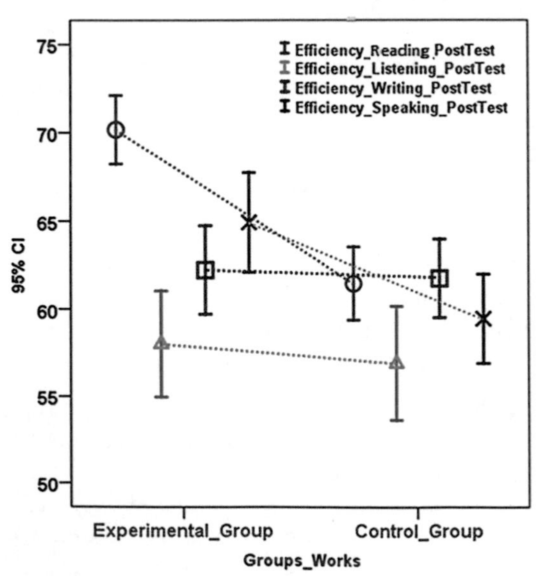

Fig. 1. Error chart for the participant groups in the research at post-test

From the above chart, one can support the view that there is a statistically significant difference between groups at post-test stage for reading and speaking skills, as the corresponding overlaps are not bigger than half the mean of the average marginal error.

Furthermore, our informal conversations with the learners revealed that students were excited to work on their programming tasks which activated them to use English for an authentic and real purpose and because they needed to communicate with one another and with the teacher in order to understand the instructions on the computer and complete their programming projects which, as they admitted, liked very much. Based on our observations and note-taking it can be said that students had a positive attitude towards the use of Kodu in the lessons. Throughout the process, it was clear to us that the implementation of the programming instructions generated discussion and we believe that the pair and group work activities also helped a lot towards this as students were instructed to work with one another and of course use only English to communicate in the lessons. According to their answers students considered Kodu a rather easy environment for the programming of simple tasks. Moreover, the learners considered the use of Kodu a fun way to learn which did not bore them but excited and activated them. Based on our notes we consider that this was so because it motivated them to use English in a pleasant and creative way while exploring their capabilities as programmers. Furthermore, the use of "Kodu" had the ability to involve learners actively in the lesson, motivate them and enhance their interest in both English and Computer science courses which students, according to their answers, previously considered boring, as lessons were quite theoretical and book-centered. Therefore, we believe that Kodu made the lessons more vivid because they created a lively and authentic atmosphere which kept the learners very energetic and participative in class. Nevertheless, we must admit that there are certain limitations as it is not always easy, at least in the schools of the research location, to find a free lab available for so many hours especially when other teachers also want to use it for their own lessons.

5 Conclusions/Recommendations/Summary

In this research we attempted to investigate the extent to which the use of the programming language of Kodu can be used as a teaching tool to improve learners' language skills in the English language in a cross-curricular comparative study. Additionally, we tried to detect learners' attitude towards the use of this tool and the extent to which it can be motivating towards learning. Based on the results it can be said that Kodu has improved learners' speaking and reading skills in English. Additionally it was found that Kodu offers simple tasks which are easy to use, thus, facilitating learning and opening new horizons for the educational ICT world which combines both fun and knowledge. Its simplicity enables even learners of very low level or no knowledge in programming to handle the tasks, follow the instructions and find them motivational to complete a programming project. Most importantly however, the research revealed a few things we consider worth mentioning: it is most likely that we never forget what we are actively involved in doing, we really like and enjoy, and become passionate about. To this end, perhaps, the G1 group improvement was owed to their deep involvement and interest in the making of their own programming tasks;

that the mere existence of a computer in a classroom, which the teacher operates and uses just for the purpose of showing something on the screen, while learners have a book to guide them which they need to follow, look for information and perhaps memorize in order to "learn about the computer", or just about anything, is not a solution. What really needs to be done is that learners should use their computers to create something with the teacher being the facilitator and coordinator instead of just a provider of knowledge. Therefore, in the case of this research we believe that Kodu might have offered learners just that: that they themselves became the creators and regulators of their own knowledge. Because, in our opinion, it is only then that learners could perhaps become autonomous learners and find pleasure, interest and real purpose in learning.

Effective teaching practices are educators' concern for improving their learners' level. Teachers strive their way through to new and magic ideas and suggestions aspiring to encounter the magic solution to the class problems. Especially for the young learners of today who are digitally oriented it is always a great challenge to invent new ways and tools to attract their attention so that they become more interested and active in knowledge learning. Sometimes however, the magic solution which all of us educators try to find so hard is so simple and so close to us that we simply need to see it just by being flexible, passionate, willing to try and realize that learning can be more efficient if it is motivating, fun and as pleasant as a game. Kodu was able to attract learners' attention and make their learning more interesting turning it into a game, thus enhancing their skills performance, but mainly making learning their own personal achievement. As a next step the researchers have planned to continue exploring the capabilities of Kodu with other age groups and types of school as well.

References

1. Sharma, A., Gandhar, K., Sharma, S., Seema, S.: Role of ICT in the process of teaching and learning. J. Educ. Pract. **2**(5), 1–5 (2011)
2. Wendy, M.: Not just tools: the role of E-technologies in culture of learning. Educ. Commun. Inf. **1**(2), 229–235 (2001)
3. Korhonen, A., Malmi, L., Myllyselka, P., Scheinin, P.: Does it make a difference if students exercise on the web or in the classroom? In: Proceedings of the 7th Annual SIGCSE/SIGCUE Conference on Innovation and Technology in Computer Science Education, ITiCSE 2002, Aarhus, Denmark (2002)
4. Flecknoe, M.: How can ICT help us to improve education? Innov. Educ. Teach. Int. **39**(4), 271–280 (2002)
5. Kulik, J.: Effects of using instructional technology in elementary and secondary schools: what controlled evaluation studies say (Final Report No P10446.001). SRI International, Arlington (2003)
6. Shen, W.-M.: Autonomous Learning from the Environment Microelectronics and Computer Technology Corporation. Computer Science Press/W.H. Freeman and Company, New York (1994)

7. Jonnavithula, L., Kinshuk, D.: Exploring multimedia educational games: an aid to reinforce classroom teaching and learning. In: Uskov, V. (ed.) Proceedings of the 4th IASTED International Conference on Web-Based Education (WBE), Grindelwald, Switzerland, 21–23 February 2005, pp. 22–27. ACTA Press , Anaheim, CA, USA (2005)
8. Leutenegger, S., Edgington, J.: A games first approach to teaching introductory programming. ACM SIGCSE Bull. **39**(1), 115–118 (2007)
9. Prambudi, S.B., Sudarmilah, E., Nugroho, Y.S.: Enpowering Kodu game as a numeracy learning media for Kindergarten. Department of Informatics, Faculty of Communications and Informatics Universitas Muhammadiyah Surakarta, (2013). (it helped the children learn to count, children were interested to play motivated by the 3D graphics, it is easy (2013))
10. Shokouhi, S., Asefi, F. Sheikhi, B.: Children programming analysis; Kodu and story-telling. In: Third International Conference on Advance Information System, E-Education & Development (ICAISED 2013) – Singapore on 6–7 November (2013)
11. Touretzky, D.S., Gardner-McCune, C., Aggarwal, A.: Teaching "Lawfulness" with Kodu. In: SIGCSE 2016, 2–5 March, Memphis, TN, USA (2016)
12. Stolee, K.T., Fristoe, T.: Expressing computer science concepts through Kodu game lab. In: Proceedings of the 42nd ACM Technical Symposium on Computer Science Education, Dallas, TX, USA, 09–12 March, pp. 99–104 (2011)
13. Fristoe, T., Denner, J., MacLaurin, M., Mateas, M., Wardrip-Fruin, N.: Say it with systems: expanding Kodu's expressive power through gender-inclusive mechanics. In: FDG 2011, 29 June–1 July, Bordeaux, France, pp. 227–234 (2011)
14. Fatiu, O.A.: Kodu game lab- a tool for ensuring quality teaching-learning for pupils in primary schools: case study (school in northern Finland). Master's thesis in Education, Faculty of Education (2014)
15. Psomos, P., Kordaki, M.: Analysis of educational digital storytelling environments: the use of the "Dimension Star" model. In: Lytras, M.D., et al. (eds.) WSKS 2011, CCIS, vol. 278, pp. 317–322 (2012)
16. Fowler, A.: Enriching student learning programming through using Kodu. Paper presented at Third Annual Conference of Computing and Information Technology Research and Education New Zealand (CITRENZ 2012). In: Lopez, M., Verhaart, M. (eds.) 25th Annual Conference of the National Advisory Committee on Computing Qualifications, Christchurch, New Zealand, 8–10 October, pp. 33–39 (2012)
17. Mulder, M.: Interdisciplinarity and education: towards principles of pedagogical practice. J. Agric. Educ. Ext. **18**(5), 437–442 (2012)

On Legal Support for Engineering Activities: A New Managerial Project

Svetlana Barabanova(✉), Vasily G. Ivanov, Raushaniya Zinurova, and Maria Suntsova

Kazan National Research Technological University, Kazan, Russia
sveba@inbox.ru, vgiknitu@mail.ru, rusha1810@mail.ru,
emci2008@gmail.com

Abstract. The paper has been prepared on the basis of its authors' great experiences in working in various areas at a technical university. It represents the result of long-term research and reflections on improving the legal support for engineering activities through developing the new forms of interaction between legal experts and engineers or between teaching staff and the management of engineering universities. The authors estimate the conventional approaches to organizing the teaching of future engineers legal disciplines as non-complying with the requirements of the modern engineering activities, and propose to both revise the contents of the existing courses and modify the organizational set-up of universities. Particularly, we propose to depart from traditional dividing into socio-humanistic and technical departments and create an interdisciplinary department. Teaching legal disciplines at an engineering university is, in our opinion, necessary and actual. However, they should be taught focusing on developing legal literacy in students as future experts in engineering. This would require retraining the teachers of law, their active interacting with the representatives of engineering sciences, and conducting relevant scientific and applied research. Training of humanities majors studying at engineering universities should be transformed in a similar way: They must acquire the basics of engineering knowledge to increase their competitiveness on the labor market.

Keywords: Engineering education · Interdisciplinary approach
Law for engineers · Engineering and technical education for humanities majors

1 Law and Engineering Education: Russian Traditions

According to the Federal Law "On Education in the Russian Federation" [1], one of the principles of national education policy is the humanistic nature of education, as well as the primacy of human life and health, responsibility, legal culture, rational nature management, environmental friendliness, etc.

Because of the step-up of technology-induced issues and the increase in the amounts of incidents and disasters caused by human factors, particularly in engineering activities, it seems to be advisable to develop new approaches to arranging the activities of non-science departments at higher engineering educational institutions. For the illustrative

M. E. Auer et al. (eds.), *Teaching and Learning in a Digital World*,
Advances in Intelligent Systems and Computing 715,
https://doi.org/10.1007/978-3-319-73210-7_68

purposes, we have chosen the department of legal studies as a department that mostly complies with the tasks of this study.

Traditionally, teaching law to Russian students has been built using a very simple scheme developed as early as in 1970s, as the basics of the Soviet law were taught. Students were offered a course on the basic branches of the Russian law, usually including 18 h, or 0.5 credit points, and 18–36 h of practical lessons, the total of 1.0–1.5 credit points. According to a large number of authors, within the system of socio-humanistic education, the course of Legal Studies promotes understanding the essential and inalienable human rights and liberties that are binding over the state and which the state may not invalidate or restrict at its discretion. Training of students is focused on developing the idea of that the civil and political rights and liberties determine the meaning, contents and enforcement of law, the activities of federal and local authorities, and are guaranteed by justice. In its turn, the state is responsible for the implementation of the political, economic, social, and any other capabilities of a person, as well as for creating the conditions that ensure the good living standards and free development of individuals.

This understanding of law complies with the ideas of a democratic law-bound state, since educating the legal consciousness of a law-abiding citizen is based on giving an insight into the characteristic of law as one of the most important standards of civilized human relations, rather than its compulsory potential [2].

At all times, the course of legal studies pursued the aims of generally cultural importance, such as giving the fundamental notions of the key branches of law; developing the legal literacy of students; elaborating the positive attitude to law; and considering law as social reality. This is why, in the most of universities, it includes the issues of the modern understanding of law, the role of law and state in the life of the society, and the basics of the theory of state and law, as well as of the constitutional, civil, family, labor, administrative, criminal, and environmental law.

Federal State Educational Standards (hereinafter, the FSESes) in technical and engineering areas, regarding the contents and amounts of legal training future engineers, had a similar concept. For instance, the FSES for training bachelors in Chemical Engineering presupposed that students had to know:

- Basics of the Russian legal system and the Russian legislation;
- Basics of establishing and functioning the judicial and other law enforcement authorities;
- Legal, ethical and moral standards in vocational activities;
- Legal standards that regulate the human relations to people, society, and environment;
- Rights and obligations of a citizen; and
- Basics of labor legislation.

A graduate had to be able to use ethical and legal standards regulating the human relations to people, society, and environment; use civil and political rights when developing social projects; use and prepare regulatory and legal documents relating to his or her vocational activities, take necessary measures to restore the rights that have been infringed; exercise civil and political rights in various areas of life; and know the basics of commercial law [3].

As we can see, all the requirements above are focused on general legal knowledge and on developing a certain level of legal culture in a professional or a manager.

The new structure of the bachelor-level program grants universities significant freedom in building the contents of education, including as regards to defining the share and amounts of socio-humanistic disciplines. Therefore, very different options are possible regarding the destiny of legal studies – from full intolerance and excluding from curricula to changing the course contents taking into account the formation of a well-rounded and vocationally oriented personality. However, it is important to remember, as P.K. Engelmeyer, the founder of philosophy of technology in Russia, said, "try as you might to stuff him (engineer) with specialized knowledge, he will be an educated craftsman, until you give him a humanistic insight into the social and economic aspects of his vocation" [4, p. 12].

2 Changing Standards and System Capabilities

The new standards of engineering education in Russian universities define the following areas of activities for the graduates: Engineering, management, research, and projecting. They have to be capable of solving particular problems, such as taking measures for protecting intellectual property and research and development results as the trade secrets of enterprises. They must be capable of using the basics of legal knowledge in various activities, ready to use legal documents on quality, standardization and certification of products and the elements of economic analysis in their practical activities; organize the activities of contractors, and find and make managerial decisions in the field of labor organizing and quota setting [5].

Russian education, along with the European one, comes to understanding that the context of professional engineering activities includes stable factors, such as interdisciplinary approach to developing a solution or the need of team-oriented engineers for efficient communications and for managing the group activities. Changing in engineering activities introduce new contextual factors. These are, in particular, sustainable development that results in paradigm shift from the exploitation of natural resources to sustainability considering their potential; globalization that presupposes international competition, collaboration and mobility of engineers; entrepreneurship, etc. [6, p. 69] Obviously, supporting such processes and solving the relevant educational problems require some changes in teaching Humanities, including law studies, because everything must be subordinated to the general idea of preparing the graduate for innovative engineering activities.

However, the analysis of the existing educational editions shows that they cannot perfectly satisfy those needs.

We have analyzed a number of original teaching aids recently published for giving classes in legal studies at universities that are not focus on legal studies:

1. Vasenkov, V.A., Korneeva, I.L., and Subbotina, I.B. (2015). Pravovedeniye. Sbornik zadach i uprazhneniy [Legal Studies. Collection of Tasks and Exercises]. Moscow: Forum, Infra-M. 160 p. (In Russian) [7],

2. Isakov, V.B., ed. (2015). Igropraktikum: Opyt prepodavaniya osnov prava: Metodicheskoye posobiye [Game-Based Workshop: Experiences in Teaching the Basics of Law: Teaching Aid]. Moscow: NIU VSHE (National Research University Higher School of Economics). 2nd ed. 304 p. (In Russian). [8], and
3. Kapustin, A.Ya., ed. (2015). Pravovoye obespecheniye professionalnoy deyatelnosti: uchebnnik [Legal Support for Vocational Activities: Textbook]. 2nd ed. Moscow: Urait. 382 p. (In Russian) [9].

Unfortunately, for all innovative methods, very useful recommendations, unique concepts of teaching law, we could not find in none of them any examples or forms of anchoring the legal studies to the future activities in production, engineering and technology. The respected authors, highly experienced in both working in legal areas and teaching law, did not adapt the discipline to engineering activities.

At the same time, if we consider the problem otherwise, for example, the practice of preparing lawyers for their future vocational activities, we can easily verify that they acquire purely legal knowledge without anchoring to their possible activities at a manufacturing enterprise [10].

We hoped to find the foundations of an integrated comprehensive approach to training professionals on an interdisciplinary basis in publications dealing with the globalization of legal education [11] or with training students majoring in Customs, which, in fact, as a whole, is the interdisciplinary education. [12] However, we did not find anything that we could apply in the present study.

At the same time, among the global trends in the development of the Russian higher engineering education, we can mention its internationalization, its focusing on the best international practices. Foreign experience in studying law by engineers could hardly be applied in Russia: It either does not exist as such – our colleagues from the Arizona State University (ASU) asked us perplexedly: why does an engineer need to study law? There are legal departments! – or it is taught focusing on the basic civil and political rights and liabilities – for instance, constitutional law is studied at Spanish universities, which is certainly important to educate civics in their students, however, it does not directly relate to their future profession. Therefore, the interdisciplinary approach that is popular in the area of engineering education may be better applicable to socio-humanistic disciplines.

It seems to be reasonable and actual to properly ground the changes in approaching to teaching law at engineering universities and, therefore to transforming the traditional departments of legal studies to focus on interdisciplinary activities.

Within the Research University, the following steps are proposed to solve the problems of educating modern engineers also having the necessary background in the legal fundamentals of their future vocational activities:

– Establishing a new Department of Judicial Support for Engineering, or transforming the existing Department of Legal Studies into an interdisciplinary department involving lecturers and instructors experienced in higher engineering educational institutions or companies that monitor or supervise the safe working practices in industries, or perform other oversight functions over the activities of economic entities;

- Extending jurisprudence-based modules and topical units within the bachelor/master degree programs, with the consideration of the students' specializations;
- Developing the engineering programs of further education for students studying under the management programs, including those majoring in public administration;
- Developing an interdisciplinary master degree program at the intersection of engineering sciences and jurisprudence to educate engineers with good command of legal regulations applicable to engineering, which ensure technosphere and industrial safety, compliance with workplace safety rules, the industrial implementation of innovations, intellectual property protection, compliance, etc.;
- Advanced training or vocational retraining the legally educated teaching staff in the areas engineering education;
- Involving the department members in working with customers on contractual basis on the development of regulatory documents for industrial enterprises;
- Integrating the department members into project teams,
- Organizing at this department further training courses for the management and personnel of industrial enterprises and corporations regarding administrative, employment and economic legislation;
- Holding academic conferences regarding the matters of judicial support for engineering, and attending engineering education conferences; and
- Preparing by engineering students their interdisciplinary final theses or projects on requests from enterprises, involving the department members.

3 New Approaches and Technologies

Interesting practices can be studied as exemplified by a number of educational institutions and scientific centers. For example, a great specificity and special features are inherent to the relations in the area of subsoil use, or mining relations. They represent a whole complex, including exploration, search, and survey of mineral resources, land allocation, extraction and processing of mineral resources, and conservation of mineral resources and continental shelf protection in a country. Obviously, the problems of legal regulation of such relations cannot be studied in isolation from understanding the special features of technical and technological processes. And, vice versa, it is impossible to teach engineering students in the processes of searching, extracting and processing mineral resources without mastering the relevant regulations, standards, and other legal instructions. This is why it was Gubkin Russian State University of Oil and Gas (National Research University) that established the department of mining law in 1998 and the Institute of Mining and Energy Law in 2014 [13].

Another good example is training lawyers at the National Research Nuclear University (MEPhI). Legal regulation of using nuclear energy, i.e. nuclear law, is studied and developed at this university in close interaction of students, teaching staff, and researchers specializing in engineering, physics, technical sciences, management, and law.

The interdisciplinary approach has been successfully tested in Russia, in the area of additional vocational education. For instance, Gubkin Russian State University of Oil and Gas offers programs that are equally actual for lawyers and managers employed in oil and gas engineering sector, such as "Occupational Standards in Oil and Gas Recovery, Processing and Transporting: Compulsory Basic Legal Skills and Competences" or "Organizing Contractual Activities at Oil-and-Gas Enterprises." The latter one includes both the general issues of civil-law regulation in contractual activities and the specificity of oil-and-gas delivery contracts, as well as other topics, such as the Market, Processing, and Logistics of Gas, Gas Products, Oil, and Oil Products; Gas/Oil Major Pipeline Transport Technology; etc.

At the Institute of Additional Professional Education of Kazan National Research Technological University, lawyers, economists, pedagogues, psychologists, and teachers of specialized engineering disciplines participate in one program of additional vocational education for oil-and-gas enterprises. According to the attendees' opinions and the employers' estimates, these are such comprehensive programs that provide the most efficient development of the new competences of the personnel [14].

Consumer feedback proves the correctness of the approaches chosen. Our colleagues from the institute of law of one of the Siberian universities tell us that they are often reached out by the representatives of oil and gas companies to help them with court proceedings: Their own lawyers "get stuck" in corporate procedures and paperwork to such an extent that they do not risk to participate in any industry-specific court proceedings involving commercial claims or administrative offence cases, such as regarding oil and gas deliveries, transporting petrochemical products; violations against environmental and ecological legislation, license agreements, etc. Thus, the deep professionalization of lawyers within the core activities of oil and gas producing companies is obvious, as well as their learning the "ropes" of engineering knowledge within the oil and gas industry, etc.

4 Case History: Unexpected Innovations

We would like to draw special attention to the practices of teaching students majoring in the Engineering Chemistry of Natural Energy Sources and Carbon-Based Materials in the Foundations of Legal Studies in the English language at Kazan National Research Technological University. In fact, the students get into another educational environment loaded with the doubled or even tripled amount of information – the standard one, the data on common law, and the special aspects of regulating professional activities in developed countries.

The work resulted in developing by Professor S. Barabanova her original course of comparative legal studies for students majoring in non-legal areas. We have already described this and similar experiences in a number of our publications. [15–17] It seems to deserve a searching examination and extension to train modern engineers that have interdisciplinary knowledge and skills and are capable of working in teams when implementing various integrative projects. Such approach would also promote intercultural communications and readiness for labor mobility.

Thus, the course of legal studies, for example, for students studying petrochemistry or oil and gas engineering, although it includes conventional sub-disciplines, they must be, however, more vocationally focused. For instance, the sub-discipline of

Administrative Law presupposes studying the topics, such as the competency of executive authorities, first of all, in the area of monitoring and supervision; legal statuses as viewed by the administrative law – licensing, accreditations, and special affirmative statuses; antimonopoly regulation; methods of administrative assistance and methods of regulating public activities; administrative violations in subsoil use, ecology, industrial safety, etc.; and liabilities for violations in economic activities.

The same is the case for civil law – students study the special features of the civil conveyance of natural resources; business contracts, such as petrochemical products delivery and transportation contracts, the characteristics of the legal statuses of natural monopolies and of adhesion contracts; and the issues of the legal protection of IP assets and of using thereof in manufacturing activities.

When studying labor law, a special focus is placed on studying the special features of working under a rotation system or in special climatic conditions, or at harmful or hazardous manufactures.

Environmental law is also largely related to the characteristics of the vocation chosen. The following should be studied: Law of special use of natural resources, the requirements of environmental law for oil and gas engineering activities, legal regulations of industrial and household waste land-filling, criminal and administrative measures applicable to environmental law violators, etc. (provided that those issues have not been considered within Administrative Law and Criminal Law, respectively).

It seems to be reasonable to add to the course structure the financial and fiscal law matters relating to how the oil presence within the territory of a country influences upon its budgeting, its excising policies, etc.

Active work in this direction has resulted in elaborating the career development program named Legal Support for the Activities of Petrochemical Enterprises, intended for the lawyers and all-level managers of petrochemical companies and allowing the lawyers to get into better understanding of the engineering specifics in their employing companies, while the "non-lawyers" are given an opportunity to be trained in the basics of law for better understanding the issues occurring in the legal support of the petrochemical business activities. The program has already been highly estimated by both our colleagues – the lawyers from among university professors and by the representatives of the customer, i.e. Gazprom PJSC.

5 Interdisciplinary Approach as a Basis for Developing Humanities at an Engineering University

The interdisciplinary approach presupposes the capability of a graduate from an engineering university of comprehensive engineering activities considering the social and environmental consequences thereof and of working in interdisciplinary multiethnic teams. Unfortunately, in actual practice, the knowledge acquired are not always transformed into a capability of implementing such approach. Inn this regard, the tasks are set of creating the tools of implementing the interdisciplinary approach; first of all, creating a methodical system of teaching staff career development, developing the mechanisms of designing the technology of planning and implementing engineering educational programs and curricula considering the interdisciplinary approach, including, but not limited to, socio-humanistic disciplines.

The authors' experiences in working in various areas of the university life allows us to draw the conclusion of the necessity of serious changes and of improving the legal support for engineering activities through developing the new forms of interaction between legal experts and engineers or between teaching staff and the management of engineering universities. As shown above, the conventional approaches to organizing the teaching of future engineers legal disciplines do not comply with the requirements of the modern engineering activities. This is why we propose to both revise the contents of the existing courses and modify the organizational set-up of universities. Particularly, we propose to depart from traditional dividing into socio-humanistic and technical departments and create an interdisciplinary department. Teaching legal disciplines at an engineering university is, in our opinion, necessary and actual. However, they should be taught focusing on developing legal literacy in students as future experts in engineering. This would unconditionally require retraining the teachers of law, forming a new engineering legal consciousness, active teachers' interacting with the representatives of engineering sciences, and conducting relevant scientific and applied research. Training of humanities majors studying at engineering universities should be transformed in a similar way: They must acquire the basics of engineering knowledge to increase their competitiveness on the labor market.

Today, the interdisciplinary approach is becoming the basic principle of the existence of a comprehensive university. [18, 19] Obviously, in all the above cases, we cannot limit ourselves with mechanical adding up information and specialists. New methods are required to use interdisciplinary approach to complex phenomena. In our opinion, the pedagogical methodology will just be enriched with interdisciplinary approaches and with the contributions made by other subjects and sciences to its development.

We suppose that the topic touched upon here is equally actual for all universities and/or colleges relating to engineering education. We can see in life that the demand for "pure" lawyers, economists, etc. is often very limited. Moreover, vice versa, an engineer or a manager who is just poorly familiar with legal regulations relating to his or her vocational activities may be in danger of remaining a second-rate performer for life. This is why the experience put forth in this paper implies discussions among teaching staff interested. However, it can already be used at the institutions of higher engineering education.

References

1. Federal Law No. 273-FZ "On Education in the Russian Federation", 29 December 2012. (in Russian)
2. Komarov, S.A., Kirillov, S.I., Ognev, V.N., Pobezhimova, N.I.: Primernaya programma po pravovedenniyu. Rekomendovana Sovetom po pravovedeniyu UMO universitetov RF pod predsedatelstvom E.A. Sukhanova [Exemplary Program of Legal Studies. Recommended by the Council for Legal Studies of the UMO (Education Review Office) of the universities of Russia, presiding by E.A. Sukhanov], Moscow (2000). (in Russian)

3. Decree No. 807 of the Ministry of Education and Science of the Russian Federation "On Approving and Putting in Force the Federal State Educational Standard of higher vocational education for Program Track 240100 Chemical Engineering (Qualification/Degree: Bachelor)", 22 December 2009 (edition as of 31 May 2011). (in Russian)
4. Gorokhov, V.G., Rosin, V.M.: Vvedeniye v filosofiyu tekhniki: uchebnoye posobiye [Introduction into the Philosophy of Engineering: Teaching Aid]. Infra-M., Moscow, p. 227 (1998). (in Russian)
5. Decree No. 1005 of the Ministry of Education and Science of the Russian Federation "On Approving the Federal State Educational Standard of higher education for Program Track 18.03.01 Chemical Engineering (Bachelor Level)" dated 11 August, 2016; Decree No. 1176 of the Ministry of Education and Science of the Russian Federation "On Approving the Federal State Educational Standard of higher education for Program of Studies 18.05.01 Chemical Engineering of Energy-Saturated Materials and Products (Specialist Level)", 12 September 2016. (in Russian)
6. Crawley, E.F., et al.: Rethinking Engineering Education. The CDIO Approach; translated from English by S. Rybushkina. In: Chuchalin, A., (ed.) NIU VSHE (National Research University Higher School of Economics), Moscow, p. 504 (2015) (in Russian)
7. Vasenkov, V.A., Korneeva, I.L., Subbotina, I.B.: Pravovedeniye. Sbornik zadach i uprazhneniy [Legal Studies. Collection of Tasks and Exercises]. Forum, Infra-M., Moscow, p. 160 (2015). (in Russian)
8. Isakov, V.B. (ed.): Igropraktikum: Opyt prepodavaniya osnov prava: Metodicheskoye posobiye [Game-Based Workshop: Experiences in Teaching the Basics of Law: Teaching Aid]. 2nd edn. NIU VSHE (National Research University Higher School of Economics), Moscow, p. 304 (2015) (in Russian)
9. Kapustin, A.Ya. (ed.): Pravovoye obespecheniye professionalnoy deyatelnosti: uchebnnik [Legal Support for Vocational Activities: Textbook]. 2nd edn. Urait, Moscow, p. 382 (2015). (in Russian)
10. Сборник учебно-методических материалов по гражданскому праву. Под ред. Е.А. Суханова. М.: 2011. 316 с. (in Russian)
11. Globalizatsiya vysshego yuridicheskogo obrazovaniya: Istoriko-pravovyye aspekty formirivaniya innovatsionnogo podkhoda [Globalization of Higher Legal Education: Historical and Legal Aspects of Developing an Innovative Approach]. In: Korovyakovsky, D.G.: Monograph. Ru-Science, Moscow, p. 144 (2015). (in Russian)
12. Podgotovka spetsialistov s vysshim obrazovaniem po spetsialnosti "Tamozhennnnoe delo": rossiysky i zarubezhny opyt [Training University-Degree Professionals Majoring in Customs]. In: Korovyakovsky, D.G., et al. Monograph. Yustitsiya [Justice], Moscow, p. 342 (2016). (in Russian)
13. Melgunov, V.D.: Teoreticheskiye osnovy gornogo prava [Theoretical Fundamentals of Mining Law]. Prospekt, Moscow, p. 336 (2015). (in Russian)
14. Ivanov, V.G., Miftakhutdinova, L.T., Galikhanov, M.F., Barabanova, S.V.: Engineering Staff Development in Research University: Synergy of Traditions and Innovations. In: Engineering Education. no. 20, pp. 9–15 (2016)
15. Barabanova, S.V.: Bilingualism, Multicultural and Comparative Law in Engineering Education. In: Vysshee obrazovanie v Rossii [Higher Education in Russia]. no. 7, pp. 20–25 (2015). (in Russian)
16. Ivanov, V., Miftakhova, N., Barabanova, S., Lefterova, O.: New Components of Educational Path for a Modern Engineer. In: Proceedings of the International Conference on Interactive Collaborative Learning (ICL), 20–24 September 2015, Florence, Italy, pp. 184–187. Institute of Electrical and Electronics Engineers, Red Hook, NY (2015)

17. Barabanova, S.V., Shagieva, R.V., Gorokhova, S.S., Popova, O.V., Rozhnov, A.A., Popova, A.V.: Innovative components in the educational strategy of training the modern graduates. J. IEJME-Math. Educ. - IJESE (IJESE 16-264)
18. Gorodetskaya, I.M., Shageeva, F.T., Khramov, V.Y.: Development of cross-cultural competence of engineering students as one of the key factors of academic and labor mobility. In: Proceedings of 2015 International Conference on Interactive Collaborative Learning, ICL 2015, pp. 141–145 (2015). https://doi.org/10.1109/icl.2015.7318015 / Source Scopus
19. Shageeva, F.T., Erova, D.R., Gorodetskaya, I.M., Prikhodko, L.V.: Socio-psychological readiness for academic mobility of engineering students. In: Proceedings of 2016 International Conference on Interactive Collaborative Learning, ICL 2016, Paper ID 1743/ Source Scopus (2016)

ICT Used to Teach Geography to Primary School Children – An Alternative Teaching Approach

Despina M. Garyfallidou(✉) and George S. Ioannidis

The Science Laboratory, University of Patras, Rio, Greece
d.m.garyfallidou@gmail.com, gsioanni@upatras.gr

Abstract. Nowadays a huge number of ICT-based tools such as Google Earth, Google maps, search engines, crosswords, as well as word processors, presentation software and more can be used in Geography teaching and learning. Creating school presentations containing locational sightseeing or school magazines, be they printed or electronic, referring to monuments, famous people originating from a locality, customs etc. are also tools that can be used for this purpose. The educational trial presented herein aims to: (a) encourage pupils learn geography in a more creative and interesting way; (b) to familiarize students with basic ICT skills; (c) teach students to seek and evaluate information about specific topics; (d) and perhaps more importantly show students ways to use computers for self-education. The trial took place in two different school years and the participants 42 altogether aged 10–11 years (grade E). The school was an ordinary one in the suburbs of Athens. An impromptu computer lab was setup in classroom with rudimentary networking utilising a projector. Most students were so excited that they were willing to sacrifice part of their break to create another crossword or complete a presentation. At the end of the test, most students had acquired basic skills in using general purpose software. They learned to carefully evaluate information content found on the internet and try checking its relevance to geography. This approach has also improved students' communicative and cooperative skills.

Keywords: Teaching geography · ICT in education · Search engines
Presentation software

1 Introduction

Geography is seen as: "the study of places and the relationships between people and their environments. Geographers explore both the physical properties of Earth's surface and the human societies spread across it. They also examine how human culture interacts with the natural environment and the way that locations and places can have an impact on people. Geography seeks to understand where things are found, why they are there, and how they develop and change over time" [1]. In this very broad sense of what Geography is all about, literally any aspect of human activity or even natural occurrence can be seen as geographical, as long as it is attached to a geographical

M. E. Auer et al. (eds.), *Teaching and Learning in a Digital World*,
Advances in Intelligent Systems and Computing 715,
https://doi.org/10.1007/978-3-319-73210-7_69

location and analysed accordingly. However, it can be argued that the essence of Physical Geography is thereby diluted to a virtual elimination.

From a more balanced viewpoint, "Geography is the study of Earth's landscapes, peoples, places and environments" [2]. In this sense, Geography informs us about:

- The places and communities in which we live and work
- Our natural environments and the pressures they face
- The interconnectedness of the world and our communities within it
- How and why the world is changing, globally and locally
- How our individual and societal actions contribute to those changes
- The choices that exist in managing our world for the future
- The importance of location in business and decision-making

There is enough evidence to support the notion that basic schooling is failing to provide school-leavers the tools to begin to approach the aforementioned complex tasks. Indeed, previous research has shown that even after studying geography at school students could not locate a certain country on the map or recall any information about it [3].

It seems that some new teaching approaches on Geography are biased towards human geography stretch the curriculum by adding new material and giving emphasis to alternative thinking, thereby failing to provide the basic geographic tools. On the other hand, "in geography, there are a great number of opportunities for using computers: to add variety to existing teaching methods, to tackle difficult areas of the curriculum, to revitalize tutors' interest in their work, and to open up new learning horizons for students" [4]. Furthermore, nowadays most students use smartphones daily, and there are numerous geography-based ICT applications for such devices. Such proliferation creates as many new educational problems as it solves, radically altering the quality of students' questions to the teachers. As today's students are, anyway, projected to live using such increasingly geolocation equipped devices offering ample geographical information content, it would appear that ICT and Geography are destined to a complex yet ideal pairing together. Yet, education is ill-preparing students for such future world.

As a cure to the aforementioned predicament, a new teaching approach with a lot of ICT use was designed and tested in class. In this, students were first asked to locate certain places on the e-map and then tried to identify places of interest, be that natural or manmade ones, and also famous and important people that were born or lived there, and to subsequently use this information in order to create a magazine and a presentation.

Could ICT usage help students to learn geography better? Could it be that exposure to information about places of interest and monuments with videos and pictures of local interest, attract students' interest in Geography in general?

Despite their almost universal approval by educators, not all primary schools have computer labs, and as far as the particular school where the trial took place, this only happened the year before. Nevertheless, recent developments in computer design, as well as the proliferation of 3G and 4G mobile internet, and the increase of the number of students that own a suitable mobile device, of one form or another, places ICT-based Geography teaching under new light. It is certain that mobile devices will keep proliferating, and that internet speeds will keep rising. What is less obvious, though, is whether or not educators will be ready to keep up with times.

2 The Research

It was decided from an early stage of the present research trial that only open source applications as well freely distributed programs were to be used. There were various reasons for this, the most important of which was (a) zero cost to obtain and maintain them in the school lab and (b) absolute freedom for children to download and install them in their home computer, at the suggestion of the researchers.

The programs used were:

1. Google Maps
2. Eclipse crossword
3. Mozilla Firefox
4. LibreOffice Impress
5. LibreOffice Writer
6. Microsoft Paint

All these programs are general purpose software and were not designed for education [5], it is the teacher who converts it to educational tool, either by using it himself or by encouraging students to exercise their creativity.

The overall educational objectives of the approach tried were:

- To familiarise students with the geomorphology of Greece.
- To learn some of the different types of maps (general, thematic of various sorts e.g. historical etc.)
- Help students learn Greek towns, islands, mountains, rivers, and seas and be able to locate them in a map be that electronic or printed.
- To become proficient in searching for places of interest around any given area assigned.
- To be able to search for important people (painters, poets, heroes) coming from (or associated with) a certain geographical area.
- To use the knowledge obtained in (d) and (e) above in order to create a school magazine, and a presentation containing a virtual tour around Greece.
- To use the knowledge obtained to create crossword-puzzles.
- Finally, to use a word processor in order to create tables where words are suitably.
- "hidden".
- To offer to the students an alternative and more attractive way of learning.

The educational objectives concerning ICT education were:

- Teach word processing to the students
- To teach the basics about presentation software
- To show to the students how to use an external storage device (USB) in order to store their work and explain them the importance of keeping backup. The possibility to use the cloud for backup services was explained but not actually tried, as it

was judged that it necessitates deliberate and specially considered educational approach
- To show to the students the basics about downloading and installing a program
- To show to the students the fundamentals of web searching, focusing on providing the basic criteria by which they could evaluate on-line material

It should be noted herein that all of the above, albeit essential, are not currently included in the 5th grade ICT school curriculum.

The present teaching trial was undertaken using two different school years while the participants were 42 altogether (24 the 1st year and 18 the 2nd) aged 10–11 years (grade E) and the school was an ordinary one in the suburbs of Athens. Rather aged PCs were utilised to set up an impromptu lab in class with rudimentary networking, using a projector.

Those students who were very enthusiastic with this learning approach, continued doing so using their home computers.

3 The Teaching Approach

The teaching sequence took place using rather old computers deployed in the ordinary classroom during the Geography lesson.

During the teaching sequence, specific activities were assigned to the students. To start with, these were common for all of them e.g. the introduction to the various software applications or the computer-based learning environments they were going to use. Later, each group could choose a task from a number of activities suggested by the teacher. Special care was given to avoid identical assignments e.g. when a group of students had already decided to create a presentation about a certain place, other groups were encouraged to undertake a different one.

The students worked in groups of 2, 3 or 4. They were freely allowed to form these groups by themselves, as well as to freely change the composition of these groups for the next educational task. This resulted in uneven grouping: the “stronger” students stuck together leaving the rest to do their best. This in itself was not thought to create any problem [6].

The bigger groups was not also considered as problem.

3.1 Google Maps

In the first lesson students familiarized themselves with Google Maps.

The use of Google Earth/Maps keenly supports the so-called “four E’s” of the learning life cycle model, allowing students to engage in the lesson, explore the earth, explain what they identify, and evaluate the implications of what they are learning. The very nature of Google Earth/Maps allows students to explore the earth in a dynamic and interactive manner, helping them understand the spatial context of their locale and engage in spatially oriented learning in an entertaining and meaningful manner. Students are not required to be at school in order to use the application [7] Fig. 1.

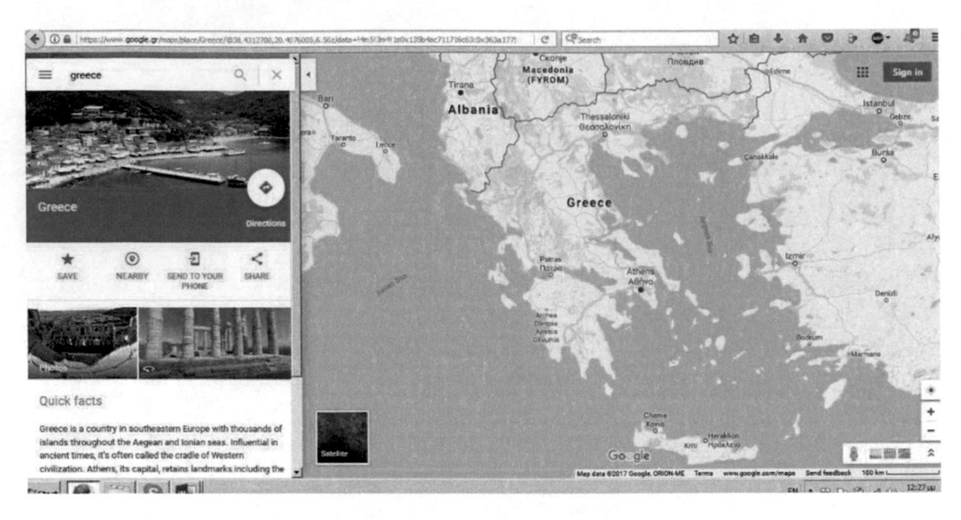

Fig. 1. Google maps - Greece

Students were asked to search for the location of cities and islands. It was observed that they searched even for their home and school. It was seen that, educationally, the most impressive aspects of Google Maps for the students were the zoom function as well as the street view, which allowed them to locate their school, their home or the place they were looking for.

Some educational observations come to mind, using this application. Students noticed that although most places could be traced in both Greek and Latin, some less known ones only appeared in Latin on the maps (e.g. very small islands such as Aggistri, or Dokos).

Also while the names of most gulfs did appear on the map, that of the majority of capes was absent. The ensuing extensive discussions concerning all these, was very fruitful for the students. They realised that it was impossible for any computer software to contain information that was not already been given to it by a human and that, consequently, errors and lapses in the information contained were inevitable. They also noticed that the "street view" function was not available everywhere, specially so for scenic walks reachable only on foot e.g. Samaria gorge, or ancient cities – places e.g. Olympia, Malvasia etc. In this case, it was explained that the real reason for this was the way these "street view" data were obtained in the first place, using cars with special devices attached on top of them. Till the next development in video data capturing and geolocation technology, that is.

The status of Google Maps as free application was explained and was much appreciated, whereas the associated URL was given so as students were able to use the application from their home computers.

3.2 Eclipse Crossword

The general-purpose freeware application "Eclipse Crossword" helps users to create crosswords, which can either be uploaded onto an interactive webpage where the user can ask for help if he/she does not know the answer or just printed out. As it happens with all software not specifically designed for education [5], it is up to the teacher to adapt it ad-hoc to education. This can be done by using it in class by the teacher, or by assigning tasks to students which, as it happened here, worked in groups of two or three.

It should be noted that the application is not operating just as a word processor, but offers significant help as it creates the 2-dimensional grid automatically doing all the cross-associations. This feature is deemed as essential for the help it offers to young children. Beforehand, students were asked to suggest words and offer clues to help readers find them.

Geography can be seen as the ideal school-subject for the playful yet educational use of crosswords, as students like to play a mental "hide and seek" game by the means of the crosswords using geography-related names. The task to search to find more details proved challenging. The task of finding the right "clues", which should be neither too obvious nor too difficult to guess the answer to were even more intriguing. Students willing to spend time searching for more details, create more difficult crosswords while the rest create ones with trivial clues. For example, the elaborate ones will offer for "Athens" the clue "the myth says that 2 major gods argued for the name of this city - the first offered an olive tree and the second a spring - the city took the name of the god that offered the olive tree". The simplistic ones would have offered as a hint "it is the capital of Greece". Although the educational game can hardly be described as a competition, the in-built human spirit to show-off newly acquired knowledge lives on and in the end education is the real winner.

All crosswords created by each group of students was eventually published in the school magazine and were uploaded as interactive web-pages in the class's web-page available, only in Greek language at http://school1.garyfallidou.org.

3.3 Searching on the Net

The object of this task, was for students to learn the art of searching for places of interest within a geographical area, and for famous people from around there, local customs etc. Students were encouraged to judge the importance of a place of interest, significance of a monument, or a person while verification was also provided by asking. Beforehand, it was explained why there is a difference between famous and important people e.g. painters, poets, novel writers as opposed to a football player who might have been popular during a certain period but has been largely forgotten a few years later.

Although the information needed for a hint in a crossword can be rather limited, when a net search was initiated, a huge amount of information revealed. This led to extended discussions, about evaluating any available information in accordance to several criteria, which have been discussed in a previous study [8].

Since students were asked to create magazine articles and presentations mostly using material found on the net, extensive discussions took place about copyright issues. It was explained to the students that when using information, be that in printed form or electronic one, they should always name their sources. It was also mentioned that they should always respect the "terms and conditions" for the use of a site.

3.4 Presentation Software

Trying to find all places of interest and important people that lived in the whole country of such a long and well documented history is an enormous task. The sheer amount of information requires a vast amount of time from the students to search, evaluate, and edit it. For this reason, Greece was divided in sub-sections and each was assigned to a group of students. And even so, an enormous amount of information was omitted.

The students were asked to collect photos, and use them to create their presentation. They could also incorporate some local music. Special care was paid so as areas with less information were assigned to less capable students, to avoid disappointments. Such groups also received more help and attention from the teacher. In the end all groups finished the work assigned to them and presented their work to the rest of the class.

3.5 Word Processing

Any word processor is suitable for creating a school magazine. The plethora of places of interest is such that it is impossible to incorporate all of them into a magazine, not even one with numerous issues. For this reason, the detailed features meant for the school-magazine were limited to places around Athens, where students could easily go with their parents. An additional challenge, in which the teaching trial seem to succeed, was to attract enough interest of students and parents alike, so as to go out of their way to visit such local places of interest. In order for the magazine to be complete and attract the readers' interest it was enriched with crosswords created by the students themselves, puzzles, jokes, dot to dot drawings, labyrinths etc.

As already mentioned students should always mention their "resources" (the pages from which they copied information) at the end of their article. Extensive discussions took place as to what type of jokes is appropriate for a school magazine.

In order to teach some more advance techniques in word-processor usage, suitable word-search puzzles were created. For this the students inserted a table. Instructions as how to set the cell height and width into a certain length were given to them. It was also explained how to insert more cells if necessary. They then placed one letter in each cell of the words they wanted to hide. At the end they filled the remaining cells with random letters. This activity was chosen mostly from the less capable (or less interested) students as it was an easier one. This allowed them to claim that they contributed to the creation of the magazine, thereby strengthening their self-esteem Fig. 2.

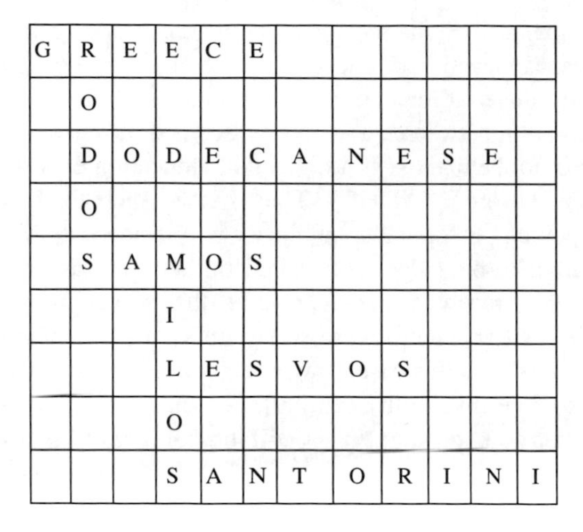

G	R	E	E	C	E						
	O										
	D	O	D	E	C	A	N	E	S	E	
	O										
	S	A	M	O	S						
			I								
			L	E	S	V	O	S			
			O								
			S	A	N	T	O	R	I	N	I

Fig. 2. A 12 by 9 table for creating word-search puzzles.

4 Educational Observations and Discussion

Students felt at ease with the new teaching approach. No problem was detected during any aspect of the task as the instructions were clear and specific. They were determined to stay in the class' computer lab in order to create one more item for the school magazine. Students were quite enthusiastic about the whole project, and indeed they were happy to stay overtime using the class' improvised computer-lab to finish their task. This agrees well with the conclusions from previous research [9]. Educational benefits included that students acquired basic computer skills like text editing (copy, paste, editing, error correction) saving and reopening their work, and use of external storage devices. They learned to locate Greek towns, islands, naming of seas etc. and came in first contact with the way landmarks and monuments are noted. They could also have a virtual tour in city centres or places of interest, using google-maps street-view feature.

This activity also helped children improve their vocabulary and their ability to write clear and very specific text, as well as improve upon their skill of self expression.

Google Maps was met with particular enthusiasm. Students familiarised themselves with the geomorphological map of the country and learned how to search for a certain place. Although the search sometimes failed for reasons explained herein, they were very impressed with the easiness they could locate a certain place.

Students learned that not all information presented on the net is equally important or even equally trustworthy and they were taught to evaluate what they saw being offered to them. Skills such as these are essential to young students, due to the insatiable proliferation of web-searches in life.

They also came to terms with the fact that all information on the net is available only because somebody has already uploaded it.

For the students themselves, the most useful gain was learning the art of learning by themselves. This was achieved very successfully by a game-style process of students creating new learning material for the rest of the class.

As far as collaboration between students is concerned, it was observed that stronger students volunteered to help those with Special educational needs, e.g. Attention Deficit Hyperactivity Disorder (ADHD). They helped them with the typing, or by correcting their spelling before the publication. Besides, they worked without demanding their names to be added in the final product. It was only after discussion with the students that received the help that the work presented appeared as co-authored. On the contrary, there were some students participating in teams while offering scant (if any) help. In this case, it was decided to omit the names from the final product. Regardless of all this, both such attitudes (in conjunction with the resulting action) helped children to understand what collaboration really means.

5 Conclusions

The new teaching approach is deemed to be very successful. It can be implemented with students working in groups or individually. For the educational trial presented herein, the students worked in small groups. Letting students create their own working-groups ensures that all groups will work properly. By working in small groups, students learned to socialise and collaborate with each other. They also learned to optimise the use of the assets available, and proceeded to evaluate their common final product. Educationally, we have reasons to believe that while "having fun" they learned more about Greece than students following the traditional (e.g. book and map) way of learning. One of the most important outcomes of this research was that students used the computer for self-learning, thereby acquiring a skill that is certain to be very useful to them in the future.

Practical educational considerations should also consider the time students needed to familiarise themselves with these new techniques. Also, special care should be taken so as the teacher observes and monitors all computer-screens. It is easy for the students to get to different sites on the net, if they are not constantly monitored.

The present study is only preliminary, awaiting for further improvement in techniques used, incorporating the latest technical developments. Meanwhile, further educational evaluation to measure the actual improvement in Geography learning is still pending.

References

1. National Geographic. https://www.nationalgeographic.org/education/what-is-geography/
2. Royal Geographical Society. http://www.rgs.org/GeographyToday/What+is+geography.htm
3. Geography - learning to make a world of difference, p. 22, February 2011, Reference no: 090224. www.ofsted.gov.uk/publications/090224
4. Shepherd, I.: Teaching geography with the computer: possibilities and problems. Middlesex Polytech. J. Geogr. High. Educ. **9**(1), 3 (1985)

5. Ioannidis, G.S., Garyfallidou, D.M.: Education using information and communication technology (ICT), and ICT education: categories methods and trends. In: Workshop ICL 2001. Kassel University Press (2001). ISBN 3-933146-67-4
6. Garyfallidou, D.M., et al.: Evaluating the combined use of video and hands-on school experiments in ICT-based science teaching: the case of atmospheric pressure. In: 11th International Conference ICL 2008, 27 p. (2008). ISBN 978-3-89958-353-3
7. Patterson, C.T.: Google Earth as a (not just) geography education tool. J. Geogr. **106**, 145–152 (2007–2008). (look in particular p. 146). ISSN 0022-1341 (Print) 1752-6868 (Online). http://www.tandfonline.com/loi/rjog20
8. Garyfallidou, D.M., Ioannidis, G.S.: Teaching geography with the use of ICT. In: Auer, M.E. (ed.) Proceedings of the 8th International Conference on Interactive Mobile Communication Technologies and Learning 2014, IMCL 2014, Thessaloniki, 13–14 November 2014, pp. 57–63 (2014). Catalogue Number: CFP14IMH-USB ISBN 978-1-4799-4743-0, ©2014-IEEE, Catalogue Number: CFP14IMH-POD ISBN 978-1-4799-4741-6
9. Garyfallidou, D.M., Ioannidis, G.S.: Design development and educational testing of a novel educational software to teach energy as a whole: the final educational results. In: International Conference ICL 2005 Ambient and Mobile Learning, 28–30 September 2005, 19 p. Kassel University Press (2005). ISBN 3-89958-136-9

Assessing the Learning Process Playing with Kahoot – A Study with Upper Secondary School Pupils Learning Electrical Circuits

Charilaos Tsihouridis[1], Dennis Vavougios[1], and George S. Ioannidis[2](✉)

[1] University of Thessaly, Volos, Greece
{hatsihour,dvavou}@uth.gr
[2] University of Patras, Patras, Greece

Abstract. The present study investigates the extent to which the popular game-based online platform of Kahoot can be used as a creative and effective tool in the teaching practice and specifically in the teaching of basic concepts of electric circuits. A comparative study was conducted for this reason with two groups of 67 learners in total, where the experimental group participated in the design of their own questions within the framework of formative assessment with the use of Kahoot, whereas the second group followed a traditional way for their assessment. According to the results, the integration of Kahoot in the teaching process improved learners' understanding of certain concepts on electric circuits, enhanced their active participation in the lesson, motivated them towards learning and constituted a creative and fun-tool to use for teaching purposes.

Keywords: ICT · Kahoot · Classroom Response Systems
Teaching electric circuits

1 Introduction

New technologies emerged in everyday education many years ago and still remain a valuable tool for instruction [1]. Their role and contribution to teaching and learning has been investigated extensively and it has been found that their use enhances motivation, class interactivity, students' active participation in the lesson, facilitation of student's learning, leading to better educational results [2]. Research findings reveal that Information and Communications Technology (ICT) plays an important role in enabling teachers and students to enhance the educational level and communicate with one another across the globe. This is because they break down borders and barriers at a faster rate than is possible in physical terms, offering authentic material and vast resources for all types of skills (at various levels) and comprise a valuable tool to enhance teaching and learning. More specifically, for teachers, ICT is a professional resource, a mode of classroom information delivery, while for learners ICT provides opportunities to communicate more effectively.

M. E. Auer et al. (eds.), *Teaching and Learning in a Digital World*,
Advances in Intelligent Systems and Computing 715,
https://doi.org/10.1007/978-3-319-73210-7_70

Within ICT, Classroom Response Systems (CRS) have been lately recognised as a distinct technology system, which allows the implementation of active and cooperative learning, creating a dynamic classroom environment where learning can be fun for all the participants [3]. According to literature, CRSs tend to cover for scarcity of traditional lectures, engage students in peer discussion, and offer attitude gains together with efficient learning outcomes [4]. Additionally CRSs present advantages like the ability they offer to teachers to display the answers of a questionnaire for all the students in the classroom to see, thus motivating them to have a challenging and fruitful discussion with one another [5]. With some of the CRSs it is also possible for students to have access to online materials for their studying and preparation for their lessons [6]. What is more, they offer an enjoyable environment which students like to engage as the majority is usually familiar with all sorts of ICT applications in their lives.

Among CRSs systems Kahoot is an on-line free platform that is known for its ability to enhance interactivity in the classroom through the making of quizzes or surveys, created by the teacher for revision or formative and final assessment purposes. This well-known CRS application and game-based on-line platform offers both a CRS function while it also maintains its core value of being a game [7]. Kahoot is used to enhance classroom participation and assess learners' cognitive level in the form of a game. Its ability to be a free online tool is an advantage and allows the implementation of quick quizzes to assess students' knowledge in real time. The quizzes of Kahoot are displayed on a computer for the whole classroom or teams or pairs, and students respond to the application downloaded on their computers, smart phones, or tablets [8]. The teacher can even add videos or pictures to accompany the questions and facilitate learning. The quizzes are answered by learners in a form of a game and students answer them in real time through an easy-to-use interface, enabling the teacher to assess their progress and the extent to which they have reached the desirable cognitive level [9]. According to literature, Kahoot has been found to be a useful learning tool creating a nice diversion during class [10]. It has also been found that as part of a multi-dimensional gamified learning approach, Kahoot has successfully achieved the pedagogical goals that had been set [11] by teachers in the classroom. Particularly, it has been reported that Kahoot has given students and teachers the opportunity to experience how game mechanics are used to make learning fun, but also has provided self-assessment opportunities through a fun and engaging atmosphere. According to other researchers, [12] Kahoot has increased students' attendance, enthusiasm, confidence, and in-class participation and at the same time has boosted students engagement, motivation and learning after having been used in the classroom.

2 Rationale of the Present Study

Over the last years, learners' ability to handle the various applications with great easiness has been remarkably improved be that for recreational or educational purposes. This gadget oriented younger generation is so skilled in ICT use that all too often their skills even surpass the ICT skills of their own teachers. Among the many ICT applications Classroom Response Systems (CRS) are considered a digital way to support learning.

The well-known game-based on-line platform Kahoot offers such a CRS function while maintaining its core value of being a game. Kahoot is used to enhance classroom participation and assess learners' cognitive level in the form of a game. Driven by the above issues the researchers decided to try the use of Kahoot as a creative tool in order to check on learners' understanding of electric circuits. The choice of the particular subject and topic was decided due to the fact that during the past years, a large number of research efforts on science teaching have been reported concerning electricity [13]. Their main interest seems to be the investigation of students' ideas (or misconceptions), the study of students' reasoning and comprehension, and the methods for overcoming any intellectual difficulties to approach scientific thinking. A long-established and most important educational research result constitutes the ascertainment that the students use alternative models, with the help of which they mediate and try to comprehend electric phenomena and everyday electrical applications [14]. These alternative ideas often remain unchanged or only partly modified, even after many years of repeated teaching at a theoretical or experimental level, throughout formal education [15]. To this end, to detect and confront them is of great interest to science education researchers, and especially so for those relating to electricity as it is important for Science and everyday life alike.

3 The Research

3.1 The Purpose of the Research

The purpose of the research was to measure the extent to which the popular game-based online platform of Kahoot can be used as a creative and effective tool in the teaching practice and specifically in the teaching of basic concepts of electric circuits. The researchers proceeded to do this by direct comparison between: (a) an experimental group of learners who used the online digital Kahoot platform as a tool for the checking of students' understanding on the concepts of electrical circuits, and (b) a control group of learners who did not use Kahoot while studying the same subject.

3.2 Research Question - The Sample

The research question was as follows: To which extent can Kahoot be integrated in the teaching process and be applied as a creative tool in the classroom in order to improve students' comprehension concerning various fundamental concepts of electric circuits (Kirchhoff's law, Ohm's law, resistive electric circuits and electric power)?

A number of 67 Upper Secondary School learners, aged 16–17 participated in the research, forming two groups: the experimental group of 41 students that used Kahoot in order to design their own questions followed by answers on electric circuits and which were used to check their understanding on these concepts and the control group of 26 learners who did not use Kahoot as a tool in the classroom.

3.3 The Research Tools

The study used both quantitative and qualitative research methods. Learners' assessment was conducted with a pre/post-test using the D.I.R.E.C.T. questionnaire (Determining and Interpreting Resistive Electric Circuits Concepts Test) (Version 1.0.) [16], which comprises 29 questions and which were answered by the experimental group through Kahoot. The test was created to assess the extent to which learners could comprehend the DC electric circuits of continuous current and detect possible difficulties or misinterpretations of concepts regarding these circuits. A wide variety of researchers who used D.I.R.E.C.T. [17, 18] describe it as a very effective assessment tool which detects the extent to which learners of all levels comprehend the electric circuits, eliciting learners' alternative ideas.

The aforementioned test was verified regarding its validity and reliability during a pilot-phase testing. The teaching objectives of the specific questionnaire used are presented grouped in various categories, in Table 1 below.

Table 1. Categories of teaching objectives for the taught subject

Categories of teaching objectives taught	Question number	Teaching objectives
O1. Physical aspects of DC circuits	10, 19, 27, 9, 18, 27, 5, 14, 23, 4, 13, 22	To understand the function of the elements of the circuit and how an element is connected to the electric circuit
O2. Potential difference	12, 13, 16	To understand and apply the concept of potential difference in electric circuits
O3. Energy – Electric power	7, 14	To apply the concept of power. To apply a conceptual understanding of conservation of energy including Kirchhoff's loop rule and the battery as a source of energy
O4. Electric DC	5, 9, 11, 15	To understand and apply the maintenance of electric circuit in various electric circuits. To understand and apply Ohm's law

In addition to the questionnaire, informal conversations were employed a posteriori to clarify and ensure firmness of opinion as regards the use of Kahoot as a tool by the students and give the opportunity to the researchers to have a clearer picture of the educational research issues.

3.4 Research Stages

The intervention took place in five successive phases and in nine teaching hours. More specifically:

First phase: During the first phase (one hour) both groups of learners' alternative ideas (experimental and control) were detected with a pre D.I.R.E.C.T. test which was appropriately adapted to their level.

Second phase: The second phase lasted for four hours of two sessions per two hours each. In this, both groups were taught a particular unit of their science course, regarding concepts that referred to the features of DC circuits, electric current, potential difference and electric power. The teaching was realized following the same teaching process and methodology for both groups.

Third phase: This phase lasted for two hours. Before the beginning of this phase and in order for the teacher to check on the understanding of the students on the taught concepts he first asked both groups to revise the material. Then, during the first hour of this phase the control group worked on worksheets with exercises and questions given by their teacher and on the second hour there was a conversation in the classroom regarding the concepts that had not been understood by the learners as a form of feedback. The experimental group however, before the beginning of this phase, was asked to create their own questionnaire: each student was asked to design three to five questions of his/her own, accompanied by the answers and give them to the teacher who went through them, omitted the questions that were similar or the same, checked their accuracy and coherence and uploaded them on the Kahoot platform. Then, during the first hour of the third phase students answered the questions that they themselves had already designed and which were displayed on Kahoot in a random order, using their mobile phones to give the answers and competing with one another in a form of a game. During the second hour, feedback was given by the teacher who had a clear and immediate picture of each one of the students' answers on his screen and therefore, was able to focus on the exact vague points and misunderstanding of each learner.

Fourth phase: During the fourth phase (one hour) semi-structured conversations were conducted with the learners for the evaluation of the whole process.

Fifth phase: Two weeks later, all learners took the same test (one hour) as a post-test in order to detect any changes on their initial ideas regarding the teaching subject.

4 Results

4.1 Statistical Analysis-Data Analysis

The use of the appropriate checking criterion (parametric or not) between research hypotheses depends mainly on the plan of the research, the commitment of the level of data, and the type of the indices of the measurement of the variables. To analyse the data obtained presently, the IBM-SPSS statistical package was used, and a t-test for independent and another for dependent samples were performed. For the purpose of the present study, the level of significance was set at 5%. The research hypotheses are:

- H0: Null hypothesis: The participant groups of learners display the same performance after the teaching intervention.
- H1: Alternative hypothesis: The participant groups of learners have displayed different performance between them, after the teaching intervention.

It should be noted that, in H1, there is no intrinsic attempt to predict which group displays the best or worst performance. Therefore, a two-sided checking of hypotheses is formulated. The results are presented below.

4.2 Discussion of the Results

It is reminded that the participant groups were two: group G1 consisting of 41 learners working with Kahoot; group G2 comprising 26 students who did not use Kahoot. The comparison process of the two groups' performance entails four basic stages of checking. (a) Checking per group and between groups, regarding pre and post instructive aim O1. (b) As above but for aim O2. (c) As above but now for aim O3. (d) As above but for aim O4.

Before the teaching interventions (pre-test) the results of the pre-test for independent samples as regards their average overall performance showed non-significant statistical results [(t (65) = 0.259, p = 0.796], a fact which implies the equivalence of the groups regarding their initial cognitive level and the teaching subject. However, after the intervention (post-test), the data revealed that the learners who played with Kahoot had better results (59.93% ± 10.09%) in comparison to those who did not use Kahoot for the checking of their understanding (51.72% ± 8.44%). The statistical significant difference between the two groups is (t (65) = 3.329, p = 0.001). More specifically, as regards the teaching sub-targets the t-test showed a statistical significant difference in the teaching target regarding the Physical aspects of DC between groups (t (65) = 2.330, p = 0.023) as well as on the teaching target concerning the concept of Energy - Electric Power (t (65) = 3.245, p = 0.002), whereas there is no statistically significant difference in the teaching target concerning the potential difference (t (65) = 0.951, p = 0.345) and Electric DC (t (65) = 0.859, p = 0.380). This allows us to assume that the use of the game-based online tool of Kahoot has improved students' learning regarding the specific topic taught and the specific age group addressed as opposed to the group which worked traditionally. In more detail:

Checking groups G1 and G2 before the teaching intervention (Pre – testing) for all individual teaching objectives.

The following Table 2 represents the t – test of independent samples.

Table 2. Results of t-test for independent samples at PRE-testing level for all individual teaching objectives

	Mean difference	Std. error difference	t	Sig. (2-tailed)
PERFORMANCE_PRE - pre-test	−.823	3.177	−.259	.796
Physical aspects of DC circuits- pre-test	−5.679	3.712	−1.530	.131
Potential difference - voltage-pre-test	−2.833	4.659	−.608	.545
Energy-electric power-pre-test	3.599	5.597	.643	.522
Electric DC- pre-test	4.784	5.415	.883	.380

This is based on the t-test, obtained for all individual teaching objectives which correspond to a (pre-determined) non-significant statistical ($p > 0.05$) result. This leads to the acceptance of the null hypothesis, meaning that the performance of learners in group G1 does not differ from that of the learners in group G2, before the teaching intervention (μ01TOTAL = μ02TOTAL), for each individual teaching objective. One can therefore proceed with the rest of the comparisons. We can reach the same conclusion using a chart (error chart 1) with the intervals of confidence set at 95% of the mean of each group's performance (Fig. 1).

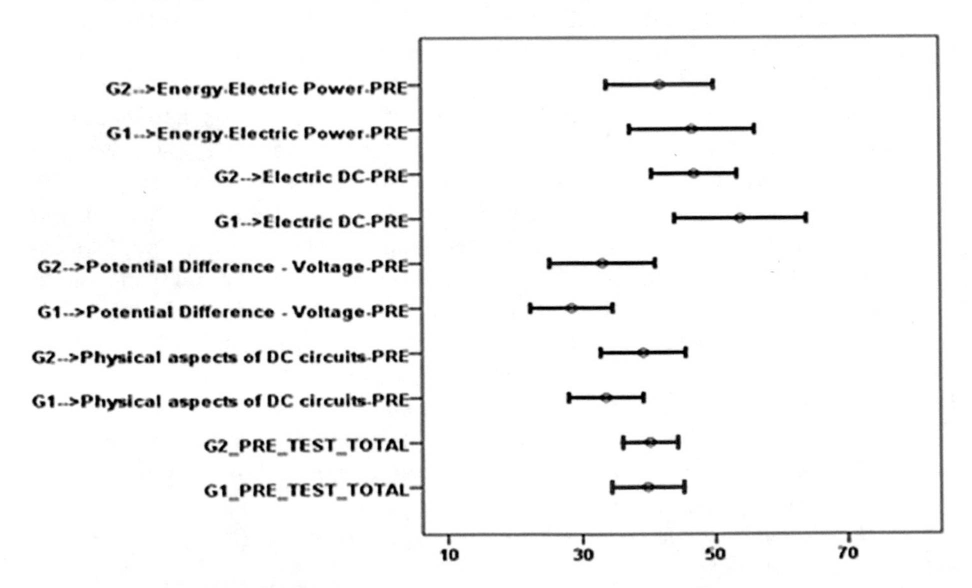

Fig. 1. Error chart for the participant groups in the research at pre-test

From the above diagrams, one can deduce that there is no statistically significant difference between groups at the pre-test stage, as the corresponding overlaps are larger than half of the mean error margin. This means that, since the two groups were indistinguishable, all further testing is therefore valid.

Checking groups G1 and G2 after the teaching intervention (Post – testing) for all individual teaching objectives.

The following Table 3 represents the t – test of independent samples

Table 3. The table with the descriptive indexes of the dependent variable (performance) at the two conditions of the independent variable at post-testing stage

	N	Mean	Std. deviation	Std. error
G1: experimental group	41	59.63	10.09	1.57
G2: control group	26	51.72	8.44	1.65

After the intervention the results of the post-test revealed that the learners who used Kahoot (G1) had somewhat better results (59.63% ± 10.09%) than those who did not (G2) (51.72% ± 8.44%) (Table 4).

Table 4. Results of t-test for independent samples at POST-testing for all individual teaching objectives

	Mean difference	Std. error difference	t	Sig. (2-tailed)
PERFORMANCE_Post-test	7.906	2.374	3.329	0.001
Physical aspects of DC circuits- post -test	8.068	3.462	2.330	0.023
Potential difference - voltage- post -test	5.196	5.465	0.951	0.345
Energy-electric power- post -test	4.221	4.913	3.245	0.002
Electric DC- post -test	16.153	4.976	0.859	0.380

Further statistical analysis using the t-test of the results, showed a statistically significant difference at the instructive goal concerning the Physical aspects of DC circuits – Post-test, i.e. [(t(65) = 3.329, p = 0.001] between the groups and in the instructive goal concerning the Energy-Electric Power- Post-test, i.e. [t(65) = 3.245, p = 0.001. This allows one to assume that the use of Kahoot has improved learning of the specific concepts, and for this specific age group, when compared to the group who worked in a traditional way (Fig. 2).

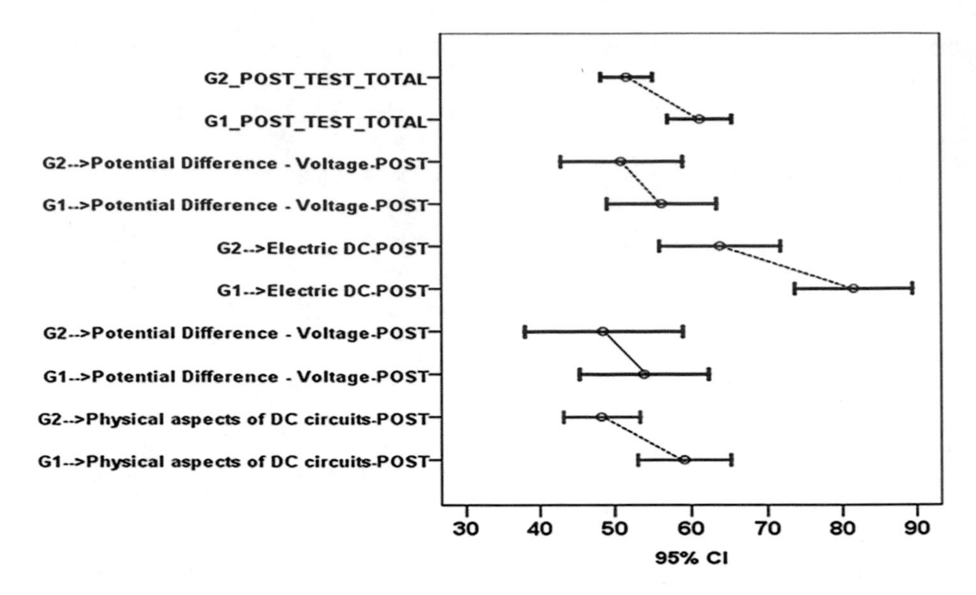

Fig. 2. Error chart for the participant groups in the research at post–test

From the above chart, one can support the view that there appears to be an overall statistically significant difference between groups at post-test stage for the concepts of electric DC and Physical aspects of DC circuits, as the corresponding overlaps are not bigger than half the mean of the average margin of error. It is worth mentioning that our informal conversations with the students after the teaching interventions revealed that their evident enthusiasm for the use of Kahoot throughout the whole process. Reasons given included the motivating features of the tool, its ability to activate them, the possibility they had to compete with one another, the possibility they had to work on the making of their own quizzes, and the interactivity features of the tool.

5 Conclusions/Recommentions/Summary

In this research the aim was to detect the extent to which the games-based learning system of Kahoot can be used as a creative and effective tool for teaching and learning purposes. For this reason Kahoot was used for the checking of students' understanding on concepts of electric circuits. Two groups participated in the research and were compared with one another with one of them using Kahoot. The analysis of students' answers at pre-test of both experimental and control groups before the teaching intervention revealed that there was no statistically significant difference in all four teaching targets of the test D.I.R.E.C.T. (O1. Physical aspects of DC, O2. Potential difference-Voltage, O3. Energy – Electric Power O4. Electric DC) showing that the participant groups initially had a similar cognitive level in electric circuits. Additionally, a big percentage of the students had misconceptions regarding electric concepts, potential difference, and intensity of electric current and electrical resistive circuits, which complicated their scientific comprehension.

However, after the teaching intervention, and for the experimental group who used Kahoot as a tool for the checking of the specific concepts understanding, a statistical significant difference was found in the teaching target regarding the Physical aspects of DC between groups ($t(65) = 2.330$, $p = 0.023$) as well as on the teaching target concerning the concept of Energy - Electric Power ($t(65) = 3.245$, $p = 0.002$).

In this research a trial was attempted, namely to integrate the game-based Kahoot platform in the teaching process, as a simple, creative, and easy tool for the purposes of completing a live questionnaire, aiming to collect efficiently and directly the learning results. According to the data, the use of the specific tool has enhanced learners' interest in the teaching, has improved their understanding of the concepts taught, has motivated them to actively participate in the teaching process and thus, achieving better educational results. Students were enthusiastic while using it, they found its use a fun process, were not bored, and in a way they were enticed (coaxed, even) to study more thoroughly, as they had to create their own questions for all the class to answer. According to the students, their active engagement in constructing their own questions to be uploaded on Kahoot to answer as part of a game and class competition, was a very exciting and innovative process. This excitement was obvious in their persistent wish to use the tool in other lessons as well. The use of Kahoot as an additional teaching tool also had pedagogical implications: on one hand it boosted learners' self-esteem since they could always provide the right answer when their classmates got it wrong (as they had studied all about it before they were able to pose the question in the first place). Furthermore, this gave them the opportunity to change roles with the teacher, who was now the coordinator and facilitator, whereas the learners were the designers of their own questionnaire, thus enhancing their creativity, while being the feedback providers to those with wrong answers. What is more, the teacher had the opportunity to have direct feedback and immediate access to each one of the students' answers, thus, having a clear picture of their problems and focusing on their exact difficulties and misunderstandings. Most of all however, the use of Kahoot has proved to assist learners' learning, and to introduce them into one of the most important life skills: the skill of learning how to learn become autonomous and self-regulated students, and citizens to be. It is true that teachers are surrounded by hundreds of proposed ideas, tools and methods to help their work in the classroom, forever hoping to find the magic solution in order to arouse their learner's interest and reach the desirable outcomes. However, it is in the hands of the teachers to choose and use any tool the best way possible for their own needs and aims, that will serve their own specific purposes and this differentiation is the real magic of it. The researchers are planning to continue exploring the use of Kahoot in other subjects as well and/or for other types of assessment with students, perhaps working in pairs, or forming groups.

References

1. Sharma, A., Gandhar, K., Sharma, S., Seema, S.: Role of ICT in the process of teaching and learning. J. Educ. Pract. **2**(5), 1–5 (2011)
2. Wendy, M.: Not just tools: the role of E-technologies in culture of learning. Educ. Commun. Inf. **1**(2), 229–235 (2001)

3. Lucke, T., Keyssner, U., Dunn, P.: The use of a classroom response system to more effectively flip the classroom. In: IEEE Frontiers in Education Conference, October 2013, pp. 491–495 (2013)
4. Duncan, D.: Clickers in the Classroom. Pearson, San Francisco (2004)
5. Owusu, A., Weatherby, N., Otto, S., Kang, M.: Validation of a classroom response system for use with a health risk assessment survey. Poster session at the 2007 AAHPERD National Convention and Exposition, Baltimore, Maryland (2007)
6. Abramson, D., Pietroszek, K., Chinaei, L., Lank, E., Terry, M.: Classroom response systems in higher education: meeting user needs with NetClick. In: IEEE Global Engineering Education Conference (EDUCON), March 2013, pp. 840–846 (2013)
7. Collins, K.: Kahoot! is gamifying the classroom (2015). http://www.wired.co.uk/article/kahoot-gaming-education-platform-norway
8. Diaz, C., Trejo, C.: Kahoot: The Student-Teacher Interactive Classroom Tool (2015)
9. Meijen, C.: Kahoot: using a game based classroom response system in teaching. School of Sport and Exercise Science, University of Kent (2015)
10. Sunde, M.T., Underdal, A.G.: Investigating QoE in a cloud-based classroom response system, a real-life longitudinal and cross-sectional study of Kahoot. Master of Science in Communication Technology, Norwegian University of Science and Technology, Norway (2014)
11. Fotaris, P., Mastoras, T., Leinfellner, R., Rosunally, Y.: Climbing up the leaderboard: an empirical study of applying gamification techniques to a computer programming class. Electron. J. e-Learn. **14**(2), 94–110 (2016)
12. Wang, A.I.: The wear out effect of a game-based student response system. Comput. Educ. **82**(C), 1–24 (2015)
13. Driver, R.: Students' conceptions and the learning of science. Int. J. Sci. Educ. **11**(5), 481–490 (1989)
14. Shipstone, D.M.: A study of children's understanding of electricity in simple DC circuits. Eur. J. Sci. Educ. **6**(2), 185–198 (1984)
15. Psillos, D., Koumaras, P., Valassiades, O.: Pupils' representations of electric current before, during, and after instruction on DC circuits. Res. Sci. Technol. Educ. **5**(2), 185–199 (1987)
16. Engelhardt, P.V., Beichner, R.J.: Students' understanding of direct current resistive electrical circuits. Am. J. Phys. **72**(1), 98–115 (2003)
17. Baser, M.: Effects on conceptual change and traditional confirmatory simulations on pre-service teachers' understanding of direct current circuits. J. Sci. Educ. Technol. **15**(5), 367–381 (2006)
18. Moore, C.J., Rubbo, L.J.: Scientific reasoning abilities of non-science majors in physics-based courses. Phys. Rev. STPER, **8**(1), 1–18 (2011)

Concept Proposal for Integrating Awareness of Sustainability Through Student Assignments

Monika Dávideková(✉) and Michal Greguš ml.

Faculty of Management, Comenius University, Bratislava, Slovakia
{monika.davidekova,michal.gregusml}@fm.uniba.sk

Abstract. Globalization has changed the economy of the whole world: today, products are being transported all over the globe to arrive at the place of sale whereas before many articles, food and drink products in particular, were mostly supplied by nearby producers. Goods often cover distances over the whole continent or even across oceans by travelling to the destined store. This logistic network creates jobs and commercial connections by transporting the product supply to its demand. It enables the availability of various articles everywhere at any season of the year not obtainable otherwise. However, the transport with all its attributes are impacting the nature. These adverse effects are often not included in the calculated end price – occurrence called tragedy of commons. The aim of this paper is to propose a concept of assignments in a course intending to enhance students' awareness of the impact of industrial globalization on nature to include sustainability in their consumer behavior.

Keywords: Teaching sustainability · Assignment concept model · Globalization Impact of industry on nature · Consumer behavior

1 Introduction

Since the earliest signs of human existence, people are interacting with each other in communication and trade. When trading (liberal trade), individuals can specialize in sectors in which they enjoy comparative or competitive advantage to the greatest extend [1] in order to maximize collective assets and gains. In current global environment raw materials necessary for production of a good are extracted in various places. These are then carried to the place of production and through assembling and mixing diverse components together goods are produced in another place. These products are subsequently transported to numerous places of sale to reach customers around the world. The global marketplace has become a vast hunting ground for the best deals [2] and lowest price. Consumers look for the product of their interest in local stores and search online through several web pages to find the best deal – to get it for the lowest price possible, not taking other aspects like for example nature pollution into consideration. However, even if there are often substitutional products for the same price and of the same quality, but from other places of origin or delivered by diverse suppliers, the local products are not selected. Consumers often place emphasis on the attractiveness and

M. E. Auer et al. (eds.), *Teaching and Learning in a Digital World*,
Advances in Intelligent Systems and Computing 715,
https://doi.org/10.1007/978-3-319-73210-7_71

appearance of the packaging [3]. The availability of all these products and articles is enabled by transport industry and logistic developed due to globalization and provided through interconnected global trade network.

On one hand, transport has always been and remained one of the main driving forces in the economic development of any country and plays a crucial role in economic development. On the other hand, the growth of transport has had a significant impact on congestion, safety and pollution [4]. The main source of energy in transport industry represents the fossil fuel burning that together with increased deforestation represent the main drivers for the continuously increasing level of carbon dioxide and other pollutants in the atmosphere for the last 55 years [5]. Besides, the road transport and the water transport are causing irreversible damage to the nature, too. Ships in tropical waters have been associated with severe long-term damages to coral reefs, seagrass and endogenous animals [6]. Among modes of transport (road, sea, rail, air), aviation counts for the one with the highest carbon intensity [7] as it burns huge volumes of fossil fuel in turbine engines.

Shopping and logistic have many things in common. One of those is the packaging industry producing plastics for transportation of products. The plastic pollution became ubiquitous not only overland, but also in the marine environment [8], representing a great danger to animals, and demanding very long time to decompose.

Globalization is not only about transport and shopping. It also creates jobs for workforce across the planet enabled by ICT. For cost reduction, many products are being produced in poor and developing countries with low cost labor: ICT industry, clothing industry, packing covers, diverse appliances, chemicals, etc. In those countries child labor is often being used [9], where children work in hard working conditions whereas the young generation in developed countries spends a lot of time checking their social network profiles and chatting [10]. Adolescents and young adults are very often unaware of wealth and good conditions in which they live and take them for granted.

The trade and business world is mostly operating with focusing on maximizing economic gains and profit whereas the "real price" paid by the society and the Earth is often not included in the monetary value expressed in numbers and currency. The price for production, transport, sale and recycling is paid also by surroundings: the nature and other present and future human generations across the world that is not always visible at a glance. To think about all those impacts, more information is necessary.

This paper provides a proof of concept for student assignments dealing with globalization and its impacts on nature that are not fully obvious to an average consumer at time of shopping. It aims to draw attention of young generation towards the implicit side effects of globalization intending to exploit sustainability in our consuming behavior. This paper outlines possible assignments for students. The main idea is that during processing and having elaborated the assigned tasks, students may better evaluate their needs and demands for a product or service by comparing the price for the offered product or service to be paid by them and the price paid by the nature through the consequences of globalization on one hand and the perceived real need on the other hand.

This paper is organized as follows: next section suggests information to be included in proposed assignments. Section 3 provides a possible use case scenario. Conclusion outlines areas of future research.

2 Assignment Composition

During diverse courses of academic study, students elaborate various assignments dealing with the topic of the course. Assignments proposed in this paper deal with economic aspects as well as aspects of globalization, transport, packaging, product life cycle, recycling etc. impacting nature and society. There are several goals to be rcached by these assignments, not limited to:

- Awareness of CO_2 pollution and O_2 consumption by transport of given product to the place of sale in comparison with the ability of the nature to obtain O_2 from CO_2 again and the consumption of O_2 necessary for breathing of humans and animals. In other words, how many hours of life or how many human lives are being given in return for unnecessary transportation.
- Awareness of the amount of plastics and waste used for the packing of products and the time necessary for the decomposition of packings; the increasing amount of plastic polluting the world and the continuously raising occurrences of damages to life beings (animals, human, earth, etc.) due to it. All dead animals represent not only animal loss, but also extinction danger to endogenous animal species, loss of possible source of food for people and/or other animals, etc. This part may include also the calculation of such loss in the number of food portions that could feed one person and not only the losses expressed in numbers and currency. At the same time, such assignment can contain calculations describing the costs of collecting and recycling the packaging pollution that is mostly not done. Students shall elaborate possible alternatives of packaging utilizing easily recyclable materials or resources that are easily decomposable in the nature including economic calculations that may be induced if replacing the current plastic packaging material with alternatives they would propose.
- Awareness of life conditions in the production country. This part should cover the comparing overview of hourly wage of that country compared to the local salaries. It shall be calculated as the time a person must work in order to purchase the particular product. It aims to increase the awareness of the value of human time and the spare time a human being has at disposal in various countries. This part shall deal with the level of child labor used in the production country to estimate statistically how many products might have been produced by a child to show that instead of enjoying his/her free time playing childish games with friends, the child has to work. This may provide insights supporting the development of tolerance towards foreigners coming from other countries including poor and developed states.
- Awareness of available substitutions being sold at the same store or in the same city location. The enumeration of alternatives enhances the awareness towards unnecessary overused transportation as it is not always necessary to get products transported across the world when there is already a producer nearby offering substitute products

at almost the same price. This part of the assignment shall demonstrate that the consumer behavior creates the demand and that only with high demand the supply raises.

- Draw their attention towards collecting information on ingredient quality and origin e.g. for food and beverages.
- Support and develop the ability to search for reliable sources, to work with scientific databases and to conduct research.

There are various aspects that may be covered in such assignments designed for groups or as individual student tasks. The above mentioned list denotes only guiding aspects intending to raise the sustainability awareness of students during the elaboration of such assignments.

3 Use Case Scenario

For the convenience of the reader, this section provides a typical use case scenario in an academic course.

A lecturer of a course introduces assignments to students at the start of the course and assigns specific articles to each student or student group so, that each elaborating unit focuses on unique product. Eventually, the assignment and picking of products can be made on "first come, first served" principle by creating an open pool where all elaborating units enter their good of interest. As a rule, each unit may register only for one item. This might be done via pooling tools e.g. Doodle. Items may denote e.g. various food or beverage products that can be bought at local supermarket. The assignments are possible also for analyzing cloths, appliances etc. However, food industry usually covers the highest number of substitutions for a product including fruits and vegetables that are usually transported and produced all around the world.

Each elaborating unit visits local store and identifies articles (e.g. various brands of yoghurt) of the given item group (e.g. yoghurt) offered and sold in given local store. Then the unit analyzes the places of production/origin for particular good (e.g. $product_1$ produced in $city_1$ $country_1$... $product_n$ produced in $city_n$ $country_n$) including price differences of these products. After that, it selects two products, one local and one delivered from far away, to conduct a calculation. The current list and the selection are sent to the lecturer for approval who reads the conducted analysis and the proposed choice and approves or corrects it in case of need.

The elaborating unit firstly calculates the differences in the transport for those two products by assuming the shortest possible route that is available for cargo transport e.g. using Google Maps. Then it calculates the volume of O_2 consumed and CO_2 produced by the transporting vehicle (truck, ship, train, or plane) by using the internet and scientific databases to search for those data. Next important part of this step is the calculation of hours a tree, flower or other plant needs to produce oxygen by turning the CO_2 produced during transport into O_2 back again. It shall also express this amount of oxygen in days/ hours an average human consumes for breathing by producing equal volumes of O_2 and CO_2. This comparison might be expressed in a percentage of an average human life for better demonstrativeness. This serves for better imagination and comprehension of the volumes and numbers.

During the second step, the elaborating team may analyze the packaging of the product, identify the time needed for its decomposition in the nature and possibilities of reuse and recycling. This analysis shall contain the level of pollution in nature e.g. in case of plastics. This part of the assignment deals with the proposal of alternative packaging solutions and their calculations. Again, the data may be found in the Internet including scientific sources to obtain the necessary information.

The following step in the assignment is the comparison of wages in local country/ city and the location of production/origin to analyze the differences in living conditions of people including their working and spare time life balancc. Important aspect deals with thc comparison of time a person has to work to purchase the product/good. The analysis of living conditions covers also the possible inclusion of child labor in the production country. It shall assume the probability that the given product might have been produced by a child. This may increase the awareness and social tolerance towards the country wealth or poverty often taken for granted and influence living conditions.

One of the last tasks in the assignment deals with the comparison of available alternatives and the price differences of available products that may equally satisfy the consumer needs. At this point, the difference in price of products calculated for the two products selected at the beginning shall be expressed not only in the purchasing "price" but also by the "price" paid by the nature identified, analyzed and calculated in previous stages of the assignment. The elaborators shall evaluate the price difference in the end products and the impacts on the nature and society that are not included in the purchasing price.

At the end of the course, each elaborating unit presents their findings to the whole group and writes a project describing all used sources of information.

4 Conclusions

The development of transport industry and the emergence of ICT have been the main drivers of globalization. Global collaboration in trade, science and other areas are driving the development of the society and ensure the availability of products from few locations of production to all consumers spread across the whole world. Everyone deals with products from all over the world during shopping and purchases in their everyday life. A consumer very often buys a good from other continent even if there are local producers of substitutional products. The purchasing price of the good not always includes everything used for its production and delivery. The supply follows the demand and the consumer decides which good he/she purchases, therefore each single person can decide on how he/she can contribute to sustainable trade by being aware of the price paid by the nature and the society to enable the production, delivery and recycling of a particular good including its packing.

This paper has proposed a concept of possible assignments for students in academic courses that aim and may lead to increase of their awareness regarding the impacts of globalization on nature and society. The proposed concept outlined possible aspects that might be included in such assignments to be elaborated by students by utilizing available information sources (internet, books, scientific databases, etc.). This can also support the development of their skills in the field of conducting research.

Acknowledgements. The support of the Faculty of Management, Comenius University in Bratislava, Slovakia is gratefully acknowledged.

References

1. Grossman, G.M., Krueger, A.B.: Environmental impacts of a North American free trade agreement (No. w3914). National Bureau of Economic Research (1991)
2. Roy, S.K., Charaborti, R.: Case study 5: Amazon.in: surviving in a jungle. Services marketing cases in emerging markets. In: An Asian Perspective, Part 1, pp. 45–59. Springer International Publishing, New York (2017)
3. Mullan, M., McDowell, D.: Modified atmosphere packaging. In: Coles, R., McDowell, D., Kirwan, M.J. (eds.) Food Packaging Technology, vol. 5, pp. 303–339. CRC Press, Oxford (2003)
4. Mačiulis, A., Vasiliauskas, A.V., Jakubauskas, G.: The impact of transport on the competitiveness of national economy. Transport **24**(2), 93–99 (2009)
5. Baker, D.F., Law, R.M., Gurney, K.R., Rayner, P., Peylin, P., Denning, A.S., Bousquet, P., Bruhwiler, L., Chen, Y.-H., Ciais, P., Fung, I.Y., Heimann, M., John, J., Maki, T., Maksyutov, S., Masarie, K., Prather, M., Pak, B., Taguchi, S., Zhu, Z.: TransCom 3 inversion intercomparison: impact of transport model errors on the interannual variability of regional CO2 fluxes, 1988–2003. Glob. Biogeochem. Cycles **20**(1), 1–17 (2006). Article no. GB1002
6. Davenport, J., Davenport, J.L.: The impact of tourism and personal leisure transport on coastal environments: a review. Estuar. Coast. Shelf Sci. **67**(1), 280–292 (2006)
7. Corbett, J., Winebrake, J.: Sustainable goods movement: environmental implications of trucks, trains, ships, and planes. EM, A & WMA's Magazine for Environmental Managers, November 2007, p. 8 (2007)
8. Eriksen, M., Lebreton, L.C., Carson, H.S., Thiel, M., Moore, C.J., Borerro, J.C., Galgani, F., Ryan, P.G., Reisser, J.: Plastic pollution in the world's oceans: more than 5 trillion plastic pieces weighing over 250,000 tons afloat at sea. PLoS ONE **9**(12), e111913 (2014)
9. Webbink, E., Smits, J., de Jong, E.: Hidden child labor: determinants of housework and family business work of children in 16 developing countries. World Dev. **40**(3), 631–642 (2012)
10. Porath, S.: Text messaging and teenagers: a review of the literature. J. Res. Cent. Educ. Technol. (RCET) **7**(2), 86–99 (2011)

Interactive Teaching Methods as Human Factors Management Tool in Dangerous Goods Transport on Roads

Jelizaveta Janno(✉) and Ott Koppel

School of Engineering, Tallinn University of Technology, Tallinn, Estonia
jelizaveta@tktk.ee, ott.koppel@ttu.ee

Abstract. This paper studies the methodological essence of ADR regulations training courses for drivers and safety advisers. The aim of research is to advance existing teacher-centred course model in Estonia with learner-centred methods that best suit specific objectives and meet expected learning outcomes. In Estonia, ADR regulations training courses are formed based on teacher-centred course design mainly. This methodological approach is outdated as the concept of learner is changing rapidly. The aim of this research is to make study based proposals, what kind of interactive methodological approach training course model meets the best trainees' expectations in Estonia.

The paper presents a combined development research strategy based on studies regarding ADR regulations training courses in Estonia as well as on analysis of teaching methods applied in professional training of adults. Data collecting on learners' attitude and preferences regarding current methodological format of courses is collected by implementing questionnaires with structured questions from consignors/consignees, freight forwarders carrier companies and drivers. Based on learners' needs and expectations, different interactive teaching methods are examined. Implementing methodology of qualitative comparison analysis (QCA) combination of best suitable teaching methods are identified.

Theoretical outcomes represent detailed review of existing ADR training courses system, training opportunities and so far implemented methods. Empirical outcomes focus on introducing suitable interactive teaching methods within the existing format of ADR regulations training courses. Finally developed ADR training course model with a new learner-centred methodological approach considers all major parties involved into transportation chain of dangerous goods. Further researches related to this issue include discussions with ADR training courses providers and introducing an actual action plan regarding the implementation of new interactive methodological approach of ADR regulations training courses in Estonia. There is also a need for measuring exact impact of new methodological approach on operational risk management.

Keywords: ADR regulations training courses · Interactive teaching methods
Qualitative comparison analysis

M. E. Auer et al. (eds.), *Teaching and Learning in a Digital World*,
Advances in Intelligent Systems and Computing 715,
https://doi.org/10.1007/978-3-319-73210-7_72

1 Introduction

The transportation of dangerous goods (DG) by road involves always risks. If substances are mishandled, injury and property damage risks are increased. From the perspective of road transport this concerns primarily main parties of a transportation chain, *i.e.* consignors/consignees and carrier companies (including drivers), but also freight forwarders, and third parties. A transport containing DG can have a serious impact on the environment if an accident occurs and these often incur a higher cost for the society than non-dangerous goods accidents. This is one reason why it is very important to focus on improving the efficiency and security for DG transport and avoid potential accidents [21].

The content of ADR regulations training is regulated by The European Agreement concerning the International Carriage of Dangerous Goods by Road (ADR). Effective training may affect the safety aspects in peculiar transportations, such as the one of dangerous goods transport (DGT) by road. The role of ADR regulations training courses has an essential impact on the human factors aspect that reveals during DG handling and transportation processes as the human factors are crucial why accidents occur within a transportation chain. Training may not only include regulations, technical and procedural aspects, but also important psychophysical aspects such as how to manage fatigue [3, 19]. The provider of training may be different according to national legislations. It can be the role of the employer (in the US and Canada) to ensure appropriate truck-driver training for the transportation of DG. In Sweden and the Netherlands, as well as in Estonia, a competent national authority must accredit training institutions or trainers and monitor the examination of truck drivers [13]. However, all training system approaches pursue the same goal: to ensure appropriate training and prevent the accidental release of DG during transportation. By implementing specific interactive teaching methods remarkable improvement of course participants' learning can be achieved. Moreover, operational risks related to human factors' issues can be reduced within entire transportation chain of DG.

Problem discussed in scope of this paper is a part of a broader study and refers to outdated methodological approach in carrying out ADR trainings in Estonia, both for drivers and safety advisers. Based on conducted survey research among representatives of different parties of a DG transportation chain in Estonia, best suitable interactive teaching methods are studied. Results can be further implemented in ADR regulations training course model development to be an effective human factor management tool. All this will contribute to improved security and efficiency of DGT by road.

2 ADR Regulations Training Courses

2.1 Literature Review

The global trend of increasing traffic due to globalization leads to a higher number of DGT [8]. Several studies focused especially on the critical analysis of ADR implementation concepts in European countries [*Ibid.*]. Chances and challenges coming along with the ADR ratification were illustrated and the concept/recommended procedures of how to train involved people in the framework of DG was developed on a basis of deep analysis and critics of current training methods.

Specific models, methods and technologies have been also studied in scope of support the training of drivers involved in the transport of DG [5]. Italian developed online training environment (TIP – Transport Integrated Platform) is addressed to operators in the transport sector and combines classroom based training with online self-learning possibilities on a distance. The platform has been continuously upgraded with innovative tools and presents a component of blended learning model where online digital media meets with traditional classroom methods [5, 22]. Implementing blended learning methodology within classes keeps students active not allowing them disconnect from the subject. This leads to a better attitude to improve learners' individual thinking and writing, motivating them for further study and development of new thinking skills [9, 14].

Training of safety and DG topics is very essential for a risk and accident minimization in the handling of DG and their transports. According to previous research studies on DGT the awareness of different parties of transportation chain in Estonia there is a lack of professional knowledge among personnel on the national level [11]. According to comparative analysis of teaching methods of ADR driver training courses of France, the Netherlands and Estonia, remarkable differences were identified [12]. In Estonia a significant lack of learning tools and no ARD based activities to endorse training courses and to increase the proportion of practice are so far in use [*Ibid.*].

Human related risk preventive mean lies in efficient personnel training. In following parts of this paper the methodology of QCA is implemented in order to analyse specific methods as cases due to set of relations and assess their consistency. Existing teacher-centred ADR training model will be completed with appropriate suggestions regarding learner-centred interactive teaching methods that best suit specific objectives and meet expected learning outcomes.

2.2 Background

As DG and their transport need special handling and attention due to their risk for the environment and health of people, the training of any persons having to deal with those goods is very essential for a safe processing [10]. Common legal requirements (ADR) states in details that drivers when transporting DG (with small exceptions) shall undergo training in the form of a course approved by the competent authority. Concerning chapter 1.3 of the ADR, every employee, which has to commit the duties of DG regulations, needs to be specifically trained [1]. Other parties involved within operations with DG can be: manufacturer or owner of DG, owner of tank containers, persons carrying out forwarder duties, persons writing and preparing transport documents, persons working for the DG receiving, persons committing packaging procedures, filling personnel of tanks, vehicle drivers, who do not need an ADR certificate, persons carrying out carrier and vehicle owner duties [2, 15].

Persons mentioned above often carry obligations of dangerous goods safety advisers (DGSA) as they are involved in operations with DG in road transportation. A DGSA is a consultant or an owner or employee of an organization appointed by an organization that transports, loads or unloads DG in the European Union and other countries [20]. There is no specific classification regarding DGSA courses generally. However ADR driver training courses can be classified according to two aspects. See Fig. 1 which visualises the content of training programs and training courses, highlighting common and distinctive elements of ADR driver training courses.

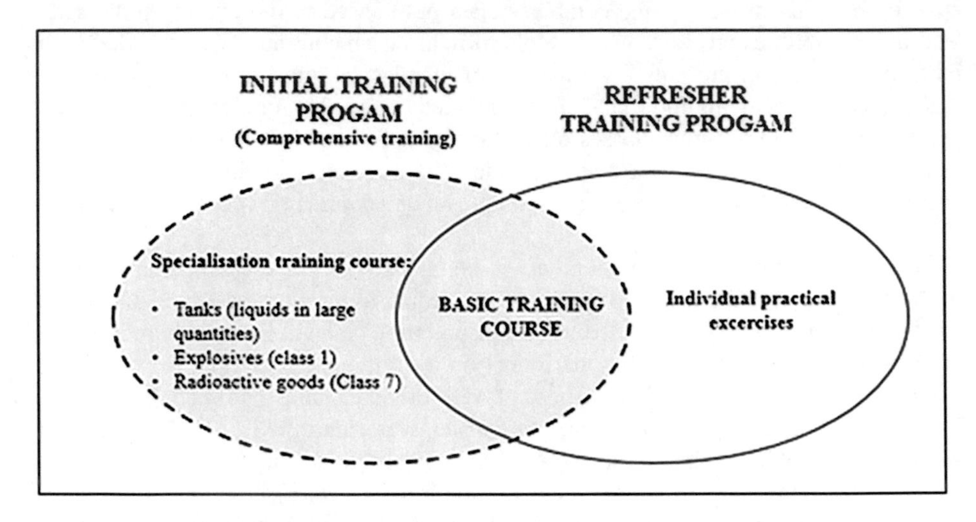

Fig. 1. Content of ADR driver training programs Source: [12]; adapted by authors

Firstly, training programs are identified on the basis of the level of the training program (initial or refresher training program) and secondly, training courses within programs are divided according to specificity (basic or specialisation training course). The minimum duration of theoretical element of each initial training course or part of the comprehensive training course are set according to common legal requirements. The overall duration of the comprehensive training course may be determined by the competent authority, which shall maintain the duration of the basic training course and the specialization training course for tanks, but may supplement it with shortened specialization training courses for Class 1 (explosives) and Class 7 (radioactive materials) [16]. Refresher trainings have to be undertaken by drivers (as well as by DGSAs) at regular intervals in every 5 years. As the form of training program is defined by compulsory topics and minimum learning hours only, it is free to choose the methodological approach to conduct the training itself [12].

2.3 Problem Description

The primary purpose of teaching at any level of education is to bring a fundamental change in the learner [23]. Due to the high risk of DG there is a must to learn before doing in the content of ensuring safety. The ADR implementation and the knowledge transfer concerning DG is complex. In Estonia, ADR regulations training courses are formed based on teacher-centred course design mainly, *i.e.* learning activity is performed during classroom lectures supported by slideshow presentation. ADR regulations training courses are mostly in-class and theoretical proceedings, even in cases, where a practical example would be considered necessary, as in the case of fire confronting and first aid issues. In most cases, in-class training is followed with the use of books, issued by the training organisations/companies, slide presentations and internal tests [12].

Today this methodological approach is clearly outdated as the concept of learner with its needs is changing rapidly. Moreover, existing learning form does not meet efficient risk management within the transportation chain that is evolving more complex due to the number of parties involved as well as due to additional risks concerned new DG and their danger characteristics. The aim of present paper is to perform the analysis and identification of teaching methods suitable to be integrated into existing ADR professional training courses in Estonia with the scope to increase the proportion of practice and thereby to minimize operational risks related to human factors in further studies.

3 Methodology

3.1 Data Collection

A research design is the set of methods and procedures used in collecting and analysing measures of the variables specified in the research problem research study [7]. The research design of this study is defined by the research problem according to which the methodological approach of ADR regulations training courses in Estonia is outdated as the concept of learner is changing rapidly. In scope of this paper data collecting on learners' attitude regarding current format of courses is collected from all main parties who operate with DG on a daily basis, *i.e.* consignor/consignee, freight forwarder and carrier company. Respondents were divided into clusters according to the type of ADR regulation training course type which is aimed at them. Clustering was performed as following:

1. CLUSTER 1 (truck drivers; ADR driver training course),
2. CLUSTER 2 (consignors/consignees, freight forwarders, carrier companies, other participants; ADR DGSA training course).

Truck drivers have been separated from carrier role in order to identify their preferences individually. The main objective is to understand attitudes and preferences by clusters toward specific teaching methods respectively. The essence of specific methods that were focused on were explained to respondents. A structured questionnaire with close-ended ordinal-scale questions has been prepared as main data collecting form, where respondents were asked to decide where they fit along a scale continuum regarding the use of particular teaching method within ADR training classes.

3.2 Data Analysis

Implementing methodology of qualitative comparison analysis (QCA) combinations of suitable teaching methods are identified that are effective both in scope of operational risk management as well as from the perspective of learner's needs and expectations. QCA is a means of analysing the causal contribution of different conditions (*e.g.* aspects of an intervention and the wider context) to an outcome of interest [17]. QCA starts with the documentation of the different configurations of conditions associated with each case of an observed outcome [18]. These are then subject to a minimisation procedure that identifies the simplest set of conditions that can account all the observed outcomes, as well as their absence. Results are typically represented in statements expressed in ordinary language or as Boolean algebra. According to formula (1) expressed in Boolean notation combination of Condition A AND (*) condition B OR (+) a combination of condition C AND (*) condition D will lead to an OUTCOME (→) E [*Ibid.*].

$$A^*B + C^*D \rightarrow E \quad (1)$$

The paper presents a combined development research strategy based on studies regarding ADR regulations training courses in Estonia as well as on analysis of teaching methods applied in professional training of adults.

4 Results

The data collecting on learners' attitude and preferences concerning methodological format of courses was performed during the period from February 3–May 3, 2017. The online survey was prepared using *Google Forms* both in Estonian and in Russian. The distribution of the questionnaire was provided via email invitations (60 companies that work with DG on a daily basis) and social media channels addressed directly to speciality-focused groups (*e.g.* Estonian truck drivers with estimated number of 1800 ADR licenced drivers). Altogether 189 replies were gathered (CLUSTER 1–151 respondents, CLUSTER 2–38 respondents). On the basis of theory the sample must represent the population as well as possible. Current sub-samples are not statistically representative enough to draw accurate conclusions concerning population.[1] To ensure the representativeness, the sub-samplings were formatted in a non-probability sampling technique where the samples are gathered in a process that does not give all the individuals in the population equal chances of being selected [4]. In scope of this study samplings are also qualified as purposive samplings where subjects are chosen to be part of the sample with a specific purpose in mind that sufficient to draw objective conclusions concerning methodological approach of some subjects are more fit for the research compared to other individuals [*Ibid.*]. This is ARD regulations training courses, but is insufficient to give an accurate picture of attitudes and preferences of all DG transportation chain participants in details.

Within the structured questionnaire interactive teaching methods were firstly explained thoroughly and then proposed to be evaluated in contrast to main existing methodological approach today - classroom lecturing with the support of slideshow.

[1] According to the statistics during the period from 2012–2016 (*i.e.* currently valid certificates) the total number of issued ADR driver licenses in Estonia was 30 539 and the number of issued DGSA training certificates during the same period 118 [6, 24].

These methods were selected into the study mainly based on the practice of other countries (*i.e.* France, the Netherlands). See Tables 1 and 2 that present respondents' attitude and preferences by clusters concerning different methods that learners have experienced or are willing to undergo when taking ADR regulations training courses. Results are given in number of respondents and in percentage share of total cluster.

Table 1. Teaching methods evaluation (CLUSTER 1)

Teaching/learning method (Category)	Evaluation scale				
	1 (most inefficient)	2	3	4	5 (most efficient)
E-learning on a distance (A)	54 (36%)	57 (38%)	28 (18%)	6 (4%)	6 (4%)
Peer-learning (B)	29 (19%)	19 (13%)	73 (48%)	21 (14%)	9 (6%)
Practical tasks (C)	28 (19%)	17 (11%)	19 (13%)	40 (26%)	47 (31%)
Solving case studies in groups (D)	23 (15%)	27 (18%)	26 (17%)	35 (23%)	40 (27%)
Watching, analysing teaching videos (E)	28 (19%)	9 (6%)	20 (13%)	48 (32%)	46 (30%)
Reading individually materials (F)	29 (19%)	38 (25%)	34 (23%)	27 (18%)	23 (15%)
Listening to lectures with assistance of slide presentations (G)	19 (13%)	12 (8%)	34 (22%)	71 (47%)	15 (10%)

Source: Authors

Table 2. Teaching methods evaluation (CLUSTER 2)

Teaching/learning method (Category)	Evaluation scale				
	1 (most inefficient)	2	3	4	5 (most efficient)
E-learning on a distance (A)	5 (13%)	10 (26%)	15 (40%)	3 (8%)	5 (13%)
Peer-learning (B)	4 (11%)	7 (18%)	10 (26%)	12 (32%)	5 (13%)
Practical tasks (C)	5 (13%)	3 (8%)	12 (32%)	10 (26%)	8 (21%)
Solving case studies in groups (D)	3 (8%)	6 (16%)	7 (18%)	10 (26%)	12 (32%)
Watching, analysing teaching videos (E)	4 (11%)	6 (16%)	10 (26%)	8 (21%)	10 (26%)
Reading individually materials (F)	20 (52%)	7 (18%)	4 (11%)	4 (11%)	3 (8%)
Listening to lectures with assistance of slide presentations (G)	16 (42%)	5 (13%)	6 (16%)	8 (21%)	3 (8%)

Source: Authors

By implementing QCA methodology best suitable combinations of teaching methods were studied. As learners' operational risks within DG transportation chain differ, as well as expectations toward training courses, two separate truth tables were formed. According to methodological approach categorical variables (conditions) were defined as following: e-learning on a distance (A), peer-learning (B), practical tasks (C), solving case studies in groups (D), *etc.* As a result combinations of conditions A–G were combined that would lead to outcome. Effective methodological approach (outcome W) for ADR regulations training courses for drivers (W1 for CLUSTER 1) and DGSAs (W2 for CLUSTER 2) in Estonia are expressed in Boolean notation below in form of formulas (2) and (3).

$$(C^{*}D^{*}F + B^{*}E^{*}G) - A \rightarrow W1 \tag{2}$$

$$E^*(D^*A + B^*C^*G) - F \rightarrow W2 \quad (3)$$

The results underline that methodological approach differs by learners' category. Empirical results indicates that classical lecturing with the support of slide presentation is still adequate and suitable teaching method concerning drivers training. Learner-centred interactive methods are expected to be implemented within a classroom lessons and individual theoretical learning is clearly outdated with regards to DGSAs training. Hence, interactive methods differ greatly on a national level. Well implemented blended learning methodological approach on example of Italy (TIP) is not suitable for Estonia's case according to results of this study. This leads to the standpoint that the attitude towards possible use of blended learning methodology at this point is clearly underestimated by trainees within ADR regulations training courses. In scope of further research the focus is to study what should be done in order to improve learners' attitude towards interactive teaching methods within ADR regulations training courses system in Estonia and to evaluate the impact of this methodological approach on operational risk management within the transportation chain of DG.

5 Conclusions

There are many prescriptions, which need to be followed by different parties within the transportation chain of DG in order to ensure safe transport and handling operations as well as to minimize operational risks related to human factors. The change in existing teaching practice today regarding ADR training courses is necessary due to many aspects. Due to continuously increasing number of the possible harm to the health of people and the environment in general, it is very important that all parties being involved are trained accordingly.

The implementation of interactive teaching methods focuses on learner during the process allowing training participant to acquire learning outcomes more efficiently. Moreover, it is important to highlight the fact that the first step when developing a new training course framework is to aware all parties involved regarding deficiency of a system. Next action is the model advancing phase which is finally followed by its partial or full implementation in practice. In scope of this paper finally developed ADR training course models propose learner-centred methodological approach with combinations of classical and interactive methods. Further research related to this issue has to consistently keep up with changes and consider new possible operational risks within the transportation chain of DG as well as with changing learner concept.

References

1. ADR. European Agreement Concerning the International Carriage of Dangerous Goods by Road (2017). http://www.unece.org/trans/danger/publi/adr/adr2017/17contentse0.html. Accessed 30 Mar 2017
2. Arnold, D., Isermann, H., Kuhn, A., Tempelmeier, H., Furmans, K.: Handbuch Logistik, vol. 3. Springer, Heidelberg (2008)

3. Arnold, P.K., Hartley, L.R.: Policies and practices of transport companies that promote or hinder the management of driver fatigue. Transp. Res. Part F Traffic Psychol. Behav. **4**(1), 1–17 (2001)
4. Babbie, E.: The Practice of Social Research. Wadsworth Publishing, Belmont (2010)
5. Benza, M., Briata, S., D'Incà, M., Pizzorni, D., Ratto, C., Rovatti; M., Sacile, R.: Models, methods and technologies to support the training of drivers involved in the transport of dangerous goods. In: Proceedings: CISAP4 4th International Conference on Safety & Environment in Process Industry (2010). http://www.aidic.it/CISAP4/webpapers/66Benza.pdf. Accessed 17 Apr 2017
6. Estonian Road Administration. ADR training of drivers. Statistics (2016). https://www.mnt.ee/et/ametist/statistika/juhiload. Accessed 9 May 2017
7. Ghauri, P., Grønhaug, K.: Research Methods in Business Studies: A Practical Guide, 2nd edn. Pearson Education Limited, Financial Times Prentice Hall, London (2002)
8. Gusik, V., Klumpp, M., Westphal, C.: International Comparison of Dangerous Goods Transport and Training Schemes, ild Schriftenreihe Logistikforschung Band 23. Institut für Logistik- & Dienstleistungsmanagement. FOM University of Applied Sciences (2012)
9. Hoffmann, M.H.W.: Fairly certifying competences, objectively assessing creativity. In: Proceedings of 2011 IEEE Global Engineering Education Conference (EDUCON 2011), pp. 270–277 (2011)
10. Klaus, P., Krieger, W.: Gabler Lexikon Logistik: Management logistischer Netzwerke und Flüsse, vol. 4. Springer Fachmedien Wiesbaden (2008)
11. Krasjukova, J.: Perception of dangerous goods in business activity. J. Int. Sci. Publ. Econ. Bus. **5**(2), 234–257 (2011)
12. Krasjukova, J.: Practical output of dangerous goods training on example of Estonia's carriers. In: The 24th Annual Nordic Logistics Research Network Conference (NOFOMA 2012). The University of Turku, Turku University Press (2012)
13. Kuncyté, R., Laberge-Nadeau, C., Crainic, T.G., Read, J.A.: Organization of truck driver training for the transportation of dangerous goods in Europe and North America. Accid. Anal. Prev. **35**, 191–200 (2003)
14. Llobregat-Gómez, N., Mínguez, F., Rosello, M.-D., Sánchez Ruiz, L.M.: Work in progress: blended learning activities development. In: Proceedings of ICL2015 International Conference on Interactive Collaborative Learning (ICL), pp. 79–81 (2015)
15. Matthes, G.: Schulung/Unterweisung nach § 6 GbV und Kapitel 1.3 ADR/RID/IMDG-Code, 7. Mitarbeiterschulung Gefahrgut. ecomed Sicherheit, Landsberg/Lech2008 (2008)
16. Ministry of Economic Affairs and Communications. Qualification requirements, training rules and the training course curriculum for driver carrying dangerous goods. Regulation of Republic of Estonia No. 37 (2013). https://www.riigiteataja.ee/akt/114062016007. Accessed 15 Apr 2017
17. Ragin, C.C.: What is Qualitative Comparative Analysis? NCRM Research Methods Festival 2008 (2008). http://eprints.ncrm.ac.uk/250/1/What_is_QCA.pdf. Accessed 3 May 2017
18. Ragin, C.C., Rihoux, B.: Configurational Comparative Methods: Qualitative Comparative Analysis (QCA) and Related Techniques. Sage, London and Thousand Oaks (2008)
19. Samuel, C., Keren, N., Shelley, M.C., Freeman, S.A.: Frequency analysis of hazardous material transportation incidents as a function of distance from origin to incident location. J. Loss Prevention Process Ind. **22**, 783–790 (2009)
20. Scottish Qualifications Authority, DGSA Administration. Dangerous Goods Safety Advisers, Scottish (2017). http://www.dgsafetyadvisers.org.uk/DGSA/Home/About_DGSA. Accessed 29 Apr 2017

21. Svensson, C.-J., Wang, X.: Secure and Efficient Intermodal Dangerous Goods Transport. Master Degree Project No. 2009:56, Economics and Law, University of Gothenburg School of Business (2009)
22. Staker, H., Horn, M.B.: Classifying K–12 Blended Learning. Innosight Institute (2012). http://www.innosightinstitute.org/innosight/wp-content/uploads/2012/05/Classifying-K-12-blended-learning2.pdf. Accessed 20 Apr 2017
23. Tebabal, A., Kahssay, G.: The effects of student-centered approach in improving students' graphical interpretation skills and conceptual understanding of kinematical motion. Lat. Am. J. Phys. Educ. **5**(2), 374–381 (2011)
24. TTK UAS Open University. DGSA training. Statistics (2017)

The IntersTICES-Type Activity (ITA) and Its Impact on Pre-service Teacher Trainers' e-Learning Culture

Genny Villa(✉)

Université de Montréal, Montreal, Canada
genny.villa@umontreal.ca

Abstract. In today's knowledge-based society, Information and communication technology (ICT) has become a must for teachers and learners. Their pedagogical potential is immense. However, when we consider the common pedagogical practices reported by the research, as well as the small impact observed on learning despite significant material investments, we can only question the reasons for such a situation. Research shows that there is a gap between how people use new technologies in their everyday lives and how they are integrated –or not in the classroom. This gap is still causing distress among teachers and even dropping-outs from the profession. One of the causes identified in the literature is the training of teachers. Professional development activities do not train teacher trainers on their e-learning culture. These training activities focus mainly on ICT use. There is lack of actual inclusion of reflective exercises regarding ICT integration.

Keywords: Training intervention · Participants' e-learning culture
ICT integration

1 Introduction

Research studies (e.g. [1]) show that there is a gap between how people use new technologies in their everyday lives and how they are integrated –or not in the classroom. This gap is still causing distress among teachers and an even higher likelihood of teachers quitting the profession [2, 3]. The lack of fluency in these areas makes teachers vulnerable, hinders their effectiveness in the classroom, and rises a questioning about their professional commitment [4, 5].

Teacher trainers' e-learning culture defined by [6], as the teachers' representations regarding ICT, their attitudes, and their skills and resources, is crucial as it shapes their practice. Surprisingly, professional development activities do not train teacher trainers on their e-learning culture. Most of the training activities focus on ICT use and there is no actual inclusion of reflective exercises regarding ICT integration.

M. E. Auer et al. (eds.), *Teaching and Learning in a Digital World*,
Advances in Intelligent Systems and Computing 715,
https://doi.org/10.1007/978-3-319-73210-7_73

2 Method

2.1 Goal and Specific Objectives

Having identified the problem-lack of (rather general) understanding regarding the need of helping teachers become aware of and develop their e-learning culture, and of effective training interventions to facilitate this type of training – and as fruit of an extensive in-depth literature review, we opted to work with the IntersTICES model [7] (Fig. 1).

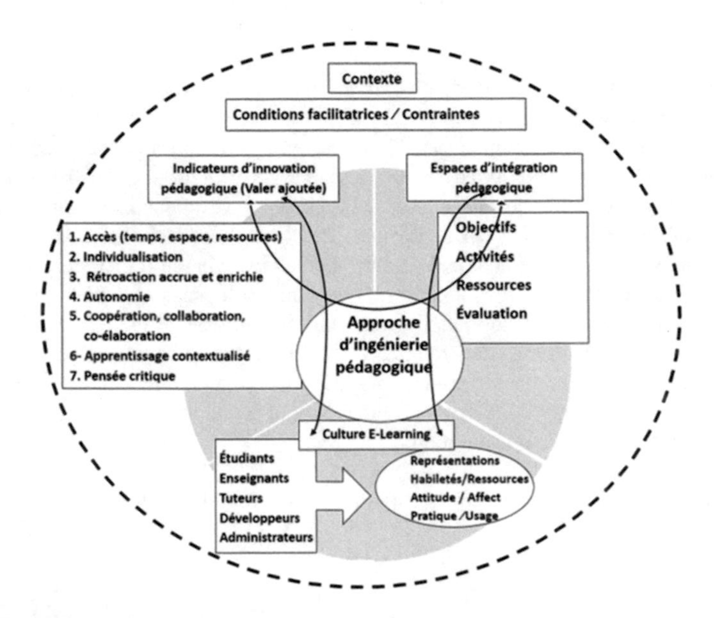

Fig. 1. The intersTICES model adapted from Peraya and Viens [13]

This model, comprising human factors and a dynamic process of interaction with the people, has already proven effective to be used when aiming at providing researchers with the appropriate means to effectively plan and intervene for a successful adoption of ICT.

Based on the specific question guiding our research: "How can the IntersTICES model [7] be used to support pre-service teacher trainers in the design of activities to successfully integrate ICT into their teaching practice, and enable them to get the very most out of the pedagogical added value of ICT", we set our goal: To help pre-service teacher trainers develop their e-learning culture; and two specific objectives: (1) to operationalize IntersTICES through a training intervention to foster the development of participant pre-service teacher trainers' e-learning culture, and (2) to analyse the impact of this intervention on participant pre-service teacher trainers' e-learning culture, and on their intention to integrate ICT in their teaching practice.

2.2 Approach

To develop a training intervention aiming at achieving our goal and specific objectives, we followed the steps of the cyclical nature of the typical Action Research (AR) proposed by [8]. This allowed us to: (1) have clearer objectives; (2) have a strategic view of what we wanted to achieve; (3) articulate various theoretical elements such as the guiding principles to be considered when working with teachers/adult learners based on [9, 10] as well as some principles also identified in our review of literature regarding the InterSTICES model [7], and the Unified Theory of Acceptance and Use of Technology (UTAUT) [11]; and (4) better coordinate the actions to be undertaken. We explored ways of operationalizing the InterSTICES model, and develop a methodology to intervene in order to accomplish this operationalization [12].

We decided to try-out the training intervention, with a population responding to the following characteristics: (1) Being pre-service teacher trainers/lecturers working in initial teacher training programs; (2) teaching subjects different from ICT at the undergraduate level; and (3) not having experience integrating ICT systematically into their teaching practice.

Our approach included allotting time to teachers for maturation processes and reflection. This implied having more than one personal encounter with participants - three times, as a minimum requirement- to be effective:

The first one, for carrying out the introductory training meeting where we presented InterSTICES along with its three main dimensions, and the pedagogical engineering approach (PEA) that articulates them to foster a thorough understanding of the model. Participants were asked to choose the pedagogical added value (PAV) they wanted to look for through the integration of an ICT tool into their activity.

The second meeting, to start exploring the ICT tool to be integrated into the activity they chose, to achieve their targeted objective(s), and to respond to their needs.

The third one, to reinforce and support them in their attempts to take ownership of the tool and of the appropriate strategy associated with their activity. During these individual meetings, the systemic and pedagogical engineering approach was also put into place to support participant teacher trainers through the whole process of integrating ICT in their chosen activity. As such, they were offered optional personal support to look for a specific pedagogical added value, to discuss alternative forms of evaluation, and to elucidate advantages and implications for them and their students.

Following [9]'s recommendation on including a discussion about the process, namely, the evolving nature of events undertaken by and with the actors within a setting, we prepared a thorough description (see below) of the 7 main steps undertaken during these meetings for personal support to make the development of the intervention more systematic and clearer in terms of presentation of the procedure to be undertaken.

Systematic Approach Implemented to Support Participant Teacher Trainers The seven steps undertaken to support participant teacher trainers during the production of activities integrating ICT [13] once their activity and ICT pedagogical added value chosen, were implemented as follows:

(1) Meeting of each participant with the researcher/trainer to reflect on and have a pedagogical discussion about every aspect conducive to its implementation. This

discussion was facilitated by exploring the IntersTICES dimension of *Spaces of pedagogical integration* (i.e. objectives, activities, resources and evaluation) that takes into account the specific contexts, and the internal and external coherence;
(2) Identification of facilitating conditions as well as of constraints, as an essential requirement to ensure allocation of required resources, successful implementation of the activity and achievement of their targeted objective(s);
(3) Identification of appropriate tool(s) that would facilitate achievement of their activity's targeted objective(s);
(4) Demonstration/modelling and co-creation of tool(s)/instruments e.g. Using *Google Form*, design of a diagnostic test to identify students' prior knowledge about a school subject matter; using *screencast-O-matic* or *Camtasia* to create on-screen videos for remedial tutorials that would allow students to acquire identified lacking pre-requisites autonomously;
(5) Complementing and enhancing mastering of the tool(s) by using YouTube video tutorials;
(6) Further discussion and reflection to try out and refine the resulting instrument(s);
(7) Discussion regarding alternative forms of evaluation, along with benefits, advantages and implications for teacher trainers and their students, e.g. Tips for designing peer evaluations by using rubrics.

The data gathered in the researcher's journal was intended to document the process and procedures we followed to operationalize the model and achieve objective 1. The pre- and post-intervention surveys, as well as semi-structured individual interviews provided data to analyze the impact of this intervention on participant pre-service teacher trainers' e-learning culture, and on their intention to integrate ICT in their teaching practice. To analyze these data we used Inspiration and QDA Miner software.

3 Actual Outcomes

3.1 Identified Guiding Principles

We clearly identified two general and overarching principles, which have to be considered when implementing a successful teachers' training intervention to integrate ICT. These principles, supported by the pedagogical engineering approach facilitated by IntersTICES, are as follows:

(1) Undertake a systemic and systematic procedure that supports carrying out more specific analyses of the needs and context;
(2) Take into account the actors' e-learning culture. For example, when aiming at integrating an ICT tool into a teaching activity, one can wonder whether the teachers have a good understanding regarding what it takes in terms of knowledge, skills, and resources to do so, or whether they require some personal support. Do they take a positive attitude towards this integration? If not, which resources should be incorporated into the pedagogical strategy when planning the training intervention?

Two other considerations need to be acknowledged:

(1) Focus on teachers' own projects; and
(2) Avoid long, drawn-out activities.

3.2 A Reflective and Critical Exercise Implied in the Four Stages of the Training Intervention

We embarked, then, on a reflective and critical exercise regarding the training intervention procedure that takes place over four main stages:

Stage 1: Undertake the analyses of thc needs and context (e.g. know the objectives; know the constraints);
Stage 2: Focus on the pedagogical added value of certain uses of ICT;
Stage 3: Consider the actors' e-learning culture. (See aforementioned example in 3.1., 2) above)
Stage 4: Foster internal consistency. This stage encompasses the pedagogical choice, a kind of pedagogical design. This involves identifying a goal; getting things implemented; handling resources carefully and according to identified needs;

In short, we strongly recommend making explicit the pedagogical added value sought; questioning the actors' e-learning culture and, only then, undertaking the pedagogical design. Often, instructional designers/teachers conduct the analysis of needs and context, as a first step, followed by the preparation of the pedagogical design (see stage 4 above) without considering the above mentioned stages 2 and 3.

Furthermore, to implement this training intervention, we took into account the guiding principles of adult learning and the IntersTICES model's mentioned above as well as notes taken while discussion and exchange of ideas with experts regarding appropriate strategies to do so.

Since we were working in an Action Research context, there were certain elements that had to be considered and respected, namely, being both participant/actor AND researcher required ongoing validation – with expert(s) and literature review - of any decision regarding, for example, the key items shaping the design of the training intervention, the training intervention itself, and its implementation. These iterations for appropriateness of decisions taken, facilitated refining and monitoring of any (planned) action, while the pedagogical engineering approach we implemented ensured systematic rigor of the whole process.

The ICT tool introduced, learned and integrated - following the strategy suggested in the *IntersTICES-Type Activity* in Table 1, below - to respond to a specific need, may have a (noticeable) impact on the teacher trainers' e-learner culture while facilitating its development taking into account on which level they may be. Every time teacher trainers need to learn about a new ICT tool –one at a time– to address a specific need, they are developing their ICT competency.

Table 1. An IntersTICES-type activity – stages and training strategies that facilitate the development of teacher trainers e-learning culture [13]

Training intervention	Training strategies
Introductory intervention (group or individual)	- Present the three dimensions of the Interstices Model and the UTAUT - Select activity into which integrate ICT tool, decide on what PAV to loot for to address in identified need or achieve a targeted objective
Follow-up & personal support customized to respond to particular needs *N.B. Keep in mind that teacher (trainers) are willing to learn about and use a tool thai respond to (one need at a time) of their identified needs, as well as one tool at a time*	1. Discuss about pedagogical added value (PAV) sought, explore tool that allows addressing need - using videos/tutorials available on YouTube
	2. Assess ease of use of tool, and its feasibility to address specific identified need
	3. Introduce (another) ICT tool e.g. *Google Form,* to the already most-used three basic ones (i.e. Word, PPT, email)
	4. Explain pedagogical added value (PAV) of suggested new tool e.g. *Google Form,* and encourage reflection about PAV
	5. Foster awareness of facilitating conditions and/or constraints regarding specific ICT tool(s) integration
	6. Support first steps toward learning about & mastering of the tool
	7. Present, show and demonstrate step by step how to use the tool through a specific and simple example of application, e.g. build a short survey from scratch, including at least two or three types of questions; choice of background or theme; sharing options
	8. Foster awareness of the teaching/learning context
	9. Encourage reflection on benefit of including some ICT PAV (initiation)
	10. Promote ability to link content, pedagogy & technology
	11. Encourage in-depth reflection on benefit of seeking specific ICT PAV to enhance teaching approach designed to achieve targeted objectives
	12. Foster awareness of specificities of the tool, its suitability and potential for integration in a chosen activity and context
	13. Provide scaffolding during actual design of activity integrating ICT tool. e.g. Use Interstices (Indicators of PAV+ Spaces of Pedagogical Integration) and the *Guidelines* designed for backing the choice of activities to support this design and implementation
	14. Enhance the capacity to transfer
	15. Foster actual practice presenting some form of (internalized) adoption of activities integrating the ICT tool looking for PAV
Post-intervention (group) Interview (individual)	- Share experiences/best practices - Share useful tips and clues

3.3 Increased Awareness and Self-efficacy

Being supported while having to learn about and being able to use the ICT tool in an activity of their own, fosters in teachers growing feelings of self-efficacy, which in turn can nurture a positive attitude towards ICT use, and increase their intentions to (actually) integrate ICT in their teaching practice [13, 14]. This whole process resulting in a developed and more comprehensive e-learning culture.

Findings suggest that the operationalization of IntersTICES via an interactive training intervention can provide teacher trainers with an opportunity for reflection and awareness about their personal representations regarding every aspect of their e-learning culture. Therefore, helping teacher trainers to have a clear and pedagogical rationale (e.g. the pedagogical added value for them and their students [15] for integrating technology in their practice kept them motivated to actually start using ICT and eager to incorporate it in their teaching.

All participant teacher trainers expressed that to start using technology in their courses, they would require training on how to use the tool, time to discuss, explore and actually start using it, as well as material resources (software) and some kind of mentoring support. These facilitating conditions would allow them to engage in pedagogical discussions to explore, get to know and align the affordances of the tool with the appropriate pedagogy strategies for achieving their targeted objectives.

Results also suggest that our intervention clearly made participant teacher trainers become (more) aware of the existence of ICT tools in terms of diversity, potential and scope. Their newly developed awareness entails an increase in intention to integrate ICT in their pedagogical activities, in their motivation to try new things out, and in introducing new strategies seeking for the pedagogical added value of ICT.

Besides, notes from pedagogical conversations during follow-up and personal-support sessions with participant teacher trainers, suggest that they are (more) aware of the reflection process in which they (have to) get involved when planning for integrating ICT in their chosen activity.

We may infer that the IntersTICES-type activity is the set of strategies more closely related to the field we are interested in: Empowering teacher trainers to develop their e-learning culture, while helping them integrating ICT in their teaching practice [16]. It may provide educators with a sound pedagogical foundation, as well as practical skills to meaningfully integrate ICT into their teaching practice [17]. Even though follow-up and personal support were considered as essential by participant teacher trainers themselves, these strategies are to be handled with care, since depending on teacher trainers' needs and profile, some of these strategies may not be appropriate.

4 Conclusions

The originality of this research resulted in the methodology used to determine the training-intervention specifications necessary to operationalize the IntersTICES model to carry-out our intervention. In fact, this methodology forced us to explicitly articulate IntersTICES to better implement it afterwards. The iterations followed to validate decisions for all actions planned and taken, ensured credibility of results -fundamental for Action Research (AR), which along with the engineering pedagogical approach implemented, contributed to a clearer identification and understanding of the different steps this type of intervention could encompass.

Analyses of the interviews and data from our other sources, suggest that teacher trainers' intentions to integrate ICT in their teaching practice are influenced by their newly-developed e-learning culture; i.e. by their improved representations of the pedagogical added-value of technology for their students, their course, and for them; their awareness regarding availability of resources; their changed attitude regarding time required to learn how to use and master the tool, and their improved self-efficacy beliefs. These changes could be thought as actually being due to the impact of the IntersTICES-Type Activity on participants' e-learning culture.

Keeping in mind that "teachers tend to teach in the way they were taught" [18], teacher-training programs should promote the development of teacher trainers' e-learning culture that will allow them to model and demonstrate effective ways and/or strategies to integrate ICT pedagogically founded [19]. This would better prepare pre-service teachers for effective use and integration of ICT in their future K-12 classrooms.

The IntersTICES-Type Activity resulting from operationalizing IntersTICES might adequately support teacher trainers during the exploration, learning and utilization of ICT-tools.

However, although skills training and follow-up support are necessary to initiate teacher trainers and future teachers in the use and integration of ICT in their teaching practice, it is essential that administrators also recognize that teachers' preparedness to achieve higher levels of integration are affected by other important factors (e.g. teacher's time required for attending ICT training sessions; for planning lessons integrating ICT, etc.). Administrators may then find it worthwhile to provide incentives for teachers to embark in ICT integration at classroom level [13].

References

1. Villeneuve, S.: L'évaluation de la compétence professionnelle des futurs maitres du Québec à intégrer les technologies de l'information et des communications (TIC): maitrise et usages. Doctoral dissertation. Université de Montréal, Canada (2011)
2. Steering Committee for Teacher Ed., Quebec - CAPFE (2002). http://www.capfe.gouv.qc.ca. Accessed 11 Feb 2013
3. OECD: Reviews of National Policies for Education. Netherlands 2016. Foundations for the future. OECD Publishing (2016)
4. The Inspectorate of Education of the Netherlands (2015). http://www.oecd.org/edu/EDUCATION%20POLICY%20OUTLOOK_NETHERLANDS_EN%20.pdf. Accessed 18 Mar 2016
5. Schleicher, A.: Teaching Excellence through Professional Learning and Policy Reform: Lessons from around the World. OECD Publishing, Paris (2016). https://doi.org/10.1787/9789264252059-en. International Summit on the Teaching Profession
6. Viens, J., Renaud, L.: La complexité de l'implantation de l'approche socioconstructiviste et de l'intégration des TIC. Éducation Canada **41**(3), 20–26 (2001)

7. Peraya, D., Viens, J.: Culture des acteurs et modèles d'intervention dans l'innovation technopédagogique. Revue Internationale des Technologies en Pédagogie Universitaire, Conférence des recteurs et principaux des universités du Québec [CREPUQ], **2**(1), 7–19 (2005)
8. Kemmis, S., McTaggart, R.: The Action Research Planner. Deakin University Press, Geelong (1981)
9. Knowles, M.S.: Andragogy in Action: Applying Modern Principles of Adult Learning. Jossey-Bass, San Francisco (1984)
10. Kearsley, G.: Andragogy (Malcolm Knowles). The theory into practice database (2010). http://tip.psychology.org. Accessed 04 July 2014
11. Venkatesh, V., Morris, M.G., Davis, G.B., Davis, F.D.: User acceptance of information technology: toward a unified view. MIS Q. **27**(3), 425–478 (2003)
12. Miles, M.B., Huberman, M.: Qualitative Data Analysis: An Expanded Sourcebook. Sage Publications, Thousand Oaks (1994)
13. Villa, G.: E-learning culture: operationalization of a systemic model to support ICT-integration in pre-service teacher trainers' practice. Doctoral dissertation. Univ. Montréal, Canada (2016)
14. Ertmer, P.A., Ottenbreit-Leftwich, A.T., Sadik, O., Sendurur, E., Sendurur, P.: Teacher beliefs and technology integration practices: A critical relationship. J. Comput. Educ. **59**(2), 423–435 (2012)
15. Viens, J., Villa, G., Stockless, A.: «IntersTICES, intégrer la recherche dans la formation initiale et continue des enseignants afin d'améliorer les usages pédagogiques des technologies» 9ª Conferência Internacional sobre Exclusão Digital na Sociedade da Informação e do Conhecimento, SEMIME 2015, Lisbonne, 30 January 2015
16. Villa, G.: E-learning culture: from theory to practice. In: Kapur, V., Goshe, S. (eds.) Dynamic Learning Spaces in Education. Springer, Singapore (2017)
17. Ito, M., Gutiérrez, K., Livingstone, S., Penuel, B., Rhodes, J., Salen, K., Schor, J., Sefton-Green, J., Watkins, S.C.: Connected Learning: An Agenda for Research and Design. Digital Media and Learning Research Hub, Irvine (2013). http://dmlhub.net/publications/connected-learning-agenda-research-and-design. Accessed 3 Oct 2014
18. Hargreaves, A.: Teaching in the knowledge society: Education in the age of insecurity. Teachers College Press, New York (2003)
19. Fiszer, E.P.: How Teachers Learn Best. Scarecrow Education, Lanham (2004)

A Look-Back to Jump Forward: From an Ancient Innovation Culture to the Exploration of Emerging Pedagogies in Engineering

Francisca Barrios, Melanie Cornejo, Brian O'Hara(✉), and Francisco Tarazona-Vasquez

Universidad de Ingenieria y Tecnologia, Lima, Peru
{fbarrios,mcornejo,bohara,ftarazona}@utec.edu.pe

Abstract. From its inception, Universidad de Ingeniería y Tecnología (UTEC) has had the vision of causing a disruptive change in society by educating a new generation of holistic engineers. The university has recently embarked on a radical transformation of its educational model, in order to deliver its promise. A flexible curriculum provides students not only with a strong STHEAM backbone imparted in a student-centered, active-learning format, but also exposes them to real engineering challenges and promotes the acquisition of professional skills from the onset. For this radical change to be implemented successfully, UTEC has decided to design and launch a Laboratory for Educational Innovation, called Moray. Moray has been conceived as an open platform, consisting of a common space and a set of protocols through which faculty, students, staff, and experts from top universities worldwide can work interdisciplinarily and collaboratively, towards the enhancement of learning experiences in higher education.

Keywords: Educational innovation · Engineering education
Emerging pedagogies · Student-centered teaching · T-shaped engineers

1 Introduction

The founders of Universidad de Ingeniería y Tecnología (UTEC) conferred one main mandate to both the university's administrative and academic teams: to cause a disruptive change in society by educating a new generation of holistic and T-shaped engineers. The university launched in 2012 as a non-for-profit organization under the premise of giving Peruvian youth access to a world-class and future-proof educational model in Engineering, regardless of each individual's socioeconomic status, and only dependent on his or her talent, disposition, creativity and intellectual capacity.

To be able to truly live up to this vision and deliver its promise to students and society at large, the university has recently embarked on a radical transformation of its educational model. It has now instituted a more flexible and project-based curriculum, that provides students not only with a strong scientific, disciplinary and technical backbone, but also exposes them to "real life" engineering challenges, thus promoting the acquisition of professional skills from the very first day. All of the content transmission and

M. E. Auer et al. (eds.), *Teaching and Learning in a Digital World*,
Advances in Intelligent Systems and Computing 715,
https://doi.org/10.1007/978-3-319-73210-7_74

skill-building dynamics are carried out in student-centered, active-learning formats that promote student engagement and the development of intrinsic motivation. This structural change should create the conditions for the formation of the T-shaped professionals the world is in such high demand for.

Furthermore, in order to facilitate the "structural re-engineering" of the traditional engineering program and a successful implementation, UTEC has decided to launch a Laboratory for Educational Innovation named "Moray", after an archeological site in Cusco, Peru, which is considered to have been a controlled environment for hydraulic and agricultural experimentation for the Inca civilization. This paper describes the design and implementation process of this laboratory, whose main purpose is to provide faculty with the training and support before, throughout and beyond this change process.

2 Background and Motivation

The learning sciences have significantly evolved in the last few decades, highlighting the deficiencies in traditional Engineering Education [1]. Diverse entities have told Engineering Schools across the globe that they should, on the one hand, strengthen and widen the scope of the fundamental sciences; and on the other, expose students to real-world engineering challenges, develop in them effective communication and teamwork skills, and also cultivate critical and ethical thinking [2]. All of this while reducing the number of hours in the curriculum, to allow most students to graduate in time. Traditional Engineering Education and its unrevised teaching methodologies cannot, as the record shows, fulfill these great expectations [1]. Today we count on proven theories on the science of learning (i.e. how people learn in a pedagogically-sound and relevant manner) and on the science of instruction (i.e. how teachers or mentors should help students learn), based on cognitive theory and neuroscience [3]. There is now an almost unanimous agreement that people learn by doing, experiencing and reflecting on the results, whereas they absorb and retain a rather small fraction of what they see and hear in a lecture format [4].

In addition, the digitalization of information and telecommunications has revolutionized the processes of knowledge production and transference. The mission of universities is to prepare their students for a modern professional career, encouraging them to actively engage with advanced technologies and develop functional skills in relation to new media, so that they can successfully integrate themselves into a society increasingly organized around these [5, 6]. Such developments have given rise to new scenarios of academic training, contents and processes of teaching and learning [7]. As a consequence, the adoption of a student-centered educational focus has led academic institutions to redefine the dynamics of teaching and learning that had previously been confined, predominantly, to the traditional classroom [8].

In view of these advancements in the learning sciences and the digital revolution, last year, and one year before graduating the first class, UTEC's higher administration and the Board of Directors decided to embark on a root-and-branch reform to be aligned with these changes and to ensure the preparation of holistic and T-shaped engineers. This disruptive change is embodied in what is now known as UTEC's Educational Model

(UEM), which in turn aims at ensuring that our students attain five main competencies. We consider these competencies necessary to truly become the holistic engineers that Peru and the world require:

1. Deep technical, disciplinary and interdisciplinary knowledge
2. Analytical reasoning and critical thinking for complex problem solving
3. Communication and collaboration skills across disciplines and cultures
4. Ability and will to lead the innovation and change processes
5. Ethical and socially responsible thinking and doing, both on a local and global basis

In short, UEM seeks to educate global professionals, able to understand, disaggregate and solve complex problems in a creative, innovative and ethical way; and communicate these solutions effectively. The model is highly student-centered and aims at awakening and empowering students' intrinsic motivation to deeply learn and arise with new solutions to the world's most pressing challenges. At UTEC, we believe that learning in a transdisciplinary and contextualized fashion and in connection with the real world, alongside the most important local and global industries, is a richer and more effective means for developing our students into adaptable, lifelong, flexible learners and change agents.

In this context, Moray has been thought-out as an open platform, consisting of both a common space and a set of flexible protocols through which faculty, students, staff and experts from top universities around the world can work interdisciplinarily and in a collaborative manner, towards the enhancement of teaching and learning dynamics and experiences both inside and outside UTEC's classrooms, fully embracing and harnessing the digitalization of the educational world. Moray crystallizes the innovative culture embedded in UTEC's DNA. This culture of innovation was, as mentioned before, first given as a mandate by UTEC's founders. It was later incorporated as one of three constitutive pillars – research, entrepreneurship, and educational innovation- during the university's design phase. Currently, it lies at the very core of the new educational model. This culture promotes experimentation, fosters a code of radical openness, is data-driven, and understands "failure" as something to be learned from and as an important step towards continuous improvement. Just like the Incas utilized Moray and its terraced depressions in concentric circles as an experimentation and testing station, the new curricular reform within UTEC has in the Educational Innovation Laboratory a platform from which to launch and monitor new initiatives. These initiatives can range from implementing technology-enhanced learning experiences to redesigning the underlying skill-building and knowledge-transfer activities, all of these with the objective of creating deep learning experiences in our students.

3 Design Process

The first step in the design process consisted of an initial (and mostly theoretical) benchmarking analysis, which allowed us to determine the key success factors for constituting an Educational Innovation Laboratory. Ten countries -USA, UK, Finland, Australia, Singapore, Korea, Uruguay, Colombia, Chile and Mexico- were covered in this analysis.

From the observations, it soon became clear that Moray had to become an overlay of physical and virtual spaces, serving as a focal point for the different actors and stakeholders of UEM. It had to be based on the following principles:

1. Be founded on a co-working and collaboration philosophy, a "radical openness" code of conduct and be staffed with a multi-disciplinary team, in order to eradicate blind spots and create communities and networks beyond the physical space or the confinement of the university.
2. Be user-centered (in this case, student-centered) and ensure the engagement of all actors within the ecosystem. The students should be co-creators of the experiences and changes.
3. Behave as an incubator or accelerator of educational projects, implementing a culture of rapid prototyping, testing, iteration and continuous improvement; always aiming to scale and replicate successful projects in other -larger- contexts.
4. Be research-based and data-driven as well as prone to incorporating new technologies and digital media into projects in order to communicate more efficiently, catalyze otherwise tedious processes, capture as much data as possible and produce relevant information models to guide the ulterior decision-making processes.
5. Provide constant mentoring and feedback to establish best practices that could later be effectively communicated to and shared with diverse audiences.

Taking these key principles into account, the following sequence of steps was pursued for the implementation of Moray:

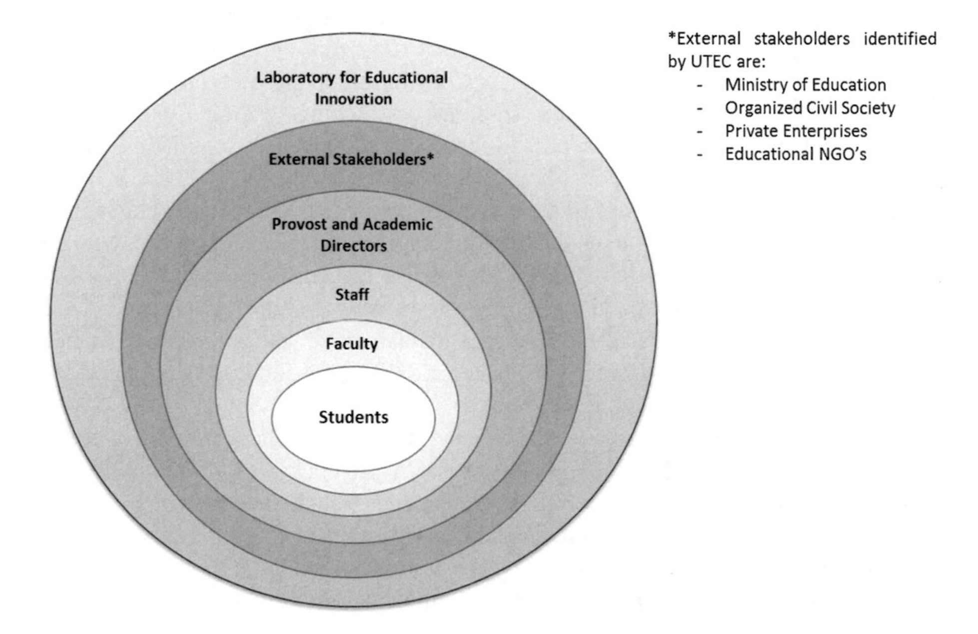

Fig. 1. Onion Map showing the different stakeholders that are being brought together by the Laboratory for Educational Innovation. At the heart of the map are UTEC students, whose experience of the new UEM is critical when it comes to orienting the Laboratory's efforts.

1. Identification of main stakeholders and possible contributors, participants and collaborators. The first step consisted of appointing the Director for the Laboratory, who would plan and supervise the initial and all subsequent processes in the establishment and governance of Moray. Moray is now being led by the Director of Educational Innovation and Quality at UTEC, and supported by two Coordinators (for Educational Quality and Digital Innovation) and two interns. In addition to the permanent staff, the champions (i.e. change leaders) and ambassadors within the organization were identified early on in the curricular transformation process, sent to specific "train-the-trainers" workshops and are now an integral part of the Laboratory. Furthermore, UTEC's Provost and the Heads of Departments have contributed to the design and launch, and now collaborate with Moray on a regular basis (Fig. 1).
2. *In-situ* contrasting of benchmarking analysis. After having carried out the benchmarking analysis based on available literature, an important step was to corroborate our theoretical findings on-site. Thus, Moray's Director and UTEC's Provost embarked on a journey that allowed for the exchange of ideas with the leaders of the most reputable Teaching and Learning Laboratories around the world (USA, UK, Denmark, Sweden, Finland, and Netherlands) and also hosted such leaders at UTEC. The main goal of these visits was to procure intensive knowledge-transfer dynamics and a bi-directional sharing of good practices with these experts.
3. Development of a strategic plan, with clear objectives, prioritized tasks, responsibilities and success metrics (Fig. 2).
   This strategic planning process had to take into account the following aspects:
   (a) Main activities: the initial focus areas would determine the activities to be developed and would take into account existing gaps and available resources. These would range from pedagogical workshops, rapid prototyping with sharebacks, presentations, one-on-one sessions, co-teaching dynamics, peer reviews, faculty exchange programs, content production (such as consolidating a problems/challenges bank), portfolio building, among others.
   (b) Personnel: the focus areas of the personnel would be based on the main educational themes and problems that the Laboratory would be focusing on with the option of calling in external talent for pedagogical consultancies.
   (c) Space: the focus areas and main functions of the Laboratory would also determine the type of space needed. Aspects to take into consideration here include location, flexibility of the given space, and the atmosphere to be created with the furniture, equipment and lighting.
   (d) Budget: the budget should be divided into four main pillars: personnel, equipment and tools, space, and operating costs by project.
4. Selecting and launching pilot project(s) that should evaluate the possible benefit to the final user and describe the concrete strategy to attain an objective. In this phase, it was important to ensure that the innovation served as enough of a proof of concept to merit the scale-up; otherwise, more pilot projects or prototypes would be necessary.

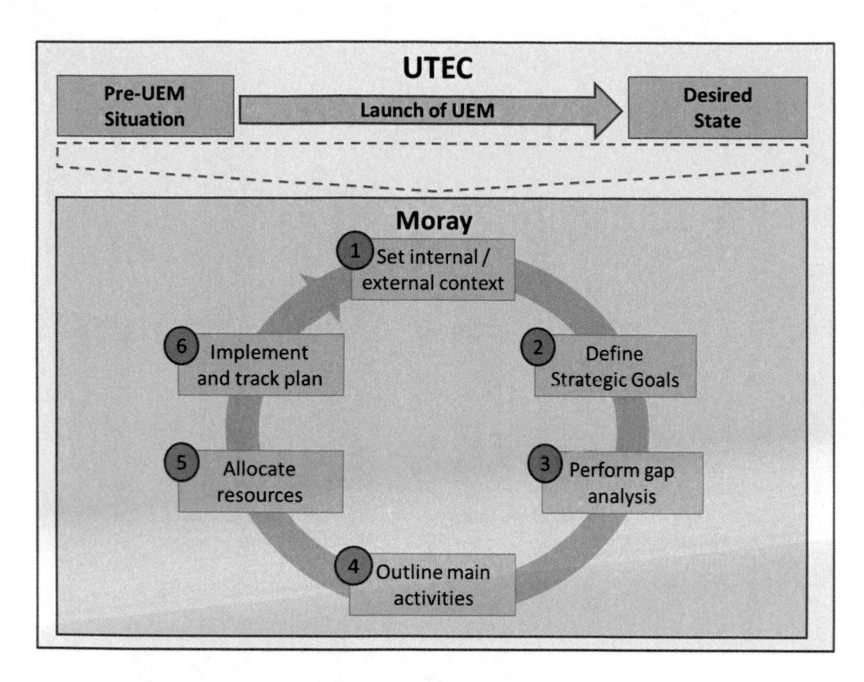

Fig. 2. Moray's strategic planning process as key support for the implementation of UEM. The benchmarking analysis and internal diagnosis (step 1) led to the definition of goals (step 2) and a follow-on gap analysis (step 3). The initial main activities (i.e. pilot projects) were focused on bridging these gaps (step 4) and were planned carefully by allocating the necessary resources to attain quick wins (step 5). The final step (6) consisted of implementing the initial plan and tracking the impact and results on the implementation of UEM. The process is iterative for continuous improvement of Moray and UEM at large.

4 Preliminary Results

To mention a few pilot projects, each faculty member has already been tasked with revising and redesigning his/her course, taking into account and introducing the newest tools and methodologies to foster student engagement, motivation, knowledge-transfer and long-term retention. In addition, UTEC's higher academic administration has created multi-disciplinary task-forces for the development or re-design of the different core courses (i.e. calculus, physics, chemistry, computer science, and communications), to ensure that foundational needs are being met. These faculty members gather at Moray every Friday for the "F3 - Faculty Feedback Fridays" meetings, a roundtable discussion and feedback session that aims at aligning on the overall objectives, iterating and improving these critical courses on the go.

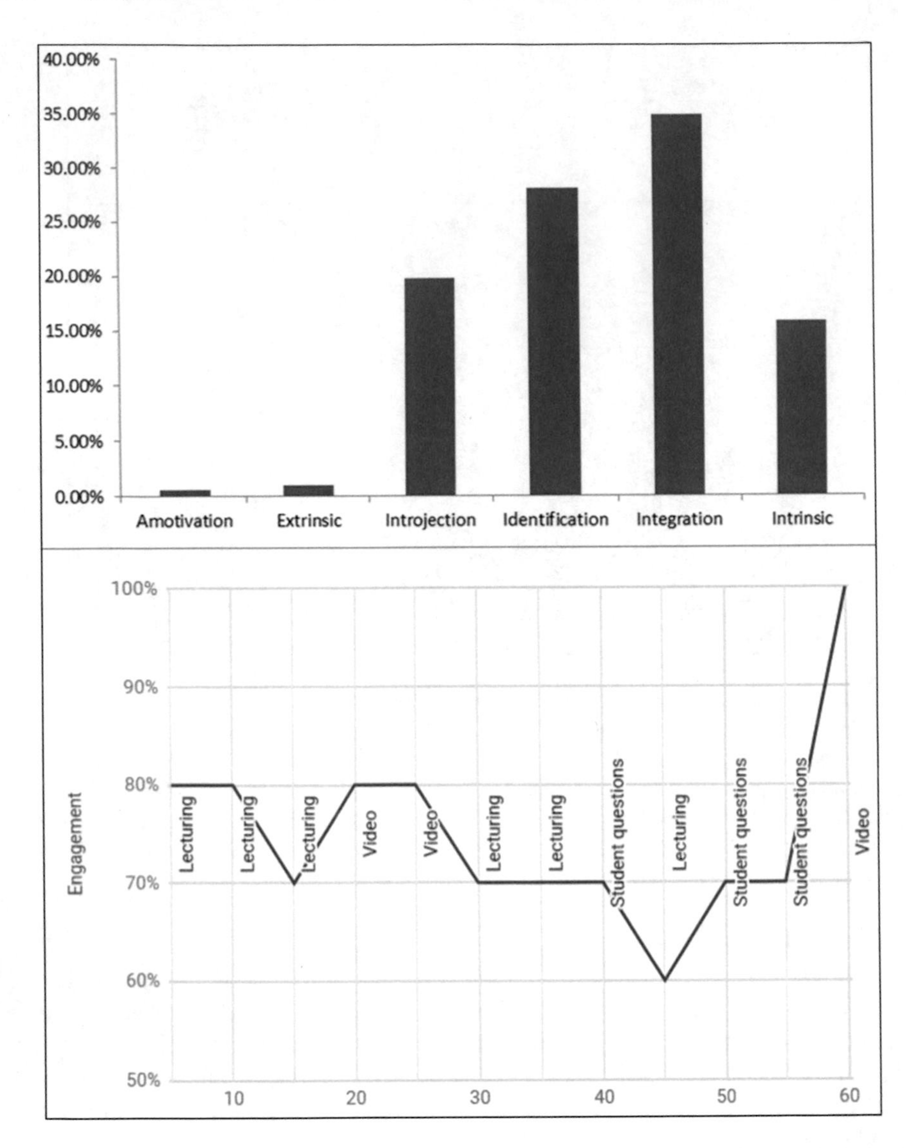

Fig. 3. Samples of data dashboards illustrating the level of achievement of two UEM-specific goals: student motivation, as accounted for by the students themselves through a student perception survey (top graph), and student engagement during classes as per an adaptation of The Classroom Observation Protocol for Undergraduate STEM [9] (COPUS; bottom graph).

The Laboratory has also launched a "Fellows Program" that chooses and supports 3-4 faculty members throughout a semester to backwards-design and innovate either the entirety of their courses or a significant portion of them, based on hand-picked literature that is read and discussed in an applied and reflective manner. Additionally, Moray is conducting several data-gathering assessments and surveys with both students and faculty. For instance, it is disseminating a student survey at two points during the term, to measure motivation, relatedness, autonomy, perception of competence and overall satisfaction with each course (Fig. 3). A rigorous classroom observation protocol has also been implemented to allow the Laboratory to capture the level of student engagement during classes, which is regarded as a direct reflection of the teacher's ability to implement emerging pedagogies successfully. In addition, Moray's staff is holding regular focus groups with faculty, staff and students, to identify areas of development and continuous improvement for UTEC as a whole.

Moray has already organized two main faculty-wide, week-long teaching and learning workshops to prepare each and every member of the academic department for the new educational approach. The faculty will continue to have several training opportunities offered by external experts in the field of emerging pedagogies, as well as receive scientific literature, research papers and other tools and resources each week, which can be reviewed individually and later discussed at Moray's Journal Club. Finally, UTEC's faculty is also encouraged to work across-fields at Moray, to devise new transdisciplinary content and explore co-teaching opportunities. Some of these initiatives have already been rolled out; examples include courses like Art and Technology, Geopolitics of Water, and Global Challenges. These courses are meant to be a platform for students to approach a specific topic from different disciplinary perspectives while making sense of these diverse approaches as a whole.

The Laboratory is also designing new classroom spaces that are more suited to promote collaboration and creative thinking than a traditional classroom format, replacing the layout, equipment, and furniture which correspond to the "sage on the stage" model, with a new set-up that instead sparks a "guide on the side" take on didactic practices [10]. In addition, Moray has also been involved in the re-shaping and re-tasking of some areas within our campus to facilitate the "design-build-implement-test" process in proper engineering workspaces [11]. Accordingly, UTEC now has a large workspace, called UTEC Garage, where students gather, exchange ideas, think them through in a hands-on approach and test them in a safe environment (Fig. 4). Likewise, UTEC's FabLab allows students to 3D-print and prototype any idea very quickly, under the guidance of experts in relevant fields. More advanced ideas that have been matured over several semesters will have the chance to be incubated in UTEC's very own business accelerator, UTEC Ventures. The university also holds over 30 laboratories with different field-specific resources and equipment that are open to students and faculty for research projects or prototyping exercises. All these spaces are thought to be spatial extensions of Moray and work in close interaction with this Laboratory.

Finally, and as an umbrella program to all these initiatives, Moray has adapted and is piloting a framework for the evaluation of teaching achievement that has been put forth as part of a study commissioned by the Royal Academy of Engineering (RAE) to a leading pedagogical consultancy. This Career Framework for University Teaching [12], proposes a standardized and transparent -therefore, portable- method for evaluating and evidencing teaching achievement. The metrics proposed by the Framework have been designed with input from sixteen universities from around the globe -one of them being UTEC- and is already being used here to evaluate teaching performance at different stages: appointment, promotion and professional development. At Moray (and UTEC at large), we view this change in mindset and focus on teaching and learning (as

Fig. 4. UTEC Garage as a MakerSpace lying at the heart of Moray. This 250 m^2-large space is open around the clock to allow students, faculty and external people to gather in groups and develop prototypes of different fidelities for their diverse projects. It is also a space for transdisciplinary workshops and roundtable discussions

opposed to mainly on research performance) as instrumental for the success of the new educational model.

5 Conclusions and Future Work

UTEC has embarked on a major transformation of its education model, to fulfill its promise of causing a disruptive change in society through the education of a new generation of engineers. This new model is based on two main, interrelated pillars: a novel curricular structure and a teaching and learning–focused Faculty Development Model. We truly believe that the early adoption, enthusiasm and active participation of the faculty in this transformation have been catalyzed by the fact that the Faculty Development Model was designed and is being launched simultaneously with UEM. It helped faculty understand and make sense of the process and allowed them to grasp the need and the sense of urgency for a shift from a highly research-focused to a holistic perspective, which truly values teaching/mentoring – and rewards them accordingly. One of the main early findings from our design and implementation process was that faculty needs to be involved from the initial stages, be part of the change process and be supported throughout. That is how the idea of Moray, UTEC's own Educational Innovation Laboratory, was born. Having both a physical and digital space, dedicated personnel and a clear strategic plan for the implementation and change management process underlying the UEM, clearly helped all the involved parties align with the overall goal.

Today, Moray is already fulfilling the promise of serving as an open platform for UTEC and the higher education community to design and test innovative pedagogical initiatives in order to improve engineering education in Peru. There are five foundational competencies that have shaped the new educational model at the university. How to help students develop these throughout their experience at UTEC has been the guiding question and the connecting thread to which this model is the answer. Every course targets specific skills that are meant to percolate to a variety of other topics and that are meant to be exercised by students and translated into a wide range of contexts of practical application. Therefore, we expect students to be able to connect the dots between their engineering education and the complexity of the world they will face as professionals, which will demand from them the ability to navigate scenarios where leadership, ethical behavior, and teamwork, among other skills, will be crucial. Faculty at UTEC are the key players for this model to be implemented successfully and Moray should continue its path toward becoming the underlying support system for them to feel backed, motivated and empowered throughout and beyond this process.

A challenge that is yet to be resolved is finding the appropriate physical location for Moray. We have a temporary space at UTEC, but it doesn't present the necessary features we believe will help spark the disruptive atmosphere that is so necessary in an innovation laboratory. Furthermore, additional work is yet to be done regarding metrics to measure Moray's impact on the change management process and on the implementation of the educational model. To this means, the Laboratory needs to continue conducting focus groups, interviews and surveys with both students and faculty, to assess the level of satisfaction with and support perceived from Moray through the preparation, launch and implementation of UEM. Also, other ICT-driven processes such as the integration of additional tools and resources (our recently purchased smart-boards, VR equipment, 360° gear, augmented reality modules and "Clickers"), the path toward a blended or inverted model of instruction, and a more powerful drill-down on Business Intelligence tools to increase student retention rates need to be assessed by Moray. All of this in order to make sure technology is properly leveraged so as to enable student-centered learning. Finally, Moray's role in integrating the initiatives of UTEC's Finance and IT departments to think about and prepare for the future of education in new media and digital formats needs to be further assessed.

References

1. Besterfield-Sacre, M., Cox, M.F., Borrego, M., Beddoes, K., Zhu, J.: J. Eng. Educ. **103**, 193 (2014)
2. Felder, R.M.: **34**, 238 (2000)
3. Ambrose, S.A., Lovett, M., Bridges, M.W., Dipietro, M., Norman, M.K.: How Learning Works: Seven Research-Based Principles for Smart Teaching, San Francisco, CA (2010)
4. Felder, R.M., Woods, D.R., Stice, J.E., Rugarcia, A.: Chem. Eng. Educ. **34**, 26 (2000)
5. Claro, M.: La Incorporación de Tecnologías Digitales En Educación. Modelos de Identificación de Buenas Prácticas, Santiago de Chile (2010)
6. Khazaal, H.F.: J. Coll. Teach. Learn. **12**, 1 (2015)
7. Correa, J.M., De Pablos, J.: Rev. Psicodidáctica **14**, 133 (2009)

8. Santiago Campión, R., Navaridas Nalda, F., Andía Celaya, L.A.: Estud. Sobre Educ. **30**, 145 (2016)
9. Smith, M.K., Jones, F.H.M., Gilbert, S.L., Wieman, C.E.: CBE-Life Sci. Educ. **12**, 618 (2013)
10. King, A.: Coll. Teach. **41**, 30 (1993)
11. Crawley, E.F., Malmqvist, J., Östlund, S., Brodeur, D.R., Edström, K.: Rethinking Engineering Education: The CDIO Approach. Springer, Cham (2014)
12. Graham, R.: Template for Evaluating Teaching Achievement, London (2016)

Student Performance and Learning Experience in MOOCs: The Possibilities of Interactive Activity-Based Online Learning Materials

János Ollé[1] and Žolt Namestovski[2(✉)]

[1] Eszterházy Károly University, Eger, Hungary
olle.janos@uni-eszterhazy.hu
[2] University of Novi Sad, Subotica, Serbia
zsolt.namesztovszki@magister.uns.ac.rs

Abstract. The most common criticisms against the effectiveness of MOOCs usually point out the high attrition rates and the low level of learning effectiveness. Open courses generally require self-regulation (self-regulated learning), task awareness and learning methodology at a level most students cannot achieve. To increase the effectiveness of open courses, the planning process should focus on the stronger control over students' activities; another important factor is that the learning materials should be able to ensure the continuous activity of the learners. The application of appropriate activity-based instructional design solutions can largely increase the effectiveness of open courses. Our research focuses on the possibilities of activity-based learning material development for online education purposes.

Keywords: Activity-based online learning material · Instructional design · MOOC

1 Introduction and Background

Networked environments offer new scope for presenting activity based courses, in which activities and reflection form the central backbone of course pedagogy. Such courses promise an enriching approach to study, but there are also challenges for the design of assessment (Macdonald and Twining 2002).

Massive open online courses (MOOCs) are the latest revolution in online teaching and learning (Liyanagunawardena et al. 2013). These academic courses are available to the general public worldwide; there are no preconditions; and they are usually free of charge (Johnson et al. 2013; Allen and Seaman 2014; Adams and Williams 2013; Fini 2009; Stewart 2013).

The field of open and distributed learning has experienced a surge of media coverage and public interest in the last several years, largely focusing on the phenomenon of massive open online courses (MOOCs). The term MOOC has been used to describe a diverse set of approaches and rationales for offering large-scale online learning experiences. MOOCs have been delivered using both centralized platforms and services

M. E. Auer et al. (eds.), *Teaching and Learning in a Digital World*,
Advances in Intelligent Systems and Computing 715,
https://doi.org/10.1007/978-3-319-73210-7_75

including learning management systems (LMSs) and decentralized networks based on aggregations of blog sites and social media feeds. MOOCs have been designed to support university curricula, academic scholarship, community outreach, professional development, and corporate training applications (Anders 2015).

A primary criticism of MOOCs is that their completion rate is very low, approximately 10% (Wilkowski et al. 2014). According to other authors (Jordan 2014) the majority of courses have been found to have completion rates of less than 10%.

Since video combines many types of data (images, motion, sounds, text) in a complementary fashion, learning can be adjusted more easily than with other tools to the diverse learning styles and individual learning pace of students. With video, the learner has more control over the information they receive and an additional opportunity for deeper learning by being able to stop, rewind, fast-forward, and replay the content as many times as needed (Greenberg and Zanetis 2012).

In the case of activity-based learning materials, the learning content is presented in a different logical sequence, while the tasks and expected activities of students are elaborately planned in advance for each content element. The close control of the learning process and course content ensures continuous activity and learning experience. Our research aims to examine student performance and learning experience by comparing a traditional teacher's presentation extended with interactive exercises and an experimental activity-based online learning material.

2 Research Problem and Hypotheses

The research - which was carried out within the frameworks of a six-week long MOOC (Instructional design, digital content development) - took place during week 4 and week 6 of an online course. During the two weeks of the experiment, the educational content of the MOOC focused on the motivation, activation and reassurance processes and the possible online tools and services used for learning material development. The 107 members of the course were divided in the learning management system (LMS) 'Schoology' into an experimental and a control group. The control group was using recorded presentations and interactive exercises for learning, while in the case of the experimental group, the activity–based experimental learning material was used for learning and completing the requirements of the same module.

After the test period, each person's learning effectiveness and learning experience was measured, both with regards to the two thematic priorities and other learning contents. The activity of the students in the online environment and their prior expectations towards the course were also taken into account during the analysis.

Within the frameworks of the experiment, the following research problems and hypotheses were set out:

H1: Students learning with interactive and activity-based learning materials will have significantly higher level of activity and motivation. As the experimental learning material attracts the attention more effectively, students will make better use of their time spent on learning. By providing continuous activity, the experimental learning material will enhance the learning experience and effectiveness even when no extra exercises are handed out to the online learning community. According to our hypothesis, at Modules 4 and 6 the measured performance of the experimental group will be significantly better than that of the control group.

H2: When learning with interactive and activity-based learning materials, the students will use their learning time more effectively. Hence, by using the experimental learning material, students will be able to achieve the same performance level within less time. Our hypothesis suggests that within the control group, there is a strong positive correlation between the amount of time spent in the online environment and the measured performance achieved at Modules 4 and 6, while in the case of the experimental group, there is no significant correlation.

3 Participants and Data Collection

The total number of course participants was 107. 41% of the participants were between the ages of 40–49 yrs; 24% of them represented the age group of 50–59 yrs; 19% fell into the age group of 30–39 yrs, while only 16% of them were under the age of 30. The majority of the participants were experienced teachers of senior age. 95% of the participants were teachers; 56% of them came from primary and 38% from secondary education. Kindergarten educators and higher education teachers were also enrolled in the course. 99% of the group declared to have previous experience in presentation making; 35% of them claimed to have made educational video recordings before and 33% said that they had experience in online learning environment development. Former experiences in MOOC participation were surprisingly high within the group: 54% of the participants had attended and completed online courses before. The content of the course was published and made accessible for all participants in advance of application; however, when asked about their anticipations, only around 10% of the participants mentioned expectations formerly planned within the frameworks of the course. The activity of the participants was tracked by the online learning environment, and the students had to take additional knowledge level measuring tests for Modules 4 and 6. At the beginning of the course, participants had to fill a questionnaire regarding their previous experiences and expectations towards the course. Another questionnaire was filled at the end of the course, inquiring about their learning experience and opinion on the course. 16% of the participants had completed the knowledge test of the experimental modules at the end of week 4 and 6. The attrition rate of the course was slightly better than the usual average of 90%, which is an important result, taking into account the significant difference between the participants' prior expectations and the actual content of the course. We have no precise information on the amount of time spent on learning outside the online learning environment.

The course ended on 19th May 2017. The assessment of the submitted assignments, the evaluation of students' performance and the implementation of the questionnaires on students' opinion and experiences are in progress. Based on the data currently being collected, further hypotheses will be examined in the upcoming two weeks.

4 Interpretation of the Research Results

H1: According to the results of the completed performance measurement tests, there is no significant difference between the performance of the experimental and the control group in the case of Module 4 ($t = -0.02$, $p = 0.98$). In the case of Module 6, the test results revealed no significant difference between the performance of the experimental and the control group either ($t = -1.15$, $p = 0.28$). Hence, our first hypothesis was refuted. The use of the experimental learning material resulted in better, but not significantly better student performance. We need to find another of interactive and activity-based solution and we also need to increase the activity of students within the online environment in order to enhance their performance. As the experimental solution resulted in a higher level of difference between the experimental and control group in the case of Module 6, the technique used in that module is advised to be considered for further development.

H2: In the case of the control group, there is a significant negative correlation ($r = -0.68$, $p = 0.02$), between the amount of time spent in the online learning environment and the performance results at the end of Module 4; in the case of Module 6, there is also a significant negative correlation ($r = -0.77$, $p = 0.02$) between the amounts of time spent in the online learning environment and the measured performance of students. The analysis of the data regarding the experimental group revealed no correlation between the time spent in the online learning environment and measured performance, neither in the case of Module 4 or Module 6 ($r = 0.64$, $p = 0.24$, and $r = -0.89$, $p = 0.07$). The surprising results in both groups call for a more detailed examination and analysis of the time spent within the online learning environment. Based on the findings, the more time spent on learning will not necessarily result in better performance. It is important to note the independence of learning time and performance in the case of the control group; however, the negative correlation in the case of Module 6 requires further analyses. The general objective of experimental learning material development is to create activity-based interactive online environments, where similar performance results to traditional learning environments can be achieved with lower time expenditure.

5 Conclusions

Our research has revealed differences between the control and the experimental group in terms of students' achievements and learning experience. The achievement and learning experience of participants were also affected by their personal prior expectations towards the course. Based on the results of our research, we can determine specific students groups, where activity-based learning materials and increased activity offer an ideal solution.

The research results can be utilised in online instructional design and can also serve as a base of further activity-based experimental learning material development. Based on the findings, further research is suggested to explore the possibilities of enhancing learning experience by the conscious planning of collaborative student activity.

Acknowledgment. The first author's research was supported by the grant EFOP-3.6.1-16-2016-00001 ("Complex improvement of research capacities and services at Eszterhazy Karoly University").

References

Allen, E., Seaman, J.: Grade change: tracking online education in the United States. Babson Survey Research Group and Quahog Research Group, LLC (2014)

Anders, A.: Theories and applications of massive online open courses (MOOCs): the case for hybrid design. Int. Rev. Res. Open Distance Learn. **16**(6), 39–61 (2015)

Fini, A.: The technological dimension of a massive open online course: the case of the CCK08 course tools. Int. Rev. Res. Open Distance Learn. **10**(5), 74–96 (2009)

Greenberg, A.D., Zanetis, J.: The Impact of Broadcast and Streaming Video in Education. Report commissioned by Cisco Systems Inc. to Wainhouse Research, LLC. Ainhouse Research (2012)

Johnson, L., Adams Becker, S., Cummins, M., Estrada, V., Freeman, A., Ludgate, H.: NMC Horizon Report: 2013 Higher Education Edition. The New Media Consortium, Austin (2013)

Jordan, K.: Initial trends in enrolment and completion of massive open online courses. Int. Rev. Res. Open Distance Learn. **15**, 133–160 (2014)

Liyanagunawardena, T.R., Adams, A.A., Williams, S.A.: MOOCs: a systematic study of the published literature 2008–2012. Int. Rev. Res. Open Distance Learn. **14**(3), 202–227 (2013)

Macdonald, J., Twining, P.: Assessing activity–based learning for a networked course. Br. J. Educ. Technol. **33**(5), 603–618 (2002)

Stewart, B.: Massiveness + openness = new literacies of participation? MERLOT J. Online Learn. Teach. **9**(2), 228–238 (2013). http://jolt.merlot.org/vol9no2/stewart_bonnie_0613.htm

Wilkowski, J., Deutsch, A., Russell, M.D.: Student skill and goal achievement in the mapping with Google MOOC. In: L@S 2014 Proceedings of the First ACM Conference on Learning @ Scale Conference, pp. 3–10. ACM, New York (2014)

Work in Progress: An Investigation on the Use of SCORM Based Pre-class Activities in the Flipped Classrooms

R. D. Senthilkumar(✉)

Department of Math and Applied Sciences, Middle East College, PB 79, KOM, PC. 124 Al Rusayl, Muscat, Oman
senthil@mec.edu.om

Abstract. An investigation has been made to study the usefulness of SCORM based education in flipped classroom through a quasi-experimental design. The coursework assessments are used as a measuring instrument and the coursework assessment scores are statistically analysed. The results of Levene's test, t-test and Mann-Whitney test reveal that mean of coursework assessment scores of the experimental and control groups are statistically significant. This confirms the Alternate Hypothesis that students who studied using SCORM based lessons (experimental group) attained higher learning compared to the students who did not use the SCORM based lessons (control group).

Keywords: Flipped classroom · SCORM · Higher education
Physics education

1 Introduction

In recent days, higher education has been changing in so many ways and these all changes are pointing in one direction to increase educational opportunities for all students. Our students who are digital native also expect learning to be more flexible, individualized/peer collaborative, learning styles based and to work at their own pace beyond the classroom walls. Therefore, to meet the students' needs, education is more interconnected with Information and communication technologies (ICTs) [1]. Hence, education at all levels can no longer be assimilated to a group of students in a classroom following a teacher with a textbook. Further, class styles are also changing due to the flipped classroom which has been recognised as an innovative and effective instructional strategy [2]. In the flipped classes, educators assign a lecture (especially as digital content comprising videos, simulations, study materials, quizzes, etc.) for pre-class activities, which is a kind of homework, and students work collaboratively and apply knowledge to solve problems during the class time (face-to-face class). However, most of the educators are struggling for the creation and deployment of quality e-learning content and to track results from the use of the content [3]. These problems can be resolved by preparing the digital lessons in the Sharable Content Object Reference Model (SCORM) format and delivering to students through Learning Management System (LMS).

M. E. Auer et al. (eds.), *Teaching and Learning in a Digital World*,
Advances in Intelligent Systems and Computing 715,
https://doi.org/10.1007/978-3-319-73210-7_76

The SCORM is basically a comprehensive suite of eLearning standards to enable durability, portability, accessibility, interoperability and reusability of the eLearning content. SCORM was an initiative by US government in 1997 and developed by Advance Distributed Learning [4] with the main objective to produce course content (SCO's or shareable content Object), which is to work on any LMS.

This paper describes an investigative study on the effectiveness of SCORM based pre-class lessons/activities on students' academic progression in the physics module offering to the first year engineering students of the Middle East College (MEC), Oman.

2 Aim of This Study

Based on the MEC's teaching and learning strategy, all the traditional classes are flipped. Although the flipped teaching strategy is an effective and innovative approach, its success completely depends on the planning of lessons, designing the study materials and activities for the pre-classes, face-to-face classes and post-classes. Generally, educators use some narrow study materials and short videos along with quizzes for the pre-class activities. But, before the face-to-face class, there is no proper mechanism to track the students' performance in their pre-class activities which is a major drawback in the flipped classrooms. Hence, to overcome this drawback the pre-class lessons are designed in the SCORM format which enables to track students' performance immediately after the completion of pre-class activities.

The aim of this study is to investigate the effectiveness of SCORM based pre-class activities in the flipped classroom and its impact on students' learning at MEC, Oman with the following research question: "What is the difference in learning between the experimental group (EG) which does use SCORM based pre-class activities and the control group (CG) that does not use SCORM based pre-class activities?" And this study is planned to conduct in two phases (first phase in Spring 2017 semester and the second phase in Fall 2017 semester) in order to get the concrete inference.

2.1 Hypotheses

The working hypotheses tested for the above research question is as follows:

Null Hypothesis (H_0). There is no difference in learning attained by the experimental group (who had SCORM based activities) compared to the control group (who didn't have SCORM based activities).

Alternative Hypothesis (H_a). There is a difference in learning attained by the experimental group (who had SCORM based activities) compared to the control group (who didn't have SCORM based activities).

3 Methodology

In this study, a quasi-experimental design is used to investigate the usefulness of SCORM based pre-class activities on students' academic progression.

First year engineering students pursuing physics module at MEC during the spring 2017 semester are formed into two groups, namely Experimental Group (EG, n = 12 as registered in Session A) and Control Group (CG, n = 13 as registered in session B). Pre-class activities prepared in SCORM packages are used for the students of EG and the normal lessons (narrow study materials and short videos) are used for the students of CG. SCORM packages are prepared using the SoftChalk software and uploaded in Moodle at least one week prior to the scheduled class. Students' performance in the SCORM based pre-class activities are monitored from Moodle and then face-to-face classes are designed for the EG, while the predesigned face-to-face class activities are used for the CG. In order to minimize the effect of the tutor's teaching skills as a variable, the same tutor taught both groups of students. The coursework assessments (Closed Book Test (CBT), Quiz and Lab Test) and final exam are being used as a measuring instrument to quantify the students learning. Both the groups were taught for 2 h per week over a period of 14 weeks and CBT, Quiz and Lab Test were administered in week 6, week 11 and week 14 respectively, for both EG and CG. As the final exam for both the groups will be conducted in week 16, results of the final exam will be included in the second phase of this study.

3.1 Statistical Analysis

As an interim data analysis, the CBT, Quiz and Lab Test scores are subjected to a descriptive statistical analysis using the Shapiro-Wilk test to verify the normal distribution of the sample. Research hypotheses are tested by analysing the coursework results by the Levene's test and Mann-Whitney U test using the IBM SPSS package.

4 Results and Discussion

Table 1 shows the descriptive statistics for the coursework assessment scores obtained by students in this study along with the difference in mean scores (ΔL) between the EG and CG.

Table 1. Descriptive statistics of closed book test, quiz and lab test scores

Parameters	CBT		Quiz		Lab test	
	EG	CG	EG	CG	EG	CG
Mean score	79.72	59.47	83.08	67.18	84.48	71.19
Median	85.50	60.53	86.67	66.67	90.00	68.50
Std. deviation	11.46	21.62	10.51	17.93	17.12	14.85
Minimum	54.66	27.00	66.67	36.67	37.50	45.00
Maximum	91.00	94.00	93.33	96.67	100.00	95.00
$\Delta Learning(\Delta L)$	20.25%		15.90%		13.29%	

Table 1 demonstrates that the mean scores of EG achieved from CBT, Quiz and Lab Test are significantly greater than CG scores which support the Alternative Hypothesis.

Table 2 presents the results of Shapiro-Wilk and Kolmogorov-Smirnov tests to verify the normal distribution of the sample collected from the coursework scores. The p-values (significance) of the Shapiro-Wilk Test is considered for assessing normality of the sample as this test is more appropriate for small sample sizes (<50 samples). The p-values for the CBT scores of EG and CG are considerably greater than 0.05 which reveal that CBT scores of EG and CG are normally distributed. The Shapiro-Wilk test allows us to accept the normality of CG scores in Quiz and Lab Test as $(p > 0.05)$ but the EG scores in Quiz and Lab Test are not distributed as a normal distribution since p-values <0.05. Therefore for the further analysis of research hypotheses, Independent Student's t-test is used for the comparison of learning between EG and CG scores obtained in CBT, and Mann-Whitney test [5] is used to analyse the difference in EG and CG scores obtained in Quiz and Lab Test.

Table 2. Normality test results

Coursework assessments	Class	Kolmogorov-Smirnov			Shapiro-Wilk		
		Statistic	df	Sig.	Statistic	df	Sig.
CBT scores	EG	.261	12	.024	.868	12	.061
	CG	.150	13	.200	.945	13	.521
Quiz scores	EG	.239	12	.057	.824	12	.018
	CG	.130	13	.200	.967	13	.856
Lab test scores	EG	.266	12	.019	.755	12	.003
	CG	.114	13	.200	.975	13	.947

Table 3 represents the Levene's test result which is in the expected direction as the mean difference between EG and CG is 20.255 and significant as $p < 0.05$. This reveals that there is a significant changes in the mean of CBT scores of E.G and C.G. Further, the two-tailed p-value from the t-test is 0.008 $(p < 0.05)$.

Table 3. Independent samples test results of CBT scores

	Levene's Test for equality of variances		t-test for equality of means						
	F	Sig.	t	df	Sig. (2-tailed)	Mean difference	Std. error difference	95% confidence interval of the difference	
								Lower	Upper
A	6.436	0.018	2.903	23	.008	20.255	6.978	5.819	34.691
B			2.971	18.606	.008	20.255	6.818	5.965	34.546

A - Equal variances assumed; B - Equal variances not assumed

The results of the Mann–Whitney test are presented in Tables 4 and 5.

Table 4. Mann–Whitney U test resutls of quiz and lab test scores

	Class	N	Mean rank	Sum of ranks
Quiz	EG	12	16.54	198.50
	CG	13	9.73	126.50
Lab test	EG	12	16.75	201.00
	CG	13	9.54	124.00

Table 5. Mann–Whitney test for a 5% significance level for quiz and lab test

	Quiz	Lab
Mann-Whitney U	35.500	33.000
Wilcoxon W	126.500	124.000
Z	−2.322	−2.450
Asymp. Sig. (2-tailed)	0.020	0.014
Exact Sig. [2(1-tailed Sig.)]	0.019	0.014

Table 4 shows that the EG have an average rank of 16.54, while the CG have an average rank of 9.73 in Quiz. It could also be seen that EG have an average rank of 16.75, while CG have an average rank of 9.54 in Lab Test. According to the researchers, theory should become a useful and meaningful tool for understanding practical, rather than an end in itself [6]. The SCORM based lessons also develop skills for performing well even in the laboratory classes as the score of EG found higher than CG in Lab Test. The results of the Mann–Whitney test (Table 5) are in the expected direction and are significant ($p < 0.05$), showing that the differences in mean scores of the experimental groups with respect to their respective control groups are significant (i.e., higher scores obtained by EG is due to the use of SCORM based teaching strategy and not by a chance).

The results of all statistical analyses demonstrate the effectiveness of the use of SCORM based lessons in flipped classrooms and indicate that the acceptance of Alternative Hypothesis: "There is a difference in learning attained by the students of EG compared to the students of CG".

According to Wade and Moje [7], the teacher's role is not to be the authority to transmit a prescribed and fixed body of knowledge to students rather to foster conditions in which students are encouraged to construct knowledge themselves. In accordance to this statement, in this study author has made SCORM based lessons which foster student-led learning in both individual and collaborative model.

5 Conclusion

The effectiveness of the use of SCORM based lessons in flipped classroom, by determining the amount of learning attained by of first year engineering students studying physics module at the Middle East College, Oman is studied. The interim data analysis results confirm the Alternative Hypothesis that there is a difference in learning attained by EG compared to CG. It is found that the use of SCORM based lessons in education increased the learning in EG by 20.25%, 15.90% and 13.29% respectively, in CBT, Quiz and Lab Test. This may be due to the face-to-face class activities which are designed based on students' performance in the pre-class activities. It is demonstrated that it is viable to track student's performance on their pre-class activities in SCORM based learning and to promote the student-led learning by which students' academic progression can be enhanced. Also, the infusion of SCORM based lessons in education can provide a paradigm shift from teaching to learning. The second phase of this research study is in progress.

References

1. Assar, S.: Information and communications technology in education. In: International Encyclopedia of the Social and Behavioral Sciences, 2nd edn., pp. 66–71 (2015)
2. Hwang, G.J., Lai, C.L., Wang, S.Y.: Seamless flipped learning: a mobile technology enhanced flipped classroom with effective learning strategies. J. Comput. Edu. **2**(4), 449–473 (2015)
3. Osten, C.: In the Eye of the SCORM (2007). <Ostyn Consulting, PO Box 2362, Kirkland, WA 98083-2362, USA>. [Accessed 14 June 2017]
4. Fletcher, J.D., Tobias, S., Wisher, R.A.: Learning anytime, anywhere: advanced distributed learning and the changing face of education. Edu. Res. **36**(2), 96–102 (2007)
5. Mann, H.B., Whitney, D.R.: On a test of whether one of two random variables is stochastically larger than the other. Ann. Math. Stat. **18**(1), 50–60 (1947). http://www.jstor.org/stable/2236101
6. Korthagen, F.A.J., Kessels, J., Koster, B., Lagerwerf, B., Wubbels, T.: Linking Practice and Theory: The Pedagogy of Realistic Teacher Education. Lawrence Erlbaum Associates, Mahwah (2001)
7. Wade, S.E., Moje, E.B.: An introduction to the case pedagogies to teachers educators. In: Wade, S.E. (ed.) Preparing Teachers for Inclusive Education, p. 8, LEA Publishers, London (2000)

Collaborative Learning Supported with Mediawiki Platform in Technical University Environment

Ján Mojžiš[1(✉)], Štefan Balogh[2], Michal Ásványi[2], and Ivana Budinská[1]

[1] Institute of Informatics, Slovak Academy of Sciences, Bratislava, Slovakia
{jan.mojzis,ivana.budinska}@savba.sk

[2] Faculty of Electrical Engineering, Slovak Technical University in Bratislava, Bratislava, Slovakia
stefan.balogh@stuba.sk, michalasva@gmail.com

Abstract. According to several studies, student collaboration can be helpful in learning process. In this paper, we propose collaborative learning for students of a technical university as part of courses on "Computer criminality". We have prepared a base platform, MediaWiki, to support the collaborative approach. The MediaWiki is a free open source platform originally used for Wikipedia. We also prepared a questionnaire for students about their opinion on using collaborative learning as part of the school courses. We assume, based on the responses given by the students, that they prefer knowledge sharing among other collaborative learning approaches. The paper summarizes findings about students' attitude to collaborative learning and suggests some improvements for the design and development of a collaborative learning tool. A case study - a course on Computer criminality, is briefly described. Preliminary results indicate, that students can highly benefit from collaboration.

Keywords: MediaWiki · Survey · Student collaboration · Constructivist learning School · Faculty · Course

1 Introduction

Traditional learning process at school originates in school attendance. In university environment, there are lectures and practical classes. During lectures, students in a classroom listen to lecturers. During practical classes, students in a classroom, perform exercises of various kinds, carried by their teachers. Often their duties include preparation of projects, seminar or semestral works. Once written, students present and defend them in front of the teacher and classmates in a classroom. Other duties include continual study, searching for resources (often on the Web or in a school library). Although, for a student, the teacher is his primary source of knowledge, his fellow students may be an additional source. For example, question sections and discussions during presentations, simple exchange of opinions. Learning through interactions and collaboration is the subject of Constructivism theory of learning.

M. E. Auer et al. (eds.), *Teaching and Learning in a Digital World*,
Advances in Intelligent Systems and Computing 715,
https://doi.org/10.1007/978-3-319-73210-7_77

For the constructivist learning, perhaps the most relevant aspect is a student-centered learning. According to the constructivist learning paradigm, students have to communicate with other students actively [1]. Based on this theory, learning occurs by social interaction and communication with others [2]. We, thus, would like to advance on Constructivist learning, support student interaction and contribute to the learning process with an innovative approach.

2 Motivation

There are very specific aspects to teaching subjects such as "Computer Criminality". It is a dynamically developing area. There exists a huge amount of online resources, but there is a lack of relevant classical course books or other reliable study materials.

Students in a local faculty already produce their schoolwork (seminar, semestral works, projects) in order to gain credits. They present their works in front of the teacher and their fellow students. They create electronic documents (like .PDF, .DOC or .PPT) as a result of their school works.

Materials produced by students are revised by teachers and commented on by fellow students. Those documents can become a good resource of knowledge for other students. We would make such documents accessible for a wider audience by publishing them on the web. The role of the teacher is very important. The teacher moderates and regulates the discussions. Even though the teacher is not supposed to be the one who knows everything, he should be the authority who decides about correctness of arguments. His function is to stop any unreliable comments and to ask for evidence of claims in students work and/or comments.

The goal is to introduce students to a process of constructivist learning. Introducing online collaborative learning platform could spare the teacher valuable time, which he can use to prepare better lectures or perform research.

According to Duffy and Johanssen [3], learners with different skills and backgrounds should collaborate in tasks and discussions to come to a shared understanding of the truth in a specific field. Constructivism is a relevant subject in learning, proving its influence in Computer science and machine learning systems [6].

3 Related Work

We can already find successful application of constructivist learning approach in school environment well documented by Tagliaferri [4] and Chu [5], where students used online collaborative tools to contribute and collaborate during school projects work and learning.

4 Preliminary Results

4.1 MediaWiki Page

We have set up a web page with a general knowledge-base for the subject "Computer Criminality". This course is periodically studied each school year in summer semester.

There are three main topics - Spam and phishing, Social engineering and Ransomware presented together with world statistics on computer criminality on the wiki pages. Students publish their documents, notes or articles on the wiki. The access for editing is restricted for the course attendees. Students are assigned roles and they have access rights according to their login credentials. However, the content of the pages is visible for public.

The page can be seen at the URL
http://147.175.98.31/wiki/index.php/Hlavn%C3%A1_str%C3%A1nka

4.2 A Survey

We prepared a survey in order to find out the actual opinion of the students, to know how they assess the possible collaboration, whether they are willing to contribute to the learning process in such a way. The survey consists of 7 questions. The questions are listed below, followed by charts of answers.

- Question 1: If your work (projects, notes, seminar works) is to be used by other students, will you contribute?
- Question 2: Is student knowledge sharing appropriate?
- Question 3: Is it appropriate to discuss the topic of collaborative learning t and to research, how students can contribute to a learning process?
- Question 4: Can MediaWiki be used as a support for collaborative learning?
- Question 5: If you are about to use MediaWiki and contribute to the content of your course/subject, what type of role would you chose?
- Question 6: If you and other students use MediaWiki, to whom should its content be accessible?
- Question 7: If MediaWiki should be used for collaborative learning, as a place where students can share their knowledge, what should be the students' contribution?

An online questionnaire in Slovak can be seen at the URL
https://drive.google.com/open?id=116Vwe5F1q9NQPXLsJYq-cYRllpsX7M0jwfLJ1g_i5-ZM

We have received answers from 57 students, in total.

According to the survey, the faculty students are open towards knowledge sharing by the web page. They welcome new possibilities and advancement in the learning process at school. The answers are below (Figs. 1, 2, 3, 4, 5, 6 and 7).

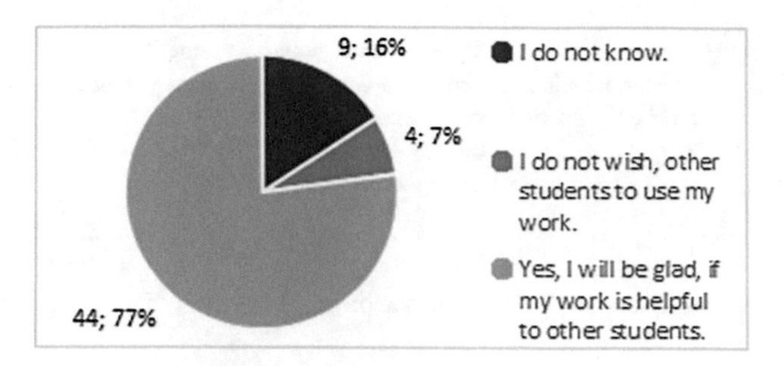

Fig. 1. Answers to the question 1

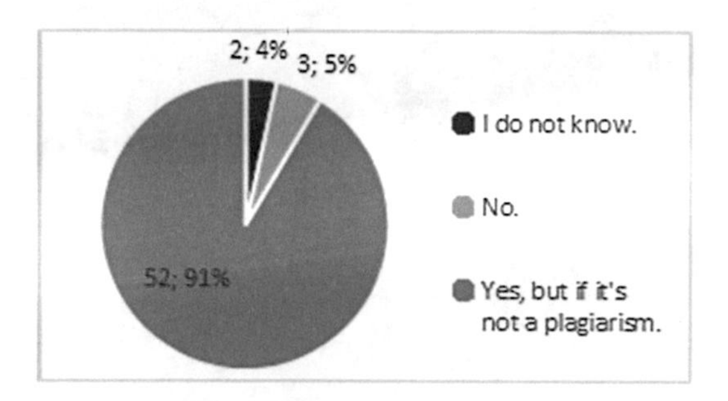

Fig. 2. Answers to the question 2

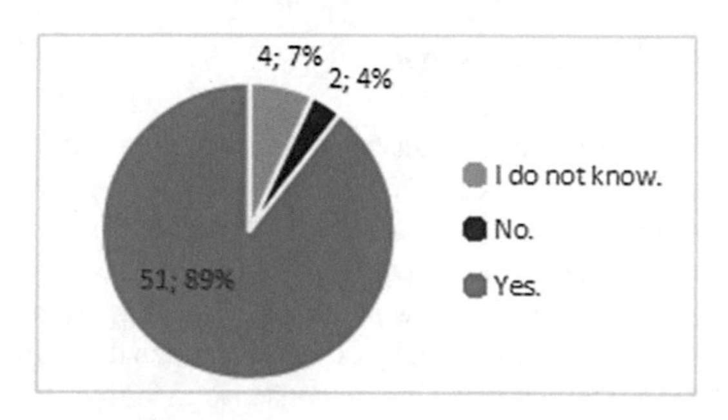

Fig. 3. Answers to the question 3

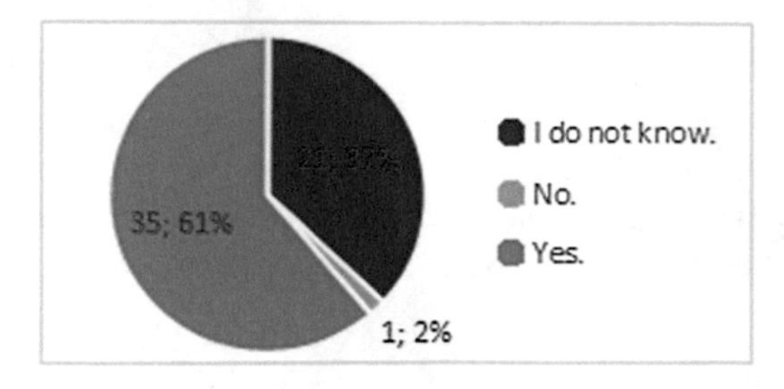

Fig. 4. Answers to the question 4

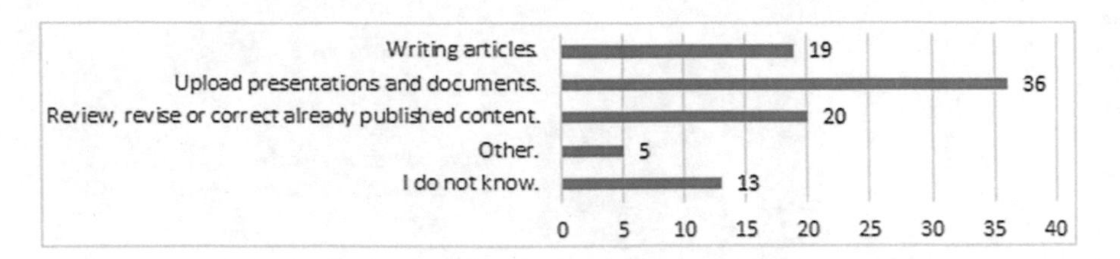

Fig. 5. Answers to the question 5. The most favorite answer was "Upload presentations and documents."

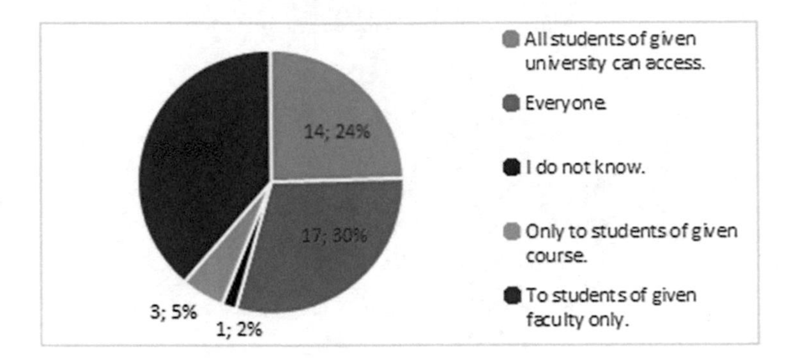

Fig. 6. Answers to question 6

5 Conclusion and Future Work

We have created MediaWiki pages as a base platform for student collaboration. It contains information about the particular subject. A survey, students had filled, indicates support for our objective - join students for collaborative learning. Students could contribute with their presentations and notes (Fig. 7). In most cases, students are willing to share their knowledge with their fellow students. According to the students, knowledge sharing is appropriate (Fig. 2). Also, it is suitable to open the collaborative learning topic for discussion (Fig. 3). Interesting are responses to the 6th question (Fig. 6). Most of them would grant access to the content of MediaWiki to all students of at least a given faculty. Being still in the work-in-progress stage, means, that collaboration would fully start in the summer semester of 2017/2018.

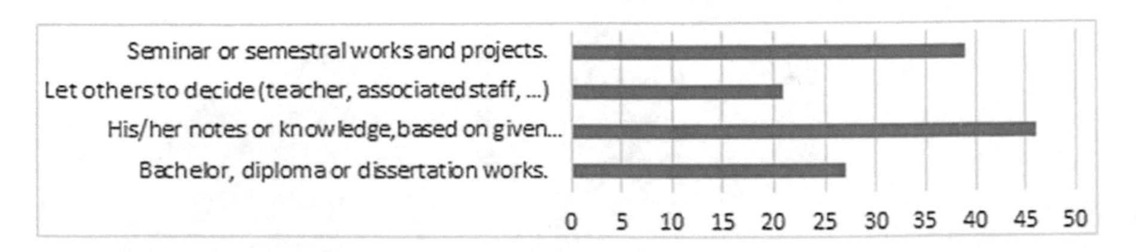

Fig. 7. Answers to the question 7. The most favorite answer was "His/her notes or knowledge, based on given course/theme/subject."

The pages would have a responsible design and in such a way it would support a ubiquitous learning that is attractive for young students.

Acknowledgment. This work has been supported by the Scientific Grant Agency - VEGA under grant No. 2/0154/16.

References

1. Harasim, L.: On-Line Education: Perspectives on a New Medium. PraegedGreenwood, New York (1990)
2. Vygotsky, L.S.: Mind in Society. Harvard University Press, Cambridge (1978)
3. Duffy, T.M., Jonassen, D.H. (eds.): Constructivism and the Technology of Instruction: A Conversation. Routledge (2013)
4. Tagliaferri, L.: Open-access student-centered learning. In: International Conference on Interactive Collaborative Learning, pp. 605–619. Springer, Cham, September 2016
5. Chu, S.K.W., Kennedy, D., Mak, Y.K.: MediaWiki and Google Docs as online collaboration tools for group project co-construction. In: Proceedings of the 2009 International Conference on Knowledge Management (2009)
6. Sarkar, A.: Constructivist design for interactive machine learning. In: Proceedings of the 2016 CHI Conference Extended Abstracts on Human Factors in Computing Systems, pp. 1467–1475. ACM, May 2016

Openness of Academic Staff for Educational Innovation in Hungarian HEIs

Maria Kocsis Baan[1(✉)], Edina Espán[2], and Adrienn Nehézy[2]

[1] Hungarian e-University Network, Miskolc, Hungary
m.kocsis.baan@uni-miskolc.hu
[2] University of Miskolc, Miskolc, Hungary
{espan.edina,nehezy.adrienn}@uni-miskolc.hu

Abstract. Dated from 2016 the Digital Educational Strategy of Hungary [1] concluded, that adaptation of advanced pedagogic methodology and ICT is generally at low level in HEIs, its progress is slow and very much uneven not only regarding different institutions but even within the same faculty or department. Working for the e-Learning Centre at a provincial university of Hungary, we launched a comprehensive survey on the openness of our colleagues for educational innovation. Teaching staff at the six different faculties at our university were asked to fill in a questionnaire, asking them about their digital competencies and habits, about the different teaching/learning tools they know/apply. Answers were also analyzed to compare the teaching practice of different faculties and different cohort of teachers. In our poster, we wish to give detailed analysis of our first findings by attractive infographics.

Keywords: e-Learning · Learning tools · ICT for learning and teaching

1 Needs for a Change in Higher Education

In the recent decades, modernization of Higher Education is considered as a key element in competitiveness of economy and sustainability of development. New demands – such as meeting the needs of internationalisation and globalisation, widening the mission and strategy of HEIs for playing an important role in knowledge economy, focusing on specific needs of Lifelong learning – also require new approaches and new tools. Integration of ICT not only changes how we teach and how we learn, but also opens up unique possibilities for answering these challenges, creating networked communities of all stakeholders of the globalised education market.

Investing and innovating in education and training are regarded as key priority areas in strategic plans of the European Union and its member states. Several EU-supported programs and projects offered outstanding opportunities for HEIs for taking part in educational innovation [2] in such a challenging period, when ICT has become the driving force of a very dynamic development, but also requiring considerable investment in infrastructure. As one of the largest provincial universities of Hungary, University of Miskolc (UoM) participated in several of such projects in the last two decades. However,

M. E. Auer et al. (eds.), *Teaching and Learning in a Digital World*,
Advances in Intelligent Systems and Computing 715,
https://doi.org/10.1007/978-3-319-73210-7_78

in spite of being among the front-runners in national and European ODL networks [3], educational innovation has stuck at project level and has not been forced and stabilized by institutional integration.

The situation is very similar in the majority of the Hungarian and European HEIs. Dated from 2016 the Digital Educational Strategy of Hungary [1] reviewed the present state-of-art at all level of the educational system. Among others it concluded, that regarding HEIs, adaptation of advanced pedagogic methodology and ICT is generally at low level, its progress is slow and sporadic, very much uneven not only regarding different institutions but even within the same faculty or department. Although in the recent years several huge national projects supported the development of digital content for courses in all kinds of professional fields, even these open-access course materials have not been implemented successfully in the daily practice of neither the developing, nor the other institutions. Similar experiences are generally mentioned regarding the impact of the former Life Long Learning and eLearning programs. Although very positive impacts are reported at micro level, i.e. for participating institutions and individuals, however these results have led to moderated strategic effects at national education systems and in European dimensions. In spite of the wide range of innovative ICT-based content developed in several subjects, applying enriched pedagogic practices, it is still a general conclusion, that more efforts are needed on a strategic level, in order to upscale the valuable results obtained in projects, small-scale research and pilot-experiments. Further improvement of teaching quality needs developing and implementing innovative pedagogies, not only “digitalisation”.

2 Changing Role of Academic Staff

For several centuries universities fulfilled their two major missions – i.e. being the sources of knowledge and providing education on the highest level – which strongly and mutually strengthen each other. Role of educators in this period has not changed dramatically – development was relatively slow and access to professional information was limited, so in this traditional educational model academics were responsible for collecting and redistributing professional knowledge in a structured system of the given professional field. But by now, these traditional missions have been complemented with a new function: universities are becoming more and more important actors of the knowledge economy. It is not only their ambition, but also a need, as governmental support can hardly cover the operational and development costs of institutions. Also, individual career of academics strongly depend on measurable bibliometric data to be achieved in scientific research. These tendencies seem to cause radical shift of attention from education to research. Unbalanced recognition and respect may not only cause short term malfunctioning, but even more serious problems: failing to meet the changing needs of teaching the “digital” generation, failing to find proper answers for the radical changes and new challenges in education.

Methodology of teaching and motivation for improving pedagogic skills of academics is far from the top of priority lists [4]: students are taught by professionals without being taught in pedagogical theory and practice, applying the same methodology as they themselves were taught - while extended content and complexity should require more effective, modern pedagogical approaches and tools. Academics have to recognise: we cannot "transfer" all the knowledge, what our students will need in their life-time as engineers – instead, we should provide them solid fundamental knowledge on the basics, and we should equip them with skills of finding and critically evaluating, filtering the enormous mass of information, and integrating them into their own problem-solving competencies [5].

Development of attractive, visually rich and pedagogically well-established learning materials is obviously needs considerable investment – even if the efficiency will prove its added values, several of the institutions may consider it as luxury expenses. However, availability of more and more open educational resources may offer cost-effective, versatile solutions for innovation. Gateway portals give easy access to learning objects under Creative Common licensing, to be remixed, refined – without costs and under transparent preconditions of copyright issues.

Sharing open education resources (OER) and courses (OCW), launching massive on-line open courses (MOOCs) have gradually broken the ice: in contrast with the "top-down" approach of strategic planning, individual scale of innovation in education has become feasible, due to the easy-access and user-friend WEB2.0 tools. Lecturers may create and share their content upon their own decision, using open tools and platforms, in a no-cost model. Paradigm shift is being implemented in a "bottom-up" approach, modernisation of HE is driven by the "capillarity effect" of individual efforts of motivated academics.

In order to facilitate the learning-centred, learning-intensive education, the Digital Educational Strategy document summarised the most urgent operational interventions to be done in the near future - among others, developing the digital skills and methodological training of academic staff, as well as establishment of motivating mechanisms to encourage them for using modern pedagogical tools.

3 Survey on the Digital Competences and Openness for Educational Innovation of Teachers

Working for the e-Learning Centre at one of the largest provincial university of Hungary, we decided to launch a comprehensive survey on the openness of our colleagues for educational innovation. Teaching staff at the different faculties at our university were asked to fill in a questionnaire, asking them about their digital competencies and habits, about the different teaching/learning tools they know and apply for educational purposes. The idea and some elements of the survey originated from the experiences of different top lists of learning tools, published on the web – among them the Top200 Tools for Learning 2016, published by the Centre for Learning and Performance Technologies (C4LPT) in early October 2016 proved to be the most appropriate for our purposes [6]. The C4LPT is one of the world's most visited learning sites on the Web, counting over 2.4 million visits viewing over 10.5 million pages in the last year. Since 2007 top lists of the most popular learning tools are published, accompanied by very useful information about these tools. In 2016, the Top200 list was compiled from the votes of contributors coming from 64 countries, working as: trainers/instructors (17%), university/college/adult education teachers (15%), instructional designers/e-learning developers (16%), consultants/advisors (12%), 11% L&D managers/specialists (12%), etc., counting a total of 1238 learning professionals. The results are published in versatile forms – for us, "Best of Breed 2016" presentation seemed to be the most useful, as in this list all tools are classified into a number of categories.

3.1 Methodology of the Online Survey

Before launching the survey in the Google Form, we organized a workshop for our most active e-learning user academics and tested some selected element of the survey. The questionnaire consisted of 4 chapters, in the first part we asked our colleagues about their faculty and age, also whether they have received any kind of pedagogical training. In the next section they were asked about the most intensively used ICT tools in education – in this stage we did not influence their answers by giving them any list. In the third chapter 10–12 learning tools were listed in all main categories of the "Best of Breed" collection, i.e. selected items from **Instructional, Content development, Social** and **Personal tools.** Finally, the last chapter was dedicated to the needs analysis, what type of support do our teachers need to improve their competencies in using advanced ICT solutions for their teaching practice.

Table 1 shows the number of academic staff at our different faculties and the number of answers we received. As an average, a bit more than the quarter of our teaching staff filled in the online form. No higher openness was shown by the academics of engineering faculties for participating in the survey, than those faculties focusing on humanities and social sciences. Nearly the same questionnaire was filled in by the same number of teachers from the public education sector (learners of a postgraduate degree course)

Table 1. Statistical data of academic staff and answers in the survey

Faculty	Total no of academic staff	No of answers	Percentages of answers, %
Faculty of Earth Science and Engineering	91	16	17.6
Faculty of Materials Science and Engineering	55	20	36.4
Faculty of Mechanical Engineering and Informatics	178	48	27.0
Faculty of Law	65	17	26.2
Faculty of Economics	77	28	36.4
Faculty of Arts	90	20	22.2
Faculty of Healthcare	50	13	26.0
Bartók Béla Music Institute	26	1	3.8
Others (Foreign Language, Sport, etc.)	34	7	20.6
Total	666	170	25.5

offering the possibility for making comparison in diffusion of digital innovation regarding different layers of teachers of the Hungarian education system.

3.2 Results of the Survey

Detailed analysis of the survey results has led to some expected but also some surprising statements and are summarised in informative infografics. Figure 1 shows the Top10 list of our teaching staff, based on the answers of two cohort groups of digital immigrants (born before 1976) and digital native (Y and Z) generation. In each column, numbers show the percentages of users of the given tool within the same age-group while the colour of the column refers to the "Best of Breed" categories. Red numbers in the double-circles indicate the position of the tool in the international TOP200 list. As the diagram shows, there is no significant differences in the most popular tools regarding the age of teachers, although the percentages of users are a bit higher in the digital native group. Comparing the top list of Hungarian teachers with the international list, more significant differences can be recognised: some tools having prestigious position in the Top200 list – typically in Social Tools category - are much less important, or even perfectly missing from tool-kit of our academics, as shown in Table 2.

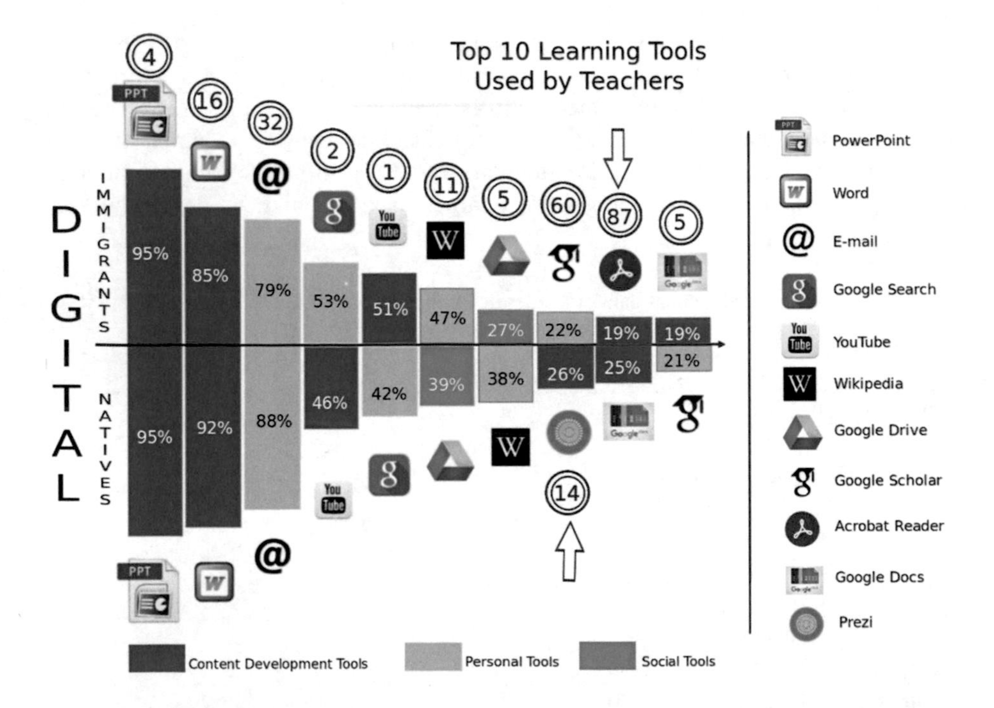

Fig. 1. Comparison of Top10 list of digital immigrant and digital native age-group of the academics at University of Miskolc, also indicating Top200 position of the given tools.

Table 2. Comparison of learning tools showing great differences in position of TOP lists

Tool	Category/ subcategory	Position in TOP200	Position in UoM	Percentage of users at UoM, %
Twitter	Social/public social networks	3	44	<1
Facebook	Social/public social networks	6	16	12
Skype	Social/group messaging apps, group video tools	7	15	13
LinkedIn	Social/public social networks	8	30	3
Word Press	Content dev./ blogging/website	9	52	<1
DropBox	Social/file synchr. & sharing	10	14	17

For the future we plan to widen our survey to the majority of the Hungarian universities, as part of a series of comprehensive studies on e-learning to be developed in collaboration with the Hungarian e-University Network.

4 Summary

Wide scale, but critical and responsible use of ICT for educational purposes can be regarded as a main driving force for the modernization of HE, supporting the overall educational reforms and extension of LLL concept. WEB2.0 technologies and open access of versatile content and tools have made individual scale of innovation feasible, teachers may create and share their content upon their own decision, using open tools and platforms, in a no-cost model. Paradigm shift is being implemented in a "bottom-up" approach, digital innovation is driven by the "capillarity effect" of individual efforts of motivated academics.

References

1. Digital Educational Strategy of Hungary (2016). (in Hungarian). http://www.kormany.hu/
2. Kocsis Baán, M.: Innovative E-learning solutions "Password" of CEE region for entering the European higher education space. Learn. Technol. Newsl. **9**(1), 2023 (2007). http://lttf.ieee.org/issues/january2007/lt_january2007.pdf
3. Kocsis Baán, M.: Multilingual E-learning programmes for engineering education. In: Proceedings of the SEFI and IGIP Joint Annual Conference on Joining Forces in Engineering Education Towards Excellence (2007). ISBN 978 963 661 772 1
4. McAleese, M., et al.: Improving the quality of teaching and learning in Europe's higher education institutions (2013). http://ec.europa.eu/dgs/education_culture/repository/education/library/reports/modernisation_en.pdf

5. Auer, M.E.: Present and Future Challenges in Engineering Education and the Strategies of IGIP. International Society for Engineering Education (IGIP) (2013). http://www.asee.org/public/conferences/27/papers/8381/download
6. Hart, J.: Top Tools for Learning 2016 (2016). http://c4lpt.co.uk/top100tools/

A Remote Mode Master Degree Program in Sustainable Energy Engineering: Student Perception and Future Direction

Udalamattha Gamage Kithsiri[1,2](✉), Ambaga Pathirage Thanushka Sandaruwan Peiris[1,2], Tharanga Wickramarathna[1,2], Kumudu Amarawardhana[1,2], Ruchira Abeyweera[1,2], Nihal N. Senanayake[1], Jeevan Jayasuriya[2,3], and Torsten H. Fransson[2,4]

[1] The Open University of Sri Lanka, Colombo, Sri Lanka
kithsirigamage@yahoo.com
[2] KTH Royal Institute of Technology, Stockholm, Sweden
[3] EIT InnoEnergy, Stockholm, Sweden
[4] EIT InnoEnergy, Eindhoven, Netherlands

Abstract. Remote mode higher education at postgraduate level is becoming popular among students because of flexible learning opportunity and the accessibility to study programs offered by renowned universities in the world. Fast development of internet facilities and learner management systems along with the development of remote educational pedagogy have been the driving force behind the acceptance and development of distant mode study programs. The success of such a study programs is largely affected by several factors that are unique to the university that offers the study program and the demography of participants as well as infrastructure and the student support available at the receiving end.

In the present study, the successes and the drawbacks as perceived by the participants of a distant master study program are evaluated. The study program considered was the Sustainable Energy Engineering Worldwide (SEEW) master degree program which was offered by the Royal Institute of Technology (KTH) in Sweden to students in Sri Lanka (Apart from Sri Lanka, SEEW was offered by KTH to some other countries; Zimbabwe, Ethiopia, Mauritius). The objective of offering the SEEW master program was to assist the developing nations to build up human resources with expertise in sustainable energy generation and utilization, hence contributing to national development. As such the program also generally contributes to global efforts of alleviating unfavourable environmental impacts connected with power generation and utilization. The SEEW master program consisted of 120 ECTS (ECTS: European Credit Transfer System) and the courses were offered over three semesters followed by a research project of 30 ECTS during the fourth semester. Lectures were delivered synchronous with the parallel KTH on-campus study program in real time through internet with the support of a learner management system. The students were attached to the Open University of Sri Lanka (OUSL) for providing

The original version of this chapter was revised: second author's name has been updated. The erratum to this chapter is available at https://doi.org/10.1007/978-3-319-73210-7_109

M. E. Auer et al. (eds.), *Teaching and Learning in a Digital World*,
Advances in Intelligent Systems and Computing 715,
https://doi.org/10.1007/978-3-319-73210-7_79

academic support where necessary and for the supervision of written and online examinations. The first enrolment consisted of 21 students in intake 2008 and the program was conducted with varying student numbers until the intake 2010. A total of 72 students have successfully completed the SEEW program and they are at presently employed in key organizations in the energy sector as well as in national universities in Sri Lanka.

The paper focusses on eight key areas that the students have identified as vital for success for this type of programs. These key areas are the effectiveness of web tools used, standard of teaching, standard of course content, examination procedures, online assessment, thesis projects, benefit to the students, and benefits to facilitating university. In the study 36 students responded to survey and overall rating of the program successfulness was identified as 72%.

Keywords: Distant mode education · Sustainable energy engineering Master degree · KTH · OUSL

Nomenclature

SEE	Sustainable Energy Engineering
SEEW	Sustainable Energy Engineering Worldwide
ECTS	European Credit Transfer System
KTH	Royal Institute of Technology
OUSL	The Open University of Sri Lanka

1 Introduction

Technological evolution has immensely improved and updated most products and services over the recent years. Drastic evolution of automobiles, aero planes, telephones and services such as medicine can be quoted as examples. However, a comparative analysis would show that education, being a crucial service of the society, has not evolved in a comparable scale. Education has however not stood still. Effective educational aspects have been absorbed into the system, which has transformed the teacher's role to be a knowledge facilitator instead of being an information/knowledge provider. Furthermore, 'online education' became a buzzword around a decade ago. It is also becoming more than a cheaper and less time-consuming alternative to traditional education system. However, advances in e-learning as well as shortcomings in conventional class room education are not familiar in the Sri Lankan context and thus make it a quite challenging process to absorb its total benefits and ensure student success.

In general, convenience and flexibility are, from a student perspective, the most interesting features of online education models. Whilst been engaged in a full-time occupation or with some other general barriers such as parental obligations, online education is still possible to be taken up by any motivated individual. For example, if an online course expects a student to have self-studies for about 40 h per week, student has the choice of distributing these 40 h in the week according to his convenience. However, this demands a lot of self-commitment towards the program to achieve the targets. Therefore, online courses cater a limited group of students who are self-motivated to pursue such courses. Online education programs are delivered to the students over the world through different digital platforms. This naturally develops the digital skills of those who participate in online programs. Since online programs do not required high cost infrastructure, it can be provided at a lesser cost. The resource personnel will be the world best as it is an online program making the quality of the program is unparalleled or even better than that of traditional educational program. In spite of all the advantages, one of the main drawbacks of online education programs is the non-availability of physical interaction among trainers as well as trainees, which lays a strong foundation to have a network during their careers later.

Some motivating factors for enrolling in online programs are career advancement, job performance improvement, convenience and obtaining a degree from a reputed institution. The strengths of such programs lie in the nature of the curriculum, availability of networking opportunities and the administrative and technical support, whereas participants' perceptions on quality of interaction with the faculty and instructional methods are mostly responsible for the weaknesses of such programs according to perceptions of students [1]. Online portfolio is a useful tool if active participation from both trainers and trainees is available. A study has revealed that lack of time, trainer support, proper introduction and personal motivation and inappropriate IT facilities were the key obstacles in an online program [2]. In another study, a structural equation modelling has been used to examine the facts that determine students' satisfaction on online courses [3]. They had used independent variables (course structure, instructor feedback, self-motivation, learning style, interaction and instructor facilitation) as potential determinants of online programs and the results indicated that online education could be a superior mode of instruction, if it is targeted to learners with specific learning styles and with timely, meaningful instructor feedback of various types.

The research literature on online programs establishes that interaction is important for the successfulness of the course. A study has carried out to ascertain the performance level of students in relationship to their quality and quantity of interaction [4]. The data on multiple independent variables (depth interaction and presence) and dependent variable (measures on performance) have been subjected to their analysis. The results of this study established a strong relationship between students' perceptions of the quality and quantity of their interaction and their perceived performance in an online program. Another study examined the skill acquisition of students enrolled in an on-campus and online introductory counseling skills course and revealed that there was no difference between students' skill acquisition in either course format [5].

In the present paper, a detailed analysis was carried out to, based on students' perception, determine the virtues and deficiencies of an online master degree program, SEEW.

1.1 Study Program

The SEEW master program is a two-year study program conducted in synchronous with an on-campus study program conducted by KTH, Sweden. This program was introduced to Sri Lanka in 2007 by KTH, Sweden, whilst the Open University of Sri Lanka played the role of the facilitating university. Since then 127 Sri Lankan students got themselves registered on the program and 72 of them successfully completed the same [5].

This program had two specializations, Sustainable Power Generation and Sustainable Energy Utilization. The uttermost objective of the program was to build world recognized resource personnel for sustainable power generation and sustainable energy utilization in countries like Sri Lanka.

The term "sustainable energy engineering" comprises a wide array of practices, policies and technologies (conventional and renewable/alternative) aimed at providing energy at the least financial, environment and social cost. To conduct a sustainable energy engineering program in master level in distant base, it is required to provide adequate amount of internet facilities for the students, lecturers, and program facilitators. In the distant learning process, it is required to provide live on-line lecture session that provide opportunity to link student- teacher directly. Several web based tools were used to facilitate the SEEW master program, such as;

- Centra[1] - Centra provided an online-platform to web-cast the lectures, where students can follow the lectures live on-line, could raise questions from the lecturers, being miles away from the lecture room. Further, recorded lectures could be downloaded for later learning activities.
- Bilda[2] - Blida facilitated the students to download the learning material and upload the assignments. Further it was used to carry out the on-line exercises, which were automatically corrected. The submission of assignments/exercises could be done through Bilda without being physically in the university.
- CompEdu[3] - CompEdu is a repository, a collection of well-structured on-line resources on sustainable energy engineering module authored and administrated by the department of Energy Technology, KTH.
- Web mail - KTH web mail address was given to all the students individually, so that KTH staff could communicate with the students more effectively.
- On-line library – The KTH on-line library has a collection of e-books and technical papers on energy engineering subjects. Students can easily access and download a copy without being physically in the library.
- On-line laboratory - The laboratory experiments were done at KTH laboratory by an instructor explaining all the procedures and steps. Once the recorded video and the set of data were given to the SEEW students, they could complete the rest of the experiment as the on-campus students. Thereby, the learning outcomes of each module were achieved for SEEW students without compromising the laboratory exposure for them.

The SEEW program consists of 90 ECTS from 14–16 of specialized modules and 30 ECTS from the research component. The modules of the master program were delivered over three semesters and the research was done in fourth semester. The research was intended to be in the field of sustainable energy engineering, enriched with novel ideas & analytical thinking and with constructive and useful outcomes. The structure of SEEW program and the aforementioned online facilities along with the

[1] Adobe Connect Meetings - www.adobe.com/products/adobeconnect/meetings.html.

[2] *Bilda* - https://bilda.kth.se/.

[3] *CompEdu* - http://www.energy.kth.se/compedu/webcompedu/.

service of the facilitators ensured that the SEEW program provides the knowledge and experience equivalent to that of the on-campus MSc degree program. KTH offered the degree certificate for SEEW students on successful completion of 120 ECTS mentioning the mode of education as online. With this background, this paper aims to critically analyse the perceptions of the students towards online education, by taking the SEEW as a model.

2 Methodology

An online survey was carried out through a questionnaire primarily to obtain the views of the participants, followed by interviews to ensure that their perceptions are correctly reflected in the study. The questionnaire, covering the following eight key aspects of the program, was administered to all students who have successfully completed and to those who are yet to complete some of the courses of the program.

1. User friendliness, accessibility and quality of web tools used in the program
2. Quality of online teaching
3. Standard of course content
4. Standard of online examinations, assignments and laboratory activities
5. Pass rate of examinations
6. Standard and relevance of thesis projects
7. Benefits of the programme
8. Benefits to facilitating institute and staff.

3 Analysis of Result and Discussion

Total of 36 students responded to the survey and 5 of them have not completed the program. Responses were analysed using the diverging Stacked Bar charts methods under two categories; views of a general students and views of students who have not completed the MSc program. Questionnaire addresses these two groups of students by covering 30 aspects of the program focused to the general students, while other aspects specifically address the issues of the remaining group.

3.1 General Student View of Program

Overall result of the survey was obtained considering eight major aspects of the program as shown in Fig. 1. Most of the aspects of the program were responded positively and nature of the course, methods conducted, facilities provided and output of program were agreed. Overall rating of the successiveness of the program was derived as 72% out of which 30% of the responds were under the category of 'strongly agree'.

Table 1. Key to the chart

A	User friendliness, accessibility and quality of web tools used in the program
B	Quality of online teaching
C	Standard of course content
D	Standard of online examinations, assignments and laboratory activities
E	Pass rate of examinations
F	Standard and relevance of thesis projects
G	Benefits of the programme
H	Service of facilitating institute and staff
T	Overall rating for the programme

The details of survey are in Annexure I.

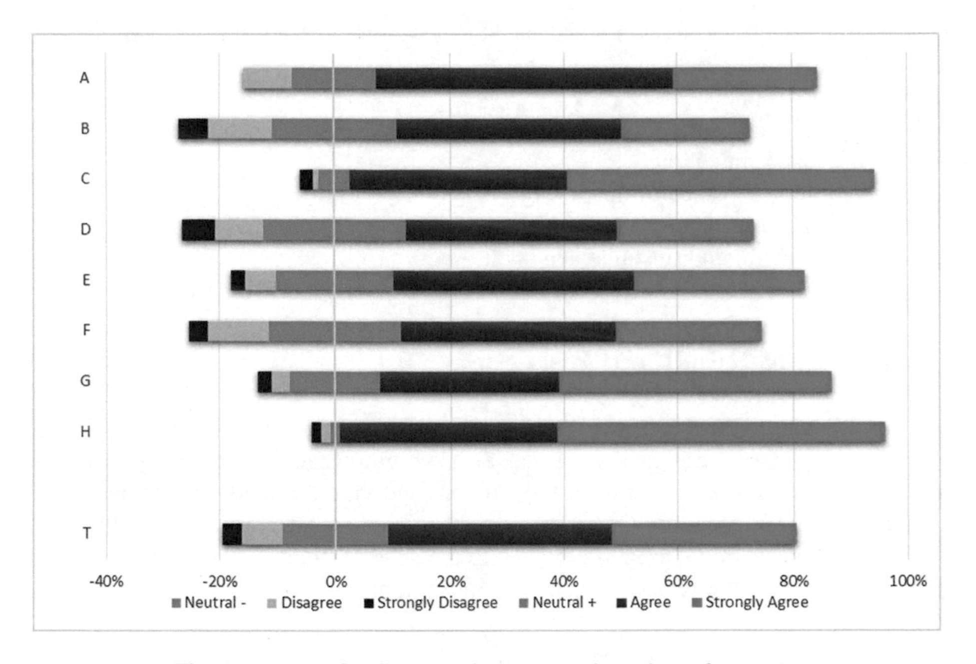

Fig. 1. Result of online questioner on student view of program

User friendliness, accessibility and quality of web tools used in the program (A)
User friendly web accessibility and quality of web tools used in study program were strong pillars in the successiveness of the program. Bilda, Centra and Compedu web tools were subjected into survey and Bilda web interface which was facilitating to online access to lectures, assignments and online examinations was highly accepted Further other web tools also recognized as useful by the majority of participants of the survey.

Quality of online teaching (B)
This was evaluated under the eight sub topics and all responses are positive. Standards of lectures were highly rated by more than 96% while quality of recorded lectures, online teaching and downloading of lecture also highly accepted. Direct interaction with lecturer and response time were identified to be at a lower scale.

Standards of course content (C)
Specialization of subjects, depth of the content of each subjects and level of provided course materials were evaluated and the responses show that the students are satisfied on all these aspects.

Standards of online examinations, assignments and laboratory activities (D)
Successiveness of online assignments, online examinations and online laboratory activities were strongly accepted by the majority. However, assignments and lab activities were identified as difficult. Further, it was realized that, delays in obtaining credits for submitted assignments and lab activities has resulted in delaying the completion of specific module.

Pass rate of examinations (E)
Level and standards of examination was positively responded by majority of the respondents, while the participants were satisfied with written examinations and result issuing time frame. A difficulty during repeat examinations and insufficient time gap between examinations were highlighted as a less responded but was above 50%.

Standards and relevance of thesis projects (F)
Most of the projects were very relevant to students working environment and local supervision of thesis projects were highly admired by the students. The role of the university (KTH) supervisor and assistance extended in selecting a suitable thesis project were highly appreciated. A few has experienced challenges in this aspect.

Benefits of the programme (G)
Output of the thesis project and knowledge and exposure gained through program were accepted to be highly related which uplift of the career. Further, this academic qualification has become a path for wage incentives for the majority of the graduates.

Service of facilitating institute and staff (H)
Service of facilitators in assisting students and service of facilitating institute were highly accepted by majority (both completed and not completed) of past students.

Total (T)
Overall assessment of the program was measured considering eight aspects described in Table 1. Each aspect was accepted in high level and recorded more than sixty percentage of successiveness. Service of facilitating institute and staff was highly recognized while standard of course content was put in much acceptable level.

3.2 'Not Completed Students' View of Program

Possible drawbacks of program which caused to unsuccessful were examined during uncompleted students' survey. Majority of them disagreed to most selected drawbacks.

Table 2. Key to the chart

A	The course content was difficult to follow as a part time student
B	Time period that could allocate for the studies was not sufficient
C	Assignments and practices were difficult
D	Assignments and practices couldn't complete before due date and it caused delaying the program completion
E	Difficulties in resetting the examination caused delaying the program completion
F	Delaying the getting credits for assignments caused delaying the program completion
G	Selecting thesis topic was difficult
H	Time period and facilities that could allocate for the t was not sufficient
I	Lack of supervision got for the thesis study caused delaying the program completion
T	Overall for not completed students

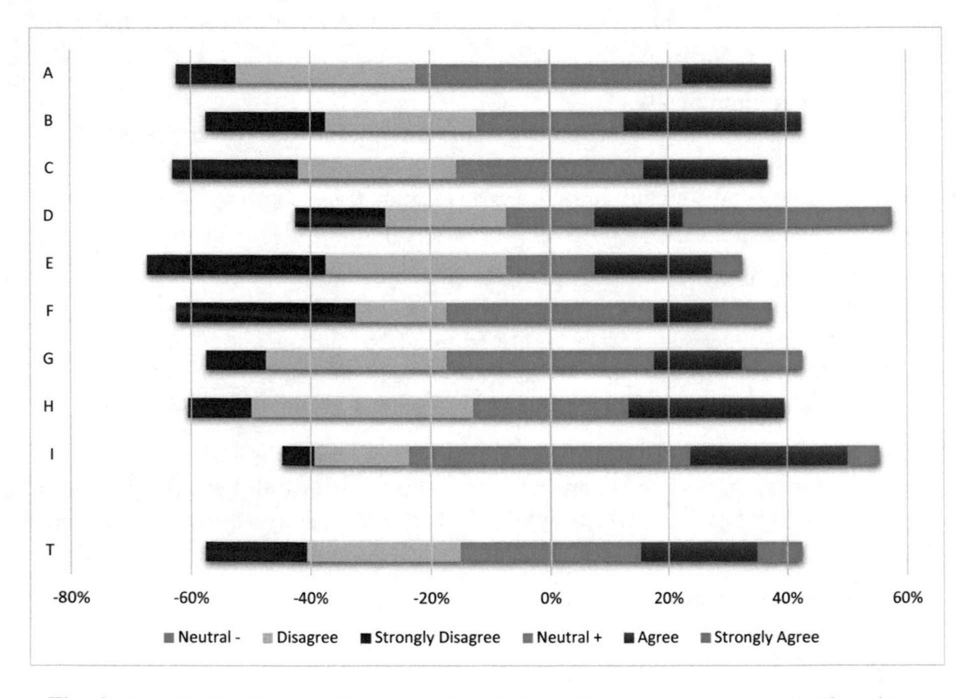

Fig. 2. Result of online questioner on student view of program - not completed students.

This means they agreed to the way the program was conducted even though they couldn't complete the program out of one reason or another. The few reasons were identified by interviewing the students who didn't complete the program, they are family commitments, working at remote locations, personnel attitudes like couldn't take much stress…etc. Only one drawback which was "Assignments and practices couldn't complete before due date and it caused delaying the program completion" (D) was accepted around 50% of domain. Therefore that could be identified as the point that should take action to improve programme structure (Table 2 and Fig. 2).

4 Conclusions

This study discussed the student view of distance based master degree program in Sustainable Energy Engineering which was offered by Royal Institute of technology, Sweden to Sri Lanka during the intakes 2008 to 2011. This was a 120 ECT credit full time program which opened real time access to Sri Lankan student through facilitated web tools. Primary objective of this distance based program is to empower Sri Lankan professionals who are working in energy sector. Seventy two students have successfully completed the program and most of them are from industry.

Scope of local facilitating institute and facilitators was very clear and level of facilitating was a decent (around 95% of participates agreed to this). Accessibility into class room lectures through provided web tools found to be high quality and user friendly (76% of domain agreed). Course content of the program was identified as well organized, advanced and very practical (93% of domain agreed). Program output has brought many benefits to the industry and to the individuals. It was found that very Pass rate of examinations held during the program (more than 70% of participates agreed to this). Few difficulties were identified during online teaching, online exams and online assignments but were in much acceptable range. The thesis projects carried out during final stage of program were industrial and output had given big contribution into local industrial development. Overall result of the survey expressed that offered program was very much success in all aspects considered in students' point of view (more than 70% of participates (both completed and uncompleted) agreed to this). This distance based educational concept is advanced, reliable and can hold big share in modern global education. It is much essential to introduce such a program with improvements.

As the enrolment to the SEEW Program was closed in 2015, an alternative program, known as MSc SELECT ("Environomical Pathways towards Sustainable Energy") is now discussed with the KTH and EIT-InnoEnergy. EIT-InnoEnergy is a European company dedicated to promoting innovation, entrepreneurship and education in the sustainable energy field by bringing together academics, businesses and research institutes. Two partner Universities of EIT-InnoEnergy, namely Royal Institute of Technology (KTH) of Sweden and the Universitat Politècnica de Catalunya (UPC) of Spain have designed MSc SELECT as a remote study program that allows students to complete Year 1 while remaining in their home countries. The OUSL will facilitate the offering of the courses in Year 1 and during Year 2, all students will study in one of the Consortium Universities in Europe. For the intake 2016 six students are registered at KTH and UPC, and they are successfully following year 1 courses with support of the OUSL.

Acknowledgment. The authors express their sincere gratitude to all past students who responded to survey. Also extend gratitude to all staff involved from Royal Institute of Technology, Sweden and Open University of Sri Lanka. The authors also express their gratitude all other persons behind to offering this program into Sri Lanka, and to InnoEnergy for the financially support to present the paper.

Appendix

Annexure I	
A	Bilda web tool was very user friendly
	Centra web tool was very user friendly
	Compedu web tool was very user friendly
B	Quality of online teaching
	The standard of the lectures was acceptable
	Interaction with the lecturer was good
	Online teaching was successful
	On campus teaching by local experts may better than online teaching by foreign experts
	The course content was difficult to follow as a part time student
	Time period that could allocate for the studies was not sufficient
	Downloading the lecture materials and uploading assignment was not a problem
	Quality of recorded lectures were good
C	The specialization subject coverage within the MSc degree program was adequate
	The depth of the syllabus of each subject was adequate subject is adequate
	The provided course materials such as presentations, recommended literature were adequate
D	Online lab sessions are manageable & useful
	No difficulties were faced in online examinations conducted
	Assignments and practices was difficult
	Assignments and practices couldn't complete before due date and it caused delaying the program completion
	Delaying the getting credits for assignments caused delaying the program completion
E	Level of examination has very good standard
	Facing repeat examination could be done without difficulties
	Time gap between examination and repeat examination was tolerable
	Issuing results of assignment was done in tolerable time frame
	Issuing results of examinations was done in tolerable time frame
	No difficulties were faced in written examination conducted on campus
	Difficulties in resetting the examination caused delaying the program completion
F	Assistance got for selecting thesis topic was good in position
	Supervision got from KTH for the thesis project was in good in position
	Supervision got from facilitator/local supervisor for the thesis project was in good in position
	Selecting thesis topic was difficult
	Lack of supervision got for the thesis study caused delaying the program completion
	Time period and facilities that could allocate for the thesis was not sufficient

(*continued*)

(*continued*)

Annexure I	
G	MSc thesis project was related to your career
	Knowledge gained from the MSc was directly help your career
	MSc in Sustainable Energy Engineering boost your career
	MSc in Sustainable Energy Engineering helped to increase your income
H	Service of the facilitator towards students was in good position
	Service of the KTH or other EU university was in good position

References

1. Cannon, M.M., Umble, K.E., Steckler, A., Shay, S.: "We're living what we're learning": student perspectives in distance learning degree and certificate programs in public health. J. Public Health Manag. Pract. **7**(1), 49–59 (2001)
2. Phipps, R.A., Wellman, J.V., Merisotis, J.P.: Assuring Quality in Distance Learning: A Preliminary Review. Council for Higher Education Accreditation, Washington DC (1998)
3. Beaudoin, M.: Learning or lurking?: Tracking the "invisible" online student. Internet High. Educ. **5**(2), 147–155 (2002)
4. Kjaer, N.K., Maagaard, R., Wied, S.: Using an online portfolio in postgraduate training. Med. Teach. **28**(8), 708–712 (2006)
5. Abeyweera, R., Senanayake, N.S., Senaratne, C., Jayasuriya, J., Fransson, T.H.: Capacity building through a web based master degree programme in sustainable energy engineering. In: IEEE Global Engineering Education Conference (EDUCON), Athens, Greece (2017)

An Empirical Study on the Use of Gamification on IT Courses at Higher Education

Balázs Barna and Szabina Fodor(✉)

Corvinus University of Budapest, Budapest, Hungary
contact@balazsbarna.hu, szabina.fodor@uni-corvinus.hu

Abstract. The aim of this work is to evaluate the effectiveness of a gamification platform during an IT course at Corvinus University of Budapest. A total of more than 2500 students attended the course during 2015 and 2016. We used a gamification environment within the Moodle e-learning platform during the course. Gamification steps included a reward system, alternative learning paths, various feedback options and social interaction platforms. Course quality was assessed based on students' willingness to participate in voluntary on-line tests, completion and results of final exams, as well as results of student satisfaction surveys. Our results indicate that gamification is able to improve IT course quality though it cannot solve all possible problems arising during such courses.

Keywords: Gamification · Blended learning · Information technology education Case study

1 Introduction

Gamification can be defined as the use of game elements and game design techniques in non-gaming environments [1]. Gamification is being increasingly used to promote the engagement and motivation of the involved individuals in business, education, health industry, societal responsibilities and many other fields of everyday life.

Gamification has recently become one of the most popular strategies to improve the methodology of promoting motivation and engagement in education. The popularity of using gamification in education is understandable as there are obvious similarities between games and the classroom. Game players work to achieve specific goals and win, while classroom students work to achieve specific learning objectives and do well academically; game players progress from level to level based on performance, classroom students must pass prerequisite courses and understand the given subjects before progressing academically [2].

Gamifying educational environments and teaching processes have massive potential as the lack of student motivation is a constantly recurring problem and an increasing number of students cannot complete the school. Although gamification could easily be applied to education applications, there are a number of challenges such as the different attitudes of individuals to gaming (especially in older adults), the involvement of game developers in educational activities, or the application of gaming principles in more

M. E. Auer et al. (eds.), *Teaching and Learning in a Digital World*,
Advances in Intelligent Systems and Computing 715,
https://doi.org/10.1007/978-3-319-73210-7_80

difficult educational tasks. Obviously, gamification should not be considered as a magic bullet that will always and consistently solve all educational problems but rather as one of many tools available to improve education efficiency. Given that gamification is a relatively new tool, its cost-benefit ratio has yet to be assessed and compared with other available tools.

If we can increase the engagement of students then this can have a significant positive impact on the effectiveness of education. Gamification can be a good tool to accomplish this by adding alternatives, personalize the curriculum.

One of the most important factors of gamification is that gamified applications mostly use on-line, digital technologies. Importantly, current students belong to Generation Y (people born between the mid-1980s and the mid-1990s) and Generation Z (born between the mid-1990s and the early 2000s), and they have radical differences in learning and gaining knowledge from previous generations. Members of these generations use internet and social media often and securely as it was part of their life and socialization from the beginning [7, 8]. Another major trait of the members of Generation Z is the decreased skill to pay attention [9] which could also be addressed by gamification as the learning process is divided into small pieces and the motivation is also expected to improve by small positive reinforcements [7].

The impact of gamification can be even bigger if we gather information from gamified platforms and analyze them. Results of such analyses allow us to adapt the educational gamified proposal to learners' special needs and pace in learning.

Here we describe an approach to use gamification as a way of teaching computing at university level to students of economic sciences. In 2015, we gamified a university course in Information Technology using an online learning platform, the Moodle (Modular Object-Oriented Dynamic Learning Environment) [3] system. The data received from Moodle platform such as achieved points, failure rate, course evaluations and students' feedback were analyzed. The course was then improved based on the findings of the study. In total, over 2500 students have been educated using gamification. Although gamification cannot solve every problems (a course requires good quality content and proper teaching skills as well), but our results suggest that gamification can lead to better course experience for the students and to better overall course outcomes.

2 Gamified Course: Information Technology

Information Technology is a compulsory, half-year undergraduate course for every full-time or part-time student with Business and Economics majors at Corvinus University of Budapest. The course is taught in blended learning format with gamification elements for everyone except for Business Informatics major since the fall of 2013, due to the large number of students (the course is taken by more than 1000 students a year). The blended form means that it combines face-to-face instruction with computer-mediated instruction [13]. In the examined courses the students attend seven lectures in a classroom and they obtain a practical material to be processed independently every week on the online educational platform of the course. The evaluation of students is based on points earned from weekly tests, assignments, and the final exam [12].

On average, approx. 1500 students participated in the research during each semester, including 1100–1400 full-time and approx. 200 part-time students (see Table 1).

Table 1. Attendance of the course each year

Year	Attendance		
	Full-time	Part-time	Total
2015	1427	189	1616
2016	1127	194	1321

2.1 The Gamification Platform: Moodle

We chose to apply gamification to Moodle as our Information Technology course already used this platform. Moodle is an open-source, PHP based education framework that allows creating a customizable learning environment. Moodle has different features such as grading and online tests support system, forums, and file-management capabilities.

2.2 Core Elements of IT Course Gamification

In this section we describe the elements used in the gamification of the IT course.

Reward system: points, badges, levels
Points are the basis for most gamified projects. Students get points when they take the right actions in the right way, by means of which they move levels, etc. Collecting points will cause a continuous gamification experience.

Alternative learning paths
The curriculum is made up of four modules, and the undergraduates should reach a minimum level of each module to complete the study period of the course. Each module has weekly tests, as well as minor (so-called 'life-belt') tests. The weekly tests can be completed two times, and the better result counts. The minor tests have not got any limitations as for filling, but only the modules' minimum level can be reached with them. The required minimum level can be achieved with both types of tests, however, achieving the modules at top level rewards the student with a badge which gives an extra point to the final exam. [10]. For each semester (14 weeks), a total of 13 weekly tests and 28 minor ('life-belt') tests could be completed.

Apart from the tests, optional assignments can be completed. If the assignment is accepted, the student gains another badge with a surplus point. If a student collects all of the badges (maximum one badge can be missing), he/she receives automatically the highest grade without taking the final exam.

Instant feedback
The students are constantly receiving feedback on their performance including their score and rating for each module, along with textual information on the further options.

Feedback themes
Students were offered seven different options to choose how they would receive their feedback. Besides the default, simple scoring theme (levels, badges, etc.), students could choose to receive their feedback as they would in a popular fiction environment (e.g. as from Professor Dumbledore in Harry Potter's Hogwarts school), through a sports theme (e.g. in a Judo learning environment) or according to a business ranking system (different ranks in a commercial enterprise). Note that such optional feedback themes have only been offered during the 2015, but not during the 2016 semester (see below).

Social interactions: forums and chats
Forum and chat are Moodle activities that allow students' interaction. Teachers can also answer questions that students ask in forums.

2.3 Changes During the Research Period

There have been a few smaller changes during the two-year research period (i.e. between 2015 and 2016) that should be mentioned:

- Apart from optional consultancy opportunities, there was no systematic direct personal interaction between students and teachers during the first year (2015). In the second year (2016), two mandatory computer lab-work lessons were provided, where students could ask questions from the instructors and they worked together in groups to deepen their knowledge.
- Since there was little interest in using the different feedback themes during the 2015 semester (see below), we decided to abandon that feature and only offer a single, basic feedback theme during 2016.

3 The Effectiveness of Gamification

The effectiveness of gamification during the IT course during the two-year research period is summarized below.

3.1 Impact of Gamification on Participation and Course Completion

The effectiveness of the engagement factor of the course is measured by the level of activity, and the ratio of failures in the mid-term section of the course.

Feedback themes
Students had six different options to choose how they would receive feedback about their performance. If they did not actively select any of the custom options, they remained in the default (basic) theme. The list of themes and the number of students choosing them is shown in Table 2. As shown in the table, only 14% of the students actively selected one of the custom themes, whereas 86% of them remained in the default theme. Furthermore, some of the themes have been selected by a very low number of students. These results indicated that there was little interest in selecting custom

feedback themes in the 2015 course period. Therefore, this option was discontinued and only the default theme was offered in 2016.

Table 2. Selection of feedback themes in the 2015 semester

Feedback theme	Actively selected?	Full-time students	Part-time students	Total	Total (%)
Basic	No (default)	1227	165	1392	86.14%
Basic	Yes	48	6	54	3.34%
Dumbledore		85	4	89	5.51%
Materials		15	3	18	1.11%
Duck		18	3	21	1.30%
Business		14	7	21	1.30%
Judo		20	1	21	1.30%
Total		1427	189	1616	100%

Weekly tests

The activity of students in the course was examined by their willingness to fill out the tests.

Each weekly test could be filled out at most twice, and from the two results, the better one was counted towards the final evaluation. It is important to note that it was not mandatory to fill out these tests, but it helped to earn part of the points that could be gained through the study period of the semester [10].

In 2015 a slight gap could be realized between full-time and part-time students (Fig. 1A). On average, 65.4% of full-time students filled out the tests at least once (with the deviation of 1.8%). On the other hand, the weekly tests among part-time students were not as popular as among full-time ones. Even the first test reached only 57.2% of part-time students, and only 37.6% of them tried to solve the test on the 8th week. (The average fill-out rate was 57.2% with a deviation of 8.1%).

The course in 2016 had a better reach among both types of students, and a substantial gap could be realized between full-time and part-time students. As for the full-time participants, the average fill-out rate was 89.7% (dev. 4.0%), and among part-time students 66.97% of them completed the test on an average week (dev. 9.2%). Though the ratio shows an almost constantly decreasing trend, at part-time students the lowest fill-rate of 2016 is about the same level as the average rate in 2015 (Fig. 1B).

On average, approx. 60% of those students who decided to fill out the tests took the opportunity to use the second chance, and there is no significant difference between the two types of training [10]. This applies to the courses of both years.

Minor 'life-belt' tests

In contrast to weekly tests, the minor tests can be completed unlimited times, but the value of one test is at most only a quarter of the value of a weekly test, though the minimum level of each module can also be achieved only by them. [11]

The willingness to fill out the minor tests is lower than the rate of weekly tests, but 2016 had more positive results compared to 2015. Among full-time students of 2015, 40–55% of the students filled out the test at least once. Examining the part-time

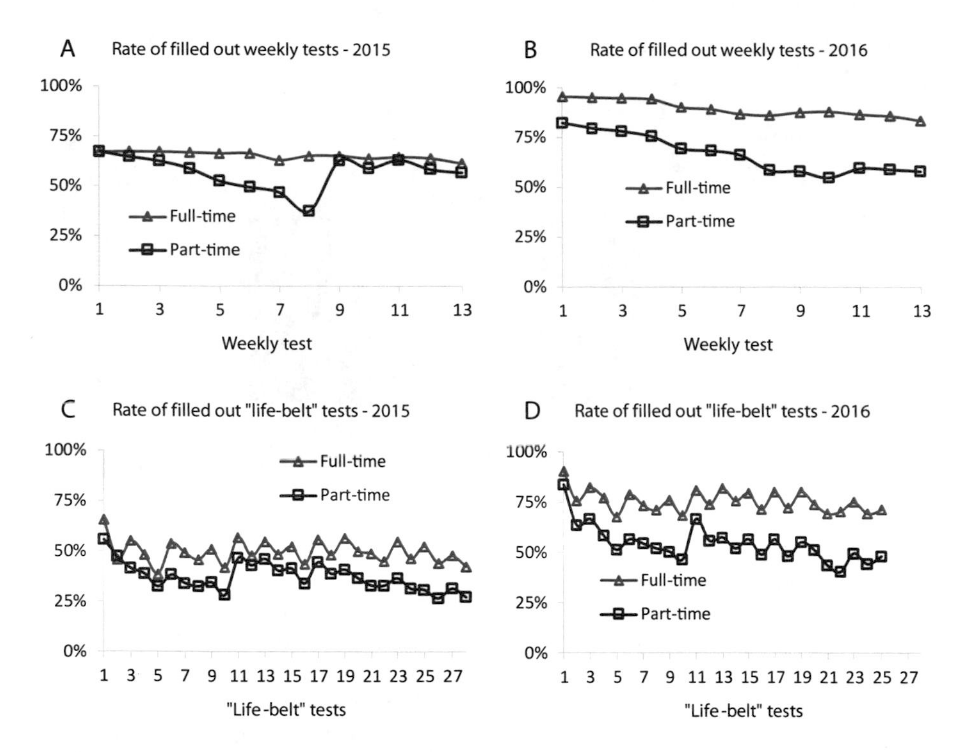

Fig. 1. Proportion of students who filled out the weekly and 'life-belt' tests at least once

participants, their completion rate had a wider fluctuation with a lower average level, it changed between 25% and 45% (Fig. 1C).

In 2016, the aforementioned completion rate among full-time students remained between 65% and 85%, while by part-time students the rate changed between 40-65% (Fig. 1D).

As the increased willingness can be realized in the aspect of weekly tests, the same rise can be seen in the case of minor tests as well.

Mid-term performance

According to the syllabus of the course, the final grade consists of the weighted average of the mid-term performance (30% weight) and the final exam (70% weight). From the mid-term, the gained points are counted only if the participant reaches the minimum level of all four modules. If even one of them is below the required minimum level, the student receives 0 point for his/her mid-term performance.

Another aspect of examining the engagement and activity of students is to check the aforementioned rate of earning 0 point on the mid-term part of the course.

In 2015, a relatively high portion of students could not reach all the required levels of the four modules. Among full-time students, 34.5% of them failed to gather points, and viewing the part-time participants, this rate rose up to 43.4%. In the following year, the rate of failure of part-time people was almost the same, with 46.9%, but a high reduction could be realized among full-time students, their rates reduced to 13.6% (Fig. 2).

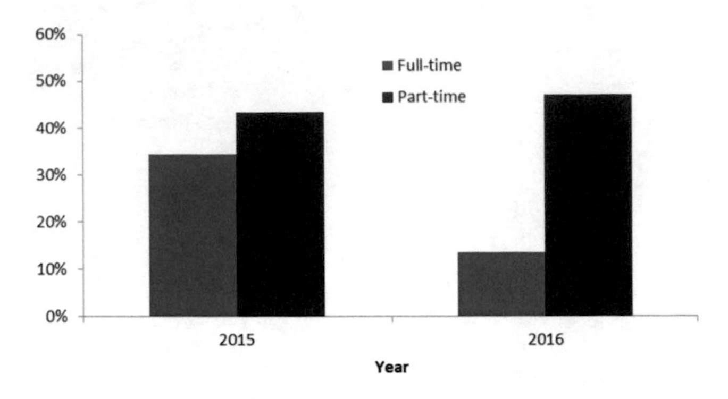

Fig. 2. Rate of failed mid-term performance by student types

3.2 Student Satisfaction

In order to measure the satisfaction of students, the result of 'Professor and course evaluation' system has been used. In this questionnaire, the students are asked to answer several questions with option between '1' and '5', where '1' means the least true statement or least amount of work, and '5' means the full approval with statement or the most amount of work.

As mentioned in Sect. 2, the examined Information Technology courses are enhanced with blended learning and gamification methods except for students attending the major of Business Informatics which course does not use gamification elements. This subject has the same syllabus as the gamified ones, therefore we could compare the courses with different teaching principles as to how they perform in the aspect of student satisfaction, and the results can be seen in Table 3.

Table 3. Result of questionnaire about student satisfaction

Question	Avg. score at gamified courses (out of 5)	Avg. score at non-gamified course (out of 5)	Percentage difference (in the view of gamified courses)
We are dealing with useful things	4.08	3.71	9.92%
My teachers evaluate my performance as I do	3.97	3.79	4.87%
The course and the related tasks are enjoyable for me	2.92	3.03	-3.54%

As for realizing the usefulness of the learned knowledge, the courses with gamification elements scored 9.9% higher average valuation, and the students scored 4.9%

higher on self-evaluation ('My teachers evaluate my performance as I do.'). On the other hand, the participants did not feel the mid-term exercises as enjoyable as the students of non-gamified course, there is a 3.5% difference between the average scoring in favor of the non-gamified course.

4 Conclusion and Future Work

In response to an increasing need to find new techniques for teaching academic-level information technology, we propose in this work to use gamification to improve student participation and success in higher education. Gamification may improve various aspects of higher education such as student engagement and motivation, interaction between students and teachers, providing regular feedback to students, and optimal use of available infrastructure and human resources.

Though our study was able to address certain aspects of the effect of gamification in IT education, a weakness of this study is that there is not any direct comparison between results in a gamified vs. a highly similar, yet non-gamified course. Our future plans include splitting the >1200 students/year into two or more smaller groups and test the effect of gamification between those groups that are in the same age and overall study environment. Such better-controlled experiments would be able to reveal fine differences in a very precise manner on the effect of gamification in a university education setting.

Acknowledgement. The authors would like to thank Gabriella Baksa-Haskó for her continuous help and support.

References

1. Werbach, K., Hunter, D.: For the Win: How Game Thinking can Revolutionize Your Business. Wharton Digital Press, Philadelphia (2012)
2. Jackson, M.: Gamification in Education: A Literature Review (2016). http://www.westpoint.edu/cfe/Literature/MJackson_16.pdf
3. Moodle: Community Moodle. http://moodle.org
4. Yahr, M.A., Schimmel, K.: Comparing current students to a pre-Millennial generation: Are they really different? J. Appl. Res. High. Educ. (2013)
5. Training Millennials: Engaging Generation Y in Training. http://www.growthengineering.co.uk/training-millennials-generation-y/
6. Kapp, K.M.: The Gamification of learning and instruction: Game-based methods and strategies for training and education. Pfeiffer, San Francisco (2012)
7. Horovitz, B.: After Gen X, Millennials, what should next generation be? USA Today (2012)
8. Bíró, G.I.: Didactics 2.0: a pedagogical analysis of gamification theory from a comparative perspective with a special view to the components of learning. Proc.-Soc. Behav. Sci. **141**, 148–151 (2014)
9. Fromann, R.: Gamification: épülőben a Homo Ludens társadalma? ELTE TÁTK, Szociológia Doktori Iskola (2012). http://www.jatekkutatas.hu/publikacio_html_files/publikaciok-gamification.pdf
10. Barna, B.: Gamification in education. SEFBIS J. **10**(1) (2014)

11. Barna, B., Fodor, S.: Gamification és közgazdászképzés – Játszani is enged? DIBIZ: Digit. Bus. **1**(4), 34–36 (2015)
12. Baksa-Haskó, G.: Efficiency over 1000 students – the evolution of an on-line course: from e-learning to flipped classroom. In: Auer, M.E., Guralnick, D., Uhomoibhi, J. (eds.) Interactive Collaborative Learning: Proceedings of the 19th ICL Conference. Springer, Cham, pp. 237–245 (2017)
13. Bonk, J.C., Graham, C.R.: The Handbook of Blended Learning. Pfeiffer, San Francisco (2011)

Academic-Industry Partnerships

Research Oriented Learning in a Research Association – Evaluated in a Maturity Model

Kerstin Haas and Jürgen Mottok(✉)

OTH Regensburg, 93053 Regensburg, Germany
{Kerstin.Haas,Juergen.Mottok}@oth-regensburg.de

Abstract. Research-oriented learning provides students the opportunity to develop their research competences by experiencing research practice, this often happens in the surrounding of research associations with different universities and companies. This paper introduces a two-step approach to evaluate the research-oriented learning within a research association. First we conduct the evaluation of the research environment with the instrument of the adapted Collaboration Maturity Model (Col-MM) to see whether the collaboration network and the management is able to support the students in their learning process. Additionally, we take into account the evaluation of the students' research competence. This approach targets the assessment of the starting conditions of the students and to compare their performance level until the end of the research association project phase. These two evaluation phases provide the potential to create an ideal research environment and consequently enable the students to develop and improve their research competence.

Keywords: Research-oriented learning · Research competence
Research association

1 Introduction

Research-oriented learning often takes place in the surrounding of a research association in a university environment. Condition precedent for successful research-oriented learning in a research association with different universities and companies is a professional research environment. This environment needs to enable the students to apply and create knowledge and to improve their research competence. For that reason it is relevant within a research association to evaluate and continuously improve the collaboration and management processes. Another condition is the research competence of the students. This means in an optimal research-oriented learning environment the students start already with research competence and are then able to improve this performance. Aim of the paper is to assess the development of the research competence to ensure that the students can improve their research competence in this setting and if necessary measures for improvement must be taken.

Consequently we introduce an evaluation in two steps. One step is the evaluation of the research competence and its development during the whole project duration and another step is the evaluation of the provided research environment such as

M. E. Auer et al. (eds.), *Teaching and Learning in a Digital World*,
Advances in Intelligent Systems and Computing 715,
https://doi.org/10.1007/978-3-319-73210-7_81

collaboration network and management in the research association. These two-step approach has been in a first step implemented in a current research association.

1.1 Research-Oriented Learning

Research-oriented learning is a concept of learning that provides students with the opportunity to develop their research competences by experiencing research practice in a university environment. The idea is that students participate at current research projects and acquire research competence. Within this learning concept the students have the possibility to become familiar with research, expand their level of knowledge, increase their network of contacts, work with others disciplines and strengthen their general competences. This research competence includes a broad definition of research abilities as the ability to review the state of research, methodological skills, skills in reflecting on research findings and communication skills [1–3].

1.2 Research-Oriented Learning in a Research Association

Promoting young researchers is a huge challenge in research associations, but this is often not an easy task. If conducted in an adequate way research-oriented learning in a research association is a huge opportunity for students. Additionally to the normal research work the students have to fulfill other duties and so have the chance to get a deeper insight into research projects and the transdisciplinary cooperation [4]. In this setting the students have the chance to conduct research in real life. In a transdisciplinary research association with different universities and companies the students have the chance to conduct research with practitioners and thus have their first interaction with the industry and consequently with practical users. Furthermore this gives them the chance to increase their network of contacts in the industry.

2 Development of the Evaluation Instrument

In the first approach we have evaluated the research environment itself. For this the Collaboration Maturity Model of Boughzala and de Vreede [5] has been chosen and adapted to measure the collaboration network and management in a research association. This model is introduced in Sect. 2.2.

The Collaboration Maturity Model allows the assessment of the research environment, but the next question that needs to be answered is, how we can measure whether the research-oriented learning really takes place, this means did the students acquire research competence. Consequently the second step is to assess the research-oriented learning conditions in a research association. To obtain this information an adapted version of the evaluation according the RMRC-K-model of Thiel and Böttcher has been added to the evaluation. This model is introduced in the following chapter.

2.1 The RMRC-K-model

Thiel and Böttcher developed a competence model which defines the following five competence categories:

- Skills in reviewing the state of research: systematically reviewing the state of research regarding a specific topic, identify gaps or unaddressed questions
- Methodological skills: systematic planning and preparation of the research process, such as formulating specific research questions and identifying the next steps to address this question, this includes a selection and application of methods appropriate to the research question
- Skills in reflecting on research findings: theorctically and methodologically reflecting on results, reflecting on scientific and practical research
- Communication skills: presenting (oral and written) research findings
- Content knowledge: knowledge of central theoretical constructs, methods and disciplinary standards

Based on these categories they have designed a self-assessment questionnaire, to evaluate the students' research competence. The goal of this model is modeling and assessing students' acquisition of research competence [1].

For the evaluation of the research-oriented learning we have chosen this model as it is covering all relevant categories to measure the development of competence in a research association. By adapting this model to our evaluation approach we can identify at which points the students can develop competence and where is need for improvement.

2.2 Collaboration Maturity Model (Col-MM)

Maturity models have been developed to assess organizations. Consequently these models are used as an evaluative and comparative basis for improvement. Maturity models define the development of an organization over time. The goal is to assess the maturity of an organization and to identify the issues that are required to increase the maturity of these processes [6, 7].

We have chosen the Collaboration Maturity Model (Col-MM) developed by Boughzala and de Vreede among the different maturity model as it is intended to be sufficiently generic to be applied to any type of collaboration, this means it is not limited to a particular setting. It is suitable for conducting self-assessments. According to Defila and di Giulio evaluation of research could or even should always include a self-assessment by the scientist involved [8]. Furthermore this model focuses on the collaboration network and management. This focus also fits for a research association.

In this model the maturity is evaluated in a four point scale with '4' representing the highest level of maturity. The following areas of concern are taken into account:

- Collaboration characteristics: This covers descriptive features and attributes of the collaboration.
- Collaboration management: This covers the way in which collaboration processes and activities are managed.

- Collaboration process: This covers how individual collaboration members perform collaboration on a day to day basis.
- Information and knowledge integration: This covers how individual collaboration members manage the information and knowledge required for productive collaboration [5].

2.3 Evaluation – Questionnaire Design

The first part of the questionnaire evaluates the maturity of the research organization regarding collaboration and management in four maturity levels. The respondents answer 27 questions on four scale responds. Additionally a few open-end questions are added to gain a deeper insight especially regarding improvement statements and to receive a better answer to the question why.

This part of the questionnaire to evaluate the research environment is split into the following six chapters:

The first part of the questionnaire inquires after the association activities as networking and publications. Research objective is to assess the conditions for students to reflect their research within and beyond the research association.

As each project member should be aware of the objectives and the targets of the research association the main objective of the second chapter is on the one hand the evaluation of the project targets, whether they are known, understandable and accessible. On the other hand the features and attributes of the collaboration within the research association are evaluated.

The management needs to provide a motivating environment in which joint results and products can be developed. Therefore the third part of the questionnaire targets the evaluation of the management of the research association. This chapter includes criteria such as the management style and the process of decision making and rewarding.

In the fourth part of the questionnaire the respondents can assess the collaboration itself, by answering questions concerning collaboration framework, conflict management and resources sharing.

Main condition for collaborative research is the information and knowledge sharing. The fifth chapter targets answering the question on which maturity level the information is collected, structured and accessed and knowledge is validated, reused and created.

In the sixth chapter the tools and equipment provided by the project management will be evaluated regarding the points frequency of use and satisfaction.

In the second part of the questionnaire students' research competence is evaluated by asking 12 Likert scale items. All items required a response ranging from fully disagree (1) to fully agree (6). For each category (skills in reviewing the state of research, methodological skills, skills in reflecting on research findings, communication skills, content knowledge) two or three questions were asked.

3 The Application

Currently the study is conducted during the whole research association project duration of three years. The evaluation is conducted in a research association with 33 project members. 9 months after project start the first wave with the goal to evaluate the research environment has been executed. 19 months after project start the second wave has been conducted with additionally the evaluation of the students' research competence. At the end of the project duration the third wave will be executed. All project members from project leaders to industry partners to students are invited to take part to gain an overall insight about the research association from every perspective. The students such as master and PhD students additionally answer the part of the survey with the research competence self-evaluation. The survey is conducted as an online questionnaire.

3.1 First Findings

This chapter introduces some examples of the findings.

Analysis of the collaboration and management. In general the members of the research association evaluate the collaboration network and management in the first wave on a medium to high level. The second wave of the evaluation has shown that the project members see an improvement in almost all areas. The following graphic shows exemplarily the result of the chapter 'information and knowledge integration' (Fig. 1).

This category is evaluated with an average maturity of 2 to 3 in the first wave. As the most positive aspect in this category is seen the point 'exchange of information among students', whereas the lowest level reached the points 'knowledge validation' and 'knowledge reusage'. After the first evaluation steps were set up by the management to improve the knowledge reusage for example by organizing regular meetings with all single project members and additionally the infrastructure to safe and share knowledge has been improved. The students got the chance to take part at conferences in their specific area to improve and reflect their knowledge and giving them the chance to write papers together with other project members. These measures improve the perception of the maturity in this category, as the analysis of the second wave has shown.

Analysis of the students' research competence. In the second part of the evaluation the students see already their research competence well defined and rate most of the statements on a medium to high level.

The following table shows that the students perceive their research skills in the area of content knowledge already at this early stage well developed. A need for improvement is seen by the students mainly in the area of methodological skills. Even 27% responded that this category does not really apply. There is a need of improvement in the area such as formulating research questions/hypotheses, what is necessary to address the research question or applying different research methods. Consequently they perceive it in some way as difficult to plan and prepare the research process and to select and apply methods (Fig. 2).

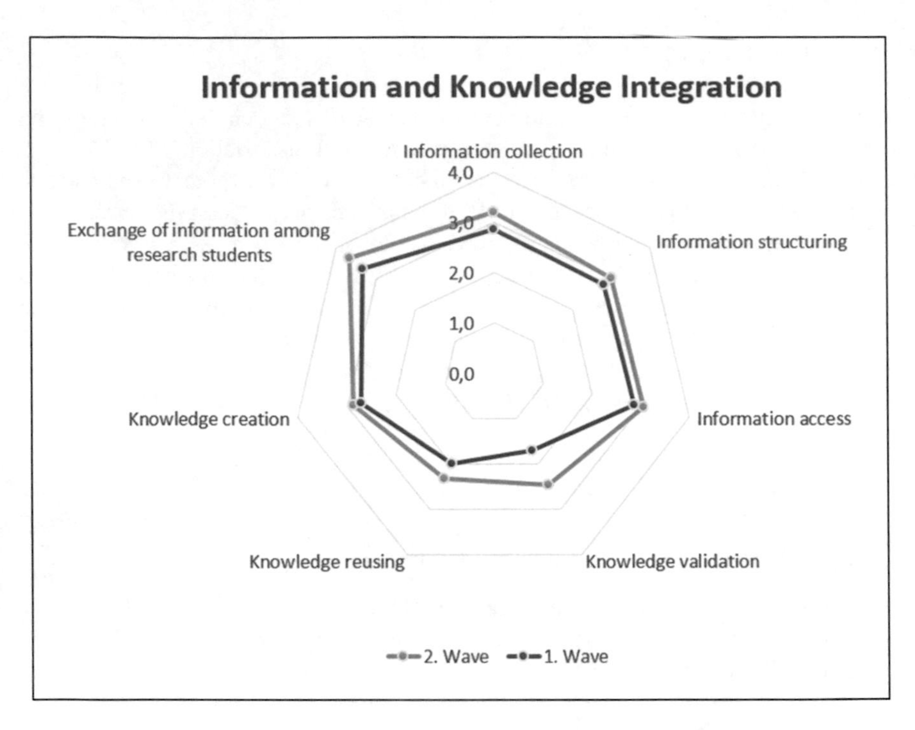

Fig. 1. Maturity of information and knowledge integration.

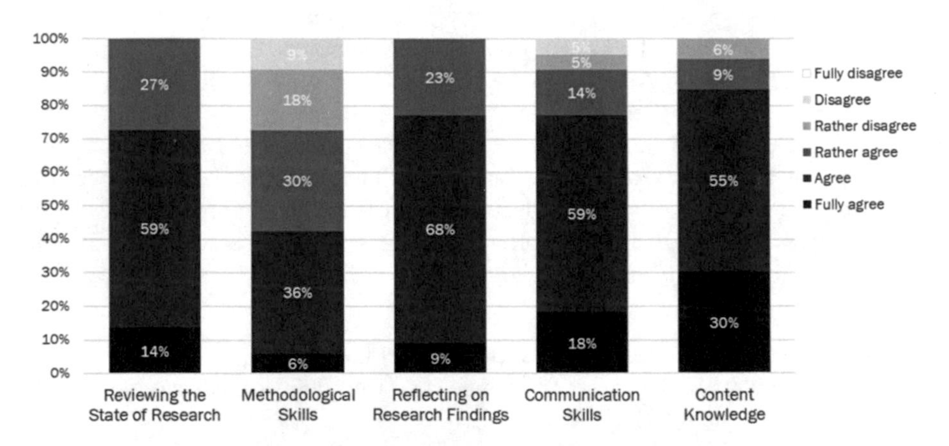

Fig. 2. Self-assessment research competence.

4 Discussion and Future Work

This paper introduces a methodology to evaluate research-oriented learning in a research association by combining two models for assessment. One evaluation step is conducted based on the Collaboration Maturity Model to assess the research environment and a further evaluation step is conducted adapted from the RMRC-K-model to assess the development of students' research competence. This two-step approach allows the assessment of the research-oriented learning from two perspectives and consequently provide an overall assessment.

The first execution of the evaluation set-up has shown that it is possible to gain feedback regarding the environment and expose this way the needs for improvement. In a further step we will include a qualitative survey (in-depth interviews or focus groups) including a content analysis with a limited number of respondents to gain deeper insights. We think this kind of analysis will allow an even more detailed information where are the needs for improvement.

Regarding the acquisition of research competence these results show that the competences can be monitored and that this offer the possibility to specifically tailor to students' needs. In a further step to gain more detailed information we will work with the statistical data collected in this survey.

References

1. Böttcher, F., Thiel, F.: Evaluating research-oriented teaching: a new instrument to assess university students' research competences. High. Educ. 1–20, (2017). https://doi.org/10.1007/s10734-017-0128-y
2. Aholaakko, T., Komulainen, K., Majakulma, A., Niinistö-Sivuranta, S.: Crossing Borders and Creating Future Competences. Laurea Publications, Helsinki (2016)
3. Europa-Universität Viadrina Frankfurt: What is research-oriented learning. https://www.europa-uni.de/en/struktur/zsfl/Hintergrundinformation/Forschendes-Lernen/index.html. Accessed 15 Apr 2017
4. Defila, R., Di Giulio, A., Scheuermann, M.: Forschungsverbundmanagement. Handbuch für die Gestaltung inter- und transdisziplinärer Projekte. Vdf Hochschulverlag (2006)
5. Boughzala, I., de Vreede, G.: Evaluating team collaboration quality: the development and field application of a collaboration maturity model. J. Manag. Inf. Syst. **32**, 129–157 (2015)
6. Magdaleno, A.M., Araujo, R., Werner, C.M.L.: A roadmap to the collaboration maturity model (CollabMM) evolution (2011)
7. de Bruin, T., Rosemann, M., Freeze, R., Kulkarni, U.: Understanding the main phases of developing a maturity assessment model. Paper in 16th Australasian Conference on Information Systems (2005)
8. Defila, R., Di Giulio, A.: Panorama Sondernummer 99: transdisziplinarität evaluieren – aber wie? In: International Transdisciplinarity Conference (2000)

Preparing Future Career Ready Professionals: A Portfolio Process to Develop Critical Thinking Using Digital Learning and Teaching

Jennifer Rowley[1(✉)], Jennifer Munday[2], and Patsie Polly[3]

[1] The University of Sydney, Sydney, Australia
jennifer.rowley@sydney.edu.au
[2] Charles Sturt University, Bathurst, Australia
jmunday@csu.edu.au
[3] The University of New South Wales, Sydney, Australia
patsie.polly@unsw.edu.au

Abstract. It is uncommon to learn, work and grow up in a world without technology and so it is pertinent to explore exactly how digital learning assists in preparing our future workforce. Employers often note that new graduates (or the 'new' employees) lack critical thinking and problem solving skills and have poor communication, as their experience is often restricted to a worldview limited by a digital communication lens. This brief paper argues, therefore, that the need for developing critical thinking skills has never been more apparent as graduates/new workers transition to employment. The conference theme is Teaching and Learning in a Digital World and this paper investigates the what, how and why technology benefits learning. The Rowley and Munday (2014) Sense of Self model underpins the theoretical paradigm that encourages individuals to experience digital technologies to support critical thinking by exploring their own self-efficacy as a learner. Authentic learning is discussed via a simulated learner setting where individuals create evidence of their own professional or personal identity (or a hybrid of both) and discuss how utilising a reflexive process to create a portfolio (as both a learning and teaching tool) can support improvements in critical thinking. The paper concludes with strategies for exploring how we encourage students and/or new employees to reflect on the what, how, why and who of themselves through a critical thinking process associated with self-reflection and portfolio creation. Understanding the progressive ways portfolio process and products can be used to develop a professional identity through encouraging students to reflect and connect themselves to multi faceted professional identities benefits life long learning for future work readiness.

Keywords: Identity development · ePortfolios · Reflective practice

M. E. Auer et al. (eds.), *Teaching and Learning in a Digital World*,
Advances in Intelligent Systems and Computing 715,
https://doi.org/10.1007/978-3-319-73210-7_82

1 Context

1.1 Introduction

An ePortfolio process of teaching across a range of educational contexts is reported to have an impact on how students learn and successfully transition to professional in different discipline areas. As a means of providing an effective way to encourage students' cognitive and motivational learning, the ePortfolio is robust in its process and outcomes. The student created ePortfolio product – whether a showcase, a presentation, a story of their learning or a collection of artefacts - has been attributed to the development of identity and praised for its use as a learning and teaching tool. This has led to identification of a way of thinking defined by Stanford University as *ePortfolio thinking* where the use of 'reflective practice guides the effective use of learning portfolios … (using) experiential learning, metacognition, reflective and critical thinking' [1] (n.p). Many years of research by the authors of this paper provide evidence that a teacher's effective use of ePortfolio can dramatically contribute to the development of what we term "soft" skills such as decision-making, problem solving, self-realisation, reflection and independent thought - termed 'soft' because these are not skills explicitly taught in educational contexts yet we often measure the success of these attributes being recognisable in graduates, for example.

1.2 Background and Literature

The 'Sense of Self' model [2] underpins the theoretical paradigm of this paper. Please see Fig. 1.

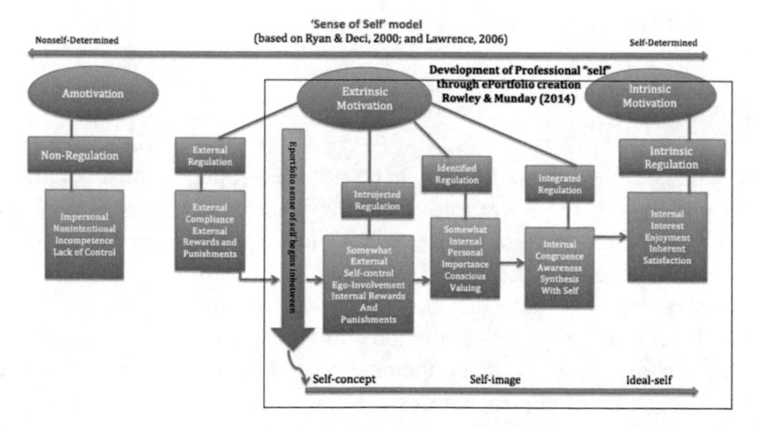

Fig. 1. 'Sense of Self' model is designed to indicate both the Self-Determination Theory by Ryan and Deci [3] and Lawrence's Self-Concept [4]. The authors note "that students producing ePortfolios embedded in Higher Education degree programs tend to move through the descriptors in the bounded rectangle, from extrinsic motivation to intrinsic motivation" [5] (p. 55).

Literature surrounding the study of ePortfolios covers a range of conceptual areas including the role of ePortfolio in encouragement of student reflection on learning and its potential for professional identity development and definition. Reflection is often a difficult task for students and maintaining engagement is challenging due to misconceptions of what reflection actually entails. However, research shows that aligning preparation for professional practice and individual development to reflection often produces positive results [6]. In demonstrating to graduates the value of "future change through the reflective process" [2] (p. 83) reflective practice becomes a desired outcome for learners as teachers become more aware of ePortfolio being instrumental in assisting students and graduates in undertaking more explicit reflective practice [7]. Brandes and Boskic [8] posited merging both the value of reflection with reflective narratives in ePortfolios and more recently, Harring and Luo [9] described case studies about the kinds of reflective ePortfolio questions they had provided to students during their undergraduate degrees. They concluded that carefully scaffolding the way to reflect is essential if you want students to produce meaningful narratives that use the 'story' to demonstrate learning.

Oner and Adadam [10] reported that handing over control to students and making them responsible for their learning increased both intrinsic motivation and active thinking – providing positive outcomes to their attitude toward professional practice. In fact, research by Rowley and Bennett [11] investigated the potential of ePortfolios to help students form professional and future identities as ePortfolios are deemed to be an appropriate medium for student (or a teacher) to negotiate and construct an online identity that relates to their future selves [2]. The reflective nature of ePortfolio assists students to view themselves and their work through a different lens that could be determined to be similar to a "student portrait" [12] (p. 8). Creating an ePortfolio during undergraduate studies enhances students' self-efficacy and reflective thinking providing evidence of impact on future career readiness.

2 Purpose

This short paper argues, therefore, the need for developing critical thinking skills through reflection and how/why reflective writing is necessary for graduates and those who are new to the world of work to take on the challenge of successful transitioning to employment. The conference theme is *Teaching and Learning in a Digital World* and hence this paper explores the what, how and why technology benefits learning as students use the ePortfolio process to demonstrate their development of skills required for the profession, as well as showcase their successful learning and achievements [13]. The process of curation of evidence necessary for an effective ePortfolio also requires deep reflective thought as authors are compelled to think about the creative ways of presenting themselves to an unknown audience (for example, in the case of a job application).

2.1 Research Questions

The two research questions framing the project were:

(1) In what ways do students require a scaffolded approach in transitioning to professional practice?
(2) What effect does creating an ePortfolio have on new graduates sense of self?

3 Approach

The session encourages participants to experience digitally assisted learning and teaching to support critical thinking by exploring their own self-efficacy as a learner. Participants experience authentic learning via a simulated learner setting where they create evidence of their own professional or personal identity and discuss how utilising a reflexive process to create a portfolio (as both a learning ad teaching tool) can support improvements in critical thinking to support outcomes that benefit life long learning for future work readiness. The approach is used to demonstrate the success of students thinking about their 'ideal selves' as future professional practitioners.

4 Actual Outcomes

Conference participants are first asked to take a photograph of something literal, metaphorical or symbolic that represents a facet of themselves before exploring individually and in small groups how that photograph is representative of their professional or personal 'self'. The process that follows demonstrates how we can encourage our students as emerging professionals and future new employees to reflect on the what, how, why and who of themselves. This is a practical example of a critical thinking process associated with self-reflection and portfolio creation. The creators of ePortfolios to see their own 'selves' through the narrative language can be regarded as the progression from a self-image to an ideal or professional self and can be enhanced by thoughtful and purposeful pedagogy [2].

4.1 Visual Images Assist in Reflective Narratives

Exploring the use of visual images and metaphors in identity building is well documented in the research literature [14]. This aim of this short paper, therefore, was to argue that visual images help people to discuss and explore future professional traits. Educators who are charged with the responsibility of ensuring students are able to use the graduate attributes for an active reflection of their studies and knowledge may see the discussion presented here as a possible pivot point in professional learning for the training of future professionals and the industries they seek to work in.

5 Conclusions Recommendations Summary

ePortfolios are robust in their ability to support different aspects of learning, assist students to address professional attributes required for future careers post- graduation and influence the development of students' digital literacy skills.

Educators see the potential for promotion of metacognition and critical thinking, whilst encouraging the development of reflective practice. The outcome of the conference session, therefore, is a series of participant generated discussion and creation of evidence of how learner professional identity intersects with developing a learning portfolio. The approach, aimed at using images to assist students, new graduates and new employees to develop a sense of 'self' encourages educators (both in formal educational settings and in the workplace) to improve critical thinking, reflection and creating the reflective narrative. Exploring innovative ways that the portfolio process and products can be used to develop a professional identity through encouraging students to reflect and connect themselves to multi faceted professional identities provides authentic learner engagement.

The overlay of two models, one illustrating Self-Determination theory [3] and the other the model of ideal self [4] to create a 'sense of self' model (Fig. 1) acknowledges the importance of the students' sense of self in ePortfolio creation, and why ePortfolio curriculum design is important [5].

Acknowledgment. My co-author of the Sense of Self Model, Dr Jennifer Munday, and second co- author, Associate Professor Patsie Polly are acknowledged for contributions to the model and the original workshop concept for this practical application of the model.

References

1. Stanford University: Folio thinking and ePortfolios at Stanford (2012). Accessed https://stanford.digication.com/foliothinking/Welcome//
2. Rowley, J., Munday, J.: A 'sense of self' through reflective thinking in ePortfolios. Int. J. Humanit. Soc. Sci. Educ. (IJHSSE) **1**(7), 78–85 (2014). https://www.arcjournals.org/pdfs/ijhsse/v1-i7/9.pdf
3. Ryan, R., Deci, E.: Self-determination theory and the facilitation of intrinsic motivation, social development, and well-being. Am. Psychol. **55**(1), 68–78 (2000)
4. Lawrence, D.: Enhancing Self-Esteem in the Classroom. SAGE Publications, London (2006)
5. Munday, J., Rowley, J., Polly, P.: The use of visual images in building professional "self" identities. Int. J. ePortfolio **7**(1), 53–65 (2017)
6. Hampe, N., Lewis, H.: E-portfolios support continuing professional development for librarians. Aust. Libr. J. **62**(1), 3–14 (2013)
7. Wakimoto, D., Lewis, D.: Graduate student perceptions of ePortfolios: uses for reflection development, and assessment. Internet High. Educ. **21**, 53–55 (2014)
8. Brandes, G., Boskic, N.: Eportfolios: from description to analysis. Int. Rev. Res. Open Distrib. Learn. **9**(2) (2008). Accessed http://www.irrodl.org/index.php/irrodl/article/view/502/1041
9. Harring, K., Luo, T.: Eportfolios: supporting reflection and deep learning in high-impact practices. Peer Rev. Assoc. Am. Coll. Univ. **18**(3), 9–12 (2016)

10. Oner, D., Adadan, E.: Are integrated portfolio systems the answer? An evaluation of a web based portfolio system to improve preservice teachers' reflective thinking skills. J. Comput. High. Educ. **28**, 236–260 (2016). https://doi.org/10.1007/s12528-016-9108-y
11. Rowley, J., Bennett, D.: Technology, identity and the creative artist. In: Carter, H., Gosper, M., Hedberg, J. (eds.) Electric Dreams: Proceedings of Ascilite Conference, Sydney, pp. 775–779 (2013)
12. Bennett, D., Rowley, J., Dunbar-Hall, P., Hitchcock, M., Blom, D.: Electronic portfolios and learner identity: an ePortfolio case study in music and writing. J. Furth. High. Educ. **40**(1), 107–124 (2016). https://doi.org/10.1080/0309877X.2014.895306
13. Stefani, L., Mason, R., Pegler, C.: The Educational Potential of e-Portfolios. Routledge, London (2007)
14. Bailey, N., Van Harken, E.: Visual images as tools of teacher inquiry. J. Teach. Educ. **65**(3), 241–260 (2014)

A Creative Ecosystem to Improve the Students Adaptation to Current Trends in IT Companies

František Babič(✉) and Vladimír Gašpar

Department of Cybernetics and Artificial Intelligence, Faculty of Electrical Engineering and Informatics, Technical University of Košice, Košice, Slovakia
{frantisek.babic,vladimir.gaspar}@tuke.sk

Abstract. An intellectual capital represents all resources that determine the value of an organization, and the competitiveness of an enterprise. We would like to develop the intellectual capital of our students to simplify their adaptation to ways of working in IT companies. This paper presents a proposal to design a creative ecosystem for helping students to become more prepared for the real-life work on IT projects. We started with the SCRUM methodology; next, we applied some selected methods of active learning and continued with supporting software tools Slack and Sli.do. Our motivation was to achieve better students' performance, timely delivery of the results and a high employment rate of our graduates. We discussed our preliminary results with participated IT companies and based on their positive feedback, we agreed on extended collaborations, e.g. invited lectures, interactive workshops or competitions for students.

Keywords: Ecosystem · Proactivity · Creativity

1 Introduction

In a fast-changing knowledge economy, 21st-century digital skills support competitiveness and innovation capacity. Authors in [1] identified seven core digital skills important for 21st-century: technical, information management, communication, collaboration, creativity, critical thinking, and problem-solving. Creativity appears to be one of the key characteristics, which development we can stimulate in the education process. A proactivity person is an assertive person who does things instead of waiting for it to be done. This characteristic can be simply identified during teamwork. It is important to form a group with fellow students to join resources and ideas along with encouragement each other.

Business Informatics is a computer science study program, which offers a combination of informatics and economic knowledge in all three levels (bachelor, master, PhD.) as the only one in Slovakia. The bachelors are familiar with the importance; role and changing the business value of Information and communication technologies, from standard software for office automation, through various Enterprise information systems, Management information systems, and Decision support systems up to the Business Intelligence tools. They are able to participate in implementation and operation

M. E. Auer et al. (eds.), *Teaching and Learning in a Digital World*,
Advances in Intelligent Systems and Computing 715,
https://doi.org/10.1007/978-3-319-73210-7_83

of various types software products, from simple web apps to complex Enterprise resource planning systems.

1.1 Related Work

Mohd Daud et al. emphasized the need to focus on knowledge of teachers about the teaching of creativity, not to the application of creativity by teachers; students who are shy and do not want to show their creativity [2]. Stojanova in her work [3] emphasized the ultimate goal of a modern educational system should be the development of independent, free, tolerant and creative young people that would satisfy their needs, but also the needs of modern society in which creativity is the basis for development. Since 2000, several engineering programs in Hong Kong have introduced design subjects aimed at nurturing students' creative capabilities [5]. Jackson understands the creativity an integral part of being a practitioner in the discipline, be it in academic research and scholarship or some other form of professional practice [6]. But being creative means different things in these different contexts. Kim and his colleagues analyzed the creativity from the point of personal and interpersonal relationships [4]. A collective of authors from the University of Auckland defined metrics of proactivity including proactive personality, confidence to perform proactive learning behaviors and frequency of proactive behaviors [7].

John Thomas explains that PBL requires complex tasks, based on challenging questions or problems, that involve students in design, problem-solving, decision making, or investigative activities; give students the opportunity to work relatively autonomously over extended periods of time; and culminate in realistic products or presentations [2]. Within the PBL, students obtained useful, real-world content knowledge potentially usable for their future tasks [11] and benefitted from improved critical thinking and problem-solving skills [3].

A paper of Mahnič [19] offers a review of studies available in Scopus database devoted to the application of Scrum to teaching. Santos et al. present some lessons learned resulted from adopting some Scrum practices in an R&D project and simultaneously using UML models, which are not commonly used in agile processes [20]. Some guidelines how to start with teaching Scrum at the university level are described in work of May et al. [21].

Bonwell and Eison defined an active learning as instructional activities involving students in doing things and thinking about what they are doing [10]. The relevant approaches focus more on developing students' skills than on transmitting information and require that students do something (read, discuss, or write) [9]. Freeman et al. realized a review of 225 existing studies focused on a comparison between some active learning versus traditional lecturing [11]. The students in traditional lectures were 1.5 times more likely to fail than students in courses with active learning. A similar approach is described in work of Ruiz-Primo et al. [12]. They found that inclusion of the active learning approaches improved student outcomes in consideration with some specifics like many of the studies did not control for pre-existing knowledge and abilities.

Barata et al. explored how gamification can be applied to education to improve student engagement. Their results showed significant improvements in attention to reference materials, online participation, and proactivity [8]. Nah et al. performed a review of literature focused on gamification in the education and learning context [15]. They identified some common game design elements like points, levels, badges, leaderboards, feedback, etc. Similar approach used a collective of authors in [16]. They confirmed that most studies focus on investigating how gamification can be used to motivate students, improve their skills, and maximize learning.

Students' perception of using a combination of smartphone and pooling application describes the work [18]. The results showed that mobile polling reduces graduate students' anxiety and improve their outcomes.

2 Creative Ecosystem

We would like to develop iteratively a creative ecosystem that will help us to improve the students' adaptation to current trends in IT companies in an effective way, This ecosystem will cover mainly master and bachelor degree with respect to particular requirements and expectations of the included domains, e.g. software development, mobile applications, project management and data analytics.

2.1 Proposal

The proposed creative ecosystem includes several innovative methods like project-based learning, AGILE methodology or active learning. We deploy these methods into specific subjects based on identified requirements and collected experiences. The first level covered a deployment of project-based learning in most subjects. The students are confronted with clearly specified goals and deadlines. The collaboration requires choosing the right team members. We would like to create an abilities wall containing information about offered skills and experiences from cooperation. The second level improved the previous one with the AGILE methodology representing a straightforward and meaningful way of managing and leading students' development projects. The third level covers carefully selected techniques of active learning like brainwriting, group decision making or retrieval practice. In addition, we support all three levels with supporting software tools designed for interaction with students or collaborative work.

We expect a better performance of the students, result's delivery within specified deadlines, creative and innovative software applications or higher percentage of the employed graduates. We evaluated initial outcomes by the questionnaires or input and output testing.

Scrum is an iterative and incremental agile framework for managing software product development. It was formalized originally for software development projects, but now can be used for any complex, innovative scope of work. Scrum was formally introduced in 1995 by Sutherland and Schwaber [13].

We selected two courses in bachelor study to test the Scrum potential: Analysis and design of information systems (ADIS), Introduction to business informatics (IBI). The ADIS offers basic theoretical and practical knowledge about software development life cycle. We focus mainly on analytical and design phases, i.e. from problem definition, through state of the art, user requirements identification and collection to technical specification containing user scenarios, UML diagrams, architecture, and mock-ups. The students have the opportunity to divide into pairs with own assignment focused on any software product. The IBI dealt with the basic concepts in business information systems domain, in particular, the concept of information system and methodology to deal with business challenges. The topic of the assignment is a response to an opportunity, threat, social or economic problem resulting in a new product, service, or in-house innovation. Both courses focus on the development of students' ability to apply creatively knowledge gained in lectures, seminars, and tutorials through project work assignments. Our motivation was to improve quality, collaboration, systematic work, communication and presentation skills.

From active learning domain, we started with two techniques. Retrieval practice represents a pause for two or three minutes every 15 min, asking students some questions from the previous class segment. Based on work Brame and Biel [14], this technique improves long-term memory and ability to learn subsequent materials. Case-based learning confronts students with a real situation requiring a use of their knowledge to find a suitable solution. Students should collect all information about the situation and consider an impact or implication of their decision.

This year we tried for a first time to organize a competition in Internet searching for students called Google war. Group of students living in the university dormitory in Košice created a concept of this competition originally. It aimed at entering correctly keywords representing the most powerful weapon in this war. We prepared three levels competition for courses IBI and Business informatics in practice (BIP).

Slack is an online communication and collaboration tool offering an opportunity to create topic-specific channels devoted to the particular teams or users. We used this tool to manage the master and bachelor thesis. The centralized on-line forum is available for general chatting and message exchange. The private channels are assigned to the individual students to brainstorm ideas, share files, ask questions or arrange meeting times and deadlines. From a teacher point of view, Slack creates an effective platform to manage the student progress and to identify his learning paths.

Sli.do is a web-based tool for real-time audience interaction through interactive Q&A and live polls. We used this tool to increase a level of interaction with our students by overcoming their shame and fear of asking. We used it for feedback collection, asking questions during lectures or voting for the best student's outcomes. A combination of smartphones and pooling provides a good possibility how to keep the student's attention.

2.2 Experiences

The student's feedback on Scrum was mostly positive. The collaborative character motivated them to meet the objectives and proposed milestones. Scrum has fostered a development of their creativity, communication and presentations skills. They were satisfied with a regular work during the semester and with a possibility to use something from the practice. The students considered a Scrum task board as a clear and simple method for collaborative creation. On the other hand, some of them expressed negative experiences with the work of classmates in their Scrum team. Some students proposed to replace the traditional paper form of the task board with electronic one or more intensive involvement of the teacher in Sprint management. Based on our experience, we can confirm the finding of Lorenzo et al. [17], i.e. an inclusion of active engagement techniques benefited all students but had the greatest impact on female students' performance. It is important for us because we have a high percentage of the female in each year (more than 30%) despite the fact that Business Informatics is a computer science program.

The use of Slack has brought the positive experiences for the teachers and students too. We were able to monitor simply the progress and consult the emerging problems or new ideas. Slack enables to store all messages and shared objects, what helped us with the final evaluation and retrospective. We present some statistics of one team (5 students, 1 teacher) from the last 2 weeks before bachelor thesis delivery: 176 messages (that's 11 fewer than the week before) and uploaded 31 files (that's 13 more than the week before), 259 messages (83) and uploaded 40 files (9).

Sli.do resulted in more intense interaction with students. For example, during a lecture, we asked three questions from related part of the theme. The students had a short time to consider the right question or prepare a joint solution by groups. We used their answers for attendance verification, knowledge evaluation and motivation support. In the case of invited lectures, Sli.do has brought many interesting questions from the students, and it was necessary to devote more time to them. One time, the lecturers were willing to come back again and answer the unanswered questions. The pooling

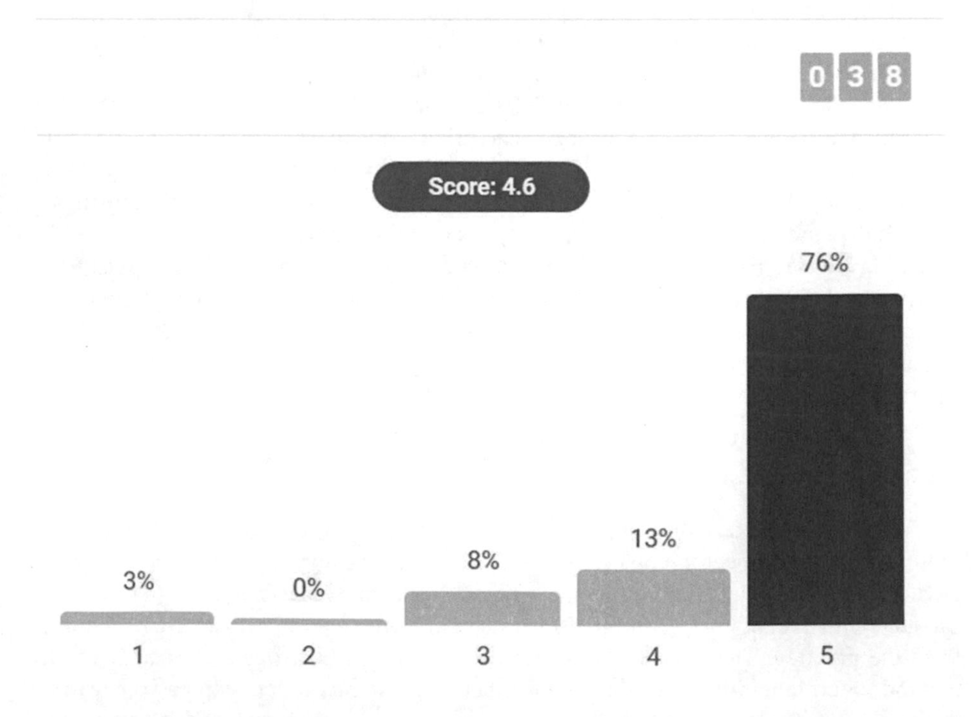

Fig. 1. The students' feedback on invited lecture (from one to five stars).

represented an important feedback for the lecturers about their presentation and communication skills (Fig. 1).

We gradually introduce the appropriate elements of gamification in our education process. Our first Google searching competition has brought interesting results like not all students were motivated to move to a higher level or some have not found the right answers before the time limit. This situation was caused by a relatively poor reading comprehension. The students focused on the wrong part of the question or did not understand the overall context. We organized six different rounds, and some winners positively surprised us. Based on collected feedback we will change the construction of some questions.

3 Conclusion

This paper describes three levels of improvement in our teaching process aimed to improve the student's adaptation to current trends in IT companies. We propose a creative ecosystem including not only new material equipment but main changes in used teaching techniques to emulate typical IT work environment in a university space. Some preliminary results showed a positive satisfaction of the students with applied changes but also identified some complication with their adaptation. The crucial factor represents their motivation, which can be supported not only with the points but also with the possibility to meet experts, to solve real tasks and to increase the price on the labour market. We discussed these results with participated IT companies and based on their positive feedback we agreed on further mutual interactions, e.g. invited lectures, interactive workshops or competitions for students.

In the future, we will continue to work on our ecosystem. We will evaluate the next iterations to provide more advantages for our students and to reduce the disadvantages. We will aim at education of data analytics [22–25], i.e. if is it possible to combine Scrum and CRISP-DM representing most widely used methodology for data mining processes. We are inspired by a combination of User Experience and Agile development resulted in Design Sprints proposed by Google or Lean UX.

Acknowledgment. The work presented in this paper was partially supported by the Cultural and Educational Grant Agency of the Ministry of Education and Academy of Science of the Slovak Republic under grants no. 025TUKE-4/2015 and no. 05TUKE-4/2017.

References

1. Van Laar, E., van Deursen, A., van Dijk, J., de Haan, J.: The relation between 21st-century skills and digital skills: a systematic literature review. Comput. Hum. Behav. **72**, 577–588 (2017)
2. Mohd Daud, A., Omar, J., Turiman, P., Osman, K.: Creativity in science education. UKM teaching and learning congress 2011. Procedia – Soc. Behav. Sci. **59**, 467–474 (2012)
3. Stojanova, B.: Development of creativity as a basic task of the modern educational system. Procedia Soc. Behav. Sci. **2**, 3395–3400 (2010)

4. Kim, S., Moon, W., Kim, W., Park, S., Moon, I.: Is it possible to improve creativity? If yes, how do we do it? In: 21st European Symposium on Computer Aided Process Engineering, pp. 1130–1134 (2011)
5. Siu, K.W.M.: Promoting creativity in engineering programmes: difficulties and opportunities. WCES 2012, Procedia – Soc. Behav. Sci. **46**, 5290–5295 (2012)
6. Jackson, N.: Making higher education a more creative place. J. Enhanc. Learn. Teach. **2**(1), 14–25 (2005)
7. Geertshuis, S., Jung, M., Cooper-Thomas, H.: Preparing students for higher education: the role of proactivity. Int. J. Teach. Learn. High. Educ. **26**(2), 157–169 (2014)
8. Barata, G., Gama, S., Jorge, J., Gonçalves, D.: Improving participation and learning with gamification. In: Proceedings of the First International Conference on Gameful Design, Research, and Applications, Toronto, Canada, pp. 10–17 (2013)
9. Brame, C.: Active learning. Vanderbilt University Center for Teaching (2016)
10. Bonwell, C.C., Eison, J.A.: Active Learning: creating excitement in the classroom. ASH#-ERIC higher education Report No. 1, School of Education and Human Development. The George Washington University, Washington, D.C. (1991)
11. Freeman, S., Eddy, S.L., McDonough, M., Smith, M.K., Okoroafor, N., Jordt, H., Wenderoth, M.P.: Active learning increases student performance in science, engineering, and mathematics. Proc. Natl. Acad. Sci. USA **111**, 8410–8415 (2014)
12. Ruiz-Primo, M.A., Briggs, D., Iverson, H., Talbot, R., Shepard, L.A.: Impact of undergraduate science course innovations on learning. Science **331**, 1269–1270 (2011)
13. Sutherland, J., Schwaber, K.: Scrum development process. In: OOPSLA Business Object Design and Implementation Workshop, Object-Oriented Programming, Systems and Applications (OOPSLA) Convention, Austin, Texas, USA (1995)
14. Brame, C.J., Biel, R.: Test-enhanced learning: the potential for testing to promote greater learning in undergraduate science courses. CBE Life Sci. Educ. **14**, 1–12 (2015)
15. Nah, F.F., Zeng, Q., Telaprolu V.R., Ayyappa, A.P., Eschenbrenner, B.: Gamification of Education: a review of literature. In: Nah, F.F.H. (eds.) HCI in Business. HCIB 2014, Lecture Notes in Computer Science, vol. 8527, pp. 401–409. Springer (2014)
16. Bordes, S.S., Durelli, V.H.S., Reis, H.M., Isotani, S.: A systematic mapping on gamification applied to education. In: SAC 2014 Proceedings of the 29th Annual ACM Symposium on Applied Computing, Gyeongju, Republic of Korea, pp. 216–222 (2014)
17. Lorenzo, M., Crouch, C.H., Mazur, E.: Reducing the gender gap in the physics classroom. Am. J. Phys. **74**, 118–122 (2006)
18. Sun, J.C.-Y.: Influence of polling technologies on student engagement: an analysis of student motivation, academic performance, and brainwave data. Comput. Educ. **72**, 80–89 (2014)
19. Mahnič, V.: Scrum in software engineering courses: an outline of the literature. Glob. J. Eng. Educ. **17**(2), 77–83 (2015)
20. Santos, N., Fernandes, J.M., Carvalho, M.S., Silva, P.V., Fernandes, F.A., Rebelo, M.P., Barbosa, D., Maia, P., Couto, M., Machado, R.J.: Using scrum together with UML models: a collaborative university-industry R&D software project. In: Proceeding of ICCSA 2016, LNCS, vol. 9789, Springer (2016)
21. May, J., York, J., Lending, D.: Play ball: bringing scrum into the classroom. J. Inf. Syst. Educ. **27**(2), 87–92 (2016)
22. Sarnovský, M., Paralič, J.: Teaching big data analysis at technical university in Kosice in business information systems study program. In: Proceedings of 13th International Conference on Emerging eLearning Technologies and Applications (ICETA), pp. 325–330. IEEE (2015)

23. Sarnovský, M.: Design and implementation of the cloud based application for text mining tasks. Data Min. Knowl. Eng. **6**(6), 261–264 (2014)
24. Muchová, M., Paralič, J., Jančuš, M.: An approach to support education of data mining algorithm. In: Proceedings of IEEE 15th International Symposium on Applied Machine Intelligence and Informatics (SAMI), Herľany, Slovakia, pp. 93–98 (2017)
25. Smatana, M., Paralič, J., Butka, P.: Topic modelling over text streams from social media. In: Text, Speech and Dialogue, LNCS, vol. 9924, pp. 163–172 (2016)

Development of the Intelligent System of Engineering Education for Corporate Use in the University and Enterprises

Alexander Afanasyev(✉), Nikolay Voit, Irina Ionova, Maria Ukhanova, and Vyacheslav Yepifanov

Ulyanovsk State Technical University, Ulyanovsk, Russia
{a.afanasev,n.voit}@ulstu.ru, epira@mail.ru, mari-u@inbox.ru, v.epifanov73@mail.ru

Abstract. In order to increase an effectiveness of the practice-oriented and dual training, the joint computer-based corporate training systems for educational institutions' students and companies' employees shall be developed. In practice of the Russian education this activity is carried out via the basic departments created by universities and colleges at industrial works and companies. Such projects use modern computer-based training systems as a software information base.

However, the modern training systems do not take into account the specifics of training in project activities, such systems are not integrated with CAD packages and project repositories, and the designers' project activity is not assessed.

A goal of this research work is to enhance the competences of trainees (students and employees) through the development and implementation of methods, models and tools for project solutions' analysis, and through the formation of personalized training on basis of the uniform intelligent project repository.

Keywords: Intelligent corporate training system
CAD mechanical engineering · Individual training path

1 Introduction

In order to increase an effectiveness of the practice-oriented and dual training, the joint computer-based corporate training systems for educational institutions' students and companies' employees shall be developed. In practice of the Russian education this activity is carried out via the basic departments created by universities and colleges at industrial works and companies.

Such projects use modern computer-based training systems as a software information base. In the current world practice of these systems' development an adaptive approach is dominating connected with the automatic generation of an individual training path. The subject-oriented (for a specific employer) competences formation based on mastering his/her experience and best practices is a key task of implementing programmed engineering training. However, the modern training systems do not take into account the specifics of training in project activities, such systems are not

M. E. Auer et al. (eds.), *Teaching and Learning in a Digital World*,
Advances in Intelligent Systems and Computing 715,
https://doi.org/10.1007/978-3-319-73210-7_84

integrated with CAD packages and project repositories, and the designers' project activity is not assessed.

2 Current Approaches and Systems' Analysis

The following summarizes the main methods and systems' analysis used in adaptive training.

The domain model is considered as a set of training courses in [1]. In order to construct the optimal training path, a genetic algorithm is used, where the genotype is a sequence of selected training courses. The optimal initial population size is chosen as 50. A larger population size will increase the probability of finding a high quality solution, but it leads to significant delays in time. The fitness functions take into account the pre-test results, the degree of training materials' relationship, the complexity and training materials. The roulette method is used as a method of the selection. The mutation operations are performed through changing the sequence of curricula. This approach does not provide a dynamic change in the training and project material.

The research work [2] describes the system of automatic generation of personal training recommendations. The domain model represents a variety of unstructured educational materials that are automatically generated from the Internet applying indexing and text mining techniques based on the Nutch software. The learner's model contains a weighted sequence of training elements that are of interest to him/her. Learners are grouped on the basis of similarities and dissimilarities of their preferences, for which different clustering techniques can be applied. The recommendations are based on the learner's expressed interests through the visited Web pages analysis, and the results are ranked according to the TF-IDF measure of the content similarity. This approach can be used for informal learning, the use of its techniques for formal education systems (such systems are the most effective in terms of quality and training time) is problematic. In addition, it does not assess the quality of the Internet sources' content.

In the research work [3] the domain model contains three components: course content organized into a content tree (course, chapters, sections and subsections), educational materials and tests. The learner's model is an overlay model, which is initialized after the pre-test. The Dempster-Shafer (DS) theory is applied to analyze the test results. The student's knowledge level is classified into 3 groups: low, intermediate, and master level. Adaptation occurs after the test, the system calculates the student's knowledge level and updates his/her model. The disadvantage of this theory is a weak adaptation.

The research work [4] represents the domain model as a graph in which the vertices are the concepts of the subject domain. The learner's model consists of three components: a learner's profile; a knowledge model; a preferred learning style. The adaptation model is built on the basis of production rules. The system uses a combination of adaptive navigation support and adaptive presentation technique.

The authors of [5] take into account the learner's thinking style. The domain model contains 12 types of materials: 8 theoretical materials and 4 practical materials. The trainee's model is described by 3 submodels: goals and preferences (which courses a trainee wants to visit, what a trainee's preferences (font, type, size, color and other parameters associated with the interface) are; thinking and learning style (contains information on the specific path of learning and the approach to learning); knowledge and execution (test results, projects, tasks). The system supports the following styles of thinking: theorist, organizer, innovator, humanitarian. Learning styles are: active, reflexive, deductive, inductive, visual, verbal, sequential, global. The learning process model builds the most appropriate learning path using adaptive navigation, content selection, and annotated references that are consistent with the learner's profile. The learning path is chosen on the basis of a formula that takes into account the thinking style, the learning style and the knowledge level.

In the GRAPPLE system [6, 7], the domain model is described through concepts and relationships. The information resources and facts can be associated with concepts. The learner's model is overlay. The adaptation model built on the basis of a set of "if-then" rules and used the domain and learner's models, decides what information and how it should be displayed (using content adaptation, presentation adaptation and link adaptation). The project developed the language called GAL (Generic Adaptation Language).

Common disadvantages of these methods and tools are: lack of techniques for integration with the CAD packages; lack of dynamic domain model filling, which does not allow the system to automatically include hybrid knowledge and enterprise's experience units in the training process; great labor intensity for developing training and practical materials; an absence of a virtual component, which makes it difficult to form practical competences for trainees.

3 Problem

The goal of this research work is to enhance the trainees' competences (students and employees) through the development and implementation of methods, models and tools for project solutions' analysis, and through the formation of personalized training on basis of a single intelligent project repository.

This approach is realized within the framework of the intelligent corporate training system, which is used jointly by the enterprise's employees and university basic department's students. The training material is formed in accordance with the employer's requirements and on the basis of the best enterprise's practices. It will increase the practical orientation and quality of training.

In order to achieve this goal, the authors have to solve the following tasks:

1. To develop the mathematical support of the adaptive training system, which allows the system to form personalized training scenarios and include an ontological domain model, a trainee's model, a scenario's model, a practical task's model, a test's model, and a scenario generation algorithm.
2. To develop the method of recommendations' forming and a designer's competence profile adjusting on basis of project operations' protocol, which is noted for the analysis of 3D-modeling of the parts performed at CAD KOMPAS.
3. To develop the software and data support of the intelligent corporate training system.

4 University – Enterprise Intelligent Corporate Training System Organization

University-Enterprise intelligent corporate training system organization is shown in Fig. 1.

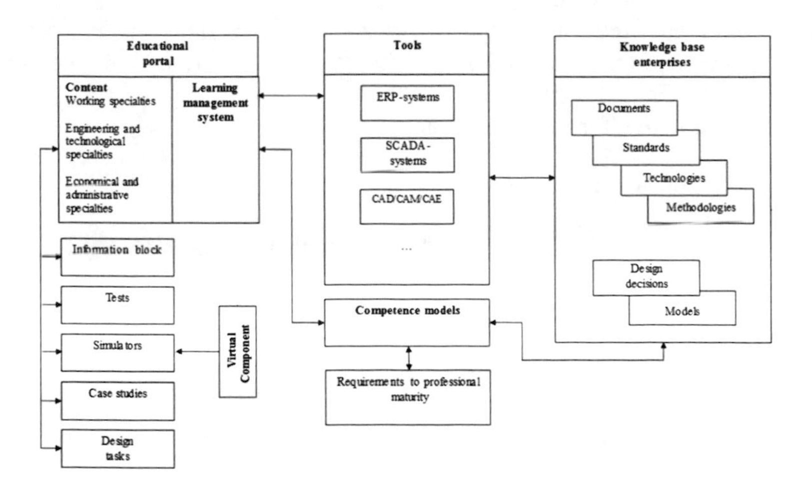

Fig. 1. University-enterprise intelligent corporate training system organization

5 Mathematical Support of the Intelligent Corporate Training System

The domain model is given as:

$$O = (PSL, T, R, F, Ax), \tag{1}$$

where $PSL = \{psl_i | i = 1..\mathrm{x}\}$ is a set of project solutions,

T is the terms of the applied domain, which the ontology describes.

A set of terms is defined as:

$$T = \{C,\ In\}, \tag{2}$$

where $C = \{A, P, D, GOAL, COMP\}$ is a set of ontology classes (*A* – knowledge atoms, *P* – concepts, *D* – training material, *GOAL* – training purpose, *COMP* – competences). The class "training material" has a slot "is a Reference" with "*true*" or "*false*" values,

In is a set of objects of ontology classes.

R is a set of relations between ontology objects:

$$R = \{R_{learn},\ R_{part},\ R_{next}\}, \tag{3}$$

where R_{learn} is the binary relation “is_studied_in” which has the semantics “connected_to” and connects the objects of the ontology classes (“Atom”, “Concept”) to the objects of the “Training Material” class,

R_{part} is the binary relation “is_a_part_of” that has the semantics “part_of” and connects the objects of the ontology classes (“Atom”, “Concept”) to the objects of the “Concept”, “Purpose of Learning” classes,

R_{next} is the binary relation “is_trained_after”, which has the semantics “after_of” and connects the objects of the ontology classes (“Atom”, “Concept”) to objects of the “Concept” and “Atom” classes.

A set of interpretive functions is defined as:

$$F = \{Fatom\_op, Fpsl\_a, Fedu, Fdefine, Fsimilar, T\}, \tag{4}$$

where $Fatom\_op$: $A \rightarrow \{Operation\}$ is a function of the “Atom” class object mapping into a set of project solution operations,

Fpsl_a : PSL $\rightarrow$ {A} is a function of the project solution mapping into a set of the “Atom” class objects,

Fedu: {A} $\rightarrow$ {D} is a function for constructing an ordered set of training materials for studying certain knowledge atoms,

Fdefine: P $\rightarrow$ {D} is a function of finding training materials that describe a certain concept,

Fsimilar: D $\rightarrow$ {D} is a function of finding the most similar training materials,

T: D $\rightarrow$ Q^+ is the didactic material complexity.

A set of axioms is defined as:

$$Ax = \{AxAHP, AxAHL, AxAHD, AxPAfP, AxPAfA, AxAAfP\}, \tag{5}$$

where $AxAHP$ – “atoms are related to concepts”, if the atom Y is related to the concept of X, which is related to the concept of Z, then the atom Y is related to the concept of Z as SWRL:

$$\begin{aligned} &Concept\ (?x) \wedge Atom\ (?y) \wedge Concept\ (?z) \wedge is\_related\_to\ (?y, ?x) \wedge \\ &is\_related\_to\ (?y, ?z)\ \rightarrow\ is\_related\_to\ (?y, ?z). \end{aligned} \tag{6}$$

$AxAHL$ – “atoms are related to training purposes”, if the atom Y is related to the concept of X, which is related to the training purpose of Z, then the atom Y is related to the training purpose of Z as SWRL:

$$\begin{aligned} &Concept\ (?x) \wedge Atom\ (?y) \wedge training\,purpose\ (?z) \wedge is\_related\_to\ (?y, ?x) \wedge \\ &is\_related\_to\ (?y, ?z)\ \rightarrow \wedge\ is\_related\_to\ (?y, ?z). \end{aligned} \tag{7}$$

$AxAHD$ – “atoms are related to training materials”, if the atom Y is related to the concept of X, which is studied in the training material Z, then the atom Y is studied in the training material Z as SWRL:

$$Concept\ (?x) \wedge Atom\ (?y) \wedge training\ material\ (?z) \wedge is\_related\_to\ (?y, ?x) \wedge is\_studied\_in\ (?y, ?z) \rightarrow is\ studied\ in\ (?y, ?z). \quad (8)$$

AxPAfP – “atoms are studied after atoms”, if the atom Y is related to the concept of X, which is studied after the concept of Z, and the atom C is related to Z, then the atom Y is studied after C as SWRL:

$$Atom\ (?y) \wedge Concept\ (?x) \wedge Concept\ (?z) \wedge Atom\ (?c) \wedge is\_related\_to\ (?y, ?x) \wedge is\_related\_to\ (?c, ?z) \wedge is\_studied\_after\ (?x, ?z) \rightarrow is\_studied\_after\ (?y, ?c). \quad (9)$$

AxPAfA – “concepts are studied after atoms”, if the atom Y is related to the concept of X, which is studied after the atom C, then the atom Y is studied after C as SWRL:

$$Atom\ (?y) \wedge Concept\ (?x) \wedge Atom\ (?c) \wedge is\_related\_to\ (?y, ?x) \wedge is\_studied\_after\ (?x, ?c) \rightarrow is\_studied\_after\ (?y, ?c). \quad (10)$$

AxAAfP – “atoms are studied after concepts”, if the atom Y is studied after the concept of X and the atom C is related to X, then the atom Y is studied after C as SWRL:

$$Atom\ (?y) \wedge Concept\ (?x) \wedge Atom\ (?c) \wedge is\_related\_to\ (?c, ?x) \wedge is\_studied\_after\ (?y, ?x) \rightarrow is\_studied\_after\ (?y, ?c). \quad (11)$$

The model of the trained engineer is defined as:

$$U = (UA,\ P,\ C,\ P\_AB,\ C\_AB,\ P\_S,\ C\_S,\ ACT\_G,\ ACT\_C), \quad (12)$$

where UA $= \{a_i \in \mathrm{A} | i = 1..\mathrm{nua}\}$ is a subset of the knowledge atoms that are required to be studied,

P $=$ A $\rightarrow$ $[0..1]$ is a degree of the A atom knowledge,

C $=$ A $\rightarrow$ N is the number of control measurements of the A atom knowledge,

P_AB $=$ A $\rightarrow$ $[0..1]$ is a degree of proficiency in the A atom abilities,

C_AB $=$ A $\rightarrow$ N is a number of control measurements of the A atom abilities,

P_S $=$ A $\rightarrow$ $[0..\infty]$ is a degree of proficiency in the atom A skill,

C_S $=$ A $\rightarrow$ N is the average time for performing operations for atom A,

ACT_G $=$ A $\rightarrow$ N is the total number of actions for the reference use of the A atom skill,

ACT_C $=$ A $\rightarrow$ N is the total number of actions for the current use of the A atom skill.

The test model is represented as:

$$TZ = (q, ANS, TA, TAQ), \tag{13}$$

where $q =$ is a question,

$ANS = \{ans_i | i = 1..n\}$ is a set of answers,

$TA = \{a_i \in A | i = 1..k\}$ is a subset of knowledge atoms that evaluate tests,

$TAQ = \{q_i \in ANS | i = 1..m\}$ is a set of correct answers.

The test execution is represented as:

$$UTZ = (U, TZ, QQ), \tag{14}$$

where U is the designer's profile,

TZ is a test,

$QQ = \{qq_i \in ANS \mid i = 1..k\}$ is a subset of the answers selected by a trained designer.

The scenario model is shown as:

$$ST = (A, P, D, STZ, SPTZ, PERIOD, Fd\_d, Fphase\_tz, Fphase\_ptz, Fphase\_d, Fphase\_first, Fnext\_phase), \tag{15}$$

where $A = \{a_i | i = 1..n\}$ is a set of knowledge atoms,

$P = \{p_i | i = 1..j\}$ is a set of concepts,

$D = \{d_i | i = 1..k\}$ is a set of training materials,

$STZ = \{tz_i | i = 1..m\}$ is a set of tests,

$SPTZ = \{ptz_i | i = 1..l\}$ is a set of practical tasks,

$PHASE = \{pz_i | i = 1..h\}$ is a set of the scenario's stages,

$Fd\_d = D \rightarrow D$ is a training material successor function,

$Fphase\_d = PHASE \rightarrow D*$ is a membership function of a training material subset to a certain stage of the scenario,

$Fphase\_tz = PHASE \rightarrow TZ*$ is a membership function of a test subset to a certain stage of the scenario,

$Fphase\_ptz = PHASE \rightarrow PTZ$ is a membership function of a practical task to a certain stage of the scenario,

$Fnext\_phase = PHASE \rightarrow PHASE$ is a scenario's stage successor function,

$Fphase\_first = PHASE \rightarrow D$ is a function of the first training material in a certain stage of the scenario.

The operation model is presented as:

$$Operation = (ID, type, number, pvo), \tag{16}$$

where ID is the operation's unique identifier,

$type \in TypeOperation$ is an operation type,

$number$ is an operation number,

pvo is a set of operation parameters with a value.

The operation parameter model with a value is defined as:

$$PVO = (key, value), \quad (17)$$

where $key \in ParamKey$ is a parameter's name,

$value \in ParamValue$ is a parameter's value.

The model of the initial data for the recommendation's formation is defined as:

$$S = (Operations, Rules, A, F\_atom), \quad (18)$$

where $Operations = \{o \in Operation\}$ is a set of project operations,

$Rules = \{r_i | i = 1..k\}$ is a set of rules for searching and replacing non-optimal project operations,

$A = \{a_i | i = 1..n\}$ is a set of knowledge atoms,

$F\_atom = Operation \rightarrow A$ is a function of mapping the operation into the knowledge's atoms.

The rule model is:

$$Rule = (tmpl, result), \quad (19)$$

where $tmpl = \{t_i | i = 1..k\}$ is a first-order logic formula for searching in the protocol of project operationsIO,

$result = \{res_i | i = 1..n\}$, res $= (C$, key, value) is a set of optimal project operations defined as ordered triples (operation code, operation parameter or a set of operation parameters, operation parameter value or a set of values), where the code is a constant, and the operation parameter and the operation parameter value are first-order logic formula.

For the first-order logic formula, the authors give the symbol's alphabet:

(a) object variables:

- X = {xp ∈ TypeOperation} is a set of operations,
- P = {p ∈ ParamKey} is a set of operation parameter keys,
- T = {t ∈ ParamValue} is a set of operation parameter values;

(b) symbols of logical operations are: $\neg, \wedge, \vee, \rightarrow, \leftrightarrow$;
(c) quantifiers: $\forall, \exists$;
(d) auxiliary symbols: ≪()≫, ≪{}≫ – brackets; ≪,≫ – comma;
(e) terms:

- type = Operations → TypeOperation – defines the operation's type,
- code = Operations → N is a definition of the operation's code,
- number = Operations → N – defines the operation's number,
- param = Operations × ParamKey → ParamValue is a definition of the operation parameter value,
- param_start = Operations × ParamKey → ParamValue is a default value of the operation parameter,

- arithmetic operations: $R \times R \rightarrow R$ - multiplication, addition, subtraction, division;

(f) predicates:

- 2-ary functions: more (>), less (<), equal (=), not equal to ($\neq$), $\in$, eq_op – equality of the operation results.

The recommendation model is defined as:

$$R = (op\_before, op\_after, F\_steps, F\_message), \tag{20}$$

where op_before $\subset$ Operations is a set of non-optimal project operations involved in the recommendation's formation,

Op_after $\subset$ Operations is a set of recommended project operations,

F_steps = Operations $\rightarrow$ N is a function for calculating the action number for constructing project operations,

F_message = Operations $\times$ Operations $\rightarrow$ Text is a function for the recommendation's text formation.

6 Personalized Training Scenario Formation Algorithm

The personalized training scenario formation algorithm is described below.

1. Choose a learning scenario.
2. Download the next stage of the selected scenario. If the previous stage is absence, the first scenario's stage is activated.
3. A trainee consistently studies the training material within the framework of the stage, including additional information assets of an industrial enterprise.
4. Perform the tests, which correct the trained engineer's knowledge level.
5. Check the skill's level on the basis of project tasks.
6. Assess the knowledge and skills' level. If the level is considered to be satisfactory, the trained engineer should go to the next scenario's stage (to the second step). If there is no the second scenario's stage, the training process is considered to be completed. If knowledge and skills' level is considered to be unsatisfactory, the individual training path is synthesized at the seventh step.
7. Generate the adaptive training path. The knowledge elements that a trainee studied at this stage are selected, but their level of knowledge and skills is less than 0.5.
8. Fill a variety of training materials that have a "part-whole" relationship with selected knowledge elements. Training materials are selected in order to minimize the total training time and the number of repetitions of the material already presented. Also, many training materials are arranged in accordance with the relation of the knowledge element order.
9. Form the minimum set of tests and project tasks that check the selected set of knowledge elements.

10. When the individual training stage is completed, the control test is performed. If the results of the trained engineer do not meet the expected, then the repeated training path is created. Its difference from the previous training path is in the fact that a lot of training materials are selected so that they the least repeat the previous path.

Consider the algorithm for the recommendation formation.

1. Start of the trainee's work with a project.
2. If the project is new, go to step 5.
3. Generate operations based on the existing project.
4. Add an operation to a sequence of operations.
5. Read trainee's control actions.
6. Generate operations based on the trainer's action.
7. Add an operation to a sequence of operations.
8. Form the project state on the basis of an operation's sequence.
9. Add the project status to the project status sequence.
10. Search for a rule that corresponds to an operation's sequence.
11. If the rule is not found, go to step 13.
12. Add a recommendation to an individual trainee's list and display it.
13. Search for a rule that corresponds to a state's sequence.
14. If the rule is not found, go to step 16.
15. Add a recommendation to the individual trainee's list and display it.
16. If the project is not finished, go to step 5.
17. Exit.

Here is the algorithm for the trainee's profile correction.

1. Obtain a formed recommendation based on the trainee's profile.
2. Make a list of the project operations involved in the rules when formulating recommendations.
3. Obtain a list of knowledge elements associated with project operations.
4. Reduce the skills' level for these knowledge elements.

7 Realization

The training block was built according to the classical three-tier architecture: a client, an application server and a database server.

Java Platform, Standard Edition technology is used as a development platform. This technology supports built-in client-server applications (RMI and ISOAP technologies).

MySQL is used as the database server. The system is scalable, allows for functionality expansion by adding components (plugins). Plugins allow client and server's components to be extended by providing a client and server API.

In order to implement the client-server technology at Java SE, the technology of Web Services is chosen. The SOAP protocol is used to exchange messages between Web services.

8 Experiments Results

The experiment in training was performed among 18 students of the Ulyanovsk State Technical University. In order to assess the initial skills' level, these students passed a pre-test, as a result of which two groups of 8 and 10 students were organized. Their average skill's level was 0.5. The groups were trained on the materials of CAD KOMPAS-3D. The first group was trained based on a linear scenario; the second group was trained in accordance with an adaptive scenario. After training, the task was to build an assembly "Pump cover" of 471 elements. The experiment's results are shown in Table 1. When applying the adaptive training scenario, the average level of students' skills is 20% higher compared with the linear training scenario due to the more detailed training. This resulted to 17% increase in time required for preparation.

Table 1. Comparison of linear and adaptive learning scenarios

	Average level of skills before training	Average learning time, h	The average level of skills after training	Average time for building an assembly, h	Total elapsed time, h
Linear scenario	0.5	1.65	0.745	2.8	4.5
Adaptive scenario	0.5	2	0.9	1.9	3.9
Gain		−17%	20%	50%	20%

9 Conclusion and Further Research Directions

The developed corporate training environment is used both by employees of an enterprise and by students of the university's basic department. As a rule, after training at the basic department, graduates are hired to work at the enterprise. Use of the theoretical and practical materials associated with the specific employer's competences as the educational objects allows graduates to shorten adaptation terms, contributes to their active career growth, and increases their motivation.

The developed mathematical support of the training system allows it to form personified scenarios.

The virtual component of the training system is represented by radio devices' assembler, adjuster and a radio equipment regulator's virtual workplaces implemented at both OpenSim and Unity.

Further directions of the research work are related to the expert system's development for the trainees' activities assessment when using CAD packages, in particular mechanical engineering based on both the KOMPAS-3D package and virtual industrial worlds.

Acknowledgement. The study was carried out with the financial support of the Russian Fund of Basic Research and the Government of the Ulyanovsk region, project No. 16-47-732152. The research is supported by a grant from the Ministry of Education and Science of the Russian Federation, project No. 2.1615.2017/4.6.

References

1. Huang, M.J., Huang, H.S., Chen, M.Y.: Constructing a personalized e-learning system based on genetic algorithm and case-based reasoning approach. Expert Syst. Appl. **33**, 551–564 (2007)
2. Khribi, M.K., Jemni, M., Nasraoui, O.: Automatic recommendations for e-learning personalization based on web usage mining techniques and information retrieval. Educ. Technol. Soc. **12**(4), 30–42 (2009)
3. Esichaikul, V., Lamnoi, S., Bechter, C.: Student modelling in adaptive e-learning systems. Knowl. Manag. E-Learn.: Int. J. **3**, 342–355 (2011)
4. Mustafa, Y.E.A., Sharif, S.M.: An approach to adaptive e-learning hypermedia system based on learning styles (AEHS-LS): implementation and evaluation. Int. J. Libr. Inf. Sci. **3**(1), 15–28 (2011)
5. Mahnane, L., Tayeb Laskri, M., Trigano, P.: A model of adaptive e-learning hypermedia system based on thinking and learning styles. IJMUE **8**(3), 339–350 (2013)
6. Ploum, E.: Authoring of adaptation in the GRAPPLE project. Master thesis, Eindhoven University of Technology (2009)
7. De Bra, P., Smits, D., Van der Sluijs, K., Cristea, A.I., Hendrix, M.: GRAPPLE: personalization and adaptation in learning management systems. In: Proceedings of the World Conference EdMedia, pp. 3029–3038 (2010)

Innopolis University, a Center of a Newly-Developed IT Hub in Russia: The Results of Four Years of Academic Operation

Tanya Stanko, Oksana Zhirosh(✉), and Petr Grachev

Department of Education, Innopolis University, Innopolis, Russia
o.zhirosh@innopolis.ru

Abstract. In 2013, a new IT university was launched in the Republic of Tatarstan, Russia, to contribute to the improvement of Computer Science education in Russia. The project is being supervised by the Ministry of Communications of the Russian Federation, and by the government of the Republic of Tatarstan. At ICL2013 in Kazan, Innopolis University (IU) announced its development plans, based on research findings [2], regarding instructional approach and curriculum development [3]. This paper aims to report on the current situation at Innopolis University, which is now in its 5th year of operation and is about to produce its first BSc graduates. The paper also presents Innopolis University's plans in terms of partnerships with academia and industry, faculty profile, curriculum development, instructional methods, student profile, student recruitment practices.

Keywords: Pilot projects · Curriculum development
Computer Science education

1 Introduction

According to the strategic Concept 2020 plan [1], the government of Russia plans to double annual productivity growth. With technology being one of the major resources for achieving this goal, the Republic of Tatarstan in Russia is planning to increase the share of the Information and Communication Technology (ICT) sector in its GDP from 3.5 to 7%. To achieve the goals, a new city of Innopolis is being built in Tatarstan, Russia. The newly established Innopolis University at its heart will be a source of qualified IT personnel for the city's resident companies and a center of research. The project was launched by the President of the Republic of Tatarstan, Rustam Minnikhanov, and the Minister of Information Technologies and Communication of the Russian Federation, Nikolay Nikiforov, in 2010.

In the paper presented at ICL2013 [3], four major priorities were identified for the then newly established university:

1. Develop practical degree and non-degree programs that address industry's needs. The curriculum should be characterized by practicality, project-orientation and

M. E. Auer et al. (eds.), *Teaching and Learning in a Digital World*,
Advances in Intelligent Systems and Computing 715,
https://doi.org/10.1007/978-3-319-73210-7_85

industry-focus, soft skills, and flexibility. Four distinct BSc programs using a common core and four years of study were being planned: Computer Science, Information Security, Database Engineering, Graphics and Game Development.
2. Build a robust research infrastructure of international caliber.
3. Create an innovation ecosystem including an IT park, an incubation/accelerator and the training to support it.
4. Create a pipeline of school graduates who are interested in and well-prepared for careers in IT.

In the subsequent five sections of the paper, we will present the state of play for each of the goals stated.

2 General Information

Innopolis University is a part of a newly built innovative city of Innopolis, in the Republic of Tatarstan, Russia. The city of Innopolis is about 60 min' drive from the capital of Tatarstan Kazan, and Kazan is about 60 min flight east of Moscow. In 2017, about 2000 citizens live in Innopolis, with the prospect of growth up to 150000 citizens.

The University leadership team includes a Director, a Rector, 4 Vice-Rectors, 3 Directors of the Institutes, and a Dean. The multinational faculty consists of 19 distinguished professors and lecturers from Russia, Western and Eastern Europe, North America and Asia. There is a body of administrative staff at the University.

Innopolis University offers Computer Science degree programs. There are about 600 undergraduate, graduate and PhD students, with 15% of them female (Table 1).

Table 1. Number of students, faculty and administrative staff by year [6].

	2013	2014	2015	2016
Number of students	14	50	322	635
Full time professors	2	3	12	19
Visiting professors	–	6	22	47
Administrative staff	30	49	93	124

The University targets intellectually advanced youth – the average enrollment rate is about 3%. The tuition fee for undergraduate programs is about 20000 USD per year, and it is being covered by the scholarships provided by the partner companies. Accommodation fees and the meal plan are covered by the monthly allowance. The amount of this allowance depends on students' performance. Upon graduation, IU students are bound to work 1–1.5 years for IU partner companies in the city of Innopolis, which means guaranteed job placement for the students.

There is one intake for both undergraduate and graduate programs, in the fall semester. The undergraduate program lasts 4 years. Each academic year consists of 2 semesters (Fall and Spring), 15 weeks each semester, with midterm and end of the term assessments. Graduate academic programs are 1 year long and consist of 2 or 3 semesters: Fall, Spring or Fall, Spring and Summer. The language of instruction is English.

3 Curriculum Development

The target is to create a practical curriculum in close collaboration with the industry partners, so that students obtain the knowledge and skills that modern industry requires from IT graduates. IU aims to be at the leading edge of research, entrepreneurship and soft skills development.

IU collaborates with about 100 industry partners. Industry partners are involved in curriculum development, course delivery, projects, thesis assessment, mentorship for projects, and ensuring internship placements for the students.

Currently Innopolis University offers 4-years Bachelor programs and 1-year intensive Master programs, with 240 ECTS for BSc, and 120 ECTS for MSc programs.

BSc degree tracks are Artificial Intelligence and Robotics, Software Engineering, and Data Science, with the first two years of general instruction in computer science, and selection following a track during the 3rd and 4th years. Minors are being offered in Game Development, Information Security, and Financial Technology. Master degree programs are Data Science, Software Engineering, AI and Robotics, Security and Network Engineering. For the list of courses delivered in the 2016/2017 academic year refer to Appendix 1. In the 2016/17 academic year IU offers 149 courses (34 in 2015/2016): 65 core courses and 63 technical elective courses and 21 Entrepreneurship and Communication elective courses.

In terms of the forms of course delivery, along with traditional 15-week courses with 4–6 h frontal classes per each week, IU introduced so-called "semi-intensive courses". This type of course is led by academia- or industry-acknowledged professionals, who visit IU 3–4 times per semester and deliver several intensive sessions two to three days each, usually over a weekend. This design was chosen to attract distinguished professors and professionals, who cannot join the institution on a full-time basis. The format was thought to increase the costs due to additional travel expenses, however, the overall course cost, estimated by board, does not exceed that of the regular courses. The efficiency and effectiveness of these course types is being analyzed, and will be reported in a separate study.

At the request of industry partner companies, the Communication and Entrepreneurship elective courses Module is an essential part of the curriculum. Thus, during 4 years of undergraduate study students complete at least 2 courses in Communication and 2 courses in Entrepreneurship, and during the graduate program, at least one course of each focus.

4 Research Infrastructure and Innovation Ecosystem

The second major target is to build the research infrastructure of international caliber. Students get practical experience in labs and research centers. 11 laboratories operate at the university, within the 3 institutes [6] (Table 2):

3 research centers are established at IU: the Center for Information Security, the Center for Modelling and Analysis of Big Data in Finance and Economy, the Center for Automation of Business Processes. The research centers are being funded by means of grants received by the university and they are meant for dealing with applied tasks.

Table 2. Innopolis University Institutes and Laboratories [6].

Institute of Technologies and Software Development	Institute of Information Systems	Institute of Robotics
Industrial Production of Software Lab	Lab of Networks and Cybersecurity	Intelligent Robotic Systems Lab
Software Engineering Lab	Artificial Intelligence in Game Development Lab	Cognitive Robotic and Systems Lab
Software Design, Models and Architectures Lab	Cyber-Physical Systems Lab	Machine Learning and Knowledge Representation Lab
	Intelligent Transport Systems Lab	
	Network Science and Information Technology Lab	

Such centers hire staff mainly from outside the university. IU is also a partner in the *CERN openlab* project.

There is significant growth in IU research activity and the number of IU publications since the beginning of operation (Table 3).

Table 3. Publications by Innopolis University faculty [6].

	2013	2014	2015	2016	2017, ½ year	2013–2017
Scopus	1	5	41	71	22	140
WoS	–	3	33	62	9	107
Google scholar	1	10	74	133	70	288

Innopolis University actively collaborates with other local and overseas universities. Some Russian partner universities are: Kazan Federal University, Moscow Institute of Physics and Technology, Saint Petersburg State University, National Research University Higher School of Economics, Russian Academy of Sciences, and National Research Nuclear University MEPhI. Some overseas partner universities are: Eigenossische Technische Hochschule Zurich, Politecnico di Milano, University of the West of Scotland, University of Massachusetts Lowell, University of Manchester, University of Cambridge, Korea Advanced Institute of Science & Technology, Pennsylvania State University, Korea University of Technology and Education, and University of Alberta.

IU faculty report and get published the results of their research in 27 countries, which proves IU solid integration into the international academic research community. Some of the countries are: United States, Italy, South Korea, Switzerland, Canada, France, United Kingdom, Saudi Arabia, Spain, United Arab Emirates, China, Denmark, Germany, Pakistan, Colombia, India, Japan, Australia, Brazil, Egypt, Iran, Malaysia, Netherlands, Oman, Singapore, Thailand, Scotland etc.

IU is a purely IT university, faculty research interests extend beyond computer science. Research papers are interdisciplinary and, according to Scopus data, their research interests are within the following knowledge domains (Table 4):

Table 4. Number of papers of Innopolis University faculty by knowledge domain, Scopus[a]

Computer Science	101	Medicine	4
Engineering	41	Biochemistry Genetics and Molecular Biology	3
Mathematics	37	Business Management and Accounting	3
Decision Sciences	13	Physics and Astronomy	3
Social Sciences	7	Agricultural and Biological Sciences	2
Materials Science	6	Energy	1
Chemistry	5	Multidisciplinary	1
Chemical Engineering	4	Neuroscience	1
Economics Econometrics and Finance	4	Psychology	2

[a]https://www.scopus.com/term/analyzer.uri?sid=BAA7B70C4ABB618332B2681E216E137B.wsnAw8kcdt7IPYLO0V48gA%3a1240&origin=resultslist&src=s&s=AF-ID%2860105869%29&sort=plf-f&sdt=a&sot=aff&sl=15&count=140&analyzeResults=Analyze+results&txGid=BAA7B70C4ABB618332B2681E216E137B.wsnAw8kcdt7IPYLO0V48gA%3a130#lvl1Tab6

The University has developed programs of advanced professional training for employees of the Innopolis resident IT companies. Besides, it organizes research and development projects in joint scientific research centers. The IT Business Module, a training program for the partner companies' employees, was developed by IU. The program includes 12 modules, and offers 86 courses delivered in the format of either information sessions, or customized courses, or individual programs. 3 partner companies have so far completed their employees training in this program.

The Centre for Specialized IT Training is another example of training support. In 2016/2017, the center has provided training for 90 candidates out of more than 5000 applicants.

5 Admissions

The ultimate target for the University is to produce highly skilled professionals, capable of creating high quality IT products for the local and international market. Hence, the university targets intellectually advanced high school graduates for BSc and experienced young professionals for MSc programs. IU encourages their BSc graduates to gain one and a half to two years of industrial experience before pursuing their Master's degree.

For attracting such students, the University hosts and participates in various IT-related events like hackathons, summer camps for high school pupils and university students, IT contests and Olympiads of national and international level in collaboration with partner universities and the Ministry of Education of the Russian Federation. The University also hosts some educational and publicity events for the public-school teachers. Innopolis lyceum and school are also source of candidates for BSc programs. Internet marketing, with a focus on social networks is utilized by the university for attracting the target youth.

The selection procedure in 2017 admission campaign is two-fold. The first selection round requires candidates to submit their application, motivation letter, portfolio of achievements and education certificates. As a part of the first selection round undergraduate candidates are required to take three online tests: English language, math, and IT basics. The second selection round requires personal presence of the candidates either at Innopolis University or at a partner venue, and it consists of 5 parts: a test in mathematics and informatics; solving programming tasks; an interview with a faculty member for further assessment of IT knowledge; an English language test; business case for evaluations of soft skills and psychological risks.

The university enrolls 3% of the applicants. With such multi-level selection procedure, the University aims to ensure the high quality of the students and minimize the dropout rate. With the prospect of significant growth in the number of students, the University is working on identifying academic success predictors and eliminating insignificant steps from the selection procedure [4, 5].

6 Challenges and Other Practices

Here are some challenges that we encountered and some solutions we came up with.

Student recruitment: during 2013–2017, we were still experiencing the consequences of the declining birth rate. Another challenge with regards to the number of potential students, for a new university, is competition with well-established universities in Russia. Hence, targeting international students.

For proper selection of the students, we developed our own procedure, including a tool for measuring soft skills which are not being measured by high school. We measure the efficiency of our admission criteria and their impact on students' performance on a regular basis, and amend procedures accordingly.

Students dropouts: one of the reasons for student dropouts is that instructional format at IU differs significantly from the high school one, as well as from the format utilized in other Russian higher education institutions. Students have difficulties adapting to a situation in which they need to take responsibility for their own decisions, work a lot in groups, manage their time on their own. Another reason for dropouts: having been exposed to a big extent to real industry experience, some students change their perception of the profession and decide to pursue a different degree.

Faculty recruitment: the general belief that it is impossible to attract world class faculty to a new project in Russia proved to be false. However, the unstable Russian currency rate, as well as political constrains, have a negative effect on the candidates and current faculty members. Russian visa and work permit procedure is time and effort consuming. Another difficulty for the faculty is living in a foreign language environment. A faculty support office is established at IU in order to assist the faculty members.

Gender diversity among the faculty at IU is quite low, IU being a purely Computer Science university.

Curriculum development: the IU curriculum should meet local and international standards. Currently, Russian higher education policies are becoming more liberal with more freedom and fewer restrictions on the educational institutions. IU takes advantage of the utilizing best practices of both Russian and international engineering education. However, IU is not modeling itself on any successful institution, but instead building its own unique system.

Funding: at the moment of writing the paper, IU relies solely on industry sponsorship for funding.

Quality assurance: an education quality assurance system, which assesses the quality of course delivery has been launched at IU. Another quality assurance procedure is yearly professional evaluation of the faculty. Both procedures were initiated from the very start of the university. They constitute a working system of regular feedback, which allows us to control our performance, make personnel decisions, and make sure that our strategic indicators demonstrate constant growth.

Faculty professional development: IU utilizes several practices for faculty professional development. In 2014–2015 academic year 11 professors completed internships and other training programs in partner universities, including those ranked top 100. Some examples of such partner universities are: IT University of Copenhagen, University of Amsterdam, Dublin City University, Technical University of Denmark, Carnegie Mellon University, Politecnico di Milano, and the Swiss Federal Institute of Technology Zurich (ETH Zurich). In 2016–2017 academic year 14 professors completed the International Engineering Educator Certification Program. In the same year one professor completed 2 months industrial internship in one of the major IT companies in the region, ICL. An ISW training program was begun in the university in 2017, with 3 professors and 2 junior instructors certified as Instructional Skills Work-shop Facilitators. In 2017 IU commenced the Shadowing Program, an Orientation Program for the faculty and a three-fold training program for teaching assistants: Orientation, Induction (ISW), and Professional Development. The last one includes mentor and peer observations and workshops.

7 Conclusion

Innopolis University is pursuing its mission to form a generation of IT experts who will take the Russian IT industry to a new competitive level [6]. During its first four years the University encountered several challenges, and had to offer practices and solutions to meet its major strategic goals.

The next 3 years the university will undergo significant growth in the number of students and faculty. Among other aspirations, the university also aims to introduce new undergraduate tracks and graduate programs, increase the number of PhD students and gender diversity, and become a Grand Challenge Scholars Program pioneer in Russia. The university aims to get accreditation and become part of the university ranking systems after 2019.

Appendix 1

List of courses delivered at Innopolis University in 2016/2017 academic year

Core courses

1. Advanced Databases
2. Advanced security
3. Advanced Statistics, Dynamics Programming and Stochastic Control
4. Analysis of software artifacts
5. Architectures for software systems
6. Calculus
7. Classical Internet applications
8. Communication for software engineers I
9. Communication for software engineers II
10. Computer architecture for BSc1
11. Computer architecture for BSc2
12. Control Theory
13. Cybercrime and forensics
14. Data modeling and databases for BSc3
15. Data modeling and databases for MSc
16. Data structures & algorithms for BSc1
17. Data structures & algorithms for BSc3
18. Data structures & algorithms for MSc
19. Discrete math/Logic
20. Discrete math/Logic for BSc1
21. Discrete math/Logic for BSc3
22. Distributed systems
23. Dynamics of nonlinear systems
24. English
25. Essential skills
26. Graph theory for DS
27. Industrial Robotics
28. Information theory
29. Intelligent mobile robotics
30. Introduction to AI
31. introduction to personal software process
32. Introduction to programming for BcS3
33. Introduction to programming for MSc
34. Introduction to programming I
35. Introduction to programming II
36. Large installation administration
37. Life safety
38. Linear algebra
39. Managing software development
40. Methods: deciding what to design
41. Models of software systems

42. MSIT project I
43. MSIT project II
44. Networks
45. Offensive technologies
46. Operating systems
47. Operating systems & networks for BSc3
48. Operating systems & networks for MSc
49. Parallel/Multicore Algorithms
50. Pattern Recognition and Machine Learning
51. Philosophy
52. Probability & statistics for BSc1
53. Probability & statistics for BSc2
54. Programming paradigms
55. Randomized Algorithm
56. Research project 1
57. Security of systems and networks
58. Security research project 2
59. Software Architecture for BSc3
60. Software Architecture for MSc
61. Software project
62. Sport
63. Theoretical computer science for BSc2
64. Theoretical computer science for BSc3
65. Theoretical computer science for MSc.

Elective courses

Technical

1. Advanced algorithms for data science
2. Advanced Data Structures and Algorithms
3. Advanced Programming in Java with Map Reduce
4. Advanced System Programming
5. Agility in practice through Scrum
6. Algorithms of machine learning
7. Architecture and mobile application development featuring Sailfish OS
8. Artificial Cognitive Systems
9. Automated Software Testing
10. Cloud Computing
11. Cognitive interaction with robots
12. Combinatorics and Graph Theory
13. Computational geometry
14. Computer architecture and organization
15. Computer Architecture for a System Developer
16. Computer Convergence Application
17. Computer Modeling
18. Computer networks
19. Computer Science Project

20. Computer Vision
21. Cyber Security
22. Data analysis and knowledge discovery
23. Data Mining
24. Dependability and performance of computer systems
25. Development and testing on C#
26. Features of Linux Kernel and libc Implementation for Elbrus Microprocessor Architecture
27. Formal methods and software verification
28. Formal Software Development
29. Game Theory and Its Application
30. Human and animal brain representation in neurosciences
31. Human Computer Interaction Design
32. Human-Computer Interaction Design
33. Independent Study
34. Information Retrieval
35. Innovative agile software development methodology for high reliability & mission critical applications
36. Intelligent Robots and Systems
37. Introduction to Computer Network
38. It biomedical instrumentation
39. Kernel based machine learning and multivariate modeling
40. Lean Software Development
41. Learning Analytics and Educational Data Mining
42. Mathematical Foundation of Big Data Science
43. Mathematical Foundations of Big Data Science II
44. Molecular and Cellular Biology
45. Natural Language Processing and Machine Learning
46. Network Programming
47. Normative and dynamic virtual worlds
48. Object Oriented Programming in C++
49. Optimization methods and applications
50. Parallelism
51. Pattern Oriented Design
52. Procedural Content Generation in Games
53. Product design
54. Programming correct real-time embedded systems
55. Security Principles and Applications
56. Software Engineering and Development
57. Software Project Management
58. Startup Entrepreneurship (Zero to Hero)
59. System Programming
60. Total virtualization
61. Transport Modelling
62. User Experience and User Interface Design Fundamentals
63. Web Game Development.

Communication electives

1. Business Communication
2. Career Development for Aspiring Entrepreneurs
3. Communication for Startups: From Bootstrap to Global Markets
4. Communication Skills
5. Communication Skills
6. Creating a culture of excellence
7. Information Technology, Ethics, and Social Responsibility
8. International Business: Legal Essentials
9. Introduction to Career Development
10. Introduction to communication
11. Technical Communication.

Entrepreneurship electives

1. Digital Darwinism and Wheel of Disruption
2. eSports industry: game design and marketing
3. Fearless Ideas
4. ICT Innovation and Entrepreneurship
5. Innovation Value Management
6. International Trade Law
7. Introduction to IT Entrepreneurship
8. Lean startup methodology
9. Performance Management and Measurement for R&D
10. Strategic management of modern technologies and innovation.

References

1. Kuchins, A.C., Beavin, A., Bryndza, A.: Russia's 2020 Strategic Economic Goals and the Role of International Integration. Center for Strategic and International Studies, Washington D.C., July 2008
2. Taran, G., et al.: Assessment Report, A Partnership Between the Republic of Tatarstan, Carnegie Mellon University and iCarnegie Global Learning, iCarnegie, Pittsburgh, PA, 21 December 2012
3. Kondratyev, D., et al.: Innopolis University – a New IT Resource for Russia (2013)
4. Stanko, T., et al.: Case study: the unified state exam and other admission tests as a predictor of academic performance at the IT university. In: ASEE Annual Conference and Exposition (2016)
5. Stanko, T., et al.: On possibility of prediction of academic performance and potential improvements of admission campaign at IT university. In: EDUCON Annual Conference (2017)
6. Innopolis University annual report (2016)

A Design Framework for Interdisciplinary Communities of Practice Towards STEM Learning in 2nd Level Education

Kieran Delaney[1(✉)], Michelle O'Keeffe[1], and Olga Fragou[2]

[1] Cork Institute of Technology, Cork, Ireland
Kieran.delaney@cit.ie
[2] Computer Technology Institute, Patras, Greece

Abstract. Modern societies need young people who are able to think creatively, to collaborate well across multiple disciplines and to use new scientific, technical and engineering knowledge to achieve effective results. This brings a requirement for better teaching and learning approaches that can operate through a real world perspective where complex systems and problems are all around us. This paper describes the development of a framework designed to empower 2nd level teachers to achieve this by using Ubiquitous, Mobile, and Internet of Things technologies to enhance their approaches to teaching engineering and science subjects to their students. We provide a methodology and design approach towards forming and developing a Community of Practice (CoP), as a knowledge management schema for this and we describe exemplars from the engineering domain to illustrate the approach.

Keywords: Knowledge management · Community of practice
Design methodology · Collaborative learning · Project based learning
Real world experiences · Technical teacher training
Academic-industry partnerships · K-12 and pre-college programs

1 Introduction

The rapid emergence of new technology-driven trends is creating many new opportunities for innovation across all types of industry domains. These are powerful tools; new areas, such as Ubiquitous Computing, Mobile Computing, and the Internet of Things (which we describe as UMI technologies) dramatically broaden the scope of what is possible in teaching and learning. However, there are also significant tensions; while the technologies are more accessible than ever before, it requires that we develop a collective ability to work together across a wide range of disciplines. These are sustained effects, where people with different skills will need to learn work with the uncomfortably different practices of others in order to reach solutions that transcend the limitations of individual domains of expertise. We must determine how to address this in a manner that recognizes the need to engage these multiple disciplines, while also enabling us to discover what learning philosophies are most effective at scale and to understand how to build communities that are ultimately sustainable. The goal of the design framework presented in this paper is to provide a foundational 'architecture' for

M. E. Auer et al. (eds.), *Teaching and Learning in a Digital World*,
Advances in Intelligent Systems and Computing 715,
https://doi.org/10.1007/978-3-319-73210-7_86

this process. Using UMI technologies as educational means to support teaching and learning practices in European countries is a key opportunity, especially in the context of 2nd level education (14–16 year olds) where students interact with many teachers, study a variety of subjects and this shapes their perspective towards careers.

2 Objectives

Our overall goal is to research how the implementation of a CoPs format can more effectively enable the use of UMI technologies in science and engineering education and to discover how this may be practically implemented and replicated. By carefully exploiting state of the art technologies in order to design educational tools and activities, we aim to

- design, develop and offer novel educational services by implementing innovative pedagogies so as to enhance students' and teachers' creativity, socialization and scientific citizenship
- design and develop an integrated yet open fully integrated training environment for 14–16 years old students

By creating CoPs, we can explore novel models for interdisciplinary collaborations between experts in teaching, in technology and in industry domains that are underpinned by engineering and scientific knowledge. In this paper we describe the design process to build the resilient framework required to engineering this, which includes new approaches, models and tools for discovery and learning, models of interdisciplinary collaboration and problem-solving and approaches to engagement that promote iterative co-design, all of which is built upon learning through addressing real world challenges and experiences.

3 Defining the UMI Domain

'*Ubiquitous computing*' as a term, was introduced by Weiser [1] as a way to describe the concept of computers fading into the background of everyday life, and being enhanced by digitized artefacts that would seamlessly blend with human environments and activities *Mobile technologies* have become almost ubiquitous to modern society particularly in the consumer market; mobile phones and tablets have become commonplace, familiar technologies to everyone. The *Internet of Things (IoT)* links computing artefacts together with everyday things in an invisible network, allow us to explore how the capabilities of these technologies can be used in new and innovative ways to solve issues in society. All of this is fostering a culture of 'making'. Computer prototyping equipment can be used to easily create new UMI systems, applications and artefacts. UDOO boards [2], see Fig. 1, are one example of an open-source electronic prototyping platform that can be used to populate such systems and environments; these solutions include sensors and actuators to collect and present data and to enable user control and interaction. In addition to these, connective technologies such as WiFi and Bluetooth equipment, allow objects within the network to 'speak' to each other to create the necessary networked environment.

Fig. 1. The UDOO Neo board, hardware platform for engineering new UMI applications [2]

Key Trends

Educational organizations begin to leverage solutions like cloud computing and radio frequency identification (RFID) across an IoT platform, they are able to capture, manage and analyze Big Data. This insight provides stakeholders with a real time view of students, staff and assets. It is this asset intelligence that enables institutions to make more informed decisions in an effort to improve student learning experiences, operational efficiency and educational settings security. UMI technologies are based on the connection of people, processes, devices and data enhancing the value and volume of information we can collect, allowing educators and administrators to turn data into actionable insight. In a hyper connected world there is a pressure to prepare students for an increasingly competitive workplace: through the use of UMI technologies institutions can improve educational outcomes by providing richer learning experiences and by gaining real-time, actionable insight into student performance.

These technologies impose the need for an education system that empowers new generation citizens who understand these technologies, being active citizens and trained to apply right the information they collect. The use of wireless devices enables online lesson plans to have the potential to feature highly engaging interactive content while assessments can become more seamless, less manual and time intensive: when connected to the cloud these technologies can collect data on student performance which can then be used to improve lesson plans. Creating networks of institutions that actively incorporate technology into learning, of teachers who share data and creating engaging educational content is a challenge that is important for modern education.

4 Communities of Practice

Lave and Wenger [3] first used the term Communities of Practice when exploring the notion of legitimate peripheral participation. Legitimate peripheral participation (LPP) describes how newcomers join a COP, become experienced members and eventually old timers of a COP. Lave and Wenger illustrate their theory through observations of different apprenticeships. Initially people join the community and learn at the periphery. As they become more competent, they become more involved in the main activities of the community. They move from legitimate peripheral participation to full participation [3, p. 37]. Wenger [4] extended the term and defined COP's as "groups of people who share a concern or a passion for something they do and learn how to do it better as they interact regularly" [4, p. 1]. Brown and Duguid [5] describe COP's as interstitial communities that exist in the 'gaps' between work as defined, and the tasks that need to be done. They use the term to describe groups that are (1) fluid and dynamic "... constantly adapting to changing membership and changing circumstances" [5, p. 41] and (2) emergent "That is to say their shape and membership emerges in the process of activity, as opposed to being created to carry out a task" and most crucially [5, p. 49] (3) exists, "... outside the organization's limited core world view" [5, p. 51]. Hildreth et al. [6] define a COP as "a group of professionals informally bound to one another through exposure to a common class of problems, common pursuit of solutions and thereby embodying a store of knowledge" [6, p. 3].

Amin and Roberts [7] conducted a survey of current literature and found different interpretations of COPs in use. They note that Brown and Duguid [5] have referred to "networks of practice" to describe relations among group members which are significantly looser than those in a COP. Knorr Cetina [8] Gittelman [9] and Haas [10] have described communities of specialized knowledge workers as epistemic communities. Fischer [11] has described groups of stakeholders from different COPs brought together to resolve a specific problem as communities of interest. Lindkvist [12] has used the term Collectivities of Practice to refer to temporary groups of project teams concerned with knowledge creation and exchange. Gherardi [13] proposed the term Community of Practitioners in order to place the emphasis on "practice" rather than community, in turn also redefining community as "an effect, a performance, realized through the discursive practices of its members."

Wenger et al. [14] defined COPs as "groups of people who share a concern, a set of problems or a passion about a topic, and who deepen their knowledge and expertise in this area by interacting on an ongoing basis" [14, p. 27]. In later work Wenger and Snyder [15] describe COPs as "...groups of people informally bound together by shared expertise and passion for a joint enterprise [which can] drive strategy, generate new lines of business, solve problems, promote the spread of best practices, develop professional skills and help companies to recruit and retain talent.." [15, pp. 139–140], while Snyder and Briggs [16] define COPs as structures that "steward the knowledge assets of organizations and society." According to them, they operate as "social learning systems" where practitioners connect to "solve problems, share ideas, set standards, build tools, and develop relationships with peers and stakeholders."

Hoadley [17] established two definition of communities of practice – the feature-based definition and the process-based definition. The feature-based definition is described as "a community that shares practices, where learning is a relational property of individuals in context and in interaction with one another." [17, p. 288], while the process-based definition describes a constant process where legitimate peripheral participation takes place. Learners enter a community and gradually take up its practices, over time they take up more and more of the group membership and centrality, and more and more of the central practices of the group [17, p. 290]. In Cleary's [18] research, she discovered that "communities of practice provide support, enable cross-functional interactions, foster new ideas and enable technical communicators to network." Communities of practice are important to technical communicators because they help to realize knowledge transfer, to facilitate communication and to enact changes [18, p. 128].

Gherardi considers a COP to be different to a 'neighborhood community' or a 'community of interest' in that its members are practitioners [13]. They are actively engaged in a particular discipline or profession. It is not enough just to be interested in a topic, members must require some form of expertise or specialization in the shared domain. Wenger stated [19] that the following characteristics are crucial to a community of practice and these are what differentiate it from a normal community:

- **Domain:** defined by a shared domain of interest; it is not merely a network of people or a group of friends, membership implies a commitment to the domain.
- **Community:** members engage in joint activities, discussions, help each other, share information; build relationships that enable them to learn from each other.
- **Practice:** members are practitioners, sharing a common practice and developing a repertoire of resources, including experiences, stories, tools, ways of addressing recurring problems, etc.; this type of interaction needs to be developed over time and the interaction needs to be sustained.

The added value of CoPs is to promote innovation, develop social capital, facilitate and spread knowledge. Using an approach to build CoPs, we aim to deliver meta-level solutions to link schools, communities, business and third-level initiatives together to foster learning models that strongly broaden impacts to the entire school population. In this paper we describe a key stage in this process, which is the formative design framework for these CoPs, including operational (i.e. models for user engagement) and infrastructural requirements, which has been built from a synthesis of real education scenarios in organizations across six countries in Europe and a series of industry trends analyses within the UMI, science and engineering domains.

5 Building the Design Framework

New technology is often rejected, so we adopt a research approach where CoPs participants (actors) interact with educational material and technologies to co-create learning and teaching structures, processes, products and activities. In this context adopting a 'making' culture within the CoPs is expected to provide productive pathways for building a tangible bridge between theory and action so as to foster pilots that enable exploration and validation of technical, pedagogical and community-based innovation approaches. Our approach is entrepreneurial and multidisciplinary to help raise young boys' and girls' motivation in science education and to increase their prospects in choosing careers in engineering, technology and science. The approach is structured around the following design principles:

- UMI applications to form an integrated learning environment for European youngsters aged 14–16.
- Knowledge management schemas such as CoPs to explore domain, practice and community important elements of CoPs.
- Emphasis on Outcome Based Learning (OBL), social constructionism and instructional design principles as means to create the learning ecosystem.
- Focus on problem solving, citizenship, leadership, work based learning and communication skills.
- Emphasis on students' setting their personal goals and monitoring their progress through milestone activities.

Brown and Duguid [5] identified three life phases which make up the stages of community development. These include: [a] Formation (potential for CoP is identified, networks discovered, relationships formed), Integration (new member admitted, unique tools and methods developed, new ideas welcomed) and [b] Transformation (community fades away, closes, merges with another community or becomes formally institutionalized). This may be mapped to Wenger's life-phases in the following way (Table 1):

Table 1. Summary of CoPs life stages

Phase	Project development stages	Towards....
Formation (potential for CoP is identified, networks discovered, relationships formed)	Special Interest Groups to establish needs, build engagement and initiate design-based research to determine learning practices	Mutual Engagement: establish norms and build collaborative relationships; engage in common actions or ideas
Integration (new member admitted, unique tools and methods developed, new ideas welcomed	Definition of learning practices and establishment of first CoP groupings to complete a first set of local pilot learning programmes	Joint Enterprise: Through the member's interaction, they develop a shared understanding of what binds them together
Transformation (community fades away, closes, merges with another community or becomes formally institutionalized)	Synthesis of promising learning approaches to be applied within broader pilots connecting local pilot activities, formalising CoP structure, process and capacity	Shared Repertoire: the community produces a set of communal resources which is called their "shared repertoire"

Phases of Development

Overall, the phases of development are the following: (a) acquiring knowledge and creating engagement, (b) applying design-based research with the actors directly involved to derive learning modules, (c) conducting local pilots in educational environments across Europe (for example, in Greece, Ireland, Italy, Norway, Finland) to validate the learning modules and (d) conducting broader pilots that replicate learning activities and grow synergies between the original local pilots and activate a CoP with shared, virtually connected infrastructure that supports ongoing use of a range of research tools/methods, educational tools/materials and communications/networking tools (See Fig. 2).

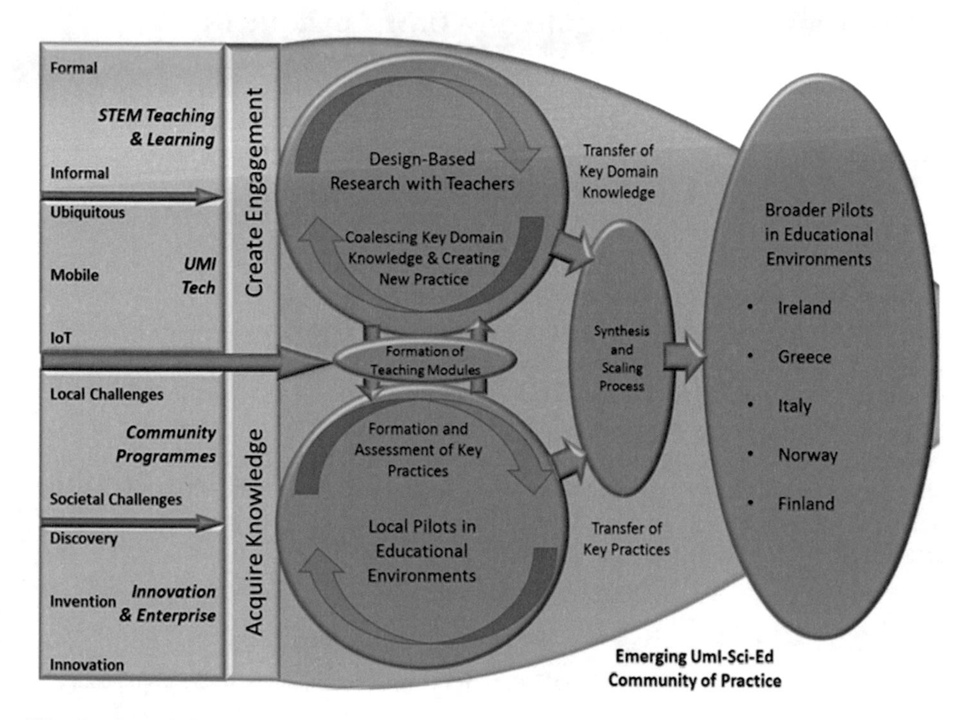

Fig. 2. Overall framework showing the formation of a CoP through engagement, design-based research and piloting of educational activities locally and then at scale

More specifically, engineering a CoP using UMI technologies for education requires a convergence between available UMI technologies and experts, STEM educators and relevant pedagogical approaches. The research process to make an effective design for this, outlined in the schematic, is a complex process of engagement based on: (a) strongly iterative procedures using a design-based research methodology; (b) shaping of artefacts and processes to promote important educational content, including educational scenarios and models of interaction; (c) use of important pillars of CoPs structure (such as the definition of priorities in the technology, engineering and science domains and their influence on specific learning practices) to further expand the knowledge and socialization schema, built upon high levels of engagement and collaboration between stakeholders in the emerging community; (d) select ethnographic practices based on mostly qualitative data collection (semi-structured interviews, focus groups, surveys and content analyses).

An educational infrastructure is also composed, including hardware and software technologies to apply UMI teaching activities, online platforms to share educational material, support assessment, archive member projects. A number of CoPs toolkits will also emerge to enable educational, research and communication tasks and activities within the CoP.

6 Designing a CoP for Engineering UMI Applications

The framework was used to enable the design of a set of local pilots for a region focusing upon applying UMI technologies through project-based learning in real world application domains. To do this, it is necessary to evaluate the potential of existing initiatives, determine models for integrating UMI teaching and learning and develop designs for a new CoP and Piloting activities that promote practical interdisciplinary activities.

Acquiring Knowledge and Creating Engagement
Many successful examples exist of project-based collaborations built by multi-disciplinary student teams. Children in coding clubs (e.g. CoderDojo) mentor each other using programs like Scratch and have subsequently self-organized to build teams to compete in robotics competitions. This existing eco-system of varied and unconnected initiatives and programs provides a platform for building networks of individuals and organizations that can combine forces to engage students in UMI, STEM and open new opportunities to industry. The following provides a brief description of three such initiatives (Table 2):

Table 2. Example collaborations between stakeholders

Community initiative – with educators and industry	CoderDojo: an international initiative where volunteers work together to teach school students coding as an extracurricular activity
Educational stakeholder – with industry	Enterprise Camps: week-long initiatives for second-level students aged 14–16; promote entrepreneurship in young people; interact with entrepreneurs; receive guidance for their own projects
Industry stakeholder – with educators	Robotics competitions run by industry where secondary school students learn to design and build a robot. STEM concepts are explored and the building of the robot is facilitated by the teacher and the industry stakeholder involved

This phase assembled a strong knowledge-base of such initiatives and commenced a range of different activities to engage with the stakeholders driving these programmes; this included desk reviews, surveys and interviews with actors, observation of the learning activities and other analyses. A broad sketch of the CoP design derived from this has the following form (Table 3):

Table 3. CoP description on engineering UMI applications

Domain	Community	Practice
UMI Technologies (U) - hardware, software, sensors, actuators; (M) - smart phones, tablets; (IoT) - Arduino platform, Raspberry PI, UDOO Kit, IoT test beds etc. STEM subjects + Social Sciences (Design, needs analysis) 21st century skills – creative thinking, collaboration, communication of concepts, critical analysis and reflection Themes - Energy & Water – Physics + Engineering for utilities (U + IoT), Environmental Sciences (IoT + M), Wellbeing & Health (M + U)	Educational Stakeholders Teachers – *mentoring students* Students – *working in teams* Parents – *mentoring students* Boards of Management – *mentoring students* Education experts – *mentoring on pedagogy* "Industry" Stakeholders Community Groups & NGOs – *add knowledge on relevant social challenges* Industry – *add knowledge on commercial needs and business models* Entrepreneurs, small start-up companies – *mentoring on design, need analysis* Research/STEM Organizations STEM experts – *mentoring* UMI experts – *mentoring*	Learning Approaches *Problem-based learning* *Project-based learning* *Action learning* *Adventure-based learning* *Service learning*

Design-Based Research and Assessment

To refine the focus of the CoP design template a series of tasks and assessments are undertaken, including workshops with stakeholders incorporating co-design activities as well as needs analyses and ideation. A portfolio of educational scenarios in the UMI and STEM domains was assembled and a portfolio of descriptions of real industry projects in the region was also compiled. Thematic analyses were undertaken on both to define major trends and themes relevant to the definition of the CoP domain and, thus, the design of the local pilots (See Figs. 3 and 4).

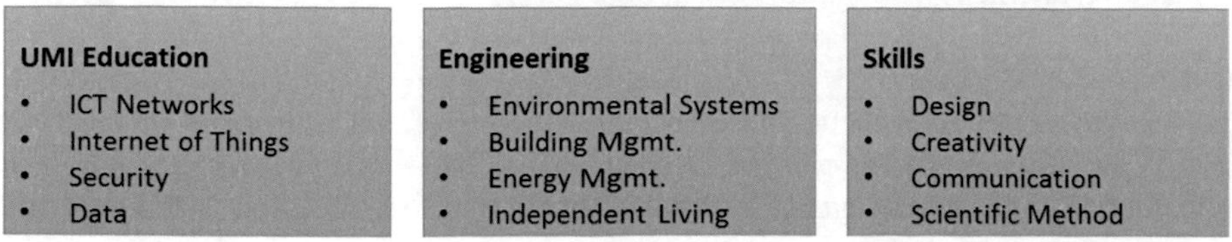

UMI Education
- ICT Networks
- Internet of Things
- Security
- Data

Engineering
- Environmental Systems
- Building Mgmt.
- Energy Mgmt.
- Independent Living

Skills
- Design
- Creativity
- Communication
- Scientific Method

Fig. 3. Learning modules on UMI technologies, applications and other required skills

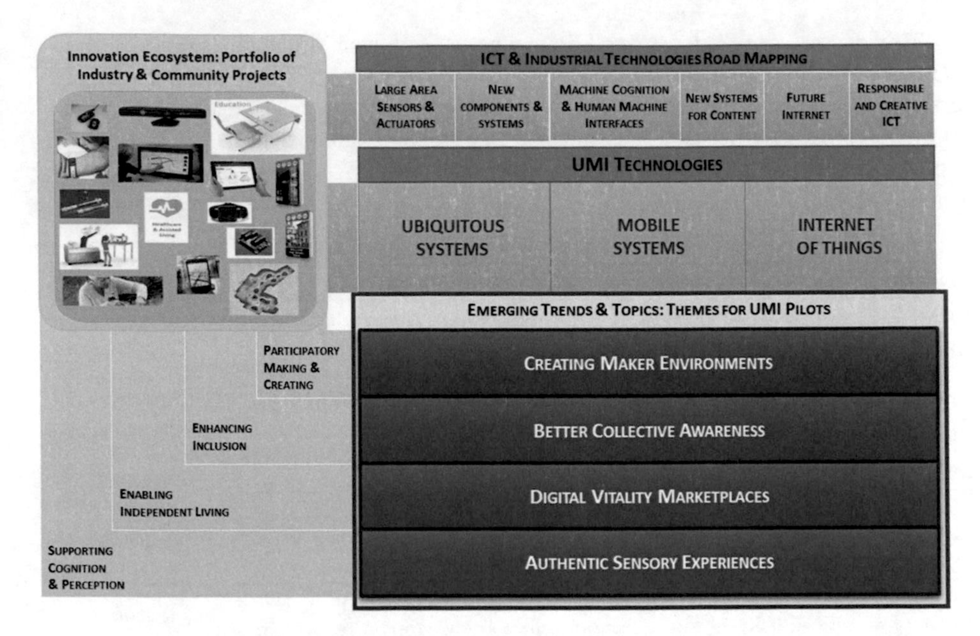

Fig. 4. Trends relating to the application of UMI technologies and skills in industry projects

Summary of Findings

- UMI technology educational modules are increasingly available for the development of basic and advanced skills; many informal learning opportunities exist;
- UMI technologies are under-utilized in teaching engineering applications;
- Programmes to teach the real world application of UMI technologies tend to have a deficit of application expertise and interdisciplinary skills;
- Learning approaches that use problem-based techniques offer accessible models for addressing this but there are challenges in assessment of learning outcomes;
- The UMI marketplace moves very quickly, making knowledge of UMI trends an important aspect of learning;
- The requirement for significant applied UMI Skills is growing rapidly with an increasing emphasis upon systems-level thinking;
- UMI marketplace innovation places a strong emphasis upon addressing societal challenges;
- Significant user engagement is a key requirement of many emerging UMI applications, even in 'traditional' engineering domains, making acquisition of collaborative & interdisciplinary skills essential.

Designing Local Pilots

The results of the assessment phase support a refinement of the focus of the CoP piloting activities. In this case it builds an emphasis on learning approaches that promote interdisciplinary engagement, access creativity and innovation in engineering domains, and build understanding of how to use UMI technologies in real world applications. The pilot designs merge the UMI teaching and learning approaches with existing educational initiatives and active stakeholders who recognize the benefits of the synthesis the forming a CoP has the potential to develop. The CoPs pilot designs are now framed through the following initiatives:

- **UmI Investigator/Creator Courses:** designing and making with UmI technologies; inventing new uses and discovering applications for UMI technology;
- **Enterprise Creation using UmI**: build innovation and entrepreneurial skills, use UmI toolkits to investigate societal challenges;
- **Workplace Learning**: embed students in real-world projects based in industry and research organizations; work in teams and solve problems using UmI toolkits.

7 Conclusions

A design framework for building Communities of Practice (CoPs) using UMI technologies as enablers that empower teaching of engineering and science topics at 2nd level has been developed. This framework uses foundational data and information from a range of educators and industry sources to help define formative designs for engineering the CoPs. It addresses key requirements for building active, resilient stakeholder engagement for the emerging community, for creating syntheses of domain subjects and supporting expertise to form an effective CoP identity, and for focusing on the most relevant educational practices for this CoP as well as the suites of tools and cyber-physical infrastructure required to deliver and evolve the CoP activities.

Acknowledgements. The UMI-Sci-Ed project has received funding from the European Union's Horizon 2020 research and innovation programme under grant agreement No. 710583.

References

1. Weiser, M.: The computer for the 21st century. SIGMOBILE Mob. Comput. Commun. Rev. **3**(3), 3–11 (1999)
2. UDOO. UDOO (2016). http://www.udoo.org/. Accessed 24 Nov 2016
3. Lave, J., Wenger, E.: Situated Learning: Legitimate Peripheral Participation. Cambridge University Press, Cambridge (1991)
4. Wenger, E.: Community of Practice: A Brief Introduction. Cambridge University Press, Cambridge (1998)
5. Brown, J.S., Duguid, P.: Organizational learning and communities-of-practice: toward a unified view of working, learning, and innovation. Organ. Sci. **2**(1), 40–57 (1991). http://dx.doi.org/10.1287/orsc.2.1.40
6. Hildreth, P., Kimble, C., Wright, P.: Communities of practice: going virtual. In: Knowledge Management and Business Model Innovation (Sachs 1995), pp. 220–234 (2001)
7. Amin, A., Roberts, J.: Knowing in action: beyond communities of practice. Res. Policy, **37**, 353–369 (2008). http://dx.doi.org/10.1016/j.respol.2007.11.003

8. Knorr Cetina, K.: Epistemic Cultures: How the Sciences Make Sense. Chicago University Press, Chicago (1999)
9. Gittelman, M.: Does geography matter for science-based firms? Epistemic communities and the geography of research and patenting in biotechnology. Organ. Sci. **18**, 724–741 (2007)
10. Haas, P.: Introduction: epistemic communities and international policy coordination. Int. Organ. **46**(1), 1–37 (1992)
11. Fischer, G.: Communities of interest: learning through the interaction of multiple knowledge systems. In: Bjornestad, S., Moe, R., Morch, A., Opdahl, A. (eds.) Proceedings of the 24th IRIS Conference, Ulvik, Department of Information Science, Bergen, Norway, pp. 1–14 (2001)
12. Lindkvist, L.: Knowledge communities and knowledge collectivities: a typology of knowledge work in groups. J. Manag. Stud. **42**(6), 1189–1210 (2005)
13. Gherardi, S.: Organizational Knowledge: The Texture of Workplace Learning. Blackwell Publishing, Oxford (2006)
14. Wenger, E., McDermott, R., Snyder, W.: Cultivating Communities of Practice. Harvard Business School Press, Boston (2002)
15. Wenger, E., Snyder, W.M.: Communities of Practice: Facing Complexity in Government. Paper presented at the Knowledge Management in Government Conference, Washington, D. C., 14 April 2003
16. Snyder, W.M., Briggs, X.d.S.: Communities of Practice: A New Tool for Government Managers. The IBM Center for the Business of Government, Arlington (2003)
17. Hoadley, C.: What is a community of practice and how can we support it? Theor. Found. Learn. **814**, 287–300 (2012)
18. Cleary, Y.: Community of practice and professionalization perspectives on technical communication in Ireland. IEEE Trans. Prof. Commun. **59**(2), 126–139 (2016)
19. Wenger, E.: Communities of Practice: Learning, Meaning, and Identity. Cambridge University Press, Cambridge, New York (1998)

Adaptive and Intuitive Environments

An Approach for Ontologically Supporting of Space Orientation Teaching of Blind Pupils Using Virtual Sound Reality

Dariusz Mikulowski(✉)

Faculty of Sciences, Siedlce University of Natural Sciences and Humanities, Siedlce, Poland
dariusz.mikulowski@ii.uph.edu.pl

Abstract. The blind encounter great difficulties in independent movement in an open environment. There are various systems that support their orientation usually based on devices such as suitably adapted GPS, cameras, RFIDs and their respective software. But there are no computer-assisted systems that would help with teaching blind children to move without exposing them to danger. It seems that a good idea for constructing such an assistant system would be to replace a physical reality with the help of suitable computer software in which the semantic technologies are used.

Keywords: Blind pupils · Obstacle hearing
Ontological support of orientation

1 Introduction

There is no doubt that blind persons encounter great difficulties in independent movement in an open environment. There are various systems that support the orientation of such people, which are usually based on suitably adapted GPS devices and their respective software. But there are no computer-assisted systems that would help with teaching blind children to move without exposing them to danger.

In order to be able to move on their own, blind people must master skills of recognizing obstacles, hearing their surroundings, recognizing dangerous elements of space such as moving cars. They do this by perceiving the auditory sensations coming from the environment. To be familiar with these skills they must complete a special course known as the space orientation course. Currently such courses are conducted using traditional methods i.e. trainer assistance during movement or showing mock-ups of stops, curbs, buses, etc., to the student. It seems that a good idea to solve this problem would be to replace such physical reality with the help of suitable computer software and adequate hardware equipment. Such virtual environment, on one hand, would make it much easier and more attractive to teach blind children to move, and on the other, would be a cheaper method than the current one that is applied now.

M. E. Auer et al. (eds.), *Teaching and Learning in a Digital World*,
Advances in Intelligent Systems and Computing 715,
https://doi.org/10.1007/978-3-319-73210-7_87

2 The Related Works

A new possibilities have emerged in terms of navigation aid for the blind with the advent of such devices as self-developed mobile phones, GPS receivers, cameras and RFID, radio labels. As we can observe, current research in terms of navigational aid for the blind concentrates in several directions. One of them is the use of odometry, RFID labels and cameras to provide navigational assistance. In his PHD Mekhalfi [1] describes a system working in public buildings, which also features obstacle detection. It uses a laser-like camera for navigation and a speech synthesizer and speech recognition engine to communicate with the user. All this equipment and software are built into a separate small device which the user can carry with him. De Lucena and Simoes [2] propose an indoor navigation system with a map consisting of RFID radio markers located in the building. The camera, RFID labels, sonar and odometry are used as devices. A similar project was presented in Sammoud and Alrjoub [3]. The authors propose a Mobile blind Navigation System Using RFID mobile device, RFID tags and Reader, GPS, text to speech, voice Recognition, and Wi-Fi. The system can work both inside and outside the building, for example in small areas such as an academic campus.

There is hope also associated with the use of stereoscopic cameras giving a three dimensional image. Jelonkiewicz [4] and Kozik [5] propose a stereo camera system that recognizes familiar faces, obstacles, and subtitles. In this solution, as in the work proposed here, a special ontology is used to help identify and eliminate dangerous obstacles. Similarly Tang et al. [6] propose an approach to construct a 3D map generated from a stereo camera system. In this project, the idea that objects in the city have a lot of flat surfaces is applied.

Let us assume here, that most of the approaches mentioned above attempt to solve the problem of navigating a blind user in a real world environment. Almost none of the solutions address the problem of teaching a blind child to move independently so that it is easier for him to travel in the real world. An exception may be a project proposed by Maidenbaum and Amedi [7]. They describe a set of educational games that they used to teach blind children to move in an open space environment. During this game, a special white cane equipped with an account sensor and distances are used by the child player. This approach is similar to that proposed in this paper in the sense that it observes the important issue of good teaching of the independent movement of the blind child before it is released into the open real environment. A solution similar to the problem of navigation of the blind can also be considered as an aforementioned project described by Kozik [5]. The similarity is the use of ontology to construct a map of objects. We would like to combine these two elements and construct a technology that allows the teaching of blind children to move around in real-world situations using a set of specially created ontologies.

3 A Concept of Virtual Audio Reality

It seems that semantic technologies, in particular, the usage of founding ontology and collection of specially created domain ontologies, could help with building such virtual audio reality. The construction of such reality should begin with the observation that sighted people perceive the world based mainly on graphical information, while blind people focus on different soundstages. For this reason, it seems that it is necessary to provide a way to automatically process a fragment of reality expressed graphically into its sound equivalent. To do this, an ontology that gathers the concepts needed to make a graphical model of the reality fragment is needed. In the next step, the same fragment of reality should be expressed in terms of sound objects understood by the blind pupil. Finally, the element that binds these two different representations should be established. It would be a founding ontology with general concepts, common to all reality, and a requesting machine with a set of rules that automatically converts one representation of the world, graphical objects, to another - sound objects. In this article, we will focus on two of these elements, namely general founding ontology and its sound representation domain ontology.

3.1 How This Virtual Reality Works

Let us imagine, that a trainer, during an orientation course, wants to teach a blind pupil to safely transit from home to school. Let us suppose that a route is split into two fragments of a street and one intersection. Lets also suppose that the trainer and student have access to a semantic ontology based simulation system.

1. At first, the trainer is watching details of a route and creates a 3-dimensional model of this, using an appropriate application. The model is stored in the system as an instance of graphic object ontology. It is a collection of graphical objects and the basic relationships established between them. There are objects such as a beginning object - home door, fragments of street1, junction, fragments of street2 and final object - school door. The schematic structure of this model is shown in Fig. 1.
2. Each of the graphic objects of a model should be now decomposed to smaller basic objects and relations relative to a student's position, who will be traveling along the route patch. Such atomic objects will be, for example: the wall, the pavement, the carriageway, and the car. They will be connected with relationships such as `is-on-left`, `is-on-right`, `is-under-feeds` and so one with the object representing a user. A schematic example of such decomposition is shown in Fig. 2.
3. Then each of these basic objects is processed into its audible representation that is understandable by the blind. For example, a basic object car and the relationship `is-on-right` are transformed into a car soundtrack of 2 s duration that will be played in the right stereo channel running from behind to the front.

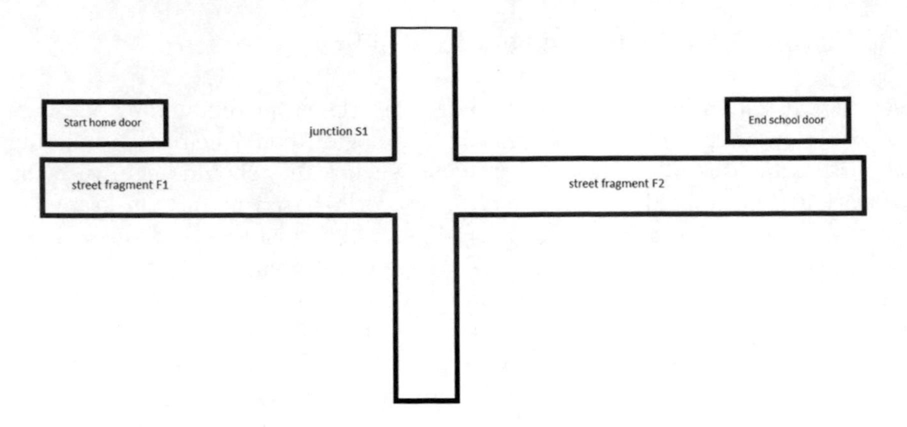

Fig. 1. General decomposition of route

4. Finally, all of these decoded sound representations are combined into an audio model of the entire route that can be played for the user. The trainer may additionally establish a blunt play of the route and add warnings for the student, needed for example, when crossing the intersection.
5. For training purposes, the student enters a special cabin equipped with an audio kit with 4 or more speakers, a microphone, and a camera. He also gets a special electronic dagger for his white stick. Its effect is that when a student moves in front of him and moves down or sideways to his hand, a pulse (like the vibration of a mobile phone) of appropriate intensity and length is generated, while the sound is also generated from the speaker. In this way it is simulated to strike a real cane on the sidewalk, in the lawn or another object, which means that the user moves around the route.
6. While the sequence of sounds representing the route is being played, the trainer has information about where the route is physically located. In this way, he can give directions to the student about obstacles in the area and other explanations needed to complete the route, and in extreme cases, to stop playing or repeat playback fragment of a route.

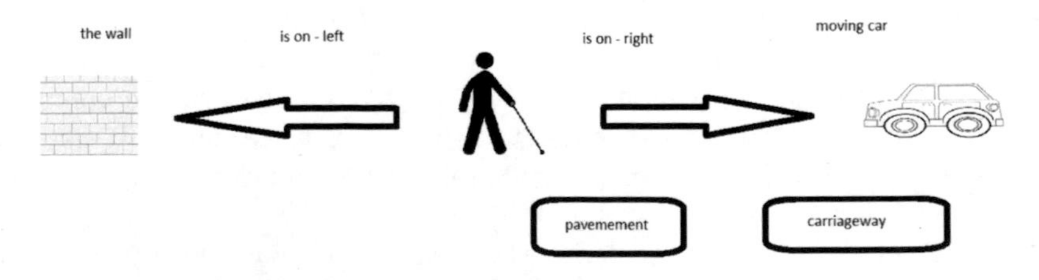

Fig. 2. Decomposition of object into atomic objects

3.2 A General Founding Object Ontology

The key component needed to create a virtual sound reality that we briefly described above, should be a general founding ontology that collects the concepts needed to describe objects surrounding a blind user. It should contain classes and their properties that will allow to create models of specific parts of that reality.

It seems that the primary class in such an ontology might be the class that represents the city. It may have properties such as name, size, communication types, streets, junctions, squares, parks, etc. In turn, another class that represents the fragment of a street is needed. It can have properties such as the sidewalk, the wall, the lawn, the carriageway, the moving cars and so on. Likewise, a class representing a junction is also required. It may have properties such as transitions and the traffic lights.

It is also important that the objects in the model of reality should be related to each other. However, in order to guide the user along the route, the relations between objects at the atomic level must directly relate to the current position of the user object in the model. These will be relationships such as: `is-on-left`, `is-on-right`, `is-in-frontof`, etc. More general relationships like `is-on-west`, `is-on-south`, etc. will also be needed to create a larger fragment of reality at an upper level.

But how is this founding ontology used to create a fragment of virtual sound reality? As shown in Fig. 1, a fragment of a route that is explored by a blind user is decomposed into several objects. When a user begins to tap through a route, the special object that represents him is placed in the proper place of the model. It means that user object is in direct relation to the nearest object in this model. For example, if he is moving from the starting object i.e. his home door to the intersection, his representative object is in relation to the sub-objects of the street1 fragment - F1. It is in the relationship `is-on-left` to the wall of the building, `is-under-feeds` to the pavement, and `is-on-right` to the roadway and `is-on-right` to the moving car as it is shown in Fig. 2. When the user continues his movement along the route, his relations with objects are continuously changing according to the direction of his movement. For example, if the user is moving west, he is in `is-moving-west` relation to the F1 street fragment. If we also know that the junction is in `is-on-west` relation to the segment of F1 Street, then we know that the user will soon enter the junction because it is in front of him. We also know that the junction has sub-objects such as transitions, traffic lights, etc., and these objects are at that junction in `is-on-west`, and so on. Thus, when the user intersects at this junction, their relationship to the sub objects of the junction must be updated. We can describe this situation by the following formula:

```
u1 is-in f1 and u1 is-moving-west f1 and
s1 is-on-west f1 and p1 is-on-south S1
then
u1 is-in s1 and p1 is-on-left u1
```

This above formula means that if user u1 are on street segment F1 and the junction s1 is to the west of street F1 and the user u1 moves west and pedestrian crossing p1 is south of junction s1, then upon entering this intersection, the user u1 will be at the junction of s1 and the transition p1 will be to his left. By making such inferences continuously while the user object is navigating through the reality model, we can dynamically describe the current user situation using proper formulas consists of terms describing his relationships with objects of a model. For such a dynamic description, a founding ontology is needed.

3.3 Sound Object Ontology

It is clear, that the general founding ontology is not sufficient enough to describe the reality of an environment to the user in a way that would be well understandable to him. For this reason, a more specialized ontology contain sound objects that are in a sense a reflection of atomic objects of the founding ontology is needed. These may include objects such as the sound of a passing car, the sound of footsteps from the wall, the sound of light signals, the sound of other people's steps and so on.

As mentioned above, when a user navigates a route, the object representing it is in relation to the nearest atomic objects in the model. These closest objects and relationships must be processed into sound signals that will be played during training. For example, a basic object car and the relationship is on right are transformed into a car soundtrack of duration 2 s that is played in the right stereo channel running from behind to the front of user. In this way, a fragment of reality constructed by the trainer in the form of a graphical model, is firstly processed into a model in the founding ontology and then decomposed into a representation of sound objects in the sound ontology instance.

For formality and good consistency purposes, all classes and relationships of these ontologies can be expressed formally in one of the known ontology description languages for example OWL.

4 Summary

Although there are systems for computer aided spatial orientation of the blind, they focus on helping people who already have the appropriate skills and experience to move independently. The teaching of children of spatial orientation is currently being carried out using traditional methods. It seems that the virtual reality based on ontologies described above could greatly facilitate and hasten such teaching process in a safe and cheaper way. By developing virtual reality snippets for the blind, a possibility of the creation of software for learning spatial orientation of the blind will be also possible. But also other solutions such as simulators or educational games that affect the imagination of a blind pupil would be possible to construct.

The ontology of sound objects mentioned above and its implementation are currently being created. The next step will be to create an ontology for the graphical objects and the conversion engine and a set of rules for it.

References

1. Mekhalfi, M.L.: Recovering the Sight to blind People in indoor Environments with smart Technologies. PhD thesis, University of Trento (2016)
2. Simoes, W.C.S.S., de Lucena Jr., V.F.: Hybrid indoor navigation assistant for visually impaired people based on fusion of proximity method and pattern recognition algorithm. In: IEEE 6th International Conference on Consumer Electronics - Berlin (ICCE-BERLIN), 05–07 September 2016, Berlin, Germany (2016)
3. Sammouda, R., AlRjoub, A.: Mobile blind navigation system using RFID. In: 2015 Global Summit on Computer Information Technology (GSCIT), 11–13 June, Sousse, Tunisia (2015)
4. Jelonkiewicz, J., Laskowski, L.: System for independent living - new opportunity for visually impaired. In: 11th International Conference on Artificial Intelligence and Soft Computing (ICAISC), Artificial Intelligence and Soft Computing, Part II. LNAI, 29 April–03 May 2012, Zakopane, Poland (2012)
5. Kozik, R., Burduk, R., Kurzynski, M., Wozniak, M., Zolnierek, A.: Stereovision system for visually impaired. In: Computer Recognition Systems, vol. 4. AISC (2011)
6. Tang, H., Zhu, Z.: A segmentation-based stereovision approach for assisting visually impaired people. In: 13th International Conference on Computers Helping People with Special Needs (ICCHP). LNCS, 11–13 July 2012, Linz, Austria (2012)
7. Maidenbaum, S., Amedi, A.: Blind in a virtual world: mobility-training virtual reality games for users who are blind. In: IEEE Virtual Reality Conference (VR) Proceedings of the IEEE Virtual Reality Annual International Symposium, 23–27 March, 2015, Arles, France (2015)

Implementation of Mobile Testing System for Control of Students' Educational Outcomes

Tetiana Bondarenko and Oleksandr Kupriyanov(✉)

Ukrainian Engineering Pedagogics Academy, Kharkiv, Ukraine
{bondarenko_tc,a_kupriyanov}@uipa.edu.ua

Abstract. This paper deals with the problem of automated testing of students in the absence of a computer room. A computer system that meets the basic requirements for the simultaneous testing of groups of students is mobile and inexpensive due to its single-computer-based design has been developed and described. The system has sufficient options to conduct testing and questioning of students.

Keywords: Computer-aided testing · Computer system · Database · Student Mobile system · Response system

1 Problem Statement

The reformation of the educational system involves the creation of a highly effective mechanism for educational quality assurance, including quality of students' educational achievements. However, it is impossible to achieve significant outcomes without an appropriate control aimed at further correction of the educational process.

For professional and technical educational institutions, the task of quality control of students' outcomes is complicated with the necessity to check both theoretic and practical training of students, which significantly broadens a general scope of control measures of educational outcomes and respectively complicates the control procedure itself. Application of conventional control measures, in particular, observation and oral and written questionnaire has serious drawbacks, among which complexity of the procedure in frontal control and a large scope of works in processing the results shall be noted.

In this connection computer testing has become extensively used as a form of knowledge control, since it combines the advantages of traditional testing system (rapid assessment of learning the studied material, an increased level of objectivity of knowledge assessment, minimum time spent to obtain reliable results of monitoring, improve control activity of the teacher by increasing the monitoring frequency and regularity) and computer system (standardized procedure, automated processing of results, the possibility of accumulation of test results with their further analysis in different assessments).

A significant shortcoming of the computerized test control systems is the requirements to the hardware. The number of computers must not be less than the number of those tested. It is naturally that the entire learning process in most subjects, in which

M. E. Auer et al. (eds.), *Teaching and Learning in a Digital World*,
Advances in Intelligent Systems and Computing 715,
https://doi.org/10.1007/978-3-319-73210-7_88

testing takes a small part of the lesson time, cannot be transferred into the computer class. As a solution to this situation, a mobile testing system (MTS) can be suggested. Using this system for testing requires only one computer the students remotely connect to. Test questions are displayed on the screen, and each student enters the answers from his/her own panel.

2 Analysis of Recent Research and Publications

The systems used to get feedback on the educational quality are divided into two groups: on-line knowledge control systems and computer-aided testing systems. Onc cannot say that it is diametrically opposite approaches, however, they have different focuses when in use.

On-line knowledge control systems [1–7] are equipped with relatively simple data input panels, connected to a computer. The teacher can display the test questions and the audience answers them using their remote panels. A specialized software will remember the response of each student and analyze it. The system is mobile and comprises a PC and panels for students. Students can answer questions using just a few buttons. The system allows maintaining estimate accounting and is easily transportable. The cost of such 24-panel systems starts from 600 USD.

Computer-aided testing systems run on personal computers [8–12]. They can be used not only to generate tests but also to organize 32 tests either in the computer room or in a local network and the Internet. Computer-aided testing systems support multiple types of questions, allow not just entering the correct answer, but also structuring it. Their drawback is the need for a computer room in case of the frontal test control. This limits the possibility of their use in the daily practice of non-computer-based subjects.

Analysis of existing computer systems for the educational control showed that most of their number is designed for testing in computer rooms. A small number of mobile feedback systems have some drawbacks: high price and dependence of the system readiness on the state of batteries of each student's panel.

3 Statement of Basic Research Material

Analysis of the list of hardware needed to implement MTS showed that the most cost-effective variant of the system design is the use of a digital HID-keyboard as remote input panels. To connect the required number of panels to a PC the USB-hubs are used. Therefore, to operate MTS, a PC, the required number of HID-keyboards, hubs, and a projector for a demonstration of test questions are needed. The connection procedure of the MTS components is presented in Fig. 1.

The developed MTS software consists of two units: the unit for filling and maintaining the MTS database and the testing unit. The system database contains tables Groups, Students, Lessons, and Tests. The table Groups stores the list of codes for the groups to be tested. The table Students contains the names of students and their group codes.

Fig. 1. MTS component connection procedure

The system database supports storage of different types of tests in text and graphical form. Tests can be grouped into classes. For this purpose, a list of lessons is stored in a special table. Thus, the teacher has the opportunity to carry out several tests on different topics during one lesson. In the case of planning to use the system for several subjects, one can group the tests by subjects instead of by lessons.

To maintain the system database, the procedures for creating and editing tests, lessons, lists, groups, students, system settings, viewing database content and test results have been developed. The program has an option to create different types of questions, the maximum number of answers — 10 (according to the number of keys on the keyboard), and the number of issues can exceed 100. There is an autosave option for tests during their creation or editing.

The testing process takes place in three stages:

- registration of students,
- testing,
- statistics displaying.

Registration of students (Keyboard registration mode on the main screen) is needed to assign the appropriate panel to the student to exercise further control the accumulation of answers in the database. The registration option also makes it possible to record student attendance. Thus, if to conduct tests at each lesson, the teacher will eventually have a picture of class attendance for each student throughout the entire training period. During the registration of students, a numbered list of names of students is shown on the displays, and they, in turn, press the button on the remote control panel with the number of their names and assign the panel to them.

Using the HID-keyboards, students enter the numbers of correct answers to the test questions. When choosing a response (pressing the corresponding key on HID-keyboard) the program will change the background of the student's panel number on the screen depending on the test configuration. Thus, the teacher can monitor the testing screen. After the answer background changed, a student cannot change his/her answer. The system stores the pressed keys of every student, and processes the results of such keys at the end of the test and converts them into an ordinary system of evaluation taking into account weight coefficients for each answer in accordance with the data that were entered during the test creation.

After the test completion, data are added to the database. Information on the test results is stored in the database without the possibility of its removing or editing with a software component of the system. To display the test statistics, the system has menu Results on its main screen. The system loads test results by groups of students and by test dates. The system has windows for convenient display of results according to the set parameters and generally throughout the database. Figure 2 shows a group of students operating the MTS.

Fig. 2. Working with testing control system

4 Recommendations for the Use of Mobile Testing System

Using MTS changes the approach to testing students. One of these opportunities is the use of MTS in a rapid survey mode. During the lesson, the teacher can conduct a short survey of students (2–3 questions) to found out their level of mastering the material. The presentation of material under continuous frontal control activates attention of students and gives them motivation throughout the lesson. Using the test results during the lesson, the teacher gets a complete picture of the material assimilation by each student and the group in general. All data are stored in a system database that provides an objective assessment of each student throughout the school year.

The panel on the table does not interfere or distract students as opposed to computers and mobile devices. At any time, the teacher can ask questions to and get the answer from all students at once. All teachers know how difficult it is to get an answer to a simple question at first glance: "Is everything clear?". The reasons are multiple: students' shyness, unwillingness to look worse than others, etc. Using MTS, the teacher who poses a question asks students to enter a response in their remote panels: 1 - yes, 0 - no. Each student enters answers individually and independently. The teacher gets a prompt response of the system that shows the percentage of students who understood the material.

The following algorithm of MTS-aided classes can be suggested:

1. Teaching of a portion of educational material according to the lesson plan.
2. Finding out of the level of mastering of the given material by the students (the percentage of students who understood the material and have no questions) with the help of MTS.
3. A more detailed explanation of the theoretical and practical provisions in case of insufficient level of mastering of the given material.
4. Quick test (3–5 questions) in the materials of the given portion with the use of MTS.
5. Analysis of typical mistakes made by students during the quick testing.
6. A repeat of par. 1–5 for each portion of the educational material to be given during the lesson.
7. The final test at the end of the lesson using MTS in all portions of the educational material given during the lesson (max 10 questions).
8. Summing up of the test results throughout the lesson and encouragement of the best students.
9. Individual discussion with the students having the lowest score according to the results of all tests.

5 Conclusions

On basis of single PC, a budget mobile computer system of test control has been developed, which can be applied in any classroom for simultaneous testing of the whole group of students. The system support development of own tests and their editing, the formation of the main test elements, computer-aided testing process and check of results, accumulation of test results on different types of questions, displaying test statistics.

A significant advantage of the proposed system as compared to the existing computer testing systems is its portability and relatively low price. The system mobility enhances the efficiency of teacher's control activities due to that MTS can be deployed at any time and in any room. At the same time, it retains all the advantages of computer-aided testing with increasing frequency and regularity of control, and the system software makes it possible to implement effective and quality testing of students. The system is cost-effective enough through the use of inexpensive components. In the presence of a projector and PC the cost of additional equipment for testing of a group of 20 students is about 100 USD, which is significantly lower than the cost of similar mobile testing systems, and the need to use only one computer instead of the computer room significantly reduces the system cost as compared to traditional computer testing systems. The mobility of the testing control system offers great opportunities to improve theoretical and practical training of students. We have examined only some of these possibilities and hope that the widespread introduction of this system will open multiple options to use it.

References

1. Audience response. https://en.wikipedia.org/wiki/Audience_response
2. Byrni, R.: Seven Good Student Response Systems That Work On All Devices (2014). http://www.freetech4teachers.com/2014/03/seven-good-student-response-systems.html#.WRi8jevyjGg
3. MQlicker: Free Audience Response System. https://www.mqlicker.com/product.html
4. OMBEA: Student Response System. http://www.ombea.com/gb/solutions/ombea-response?gclid=CJb684ya8NMCFcOoGAodCpIAvw
5. Affordable student response system. https://www.polleverywhere.com
6. SMART Response System. http://education.smarttech.com/en
7. Student Response Systems. http://turningtechnologies.co.uk/software/
8. Schoolhouse Test: Easy test maker for teachers. https://www.schoolhousetech.com/test
9. Mentimeter: Interact and vote with smartphones during presentations. https://www.mentimeter.com/features
10. SurveyMonkey. https://www.surveymonkey.com
11. Adit Testdesk: Exams, Tests and Quizzes Made Easy. http://www.aditsoftware.com
12. Moodle_Mobile. https://docs.moodle.org/30/en/Moodle_Mobile

Cloud Monitoring of Students' Educational Outcomes on Basis of Use of BYOD Concept

Denys Kovalenko(✉) and Tetiana Bondarenko

Ukrainian Engineering Pedagogics Academy, Kharkiv, Ukraine
{kovalenko_denys,bondarenko_tc}@uipa.edu.ua

Abstract. This paper proposes the technology of using own mobile devices based on BYOD concept for testing educational outcomes of students. The technology is based on Google cloud services, which provide a comprehensive support to the testing system from the creation of appropriate forms and storage of results in cloud data storage to the processing of test results and management of the testing system through the use of Google-Calendar service. The stages and types of testing of the educational outcomes based on Google search services using BYOD concept are described. It is noted that the use of BYOD concept extends the testing frameworks in space and in time, makes the testing procedure more flexible and systematic, adds elements of gameplay to it.

Keywords: Control of knowledge · Computer-aided testing
Cloud technologies · Google Docs · BYOD · Knowledge updating

1 Problem Statement

A significant shortcoming of the traditional computerized test control systems is the requirements to the hardware. The number of computers must not be less than the number of those tested. It is naturally that the entire learning process in most subjects, in which testing takes a small part of the lesson time, seems impractical to be transferred into the computer class. As a solution to this situation can be the use of own mobile devices of the students.

The Open University of Great Britain [1] in its report singles out BYOD (Bring Your Own Device) concept among the major innovations that can cause global changes in education. The traditional educational environment becomes wider through social networks and open educational resources. There is a unique opportunity to combine full-time and extramural education in the educational practice. The teachers instead of their former role as a source of information and knowledge get a new one—administrator of students provided with access to network resources.

The author of the article [2] lists the additional advantages of this very simple concept:

- firstly, it increases the amount of devices in educational institution that can be used to enhance learning;

M. E. Auer et al. (eds.), *Teaching and Learning in a Digital World*,
Advances in Intelligent Systems and Computing 715,
https://doi.org/10.1007/978-3-319-73210-7_89

- secondly, it avoids unnecessary spending on hardware resources, and this finance can then be re-directed to other areas of ICT development within the educational institution;
- thirdly, it avoids the 'doubling' or sometimes 'tripling' up on devices, where a computer is redundant for much of the day because it is either at educational institution, at home or hidden in your pocket.

In our opinion, one of the most effective ways to use their own mobile devices is the computer-aided testing of students' educational outcomes. Computer-aided testing as a form of knowledge control combines the advantages of traditional testing system and benefits of a computer system.

2 Analysis of Recent Research and Publications

In the article [3] the features of the BYOD trend in engineering education are described. Research results on various aspects of BYOD in one of the largest technical university of Russia - Bauman Moscow State Technical University - are summarized, but the question of the using of this concept for assessment of students' academic achievements don't discussed.

Today, there are a number of software products based on cloud technologies that allow running BYOD concept for testing the educational outcomes. They include mQlicker service [4], Mentimeter cloud services [5], SMART Response VE system [6], SurveyMonkey service [7], Anketolog system [8], Webanketa service [9], and Hearne Software Survey System [10].

Drawbacks of the above services are that they are designed primarily for compiling questionnaires and surveys. Moreover, the announced full package of options of these systems is implemented only in the paid version, and in the case of using free services, their opportunities are significantly limited.

Another possibility of implementing BYOD concept to test the students' educational outcomes is built into the mobile versions of distance learning systems. For example, an official application Moodle Mobile [11] supports Moodle websites, configured for operation with the app. After setting the Moodle website one can use this application to organize testing of students including using BYOD concept. However, the implementation of such projects requires considerable financial costs and human resources.

We think that one of the prerequisites for full implementation of computer-aided testing in the educational process is to create systems adapted to different levels of tasks and scales of education institutions.

Objective of the study is to suggest a budget version of a system of mobile testing of educational outcomes of students based on Google cloud services, which provide a comprehensive support to the testing system from the creation of appropriate forms and storage of results in cloud data storage to processing of test results and management of the testing system through the use of Google-Calendar service.

3 Statement of Basic Material and the Substantiation of the Obtained Results

The use of BYOD-based mobile devices allows implementing the information-educational environment of mobile learning. Mobile devices include smart phones, communicators, tablets, etc. These devices have the International Mobile Equipment Identifier (IMEI) and run under the operating system (for example, MaciOS, Android, Windows Phone, etc.), support the mobile networks and Wi-Fi. Creation of the information-educational environment of mobile learning based on BYOD concept makes it possible to ensure control over the educational process not only in but also outside the computer classrooms.

Figure 1 shows the structure of the Google-based mobile testing system. The system includes the following components:

- test generation subsystem based on Google Forms service;
- Google Drive cloud storage for storing test forms and their results in the result database;
- subsystem for testing processing in electronic Google Tables;
- testing system management subsystem based on Google Calendar, necessary for planning and coordinating all activities related to testing of initial outcomes.

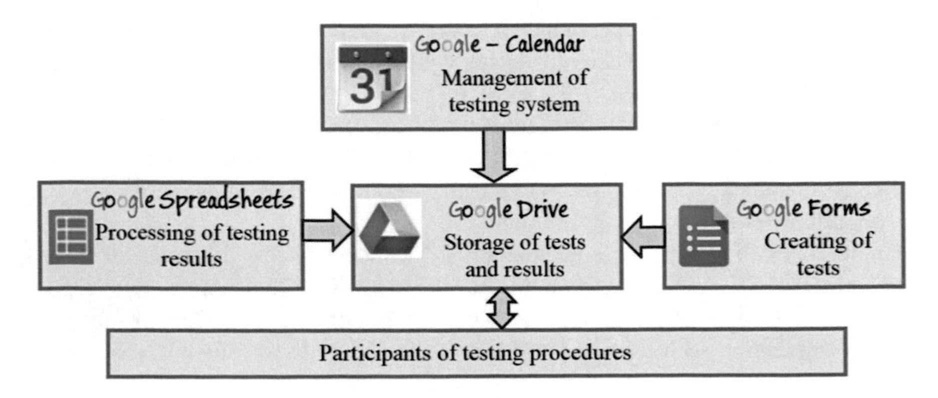

Fig. 1. Google-based mobile testing system

Testing of educational outcomes based on Google search services using BYOD concept includes the following steps:

- test preparation;
- test development in Google form;
- registration of the test takers;
- generation of Google Calendar with the events of test management and provision of access of the test takers to such calendar;
- the connection of the mobile devices of the participants to the created Google Calendar;

- testing;
- processing of test results in electronic Google tables.

The development of tests in Google form is a simple process. Eight possible types of questions ensure check of the assimilation of almost all the material learnt. Types of questions in Google form: one from the list, a few from the list, text, scale, grid, drop-down list, date, and time. At the final stage of creation, one needs to set the form on the following parameters:

- show progress at the bottom of the page;
- one response per one participant only (requires login to the account);
- mix the questions (questions for each participant are in a random order).

After the creation of test questions, the created Google form is automatically stored on your Google Drive account. The created form can be pasted into a website or blog. To do this, save it first, and then at the top of the editing window click *More actions* button, select *Paste* in the open menu.

If there is a significant number of tests in subjects, use Google Calendar cloud service for their convenient administration. Google Calendar supports synchronization with mobile devices. Reminders are sent via email or SMS. Google Calendar can be used to schedule work related to solving the problem of testing. For each testing, Google Calendar creates two events with the presetting of alerts through SMS messages and/or emails and/or messages via social networks. One message contains a reminder of what happens at the appointed time of testing on a given topic, the second at the appointed time will be sent to users of the testing system with reference to the form of the test in the previously announced subject. In addition, one of the components of Google-Calendar is a "task list", which determines the future and current tasks, for which the user can set priority. This is especially useful for scheduling the preparation of tests.

In order to combine all events associated with testing in one calendar in Google services, create a new calendar in *My calendars* menu, for example, called "Testing". Calendar "Testing" with events of testing is shown in Fig. 2.

To send the notifications of events to the participants the latter need to share the calendar "Testing" (specify a list of email addresses of users in Share calendar menu). To receive SMS notification, each participant must independently connect his/her mobile phone to calendar "Testing".

After creating events related to testing in Google Calendar "Testing", send the messages to the participants (a reminder of the time of the testing and the forms with tests). In a reply SMS with the reminder, the participant should click *Accept invitation* button and the organizer of the event will receive notification in the calendar "Testing" of participant's consent to undergo testing.

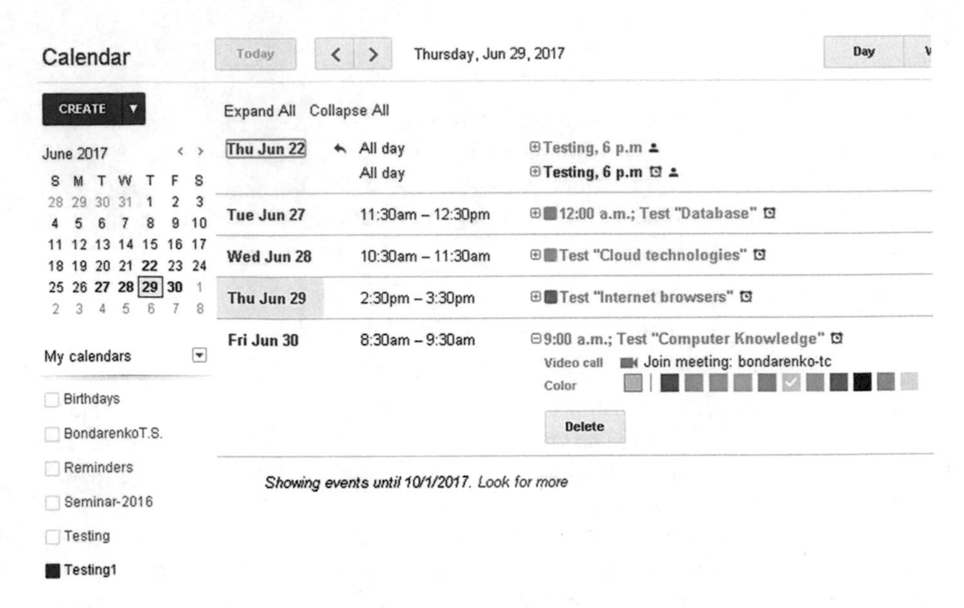

Fig. 2. Calendar "Testing" with events of testing

Messages with a link to the form with the test come to the e-mail of the participants (Fig. 3).

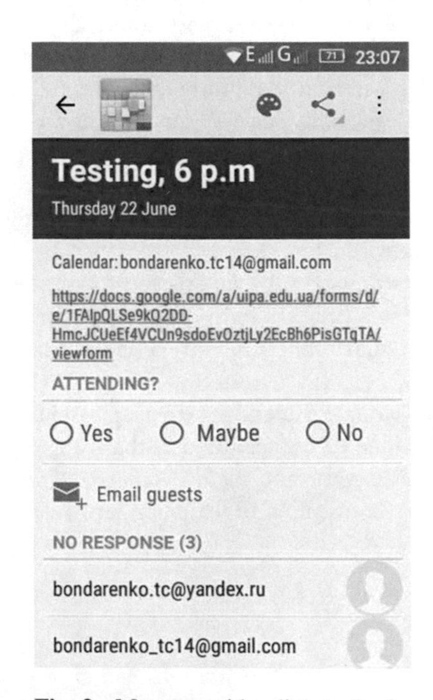

Fig. 3. Message with a link to the form with the test

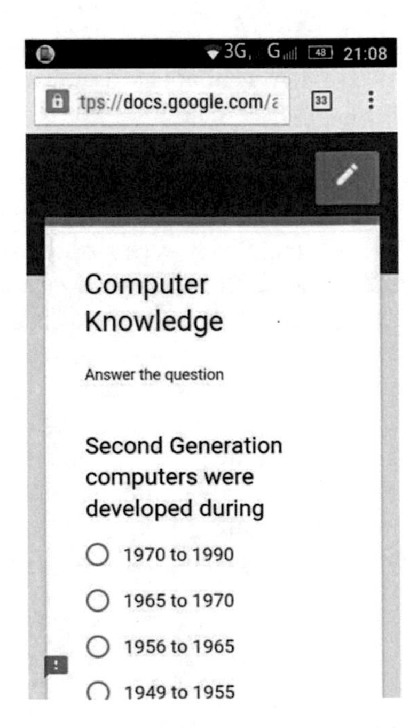

Fig. 4. A testing form on smartphone

By clicking the received link, the mobile device opens the created form and the participant consistently answers the test questions contained in it (Fig. 4).

After filling the form press *Send* button. The test taker enters the testing result in the electronic Google table in the cloud storage. The table appears in Docs.Google file manager; its name is taken from the form name plus the word 'response'.

Answers obtained through the form can be viewed in four ways:

- as a summary;
- as individual user's responses;
- as a table;
- as CSV file.

Using the tool *Summary*, a tab opens with a chart on every question of the form. In addition, next to the chart, the results will be presented in the form of numbers and percent (Fig. 5).

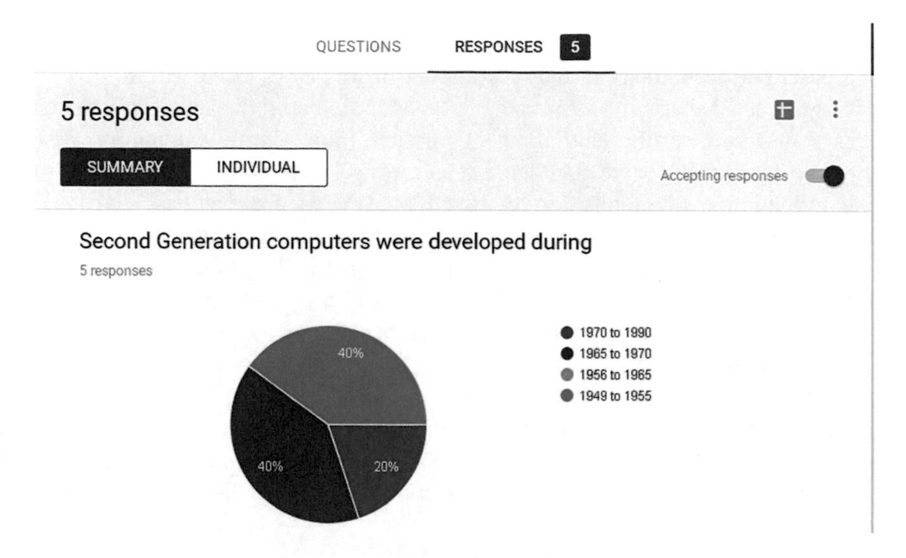

Fig. 5. View of answers as a summary

Viewing responses of individual participants makes sense if there is a small number of both questions and test takers because in this case, a sufficiently large number of forms will have to be reviewed.

Test results in a table form contain information about the answers to test questions for each student. The table is attached to *Entry Timestamp* column that is also a data set "index". The information is sorted in this field by default, so we get a chronological list of all entries. Despite the fact that to fill the test forms the participants need to login to their accounts, information about the person who filled the form is not reflected in the test results. Therefore, the form should include questions of Text type to fill in the names and surname of the person undertaking the testing.

During data processing, the electronic table allows easily re-sorting information by any column, not breaking the entries. Google table (as well as Excel table) has a set of options for statistical data processing. If desired, the table can be exported to MS Excel for processing.

In contrast, for example, to the Moodle-based distance learning system, which carries out processing of test results automatically by means of the system, in this case, the test developer has to individually perform the processing of test results.

Test results are sent using shared access of the users to the files stored on Google Drive.

We should have a closer look at possible options for the use of testing control system based on BYOD concept. We can suggest the following schemes of lessons with such testing control:

- initial assessment on the subjects of the previous lesson;
- 2–3-step continuous assessment during the lesson to determine the level of mastering the material learnt;
- final assessment covering all the material learnt during the lesson.

The described technology of the testing control was approbated within the subject "The Integrity and Security of Information". Its Moodle-based distance learning course involves four test controls, held in the computer room during laboratory lessons. Another two assessments were conducted at lectures prior to the use of this technology.

The introduction of mobile testing based on BYOD concept has increased the number of test controls up to six. Four of these controls were conducted at lectures and two were conducted outside of classes as an experiment. By setting the option "Mix questions" the students could not use the answers of other participants during testing.

As mentioned above, despite all the positive benefits of computer-aided testing its widespread use has been constrained due to lack of the necessary technical base. The BYOD concept removes this limitation, and the computer-aided testing of the educational outcomes can be conducted both in the classroom and beyond it by using student's own mobile devices. This approach allows implementing the principle of systematic test control at all stages of the didactic process from the initial perception of knowledge to its application in practice.

4 Conclusions

The study "Values and Interests of Students" [12] puts the Internet first among the priorities of young people. However, as the main interest of the youth, we would like to note their interest in everything new. In this regard, a new testing technology, which uses their favorite mobile device and no less favorite social networks, where they receive an invitation to pass another test control, has aroused a considerable interest on the part of those being tested.

Despite some complications (standardization of mobile devices, technical and pedagogical problems of introducing the mobile devices in the learning process, etc.), the use of BYOD concept makes broader the testing procedure in space and in time, makes the testing procedure more flexible and systematic, adds elements of gameplay to it. The ability to organize whenever and wherever a systematic control of educational outcomes on the basis of BYOD concept facilitates updating the knowledge of those being tested.

Using the considered concept based on cloud services of Google search engine in testing has a number of advantages as compared to similar software services. First of all, this is a comprehensive support to the testing system from the creation of appropriate forms and storage of results in cloud data storage to the processing of test results and management of the testing system through the use of Google-Calendar service. Second, it provides the ability to create a cost-effective testing system through the free use of free Google services.

References

1. Sharples, M., Adams, A., Ferguson, R., Gaved, M., McAndrew, P., Rienties, B., Weller, M., Whitelock, D.: Innovating Pedagogy 2014: Open University Innovation Report 3. Milton Keynes: The Open University (2014). 43
2. Bray, O.: Empowering Learning with BYOD. Bloxx, p. 18 (2013). http://www.bettshow.com/library_10/1456146_assocPDF.pdf
3. Shahnov, V., Zinchenko, L., Rezchikova, E., Glushko, A., Sergeeva, N.: Peuliarities of BYOD Trend in Engineering Education. - Educational technologies and society, # 4, vol. 19, pp. 334–346 (2016). https://cyberleninka.ru/article/v/osobennosti-tendentsii-byod-v-inzhenernom-obrazovanii
4. MQlicker - Free Audience Response System. https://www.mqlicker.com/
5. Mentimeter - cloud-based tool for interact with your audience in real-time. https://www.mentimeter.com
6. SMART Response System. SMART response 2 (Beta Version). http://education.smarttech.com/en
7. SurveyMonkey. https://www.surveymonkey.com
8. Anketolog – servis onlayn-oprosov. https://anketolog.ru
9. Webanketa. Information/ Businesses and organizations. http://webanketa.com/ru/info/service/factory
10. Hearne Software Survey System: Editions and Addons. https://www.hearne.software/Software/Survey-System/Editions
11. Moodle_Mobile. https://docs.moodle.org/30/en/Moodle_Mobile
12. Menshikova, N.A.: Student's Value Orientations and Interests. Ivanovo State University Shuya, Russia (2015). http://www.scienceforum.ru/2015/pdf/14522.pdf

A Concept for an Intelligent Tutoring System to Support Individual Learning Paths in Software Development Courses

Veronika Thurner(✉), Philipp Chavaroche,
Axel Böttcher, and Daniela Zehetmeier

Department of Computer Science and Mathematics,
Munich University of Applied Sciences, Lothstraße 64, 80335 Munich, Germany
{veronika.thurner,axel.boettcher,daniela.zehetmeier}@hm.edu,
PhilippChavaroche@gmx.de
http://www.cs.hm.edu/

Abstract. Freshmen students of computer science usually are a highly heterogeneous set, especially regarding their initial programming skills, which range from none to professional. As well, students are unequally equipped with essential base competencies, and differ in their respective pace of learning and the amount of practice they need to get new skills under their belt. As a consequence, every student requires an individual learning path to meet his or her specific needs. To ensure an efficient learning progress, it is crucial to select additional exercises for the students' self-study phases appropriately, so that they really meet the individual student's current need. Therefore, we developed a concept and implemented a prototype for a tutoring system that supports individual learning paths by providing each student with exercises that specifically address his or her specific needs and take existing skills and competencies into account.

Keywords: Intelligent tutoring system · Individual learning path
Competence level

1 Introduction

Experience shows that freshmen students of computer science usually are a highly heterogeneous set, especially regarding their initial programming skills, which range from "none" to "professional" [3]. As well, students are unequally equipped with essential base competencies. These comprise not only cognitive skills such as analytical, abstract or logical thinking, but personal skills such as perseverance or self discipline, as well as social skills, e.g. the abilities to communicate, work in a team or deal with criticism. Furthermore, they differ in their respective pace of learning and the amount of practice they need to get new skills under their belt. As a consequence, every student requires an individual learning path to meet his or her specific needs.

M. E. Auer et al. (eds.), *Teaching and Learning in a Digital World*,
Advances in Intelligent Systems and Computing 715,
https://doi.org/10.1007/978-3-319-73210-7_90

However, university courses are mostly designed as educational mass production, where one lecturer faces tens or even hundreds of students at a time. As a consequence, it is a tough challenge for lecturers to foster their students individually. Moreover, class time is usually scarce – and often too short to provide students with sufficient practice. Therefore, many students need to do additional exercises during their self-study phases. To ensure efficient learning progress, it is crucial to select these additional exercises appropriately, so that they really meet the individual student's current need and help to reliably produce the desired learning outcome.

2 Goal of this Contribution

Established e-learning systems such as Moodle (https://moodle.com) or Lon Capa (http://www.lon-capa.org/) provide a platform through which the lecturer may publish assignments and supplementary exercises on which students can sharpen their skills. However, they do not support the student in identifying from the masses of available material, those exercises that he or she should tackle next, to gain the best possible benefit based on the student's current knowledge, skill level and learning requirements.

Therefore, to improve the situation we developed a concept and implemented a prototype for a tutoring system that supports individual learning paths by providing each student with exercises that specifically address his or her specific needs and take existing skills and competencies into account. Note that by competence we denote the combination of a technical content and an action that the student can perform with respect to this technical content, following the revised Bloom taxonomy of educational objectives [1].

As a first example, we apply this concept to an introductory class on software development that is taught in the freshman year of a computer science bachelor programme at Munich University of Applied Sciences. This software development class comprises four hours of lecture and two hours of lab session per week, rounded off with plenty of self-study time if needed.

3 Approach for Supporting Individual Learning Paths

To achieve this, as a first step we define learning objectives for the topics dealt with in class. Secondly, we specify a syllabus that defines the order in which technical content is presented. Then, we define exercises that address the different topics on the desired skill levels.

On this basis, we provide students with exercises, keeping track of each student's performance, which we use to systematically suggest which exercise a student should tackle as a next step on his or her individual learning path.

3.1 Define Learning Objectives

As a first step towards supporting individual learning paths, we define learning objectives for the topics dealt with in class. Within these learning objectives, we differentiate the aspired competencies according to the six cognitive levels (1) remember, (2) understand, (3) apply, (4) analyse, (5) evaluate and (6) create that are described by Bloom's revised taxonomy [1]. Each learning objective combines a technical topic with one of these cognitive levels.

Note that a skill is not necessarily developed in the exact order of these levels. For example, many students that already possess some programming experience are able to correctly use concepts and develop larger programs on their own (level (6) create), but are sloppy in their usage of terms and definitions when talking about their solutions (level (1)).

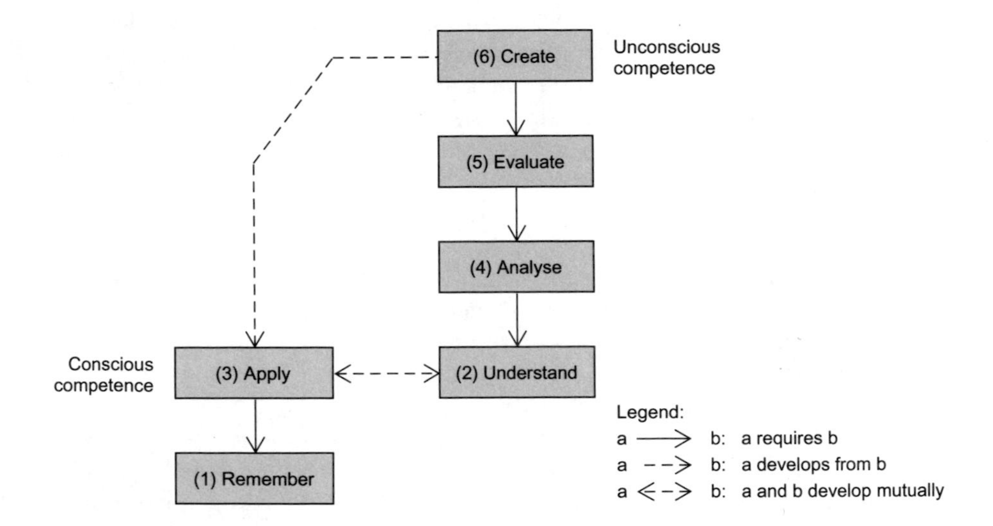

Fig. 1. Bloom levels and their relations with respect to their development

Figure 1 illustrates the interdependencies of skill development for a single technical content across different Bloom levels. Skill levels in the left column ((1) remember and (3) apply) usually can be achieved by a sufficient amount of diligence and hard work, but do not necessarily require outstanding intellectual effort. In contrast to this, skill levels in the right column require the ample application of cognitive and methodical competencies such as abstract, analytic and critical thinking as well as creativity, among others.

Level (3) apply corresponds to a conscious competence (according to the four stages of competence development [2]), where students are explicitly made aware of which steps they have to execute to achieve a desired result. In contrast to this, level (6) create denotes an unconscious competence, where experts intuitively act in a way that is adequate to solve the task at hand.

Especially in the STEM-area, professional practice involves problem solving skills in the respective area of expertise, which usually corresponds to cognitive level (6) create. However, skills on *all* cognitive levels are required in professional practice. For example, settings such as pair programming or code reviews require the ability to correctly communicate about programming concepts (level (1)). Similarly, although it goes without saying that a program needs to deliver the intended results, this in itself is not sufficient to make this program a high quality artifact. Rather, the program must fulfill established quality criteria, such as maintainability or efficiency. Reading and understanding code that has been developed by others and assessing its quality requires analysis (level (4)) and evaluation skills (level (5)).

3.2 Outline a Syllabus

As a next step, we outline a syllabus that defines the order in which topics are dealt with in class. This order is based on dependencies within the technical content. (For example, Boolean values must be introduced at least in a rudimentary way before if-statements come up.) In addition, didactic considerations influence the order in which topics are covered in class. (E.g. to ensure that students accept unit testing and the test first approach as a matter of course, we introduce unit testing in an early phase of the course.) The resulting order of topics is represented by a tree-structured collection, where the tree's directed edges indicate prerequisites of a topic as specified by the lecturer.

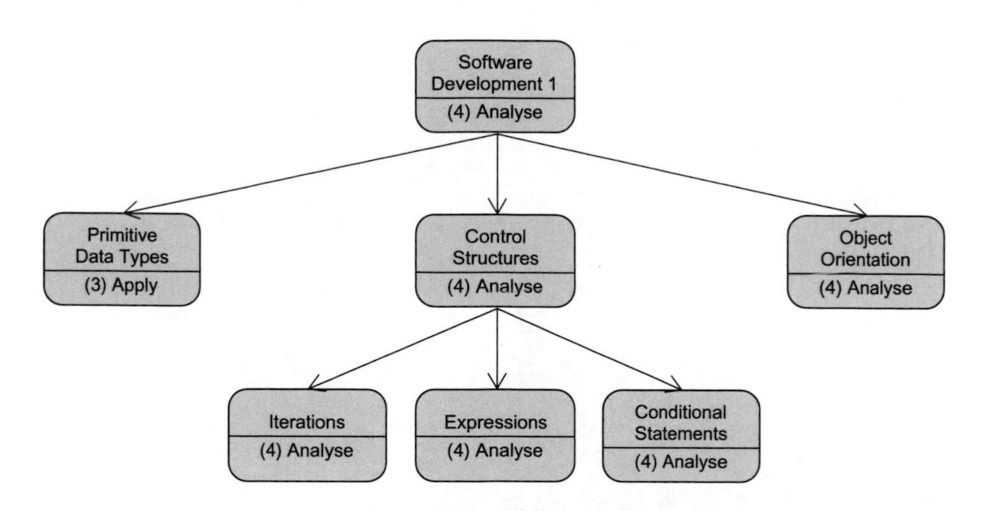

Fig. 2. Example syllabus

Figure 2 shows an example for technical competencies that are to be developed in an introductory class on software development. Competencies are organized into an ordered tree. Learning objectives are depicted in the leaves of the tree, whereas inner nodes merely help to structure the different competencies.

To bring learning objectives into a sequential order that the lecturer will follow in class, the tree is traversed in postoder, i.e. from left to right and bottom to top.

3.3 Generate Appropriate Exercises

Having achieved this, the lecturer and supportive staff specify exercises and questions, each of which addresses a certain topic on a particular cognitive level, and enter them into the tutoring system. Figure 3 sketches the data model we employ for representing exercises in our tutoring system.

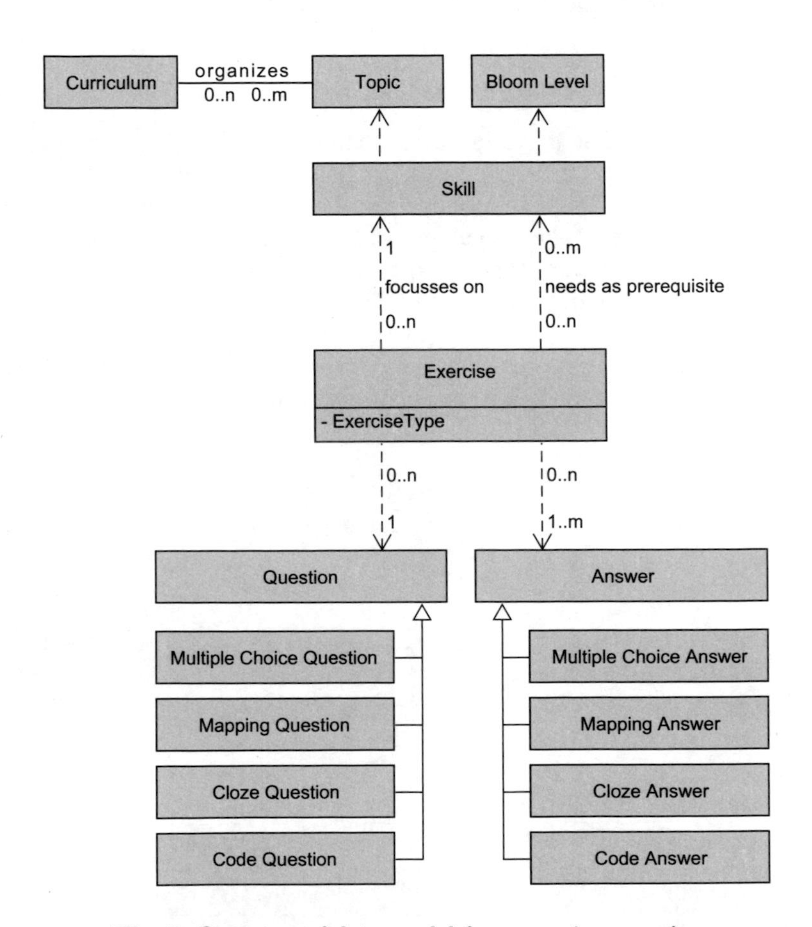

Fig. 3. Overview of data model for managing exercises

To ensure optimum support of the students' self study time through the tutoring system, exercises should be specified in a way that allows for automatic evaluation of the student's answers. To this end, exercise types in our tutoring system are currently restricted to multiple choice, mapping, cloze and code.

For all exercise types, it is possible to predefine a set of possible answers, rather than just a single answer. Obviously, exercise types multiple choice and mapping require several answer options, so that students are required to choose (or map) the correct answer or answers from the given set.

In exercises of type cloze, students are required to enter a specific word, such as the name of a specific concept. To allow for typical variations in spelling, the staff may specify a variety of acceptable spellings, each of which will be evaluated to true.

Currently, the code snippets that are expected as answers to exercises of type code are explicitly provided by the staff as well, in different varieties, similarly to the answers for exercise type cloze. Of course, it would be desirable to connect the tutoring system to a compile- and testing-environment. However, the tool does not yet offer this possibility.

Topic and cognitive level that are focussed on by an exercise are stored as attributes of the exercise. Similarly, prerequisites that are necessary to solve an exercise are stored there as well, as a set of tuples (topic, level).

3.4 Monitor the Student's Performance and Suggest Exercises

When a student uses the tutoring system for supporting his or her self-study efforts, the system keeps track of the student's individual skill profile as well as of detected knowledge gaps.

Figure 4 sketches the data model we employ for managing a student's skills and knowledge gaps in our tutoring system, and for suggesting appropriate exercises.

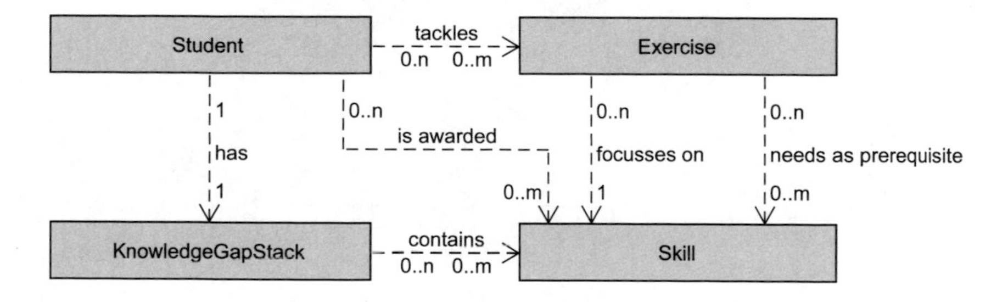

Fig. 4. Overview of data model for managing a student's skills

If the student solves an exercise correctly, the system adds the topic and level addressed by the exercise to the student's skill profile and moves along the specified syllabus on to more advanced exercises. However, if the student provides a wrong solution to an exercise, the topic and level focussed by the exercise as well as all its prerequisites are knowledge gap candidates that might have caused the incorrect answer. In this case, following certain heuristics such as the kind and number of prerequisites, the system automatically selects follow-up questions that close in on the skill that is missing.

Generally, when knowledge gaps were detected for a student, the tutoring system will focus on closing these existing knowledge gaps before moving on to new skills. Therefore, as long as the knowledge gap stack is not empty, the system will choose the most basic competence, i.e. the one with the least prerequisites, of the competencies in the knowledge gap storage. In case the knowledge gap stack should be empty, the system will focus on the competence that "comes next" according to the selected curriculum (or the default curriculum, if no specific curriculum has been selected). For the competence that has thus been selected, the system then suggests an exercise that specifically adresses this competence. Figure 5 illustrates the algorithm for choosing the next exercise.

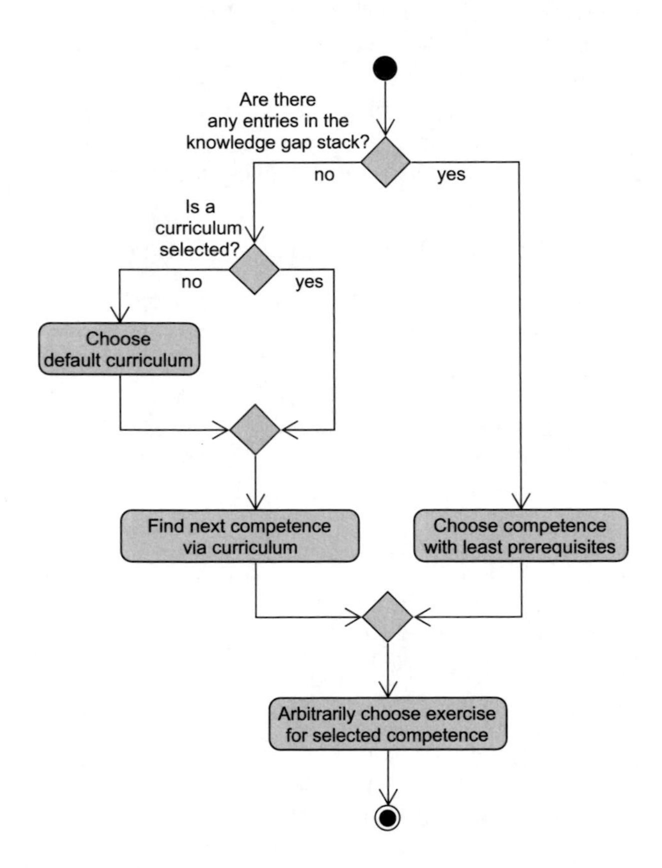

Fig. 5. Algorithm for choosing the next exercise

Experience shows that students greatly differ with respect to their speed of learning. While some lucky students achieve a certain skill by applying it just once, others have to involve a higher amount of practice on the same skill to have it securely under their belt. Therefore, the system parameterizes the student's individual learning speed.

As a default, a student needs to correctly solve three exercises to be awarded the skill these exercises focus on. However, if the student's learning protocol shows that generally he or she requires more (or less) than three exercises to obtain a certain competence, this default value can be increased (or reduced) accordingly.

3.5 Calibrate Initial Parameter Values

When a student uses the system for the first time, the system initially has no information on the student's existing competencies, since there are no tasks yet that the student has answered either correctly or incorrectly. In particular, both the storage of previous errors (i.e. the knowledge gap stack) and the set of awarded competencies are still empty. As a consequence, the system cannot suggest an exercise that adresses the student's specific need.

One strategy for dealing with new students is to have them work on the "regular" system right from the start, beginning with really simple tasks that do not require any prerequisite skills. This ensures an easy start and a quick sense of achievement, which is motivating for students that are novices to the subject at hand. However, there is a significant risk that students which already possess some knowledge will get bored quickly if tasks are too simple.

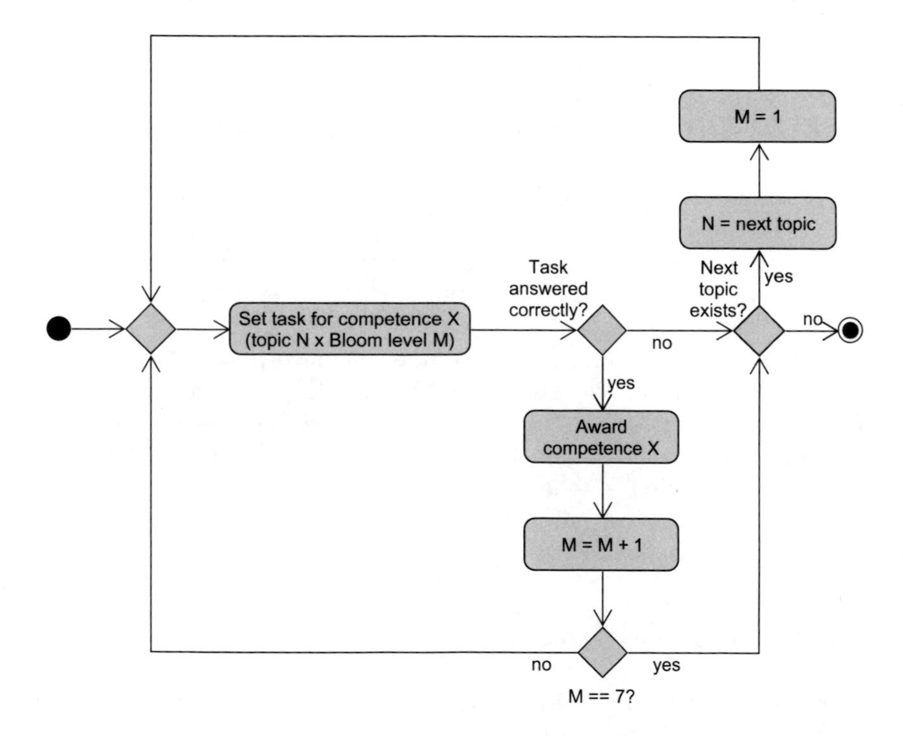

Fig. 6. Introductory test for calibrating parameters

Therefore, we suggest an introductory test for quickly calibrating parameters for each individual student (see Fig. 6). Whereas the regular system involves a certain repetetiveness to ensure that students develop an adequate dexterity in the different competencies, the introductory test aims at quickly (but just roughly) identifying areas where the student needs to invest some effort. Therefore, the introductory test encompasses just one single task for each competence, i.e. for each combination of topic and skill level.

For each topic, tasks are run through starting from level (1) remember, until level (6) create is successfully accomplished – or until the student provides a wrong answer. At the first wrong answer, the topic is dropped entirely and the introductory test moves on to the next topic, on which it will start again on level (1).

Note that the system merely awards competencies but does not fill the knowledge gap stack. Thus it focusses on the students' initial skills, rather than on whatever students don't know yet.

Figure 6 visualizes the algorithm that the introductory test employs for initially calibrating the parameters for a new student that uses the tutoring system for the first time.

4 Results

So far, the system supports only a limited set of question types, e.g. single choice (one out of n alternatives is correct), multiple choice (m out of n alternatives are correct), cloze (asking for text fragments) or code (small code snippets have to be inserted).

Most of the exercises provided so far address skills on the lower cognitive levels, i.e. (1) remember, (2) understand and (3) apply. With the available exercise types, it would be possible to address skills up to level (5). However, specifying exercises that address higher skill levels but can still be evaluated automatically is a time consuming task on which the staff is still working.

The algorithm for selecting follow-up questions incorporates heuristics that are parameterized to a certain extent, in order to reflect the students' individual need for practice. For example, one parameter specifies how many exercises on a specific (topic, level) combination the student must answer correctly before the (topic, level) combination is added to the student's individual skill profile.

All in all, initial tests confirm that the tutoring system is able to select appropriate exercises, on which students can focus in their self study phase.

5 Conclusions and Future Work

Providing students with an individually appropriate amount of exercises that address the students' specific needs is crucial for establishing individual learning paths that are both effective and efficient.

The tutoring system presented here introduces a concept for selecting exercises that incorporates both the syllabus of a class and the individual student's skill profile as well as detected knowledge gaps. Thus it helps students to focus their study and practice efforts on those skills that are already within reasonable reach, but where they still need to gain proficiency.

To further enhance the benefit of the tutoring system, future versions should provide exercise types that allow for adressing higher skill levels such as (4) analyse, (5) evaluate or (6) create. One possibility would be ranking tasks, as proposed in [4], to address level (5).

For automatically evaluating exercises that address skill level (6) create, it would be helpful to enhance the tutoring system by connecting it to a compile- and testing-environment, which allows for running the student's solution against a predefined set of unit tests.

References

1. Anderson, L.W., Krathwohl, D.R., Bloom, B.S.: A taxonomy for learning, teaching, and assessing. A Revision of Bloom's Taxonomy of Educational Objectives, 1st edn. Longman, New York (2001)
2. Burch, N.: Learning a New Skill is Easier Said than Done (1970). http://www.gordontraining.com/free-workplace-articles/learning-a-new-skill-is-easier-said-than-done
3. Santos, A., Gomes, A., Mendes, A.: A taxonomy of exercises to support individual learning paths in initial programming learning. In: 2013 IEEE Frontiers in Education Conference (FIE), pp. 87–93, October 2013
4. Tao, Y., Liu, G., Mottok, J., Hackenberg, R., Hagel, G.: Ranking task activity in teaching software engineering. In: 2016 IEEE Global Engineering Education Conference (EDUCON), pp. 1023–1027, April 2016

Problems and Educational Responses in the Age of Digital Technologies

Educational Robotics like Pedagogical Philosophy

Anita Gramigna and Giorgio Poletti(✉)

University of Ferrara, Ferrara, Italy
{anita.gramigna,giorgio.poletti}@unife.it

Abstract. According to recent research [1, 2], the latest generation technology causes disturbing forms of dependence on children, young people and adolescents, as well as increasing concentration and abstinence problems with obvious cognitive consequences and school behaviors. Moreover, subtle forms of solipsism would affect our young people; young people would try to solve such solipsism in virtual relationships that are consumed predominantly in social networks.

Our thesis is that educative and cognitive training of new technologies implies knowledge of the thinking that supports its use, which in turn is the basis of a young person's creative action and effective learning. We have identified the operational strategies in Educational Robotics. This article represents a reflection based on the work done with the teachers who participated in the course on Educational Robotics held at the Summer School organized by the Epistemology Laboratory of Euresis Training at the University of Ferrara and by the Robocup jr Italia State Network in Stresa (Italy) in August Of 2016. *Chapters* 1 *and* 3 *are assigned to Prof. Anita Gramigna and Chaps.* 2 *and* 4 *to Dr. Poletti.*

Keywords: Robotics · Learning · Education

1 Introduction: Digital Lovers

The fascination that digital technologies have on children, teenagers and young people is becoming total. As evidenced by the reports of elementary, middle and higher school teachers, but also from the daily observation that each of us can do in university classrooms, media Even at the restaurant, as is dramatically evident from car accidents that saw young stars unable to break away from the smartphone even while driving.

The little ones play with tablets, cell phones and video games, experimenting perhaps with the feeling of the extraordinary and wonderful we had experienced with traditional fairy tales and cartoons. Technology is introducing epochal changes, both in social dynamics and in knowledge-building processes, and finally, in the elaboration of thought. However, that is not all, the consequences are also evident in our children's behavior and therefore in ethical choices.

M. E. Auer et al. (eds.), *Teaching and Learning in a Digital World*,
Advances in Intelligent Systems and Computing 715,
https://doi.org/10.1007/978-3-319-73210-7_91

According to a study conducted by the Children's Health Center led by Professor Tamburlini, a baby in five in Italy will contact the cell phone in the first year of life. At this stage of development, the brain of children is extremely plastic; as a result, both the risks and benefits of any environmental exposure are maximized [3].

Between 3 and 5 years of age, 80% of small hands are now able to use mama and dad's mobile phones [4]. In the second year of life, 60% of parents are using their cell-phone or smartphone to their children and in the range of 2 to 5 years, this percentage will gradually rise to 80%. Over 50% keeps the phone switched on day and night. Already in 9–10 years, 26% have their own laptop, 11% a latest generation cell phone and 4% a personal tablet. Thirty percent of parents would resort to this expedient in the first year of age, to reach over 70% after two years. The percentage remains constant in the range of 3 to 5 years.

Finally, 50% of teenagers spend from 3 to 6 h each day with the smartphone in hand, 16% between 7 and 10, and 10% quietly over 10 h a day. According to some studies, the effects of this massive exposure would be observable at school age with the appearance of socio-emotional disorders, with increased aggressive behaviors, reduced attention and mental flexibility. We are convinced that a fundamental element of the fascination that the technology world exerts on young people is linked to the processes of building their identity. Their subjectivity compares with the approaches of the latest generation technology as it represents the figure of definition of contemporary society.

The boys define, through the possession of technological goods, their status as well as their membership of a culture and the level of their technical expertise.

At university, the demand for fast learning, pragmatic knowledge, and a tiring study is revealed in the widespread pretense of receiving concrete lessons, even when, as in our case, this is a course of study that should form not only aspiring educators but also future philosophers. The massive exposure to digital instruments has contributed to forge anthropology and hence the mentality of our young people on the value of operational effectiveness [5].

This means that the sense attributed to learning and, in general, to study depends on satisfying needs and desires, and stresses of each and every other, that are increasingly bound to the concreteness and immediacy of the results.

There is a true dependence that becomes more massive with the passing of time and that marks the changing of youth behaviors, their relational style, cognitive approaches, cultural consumption, and which would be the source of an increasingly marked Lack of empathic skills. Advice to parents and educators on the widespread use of new digital media, on adult surveillance, and the limitation of time spent in exchange for good readings and serene conversations in the family are as sensational as ineffective because of a sort of fascination, which links children, young people and the young to the technological medium in a pathological way.

Fight the risks to which digital technology exposes our kids with the same technology: this is our proposal. We are convinced that thinking strategies and building intelligence in the environment of robotic systems can help shorten the distance between the boy's lives, including the playful and social ones, and the disciplinary knowledge and, in all of this, leverage on his participation in building-exploration of knowledge.

The answer to what is called new postmodern drugs, or addicts, is to try to clarify the primitive implications of knowledge shaken by the latest generation technology, namely, to understand and help explain what is needed. Nevertheless, given the pragmatic tension that animates our young people's educational needs and their criticalness, to understand what knowledge is, and in particular the knowledge transmitted by digital technologies, it is good to start from what it needs. In fact, due to the intense use of technology both in the field of play and in communication, youth mentality is highly conditioned by a kind of technical rationality. Our reason follows more and more logical "*techniques*" [6].

In our practice of teaching at the University and in research at school we have been able to observe that the attribution of value today is more than ever pragmatism. We believe that it is possible to begin the operational effectiveness of "*what is needed*" to clarify cognitive processes and that "active thing" and finally to understand what it is.

In our opinion, what "serves" is to consider the study as a practical relational model for lived-in life. It is from these shared considerations that, with our fellow teachers, we have tried to elaborate good questions more than to provide answers. Our attempt was to explore with them the possibility of an operational definition of knowledge built with robotic systems. It was to help colleagues to work on educational proposals; didactic offers that offer work materials that also meet the student's need for concreteness.

In fact, the teachers showed a striking discomfort to their students' cognitive behaviors. It is evident that the training needs of our students are linked to an imaginary and, above all, an anthropology that is very different from a few decades ago.

For them, as we know, it was getting to get a clear understanding of the phenomenon studied. So the "*how*" to get there was one with the "*what it is*" and with its own functionality or, in fact, more concretely, with what "*serves*". To know was to come to a meaning both subjective and clear of a problem or an object. The usefulness of the thing studied was contained in the sense that they attributed it, but did not replace it. Today things are very different. Hence, on the one hand, the sense of loss that becomes increasingly strong in the school, on the other hand, the gap between school teaching and the real world. Compared to our students, we faced the study by attributing a very different value to knowledge. Because our imagination, our mentality, our anthropology was and is different.

It is now necessary for an epistemological clarification that is the cornerstone of our arguments: we believe that knowledge, all knowledge, requires an epistemological preparation because it refers not so much to the reception of the information we think we owe to our Students or fellow teachers, as far as it's processing is concerned. Knowledge is, in our perspective, content and method, as it is phenomenal, that is, and it deals with problems and uses technologies, and is procedural, because it also involves methods, strategies and tactics.

It is, at the same time, "how" and "what", because it is the phenomenon and movement of its own construction. For this reason, the use of digital technologies cannot be exhausted in a pure economic recipe. Nevertheless, the study of knowledge is a practical science because knowledge allows us to achieve goals whose concrete is tangible, often beyond the immediate conveniences.

In fact, our purpose was, in fact, to discuss what knowledge is - whether for educational purposes from "what is needed" - to clarify how it works, how it is built, to what causes it can be connected, what processes it triggers, In short: how to deal with life in its concreteness. Robotic systems allow, in a clear and immediate way, the knowledge to be realized. This, through the teacher's action, must foster access to abstract knowledge. **As?** With a structured approach, that knows how to adapt to the wide variety of content and patterns of training thought [7].

2 The Operational Proposal

We started to do with the teachers who attended the meetings of Stresa; the teachers present are fervent supporters of Educational Robotics. We have tried to clarify the definition of educational Robotics that each teacher had matured over his or her experience to reflect on the structure, evolutionary paths, and the worldview that this fascinating world carries.

We have agreed that understanding the nature of what we are talking about will help us had better develop applications in the field of education, because we acquire cognition of the cause, understand the nature of processes that affect our thinking in building knowledge.

The teachers asked us to provide them with pedagogical coordinates on the subject. In short, they knew how to work with robotic classroom systems but wanted to deepen their reflection on emotional dynamics, cognitive approaches, metacognitive value that these marvelous artifacts could trigger in order to fully utilize their educational potential.

The first goal was to gain a definition, both dynamic and provisional, that connotes the deep meaning of knowledge and that it is oriented around the theme and the problems that the digital apparatus also poses us emotionally and behaviorally, inevitably, ethical. Indeed, this pragmatic, so pervasive propensity of our time represents both a moral posture and a cognitive approach; which has obvious consequences on the level of education.

At the beginning of the course, which lasted three days we asked them to develop a definition of the Educational Robotics from the use that each of them, starting from their own subject area. The purpose was to build together a transitional definition that would help our interlocutors to develop an autonomous thought about the great problem of robotic systems. It was to formulate a theoretical construct from the concreteness of their teaching experience; a constructor that was at a transitional time, that is available for subsequent adjustments, but also orienting, for epistemologically more conscious use of classroom robots.

Such a definition would therefore have to assume, in itself or better, in itself a formative value, not so much in providing a certain and tangible answer, as in soliciting our interlocutors to trace and build cultural and methodological tools to orient themselves In an autonomous and critical way around robotics. After the first loss, questions and requests for clarification, colleagues in their group had children who had socio-relational problems, they observed that building robotic systems in the class facilitated integration and collaboration even with those students who were traditionally inactive and participatory.

Other teachers, those who had experienced disciplinary problems among their students, noted that the interest generated by the robotics provided a more attentive and orderly attitude. Finally, professors who taught science disciplines estimated that they had been able to convey some of the important contents of their discipline more effectively. We could see that the definition teachers wanted to set up for Robotic Education started from the specificity of each's experience, but struggled to trigger strategic hypotheses that would guide their didactics.

In addition, here is a reflection on the metacognitive sense of learning built with robotic systems. It was crucial to be able to transfer content and strategies learned by working with the robot in other contexts, which concerned both the school discipline for the learning envisaged in the ministerial program and a more aware management of the emotional relationships that conveyed the working groups.

At the end of the course, the definitions became much more significant and complex. We include these definitions.

3 Collected Definitions of Educational Robotics

E. "Educational Robotics is that part of the robotics that allows us, through the playful element and learning for discovery, to understand how the mind works and how it builds the knowledge. With a real instrument, thought is deconstructed and re-assembled, enabling pupils to be active builders in knowledge. The task is divided into classes of problems and with the help of the companions one learns to attribute to the error an educational value considering it not as failure, but as a tool for refinement. The role of the teacher qualifies in favor of this kind of reflection and in facilitating the identification of the process and knowledge used (to solve the problem)."

G. "Educational Robotics is the conjunction between student and teacher, and it is the discipline that building, planning and ***playing*** with artifacts, fosters knowledge building and learning. Educational Robotics, in fact, motivates, connects (disciplines and pupils) and contributes to the discovery and construction of self; In fact, thanks to its playful aspect, it allows stimulating attention, motivation, thus improves storage, and develops a logical thinking and a greater design aptitude. It also allows co-operation, self-discovery, hidden skills or attitudes to increase self-esteem and the "creation" of a univocal link with the teacher. It allows, finally, the connection of knowledge, interdisciplinary and skills development."

L. A and J. "Robotics Educative is the development and use of robots for educational purposes, for teaching and learning. However, among those who teach and learn, they are definitely digital native students, to be more in the advantage of recognizing and using innovative technologies to learn how to learn.

The educational-didactic aims that the lecturer places for this research-action laboratory are as follows:

1. **Help the student** understand how his knowledge works, understand the critical elements and then where to intervene to improve and enhance it.

2. **Build cross-disciplinary and interdisciplinary skills** and transfer them from field to field.
3. **Encourage fun activities** not only between students but also between students and teachers.
4. **Use cooperative work** for problem solving situations.
5. **Promote manipulation activities** to strengthen synaptic structures (intelligence in and with hands).

The robotics mediated teaching-learning allows the student to:

1. **See the steps of creating your model** and benefit from immediate feedback on your progress.
2. **Use your own imagination** and be active in designing and modeling the model.
3. **Introduce your project to other students** by sharing suggestions and opinions.
4. **Build your own knowledge** based on the knowledge you already have, stimulating, with the help of the teacher, new learning when these become necessary.

Expose your project to other students by sharing suggestions and opinions. Build your own knowledge based on the knowledge you already have, stimulating, with the help of the teacher, new learning when these become necessary."
F. "Educational Robotics serve to foster:

1. Interdisciplinary Teaching.
2. Learning for Discovery.
3. Continuous problem solving situations.
4. Laboratory activity.

Educational Robotics can become the easiest way to create an innovative, creative and entertaining learning environment, thanks to the great involvement of the students:

1. Learn to learn.
2. Teamwork.
3. Grounds.
4. Communication.
5. Empowerment.

Educational robotics are interdisciplinary. In fact, at least the following disciplines are involved:

1. Mathematics.
2. Science.
3. Informatics technology.
4. You can assign specific tasks to each student /student and even to those who have difficulty.

The elements that characterize the robotics are:

1. The natural appeal that the robots exert on the boys.
2. Making the learning process more enjoyable and fulfilling, allowing you to build a stimulating path, perfect to motivate even less students in the school context.

3. However, programming is no longer a fine discipline, but also for the learning of different subjects.

Educational Robotics: What Is It?

1. The development and use of learning environments based on robotic technologies. Such environments usually consist of: robot + software + curricular material.
2. A discipline widely acknowledged and appreciated in the context of school dispersion and juvenile inclusion.
3. Using a constructivist didactic or *learning by doing* and experimenting, in fact, by trying and rethinking students, they realize mistakes and correct them by becoming the protagonists of their learning!

Summing
Educational robotics are NOT:

1. It is not the teaching of robotics.
2. It's not the teaching of programming applied to robots.
3. It is not the study of how the android works.

The educational robotics is:

1. Teaching with robotics.
2. Cross-disciplinary and multidisciplinary discipline.
3. Students passing passive users to active subjects."

V. "To explain what is the educational robotics I must first say what is not really: robotics teaching, programming applied to robots, studying how an android works… it's not just that, it's a lot more. The development and use of robotic learning-based learning environments where the child learns more effectively while engaging in a creative process, promoting the taking of an active role in building something that is motivating and interesting from the His point of view. Children learn to analyze problems that do not have a predetermined response and allow them to develop new and creative solutions. This is through a process of experimentation and modeling in which students manipulate external reality, analyze observations made in their internal models, or modify their mental models in order to make them compatible with our observations. This process is influenced by the availability of concrete objects that facilitate the development of specific learning. These objects can be examined, manipulated, displayed, discussed and admired. Learning processes, strategies, information that pupils encounter and use, working and playing with robotic systems are organized not only on an abstract chronological pattern, as is the case with books; the android puts in place a simulation, physical actions, like their symbolic content, are organized on an operational logic; are less abstract, less liberated from a significant socio-anthropological context. The robot is a product of knowledge that is realized during its construction; the likelihood of the actions that the robot does makes it possible to favor an environment of consistency and practical use. Educational Robotics allow the teacher to become a facilitator of the learning process that, instead of or in addition to, illustrates concepts and ideas, proposes to students the recipes on building objects or realizing paths, which he learns with them,

Participating in the manufacturing process. Interaction with autonomous or semi-autonomous programmable machines facilitates listening, self-esteem; it allows interdisciplinary experiences and relational skills to improve (*small group work in problem solving situations*), establishing solid links in the social context and the environment."

L. "Educational Robotics is a didactic methodology that involves the creation of working groups and which, through the design, construction, manipulation of programmable and automated systems, allows each individual to enter the knowledge world as a deep exploration of the meaning Of the concepts, actions and relationships that bind them. Using as a starting stimulus, the playful aspect and challenge concept, seen as a creative stimulus, is able to capture the attention of the boys and, through manipulation of objects, to involve them in the process of self-learning concepts. In fact, for the achievement of the didactic goal, robotics require knowledge not only of the technical characteristics of the components that are used for the creation or programming of the robot, but also of the physical and conceptual relationships that bind the various components between them. This cognitive process is mainly acquired experimentally with knowledge already belonging to the cultural baggage of the children, who are asked to design and implement, through the construction of a technological artifact, a strategy that will solve a given problem Then extend it to a class of problems. This methodology allows students to acquire the ability to establish relationships between theoretical and practical concepts. Boys also become able to apply their knowledge to different aspects of reality by adapting them to the context. Applying the acquired knowledge makes it possible for the student to improve the strategy in relation to failures, thus enabling them to learn how to self-evaluate their learning and cognitive processes; This process allows to improve the ability of attention and concentration, to acquire the ability to identify relationships and relationships between similar concepts. The methodology in this context also allows formulating and documenting logical processes that allow addressing and solving problems of a different nature. In addition to solving problems, the boys are able to establish an emotional relationship with each other and with the teacher who allows an exchange of knowledge; an exchange of knowledge related not only to didactic topics but also to life experiences that help to shape their personality."

M. "Teaching strategy that uses a playful environment in which the pupil builds and program robots, developing problem solving skills and problem setting, creativity and relational, through group work. This process allows us to build a new approach to life. In this process the teacher does not teach, but becomes, together with the pupils, a builder of knowledge. The advantage of robotics is to relate what the pupil designed and programmed with an artifact. The robot moves in the real environment and does things that have been programmed by the pupil through a process of trial and error and discovery that is typical of the game, where creativity and emotion are the main ingredients that make the learning."

4 Conclusion

Why is robotics, when it is educational, helping solve the criticalities of a massive and unattractive use of digital technologies? In the light of the above, some concepts can be enucleated.

Youngsters as adherents to the contemporary world experience the processes of building, organizing, disseminating and transforming knowledge through robotic systems. In this way, the kids are part of their self-determination and, just as rewarding, the activities of using tablets and other digital technologies. In the case of robotics, since the role of students is active and creative, the exercise of thought puts in place reflection processes that obstruct passive use and blind dependence on the technological medium. Integrated group work helps young people mature respect for rules, consideration of the value of the role specificity, the difference in viewpoints. These factors help develop those relational skills that activate an emotional awareness and that can prevent aggressive behavior and exclude them.

The conditions, in turn, pose the problem of the verifiability of robot-activated processes, which, once again, forces the boy to think and review all the steps of the procedures that are in place (e.g., when and in what degree does acquired knowledge have criteria of truth, certainty and effectiveness?) [8].

Choosing the information that robotic experience suggests, their interpretation and their placement within a cognitive system, if guided, helps students to develop their own theory of knowledge and mature.

The relationship of these processes to the pupil's self-consciousness - that is, with conscious and non-perceived perception, having their own cognitive field of acquisition, processing, invention - allows us to formulate a diagnosis of both cognitive difficulties and potentialities;

Teaching with educational robotics then helps teachers to reflect critically on the tools for controlling the fundamentals of various school disciplines: specific language, field of study and application, peculiarities of content, methodology, procedures, theoretical background, consequentiality, verifications, tools and consistency of meaningful and procedural relationships that lie with each other. Robotics facilitate transversally and epistemological contaminations between different disciplines: metaphors transfer, use of narrative segments from other domains, and methodological ideas. In short: it promotes intercultural scientific competence. We mean competence, a basic knowledge that activates a series of acquisitions and therefore, has a metacognitive value.

Since the social image of knowledge as an "encyclopedia" has been replaced by that of "context", knowledge built with robotic systems is more adhering to the anthropology of contemporary formation and, consequently, more easily recognizable, in the sense and in the Use, by the students. The social image of knowledge as an encyclopedia, in fact, implied a precognitive and accumulative behavior for sectors. The second, context-related, enhances the heuristic and strategic function of each learning, as well as the metacognitive sense of procedures, codes, and approaches. In this second perspective, which is most favored by the epistemologically conscious use of robotics at school, the plurality of points of view, languages, theoretical constructs is fundamental. A consequence of this new way of understanding knowledge is that many conceptualizations come to disciplinary fields or areas of experimental research that are very different from those in which they are germinated. For this, it is important to know the dynamic semantics of robot-driven knowledge, i.e., knowing processes and mechanisms, to be able to build "other" knowledge and transfer skills from different fields and times. For a long time we have concluded that, life stories have a strong educational dimension, both for those who tell it and for those who listen to it. Kilpatrick's "*method of projects*" is considered essential for its operational strength, which leads the teacher to move forward to leave room for his students in research. In many words, many of the trained teachers, reflecting on their own laboratory experience, used expressions such as "let the students discover" the good of the ongoing process, put them in the condition of capturing the fascination of the invasion, with the astonishment that conveyed Spirit of cognitive research since ancient times, as evidenced, among others, by Plato and Aristotle. These teachers are thought, Socratic, as non-teachers and have as their purpose the stimulus for the creativity and the concrete understanding of their pupils. The laboratory size of the rest, at all ages, favors the gradual abandonment of that narcissistic that seems to contaminate as a contagious disease the existence of current children, boys and adults. Working together in a robotic project implies providing the group with its own resources, without calculating and misleading doubts about their own protagonist, which in the eventuality will be sanctioned by others. It is the same ongoing practice of the project to select the good attitudes and behaviors appropriate to the "product". In addition, this is worthy of some insight: it is not just a more or less successful artifact, but also rather a cognitive process that translates into meaningful symbolic object. In its complex physicality, it is enclosed by the efforts of many people who are seeing the realization of a working hypothesis, which can then be turned backward and consciously framed in positive outcomes, as in any eventual errors. The philosopher Anassagora recognizes human development through experience, memory, knowledge and technique. However, these would not be possible without the skilled use of hands, which in essence behave as knowledgeable tools and civilization producers. The same thought without their incessant contribution would be poor. In addition, thanks to his hands, the philosopher Aristotle reminds us that we differentiate ourselves from other animals and, instead of adapting ourselves to the environment, we adapt to our living needs.

Robotics today seem to be the most authentic symbol of our specificity: it is not a mere technique to survive in a global world torn by unmistakable local conflicts; But a new humanism that prefigures the awareness of a complex universe. Not only a sign of anxiety without certain answers, but also opportunities to look for new ways open to the future, as in any intelligent game that is not afraid to bet on man and his talents.

References

1. Chassiakos, R., Radesky, J., Christakis, D., et al.: AAP council on communications and media: children and adolescents and digital media. Paediatrics, **138**(5) (2016)
2. Chiong, C., Shuler, C.: Learning: is there an app for that? Investigations of Young Children's Usage and Learning with Mobile Devices and Apps. The Joan Ganz Cooney Centerat Sesame Workshop 2010, New York, NY (2010)

3. Shonkoff, J.P.: Building a new biodevelopmental framework to guide the future of early childhood Policy. Child Dev. **81**(1), 357–367 (2010)
4. Dusi, E.: Bambini, già a un anno con il cellulare. In: La Repubblica, 5 gennaio (2017)
5. Queraltò, R.: La estrategia de Ulisses o ética para una sociedad tecnologica, Sevilla. Doss Ediciones (2018)
6. Broers, A.: The Triumph of Technology. Cambridge University Press, Cambridge (2005)
7. Piro, G.: La robotica educativa: luci e ombre nel panorama europeo e italiano. In: Pedagogika.It, a. XXI, no. 1, pp. 8–18 (2017)
8. Reale, G. (ed.): Plato, Teeteto, in Tutti gli scritti. Milano, Bompiani (2000)

Model of M-Learning by Multimedia Content Delivery from mCloud to Mobile Devices

Danco Davcev[1], Goran Jakimovski[2(✉)], and Snezana Scepanovic[3]

[1] Faculty of Computer Science and Engineering, University Ss Cyril and Methodius, Skopje, Republic of Macedonia
danco.davcev@finki.ukim.mk

[2] Faculty of Electrical Engineering and Information Technologies, University Ss Cyril and Methodius, Skopje, Republic of Macedonia
goranj@feit.ukim.edu.mk

[3] Mediteranian University, Podgorica, Montenegro
snezana.scepanovic@unimediteran.net

Abstract. Integration of mobile devices in the multimedia delivery systems provides the users (learners or business people) with access to multimedia content outside the classroom (workplace). It allows them to easily store, record and deliver multimedia content in real-time. Process of delivering multimedia content to the users requires more computational resources than mobile device can provide. In order to provide users with multimedia content that is suitable for their mobile devices and according to their needs we introduce the mobile cloud (mCloud) computing environment as paradigm that is ideal to overcome these problems. The proposed interactive mCloud system should provide high scale collaboration and interaction between the users. The system is designed for delivery of multimedia learning content to the users according to the user's cognitive style. The content is adapted according to the context-aware network conditions. In this paper, we provide some experimental results based on experience of 90 users (students) that participated in the m-learning cooperative process. After completing the course of Database Systems, the users estimated the Quality of Experience (QoE) by using a Learning Scenario Questionnaire.

Keywords: Learning · Multimedia content · Mobile cloud computing
User profile · QoE

1 Introduction

Throughout the traditional learning, educational process takes place in classroom, where professors and students are meeting face to face at the same time and in the same place. During the lectures professors present the learning material and interact with the students through sequence of questions and answers, which allows them to give rough assessment on level of achieved student's knowledge. The learning materials are limited with certain multimedia objects that the professors have previously arranged and it is not possible to adapt that content to individual student's learning requirements and cognitive style.

M. E. Auer et al. (eds.), *Teaching and Learning in a Digital World*,
Advances in Intelligent Systems and Computing 715,
https://doi.org/10.1007/978-3-319-73210-7_92

Introduction of the mobile and online learning system has provided an educational environment that can be accessed from anywhere and anytime and adapts to the student's learning process and keeps the continuity of life-long learning.

Mobile devices as service platforms for distance learning are considered as excellent educational tool for sharing multimedia learning content between learners and teachers. Particular benefits of mobile learning systems are easy portability, real-time learning, interaction and collaboration in the process of m-learning [4]. However, mobile devices, in order to provide these advantages, are facing certain limitation, computational capacity and limited battery power. Existing mobile learning systems, which are based on traditional teaching and assessment principles, provide one and the same learning content to every student [5]. Certain progress can be observed with the introduction of the interactive mobile live video learning system in a Cloud Computing environment [6]. In paper [6], instructors' video presentation was captured and stored on a private Cloud and students using GPRS/WiFi connectivity on their mobile device are able to progressively download and/or play the video. Existing M-Learning environments still experience diverse technological and Quality of Service (QoS) problems, such as: delivery of different kind of multimedia materials and adaptation of the learning material to individual student needs. Another important challenge is the real-time interaction between students and the environment. This is difficult to achieve because of the context-aware bandwidth limitations.

Main advantages of the cloud computing in general are: centralized storage, memory and data processing providing increased capacity for multimedia distribution. The existing intelligent multimedia delivery, based on Cloud computing infrastructure called EVE (Elastic Video Endpoint), provides dynamic provisioning of multimedia content [1]. Similarly, there is positive improvement in M-Learning systems with the emergence of EBTIC's international iCampus (intelligent campus) initiative, which is delivering customized and adapted learning to individuals via mobile devices [7]. Its intelligent engine uses the Learning-Assessment-Communication-Analysis (LACA) model. The dynamic engine generates content based on the assessment of learners' effectiveness and outcomes, rather than time spent on learning [7]. Benefit of delivery of multimedia content in cloud computing for the m-learning has already provided enhanced learning and more transparency and collaboration in the education [2, 3].

In this paper, we propose a system for delivery of multimedia learning content to the users according to the user's cognitive style. The content is adapted according to the context-aware network conditions. We provide some experimental results based on experience of 90 students that participated in the M-Learning cooperative process. After completing the course of Database Design Systems, the users estimated the Quality of Experience (QoE) by using an adapted Learning Scenario Questionnaire, [8]. The paper is organized as follows. Section 2 describes the proposed model of m-learning system for multimedia delivery in mCloud computing environment. Section 3 describes experienced quality of m-learning evaluation and Sect. 4 concludes the paper.

2 Model of M-Learning System for Multimedia Delivery in mCloud Computing Environment

Considering the benefits of the Cloud Computing, our proposed model for multimedia delivery in mCloud goes beyond the existing solutions by providing adaptive delivery of multimedia learning content. Our system architecture is given in Fig. 1.

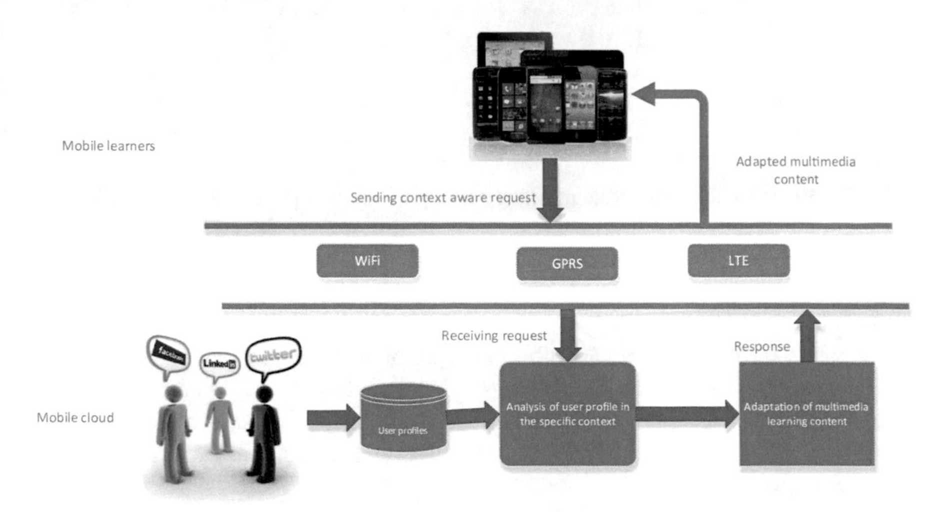

Fig. 1. Architecture of multimedia delivery in mCloud computing environment

All of the requests from user's mobile devices and heavy-duty processing tasks will be executed in the mCloud and the response is the appropriate type of multimedia learning content. The main role of m-Cloud computing is to offload and reduce the workload of mobile devices by exploiting the remote multimedia processing resources in the mCloud.

The user's cognitive style estimation is provided by the interaction with the user's social network profiles. In order to prepare adapted learning material for the student, our system continuously monitors the network's capacity and adapts the content to the appropriate format. This means that the content will be aware of congestions and be adapted accordingly by the mCloud. The multimedia content is sent to the content adaptation engine (Fig. 2) and delivered to the mobile device.

In order to increase the quality of delivering the multimedia content, current systems try to improve the network parameters such as delay, jitter and packet loss, as they are important parameters to be considered when using the Quality of Service (QoS) metric.

Delivery of multimedia content is highly dynamic and the innovative mobile technologies that are introduced can increase the capabilities. When new media content is offered by the teacher, students can change their preferences (ways that they want to learn), which makes a feedback loop. In order to have an efficient learning system that will increase Quality of Learning (QoL), it is necessary to continuously adapt the feedback from the students. This adapted feedback is kept and used by the social networks

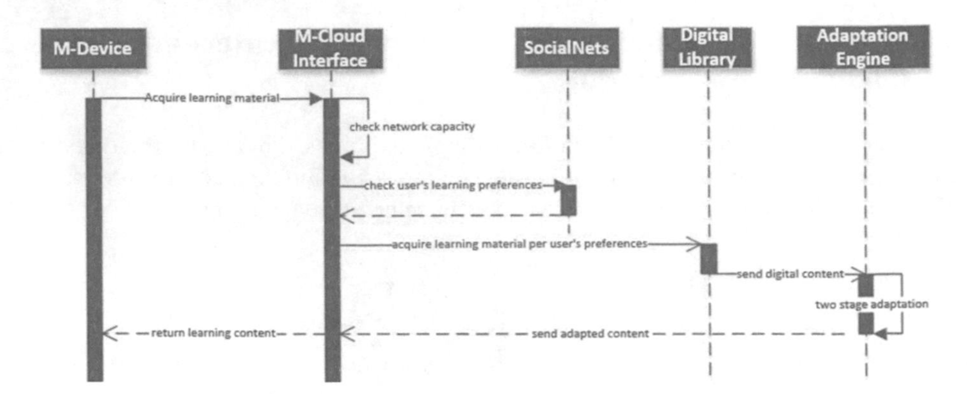

Fig. 2. Sequence diagram of the message passing between objects of the architecture

(SocialNets) to update user's profile (Fig. 2).This means that the SocialNets object in Fig. 2 is continuously adapted by the students (users). The main characteristics of the user's profile belong to three groups: audio, visual and mixed learners. This can be used for more precise profiling. There are different types of user profiling, as described in [9], where they use Ontology Web Language (OWL) and RDF/XML format to get information about the user and his preferences. Ontologies can be used to define the group in which a user profile belongs, whether it is individual, canonical, explicit, long-term or short-term grouping.

In our case, a user's profile can belong to more than one profiling group, so OWL can help us in defining and describing the links between different objects (users, network characteristics, type of access and etc.) in the system. As described earlier, M-learning content is adapted in two steps, one step using user's preferences in learning and the other step using the network capacity (network type). The first step is modeled using long-term individual

```
<?xml version="1.0"?>
<rdf:RDF xmlns:rdf="http://www/w3.org/1999/02/22-rdf-syntax-ns#"
    xmlns:rdfs="http://www.w3.org/2000/01/rdf-schema#"
    xmlns:mlearn="http://www.m-learn.org.mk/students#"
    xmlns:base="http://www.m-learn.org.mk/students">
<rdf:Description rdf:ID="Net1">
    <mlearn:hasName>WiFi</mlearn:hasName>
    <mlearn:typeData>All</mlearn:typeData>
</rdf:Description>
<rdf:Description rdf:ID="student1">
    <mlearn:hasName>Aleksandar Aleksandrovski </mlearn:hasname>
    <mlearn:yearOfStudy>3</mlearn:yearOfStudy>
    <mlearn:mlearnerType>visualizer</mlearn:mlearnerType>
</rdf:Description>
<rdf:Description rdf:access="172346350">
<rdf:whoAccessed rdf:resource="#student1"/>
<rdf:netType rdf:resource="#Net1"/>
<rdf:dateTimeAccess>2016-03-12T10:07:12.114z</rdf:dateTimeAccess>
<rdf:subject rdf:resource"#dataBaseIntro" rdf:lesson="03"/>
</rdf:Description>
```

Fig. 3. RDF/XML example of the system

profiling, whereas the second step is modeled using short-term individual profiling. In Fig. 3 we show an example for the student Aleksandar Aleksandrovski, who is in the third year of study and prefers visual content (visualizer). Next, we define the network access of the student, which is "#Net1", the date of access, which is "2016-03-12T10:07:12.114z" and the subject which is accessed is "#dataBaseIntro".

3 Experienced Quality of M-Learning Evaluation

This paper focuses on the real-time interaction that provides collaborative and adaptive learning environment for the students, which takes into account context-aware conditions and their different cognitive perception for the multimedia content. Mayer in [8] has described multimedia learning as a field which has matured over the past decade; it proposes that M-Learning courses should be based on a cognitive theory. The proposed model has an open issue, which is: how to measure the cognitive dimension to the individual student, in order to deliver multimedia M-Learning material that corresponds to the face-to-face knowledge communication. User Interaction (UI) between the participants is achieved by using the social network interaction forum.

However, QoS parameters are, in general, measuring the accuracy of networked data delivery that is not sufficient to describing the actual experience of the user. Quality of Experience (QoE) is used as an overall acceptability of an application or service, as perceived subjectively by the end-user and represents multidimensional subjective concept that is not easy to evaluate [1]. The relationship between QoE and QoS is non-trivial. It is a real research challenge to investigate and analyze what additional factors are influencing the user's perception of quality for delivery of multimedia content in Cloud computing environment. The research presented in this paper is the first step in that direction.

We propose an adapted questionnaire for our research area of Database systems which is based on the Learning Scenario Questionnaire adapted from [8]. By using this questionnaire in our experiments with 90 users with mobile devices connected to the multimedia learning content in mCloud environment, we have provided our experimental M-Learning results. The proposed QoE model for estimating the multidimensional metric has been demonstrated to the students using the university distance learning web 2.0 mCloud portal. Students have been enrolled to learn the course Database Systems, which covers the Extended Entity Relationship (ER) and UML modelling as well as SQL query data manipulation class. After completing the course of Database Systems, the 90 users estimated the Quality of Experience (QoE) by using a Learning Scenario Questionnaire for estimating preferences in five learning situations, which is presented in Fig. 4.

Learning Scenario Questionnaire:

Q1) Which format do you prefer for describing the Database system?
1) A paragraph (text) description
2) A graphic diagram description
3) A video sequence combining graphics and text

Q2)Which format do you prefer for learning SQL and DB querying?
1) An essay describing what happens with each SQL command
2) A labeled diagram showing the status of each execution
3) A diagram accompanied with text description

Q3)Which format do you prefer for describing the modeling process of an EER and/or UML?
1) Audio directions that include directions that show you what to draw first
2) A map showing how to draw the entities, relations and how they are connected
3) Video directions that describe the process step by step by drawing and connecting

Q4) Which format do you prefer for learning data aggregation with SQL statements?
1) A textual list of words
2) A diagram showing the steps
3) A moving presentation accompanied with example tables and the result from the aggregation

Q5)Which format do you prefer for describing the results of an executed SQL query?
1) A list of resulting tables
2) A graphical chart, diagram or pie
3) A step by step graphical representation of the result with text

Fig. 4. Learning scenario questionnaire

We provided experimental results of using this system by the group of 90 users that participated in the M-Learning process. This evaluation has been conducted by using the discussion forum, which allowed each user independently to answer the questions from the Learning Scenario Questionnaire. The mCloud learning portal is consisted of social network forum that provides interactive communication that was used to conduct quality evaluation and receive feedback from the students. Complete review of the results from the completed Learning Scenario Questionnaire is given in Fig. 5.

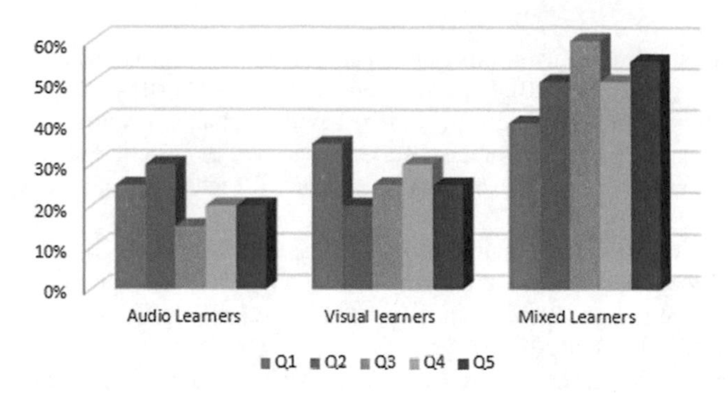

Fig. 5. Results from Learning Scenario Questionnaire

These results confirm that majority of the students are mixed learners; they prefer to receive graphical/image accompanied with a textual description of the image. The peak which corresponds to question 3 from the Learning Scenario Questionnaire emphasizes that the mixed learners prefer to receive visual instructions in form of ER and UML diagrams that is mixed with audio or text description. Similar preferences for mixed learning objects (in connection with the question 2 from the Learning Scenario Questionnaire) highlight the user's preference of labelled diagrams.

The proposed m-learning model for multimedia content delivery from mCloud to mobile devices uses the benefits of the cloud computing technology to deliver the learning materials. It enables the students to learn their topics of interest transparently and immediately using various devices whenever and wherever they want.

4 Conclusion

In this paper we have proposed a model of m-learning by multimedia content delivery from mCloud to mobile devices. The proposed interactive mCloud system should provide high scale collaboration and interaction between the professor and students (students and business people) in direction of increasing the quality of learning and business cooperation. Our system is designed for delivery of multimedia learning content to the users according to the user's cognitive style. The content is adapted according to the context-aware network conditions.

We provided experimental results of using this system by a group of 90 users participated in the m-learning process. After completing the course of Database Systems the users estimated the Quality of Experience (QoE) by using a Learning Scenario Questionnaire. On the basis of the experimental results, we concluded that cognitive style and user interaction, with hands on practice, provides increased educational benefit.

References

1. Agboma, F., Liotta, A.: Quality of experience management in mobile content delivery systems, pp. 85–98. Springer 2010 (2010). https://doi.org/10.1007/s11235-010-9355-6
2. Alabbadi, M.M.: Mobile learning (mLearning) based on cloud computing: mLearning as a service (mLaaS). In: Proceedings of the UBICOMM 2011: The Fifth International Conference on Mobile Ubiquitous Computing, Systems, Services and Technologies, IARIA, pp. 296–302 (2011)
3. Blair, A., et al.: Cloud based dynamically provisioned multimedia delivery: an elastic video endpoint. In: Proceedings of the Third International Conference on Cloud Computing, GRIDs, and Virtualization, IARIA, Cloud Computing 2012, 22–27 July 2012, Nice, France, pp. 260–265 (2012). ISBN: 978-1-61208-216-5
4. Chen, S., Lin, M., Zhang, H.: Research of mobile learning system based on cloud computing. In: Proceedings of the International Conference on e-Education, Entertainment and e-Management ICEEE, 27–29 December 2011, pp. 121–123 (2011). ISBN: 978-1-4577-1381-1
5. Hirsch, B., Ng, J.W.P.: Education Beyond the Cloud: Anytime-anywhere learning in a smart campus environment. In: Proceedings of the 6th International Conference of Internet Technology and Secured Transactions, 11–14 December 2011, Abu Dhabi, United Arab Emirates, pp. 718–723 (2011). ISBN: 9781457708848
6. Saranya, M., Vijayalakshmi, M.: Interactive mobile live video learning system in cloud environment. In: Proceedings of the IEEE-International Conference on Recent Trends in Information Technology, ICRTIT, June 2011, pp. 673–677 (2011). ISBN: 978-1-4577-0588-5
7. Mayer, R.E.: Multimedia Learning. Cambridge University Press, New York (2001)
8. Mayer, R.E., Massa, L.J.: Three facets of visual and verbal learners: Cognitive ability, cognitive style and learning preference. J. Educ. Psychol. **95**, 833–846 (2003). https://doi.org/10.1037/0022-0663.95.4.833
9. Pena, P., Hoyo, R., Vea-Murguia, J.: Collective knowledge ontology user profiling for twitter. In: International Conference for WI and IAT, Atlanta, USA (2013)

Assess and Enhancing Attention in Learning Activities

Dalila Durães[1,3(✉)], Javier Bajo[1], and Paulo Novais[2]

[1] Department of Artificial Intelligence, Technical University of Madrid, Madrid, Spain
d.alves@alumnos.upm.es, jbajo@fi.upm.es
[2] Algoritmi Center, Minho University, Braga, Portugal
pjon@di.uminho.pt
[3] CIICESI, ESTGF, Polytechnic Institute of Porto, Felgueiras, Portugal

Abstract. The rapid progress of technologies has enabled the development of innovative environment in learning activities when the student used computer devices with access to Internet. The goal of this paper is to propose an ambient intelligent (AmI) system, directed at the teacher that indicates the level of attention of the students in the class when it requires the use of the computer connected to the Internet. This AmI system captures, measures, and supervises the interaction of each student with the computer (or laptop) and indicates the level of attention of students in the activities proposed by the teacher. When the teacher has big class, he/she can visualize in real time the level of engagement of the students in the proposed activities and act accordingly when necessary. Measurements of attention level are obtained by a proposed model, and user for training a decision support system that in a real scenario makes recommendations for the teachers so as to prevent undesirable behaviour and change the learning styles.

Keywords: Ambient intelligent system · Learning activities
Attentiveness · Learning styles · Innovative environment

1 Introduction

The rapid progress of wireless communication and sensing technologies has enabled the development of innovative environment in learning activities. For this reason making learning systems innovative and smart has been the objective of many researchers in both the fields of computer science and education [1].

In the field of computer science an innovative environment is a digitally augmented physical world where sensor-enabled and networked devices work continuously and collaboratively to make lives of inhabitants more comfortable. Indeed, tremendous advances in smart devices, wireless mobile communications, sensor networks, pervasive computing, machine learning, robotics, middleware and agent technologies, and human computer interfaces have made the dream of smart and innovative environments a reality.

M. E. Auer et al. (eds.), *Teaching and Learning in a Digital World*,
Advances in Intelligent Systems and Computing 715,
https://doi.org/10.1007/978-3-319-73210-7_93

In the field of education learning theories provide insights into the very complex processes and factors that influence learning and give precious information to be used in designing instruction that will produce optimum results. The learning models are designed in order to supply to the students with practice, evaluation and improvement procedures which will adjust the model [2]. The teaching process first requires that the instructor creates a pedagogical design of the objectives and determines the content to be taught. Second, a pre-assessment is used to determine learning abilities. Third, pedagogical procedures are used when teaching is initiated. Finally, assessment is applied to determine what learners have achieved, and, according to the assessment results, instructors should use feedback to determine the cause of ineffective instruction [2, 3].

It's crucial to improve the learning process and to mitigate problems that might occur in an environment with learning technologies. Besides, each student has its own particular way of assimilating knowledge, that is, his learning style. Learning styles specify a student's own way of learning. Someone that has a specific learning style can have difficulties when submitted to another learning style. When the given instruction style matches the student's learning style, the process is maximized which guarantees that the student learns more and more easily. Attention is a very complex process through which one individual is able to continuously analyze a spectrum of stimuli and, in a sufficiently short amount of time, chose one to focus on. In most of us, who can only focus on a very reduced group of stimuli at a time, this implies ignoring other perceivable stimuli and information.

However, for various reasons, students may not be predisposed to learning. In this sense, and in bigger classes, it is important that the teacher has instruments to point out potential distractions (namely in what concerns the applications being used by the students) that may indicate a lack of predisposition to learning.

This paper deals with the issue of AmI system, with the aim of proving a non-intrusive and non-invasive smart environment, reliable and easy tool that can be used freely in schools, without changing or interfering with the established working routines of students.

This paper is organized as follows. In the next Section the state of art of attention concept, learning styles, and AmI system where scientific literature is reviewed. Section 3 contains the proposed architecture, Sect. 4 presented the methodology applied with data acquisition and in Sect. 5 results are presented. Finally, in Sect. 6 some discussions and conclusions of this work are presented.

2 State of Art

In educational environments, attention is considered a fundamental factor in the evolution and success of the student. If the student is not concentrating and paying attention to what is being taught, he will not capture information that is being provided and consequently the academic course will be compromised.

Attention is a resource that allows the human being to be focused on a situation and to be able to ignore non-priority information. As happens with performance, several factors can influence attention, like, stress, mental fatigue, anxiety, emotion, new environments, and human health. Besides these factors, the advancement of technology has been a real problem which has increased the lack of attention. With the emergence of the smartphone that provides new and varied information in real time and new ways of communication, people's attention is easily captured and the task that was meant to be done is left out [4, 5].

2.1 Attention Concept

In educational environments, attention is considered a fundamental factor in the evolution and success of the student. If the student is not concentrating and paying attention to what is being taught, he will not capture information that is being provided and consequently the academic course will be compromised.

Generally there is no universally accepted definition of attention because there is a diversity of disciplines that are focused on it. In the past, only psychologists studied attention, however in present days attention is highly important for other fields like philosophy, chemistry, anatomy, and even computational science [6].

The concept of attention may be defined as the transforming of a huge acquired unstructured data set into a smaller structured one where the main information is preserved. In Computer Science, attention means that there is a filtering input space that selects the most important data in processing and this is a key mechanism of behavioral control for tasks, which is related to planning, decision making, and preventing new situations, however there are limited computation capabilities [6, 7].

Attention means focusing on clear thinking, among one of several subjects or objects that may capture mind simultaneously. Attention implies the concentration of mental powers upon an object by close or careful observing or listening, which is the ability or power to concentrate mentally.

2.2 Learning Styles

In order to maximize the learning is also important to consider the concept of learning styles. A learning style is the method that allows an individual to learn best. Different people learn in different ways, each one preferring a different learning style.

Learning style not only specifies how a student learns and likes to learn, but it can also help a teacher to adapt to individual students, so that they might learn successfully. When the teacher's methodologies do not support a specific learning style, the student will find it more difficult to learn and acquire knowledge. Everyone has a mix of learning styles, but some people may find that they have a dominant style of learning. Others may find that they have different learning styles in different circumstances.

Learning styles can be defined as cognitive, affective, and physiological features that serve as relatively stable indicators of how learners perceive interaction and respond to their learning environments [8].

There are several models developed by several authors that try to represent the way people learn [9]. Previous research suggests that, in the context of learning activities, different learning styles can influence learning performance [10, 11]. Learning styles are considered one of the more important factors influencing learning [12].

Some researchers have argued that learning style is also a suitable indicator of potential learning success because it provides information about individual differences in learning preferences and information-processing [10, 13].

However the field of learning styles is a very controversial field, because there are some authors that consider that scientific support for learning styles theories is lacking [14].

2.3 AmI System

Important features of AmI system or smart environments are that they possess a degree of autonomy, adapt themselves to changing environments, and communicate with humans in a natural way. An AmI system is a tool in which technology is embedded, hidden in the background, sensitive, adaptive and responsive to the presence of people and objects. This system also preserves security and privacy while using information when needed and with an appropriate context [15].

In our case we proposed an AmI system that aim to support the student learning process. This system is adaptability, which means that consider the student's knowledge, background, interest, goals, targets and/or choices.

3 The Proposed Architecture

Once information about the individual's attention exists in these terms, it is possible to start monitoring attentiveness in real-time and without the need for any explicit or conscious interaction. This makes this approach especially suited to be used in learning activities in which students use computers, as it requires no change in their working routines. This is the main advantage of this work, especially when compared to more traditional approaches that still rely on questionnaires (with issues concerning wording or question construction), special hardware (that has additional costs and is frequently intrusive) or the availability of human experts.

The present work adds a new feature to this previously existing framework, by providing the learning styles theory, where the applications of different type of exercises obtained different results of level of attentiveness. It constitutes a much more precise and reliable mechanism for attention monitoring, while maintaining all the advantages of the existing system: nonintrusive, lightweight, and transparent.

This work was detail in [16], but briefly as show in Fig. 1, it is possible to collected data that describes the interaction with both the mouse and the keyboard in the devices in which students work have software that generates raw data, which store the raw data locally until it is synchronized with the web server in the cloud. The Mouse and Keyboard collected data captured information describing the behavioral patterns of the students', and receiving data from events mouse and keyword students'. This layer encodes each event with the corresponding necessary information (e.g. timestamp, coordinates, type of click, key pressed). These data are further processed, stored and then used to calculate the values of the behavioral biometrics. Mouse movements can also help predict the state of mind of the user, as well as keyboard usage patterns.

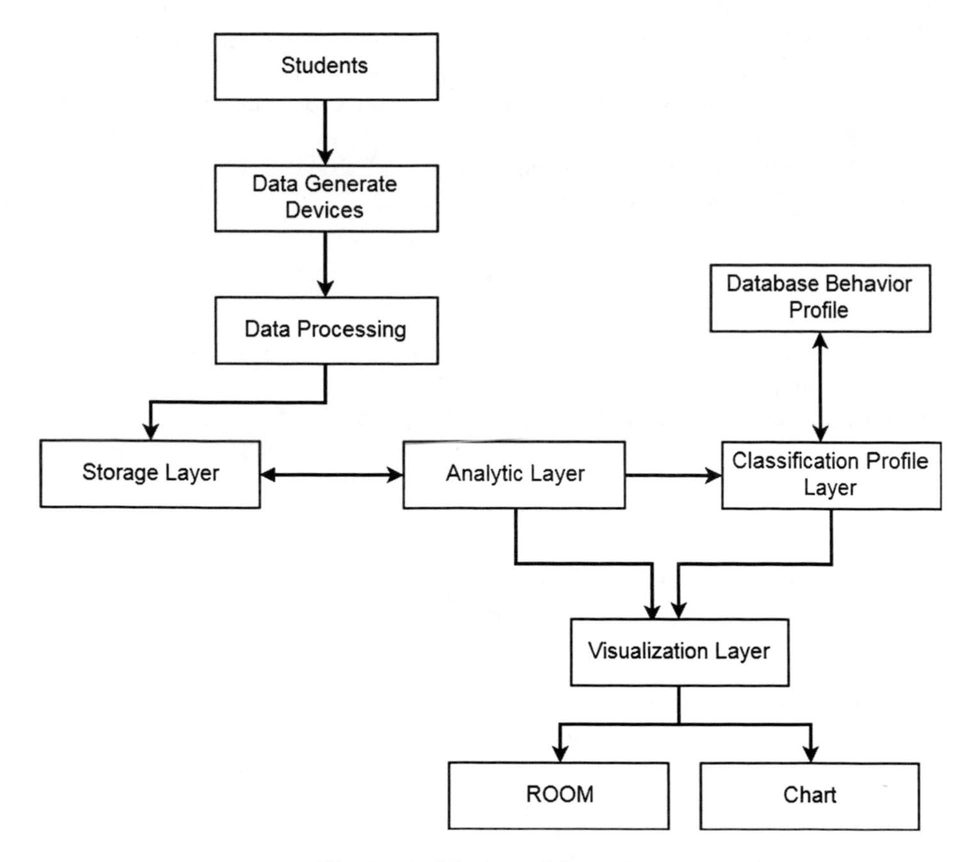

Fig. 1. Architecture of the system.

After the raw data was stored in a data store engine, the analytic layer is responsible to process the data received from the storage layer in order to be evaluated those data according to the metrics presented. It's important that in this process some values are filtered to eliminate possible negative effects on the analysis (e.g. a key pressed for more than a certain amount of time). The system receives this information in real-time and calculates, at regular intervals, an estimation of the general level of performance and attention of each student.

In the classification profile layer the indicators are interpreted. Based on data from the attentiveness indicators and building the meta-data that will support decision-making. When the system has a sufficiently large dataset that allows making classifications with precision, it will classify the inputs received into different attention levels in real-time, creating each student learning profile. With these results it is possible to obtain a profile of the learning style.

Finally, the actual students' attention information is displayed in the visualization layer, and can be used to personalize instruction according to the specific student, enabling the teacher to act differently with different students, and also to act differently with the same student, according to his/her past and present level of attention.

In visualization layer it's possible to obtain some graphical modules that allow showing the information in an intuitive way to the user. These graphical modules stand out the user interface that allows the teacher to control the application. This graphical module user interface is composed by a module that allow the creation of charts (CHART) and the layer that allows the creation of virtual classrooms (ROOM) so that the teacher may view intuitively the students behavior.

4 Methodology

This work was applied on a vocational course while performing an activity based on Adobe Photoshop at the high school of Caldas das Taipas, Guimarães, Portugal. We want to determine how the class reacts during the lessons and the effect on mouse and keyboard dynamics, and attention level.

For this purpose one group of 22 (9 girls and 13 boys) students were selected to participate in this experience. Their average age is 17.6 years old (SD = 1.4 years). The experiment was applied in four different lessons, where they have access to an individual computer and 100 min to complete the task. Students received, at the beginning of the lessons, all necessary data with the goals of the task.

To quantify attentiveness the following methodology was followed. Apart from capturing the interaction of the students with the computer, the monitoring system also registers the applications with which students are interacting. Attention is calculated at regular intervals, as configured by the teacher (e.g. five minutes). The teacher may also want to assess, in real-time or a posteriori, the evolution of attention of the whole class.

In order to determine the learning style of each student, four different exercises were applied in four different days where the room had similar conditions in terms of lightning, temperature and humidity. The exercises applied were the following: on the first day a video exercise without audio; on the second day, an exercise only with images; on third day, an exercise only with text; and on the fourth day an exercise only with audio. In the end of each class, the exercise was saved in order to be assessed by the teacher.

4.1 Features Extraction

The process of feature extraction starts with the acquisition of interaction events, which is carried out by a specifically developed application that is installed in each of the computers, laptops or tablets. The first stage in the life cycle of the proposed system takes place in the data generating devices, which was designed and implemented using a logger application.

The data collected by the logger application characterizing the students' interaction patterns is aggregated in a server to which the logger application connects after the student logs in. The privacy of the students is ensured, since the necessary data that is collected in the registration process are an ID that does not identify the student, password, and gender. Furthermore, the privacy issues of the system are assured, since the teacher will only have access to the final results on the level of attention.

The Mouse and Keyboard Sensing layers are responsible for capturing information describing the behavioral patterns of the students while interacting with the peripherals [17].

5 Results

During the lessons the monitoring system was used to assess the interaction of the students with the computer and to quantify their level of attentiveness as well.

On each lesson the level of attention of each student was quantified. However, at the beginning it is necessary that the teacher define the task-related applications that the students will use during the class. For that he/she uses a graphical interface to set rules such as "starts with Photoshop" or "Contains the word Photoshop" which are then translated to regular expressions that are used by the algorithm to determine which applications are and are not work-related [16]. In this sense it is necessary to measure the amount of time in each interval, which the student spent interacting with task-related applications. By default, applications that are not considered task-related are marked as "others" and count negatively towards the quantification of attention. The teacher may also determine the regular intervals at which attention is calculated.

Figure 2 shows the output of the evaluation of attention worked task-related of a number of students in the four different lessons.

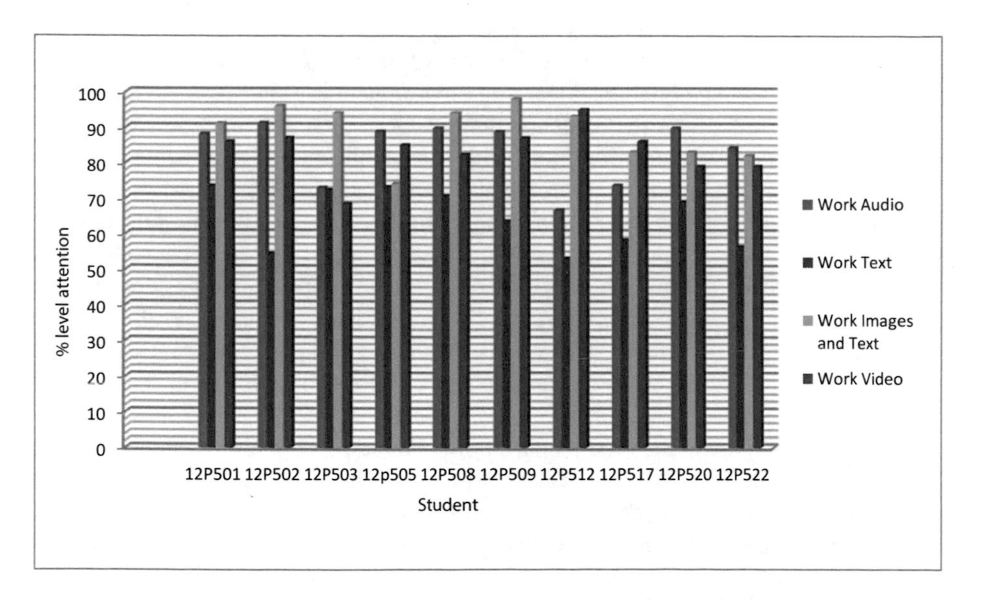

Fig. 2. Detail of evaluation of attention of the students in the four different lessons.

The first lesson was a video exercise without audio; the second lesson, was an exercise only with images; the third lesson was an exercise only with text; and the fourth lesson was an exercise only with audio.

In each of these lessons it's analyzed the level of attention, which is measuring with the work task-related. This is an example of a set of students, but the teacher had access to all students and the global of the class, which allows the teacher to assess the temporal evolution of attention. These results consider the entire length of a class and give the percentage of time spent in task-related or other applications, for each student.

6 Discussions and Conclusions

The main goal of this paper was to present an AmI system approach that analyzed the interaction of student's in learning activities using technologies connected to the Internet. In this case, a specific subject was focused (Adobe Photoshop) and it was analyzed in four different lessons, using four different learning style approach. For this case it was observed the performance of the class and each student. An example of the results of one student was showed and we can observe that these students for the same subject react differently depend on the learning style applied.

In general, the lesson that obtained better results is the lesson where it was applied Text and Images. However the student 12P505, have a better result for the lesson where the exercise was video. In the opposite, the lesson that the level of attention was worst was the exercise where it was applied only text. Although, the student 12P503 have the worse result in the lesson where the exercise applied was video.

These indicators are useful to predict the behavior of a student and identify potential problems in the course of their learning.

AmI system make possible the enhanced of learning/teaching processes. The architecture of an ambient intelligent learning environment is proposed to address these issues, especially to monitor the students' attentiveness students in learning activities, through the use of a developed log tool. With this architecture it is possible to detect those factors dynamically and non-intrusively, making it possible to foresee negative situations, and taking actions to mitigate them. In this case the door is then open to allow to analyze students' profile, taking into account their individual characteristics, learning styles, and to propose new strategies and actions, minimizing issues such as stress, anxiety, and new environments, which can influence students' results and are closely related to the occurrence of conflicts. Moreover it's possible to maximize the performance and attentiveness since the teacher is informed about the behavioral of each student.

Enlarging this study to the use of smartphones and tablets, taking advantage of their new features such as several incorporated sensors, and high resolution cameras, is the next step that may allow a wider characterization of the student, making it possible to enhance the learning experience, though better recommendation and personalization.

Acknowledgements. This work has been supported by COMPETE: POCI-01-0145-FEDER-007043 and FCT – Fundação para a Ciência e Tecnologia within the Project Scope: UID/CEC/00319/2013.

References

1. Hwang, G.J., Chang, H.F.: A formative assessment-based mobile learning approach to improving the learning attitudes and achievements of students. Comput. Educ. **56**(4), 1023–1031 (2011)
2. Eggen, P., Kauchak, D.P.: Educational Psychology: Classroom Connections. Merrill, New York (1992)
3. Hopkins, K.D.: Educational and Psychological Measurement and Evaluation, 8th edn. Allyn and Bacon, Boston (1998)
4. Pimenta, A., Carneiro, D., Novais, P., Neves, J.: Detection of distraction and fatigue in groups through the analysis of international patterns with computers. In: Intelligent Distributed Computing VII, pp. 29–39. Springer International Publishing (2015)
5. Mancas, M.: Attention in Computer Science - Part 1. News and insights from EAI community (2015). http://blog.eai.eu/attention-in-computer-science-part-1/. Accessed 31 Dec 2016
6. Mancas, M.: Computational Attention Toward Attentive Computers. Press Universitaire de Louvain (2007)
7. Tamiz, M., Karami, M., Mehorabi, I., Gidary, S.S.: A novel attention control modeling method for sursor selection based on fuzzy neural network learning. In: First RSI/ISM International Conference on Robotics and Mechatronics (ICRoM) (2013)
8. Keefe, J.W.: Learning style: an overview. In: NASSP's Student Learning Styles: Diagnosing and Prescribing Programs, pp. 1–17 (1979)
9. Morgan, R., Baker, F.: VARK analysis and recommendations for educators. In: McBride, R., Searson, M. (eds.) Proceedings of Society for Information Technology & Teacher Education International Conference 2013, pp. 1381–1385. Association for the Advancement of Computing in Education (AACE), Chesapeake (2013)
10. Smith, P.L., Ragan, T.J.: Instructional Design. Wiley, New York (1999)
11. Bybee, R.W., et al.: Science and Technology Education for the Elementary Years: Frameworks for Curriculum and Instruction. The National Center for Improving Science Education (1989)
12. Ford, N., Chen, S.Y.: Individual differences, hypermedia navigation and learning: an empirical study. J. Educ. Multimed. Hypermedia **9**(4), 281–311 (2000)
13. Kolb, D.: Experiential Learning: Experience as the Source of Learning and Development. Prentice-Hall Inc., New Jersey (1984)
14. Willingham, D.T., Hughes, E.M., Dobolyi, D.G.: The scientific status of learning styles theories. Teach. Psychol. **42**(3), 266–271 (2015)
15. Weber, W., Rabaey, J.M., Aarts, E.: Ambient intelligence, pp. 1–2. Springer (2005)
16. Durães, D., et al.: Detection of behavioral patterns for increasing attentiveness level. In: International Conference on Intelligent Systems Design and Applications. Springer, Cham (2016)
17. Durães, D., Jiménez, A., Bajo, J., Novais, P.: Monitoring level attention approach in learning activities. Advances in Intelligent System and Computing, vol. 478, pp. 33–40. Springer, Cham (2016)

The General Model of the Cloud-Based Learning and Research Environment of Educational Personnel Training

Mariya Shyshkina(✉)

Institute of Information Technologies and Learning Tools of NAES of Ukraine, 9 M.Berlynskoho St., Kyiv, Ukraine
shyshkina@iitlt.gov.ua

Abstract. The article highlights the promising ways of providing access to the cloud-based learning and research software in higher educational institutions. It is emphasized that the cloud computing services implementation is the actual trend of modern ICT pedagogical systems development. The analysis and evaluation of existing experience and educational research of different types of software packages use are proposed. The general model of formation and development of the cloud-based learning and research environment of educational personnel training is substantiated. The reasonable ways of methods selection on the basis of the proposed model are considered and the prospects for their use in educational systems of higher education are described. The analysis and assessment of the prospects of the cloud-based educational and research environment development is fulfilled.

Keywords: Cloud computing · Cloud-based learning environment
Cloud services · Design · Model · Openness · Flexibility

1 Introduction

1.1 The Problem Statement

Nowadays, innovative technological solutions for learning environment organization and design using cloud computing (CC) and ICT outsourcing have shown promise and usefulness [1–4]. So, the modelling and analysis of the processes of the cloud-based learning environment formation, elaboration and deployment in view of the current tendencies of ICT advance have come to the fore. Insufficient attention to these issues may have the negative impact on the level of ICT competence of lecturers and academic staff, organization of their educational and research activities. There is a need to produce some general framework of learning environment formation and design that may be detailed through the system of the models of its parts so as to elaborate and evaluate different CC based learning components, to consider and compare the most advisable methods of their design and delivery.

M. E. Auer et al. (eds.), *Teaching and Learning in a Digital World*,
Advances in Intelligent Systems and Computing 715,
https://doi.org/10.1007/978-3-319-73210-7_94

1.2 The State of the Art

According to the recent research [2, 4, 6, 9, 14] the problems of cloud technologies implementing in higher educational institutions to provide software access, support collaborative learning, research and educational activities, exchange experience and also project development are especially challenging. The formation of the cloud-based learning environment is recognized as a priority by the international educational community [7], and is now being intensively developed in different areas of education [6, 10, 14].

According to IDC, at the end of 2016, two-thirds of IT executive directors of the 500 largest companies (rated Financial Times) reported that they considered digital transformation as the main objective of corporate strategy. Under the "digital transformation" the gradual transfer of IT infrastructure to the corporate cloud and then partly or completely to the external provider cloud to create more efficient business, accelerate the process of innovation, to achieve cost reduction, improve production safety, etc. is implied [5, 13].

Growing demand for cloud services is reflected by the following figures: the share of those companies in the world that are using the cloud solutions increased from 26% in 2015 to 59% in 2016, representing 127% of annual growth; while in Ukraine this figure of annual growth was 26% (an increase from 38% in 2015 to 48% in 2016) - according to the survey IDC, that was conducted in 2016 with the number of respondents n = 11 350 from all over the world [5, 13].

Among the current issues there are those concerning existing approaches and models for electronic educational resources delivery within the cloud-based setting; the methodology of CC-based learning and research university environment design; evaluation of current experience of cloud-based models and components use [2, 6, 8, 9, 14]. This brings the problem of educational personnel training to the forefront.

1.3 The Purpose of the Article

The *main purpose* of the article is to define the general model of the cloud-based learning and research university environment for educational personnel training and consider the possible ways and techniques of its use and application within the pedagogical systems of higher education. The main idea is in the hypothesis that design and development of learning and research environment due to the proposed approach will result in positive effect of better access to electronic educational resources and more efficient use of ICT and also rising of the ICT competence of educational personnel.

1.4 The Research Methods

The research method involved analysing the current research (including the domestic and foreign experience of the application of cloud-based learning services to reveal the concept of the investigation and research indicators), examining existing models and approaches, technological solutions and psychological and pedagogical assumptions about better ways of introducing innovative technology so as to consider and elaborate the general model and special methodical system of educational personnel training with the use of the cloud-based components on the basis of Microsoft Office 365, AWS, SageMathCloud. To measure the efficiency of the proposed approach the pedagogical experiment was undertaken. The special indicators to reveal ICT competence of educational personnel trained according to the proposed methodical system within the cloud-based learning environment were used.

2 Presenting the Main Material

The use of ICT affects the content, methods and organizational forms of learning and managing educational and research activities that require new approaches to learning environment arrangement [1, 2]. Therefore, the formation of modern cloud-based systems for supporting learning and research activities should be based on appropriate innovative models and methodology that can ensure a harmonious combination and embedding of various networking tools into the educational environment of higher education institution [1, 2, 7, 9].

Taking into account the stages and components of the learning environment design process the general model of this environment formation and development is to be considered (Fig. 1).

The *target* component of this model reveals the goals and objectives of the *CC-based learning and research environment* design such as ICT-competent specialist formation, wider access to ICT; providing the use of the most advanced tools and technologies in education and research and others.

The learning and research process is realized within some *Pedagogical System* (PS) being the component of this model. The CC-based environment is designed so as to support the specific aims achievement within this system. For this purpose the certain *functions of PS* should be revealed, among them there are such as learning, research, development, education, control. To fulfill these functions the necessary educational preconditions are to be provided with the use of the CC-based learning environment tools and services.

For the CC-based Learning Environment (LE) would provide the PS functions implementation the specific services of LE are arranged to support the *functions of LE* such as collection, accumulation, storage, input, representation, manipulation and reorganization of data, learning tools management, measurement, communication, maintaining domain Electronic Educational Resources (EER) and others. To deploy the relevant environment components the special *methodical system* is elaborated basing on certain methodological principles, methods and approaches such as open education principles and also the principles being specific only for the cloud-based learning environment in particular adaptability; services personalization; infrastructure unification; full-scale interactivity; flexibility and scalability; consolidation of data and resources and others; the problem-oriented, the person-oriented, operational and processing, holistic and other conceptual approaches are used.

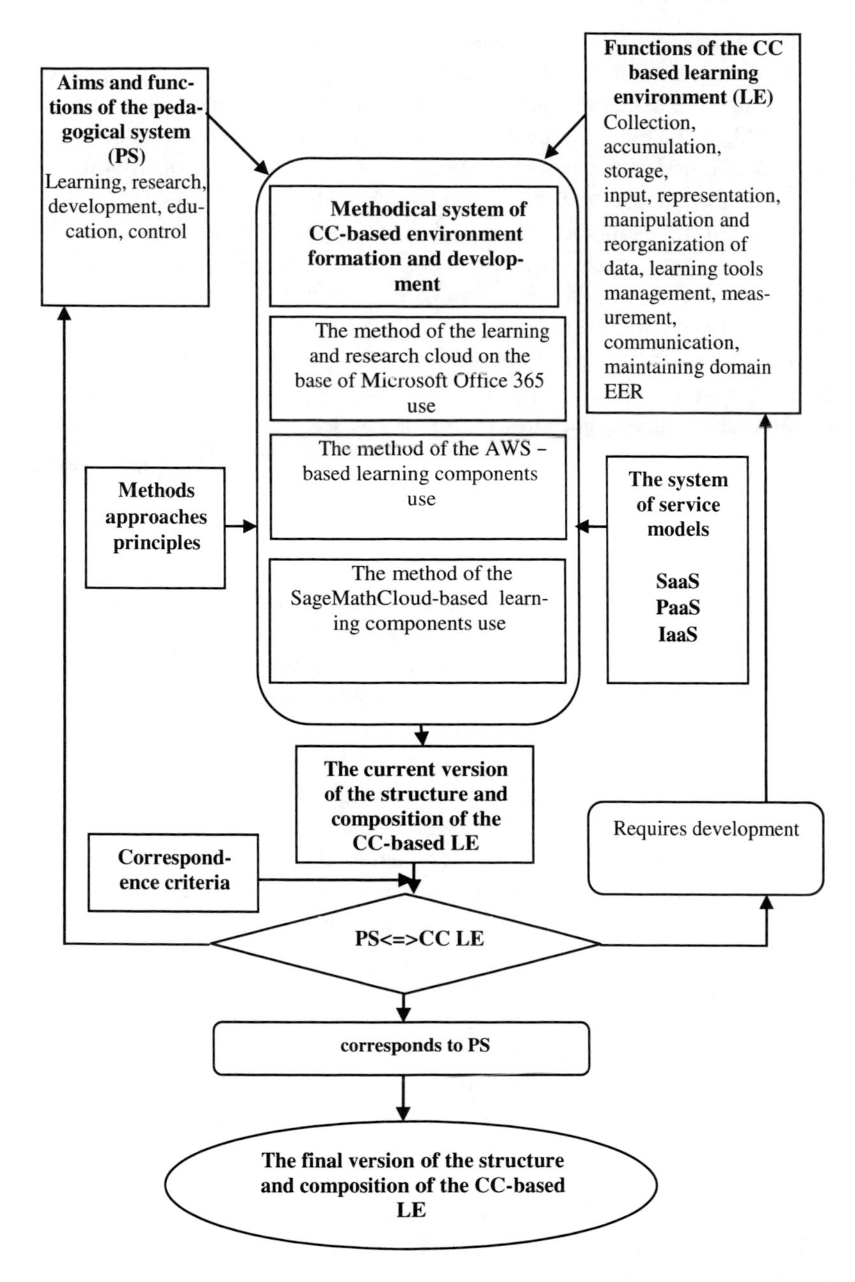

Fig. 1. The general model of formation and development of the CC-based learning and research environment of educational personnel training.

The methodical system includes the number of special developed methods that may be realized at 3 levels due to the main types of CC service models such as SaaS (software-as-a-service); PaaS (platform-as-a-service); IaaS (infrastructure-as-a-service) [9].

Conceptual principles, and also the basic characteristics and service models of CC appear to be the system creative factor that brings together the separate methods into the unite system. Due to this the method of the learning and research cloud on the base of Microsoft Office 365 use, the method of the learning components on the basis of AWS use [9, 10], the method of the learning components on the base of SageMathCloud use [6] are elaborated as the methodical system components.

On the basis of the methodical system of the CC-based LE formation and development the current version of its structure and composition is built then this is checked up as for the correspondence to the most complete realization of the PS functions due to the determined criteria. In case of compliance achievement the environment is successfully created. If the structure and composition of LE don't correspond to the functions of PS the environment development is required. For this purpose the return to the review of the environment structure and composition is needed.

The peculiarities of methods selection for the formation and development of the cloud-based learning environment. The cloud-based learning environment design basing of the general model of its formation and development involves the methodical system that covers several methods of the learning and research components use that are united by the system creating factor such as the cloud-based approach with the most important basic features and service models. The cloud technologies are to provide the key features of the learning environment that are openness and flexibility. If the tasks of environment development are changed it is possible to modify adequately the tools features and overall composition of the environment and also to modernize the methods of its use. So the structure and the composition can be aligned with the planned development goals, new challenges appeared or appearing in future.

So the methods of the learning and research components use may be selected and elaborated depending on the service model chosen for the environment deployment. For example using the SaaS model the services of the general purpose and also the specialized services may be provided. It may be used to support the learning or research activity regardless of the domain by means of documents, electronic tables, databases, files repositories, sites elaborating and processing and also by means of communication services [4, 7]. On the basis of the general use services the method of the Microsoft Office 365 learning and research cloud use was elaborated.

The specialized services may cover the remote learning laboratories and software tools for programming, design, modeling, data processing, research and others. The clear example of such kind of tools is the SageMathCloud mathematical software [6]. The method of the SageMathCloud learning components use was elaborated within this study. In this case the specialized software is provided as the turnkey solution it does not require deployment on the equipment of a user but still can not be modified in realization proposed by the provider [6].

As opposed to it the IaaS or PaaS model is to provide the software access of whatever kind deployed on the cloud servers of the provider by the user. The method of the learning components use within this kind of model was elaborated on the base of AWS (Amazon Web Services) [9, 10]. This approach may be used to provide access to different kinds of learning services deployed in the cloud. For example the learning component with the Maxima system use was elaborated and researched [10].

The Results of Experimental Work. The aim of pedagogical experiment was to justify empirically the proposed theoretical assumptions and methodical system to be really effective.

At the ascertaining stage of the experiment (2012) some problems of ICT competencies of lecturers and academic staff taking part in the experiment as for using CC technologies in learning and research were found. The proportion of those having the high levels of the relevant ICT competencies was 14,65% in the experimental group (58 members) and 12,65% in the control group (60 members). The participants would like to improve their knowledge through greater use of cloud-based tools in their professional activities.

On the formative stage (2012–2014) the different components of the methodical system were introduced in the process of learning and research activities of the experimental group. Comparing the levels of the relevant ICT competencies in the early formative stage and at the end of the experiment gives the rise of the proportion of research and lecturers staff with the high levels of relevant ICT competencies (29,5% in the control group and 48,05% in the experimental group). It shows the statistically relevant increase proved by the angular Fisher criteria ($\varphi_{exp} = 2{,}04 > \varphi_{0{,}05} = 1{,}64$).

Analysis of the formative stage of pedagogical experiment showed that the distribution of the levels of ICT competences in the experimental and the control groups of teaching staff have statistically significant differences due to the introduction of the developed technique, which confirms the hypothesis of the study.

The advantage of the approach is the possibility to compare the different ways to implement resources with regard to the learning infrastructure. Future research in this area should consider different types of resources and environments.

3 Conclusions

The general model of the university cloud-based learning and research environment formation and design proved to be a reasonable framework to deliver and research the cloud-based learning resources and components. The ways of methods selection on the basis of the proposed model and the prospects for their use within the learning systems of higher education are considered. Among them there are the methods of cloud-based components design on the basis of Microsoft Office 365, AWS, SageMathCloud to support learning and research processes. The recent tendencies of CC development are considered and evaluated in view of the cloud-based learning and research environment creation.

References

1. Bykov, V.: ICT outsourcing and new functions of ICT departments of educational and scientific institutions. Inf. Technol. Learn. Tools, **30**(4) (2012). http://journal.iitta.gov.ua/index.php/itlt/article/view/717. (in Ukrainian)
2. Bykov, V., Shyshkina, M.: Emerging technologies for personnel training for IT industry in Ukraine. In: 2014 International Conference on Interactive Collaborative Learning (ICL), pp. 945–949. IEEE (2014)
3. Doelitzscher, F., Sulistio, A., Reich, C., Kuijs, H., Wolf, D.: Private cloud for collaboration and e-Learning services: from IaaS to SaaS. Computing **91**, 23–42 (2011)
4. Lakshminarayanan, R., Kumar, B., Raju, M.: Cloud computing benefits for educational institutions. In: Second International Conference of the Omani Society for Educational Technology, Cornell University Library, Muscat (2013). http://arxiv.org/ftp/arxiv/papers/1305/1305.2616.pdf
5. Middleton, S.G.: Desktop PC Residual Value Forecast. Industry Development and Models. In: IDC US41663116, August 2016, 15 p. (2016)
6. Popel, M., Shokalyuk, S., Shyshkina, M.: The learning technique of the SageMathCloud use for students collaboration support. In: ICT in Education, Research and Industrial Applications: Integration, Garmonization and Knowledge Transfer, CEUR-WS.org, vol. 1844, pp. 327–339 (2017)
7. Shyshkina, M.P., Popel, M.V.: The cloud-based learning environment of educational institutions: the current state and research prospects. J. Inf. Technol. Learn. Tools **37**(5), 66–80 (2014). https://journal.iitta.gov.ua/index.php/itlt/article/view/1087/829. (in Ukrainian)
8. Shyshkina, M.: Emerging technologies for training of ICT-skilled educational personnel. In: Ermolayev, V., Mayr, H.C., Nikitchenko, M., Spivakovsky, A., Zholtkevych, G. (eds.) ICT in Education, Research and Industrial Applications. Communications in Computer and Information Science, vol. 412, pp. 274–284. Springer, Heidelberg (2013)
9. Shyshkina, M.: The hybrid service model of electronic resources access in the cloud-based learning environment. In: CEUR Workshop Proceedings, vol. 1356, pp. 295–310 (2015)
10. Shyshkina, M.P., Kohut, U.P., Bezverbnyy, I.A.: Formation of professional competence of computer science bachelors in the cloud based environment of the Pedagogical University. J. Probl. Mod. Teach. Train. **9**(2), 136–146 (2014). (in Ukrainian)
11. Shyshkina, M.P., Spirin, O.M., Zaporozhchenko, Y.G.: Problems of informatization of education in Ukraine in the context of development of research of ICT-based tools quality estimation. J. Inf. Technol. Learn. Tools **27**(1) (2012). http://journal.iitta.gov.ua/index.php/itlt/article/view/632/483. (in Ukrainian)
12. Smith, A., Bhogal, J., Sharma, M.: Cloud computing: adoption considerations for business and education. In: 2014 International Conference on Future Internet of Things and Cloud (FiCloud) (2014)
13. Turner, M.J.: CloudView Survey 2016: U.S. Cloud Users Invest in Mature Cloud Systems Management Processes and Tools. In: IDC Survey Spotlight US40977916, January 2016
14. Vaquero, L.M.: EduCloud: PaaS versus IaaS cloud usage for an advanced computer science course. IEEE Trans. Educ. **54**(4), 590–598 (2011)

Project Based Learning and Real World Experiences

Introducing Project Based Learning into Traditional Russian Engineering Education

Phillip A. Sanger[1(✉)], Irina V. Pavlova[2], Farida T. Shageeva[2], Olga Y. Khatsrinova[2], and Vasily G. Ivanov[2]

[1] Purdue Polytechnic, Purdue University, West Lafayette, USA
psanger@purdue.edu
[2] Kazan National Research Technological University, Kazan, Tatarstan, Russian Federation
ipavlova@list.ru, faridash@bk.ru, khatsrinovao@mail.ru, idpokgtu@hotbox.ru

Abstract. An extended Algarysh grant from the Republic of Tatarstan, Russian Federation was awarded to KNRTU. The goal of the project was to introduce PBL to the faculty, to develop a PBL training manual for interested faculty and to apply the PBL approach to several classes of students at the university as a demonstration of its effectiveness. This paper gives the structure of the classes, describes the project management lecture content and the related practicum, describes the projects that were undertaken during this program and finally evaluates the impact of the program using student survey and open-ended comments. For the undergraduate and master students, the experience was unique in their education, as many had never worked in teams on a project. Their feedback was highly positive. In general, faculty acceptance was limited to the younger faculty with senior faculty being resistant to the new techniques of PBL.

Keywords: Project based learning · PBL · Teaming

1 Background

1.1 Challenges Facing the Russian Educational System

Russia's desire to take its place in the international labor market cannot be realized without advanced training in the field of engineering and technology. Graduates must be able to solve problems taking into account not only the current needs of industry but also the long-term needs of society as a whole. This implies the need for forming engineers who are able to formulate and solve new problems and offer innovative engineering solutions. The most common pedagogical approach is traditional verbal lecture reinforced by homework and exams. This approach results in clean and unambiguous answers and is easily assessed and evaluated. However this approach does not facilitate the application of the specific content to real world situations which unfortunately tends to be messy, complicated and requiring a great deal of judgement in arriving at a solution. Furthermore, many of the faculty have not had industry

M. E. Auer et al. (eds.), *Teaching and Learning in a Digital World*,
Advances in Intelligent Systems and Computing 715,
https://doi.org/10.1007/978-3-319-73210-7_95

experience and have only practiced engineering within the academic research domain. Specifically, the Republic of Tatarstan is experiencing a serious need for qualified engineering personnel as a necessary resource for economic development. Being "qualified" within the context of economic development requires more than knowledge. It requires the ability to apply that knowledge. Prior to implementing the Bologna process, the transition from academic pure knowledge to practical applied engineering took place in the "specialist" degree in which industry was heavily involved and assisted in the educational process with extensive internships and work experience. Verbal feedback from local industry supports the feeling that several key skills have gotten lost in the transition such as the ability to apply knowledge to complex technical and social situations and the ability to communicate effectively across age groups, technical areas and cultures.

1.2 The Algarysh Grant in the Republic of Tatarstan

To spur this needed transformation in the Republic of Tatarstan of the Russian Federation, the Tatarstan government created the Algarysh grant program in which internationally recognized experts were invited to regional universities to share best practices in a wide range of subject areas. Among the subjects included was engineering education. The structure of the Algarysh grant is an extended, on-site visitation of these experts to participate in effective technology and knowledge transfer. Activities include research, lectures to a broad range of audiences and active engagement with the full range of students and faculty.

In 2016, an extended Algarysh grant from the government of Tatarstan was awarded to KNRTU. The goal of the grant was to introduce PBL to its faculty, to develop a PBL training manual for interested faculty and to apply the PBL approach to several classes of students at the university as a demonstration of its effectiveness.

2 The Approach for Introducing Project Based Learning (PBL) to Students of Different Levels and Groups of Faculty of KNRTU

2.1 About Project Based Learning

The incorporation of real world projects into technical engineering curricula provides a unique and invaluable enhancement to the educational experience. This inclusion of projects into the curricula is often referred to as Project Based Learning (PBL). PBL is one of the modern technologies that universities in many parts of the world are adopting to develop engineering graduates capable of being the practical, application oriented engineers needed in industry. This pedagogical approach is well established and has been reviewed extensively [1–6].

PBL is being implemented in a variety of different ways depending on the curriculum and the surrounding economic climate. Essential characteristic of projects within PBL are that the projects are central to the course, not peripheral, that they are focused on a driving question, that they require transforming acquired knowledge, that they are largely student controlled, and finally the project address are real world problems [1].

Numerous studies have shown that project based learning (PBL) can be an effective way to deepen students' understanding of engineering principles and apply these principles in engineering design [1]. Additionally PBL can also help students develop essential interpersonal and communication skills (so called "soft" skills) which are vital to being effective in the global multi-cultural engineering community. The challenge then becomes to develop an effective training system for incorporating PBL in an established, very traditional system of engineering education.

Within the context of the Tatarstan need, the goal is to produce applied engineers and technologists who are innovative, entrepreneurial and self-directed learners. These descriptive words mean many things depending on the source. In the context of the Algarysh grants, these terms have the following meaning:

Innovation: problem solving without boundaries,
Entrepreneurial: transforming innovation into economic good, and
Self-directed learners: having the confidence to learn something new.

Typically, projects within a PBL curriculum require essential content knowledge and the ability to apply that knowledge. These projects create a structure in which inquiry is demanded and the desire to create something new is invited. A full set of personal skills is essential: critical thinking, problem solving, collaboration (teaming skills), and the full range of communications skills including listening, writing and several varieties of oral communication. The project provides a context for introducing the need for project management skills, listening to client and stakeholder requirements, teaming and interpersonal skills, conflict management skills, problem solving, creativity and intuition and even, attention to aesthetics.

2.2 Instructional Framework at Work with Students and Faculty

Four groups of participants were included in this effort to introduce PBL techniques to KNRTU:

(1) a group of faculty in the field of engineering education and communication sciences,
(2) a class of students pursuing a masters in translation (note that many of these students were teaching faculty in English),
(3) a class of students pursuing continuous professional education at the Master's level while working full time, and
(4) a class of undergraduate 3rd year students in the field of professional learning.

Due to the limitation of the instructor the language in the classroom was predominantly English often with the assistance of translators. A series of PBL oriented lectures, exercises and videos were given over a number weeks including project management, project planning, failure modes and effects analysis, teaming skills and conflict management. In parallel to these classes, two of the classes were divided into teams to work on projects with an eight week duration. The topics for the project and the formation of the teams were student lead and directed. As will be seen later, the project goals were largely aimed at improving society and their community. Consistent with the PBL approach, many of the lecture areas were reinforced by being applied to their projects. For example each team was asked to form a Gantt chart to organize their project and monitor progress. For all these students, this was their first experience with working on a team project where decisions had to be made as a group and successful completion included a mutual reliance of each other. The lack of earlier team project experience was a surprise to the instructor.

Some of the early classes were dedicated to introductory exercises and videos introducing the ideas and techniques surrounding innovation, brainstorming and new product development. The ABC video, "The Deep Dive" [7] was shown in three parts with discussion and reflection done in teams between each of the parts. This video documents the process of developing a better shopping cart. The themes in the video reinforce the ideas of PBL: researching the problem thoroughly, brainstorming and the stimulation of outlandish ideas, and down selecting to the best idea and the importance of early experimentation in the development process. The timeline of the "shopping cart" project was used as a guideline for budgeting their time in their own projects: 20% thoroughly researching the problem, 20% ideation, idea development and concept exploration, 20% generating the final design, 20% fabrication and 20% testing.

A second, well-known PBL exercise, the skyscraper project, was used to explore leadership and group decision making issues [8]. This project requires building a tall structure using foam blocks and sharp pencils that must meet firm testing requirements. A sample of the results are shown in Fig. 1. During this three hour exercise, it was interesting to observe how natural leaders gradually emerged. Particularly in the younger 3rd year class, these leaders were initially uncomfortable with this new role but they were quite successful in performing this role and gaining the confidence of their group.

Fig. 1. Students engaged in the Skyscraper project

Several TED videos were used to stimulate discussion on inter-personal issues such as motivation and leadership. A flipped classroom approach was used for these videos but the instructor also chose to reinforce the message of the videos in class with rerunning the videos and discussion and see whether the videos were effective. Note that the videos were in English but with Russian subtitles. The first video on motivation [9] focused on the role that autonomy plays in getting and keeping humans motivated. It was noticed that the idea of autonomy and independence did not seem to fit in the Russian culture although the students seemed to acclimate to the idea quickly.

The second TED video focused on leadership being built on a belief in something greater, something better, making a change in people's lives [10]. Examples that are given were the Wright brothers, Martin Luther King and the success of Apple. This video was selected to reinforce the students' choices for projects in that most of them were about improving their world.

2.3 And Now the Projects.....

Two of the classes were able to complete projects during the class. As mentioned before, the ideas for the projects came from the students themselves and the students were allowed to form their own teams. The latter freedom resulted in un-even team sizes but the instructor felt that this size variation did not interfere with the learning outcomes. The topics selected were oriented toward solving societal or community problems but also included topics that came from their professional focus such as safety. In the BS 3rd year student class, the four projects were: (1) arranging professional speakers to come and talk to students, (2) making available sign language for TV programs, (3) development of a Russian slang translator, and, lastly, (4) arranging for nature trips for young students. In the MS class of Working Professionals, projects included a tourist website for the city of Kazan, a recycling trash compactor, an electric ice scraper, an enhanced fire extinguisher tool and an integrated head protection system. Keep in mind that the role of the projects was to develop teaming skills and to exercise the project management tools presented in class. Clearly, the scope of some of these projects was too large to be reasonably accomplished in the eight week time frame and beyond the expertise within the teams. Part of the learning was to better assess scope and skills needed in projects.

An important component of the communication skills of the classes was the final presentation that was made to a combined audience of students and visiting faculty. For some of the students, preparing for and presenting their project was a first time experience and an important learning outcome. Preparing the material in an interesting but professional way using slide making software was challenging.

3 Assessment of Learning Outcomes

3.1 Attitude Survey

At the end of the nine week course, each of the classes in which projects were undertaken was given a survey of ten questions with an additional opportunity for open ended feedback. Questions 3, 4, 7 and 9 explore feelings toward team dynamics and team satisfaction. Questions 5 and 6 explore project planning and time management (Table 1).

Table 1. PBL survey questions

1	If I did the project again, I would spend more time doing research about the project background
2	I was able to apply my computer science knowledge to this project
3	I was very pleased to work with my teammates
4	All of my teammates did their fair share of the work
5	Working to a deadline was difficult
6	Our team created a schedule for the tasks to do
7	Making a decision was difficult
8	Our team had to learn about topics that we had not had in class
9	Helping and supporting my teammates was fun
10	Afterwards I used skills obtained during the work on this project in other activities

The remaining questions look into usage of their new and existing skills within the context of a project. Students were asked to rate each question on a scale of 6 with 1 being strongly disagree and 6 being strongly agree. Note that questions 5 and 6 are framed in the negative tone and reverse scored.

3.2 Analysis of the Survey

The results of the survey are presented numerically in Table 2 and graphically in Fig. 2. In general, both groups found that working to deadlines was not difficult but neither class developed effective schedules that they used. This situation is not unusual as the instructor has experienced this before in the United States. It is only when the students are put in a situation that they must report progress to superiors or to customers that a work breakdown structure and a schedule is found to be necessary and

Table 2. Survey data from PBL experience with classes of MS and BS degree students

	MS working Prof.		3rd year BS		Difference	Effect size	T-value	p-value
Count	21		19					
Item	Mean	St Dev	Mean	St Dev				
1	4.333	0.66	3.789	0.92	0.544	2.1339	6.7396	0.000
2	5.000	1.10	3.263	1.41	1.737	4.3221	13.6507	0.000
3	5.095	1.00	5.737	0.45	−0.642	−2.6655	−8.4186	0.000
4	5.000	1.10	5.474	0.77	−0.474	−1.5919	−5 0279	0.000
5	3.762	1.37	3.737	0.93	0.025	0.0680	0.2147	0.387
6	4.095	1.09	4.158	1.21	−0.063	−0.1710	−0.5401	0.341
7	3.571	1.54	3.684	1.20	−0.113	−0.2597	−0.8203	0.281
8	5.048	0.97	4.737	1.33	0.311	0.8373	2.6444	0.015
9	5.238	0.70	5.474	0.61	−0.236	−1.1354	−3.5859	0.001
10	5.333	0.58	5.263	0.73	0.070	0.3338	1.0543	0.226

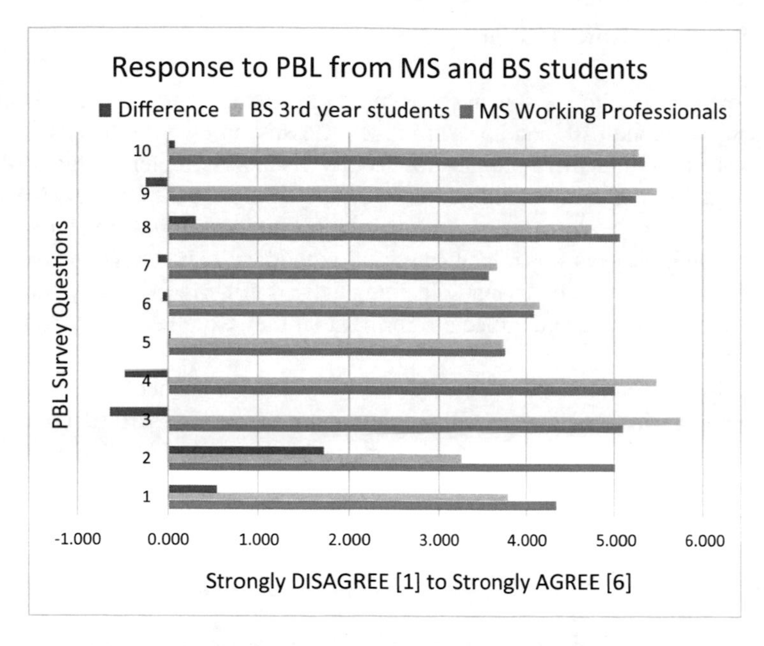

Fig. 2. BS students responded differently than MS students

often, demanded by the client. Neither class found it difficult to get agreement on choices that they needed to make. However in the open comments, several students identified reaching decisions was difficult. So this answer is misleading and highly dependent on the team and the project.

There were several areas of strong differences between the two groups. The class of working professionals appreciated the need for more research in the problem before launching into work (question 1). This group also came into the class with more IT skills and were able to apply them more to the project (question 2).

The younger students expressed a higher enthusiasm for the teaming aspects of the project (questions 3 and 9). This is not surprising since this was a first experience for many of them and the pressure for results was reduced.

The overwhelming feedback from the students was very positive. For most of the students, this was the first team project that they had encountered in their education. Most challenging for them (as per their comments) was time management, arriving at consensus and establishing functional roles within the team. As noted above, there was significant variation between teams and projects. Many of the projects required the team to learn a new skill or get familiar with a new technology, a task which they found to be challenging. The preparation and giving of a formal presentation was a first time experience for some of the students.

4 Faculty Adoption of PBL

While the interaction with the students was highly positive, the true test is whether the faculty chose to pursue PBL and apply it in their classroom. Overall the answer to that question was no. While there are several faculty members implementing PBL as a result of these lectures and seminars, they remain a very small number. No formal survey was done to explore the reasons for this attitude but informal discussions suggest that doing projects was not an activity that the faculty had personal experience with and there was a fear of trying something new. Furthermore, many did not see anything wrong with the standard lecture format that they experienced as students and they had survived.

Despite this powerful inertia, several faculty members launched efforts in PBL. In another paper in this conference (Khasanova and Sanger) describes collaborative projects on computerized psychometry instrument development that affected positively engineering students' engagement. The conclusion from this paper was that these students were not introduced to the PM tools that would have helped them with time management and meeting deadlines. The handbook that was written in this Algarysh grant can provide support to this and other instructors implementing PBL in their courses.

Additionally several of the instructors from the Professional Language department are trying to create a group of projects that would spread over several courses as the students improve their language skills. In many parts of Russia, the museums and places of interest do not have signage and texts in any language except Russian. Tourists are limited in what they learn about the culture. The concept is to develop text, audio files, and multi-media files that could be available to tourists in signage as well as in audio headsets and smart phone apps. This project would help students apply their writing and oral skills to real situations and benefit from constructive criticism from their teammates which often is more accepted than criticism from professors. This approach could be applied to touring of technical labs at the university further expanding these projects and the students' language proficiency into the technical jargon and terminology of relevant engineering disciplines.

Acknowledgment. The authors are pleased to recognize and express their appreciation for the support of the Republic of Tatarstan through the Algarysh grant who understood the value of Project Based Learning in giving students the real world experience that prepares them for the real world. And to all the students and faculty that participated and supported the PBL effort.

References

1. Thomas, J.W.: A Review of Research on Project-Based Learning (2000). http://w.newtechnetwork.org/sites/default/files/news/pbl_research2.pdf
2. Helle, L., Päivi, T., Erkki, O.: Project-based learning in post-secondary education–theory, practice and rubber sling shots. High. Educ. **51**(2), 287–314 (2006)
3. Bell, S.: Project-based learning for the 21st century: skills for the future. Clgh. **83**(2), 39–43 (2010)

4. Northwood, M.D., Northwood, D.O., Northwood, M.G.: PBL: from the health sciences to engineering to value-added in the workplace. Glob. J. Eng. Educ. **7**(2), 157–164 (2003)
5. Pavlova, I.V., Sanger, P.A.: Applying Andragogy to promote active learning in adult education in Russia. Int. J. Eng. Pedag. **6**(4), 41–44 (2016)
6. Mills, J.E., Treagust, D.F.: Engineering education – is problem-based or project-based learning the answer. Australasian J. Eng. Educ. **3**(2), 2–16 (2003)
7. ABC News video called: The Deep Dive in three parts. Part 1 http://www.youtube.com/watch?v=JkHOxyafGpE, Part 2 http://www.youtube.com/watch?v=pVZ8pmkg1do and Part 3 http://www.youtube.com/watch?v=nyugyrCQTuw
8. http://www.cdio.org/knowledge-library/documents/skyscraper-exercise
9. https://www.youtube.com/watch?v=rrkrvAUbU9Y&list=PL70DEC2B0568B5469&index=11
10. https://www.youtube.com/watch?v=qp0HIF3SfI4&index=7&list=PL70DEC2B0568B5469
11. Khasanova, G.F., Sanger, P.A.: Collaborative Project-based Learning in Training of Engineering Students. IGIP, Budapest, Hungary (2017). (to be published)

Recoil – Measurement, Simulation and Analysis

A Study Performed by Students for Students

Monika Grasser[1,2](✉), Mayer Florian[1,2], Hanzlovich Christian[1], Moschner Gerald[1], and Silke Bergmoser[1]

[1] EUREGIO HTBLVA Ferlach, Ferlach, Austria
monika.grasser@htl-ferlach.at
[2] SciReAs, Ferlach, Austria
florian.mayer@scireas.org

Abstract. Education of engineers at Higher Technical Colleges is completed after five years with a final degree that requires the realization of a diploma thesis. In the presented paper the development of a recoil measurement device and first measurements are presented based on students work within diploma theses and on the study of teachers who supported the diploma theses within SciReAs, the scientific research association at the EUREGIO HTBLVA Ferlach. As the strength of recoil is an important property in security and weapon engineering, it is of huge interest of the Higher Technical College EUREGIO HTBLVA Ferlach to be able to perform and analyze these kind of measurements. Furthermore, it is planned to study the behavior of the human body on the occurring recoil when shooting a hunting rifle or a sports gun.

Keywords: Recoil measurement · Security engineering · Data analysis

1 Introduction

The aim of the paper is to present the development of a measurement device for recoil force and to discuss first results of recoil measurements. The recoil is specific for a weapon itself based on its material, mechanical behavior and on the ammunition used. Besides the felt impact of the recoil on the body of a shooter depends on the shooter himself, on his mass and on his specific handling of the shotgun. The recoil force is based on the momentum transfer induced by the acceleration of the bullet and the occurring gases to the human body during firing. For industry, it is of interest, how changes in the weapon system, for example by applying different materials or mechanical adaptations, change the recoil behavior of the weapon system and the "felt recoil" for the shooter. Based on this demand, a measurement device has been developed within three diploma theses. The first measurement set up was based on an apparatus that allowed the fixed firearm to move damped by springs or gas springs. As it is an important to model the behavior of the human body this was developed within the diploma theses 2014/2015 [1] and 2015/2016 [2]. Based on the defined demands, the apparatus has to enable a movement that is modelling the moving human body

M. E. Auer et al. (eds.), *Teaching and Learning in a Digital World*,
Advances in Intelligent Systems and Computing 715,
https://doi.org/10.1007/978-3-319-73210-7_96

during firing. Additionally, the measurement device should be adaptable for different masses and bodies. Parts of the results of the first two diploma theses are published in [3]. Within the diploma thesis 2016/2017 [4] the apparatus was further developed to specify the recoil force directly at the butt plate. This measurement device will be used in future in laboratory to study the recoil based on real time measurements, where data acquisition forces the students to develop skills for working with a huge amount of data which is a part of the focus of industry 4.0.

2 Development of the Measurement Device

The development of the recoil measurement device is divided in three stages: (1) discussion of performed research, (2) description of the first recoil measurement device 2014–2016, and (3) latest development of the recoil measurement device and first measurement results.

2.1 Development of the Measurement Device – State of the Art

Defining the recoil of a firearm can be done by using specified tables as for example published on the web [5, 6], by applying an implemented calculator in the web [7], or by detecting the recoil force during the shooting process. In the web, there are values available for the recoil of different rifles depending on the cartridge, the rifle weight, the recoil energy, and the recoil velocity of a defined system [4]. Anyhow, differences may occur, for example due to the usage of different loads for a given bullet [5]. Möller offers a calculation model for the recoil by an implemented calculator on his webpage [6] and he describes a measurement set up to detect the maximum of the recoil displacement by a "recoil pendulum". Matthew Hall developed a measuring device for recoil of sporting arms [7]. He describes a measurement device to detect the recoil force in vertical and horizontal direction.

2.2 Development of the Measurement Device – Part 1

The aim of the development of the recoil measurement device is given by three requirements. (i) The measurement device has to be able to model the recoil depending on the shooter. (ii) It should be possible, to adapt the apparatus for different firearms, and (iii) the momentum in the vertical direction has to be suspended which is the main difference to the recoil apparatus of Hall [7]. In Fig. 1 the 3D model of the developed apparatus and a 2D schematic drawing are displayed. The apparatus consists of a retaining plate in the front and in the back. These two plates are fixed and connected by a carrier plate and by two steel pipes, positioned symmetrically. Within the Teflon pipes, steel pipes are used to guarantee the necessary small rigidity of the measurement set up. Springs and pneumatic springs are used to model the mechanical behavior of the human body. It is possible to include different springs to take into account varying mass and different performance of diverging human bodies. For the detection of the recoil force and the displacement, an inductive displacement sensor and a load cell purchased from HBM Hottinger Baldwin Messtechnik GmbH is applied. Based on this setup, two measurement possibilities are given: (i) measuring the displacement (length of the backwards movement) of the rifle and the according time and (ii) measuring the force noticed by the system. The data acquisition is performed with the MX410 of HBM which is a four channel amplifier Quantum X. The data analyses is done with the software Catman Easy.

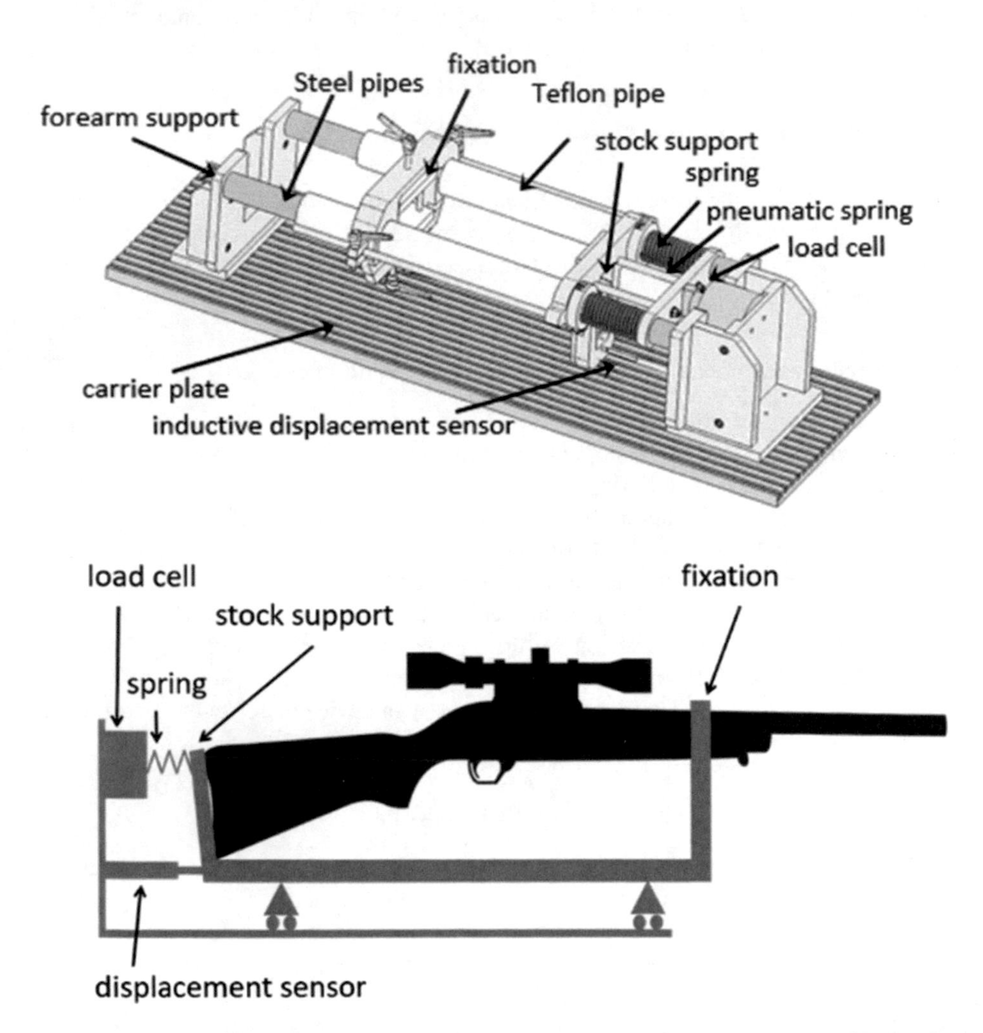

Fig. 1. (A) Upper picture: 3D model of the developed apparatus, (B) lower picture: 2D drawing (Source: Grasser [3])

2.3 Development of the Measurement Device – Part 2

The results of the measurements performed with the measurement device as described in Part 1 showed, that the application of springs in front of the load cell according to the line of action of the recoil force induces a sigmoidal behavior of the signal. Therefore

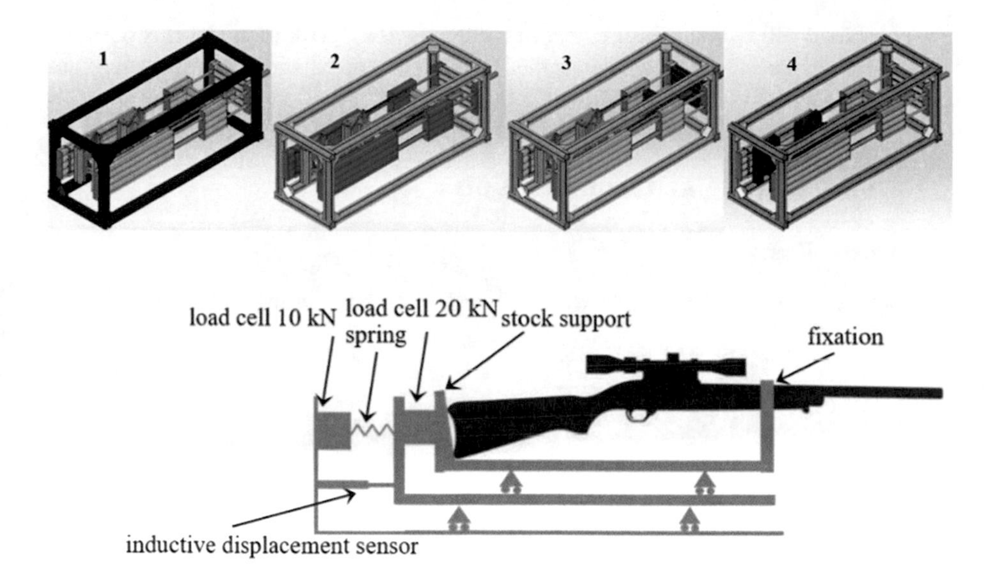

Fig. 2. (A) 3D model developed apparatus, (B) 2D schematic drawing

the next step in the development of the measurement device was to implement a situation, where the recoil force can be detected directly at the butt plate of the shotgun. The requirements are that the measurement device has to be able to move and the recoil force has to be detected in the moment, it occurs. Additionally, it should be possible, to adapt the apparatus for different kind of rifles and the momentum in the vertical direction has to be suspended. Based on these conditions, it was decided to build a new measurement set up as shown in Fig. 2. Here, the 3D model (Fig. 2A, 1–4) and the 2D schematic drawing (Fig. 2B) are displayed. For the new construction mainly ITEM profiles are applied. The blue color shows the inner movable part of the measurement device (Fig. 2A, 4).

It includes the fixation for the rifle and the load cell with a measurement range up to 20 kN and the connection to the outer load cell. When the shot breaks, the colored part in Fig. 2A (4) is accelerated backwards together with the colored part in Fig. 2A (2) to

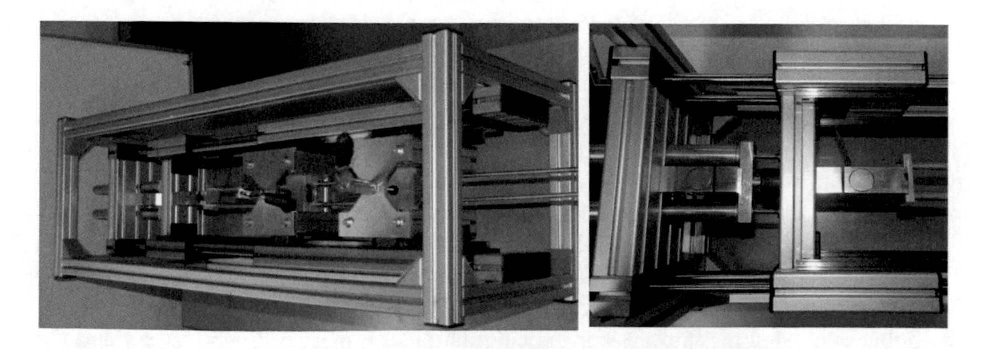

Fig. 3. Recoil apparatus, left: whole apparatus, right: load cells, top view

the second load cell with a measurement range up to 10 kN. Additionally, it is possible to detect the displacement by a displacement sensor. Figure 3 shows the measurement device Part 2 with the two applied load cells.

3 Recoil – Theory and Computation

3.1 Kinetic Energy

The kinetic energy that is induced by a backwards moving system can by described by Eq. (1) [7].

$$E_{kin} = \frac{1}{2} m_M * v_M^2 \tag{1}$$

where E_{kin} is the kinetic energy of the moving system, m_M is the mass of the firearm (in this case of the moving system), and v_M is the residual velocity of the firearm when moving backwards (in this case of the moving system). In all practical processes, parts of the kinetic energy of a moving system are transferred to inner energy U by friction and other effects. Based on this fact, in the following maximum energy and maximum force are taken into account. The preload force of the springs F_P is calculated based on a defined displacement of the springs as described in part 1.

3.2 Conservation of Momentum

The physical recoil is calculated based on velocity and mass of the involved frame work. Therefore, the important quantity in calculation is the conservation equation of momentum as given in Eqs. (2) and (3). It is applied for calculation of recoil force as given in Eqs. (4) and (5) [8].

$$p_1 = p_2 \tag{2}$$

$$m_1 * v_1 = m_2 * v_2 \tag{3}$$

$$m_B * v_B + m_G * v_G = m_M * v_M \tag{4}$$

$$m_B * v_B + m_G * v_G = F * \Delta t \tag{5}$$

where p_1 is the inducing momentum of the bullet when leaving the muzzle, p_2 the responding momentum after the shot mentioned as backwards motion of the weapon system, m_1 the mass of the moving body for p_1, v_1 the velocity of m_1, m_2 the mass responding in p_2, v_2 the velocity of m_2, m_B the mass of the bullet, v_B the velocity of the bullet, m_G the mass of the gun powder, v_G the velocity of the gun powder, m_M the mass of the backwards moving parts, v_M the velocity of the backwards moving system, F the recoil force, and Δt the time that the system takes for the backwards movement. p_1 is the induced momentum at the beginning, which is the momentum induced by the shot when the bullet is passing the muzzle. For the calculation, the mass of the bullet m_B and the velocity of the bullet v_B and the mass m_G and the velocity v_G of the gunpowder are applied. v_G is approximated by half of v_0 [8]. Here v_0 is assumed to be v_1 when the bullet is leaving the barrel and p_2 is the momentum of the backward moving weapon system and fixation, depending on the mass of the weapon system m_M and the velocity of the backwards moving system v_M. F is representing the repulsive force lasting over time Δt.

3.3 Calculation Energy and Recoil Force

Combining Eqs. (1) and (4), Eq. (6) leads to the calculation of the kinetic recoil energy.

$$\mathrm{E_{kin}} = \frac{(\mathrm{m_B} * \mathrm{v_B} + \mathrm{m_G} * \mathrm{v_G})^2}{2 * \mathrm{m_M}} \tag{6}$$

The recoil energy is directly proportional to the projectile mass and projectile velocity but inversely proportional to the mass of the firearm [8]. E_{kin} is used to calculate the maximum recoil force by applying preload force and maximum displacement and ignoring friction or other effects. Based on this, the maximum recoil force F_m is calculated by Eqs. (9) and (10) [8].

$$E_{kin} = \frac{1}{2}\left(F_m + F_p\right) * x_r \tag{7}$$

$$x_r = \left(F_m - F_p\right)/R \tag{8}$$

Where F_{m} is the maximum recoil force, F_p the preload force of the spring, x_r the maximum displacement caused by recoil, and R the average spring rate.

Maximum recoil force F_m and displacement x_r are calculated by Eqs. (9), (10) and (1)

$$F_m = \sqrt{2 * R * E_{kin} + F_p^2} \tag{9}$$

$$x_r = \frac{2E_{kin}}{F_p + F_m} \tag{10}$$

Based on this theory a maximum recoil force for the investigated system with recoil measurement device 1 of F_p = 60 N is calculated. Therefore the load cell S2M 200 N with a class of accuracy of 0.02 is applied for the measurements performed with measurement device 1 [3].

4 Velocity Simulation of the Gas Around the Bullet

The velocity change of the bullet over time has a huge impact on the recoil behavior of a specific gun. In addition, the flow field around the bullet at the moment it leaves the barrel is thought to be able to influence the recoil behavior remarkable. Therefore, different muzzle breaks are applied to model the recoil force over time by manipulating the flow field around the bullet. Based on this fact, a first attempt is presented to model the flow field surrounding the bullet applying the CAD Software Solid Works with the CFD application ("SolidWorks flow simulation"). The calculation uses the boundary condition of a velocity inlet with 1000 m/s. This velocity is about 100 m/s higher than the muzzle velocity of the flying bullet. A pressure outlet is defined at the other three boundaries of the surrounding cylindrical domain and the flow und pressure field is computed. The simulation is performed for a steady position of the bullet when moving through the barrel. After the first simulation, the bullet is positioned 1 mm further in moving direction towards the compensator and the next simulation is performed. For each simulation the same boundary conditions and displacement are applied till the end of the compensator is reached. Therefore, each simulation series consists of 32 simulations. Figure 4A shows the velocity distribution as a result of the performed simulation at the beginning of the study and Fig. 4B the velocity field at a position of the bullet 4 mm before the end of the muzzle break. The simulation results are thought to give additional information for the evaluation of the measured recoil force for future work. For example, the projected area of the compensator holes can be used in combination with the pressure field to calculate the impact of the gas flow on the recoil. As this simulation is a first trial, the simulation model did not take into account influences like: mass of the gun powder, temperature changes, rotation and translational motion of the bullet and deviation of the form of the barrel from a cylinder etc.

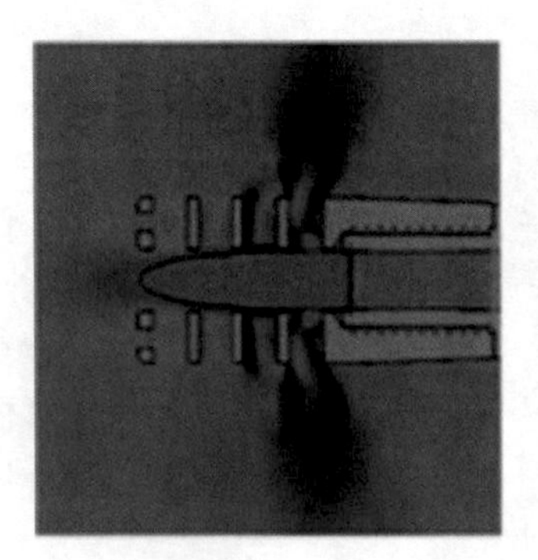
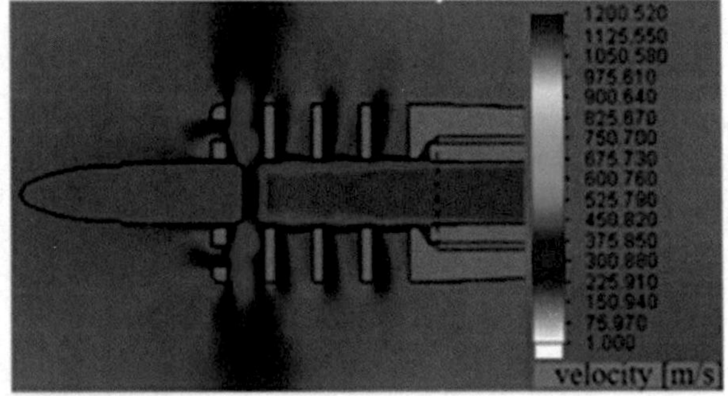

Fig. 4. Velocity field. (A) Simulation 28 mm, (B) 4 mm from the end of the muzzle break.

5 Measurement Results

The first data set presented in Part 1 is based on measurements with a shotgun, caliber 20 with slug Brenneke caliber 20/76-24 g/370 grain and the presented measurement results with measurement device in Part 2 are performed with a shotgun, Rößler Titan 6 caliber 300 Win.Mag with ammunition Sellier & Bellot.

5.1 Measurement Data - Part 1

Figure 5A shows typical measurement results for displacement Δx for seven shots. In addition to the time dependent graph, the values of maximum displacement and of maximum recoil force are detected. These measured values are compared with the calculated ones [3]. The measurement given in Fig. 5A (1, 2) is performed with two springs with a spring rate R = 0.305 N/mm and a preload displacement of 63 mm. The seven graphs show a good reproducibility at the beginning of the measurement (1, 2). Anyhow, the deviation at the maxima of the displacement (3) lead to almost parallel graphs while the system is moving forward (4) induced by the spring force. After about 0.6 s, the system reaches the initial point again (5). The average of the maxima is 40.93 mm with a range of 7.92 mm. The average of the according muzzle velocity is 431.02 m/s with a range of 10.7 m/s. Figure 5A and B show a comparison of one measured displacement curve with an expected curve based on [8] for a recoil measurement. The first (Fig. 5B, 1) peak is not expected by [8]. It is thought to appear due to the trigger mechanism. However, the effect of buffering is detected (Fig. 5B, 2). The average of the maxima of the recoil force in this measurement series is 58.74 N with a range of 21.27 N. The calculated values of the maximum displacement and maximum recoil force as given in [3], $x_r = 0.065\,\mathrm{m}$; $F_m = 59\,\mathrm{N}$, show a deviation of $\Delta x_r = 0.0205\,\mathrm{m}$ and of $\Delta F_m = 11,7\,\mathrm{N}$ compared to the measured maximum displacement $x_r = 0.0445\,\mathrm{m}$ and force $F_m = 70.7\,\mathrm{N}$ [3].

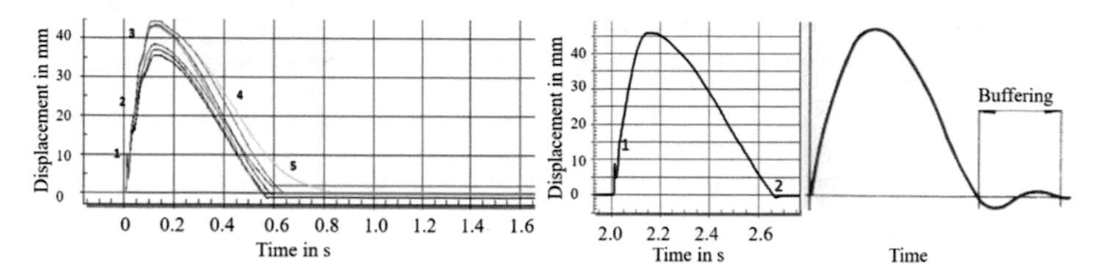

Fig. 5. (A) Repeatability of horizontal displacement. (B) Function of displacement versus time measured, maximum value 44.9 mm; (C) Release travel diagram - displacement versus time of a shotgun with fixed zero line – published by (Source: [8] p. 429).

5.2 Measurement Data – Part 2

Figure 6 shows typical measurement results for recoil force F and Δt of the inner load cell (F_1) and the outer load cell (F_2). In addition to the time dependent graph, the values of maximum recoil force F_{m1}, F_{m2}, and the muzzle velocity are detected. These measured values are compared with the accepted load during shooting by a human with a mounted load cell (max. 10 kN) in the gun stock (Part 2). The results of three different measurements are presented, each including six measurements:

- Case 1: without muzzle break (m.b.) using measurement device - Part 2 (Fig. 6)
- Case 2A: without muzzle break without measurement device - Part 2 (Fig. 7A)
- Case 2B: with muzzle break without measurement device 2- Part 2 (Fig. 7B)

Table 1 shows the average values of the detected maximum forces F_{m1} and F_{m2} for the two load cells in comparision with the detected F_{m1} when shooting out of the shoulder. Load cell 1 measures the force directly at the stock and load cell 2 measures the force induced by the backwards moving damped system. Figure 6 shows the measurement data of F_1 and of F_2 for Case 1. The measurement results show a definded peak for F_1 and the reproducibility of the six shots. The measurement data of F_2 is complicated. The reproducibility is worse which has to be investigated further on. Figure 7 shows the measurement data of F_1 for Case 2A and Case 2B. The measurement results show a small range for F_1 even when the measurements are performed without fixation.

Case 2A and Case 2B show a remarkable difference in the second peek of the force due to the application of the muzzle break but no reduction of the maximum force although the felt recoil force is reduced.

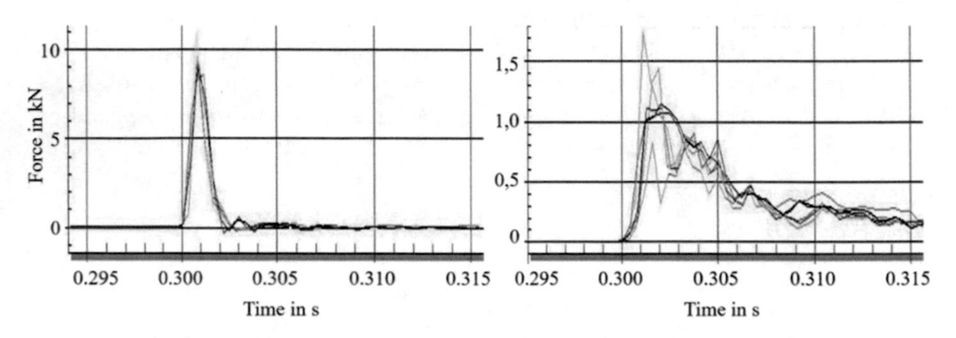

Fig. 6. Recoil force inner load cell F_1 (A), outer load cell F_2 (B), ammunition Sellier & Bellot

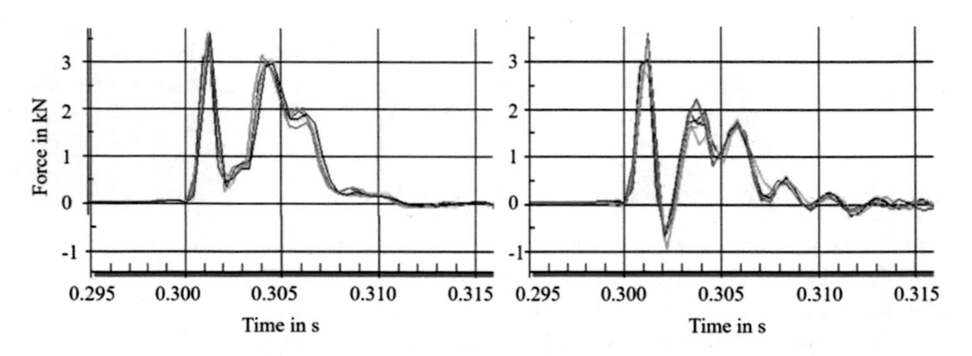

Fig. 7. Recoil F_1 Case 2A without muzzle break, (A), recoil force F_1 Case 2B with muzzle break (B)

Table 1. Average values of the detected maximum forces F_{m1} (load cell 1), F_{m2} (load cell 2)

Average value	F_{m1} [kN]	F_{m2} [kN]	v_m [m/s]
Case 1	14.372	0.910	918.92
Case 2A	*3.520*		*915.95*
Case 2B	*3.368*		*918.75*

6 Conclusions

An apparatus for horizontal recoil measurement of a firearm by suspending dynamic torque is constructed and measurements are presented by a group of students from the Higher Technical College EUREGIO HTBLVA Ferlach. The development of the measurement device was performed during three diploma theses. The developed measuring apparatus for horizontal recoil allows the detection of the movement of the system and the measurement of the force occurring based on recoil during a shot. For the detection of the deflection of a rifle during recoil, an inductive displacement sensor and two load cells are installed. The vertical movement of the rifle is minimized by the used symmetrical fixation in both cases. Hence, the results of the first apparatus are impressed by the applied springs. First results show good repeatability of the measurement device. Additionally, calculation and simulation for the recoil force and the velocity field are presented. A simulation of the velocity of the produced gases around the bullet in the barrel was performed. The recoil measurement is installed in the laboratory lecture as measurement application combining calculation, simulation and measurement on the one hand for research of the present diploma thesis and on the other hand for lecturing. With that, the results of this study are available for students in future. As the application of simulation will be involved further in construction lectures and theory, praxis and numerical simulation are combined which enforces abilities in data management and Industry 4.0.

Acknowledgements. We want to acknowledge the EUREGIO HTBLVA Ferlach for providing the infrastructure for the study and the involved students for their work.

References

1. Gallob, S., Hechtl, L., Prosic, F.: Development of a Curbed Recoil Model. EUREGIO HTBLVA Ferlach, Ferlach (2015)
2. Kullnig, D., Rainer, D., Schrödl, M.G., Sükar, D.: Recoil Measurement – Testing, Evaluation and Optimization. EUREGIO HTBLVA Ferlach, Ferlach (2016)
3. Grasser, M., Mayer, F.: Recoil measurement - application in security and weapon engineering. In: Conference Innovative Metrology, Institute of Electrical Measurements, Johannes Kepler University Linz, pp. 12–17, Linz (2017)
4. Ganner, F., Kloger, G., Plattner, R., Steiner, S., Tschebull, G.: Muzzle Break Optimization for Röwa Titan, vol. 6. EUREGIO HTBLVA Ferlach, Ferlach (2017)
5. Hawks, C.: Rifle Recoil Table. http://www.chuckhawks.com/recoil_table.htm
6. Möller, L.: Rückstoß (Recoil). http://lutzwmoeller.net/Bremse/Rueckstoss/Rueckstoss.php
7. Hall, M.J.: Measuring felt recoil of sporting arms. Int. J. Impact Eng. **35**, 540–548 (2008)
8. Apotheloz, R., Beisken, H., Bossel, R., Brack, K., Brandenberger, E., Burkhardt, K., Cremosnik, G., Daume, D., Diewald, G., Dietz, W., Freymond, P., Flückiger, H., Frommer, H., Groebsch, W., Herter, R., Hottinger, K., Isenring, U., Kellenberger, A., Keller, F., Krüger, H., Mamie, J., Meier, H., Rochat, E., Sauer, W., Saugy, K., Schneider, K., Spalding, W., Spring, F., Stadelmann, R., Stoll, W.W., Strüby, E., Tschui, F., Werlen, A., Wildi, G., Wyss, D.: Oerlikon Taschenbuch. Werkzeugmaschinenfabrik Oerlikon-Bührle AG, Zürich, pp. 426–430 (1981)

Topic Oriented Mixed-Method System (TOMMS)

A Holistic System to Integrate English Language Classes into English-Medium Instruction Courses in European Higher Education

Mathew Docherty(✉) and Kurt Gaubinger

School of Engineering, University of Applied Sciences Upper Austria,
Stelzhamerstraße 23, 4600 Wels, Austria
mathew.docherty@fh-wels.at

Abstract. This paper addresses the issue of English-Medium Instruction (EMI) in European higher education, which has been referred to as 'the language of higher education in Europe' [1]. It outlines the goals of this new phenomenon, analyses the current situation and evaluates potential limitations. It investigates how, using team-teaching, inverted classrooms and active classroom techniques, potential hurdles may be reduced. How, through careful planning and coordination, students and teachers alike, can be prepared for the new challenge of teaching and learning content through English. It goes on to outline a systematic approach, coined by the authors as the Topic Oriented Mixed-Method System (TOMMS), where each topic is initially addressed in a contact lesson with a Language Teaching Expert (LTE), to prepare students, it then introduces them to the theory in self-study sessions, before applying the theory, under guidance of the Content Expert (CE), in tasks and projects. This paper provides a holistic mixed-methods pedagogical approach to address the issue of knowledge transfer in EMI programmes using state of the art methodologies to allow the LTE to support the CE in planning and teaching of content. It offers higher education institutes (HEIs), who may be averse to teaching content courses in English, an effective system to implement EMI and, for those that already have, it offers ways to improve the efficiency of them.

Keywords: English as a Medium of Instruction (EMI) · Inverted classroom Blended learning · Active classroom · Web-based learning · Project-based learning Team-teaching

1 Introduction

English as a Medium of Instruction (EMI) is well described as "The use of the English language to teach academic subjects in countries or jurisdictions where the first language (L1) of the majority of the population is not English" [2]. This growing global phenomenon [1] is employed in a quarter of all universities across Europe [3] and the spread of English is said to be "inseparable from globalization" [4]. However, is this medium of

M. E. Auer et al. (eds.), *Teaching and Learning in a Digital World*,
Advances in Intelligent Systems and Computing 715,
https://doi.org/10.1007/978-3-319-73210-7_97

content transfer really as efficient and effective as teaching in the native tongue? Here the data is unclear as there is a lack of empirical studies [5].

Can content teachers be expected to achieve the equivalent level of knowledge transfer in a foreign language? There is conflicting evidence from studies [6–8] as to whether it does indeed lead to a reduction in academic outcomes, but experts agree that further research is urgently required [5]. Also, there is a general perception that implementing EMI without the required support 'may actually hinder students' acquisition of the subject matter being taught' [8].

Another issue is whether students can really be expected to learn language indirectly from EMI content lessons without the support of a language specialist? This problem was anticipated for in the early stages EMI, notably at the Maastricht University in the mid-1980s, where the importance of involving the language teacher in the planning and implementation of the content courses was well known. As stated by Wilkinson, in his chapter in English-Medium Instruction at Universities [5], content teachers would discuss each lesson with the English teacher, before and after, to ensure that pedagogical goals were being met. Wilkinson goes on to explain that content lessons were monitored by an English teacher who would offer feedback and help students at the end of content lessons. This close coordination of the two departments allowed the teachers to tweak the lessons to better support each other and was a contributing factor to the programme's success (Ibid).

It is the authors' contention that EMI courses are now being used in European higher educational institutes as a matter of course, but the amount of care and consideration that goes into the planning of syllabi, materials and methodologies is underestimated and, therefore, not performed optimally. In this respect, it is the aim of this paper to answer the following question:

How to enhance knowledge transfer on EMI courses through improved interdisciplinary synergy by implementing state-of-the-art pedagogical approaches and methods?

Derived from this aim, this paper starts with a literature review addressing the aims of EMI, the current situation in Europe as well as critical issues. Based on this knowledge the authors have developed a framework which can act as a reference model for topic-based cooperation of language and content teacher using state-of-the-art pedagogical methods. Subsequently, an exemplary practical implementation of this model is described. At the end of this paper, the authors discuss the practical implications and limitations of the framework and address aspects for further research.

2 English-Medium Instruction

2.1 Drivers and Aims

The EMI approach of integrated language and content transfer has shown exponential growth in European universities over the past twenty-five years. The development, driven by internationalisation [9], can cause difficulties for indigenous students due to their language capabilities, as well as staff, due to inadequate language skills and unwillingness to teach through English [1].

The aims of EMI in European universities are 'enhanced employability for domestic graduates, institutional prestige and greater success in attracting research and development funding' [1]. EMI is intended to increase the global employability of students, through improved linguistic competence in the subject matter, and to improve English language proficiency [10], however, EMI sets no requirements or guidelines regarding English language skills. It may be assumed, that English-language learning will be a by-product of learning the content, but complications of content knowledge transfer must also be considered. Dearden states, "how are students supposed to understand lectures and classes if the EMI teacher does not help [them] with their knowledge of English" [2].

2.2 Current Situation

English-taught programmes (ETPs) in European universities have increased, according to studies by Maiworm and Wächter [3, 9, 11], from 725 in 2001 to 8089 in 2014, with 27% of higher education institutions (HEIs) now offering at least one ETP. Nordic regions are leading the way in the march with 61%, Central West Europe has 44.5% of HEIs offering ETPs, and nearly six percent of all programmes in HEIs in Europe are English-taught programmes [3].

Studies have shown that the level of English language proficiency of students and staff was shown to be perceived as adequate but potentials of English language diversity of the students viewed as a problem [3]. The issue, in the authors' opinion, seems not to be whether the students and teachers meet basic language requirements but more that they are reduced in their capacity to express themselves as they would in their native tongue.

A further issue is the amount of language support classes or rather lack of them. Wächter and Maiworm's study [3] showed that around one-third of the HEIs in the study provided no language training at all for students on ETPs and that it was more common to be offered language support on Bachelor programmes in comparison to Master or PHD programmes.

The statistics regarding English language training for academic staff were no more reassuring, finding that even though half of master programmes, and nearly three-quarters of Bachelor programmes, stated English proficiency as an important selection criterion for the recruitment of new academic staff, mandatory English courses for academic staff are rather rare (19% for Bachelor and 11% for Master programmes) [3].

The size of the HEI is a major factor in the decision to offer ETPs, with over 80% of larger HEIs, of 10,000 students and more, offering at least one course, but only 14% of small HEIs having at least one programme on offer [3]. This, as the authors state, is unsurprising as the larger the institute the more programmes generally on offer. However, it does illustrate a need to implement such programmes in these smaller HEIs, therefore, allowing them to contribute to the contingent offering programmes for international ERASMUS students.

So, which students are taking advantage of the EMI courses, indigenous or foreign students? According to the most recent study [3], just over half are foreign students, which shows that the initial aim of internationalisation has been successful. However, upon further analysis, it can be seen that the number of foreign students on Master programmes (57%) is markedly higher than bachelor programmes (39%) (Ibid.). Wächter and Maiworm summarise, that "This finding supports the assumption that Bachelor programmes more often serve as a means to make domestic students fit for the global market while Master programmes more often suit to attract foreign students as top talents for the own labour markets, as fee payers, etc." (Ibid. p. 83).

2.3 Critical Issues of EMI

The rise of English as a global Lingua Franca, although long predicted [1] and widely practised, still faces many hurdles in education. Dafouz and Nunez, in Doiz, Lasabaster and Sierra, warn of possible implications for pedagogical practices and learning outcomes [5]. Coleman states, that even if content teachers have adequate English language proficiency skills, they may well lack the knowledge of the demands of higher level education through it [1]. Dearden goes on to say that "EMI teachers are not, or at least do not see themselves as, language teachers" [2] and Sowden points out another issue, that both teachers and students are forced to 'embrace and foster a variety of English which up to now they have learnt to treat as inferior and by doing so risk undermining their academic self-image' [12].

Some HEIs are reluctant to introduce EMI courses, due to inadequate English skills of teachers and the resulting resistance to teaching in this medium [3], as they may perceive themselves as not fully capable of achieving the required language level [13]. Studies [Arantegui et al. and Dafuz et al. in, 5] have shown that although lecturers strongly supported introducing subjects using EMI they did not themselves feel prepared enough to teach it.

It may also be argued, in the age of EMI, that classical English lessons are superfluous, which could lead to a reduction in teaching hours for English as a second language (ESL) teachers. However, even if English language lessons are still offered, they may not necessarily be aimed at supporting the EMI classes, or the students' explicit requirements surrounding them, but rather at generic English language skills.

Although EMI has been in practice across HEIs in Europe for over 30 years the amount of empirical research to prove its effectiveness is lacking [5]. Much research is still needed in order to ascertain how much language is actually being learnt, and if there is a compromise in the level of academic content conveyed. However, research in contexts of immigration shows evidence that 'this type of learning is not very effective' [Shohamy in, 5].

3 Topic Oriented Mixed-Method System - TOMMS

3.1 Aims

In order to address the issue of effective and efficient knowledge transfer in EMI courses, the authors have derived the following criteria, which is to be taken into consideration when selecting methodologies and approaches: increased levels of cognitive work in the classroom, interdisciplinary synergies, cost efficiency of classes, improved course flexibility and increased individual learning.

3.2 Methodologies Applied

The mixed-methods pedagogical concept developed is a building-block system and has been coined the Topic Oriented Mixed-Method System (TOMMS), it applies inverted classroom, blended learning, active learning and project-based learning, as well as team-teaching, to improve learning effectiveness and efficiency.

Inverted Classroom. This methodology was chosen to fulfil the criteria of improved course-flexibility, increased individual learning and cost efficiency of classes.

An inverted classroom approach will be used to introduce theory, as distance E-learning [14], prior to the session with the teacher. In this way, all the students will come with a basic understanding of the topic, from previewing the theory at home, allowing them to be able to process the information at their own pace [15], and evens out potential disadvantages and misunderstandings due to language deficiencies.

King [16] first broached the idea of changing the learning process from the traditional transmittal model, whereby knowledge is seen purely as generic input, to a constructivist one, where the learner's existing knowledge is used to help them understand new materials, in the article "From sage on the stage to guide on the side". With it came the notion of changing the teacher's role from central figure to support person, which was later developed by Lage et al. [17] into the inverted, or flipped, classroom approach. Here out-of-class asynchronous theory is combined with synchronous classroom tasks and activities, and the learner's role goes from passive observer to active participant.

Decentralised teaching methods, as opposed to frontal teaching and rote learning, are relatively commonplace in English language classrooms and aim to increase motivation through participation [18]. Additionally, this methodology can lead to reduced costs which are of relevance to counteract team teaching methods employed.

Blended learning. Blended learning can be more successful than a non-blended approach and aims to support effective and efficient knowledge transfer by obtaining "a perfect blend between face-to-face learning done in the classroom by teachers and online learning experience done outside the classroom as a complement" [19]. In a study by Thompson [20], learners showed improved accuracy of 30% and an increased learning tempo of 40% in comparison to a non-blended approach control group. A Meta-Analysis and Review of Online Learning Studies, by the U.S. Department of Education [21], into face-to-face instruction compared to blended learning in K-12 students, showed that blended learning improved student performance over face-to-face and also over purely online learning techniques.

Active Learning. As sections of the theory are moved from the classroom to web-based environments it is important to implement new activities into the contact teaching time. This, done through active learning, aims to actively involve students in the learning process. It uses Bloom's taxonomy model to progress from lower order thinking tasks to higher level cognitive work. For the purposes of this paper, a more real-world problem-solving approach will be taken and methods such as task-based and project-based learning will be suggested.

Student centred learning (SCL), which has always been seen as conducive to learning in EMI classes, reduces pressure on teachers to produce English language content at an equivalent level to that of their L1 [5]. SCL allows the student more scope and responsibility by implementing methods such as task-based learning (TBL) and problem-based learning (PBL), whereby the learners are required to use language skills in order to perform tasks and solve problems. Such higher–level cognitive activities, according to Bloom et al. [22], increase learner motivation, engagement and understanding, and therefore ultimately student success.

This higher-level cognitive work will be done in face-to-face sessions. The theory is applied through active learning [23], under supervision of the teacher, or teachers where team-teaching is employed, and will typically be done in groups, allowing for focus on their individual needs, in order to increase student engagement and promote understanding of the core content [24]. Research also suggests that this active participation, whereby students are engaged in solving problems through discussion and collaboration, improves knowledge understanding and therefore retention [25].

Project-based Learning. This methodology was selected to increase the amount of individual learning in the classroom. Project-based and task-based learning are both student-centred learning approaches, where content is taught through problem-solving and critical thinking, instead of rote learning, to engage students in real-world situations [24]. It is often employed in language teaching as a top-down method, that exposes learners to language in its full natural form, before going on to focus on specific aspects of it, this is called a "meaning-based approach" by Willis and Willis [26]. It uses active learning methods to apply learnt theory, in interactive situations [27], and "requires learners to arrive at an outcome from given information, through some process of thought, and which allows teachers to control and regulate that process" [28]. In regards to language learning is has more recently been defined as "a goal-oriented communicative activity with a specific outcome where the emphasis is on exchanging meanings not producing specific language forms" [29]. In addition, this methodology again supports higher-level cognitive work and interdisciplinary projects.

Team-Teaching. Cots, in Doiz et al. [5], suggests that the neglect of focus on language in EMI may in part be due to the fact that the EMI instructors are lacking in language training as well as language proficiency (Ibid., p. 117). Tandem teaching, with a Content Expert (CE) and a Language Teaching Expert (LTE) working together, may help to remedy this by creating interdisciplinary synergies and is a cornerstone of TOMMS. However, it must be stressed that this collaboration must exceed the teaching task alone and should also include syllabus creation, material design and the planning of classroom activities.

3.3 Language Acquisition Approaches

The model of Second Language Acquisition (SLA) that TOMMS is based on is the Input Hypothesis, this states "that in order for acquirers to progress to the next stage of the acquisition of the target language, they need to understand input language that includes a structure that is part of the next stage" [30]. Such comprehensible input, or input at i+1, was originally described by Krashen as "when we understand language that contains structure that is "a little beyond" where we are now" [31].

This comprehensible input approach is further developed by Long, in his Interaction Hypothesis [32], which states that the effectiveness of comprehensible input is greatly increased when learners have to negotiate for meaning. This can be understood in the form of classroom tasks whereby the students are forced to communicate in English whilst debating topics and discussing their opinions.

In TOMMS, language teaching experts work closely with content experts, to develop materials in a contextualised framework, which introduces new topics in order to extrinsically motivate students, whilst stimulating interaction, and encouraging learning skills and strategies. The teachers prepare materials in such a way that when the students come to the lessons they will encounter new concepts as comprehensible input in low-stress situations. This means, that they are prepared with tools and vocabulary to be able to conceptualise the idea and learn language simultaneously. They are also supported, whilst they work on group tasks and problems, and implement the target language in interactive communication.

3.4 Conceptual Framework

By implementing the above-detailed methodologies and approaches, into an existing EMI content lesson, the building-block strategy called Topic Oriented Mixed-Method System (TOMMS) has been developed and coined.

TOMMS ensures that students are initially exposed to new concepts in a physical language lesson, taught by an LTE, who supports the students' needs in regards to new language constructions, techniques to manage complex concepts in English as well as introducing the topics themselves. This initial lesson is followed by a self-study session where, through blended learning, the principles and theories behind the concept are further explained. Thirdly, the students attend a physical EMI lesson where the CE, possibly aided by the LTE in a team-teaching approach, guides and supports the students, while they apply the learned concepts, in interdisciplinary long-term projects. It is in this phase, that the students are instructed, under the premise of the active classroom, in the implementation of the principles, in what Bonwell and Eison [23] call meaningful activities, that require students to think about what they are doing. Finally, students are required to work independently, at home, on creative output. These individual parts, content lesson LTE, self-study, content lesson CE and creative output, complete one block and cover one topic, they are then repeated for each topic in the curriculum. The individual tasks and projects merge, to form an interdisciplinary long-term project, that runs in parallel, therefore, allowing the students to combine the individual topics, concepts and theories into one solution-based example (see Fig. 1).

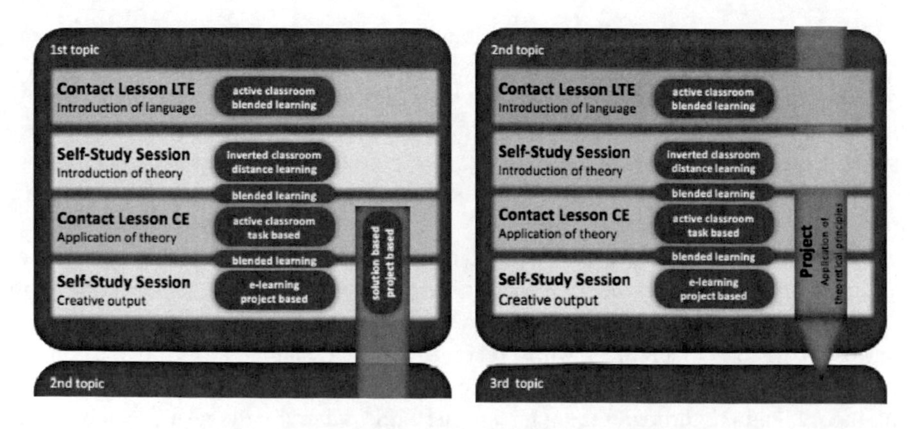

Fig. 1. Conceptual framework TOMMS

3.5 Implementation

In this paper, the syllabus, methodological planning and materials creation phases are not detailed due to space limitations, but were performed through the close collaboration of the LTE and the CE.

The conceptual framework was implemented in the course Innovation Management, in the degree programme Automotive Mechatronic and Management (AMM), in order to create synergies across the disciplines and competence areas of each expert. Originally this course did not include any ESL quota, it was, therefore, necessary to reduce the amount of CE teaching time, through blended learning and self-study, in order to be able to allocate teaching time to the LTE. It is, therefore, cost neutral and includes the equivalent amount to teaching hours, it is however more effective and efficient as both lessons now support, and build upon, each other and the finished product should ideally be more than just a sum of its constituting parts.

Figure 2 is an excerpt from the AMM syllabus showing the individual lessons and self-study session, of one topic from a total of five, illustrating the interdisciplinary synergies.

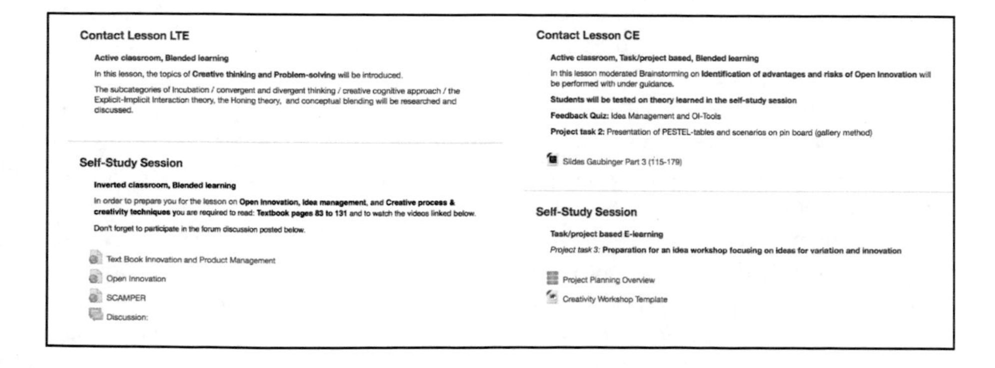

Fig. 2. Implementation of TOMMS (excerpt)

Initially, students are introduced to the main topics, creative thinking and problem-solving with language pertaining to them, in the contact lesson with the LTE. Students work in groups, on an online research task, where they are given different subcategories to investigate and then are required to explain their findings to the others. In this way, all students are active all of the time, whilst completing specially designed activities. This is followed by a reading and writing task, to expose them to language structures, required for the content class.

Subsequently, students are given multi-media materials, in a self-study session, which introduces the main content theory to them at their own pace before coming to the physical content lesson with the CE where these will be applied to real-world scenarios, in the form of a project, whilst receiving individual guidance from the content teacher. In addition, some time is taken in the contact lesson to check the understanding of the theory learned, through digital testing methods, whereby the results can be evaluated immediately and any deficiencies addressed. Students are also required to present the findings of a previous task (PESTEL tables) to their peers, which leads into a class discussion, on similarities and differences of various problem-solving skills, previously learnt.

Finally, students continue to apply the theory learned into their allocated projects, via a self-study session, where support is offered through chat forums and in person at the university.

4 Discussion

The English-Medium Instruction approach, with the supposition that language learning will be a by-product of the content lessons [6], is seen by the authors as credulous and the probability that effective transfer of content knowledge will be reduced, due to language obstacles, must be considered. Some of the basic principles and practices, as discussed in the introduction, that were initially employed in the Maastricht university, have been forgotten and this has led to ineffective practices where content teachers are unsupported and face issues for which they are not equipped to deal with.

Additionally, the linguistic objectives of the EMI courses need to be clearly defined and the teacher needs to be able to support the students with their knowledge of English [2]. TOMMS aims to illustrate how this can be effectively implemented, with language lessons introducing content, and preceding the content lessons themselves, thus creating the core competencies and skill sets needed to tackle academic materials in a foreign language. Again, the importance of the content expert and the language teaching expert working together in the creation of syllabus and materials, plus the continued collaboration during the teaching semester, cannot be understated as one of the underpinning concepts of TOMMS.

One of the major issues of English-taught programmes is the variation of academic ability of students [3], although not solely due to the new medium of instruction this variation may be emphasised (Ibid.) by it. The use of inverted classroom techniques is vital in order to prepare students for new academic topics, by supplying materials, which they can watch at their own pace, using the techniques taught to them. This allows for more effective knowledge transfer, and 'levels the playing field' of student ability before they attend a contact lesson.

There is a risk that students will not participate in activities or learn sufficient content [25] from the online activates set for self-study. As with all out-of-class activities, it is essential to check understanding at intervals, to reduce the possibility of students misunderstanding and falling behind. Here it is suggested to 'test' student's knowledge briefly, after each self-study block, during the contact lesson. This could be formal paper tests or quick opinion poll questions, through web-based apps, to get a general feeling of understanding. The teacher then has the possibility to re-address areas deemed necessary or to expand on the basic premise through active learning higher order tasks.

Teachers may lack the necessary skills to be able to implement the methodologies and tools described as well as possibly having a low affinity for them. It is the intention of this concept that teachers are guided, on a one-to-one basis, during course planning, by a methodological expert, and are able to choose the tools that they see as appropriate for them. They are then aided in implementing and setting them up. In this manner, it is hoped, that insecurities and aversions be reduced. Another issue is the involved development costs and time in creating new materials and applying such methods, which may even require dedicated members of staff initially. This is, in the author's opinion, unavoidable and must be accepted if the HEI is serious about developing itself in EMI, and therefore reaping the rewards that go with it.

The TOMM system is currently being implemented in one study programme in the university of applied sciences in Wels, Austria. Upon completion of the trial phase, predicted to be in 2018, it will be rolled out across more study programmes and disciplines. However, the actual effects on, and improvements in, knowledge transfer efficiency and accuracy have not yet been empirically measured.

This system requires at least one of the teachers to be the teaching expert, with deep pedagogical knowledge. In our scenarios, this is the language teaching expert but it must be noted that it could quite as well be the content teaching expert who plays this role, and leads in the creation of materials and methods, for the topics.

Although, in this example, the course where TOMMS was implemented was an EMI programme without language support quota, the framework can just as well be used to integrate already existing language lessons, and in this manner create better synergy, through the integration of language teachers into content courses.

References

1. Coleman, J.A.: English-medium teaching in European Higher Education. Lang. Teach. **39**, 1–14 (2006)
2. Dearden, J.: English as a medium of instruction – a growing global phenomenon, pp. 1–8. Going Glob 2014, Interim Report, Oxford Dep. Educ. Univ., Oxford (2014)
3. Wächter, B., Maiworm, F.: English-Taught Programmes in European Higher Education: The State of Play in 2014 (2014)
4. Hüppauf, B.: Globalization: Threats and Opportunities, Globalization and the Future of German (2004)

5. Doiz, A., Lasagabaster, D., Sierra, J.M.: English-Medium Instruction at Universities: Global Challenges. Multilingual matters, Bristol (2012)
6. Dafouz, E., Camacho-Miñano, M.M.: Exploring the impact of English-medium instruction on university student academic achievement: the case of accounting. Engl. Specif. Purp. **44**, 57–67 (2016)
7. Sert, N.: The language of instruction dilemma in the Turkish context. System **36**, 156–171 (2008). https://doi.org/10.1016/j.system.2007.11.006
8. Byun, K., Chu, H., Kim, M., et al.: English-medium teaching in Korean higher education: policy debates and reality. High. Educ. **62**, 431–449 (2011). https://doi.org/10.1007/s10734-010-9397-4
9. Maiworm, F., Wächter, B.: English-language-taught degree programmes in European higher education: Trends and success factors. ACA Papers on International Cooperation in Education. Lemmens, Bonn, Germany (2002)
10. Tatzl, D.: English-medium masters? programmes at an Austrian university of applied sciences: Attitudes, experiences and challenges. J. Engl. Acad. Purp. **10**, 252–270 (2011). https://doi.org/10.1016/j.jeap.2011.08.003
11. Wachter, B., Maiworm, F.: English-Taught Programmes in European Higher Education: The Picture in 2007 (2008)
12. Sowden, C.: Elf on a mushroom: the overnight growth in English as a Lingua Franca. ELT J. **66**, 89–96 (2012). https://doi.org/10.1093/elt/ccr024
13. Spolsky, B.: Language Policy. Cambridge University Press, Cambridge (2004)
14. Shepard, J.: An e-recipe for success. EL Gaz. 312, 5 December (2005)
15. Rotellar, C., Cain, J.: Research, perspectives, and recommendations on implementing the flipped classroom. Am. J. Pharm. Educ. **80**, 1–9 (2016). https://doi.org/10.5688/ajpe80234
16. King, A.: From sage on the stage to guide on the side. Source Coll. Teach. **41**, 1–7 (2008)
17. Lage, M.J., Platt, G.J., Treglia, M.: Inverting the classroom: a gateway to creating an inclusive learning environment. J. Econ. Educ. **31**, 30–43 (2000). https://doi.org/10.1080/00220480009596759
18. Harmer, J.: How to Teach English (Second Edition). ELT J. **62**, 313–316 (2007). https://doi.org/10.1093/elt/ccn029
19. Lim, D.H., Morris, M.L., Kupritz, V.W.: Online vs. blended learning: differences in instructional outcomes and learner satisfaction. J. Asynchronous Learn. Netw. **11**, 27–42 (2007)
20. Thompson, I.:Job impact study: the next generation of corporate learning. JobImpact. pdf (2002). 07 Octubre 2003
21. Means, B., Toyama, Y., Murphy, R., et al.: Evaluation of evidence-based practices in online learning: a meta-analysis and review of online learning studies. US Dep Educ, vol. 94 (2010). https://doi.org/10.1016/j.chb.2005.10.002
22. Bloom, B.S., Englehard, M.D., Furst, E.J., et al.: Taxonomy of Educational Objectives: The Classification of Educational Goals: Handbook I Cognitive Domain, vol. 16, p. 207, New York (1956)
23. Bonwell, C.C., Eison, J.A.: Active Learning: Creating Excitement in the Classroom. 1991 ASHE-ERIC Higher Education Reports. ERIC (1991)
24. Laur, D.: Authentic Learning Experiences: A Real-World Approach to Project-Based Learning. Routledge, New York (2013)
25. Chickering, A.W., Gamson, Z.F.: Seven principles for good practice in undergraduate education. AAHE Bull. **3**, 7 (1987). https://doi.org/10.1016/0307-4412(89)90094-0
26. Willis, D., Willis, J.: Doing task-based teaching. Tesl-Ej **12**, 173–176 (2008). https://doi.org/10.1093/elt/ccm083

27. Nunan, D.: Task-Based Language Teaching, pp. 1–15 (2004). https://doi.org/10.1017/CBO9780511667336
28. Prabhu, N.S.: Second Language Pedagogy. Oxford University Press, Oxford (1987)
29. Willis, J.: A flexible framework for task-based learning. In: Challenge and Change in Language Teaching, pp. 52–62 (1996)
30. Krashen, S., Terrell, T.D.: The Natural Approach: Language Acquisition in the Classroom. Pergamon, Oxford (1983)
31. Krashen, S.D.: Principles and practice in second language acquisition. Mod. Lang. J. (1982). https://doi.org/10.2307/328293
32. Long, M.: The role of the linguistic environment in second language acquisition. In: Handbook Language Acquisition, pp 413–468 (1996)

Developing Assessment for a Creative Competition

Zsuzsa Pluhár(✉)

Faculty of Informatics, Eötvös Loránd University, Budapest, Hungary
pluharzs@inf.elte.hu

Abstract. The significant role of ICT in everyday life could change concepts about skills, education and learning.

The digital fluency means not more to use of ICT tools (browsing, chatting, interacting etc.), but also to be able to design and create something new with the possibility of new media and to be more than an ordinary user, to be a creative creator. Create a program is one of the possibilities that supports parts of computational thinking and helps express yourself, explore the range of computers and yourself, involve external representation of problem solving processes, and to reflect on your own thinking – and even to think about thinking itself.

Men majorly dominate the IT sector. On average 30% of the tech jobs around the globe are filled with women, but in Europe this number is even lower, only 7%. The reasons behind the decaying numbers root in our culture too. To meet more women on this field we have to change the mindset of the kids – attitudes of boys and girls, too.

The work presented in this study focuses specifically on the improvement of developing and organizing a competition in Scratch for creative groupwork for girls, called Scratchmeccs (ScratchMatch). 2017 was the second year that we could organized this event. We used background questionnaire and would like to study the workflows in groups and what influence the successful groups.

Keywords: ICT · Creative programming · IT · Scratch

1 Introduction

The use of information and communication technologies (ICT) in the 21st century and the need to develop relevant skills to participate effectively in the digital age [1] have increased.

The significant role of ICT in everyday life could change concepts about skills, education and learning [2, 3].

1.1 Computational Thinking and Creative Programming

The change in definitions of digital skills and fluency from digital literacy to computational thinking (CT) – a fundamental skill for everyone, not just for computer scientists – [4] shows the changing of concepts not only about the ICT influences and possibilities but about the role of users, teachers and students, as well [2].

M. E. Auer et al. (eds.), *Teaching and Learning in a Digital World*,
Advances in Intelligent Systems and Computing 715,
https://doi.org/10.1007/978-3-319-73210-7_98

"It is the mission of education to adequately supply students not only with factual knowledge and domain-specific problem solving strategies (which are crucial in and of themselves as well) but also with a broader set of skills required in today's societies" [5]. It means to support changes in roles and the formal and the informal learning environments, too.

The term "user" no longer means only to use ICT tools (browsing, chatting, interacting etc.), but also to be able to design and create something new with the possibility of new media and to be more than an ordinary user, to be a creative creator [6].

One possible tool to support more dimension of computational thinking and our aims in educational changing is to create a program [4]. It is not only coding – it is a process from the wording of the first idea, by planning and developing to create a new product. It helps to express yourself, explore the range of computers and yourself, involve external representation of problem solving processes, and to reflect on your own thinking – and even to think about thinking itself [7].

1.2 Constructionism and the Scratch

We can find the main idea to support "creative creators" by Papert who is "looking at children as the active builders of their own intellectual structures'' [8, p. 19].

The "learning-by-making" or "learning-by-doing" philosophy could give the theoretical background of logo and his constructionism. They have influenced the direction of educational reform and the roles of technology in education [9].

Nowadays, we can find more environments which support this philosophy in programming. Many of them are block-program languages, which means you can manipulate programming blocks with the drag and drop technique and you don't need to attend to syntax and spelling.

With them, kids can make animations, interactive stories or games without thinking about the complexity of program languages, and they can learn to think about algorithms without speak about programming specifications.

One of these languages is Scratch, developed by the Lifelong Kindergarten Group at MIT.

Scratch is not only a programming language – it is an educational environment where you can:

(1) "write" a program in a block-based language;
(2) be a part of a large community – get and give help, see others ideas or solutions for a problem.

An important aspect to using a first programming language is not only the usability and easiness (Papert's "low floor" and "high ceiling"), but to have support and the possibility to go on a higher level of programming.

Guzdial [19] believes that "languages need wide walls (supporting many different types of projects, so that people with different interests and learning styles can all become engaged)" [10, p. 63].

Scratch community gives these educational aspects, too.

2 Girls in IT

In last few years the imbalance between the genders in CS has received more focus.

Women in the IT sector are underrepresented [11], they are missing the possibilities of an IT career, and "computer science is missing out on female perspectives — a fact that can have negative consequences for society, as evidenced by the negative outcomes attributed to all-male design teams (Margolis and Fisher, 2002)" [11, p. 1045].

The main influence factor is based on the stereotypes, that girls and women are bad at CS, and the technical jobs are still considered as professions for men. More studies have found that these stereotypes – the gender differences in attitudes about technology – are formed in earlier age groups [12] and based on the Western socialization. It is influenced by importance of role models, access to and experience with computers, teacher methods and computer environments [13–16].

These stereotypes can be changed. But "the task of changing the outcomes of women's education in computer technologies is more complicated than simply teaching them how to use computers. … It is also necessary to change how the women (and the men around them) understand and talk about the presence and competence of women" [17].

3 Scratchmeccs

A non-profit organization (Skool) and the Faculty of Informatics at Eötvös Loránd University prepared a new competition, called Scratchmeccs ("meccs" means "match" in Hungarian). It is not exclusively for girls who already have some knowledge in programming. On the contrary, we want to inspire all girls who have an interest in technology to apply, even if they are completely new on the ICT field.

Girls between ages 12 and 16 can enter the competition in teams of three. There they need to plan, develop, and submit their projects in Scratch [10]. The project can be "anything": an animation, game, or a freestyle work.

Our main aim was to show (not only girls), that

(1) they can be creative on a computer, in ICT work and expressing themselves easily;
(2) they can work together – in a group; they don't need to be programmer, designer and musician in one, but they have possibilities to organize work and divide a project into parts;
(3) they have place in the CS world.

Our group at ELTE had (and has) two main parts in the preparation:

(1) to develop the evaluation methodology: how a creative work can be rated limiting the subjective influence in evaluation;
(2) to do research on how the idea works, what can influence the success on a creative competition.

The competition has three rounds: (1) a pre-screening, where several volunteers scored the projects with the help of our scale (see caption "The evaluation of projects").

Then (2), a five-member jury used our scale for evaluating the first 20 projects selected in pre-screening. For the jury, we invited generally recognized experts in music, graphics (design), education and game developing, as well.

(3) After the first rounds, 10 groups with the highest scores had to present how they organized their work, how they worked, some facts about ideas, group management and the project. The same jury scored for the groups.

To help the girls participating, we prepared tutorials and online courses about Scratch. Skool organized courses in not only getting to know Scratch in general, but also longer ones-, for programs specified in this competition.

4 Method and Results

4.1 The Method

The evaluation of projects

For the evaluation process, we developed a complex scale for topics as engagement (playability, playing experience), artwork (design), development (coding), complexity, completeness, usability, originality (creativity) and media appearance.

We weighted for subcategories. All subcategories were divided into three main parts – with a short description and more scaled scoring points.

This scale was used in two parts of the evaluation process. First in pre-screening, where volunteers scored the projects. One volunteer evaluated several projects and one project was evaluated by several volunteers – but not always the same ones. The first result was the average of scores. The five-member professional jury then evaluated the first 20 projects.

In the statistics about correlations, we made calculations only with these 20 projects and the scoring by the five-member jury.

Background questionnaire

We developed a background questionnaire to ask girls about motivations, attitudes and previous-knowledge which we thought could be important for this competition. For attitudes and knowledge, we used a five-level Likert scale and sometimes we lessened the "other" possibilities.

We asked about knowledge of program languages as Comenius and Imagine Logo, Kodu and C++ because these are the most used languages in Hungarian schools. We asked about Scratch – experiences, usage and attitudes.

We have some facts about workflows only from the second round – the presentation.

Because girls uploaded their projects after their work, it was not possible to use the questionnaire before and after the workflow, and it is not possible to study and measure the changes of thinking and knowledge in Scratch or programming. But with the questionnaire, we want to find influence-factors to be successful in a creative competition with group work.

4.2 Results

The first competition was in 2015/2016. We had 24 groups (72 girls). The second competition was this year (2016/2017), where we had 33 groups (99 girls). In the statistics about correlations, we calculate and write only about the first 20 projects in each and so about 60 girls in each.

There were 10 teams each year who registered themselves but did not send projects. They could not end the work, mostly because of the problem in cooperation.

Projects

The projects have mostly a basic story where there are some exercise games to play or a problem to solve.

In 2016, the additional games were mostly logic games. The game ideas were more specialized than those in 2017. This year, we had more "standard" projects – more mazes, drag-and-drop games from tutorials, and fewer games with new ideas or tricks. Every basic story could be found each year in all projects.

The topics (the basic story) of projects were very different – from the pink girl-world to the deep feeling about the meaning of life (Figs. 1 and 2).

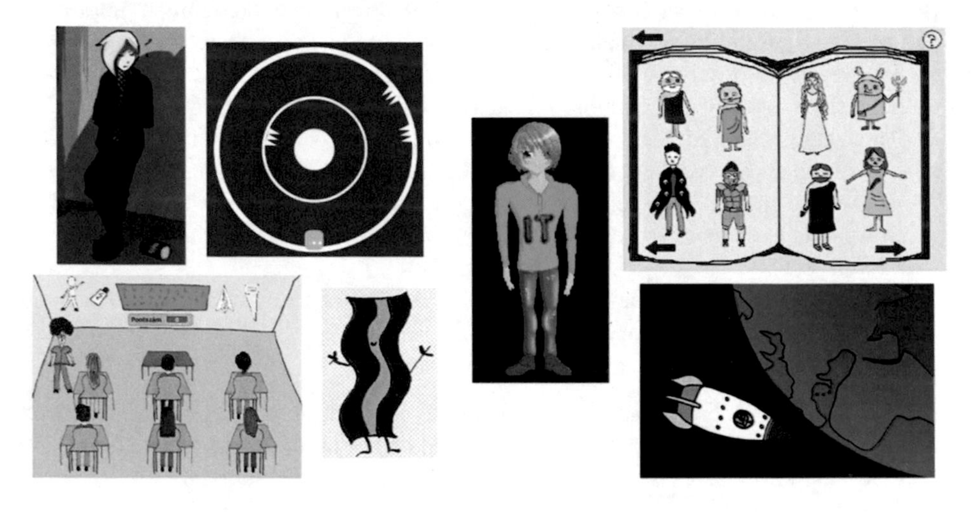

Fig. 1. Pictures about the winner's project in 2016

Fig. 2. Pictures about the winner's project in 2017

Girls mostly used their own pictures and music and really worked as designers, musicians and programmers. From presentations, we know they thought the hardest work was to come up with the main idea about topic, making the plans and working together.

Background questionnaire

Both years the girls were between 12 and 16. The average age was 14, but within the groups, at the greatest difference was one year. Girls in one group were not always from the same school or class; those were relatives or friends from camp.

In 2016, most girls had not participated in ICT competitions before (90%) and had no possibilities for afterschool activity in ICT (54%). In 2017, 60% had participated in one or more ICT competition before: 36% in a school-competition in programming, 22% in Bebras[1] and only 2% in Scratchmeccs the previous year. Nearly third (31%) had visited after-school activities once a week and 36% had no possibility for such.

Most girls had never heard about or used kodu or C++ yet both years. The most popular languages were Comenius Logo and Imagine Logo – they had used it in the school, but only 36% a few times, and 44% seriously (Fig. 3).

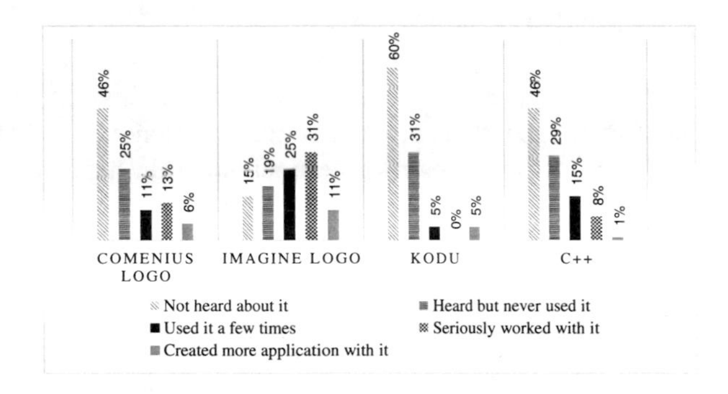

Fig. 3. How well do the participants know program languages.

The girls had heard about html, the LEGO Mindstorms control language, JavaScript and Python and in 2017 they had written C#, too.

In 2016, 30.6% had begun programming only for the competition. This number decreased in 2017: 27% had never programmed before. Few (2%) had more than two years' programming experience both years.

They had mostly used Scratch in school (33%), in after-school activities (7%) or only for their own fun (19%), but 26% had never used Scratch before the competition.

The attitude to Scratch was very good: 21% (2016: 38%) wrote that they could work with Scratch without assistance and 41% (2016: 41,7%) needed only a little bit. Only 6% (2016: less than 1%) reported problems with Scratch.

Correlations

The rates for engagement show a strong positive correlation with most other rates: with art and design ($r = 0.719$, $p < 0.01$), with development, coding ($r = 0.793$, $p < 0.01$), with complexity ($r = 0.869$, $p < 0.01$) and with creativity ($r = 0.908$, $p < 0.01$). A lower correlation was found between usability and creativity (0.375, $p < 0.05$), or development ($r = 0.432$, $p < 0.01$) and a low (the lowest) correlation between usability and art, design ($r = 0.299$, $p < 0.01$).

There was a very strong positive correlation between complexity and art, design ($r = 0.805$, $p < 0.01$) and between creativity and complexity ($r = 0.886$, $p < 0.01$).

We did not find a significant relationship between age and scores neither in sum nor in subcategories.

[1] http://e-hod.elte.hu; http://bebras.org.

There was no correlation between scores and previous knowledge-factors in programming and between scores and the facts how easy (or difficult) find participants Scratch.

There was no relationship between project topics and age. The school and the school environment influenced the topics the most.

An independent samples t test was conducted to compare the scores in 2015 and 2016. There was no significant difference between the two years.

5 Conclusion and Future Directions

Our results indicate that the main idea and the used tool fit our aims, and that motivated girls are creative, but a motivated teacher (or the Scratch community) in the background is helpful. Overall, the hardest point for girls in competition is the teamwork, to organize developing and implementing the projects.

The results confirm that successful participation in a creative competition is independent of previous knowledge of programming or knowing program languages.

Our next goals are

(1) to motivate more girls to participate;
(2) to give more support and possibilities to prepare teams for the competition;
(3) to open the competition to younger girls, as well.

We would like to be in communication with the girls who participated in Scratchmeccs and to follow them in the ICT world. They can set a good role model for other girls, as well [16, 18].

This year started Scratchmeccs in two other countries: in Poland and Slovakia. We hope to organize research next years over countries.

References

1. Fraillon, J., Ainley, J., Schulz, W., Friedman, T., Gebhardt, E.: Preparing for Life in a Digital Age. Springer International Publishing (2013)
2. Csapó, C.: A tudáskoncepció változása: nemzetközi tendenciák és a hazai helyzet. Új Pedagógia Szemle, vol. 2, pp. 38–45 (2002)
3. Molnár, G., Kárpáti, A.: Informatikai műveltség. In: Csapó, C., (ed.). Mérlegen a magyar iskola. Nemzeti Tankönyvkiadó, Budapest, pp. 441–476 (2012)
4. Wing, J.: Computational thinking. Commun. ACM **49**(3), 33–35 (2006)
5. Greiff, S., Wüstenberg, S., Csapó, B., Demetriou, A., Hautamäki, J., Graesser, A.C., Martin, R.: Domain-general problem solving skills and education in the 21st century. Educ. Res. Rev. **13**, 74–83 (2014)
6. Resnick, M.: Sowing the seeds for a more creative society. learning and leading with technology, pp. 18–22, December (2007)
7. Disessa, A.: Changing Minds: Computers, Learning, and Literacy. MIT Press, Cambridge (2000)
8. Papert, S.: Mindstorms: Children, Computers, and Powerful Ideas. Basic Books, New York (1993)

9. Kafai, Y., Resnick, M. (eds.): Constructionism in Practice: Designing, Thinking, and Learning in a Digital World. Lawrence Erlbaum, New Jersey (1996)
10. Resnick, M., Maloney, J., Monroy-Hernandez, A., Rusk, N., Eastmond, E., Brennan, K., Millner, A., Rosenbaum, E., Silver, J., Silverman, B., Kafai, Y.: Scratch: programming for all. Commun. ACM 52(11), pp. 60–67 (2009). http://web.media.mit.edu/~mres/papers/Scratch-CACM-final.pdf
11. Cheryan, S., Plaut, V.C., Davies, P.G., Steele, C.M.: Ambient belonging: how stereotypical cures impact gender participation in computer science. J. Pers. Soc. Psychol. **97**(4), 1045–1060 (2009)
12. Fletcher-Flinn, C.M., Suddendorf, T.: Computer attitudes, gender and exploratory behavior: a developmental study. J. Educ. Comput. Res. **15**(4), 369–392 (1996)
13. Klawe, M., Leveson, N.: Women in computing: where are we now? Commun. ACM **38**(1), 29–35 (1995)
14. Funk, J., Buchman, D.D.: Children's perceptions of gender differences in social approval for playing electronic games. Sex Roles **35**(3/4), 219–231 (1996)
15. Davies, A.R., Klawe, M., Ng, M., Nyhus, C., Sullivan, H.: Gender issues in computer science education. In: Proceedings of the National Institute of Science Education Forum, Detroit. Accessed: May 2017. http://www.wcer.wisc.edu/nise/News_Activities/Forums/Klawepaper.htm
16. Sanders, J.: Girls and technology: villains wanted. In: Rosser, S. (ed.) Teaching the Majority: Breaking the Gender Barrier in Science, Mathematics, and Engineering, pp. 147–159. Teachers College Press, New York (1995)
17. Henwood, F.: Exceptional women? gender and technology in U.K. higher education. IEEE Technol. Soc. Mag. **18**(7), 21–27 (1999)
18. Teague, J.: Women in computing: what brings them to it, what keeps them in it? SIGCSE Bull. **34**, 147–158 (2002)
19. Guzdial, M.: Programming environments for novices. In: Fincher, S., Petre, M., (eds.) Computer Science Education Research. Taylor & Francis, pp. 127–154 (2004)

Modbus Protocol as Gateway Between Different Fieldbus Devices - a Didactic Approach

Armando Cordeiro[1], Paulo Costa[1], Vitor Fernão Pires[2,3](✉), and Daniel Foito[2]

[1] ADEEEA – Área, Departamentalde Engenharia Eletrotécnica, Energia e Automação, ISEL – Instituto Superior de Engenharia de Lisboa, Rua Conselheiro Emídio Navarro 1, 1959-007 Lisbon, Portugal
acordeiro@deea.isel.pt, pauloarcosta81@gmail.com
[2] ESTSetúbal-Instituto Politécnico Setúbal Campus do IPS, Estefanilha, 2914-761 Setúbal, Portugal
{vitor.pires,daniel.foito}@estsetubal.ips.pt
[3] INESC-ID Lisboa, Rua Alves Redol, 9, 1000-029 Lisbon, Portugal

Abstract. This paper is dedicated to improve the skills of Electric Engineering students about Industrial Automation. This is done creating a set of practical exercises in the laboratory that allow students to interact with didactic and industrial equipment understanding the compatibility problems and give students the tools, the knowledge and even some creative freedom to solve problems that happen in real situations. In this case it is focused on the interconnection of equipment from different manufacturers using the Modbus protocol. To allow to the students this practical study, it was developed a solution based on the Modbus RTU protocol directly implemented in the program memory of both S7-200 (using the Freeport configuration mode) and S7-300 PLCs. The solution allows a gateway between Modbus-TCP (over a Master TSX PLC from Schneider-Electric) and Profibus-DP (over a Master S7-300 PLC from Siemens).

1 Introduction

Industrial Automation is a multidisciplinary area that requires experience and competences in control, energy, electronics, robotics and computer engineering, among others [1, 2]. Despite the existing regulations and standards, many professionals in this area have already experienced the difficulty of interconnecting equipment from different manufacturers. This problem may arise during the upgrade or modification of industrial facilities where manufactures propose new products with new technologies but unfortunately not always compatible with already existing devices. This is also a common issue in fieldbus devices for real-time DCS (Distributed Control Systems) were the development of various devices conforming to different standards, such as CAN, Profibus, Interbus, WorldFIP, etc., usually originates compatibility problems [3, 4]. Fortunately, the international standard committees had the merit of restraining the different solutions within a limited range, thus addressing users, device manufacturers, and integration experts toward a limited number of alternative choices among

M. E. Auer et al. (eds.), *Teaching and Learning in a Digital World*,
Advances in Intelligent Systems and Computing 715,
https://doi.org/10.1007/978-3-319-73210-7_99

components and solutions [3]. The easiest way to solve this problem in a real application is to purchase a specific communication interface when it is available. Nevertheless, it is not always possible to carry out this acquisition, due to limited financial resources or just because the interface is not available/compatible. Thus, it is necessary to give training to future engineers to solve problems of this nature. The Electrical Engineering Department at the Instituto Superior de Engenharia de Lisboa (ISEL) has a long tradition teaching fieldbus technologies for Industrial Automation. The knowledge and experience of teachers about fieldbus technologies is largely due to the close relationship of this institute with industries of electrical equipment and services. In terms of education about theory of fieldbus systems students are encouraged to learn different aspects, namely:

- Principles of communication theory (communication techniques, codification, modulations, topologies, data rates, media access methods, etc.);
- Fieldbuses for Industrial Automation;
- Commonly used Fieldbus standards (physical layer standard, media supported, maximum nodes, deterministic, intrinsically safe, maximum segment length, etc.);
- Describe the network processes that use protocols and how they utilize them;
- Identify the functions of protocols and protocol stacks;
- Map specific protocols to the appropriate OSI layer.

In order to consolidate theoretical knowledge about fieldbus systems several practical experiments are normally proposed. Some experiments are dedicated to simple RS-232 communications through half-duplex transmissions using Modbus RTU. Others are dedicated to more complex configuration of fieldbus communication devices such as Profibus-DP using RS-485 or WorldFip using IEC61158 as physical layer standard. Usually, students who attend the Master degree in Electrical Engineering and want a specific expertise at Industrial Automation it is normally proposed a more advanced practical experiment such as described in this paper.

The proposed paper presents a solution to interconnecting PLCs with fieldbus interfaces and SCADA systems from different manufactures using the Modbus RTU protocol as gateway between these devices. Despite some limitations, it shows the potential of using open protocols, such as Modbus RTU, implemented in software directly in the program memory of the PLCs. It is necessary to keep in mind that the main objective is not to create an optimal or better solution than those ones proposed by manufacturers, because this would be an almost impossible task, but to provide advanced training for students to master aspects related to interrupts, timeouts, data synchronization, watchdogs, subroutines, direct and indirect addressing, buffering, among other aspects of fieldbus communications.

2 Equipment Description

The industrial fieldbus devices of the Automation Laboratory come from different manufacturers and have consequently distinct interfaces (RS-232 and RS-485), net-works and communication protocols, such as Modbus TCP [11, 12, 14], Modbus RTU [7, 10, 11, 13], Profibus-DP [5] and AS-I [6], among other devices that are not directly related with the presented paper. This section provides an overview of the main devices used in the present work.

2.1 TSX57103/302 Premium Controllers

The TSX57103 and TSX57302 CPU controllers from Schneider-Electric are modular controllers supplied with 24VDC. Each CPU has by default a serial communication port (TER) using UNI-TELWAY protocol for PLC programming and a RS-232 serial interface module (TSX SCP 111) to communicate with other devices using the Modbus RTU protocol. Each controller also has an Ethernet communication module (TSX ETY 110). Beside the digital I/O interfaces (TSX DEY16D2 and TSX DSY16R5), the TSX57302 CPU has also an additional Modbus interface (TSX SCY21601) and is the Master of a two fieldbus devices: a power meter and a protection relay both from Schneider-Electric (PM500 and SEPAM series 40). The TSX57103 CPU controller is also operating as a Master (through the TSX SCP 111 interface) with other fieldbus device, the S7-200 (slave) from Siemens.

2.2 S7-200 Controller

The S7-200 is a compact controller from Siemens supplied with 230VAC and has a 215-2DP CPU [18]. This controller has two communication Ports, Port 0 and Port 1. Since this controller does not support the Modbus RTU protocol it was used in Freeport mode to communicate with the TSX57103 CPU controller. The port 0 was configured in the programming software (STEP7 Microwin) to receive data from Modbus RTU messages sent by the TSX57103. Since the S7-200 has RS-485 interfaces and the TSX SCP 111 module installed in the TSX57103 has an RS-232 interface it was necessary to use an appropriate RS-232/RS-485 cable converter and perform some changes in the pinout. The S7-200 is also connected as a slave to a PROFIBUS-DP network through Port 1 where other PLC (S7-300) is the master.

2.3 S7-300 Controller

The S7-300 is a modular controller from Siemens supplied with 24VDC and has a 315-2DP CPU. It has two communication ports: X1 with MPI (Multi-Point Interface) for programming the PLC and X2 for Profibus-DP network communication [18]. The S7-300 is the master of the PROFIBUS-DP network and has three slaves: the S7-200 (215-2DP), the gateway DP/AS-i Link 20 and a MicroMaster VFD which is a variable speed electrical drive [19].

3 Proposed Solution

The proposed solution to integrate fieldbus devices from different manufactures is presented in Fig. 1. This solution took into account the following aspects/restrictions:

- Minimal changes in the physical layout (mainly because the devices are used by other students in other laboratorial experiments);

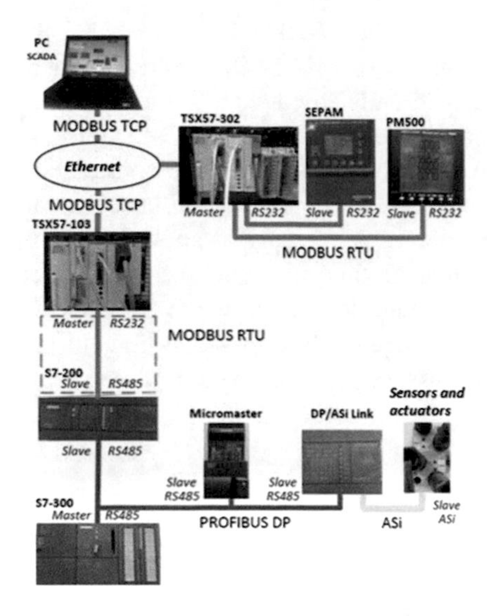

Fig. 1. Proposed solution to integrate fieldbus devices from different manufactures

- Create a solution as close as possible to a real scenario.
- No investments in new equipment or interfaces;

As a result of these restrictions, the solution was divided in two main components: hardware and software. Concerning hardware the only possible solution to create and connect the SCADA system was using the Ethernet interfaces available in both TSX57 controllers. The physical location of the devices dictated that the connection of the TSX57103 to the PROFIBUS-DP network had to be made through the S7-200 (working as a gateway through software).

Concerning software, the S7-200 had to be programmed (Port 0 in Freeport mode) to support the MODBUS RTU protocol sent by the TSX57103 CPU (through the TSX SCP 111 module), and had to transmit data messages to other devices present in the PROFIBUS-DP network if needed. To send messages between different networks, they must be sent to the master of the network which as the capability of, after identifying the slave address, transmitting messages. The difficulty of this solution lies in the fact that the S7-200 is a slave device in both networks which means that it can receive messages from the TSX57103 but cannot send directly the information to the S7-300 (the master of the PROFIBUS-DP network) or to other slave devices.

To solve this problem it was necessary to create and manage memory buffers in the S7-200 CPU. This solution allowed to store the "received messages" and create conditions to "send messages" to other network if necessary (depending of the origin and destination of the message). The most critical scenario happens when the messages sent to S7200 CPU have another destination (from TSX57103 to other device in the Profibus-DP or the opposite as result of the first action). In this case it was necessary to create a polling cycle in the PLC program of the S7-300 CPU to read the information stored in the memory buffers of the S7-200 CPU to realize if there are messages to any device of the PROFIBUS-DP network. If the S7-300 reads that a new message is present in the S7-200 and the destination is any device of the Profibus-DP network, processes the message executing the command and writing the response back to the S7-200 again. The S7-200 does not have the ability to initiate the communication with other devices but can reply to a message received, in this case it will use the data written in the memory buffers by the S7-300 to build a new message as answer to the original message sent by the TSX57103. An example of this procedure is exemplified in Fig. 2(a) where a message sent from SCADA requests an input value from a sensor in the AS-i network. This procedure has to be managed before message timeout (100 ms by default in the TSX57103). Figure 2(b) illustrates the approximate time involved in this procedure.

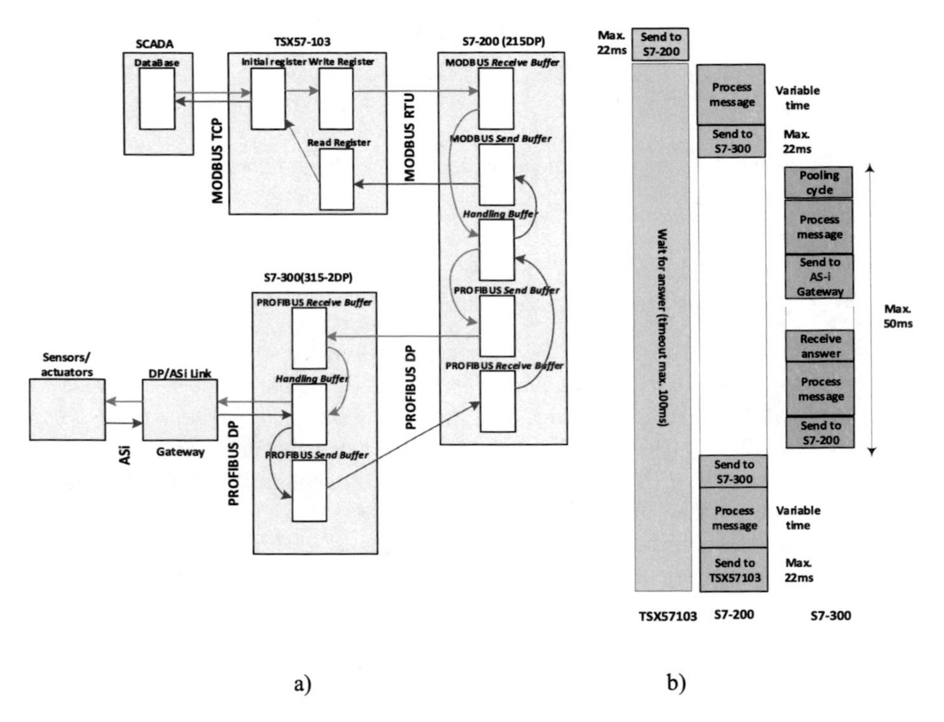

Fig. 2. Procedure example of message sent from SCADA software to a sensor in the AS-i network. (a) Example that demonstrates the management of the memory buffers; (b) Time necessary to process the message from TSX57103 to an AS-i device placed in the Profibus-DP network.

3.1 Details About the S7-200 Program

The developed program for the S7-200 is composed by two parts: the generic control program and the message processing program, both running in different subrotines. Because the Modbus RTU protocol was implemented in the software program of the S7-200 and is running along with other generic control program, the code must be optimized to process fast enough at least two messages (message and answer) before the message timeout of the TSX57103 (100 ms by default and without possibility of change). The Modbus RTU protocol was developed according with Modbus Application Protocol Specification [8]. Some aspects related with the configuration of the communications in the S7-200 program are presented next.

The first step of configuration consists of set the Port 0 to operate in Freeport mode, sending specific values to special memory bytes. This is done in sub-routine 0, which is only processed on the first cycle after a CPU reboot. When a new message is received in Port 0 the first action is always to check the message integrity by performing a 16 bit CRC calculation. The result is compared with the received CRC to validate the message. The second action is to identify the slave address, checking if it is a broadcast message, a message to the S7-200 or for any other equipment. If the message is to S7-200 (direct address) it will be processed immediately. When the address belongs to a device operating in the PROFIBUS-DP network (indirect address) the message will be transferred to the "PROFIBUS Send Buffer" (see Fig. 2(a)) and the S7-300 PLC will be responsible for reading this buffer. In case of any error, the program will send defined error messages according with Modbus specifications.

The program was tested with write and read bits to the peripherals in the AS-i network and read/write 16 bit registers in the MicroMaster VFD and S7-300 with success (response time < 100 ms). For very long and not optimized generic control program is not always possible to succeed in communications.

3.2 Communication Between the S7-200 and the S7-300

The communication between the S7-200 (slave) and S7-300 (master) is done directly through the PROFIBUS-DP network. Both devices have a specific RS-485 to this effect and this network is configured using the programming software (STEP7 Manager). The master sends and receives messages using instruction blocks already available in the software. As mentioned before, the S7-200 is a slave device in both networks and cannot send automatically messages available in the "PROFIBUS send buffer". The S7-300, which is the master of the PROFIBUS-DP network, is responsible for reading this buffer, using a polling cycle in PLC program. The S7-300 after processing the message writes the response in the S7-200 "PROFIBUS Receive buffer" and the S7-200, after detecting that a new response message has been sent back, transfers the data to the "MODBUS Send buffer" and complete the message to send now to the TSX57103.

3.3 Communication Between the TSX57103 and the S7-200

As described earlier, the TSX57103 PLC has available a TSX SCP 111 (Modbus RTU PCMCIA RS-232 interface) and the S7-200 uses Port 0 with a RS-485 interface in Freeport mode to communicate. To achieve hardware compatibility between both devices it was necessary to connect the cables from different manufactures. A direct RS-232 cable was connected to a PPI *Multi-Master* Cable (which has a RS-485/RS-232 converter). Because cables have different pinout it was necessary to create a pinout inversion box. Figure 3 presents the physical connection between TSX57103 and S7-200.

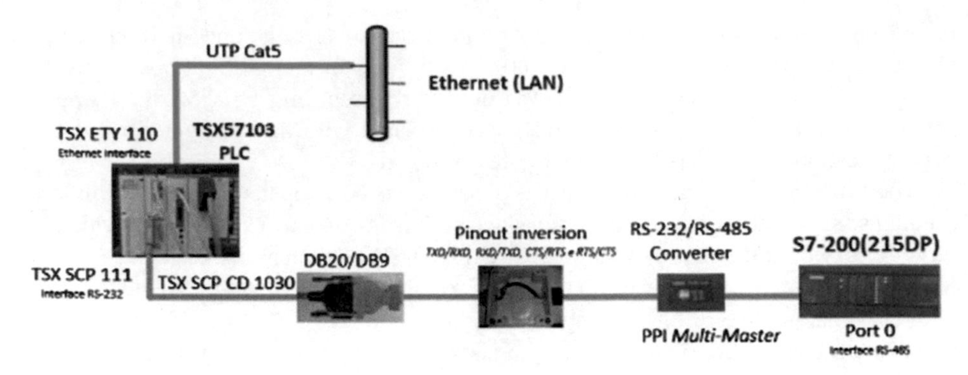

Fig. 3. Physical connection between PLCs: TSX57103 and S7-200.

Another relevant aspect of this solution is that the S7-200 program can only process one received Modbus RTU message at a time (message plus response) before processing another. On the other hand, the TSX57103 CPU uses a dedicated communications interface capable of processing eight messages sequencialy. Since the TSX57103 CPU is able to send messages faster than the processing time of the S7-200 it was necessary to create a control structure to implement a delay cycle in the TSX57103 CPU.

3.4 Communication Between SCADA Software and TSX57's

The communication between SCADA software and the TSX57103/302 PLCs was established through the Ethernet LAN of the campus. This solution is not the most desirable since it has not the same requirements (high reliability, safety critical, deterministic behavior, etc.) of a dedicated Industrial Ethernet network, but can be considered acceptable for academic purposes. Because the TSX57103/302 PLCs have Ethernet communication interfaces (TSX ETY 110) it is quite simple to perform communication between the PC hosting the SCADA software and the PLCs [15–17]. The adopted protocol in this connection was the Modbus TCP [9]. Two types of messages where considered, direct messages to the TSX57103/302 PLCs and indirect messages if destined to any other equipment. Direct messages are processed directly on the TSX57 controllers and the instructions executed directly. Indirect messages consist of writing specific memory registers with data, which will be sent to remote devices using MODBUS RTU protocol through the TSX SCP 111 interface (see Fig. 4).

The addresses of the devices were divided according with three different zones. Addresses 00hex, 10hex, and 20hex were used to send broadcast messages to the respective zone. Devices with addresses from x1 to x9 (where x represents 0, 1 or 2) are allowed for each zone. For example the S7-200 it is in zone 2 and is the device number 5 and so it have the address 25hex (slave address 5 on the PROFIBUS-DP network).

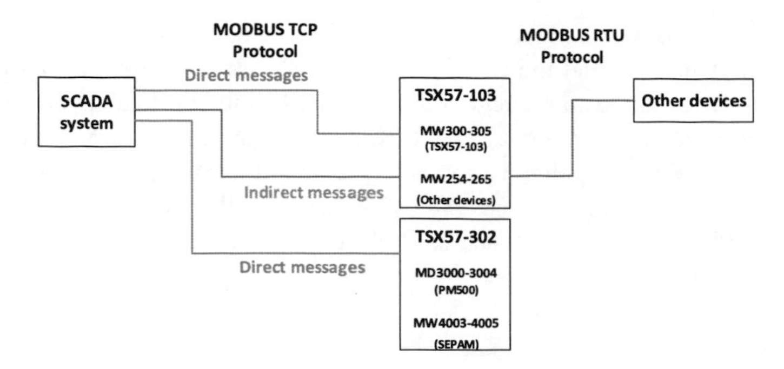

Fig. 4. Destination messages from the SCADA system.

3.5 SCADA System

The solution for SCADA system is quite simples and is based on a PC running the MOVICON SCADA software on a windows operating system. It were not developed specific layouts to simulate industrial processes. The layouts of the SCADA software were designed to test the operation of the fieldbus devices using automatic commands (scripts) or using manual inputs through a control panel. The SCADA solution developed allows to send and receive messages to any device available on the existing networks. Figure 5 shows the main layout of the SCADA software for this purpose. In this layout it is possible to choose que device and select in other layouts (with certain limitations) the variables that we want to read and write and the obtained communication result.

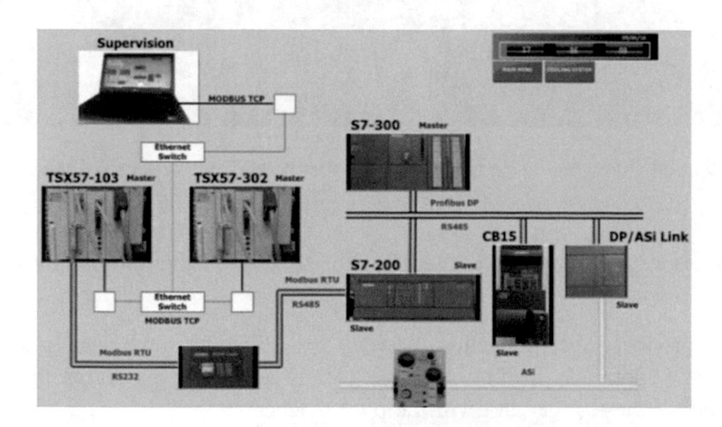

Fig. 5. Main layout of the SCADA software developed.

4 Laboratory Experiments and Evaluation of This Didactic Approach

With the developed system it is possible to perform new laboratory experiments, helping students to understand several aspects of fieldbus communications. Figure 6 shows a partial view of the developed system in the Automation Laboratory. In this picture is possible to see at right the SCADA system developed, in the middle the Siemens fieldbus devices and the Schneider-Electric devices at right. Using this new experiment students can test an integrated solution of different fieldbus devices using the well-known Modbus protocol. The PLCs programs are available and can be reprogramed by students any time to add new features if necessary. Using this system, students can also perform laboratory exercises in which they will learn how to configure different fieldbus networks, such as Profibus-DP, AS-i and WorldFIP. This kind of laboratory experiments also provides a good background to take the next steps into the future of automation: the Industry 4.0.

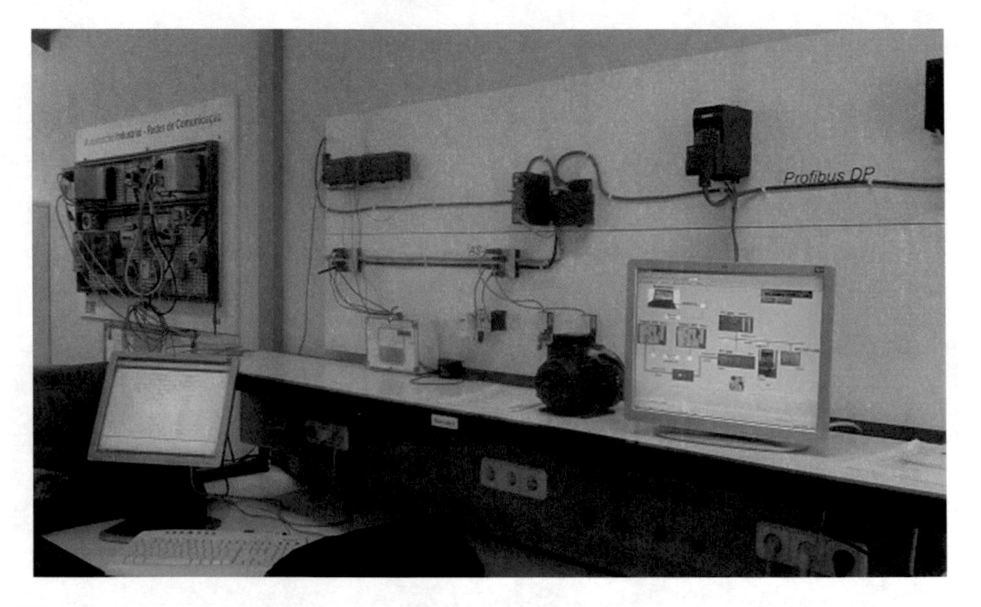

Fig. 6. Photograph showing a partial view of the developed system in the Automation Laboratory.

Several practical exercises are available to students in the automation laboratory according with a certain educational strategy, namely:

- First it is proposed the connection between two PLCs from the same manufacture. As expected, normally in these exercises there are minor problems they have to solve reading the manual and with help of teachers;
- Second, it is proposed the connection between two PLCs from different manufactures. This exercise is more complicated and it is requested to the students to study the service manual of both devices and other documents and produce a report in order to solve the problem. Each group of students presents the main topics of the report during the class and discuss the solution with colleagues and teachers;
- In the next step and after the discussion (and eventually after some modifications), students try to connect the devices according with the study presented in the report;
- Finally the students will test the solution developed by the teachers to compare the results.

Usually this kind of laboratory experiments is performed by Master's students of Industrial Automation. Through the Master course, students are normally seeking for new opportunities to increase their knowledge in this area. It is at this stage of their lives that usually, after having had contact with the first job as a technical engineer, that they assume a more active role within school and society. It is remarkable to see students during this stage trying to do work with a well-defined objective and looking for answers by themselves with self-confidence. This psychological aspects give students additional experience to face the challenge of working in the real world. It is necessary to understand that the role of engineering schools should not be only dedicated to train engineers since the real world goes far beyond from math and physics calculations. This is a role that teachers from any area should never forget. Students who have already used this system have recognized that it is a useful tool that helps to understand important aspects related to communications in an industrial environment, such as the diversity of communication interfaces, transmission speed versus cable length, manage communication errors, determinism in the delivery of data between devices, different network access rules (master-slave, spontaneous, token ring), among other aspects. The results show that students who perform this kind of laboratory experiments (didactic but using real devices) have increased their knowledge and interest in industrial automation and are better prepared for their professional life in this area. The increase of knowledge through laboratory experiments is also very rewarding for teachers and the feedback from employers is very good. This way of teaching seems to be a good complement to theoretical classes of Industrial Automation.

5 Conclusions

The objective of the present study was to create a solution to integrate several devices from different manufactures without purchase specific hardware interfaces, providing training to future Electric Engineers to solve problems of this nature. To achieve this objective it was necessary to study and understand the devices and the protocols and create a specific solution to deal with the compatibility problems. As expected, the most complex problem to solve was the communication between the MODBUS RTU protocol used by the TSX57103 controller and the PROFIBUS-DP network used by the S7 controllers. It was necessary to program the S7-200 PLC to behave as a gateway between these two networks. This was difficult because the S7-200 is a slave device in both networks. To solve this problem were used memory buffers in S7 PLCs combined with indirect readings cycles to detect the presence of new messages. Also the generic control programs of the S7 PLCs (with main concern in the S7 200) must be optimized to avoid messages timeout. For very long and not optimized generic control program is not always possible to succeed in communications. This is the one of the main disadvantage of the proposed solution. Students who participate in this kind of experience are better prepared for their professional life with regard to industrial automation.

References

1. Vyatkin, V.: Software engineering in industrial automation: state-of-the-art review. IEEE Trans. Ind. Inform. **9**(3), 1234–1249 (2013)
2. Efe, M.Ö.: Fractional order systems in industrial automation - a survey. IEEE Trans. Ind. Inform. **7**(4), 582–591 (2011)
3. Benzi, F., Buja, G., Felser, M.: Communication architectures for electrical drives. IEEE Trans. Ind. Inform. **1**(1), 47–53 (2005)
4. Felser, M., Sauter, T.: The fieldbus war: history or short break between battles? In: Proceedings of IEEE International Workshop on Factory Communication System, pp. 73–80 (2002)
5. Weigmann, J., Kilian, G.: Decentralization with Profibus-DP: Architecture and Fundamentals, Configuration and use with SIMATIC S7. Publicis MCD, Verlag (2000). ISBN 3-89578-144-4
6. Švéda, M., Vrba, R.: Actuator-sensor-interface interconnectivity. Elsevier-Control Eng. Pract. **7**(1), 95–100 (1999)
7. "Modicon Modbus Protocol Reference Guide", PI–MBUS–300 Rev. J, 1996, MODICON, Inc., Industrial Automation Systems, One High Street North Andover, Massachusetts 01845. http://www.Modbus.org
8. Lemay, A., Fernandez, J.M., Knight, S.: A modbus command and control channel. In: Proceedings of IEEE Annual Systems Conference (SysCon), Orlando, Florida, USA (2006)
9. Al-Dalky, R., Abduljaleel, O., Salah, K., Otrok, H., Al-Qutayri, M.: A Modbus traffic generator for evaluating the security of SCADA systems. In: Proceedings of 9th IEEE International Symposium on Communication Systems, Networks & Digital Signal Processing (CSNDSP), 23–25 July 2014, Manchester, UK (2014)
10. Bonganay, A.C.D., Magno, J.C., Marcellana, A.G., Morante, J.M.E., Perez, N.G.: Automated electric meter reading and monitoring system using ZigBee-integrated raspberry Pi single board computer via Modbus. In: Proceedings of IEEE Students' Conference on Electrical, Electronics and Computer Science (SCEECS), 1–2 March 2014, Bhopal, India (2014)
11. Tamboli, S., Rawale, M., Thoraiet, R., Agashe, S.: Implementation of Modbus RTU and Modbus TCP communication using Siemens S7-1200 PLC for batch process. In: Proceedings of IEEE International Conference on Smart Technologies and Management for Computing, Communication, Controls, Energy and Materials (ICSTM), 6–8 May 2015, Chennai, India (2015)
12. Liu, Q., Li, Y.: Modbus/TCP based network control system for water process in the firepower plant. In: Proceedings of 6th World Congress on Intelligent Control and Automation (WCICA), vol.1, pp. 432–435 (2006)
13. "Introduction to Modbus Serial and Modbus TCP", vol. 9, Issue 5 Sept-Oct.2008, the Extention, a technical supplement to control network, 2008 Contemporary Control Systems, Inc
14. Modbus Messaging on TCP/IP implementation Guide V1.0b (2006). http://www.Modbus-IDA.org
15. Daneels, A., Salter, W.: What is SCADA? In: Proceedings of International Conference on Accelerator and Large Experimental Physics Control Systems, Trieste, Italy (1999)
16. Amy, L.T.: Automation Systems for Control and Data Acquisition - Resources for measurement and control series, 1st ed. ISA, 1992. ISBN 13: 9781556177798 (1992)

17. Clarke, G., Reynders, D., Wright, E.: Practical Modern SCADA Protocols: DNP3, IEC60870.5 and Related Systems, Elsevier, 1st ed. (2004)
18. SIEMENS, "SIMATIC S7-200 Programmable Controller System Manual," 2008; SIEMENS, "SIMATIC S7-300 CPU 31xC and CPU 31x: Technical specifications," (2011)
19. SIEMENS, "MICROMASTER 420 0.12 kW - 11 kW Operating Instructions," (2006)

Industrial Automation Self-learning Through the Development of Didactic Industrial Processes

Armando Cordeiro[1], Manuel Abraços[1], Luis Monteiro[1], Euclides Andrade[1], Vitor Fernão Pires[2,3], and Daniel Foito[2(✉)]

[1] ADEEEA – Área Departamental de Engenharia Eletrotécnica, Energia e Automação, ISEL – Instituto Superior de Engenharia de Lisboa, Rua Conselheiro Emídio Navarro 1, 1959-007 Lisbon, Portugal
acordeiro@deea.isel.pt

[2] EST Setúbal-Instituto Politécnico Setúbal, Campus do IPS, Estefanilha, 2914-761 Setúbal, Portugal
{vitor.pires,daniel.foito}@estsetubal.ips.pt

[3] INESC-ID Lisboa, Rua Alves Redol, 9, 1000-029 Lisbon, Portugal

Abstract. Teaching industrial automation is a complex mission. The classical approach is based on lectures and laboratories assisted by teachers. Nevertheless, teaching industrial automation is not easy because this multidisciplinary area requires knowledge in control, energy, electronics, robotics and computer engineering, among others. In this way, this paper presents an approach to teach Industrial Automation based on a self-learning strategy. Instead of using only the classical approach where they use didactic back boxes with a particular system, students must also develop a research work and a didactic automation prototype. Since the approach is based in the autonomy and self-learning, the evaluation of this methodology indicates that students increase the interest about industrial automation and clarify important aspects of assembly, commissioning, parameterization and programming of electric and electronic devices. Additionally, this methodology seems to increase their self-confidence, apart from the necessary background to face the challenge of working in the real world. The feedback from employers is also very satisfactory.

1 Introduction

This paper is dedicated to improve the motivation and skills of Electric Engineering students about Industrial Automation. This specific area of Electrical Engineering courses is sometimes a complex task for teachers and students. Teaching Industrial Automation is not easy since this multidisciplinary area requires knowledge in control, energy, power and signal electronics, robotics and computer engineering, among others [1, 2]. As other areas with the same level of difficulty, this one also requires an innovative approach in order to be successful in transmitting knowledge. In fact, in the area of electrical engineering, several approaches based on didactic equipment or software tools have been used [3–6]. Due to the importance of the practical experience, internet also

M. E. Auer et al. (eds.), *Teaching and Learning in a Digital World*,
Advances in Intelligent Systems and Computing 715,
https://doi.org/10.1007/978-3-319-73210-7_100

allowed students to have access to laboratories and experiments at long distance. Several interesting and helpful solutions of remote laboratories have been proposed in literature [7–16]. Unfortunately they are dedicated to specific applications or to access remotely to a very specific experiment of power electronics, digital circuit or automation/robotics, and not meet all the necessary aspects for a correct learning, requiring in-class support.

The Electrical Engineering Department at the Instituto Superior de Engenharia de Lisboa (ISEL) and Setúbal School of Technology of Polytechnic Institute of Setúbal has a long tradition teaching Industrial Automation, Energy Production and Energy Management. The knowledge and experience of teachers is largely due to the close relationship of these institutes with industries of electrical equipment and services. In this way, this paper presents an approach to teach Industrial Automation based on students self-learning. In order to motivate students, several visits are made to local companies with a strong automation component in their industrial processes. Some protocols made with these companies will allow best performing students to get a temporary work contract.

Instead of using the classical approach based on lectures and laboratories exercises (using commercial didactic equipment) assisted by teachers and where students have a passive role, a different approach was tested. So, in the Master course was adopted an approach in which the role of students and teachers are changed up to a certain extent. According to this, students must propose a specific project that can result in a didactic industrial equipment. After the discussion of the proposed project with teachers and other students, they must implement it. As a result it is expected that through the development of didactic equipment with industrial devices from several manufactures, students clearly understand the functionalities of specific devices and be able to handling and commissioning them as they should do in real scenarios. Usually this kind of projects is developed by a group of Master's degree students of Industrial Automation with some help of teachers. At the same time this equipment will be used in exhibitions and workshops were students must explain the operational detail of the didactic equipment to other students. This increases the self-confidence and motivation of students and gives to them the necessary background to face the challenge of working in the real world. It is necessary to understand that the role of engineering schools should not be only dedicated to train engineers since the real world goes far beyond from math and physics calculations. This is a role that teachers from any area should never forget. The proposed self-learning system is also able to improve students relationships with other people, communication skills (talk, listen and understand others), increasing trust and leadership. On other hand, Bachelor students were also involved in this process. Due to this an interesting completion between them was created.

2 Industrial Automation Previous Knowledge

Students attending Electrical Engineering at ISEL are normally performing their Bachelor or Master degrees. According with the proposed solution, students who attend classes of Industrial Automation in the Master degree are encouraged to develop a specific didactic industrial equipment. To develop such equipment is necessary to first provide some contents about industrial automation in the Bachelor course, namely:

- Understand the principle of operation of some of the main devices and equipment used in automation, including DC and AC motors, sensors and actuator, mechanical and solid state devices;
- Understand the main failure modes and the respective detection and protection equipment;
- Basic concepts related to schematic rules and graphic symbols for control and power diagrams using electrical wires;
- Provide students basic concepts about I/O interfaces, programming PLCs (Programmable Logical Controller) according with the adopted languages of the IEC61131-3 and SCADA (Supervisory Control And Data Acquisition);
- Introduce the PLC programming software to be used in the laboratory classes and perform some application examples;

These contents are conducted over 30 weeks in two semesters at the end of the Bachelor course separated in two different classes: Automation I (beginner) and Automation II (intermediate). For students of the Master degree without any background in Automation, (as happens sometimes with students from other schools or from other areas), they must attend these two classes of the Bachelor course. During these 30 weeks, now the students have access to laboratory classes in which they must use didactic equipment (composed by real devices, instead of a programmable black box) developed by Master Students. Students are also divided by groups, and for each group are assigned a workstation in the laboratory. They carry out their practical work in the automation laboratory with the support and supervision of teachers. Each workstation has available all the practical tools for connect the devices (wires, connectors, pliers, screwdrivers, etc.), a PLC and a computer with both SCADA and PLC programming software. The teacher distributes for each group a working document with the proposed practical exercises along with a list of input and output variables to use in the PLC. Prior to programming, physical connections must be made to input and output interfaces. This task is also performed by students.

The content of the Automation classes are available in Moodle for all students. There, they can have access to slides, quizzes, laboratory manuals, reference books and web links. One of the great problems of teaching automation, is that in a world that increasingly is used online tools and computer programs, there is a lack of advanced simulation tools that can help students to understand concepts about automation in contrast to power electronics, digital systems or computer engineering where is possible to find many computer programs (e.g. Matlab/Simulink, Pspice, Logisim, Sapwin, KTechLab, Micro-CAP, Oregano, etc.). There are only few and limited simulation packages for automation, (e.g. FluidSim and CadeSimu).

At the end of each laboratory exercise, Bachelor students must do a written report. At the end of the semester, each group will have a meeting with teachers, which discuss the report with students, make questions about each specific exercise and show what could be improved. For the evaluation of Bachelor students, it is considered an examination with theoretical questions (60%) and some exercises about the laboratory component (40%). Regarding the laboratory component, students are also evaluated by their capability in realize the practical work, by the quality of the report and by the oral answers regarding the laboratory report.

The implementation of the proposed approach in the automation Bachelor course, in which was implemented laboratory exercises using didactic equipment developed by other students, showed that many positive aspects can be achieved, as well a good background that will give to the students the necessary confidence to take the next step in their learning process.

3 Self-learning Approach

Despite the good results achieved by students of the Bachelor course in Electrical Engineering, proven by the high rate of employability, these three years are usually not enough to consolidate all the crucial knowledge about automation. Thus, through the Master course, students are normally seeking for new opportunities to increase their knowledge in this area as well a better job in the near future. It is at this stage of their lives that, after having had contact with the first job as a technical engineer, that they assume a more active role within school and society. It is remarkable to see students during this stage to try to do work with a well-defined objective and looking for answers by themselves with self-confidence.

In the Master degree it is proposed an advanced course of industrial automation using a self-learning approach. So, instead of using the classing teaching/learning method, it is used a different method, where students play a more active role. In the classical approach the teaching process is based on lectures and laboratories exercises using equipment as black boxes. In such equipment, students only see some input and output terminals that simulate a specific system (Fig. 1). Although they acquire some programming skills, this black boxes are normally far from reality concerning certain limitations and complexity of real systems, such as the appropriate sensors and signal conditioning. As mentioned before, in this method the role of students is very passive, since there is a guide where is explained to the students what they should do.

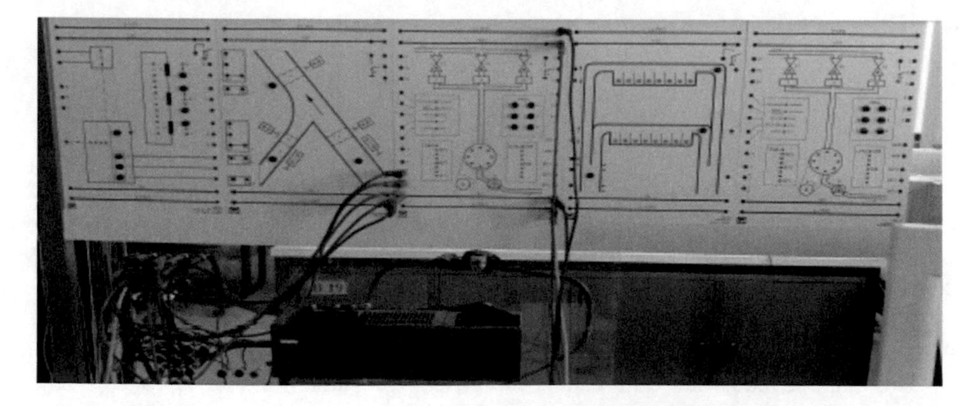

Fig. 1. Photograph of a didactic automation equipment based on classical black boxes.

Taking all this questions into consideration, Master students must develop the following tasks:

- Choose a specific automation or robotics theme and a research work about it, exploring different possibilities and solutions;
- Discuss the proposed project with teachers and other students;
- A laboratory project where students must develop a didactic industrial equipment based on the research work. This equipment should be as close as possible from a real scenario and using real industrial equipment;
- Be able to handling and commissioning them as they should do in real scenarios;
- The developed equipment must be presented in exhibitions and workshops were students must explain the operational detail of the didactic equipment to other students.

These tasks are usually made by a group of two or three students according with the extension of project. This has the advantage to give students team working capability. Most students are not prepared to work in group and are not able to manage conflicts. The teacher must be a moderator and adviser in this situations and a supervisor of all the stages of such projects. On the other hand, the success of finishing the project is bigger when they work together. They should give each other mutual support and respect in order to finish the project.

When the projects are finished, Master students are invited to explain the operational detail of the didactic equipment to other students in exhibitions and workshops organized by the department of Electrical Engineering. In this events are also invited companies that provide some of the industrial equipment used in the projects to discuss problems and solutions. Also, in these events, new suggestions for future projects usually appear, and are a good opportunity for establish new partnerships with companies with advantages for all.

4 Case Study

In this section is described a project developed by students using the self-learning methodology. The development of this project was only possible due to partnerships with some companies that produce or sell industrial equipment which provided a significant part of the equipment that is used. In Fig. 2 is present a photograph of a didactic automation equipment made by Master's students of Electrical Engineering at ISEL according with the adopted self-learning methodology. This didactic equipment was developed taking into account several aspects that compose usually any automatic system. This equipment was designed to simulate a batch process for mechanical parts. A brief description is presented next.

This didactic industrial equipment is composed by a garner (similar to a silo) where the mechanical parts (see Fig. 3(a)) are placed manually. When the operator initiates the process (there is a manual or automatic operation cycle which can be selected by the operator), a mechanical part is pushed by a pneumatic cylinder to a conveyor belt (TTP) as can be seen by Fig. 3(b).

Fig. 2. Photograph of a didactic automation equipment developed by Master's students according with the self-learning methodology adopted.

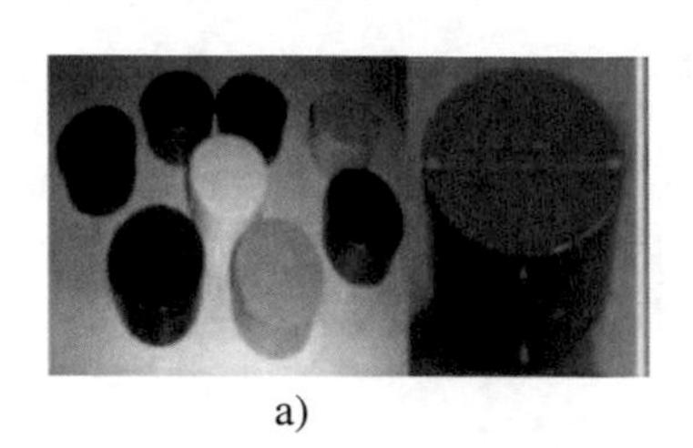

a)

b)

Fig. 3. Photograph of the: (a) mechanical parts to be processed by the didactic equipment (6 cm diameter × 3 cm height); (b) garner with a mechanical part to be processed by the machine.

After this initial step, the mechanical part is detected by a photoelectric sensor which will be responsible to send information to the PLC. This initiates the batch process of the mechanical part, opening the door of the heating and cooling chamber (CAA). Figure 4(a) shows a photograph of this chamber where is possible to see some associated devices inside and outside the chamber. The doors of this chamber are opened and closed by pneumatic cylinders. The mechanical part will be sent to the center of this chamber using the conveyor belt. When the mechanical part arrives to the center of the chamber, an inductive detector (SC) will stop the linear movement of the conveyor belt and start the heating process (which is made through the connecting of a set of heating resistors (RA) and a ventilator (TM) placed inside the chamber). The temperature in this chamber will be controlled by a temperature controller placed bellow the HMI (Human-Machine Interface) touch screen (see Fig. 4(b)). Temperature indicators are also presented in this front panel of a small electrical board. The temperature sensors are PTC thermistors (Positive Temperature Coefficient). Due to safety reasons this chamber has available an emergency push-button and two safety locks in the front door, avoiding the access to the interior. This didactic industrial equipment will stop immediately if any of these components are activated.

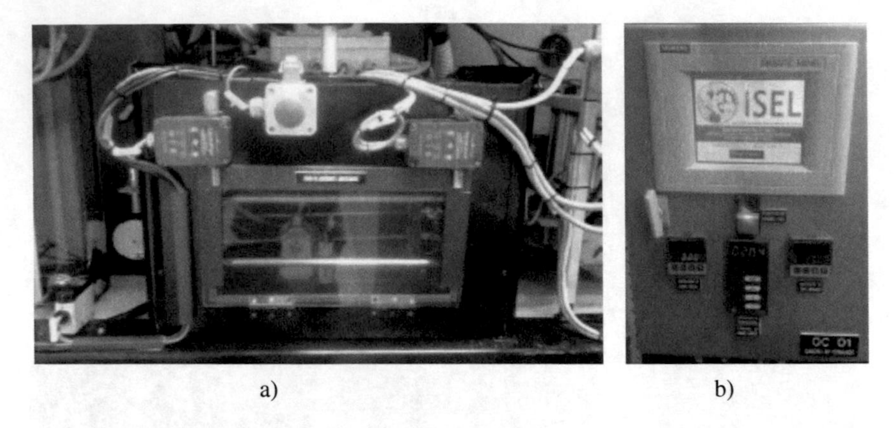

a) b)

Fig. 4. Photograph of the: (a) chamber dedicated to heating and cooling the mechanical parts; (b) electrical board with the touch panel, temperature indicators and controller.

After reaching the desired temperature, the system starts the cooling cycle through the use of a ventilator. To accelerate this process, it is possible to use a heat exchanger, transferring water in a closed loop between two plastic tanks (Dep. A and Dep. B) that is available in the system (see Fig. 5(a)). The water in circulation is generated by a centrifugal pump (BA) placed in the lower part of this equipment. Inside the chamber, the water flows in several pipes mounted around the interior walls (see Fig. 5(b)).

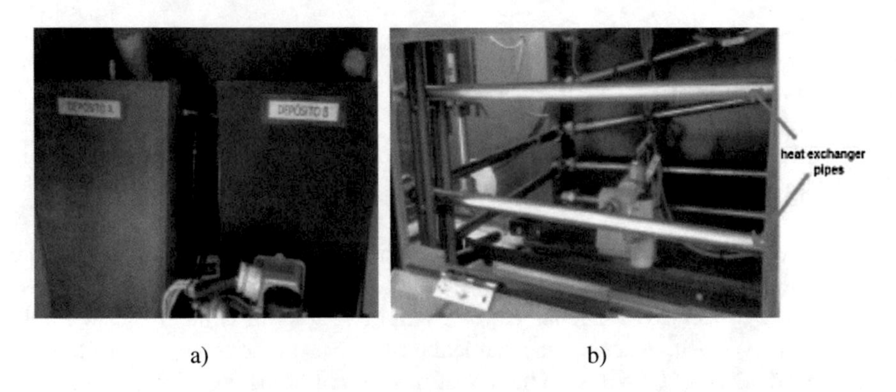

a) b)

Fig. 5. Photograph of: (a) both plastic tanks. They are used by the heat exchanger, transferring water in a closed loop between the tanks and; (b) the cooling tower used to cool down the mechanical part to the room temperature.

At the end of this process, the output door of the chamber will open and the mechanical part will be sent by the conveyor belt to a cooling tower (TV) which represents the final stage of this didactic system. This tower is composed by a pneumatic cylinder, which moves the tower up and down to lock the part inside of it, allocating another ventilator. This ventilator will cool down the mechanical part up to the room temperature. In Fig. 6 it can be seen a photograph of the cooling tower. Another two PTC thermistors (PTC2 and PTC3) are used to measure the temperature inside and outside the tower during this final cooling process. An overview of all this system can be seen in the diagram of Fig. 7. This diagram gives a better idea of the developed equipment.

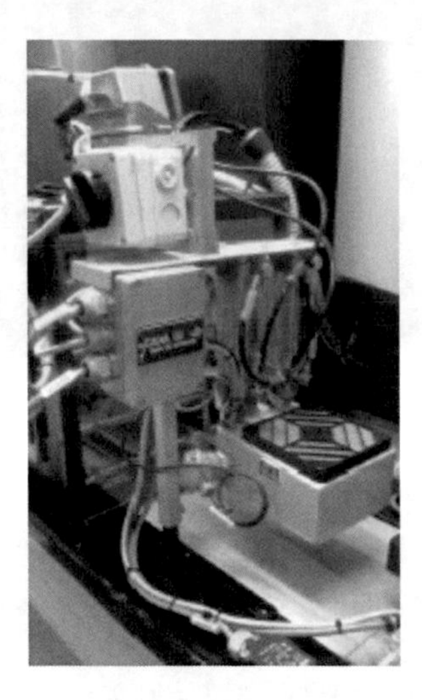

Fig. 6. Photograph of the cooling tower used to cool down the mechanical part to the room temperature

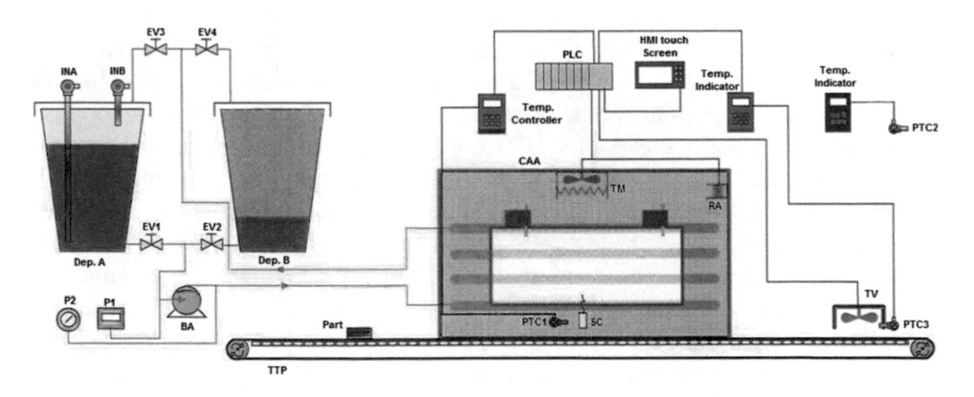

Fig. 7. General diagram of the developed equipment.

A supervisory panel was also created to identify any failure in the process, as well for a better understanding of all the stages of this didactic equipment. This panel can be seen in Fig. 8(a). This didactic equipment is controlled by several devices that are placed inside of an electrical board behind the supervision panel. In Fig. 8(b) is presented the PLC, a variable speed electrical drive, several relays, contactors, transformers and power supplies, among other devices, installed similarly to that found in a real industrial environment.

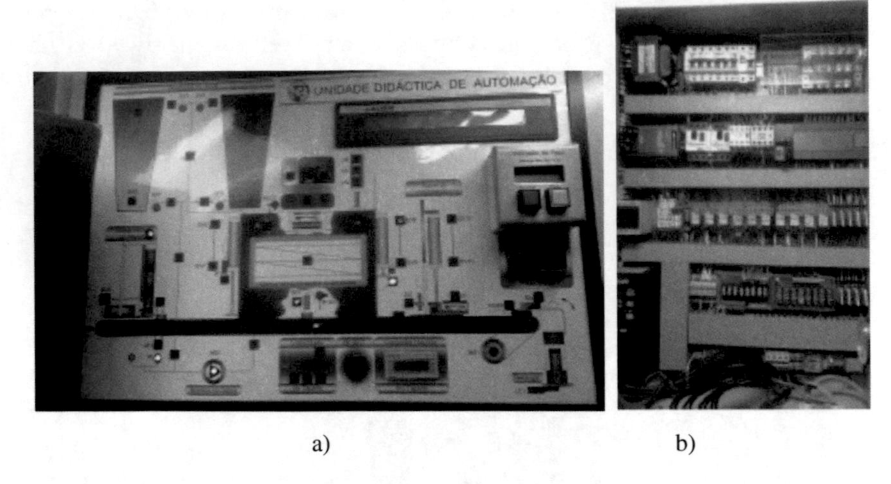

Fig. 8. Photograph of the: (a) supervisory panel placed in the door of the electrical board and; (b) interior of the main electrical board.

5 Evaluation of Achieved Results

This type of projects began in the year 2012 and has required a great concerted effort among all the students that are involved in terms of research and practical work. More than 150 students accepted this challenge. Until now, all of them completed the Master course in Industrial Automation with very good results. Other students decide to do other projects in the area of energy and are not included in this evaluation. Industrial Automation students must present an interim and a final evaluation report based on the developed work. The interim evaluation report makes it possible to assess the progress of work. They should also create a final PowerPoint presentation and discuss the results with colleagues and teachers without any additional pressure. The experience reveals that they seem to feel comfortable explaining ideas and concepts and sharing experiences with other colleagues (which have done other projects). The defined time for presentations is about 30 min. After this presentation, all students must maintain an active role and interact with other colleagues, asking questions, propose new solutions and discuss ideas. If necessary the teacher make also some questions, most of the time with the purpose of create interest and discussions about the developed equipment, since the teacher at this stage knows the work very well and all its participants. There students must also present their work to the Bachelor students. On other hand, when the Bachelor students use the developed didactic industrial equipment, they must give feedback to the Master students that developed that system. In this way, this method seeks to achieve the main following objectives:

- Motivate students to study by themselves subjects in which they are unfamiliar in automation;
- Increase the experience of operating new equipment;

- Improve the writing of reports using the correct language and in a reduced number of pages (expressing ideas using few words is a difficult exercise);
- Increase the self-confidence in oral presentations;
- Improve the technical language and the ability to expose and discuss a point of view;
- Improve the way students relate to each other, to achieve a common goal.

For the evaluation there are five different criteria: quality of the presentation, security in answering the questions, quality of the final report, developed work during the semester and the evaluation given by other students (secret evaluation). Almost all students complete the proposed tasks with high grades since the developed didactic equipment presents normally high quality. The experience of teachers is that students have strong knowledge about industrial automation at the end of these projects. An online survey is available for students to send comments, suggestions and evaluate the physical laboratories (and available equipment) and teachers concerning punctuality, quality, behavior in classroom and laboratories, support given to students, etc.

This type of educational project is also subject to an annual evaluation. This evaluation is carried out by a committee created by the course director. For that evaluation is necessary to send all the reports (students and teachers), surveys and documents with the contents produced by the automation teachers. The results of surveys carried out to various employers in order to have feedback from the knowledge of former students are also included. This and other projects among other subjects are evaluated from time to time by an independent accreditation body. In Portugal this entity is A3ES (www.a3es.pt). The results of this evaluation is favorable and indicate that students who participate in this kind of experiences are better prepared for their professional life with regard to industrial automation. The reality of numbers show that the employability rate of these students is very high and many of them already hold important positions in many companies in the region.

6 Conclusions

Industrial Automation is a very important course for an Electric Engineer. Classical approach with lectures, practical exercises and laboratories show that many students have difficulties in learning Industrial Automation since students play a very passive role. In order to overcome this problem, the Electrical Engineering Department at the Instituto Superior de Engenharia de Lisboa (ISEL) has modified the Industrial Automation course of the Master degree since 2012 introducing a different teaching approach. In this course the teaching method is based on students self-learning.

In the Master course of Industrial Automation, students will develop a research work which must lead to a didactic industrial equipment. All the didactic equipment's are normally used in exhibitions and workshops were students must explain the operational detail of the didactic equipment to other students and invited companies.

The objectives of this methodology are mainly to motivate students to study by themselves subjects about automation, increase the experience of operating new equipment's, improve the writing of technical report, increase the self-confidence in oral presentations and the ability to expose and discuss their point of view. This methodology also provides proximity between teacher and students. The obtained results show an increase in student's interest about the subjects of automation and a general satisfaction with the method in which they are evaluated. Through the implementation of this new approach, teachers verify that students acquired strong knowledge about industrial automation. The main disadvantage of this methodology is that requires the necessary financial resources to buy all the necessary equipment. Most of the times this is only possible due to partnerships with certain manufactures.

References

1. Vyatkin, V.: Software engineering in industrial automation: state-of-the-art review. IEEE Trans. Ind. Inform. **9**(3), 1234–1249 (2013)
2. Efe, M.Ö.: Fractional order systems in industrial automation - a survey. IEEE Trans. Ind. Inform. **7**(4), 582–591 (2011)
3. Cristolţean, D-C., Silea, I.: Didactic equipment for studying automation and applied informatics. In: 12th IEEE International Symposium on IEEE Transaction on Industry Informatics Electronics and Telecommunications, ISETC 2016, Timisoara, Romania, 27–28 October 2016
4. Parkin, R.M.: The mechatronics workbench. Eng. Sci. Educ. J. **11**(1), 36–40 (2002)
5. Palma, J.P., Antonio, F.A., Virtuoso, V.F.: A didactic configurable converter for training on inverter and chopper topologies. In: 8th IEEE Mediterranean Electrotechnical Conference (MELECON 1996), Bary, Italy, 13–16 May 1996
6. Cunha, B.G.P., et al.: DidacTronic: a low-cost and portable didactic lab for electronics: kit for digital and analog electronic circuits. In: Proceedings of Global Humanitarian Technology Conference (GHTC), Seattle, WA, USA, 13–16 October 2016
7. Esteves, L., Pires, V.F.: WPEC - a new web tool for the power electronics learning. In: 31st Annual Conference of the IEEE Industrial Electronics Society (IECON 2005), pp. 2152–2155 (2005)
8. Drofenik, U., Kolar, J.W., Van Duijsen, P.J., Bauer, P.: New web-based interactive e-learning in power electronics and electrical machines. In: Proceedings of the IEEE Industry Applications Conference, 36th Annual Meeting (IAS 2001), vol. 3, pp. 1858–1865
9. Sziebig, G., Korondi, P., Suto, Z., Stumpf, P., Jardan, R.K., Nagy, I.: Integrated E-learning projects in the European Union. In: Proceedings of the Annual Conference of IEEE Industrial Electronics, pp. 3524–3529 (2008)
10. Pires, V.F., Martins, L.S., Amaral, T.G., Marçal, R., Rodrigues, R., Crisóstomo, M.M.: Distance learning power system protection based on testing protective relays. IEEE Trans. Industrial Electron. **55**(6), 2433–2438 (2008)
11. Rojko, A., Hercog, D., Jezernik, K.: Power engineering and motion control web laboratory: design, implementation, and evaluation of mechatronics course. IEEE Trans. Ind. Electron. **57**(10), 3343–3354 (2010)
12. Hercog, D., Gergic, B., Uran, S., Jezernik, K.: A DSP-based remote control laboratory. IEEE Trans. Ind. Electron. **54**(6), 3057–3068 (2007)
13. Costas-Perez, L., Lago, D., Farina, J., Rodriguez-Andina, J.J.: Optimization of an industrial sensor and data acquisition laboratory through time sharing and remote access. IEEE Trans. Ind. Electron. **55**(6), 2397–2404 (2008)
14. Aydogmus, Z.: A web-based remote access laboratory using SCADA. IEEE Trans. Educ. **52**(1), 126–132 (2011)

15. Chang, W.F., Wu, Y.C., Chiu, C.W.: Development of a web-based remote load supervision and control system. Elect. Power Energy Syst. **28**, 401–407 (2006)
16. Ko, C.C., Chen, B.M., Chen, J., Zhuang, Y., Tan, K.C.: Development of a web-based laboratory for control experiments on a coupled tank apparatus. IEEE Trans. Educ. **44**(1), 76–86 (2001)

Escargot Nursery – An EPS@ISEP 2017 Project

Lauri Borghuis, Benjamin Calon, John MacLean, Juliette Portefaix, Ramon Quero, Abel Duarte, Benedita Malheiro, Cristina Ribeiro, Fernando Ferreira, Manuel F. Silva(✉), Paulo Ferreira, and Pedro Guedes

ISEP/PPorto – School of Engineering, Polytechnic of Porto, Porto, Portugal
epsatisep@gmail.com
http://www.eps2017-wiki1.dee.isep.ipp.pt/

Abstract. This paper presents the development of an Escargot Nursery by a multinational and multidisciplinary team of 3rd year undergraduates within the framework of EPS@ISEP – the European Project Semester (EPS) offered by the Instituto Superior de Engenharia do Porto (ISEP). The challenge was to design, develop and test a snail farm compliant with the applicable EU directives and the given budget. The Team, motivated by the desire to solve this multidisciplinary problem, embarked on an active learning journey, involving scientific, technical, marketing, sustainable and ethical development studies, brainstorming and decision-making. Based on this project-based learning approach, the Team identified the lack of innovative domestic snail farm products and, consequently, proposed the development of "EscarGO", a stylish solution for the domestic market. The paper details the proposed design and control system, including materials, components and technologies. This learning experience, which was focussed on the development of multicultural communication, multidisciplinary teamwork, problem-solving and decision-making competencies in students, produced as a tangible evidence the proof of concept prototype of "EscarGO", an Escargot Nursery designed for families to easily grow snails at home.

Keywords: Collaborative learning · Project based learning
Escargot Nursery · Snail framing · Home · Technology · Education
Curtain system

1 Introduction

The European Project Semester (EPS) is a multicultural, multidisciplinary teamwork and project-based learning framework provided by 18 European engineering schools. In Portugal, EPS is offered by Instituto Superior de Engenharia do Porto (ISEP) since 2011 to 3rd and 4th year engineering, business and product design undergraduates. In the spring of 2017, an EPS@ISEP team composed of Lauri Borghuis, a Biology and Medical Laboratory student from the Netherlands, Benjamin Calon, a Product Development student from Belgium, John MacLean, a Mechanical Electronic Systems Engineering student from Scotland, Juliette Portefaix, a General Engineering student from France, and

M. E. Auer et al. (eds.), *Teaching and Learning in a Digital World*,
Advances in Intelligent Systems and Computing 715,
https://doi.org/10.1007/978-3-319-73210-7_101

Ramon Quero, an Engineering and Architecture student from Spain, chose to develop an Escargot Nursery. This group of five students, from different national and scientific backgrounds, joined efforts around this project-based learning experience, exchanging and gaining knowledge together.

Nowadays, while many tend to live disentangled from the natural habitat, others are in pursuit of natural processes and experiences. On one hand, the digital revolution, which improved our communication channels through social media, mobile phones and video conferencing, also isolated people from a real social life [8]. On the other hand, more people are aware of the use of genetically modified organisms and, consequently, want to know the origins and growth processes of their food. Genetic modification is being used to improve the colour, smell and taste of food, trying to make it more attractive and durable in terms of the shelf life. However, there is not enough scientific knowledge regarding the long term side effects of genetically modified food on people [15].

The Team saw this project as an opportunity to contribute to the mitigation of both problems by deciding to build a unique and innovative product to help people produce their own snails at home for educational and consumption purposes. Specifically, the focus was on the design of an educational product mainly targeted for families with children. This would help children relate and establish bonds with nature, while developing autonomy, responsibility and an interest in science. To create a new and fun way of producing food, the Team identified the need to include technology and create a comfortable habitat for the snails, allowing the end user to grow snails for food or as pets.

The Team performed a series of background scientific, technical, marketing, ethical and sustainable development studies to specify the requirements, design and control system of "EscarGO". Based on this research, it decided to: (*i*) farm the *Cornu aspersum*, one of the most commonly consumed breed of snails, which take six months to grow to their optimal size [4]; (*ii*) design a terrarium to house up to fifty snails, producing two meals a year for a family of four; and (*iii*) control the light, humidity and temperature in the terrarium. With "EscarGO" the Team hopes to contribute to the education of children, bring families together by anchoring members to nature, rather than the digital world, and reduce the carbon footprint of food by decreasing the distance travelled. Moreover, the educational purpose would be the main selling focus of the product.

Given these ideas, in the sequel Sect. 2 covers the State of the Art on escargot farming, Sect. 3 describes the proposed solution, Sect. 4 the tests and results and, finally, Sect. 5 draws the project conclusions.

2 Escargot Farming

This section summarises the different background studies conducted by the Team in order to reach the proposed solution.

2.1 Related Work

There are several snail farms available on the market. Since most of the home-use competitors of the "EscarGO" were not designed for snails, a comparison between large scale snail farming solutions was made. The Team considered this comparison relevant to the development of the product, due to the lack of technologies used in the products for domestic use. These technologies were dedicated to the production of a much larger number of snails, whereas this project is designed for a much smaller number and for domestic use.

After this comparison, the Team decided to choose the species that seemed the most relevant. The goal was to adapt the product to this particular species. So, the Team decided to use the *Cornu aspersum* because this is one of the most common snail breeds and the most consumed in France, the main market target. The *Cornu aspersum* species belongs to the Gastropoda class and they prefer an undisturbed habitat with adequate high moisture level and good food supply. This species needs a specific habitat. First, these snails require a temperature between 15 °C and 25 °C, with an optimal temperature of 21 °C [3]. The humidity level is essential for the activity of the snails. They are more comfortable with humidity levels from 75% to 90%. For an optimal reproduction and breeding process, the snails require 16 h/d of light [5,10]. Finally, the snail population density must be considered since too many snails have a negative impact in their successful growth and breeding. The recommended density for *Cornu aspersum* is 1.0 kg/m^2 to 1.5 kg/m^2. Since an adult snail weights approximately 10 g, it was possible to have up to 100 snail/m^2 [7,10].

Based on this study, the Team chose the *Cornu aspersum* breed and learned that it is sensitive to humidity, temperature, light and population density. Consequently, the Team derived the following requirements: (*i*) breed up to 50 snails; (*ii*) design a terrarium with a dimension of 400 mm × 300 mm × 375 mm; and (*iii*) include a light, humidity and temperature control system to make the product user-friendly. It should be kept on mind that when the user touches the snails, for hygiene purposes, it is necessary to wash their hands.

2.2 Marketing

During the marketing study, the Team worked on logos and commercial names and decided to launch "EscarGO" in France, where the cultural barriers related to the consumption of snails seem more diluted. The product should be sold on the Internet, targeting the gourmet costumer, who wants to grow snails at home for self-consumption, and the parent costumer, who wants an educational and recreational product. Despite the lack of competitors at the time of the study, the Team determined the need to keep the production costs low enough to generate profit and still be able to sell the product at a competitive price of 50 € to 70 €. In terms of marketing plan, the Team concluded that "EscarGO" should be a domestic product and, therefore, its dimensions should not be bigger than any other home-size product, *e.g.*, a microwave, while being able to host 50 snails.

2.3 Eco-efficiency Measures for Sustainability

Concerning the sustainability of the product, its design must be simple to reduce the environmental impact. The Team tried to create a low impact system by choosing low impact materials and recycling as much as possible. The structure of the final product should be in polypropylene (PP) since it is resistant to the growth of bacteria and has a lower impact on the environment when compared with other plastics. Due to budget constraints, the Team chose to use Polyvinyl Chloride (PVC) for the structure of the prototype. Finally, the Team decided to use the curtain method in order to increase the livable surface area for the snails, while keeping the dimensions smaller. This method consists of two curtains, made of Nylon mesh, allowing to host more snails in a smaller space.

2.4 Ethical and Deontological Concerns

The Team adopted the French ethics charter for engineers drafted by CNISF, since the target is the French market, and made every effort to create a safe and sustainable Escargot Nursery, *i.e.*, with a minimal impact on the environment regarding to the environmental ethics [1,2,11].

From the marketing and sales ethics perspective, the Team considered the need to build a positive image of the company, define fair prices and develop a trust relationship with customers, *i.e.*, never lie about the benefits of the product. Regarding liability, the Team, while being responsible for the product and for any injuries it might cause, assumed its responsibility towards supervisors and customers, *i.e.*, if something unfortunate happens, the responsibility is with the Team [14]. Finally, the Team discussed the need to be ethical during the construction of "EscarGO" to ensure that the product meets all ethical requirements and the applicable European directives.

3 EscarGO

3.1 Requirements

The Team started by specifying the requirements for the project. The requirements imposed in the project proposal were using sustainable materials, using low cost hardware solutions, sticking to the budget (100.00 €), complying with the relevant European Union (EU) Directives, using of the International System of Units and using open source software and technologies.

Additionally, the Escargot Nursery had to meet other requirements, namely an aesthetically pleasing design, as the product would be on display, so the Team wanted the product to be an attractive appliance. For the electronic aspect of the nursery, the Team wanted the product to be as fully automated as possible, with little need for human interaction. The product had to be able to set and display the temperature, humidity and light, while using as little power as possible.

3.2 Functionalities

The Escargot Nursery is expected to achieve certain functions. It has to keep the climate inside at a comfortable level for the snails. For this, the humidity needs to be controlled, so a liquid spray system needs to be used to keep the soil moist, the temperature also needs to be kept within the safe range, i.e. between 15 °C and 25 °C. Lighting in the form of LED needs to be controlled to ensure the snails have enough light to thrive. The system measures the temperature and humidity inside the nursery and measures the light level outside the nursery. The program stored in the Arduino board gathers information and controls each output, to ensure automatic climate control.

3.3 Design of the Structure

The final design of the "EscarGO" is minimalistic yet functional, with dimensions of 400 mm × 300 mm × 375 mm. Ease of use was one of the most important design motivators of "EscarGO".

The housing is made of black and white PP, and the front and rear cover have a transparent area made of Polymethylmethacrylate (PMMA). The black PP in the front covers a display with relevant information about temperature, humidity and light. On the top there are two removable plates: one gives access to the living environment of the snails and one to the water supply. The water supply and the curtains are easily removable for feeding and maintenance. The right side has openings to check the water level and vents for the fan and both sides are provided with sunken handles so the product can be moved easily.

Inside the nursery there are two compartments. One large compartment hosts the snail habitat. It is equipped with planting, curtains, soil and small rocks to act as natural heat regulators. The curtains are made from Nylon mesh, so they are easy to clean. The tube under the soil helps to keep the humidity at a certain level. The LED lighting gives additional light when needed. All electronics are kept to the right side of the product in a smaller compartment, next to the water tank. These components are an Arduino Uno, a fan, a heating device, an actuator and sensors. The Arduino controls all processes so the snails can live in optimal conditions. The fan blows air into the nursery which can also be heated by the heating device. The actuator releases water from the tank to the tubing when the Arduino sends a signal. The sensors give all the information needed to the Arduino board. In Fig. 1 the 3D model of the product is shown with all the components of "EscarGO". The "EscarGO" will be assembled in a factory so the customers do not have to built the product.

3.4 Control System

Figure 2 presents a BlackBox schematic diagram describing the main functions of the "EscarGO" control system.

An Arduino Uno is used as the controller, as this complies with the requirement to use open source software. Arduino was mainly chosen because the Team did not have much experience with designing electronic systems, and little coding experience, so the variety and amount of tutorials that are on-line that use Arduino Uno made it the most attractive prospect.

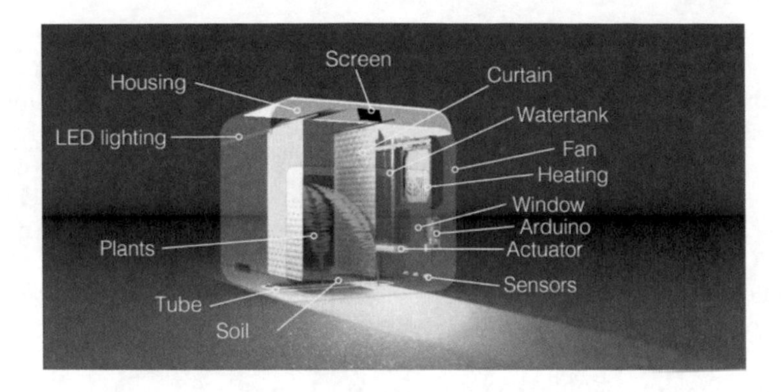

Fig. 1. 3D model of the "EscarGO" with all the components

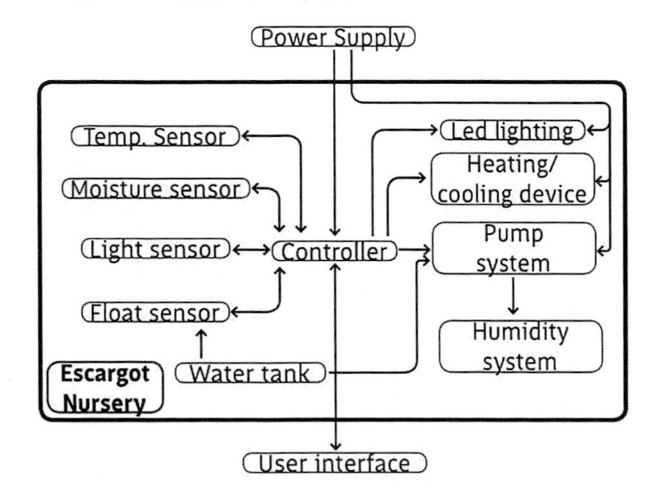

Fig. 2. The BlackBox schematic

The control system requirements are as follows. The enclosure needs to be able to maintain a comfortable temperature for the snails, without requiring much energy. It is recommended that the enclosure stays inside the home. Proposed is a heater element that will turn on if the temperature drops below 15 °C and a cooling fan that will turn on if the temperature rises above 25 °C. The reason for this is that the market research suggests that people want a system as simple as possible, that rarely uses any energy. The final product uses power resistors as heating elements, and a fan to cool the air and provide air movement.

Humidity is another aspect that needs to be controlled. A moisture sensor is inserted into the enclosure, and there is a sprinkler hose pipe inside to release water if the environment is not humid enough. These have to be short bursts since over watering, or flooding, might drown the snails, and also there is a possibility of a small water tank on the system for the humidity control so that the tank does not need to be fed with a constant water supply. The Team decided to use a combined temperature and humidity sensor, DHT22. This is meant to reduce the amount of components and keep the electronics as compact as possible, and with as little intrusion into the space as possible.

Another requirement of the system was to display the temperature and humidity on a small LCD screen. Humidity and temperature are easily controlled using Arduino boards and software, which is a great benefit of using Arduino. One of the concerns with the project was that because there needed to be a humid environment for the snails, and they also required oxygen, there was the issue of dampness and humidity getting into the room where the terrarium is stored. This needs to be carefully controlled and monitored because dampness can cause damage to the room around the enclosure.

4 Tests, Results and Discussion

4.1 Tests

To evaluate the work and to make the product as safe as possible, the Team performed functional tests and soil tests. The functional tests gave an insight into whether the Escargot Nursery complied with the requirements, and was ready to be produced and released onto the market. The soil tests focussed on the determination of the most appropriate soil conditions for growing snails.

Functional Tests were concerned with the verification of the correct operation of the control system and included testing:

Temperature & Humidity Sensor: by lowering the temperature (ice pack), raising the temperature (hair dryer) and the humidity (warm damp cloth);
Light Sensor: by covering the sensor to see if the lights switch on automatically;
Fan: by rising the temperature above 25 °C (hair dryer) to check if the fan starts;
Heating Resistor: by lowering the temperature inside "EscarGO" below 15 °C (ice pack) to see if the heater activates;
Liquid Crystal Display (LCD): by verifying if the LCD displays the correct temperature, light level and relative humidity;
Water Tank: by checking that there are no water leakages and the water tank moistens the soil.

Relative Humidity Soil Tests focussed on choosing the most appropriate soil conditions for the snails. Moisture along with available calcium content are two extremely important environmental factors that dictate the health of molluscan fauna such as snails [6,9,13]. In order to keep the soil moist, two different strategies were used in the present work, namely the addition to the soil of calcium alginate microspheres or sodium polyacrylate particles. Alginate is a natural biodegradable polymer extracted from brown algae that forms hydrogels under mild conditions, in the presence of divalent cations, such as calcium. Sodium polyacrylate is a superabsorbent polymer that has the ability to absorb as much as 200 to 300 times its mass in water. It is frequently used in agriculture since it can absorb water when it rains and release it when needed [12]. The preparation of these experiments involved, in the case of the:

Alginate Microspheres, using a 1 ml syringe, a 3% (w/v) Na-alginate (Pronova Biopolymers) solution was extruded dropwise into a 0.1 mol $CaCl_2$ cross-linking solution, where spherical-shaped particles instantaneously formed and were allowed to harden for 30 min. At completion of the gelling period the microspheres were recovered and rinsed in water in order to remove the excess $CaCl_2$ [9,13].

Sodium Polyacrylate Particles, using 2 g of sodium polyacrylate crystals were added to 1000 ml of water and let them swelling for 72 h.

In order to investigate the capacity of the two materials to release water and keep the soil humid, humidity tests were performed. Three soil boxes were prepared:

- Box 1: soil (used as control);
- Box 2: soil with calcium alginate microspheres;
- Box 3: soil with polyacrylate hydrated particles.

The same ratio of soil mass and calcium alginate microspheres or polyacrylate particles mass were used (700 g of soil to 82.81 g of microspheres or particles). A humidity sensor was placed inside each of the boxes, connected to an Arduino, in order to collect the relative humidity soil changes over 72 h. The soil boxes were covered with a perspex box positioned to allow the circulation of air.

4.2 Results

The team tested all electronic components separately, to be sure that all components worked, and then combined them to test the whole system. Table 1 displays the results of the electronic components tests.

Table 1. Electronic components tests and results

Part	Test	Outcome	Solution
DHT22	Ice	Humidity rise and temperature fall	
DHT22	Hair dryer	Humidity and temperature rise	
Fan control	Air circulation	Fan started with $T \geq 25\,^\circ C$	
TSL2561	Daylight illuminance	Wrong reading (65 536 lx)	Rewire
LED	Operation	LED on when powered with 12 V	
Lighting	Light dimness	LED on with illuminance $\leq$ 50 lx	
System	Breadboard assembly	Operational	
System	Soldered assembly	TSL2561 not powered	Rewire and solder

The Team did both soil tests at the same time, with the same conditions for all boxes, being the temperature 20 °C. The three boxes had 700 g of soil and 10 ml of water. One was the box with 82.81 g of microspheres, the second box was filled with 82.81 g of poly-acrylate and the third was the control one. In each box was a soil moisture sensor, that measured the soil moisture content, connected to the Arduino and an SD card module which registered and stored the data. Figure 3 depicts the results obtained.

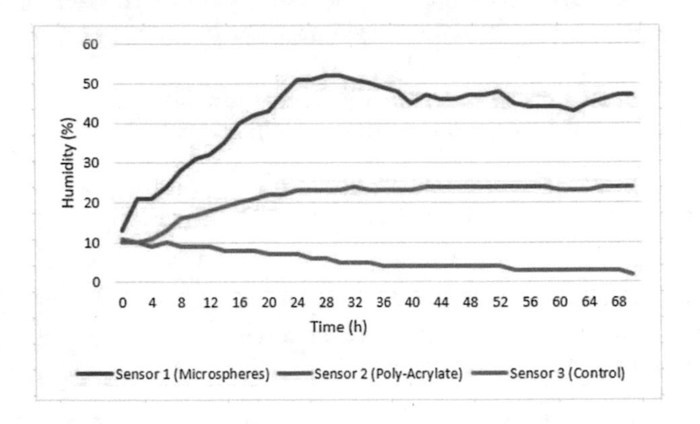

Fig. 3. Results of the soil tests

The relative humidity of the soil (Box 1) decreased during the duration of the test as opposed to what happened with the presence of calcium alginate microspheres or polyacrylate hydrated particles. For both materials, there is a burst of water release during the first 24 h of the test and then the humidity stabilises. The quantity of water released was higher for the calcium alginate microspheres in comparison to the polyacrylate particles. The amount of material to be added to the soil must be adapted to the desired humidity level.

The calcium alginate microsphere solution is the material of choice to keep the soil moist, not only because the humidity level was the highest attained during the whole test, but also because as they degrade, they release calcium into the soil, and calcium is very important for the snails health. Snail shell is made of calcium carbonate and keeps growing as long as the snail grows. In this particular application, microspheres can act simultaneously as a water and calcium reservoir. Additionally, alginate microspheres can also be used as a controlled-release product of other substances that are identified as necessary for snail's development, such as, for instance, vitamins.

Figure 4 displays the final prototype, which includes all the electronics for the optimal living conditions of the snails.

4.3 Discussion

The objectives of the Team were to design and build a unique and innovative product that would help people to produce their own snails at home, whether for

Fig. 4. Final prototype

recreational purposes or consumption. The main objective was to bring families together and educate children about animals and food.

The design of the "EscarGO" needed to be simple but attractive, without too much technology, since it is intended to be placed in a house kitchen or living room. From the results of the marketing study (where a survey was conducted), users want a simple product, so the Team took this in account while developing the prototype. It also had to be sustainable and protect the environment. The Team tried to reduce its environmental impact, by creating a low impact system and using low impact materials. An innovative sustainable tool used, was the curtain system. With the curtain system, the service intensity was increased. Increased hygiene was also a positive side effect of the curtain method. For this project, the Team used two curtains. This way the same amount of energy was used, but a larger number of snails could be kept in the enclosure.

There were some problems with the development of the control system during the project due to some bad connections and since no Team element had previously worked with Arduino. Nevertheless, the Team solved everything and could do almost all tests. There were also some time constraints and, for this reason, it was not possible to test the water tank and the heating element could not be added on time for the prototype but will be used in the final product. While the Team was developing the electronic system, they discovered a more ingenious solution. For the prototype it is necessary to connect the Arduino board to a computer at all times in order to operate the program. This is because the time has to be synchronized with the computer, and the "EscarGO" looses the time data if switched off. This solution is not optimal neither practical. In order to avoid the mandatory connection to a computer, the Team thought of including a Real Time Clock (RTC), but it did not have time to buy it for the prototype.

5 Ideas for Future Developments

For future development of the product, the Team considered different uses for the "EscarGO". One alternative is using the product to breed insects. Alternatively, the product can also be changed to suit small reptiles, and developed further to house snakes. There are already plenty of those on the market, but the Escargot Nursery concept is a much more attractive design.

Another improvement that the Team considered is adding controls so that the humidity and temperature can be changed to suit different insects or reptiles. Controls for the time function, such as a RTC, could also be included in the final product, but for the prototype the Team was limited by cost requirements so these details couldn't be added.

In the prototype, there is a water tank included to keep the soil moist. The team did some tests with poly-acrylate and microspheres. The microspheres are an improved solution compared with the water tank so, for the future, the Team wants to include the microspheres in addition to the water tank. These are not harmful for the snails and the family can make these microspheres together to make the family bond stronger.

Another test that needs to be completed is testing if the snails are comfortable within the prototype. This is necessary since they need to live in the "EscarGO". The team did not have time to test the prototype with the snails because, to monitor the snails growth and quality of life, would require a minimum of 3 months of living within the Escargot Nursery.

6 Conclusions

The EPS programme provided an innovative educational experience for the team members, in terms of the learning approach, drawing knowledge from different fields in a series of subjects, as well as from the different attributes and cultures of each member of the group.

In this paper a short overview was given about the development of the "EscarGO", a user friendly home Escargot Nursery. The goal was to make a product that brings families together and educates children. Different fields were taken into account while designing the product, such as sustainability, ethics and marketing. Along with the requirements definition, a solution was proposed. Further details were given, concerning the product's characteristics related to materials, systems and functionalities. Different techniques for the control system were tested, to make sure the proposed solutions work as expected, and distinct innovative techniques for keeping the soil moisture were considered and were tested. The tangible result was a product, the "EscarGO", but more importantly was the learning experience of the Team.

Acknowledgements. This work is financed by the ERDF – European Regional Development Fund through the Operational Programme for Competitiveness and Internationalisation – COMPETE 2020 Programme within project POCI-01-0145-FEDER-006961, and by National Funds through the FCT – Fundação para a Ciência e a Tecnologia (Portuguese Foundation for Science and Technology) as part of project UID/EEA/50014/2013.

References

1. Association des Architectes Paysagistes du Québec. L'éthique et la déontologie (2016). https://formation.aapq.org/ethique.php. Accessed 15 Mar 2017
2. Conseil National des Ingénieurs et des Scientifiques de France. Charter d'Ethique de l'Ingénieur (2017). http://guide.ensait.fr/lib/exe/fetch.php?media=charte_ethique-cnisf.pdf. Accessed 15 Mar 2017
3. Cowie, R., Ansart, A., Madec, L., Guillier, A.: Cornu aspersum (common garden snail) (2015). http://www.cabi.org/isc/datasheet/26821. Accessed 06 Mar 2017
4. Bourgogne, C.: Quelles sont les différentes espèces d'escargots comestibles? (2016). http://www.croquebourgogne.com/index.php?rubrique=faq. Accessed 01 Mar 2017
5. Lubell, D.: Are land snails a signature for the Mesolithic-Neolithic transition? Department of Anthropology (2004)
6. FMC Biopolymer. Alginates (2003). http://www.fmcbiopolymer.com/Portals/Pharm/Content/Docs/Alginates.pdf. Accessed 15 May 2017
7. Onwuka, B., Cobbinah, J.R., Vink, A.: Snail farming, production, processing and marketing. CTA (2008)
8. Udorie, J.E.: Social media is harming the mental health of teenagers. The state has to act (2015). https://www.theguardian.com/commentisfree/2015/sep/16/social-media-mental-health-teenagers-government-pshe-lessons. Accessed 04 Mar 2017
9. Lee, K.Y., Mooney, D.J.: Alginate: properties and biomedical applications. National Institutes of Health, 1 January 2013. Accessed 15 May 2017
10. Murphy, B.: Breeding and Growing Snails Commercially in Australia. Rural Industries Research and Development Corporation (2001)
11. Ordre des Ingénieurs du Québec. Distinction entre éthique et déontologie (2001). http://gpp.oiq.qc.ca/distinction_entre_ethique_deontologie.htm. Accessed 15 Mar 2017
12. Puoci, F., Iemma, F., Spizzirri, U.G., Cirillo, G., Curcio, M., Picci, N.: Polymer in agriculture: a review. Am. J. Agric. Biol. Sci. **3**, 299–314 (2008). Accessed 15 May 2017
13. Ribeiro, C.C., Barrias, C.C., Barbosa, M.A.: Calcium phosphate-alginate microspheres as enzyme delivery matrices. Biomaterials **25**, 4363–4373 (2004). Accessed 18 May 2017
14. Stearns, D.W.: An introduction to product liability law (2001). http://www.marlerclark.com/pdfs/intro-product-liability-law.pdf. Accessed 18 Mar 2017
15. World Health Organization: Frequently asked questions on genetically modified foods (2014). http://www.who.int/foodsafety/areas_work/food-technology/faq-genetically-modified-food/en/. Accessed 04 Mar 2017

Wearable UV Meter – An EPS@ISEP 2017 Project

Elin Lönnqvist, Marion Cullié, Miquel Bermejo, Mikk Tootsi, Simone Smits, Abel Duarte, Benedita Malheiro, Cristina Ribeiro, Fernando Ferreira, Manuel F. Silva(✉), Paulo Ferreira, and Pedro Guedes

ISEP/PPorto – School of Engineering, Polytechnic of Porto, Porto, Portugal
epsatisep@gmail.com
http://www.eps2017-wiki2.dee.isep.ipp.pt/

Abstract. This paper reports the collaborative design and development of Helios, a wearable UltraViolet (UV) meter. Helios is intended to help preventing the negative effects of over-exposure to UV radiation, *e.g.*, sun burning, photo ageing, eye damage and skin cancer, as well as of under-exposure to solar radiation, *e.g.*, the risk of developing vitamin D shortage. This project-based learning experience involved five Erasmus students who participated in EPS@ISEP – the European Project Semester (EPS) at Instituto Superior de Engenharia do Porto (ISEP) – in the spring of 2017. The Team, motivated by the desire to find a solution to this problem, conducted multiple studies, including scientific, technical, sustainability, marketing, ethics and deontology analyses, and discussions to derive the requirements, design structure, functional system and list of materials and components. The result is Helios, a prototype Wearable UV Meter that can be worn as both a bracelet and a clip-on. The tangible result was the Helios prototype, but more importantly was the learning experience of the Team, as concluded from their closing statements.

Keywords: Project based learning · Collaborative learning
Ethics and engineering education
New learning models and applications · European Project Semester
Ultraviolet radiation · Wearable UV meter

1 Introduction

The European Project Semester (EPS) is a one-semester capstone project or internship programme offered to engineering, product design and business undergraduates by 18 European engineering schools. Teams are assembled according to the EPS team assembly rules, ensuring inner multiculturalism and multidisciplinary scientific backgrounds, and the Belbin teamwork profiles. Team Helios was composed of Elin Lönnqvist, an Industrial Engineering and Management student from Finland; Marion Cullié, a Packaging Engineering student from

M. E. Auer et al. (eds.), *Teaching and Learning in a Digital World*,
Advances in Intelligent Systems and Computing 715,
https://doi.org/10.1007/978-3-319-73210-7_102

France; Miquel Borras Bermejo, an Industrial, Product and Graphic Design student from Spain; Mikk Tootsi, an Electrical Engineering student from Estonia and Simone Smits, a Biology and Medical Laboratory Research student from the Netherlands. The team members chose to develop the wearable UV meter because they considered it well suited for their different skills and knowledge and expected it to be a great learning experience, both in terms of knowledge acquisition from the project and each other, but also in terms of the development of soft skills.

The UV solar radiation has positive and negative health effects. It is a natural source of vitamin D, also called serum 25(OH)D, a hormone which controls calcium levels in the blood. However, UV radiation is also a mutagen, *i.e.*, can cause mutagenic damage to the DNA. This is the case of UV photons which bypass the natural defences of the skin and melanin. By changing the skin's cellular DNA, excessive UV radiation produces genetic mutations which can lead to several skin cancers. Suntan, freckling and sunburn are familiar effects of over-exposure, along with early ageing of the skin, wrinkling and eye damage. UV radiation may also cause suppression of the immune response system [1].

The dangers of UV exposure are undeniable and public ignorance results in the increase of skin related health problems. However, not every type of UV radiation is equally harmful. UV radiation is divided into, at least, three different groups. These groups are classified according to their wavelength:

- UltraViolet A (UVA) wavelengths (320 nm to 400 nm);
- UltraViolet B (UVB) wavelengths (280 nm to 320 nm);
- UltraViolet C (UVC) wavelengths (100 nm to 280 nm).

The types of UV radiation differ in their biological activity and the extent to which they can penetrate the skin. If the wavelength is shorter, the UV radiation is more harmful. However, shorter wavelength UV radiation is less able to penetrate the skin, thus UVB does not penetrate the skin as much as UVA. However, UVB damages skin cells called keratinocytes in the basal layer of the epidermis, where most skin cancers occur. UV radiation is considered the main cause of nonmelanoma skin cancers (NMSC), including basal cell carcinoma (BCC) and squamous cell carcinoma (SCC). These cancers strike more than a million people worldwide each year [1].

This project had two complementary goals. The first was at the educational meta-level – to foster multicultural multidisciplinary teamwork, autonomous problem-solving and ethical and sustainable development practices – and the second at the design and implementation level – to develop and test a wearable UV meter based on the technical, marketing, sustainability and ethical analyses, as well as on the needs of the user. In particular, the Team's objective was to develop a user-friendly and waterproof device to be worn as a bracelet or a clip-on, latter named Helios. In the first instance, this wearable UV meter should depict an indication of the UV index (from 1 to 11) and should notify the user in case of over-exposure or under-exposure to UV radiation. Eventually, the notifications could be done by means of a mobile application running on a smartphone.

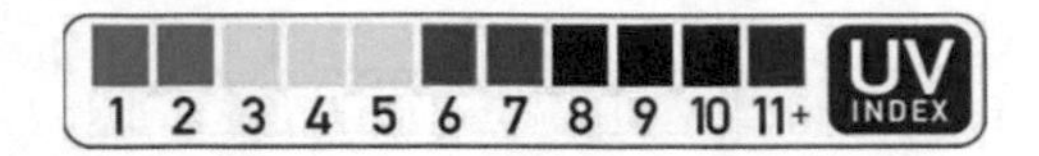

Fig. 1. UV index scale

The Team was able to work together, gather the knowledge required to address the problem, decide on the requirements, design, materials and components, assemble and test a proof of concept prototype as well as produce the set of mandatory deliverables.

Bearing these ideas on mind, the paper is divided in five parts. The first section is the introduction to the project, including the problem, goals and achievements of the Team. The second section presents the technical approaches of several devices existing on the market together with the Marketing, Sustainability and Ethical research done. The third section describes the proposed solution, including the design structure and control system architecture of the wearable UV meter. Section four contains the functional test results and a discussion on the obtained results. Finally, section five presents the conclusions of the project and suggestions for further work.

2 Measurement of UV Radiation

The intensity of the UV radiation corresponds to the energy received per second by a surface, measured in mW/cm^2, or, alternatively, to the energy received per unit area in a given period of time, measured in mJ/cm^2.

To measure the UV solar radiation, the UV meter requires a sensor capable of detecting the full spectrum of UV light. However, most UV meters measure the radiation on the spectrum of UVA and UVB, but not of UVC since UVC is filtered by the ozone layer in the atmosphere.

In 1992, Canada introduced the UV index to quantify the levels of UV radiation at the surface of the Earth for the general public. The UV index scale runs from 0 (when there is no sunlight) to 11+ (extreme). These values give an indication of the expected risk of overexposure to UV radiation and predict the UV intensity levels on a scale of 0 to 11, with 5 different colours, as presented in Fig. 1. This index was approved as a standard indicator of UV levels by the World Meteorological Organization (WMO) and World Health Organization (WHO), in 1994, and takes into account several location-based variables:

- The thickness of the ozone layer over the location (detected using satellites);
- The cloud cover over the location (clouds block UV radiation);
- The time of year at the location (UV radiation in the winter is lower than in the summer because of the angle of the sun rays);
- The altitude of the location (higher locations get more UV radiation).

2.1 Wearable UV Meters

There are several types and shapes of wearable UV meters, ranging from bracelets, clip-ons, key chains, UV cards, UV patches, body stickers and mobile phone applications. Some of these devices, commercially available, are presented in the sequel. Typically, they expect the user to provide the skin type, eye colour, hair colour and the Sun Protection Factor (SPF) usually applied.

Microsoft Band is the market leader. It is a multifunctional watch, including a Global Positioning System (GPS) receiver, UV sensor, personal trainer, sleep helper and calories tracker. The band is made of Thermoplastic Silicone Vulcanizate (TPSiV) and, inside, contains two batteries, a flexible printed PCB, a Bluetooth interface and an Amoled screen [5].

Raymio contains an accelerometer (motion sensor), a gyroscope (rotation sensor) and a UV sensor that measures UVA and UVB. Raymio is made of Thermoplastic Elastomer (TPE) with adjustable length [2]. The device measures the user time exposure and notifies when it is time to seek protection in real time.

June, by Netatmo, includes a leather bracelet with a small UV sensor and a low energy Bluetooth interface together with its companion iOS App. June gives personalized sun protection advice by determining the skin type of the user, measuring the UV exposure and calculating the recommended daily dosage [6].

SunSprite is an eco-friendly personal light tracker together with a dedicated iOS/Android compatible App. This thumb-sized flexible magnetic clasp relies on solar cells for power and on ten light-emitting diodes (LED) to display the UV exposure of the user. The device includes visible light and UV sensors, a Bluetooth interface, a custom-built polymer battery and high-efficiency solar cells [4].

CliMate, by Rooti, claims to track the humidity, UV radiation and temperature of the user environment and to communicate this information through Bluetooth to an accompanying application every 15 min. The device tags the collected data with the location and time, allowing users to generate accurate local weather maps. CliMate is made of Polyethylene Terephtalate (PET) [3].

Although the most advanced technologies found in existing UV meters are above the 100 € budget limit for this project, the Team, after analysing the competitor devices and their marketing strategies, was able to identify which features to include in Helios.

2.2 Complementary Studies

To develop the product, the Team contemplated the Marketing, Sustainability and Ethical perspectives of the development of a wearable UV meter.

Marketing Plan. Helios, for differentiation purposes, targets specifically families with children. The aim was to reduce the effects of over-exposure to UV radiation, especially the development of new skin cancers, by making children aware of this risk. The goal of the Team is to have Helios recommended by doctors and sold in pharmacies, near to the sun block cream, from 100 € to 150 €. The philosophy of the company is to be involved in this medical issue, by doing a campaign and donating 5 € per device sold to a skin cancer organisation.

Sustainable Development. The Team performed the life-cycle analysis of Helios and of the future Helios company in terms of the Environmental, Economic and the Social dimensions of sustainability. These considerations made the Team change the design of Helios, namely, choosing cork for the casing, using a rechargeable battery instead of regular batteries and adding solar cells to the final product, as well as on the creation of a sustainable and eco-efficient factory.

Ethical and Deontological Concerns. Team Helios focussed on the Engineering, Sales and Marketing and Environmental Ethics. The team chose to apply honesty, respect and high standards to every step of the production process, including choosing suppliers and components, selling and marketing the product and providing a warranty of two years.

3 Helios

To make Helios unique, the Team created a UV meter which can be worn both as a bracelet and as a clip-on, and using a UV sensor to measure and display the UVA and UVB radiation in a UV index scale. For sustainability reasons, the casing is in cork, an ecological and renewable resource.

3.1 Requirements

The Team specified the requirements of Helios based on the studies described in the previous section. Additionally, the project proposal specified the budget (100 €) and the need to chose low cost hardware solutions, use open source software and technologies and comply with the EU Directives and the International System of Units.

3.2 Architecture

As can be seen in the requirements for the Helios device development, the architecture had to be a combination of attractive, wearable and built-in low price hardware solutions. Meaning that the device had to be as small as possible conflicts with low cost hardware solutions, since these are not the smallest on the

Fig. 2. Helios device with the LED lights lighted up

Fig. 3. Wearable systems

market. For this reason, Team Helios used Computer Aided Design (CAD) programs, such as Solidworks, and rendering solutions, such as KeyShot, to develop a 3D model with these tools, including real size components, as small and elegant as possible. Figure 2 displays the initial 3D model, with all the LED lights lighted up and five semitransparent plastic parts that let the light of the LED lights go through, indicating to the user the level of UV radiation in real time. On top of the case is also included a hole to let the UV radiation reach the UV optical sensor.

For the "wearable" requirement, the Team also developed two rubber-made attachments, as depicted in Fig. 3, allowing the user to wear Helios as a watch, or hooked into a backpack, for example.

3.3 Structure

The system architecture of the device is presented in Fig. 4. As mentioned, the device has LED lights which light up, one at the time, according to the UV radiation that the sensor is recognizing. Further, a vibrating sound from the buzzer is heard when the LED is changing colour, to alert the user. The user of the device is able to wear the device and put it on and off by a button. In future versions, the user will also be able to connect the device to a smartphone using Bluetooth.

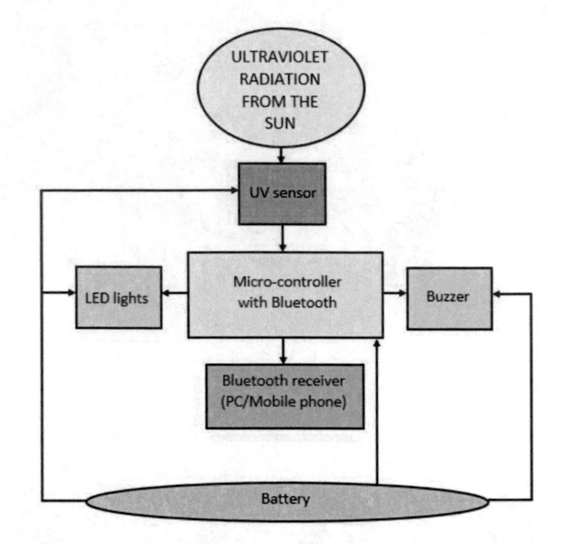

Fig. 4. System diagram

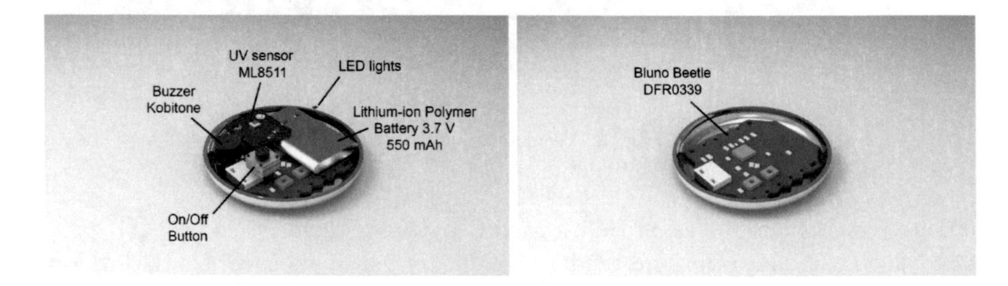

Fig. 5. Inside the Helios device

The 3D model in Fig. 5 includes the controller board port, the buzzer, the battery, the UV sensor, the on/off button and the five different LED lights inside.

To create the prototype, a controller board and a UV sensor were essential. Regarding quality, specifications, size and price, the Team chose the Bluno beetle DFR0339 and the UV sensor ML8511.

Bluno Beetle is a wearable Arduino Uno based board with the CC2540 Bluetooth 4.0 (BLE) module. It is suitable for this project because it has the exact amount of pins that are needed and also because of the low price, integrated Bluetooth, and small size. It uses low-energy technology Bluetooth 4.0. devices, the same technology which is used in most of the newest phones and computers.

The ML8511 is a low current UV sensor, which is suitable for acquiring UV intensity indoors or outdoors. The ML8511 is equipped with an internal amplifier, which converts photo-current to voltage depending on the UV intensity. This unique feature offers an easy interface to external circuits such as

Analog-to-Digital Conversion (ADC). In the power down mode, typical standby current is 0.1 μA, thus enabling a longer battery life.

A rechargeable battery was used for the prototype, namely a 3.7 V 550 mAh Lithium-ion Polymer battery. For the final device team Helios wants to charge the device battery directly with solar panels. Currently the smallest solar panel systems are quite expensive but the solar technology advances everyday and with that the price also decreases.

4 Tests and Results

Included in the prototype development phase, the Team had to perform the following functional tests:

- **UV sensor test:** The device has to be taken outside, on different days, and the UV index calculated by it has to be compared with the official reported index;
- **Software test:** All different functionalities must be tested to check if the programming code works accordingly to the requirements;
- **Waterproof test:** The device needs to be put into fresh and salt water, for at least an hour, in a 1m depth. Afterwards the device should be tested to be sure that it still works properly;
- **Wearability test:** The device will be given to different people, for a one week period, so that they use it and provide feedback;
- **Robustness test:** To make sure that the device is durable, it must be tested to check if it is strong enough and if it can be dropped without breaking.

Initially, the Team tested the controller board code with an Arduino Uno and the UV sensor to see if the programming was working properly. The first objective was to have a code that made LED lights shine according to the value that the UV sensor measured. The LED timer also worked, and it was verified that the LED lights shined for five seconds before they were turned off. The buzzer also went on and buzzed for two seconds every time the UV level changed. When the Team was sure that the code was working, the Arduino Uno was changed to the Bluno Beetle, as this was the board to be used. After several tests, summarized on Table 1, the code worked with success: the device showed UV radiation levels according to the level the weather forecast gave. However, the Team noticed that the UV sensor should be turned straight towards the Sun to show the right results.

Next the device was tested in water, with a cotton pad inside of it instead of the electronic components. This test revealed that the case is not waterproof. This result was predictable given that the groove for the rubber o-ring that should seal the two halves of the case was made by hand, so it was not perfectly smooth, and the size of the O-ring used was also deemed to small.

Afterwards, the Team made the wearability test (shorter than the one week period intended, given time constraints) with a bracelet and a clip on. The Team chose a bracelet with Velcro, so the size can be easily adjusted. Thanks to

Table 1. UV sensor test

Timer [minute]	UV intensity [mW/cm^2]	UV level group	LED	Buzzer	Comments
1	2.22	<3	Green	On	First value
2	2.21	<3	No	No	Same UV level, nothing happened
3	3.77	3–5	Yellow	On	The sensor was turned towards the Sun
4	3.85	3–5	No	No	Same UV level, nothing happened
5	3.22	3–5	Green	On	Green because the reset button was pushed at the same time
6	3.14	3–5	Yellow	On	A new loop was started after the reset button

magnets, the device stayed fixed to its accessory. However, the Team imagined a better wearable experience in a future device version – particularly smaller, because the prototype is bigger than expected.

Finally, the robustness was tested inserting a cracker in the case interior and dropping the device from 1.2 m height to see if it break. The conclusion was that the cracker did not break, so the device is robust to falls. Anyway, the Team does not recommend to repeatedly drop Helios with the electronic components inside.

Figure 6 displays 3D image of the final version of the prototype and Fig. 7 shows an image of the final built product.

Fig. 6. 3D image of the final version of the prototype, with the placement of the electronic components in it

Fig. 7. Photograph of the final developed prototype

5 Future Developments

Concerning future developments of the prototype, the Team identified that the device should be smaller, with a better organisation of the components and a plastic case (PET for example). The bracelet and the clip on should be in TPSiV because of its contact with the user skin.

Finally, in future versions it is intended to have Helios communicate with iOS and Android devices using Bluetooth. The idea is that the user can make a picture of their skin and send it to the mobile application, which will have a database with many types of skin types. Then, this mobile application notifies the user of the risk of underexposure or over-exposure to UV radiation, including personal protection advice, *e.g.*, recommend the Sun Protection Factor (SPF) level of the sun cream based on a picture of user skin and when to (re-)apply this sun screen.

6 Conclusions and Discussion on the Students Experience

The goal of the project described in this paper was to design, develop and build a user-friendly and waterproof device, which can be worn as a bracelet or a clip-on, for measuring the UV index. In the first instance, this wearable UV meter should depict an indication of the UV index (from 1 to 11) and should notify the user if he/she is over-exposed to UV radiation. Additionally, it should notify the user when he/she has been out of the sun too long and thus has a bigger chance of vitamin D shortage.

It can be concluded that developing a wearable UV meter is very important, namely to make people aware of the rising risk of getting skin cancer, because the ozone layer is decreasing due to the climate changes. The project has met most of the requirements set by the Team, and now the members can say they developed a UV meter that is wearable as a bracelet and as a clip-on. The device is showing the UV radiation level by lightning up five different LED lights and gives a sound from a buzzer when the level is changing. The two things missing in the present version of the prototype are the on and off function for the button and a mobile application for iOS and Android.

However, the global objective of this project was not only to have the students implement a final prototype, but also to make them contribute with their distinct visions of the problem to a common solution. This process is not always easy, since at this educational level students are not usually used to collaborate with colleagues from different nationalities (implying distinct cultural backgrounds) and from different backgrounds (students from engineering areas tend to think differently from students from business and product design).

It was verified that working as a team forced the members to divide tasks and have trust in each other. When people are in their professional life, they do not know with whom they will work. This was the same for these students, as they came from five different countries, have five different mother tongues and backgrounds on education. The team members learned from each other and, moreover, learned about themselves. Within the EPS, they also discovered a new country with its own culture and traditions. Participating in several events, visits and travelling around was a part of the learning progress and staying open-minded.

In the words of the students, *"I have also learned to work with people from different countries with different cultural personalities. To learn languages is important for me, and it is something that I will use in my future working life. This semester I got the ability to improve my English and learn some Portuguese... The project was interesting and I could not ask for lovelier team members... Overall I am glad I took the opportunity and went on an Erasmus...*

This semester abroad was a rewarding adventure and I enjoyed each moments. Thanks to this EPS, I worked in a really cool international team with different backgrounds as in a future professional environment. This allowed us to develop our skills and learn new knowledge... it was incredible to live in this foreign country, discover a new culture and travel around beautiful landscapes. Finally, I can say that I have grown from this experience, I improved my English, learned from this project, from the others and about myself.

During EPS I learned a lot, especially in the field of electronic engineering. It was my first time writing a code and I'm really glad that finally after many hours working on the code it was working correctly. Working in a multicultural team was a really good experience. Our team was working together perfectly! I liked Portuguese, communication and project management classes. I'm thankful for this opportunity and ISEP also I want to thank all the supervisor and teachers who really came and helped us out with our project!

I am thankful for having the opportunity to participate in the EPS at ISEP, it was an amazing experience to be in Porto learning many new things and meeting new people, while discovering Portugal and its culture. Within the EPS I improved my soft skills and broadened my knowledge... I did had a great team and I am glad we successfully made it until the end together."

Finally, and as a concluding remark, *"...team Helios was glad to live this amazing adventure, and work on this project."*

Acknowledgements. This work is financed by the ERDF – European Regional Development Fund through the Operational Programme for Competitiveness and Internationalisation – COMPETE 2020 Programme within project POCI-01-0145-FEDER-006961, and by National Funds through the FCT – Fundação para a Ciência e a Tecnologia (Portuguese Foundation for Science and Technology) as part of project UID/EEA/50014/2013.

References

1. Lucas, R., McMichael, T., Smith, W., Armstrong, B.: Solar Ultraviolet Radiation. Environmental Burden of Disease Series, No. 13 (2006)
2. Raymio.com: Raymio Sun Protection Wearable (2015). https://www.indiegogo.com/projects/raymio-sun-protection-wearable#/. Accessed 15 May 2017
3. Best Fitness Tracker Reviews. Tracking Sun Exposure (2015). http://www.bestfitnesstrackerreviews.com/sun-exposure-trackers.html. Accessed 15 May 2017
4. Sunsprite.com: Personal Life Tracker (2017). https://www.sunsprite.com/. Accessed 15 May 2017
5. Technologyreview.com: Microsoft's Wristband Would Like to Be Your Life Coach (2015). https://www.technologyreview.com/s/535956/microsofts-wristband-would-like-to-be-your-life-coach/. Accessed 15 May 2017
6. Wareable.com: Netatmo June Review (2015). https://www.wareable.com/wearable-tech/netatmo-june-review-1176. Accessed 15 May 2017

Challenge Based Learning: The Case of Sustainable Development Engineering at the Tecnologico de Monterrey, Mexico City Campus

Jorge Membrillo-Hernández(✉), Miguel de J. Ramírez-Cadena, Carlos Caballero-Valdés, Ricardo Ganem-Corvera, Rogelio Bustamante-Bello, José Antonio Benjamín-Ordoñez, and Hugo Elizalde-Siller

Escuela de Ingeniería y Ciencias, Tecnológico de Monterrey, Campus Ciudad de México, Calle Puente 222, Col. Ejidos de Huipulco, 14380 Tlalpan, Mexico City, Mexico
jmembrillo@itesm.mx

Abstract. Recently, The Tecnológico de Monterrey (ITESM) in Mexico has launched the ***Tec21*** Educational Model. It is a flexible model in its curriculum that promotes student participation in challenging and interactive learning experiences. At the undergraduate level, one of the central scopes of this model is addressing challenges by the student, to develop disciplinary and cross-disciplinary skills. Two institutional strategies have been implemented to reach the ultimate goal of the ITESM, to work in all careers under the Challenge Based Learning (CBL) system: the innovation week (*i-week*) and the innovation semester (*i-semester*). Here we report on the results of four *i-week* and one *i-semester* models implemented in 2016. The *i-semester* was carried out in conjunction with a training partner, the worldwide leader Pharmaceutical Company Boehringer Ingelheim. Thirteen Sustainable Development Engineering career students were immersed for a 14 week period into the strategies to solve real-life challenges in order to develop the contents of four different courses. Six teachers of the academic institution and four engineers from the Boehringer plant served as mentors. Continuous evaluations were carried out throughout the abilities examination and partial and final examinations were performed by both experts, from the company and from the University.

Keywords: Challenge based education · Sustainable development Engineering

1 Introduction

The Tecnologico de Monterrey in Mexico City (ITESM-CCM) began operations in 1973 in downtown Mexico City. The ITESM was the first University in Latin America that was associated with the Massachusetts Institute of Technology, Carnegie Mellon University and Yale University in the form of a consortium. According to the Academic Ranking of World Universities 2016 [1] and the AmericaEconomia Intelligence 2017 [2] the ITESM is ranked as the second best university in Mexico in general terms but

M. E. Auer et al. (eds.), *Teaching and Learning in a Digital World*,
Advances in Intelligent Systems and Computing 715,
https://doi.org/10.1007/978-3-319-73210-7_103

the best valued by the labor market due to the skills and competences acquired by the graduated students. A key role in this achievement has been played by The School of Design, Engineering and Architecture at the Graduate Student Division (EDIA), now known as the School of Engineering and Sciences (EIC), which bases its educative growth strategy on a concept that integrates the use of technology, creation and management of innovative companies, business linkage and applied research. Inside the EDIA, the Sustainable Development Engineering (IDS) program aims to prepare skilled professionals in sustainable development taking into account that this area is considered as strategic for almost all governments.

Recently, ITESM has launched the *Tec21* Educational Model, a flexible model in its curriculum that promotes student participation in challenging and interactive learning experiences. At the undergraduate level, one of the central scopes of this model is addressing challenges by the student, to develop disciplinary and cross-disciplinary skills. Challenge Based Learning (CBL) promotes the development of skills in students [3, 4]. This model exposes students to situations of uncertainty and in some cases failure tolerance in order to develop their resilience [5]. This is a concern for students in the colleges of engineering, as they are required to have the ability to think critically and solve problems as outlined in the Accreditation Board for Engineering and Technology Inc. (ABET) criteria. Besides the development of disciplinary skills, with this pedagogical approach student motivation toward learning, for their connection to the environment it is encouraged. At the same time, during the process of solving the challenge of innovation, collaboration and multidisciplinary work is encouraged [2].

Here we report on two important ITESM efforts to develop and cross-disciplinary skills in students through experiential learning experiences. The Innovation Week (*i-week*) and the Innovation Semester (*i-semester*).

2 Experimental Design

The general purpose of this research was to investigate the use of Challenge Base Learning in two formats:

Firstly, the undergraduate students of several careers at EDIA were auto enrolled in a one-week intensive period called *i-week*. Four *i-week* subjects were offered at EDIA: ELARA Challenge, PROFEPA Challenge, Ziklum Challenge, and Xochimilco Challenge. A minimum of 15 students (all undergraduate) were enrolled in each of the challenges, no classes were given during the whole week to allow the students to focus on the *i-week* activities. A minimum of three expert teachers (lecturing related subjects) were in charge of the design of the challenge and all its associated activities. The *i-week* was divided into three steps: getting involved (reading, planning the activities and determination of the schedule of actions), development (carry out the planned activities, innovating actions), and discussion and conclusions (where all students compare their results and may improve the conclusions of the others).

Secondly, a 14 weeks/4 months in duration challenge-based-education period, or *i-semester*. Thirteen IDS students were enrolled into what we called Pharmaceutical *i-semester*. Six teachers of the academic institution served tutors of the students, of which four were in charge of each one of the four courses the *i-sesmester* consisted of. This experience was carried out through the participation of a training partner who in this case was the pharmaceutical company Boehringer Ingelheim (BI).

The general purpose of the research was to investigate the use of CBL in the undergraduate students of the Sustainable Development Engineering career at EDIA of ITESM-CCM.

3 Results and Discussion

3.1 The *i-week*

Activities are published four to six weeks prior to the *i-week* and students get involved full-time in a challenging experience they have chosen. The activities during the *i-week* are aimed at: enrich training and competency profile student experiences through innovative and challenging learning, develop disciplinary and transversal competences and promote collaborative and multidisciplinary work. A total of 50,000 students in 26 professional ITESM campus as well as more than 3,000 teachers supporting the development of more than 1,800 projects were involved. Students chose one activity among the options which were offered any campus, including projects with companies or local, national or foreign organizations. At EDIA of ITESM-CCM, samples of projects offered were (each challenge was carried out with a minimum of 15 students):

1. ELARA Challenge. ELARA is a Mexican telecommunications company offering a wide range of products and services in Mexico, Latin America and the US, among which, a wide telephony, pay TV, Internet, data transfer and interconnection network are offered. The challenge consisted in designing an electrical and electronic system that would allow to bring communication to isolated communities in the country, where no electricity is present. One key step of this challenge was the implementation of solar panels to provide enough energy for all the required devices.
2. PROFEPA Challenge. It was led by personnel of the Federal Attorney for Environmental Protection (PROFEPA) where 40 students were approached to the work of monitoring and evaluation of the Attorney General. A specific challenge was to review a company for a week detecting procedures to protect the environment and compliance with current standards.
3. ZIKLUM Challenge. This challenge was carried out in conjunction with the Ziklum Company, an enterprise that recycles more than 5,000 tons of Tetra-Pak containers a year. This prevents around 150 million containers go to garbage dumps. The challenge consisted in designing new lines of treatment of Tetra-Pak containers to open new production lines.
4. XOCHIMILCO Challenge. Xochimilco is a World Heritage City declared by UNESCO in 1987, specifically due to the very productive agriculture system called Chinampa, a pre-Spanish ancient knowledge that has survived throughout the times, this system is placed on a lake that serves as a reservoir of aquatic and aerial species giving a unique and exceptional feature. However, due to the fact that Xochimilco is embedded in Mexico City, there is a great risk of losing its identity by population

and urban growth. So an awareness campaign based on knowledge is necessary. Therefore a challenge was established in finding the way to get the message across to preserve the Chinampas that give identity to the population of Xochimilco.

In all cases, the *i-week* fulfilled the goal to approach the challenged based learning technique to all the students of the campus and helped to establish the more complex strategy: the *i-semester*.

3.2 The *i-semester*

3.2.1 School Setting and Students

A research study on the teaching strategics and the impact on the learning experience was carried out. Research was conducted in the fall of 2016 with 14 weeks/4 months in duration. Thirteen IDS students (8 males and 7 females) from 2nd (Freshman, 1), 4th (Sophomore, 6), 6th (Pre junior 3), 7th (Senior, 3) semester were enrolled in a 4-course credited "*i-semester*" CBL experience. Students were grouped in 4 teams (3–3–3–4 format).

3.2.2 Instructional Design

Participant teachers were trained during the summer 2015 in a 20-h course in which the teachers discussed strategies suitable to implement teaching techniques appropriate to the CBL in order to become mentors or coaches more than teachers of a normal classroom, since the objective is to cover the subjects of the courses through the resolution of challenges. The teachers met with BI staff to determine the challenges to be solved. It is important to note that the challenges were decided on the basis of the professional skills a graduated Sustainable Development Engineer must have, therefore the following challenges were established:

(A) Comprehensive pruning and solid waste management inside the production plant of BI.
(B) Disabling dangerous category waste such as blisters and other packaging of medicines.
(C) Use of residual food oil in the BI cafeteria to make some useful fuel.
(D) Determine the amount of methane produced in the wastewater treatment plant and establish strategies for its use or disposal.

BI participated with 4 Engineers responsible of the areas where the challenges took place and one assessor that monitored all the activities; on the other hand, ITESM-CCM participated with 6 teachers (4 responsible of each course and two assessors that monitored all the activities). BI staff and the ITESM-CCM assessors had two regular meetings a week, one teachers-BI staff only and the other one in the presence of students to monitor the developments of the resolution of challenges. Students spent 4 to 6 h immersion at the BI-Plant (2 miles away from the ITESM-CCM) from Monday to Thursday and a total of six hours Friday sessions corresponding to every single course (1.5 h each) with a specific mentor. One of the properties of the challenges is the uncertainty, this feature forced the students to have at least one 4-h session a week to visit libraries, other experts or field trips to acquire more knowledge to solve the challenges.

3.2.3 Data Collection Procedure

The analyses reported herein focused on the performance of the 4 teams, two partial and one final examination of each of the four courses, three oral presentations of the developing of the resolution of the challenges (examined by both BI and ITESM-CCM staff), and two student satisfaction surveys given at the mid and at the end of the semester answered anonymously that did not count toward the grade for the *i-semester*. The courses by which this CBL *i-semester* strategy was credited were: (a) Sustainable products and services, (b) Environmental and Sustainable research project (c) Environmental management and (d) Cleaner production and industrial ecology.

3.2.4 Analysis or Performance

Students spent approximately 280 h at the BI plant and 110 h of mentoring at the school. Performance was analyzed based on exam scores and rubric-driven examination of oral presentations regarding developments of the resolution of challenges. All exams contained a maximum of 100 points. Descriptive statistics are given in Table 1. As it is shown, the standard deviations indicated that the exam scores were widely dispersed amongst the mean for all three exams but not when oral presentations were examined. It must be noted that the results shown in Table 1 are 20 to 25% higher than the traditional academic lecture courses. These results indicate that the contents of the four subjects were reviewed in full by means of the resolution of challenges and the students fulfilled to 100% the syllabus contents of every course.

Table 1. Descriptive statistics for CBL-format 4 different course (N = 13 students)

	Partial exam 1	Partial exam 2	Final exam	Oral presentation 1	Oral presentation 2	Final presentation
Mean	87.5	89.3	85.9	88.9	92.1	94.6
Median	89	92	89	92	94	96
Standard Deviation	6.07	7.12	8.12	4.15	2.6	2.14
Maximum	100	98	97	96	98	99
Minimum	72	74	75	80	90	90

3.2.5 Analysis of Experience

Students and BI-Staff experiences were analyzed through the surveys. Students were asked two open-ended questions to rate their CBL experience. Question 1 asked students to write the best features of the CBL strategy. Four themes emerged from the students, as shown in Table 2, interaction and the exposure to real-life challenges were the top two themes that were mentioned. To have a professional experience was also mentioned, it is important to note that having a training partner is difficult as many of the companies have their goals focused on the production and business, as the competition is everyday stronger, therefore it is difficult to spend time from the company's human resources in the formation of students or to establish a Challenge-based not a Project-based or Practical-based program. Question 2 asked students to write the worst features of CBL strategy, two main themes emerged from responses. The first thing to arise was the time of the course, it is important to note that the solution of the challenge is not the most important aim under CBL, the goal is to learn the contents of the four subjects throughout the solution of challenges. It is common to hear that the time is short as the students get increasingly interested in the challenge. On the other hand the nature of a challenge is the uncertainty, some methods to solve may not be always available, and it is quite a lot of work to search sources of relevant knowledge, this is in line with the fact that the time spent at the library searching for sources was also one of the themes mentioned.

Table 2. Emerging themes for the "best thing" and the "worst thing" question on CBL strategy (N = 13 students)

Rank	Theme (Best)	Theme (Worst)
1	Interaction	Short time
2	Real-life challenges	Exam preparation
3	Professional contact	No clear order on topics
4	Innovation	Too many books for consulting
5	Applied concepts	Self-learning

4 Concluding Remarks

Challenge Based Education is a key model for teaching Engineering, in the case here described, sustainable development engineering is a recently created area that emerges from a requirement of the development of many companies that need to implement solutions with new ideas coming directly form the academy. Students should be exposed to new course materials to be able to solve the real-life problems and teachers must be ready to learn the state-of-the-art tools to implement innovative solutions. This method was developed with the purpose of improving the ability of engineering students to solve new problems and transfer knowledge from one context to another.

CBL is a pedagogical technique that has been incorporated into areas of study such as science and engineering, and demands a real-world perspective because it suggests that learning involves making or acting student on a subject of study [6, 7]. CBL forces the students to be reflective and flexible thinkers who can use knowledge acquired to take action. Thus CBL triggers the interest of students by giving practical meaning to education, while developing key skills such as collaborative and multidisciplinary work, decision making, advanced communications, ethics and leadership [8].

The Tecnologico de Monterrey will implement as soon as in three years, institutional programs of Education based on challenges for all the careers, implying a great challenge for both teachers and students. The approaches mentioned in this article are two experiences that have to be taken into account for the design and programming of the following programs of study.

Acknowledgments. We thank all the members of the Mechatronics and Sustainable Development Staff at Teconologico de Monterrey, CCM for helpful discussions and to Dr. Ricardo Swain-Oropeza for his support throughout this experience.

References

1. http://www.shanghairanking.com/ARWU2016.html
2. http://rankings.americaeconomia.com/universidades-mexico-2017/
3. Educause: Seven Things You Should Know About Challenge Based Learning (2012). http://educause.edu/eli. Accessed
4. Johnson, L.F., Smith, R.S., Smythe, J.T., Varon, R.K.: Challenge Based Learning: An Approach for Our Time. The New Media Consortium, Austin, Texas (2009)
5. Gaskins, W.B., Johnson, J., Maltbie, C., Kukreti, A.R.: Changing the learning environment in the college of engineering and applied science using challenge based learning. Int. J. Eng. Ped. **5**(1), 33–41 (2015)
6. Jou, M., Hung, C.K., Lai, S.H.: Application of challenge based learning approaches in robotics education. Int. J. Tech. Eng. Educ. **7**, 1–42 (2010)
7. Santos, A.R., Sales, A., Fernandes, P., Nichols, M.: Combining challenge-based learning and scrum framework for mobile application development. In: Proceedings of the 2015 ACM Conference on Innovation and Technology in Computer Science Education, New York, USA, pp. 189–194 (2015)
8. Malmqvist, J., Rådberg, K.K., Lundqvist, U.: Comparative analysis of challenge based learning experiences. In: Proceedings of the 11th International CDIO Conference, Chengdu University of Information Technology, Chengdu, Sichuan, People's Republic of China (2015)

Design, Development and Implementation of E-learning Course for Secondary Technical and Vocational Schools of Electrical Engineering in Slovakia

Juraj Miština[1(✉)], Jana Jurinova[1], Roman Hrmo[2], and Lucia Kristofiakova[2]

[1] University of SS. Cyril and Methodius in Trnava, Trnava, Slovakia
juraj.mistina@ucm.sk
[2] DTI University in Dubnica nad Vahom, Dubnica nad Váhom, Slovakia

Abstract. Electronic course development tools for web-based training make crucial prerequisites to meet the target group instructional design requirements, and ways it has to be delivered to the trainees. The unsatisfactory situation in teaching technical subjects was an incentive to search for the innovative forms and methods of education and modernizing the educational process, and a stimulus to apply them into educational practice. The authors draw attention to the design of methodology for development of multimedia presentations with elements of e-learning. They introduce the design, development and implementation of Multimedia Educational Aid (MEA) developed on the basis of the proposed methodology, multimedia processing and Flash, PHP and XML technology. They used this tool in the educational process of secondary technical and vocational schools in Slovakia with emphasis on achieving greater clearness of abstract concepts included in the thematic unit "Electricity". Conducting research methods, they investigated experimental and control group of students using Niemierko's taxonomy of educational objectives and Simpson's taxonomy of educational objectives to verify the seat up hypothesis. The aim was to raise the students' active approach to this issue. The paper presents results achieved by natural pedagogical experiment.

Keywords: e-learning · Technical Vocational Education and Training (TVET) Multimedia Education Aid

1 Introduction

From various studies [1–4], as well as from the research results carried out by the authors [5, 6], we have found out that the more diversified and entertaining forms of learning by using interactive multimedia applications developed with the purpose to influence the specific knowledge and skills, have positive impact on learning. Students can better remember the subject matter and are more motivated to study the topic and improve knowledge. The aim of technically oriented courses is completion of a comprehensive system of knowledge as a part of general education, which shall include the conditions for a solution of pattern situations associated with the use of technology in professional,

M. E. Auer et al. (eds.), *Teaching and Learning in a Digital World*,
Advances in Intelligent Systems and Computing 715,
https://doi.org/10.1007/978-3-319-73210-7_104

social and private lives of learners. In addition to the knowledge of basic technical issues and performance acquiring the necessary skills and competencies, it covers also the training, educational and developmental content potential for the correct development of the learners' habits in relation to environment, health protection, safety and hygiene behaviour, as well as economic, social and ecological view on life and activities associated with it. The students learn what they consider useful, attractive, interesting, and that what works. Such education focussed only on technical issues is not sufficient and comprehensive. It is important for the learners to know that technology is associated with a range of contexts with a human, society and nature [7].

Electrical engineering remains one of the key areas of technology in the 21st century. The effectiveness of education is directly affected by the process of transferring information, methods and the ways the subject matter is used to explain, train, rehearse and strengthen knowledge and skills. For this reason, the diversity of ICT, multimedia, animations, simulations, etc., and their use in combination with the necessary theoretical and practical teacher's knowledge is a reasonable prospect for increasing the efficiency of educational process. A number of experiments [8–11] have confirmed the educational value of digital technologies. They confirmed that the involvement of multimedia in education contributes to clearness, greater activity and concentration of students in the classroom, as well as to better educational outcomes.

We addressed the outlined problem comprehensively, and drew our attention to the design, development and verification of methodology of the Multimedia Educational Aid (MEA) with elements of e-learning, which we can also be seen as an integrated e-module of a comprehensive e-course. After creating the MEA, the main objective was its implementation into the educational process and verification of its influence on acquired cognitive knowledge, as well as on the psychomotor skills of learners with an emphasis on achieving greater clarity of abstract concepts, growth of interest and active participation of learners in the subject.

2 Analysis of Deficiencies in Technical and Vocational Education

Restructuring national economy in the post- communist years has influenced not only the economy sector. Fluctuation of the global economic environment, as well as the economy crisis has significantly affected the educational sector. Although declared, but financially undervalued, the transition from the industrial society to the information and learning based society has had no solid foundation in the material base. In recent years, under the influence of the societal, political and global changes, the educational system in Slovakia has undergone several transformational modifications that are reflected in different areas. In terms of restructuring the curricular subjects at lower secondary schools, also in technical education as a part of general education, the significant organizational and content changes require to pay increased attention from various perspectives, especially in terms of providing schools with necessary teaching aids, without which the goals cannot be completed. Based on reviewing the results of international studies (PISA 2009–2016) as well as the research and knowledge of experts in the field, we can confirm that technical education in primary schools is currently faced with several difficulties that persist for decades and it is therefore necessary to solve them. Upon the former research carried out within the period of 2009–2016, we analysed the following weaknesses:

- low links of the learning process with the practice;
- lack of material-technical and organizational support of schools;
- quality of the courses themselves (lack of textbooks and video material, lack of visual aids, construction kits, appliances and other equipment);
- insufficient time allotment of the "technology" courses providing technical education at basic, secondary technical and vocational schools;
- lack of qualified teachers to teach technically oriented subjects;
- students' lack of interest in technical education and further education (self-learning, or lifelong learning in this area) etc.

Based on the above knowledge, we therefore pay attention to the design and development of the Multimedia Educational Aid (further referred as "MEA") and related methodological guides, created with an emphasis on flexibility and applicability of these materials in the technical preparation of pupils and students at primary, secondary technical and vocational schools and 8-year grammar schools. We follow the action plan "Learning in the Information Society" launched in 1996 by the European Commission, which highlighted the political commitment of Member States to incorporate multimedia and the Internet into the learning process. Improving the quality of education through multimedia and the Internet is one of the priorities of European cooperation [12].

3 Materials and Methods

The main role of multimedia and ICT, the development of information society, by providing access to a wealth of information processed in different ways [13] is appropriate to different learning styles of pupils. There are a number of different theories of how the human memory works, together with classifications of types of memory and learning styles. However, we have to note that these are just theories and not exact assumptions under which we could propose an e-course. They are stimuli and reference points for the whole concept of MEA, since the didactic applications should respect individual learning styles and the teaching process should be managed so that pupils would follow their own style, eventually, to combine different styles. Several studies suggest that "student learning styles are the key factor that affects learning achievement when adopting the multimedia computer assisted instruction model or (hypermedia) the web-based learning instruction model. Other studies argue that there is no significant difference. It is suggested that, the teachers taking into consideration of the preferred learning strategies for students with different learning styles when making multimedia instructional design will help students attain similar learning effects despite of their learning styles" [14].

An integral benefit of working with multimedia applications is the development of digital (computer) literacy of pupils, which is a prerequisite for their further learning and self-assertion in life. We reflected the foregoing facts and in formulating the objective of increasing the sake of clarity, attractiveness and quality of education itself, along with shaping positive attitudes of pupils to the subject of technology by MEA inclusion into the educational process within the course of "Technology". Prior the research itself and its organization, we faced the issue of the MEA and development of its components. Due to the wide range of thematic unit "Electricity", we concentrated only on a part of the logically interdependent issues: "Electrical Installation Technology". We tried to process the content of MEA in order to transform the curriculum into elements of e-learning. Identify parts of educational content suitable for virtual visualization, based on animation and simulation modelling and technology of visualization based on hypertext, video sequences, etc. The concept of the MEA structure is designed in a modular way. MEA is divided into particular topics characterizing smaller logically interconnected sections of the relevant modules. MEA covers the subject area of residential wiring, reflects the pedagogical-didactic patterns, meets the requirements for a teaching tool and includes the multimedia and e-learning elements that were combined and formed so that the pupils would achieve the expected results as efficiently as possible. The requirements for the learners' competences were specified according to the State Education Programme (ISCED 3A and 3C) for a group of study fields 26 - Electricity. The aid can be characterized as a complex, interactive and graphically friendly learning environment with suggestions and ideas that the teachers can use, combine and partially modify according to their needs [15]. The MEA content can be briefly defined as follows:

- content of teaching material in the form of text and static imaging system (vector graphics) - about 135 pages,
- fourteen interactive photo galleries (14 galleries = about 600 images),
- didactic PC "Pexeso" Memory Match Game of pairs with electricity content,
- four animations illustrating the operation of the breaker in the circuit,
- eight instructional videos of the socket and lighting circuits installation,
- seven interactive descriptive animations of sockets and switches,
- seventeen interactive didactic animations illustrating safety rules when operating electrical appliances and wiring (Fig. 1),
- interactive 3D home visualization of lighting circuits,
- proposals for practical activities (designed with respect to the low cost and simple organizational support),
- thirteen interactive tests (13 tests = about 270 questions).

4 Results and Discussion

The core set of research were all pupils of secondary technical and vocational schools providing education in the field of electrical Engineering. We have chosen a representative set of students. We applied the random selection in dividing the research sample into control and experimental groups (hereinafter referred to as CG and EG). We set up the following hypotheses:

- H: The use of MEA in the educational process will increase the effectiveness of the technical education of learners in EG, where the training is carried out by using MEA, in comparison to the traditionally taught learners in CG (without using MEA). Hypothesis H was considered to be valid, if at least one of the two following working hypotheses H1 and H2 is confirmed:
- H1: We assume that in didactic test at the end of experimental classes, the learners in EG, where teaching is carried out by using MEA, will achieve better learning outcomes in cognitive area from selected themes than learners in CG taught traditionally (without MEA).
- H2: We assume that in solving practical tasks at the end of experimental classes, the learners in EG, where teaching is carried out by using MEA, will achieve better results in psychomotor area from selected themes than learners in CG taught traditionally (without MEA).

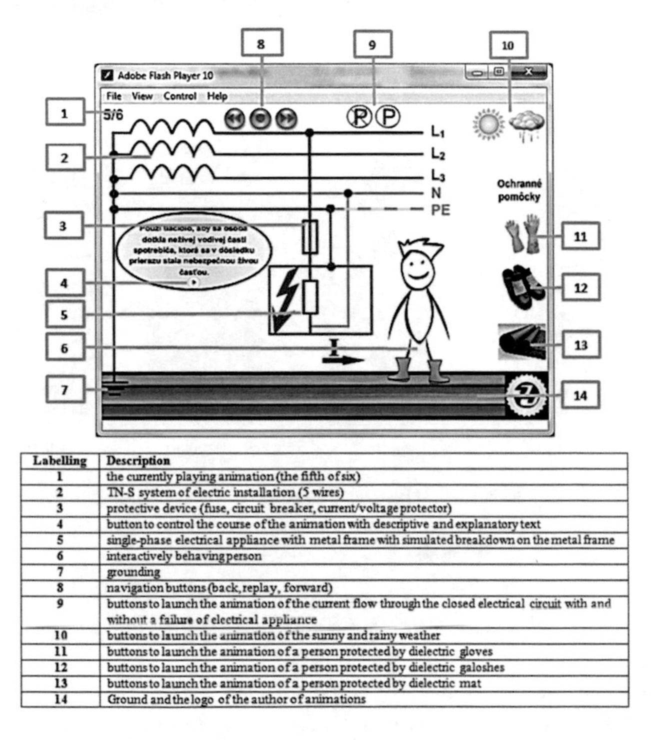

Labelling	Description
1	the currently playing animation (the fifth of six)
2	TN-S system of electric installation (5 wires)
3	protective device (fuse, circuit breaker, current/voltage protector)
4	button to control the course of the animation with descriptive and explanatory text
5	single-phase electrical appliance with metal frame with simulated breakdown on the metal frame
6	interactively behaving person
7	grounding
8	navigation buttons (back, replay, forward)
9	buttons to launch the animation of the current flow through the closed electrical circuit with and without a failure of electrical appliance
10	buttons to launch the animation of the sunny and rainy weather
11	buttons to launch the animation of a person protected by dielectric gloves
12	buttons to launch the animation of a person protected by dielectric galoshes
13	buttons to launch the animation of a person protected by dielectric mat
14	Ground and the logo of the author of animations

Fig. 1. Description of the elements of one of the seventeen screens of interactive animation. The animation of single pole touch with exposed conductive part of an appliance visualizes five different scenarios.

Niemierko's taxonomy of educational objectives was used to verify the hypothesis H1. We formulated four working hypotheses (H1.1–H1.4). These were verified by the outputs of non-standardized didactic test. To verify the hypothesis H2, we used Simpson's taxonomy of educational objectives, which consists of seven levels. We verified the results for only the first and fifth level of significance (H2.1–H2.2). Didactic test was also used to monitor some parameters of psychomotor skills (to verify the hypothesis H2.1). Learners' worksheets with guidance and examples of solutions of different types of tasks in didactic test were made. Because of elimination of the subjective assessment of teachers, evaluation sheet for the teachers with the key to the tasks solutions and with detailed instructions for the evaluation of open tasks and practical tasks validating psychomotor skills of learners was developed. Using natural pedagogical experiment as the main research method and the event analysis of the didactic test tasks categorized according to the levels Niemierko's taxonomy of educational objectives, we obtained data that clearly confirm that the EG learners achieved better results in all monitored levels, compared to students in CG. This fact is illustrated in Fig. 2 above. We determined the statistical significance of the results reported above by the nonparametric Wilcoxon rank-sum test in the Statistica program (Table 1).

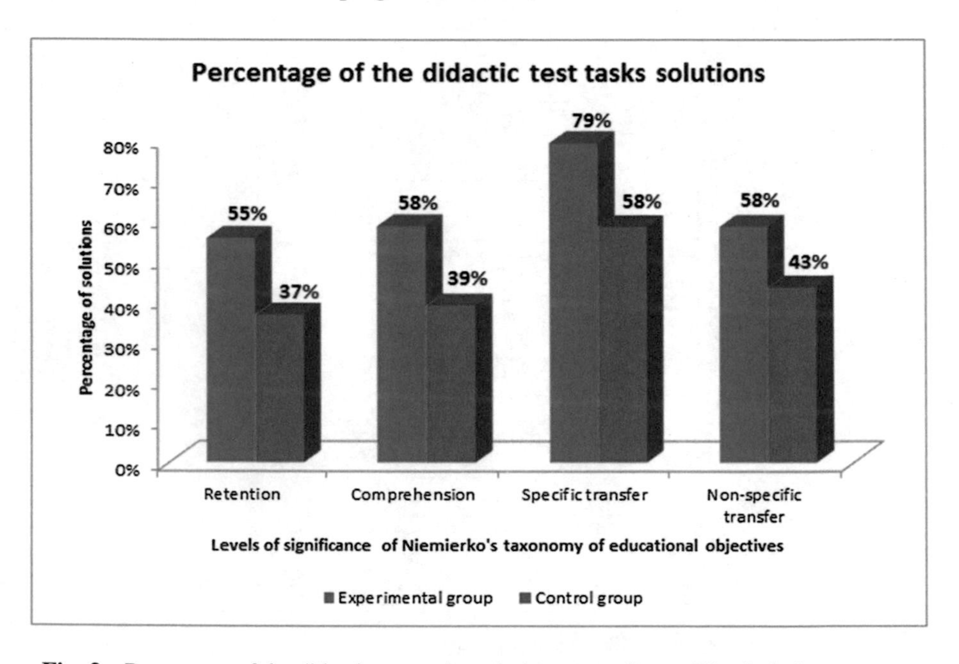

Fig. 2. Percentage of the didactic test tasks solutions according to Niemierko's taxonomy

By the statistical verification of hypotheses H1.1–H1.4, we confirmed statistical significance in all four monitored areas for the learners in EG. We assumed the validity of hypothesis H1 if the at least three of the four working hypotheses are confirmed. All hypotheses were confirmed, and therefore the hypothesis H1 is regarded as valid. The difference in the results obtained in the cognitive area is statistically significant. We can therefore conclude from these results that the use of MEA positively affects cognitive skills of learners in all monitored areas (retention, comprehension, specific transfer, non-specific transfer). For the assessment of hypothesis H2.1, we used the tasks No. 27, 28 and 29 of didactic test, which were directly related to practical activity represented by the task No. 30. On the basis of the event analysis of these tasks, we can say that EG learners achieved better results in comparison to learners in CG. This fact is presented in Fig. 3.

Table 1. Results of statistical evaluation

Hypothesis	P error probability	Z value of testing criteria	Result	Conclusion
H 1.1 (area of retention)	0,000004	4,59565	the observed features are statistically significant	the hypothesis was confirmed
H 1.2 (area of comprehension)	0,000359	3,568692	the observed features are statistically significant	the hypothesis was confirmed
H 1.3 (area of specific transfer)	0,000004	4,586946	the observed features are statistically significant	the hypothesis was confirmed
H 1.4 (area of non-specific transfer)	0,038311	2,071524	the observed features are statistically significant	the hypothesis was confirmed

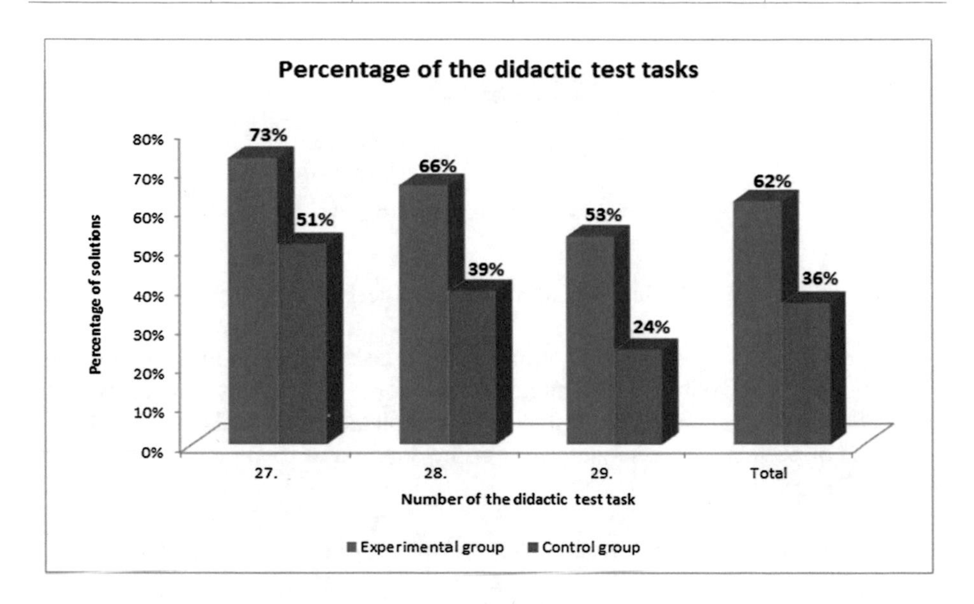

Fig. 3. Percentage of the didactic test tasks solutions

For statistical verification and assessment of the fifth level of significance by Simpson taxonomy of objectives (H2.2), which presupposes a more complex activity of the learner requiring highly coordinated motor activities, four tasks focused on practical activities were compiled. These were framed in such a way that they can be objectively scored in the same grading category.

Each learner because of timing reasons always dealt with only one task. Evaluation of practical activity consisted of assessing the functionality and correctness of electrical wiring, using pre-prepared evaluation system, and of the record total activity time. We evaluated whether learners carry out the activity quickly, without hesitation - automatically. For clarity of evaluation, guidelines for evaluation were developed. The maximum number of points for practical activity that a learner could obtain was 18 points. EG learners were more successful in comparison to learners in CG by 18%. Especially important for us is that the EG learners achieved 95% success rate. The learners in CG achieved 77% success rate (Fig. 4).

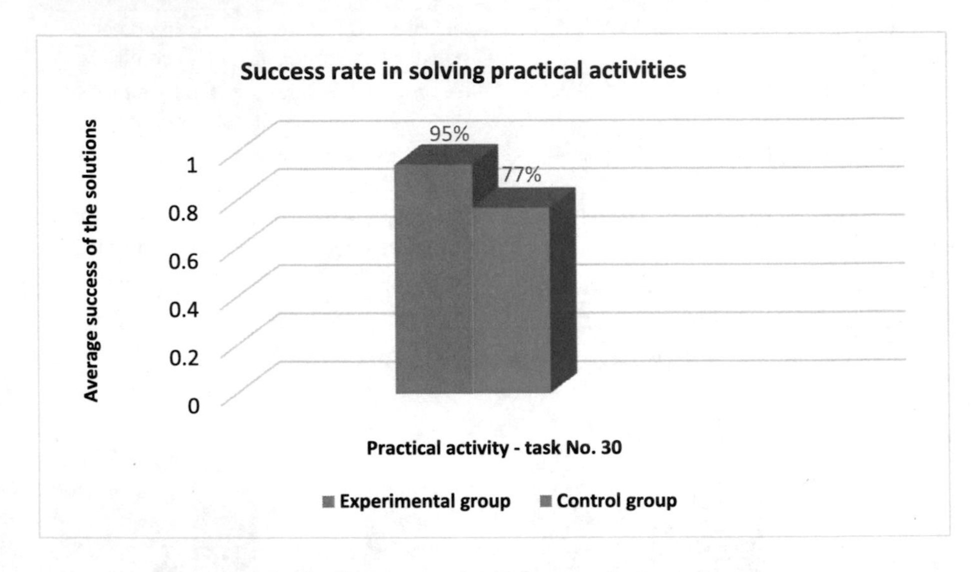

Fig. 4. Success rate in solving practical activities

The time for the implementation of the practical task was determined on the basis of pilot testing for a maximum of 45 min. However, for some students this time was not enough. So, the time limit to allow all students to complete the practical tasks was shifted to the limit of 60 min. The detailed results of the duration of the practical activity are shown in Table 2.

Table 2. Duration of practical task in the control and experimental group.

Student no.	1	2	3	4	5	6	7	8	9	10	11	12	13	14	15
EG time	25	26	26	27	28	31	32	35	35	35	36	37	37	38	38
CG time	35	38	38	39	40	40	41	42	42	42	42	42	43	43	43
Student no.	16	17	18	19	20	21	22	23	24	25	26	27	28	29	30
EG time	38	40	41	42	43	43	44	57	57						
CG time	43	44	45	45	45	53	56	60	60	60	60	60	60	60	60

From the data in the table, it can be stated that only two EG learners exceeded the limit of 45 min, unlike the CG students. This limit was not enough for ten CG students. The shortest duration - 25 min was recorded for an EG students. The shortest activity time in the control group was up to 35 min. Compared to the CG, up to seven EG students carried out the practical activity in shorter time than the shortest time in CG and the three students at the same time, i.e. 35 min. The results verified that MEA streamlined the teaching process - in particular clarification of theoretical rules. Students using MEA than have more time to implement practical activities during the lessons, compared to the pupils in the CG. Obviously, acquiring psychomotor skills requires re-training of individual tasks.

5 Conclusion

Creating multimedia teaching aids, as well as e-learning courses, is generally time-consuming as well as the technically very demanding activity. When creating this type of material, the author must dispose of a range of knowledge and competences in the fields of didactics, pedagogy, profession, IT development, programming, ICT, multimedia material processing, legislation, etc. We believe that similar devices should be developed, as well as made accessible to all schools on a centralized basis, by professionally oriented institutions, or by specialized workplaces. We cannot expect that teachers will create such materials by themselves in their free time. We assume that the positive results of our research will contribute to solving the current situation, and thus open the way to effective teaching of some teaching units in the subject “technology” that is not covered in the nationwide multimedia teaching material “the Planet of Knowledge”.

Acknowledgment. The authors gratefully acknowledge the contribution of the KEGA Grant Agency of the Slovak Republic under the KEGA Project 016UCM-4/2017 “Implementation of the project for improving the quality of graduates into practice at secondary technical and vocational schools”.

References

1. Sutopo, H.: Selection sorting algorithm visualization using flash. Int. J. Multimedia Appl. (IJMA) **3**(1), 22–35 (2011)
2. Saraiya, P.: Effective features of algorithm visualizations. Virginia Polytechnic Institute & State University, Blacksburg. Master thesis (2002). http://scholar.lib.vt.edu/theses/available/etd-08202002-132927/unrestricted/Thesis.pdf. Accessed 04 May 2017
3. Urquiza-Fuentes, J., Velázquez-Iturbide, J.A.: A survey of successful evaluations of program visualization and algorithm animation systems. ACM Trans. Comput. Educ. (TOCE) **9**(2), 1–21. (2009). Special Issue on the 5th Program Visualization Workshop. http://www.researchgate.net/publication/220094505_A_Survey_of_Successful_Evaluations_of_Program_Visualization_and_Algorithm_Animation_Systems
4. Avancena, T., Nishihara, A., Kondo, C.: Developing an algorithm learning tool for high school introductory computer science. Educ. Res. Int. 11 pp. (2015). Article ID 840217. http://www.hindawi.com/journals/edri/2015/840217/
5. Jurinová, J.: The use of multimedia in technical education. In: Cápay, M., Mesárošová, M., Palmárová, V. (eds.) DIVAI 2012: 9th International Scientific Conference on Distance Learning in Applied Informatics: Conference Proceedings, pp. 165–173. Faculty of Natural Sciences Constantine the Philosopher University in Nitra, Nitra (2012)

6. Hrmo, R., Miština, J., Krištofiaková, J.: Model for improving the quality of graduates and job applicants in European labour market. In: International Conference on Interactive Collaborative Learning, pp. 429–439. Springer, Cham (2016)
7. Smékalová, L., Němejc, K., Slavík, M.: Plasticity of evaluation of transferable competences of students of technical and natural science fields of study: a transversal research. In: Miština, J., Hrmo, R. (eds.) Key Competences and the Labour Market: Scientific Monograph, pp. 29–45. Management University Publishing House, Warsaw (2016)
8. Chmelíková, G., Mironovová, E.: Students' on-line conferencing skills trained within the Slovak/Serbian student-teacher-young researcher cooperation. In: Synergies of English for Specific Purposes and Language Learning Technologies, 1st edn., pp. 311–321. Cambridge Scholars Publishing, Newcastle upon Tyne (2017)
9. Mironovová, E., Chmelíková, G., Fedič, D.: Developing e-Learning skills within English language training in the Slovak University of Technology. In: IC4E 2010: International Conference on e-Education, e-Business, e-Management and e-Learning, Sanya, China, pp. 203–206. IEEE Computer Society, Sanya (2010)
10. Šimonová, I.: Students' assessment preferences in ESP in the smart learning. In: Smart Education and e-Learning, pp. 387–396. Springer, Berlin (2016)
11. Šimonová, I., Poulová, P., Kříž, P., Sláma, M.: Intelligent e-learning/tutoring - the flexible learning model in LMS blackboard. In: Computational Collective Intelligence - Technologies and Applications, pp. 154–163. Springer, Berlin (2015)
12. EURIDYCE 2004: Key Data on Information and Communication Technology in Schools in Europe. ©Eurydice (2004)
13. Arvanitis, K.G., Patelis, G., Papachristos, D.: Application of a model of asynchronous Web based Education (WbD) in the agricultural engineering sector. In: WSEAS Transactions on Advances in Engineering Education, vol. 9, pp. 12–22. WSEAS Press (2012)
14. Cheng, Y., Cheng, J., Chen, D.: The effect of multimedia computer assisted instruction and learning style on learning achievement. In: WSEAS Transactions on Information Science and Applications, vol. 9, pp. 24–35. WSEAS Press (2012)
15. Andres, P., Macák, T.: Unconventional methods of human capital management - key competencies as a source of competitive advantage. In: Miština, J., Hrmo, R. (eds.) Key Competences and the Labour Market: Scientific Monograph, pp. 215–265. Management University Publishing House, Warsaw (2016)
16. Bilčík, A.: Posudzovanie a hodnotenie úrovne vzdelávania. In: Krpálek, P., Krpálková, K.K. (eds.) Autoevalační kultura a kvalita vzdělávaní. Sborník recenzovaných příspěvků mezinárodní vědecké konference, pp. 21–26. Extrasystem Praha, Praha (2017)
17. Gabrhelová, G., Pasternáková, L.: The intersections of education and management, 164 p. Ste-Con, GmbH, Karlsruhe (2016)
18. Hagara, V., Ružinská, E., Jakúbek, P., Paľun, M.: Creativity and positive image in educational institutions, 202 p. Tribun EU, Brno (2015). ISBN 978-80-263-0968-0
19. Chromý, J., Sobek, M., Krpálková, K.K., Krpálek, P.: Determining aspects of electronic systems for teaching support. In: Recent Advances in Educational Methods: Proceedings of the 10th International Conference on Engineering Education (EDUCATION 2013), Proceedings of the 1st International Conference on Early Childhood Education (ECED 2013), Cambridge, UK, pp. 62–67 (2013)
20. Kučerka, D., Kučerková, M.: Analýza vzdelávania pedagogických pracovníkov na základných a stredných školách v Slovenskej republike. In: Almanach příspěvků konference Moc moudrosti a moudrost moci, České Budějovice: Vysoká škola technická a ekonomická v Českých Budějovicích, pp. 119–129 (2012)

21. Lajčin, D.: The specifics of the Manager of the private sector in the spectrum of approaches of education legislative conditions Slovakia, 109 p. Ste-Con, GmbH, Karlsruhe (2016)
22. Oberuč, J., Lajčin, D., Porubčanová, D.: Stress management in the workplace. J. Law Econ. Manag. **4**(2), 47–51 (2014). Eastern European development agency, London
23. Szököl, I.: Key competences in educating teachers. In: Edukacia-Technika-Informatyka. Rzsesow: Wydawnictwo uniwersytetu Rzeszowskiego. roč. 1(11), 249–253 (2015)
24. Tamášová, V., Barnová, S.: School climate as the determinant of the relationship between the level of students resilience and school satisfaction. In: ACTA- technologica Dubnicae, pp. 19–37, no. 1 (2011)

A Method to Design a Multi-player Scenario to Experiment Risk Management in a Digital Collaborative Learning Game: A Case of Study in Healthcare

Catherine Pons Lelardeux[1(✉)], Michel Galaup[2], David Panzoli[1], Pierre Lagarrigue[3], and Jean-Pierre Jessel[4]

[1] IRIT, University of Toulouse, INU Champollion, Serious Game Research Network, Toulouse, France
{catherine.lelardeux,david.panzoli}@univ-jfc.fr

[2] EFTS, University of Toulouse, INU Champollion, Serious Game Research Network, Toulouse, France
michel.galaup@univ-jfc.fr

[3] ICA, University of Toulouse, INU Champollion, Serious Game Research Network, Toulouse, France
pierre.lagarrigue@univ-jfc.fr

[4] IRIT, University of Toulouse, UPS, Serious Game Research Network, Toulouse, France
Jean-Pierre.Jessel@irit.fr

Abstract. In recent years, there has been an increasing interest for collaborative training in risk management. One of the critical point is to create educational and entirely controlled training environments that support industrial companies (in aviation, healthcare, nuclear...) or hospitals to train (future or not) professionals. The aim is to improve their teamwork performance making them understand the importance applying or adjusting safety recommendations. In this article, we present a method to design multi-player educational scenario for risk management in a socio-technical and dynamic context. The socio-technical situations focused in this article involve non-technical skills such as teamwork, communication, leadership, decision-making and situation awareness. The method presented here has been used to design as well regular situations as well as critical situations in which deficiencies already exist or mistakes can be freely made and fixed by the team in a controlled digital environment.

Keywords: Collaborative virtual environment · Learning game
Risk management · Communication · Healthcare training · Simulation

1 Introduction

In recent years, there has been an increasing interest for team training in risk management especially in case of socio-technical and dynamic systems.

M. E. Auer et al. (eds.), *Teaching and Learning in a Digital World*,
Advances in Intelligent Systems and Computing 715,
https://doi.org/10.1007/978-3-319-73210-7_105

On one hand, the investigations against the most tragic accidents lead most of the time politicians and managers to establish new rules, policies or recommendations to prevent new similar accidents. On the other hand, some professionals often interpret new rules and procedures as a stronger control of their activities. One of the critical components of a comprehensive strategy to improve the safety in transport industry, as well as in healthcare is to create educational and training environments that support providers to train teams to improve their performance making them understand the importance applying or adjusting safety recommendations. Several markets have already been addressed to train on emergency situations such as medical emergency [15], military intervention, bio-terrorism preparedness, nuclear emergency, chemical industrial risk [4]...

However, these works fails to consider the human factors such as communication which is listed among the main root causes of accidents. They mainly focus on scheduling or technical skills and their approaches are centered on the individual aspects of risk management.

2 Purpose and Goal

To teach and monitor non-technical skills such as teamwork, communication, leadership, decision-making and situation awareness, a digital collaborative environment has been designed to represent the socio-technical and dynamic situation in which a team will be involved [11]. The system we choose to exemplify our work is the operating theater. This digital collaborative environment represents with high fidelity its structure and its complexity. It allows controlled manipulations of the decision context and controlled information. It has been designed with game design mechanisms and interactive features to mimic professional activity. Each one plays the role of a professional with their own expertise field (such as surgeon, anesthetist, operating-nurse in an operating room). The participants must manage all together a real-life like professional and uncertain situation. These situations should be designed to make teams understand the importance of non-technical skills and their impact on the way they manage the situation. The more they communicate, the more realistic their representation of the current situation should be. As a result, the most suitable decision should be made.

In others words, the challenge relates to provide a library of controlled educational situations where students are relatively free to act and can reproduce a causal chain of events that leads to virtual accidents.

So far, however, there has been little discussion about how designing a controlled and educational scenario to train staff to manage risk with real-life like situations.

Designing educational scenarios for risk management training is particularly complex because most of the time, the causal chain of events that leads to an accident is unpredictable. It implies a large variety of contributing factors, such as human factors and technical failures, which are difficult to combine artificially. This paper aims to present a method to design such educational scenario for a virtual collaborative environment that particularly intents to train staff on risk management in a socio-technical an dynamic system. We focus particularly on risks connected to human errors such as communication default or non-suitable decision-making.

3 Approach

The online Cambridge dictionary defines a scenario as "a description of possible actions or events in the future" or "a written plan of the characters and events in a play or film".

In the field of learning games, a scenario can be considered here as a set of elements: (1) a briefing (mission): presentation of the current situation and expected objectives to reach, (2) a virtual universe: objects, furniture, documents, characters... (3) a set of actions, pieces of information, documents, furniture and objects which can be manipulated throw the universe to achieve the mission, (4) playful and educational lockers such as educational prerequisites, educational failures to avoid... (5) educational skills to develop or acquire, (6) abstract or concrete concepts which can be manipulated with interactivity throw the environment: game play elements as inventory of assets, monetary system, virtual store... and educational concepts as programming, making decision... (7) steps or levels which compose the mission, (8) educational objectives to reach (visible or not in a briefing stage) (9) a debriefing: summary of outcomes with feedback that should help the player to succeed in the future.

Firstly, a scenario proposes to the players a short storytelling of what is the actual situation and what is the expected situation at the end. Secondly, a scenario provides interactions that allow the players to achieve the mission and lockers (educational lockers or playful lockers) to prevent the player to succeed. Finally, outcomes are compared to expected objectives and results are immediately displayed at the end of a game session.

This definition particularly suggests that interactive storytelling triggers challenging opportunities in providing effective models for enforcing autonomous behaviors for characters in complex virtual environments. In other words, players should be able to be wrong, patch their errors, succeed or fail.

3.1 Challenges to Design an Interactive Scenario for Risk Management

In the case of training environment for high graduated students, the classical challenge to design an interactive scenario consists in either (1) representing with creativity but also with high fidelity the professional environment through the virtual universe or (2) providing opportunities to characters to choose what they want to do, or (3) providing interactions as part of the professional activity using objects/equipment/furniture/abstract elements arranged in the virtual universe, or (4) giving relative but controlled freedom to act in the universe in order to compare with the expected behaviors.

In the case of training in risk management, different approaches exist such as training for emergency situations, improving the ability to identify and understand a critical situation and improve the situation awareness of a critical and risky situation [5], providing technical skills training on technical equipment with or without automation [12], providing maintenance training on dangerous equipment [6], training with exceptional/rare situations, training in a safety environment without any consequences in real life [1]...

We choose here to design educational situations to make teams able to identify and understand critical situations. The library of educational situations must be composed of regular situations as well as critical situations in which deficiencies already exist or mistakes can be made and fix by the team.

3.2 Overview of Methods to Design a Scenario

Three kinds of scenario can be listed: - entirely controlled scenario: a script defines every possible paths to succeed or fail the mission. - controlled scenario with a limited but real freedom: a large number of paths are possible to succeed or fail. As a consequence, none script defines all the possible combinations but algorithms can calculate them if necessary. - entirely free scenario: machine learns from user's interactions and builds a statistically-realistic behavior. At the beginning, paths are unknown but machine learns from the user's experience over the time.

Designing an entirely controlled scenario consists in determining in details every available alternatives and their consequences in the virtual world. Such a scenario is called scripted or branching scenario. It can be graphically represented by a tree-like structure. Each arc represents a choice and each node represents the state of the world. For every state, the user has to make a choice between different alternatives. There is normally one best choice, with the other alternatives being either wrong, or not as good. This kind of scenario offers different pipelines and maintains the user in a entirely controlled session.

We choose here the second method. The main argument that can be advanced to support this choice is that it offers a real-life-like experience even though the level of freedom is limited in a virtual world comparing to real-life. Despite of limited freedom, users should be free to manage a professional-like situation in a virtual world.

4 A Method to Design an Educational Scenario for Risk Management Training

Designing scenarios based on real-life situation for risk management training consists both in (1) representing a perfect initial situation with competitive experts who made zero error before the team must manage the current situation and (2) representing an irregular situation where experts made mistakes that can lead to an incident if the errors are not tracked and fixed in time. If they are not fixed in time, problems will reveal as being part of the causal chain of events that leads to an adverse event. To that end, two categories of multi-player scenario have been designed.

The first one represents a regular situation embedded in a standardized scenario. It aims to train teams to apply safety recommendations and security process.

The second one represents an "irregular situation" embedded in a critical scenario. It aims to make team understand the interest of applying or adjusting policy safety procedure to avoid accidents. Designing such a scenario is more complex because it requires also analyzing the chronology of events before an accident and identifying the causal chain of events and their root-causes. The method described here has been inspired by the systemic method used to analyze real accidents that occurred in socio-technical and dynamic systems. Basing our thoughts on systemic methods analysis, we designed "irregular situations" both dispatching failures or errors in an initial perfect situation and providing erroneous available issues during a decision-making or inappropriate tasks in the cloud of possible tasks.

The Sect. 4.2 describes the Reason's Swiss model to represent the chain of events that leads to an accident. It helps to understand the method we use to design a scenario based on an "irregular situation".

4.1 Draw up an Educational Scenario Based on Standardized Situation

Surgical interventions were recorded with four video cameras in the operating room. Analyzing these recordings, Devreux [3] studied how professionals communicate according to the level of experience they have.

The Fig. 1 illustrates the method we used to design such a scenario.

Firstly, we decide to specify the data using the Business Process Modeling and Notation graphical representation (BPMN). "BPMN defines a Business Process Diagram (BPD), which is based on a flow-charting technique tailored for

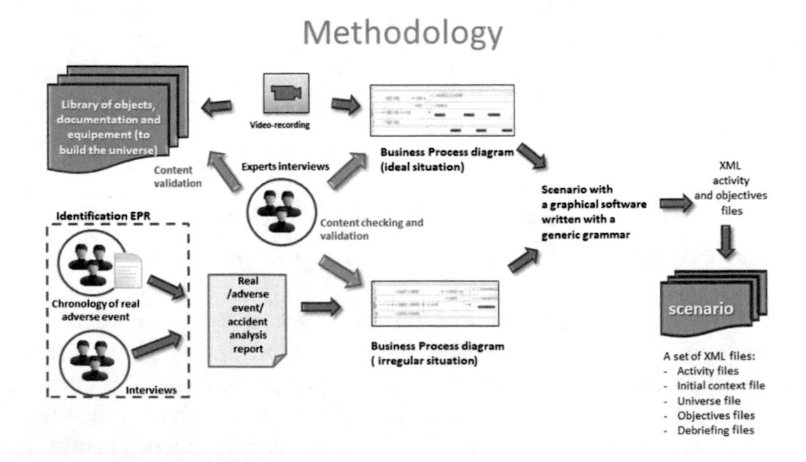

Fig. 1. Methodology used to design a risk management scenario.

creating graphical models of business process operations. A Business Process Model, then, is a network of graphical objects, which are activities (i.e., work) and the flow controls that define their order of performance" [16]. BP Diagram enables us to represent parallel and collaborative tasks. This choice was also motivated by the necessary content validation step. In other words, it is easier to make experts validate contents with a graphical representation than with text-document. However, the computerized human activity held in the BP diagrams cannot be used straightforwardly. The computer can neither understand the interactions labeled on the activity nodes of the BP diagram nor relate them with their expected impact or meaning in the environment. The BP Diagram does not enable us to anchor the tasks to virtual objects on the virtual scene. The process of anchoring the interactions into a semantic environment is another necessary step towards solving that problem. As a result, we enhance BP diagram with a specific grammar to describe an educational scenario. This grammar enables us to anchor the tasks in the virtual environment and to increase the number of possible combinations for each character's role. It uses predicates to combine sets of available tasks and pieces of information. As an illustration, the action "Wash hands" is no longer available if the action "put the surgical gloves" has already be achieved.

4.2 Draw up an Educational Scenario Based on Irregular Situation

Extensive researches into disasters such as the disintegration of space shuttle Columbia [7], typically focus on the chains of events which caused these disasters. When such accidents are analyzed more closely, organizational problems, equipment breakdowns or loss of communication accuracy are often revealed as being part of the causal chain of events.

Different methods to analyze accidents and risks after slips happened exists. Some of them are based on systemic-based technique. System-based technique methods are specially used for analyzing the causes of accidents that occurred in socio-technical systems.

Studying complex system, Reason [14] shows that most of the time, accidents result from multiple successive failures which could not have been corrected or stopped in time. Reason's model proposes that within complex systems, multiple barriers or layers exist to prevent accidents or errors. Mostly they do this very effectively, but there are always weaknesses. Among the weaknesses, a poor communication between team members is often identified as an underlying factor of near-misses or accident.

The Reason's Swiss Cheese Model represents the system as a whole. Each slice of cheese represents a barrier defense against failures and mistakes. The holes in the slices represent individual or collective weakness. The whole system is dynamic and the holes can varies in size and position on the slices as far as the situation evolves. The system can trigger accident when errors or mistakes are temporary aligned because none defense barrier can avoid the accident. As the result, when the holes in all slices are temporary aligned, they allow 'a trajectory of accident opportunity'.

Committing zero error is most of the time nearly impossible. However, it is possible to build defense barrier to detect mistakes and avoid unpredictable accident. The pursuit of greater safety is hindered by an approach that does not seek to remove the error provoking properties within the system at large. Advancing mistakes or identifying likely errors and then removing or correcting them before the accident would be a better way to improve safety.

As a consequence, designing educational real-life situation for training consists in dispatching holes in a predefined situation and providing features that make team able to act, track and correct mistakes/failures using defense barriers.

The next section describes methods used to analyze the chain of events that leaded to an accident. It helps to understand the method we use to model educational feedback at the end of a game session.

Systemic analysis methods in healthcare. The idea that not only the disease but also the actors, teams, equipment, organizations, patient's profile... can ultimately be harmful to the patient is fairly recent in medicine. This new awareness dates from the 1990s and the report "To Err is Human" [8]. The same awareness has risen in France where the National Authority for Health (HAS) requires professionals to evaluate their practices through morbidity-mortality meetings. During these meetings, professionals declare and collect data on health care related adverse events. To help practitioners through a rigorous and structured approach, the HAS recommends a systemic analysis method be used. The main two systemic analysis methods used to study near-misses or adverse events are ORION [2] and ALARM method [13].

The systemic analysis which is supported by the French National Healthcare Authority is composed of 5 defense barriers: the patient, the actors, the team, the tasks, the environment, the institution and the organization.

When we designed a scenario representing an irregular situation, predefined anomalies were dispatched and hidden throw the barriers. Designers provide large variety of actions and pieces of information to create diversion. As the consequence, the anomalies are drown in a sea of pieces of information. Furthermore, some actions or decisions might launch an uncontrollable situation. At the end, players are asked to identify what was wrong and what was right from their point of view. The professional process from systemic methods has been reversed to force the students to identify their weakness and strengths.

5 Experiment and Results

5.1 A Case of Study: The Operating Room

The educational content used to apply this method is based on real adverse events. These adverse events have common characteristics: a communication or decision making defaults have been identified as a contributing factor.

Less than 10 scenarios have been designed using the method described in Sect. 4. Training sessions have been organized using the learning game environment, which offers a library composed of standardized and critical scenarios.

The experiments were carried out with the help of medical trainers and anesthetist-nurse trainers at the University Hospital of Toulouse. They aim to control how the students apply safety recommendations in real-life like situations. Lessons had already been delivered to the students on said topics prior to the experimentation and all the students had already worked in a real operating room during a professional internship.

5.2 Results

During the experiment, computer data were logged; training sessions were video-recorded. Questions were asked the students on their game training experience through surveys (32 persons were concerned). Two scenarios were played. The first one presents a regular situation and the second one presents an irregular situation even though they seem to be similar. Both scenarios used in this experiment focuses on two adverse events: wrong patient's identity and wrong site surgery which should be avoided using the Surgical Safety Checklist recommended by the World Healthcare Organization. The two first checklist's items concerned are: "Is the patient's identity confirmed?" and "Is the patient's operating site confirmed?". The scenarios are divided into three steps: (i) Verifying patient's identity (ii) Verifying patient's surgical site (iii) Move the patient to the operating room. The scene takes place when the patient has been transferred from their room to the operating room. The mission assigned to the students team consists in preparing the patient from his arrival in pre-operating room until the end of the anesthesia procedure. The students are unaware of the hidden educational objectives that are: "Reducing the wrong patient risk applying the checklist", "Reducing the wrong side surgery risk applying the checklist" and "Adapting the security procedure to the context".

Data logged were analyzed and presented in details by Pons Lelardeux et al. [9,10]. Some results are mentioned below. Analyzing the topics of dialogue, the patient's identity was the first topic of discussion or collected information and the operating surgery site was the second one. Different strategies were observed: the first one consists in collecting pieces of information from documents whereas the second one consists in collecting information asking questions to teammates. The debate related to moving the patient to the operating room was more often triggered than the others. The analysis of computer data logged revealed that the leader's behavior during the decision making process is closely linked to the expert's behavior. Relevant arguments placed by experts were able to inflect the decision of the leader whereas the lack of a relevant argumentation was the main root-cause of a disagreement. For all teams, every deficiency (initially hidden and dispatched through the scenario) was found and exchanged between the team members. During the debriefing, they verbally expressed all the deficiencies found.

The data from the survey have been analyzed. 78% of players declare that the virtual environment is similar to a real operating theater. The Fig. 2 shows that the scenario designed using the model described in Sect. 4 gives the feeling of learning. They mainly express that they feel free (even if the communication is restricted to bubble of information exchanges) to manage a real-life like situation. 72% of students declare that they feel perform learning safety process using such an environment. This particular result matches with the main goal of this typical multiplayer digital environment.

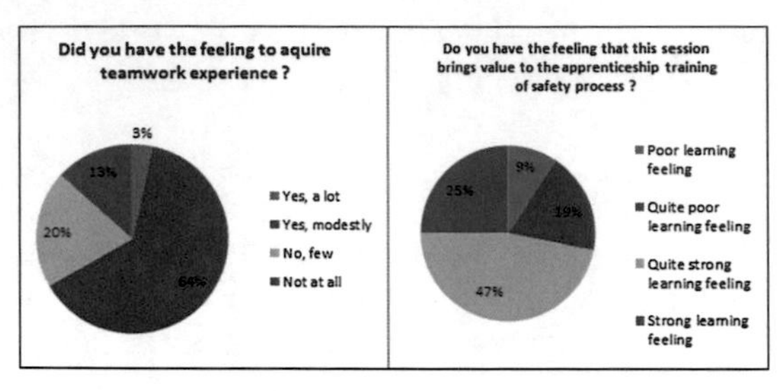

Fig. 2. Questions were asked to students about their feeling on the training session.

6 Conclusion

Training teams on risk management keeps up the interest of many companies in industry such as aviation, nuclear, healthcare as they work in dynamic and unpredictable contexts. One of the critical point is to create educational and entirely controlled training environments that support providers to train staff to improve their teamwork performance making them understand the importance applying or adjusting safety recommendations. In this article, we present a method to design multi-player educational scenario for risk management in socio-technical and dynamic context. This method has been inspired by the systemic method used to analyze real accidents. A dozen of scenarios in the healthcare field have been designed using this method. Training sessions have been organized using the learning game environment with standardized and critical scenarios designed using this method. Experiments were carried out with trainers and their students at the University Hospital of Toulouse. They aimed to analyze student's behaviors facing to a real-life like professional situation. Data analysis shows the different strategies used by the teams facing to a regular or unexpected situation. The experiments allow identifying discrepancies on what must be done if such a case happens. Most of the time, team members found contradictions and hidden anomalies. It was the starting point to trigger debates on what must be done. The results confirm that the method employed to design risk management educational scenario works. Educational scenarios designed with this method support teams to experiment with relative freedom a real-life like professional situation.

Acknowledgments. These works are part of a global national innovative IT program 3D Virtual Operating Room (http://3Dvor.univ-jfc.fr) whose partners are KTM Advance company, Novamotion company, the Serious Game Research Network and University Hospital of Toulouse (France). The steering committee is composed of Pr. P. Lagarrigue, M.D. Ph.D. V. Lubrano, M.D. Ph.D. V. Minville and C. Pons Lelardeux.

References

1. BinSubaih, A., Maddock, S., Romano, D.: Developing a serious game for police training. In: Handbook of Research on Effective Electronic Gaming in Education, pp. 451–477. IGI Global (2009)
2. Debouck, F., Rieger, E., Petit, H., Noël, G., Ravinet, L.: ORION®: a simple and effective method for systemic analysis of clinical events and precursors occurring in hospital practice. Cancer Radiothérapie: Journal De La Société Française De Radiothérapie Oncologique **16**(3), 201–208 (2012)
3. Devreux, G.: Le rôle des comportements informationnels dans la prise de conscience de la situation: Usage dans le Serious Game 3D Virutal Operating Room. Thèse de doctorat en Psychologie, Université Jean Jaurès, Toulouse, France. Ph.D. thesis, Institut National Universitaire Champollion, France (2015)
4. Edward, L., Lourdeaux, D., Lenne, D., Barthes, J.P., Burkhardt, J.M., Camus, F.: Multi-agents approach for modelling safety interventions on a SEVESO site through virtual reality. In: International Meeting "Virtual concept" (2006)
5. Frank, G., Guinn, C., Hubal, R., Pope, P., Stanford, M., Lamm-Weisel, D.: Just-talk: an application of responsive virtual human technology. In: Proceedings of the Interservice/Industry Training, Simulation and Education Conference, pp. 773–779. Citeseer (2002)
6. Gerbaud, S., Mollet, N., Arnaldi, B.: Virtual environments for training: from individual learning to collaboration with humanoids. In: Edutainment, Hong-Kong, Hong Kong SAR China, June 2007
7. Hall, J.L.: Columbia and challenger: organizational failure at NASA. Space Policy **37**, 127–133 (2016)
8. Leape, L.L., Berwick, D.M.: Five years after To Err Is Human: what have we learned? Jama **293**(19), 2384–2390 (2005)
9. Lelardeux, C.P., Panzoli, D., Lagarrigue, P., Jessel, J.P.: Making decisions in a virtual operating room. In: 2016 International Conference on Collaboration Technologies and Systems (CTS), pp. 136–142, October 2016
10. Lelardeux, C.P., Panzoli, D., Galaup, M., Minville, V., Lubrano, V., Lagarrigue, P., Jessel, J.P.: 3d real-time collaborative environment to learn teamwork and non-technical skills in the operating room. In: Interactive Collaborative Learning, pp. 143–157. Springer, Cham, September 2016
11. Pons Lelardeux, C., Panzoli, D., Lubrano, V., Minville, V., Lagarrigue, P., Jessel, J.P.: Communication system and team situation awareness in a multiplayer real-time learning environment: application to a virtual operating room. Vis. Comput. **33**(4), 489–515 (2016)
12. Puentes, A.F.: The manual flight skill of modern airline pilots. Ph.D. thesis, San Jose State University, USA, December 2011
13. Raux, M., Dupont, M., Devys, J.M.: Systemic analysis using ALARM process of two consecutive incidents during anaesthesia. Annales Francaises D'anesthesie Et De Reanimation **26**(9), 805–809 (2007)
14. Reason, J.: A Life in Error. Ashgate Pub Ltd., Farnham (2013). Ashgate edn
15. Stytz, M.R., Garcia, B.W., Godsell-Stytz, G.M., Banks, S.B.: A distributed virtual environment prototype for emergency medical procedures training. Stud. Health Technol. Inform. **39**, 473–485 (1996)
16. White, S.A.: Introduction to BPMN. IBM Cooperation 2(0) (2004)

The Role of Foreign Aid in Education and Development: Field Experiences, Examples from East Africa

Ibolya Tomory(✉)

Trefort Agoston Centre of Engineering Education, Obuda University, Budapest, Hungary
tomory.ibolya@tmpk.uni-obuda.hu

Abstract. Education is an important service which can be financed from local sources of a country. However, many countries have a large uneducated population, people who are unable to make significant progress. International organizations have been calling for different ways to support the education sector for a long time. What determines the success of a development program? This study provides an insight into some programs, pilots and the socio-cultural background of the participants. It is based on participant observation and ethnographic interviewing methodology.

Keywords: Educational development · Cultural-sensitive planning Foreign aid

1 Introduction

1.1 The Role of Foreign Aid

Foreign aid has become institutionalized since World War II and today it is a basic concept in politics of the so-called developed countries. It is also well known in economy and social relationships among nations. In many countries, especially in the so-called Third World countries donors' aid plays a key role at all level of education, even in many cases this is the only source to opening the way of learning.

UNESCO and the World Bank, for instance, have developed a joint program in 1990 involving of other UN organizations entitled "Education for All" (EFA). Primarily this world-wide program focused on formal education but today it focuses on non-formal education which conveys knowledge that is well-utilized. There have been many projects (not only in schools), which have resulted in an increase in the number of school enrollments (now it reaches up to 80%), promotes the acquisition of certain occupations and livelihoods. Primary education is still high priority, although secondary and higher education but mainly various vocational programs are increasingly important (UNICEF 2016).

M.E. Auer et al. (eds.), *Teaching and Learning in a Digital World*,
Advances in Intelligent Systems and Computing 715,
https://doi.org/10.1007/978-3-319-73210-7_106

2 Purpose

Year to year new programs have been emerging and they call attention to education for all, in many cases to support them and year after year statistics supporting encouraging progress are being produced. For the sake of achievement a series of campaigns, programs, calls, radio and TV spots are often made by famous people such as Henry Belafonte, Nelson Mandela, Angelina Joly who raise awareness to fundamental right to education and its access for more and more children and adults.

Despite all good intentions positive statistics do not always reflect reality. Tens of millions of children could still not be in school by 2015: "Based on the trends of the past five years, 57 million children would still be out of school in 2015… Estimates for 2012 indicate that about 25 million, or 43% of out-of-school children, will never go to school; the rate is 50% in sub-Saharan Africa and 57% in South and West Asia."

"Only two in three adolescents finish lower secondary school in countries with low or middle income" (UNESCO EFA 2015:7., 12.).

There are always measurable and immeasurable changes in development and one of the questions is perhaps the interpretation of underdevelopment/backwardness and development itself. These can be interpreted as the growth of the economy, the development of better abilities/skills, and the side of psychological-motivational effects. Another psychological aspect is the developmental vision itself as virtual reality rather than vision, and how promising or unacceptable it is for a given society, community, and ethnic group. Motivation is one of the cornerstones of the acquisition of the required qualification (Tomory 2015). Questions have been raised about effectiveness, needs and about its terms, conditions, and purposes.

On the other hand, this is closely related to the cultural background, so culturally sensitive planning plays a decisive role. This is especially important in some cases, such as The girls' education, as their participation in education is far behind the boys for many different reasons, even cultural reasons (UNESCO EFA 2015; Ondiek 2010). "Countries in which the net enrolment ratio for girls is generally considered to be low (below 70%)……there are 44 such countries and some have less than 40%" (UNICEF 2003).

3 Methods

In addition to studying existing literature, I have visited East-Africa (Kenya, Uganda, Tanzania, Ethiopia) several times to gather on-site data on Black-African education.

My basic working method is the specific method of cultural anthropology, participant observation. Anthropology is trying to get to know other cultures and phenomena in their entirety and this method allows to reach out to a phenomenon, not only in general but through individuals, based on first-hand experience (Lajos 2005).

The focus is on exploration and non-statistical data testing. Thus, for example, statistical data collected from interviews does not produce results (not emphasized), but rather the convergence of their content, the strengthening of information in different environments and perspectives. Cultural anthropologists rather look for different ways to study how people interact. Of course, it is possible to present the collected data, but here (I think so) the presentation of specific examples tells more than any data.

Data in this approach often mean local perceptions, statements, stories, life histories gathered by spontaneous conversations, interviews, or perceiving, tracking, and linking a phenomenon with another mosaic. My main research tool is a fieldwork diary (field notes), and audiovisual tools (depending on the situation).

Participating observation is a complex method that allows to use many methods and techniques: the anthropologist can perform continuous observations, see spontaneous situations, might conduct interviews, might also refrain from asking questions in certain circumstances to observe individuals interacting more naturally, without interruption etc. Anthropologists select a topic such as education and they do one or more research in a certain place and/or do a multi-sited research with the aim of documentation of variations in cultural forms, perspectives, phenomenons for the topic (Kottak 2002).

I have met a wide range of participants in education and development programs (schoolchildren, illiterate adults, university teachers, development staff). Meanwhile, it has become clear to me what it means in reality. I have read statistics and reports about what learning means for the people concerned and what they think about the programs.

In this way I could manage to put written sources into practical illumination, to understand the complexity of the socio-cultural system, its perspectives, the role of education and to recognize contexts that cannot be perceived in any other ways. This study is a part of this research process.

4 Effects of Foreign Aid

4.1 Types of Organizations

There are several different categories of organizations. M*ultilateral organizations* can grant non-repayable funds (UN organizations) or (the World Bank). No single nation controls them. *Bilateral organizations* organize the development projects country by country (The U. S. Agency for International Development) but each donor country has an agency. There are many *private sector organizations* in two categories: Non-governmental organizations (NGOs) and profit-seeking organizations (commercial enterprises) (Heyneman 2005).

All of these organizations consider the mission of the Millennium Development Goals (MDG) to be their mission, with the support of education in every possible way. In 2000, 189 UN Member States committed themselves to realize the eight MDGs, providing universal primary education, all by the target date of The United States is the leading nation in charitable donations to developing countries ($4 billion in 2000) (Bernard 2005).

4.2 Positive and Negative Contribution of Foreign Aid

The literature and my field experiences also confirm the fact that from a certain perspective funds are invariable put to use to create new facilities or upgrade existing ones adjusted to local needs.

Usually in-built are training activity components giving opportunities to local staff members to undertake different activities at institutions. The other beneficial component of the aid is technical assistance which means providing technical tools and training local staff and/or providing a foreign expert.

A practical example is Moi University in Kenya where the 'Margaret Thatcher Library' was built from funds of the British Government or in Tanzania where assistance by the UN was given to curriculum development at all level of education helped by various foreign professionals working for the Institute of Education, and other agencies.

Foreign aid has another value of enhancing equity and efficiency as well as optimizing output and increasing opportunities to a large number of people. Labor and capital, tools and information, knowledge and education are the main resources (inputs) in a development project (Mihály 2005).

Given the current state of Tanzanian economy, most of the schools - on their own - do not have resources to integrate ICT into education which is one of the most important steps of progress. One of the cornerstones of this solution is UNESCO's ICT program for primary and secondary education, a top priority for girls' education (UNESCO Tanzania 2014).

Various organizations have many similar initiatives for complex programs and collaborations across East Africa (Oxfam, SIDA) but more and more small NGOs have successful projects that I have seen (GOIG, Sidai Design, Maasi Association etc.).

Swedish Agency for Development Evaluation shows that "countries in Sub-Saharan Africa have experienced positive growth after seeing negative growth during the last two decades of the 20th century" (Viederpass and Andersson 2007, 16.).

During my fieldworks I met successful and less successful co-operation from the point of view of the local stakeholders. It was clear that foreign aid is not always a bedrose for local target groups, and it is not without negative effects and constraints.

Basically, foreign aid raises several problems:

- One of the hard-to-manage factors is the string of conditions attached to the aid package. My informators (work mainly on the plan and organization side) talked about the fact that the negotiations take too long and some donors have made it clear that the "price" of the intervention is a contract for some companies from their country who provide services (building materials, textbooks, training) for what the side has to be paid.
- Closely related is the often condition that goods and services can only be procured in a certain way even though it is unfavorable, as the supplier may be expensive but there is no other option.
- A less visible, but more important issue is the difficulties of differentiating between the nominal value and real value of the assistance, especially during a period of rapid inflation.

- The significance of aid in the perception of the donor and the recipient can be different. The question then arises how much target groups have been involved in the planning of programs, whether local needs have been raised beforehand?
- A controversial issue is involvement in the programs, there are several types of campaigns, and donors tend to have the fastest results that would attract more participants and shows results quickly. But sometimes quantity goes against quality and usability. It is a nice memory for me when in a small village in the Pare Hills, Tanzania, I was asked not to tell anyone that a few students do not wear all the required uniforms but there is not enough money for purchasing one for each child in each family. Figures and reality are therefore different: the ratio of school-age enrollers in the reports can change in reality.
- Moreover imported materials do not fit to local needs and do not take existing local knowledge into account, and donors do not even know it. Education often takes place within the institutional framework of a stronger, power culture coming from outside, its values and contents, and students often try to identify with things (content, thought, values, standards, communication) that are difficult for them to understand.
- Development organizations do not understand the local socio-cultural conditions they often go against the habits, customs that create tension.

These issues appear during the implementation of the programs in daily situations and also affect participants and are invisible behind the data as shown in the following examples that I often encountered as a recurring phenomenon.

4.3 Two Programs Example

I met many people in East Africa working along the roads with a sewing machine, they offered their work capacity and seemed to be looking for at least a couple of pounds per day. I got to know that with the aim of eradicating unemployment, a sewing machine program was launched in Uganda and then in other East African countries for caring of family and business start-up, which was supported by several international and smaller organizations in the spirit of initial and vocational education. "On-off Mass" large-scale general education campaign was also launched. This is a highly "inviting", "militaristic" form, giving the call as many places as possible, and more people wanting to win the benefits of participating. Tutors help participants but they are usually not local people.

The organizations argued for the program with the following: participants would be able to earn money for the family but to be able to do that they would need a sewing machine and to be taught how to make clothes. Along with that writing and reading was also taught to adult participants, men and women alike. Moreover they argued that - as educated people - they would be able to start their own businesses. Many people applied and learned the professional basics of sewing (thousands of people) but few of them were able to meet basic education requirements. At that time they promised to purchase their own sewing machines, and again they got some help by a wide range of campaigns, recruiting, which again brought the expected number of employees.

However, as a result of the massive campaign, a lot of people started sewing, but there were few orders as there is a huge competition in supply. People have been sitting along the roads, waiting for lucky days, so the chance of a better life is still unseen. As far as I know they have not been able to utilize other knowledge and general literacy because wherever they try their luck, they can hardly apply them. Many people forget, barely or cannot use the rapidly acquired knowledge. For them the only question left is where the boom is, which they have been urged to enter.

In Ethiopia SNNPR (Southern Nations, Nationalities and Peoples Region) I met a successful development program targeting girls' education. Professionals and responsible people work in different fields, combining everything with education: health, HIV/AIDS prevention, education. An important goal is to involve young children in the pre-school/school education process, to educate girls, to reduce the consumption of female circumcision and "kat" (narcotics). In all the five schools around the country, they work with curricula and textbooks developed for their own local needs. Teachers will receive special training from time to time to get acquainted with and understand the cultural background factors of the area, both to meet the needs of young people who are out of school and return, and adults involved in education.

By mobilizing the community, persuading local leaders and senior citizens - to make them understand how important it would be to create better living conditions for children and adults, - by involving local people and people who are well-versed in the culture, more credit for the program could be given and it would be easier for the ones in need to accept help. Following a broadband, long-term, successful pilot, a successful development is under way, which is the result of today: slowly leaving the bottom, young children at school, learning girls, beginners and married men return to school, parents also come and are willing to learn, many are interested in professions, - local people also wish to introduce agricultural methods in several villages, - so all in all the proportion of graduates increases.

This program was one of the few, which, for me, symbolizes the importance and effectiveness of cultural-sensitive planning. It was also a lesson for me that perhaps the most important link between preparation and implementation is the cultural-sensitivity, which means the cultural knowledge but even a great empathy and acceptance.

5 Conclusion

The above-mentioned suggests that the desirable ways in which development is carried out in terms of individual improvement and labor market opportunities are well thought-out and culture-responsive solutions have an increasing importance in development. The approach and intercultural competence of the donor is the determinant of the cultural-sensitive planning. This means the thorough preparation of the donors, the proper preparedness of the institutions and individuals for the process of change, a concrete program, and cooperation with local civil society organizations. It is indispensable to have accurate knowledge of local characteristics, to explore customs and to coordinate with the donors' ideas. The donating and their representatives should be well prepared and, alongside the different knowledge, social and cultural know-how is important. By involving local people, you can assess the content and methodological ideas and plan a thematic program/series in small units. Preparing the locals in the form of training can help to understand why and for whom it is done. The lesson of all this should be taken into account in the development of every disadvantaged, here in the integration program of the Roma minority, where similar problems can be identified.

For me, these experiences provide a personal surplus in my educational work, enabling me to develop the critical sense of students and the need for realistic orientation beyond interestingness. That is why I think multilateral formulation and up-to-date information on a given subject are important.

Several authors report similar results. They point out the luck of reform pilot programs in African and other countries, confirm that local roots need to be cultivated and their findings emphasize the importance of local ownership and cultural conditions as crucial ingredients of success (Samoff et al. 2011). So the results of this study can be generalized on other international programs. Nevertheless, I consider the texts here to be the most valid for East Africa, as I have gained field experience here.

References

Bernard, P.: EC contribution towards the MDGs. In: A nemzetközi fejlesztési együttműködés a XXI. században, Magyar ENSZ Társaság, Budapest, pp. 77–83 (2005)

Lajos, B.: A tükör két oldala, Nyitott Könyvműhely, Budapest, pp. 13–31 (2005)

Clarke, F., Ekeland, I.: Nonlinear oscillations and boundary-value problems for Hamiltonian systems. Arch. Rat. Mech. Anal. **78**, 315–333 (1982)

Heyneman, S.P.: Foreign aid to education: recent U.S. initiatives—background, risks, and prospects. Peabody J. Educ. **80**(1), 107–119 (2005). Nashville

Kottak, C.P.: Cultural Anthropology. McGraw-Hill Higher Education, Boston (2002)

Mihály, S.: The UN and the Global Development Process, HUNIDA, Budapest, pp. 55–69 (2005)

Ondiek, C.A.: The persistence of female genital mutilation (FGM) and its impact on women's access to education and empowerment: a study of Kuria district, Nyanza province, Kenya. University of South Africa, Pretoria (2010)

Samoff, J., Dembélé, M., Sebatane, E.M.: 'Going to scale': nurturing the local roots of education innovation in Africa, EdQual Working Paper No. 28, EdQual, Bristol (2011)

Tomory, I.: "Education for All" – Az oktatási kampányok típusai és hatékonyságuk a nemzetközi fejlesztésben. In: Tudás-Tanulás-Szabadság, Babes-Bolyai Tudományegyetem, Kolozsvár, pp. 273–281 (2015)

UNESCO EFA: Global Monitoring Report, Education for All 2000–2015, Achievements and Challenges, UNESCO Publishing, Paris (2015)

UNESCO Tanzania: Tanzania and UNESCO, UN, Dar es Salaam, pp. 19–53 (2014)

UNICEF: Accelerating progress in girls education: towards robust and sustainable outcomes. In: The State of the World's Children. UNICEF House, New York (2003)

UNICEF: The State of the World's Children, A fair Chance. UNICEF, New York (2016)

Viederpass, A., Andersson, P.-A.: Foreign aid, economic growth and efficiency development. SADEV report, Karlstad (2007)

Crowdsourcing Project as Part of Non-formal Education

György Molnár(✉) and Zoltán Szűts

Department of Technical Education,
Budapest University of Technology and Economics, Budapest, Hungary
molnargy@eik.bme.hu, szutszoltan@gmail.com

Abstract. Crowdsourcing is the latest revolution brought by the digital technologies of computing and communication. It is a nowadays popular way of finding services, concepts, or content by asking contributions from a large group of people, particularly from users. According to Jeff Howe, crowdsourcing generally refers to the participatory online activity of calls for individuals to voluntarily undertake a task. The key elements of a crowdsourcing project are the open call format intended for an enormous network of potential contributors. It is a revolution that brings people together and harnesses their collective intelligence. Crowdsourcing in an online, distributed problem-solving model that pulls the collective intelligence of online communities to assist explicit goals. Online communities, or crowds, are given the opportunity to answer to crowdsourcing activities requested. In crowdsourcing, there is no clear frontier between the subjects of a research and the researchers themselves. It differs from traditional outsourcing as it involves a random, volunteer crowd and not previously selected group of individuals. A crowdsourcing - along with big data and citizen science – is a key part of an important scientific, methodological and educational phenomenon. With advent of crowdsourcing, a paradigm shift can be witnessed in information procurement, transfer, storage and processing as well as in learning. In the practice, crowdsourcing forms a firm bond with the phenomenon of wisdom of the crowds and user-generated content.

Keywords: Crowdsourcing · Non-formal education · Waze · Wikipedia

1 History and Clarification of the Terminology

Creation of the Oxford English Dictionary in the 1800s is the initial example of crowdsourcing that can be used in non-formal education. Over an open call, public were asked to gather English words and their usage and show them to editors to be indexed in the dictionary [6]. Throughout history we can find open calls for solutions to solve tough tasks. Though crowdsourcing has a long history, it has really surfaced as a paradigm-shift phenomenon based on the tools of the Internet. When addressing crowdsourcing, the term fundamentally refers to crowdsourcing involving online technology [8].

M. E. Auer et al. (eds.), *Teaching and Learning in a Digital World*,
Advances in Intelligent Systems and Computing 715,
https://doi.org/10.1007/978-3-319-73210-7_107

1.1 Information Society

Authors would also like to look at the context, which made possible for crowdsourcing to become such an important phenomenon, the information society. As we all know, information society is a culture where the formation, dissemination, use, incorporation and management of information is a substantial activity.

According to György Csepeli and Gergő Prazsák, information society comes into being where broadband internet, capable of symmetrical connections is present both in physical and social space [2].

1.2 Web 2.0 and Crowdsourcing

Another phenomenon related to crowdsourcing is Web 2.0. Web 2.0 services permit users to cooperate and collaborate with each other as creators of user-generated content online, in contrast to Web sites where people are limited to the passive viewing of content. Web 2.0 services are the most known social networking sites (Facebook), blogs (Twitter, Wordpress), wikis (Wikipedia), folksonomies, video sharing sites (YouTube), hosted services and tags [7]. One of the most notable effects that mark Web 2.0 are not the services but the ways that ICTs have restructured the interaction. Government, businesses, NGOs and schools frequently the power of online communities for routine procedures [3].

1.3 Collective Intelligence and Wisdom of Crowds

Pierre Lévy defined collective intelligence as a "form of universally distributed intelligence, constantly enhanced, coordinated in real time, and resulting in the effective mobilization of skills" [5]. James Surowiecki defines the phenomenon "wisdom of crowds," as where, under the right conditions, groups of people can outperform even the best individuals or experts. Surowiecki claims that the wisdom of crowds is based on the freedom of entities in a group, the multiplicity of the group, and the combination of their individual productions rather than the averaging of their cooperative effort. With diverse individual and/or group skills, people will give varying connotations to a single detail or inquiry [9].

1.4 Crowdsourcing

Crowdsourcing is also a way of cooperation, accumulation, collaboration, agreement, and originality. A new way of doing work, but it also is a phenomenon where clusters of people can outdo individual professionals, outsiders can bring different visions to internal problems, and geographically isolated user can work together to create policies and proposals that are agreeable to most.

Crowdsourcing means access to talent: Many people crowdsource to get access to talent that they can't get in any other way. Crowdsourcing can help find individuals who have special skills. It can also bring you the talent that comes from the collected intelligence of the crowd, the ability to do things that are difficult for machines to do.

1.5 Five Categories of Crowdsourcing

Government Crowdsourcing
The subcontracted actions are those handled by government institutions that provide constant support to certain categories or whole population. This assistance might be health or wellbeing associated, emotive, material, etc.

Business Crowdsourcing
Business crowdsourcing platforms are a meeting place for institutions offering work and those looking for contracts. The businesses promoting work are characterized by offering ventures that can be fragmented down into simple tasks; these ventures, then, are complicated but not complex.

Ngo and Non-profit Crowdsourcing
Crowdsourcing enables NGOs to hire people with specialized skills on a short-term basis or to use inexpensive services that provide the skills they need.

Cultural Crowdsourcing
Cultural crowdsourcing motivate the public to take on tasks that cannot be done mechanically, in an environment where crowd's input contributes to a common, substantial goal or research interest. Crowdsourcing can be vastly effective for engaging audiences with the work and assortments of galleries, libraries, archives and museums.

Scientific Crowdsourcing and Citizen Science
Citizen science are projects conducted, in whole or in part, by amateur or nonprofessional scientists. It is also a non-formal educational method.
But citizen science is not about amateurism. Based on a survey of 320 participants in the InnoCentive "solver" community, Karim R. Lakhani and colleagues found that solvers were "highly qualified," with 65% of solvers holding doctorates and nearly 20% holding some other advanced degree, mostly degrees in the sciences [4].

2 Example of Non-formal Educational Crowdsourcing Project

Meteorologists have found great value in the log books of the British Royal Navy, which contain one of the most complete records of global weather data. The logs of the British Navy are handwritten, and no computer can scan them successfully. Examples include: 'Pistols explode,' 'Seamen abscond with dinghy,' and 'Prayer services held.' Even the best programs may not realize that these events had anything to do with the weather (Fig. 1).

For each task, the crowd is shown a single page of a ship's log. When they see something of interest on the page, they're supposed to move the cursor to that point and click. The Zooniverse interface then opens a little window, which allows the worker to transcribe the information. The boxes for weather data have specific fields for wind, temperature and barometric data.

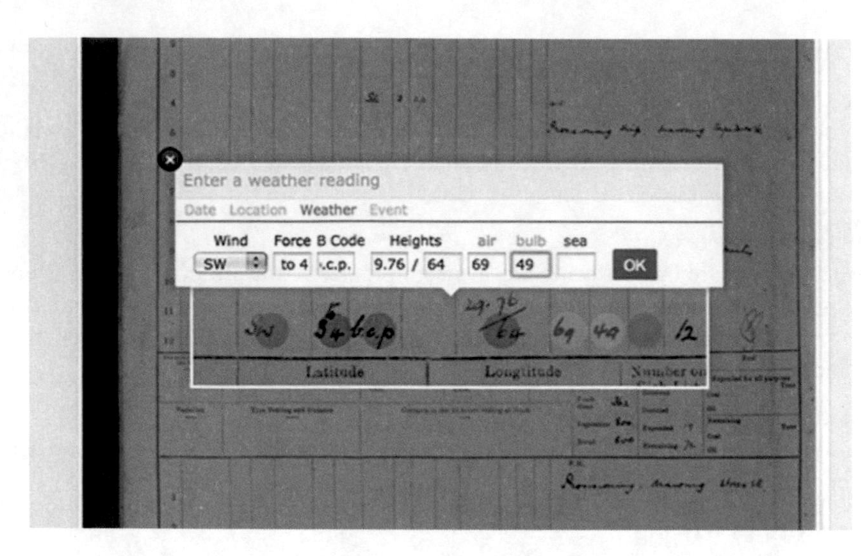

Fig. 1. Old weather

2.1 Knowing the Limits

We have to notice that not everything can be crowdsourced. You should not crowdsource in education if:

- Sometimes, crowdsourcing does not bring in the talent you need. You cannot always find the talent that you need in the crowdmarket.
- Crowdsourcing hurts quality. Crowdsourcing can easily lower the quality of your "product". This kind of problem often occurs when you do sophisticated things that need to be monitored constantly.
- Crowdsourcing makes things too complicated. If you need to give more to crowdsourcing than the benefits you get back from it, you should not crowdsource.
- Nevertheless its very rare, there is always some vandalism in crowdsourcing, for example in Wikipedia.

2.2 Non-formal Teaching and Learning – the Use of Waze and Wikipedia

There are several crowdsourcing projects that can be used in education. In the survey two of them will be enquired. Waze a GPS navigation software that works on smartphones and tablets with GPS support provides real time information on traffic. Some of this information (for example traffic jams and accidents) is provided by users, and the same time it is immediately used for process of learning. During this process drivers learn about actual traffic situation and through this knowledge, decide on their further actions. Seeing the popularity of the application, it can be suggested, that education systems ought to be more like crowd-sourced apps, especially Waze [1]. They must draw on the knowledge of those using it to adapt and become more relevant. This education should offer customisable experiences and real time application of knowledge. And the educational systems should create a platform for collaboration [6].

Wikipedia a great example of crowdsourcing, where it can be immediately used for learning. The most commonly critic is that the Wikipedia's inaccurate material is incomplete not valid. It is based on the fact that its content is not created in a traditional publishing environment, with professionals and editors. Wikipedia is being written and edited by individuals with different motives and can be authored by any internet user.

3 Empirical Survey

In this chapters the authors will introduce a survey conducted at Budapest University of Technology and Economics. As a measuring instrument we used the quantitative questionnaire survey method, using simple random sampling from a student base population. The survey was conducted in May 2017, where N = 59 evaluable answers were received. The majority of the chosen target group were correspondent vocational teachers, which means engineers and economics teachers. The results of the questionnaire test were performed using textual and diagrammatic processing using a simple descriptive statistic method. The questionnaire contained a total of 16 questions, of which the authors in answers included only more relevant ones. Most of our respondents (53%) were male and minority (47%) female students. Most of the target group were an engineer and economy teachers, this explains the gender distribution (Fig. 2).

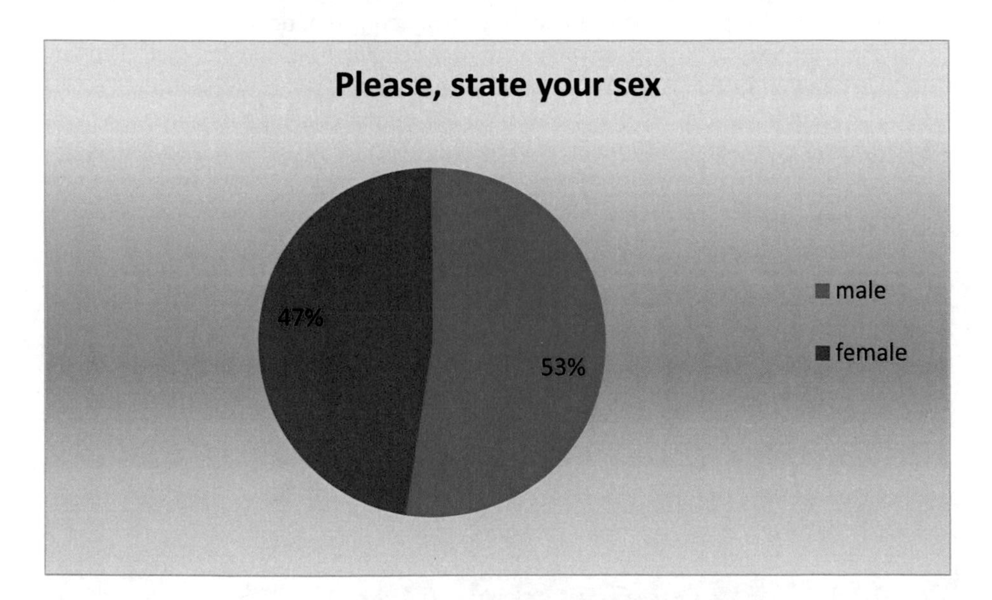

Fig. 2. Repartation of gender, own diagram

Most of the respondents, 38% were people aged 41–50, this is the X-generation, the digital immigrants. Nearly one third of the respondents, 28% were aged 31–40, they are members of the Y generation, who are more of a digital natives. 20% of respondents were aged 51–55, they belong to the baby boom generation, while 7% are 56–60 years old.

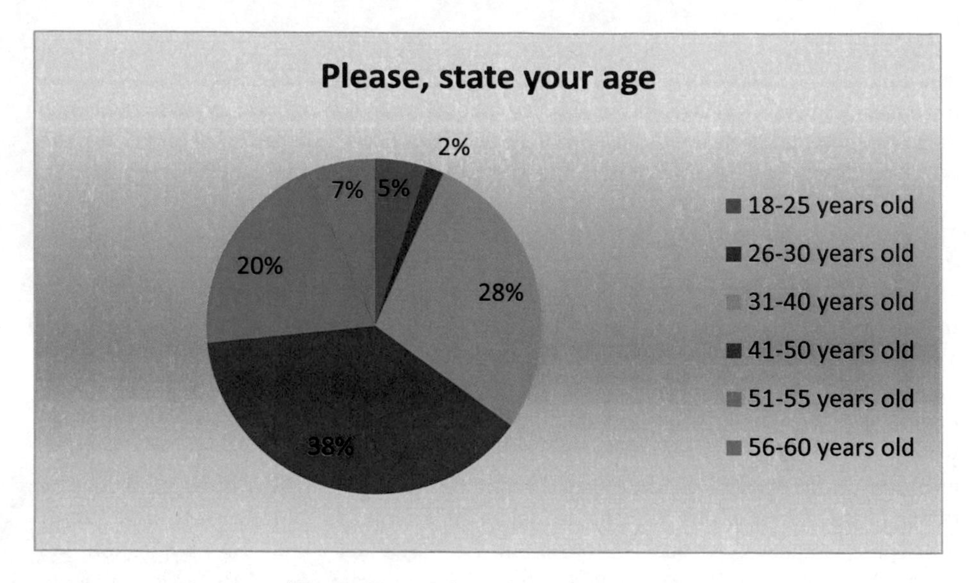

Fig. 3. Repartation of age, own diagram

Overall, a broad spectrum is shown by the age group, and the majority of the elderly respondents can be explained by fact that they attend correspondent training (Fig. 3).

All the interviewed used to download digital content from the Internet. The frequency distribution of this is shown in the following figure, where we see that the vast majority of respondents (57%) use to download of digital content every week. This is 21% download daily, and 14% monthly (Fig. 4).

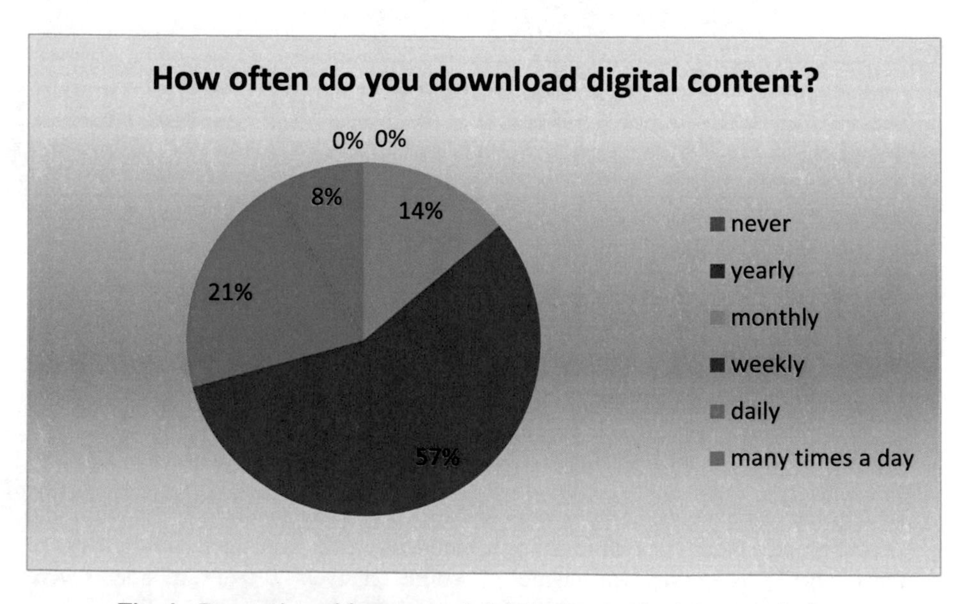

Fig. 4. Repartation of frequency of digital content materials, own diagram

Almost half of the respondents, 47% do not know the Waze mobile navigation application, which is a striking practical example for crowdsourcing. Only 35% use it, and 18% did not answer (Fig. 5).

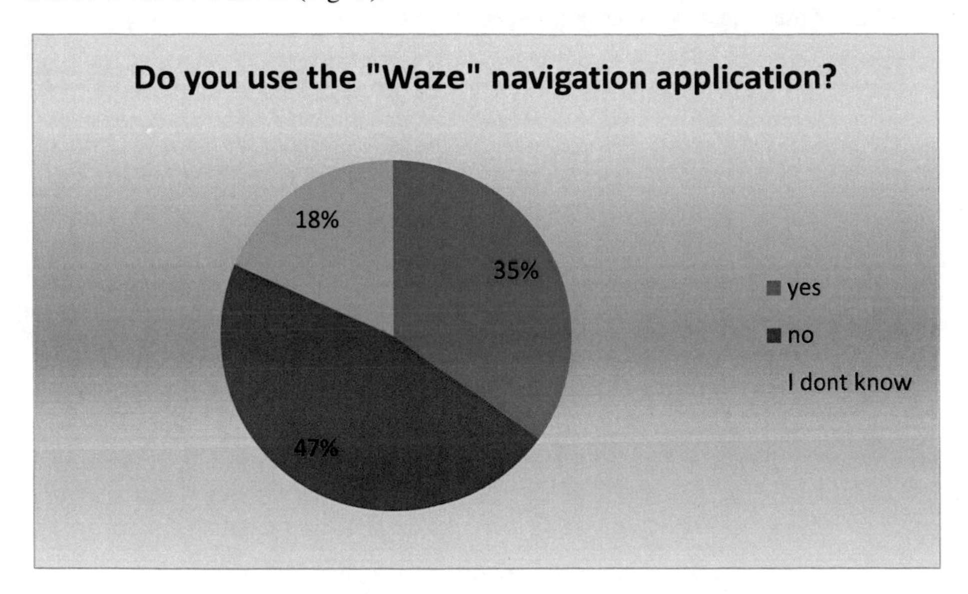

Fig. 5. Repartation of use the "Waze" navigation application, own diagram

The following diagram shows that 98% of respondents use the most popular crowdsourcing application, Wikipedia. This is a welcome fact, especially if we look at the age of affliction. There is no one who does not know it (Fig. 6).

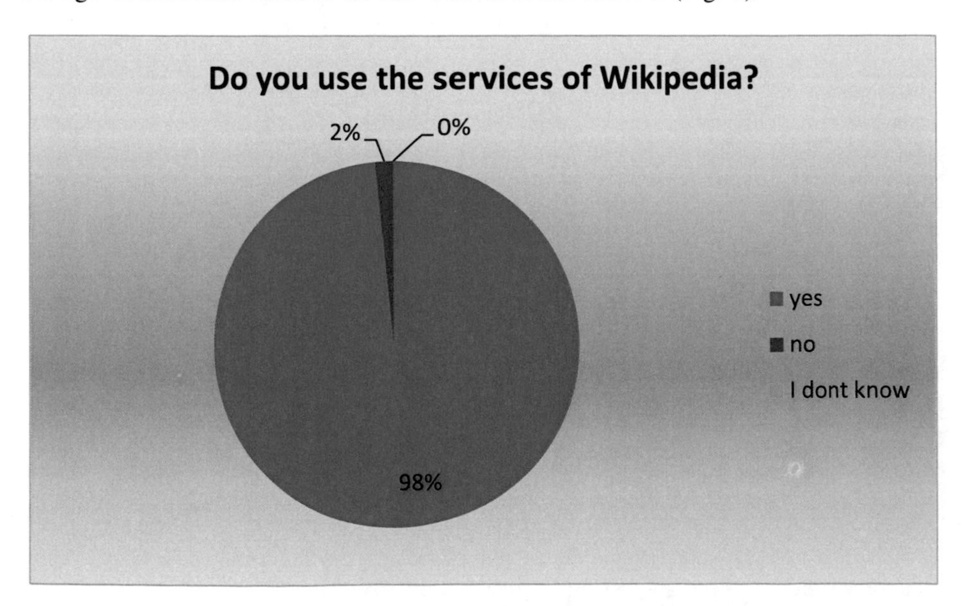

Fig. 6. Repartation of use the services of Wikipedia, own diagram

In the following diagram it is presented that, Wikipedia's services are used by 46% of the respondents per week, 25% each month, and 17% only yearly. This is essentially a sign of an advanced digital culture, but 100% of respondents use Wikipedia only as content consumers, just for collecting information (Fig. 7).

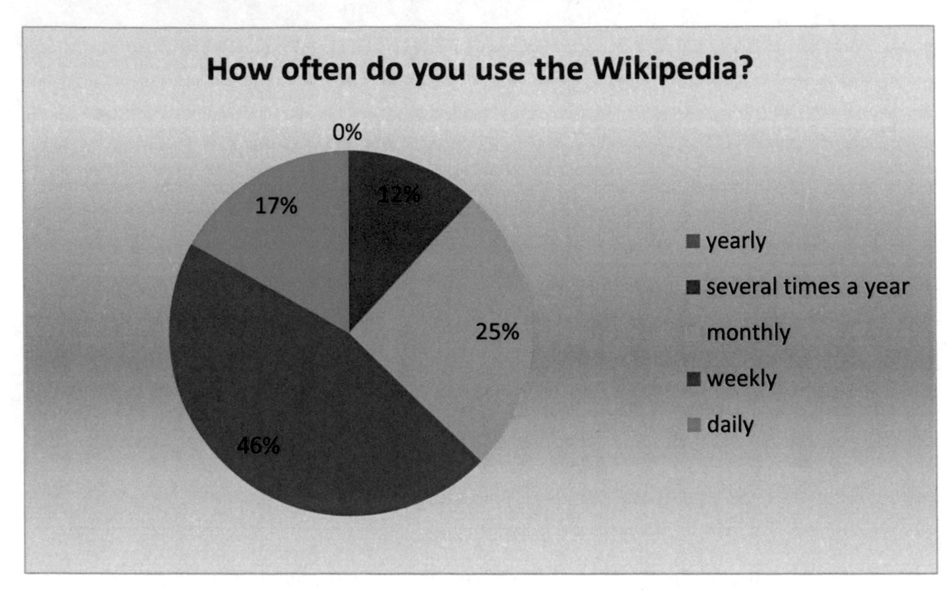

Fig. 7. Repartation of frequency of use of Wikipedia, own diagram

The following figure shows the use of Wikipedia. It shows that the majority, that is, 55 respondents use Wikipedia only as source of information. 19 people use it in education (Fig. 8).

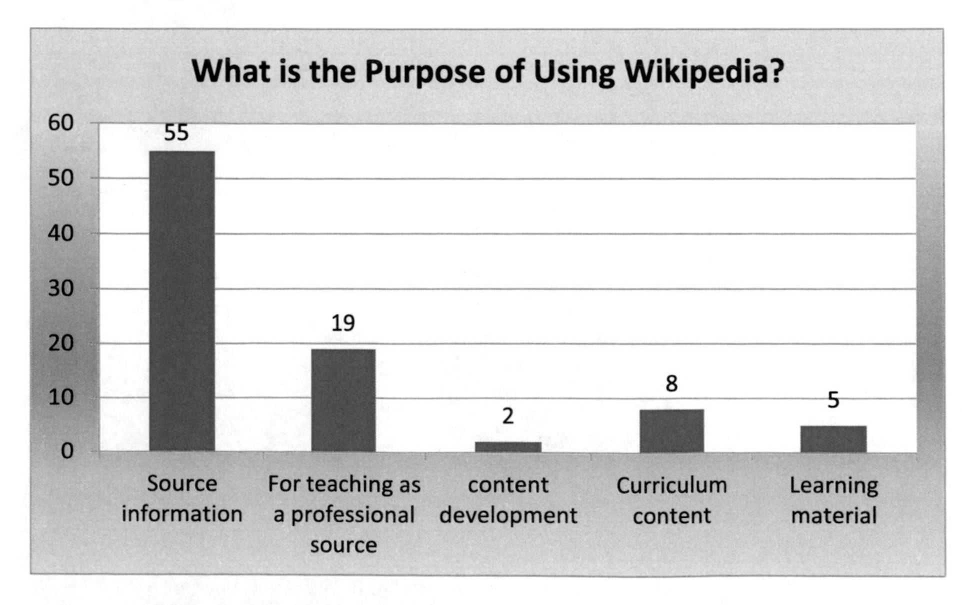

Fig. 8. Repartation of Purpose of Using Wikipedia, own diagram

More than half of the respondents, 52%, would like to share the content created in Wikipedia (Fig. 9).

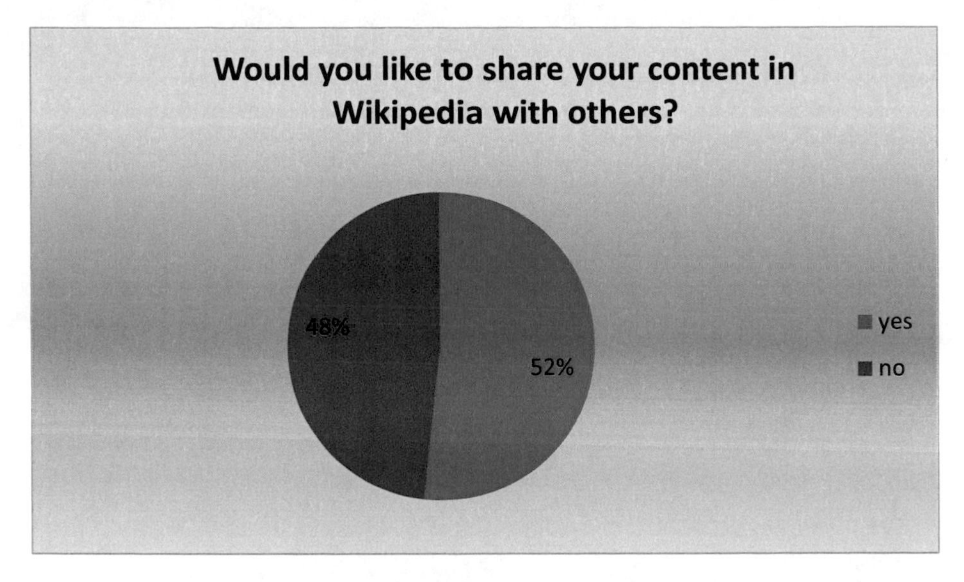

Fig. 9. Repartation of content to share in Wikipedia with others, own diagram

75% of all respondents fully, while 18% more or less believe it to be important for content produced in the Wikipedia to be accessible to the open community (Fig. 10).

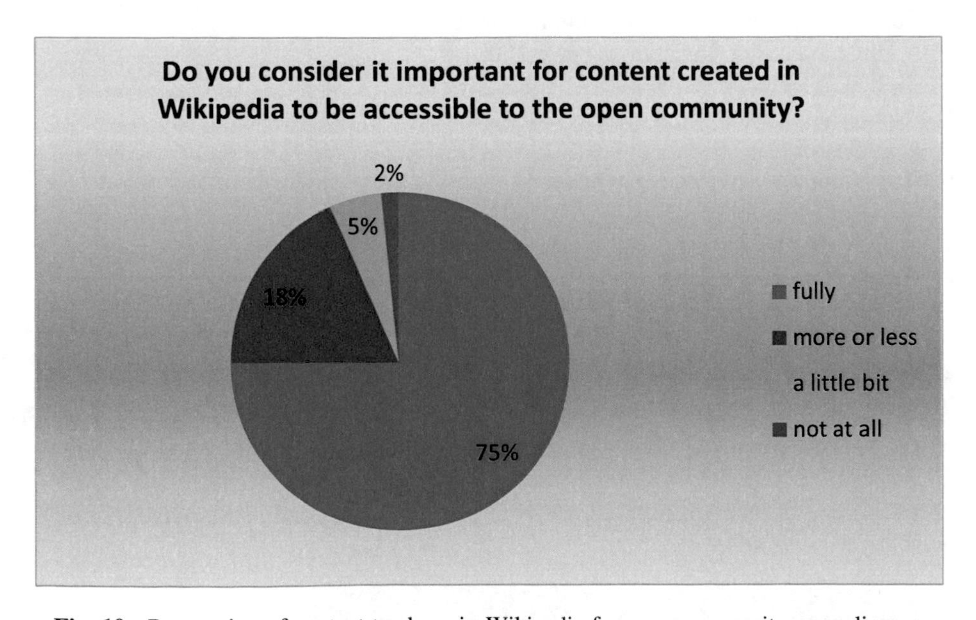

Fig. 10. Repartation of content to share in Wikipedia for open community, own diagram

Almost half of the respondents consider the Wikipedia services to be of little importance, while 38% of them consider it very important (Fig. 11).

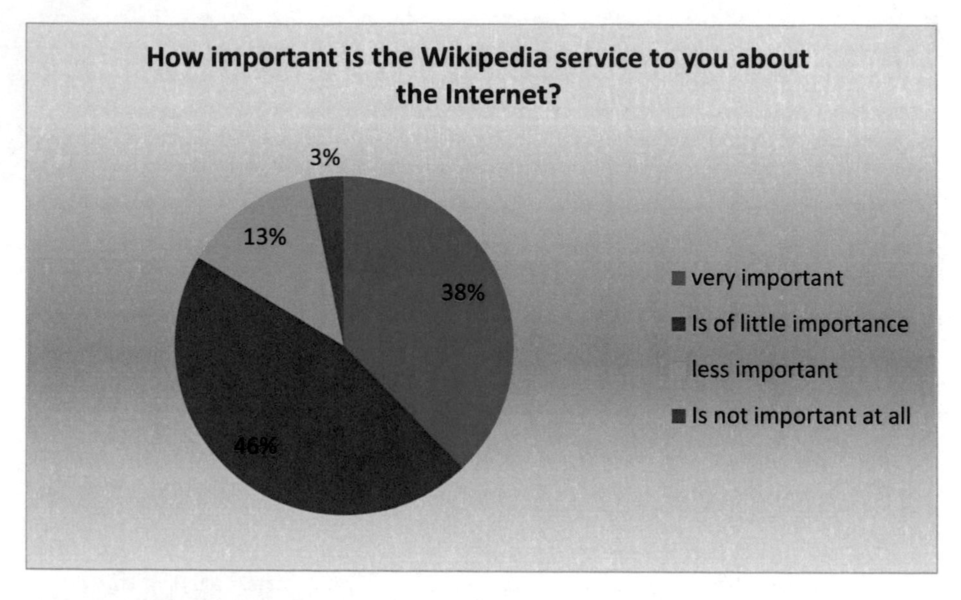

Fig. 11. Repartation of role of Wikipedia in Internet, own diagram

A Proposal for a Genealogy Research Project

- Genealogy research in crowdsourcing model
- A Hungarian genealogical tree from 1900
- Unified, open source, editable Web 2.0 database
- Dedicated editors
- A graph editor
- Data mining from offline databases
- Using oral history methods
- Users with free capacity map uncharted parts
- Word-of-mouth marketing.

4 Conclusion and Future of Crowdsourcing

What can be the future of crowdsourcing: There will be interdisciplinary collaboration between the scholars. Crowdsourcing will be part of non-formal educational system. Crowdsourcing seems a natural approach to processing big data. And maybe, professional crowd will arise.

Based on the method and the empirical study presented in this article, the difficulty of the method is that a large amount of unstructured data from the entire Hungarian online public has to be analyzed in a given time interval.

It is a help that the audio and video content is beyond the survey's horizon, so it is only necessary to concentrate on textual, Hungarian-language content. The upside of the method is that no special, costly IT infrastructure is needed. Similarly, the data comes directly from the students and it is not hindered by filters like hypothesis, questionnaire, questioner.

References

1. Benedek, A.: New educational paradigm: 2.0: items of digital learning. In: Benedek, A. (ed.) Digital Pedagogy 2.0., pp. 15–48. Typotex, Budapest (2013)
2. Csepeli, G., Prazsák, G.: Örök visszatérés? Társadalom az információs korban. Jószöveg Műhely Publisher, Budapest (2010). ISBN: 9789637052934
3. Molnár, G.: The impact of modern ICT-based teaching and learning methods in social media and networked environment. In: Turčáni, M., Balogh, Z., Munk, M., Benko, Ľ. (eds.) 11th International Scientific Conference on Distance Learning in Applied Informatics, pp. 341–351 (2016)
4. Lakhani, K.R., Panetta, J.A.: The principles of distributed innovation. Innovations **2**(3), 97–112 (2007)
5. Lévy, P.: Collective intelligence: Mankinds Emerging World in Cyberspace. Published by Plenum Trade, New York (1997)
6. Eskenazi, M., Levow, G.-A., Meng, H., Parent, G., Suendermann, D.: Crowdsourcing for Speech Processing: Applications to Data Collection, Transcription and Assessment. Wiley, Chichester (2013)
7. Siemens, G.: Connectivism: a learning theory for the digital age. Int. J. Instr. Technol. Distance Learn. **2**(1), 3–10 (2005)
8. Surowiecki, J., Silverman, M.P.: The wisdom of crowds. Am. J. Phys. **75**(2), 190–192 (2007)
9. Szűts, Z., Jinil, Y.: Recent problems of digital culture. Value of incomprehensible. Információs Társadalom **2014**(1), 109–116 (2014)

Serious Games Development as a Tool to Prevent Repetitive Strain Injuries in Hands: First Steps

Hélder Freitas[1], Filomena Soares[1,2(✉)], Vítor Carvalho[2,3], and Demetrio Matos[4,5]

[1] Department of Industrial Electronics, University of Minho, Guimarães, Portugal
a68580@alunos.uminho.pt, fsoares@dei.uminho.pt
[2] R&D Centre Algoritmi, University of Minho, Guimarães, Portugal
vcarvalho@ipca.pt
[3] School of Technology, Polytechnic Institute of Cavado and Ave, Barcelos, Portugal
[4] School of Design, Polytechnic Institute of Cavado and Ave, Barcelos, Portugal
dmatos@ipca.pt
[5] MEtRICs Research Centre, University of Minho, Guimarães, Portugal

Abstract. This paper is focused on the problem of repetitive strain injuries in hands. These injuries are mostly related to professional activities where people are subjected to a high rate of work and the performed tasks often lead to repetitive actions. The objective of this paper is to develop a serious game to prevent strain injuries in hands. The game scenarios promote warm-up and stretching off hand exercises that should be executed before and after the working period. The game is developed in Unity software, associated with a 3D sensor, Intel RealSense 3D Camera F200 for hand and movements detection. With the activity implementation in companies and establishments the employees will be able to exercise their hands, thus reducing the risk of being affected by strain injuries.

Keywords: Serious games · Unity 3D · Image processing · 3D sensor
Labor gymnastics · Hand injuries prevention · Repetitive strain
Muscle articular exercises

1 Introduction

This paper is focused on the problem of the strain injuries in hands that occur due to repetitive execution of a certain hand movement, causing an excessive use in certain muscles or joints [1]. These injuries are mostly related to professional activities, especially in people that work at high levels of industrialization, patchwork or in use of advanced technology in productive process [2], as they are subject to a high rate of work where the performed tasks often lead to repetitive actions.

The main injuries in hands caused by repetitive efforts are the carpal tunnel syndrome, tendonitis and tenosynovitis (DeQuervain disease). These injuries are

M. E. Auer et al. (eds.), *Teaching and Learning in a Digital World*,
Advances in Intelligent Systems and Computing 715,
https://doi.org/10.1007/978-3-319-73210-7_108

responsible for causing pain and functional incapacity on upper limbs, which involve tendons, muscles, joints, nerves and blood vessels [3].

In the literature there are studies on repetitive strain injuries in several fields and it was identified that one of the most used body segments were the upper limbs [4, 5]. There is a need to exercise them in order to prevent repetitive strain injuries, as well as to improve the quality life of the people in general and the working class in particular [5]. For this purpose classes of labour gymnastics were created. This consists in the performance of physical activities during working days in companies [6]. However, in addition to the recommendation of physical exercise or the labour gymnastics there is no method for the prevention of these injuries [3, 5–10].

The lack of awareness for prevention often leads the employers to believe that costs associated to prevention are unnecessary expenditure. However, in the United States of America the workers compensation costs $20 billion, and another $100 billion on lost productivity, employee turnover, and other indirect expenses [11].

The objective of this paper is to develop a sequence of serious games to prevent strain injuries in hands through stretching off and warm-up exercises. The game activities promote hands detection and their movements which are automatically detected by the proposed system. To accomplish the goal, it is first necessary to define the adequate exercises to prevent strain injuries through exercises on the wrists and fingers for each hand. Given the nature of the problem, the tests should be performed for a long period of time to detect results. To perform these tests there will be defined two groups with similar functions and workloads, where one group should perform this set of games and the other does not.

Being the wrist where all the nerves, tendons and blood vessels pass to the hand it is very important to perform an appropriate stretching. So, the stretching exercises should be performed at least twice a day, at the beginning and end of daily functions, as stated in [1]. Performing regular stretching exercises increases flexibility, and it may promote injury prevention and improve the recovery ability, in case of the worker eventually suffers from any injury [12].

The warm-up exercises activate blood circulation and warm up muscles and joints, as stated in [13], and will prepare the user for intense muscular activities [1].

The execution of these exercises are important, since the main reason for the existence of strain injuries results from the fact that the hands and wrists usually are not sufficiently exercised before performing a repetitive task [14].

This paper is organized in five sections, where Sect. 2 presents the proposed system. Section 3 refers to the games developed and the respective movements implemented. Section 4 presents the preliminary results of the detection of the hands with their tracking points in laboratory. The article finishes with the final comments.

2 Implemented System

Figure 1 presents the implemented system. The system consists of the Intel RealSense 3D Camera F200 and a personal computer. The user is placed in front of the camera to perform the warm-up and the stretching exercises to be tracked.

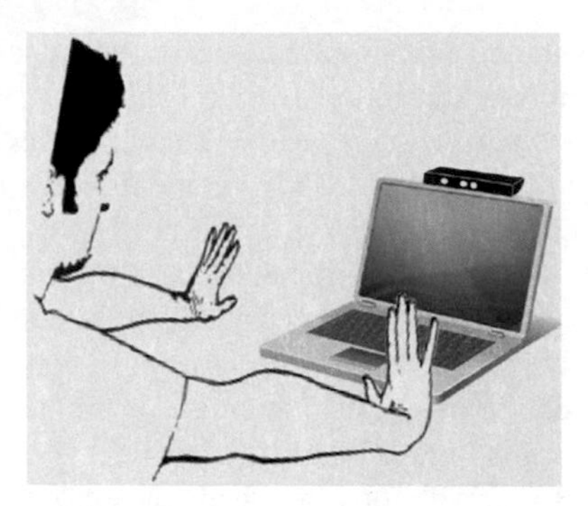

Fig. 1. Implemented System, adapted from [15]

The Intel camera is responsible for the 3D hands detection and it has a color camera, infrared laser projector and a depth camera [15]. The effective range for gesture capture is between 20 and 60 cm, as stated in [16].

The computer is responsible for the game environment, data acquisition from the camera, and processing data to recognize the movements (the warm-up and the stretching exercises) made by the player as commands in the game.

Figure 2 shows all points tracked from the hand by the camera using the image processing modules from Intel. However, on the application only the red tracking points are tracked, therefore the frame processing is faster and lighter [14].

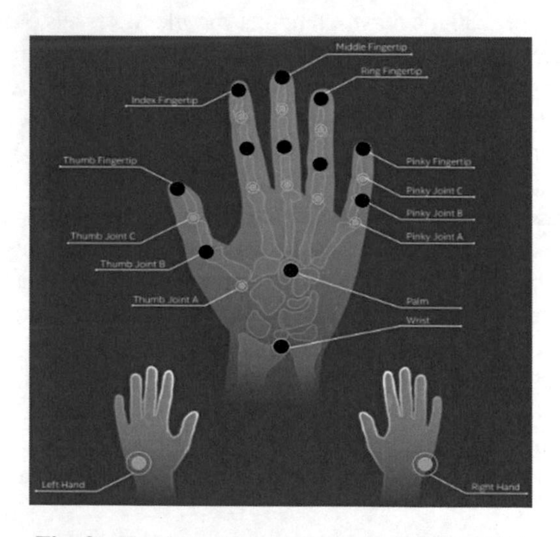

Fig. 2. Hand Tracking Points, adapted from [16]

The game application is developed in Unity 3D software, a flexible software which allows game development in 2D or 3D, compatible with C# programing. It has available several free assets and it allows to target many devices more easily, as stated in [17, 18].

3 Developed Games

In this section, there will be discussed the implementation of the games developed.

To prevent the strain injuries, as previously referred, there are two sets of exercises that can be implemented: warming up and stretching off exercises, to the wrists and fingers for each hand, as stated in [5, 6, 15, 19] - these movements were defined by a physiotherapist. So, several games were developed, where the game controls are the movements corresponding to the warm-up and stretching exercises.

The use of serious games have several benefits, because they can be played on any computer, being a low cost solution, especially in companies where the daily use of computers is given. Moreover, in relation to the classes of labour gymnastics [4, 6], it has the advantage of not being necessary to create a space in the company for the performance of classes, neither the hiring of a teacher, besides being able to be used at any time of the day, without having to join the employees. Besides that, a serious game playing can also be a stimulating and entertaining activity [20].

3.1 Ping Pong Game

The Ping Pong Game, Fig. 3, is based on the classic game Pong, one of the first computer games. The game features are two paddles and a ball; the goal is to defeat an opponent by being the first to gain 10 points; a player gets a point once the opponent misses a ball, as stated in [21].

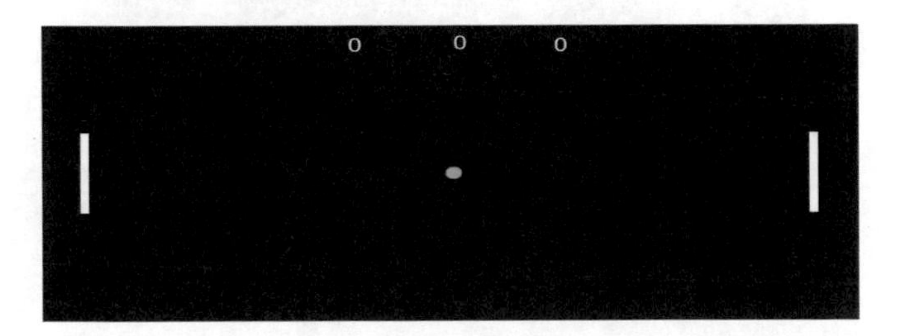

Fig. 3. Ping Pong Game

In this game, the objective is to suffer the least number of goals in one minute of play. Since, the player controls both paddles with the movement of one hand, each hand controls both paddles during 30 s.

Figure 4 represents the movement correspondent, to move the paddles, up and down, with the hand. This is the warming-up exercise for the wrist with the vertical movement.

With this game, it is possible to perform one of the several warm-up exercises, to prevent strain injuries.

To play the Ping Pong Game, the system needs the wrist point and the center point to move the paddles.

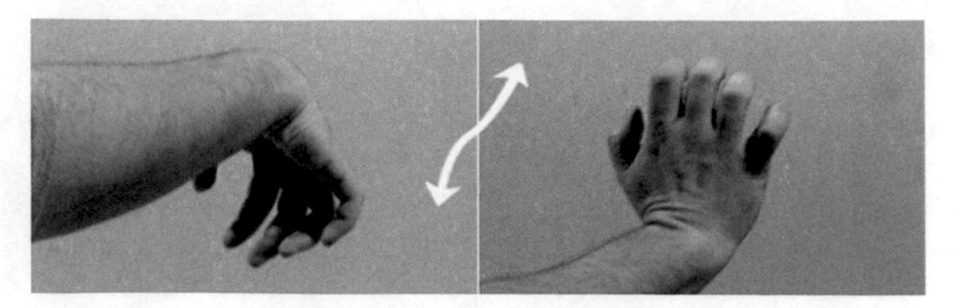

Fig. 4. Wrist warm-up exercise, vertical movement, adapted from [15]

3.2 Space Ship Game

Based on the arcade game Space Invaders [22], the Space Ship game consists on a spaceship that is traveling through space and must avoid collisions with the meteorites that are in its way, Fig. 5. The player can also destroy these meteorites with the shooting weapons implemented on the ship.

So, to move the ship from one side to another, the player must do the wrist warm-up exercise with the horizontal movement, Fig. 6. To activate the shooting system from the ship, the player must do the exercise from Fig. 7.

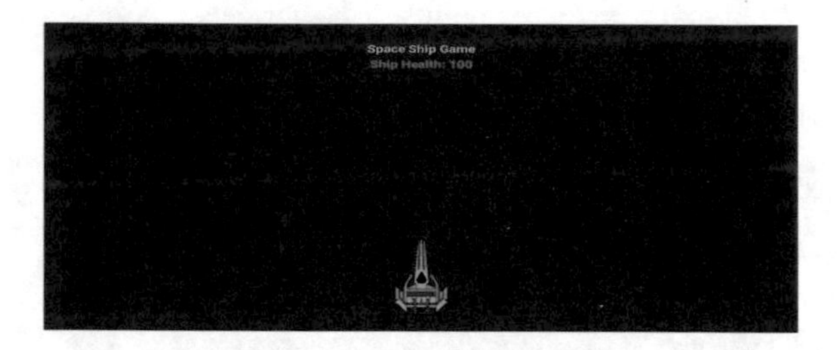

Fig. 5. Space Ship Game

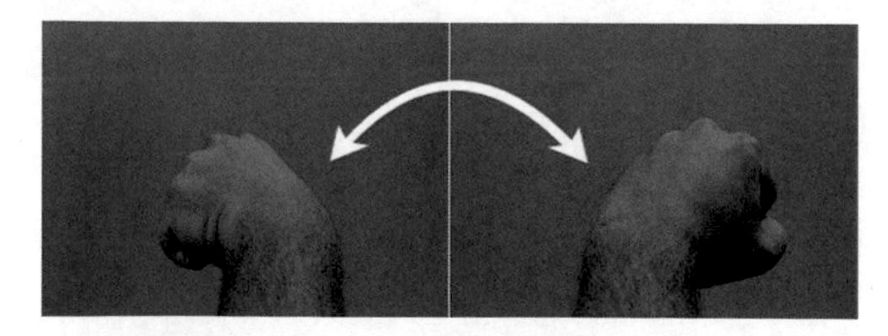

Fig. 6. Wrist warm-up exercise, horizontal movement, adapted from [15]

The player should control the ship for 30 s with each hand, to prevent destroying it. So, the player can perform both warming up exercises, one to the wrist another to the fingers, in both hands.

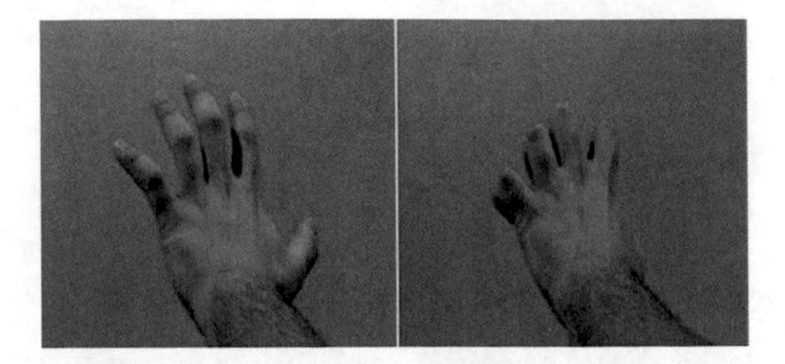

Fig. 7. Warm-up exercise, stretching and flexion of the fingers, adapted from [15]

To play the Space Ship Game, the system needs the wrist point and the center point and, also need the tracking points corresponding to the fingertips. The wrist point and the center point are responsible for the ship movement, but the fingertips are needed to use the shooting guns implemented on the ship.

3.3 Hedgehog Invasion Game

The Hedgehog Invasion Game, Fig. 8, is based on the Angry Birds game [23], the Siege Hero game [24], and the Crush the Castle game [25], where the goal is to eliminate all enemies while destroying their buildings, with as few attempts as possible.

Fig. 8. Hedgehog Invasion Game

To launch the hedgehogs, it is first necessary to close the hand in fist form and then to load the catapult. To do so, it is necessary to apply the movement shown in Fig. 9 several times. In the end, open again the hand to release the hedgehog and throw it against the construction.

As it can be seen in Fig. 8, there are two catapults, so to launch the hedgehog from the right catapult it is necessary to perform the game exercise with the right hand and to use the left catapult it is used the left hand.

The player has five shots on each catapult to progress in the game levels. There are ten levels in case the player eliminates all the enemies of the level with a single shot.

To launch the hedgehog, the system uses all the tracking points because it is first necessary to recognize that the hand is closed in a fist form. Then, it is needed to use the wrist point and the center point to load the catapult; to recognize this circular

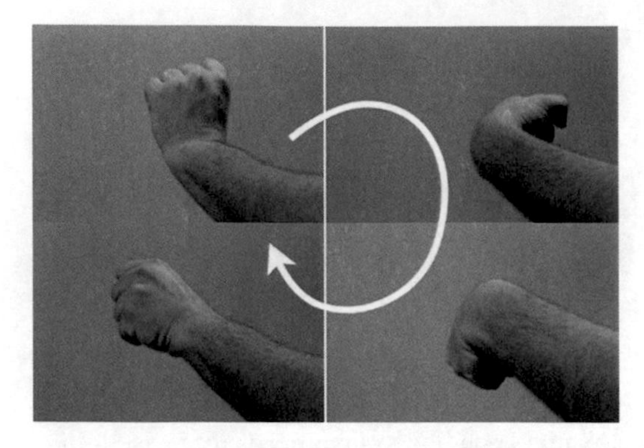

Fig. 9. Wrist warm-up exercise, wrist rotation, adapted from [15]

movement it is necessary to recognize several positions on each lap. It is necessary to use all the tracking points to recognize the movement when the hand is opened to launch the hedgehog.

These game activities cover the warming-up exercises. Figure 10, presents some examples of the stretching exercises that will be translated in a game scenario.

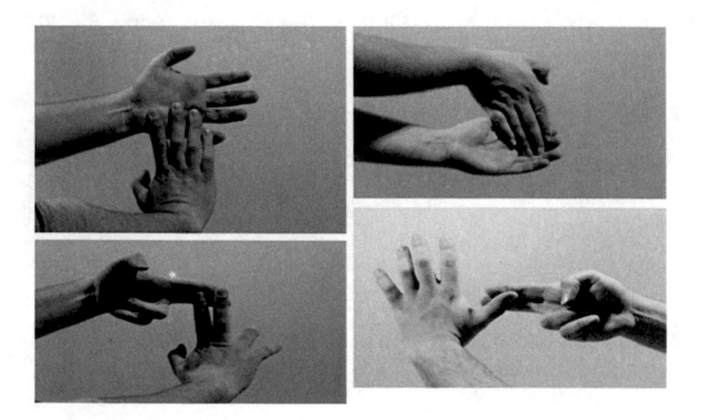

Fig. 10. Stretching exercises, adapted from [15]

4 Results

In this section there will be presented the results of the developed games. Figure 11 represents a 3D tracking result for both hands, where the hand with the blue tracking fingers represents the right hand and the other, the left hand. Moreover, the hand is represented in 3D, but during the execution of the games, the hand will be represented in 2D.

The three game scenarios were tested in laboratory environment. These preliminary tests allow evaluating system constraints and functioning. It is worth pointing out that it is still necessary to conduct more tests with all games with several people, to obtain a more consistent feedback.

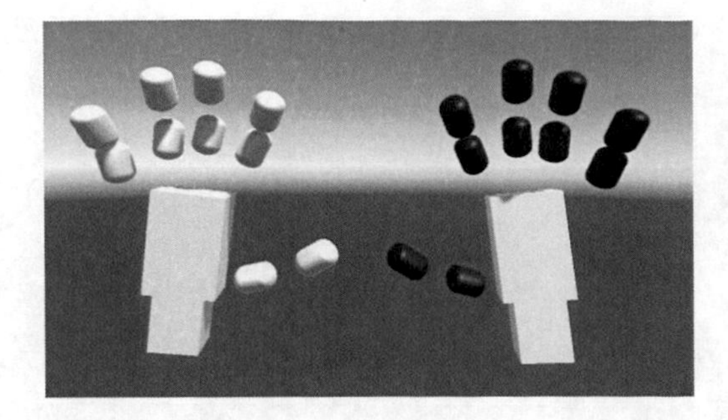

Fig. 11. Hands points tracked by Intel camera in Unity, adapted from [14]

In these first tests, the games presented a good gameplay, the commands were correctly detected and hands movements were correctly detected.

The use of both hands simultaneously (tested at the beginning) offered some detection problems, causing recognition errors. The game where this problem was most felt is in the game of Ping Pong, where one hand could control both paddles at the same time. To overcome this problem, all games were adapted to be controlled by one hand at a time. So, in both Ping Pong and Space Ship games it was necessary to create a playing time for each hand, corresponding to 30 s each.

Figure 12 shows a game scenario of Ping Pong. On this position, the tracking points are directing the paddles to go down. The central value is the game time and the score of the game is given by the remaining two numbers.

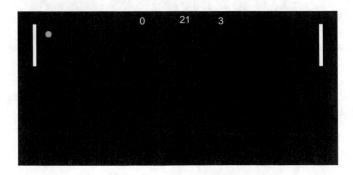

Fig. 12. Ping Pong Game scenario

In Fig. 13 it is being played the Space Ship Game and the tracking points are in position to activate the firing command, because to firing the fingertips must be closer to the center point. The player's score is written in yellow, and the ship's health in blue.

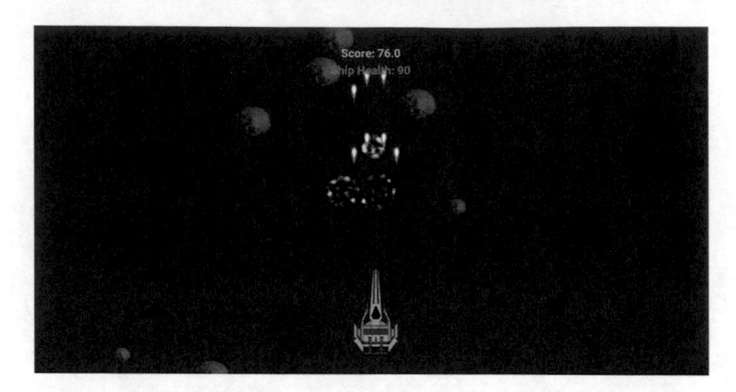

Fig. 13. Space Ship Game, shooting result

In Fig. 14, due the tracking points location, the ship is moving to the left position, to avoid collision with meteorites.

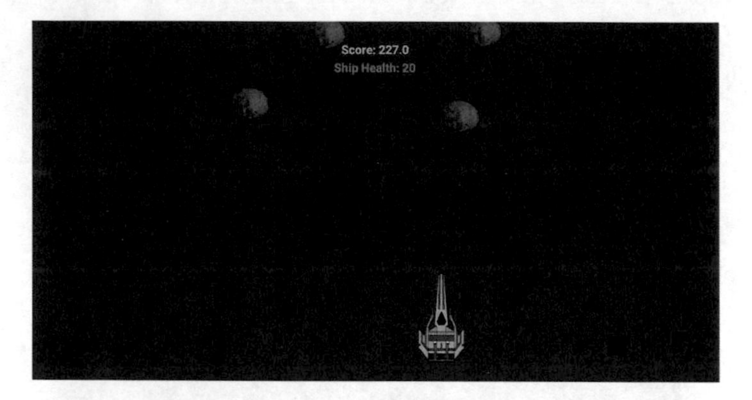

Fig. 14. Space Ship Game, movement result

In Fig. 15, the hedgehog was thrown against the enemy buildings, beginning to overthrow them.

Fig. 15. Hedgehog Invasion Game, launching result

5 Final Comments

The objective of the work presented in this paper was to design and develop a sequence of serious games where with the hands detection and their movements it is possible to prevent strain injuries through warm-up and stretching off exercises.

The games were developed, tested and optimized in laboratory environment. The next step in the research is to design and implement the stretching exercises and some other warming-up scenarios. When all the game scenarios are implemented, the overall system will go under extensive tests for validation. Moreover, there will be used a questionnaire in order to evaluate the comfort of the users in the daily work and about their game experience.

The growing concern about strain injuries caused by professional activities leads to a higher pursuit for solutions to perform labour gymnastics in companies. So, the implementation of this type of games can be an asset, as it can lead, in a more fun and motived way, to a reduction or even help to eliminate strain injuries in employees.

Acknowledgment. The authors would like to express their acknowledgments to COMPETE: POCI-01-0145-FEDER-007043 and FCT – *Fundação para a Ciência e Tecnologia* within the Project Scope: UID/CEC/00319/2013.

References

1. Moreira, J.A.: "A prevenção de Lesões por Esforço Repetitivo (LER) nas aulas de saxofone," Universidade do Minho (2015)
2. de Oliveira, E.M., Barreto, M.: Engendrando Gênero na Compreensão das Lesões Por Esforços Repetitivos. In: Saude e Sociedade, pp. 77–99 (1997)
3. Aptel, M., Aublet-Cuvelier, A., Cnockaert, J.C.: Work-related musculoskeletal disorders of the upper limb. Joint. Bone. Spine **69**(6), 546–555 (2002)
4. Ferracini, G.N., Valente, F.M.: Presence of musculoskeletal symptoms and effects of labor gymnastic in employees of the administrative sector of a public hospital. Rev. Dor **11**(3), 233–236 (2010)
5. Machado, L.: "Proposta de um conjunto de exercícios de Ginástica Laboral, como resposta às principais Lesões Músculo – Esqueléticas Relacionadas ao Trabalho," Universidade do Porto (2008)
6. Delani, D., Evangelista, R.A., Pinho, S.T., Silva, A.C.: Labor gymnastics: improving the quality of life of worker, evista Científicada. Fac. Educ. e Meio Ambient. **1**, 41–61 (2013)
7. Serranheira, F., Lopes, F., Uva, A.S.: Lesões Músculo-Esqueléticas (LME) e Trabalho: Uma associação muito frequente. In: Sociedade Portuguesa de Medicina do Trabalho (2003)
8. Silva, L.M.O.: Estudo de casos de Lesões Musculosqueléticas Relacionadas com o Trabalho dos Membros Superiores existentes numa empresa de componentes automóveis (2015)
9. Nunes, I.L., Bush, P.M.: Work-related musculoskeletal disorders assessment and prevention. Ergon. Syst. Approach **26**, 1–31 (2011)
10. Simoneau, S., St-Vincent, M., Chicoine, D.: Work-Related Musculoskeletal Disorders (WMSDs). Saint-Léonard, Québec: IRSST and the A.S.P. Métal-Électrique
11. Anliker, J.: "Carpal Tunnel Syndrome - Damaging U.S. Economy," RSI-Therapy (2005). http://www.rsi-therapy.com/Articles/a_ctsdamaginguseconomy.htm. Accessed 08 Feb 2017

12. Fragelli, T., Günther, I.: Abordagem ecológica para avaliação dos determinantes de comportamentos preventivos: proposta de inventário aplicado aos músicos. In: Borém, F. (ed.) Per Musi, vol. 25, pp. 73–84. Belo Horizonte (2012)
13. Marques, R.M.M.: Identificação dos fatoresde risco determinantes da prevalência lesões músculo-esqueleticas nos membros superiores e coluna vertebral nos múicos profissionais em Portugal (2011)
14. Aparício, L.N., Silva, A.: Postura, Dor e Perceção de Esforço na Aprendizagem do Acordeão (2014)
15. Freitas, H., Carvalho, V., Soares, F., Matos, D.: Virtual application for preventing repetitive strain injuries on hands: first insights. In: 10th International Conference on Biomedical Electronic Devices, Biostec, vol. 1, pp. 237–244 (2017)
16. IntelRealSense, "SDK Design Guidelines" (2014)
17. Unity, "Unity - Game engine, tools and multiplatform". https://unity3d.com/pt/unity. Accessed 09 Dec 2016
18. Unity Editor, "Unity - Editor." https://unity3d.com/pt/unity/editor. Accessed 09 Dec 2016
19. Longen, W.C.: "Ginástica Laboral na prevenção de LER/DORT? Um estudo reflexivo em uma linha de produção," Universidade Federal de Santa Catarina (2003)
20. Definição ou significado de jogo no Dicionário Infopédia da Língua Portuguesa com Acordo Ortográfico. https://www.infopedia.pt/dicionarios/lingua-portuguesa/jogo. Accessed 04 July 2017
21. Pong Game. http://www.ponggame.org/. Accessed 28 Mar 2017
22. SPACE INVADERS. http://www.pacxon4u.com/space-invaders/. Accessed 16 Apr 2017
23. Angry Birds. https://www.angrybirds.com/games/. Accessed 16 May 2017
24. Siege Hero. http://www.postmania.org/siege-hero-gratis-para-iphone/. Accessed 16 May 2017
25. Crush the Castle 2. http://www.1001jogos.pt/accao/crush-the-castle-2. Accessed 16 May 2017

Erratum to: A Remote Mode Master Degree Program in Sustainable Energy Engineering: Student Perception and Future Direction

Udalamattha Gamage Kithsiri,
Ambaga Pathirage Thanushka Sandaruwan Peiris,
Tharanga Wickramarathna, Kumudu Amarawardhana,
Ruchira Abeyweera, Nihal N. Senanayake, Jeevan Jayasuriya,
and Torsten H. Fransson

Erratum to:
Chapter "A Remote Mode Master Degree Program in Sustainable Energy Engineering: Student Perception and Future Direction" in: M. E. Auer et al. (eds.), *Teaching and Learning in a Digital World*, Advances in Intelligent Systems and Computing 715, https://doi.org/10.1007/978-3-319-73210-7_79

In the original version of this book, the second author's name has been updated from "Amagaha Pathirage Thaushka Sandaruwan Peiris" to "Ambaga Pathirage Thanushka Sandaruwan Peiris" in Chapter 79, which is a belated correction. The erratum chapter and the book have been updated with the change.

The updated online version of this chapter can be found at
https://doi.org/10.1007/978-3-319-73210-7_79

M. E. Auer et al. (eds.), *Teaching and Learning in a Digital World*,
Advances in Intelligent Systems and Computing 715,
https://doi.org/10.1007/978-3-319-73210-7_109

Author Index

M. E. Auer et al. (eds.), *Teaching and Learning in a Digital World*,
Advances in Intelligent Systems and Computing 715,
https://doi.org/10.1007/978-3-319-73210-7

Printed in the United States
By Bookmasters